WileyPLUS

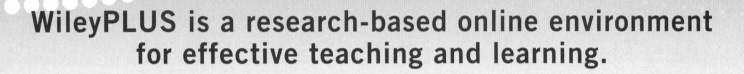

WileyPLUS is a research-based online environment for effective teaching and learning.

WileyPLUS builds students' confidence because it takes the guesswork out of studying by providing students with a clear roadmap:

- what to do
- how to do it
- if they did it right

It offers interactive resources along with a complete digital textbook that help students learn more. With *WileyPLUS*, students take more initiative so you'll have greater impact on their achievement in the classroom and beyond.

Now available for

Bb
Blackboard

For more information, visit www.wileyplus.com

WileyPLUS

ALL THE HELP, **RESOURCES,** AND PERSONAL SUPPORT YOU AND YOUR STUDENTS NEED!

www.wileyplus.com/resources

1st DAY OF CLASS ...AND BEYOND!

2-Minute Tutorials and all of the resources you and your students need to get started

WileyPLUS

Student Partner Program

Student support from an experienced student user

Wiley Faculty Network

Collaborate with your colleagues, find a mentor, attend virtual and live events, and view resources
www.WhereFacultyConnect.com

WileyPLUS

Quick Start

Pre-loaded, ready-to-use assignments and presentations created by subject matter experts

Technical Support 24/7
FAQs, online chat, and phone support
www.wileyplus.com/support

© Courtney Keating/iStockphoto

Your *WileyPLUS* Account Manager, providing personal training and support

PHYSICAL GEOLOGY

The Science of Earth, Canadian Edition

PHYSICAL GEOLOGY

The Science of Earth, Canadian Edition

Charles Fletcher

University of Hawai'i

Dan Gibson

Simon Fraser University

Kevin Ansdell

University of Saskatchewan

WILEY

VICE PRESIDENT & PUBLISHER:	Veronica Visentin
ACQUISITIONS EDITOR:	Rodney Burke
MARKETING MANAGER:	Patty Maher
EDITORIAL MANAGER:	Karen Staudinger
PRODUCTION MANAGER:	Tegan Wallace
DEVELOPMENTAL EDITOR:	Andrea Grzybowski
MEDIA EDITOR:	Channade Fenandoe
PUBLISHING SERVICES COORDINATOR:	Lynda Jess
EDITORIAL ASSISTANT:	Luisa Begani
LAYOUT:	cmPreparé
COVER DESIGN:	Joanna Vieira

Cover Photo: Courtesy of Dr. Dan Gibson, Simon Fraser University

Figure 7.30 Credits:
7.30a (left): USDA.
7.30a (right), 7.30f (right): © Paul Sanborn, Ecosystem Science and Management Program, UNBC.
7.30b (left): Department of Soil Science, University of Saskatchewan
7.30b (right), 7.30c, 7.30d (right), 7.30 (right), 7.30f (left): Maja Krzic, Faculty of Land and Food Systems/Faculty of Forestry, University of British Columbia, Vancouver.
7.30d (left): © Daniel J. Pennock, University of Saskatchewan.
7.30e (left): © Daniel J. Pennock, University of Saskatchewan.

Library and Archives Canada Cataloguing in Publication

Fletcher, Charles H., author
Introduction to physical geology : the science of earth/Charles Fletcher, University of Hawi'i, Dan Gibson, Simon Fraser University, Kevin Ansdell, University of Saskatchewan. – Canadian edition.

Includes bibliographical references and index.
ISBN 978-1-118-30082-4 (pbk.)

1. Physical geology–Textbooks. I. Gibson, Dan, 1969-, author II. Ansdell, Kevin Michael, 1961-, author III. Title. IV. Title: Physical geology.

QE28.2.F52 2013   551   C2013-906982-8

Printing and binding: Quad Graphics
Printed and bound in the United States.
1 2 3 4 5 QG 18 17 16 15 14

John Wiley & Sons Canada, Ltd.
6045 Freemont Blvd.
Mississauga, Ontario L5R 4J3

Visit our website at: www.wiley.ca

ABOUT THE AUTHORS

Charles "Chip" Fletcher is a professor (and past chair) in the Department of Geology and Geophysics at the University of Hawai'i. Chip earned his Ph.D. at the University of Delaware where he met his wife, also a geology graduate student at the time. They have raised three children in the beachside town of Kailua, Hawaii.

Chip's career has focused on coring the reefs of Hawaii to interpret the changing history of sea level, environments, and ecosystems over the great global climate shifts of the Pleistocene and Holocene Epochs. He also works closely with government agencies to improve understanding of coastal erosion, climate change, and other types of hazards on Pacific islands. Much of this research formed the basis for his book, *Living on the Shores of Hawai'i: Natural Hazards, the Environment, and Our Communities,* published by University of Hawai'i Press.

For 25 years, Professor Fletcher's physical geology classroom has been a lively place where students learn to think critically about Earth's future. Students are asked to analyze, synthesize, and evaluate relevant information and wrestle with questions such as "Why is gasoline cheaper than bottled water?", "Will we run out of economic metals?", and "What are the potential consequences of global warming where I live?" Professor Fletcher believes that students trained in the processes of science and who possess an enduring understanding of geologic principles will be able to effect change in the world. Physical geology is the science of Earth, and the preparation of tomorrow's decision makers begins in the physical geology classroom.

Dan Gibson is an associate professor in the Department of Earth Sciences at Simon Fraser University. Dan completed his Ph.D. at Carleton University, working on the tectonic evolution of the southeastern Canadian Cordillera.

Dan's research program is focused on the development and evolution of mountain belts (orogens). Dan and his students apply a multidisciplinary approach to research that includes geological mapping and structural, tectonic, petrological, and geochronological analyses. Dan's current research emphasis is on the Canadian Cordillera, a spectacular mountain belt that provides ample opportunity to learn about past and present mountain-building processes, the resources they produce, and the hazards they can pose.

Dan is also passionate about teaching, seeing it as an opportunity to instill in students a desire to learn, to enable them to embrace their curiosity, and to encourage them to think critically about the world around them. Pressures such as increasing populations, coupled with ever-dwindling resources and a warming climate, have put us in a precarious position. Now more than ever society needs to understand how Earth works and the impact that our actions and inactions have on the sustainability of the resources we require, the air we breathe, and the water we drink. Education is the key. The physical geology classroom provides a means to send forth wave upon wave of critically thinking, educated students, ready to be our ambassadors of geology and help society make better-informed decisions.

Kevin Ansdell is a professor in the Department of Geological Sciences at the University of Saskatchewan. He grew up in the United Kingdom and completed his bachelor's degree in geology at Oxford University before coming to Canada for a master of science project at the University of Alberta. After working for mining and mineral exploration companies in South Africa and Australia, he returned to Canada and completed his Ph.D. at the University of Saskatchewan, working on gold deposits in the Flin Flon area of Manitoba and their relationship to the tectonic evolution of the southern Canadian Shield.

Kevin's research program is focused on understanding the origin of hydrothermal mineral deposits and their relationship to the development and evolution of ancient mountain belts. Very similar to the approach taken by Dan and his students, Kevin's research group also applies a multidisciplinary approach to research that includes geological mapping and petrological, geochemical, and geochronological analyses. Kevin's current research is focused on the mineral deposits associated with the Trans-Hudson Orogen, a Himalayan-scale mountain belt, the core of which is preserved in the Precambrian Shield of Canada.

Kevin has taught introductory physical geology for many years, as well as more advanced geoscience courses in mineralogy, petrology, mineral deposits, and field mapping. He won the 2013 Teaching Excellence award from the Division of Science in the College of Arts and Science at the University of Saskatchewan for his undergraduate teaching initiatives. He designs the introductory course so that students interested in pursuing a degree in geology have a solid background, and all students gain an appreciation of the importance of geological processes in shaping our environment and an understanding of natural hazards and the resources we require in our daily lives. It is vital that everybody is educated about geological processes so that critical decisions concerning sustainable development and preservation of our planet are made in the future.

This is dedicated to Yolande, Rachel, and Ben, and to all my students over the years — D.G.
To my wife, Lana, and all the students I have had the pleasure of interacting with over the years — K.A.

The physical geology classroom has evolved in recent years. The traditional text and instructor lectures are now accompanied by interactive lectures, online resources, and e-books. The world has changed as well. Global warming brings new intensity to natural hazards. The human population has reached record levels, and natural resources have grown in expense and scarcity. These are global challenges requiring our best thinking and a need to know how Earth works. The physical geology classroom is where this training begins.

The study of physical geology has never been more relevant, timely, and fast-paced. However, three things have not changed:

1. Instructors want to enrich the lives of their students and help them leave the course as more responsible and informed global citizens. This requires students to develop a basic understanding of geologic processes as well as strong critical thinking skills.

2. Students need a book that is easy to read, well illustrated, relevant, and engaging.

3. Instructors need a text that focuses on basic principles with up-to-date information.

Why Publish Another Physical Geology Text?

There are many physical geology texts on the market, so why publish another one? This book has what we think are important innovations and strengths.

Innovations

- For ease of reading, we offer text in two-page segments. Students will find the book approachable, succinct, and well organized.
- Woven into each page are critical-thinking exercises linked to illustrations and photos. These exercises are guided by Bloom's taxonomy, a classification that identifies six sequential levels of building intellectual skill.
- We discuss relevant and provocative global issues such as peak oil, global warming, drought and water stress, population growth, natural hazards, and environmental management.

Physical Geology, Canadian Edition, offers the text in **two-page section segments**. Students find that a reading assignment consisting of "bite-size pieces" is easily accomplished, the content is easier to understand, and thus the assignment is more readily started and more likely to be completed. As instructors, we want students to do the reading and the work we assign. This book advances that goal with a purposeful design to facilitate a student's learning. Every two-page section concludes with an "Expand Your Thinking" question that provides further critical

thinking opportunity for students to check their conceptual understanding of the material before moving forward.

Physical Geology, Canadian Edition, is a learning system rich with critical thinking questions and activities that drive students to challenge their understanding of the world. What is critical thinking? We identify it as the use of reasoning to define the world around us by developing hypotheses and theories that can be rigorously tested. *Physical Geology: Science of the Earth, Canadian Edition* treats the **art and photo program as a critical thinking opportunity**.

It is becoming common for texts to include big two-page art and photo spreads in each chapter—we call ours "showcase art." It is also common for all art and photographs to be colour, three-dimensional, and beautiful. We achieve all this, but why stop at showing a bunch of pretty pictures? We take it one step further—we make almost every visual piece a critical thinking exercise. Our showcase art pieces and photos are part of problem-solving exercises that students can do as homework, in teams or solo, or simply as another learning opportunity. Similarly, a question with almost every art piece and photo asks students to stop and think further about what they have learned. Good questions are the driving force of creative learning. This integration of art, photos, and critical thinking promotes the growth of problem-solving skills in the context of a lifelong understanding of Earth. Students are reconnected with the natural resources and landscapes that sustain us all, and they exercise the thinking tools needed to address global challenges.

Our critical thinking approach is guided by Bloom's taxonomy (1956), a classic study that identified six sequential levels of cognitive skill. Below are the six levels in ascending order of building knowledge, comprehension, and critical thinking, and examples of how we use them:

1. **Knowledge**—This is the basic level. Bloom found that over 80 percent of test questions simply test knowledge. Students must arrange, list, recall, recognize, and relate information.

   Example: How are loose sediments lithified? What is Bowen's reaction series?

2. **Comprehension**—Knowledge should build comprehension. At this level, students must classify, describe, and explain concepts. For instance, students are asked to state a problem in their own words.

   Example: Explain why waves change shape as they enter shallow water. Why do corners and edges of rock weather faster than a flat surface? Describe the crystallization of minerals in magma of average composition.

We use knowledge and comprehension questioning often throughout the book, typically with every art piece or photo and as a warm-up to more complex activities.

3. **Application**—Students must apply what they learned in the classroom and their reading to novel problem-solving situations.

   Example: Using your understanding of mass wasting and stream behaviour, define three general guidelines for safely locating a new road in a hilly region.

4. **Analysis**—Students must calculate, analyze, categorize, and compare/contrast information.

   Example: What evidence would you look for to test the hypothesis that humans are contributing to, or the cause of, global warming?

Application and analysis are higher levels of cognition. Students are asked to exercise these skills in our "Expand Your Thinking" exercises on each two-page segment of the book.

5. **Synthesis**—Students must put parts together to form a new meaning or structure. This involves arranging information to construct new knowledge.

   Example: In some areas, climate change is causing less overall rainfall but more intense rain events. As a community leader, what new challenges do you think these trends present and how will you manage them?

6. **Evaluation**—Students must make judgements about the value of ideas or materials. We ask them to appraise, assess, choose, and predict outcomes.

   Example: What criteria would you use to assess the hazards associated with a volcano that no one had previously studied?

We have designed critical thinking exercises around showcase art and photos. These typically begin with knowledge and comprehension questions and culminate with synthesis and evaluation questions. We also ask synthesis and evaluation questions in "Expand Your Thinking" exercises.

Strengths

- Emphasizing the basic principles of geology, this text is easy to adopt into an existing syllabus.
- The writing is clear and jargon free.
- Every page has several pieces of accurate, detailed, beautiful art and photos with many Canadian examples.

Physical Geology, Canadian Edition, focuses on the **basic principles of geology** so that instructors can easily adopt the book into their existing syllabus. For example, plate tectonics is introduced early, and the entire text teaches tectonics as a unifying theory of geology. It contains chapters on each rock group and surface environment plus some new ones not normally seen in other texts—"The Geology of Canada," "Global Warming," "Glaciers and Paleoclimatology"—as well as separate chapters on "Coastal Geology" and "Marine Geology." There are also separate chapters on "Geologic Time" and "Earth's History," which include sections on evolutionary theory. Because the book has 23 chapters, instructors can choose a subset of chapters that best fits their goals.

Another strength is that *Physical Geology, Canadian Edition,* is easy to read. Throughout the review process, dozens of faculty from around the country were asked to critically review each chapter. One typical comment was "This text is well written and easy to read."

Together these characteristics address another problem commonly expressed by students: science classes tend to bury students in information. We manage this issue with our focus on basic principles, the frequent use of critical thinking exercises where students can check their understanding, and the easy-to-approach design of two-page spreads, accompanied by a well-written, non-technical text.

The content of *Physical Geology, Canadian Edition*, also reflects the fact that students are more likely to become engaged in the subject matter if they can see some connection or relevance to their everyday lives. In addition to updating the material of the original textbook, every chapter includes references to geological features in Canada or geological processes that can be observed to be operating in the Canadian environment. There are numerous photographs, maps, and statistics focused on showcasing Canadian examples, but without lessening the impact of global examples. Students must learn that important Earth processes do not know the meaning of political boundaries!

There is a new chapter on the geology of Canada, which divides the country into a number of geologically important regions. These range from the vast areas of the Canadian Precambrian Shield, which includes the oldest known rocks on Earth, to some of the most spectacular scenery in the mountainous areas in western Canada, but all students will be able to gain an appreciation of the geological history of their area. Wherever possible, links are provided to Canadian websites or other reading material for additional insight and information on Canadian geological features, resources, hazards, and environmental issues.

Pedagogical Help for Students

Each student in the classroom is capable of learning, yet none think or process information in the same way. Some are social and learn when interacting with other people. Some are analytical and learn best by problem solving. Others are spatial learners and enjoy graphics. Instructors, parents, and modern brain-based researchers understand that many different types of learners populate the classroom.

Instructors are challenged to reach all of these learners. Hence, *Physical Geology, Canadian Edition,* provides a variety of ways to deliver information.

- **A specific learning objective for every concept**—There are 23 chapters, each chapter is organized into 7 to 12 basic concepts, and each concept is delivered in two pages. With every turn of a page, a new concept with a specific learning objective appears. The learning objectives identify what the students should know, understand, or be able to accomplish as a result of their reading. Achieving each learning objective will help students study and help instructors assess and evaluate students' progress. This design facilitates learning in a busy world where our attention is challenged by texting, cellphones, social networking, and other distractions.
- **A diversity of thinking**—Every chapter has several types of review, critical thinking, and problem-solving exercises. Students are asked to contemplate geological problems by focusing on art and photos, calculating, inferring outcomes, recognizing patterns, connecting concepts, and doing other activities.

- **Relevant and provocative**—Four themes recur throughout the book to remind us that we are surrounded by dynamic and vital natural processes that need careful management. "Critical Thinking," "Geology in Our Lives," "Geology in Action," and "Earth Citizenship" features provide examples, applications, and activities for students to further their understanding.
- **Easy to read, easy to teach from**—Students will enjoy reading this book because of its simple organization and direct non-technical language. Instructors will find this book easy to adopt because it presents the basic principles of geology in a flexible array of chapters.

Themes

Four themes run throughout this text. Each one is identified by an icon.

Critical Thinking Critical thinking is our most powerful tool in achieving sustainability. Like any skill, problem solving improves with practice. This text provides instructors with numerous opportunities to encourage students to think critically about scientific problems. We don't tell students WHAT to think; we ask them TO THINK and help them learn how to do it, so they can become more informed about the planet on which they live.

Geology in Our Lives Every minute of every day we are surrounded by geology. Where do drinking water, metals, and petroleum come from? How can we reduce the threat of landslides, earthquakes, and storms? This book provides students with answers and encourages them to appreciate the impact we all have on Earth.

Geology in Action Earth is in a state of constant change, largely related to plate tectonics and physical, chemical, and biological processes in the environment. Highlighting these brings physical geology to life and provides exciting real-world examples of what students are learning.

Earth Citizenship We all have the power to influence and manage natural resources locally and globally to meet the needs of both present and future generations, and to become geologically informed, global citizens. Our goal is to give students an understanding of how their choices contribute to changing the world.

Organization

The table of contents is organized in a straightforward manner and provides for flexibility in selecting topics suited to any syllabus.

- Chapter 1 introduces plate tectonics, geologic time, scientific thinking, and the idea of natural resources and natural hazards. These concepts are all revisited throughout the text.
- Chapter 2 develops scientific understanding of the Solar System and Earth's unique status among the planets.
- Chapter 3 is an in-depth review of plate tectonics. The material presented here is referred to and developed throughout the book.
- Chapters 4 to 10 present minerals, rocks, volcanoes, weathering, and geologic resources.

- Chapters 11 and 12 discuss mountain building and earthquakes.
- Chapters 13 to 15 present geologic time and the geology of Canada.
- Chapters 16 to 23 discuss surface environments, working down the watershed from the atmosphere (global warming) to glaciers, surface water and groundwater, wind, and finally to marine geology in the last chapter.

In Closing

The role of teaching is to educate the next generation of leaders. We think physical geology should be a required course for every student on the university campus. So many geologically related problems have surfaced in the past decade that adults with even minimal training in the basic principles of geology are capable of offering important and critically needed help. Climate is changing, resources are dwindling, environments are threatened, and communities are expanding into hazardous areas—how we respond to these challenges will shape the future and fate of humankind on Earth. These grand challenges are on our doorstep now. Our greatest hope lies with the best critical thinkers who understand how Earth works.

Supplements

Physical Geology, Canadian Edition, is accompanied by a wealth of study and practice materials available in *WileyPLUS* and on our book companion site, www.wiley.com/go/fletchercanada.

- **GeoDiscoveries Media Library**—This easy-to-use website helps reinforce and illustrate key concepts from the text through the use of animations, videos, and interactive exercises. Students can use the resources for tutorials as well as self-quizzing to complement the textbook and enhance their understanding of physical geology. Easy integration of this content into course management systems and homework assignments gives instructors the opportunity to incorporate multimedia with their syllabi and with more traditional reading and writing assignments.
- **Concept Caching**—An online database of photographs allows students to explore what a physical feature looks like. Photographs and GPS coordinates are "cached" and categorized using core concepts of geography. Instructors can access the images or submit their own by visiting www.ConceptCaching.com.
- *Google Earth™ tours*—Virtual field trips allow students to discover and view geological landscapes around the world. Tours are available as .kmz files for use in Google Earth™ or other virtual Earth programs.

Student online resources include the following:

- *Self-quizzes*—chapter-based multiple-choice and fill-in-the-blank questions
- *Web resources*—useful websites to explore more about geology
- *Lecture note handouts*—key images and slides from the instructor PowerPoints so students in class can focus on the lecture, annotate figures, and add their own notes

Instructor online resources include all student resources, plus the following:

- *PowerPoint lecture slides*—chapter-oriented slides including lecture notes and text art

- *Test bank*—multiple-choice, fill-in, and essay questions available in both Word and computerized formats.
- *Instructor solutions manual*—answers to all figure, Expand Your Thinking, and Critical Thinking questions as well as the Assessing Your Knowledge self-test questions that appear at the end of each chapter.
- *Clicker questions*—a set of questions for each chapter formatted for clicker systems, which can be used during lectures to check understanding
- *Image gallery*—both art and photos from the text

WileyPLUS

WileyPLUS is an innovative, research-based, online environment for effective teaching and learning.

WileyPLUS builds students' confidence because it takes the guesswork out of studying by providing students with a clear roadmap: what to do, how to do it, if they did it right. Students will take more initiative so you'll have greater impact on their achievement in the classroom and beyond.

Acknowledgements

Physical Geology, Canadian Edition was built on the superb framework provided by the original book, *Physical Geology: The Science of Earth*. We would like to thank Chip Fletcher for his amazing work on that book. That could not have been completed without the hard work of the Wiley team in the United States, and the support and encouragement of colleagues in the Department of Geology and Geophysics at the University of Hawai'i at Manoa. We appreciate the stimulating educational environments provided by the Department of Earth Sciences in the Faculty of Science at Simon Fraser University and the Department of Geological Sciences in the College of Arts and Science at the University of Saskatchewan. Interactions with colleagues over the years have been very important in improving our teaching, but we are both still learning. Several colleagues with expertise in various fields answered our questions with patience and friendship throughout the process, thank you. We would especially like to thank Jim Monger from the Geological Survey of Canada for his fabulous work on developing the framework for the chapter "The Geology of Canada", and for writing the first draft of most of that chapter. Ivanka Mitrovic assisted in the search for some of the great photographs in the textbook. The Wiley Canada team is gratefully acknowledged for all their extraordinary effort throughout this process, from start to finish. A special thank you is extended to Andrea Grzybowski, developmental editor, for her excellent handling of the *Physical Geology, Canadian Edition*, and "almost" unending patience, even with all of our idiosyncratic tendencies and tiptoeing around deadlines. The superb copyediting by Yvonne Van Ruskenveld and proofreading by Ruth Wilson are greatly appreciated; the book is definitely better for their efforts. Thank you to Rodney Burke, who originally persuaded us to take on this project as it certainly has been a learning experience. And thanks to Patty Maher for your strategies as to how to market this new book. Many reviewers and focus groups provided valuable input: thank you all for your time and energy. Finally, and most importantly, thanks to the many students who have passed through our classrooms over the years, as we have learnt an enormous amount from those interactions.

Reviewers

Katherine Bruce, *Cambrian College*
Dante Canil, *University of Victoria*
Michael Cuggy, *University of Saskatchewan*
Keith Delaney, *University of Waterloo*
Sandy Denton, *SAIT Polytechnic*
Robbie Dunlop, *Simon Fraser University*
Steven Earle, *Vancouver Island University*
Galen Halverson, *McGill University*
Tark Hamilton, *Camosun College*
Cindy Hansen, *Simon Fraser University*
Mary Louise Hill, *Lakehead University*

Timothy Jones, *Cambrian College*
Eric Mattson, *Nipissing University*
Nancy Mckeown, *MacEwan University*
Ken Munyikwa, *Athabasca University*
Adrian Park, *University of New Brunswick, Fredericton*
Sylvie Pinard, *Mount Royal University*
Cliff Shaw, *University of New Brunswick, Fredericton*
Martin Sheppard, *SAIT Polytechnic*
Mark Smith, *Langara College*
Lisa Tutty, *University of Toronto*

BRIEF CONTENTS

Chapter 1 – AN INTRODUCTION TO GEOLOGY 2

Chapter 2 – SOLAR SYSTEM 26

Chapter 3 – PLATE TECTONICS 46

Chapter 4 – MINERALS 78

Chapter 5 – IGNEOUS ROCK 104

Chapter 6 – VOLCANOES 130

Chapter 7 – WEATHERING 160

Chapter 8 – SEDIMENTARY ROCK 190

Chapter 9 – METAMORPHIC ROCK 218

Chapter 10 – GEOLOGIC RESOURCES 242

Chapter 11 – MOUNTAIN BUILDING 270

Chapter 12 – EARTHQUAKES 302

Chapter 13 – GEOLOGIC TIME 336

Chapter 14 – EARTH'S HISTORY 364

Chapter 15 – THE GEOLOGY OF CANADA 392

Chapter 16 – GLOBAL WARMING 420

Chapter 17 – GLACIERS AND PALEOCLIMATOLOGY 450

Chapter 18 – MASS WASTING 482

Chapter 19 – SURFACE WATER 508

Chapter 20 – GROUND WATER 538

Chapter 21 – DESERTS AND WIND 566

Chapter 22 – COASTAL GEOLOGY 590

Chapter 23 – MARINE GEOLOGY 622

TABLE OF CONTENTS

Chapter 1 AN INTRODUCTION TO GEOLOGY 2

1-1 Geology Is the Scientific Study of Earth and Planetary Processes and Their Products through Time 4

1-2 Critical Thinking Based on Good Observations Allows Us to Explain the World around Us 6

1-3 Global Challenges Will Be Solved by Critical Thinkers with an Enduring Understanding of Earth 8

1-4 The Theory of Plate Tectonics Is a Product of Critical Thinking 10

1-5 Rock Is a Solid Aggregation of Minerals 14

1-6 Geologists Study Natural Processes, Known as Geologic Hazards, that Place People and Places at Risk 16

1-7 The Geologic Time Scale Summarizes Earth's History 20

Chapter 2 SOLAR SYSTEM 26

2-1 Earth's Origin Is Described by the Solar Nebula Hypothesis 28

2-2 The Sun Is a Star that Releases Energy and Builds Elements through Nuclear Fusion 30

2-3 Terrestrial Planets Are Small and Rocky, with Thin Atmospheres 32

2-4 Gas Giants Are Massive Planets with Thick Atmospheres 34

2-5 Objects in the Solar System Include the Dwarf Planets, Comets, and Asteroids 36

2-6 Earth's Interior Accumulated Heat during the Planet's Early History 40

Chapter 3 PLATE TECTONICS 46

3-1 Earth Has Three Major Layers: Core, Mantle, and Crust 48

3-2 The Core, Mantle, and Crust Have Distinct Chemical and Physical Features 50

3-3 Lithospheric Plates Carry Continents and Oceans 52

3-4 Paleomagnetism Confirms the Seafloor-Spreading Hypothesis 56

3-5 Plates Have Divergent, Convergent, and Transform Boundaries 58

3-6 Oceanic Crust Subducts at Convergent Boundaries 62

3-7 Orogenesis Occurs at Convergent Boundaries 64

3-8 Transform Boundaries Connect Two Spreading Centres 66

3-9 Earthquakes Tend to Occur at Plate Boundaries 68

3-10 Plate Movement Powers the Rock Cycle 72

Chapter 4 MINERALS 78

4-1 Minerals Are Solid Crystalline Compounds with a Definite (but Variable) Chemical Composition 80

4-2 A Rock Is a Solid Aggregate of Minerals 82

4-3 Geologists Use Physical Properties to Help Identify Minerals 84

4-4 Atoms Are the Smallest Components of Nature with the Properties of a Given Substance 86

4-5 Minerals Are Compounds of Atoms Bonded Together 88

4-6 Oxygen and Silicon Are the Two Most Abundant Elements in the Crust 90

4-7 Metallic Cations Join with Silicate Structures to Form Neutral Compounds 92

4-8 There Are Seven Common Rock-Forming Minerals 94

4-9 Most Minerals Fall into Seven Major Classes 96

Chapter 5 IGNEOUS ROCK 104

5-1 Igneous Rock Forms When Molten, or Partially Molten, Rock Solidifies 106

5-2 Igneous Rock Forms through a Process of Crystallization and Magma Differentiation 108

5-3 Bowen's Reaction Series Describes the Crystallization of Magma 110

5-4 The Texture of Igneous Rock Records Its Crystallization History 112

5-5 Igneous Rocks Are Named on the Basis of Their Texture and Composition 114

5-6 The Seven Common Types of Igneous Rock in More Detail 116

5-7 All Rocks on Earth Have Evolved from the First Igneous Rocks 118

5-8 Basalt Forms at Spreading Centres Hotspots and Subduction Zones 122

5-9 Igneous Intrusions Occur in a Variety of Sizes and Shapes 125

Chapter 6 VOLCANOES 130

6-1 A Volcano Is Any Landform that Releases Lava, Gas, or Ash or Has Done So in the Past 132

6-2 There Are Three Common Types of Magma: Basaltic, Andesitic, and Rhyolitic 134

6-3 Explosive Eruptions Are Fuelled by Violent Releases of Volcanic Gas 136

6-4 Pyroclastic Debris Is Produced by Explosive Eruptions 138

6-5 Volcanoes Can Be Classified into Six Major Types Based on Their Shape, Size, and Origin 140

6-6 Shield Volcanoes Are a Type of Central Vent Volcano 142

6-7 Stratovolcanoes and Rhyolite Caldera Complexes Are Central Vent Volcanoes 144

6-8 Large-Scale Volcanic Terrains Lack a Central Vent 148

6-9 Most Volcanoes Are Associated with Spreading Centre Volcanism, Arc Volcanism, or Intraplate Volcanism 150

6-10 Volcanic Hazards Threaten Human Communities 154

Chapter 7 WEATHERING 160

7-1 Weathering Includes Physical, Chemical, and Biological Processes 162

7-2 Physical Weathering Causes Fragmentation of Rock 164

7-3 Hydrolysis, Oxidation, and Dissolution Are Chemical Weathering Processes 166

7-4 Biological Weathering Involves Both Chemical and Physical Processes; Sedimentary Products Result from All Three Types of Weathering 170

7-5 Rocks and Minerals Can Be Ranked by Their Vulnerability to Weathering 172

7-6 The Effects of Weathering Can Produce Climate Change 174

7-7 Weathering Produces Soil 176

7-8 Soil, Spheroidal Weathering, and Natural Arches Are Products of Weathering 178

7-9 Soil Erosion Is a Significant Problem 182

7-10 There Are 10 Orders in the Canadian Soil Classification System 184

Chapter 8 SEDIMENTARY ROCK 190

8-1 Sedimentary Rock Is Formed from the Weathered and Eroded Remains of the Crust 192

8-2 There Are Three Common Types of Sediment: Clastic, Chemical, and Biogenic 194

8-3 Sediments Change as They Are Transported across Earth's Surface 196

8-4 Clastic Grains Combine with Chemical and Biogenic Sediments 198

8-5 Sediment Becomes Rock during the Sedimentary Cycle 200

8-6 There Are Eight Major Types of Clastic Sedimentary Rock 202

8-7 Some Sedimentary Rocks Are Formed by Chemical and Biogenic Processes 204

8-8 Sedimentary Rocks Preserve Evidence of Past Environments 206

8-9 Primary Sedimentary Structures Record Environmental Processes 212

Chapter 9 METAMORPHIC ROCK 218

9-1 Metamorphic Rocks Are Composed of Sedimentary, Igneous, or Metamorphic Minerals that Have Recrystallized 220

9-2 Changes in Heat and Pressure Can Cause Metamorphism 222

9-3 Chemically Active Fluids Transport Heat and Promote Recrystallization 224

9-4 Rocks Evolve through a Sequence of Metamorphic Grades 226

9-5 Foliated Texture Is Produced by Directed Stress Related to Regional Metamorphism 228

9-6 Nonfoliated Rocks May Develop during Regional or Contact Metamorphism 230

9-7 The Relationship between Mineral Assemblage and Metamorphic Grade Is Expressed by Metamorphic Facies 232

9-8 Metamorphism Is Linked to Plate Tectonics 236

Chapter 10 GEOLOGIC RESOURCES 242

10-1 The Crust Contains Metals, Building Stone, Minerals, and Sources of Energy 244

10-2 Mineral Resources Include Non-Metallic and Metallic Types 246

10-3 Ores Are Formed by Several Processes 248

10-4 Fossil Fuels, Principally Oil, Provide Most of the Energy that Powers Society 254

10-5 Oil Is Composed of Carbon that Is Derived from Buried Plankton 256

10-6 About 77 Percent of the World's Oil Has Already Been Discovered 258

10-7 Coal Is a Fossil Fuel that Is Found in Stratified Sedimentary Deposits 260

10-8 Nuclear Power Plants Provide about 17 Percent of the World's Electricity 262

10-9 Renewable Energy Accounts for More than 20 Percent of Canada's Energy Sources 264

Chapter 11 MOUNTAIN BUILDING 270

11-1 Rocks in the Crust Are Bent, Stretched, and Broken 272

11-2 Strain Takes Place in Three Stages: Elastic Deformation, Ductile Deformation, and Fracture 274

11-3 Strain in the Crust Produces Joints, Faults, and Folds 276

11-4 Dip-Slip and Strike-Slip Faults Are the Most Common Types of Faults 278

11-5 Rock Folds Are the Result of Ductile Deformation 282

11-6 Outcrop Patterns Reveal the Structure of the Crust 284

11-7 Mountain Building May Be Caused by Volcanism, Faulting, and Folding 286

11-8 Volcanic Mountains Are Formed by Volcanic Products, Not by Deformation **290**

11-9 Crustal Extension Formed the Basin and Range Province **292**

11-10 Fold-and-Thrust Belts Produce Some of the Highest and Most Structurally Complex Mountain Belts **294**

Chapter 12 EARTHQUAKES 302

12-1 An Earthquake Is a Sudden Shaking of the Crust **304**

12-2 There Are Several Types of Earthquake Hazards **306**

12-3 The Elastic Rebound Theory Explains the Origin of Earthquakes **310**

12-4 Most Earthquakes Occur at Plate Boundaries, but Intraplate Seismicity Is Also Common **312**

12-5 Divergent, Convergent, and Transform Boundaries Are the Sites of Frequent Earthquake Activity **314**

12-6 Earthquakes Produce Four Kinds of Seismic Waves **318**

12-7 Seismometers Are Instruments that Measure and Locate Earthquakes **320**

12-8 Earthquake Magnitude Is Expressed as a Whole Number and a Decimal Fraction **322**

12-9 Seismology Is the Study of Seismic Waves in Order to Improve Understanding of Earth's Interior **326**

12-10 Seismic Data Confirm the Existence of Discontinuities in Earth's Interior **328**

12-11 Seismic Tomography Uses Seismic Data to Make Cross-Sections of Earth's Interior **330**

Chapter 13 GEOLOGIC TIME 336

13-1 Geology Is the "Science of Time" **338**

13-2 Earth History Is a Sequence of Geologic Events **340**

13-3 Seven Stratigraphic Principles Are Used in Relative Dating **342**

13-4 Relative Dating Determines the Order of Geologic Events **344**

13-5 Isotopic Dating Uses Radioactive Decay to Estimate the Age of Geologic Samples **346**

13-6 Geologists Select an Appropriate Radioisotope when Dating a Sample **348**

13-7 Accurate Dating Requires Understanding Various Sources of Uncertainty **352**

13-8 Potassium-Argon and Carbon Provide Important Isotopic Clocks **354**

13-9 Earth's Age Is Measured Using Several Independent Observations **356**

Chapter 14 EARTH'S HISTORY 364

14-1 Earth's History Has Been Unveiled by Scientists Applying the Tools of Critical Thinking **366**

14-2 Fossils Preserve a Record of Past Life **368**

14-3 There Are Several Lines of Evidence for Evolution **370**

14-4 Molecular Biology Provides Evidence of Evolution **372**

14-5 Mass Extinctions Influence the Evolution of Life **374**

14-6 The Geologic Time Scale Is the "Calendar" of Events in Earth's History **376**

14-7 The Archean and Proterozoic Eons Lasted from 4.0 Billion to 542 Million Years Ago **378**

14-8 In the Paleozoic Era, Complex Life Emerged and the Continents Reorganized **380**

14-9 In the Mesozoic Era, Biological Diversity Increased and Continents Reorganized **382**

14-10 Modern Mammals, Including Humans, Arose in the Cenozoic Era **386**

Chapter 15 THE GEOLOGY OF CANADA 392

15-1 Canada Can Be Divided into Six Major Geologic Provinces **394**

15-2 The Canadian Shield Contains the Oldest Rocks in the World **398**

15-3 The Continental Platform Consists of Exposed Flat-Lying Phanerozoic Rocks **400**

15-4 The Appalachian Orogen Has Been Evolving Since the End of the Precambrian **404**

15-5 The Innuitian Orogen Is Canada's Northernmost Mountain Belt **408**

15-6 The Cordilleran Orogen Is Canada's Youngest Mountain Belt **410**

15-7 Continental Shelves and Slopes Form the Margins of Canada **414**

Chapter 16 GLOBAL WARMING 420

16-1 "Global Change" Refers to Changes in Environmental Processes Affecting the Whole Earth **422**

16-2 Heat Circulation in the Atmosphere and Oceans Maintains Earth's Climate **424**

16-3 The Greenhouse Effect Is at the Heart of Earth's Climate System **426**

16-4 The Global Carbon Cycle Describes How Carbon Moves through Natural Systems **430**

16-5 Climate Modelling Improves Our Understanding of Global Change **432**

16-6 Human Activities Have Increased the Amount of Carbon Dioxide in the Atmosphere **434**

16-7 Earth's Atmospheric Temperature Has Risen by About 0.8°C in the Past 100 Years **438**

16-8 Global Warming Leads to Ocean Acidification and Warming, Melting of Glaciers, Changes in Weather, and Other Impacts **440**

16-9 Several International Efforts Are Attempting to Manage Global Warming **444**

Chapter 17 GLACIERS AND PALEOCLIMATOLOGY **450**

17-1 A Glacier Is a River of Ice **452**

17-2 As Ice Moves, It Erodes the Underlying Crust **454**

17-3 Ice Moves through the Interior of a Glacier as if on a One-Way Conveyor Belt **456**

17-4 Glacial Landforms Are Widespread and Attest to Past Episodes of Glaciation **458**

17-5 The Majority of Glaciers and Other Ice Features Are Retreating due to Global Warming **464**

17-6 The Ratio of Oxygen Isotopes in Glacial Ice and Deep-Sea Sediments Is a Proxy for Global Climate History **466**

17-7 Earth's Recent History Has Been Characterized by Cycles of Ice Ages and Interglacials **470**

17-8 During the Last Interglacial, Climate Was Warmer and Sea Level Was Higher than at Present **472**

17-9 Glacial-Interglacial Cycles Are Controlled by the Amount of Solar Radiation Reaching Earth **474**

17-10 Together, Orbital Forcing and Climate Feedbacks Produced Paleoclimates **476**

Chapter 18 MASS WASTING **482**

18-1 Mass Wasting Is the Movement of Rock and Soil Down a Slope Due to the Force of Gravity **484**

18-2 Creep, Solifluction, and Slumping Are Common Types of Mass Wasting **488**

18-3 Fast-Moving Mass-Wasting Events Tend to Be the Most Dangerous **490**

18-4 Avalanches, Lahars, and Submarine Landslides Are Special Types of Mass-Wasting Processes **492**

18-5 Several Factors Contribute to Unstable Slopes **494**

18-6 Mass-Wasting Processes Vary in Speed and Moisture Content **496**

18-7 Human Activities Are Often the Cause of Mass Wasting **498**

18-8 Research Improves Knowledge of Mass Wasting and Contributes to the Development of Mitigation Practices **502**

Chapter 19 SURFACE WATER **508**

19-1 The Hydrologic Cycle Moves Water between the Atmosphere, the Ocean, and the Crust **510**

19-2 Runoff Enters Channels that Join Other Channels to Form a Drainage System **512**

19-3 Discharge Is the Amount of Water Passing a Given Point in a Measured Period of Time **514**

19-4 Running Water Erodes Sediment **516**

19-5 There Are Three Types of Stream Channels: Straight, Meandering, and Braided **518**

19-6 Flooding Is a Natural Process in Normal Streams **520**

19-7 Streams May Develop a Graded Profile **524**

19-8 Fluvial Processes Adjust to Changes in Base Level **526**

19-9 Fluvial Sediment Builds Alluvial Fans and Deltas **528**

19-10 Water Problems Exist on a Global Scale **532**

Chapter 20 GROUND WATER **538**

20-1 Groundwater Is a Very Important Source of Fresh Water **540**

20-2 Groundwater Is Fed by Snowmelt and Rainfall in Areas of Recharge **542**

20-3 Groundwater Moves in Response to Gravity and Hydraulic Pressure **544**

20-4 Porous Media and Fractured Aquifers Hold Groundwater **548**

20-5 Groundwater Is Vulnerable to Several Sources of Pollution **550**

20-6 Common Human Activities Contaminate Groundwater **552**

20-7 Groundwater Remediation Includes Several Types of Treatment **556**

20-8 Groundwater Is Responsible for Producing Springs and Karst Topography **558**

20-9 Hydrothermal Activity and Cave Formation Are Groundwater Processes **560**

Chapter 21 DESERTS AND WIND **566**

21-1 Deserts May Be Hot or Cold, but Low Precipitation Is a Common Trait **568**

21-2 Atmospheric Moisture Circulation Determines the Location of Most Deserts **570**

21-3 Not All Deserts Lie around 30° Latitude **572**

21-4 Each Desert Has Unique Characteristics **574**

21-5 Wind Is an Important Geological Agent **576**

21-6 Sand Dunes Reflect Sediment Availability and Dominant Wind Direction **578**

21-7 Arid Landforms Are Shaped by Water **582**

21-8 Desertification Threatens All Six Inhabited Continents **584**

Chapter 22 COASTAL GEOLOGY **590**

22-1 Change Is Constantly Occurring on the Shoreline **592**

22-2 Wave Energy Is the Dominant Force Driving Natural Coastal Change **594**

22-3 Wave Refraction and Wave-Generated Currents Occur in Shallow Water **596**

22-4 Longshore Currents and Rip Currents Transport Sediment in the Surf Zone **598**

22-5 Gravity and Inertia Create Two Tides Every Day **600**

22-6 Hurricanes and Tropical Storms Cause Enormous Damage to Coastal Areas **602**

22-7 Sea-Level Rise since the Last Ice Age Has Shaped Most Coastlines and Continues to Do So **604**

22-8 Barrier Islands Migrate with Rising Sea Level **606**

22-9 Rocky Shorelines, Estuaries, and Tidal Wetlands Are Important Coastal Environments **608**

22-10 Coasts May Be Submergent or Emergent, Depositional or Erosional, or Exhibit Aspects of All Four of These Characteristics **610**

22-11 Coral Reefs Are Home to 25 Percent of All Marine Species **614**

22-12 Coastal Problems Are Growing as Populations Increase **616**

Chapter 23 MARINE GEOLOGY 622

23-1 Marine Geology Is the Study of Geologic Processes within Ocean Basins **624**

23-2 Ocean Waters Are Mixed by a Global System of Currents **626**

23-3 A Continental Shelf Is the Submerged Border of a Continent **628**

23-4 The Continental Margin Consists of the Shelf, the Slope, and the Rise **630**

23-5 Most Ocean Sediment Is Deposited on the Continental Margin **632**

23-6 Pelagic Sediment Covers the Abyssal Plains **634**

23-7 Pelagic Stratigraphy Reflects Dissolution, Dilution, and Productivity **636**

23-8 The Mid-Ocean Ridge Is the Site of Seafloor Spreading **638**

23-9 Oceanic Trenches Occur at Subduction Zones **642**

23-10 Human Impacts on the Oceans Are Global in Extent **644**

APPENDIX 651

GLOSSARY 656

INDEX 665

A Note from the Authors to the Student— How to Get the Most from Your Text

Human impacts on Earth are now global in extent. Managing these challenges requires our best thinking, based on a knowledge of how Earth works. For many of you, this course in physical geology may be your only opportunity to learn about the planet you call home. We have designed the book to help you develop a lifelong understanding of Earth and its processes. This will enhance your role as a decision maker and a steward of Earth. Let's walk through the book and examine the special features that will help make your learning more enjoyable and successful.

Chapter Opener

Every chapter opens with a critical thinking question to keep in mind as you read the chapter and a brief section called "Geology in Our Lives." These introduce you to the overall theme of the chapter and the relevance to us all. This theme is addressed again at the end of the chapter in "Let's Review Geology in Our Lives," where the ideas are revisited in terms of what you have learned in the chapter.

Chapter Contents and Learning Objectives

Each chapter is divided into sections with specific learning objectives (LOs) to help you understand what you should know or be able to accomplish after studying that section. Each LO is designed to guide and reinforce what you learn by reading the text.

Two-Page Organization

Each chapter is organized in two-page segments. This presentation allows for a full discussion of a topic in just the right amount of detail and allows you to study in succinct, digestible segments. The short, easy-to-read segments mean your reading assignments are approachable and fit in with your busy life.

Visual Learning

The book contains hundreds of extraordinary photos (many taken by the authors), illustrations, graphs, and tables, all of which were selected to enhance your learning of physical geology. But instead of being just "pretty pictures," we have added review and critical thinking questions to many of the visuals. These questions are marked with a ⑦ and ask you to really look at and think about the image being presented. By answering these questions, you will be more involved in the subject matter and have a better understanding of what you are seeing. Your instructor may use these questions as lecture starters, class discussion topics, or assign them as quiz items.

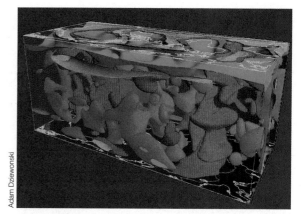

FIGURE 3.5 Earth's interior as viewed from the north. This complicated model of the mantle is based on records of the behaviour of seismic waves passing through warm (red) and cool (blue) rock. The structure of the planet is more complex than a simple sequence of uniform layers.

⑦ How will the red areas and blue areas change over time?

THE CANADIAN PRESS/Jonathan Hayward

FIGURE 1.2 On July 29, 2008, a rock slide blocked the Sea-to-Sky highway and rail line just north of Porteau Cove, B.C. This highway is the main corridor between Vancouver and Whistler and was used extensively during the 2010 Winter Olympic Games. The rock slide narrowly missed crushing a bus, only breaking some of its windows; fortunately, no one was hurt.

⑦ Explain the role that critical thinking can play in sustainably managing natural resources.

Geology in Action, Geology in Our Lives, Earth Citizenship, and Critical Thinking

The text supports four major themes to help you understand the importance and relevance of physical geology. Embedded within chapters are three types of boxes that bring these physical geology themes to life and provide examples of what you are learning. "Geology in Action" presents topics that show how Earth is in constant change and the relationships between the many forces of nature. "Geology in Our Lives" offers examples of how geology is relevant to everyone in day-to-day life and encourages you to appreciate the impact we all have on Earth. "Earth Citizenship" encourages important understanding of the role we play in the community and the world. We all have the power to influence and manage natural resources locally and globally to meet the needs of both present and future generations. An important goal of this course is to encourage you to become a geologically informed, global citizen.

GEOLOGY IN ACTION

RADIOCARBON DATING THE FIRST MAP OF NORTH AMERICA

Scholars believe that the Vinland Map (**FIGURE 13.20**) is a 15th-century map depicting Viking exploration of North America half a century before the arrival of Columbus's expedition. If genuine, the Vinland Map is one of the great documents of Western civilization; if fake, it is an amazing forgery.

According to one team of scientists, the kind of ink used on the map was made only in the 20th century. According to another, the parchment dates from the mid-1400s. Doubters argue that the parchment's age is irrelevant, while others, who believe that the analysis of the ink is flawed, have introduced evidence that many medieval documents used similar ink.

The Vinland Map shows Europe, the Mediterranean, northern Africa, Asia, and Greenland, all of which were known to travellers of the time. In the northwest Atlantic Ocean, however, it also

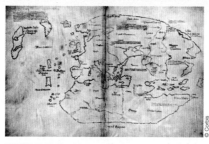

© Corbis

FIGURE 13.20 The oldest map showing North America. The map shows Europe (including Scandinavia), northern Africa, and Asia. A previously unmapped portion of North America—the region around

GEOLOGY IN OUR LIVES

BRINGING EUROPE TO A STANDSTILL

On April 15, 2010, British civil aviation authorities ordered the country's airspace closed because a cloud of ash drifting from the erupting Eyjafjallajökull (Icelandic for "island-mountain glacier") volcano in Iceland had made it too dangerous to allow air traffic (**FIGURE 6.7**). Flying through volcanic ash can abrade the cockpit windshield, making it impossible for the pilots to see out; damage communication and navigation instruments on the outside of the aircraft; and worst of all, coat the inside of jet engines with concrete-like solidified ash, causing the engine to shut down.

Within 48 hours, the ash cloud had spread across northern Europe. Three hundred and thirteen airports in England, France, Germany, and other European nations closed for about one week causing the cancellation of over 100,000 flights. This affected airline schedules around the world and resulted in some 10 million stranded air travellers at a cost estimated at $3.3 billion. As a consequence of this event, caused by what volcanologists describe as a rather modest eruption, European authorities are considering bringing all European air traffic

AP/Wide World Photos

FIGURE 6.7 The April 14, 2010 eruption of Eyjafjallajökull in Iceland brought air travel throughout Europe to a standstill for nearly a week. The cancellation of over 100,000 flights stranded some 10 million

EARTH CITIZENSHIP

OCEAN ACIDIFICATION

Acid Bath

The same process involved in dissolution, namely the reaction of CO_2 and H_2O to produce H_2CO_3 (carbonic acid), is responsible for *ocean acidification*, a growing danger from global warming that is gaining worldwide attention. Humans release CO_2 to the atmosphere—lots of it by burning coal and oil, making cement, and clearing forests to make farmland. This CO_2 traps heat and leads to global warming. (See Chapter 16, "Global Warming," for more details on the problems of global warming.) However, the ocean

absorbs approximately one-third of our CO_2 emissions, leading to another set of problems related to lowering the *pH of the water*.

The pH scale measures how acidic or *basic* a substance is. Acidic and basic are two extremes that describe chemicals, just like hot and cold are two extremes that describe temperature. Mixing acids and bases can cancel out their extreme effects, much like mixing hot and cold water can moderate water temperature. The pH ranges from 0 to 14; a pH of 7 is *neutral*, a pH less than 7 is acidic, and a pH greater than 7 is basic. Each whole number decrease of pH value is 10 times times more acidic than a pH of 5 and 100 times (10 times 10) more acidic than a pH of 6.

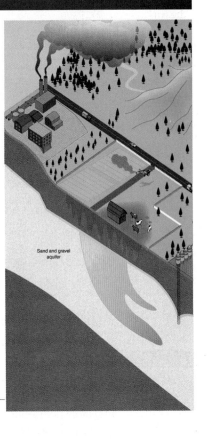

CRITICAL THINKING

YOUR DRINKING WATER

Groundwater can be contaminated by chemical products, animal and human waste, and bacteria (FIGURE 20.22). Major sources for these pollutants include storage tanks, septic systems, animal feedlots, hazardous waste disposal sites, landfills, pesticides and fertilizers, road salts, industrial chemicals, and urban runoff.

Please work with a partner and answer the following questions.

1. Make a list of the sources of contamination in Figure 20.22.
2. How many of these sources are found where you live?
3. List additional sources of groundwater contamination that are not shown in Figure 20.22.
4. The contamination plumes shown have city managers concerned. Describe a method of controlling the plumes and eventually stopping their movement.
5. Do you agree that this area has historically lacked an overarching plan for managing water? Why or why not? Describe the necessary elements of such a plan. Write a list of steps to ensure future safe drinking water in this region.
6. As mayor of the city, you know that just controlling the spread of groundwater contamination (question 4 above) is a short-term fix. What longer-term actions do you propose for sustainably managing the water resource? Write your plan in the form of a press release.
7. Participate in a class forum on managing water.

Sand and gravel aquifer

FIGURE 20.22 Potential sources of groundwater contamination.

"Critical Thinking" features provide you with the skills you need to fully understand the science and develop problem-solving ability that can be applied to many aspects of your studies and future careers. In addition to the embedded boxes, critical thinking is further developed with expanded two-page showcase activities that allow you to improve your knowledge and skills.

Expand Your Thinking

The theme of critical thinking is continued through "Expand Your Thinking" questions posed at the end of every two-page section. These questions provide a check for you to build self-confidence by making sure you have mastered the ideas in the discussion and enable you to further your understanding.

? **Expand Your Thinking**—Explain or write about the tectonic processes responsible for each of the three primary magma types.

Let's Review "Geology in Our Lives" and Study Guide

Every chapter ends with a variety of study materials for you to review and help you prepare for quizzes and exams. The "Let's Review…" section revisits the "Geology in Our Lives" feature from the chapter opening page and reviews it in terms of the knowledge you have gained after reading the chapter. The "Study Guide" provides an outline of the major sections in the chapter and summarizes the key points, while providing a review of the learning objectives for each chapter.

LET'S REVIEW "GEOLOGY IN OUR LIVES"

Now that you have finished the chapter, "Geology in Our Lives" will have taken on new meaning for you. Let us review it: The North American continent consists of two major geologic features: the craton and the surrounding orogenic belts. The craton is the stable core of the continent, consisting of the Precambrian shield and the Phanerozoic platform. The orogenic belts formed during the Phanerozoic when island arcs, oceanic rocks, and fragments of other continents accreted to the ancient continental core during plate convergence. This accretion resulted in the formation of mountain ranges.

All Canadians depend on geology. Geology influences our water quality and food supplies and provides our energy requirements and manufacturing materials. Exported minerals, metals, and hydrocarbons bring vast amounts of money into the country. Understanding our geology is a shared effort of the federal, provincial, and territorial geological surveys, along with university and industry researchers across the country. These people and their institutions serve a critically important function of providing the geologic framework needed for the discovery and management of natural resources, identification of geologic hazards, and evaluation of long-term environmental changes.

The central craton, composed of heavily metamorphosed Archean and Proterozoic crust, consists of two parts: a shield and a platform. The Canadian Shield is the largest of the geologic provinces

of Canada and the largest continuous area of Precambrian rocks on Earth. It also contains the oldest rocks on Earth. The continental platform is a stable, flat interior region covered with generally flat-lying beds of mainly Phanerozoic sedimentary rock.

There are three orogenic belts in Canada. The Appalachian Orogen, exposed in the Maritime provinces of Canada, was active mainly in the Paleozoic during the formation of Pangaea. In the far north of Canada, the Innuitian Orogen consists of rocks affected by two distinct mountain-building events, one in the Late Paleozoic and one in the Cenozoic. The third orogenic belt is the Cordilleran Orogen, consisting of the Eastern System, which includes the Rocky Mountains, the Interior System, and the Western System, which includes the mountains along the Pacific coast. These regions were formed as a result of the westward movement of the North American Plate following the breakup of Pangaea, which led to the accretion of several terranes onto the western side of North America during the Mesozoic. The western margin of Canada remains active today, whereas the eastern and northern continental margins of Canada are good examples of passive continental margins.

In the remaining chapters, we focus on modern geologic processes characterizing the major environments on Earth's surface. These chapters deal with global warming, glaciation, the hydrologic cycle, deserts, and the oceans.

STUDY GUIDE

15-1 Canada can be divided into six geologic provinces.

- Canada can be divided into six **geologic provinces** that are identified on the basis of their physiography and their geologic structure and history. They are the Canadian Shield, the **continental platform**, the Appalachian Orogen, the Innuitian Orogen, the Cordilleran Orogen, and the continental shelf and slope.
- The interior lands of Canada consist of the **craton**, which has the Precambrian rocks of the Canadian Shield at its core. These were periodically covered by shallow seas in which the Paleozoic and Mesozoic sedimentary rock of the continental platform were deposited. To the east, west, and north of the continental platform, the Appalachian, Innuitian, and Cordilleran Orogenies added new material to its edges, in a process known as continental accretion. These lands are composed of former island arcs, slivers of continental rock, and basaltic crust and marine sediments delivered by plates that converged on North America.

15-2 The Canadian Shield contains the oldest rocks in the world.

- The **Canadian Shield** is the largest of the geologic provinces and is the largest area of exposed Precambrian rocks on Earth. It contains rare rocks slightly older than 4 billion years.

- The shield contains five Archean cratons welded together during Proterozoic collisions, forming a supercontinent called **Nuna**. A later collision, called the Grenville Orogen, led to the incorporation of the Shield into a new supercontinent called **Rodinia**. This was followed by the formation of the supercontinent called **Laurentia**. Finally, the Shield was fully developed, and this represents the stable core of present-day North America.
- During the Quaternary Period, the region was glaciated by the Laurentide Ice Sheet. Erosion by the ice sheets helped to expose the Canadian Shield.

15-3 The continental platform consists of exposed flat-lying Phanerozoic rocks.

- The continental platform deposits, which are typically flat-lying, undeformed sedimentary rocks, occur in three areas. The largest area is the Interior Plains of western Canada, where the rocks form the Western Canada Sedimentary Basin. Two smaller areas are the St. Lawrence Lowland, and the area around Hudson Bay. These deposits are largely undisturbed by mountain building and are composed of flat-lying or gently dipping sedimentary strata of Paleozoic and Mesozoic age. They have buried a portion of the craton known as the "stable platform."
- The geologic history of the platform deposits reflects the rise and fall of sea level, migration of the continent through tropical regions,

KEY TERMS

Appalachian Orogen (p. 404)
Canadian Cordillera (p. 410)
Canadian Shield (p. 395)
continental platform (p. 395)
craton (p. 394)
Ellesmerian Orogeny (p. 408)

Eurekan Orogeny (p. 408)
field geology (p. 396)
geologic provinces (p. 395)
Innuitian Orogen (p. 408)
Laurentia (p. 399)
Nuna (p. 398)

Pangaea (p. 400)
Rodinia (p. 398)
Sverdrup Basin (p. 408)
terranes (p. 397)

Key Terms

A list of the key terms (boldfaced terms in the text), along with page references, provide a quick resource to highlight the most important terms in the chapter. The key terms are again defined in the glossary. This is a helpful tool to use as you prepare for quizzes and exams.

Assessing Your Knowledge, Further Research, and Online Resources

Multiple-choice questions at the end of each chapter review and assess your knowledge of the main concepts and offer you another way to check your mastery of the chapter. By answering these questions, you will know whether you have achieved the learning objectives for the chapter. These questions may serve as a self-check or be assigned by your instructor as homework.

"Further Research" assignments offer a higher level of questions or assignments to encourage additional critical thinking and ask you to bring together multiple concepts.

In "Online Resources," helpful websites are presented with ideas for investigating additional information of interest and relevance if you would like further information on a topic. These resources are useful for preparing class papers or projects.

ASSESSING YOUR KNOWLEDGE

Please answer these questions before coming to class. Identify the best answer.

1. What is a geologic province?
   a. a fault-block mountain system
   b. a change in crustal densities
   c. a condition of equilibrium
   d. a region of crust with similar physiography
   e. None of the above.

2. The rocks on the Atlantic continental shelf are
   a. highly deformed.
   b. composed largely of Paleozoic strata.
   c. composed largely of Mesozoic and Cenozoic strata.
   d. a result of the breakup of Rodinia.
   e. composed of Precambrian igneous and metamorphic rocks.

3. The regions of the Appalachian Orogen from northwest to southeast are
   a. Laurentia, Dunnage, Gander, Avalon, Meguma.
   b. Avalon, Gander, Dunnage, continental shelf.
   c. Coast Mountains, Interior System, Rocky Mountains.
   d. Laurentia, Continental Platform, Meguma, Dunnage.
   e. None of the above.

4. The craton of North America consists of the
   a. Shield and orogenic belts.
   b. orogenic belts and the continental shelf.
   c. Platform and shield.
   d. Glacial deposits and shield.
   e. Innutian Orogen and the Cordilleran Orogen.

5. Sedimentary rocks on the continental platform come from
   a. largely volcanic depositions.
   b. largely marine and coastal environments.
   c. largely deep sea environments.
   d. thick sequences of glacial sediments.
   e. None of the above.

6. The Cordilleran Orogen
   a. does not extend into Alaska.
   b. is composed of only two accreted terranes.
   c. reaches from Alaska to the tip of South America.
   d. was created by the formation of Pangaea.
   e. is Precambrian in age.

7. Canadian orogenies were largely
   a. Paleozoic in the east and north, and Mesozoic to Cenozoic in the west.
   b. related to destruction of the stable craton.
   c. the product of the formation of Eurasia.

9. The Innuitian Orogen consists of
   a. a core of Precambrian metamorphic rocks.
   b. sedimentary rocks affected by the Ellesmerian and Eurekan Orogenies.
   c. sedimentary rocks eroded from the Rocky Mountains.
   d. volcanic arc rocks.
   e. granitic intrusions.

10. The North American Plate is moving to the
    a. north.
    b. south.
    c. east.
    d. west.
    e. None of these.

11. What is an accreted terrane?
    a. rock accreted onto the edge of a plate by convergence
    b. sedimentary layers deposited in shallow interior seas
    c. faulted blocks produced by rifting
    d. new continental outlines produced by the breakup of Pangaea
    e. uplift caused by the Juan de Fuca Plate

12. The western margin of Canada is
    a. an active continental margin.
    b. a passive continental margin.
    c. the product of continental accretion.
    d. uplifted by isostatic rebound.
    e. formed by plate movement over a hotspot.

13. The geologic structure of Canada consists of
    a. a central craton and three orogenic belts.
    b. a stable orogenic belt and a craton.
    c. accreted terranes forming a craton and stable orogenic belts to the north.
    d. a central orogenic belt and a hotspot.
    e. a central carbonate platform with four Precambrian fold-and-thrust belts.

14. What is the name of the oldest supercontinent, elements of which are preserved in the Canadian Shield?
    a. Pangaea
    b. Rodinia.
    c. Nuna.
    d. Meguma.
    e. Laurentia.

15. The thickest sequences of Phanerozoic sedimentary rocks can be

FURTHER RESEARCH

1. Does orogeny happen on other planets?

2. What factors control the flooding of the craton by marine environments?

3. British Columbia has several deep valleys separated by steep mountain ranges. What geologic processes led to this type of topography?

4. What is a *jokulhlaup*?

5. Why do continents rift apart?

ONLINE RESOURCES

Explore more about the geology of Canada and North America on the following websites:

Dr. Ron Blakey, a professor of Geology at Northern Arizona University, maintains a website showing global reconstructions of paleogeography through time. Some of these are used in this chapter. Explore the best global reconstructions at:
http://jan.ucc.nau.edu/~rcb7/RCB.html

The CBC series, Geologic Journey *visits geologic locations across the country:*
www.cbc.ca/geologic/field_guide/table_of_contents.html

The book Four billion years and counting: Canada's geological heritage *presents a more detailed overview of the geology of Canada:*
www.earthsciencescanada.com/4by

Natural Resources Canada provides extensive information about the geology of Canada, as well as maps, in the Atlas of Canada:
http://atlas.nrcan.gc.ca/site/english/maps/geology.html

The U.S. Geological Survey provides "A Tapestry of Time and Terrain":
http://tapestry.usgs.gov/na-info.html

Additional animations, videos, and other online resources are available at this book's companion website:
www.wiley.com/go/fletchercanada

This companion website also has additional information about WileyPLUS and other Wiley teaching and learning resources.

CHAPTER 1
AN INTRODUCTION TO GEOLOGY

Chapter Contents and Learning Objectives (LO)

1-1 Geology is the scientific study of Earth and planetary processes and their products through time.
LO 1-1 *Describe why the science of geology is important in our daily lives.*

1-2 Critical thinking based on good observations allows us to explain the world around us.
LO 1-2 *List the five steps in the scientific method and explain each step.*

1-3 Global challenges will be solved by critical thinkers with an enduring understanding of Earth.
LO 1-3 *Describe some characteristics of the science of geology.*

1-4 The theory of plate tectonics is a product of critical thinking.
LO 1-4 *List three observations that support the theory of plate tectonics.*

1-5 Rock is a solid aggregation of minerals.
LO 1-5 *Compare and contrast the three families of rocks.*

1-6 Geologists study natural processes, known as geologic hazards, that place people and places at risk.
LO 1-6 *List geologic hazards that are capable of causing severe damage.*

1-7 The geologic time scale summarizes Earth's history.
LO 1-7 *Appreciate the extent of the geologic time scale.*

GEOLOGY IN OUR LIVES

The dramatic increase in human population over the past century has had serious consequences for the Earth we live on. Our expanding activities have affected the natural environment, and evidence suggests that we have had a significant impact on global climate. We have depleted natural resources that we depend on. In many areas, water resources are in short supply, energy costs have risen, and some mineral resources are dwindling. Communities are expanding now into areas formerly considered too dangerous to live in, and thus we are increasingly exposed to natural hazards. The sustainable economic development of the ever-growing human population is a global challenge that needs to be managed by critical thinkers with an understanding of Earth. Critical thinking is the use of reasoning to explain the world around us.

(?) What global challenges concern you as you gaze on this image of Earth?

3

1-1 Geology Is the Scientific Study of Earth and Planetary Processes and Their Products through Time

LO 1-1 *Describe why the science of geology is important in our daily lives.*

This chapter introduces you to the science of **geology**. Geology literally means "study of Earth," but geologists also study other planets in the Solar System and beyond (see Chapter 2). **Physical geology** is the study of the materials that compose Earth; the chemical, biological, and physical processes that create them; and the ways in which they are organized and distributed throughout the planet. Earth is our resource-rich, sometimes hazardous, always-changing home in space. By understanding our restless planet, we improve our ability to conserve geologic resources, avoid geologic hazards, and manage critical environments. This makes us better Earth citizens.

Physical geology is one of two broad branches of geology. The other is **historical geology**, the study of Earth's history. We introduce you to Earth's history so you will understand when and how Earth (and the Solar System) came into existence. We also explain how geologists interpret Earth's past and how they collect evidence revealing the evolution of life.

Geology is an integrative science. To figure out how Earth works, geologists must *integrate*, or combine, elements of chemistry, physics, mathematics, and biology (**FIGURE 1.1**). For example, when examining a piece of shale (a common type of rock), a geologist must think like a biologist to describe the fossils it contains, analyze its chemistry to help identify the minerals within it, and assess its material properties using principles of physics and mathematics. The information produced by this analysis enables the geologist to determine the rock's history, how it might be economically useful, and whether it represents a **sustainable resource** for humanity. A sustainable resource is any natural product used by humans that meets the needs of the present without compromising the ability of future generations to meet their own needs. In doing so, geologists rely on **critical thinking**—the use of reasoning to explain the world around us by developing hypotheses and theories that can be rigorously tested.

When the science of physical geology is distilled down to its essentials, four fundamental themes emerge. Our discussion of geology will revolve around these four themes, each identified by an icon to draw your attention to these important aspects of geology.

? CRITICAL THINKING

Critical thinking is used by everyone, every day in some capacity, and it is the most important skill that scientists and engineers possess. It is also a central component of this course. Your study of geology will provide you with training in the methods and processes of science, which focus on *problem solving,* a key element of critical thinking. You will find that the critical thinking skills learned in this course can be applied to many aspects of everyday life.

FIGURE 1.1 Geology students assist in the removal of a 49-million-year-old fish fossil from an oil shale mine in Germany.

? What does the presence of the fossilized fish tell us about the history of this rock?

Jonathan Blair/© Corbis

🌐 EARTH CITIZENSHIP

Geologists are uniquely trained to understand the potentially destructive and potentially beneficial roles of humankind in the future of Earth's environments and natural resources. This understanding is embodied in the concept of *sustainability*—meeting the needs of the present without compromising the ability of future generations to meet their own needs. Humans have the power to manage environments and ensure that natural resources remain abundant for the use of future generations. Geologists know the time scales necessary to create natural resources such as fossil fuels and groundwater, and they recognize that rates of human use of those resources greatly exceed their natural rates of renewal. It is incumbent on everyone who learns this lesson to practise lifelong Earth citizenship—for instance, through informed voting and participation in community affairs.

GEOLOGY IN OUR LIVES

Geology is relevant to everyone's day-to-day life, and we point this out at the start of each chapter. For example, the protection provided by Earth's *geomagnetic field* (magnetism that emanates from Earth's core) shields us from harmful solar radiation, making life on the planet possible; metals and energy sources are geologic products that build and power modern society; and water, essential for life, industry, and agriculture, is also a geologic product. These examples are a small sample of literally hundreds of geologic products and processes that are indispensable in our lives.

Geology enters our lives dozens of times a day, usually without our being aware of it. To appreciate this aspect of your life, it is important to become conscious of where geologic resources (water, soil, fossil fuels, metals, and others) come from, how they are extracted from the ground, and how they can be conserved. It is also a good idea to become aware of the many geologic hazards (**FIGURE 1.2**) that threaten our safety, including earthquakes, tsunamis, landslides, volcanic eruptions, flash floods, and others.

THE CANADIAN PRESS/Jonathan Hayward

FIGURE 1.2 On July 29, 2008, a rock slide blocked the Sea-to-Sky highway and rail line just north of Porteau Cove, B.C. This highway is the main corridor between Vancouver and Whistler and was used extensively during the 2010 Winter Olympic Games. The rock slide narrowly missed crushing a bus, only breaking some of its windows; fortunately, no one was hurt.

 Explain the role that critical thinking can play in sustainably managing natural resources.

GEOLOGY IN ACTION

Earth is perpetually changing. Geologic change happens because Earth is the product of interactions and relationships among rocks (the geosphere), living organisms (the biosphere), surface and groundwater (the hydrosphere), gases (the atmosphere), and the various forces affecting them (such as heat, gravity, and chemical reactions).

Examples of geology in action include *weathering* (chemical and physical reactions between rock and the environment) of Earth's **crust** (Earth's outer, rocky layer), which produces sediments and the mineral components of soil; folding and faulting of rock layers in the crust, which occurs when they are compressed or stretched; and volcanic eruptions, which are hazardous and influence climate. While the details of these processes are fascinating

and well worth learning in this course, it is also important to appreciate that Earth is the product of many complex interrelationships among all its components, many of which scientists are only beginning to understand.

In recent years, climate change, the search for new energy sources, natural hazards, and human impacts on the environment have all emerged as global issues. These concerns do not fall exclusively in the fields of chemistry or geology or biology. Managing these issues requires integrating knowledge from many fields, and the natural sciences have thus evolved to become more integrated. We highlight these integrated relationships throughout this course.

 Expand Your Thinking—Where have you encountered geology in your life today?

1-2 Critical Thinking Based on Good Observations Allows Us to Explain the World around Us

LO 1-2 *List the five steps in the scientific method and explain each step.*

The purpose of science is to find laws or theories that explain the nature of the world we observe. This is achieved by using critical thinking.

Critical thinking about the world around us begins with the use of *reasoning* (the process of drawing logical inferences) to develop a **hypothesis** (a *testable* educated guess that attempts to explain a phenomenon). A true hypothesis lends itself to being tested and to being revised or rejected if it fails the test. If it passes repeated and rigorous tests, it may become a **theory**.

A theory must be as simple and self-contained as possible and must account for all the known facts that are relevant to a subject. A theory is a hypothesis that has been aggressively tested and is generally accepted as true. Examples of successful theories include Darwin's theory of evolution, which explains how life has developed into diverse forms, and the theory of plate tectonics, which explains the organization and history of many geologic phenomena.

Critical thinking works only if carefully constructed tests are used to examine the hypothesis (or theory). A poorly designed test can lead to false conclusions. Eventually, a successful theory that, over time, has been shown always to be true can come to be considered a **natural law**, if it is useful in predicting a wide range of phenomena. The laws of thermodynamics are examples of natural laws that describe the behaviour of energy in all types of settings.

Reasoning

Typically, critical thinking about a new problem begins with *inductive reasoning*—the discovery of a general principle based on the study of specific observations. Inductive reasoning moves from specifics to generalizations by using observations to build a hypothesis that expresses a general principle.

The reasoning process goes like this: A scientist observes a phenomenon that is not explained well by current knowledge. To improve understanding of that phenomenon, the scientist uses inductive reasoning. He or she will gather specific observations and measurements of the phenomenon, begin to detect patterns and relationships, and formulate a tentative, testable hypothesis that attempts to explain these observations.

Here is an example of inductive reasoning in which specific observations lead to a hypothesis expressing a general principle:

- Observation: A robin is a bird that flies.
- Observation: A crow is a bird that flies.
- Observation: An eagle is a bird that flies.
- Hypothesis that explains all the observations: All birds fly.

This hypothesis can now be tested using *deductive reasoning*. Deductive reasoning uses generalizations to predict specific occurrences.

This kind of reasoning often is used to understand a problem that is not new. For example, our hypothesis about birds, while it has a high success rate, ultimately will be proven false by deductive reasoning, as follows:

- Hypothesis: All birds fly.
- Prediction: Robins fly (true).
- Prediction: Eagles fly (true).
- Prediction: Penguins fly (false).

Inductive reasoning can lead to deductive reasoning (**FIGURE 1.3**). That is, observations can be used to build a hypothesis (induction), and the hypothesis can be used to generate testable predictions (deduction).

Another important type of reasoning is the *principle of parsimony* (also referred to as *Occam's razor*). This principle states that among competing hypotheses, the simplest and most direct one is preferable. Parsimony assumes that an inefficient and complicated explanation of a phenomenon is not desirable if a simpler explanation would serve the same purpose. Scientists use parsimony to select the simplest explanations of nature.

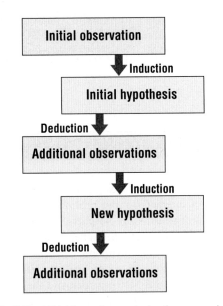

FIGURE 1.3 Critical thinking relies on inductive reasoning and deductive reasoning.

? How does inductive reasoning differ from deductive reasoning?

We all use inductive, deductive, and parsimonious reasoning several times a day. Next time you are standing on a busy street corner, see if you can identify when and how you employ all three types of reasoning in deciding when to cross the street. (Be careful!) By recognizing and naming these thinking tools, we improve the ways in which we use them to make new discoveries.

The Scientific Method

Scientists use these reasoning tools in a process known as the **scientific method**. The method has five steps (**FIGURE 1.4**).

1. **Observation.** Observe and describe a phenomenon. Observation is collecting data using your five senses.

2. **Hypothesis or model.** Using inductive reasoning and parsimony, formulate a tentative hypothesis to explain the phenomenon. In geology, a hypothesis often takes the form of a description of a natural process (known as a *model*) that predicts a particular outcome. The model can be either mathematical or verbal.

3. **Prediction.** Use the hypothesis (or model) to *predict* the existence of other phenomena or to predict the results of new observations. This is done using deductive reasoning.

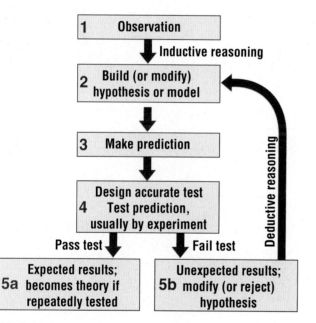

FIGURE 1.4 The scientific method.

 Why is "prediction" a significant function in the scientific method?

4. **Testing predictions.** Testing a prediction involves designing an accurate examination of the hypothesis. An *experiment* is a type of test that examines the effect on a phenomenon of changing variables in an attempt to produce a specific outcome. For instance, when studying mineral formation, a geologist may experiment with melted rock by observing how it reacts to changes in temperature and pressure. This improves understanding of how minerals form (called *crystallization*) under various conditions. However, geologists also deal with large-scale natural processes that cannot be manipulated for experimental purposes. For instance, in answering the question, "How are mountains built?" it is not possible to manipulate nature; mountains are just too large. In such cases, geologists may apply mathematics to describe the physical forces in the crust and how they cause rock to deform and break. In both cases (crystallization and mountain building), the goal is to predict the existence of a particular phenomenon and then determine if it exists.

Another approach that geologists use is to reproduce a large-scale phenomenon on a smaller scale in the controlled environment of a laboratory setting. Under such conditions, they can watch and manipulate certain aspects of the phenomenon in order to improve their understanding of its behaviour. For example, some mountains are built when the crust is compressed by opposing forces acting parallel to the surface. Geologists may use clay or other deformable materials as models of mountains to improve understanding of how rocks behave when compressed in this way. It is important to design tests that are appropriate to the hypothesis. A poorly designed experiment will not shed light on the accuracy of the hypothesis.

5. **Modifying or rejecting the hypothesis.** If predictions based on a hypothesis fail testing (Step 5b in Figure 1.4), the hypothesis may be modified. The existing hypothesis may also be rejected in favour of a new hypothesis if the results of testing show that it is inadequate to explain the phenomenon of interest. What is most important in the use of the scientific method is the predictive power of a hypothesis. That is, a well-tested hypothesis is a tool that can be used to predict the behaviour or existence of unseen natural phenomena.

Once a hypothesis has passed vigorous testing by a community of skeptical scientists (Step 5a), it may become a theory if it has broad implications and applies to a wide range of natural processes. It is often said that "theories can never be proved, only disproved." There is always the possibility that a new observation or new experimental data will conflict with a long-standing theory.

 Expand Your Thinking—What might the outcome be if a hypothesis is not testable?

1-3 Global Challenges Will Be Solved by Critical Thinkers with an Enduring Understanding of Earth

LO *1-3 Describe some characteristics of the science of geology.*

The goal of this geology course is to embed an enduring understanding of Earth in your thinking. In fact, on the opening page of every chapter is a short paragraph called "Geology in Our Lives" that describes an overarching concept expressing the enduring understanding of the chapter. Years from now when you recall your geology class, we hope that these concepts are what you remember most vividly. Keep "Geology in Our Lives" in mind as you read the chapter. For instance, here in Chapter 1, you have been introduced to *critical thinking* (Section 1-2). Critical thinking is how scientists answer questions about nature. The global challenges that come with living on a crowded planet will be managed most effectively by critical thinkers with an understanding of geology—like you.

Retaining these important ideas throughout your life will help you promote sustainable living. For instance, it is important for you to limit your consumption of goods; to recycle metals, paper, and glass; to preserve clean water; to avoid geologic hazards. Adults who are trained in geology make better decisions about the use of precious natural resources and are more knowledgeable about managing geologic hazards. Global challenges faced by a population with dwindling natural resources, dangerous geologic hazards, and growing pollution will be solved by critical thinkers with an understanding of our planet. This section presents six important concepts to begin your education as a geologist.

The Study of Geology Encompasses a Vast Range of Time and Space

Natural phenomena studied by geologists cover an immense span of time and space. They range from the megascopic (the length of the Solar System is measured in trillions of kilometres) to the microscopic (the bonding of atoms occupies an infinitesimally small space, on the order of 0.000000000001 km, which is the same as 1 nanometre). These phenomena can exist side by side (**FIGURE 1.5**). Massive planets are constructed from the tiniest molecules. Earth has existed for about 4.6 billion years, during which time there have been long periods of slow and gradual change punctuated by short, violent convulsions, such as earthquakes, volcanic eruptions, floods, and landslides. For example, a major earthquake can destroy communities in minutes.

Earth Materials Are Recycled over Time

The main subdivisions of Earth (crust, mantle, and core) started to form early in Earth's history. Since then, Earth materials have been continuously affected by chemical and physical processes that are part of a never-ending recycling machine that incessantly destroys and renews these materials. Although rocks may seem indestructible, in fact they are perpetually degraded by chemicals in the air and water, baked and contorted by forces within the crust, and melted if they are conveyed to Earth's interior. But molten rock will solidify once it returns to the crust, whereupon the process of recycling rock starts over again.

These are all steps in the great recycling machine known as the **rock cycle**. The rock cycle simultaneously destroys and renews Earth's crust. This cycle makes many of the natural resources that are the foundation of civilization: soil, water, petroleum, coal, natural gas, metals, and others. These resources form on geologic time scales of

Photo courtesy of Dr. James MacEachern, Simon Fraser University

FIGURE 1.5 In this photo, the energy from the sun streams onto individual sand grains composed of bonded oxygen and silicon atoms within the layered rock of this hoodoo in the Alberta Badlands. Sediment that formed these rocks accumulated in an ancient sea many millions of years ago during the Cretaceous Period. The rock has since been sculpted by streams carrying glacial outwash thousands of years ago, and today is being eroded by rain, wind, and freeze-thaw cycles.

 What is the evidence in this photo that the rocks are being eroded?

thousands to millions of years. Humans must carefully (sustainably) manage these resources because we use them at much faster rates than they are renewed.

Plate Tectonics Controls the Geology of Earth

Earth's crust is in constant motion. Right now the ground under your feet is moving at about the same pace as your fingernails grow. This phenomenon was not understood by geologists until only a few decades ago. The hypothesis of **plate tectonics** was first proposed in the 1960s, and it has now grown into a theory. According to this theory, Earth's crust and the uppermost part of the mantle is organized into a dozen or so pieces like those of a jigsaw puzzle; these pieces are called *plates*. At their edges, some plates are recycled into Earth's interior, some collide and crumple, and others separate, allowing molten rock to rise from below. This behaviour is called *tectonics*, and it has far-reaching implications. As plates move, they change the way Earth looks. Mountain ranges rise when plates collide, only to be worn down by weathering and *erosion* (carrying away bits of weathered rock by running water, gravity, and wind). Ocean basins open and close as continents drift apart and converge again. Nearly every aspect of Earth's surface is related to how plates interact and change over time. The theory of plate tectonics, which is invoked to help explain the content of every chapter in this text, is described further in Section 1-4 and it is the subject of Chapter 3.

The Earth System Is the Product of Interactions among the Solid Earth, Water, the Atmosphere, and Living Organisms

Earth is organized into overlapping and complex systems that influence and react to one another at a variety of scales. *Earth systems* consist of interdependent materials, such as rocks and sediments (*geosphere*), living organisms (*biosphere*), gases (*atmosphere*), and water (*hydrosphere*) that interact with each other by way of physical and chemical processes (**FIGURE 1.6**). In a broad sense, these interactions occur because nuclear energy (heat from the Sun and from Earth's interior) and gravitational energy are at work mixing the air, water, and solid earth.

For example, the atmosphere contains carbon dioxide (CO_2), some of which was erupted from volcanoes. Carbon dioxide is used by plants in photosynthesis, which yields oxygen (O_2) back to the atmosphere, allowing us and other animals to survive. Plants help to stabilize hillsides and make the organic part of soil when they decay. But soil is also composed of minerals that initially formed in the Earth's interior. The heat in Earth's interior helps to drive the movement of the immense plates that interact at their edges, generating volcanoes, earthquakes, and mountain ranges. When plates collide, one plate may recycle back into the interior of the planet. These recycled plates may melt and feed volcanoes with the molten material they erupt, as well as gases like CO_2. Plants, tectonic plates, and humans are thus inextricably linked as part of the large Earth system.

Courtesy of Dr. Dan Gibson, Simon Fraser University

FIGURE 1.6 This photo taken near Harrison Lake, British Columbia, shows a geologic system that consists of interactions among plants, sediments, rocks, gases, and moisture.

 What is the connection between the rocks in the foreground and the river?

Earth Continuously Changes

We live on an ancient and restless landscape that is changing under our feet. All forms of life have evolved partially in response to this geologic change over time. Today's Earth is the result of gradual, rapid, and catastrophic changes whose products (including minerals, rocks, topography, ecosystems, and others) have been accumulating for 4.6 billion years. Hence, our planet looked very different in the past than it does today, and it will look very different in the future. Humans, with our technological abilities, may cause changes that rival, and often exceed, the pace and extent of natural processes that change Earth. We extract materials from the crust, alter the landscape, and pollute the air and water. It is vital that we understand the effect that we have on the ever-changing Earth.

Rocks and Sediments Are Pages in the Book of Earth's History

By "reading" the evidence of past events in Earth's crust, geologists piece together the history of our restless planet. This evidence shows that Earth is very old, evolution is responsible for life's incredible diversity, continuous change is a characteristic of Earth systems, and geologic processes operate on an immense stage of time and space.

 Expand Your Thinking—Why is understanding the geologic systems significant to managing natural resources sustainably?

1-4 The Theory of Plate Tectonics Is a Product of Critical Thinking

LO *1-4 List three observations that support the theory of plate tectonics.*

No one person is responsible for developing the theory of plate tectonics. It is a still-evolving product of thousands of critical thinkers trying to make sense of the patterns of nature that surround us. Why are there continents? Why are there ocean basins and islands? Why are earthquakes common in some areas and rare in others? Why are most volcanoes found in chains? Many scientists have worked on these problems, but the contribution of a handful of researchers stand out as the most important in developing the early components of the theory of plate tectonics.

Alfred Wegener and Continental Drift

Like many observers of nature before him, *Alfred Wegener*, an early-20th-century German meteorologist, noticed the natural "fit" of the continents. Conducting his own research as well as drawing on that of others, Wegener employed inductive reasoning to propose the hypothesis known as *continental drift*. He cited observations that the edges of the continents seem to fit together, that continents separated by wide oceans have similar topographic and geologic features along opposite shores, and that fossil evidence suggests that continents were formerly connected. Based on these observations, Wegener proposed that continents move, or "drift," across Earth's surface, but he was unable to describe how this movement occurs. He hypothesized that Europe and North America, and Africa and South America were formerly joined together in a "supercontinent" that he named **Pangaea**.

Evidence for continental drift is provided by the observations that similar deposits and ages of rock span Europe and North America (**FIGURE 1.7**) and glacial deposits from a single ice source are found in the southern portions of South America, Africa, India, Australia, and Antarctica. Wegener also recognized that the presence of certain distinctive fossils of extinct plants and animals shared by different continents now separated by vast oceans is best explained if those continents were formerly joined together.

For example, the freshwater fossil reptile *Mesosaurus* is found only in the southern regions of Africa and South America, suggesting that those areas were once connected. Fossils of *Cynognathus*, a land-dwelling reptile, are found only in central Africa and central South America. Unless it could walk with ease between those lands, there is no parsimonious explanation for the far-flung fossil locations. Fossils of the fern *Glossopteris* are found spread across all the southern continents but not on any lands in the Northern Hemisphere, suggesting that a great southern supercontinent once existed.

However, despite strong supportive evidence, Wegener's continental drift hypothesis failed to gain acceptance by most scientists. Continental movement seemed impossible, and Wegener could not suggest a process by which such movement could be accomplished. Continents simply cannot plow through Earth's crust like ships at sea, and the scientific community refused to believe Wegener's conclusions unless he produced a plausible physical mechanism for moving continents. But this problem would be resolved (in the true style of

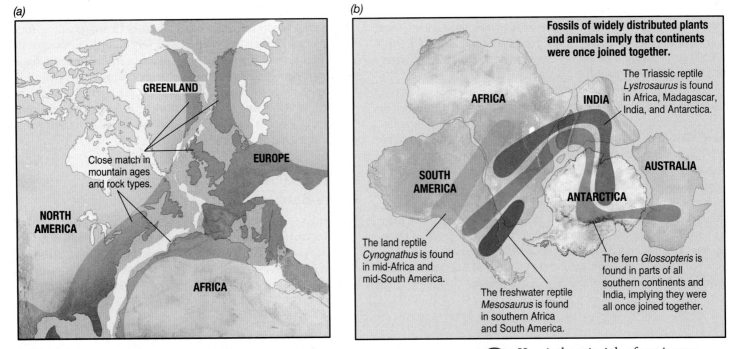

FIGURE 1.7 Wegener's evidence included (a) the "fit" between continental outlines, the similar nature of their rocks, and (b) the land-bound fossils they share.

(?) How is the principle of parsimony applied in analyzing this evidence?

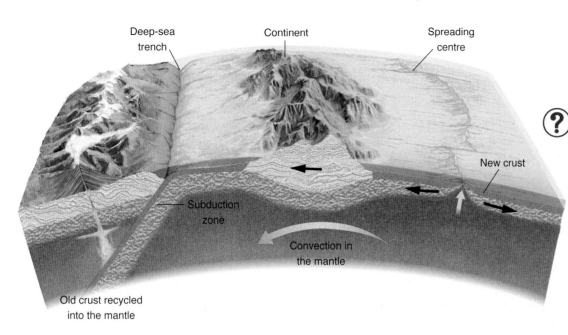

Deep-sea trench

Continent

Spreading centre

New crust

Subduction zone

Convection in the mantle

Old crust recycled into the mantle

? What do you think will happen when the continent arrives at the subduction zone?

FIGURE 1.8 Spreading centres are long, relatively narrow openings in the crust where molten rock rises from below to cool and form new sea floor. Like a conveyor belt, the sea floor moves away from these areas toward subduction zones where it is recycled into the mantle.

critical thinking) as the hypothesis was modified on the basis of new observations by later researchers.

Seafloor Spreading

Wegener's continental drift hypothesis languished for decades with few proponents. Basic laws of physics suggested that continents cannot "drift" across Earth's surface. After World War II, however, new observations led to a modification of the hypothesis and the emergence of **seafloor spreading** as an explanation of continental drift.

During the war, Captain *Harry Hess,* a geologist from Princeton University, commanded a ship that mapped the seafloor of the Pacific Ocean. He was able to show that the ocean floor is not flat and featureless, as was commonly thought. At the time, scientists assumed that ocean basins were unchanging and the ocean floor was covered in thick layers of sediment that buried the topography beneath it. Instead, Hess found sharp, rugged ridges on the sea floor. He inferred from this evidence that the sediment cover was thin, and thus the sea floor must be young. But how can the crust on an old planet be young?

In 1962, aided by the North Atlantic seafloor maps produced by Bruce Heezen and Marie Tharpe, Hess published his answer in a paper titled "History of Ocean Basins." Using deductive reasoning to establish a robust, predictive hypothesis that improved on the idea of continental drift, he proposed the important concept of seafloor spreading. Hess predicted that new sea floor is formed at places called *spreading centres* where Earth's crust tears open and molten rock from below rises and cools to form new crust. Spreading centres, he said, are marked by submerged mountainous ridges because they are regions of high heat flow and the crust is hot and buoyant. At other locations where ocean basins have long and narrow *deep-sea trenches,* the sea floor is recycled into Earth's interior at **subduction zones**. Hence, like a conveyor belt, new oceanic crust is created at a spreading centre and moves across Earth's surface until it reaches a subduction zone, where it is recycled back

into Earth's interior (**FIGURE 1.8**). Continents, rather than plowing through Earth's crust, ride embedded within these moving plates of rock.

Seafloor spreading differs from continental drift in that the movement of continents is driven by *the motion of the sea floor.* Unlike the continental drift hypothesis, Hess's concept of seafloor spreading proposed an explanation for seafloor motion, namely that plates ride on currents of hot rock circulating in Earth's interior. These currents are the product of **convection**, the transfer of heat by the movement of rock from areas of high heat to areas of low heat. In the thickest layer of Earth's interior, the **mantle** (which lies directly beneath the crust), convection is thought to occur where great volumes of hot rock cyclically migrate upward and downward as they heat and cool.

Hotspots

In 1963, the Canadian geophysicist *J. Tuzo Wilson* seized on the significance of Hess's ideas. He used deductive reasoning based on the seafloor-spreading hypothesis to predict the existence of **hotspots** that would explain mid-plate chains of volcanoes like the Hawaiian Islands (**FIGURE 1.9**). Wilson deduced that a single, stationary source of magma in the mantle periodically erupts onto the sea floor, forming an active volcano that emerges above the surface of the ocean, creating an island. As the plate shifts a volcano away from the hotspot, the volcano is cut off from the magma source and becomes extinct. Within a few hundred thousand years, a new volcano erupts onto the sea floor above the hot spot, starting the cycle again.

Since no new land is formed on the extinct volcano, it is eroded by weathering processes that include rainfall, streams, landslides, and ocean waves. As the island travels away from the hotspot, the plate beneath it cools and thins. This, along with persistent erosion of the land surface, eventually causes the island to subside below the sea and form a topographic feature called a *seamount.* This mechanism is responsible for some chains of volcanic islands found around the world.

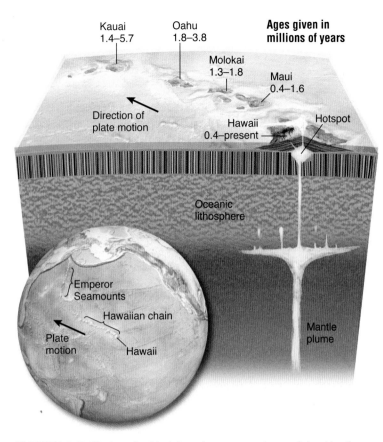

Kauai
1.4–5.7

Oahu
1.8–3.8

**Ages given in
millions of years**

Molokai
1.3–1.8

Maui
0.4–1.6

Direction of
plate motion

Hawaii
0.4–present

Hotspot

Oceanic
lithosphere

Emperor
Seamounts

Hawaiian chain

Plate
motion

Hawaii

Mantle
plume

FIGURE 1.9 Chains of mid-plate volcanoes can be explained by the existence of a single stationary source of magma, a "hotspot" in the mantle.

 What characteristic of the island ages supports hotspot theory?

The hotspot hypothesis predicts that islands farther from a hotspot are older than those closer to it, a prediction that has been supported with geologic data numerous times. For instance, the Hawaiian island of Kauai (the farthest of the main islands from the Hawaiian hotspot) is the most highly weathered and the oldest (approximately 1.4 to 5.7 million years).

Pangaea

Modern research has verified Wegener's hypothesis regarding the former existence of a supercontinent. About 255 million years ago, plates moved in such a way that all the continents merged together and formed a single supercontinent called Pangaea (from the ancient Greek words for "all lands"). Since then, plates have shifted so that Eurasia, Antarctica, Africa, South America, North America,

and Australia have broken away into separate pieces, creating the continental pattern we know today (**FIGURE 1.10**).

Rift Valley Topography

Because spreading centres experience high heat flow from Earth's interior, the crust is warm and sits higher than the surrounding sea floor. This forms a ridge called a **mid-ocean ridge** that runs for thousands of kilometres around the globe. The top of this ridge has a narrow valley, which is where the sea floor is *rifted* (opened up) and molten rock from the interior rises to cool at the surface and form new crust.

Critical thinking has helped improve our understanding of why **rift valleys** in mid-ocean ridges have different shapes. Some rift valleys are wide with numerous volcanic features and high topography with steep walls of rock over 1,000 m tall. Other rift valleys are narrow and shallow, lacking the high walls, with a gently rounded central floor and relatively smooth relief. What process causes these differences?

Geologists know these rift valleys are the site of new crust fed by molten rock from the mantle. It is reasonable to expect the rate at which magma is supplied from below to vary along a ridge and over time at any location. Hence, they hypothesize that *variations in magma supply rates* could account for differences in rift valley shape. Using deductive reasoning, researchers realized that variations in heat flow from Earth's interior could cause such differences. Mid-ocean ridges experiencing high heat flow would have warm crust forming a high, thermally buoyant central valley where spreading is rapid, the valley is narrow, and the relief is smooth. A region experiencing low heat flow where magma is supplied at a slow rate would allow the sea floor to cool, causing the crest of the ridge to subside. This tends to form deep, wide, rugged rift valleys with floors that are not arched upward. Heat flow and magma supply are related to the rate at which seafloor spreading takes place. Hence, geologists predicted that fast-spreading centres would be characterized by smooth, narrow valleys and slow-spreading centres would be characterized by rugged, wide valleys.

Indeed, this hypothesis has been supported by direct observations. Fast-spreading mid-ocean ridges move apart at 10 cm to 20 cm per year. At one well-studied spreading centre, the *East Pacific Rise,* the sea floor moves away from the rift valley in either direction as much as 14 cm each year. This place is characterized by a relatively smooth topography at the ridge crest with small, narrow troughs in the rift valley only 50 m to 300 m wide and 8 m to 15 m deep in some locations. By contrast, slow-spreading mid-ocean ridges, like the *Mid-Atlantic Ridge,* which is moving at about 2 cm to 4 cm per year, are broad and have a deep central rift valley. Some rifts are 10 km to 20 km wide and 3 km deep. Broken bedrock is widespread and creates rough topography.

 Expand Your Thinking—Drawing geologic features can help us identify what we do not understand. Draw a simple cross-section of the crust and upper mantle showing a spreading centre, a subduction zone, and mantle convection. Label as many aspects of the diagram as possible.

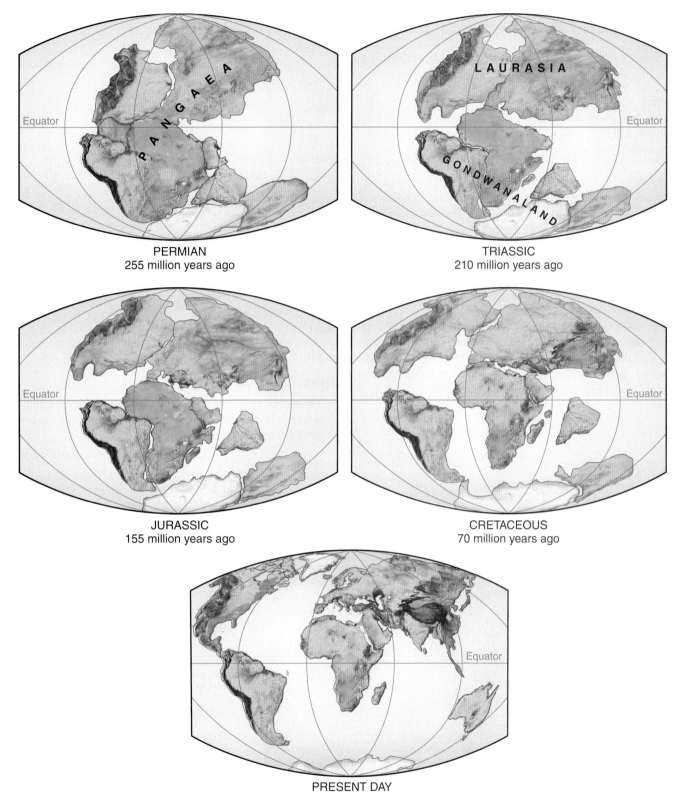

FIGURE 1.10 About 255 million years ago, continents were joined as a single land mass called Pangaea. Since that time, they have separated into the present geography we know today. Each time slice in the figure bears the name of its corresponding geologic period. The geologic time scale is discussed in Section 1-7.

 In which direction is the North American continent moving?

1-5 Rock Is a Solid Aggregation of Minerals

LO 1-5 *Compare and contrast the three families of rocks.*

It is commonly said that geology is "the study of rocks." While geology involves much more than the study of rocks, the question "What is a rock?" is a fundamental one that we need to answer before proceeding with our discussion of Earth.

Rock is a solid aggregation of minerals. *Minerals* (the subject of Chapter 4) are solid chemical compounds, or sometimes single elements, that can be seen in rocks as crystals or grains. If you hold loose sand in your hand, you are not holding a rock. If you cement those sand grains together, you have made a rock called *sandstone* (studied in Chapter 8). Look closely at the speckled surface of a rock sample (**FIGURE 1.11**). Each speck of white, black, grey, brown, red, or other colour is a separate mineral.

As you will learn in later chapters, there are many types of minerals. Hence, there are also many kinds of rock. Despite this diversity, all rocks can be grouped into three distinct families depending on how they were formed. These three families are **igneous rock**, **sedimentary rock**, and **metamorphic rock**. These rock families do not exist independently of one another. Rather, they are often found together and are recycled into new forms through the rock cycle.

Courtesy of Chip Fletcher

FIGURE 1.11 Rock is a solid aggregation of minerals. Each speck of black, pink, grey, and white is a separate mineral in this sample of *granite*.

Igneous Rock

Igneous rock (Chapter 5) is produced by the development, movement, and then cooling of molten rock, called *magma*. Igneous rock is formed in two ways (**FIGURE 1.12**):

1. Cooled magma within Earth's crust forms *intrusive* igneous rock.

2. Cooled *lava*—magma located on Earth's surface (after erupting from a volcano)—forms *extrusive* igneous rock.

Sedimentary Rock

Fragments of mineral and rock (known as *sediment*) are constantly dissolving or breaking off the crust and being moved (by gravity, wind, and water) from one place to another on Earth's surface. The process of degrading and breaking the crust through chemical reactions or mechanical processes is called *weathering* (Chapter 7). Weathering is usually followed by *erosion*, which is the process of moving sediment from its place of origin, where it was formed by weathering, to its final resting place, where it is *deposited*. You can see sediment as dust blowing in the wind or as muddy silt carried along in a stream or river (Chapter 19).

Basalt

Volcanic terrain

Rapid-cooling, fine-grained igneous rock

Crust

Slow-cooling, coarse-grained igneous rock

Mantle lithosphere

Partial melting

Asthenosphere

Photos courtesy of Chip Fletcher

Gabbro

FIGURE 1.12 Molten rock, or magma, forms in Earth's interior. Magma in contact with the ocean or atmosphere is lava and forms extrusive igneous rock, which is fine-grained (for example, basalt). Cooled magma in Earth's interior forms intrusive igneous rock, which is coarser-grained (for example, gabbro).

 Which type of rock cools faster, intrusive or extrusive?

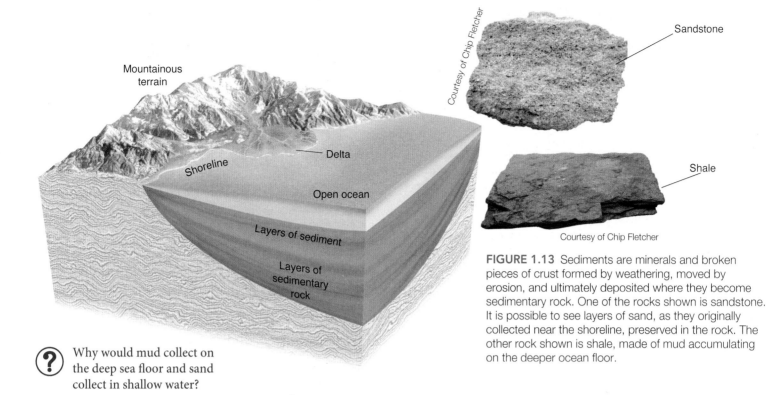

Courtesy of Chip Fletcher

FIGURE 1.13 Sediments are minerals and broken pieces of crust formed by weathering, moved by erosion, and ultimately deposited where they become sedimentary rock. One of the rocks shown is sandstone. It is possible to see layers of sand, as they originally collected near the shoreline, preserved in the rock. The other rock shown is shale, made of mud accumulating on the deeper ocean floor.

(?) Why would mud collect on the deep sea floor and sand collect in shallow water?

Over time, huge quantities of eroded sediment are carried by wind and water from the land into the oceans, where they are eventually deposited. Sedimentary rock is made of sediment that has collected and become solidified into rock by *compaction* or by *cementation* of grains (**FIGURE 1.13**). Sedimentary rock makes up a significant proportion of the exposed rock on Earth's surface.

Metamorphic Rock

The third family of rock is metamorphic rock (Chapter 9; **FIGURE 1.14**). It forms beneath Earth's surface in several ways. When rock is buried at great depths—for example, beneath mountain ranges—it is subjected to heat and pressure. This causes minerals to change and allows for new minerals to grow out of elements that formed the old ones. *Metamorphism* (change to a rock caused by heat and pressure) also occurs when magma bakes the rock surrounding it.

Both the pressure of deep burial and exposure to the heat of Earth's interior produce new conditions that lead to the growth (i.e., *recrystallization*) of *metamorphic minerals*. Metamorphic rock is made of new *recrystallized* minerals that form under relatively high temperature and pressure conditions. It is very often folded or contorted by the pressure that forms it.

FIGURE 1.14 Contorted rock is a sign of metamorphism that may have occurred in the core of a mountain range. The sample shown is gneiss.

(?) What is the significance of the contortions?

 Expand Your Thinking—Why is sedimentary rock most common on Earth's surface and igneous rock most common below the surface?

1-6 Geologists Study Natural Processes, Known as Geologic Hazards, that Place People and Places at Risk

LO 1-6 *List geologic hazards that are capable of causing severe damage.*

Geologists study dangerous natural processes, such as landslides, floods, erosion, volcanic eruptions, tsunamis, earthquakes, hurricanes, and other **geologic hazards**. In most cases, when they occur in remote areas, these processes do not affect humans. However, some geologic hazards do impact humans. Geologists monitor geologic hazards to help communities avoid casualties and reduce damage. Hazard avoidance is one type of *mitigation;* mitigation is achieved by reducing community vulnerability to a geologic hazard.

Floods

Floods (Chapter 19) are part of the natural life cycle of streams. Stream channels and their *flood plains* depend on regular flooding to maintain healthy ecosystems and to clear accumulated sediment. However, in Canada, flooding has caused over $5 billion in damage over the last century, including the extensive flooding of the Souris and Red Rivers in Manitoba and North Dakota in 2011. One reason for these losses is our tendency to build houses, farms, and even entire towns in areas that are vulnerable to flooding. Scientists think that flooding will increase due to global warming as rainfall in some areas becomes more intense.

Hurricanes and Tornadoes

Hurricanes (Chapter 22) are tropical cyclones with winds that exceed 118 km per hour. High winds are a primary cause of injuries and property damage. However, flooding resulting from storm waves, high water levels, and torrential rains also causes property damage and loss of life. Hurricanes that affect the southern and eastern parts of North America are generated in the tropical regions of the North Atlantic Ocean. The most destructive storm in the history of North America was Hurricane Katrina in 2005, which affected the Mississippi and Louisiana coasts, flooding New Orleans, killing some 1,200 people, and leaving hundreds of thousands of people homeless. North America, and especially the region termed "Tornado Alley," from the central United States to the Prairies of Canada have the most tornadoes of any part of the world. Tornadoes are a type of cyclone, and a devastating one demolished part of Goderich, Ontario, in August 2011, with winds up to 300 km per hour resulting in one death, numerous casualties, and damage in excess of $100 million.

Volcanic Eruptions

In 1991, thousands of lives were protected when government geologists evacuated towns and villages on the slopes of Mt. Pinatubo in the Philippines. Soon afterward, the volcano exploded in a massive eruption that devastated vast tracts of land in the region. There are several types of volcanoes (Chapter 6), and understanding how and why they differ is a key to understanding the threat they pose to local populations (FIGURE 1.15).

Earthquakes

When rock in the crust breaks, it may cause an *earthquake* (Chapter 12). Shock waves radiating outward from the fracture shake buildings and jolt bridges, often causing massive damage. In Haiti, more than 230,000 people died and over 1 million were left homeless when an earthquake struck on January 12, 2010. These death tolls are due not only to poorly constructed buildings that collapse and trap people but also to landslides that bury cities and towns, fires caused by ruptured gas lines and downed electrical lines, and lack of adequate rescue and medical facilities.

Tsunamis

The most damaging and deadly *tsunami* (a series of huge waves caused by sudden movement of the sea floor) in recorded history occurred on December 26, 2004, when coastlines were inundated without warning around the Indian Ocean. Known as the *Sumatra-Andaman tsunami,* it was caused by an undersea earthquake in Indonesia (the second-largest earthquake ever recorded) and resulted in the deaths of more than 230,000 people. On some shorelines, the tsunami produced water levels over 30 m high, and many coasts were swept clear of all human habitation. On March 11, 2011, a tsunami generated by a large earthquake off the east coast of Japan resulted in the loss of over 15,000 lives, caused an estimated $235 billion of damage, and led to explosions and radioactive leakage from the Fukushima nuclear power plant.

Mass Wasting

Mass wasting (Chapter 18) is one of the processes that forms valleys; it is the movement of soil and surface materials down a slope under the force of gravity. *Landslides* and other types of mass wasting

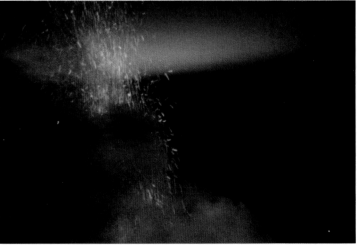

FIGURE 1.15 Mount Stromboli erupts in Italy.

 What drives the volcanic eruption into the air?

Courtesy of Dr. Glyn Williams-Jones, Simon Fraser University

are one way in which land erodes. As mentioned earlier, erosion is important in recycling sediments and nutrients through the environment. However, excessive mass wasting caused by heavy rains, earthquakes, or other triggering events constitute a major geologic hazard. Mass wasting occurs in all parts of Canada and causes about $300 million in damage each year. The worst landslide in Canada occurred in April 1903, when the mining town of Frank, Alberta, was buried and 75 lives were lost.

As well as trying to understand the Earth processes that are potentially hazardous, geologists also contribute to our understanding of natural resources that are vital for our survival. See the "Earth Citizenship" box below for examples of this problem.

Expand Your Thinking—Describe the geologic hazards and natural resources that are most common where you live.

EARTH CITIZENSHIP

NATURAL RESOURCES ARE NOT ENDLESS AND MUST BE MANAGED

Although Earth is large, its **natural resources**, which have been formed through normal geological processes, are not endless. Natural resources are materials that occur in nature and are essential or useful to humans, such as water, air, building stone, topsoil, and minerals. To ensure that future generations will have the resources they need, geologists work to understand how natural resources are formed, their abundance and distribution, and how they can be conserved.

Water
Fresh water is relatively scarce, representing only about 3 percent of all water on Earth. Of this total, almost 70 percent is unusable as it is stored in icecaps and glaciers. Although water is a renewable resource, it takes time for rain and snow to replenish the water we pump out of the ground (Chapter 20). Moreover, water-rich rock layers can be damaged by over-pumping so that they are no longer useful. Agriculture and industry account for over 90 percent of total water withdrawals, although the exact proportions vary from place to place. While the daily drinking water needs of humans are relatively small—about 4 L per person—the water required to produce the food a person consumes each day is much higher, ranging between 2,000 and 5,000 L.

Soil
Soil (Chapter 7) is the layer of minerals and organic matter that cover the land. Soil ranges in thickness from a few centimetres to several metres or more. Its main components are bits of rock and minerals, organic matter, water, and air. Researchers report that 65 percent of the soil on Earth is degraded by erosion, *desertification* (Chapter 21), and *salinization* (high salt content). Soil quality must be protected because soil provides us with food and sustains natural ecosystems. Soil is a renewable resource, but only if it is treated with natural fertilizer (such as manure) several times each year and carefully protected from erosion. Under natural conditions, it may take up to 1,000 years to form only 2.5 cm of topsoil, and this can be swept away in a single rainstorm if it is carelessly exposed to erosion. Worldwide, soil is being lost many times faster than it can be replaced naturally, so soil conservation is of vital importance.

Fossil Fuels
One of our major means of generating electricity, by burning fossil fuels (Chapter 10) such as coal, oil, and natural gas (**FIGURE 1.16**), pollutes air and water. In addition, carbon dioxide released during burning has resulted in rising carbon dioxide in the atmosphere and related changes to Earth's climate (Chapter 16).

Since the 1970s, far more oil has been consumed than has been discovered. Humans now burn 1,000 barrels of oil per second. Yet we are finding only 1 barrel of new oil for every 4 barrels that we consume. Many geologists believe that all the significant oil fields have already been discovered, and they predict that a permanent downturn in oil production will occur in less than 20 years. Oil prices are very sensitive to small changes in supply, so when the rate of production begins to decline, the combination of growing demand and falling production will bring about a significant increase in oil prices.

Minerals
Every year we use enormous quantities of rocks and minerals, including stone, gravel and sand, clays, limestone for cement, salt, and various metals, such as iron, zinc, aluminum, copper, silver, and many others. These mineral resources are used to build highways, cars, computers, CDs, buildings, and so on. But mineral resources are finite and increased demand can make them unaffordable. Geologists are responsible for finding and assessing the precious minerals we depend on and for identifying alternatives when necessary.

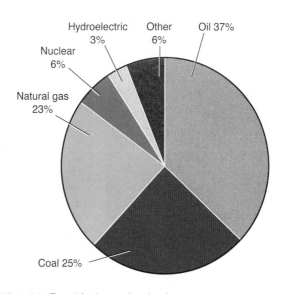

FIGURE 1.16 Fossil fuels are the dominant energy source, accounting for about 85 percent of primary energy consumption.

❓ CRITICAL THINKING

APPLYING GEOLOGY IN OUR LIVES

Throughout this chapter, we have emphasized the fact that geology is relevant in our lives. **FIGURE 1.17** depicts some common natural resources and geologic hazards. Work with a partner to answer the following:

1. Classify each image as either "Geologic Hazard" or "Natural Resource" or both.
2. What criteria would you use to rank the value of each natural resource in our lives?
3. Describe how answering question 2 led you to think of new ways to value resources.
4. For the hazards, describe how humans might decrease the damage each hazard causes in our communities.
5. Write a short description of how each image depicts the role of geology in our lives.
6. What additional geologic hazards and natural resources can you think of that are not depicted here?

Carson Ganci/Getty Images, Inc.

Louie Psihoyos/© Corbis

Construction Photography/Corbis Images

© iStockphoto.com/Ermin Gutenberger

AP/Wide World Photos

Lester Lefkowitz/Iconica/Getty Images, Inc.

© iStockphoto.com/Dragan Trifunovic

American School/The Bridgeman Art Library/Getty Images, Inc.

Science Source

Michael Hall/Photonica/Getty Images, Inc.

David Greedy/Getty Images, Inc.

© iStockphoto.com/ Sawayasu Tsuji

Ian Shive/Getty Images, Inc.

Carsten Peter/National Geographic Creative/Getty Images, Inc.

FIGURE 1.17 The role of geology in our lives.

1-7 The Geologic Time Scale Summarizes Earth's History

LO 1-7 *Appreciate the extent of the geologic time scale.*

Everyone has heard of dinosaurs, but *Tyrannosaurus rex* and its cousins are only one group of animals among hundreds of thousands (and millions of insects) that have come and gone in the course of Earth's 4.6 billion-year-history (**FIGURE 1.18**). Until the late 1700s, Earth was believed to be only a few thousand years old, and all those species were thought to have been created simultaneously. The fact that Earth is much, much older was recognized by early geologist *James Hutton* (1726–1797), among others, who based his claim on a close analysis of fossils and rock strata.

Based on his analysis, Hutton proposed one of the most important concepts in geology: the **principle of uniformitarianism**. This principle states that Earth is very old, that natural processes have been essentially uniform through time, and that the study of modern geologic processes is useful in understanding past geologic events.

Uniformitarianism is often summarized in the statement, "The present is a key to the past." In Hutton's day, this idea was a direct rejection of the belief that Earth's landscape and history were the product of the great flood described in the Bible. Today, we hold uniformitarianism to be true and know that natural disasters such as earthquakes, asteroid impacts, volcanic eruptions, and floods are normal Earth processes.

The Geologic Time Scale

Geologists have created a calendar of events that covers Earth's entire history, called the **geologic time scale**. Like any other calendar, it is divided into periods of time. However, since days, weeks, and months have very little meaning over millions and billions of years, the geologic time scale is divided into longer units of time that vary in length: *epochs* (tens of thousands to millions of years), *periods* (millions to tens of millions of years), *eras* (tens of millions to hundreds of millions of years), and *eons* (hundreds of millions to billions of years). **TABLE 1.1** provides an overview of how these time units are organized.

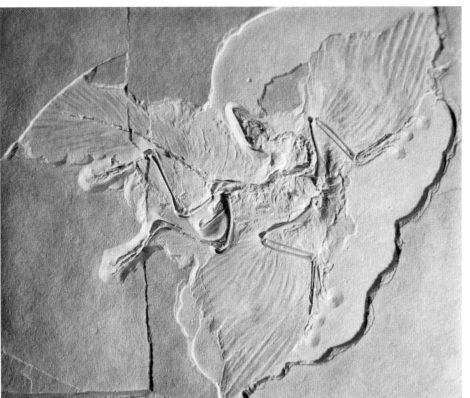

François Gohier/Photo Researchers, Inc.

(?) What evidence supports the conclusion that this was a birdlike creature?

FIGURE 1.18 *Archaeopteryx* lived approximately 150 million years ago and is considered to be the first bird. It was a transitional form between reptiles and birds.

(?) **Expand Your Thinking**—The earliest human settlements and intensive agriculture occurred about 10,000 years ago. Calculate the percentage of geologic time this represents.

TABLE 1.1 Overview of the Geologic Time Scale

Began Years Ago	Notable Events	Epochs	Periods	Eras	Eons
10 thousand	Modern epoch	Holocene	Quaternary	Cenozoic	Phanerozoic
2.6 million	Global cooling	Pleistocene			
23 million	Earliest Hominids, first apes		Neogene		
65 million	Early modern mammals		Paleogene		
	Extinction of dinosaurs Flowering plants abundant Modern sharks		Cretaceous	Mesozoic	
	First birds and lizards Dinosaurs abundant		Jurassic		
251 million	First dinosaurs First mammals and crocodilians		Triassic		
	Major extinction of many life forms Beetles and flies evolve Reef ecosystems flourish Pangaea forms		Permian	Paleozoic	
	First reptiles and coal forests		Pennsylvanian		
	First reptiles First land vertebrates Early sharks		Mississippian		
	First insects and amphibians First ferns and seed-bearing plants		Devonian		
	First land plants First jawed fishes and vascular plants		Silurian		
	First green plants and fungi on land		Ordovician		
542 million	First abundant fossils		Cambrian		
2.5 billion	Oxygenated atmosphere				Proterozoic
3.8 billion	Earliest life (single-celled algae), oldest crust				Archean
4.6 billion	Oldest mineral grain, oldest asteroids and Moon rocks				Hadean

Note: The time periods are not scaled linearly.

LET'S REVIEW "GEOLOGY IN OUR LIVES"

Now that you have finished the chapter, "Geology in Our Lives" will have taken on new meaning for you. Let us review it: The human population has dramatically increased over the past century—with consequences. Our expanding activities have damaged the natural environment. We have depleted natural resources that we depend on, and we are increasingly exposed to natural hazards. Global climate is changing; in many areas, water resources are in short supply; energy costs have risen; and some mineral resources are dwindling. Communities are expanding now into areas formerly considered too dangerous to live in, and the annual cost of natural disasters is at an all-time high. These are global challenges that need to be managed by critical thinkers with an understanding of Earth. Critical thinking is the use of reasoning to explain the world around us.

By the end of the semester, you will know a great deal about how Earth works. This means you can use your knowledge to make better decisions about conserving natural resources, avoiding geologic hazards, and managing natural environments. After you graduate, you may become a business person, a teacher, an artist, an engineer, or any of a multitude of other professions. But regardless of the career you choose, you will take with you the knowledge that we have only one planet, and humanity is its caretaker. We have only one Earth, and the best tool to keep it clean, safe, and livable is *critical thinking*. By applying the skills of critical thinking, we stand the best chance of managing the grand global challenges of our time and overcoming difficulties that may stand between us and this important goal.

This chapter has introduced you to the science of geology and the methods and processes involved in critical thinking. The stage is now set to journey into the Solar System and investigate its earliest history and how and why Earth and the other planets came into existence. These are the subjects of the next chapter, "Solar System."

STUDY GUIDE

1-1 Geology is the scientific study of Earth and planetary processes and their products through time.

- **Geology** is the scientific study of Earth and the other planets—the materials they are made of, the processes that act on those materials, and the products that are formed as a result. Geology includes the study of Earth's history and the life forms that have emerged on Earth since its origin.

- Geological knowledge contributes to the production of the materials that are used in building modern society and constructing the everyday conveniences that we rely on in our lives.

- Dynamic interactions occur among rocks, water, gases, sediments, and living plants and animals. These interactions are powered by solar energy, gravity, and heat from Earth's interior. As a result, Earth's surface is actively evolving, and natural materials are recycled and renewed across a broad range of time and space.

- Geology fosters recognition of the importance of conserving natural resources as well as the human need to use those resources in producing material goods. Carefully using and managing resources and planning for their eventual depletion and replacement by alternatives are important aspects of the practice of modern geology.

1-2 Critical thinking based on good observations allows us to explain the world around us.

- Science is conducted with the help of **critical thinking**, which requires that any explanation for a process or a phenomenon must be testable. This **scientific method** consists of the use of observations and experiments to build and refine **hypotheses** that explain phenomena.

- Scientists use several types of reasoning to solve problems, including inductive reasoning, deductive reasoning, and parsimony.

1-3 Global challenges will be solved by critical thinkers with an enduring understanding of Earth.

- An enduring understanding of Earth can shape your decision making to take into account the need to live in sustainable ways on a planet with limited resources. People who are trained in geology make better decisions about the use of precious natural resources and are more knowledgeable about avoiding geologic hazards. Global challenges faced by a population with dwindling natural resources, geologic hazards, and growing pollution will be solved by critical thinkers with an enduring understanding of Earth.

- Six concepts provide an overarching, big picture of Earth: the study of geology encompasses a vast range of time and space; Earth materials are recycled over time; **plate tectonics** controls Earth's geology; Earth's geologic systems are the products of interactions among the solid Earth, water, the atmosphere, and living organisms; Earth continuously changes; rocks and sediments are pages in the book of Earth's history.

1-4 The theory of plate tectonics is a product of critical thinking.

- Alfred Wegener deduced that the continents have shifted positions over Earth's history. His evidence included similar rock formations and fossils on continents separated by wide ocean basins. His hypothesis of continental drift was never widely accepted because it did not provide a physical process that could account for the movement of continents across Earth's surface.

- The theory of plate tectonics describes spreading centres (where new crust is manufactured), **subduction zones** (where old crust is recycled), and moving plates of crust that act like conveyor belts carrying continents and the sea floor across Earth's surface. Plate movement was originally described by Harry Hess as the result of convection in the **mantle**.

- Plate tectonics provides an explanation for chains of volcanic islands (**hotspots**), the past existence of the supercontinent **Pangaea**, and the topography of **rift valleys** at spreading centres.

1-5 Rock is a solid aggregation of minerals.

- A **rock** is a solid aggregation of minerals. All types of rock can be grouped into three distinct families according to how they formed: **igneous**, **sedimentary**, and **metamorphic**. These families do not coexist independently but can be found together and are recycled into new forms through the rock cycle.

- Igneous rock is composed of cooled molten rock. Sedimentary rock is made of sediment that collects and becomes solidified into rock. Metamorphic rock is made of new minerals that are formed under relatively high temperature and pressure conditions while the rock remains solid.

1-6 Geologists study natural processes, known as geologic hazards, that place people and places at risk.

- Geologists study and monitor landslides, floods, erosion, volcanic eruptions, earthquakes, hurricanes, and other **geologic hazards**.

- When human land use interferes with natural processes, the results may damage property and harm the environment.

- Although Earth is large, its **natural resources** are not endless. Natural resources are materials that occur in nature and are essential or useful to humans, such as water, air, building stone, topsoil, minerals, and others.

1-7 The geologic time scale summarizes Earth's history.

- Geologists have developed a **geologic time scale** that divides time into eons, eras, periods, and epochs. Each time interval has unique characteristics in terms of living organisms and geologic events that are often related to plate tectonics.

- The **principle of uniformitarianism** is a unifying concept in the geosciences that states that Earth is very old, natural processes have been uniform through time, and the study of modern geologic processes is useful in understanding past geologic events. Uniformitarianism is often summarized in the statement, "The present is a key to the past."

KEY TERMS

convection (p. 11)
critical thinking (p. 4)
crust (p. 5)
geologic hazards (p. 16)
geologic time scale (p. 20)
geology (p. 4)
historical geology (p. 4)
hotspots (p. 11)
hypothesis (p. 6)
igneous rock (p. 14)

mantle (p. 11)
mass wasting (p. 16)
metamorphic rock (p. 14)
mid-ocean ridge (p. 12)
natural law (p. 6)
natural resources (p. 17)
Pangaea (p. 10)
physical geology (p. 4)
plate tectonics (p. 9)
principle of uniformitarianism (p. 20)

rift valleys (p. 12)
rock (p. 14)
rock cycle (p. 8)
scientific method (p. 7)
seafloor spreading (p. 11)
sedimentary rock (p. 14)
subduction zones (p. 11)
sustainable resource (p. 4)
theory (p. 6)

ASSESSING YOUR KNOWLEDGE

Please answer these questions before coming to class. Identify the best answer.

1. The reason we emphasize learning the most important ideas of geology in this text is that
   a. your knowledge of geology can guide your decision making throughout your life.
   b. as an adult, it is important that you understand resource sustainability and hazard mitigation.
   c. an understanding of Earth can help you make good decisions in the voting booth.
   d. geology is important in our daily lives.
   e. All of the above.

2. Physical geology is the study of
   a. the materials that compose Earth and the ways in which they are organized and distributed throughout the planet.

   b. the history of life on Earth.
   c. the nature of how humans and animals interact and influence one another.
   d. fossils and past environments.
   e. None of the above.

3. A hypothesis must be testable because
   a. otherwise it cannot be objectively evaluated.
   b. if it is testable, then its accuracy can be independently judged.
   c. its ability to make accurate predictions can be appraised.
   d. unless it is testable, there is no way to determine if it provides useful predictions.
   e. All of the above.

4. Which of the following statements is most likely to be proven using the scientific method?
   a. My dog is better than your dog.
   b. Bacteria cause tooth decay.
   c. That is a beautiful flower.
   d. Apples taste better than bananas.
   e. All of the above.

5. The rock cycle is
   a. a concept describing how rocks are moved in running water.
   b. a concept describing the fact that rocks roll downhill.
   c. a concept proposing that rocks are naturally recycled.
   d. a concept predicting that rocks eventually return to their birthplace.
   e. None of the above.

6. Subduction occurs
   a. when one plate slides beneath another.
   b. when a plate is recycled into Earth's interior.
   c. when a plate enters the mantle at a deep-sea trench.
   d. when the sea floor is recycled beneath an overriding plate.
   e. All of the above.

7. Hess's mechanism for moving plates was
   a. continental drift.
   b. ocean currents.
   c. mantle convection.
   d. volcanic eruptions.
   e. None of the above.

8. Among Wegener's evidence for continental drift was
   a. fossils floating from one continent to another.
   b. distinctive fossils of extinct plants and animals on different continents now separated by vast oceans.
   c. the existence of former land bridges.
   d. simultaneous volcanic eruptions across the globe.
   e. All of the above.

9. Mantle convection
   a. was proposed by Hess as the driver of seafloor movement.
   b. is the product of heat flow within Earth's interior.
   c. causes rifting and seafloor movement between spreading centres and subduction zones.
   d. All of the above.

10. The process of seafloor spreading predicts that
    a. the oldest sea floor will be found at subduction zones.
    b. the sea floor moves like a conveyor belt from spreading centres toward subduction zones.
    c. there is high heat flow from Earth's interior into the crust at spreading centres.
    d. continents shift locations because they are embedded in moving plates.
    e. All of the above.

11. Pangaea is
    a. the name for ancient Greece.
    b. the name of an ancient supercontinent.
    c. the Greek name for plate tectonics.
    d. the concept that rocks are recycled at subduction zones.
    e. the portion of crust that is dry land.

12. The three rock groups are
    a. oceanic, continental, and Pangaean.
    b. weathered, eroded, and contorted.
    c. molten, crustal, and recycled.
    d. igneous, sedimentary, and metamorphic.
    e. shale, sandstone, and basalt.

13. Geologic hazards
    a. are impossible to prevent.
    b. can be eliminated as we learn more about plate tectonics.
    c. cannot always be prevented but often can be avoided.
    d. rarely result from human actions.
    e. tend to cluster in distant areas.

14. Common geologic hazards studied by geologists include
    a. oil spills, pollution, earthquakes, and landslides.
    b. landslides, floods, drownings, and shark attacks.
    c. floods, landslides, earthquakes, and volcanic eruptions.
    d. volcanic eruptions, tsunamis, potholes, and forest fires.
    e. building fires, forest fires, and volcanic eruptions.

15. The subdivisions of the geologic time scale that represent the greatest expanse of time are called
    a. epochs.
    b. eras.
    c. eons.
    d. periods.
    e. eternity.

FURTHER RESEARCH

1. Search the web for a description of or a story about a geologist. What does he or she do? Make contact by email and ask why the person became a geologist and what he or she likes about the job.

2. Why do geologists study geologic hazards if we usually cannot prevent them?

3. From your perspective, what is the reason for learning about science?

4. Why are fossils important markers for understanding geologic time?

5. Give an example of how you solved a problem using inductive reasoning. Give an example of how you used deductive reasoning.

6. Describe what it means when scientists say that a hypothesis must be testable.

7. List two natural resources. What happens when they are poorly managed? How can they be managed effectively?

ONLINE RESOURCES

Explore more about geology on the following websites:

Earth Exploration Toolbox:
serc.carleton.edu/eet/index.html

Canadian Geoscience Education Network:
http://earthsciencescanada.com/cgen

United States Geological Survey (USGS) undergraduate education page:
education.usgs.gov/undergraduate.html

SciCentral stories and videos – news stories in science:
www.scicentral.com

See the latest geology findings announced at Science Daily website:
www.sciencedaily.com/news/earth_climate/geology

Additional animations, videos, and other online resources are available at this book's companion website:
www.wiley.com/go/fletchercanada
This companion website also has additional information about WileyPLUS and other Wiley teaching and learning resources.

CHAPTER 2
SOLAR SYSTEM

Chapter Contents and Learning Objectives (LO)

2-1 Earth's origin is described by the solar nebula hypothesis.

LO 2-1 *List the terrestrial planets and the gas giants in order from the Sun.*

2-2 The Sun is a star that releases energy and builds elements through nuclear fusion.

LO 2-2 *Describe the Sun and how it works.*

2-3 Terrestrial planets are small and rocky, with thin atmospheres.

LO 2-3 *State the ways that Mercury, Venus, and Mars are different from Earth.*

2-4 Gas giants are massive planets with thick atmospheres.

LO 2-4 *Describe each of the gas giants.*

2-5 Objects in the Solar System include the dwarf planets, comets, and asteroids.

LO 2-5 *Define a dwarf planet and name the five that are currently recognized.*

2-6 Earth's interior accumulated heat during the planet's early history.

LO 2-6 *Describe the sources of heat for early Earth and the consequences of heat buildup.*

GEOLOGY IN OUR LIVES

Earth, the Sun, and other objects in the Solar System originated at the same time from the same source and have evolved in varying ways since then. The Solar System began with the collapse and condensation of a planetary nebula. Understanding Earth's place in the cosmos provides us with perspective on and appreciation for the scale of nature. Exploring the Solar System is one of humanity's grand accomplishments. Someday our planetary neighbours may provide us with useful resources.

(?) Why would planets close to the Sun, such as Earth, have thin atmospheres and those far from the Sun, such as Jupiter, have thick atmospheres?

2-1 Earth's Origin Is Described by the Solar Nebula Hypothesis

LO 2-1 *List the terrestrial planets and the gas giants in order from the Sun.*

Earth and the other planets have a common origin that is described by the **solar nebula hypothesis**. This hypothesis states that the Sun, the planets, and other objects orbiting the Sun originated at the same time from the same source through the collapse and condensation of a *planetary nebula* (a great cloud of gas formed from an exploding star) and have evolved in varying ways since that time. Support for this hypothesis is found in the observation that the planets orbit on nearly the same plane and in the same direction around a common focus, the Sun. Differences among the planets and other Solar System objects that are not explained by the solar nebula hypothesis are attributable to events that have happened since their origin.

The **Solar System** (FIGURE 2.1) consists of these objects:

- the Sun
- eight planets: Mercury, Venus, Earth, Mars, Jupiter, Saturn, Uranus, and Neptune
- five dwarf planets (Pluto, Ceres, Haumea, Makemake, and Eris)
- small Solar System bodies (including asteroids, comets, and objects in the Kuiper Belt, scattered disc, and Oort Cloud)
- a total of 162 moons orbiting the eight planets
- countless particles and interplanetary space

A *star* is a celestial body of hot gases that radiates energy produced by nuclear reactions in the interior. The Sun is a star located at the centre of the Solar System. It is huge; in fact, if you think of the Sun as a basketball, Earth would be only the size of the head of a pin—a mere speck. Also, if the basketball representing the Sun were at one end of the court, then the pinhead would be at the other end. The **Milky Way Galaxy** contains billions of other stars besides the Sun, but most of them are smaller than the Sun, which is in the top 10 percent of stars by mass. The median size of stars in our galaxy is probably less than half that of the Sun.

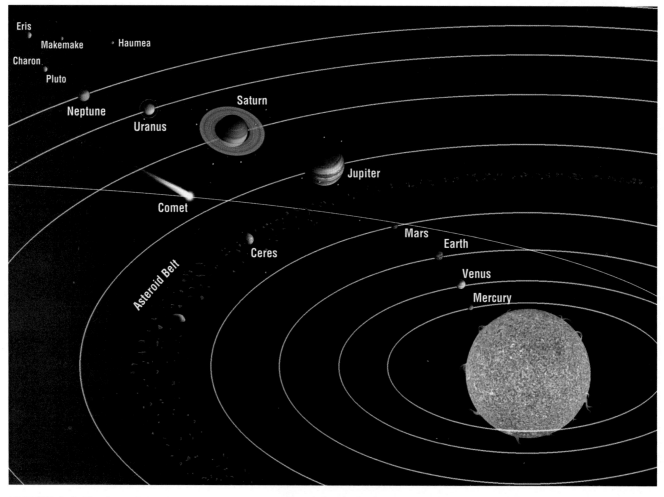

FIGURE 2.1 The Solar System.

 What is the source of the elements and compounds that compose the Solar System?

Everything in our Solar System orbits around the Sun on a path that balances the Sun's gravity with the *centrifugal force* (an outward tug like the one you feel in a car going around a curve) acting on the object. The Sun contains over 99 percent of the Solar System's mass.

The Solar Nebula Hypothesis

As shown in Figure 2.1, the planets lie nearly on a single plane, called the *ecliptic,* with the Sun at the centre. All the planets orbit the Sun in the same direction, and all the planets, except Venus, rotate, or spin on their axes, in the same direction. Venus rotates in the opposite direction, for reasons of which scientists are unsure. Perhaps it suffered a colossal collision that tipped it nearly upside down. Uranus also has a steeply tilted axis, thought to be the result of a collision with a large object early in its history. All the planets orbit the Sun in the same direction.

According to the solar nebula hypothesis, about 5 billion years ago, a swirling cloud of interstellar gas and dust called a *nebula* (**FIGURE 2.2**) began to collapse inward under the pull of gravity. Shock waves from the explosion of a nearby star, or some other disturbance, may have initiated this collapse. As the volume of the cloud decreased, its density and rate of rotation increased—a phenomenon similar to when ice skaters pull in their arms in order to spin faster. As the spinning accelerated, the cloud flattened and developed a hot core, giving birth to an infant star, our future Sun.

The cloud continued to collapse as material streamed into its centre. Climbing temperatures vaporized dust, and the resulting gases joined the other gases feeding the growing core. Astrophysicists calculate that it took less than 100,000 years for the nebula to evolve into a hot "protostar" fed by gases flowing inward.

FIGURE 2.2 The solar nebula hypothesis proposes that the Solar System originated from a cloud of interstellar gas and dust (the solar nebula). This is a photo of the Crab Nebula. It is a supernova remnant, all that remains of a tremendous stellar explosion. Observers in China and Japan recorded the supernova nearly 1,000 years ago, in 1054. NASA, ESA, J. Hester, A. Loll (ASU)

© Don Dixon/cosmographica.com

FIGURE 2.3 Planetesimal accretion.

(?) Write a short description of the planetesimal accretion process.

Simultaneously, centrifugal force prevented some gas and dust from reaching the core of the nebula. This material formed a cooler outer region where rocky particles and gases coalesced to become our system of planets. This process of planet growth is termed **planetesimal accretion**.

The original nebula was composed mostly of atoms of hydrogen (H) and helium (He), the two simplest and lightest elements, in gaseous form. It also contained heavier atoms, such as oxygen (O), carbon (C), silicon (Si), iron (Fe), and additional metals and other elements, some of them in the form of dust particles. As the new star took shape in the centre of the disc-like cloud, matter also coalesced in the outer regions of the disc. This was a critical phase in the development of the Solar System. Earth and other planets would soon form in an area of the disc that was enriched with many of the elements that would form the Earth materials we see today.

The outer region of the disc cooled rapidly, with the result that metal, rock, and ice condensed as tiny particles. These particles grew by colliding with one another, forming larger masses of rock. Gravity gave larger particles an advantage over smaller ones, and their growth accelerated as they pulled in additional solid and gaseous matter. These objects grew still larger, becoming *planetesimals.* Planetesimals, in turn, coalesced to form *planetary embryos,* a process that unfolded over a period of hundreds of thousands of years. Finally, over the course of another few million years, planetary embryos combined to form true planets (**FIGURE 2.3**).

Today, there are two major groups of planets in the Solar System: the four **terrestrial planets** (Mercury, Venus, Earth, and Mars) and the four **gas giants** (Jupiter, Saturn, Uranus, and Neptune). The asteroids orbit the Sun in a belt beyond the orbit of Mars. Made mostly of rock and iron, they are probably debris from the planetary nebula that never coalesced.

(?) **Expand Your Thinking**—Describe how the Solar System would be different if the original solar nebula did not rotate.

2-2 The Sun Is a Star that Releases Energy and Builds Elements through Nuclear Fusion

LO *2-2 Describe the Sun and how it works.*

When the universe first came into existence some 14 billion years ago, it was composed only of hydrogen, helium, and small amounts of lithium (Li), beryllium (Be), and boron (B), the first 5 elements in the periodic table. Today, it consists of 90 naturally occurring elements, all of which have been discovered here on Earth. If there were only 5 elements to begin with, where did the remaining 85 come from? They were created through **nuclear fusion** in stars. In other words, stars are the source of most of the elements that make up our world.

The extremely high temperatures and pressures within stars allow protons and neutrons to be added, or fused, to the nucleus of a hydrogen atom, thereby forming a heavier element: helium. This process generates an explosive burst of energy. Just like any engine, when the concentration of fuel, in this case hydrogen, runs low, other fuels, such as helium, must be used to power the fusion process. Helium fusion produces lithium, an even heavier element. Then, when the concentration of helium fuel runs low, lithium is used, producing carbon and oxygen—which, in turn, produce still heavier elements. This process can be likened to burning the ashes of one fire as fuel for another. In the billions of years since the origin of the universe, it is thought that nuclear fusion within stars has formed the heavier elements (elements with more protons in the nucleus), up to and including iron, which are found in nature. This is the process of **stellar nucleosynthesis**. However, this is not the end of the story.

Planetary Nebulas

Stars die, or burn out, when they use up their nuclear fuel. When a star's concentration of hydrogen is too low to support the fusion process, it begins fusing helium in its core. At this point, it will be 1,000 to 10,000 times brighter than before, becoming what is termed a *red giant,* a large, bright, unstable star whose hydrogen has run out.

Our Sun will eventually become a red giant that is likely to engulf Mercury, Venus, and perhaps Earth. Astrophysicists hypothesize that the Sun, which has now completed approximately half its life, will enter the red giant phase in 4 to 5 billion years. Red giants, like all other stars, will eventually burn out. If it is large enough, a red giant ends its life in a process in which its core collapses. The intense fusion taking place in the collapsing core causes the outer gaseous layers to be ejected explosively into space. An exploding star is called a *nova.* When a more massive star explodes, it is called a *supernova.* The elements heavier than iron are thought to be produced during these cataclysmic explosions. The material ejected by a nova or supernova is composed of newly made elements, and the resulting field of debris—the stuff of which planets are made—is a *planetary nebula* (**FIGURE 2.4**). As described earlier, it was a planetary nebula that evolved into our Solar System.

NASA, ESA, and the Hubble SM4 ERO Team

FIGURE 2.4 The Hubble Space Telescope took this image of hot gas escaping a dying star 3,800 light years away in the Scorpius constellation. This planetary nebula is known as the Butterfly Nebula. The features that resemble dainty butterfly wings are actually roiling clouds of gas heated to more than 20,000°C. The star itself, once about five times more massive than the Sun, is over 200,000°C, making it one of the hottest objects in the galaxy. The ejected gas, enriched with oxygen, nitrogen, and carbon produced by the formerly massive star, will form the stuff for future stars.

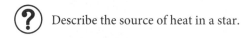 Describe the source of heat in a star.

Our Sun: A Massive Hydrogen Bomb

The Sun is the most prominent feature of our Solar System and the most massive, yet its interior structure is still somewhat of a mystery. One hundred and nine Earths would be required to equal the Sun's diameter, and its interior could hold more than 1.3 million Earths. Like Earth, the Sun is composed of several layers that define its internal structure (**FIGURE 2.5**). Unlike Earth, however, it is completely gaseous and lacks a solid surface.

The Sun's extremely dense core accounts for about 50 percent of its total mass but only about 1.5 percent of its total volume. Physical conditions inside the core are extreme. The temperature is thought to be around 16 million degrees Celsius and the pressure perhaps 250 billion times the air pressure at Earth's surface. Under these intense

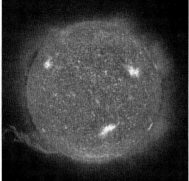

NASA Goddard Space Flight Center

 Why is the Sun getting lighter with time?

FIGURE 2.5 Energy is generated in the Sun's core where temperatures reach 16 million degrees Celsius.

produces X-rays that work their way slowly to the surface. As they move upward, the X-rays collide with subatomic particles and change direction in random ways. Each X-ray may travel only a few millimetres before it experiences another collision and sets off in a different direction. The time to complete this journey out of the Sun's interior is measured in millions of years—an incredible fact, given that X-rays travel at the speed of light. Only very hot gases emit X-rays, such as those in the Sun's core or in the corona part of its atmosphere.

The surface of the Sun, in contrast, is only about 6,000°C and emits most of the electromagnetic radiation in the visible spectrum. This is the sunlight that we see, although we have to wait another eight minutes because that is how long it takes for it to travel the 150 million kilometres from the Sun to Earth.

The Sun is essentially an enormous hydrogen bomb in a constant state of explosion. Yet it does not explode outward. Instead, its immense gravity, which pulls its mass inward toward the core, keeps the explosion in check. At the same time, the outward pressure created by the continuous nuclear reaction prevents it from collapsing under its own gravity. The Sun thus is in a state of *dynamic equilibrium* between exploding and collapsing, a balance of nature to which we owe our existence.

The Sun is clearly fundamental to life on Earth. If we wish to know the potential for life elsewhere in the universe, we first need to know how many stars there are in the universe. For the answer, see "How Many Stars Are There?" in the "Critical Thinking" box.

conditions, atoms are stripped of their electrons. As a result, the Sun's core is a mixture of protons, neutrons, nuclei, and free electrons. The temperature and pressure deep within the core is so intense that nuclear reactions take place, causing the fusion of hydrogen nuclei to form helium nuclei. Every second, 700 million tonnes of hydrogen is converted into helium. In the process, 5 million tonnes of energy is produced, which is carried toward the surface, where it is released as light and heat.

If we could see into the core, it would appear black, since none of the energy produced there is visible to our eyes. The Sun's core mainly

---

CRITICAL THINKING

HOW MANY STARS ARE THERE?

It is impossible to physically count every star in the universe (**FIGURE 2.6**). But what if you wanted to know how many stars actually exist? How would you find out? You could construct an estimate based on **extrapolation**. Extrapolation is a form of critical thinking that is commonly used to estimate things that cannot be directly observed. It consists of inferring or estimating an answer by projecting or extending a known value. Questions such as

- How many cells are in the human body?
- How many tuna are in the sea?
- How many grains of sand are on the world's beaches?

can be answered using extrapolation, as long as you have some basic information, such as the size of a cell.

So, how many stars are there? The answer to this question has been estimated using extrapolation. Astronomers at Australia National University used two telescopes to count the stars in a single strip of sky. Within that one region, some 10,000 galaxies were pinpointed, and detailed measurements of their brightness were made. These measurements were used to calculate how many stars they contained. That number was then multiplied by the number of similar-size strips it would take to cover the entire sky, then multiplied again out to the edge of the visible universe. The result is 70

sextillion, or 70,000,000,000,000,000,000,000. As it turns out, this is about 10 times more than the total number of grains of sand on all the beaches in the world!

NASA, ESA, and the Hubble SM4 ERO Team

FIGURE 2.6 How would you count the stars in the universe? This is a view into one of the more crowded places in the universe, a small region inside the globular cluster Omega Centauri, which has nearly 10 million stars. Globular clusters are ancient swarms of stars united by gravity. The stars in Omega Centauri are 10 to 12 billion years old.

Expand Your Thinking—What would the result be if the force of gravity in the Sun were greater than the release of energy by nuclear fusion?

2-3 Terrestrial Planets Are Small and Rocky, with Thin Atmospheres

LO *2-3 State the ways that Mercury, Venus, and Mars are different from Earth.*

The terrestrial planets (**FIGURE 2.7**) are relatively small and consist primarily of compounds of silicon, oxygen, iron, magnesium (Mg), and other metals [potassium (K), calcium (Ca), sodium (Na), and aluminum (Al)]. They were formed close to the developing Sun during the first 100 million years after its birth, when the heat was too great to allow them to accumulate much *water* (H_2O) or other *volatile gases* (easily evaporated gases). These gases accumulated as thin primitive atmospheres on the terrestrial planets but are found in much larger quantities farther from the Sun. They include *carbon dioxide* (CO_2), *methane* (CH_4), and *ammonia* (NH_3), all of which were abundant in the planetary nebula but too unstable to survive in great abundance close to the Sun.

At the same time, in the cooler outer portion of the Solar System, the original nebula gases froze on some planets and formed thick gaseous atmospheres on others (**FIGURE 2.8**). The accumulation of gases allowed the outer planets to acquire massive gravitational power. This gravitational force enabled them to hold onto huge quantities of helium and hydrogen, which were still present in the outer region of the condensing nebula. These outer planets thus became gas giants. Each gas giant has several moons with thick icy surfaces made mainly of water, ammonia, and carbon dioxide. Many scientists think that conditions on some of these moons might be conducive to the formation of life.

While planetesimals were forming in the outer region of the condensing solar nebula, nuclear fusion began in the nebula's core. With fusion came a release of energy that stopped the contraction of the nebula. This event marked the birth of our Sun. It let loose a fiery storm of high-energy subatomic particles and radiation called the **solar wind**. The solar wind, still present today, swept away the gases left within the nebula and stripped the inner planets of their thin primitive atmospheres. Mercury, Venus, Earth, and Mars likely became barren rocky spheres with no atmospheres and no surface water. Beyond Mars, the huge gravitational mass of the gas giants stabilized them and enabled them to retain their immense atmospheres.

Mercury

Mercury's Atmosphere: oxygen (O_2), 42 percent; sodium (Na), 29 percent; hydrogen (H_2), 22 percent; helium (He), 6 percent; and potassium (K), 0.5 percent

Five elements have been detected in minute quantities in Mercury's unstable atmosphere. Oxygen, sodium, and hydrogen are present at all times, but helium and potassium are found only at night. During the day, they are absorbed by rocks on the surface. Mercury is the closest planet to the Sun and is dominated by unrelenting and intense heat during the day. But because it lacks a robust atmosphere to trap that heat, during the night temperatures plummet. Thus, surface temperatures exceed 227°C on the sunlit side but chill to −137°C at night.

Mercury's surface resembles that of Earth's Moon because it is heavily marked by impact craters and includes areas that appear to be *lava plains*. On one side of the planet is a large impact crater called the Caloris Basin. Exactly opposite, on the other side, is an area of unusual topography called the "weird terrain," where the shock waves from the impact that created the Caloris Basin apparently converged and disturbed the surface of Mercury.

Mercury's axis of rotation does not tilt as Earth's does. Therefore, craters at Mercury's poles may be in perpetual shadow and escape the planet's staggering temperature extremes. Radar mapping of Mercury's surface in 1991 revealed unusual signals from within the polar craters that may indicate the presence of ice. The possibility of a polar icecap so close to the Sun is certainly a surprise. At present, the National Aeronautics and Space Administration (NASA) has a spacecraft, *Messenger*, orbiting the planet and collecting more information.

Venus

Venus's Atmosphere: carbon dioxide (CO_2), 96.5 percent; nitrogen (N_2), 3.5 percent

NASA

(?) This image clearly illustrates one of Earth's most distinguishing features—what is it?

FIGURE 2.7 The terrestrial planets include Mercury, Venus, Earth, and Mars.

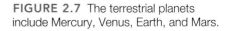

FIGURE 2.8 Rocky particles and metallic compounds formed solids in the inner portion of the condensing solar nebula. But it was too hot for hydrogen compounds to solidify. In the cooler outer region, hydrogen compounds, metals, and rocks all condensed to a solid state. The transition zone between the two regions is known informally as the frost line.

Why do scientists use the term frost line?

Venus, the second planet from the Sun, is often known as Earth's sister because it is our immediate neighbour and of all the planets is closest to Earth in size. There are also features on the surface that may have been formed by volcanoes, similar to those seen on Hawaii. Its atmosphere, however, makes Venus a radically different place. Carbon dioxide, which is very efficient at trapping heat, makes up most of the atmosphere. As a result, surface temperatures reach a hellish 477°C—even hotter than Mercury's daytime temperatures. Global winds distribute heat around the planet and, combined with the thick atmosphere, ensure that night and day temperatures are about the same. Deep white clouds composed of sulphuric acid (H_2SO_4) fill the atmosphere, with the result that acid rain continually falls on the planet's surface.

As Venus travels around the Sun, it rotates very slowly on its axis, only once every 243 Earth days. Venus's axis is very nearly perpendicular, although it may be upside down. This is because, unlike Earth, Venus does not rotate in the same direction that it travels around the Sun. Rather, it rotates in the retrograde (opposite) direction—the only planet to do so. One possible explanation for this is that it may have been knocked over by a collision with another planetesimal during the accretion phase early in its history.

Earth

Earth's Atmosphere: nitrogen (N_2), 78 percent; oxygen (O_2), 21 percent; argon (Ar), 1 percent

In many ways, Earth is unlike its neighbours in the Solar System. Its hot interior provides the energy to fuel volcanoes and drive internal convection, yet its surface is cool enough to support life. Its position relative to the Sun provides sufficient warmth to maintain liquid water, yet we still experience freezing seasons. Earth's winds are relatively calm; the

atmosphere is thin and transparent so it is easily penetrated by light; and daily extremes in temperature are comparatively modest. Earth also has a magnetic field, which is generated by motion within the molten outer core (see Chapter 3), and this helps to shield Earth's surface from the solar wind. All these conditions promote the existence of life, yet they are rare in the rest of the Solar System.

Mars

Mars's Atmosphere: carbon dioxide (CO_2), 95.3 percent; nitrogen (N_2), 2.7 percent; argon (Ar), 1.6 percent; oxygen (O_2), 0.13 percent

Mars is the most Earthlike of the planets in the Solar System. Although it is only half the size of Earth, it has many similar features, such as a nearly 24-hour day, clouds, layers of sediment, sand dunes, volcanoes, and polar icecaps that grow and recede with the changing seasons. Mars has dry river channels, along with flood plains and watersheds, indicating that water flowed across its surface in the past, perhaps from springs. Physical features closely resembling shorelines, gorges, riverbeds, and islands suggest that the planet formerly possessed great rivers, lakes, and even seas.

Scientists hypothesize that after its formation, Mars cooled enough that large volumes of water vapour collected as thick clouds in the atmosphere. The resulting rain fell onto the planet's surface, forming the numerous water-carved features we observe today. The lack of a protective ozone layer meant that water molecules were easily destroyed by sunlight, and the constituent elements probably escaped the atmosphere because of Mars's weak gravity field. Mars today has no liquid water, although frozen water is found in the polar icecaps and in the form of a shallow layer of permafrost under the ground.

 Expand Your Thinking—Why is an atmosphere important in equally heating a planet's surface?

2-4 Gas Giants Are Massive Planets with Thick Atmospheres

LO 2-4 *Describe each of the gas giants.*

The gaseous planets, or gas giants, include Jupiter, Saturn, Uranus, and Neptune (**FIGURE 2.9**).

Jupiter

> **Jupiter's Atmosphere:** hydrogen (H_2), 89.8 percent; helium (He), 10.2 percent

Jupiter is the largest planet in the Solar System, more than twice as massive as all the other planets combined. Its enormous size is due to its huge gaseous atmosphere, a holdover from the early days of the solar nebula. While the terrestrial planets lost their early atmospheres to the heat of the developing Sun, Jupiter did not. Jupiter's most famous feature is the *Great Red Spot*. Still there after more than 300 years, the spot is the largest storm in the Solar System with winds reaching 434 km per hour. Fed by heat generated by the huge planet, it may still exist because it never runs over land, where it would lose its source of energy.

Scientists disagree about Jupiter's inner structure. Some suggest that it has no solid surface. This leaves nothing to "tie down" its atmosphere, and as a result its rotation rate is not the same from one location to another. The equatorial zones rotate more rapidly than the polar regions, giving rise to the ever-present cloud banding observed in the planet's atmosphere. However, scientists are still unsure as to whether circulation in the atmosphere is confined to the upper parts, or whether the deepest parts of the atmosphere are also involved.

FIGURE 2.9 The gas giants (left to right): Jupiter, Saturn, Uranus, and Neptune.

(?) Why do different planets have different colours?

Nevertheless, it is thought that the gases increase in density toward the planet's core, essentially becoming liquid at a certain point.

The moons of Jupiter have attracted scientific interest in recent decades. Organic molecules that are believed to be necessary for the formation of life have been detected on some of those moons. Scientists study these conditions in the hope of learning more about the development of life on Earth. Jupiter's four largest moons (known as the Galilean satellites)—*Io, Europa, Ganymede,* and *Callisto*—can readily be seen from Earth. The moons are also interesting, because they are volcanically active. For example, Io has about 400 volcanoes that erupt sulphur and sulphur dioxide.

Saturn

> **Saturn's Atmosphere:** hydrogen (H_2), 96.3 percent; helium (He), 3.25 percent

Saturn is famous as the planet with the rings. These rings are composed of myriad particles of ice and rock that are thought to be the remains of ancient "moons" that have been pulled apart by Saturn's gravity. The ring system is approximately 100,000 km wide and 200 m thick. Much of our knowledge of this gas giant comes from the space probes *Voyager 1* and *Voyager 2*. These were launched by NASA in 1977 with a mission to conduct close-up studies of Jupiter and Saturn, Saturn's rings, and the larger moons of the two planets. The *Voyager* data showed that Saturn has a mass about 95 times that of Earth as well as a very rapid spin rate that has the effect of flattening the planet. Saturn is clearly oblate at the poles, the result of an outward push in the equatorial region due to rapid spinning.

Saturn's exterior is composed primarily of frozen ammonia. The interior consists mostly of hydrogen, with lesser amounts of helium and methane. Saturn is the only planet that is less dense than water (about 30 percent less). Amazingly, if you were able to place it in a freshwater bath, Saturn would float!

Uranus

> **Uranus's Atmosphere:** hydrogen (H_2), 82.5 percent; helium (He), 15.2 percent; methane (CH_4), 2.3 percent

Uranus has a unique feature: Its axis of spin is approximately parallel to the ecliptic. In other words, Uranus lies on its side and its north pole points toward the Sun for half of a Uranian year. Its south pole points toward the Sun for the other half of the year. This strange tilt produces some extreme seasonal effects. For one thing, the polar regions receive the greatest amount of sunlight, and the rest of the planet does not experience the daily heating and cooling that other planets do. Researchers hypothesize that collision with another object altered the planet's spin axis, or "tipped it over."

Neptune

> **Neptune's Atmosphere:** hydrogen (H_2), 80 percent; helium (He), 18.5 percent; methane (CH_4), 1.5 percent

Neptune is the outermost planet in our Solar System. Scientists hypothesize that the inner two-thirds of the planet are composed of a mixture of molten rock, water, liquid ammonia, and methane. The outer third is a mixture of heated gases comprising hydrogen, helium, water, and methane. It is methane that gives Neptune its blue cloud colour. Neptune has the strongest winds on any planet. Most of the winds blow westward, which is in the opposite direction from the rotation of the planet. Near one spot, winds blow up to 2,000 km an hour. Neptune has four rings, which are narrow and very faint. The rings are made up of dust particles originating from tiny meteorites smashing into Neptune's 13 known moons.

The Solar System is part of the Milky Way Galaxy, one of billions of galaxies in the universe. Read the "Geology in Action" box on "The Nature of the Universe."

GEOLOGY IN ACTION

THE NATURE OF THE UNIVERSE

Have you ever wondered whether the universe is endless? If it is not, and you were able to travel to its edge, what would you see? If there is an edge, there must be a centre, but what is in the centre of the universe? We can begin to answer these questions by *thinking critically* about a simple observation: It gets dark at night.

If the universe were infinite, eternal, and uniformly filled with stars, every point in the sky would be filled with a star radiating light and heat, and it would not get dark at night. Indeed, the glare would be overpowering, and the combined radiation from all the stars would heat Earth and all other objects in the universe to the temperature of the Sun's surface. This observation is called *Olbers' Paradox* (after astronomer *Heinrich Olbers*, 1758–1840) because in Olbers' day, it was assumed (largely for religious reasons) that the universe must be infinite, yet his observations suggested that it was finite. The solution to the paradox was proposed in 1848 by the poet Edgar Allen Poe, who stated that the night sky is dark (**FIGURE 2.10**) because *the universe is not infinite.*

Does the universe have an edge and therefore a centre? The *Doppler effect* (named for physicist *Christian Doppler*, 1803–1853) helps answer this question. This term refers to the change in frequency of energy as the source of that energy approaches or recedes. To illustrate: You can determine the direction in which an ambulance is travelling based on the pitch of its siren. The pitch rises if the ambulance is coming toward you and falls if it is travelling away from you. Inside the vehicle, the sound is constant; the driver does not notice a change in pitch. Light from a distant galaxy does the same thing. The wavelengths of light from a retreating galaxy appear to decrease in frequency (the light is "red-shifted"). The wavelengths of light from an approaching galaxy appear to increase in frequency (the light is "blue-shifted").

In 1929, astronomer *Edwin P. Hubble* (1889–1953) astounded the scientific community by announcing that other galaxies in the universe are red-shifted and that the size of the shift increases the farther away a galaxy is. That is, galaxies are moving away from

NASA and the Hubble Heritage Team (STScI/AURA)

FIGURE 2.10 If the universe were infinite and filled with stars, the sky should be brilliantly lit at all times and the radiation would heat Earth thousands of degrees. It gets dark at night because the universe is not infinite.

Earth, and more distant galaxies are moving away faster. The explanation for this phenomenon is simple but revolutionary: *The universe is expanding.*

Hubble's finding, which has been tested many times by other astronomers, indicates that the universe is expanding. What does that actually mean? For one thing, it means that no matter which galaxy we happen to be in, virtually all the other galaxies are moving away from us. In other words, it's not as if Earth (or any other location) is at the centre of the universe and everything else is receding from it. Thus, the universe has no centre and therefore no edge as such.

Expand Your Thinking—Make a list of the five most abundant gases in the atmospheres of the planets and calculate the average concentration of each gas using the data given in Sections 2-3 and 2-4.

2-5 Objects in the Solar System Include the Dwarf Planets, Comets, and Asteroids

LO *2-5 Define a dwarf planet and name the five that are currently recognized.*

In addition to the "classical" planets, the Solar System includes five dwarf planets (**FIGURE 2.11**). A *dwarf planet* is an object in the Solar System that orbits the Sun and is not a satellite of a planet or other celestial body. It must be spherical (or nearly so) in shape. This means that it is large enough so that its own gravitational field has overcome the internal rigidity of rock to shape it into a round object. Dwarf planets must also clear the neighbourhood of their orbits, meaning that they interact gravitationally with smaller bodies nearby and either accrete with them or propel them into another orbit. Under these strict requirements, the International Astronomical Union recognizes five dwarf planets: Pluto, Haumea, Makemake, Eris, and Ceres.

Pluto, Haumea, Makemake, Eris, and Ceres

Pluto's Atmosphere: nitrogen (N_2), 98 percent; water (H_2O), < 1 percent; methane (CH_4), < 1 percent; carbon monoxide (CO), < 1 percent

Pluto, Haumea, and Makemake reside in the region of the *Kuiper Belt,* a sea of icy bodies extending beyond Neptune that is one source of comets. Eris is thought to be a member of the *scattered disc* region. The scattered disc region overlaps with the Kuiper Belt, but objects within it have highly eccentric and inclined orbits, and its outer limits extend much farther from the Sun. Although there are just five dwarf

planets now, their number is expected to grow. Scientists estimate there may be many dozens of dwarf planets in the outer Solar System. But since we don't know the actual sizes or shapes of many of these faraway objects, we can't yet determine which are dwarf planets and which are not. More observations and better telescopes are likely to add to the list in coming years.

Pluto was discovered in 1930 and considered a planet until it was reclassified as a dwarf planet in 2006. Haumea, Makemake, Eris, and Ceres have been classified since then. Pluto's mass is thought to be only 0.0025 that of Earth's. Eris is slightly larger than Pluto. The mass of these bodies is closer to that of a moon than to that of a planet. Indeed, Pluto, together with its moon, Charon, is so small that together they could fit within the borders of Canada. Because Charon and Pluto are similar in size, they are really a double dwarf-planet system rather than the more typical combination of a planet with a much smaller moon.

These dwarf planets do not resemble the inner, terrestrial planets or the outer, gaseous planets in their makeup. Instead, they more closely resemble the ice moons of the outer planets, which has led some researchers to suspect that they are large icy chunks of debris left over from the formation of the Solar System.

Ceres is a dwarf planet in the asteroid belt. Although it is the largest of the asteroids (about 950 km in diameter), it is less than half the size of Pluto and Eris. Still, Ceres accounts for approximately one-third

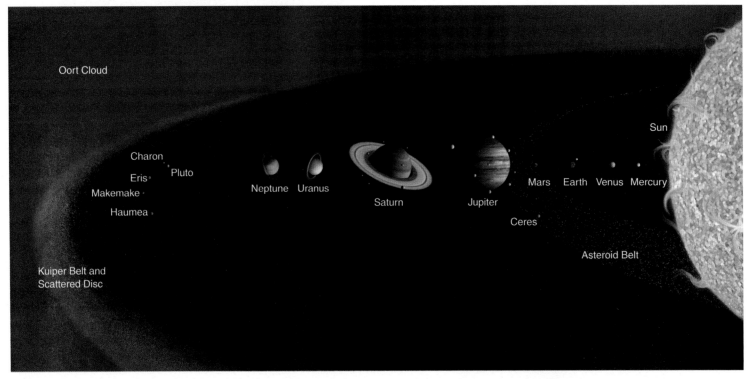

FIGURE 2.11 The dwarf planets include Eris, Makemake, Haumea, Pluto (with its moon Charon), and Ceres.

 What is a dwarf planet?

FIGURE 2.12 Comets orbiting the Sun exhibit a coma and tail.

 What causes a comet to have a blue tail and a white tail?

FIGURE 2.13 Two hemispheres of the asteroid Eros. Most asteroids reside in a belt of rocky debris between Earth and Jupiter that may be left over from the original nebular disc. The total mass of all the asteroids is less than that of our Moon.

 Where are most asteroids found?

of the belt's total mass. Ceres' physical aspects are not well understood. Researchers hypothesize that its surface is warm and that it has a tenuous atmosphere that includes frost. One study concluded that this dwarf planet has a rocky core overlain by an icy mantle possibly containing 200 million cubic kilometres of water, more than the total amount of fresh water on Earth.

Asteroids and Comets

Not all of the planetesimals ended up becoming planets. Some were made up primarily of rocky and metallic substances, and they became *asteroids*. Other planetesimals were made up primarily of ice and other frozen liquids or gases—hydrogen, helium, methane, carbon dioxide, and others—and are known as *comets* (**FIGURE 2.12**).

Because they are made of ice, comets would not last long if they inhabited the asteroid belt. Continuous exposure to the Sun's heat would cause them to waste away. Instead, most of the comets in the Solar System are found in two distant regions: the *Oort Cloud,* a vast spherical cloud of comets at the extreme outer edge of the Solar System; and the closer Kuiper Belt, a wide band of comets located beyond the orbit of Neptune. Comets range from 1 km to 10 km in diameter. In addition to ice, the nucleus, or centre, of a comet may contain frozen carbon dioxide, carbon monoxide, methane, and other easily evaporated compounds. Interspersed with this ice are tiny mineral grains (dust); together, the ice and dust in the nucleus of a comet form

a "dirty snowball," an idea first put forth by astronomer *Fred Whipple* in 1950 and verified by the spacecraft *Giotto*'s flyby of Halley's comet in 1986.

When the solar wind, which consists of a stream of energetic, charged particles, encounters the material in the nucleus of a comet, it blows it back behind the nucleus, thus creating the comet's tail (which usually extends behind the comet, away from the Sun). Early in the history of the Solar System, large numbers of comets bombarded the terrestrial planets, delivering some of the first water and important atmospheric components. These impacts also added heat, contributing to the creation of the planets' internal structure.

Comets actually have two tails: the dust tail and the gas or ion tail. The dust tail is formed of solid particles escaping from the comet along its flight path and may be slightly curved. Its whitish colour is due to reflected sunlight. The gas tail is pushed straight back from the comet by the solar wind (it always points directly away from the Sun) and has a bluish glow due to the fluorescence of ionized molecules of carbon dioxide, carbon monoxide, nitrogen, and methane.

The asteroid belt is a zone of rocky debris (**FIGURE 2.13**) between Mars and Jupiter. Asteroids are rocky, metallic objects that orbit the Sun but are too small to be considered planets. Asteroids range across the central Solar System from inside Earth's orbit to beyond Saturn's. Most, however, are contained in the region between Mars and Jupiter. Study the "Critical Thinking" box on the next page and answer the questions on "Summarizing the Solar System."

 Expand Your Thinking—If a comet had the composition of an asteroid, how would it look different when observed from Earth?

CRITICAL THINKING

SUMMARIZING THE SOLAR SYSTEM

Work with a partner to answer the following questions based on **FIGURE 2.14**.

1. Label each of the objects in this image. Identify the classical planets and dwarf planets.
2. Identify differences in the chemistry of the terrestrial planets and the gas giants. Be sure to indicate the planets that fall into each group.
3. What was the source of heat for the solar nebula? What role does heat play in organizing the Solar System?
4. Describe the stage of Solar System development known as planetesimal accretion. What major processes occurred during that time?
5. What is unique about Earth compared to the other planets? Describe differences and similarities in the chemical composition of the planets.
6. Write a timeline of events in the formation and future of the Solar System.
7. How would the Solar System be different if the entire system rotated at a slower rate?
8. Scientists classify the planets as either classical or dwarf. Use the chemical information provided in this chapter to suggest a new classification system, one based on their chemistry.

FIGURE 2.14 The Solar System.

2-6 Earth's Interior Accumulated Heat during the Planet's Early History

LO 2-6 *Describe the sources of heat for early Earth and the consequences of heat buildup.*

Most scientists agree that Earth and the Solar System were formed by planetesimal accretion and gravitational contraction of a massive nebula. Driven by gravity, the nebula contracted, and its temperature rose until it achieved nuclear fusion, forming the Sun. Energy from the Sun constitutes the solar wind, which stripped the inner planets of their primitive atmospheres of helium and hydrogen (**FIGURE 2.15**).

The solar nebula hypothesis describes the formation of the early Solar System approximately 4.6 billion years ago. This period of Earth's history is informally known among geologists as the "Hadean Eon" (named after Hades, another word for hell) because shortly after its formation, Earth's surface turned into a sea of molten rock. The Hadean Eon lasted about 800 million years, from approximately 4.6 to 3.8 billion years ago. Changes that occurred during that time set the stage for plate tectonics and the origin of life on Earth.

Condensing the Planet

Earth formed in four basic steps:

1. It began to accrete from the nebular cloud as particles smashed into each other, forming planetesimals.

2. As Earth's mass grew, so did its gravitational field, causing it to compress into a smaller, denser spherical body.

3. This compression generated heat in Earth's interior. Heat was also produced by the decay of *radioactive elements* in the interior as well as by the impacts of many thousands of extraterrestrial objects. Earth's interior began to melt. Some of the iron, which is the heaviest of the common elements, and other elements that are chemically similar to iron, such as nickel (Ni), sank toward the planet's centre and eventually condensed to form the core. Chapter 3 describes the development of the layers within Earth in more detail.

4. Molten iron moving through the planet resulted in the conversion of potential energy to kinetic energy, but some energy was also converted into heat that helped raise Earth's temperature. A deep magma ocean formed on the planet's surface.

Extraterrestrial Bombardment

Early in its history, Earth (and other planets) endured heavy impacts by asteroids and comets (**FIGURE 2.16**). Evidence in the form of overlapping craters on other planets and moons suggests that these impacts occurred in two waves, the first consisting of small to medium-size rocky asteroids, most likely from the nearby asteroid belt, and the second probably consisting of icy comets from the distant Kuiper Belt and Oort Cloud that surround the Solar System. Some meteorite debris on Earth is even thought to have come from the planet Mars when it sustained large impacts that hurtled Martian rock into space.

As you can imagine, when an asteroid the size of several city blocks slams into the planet at thousands of kilometres per hour, the collision releases a great deal of heat. In fact, just one such collision would release more heat than all the current nuclear bombs on Earth detonated at once. During the Hadean Eon, tens of thousands to millions of such violent collisions produced enormous quantities of heat, much of which was stored in the solid rock of the planet. With more impacts came more heat. Large impacts, such as the one that may have formed our Moon, were especially important in this process. In time, Earth was transformed from a cold planet into one that was warmed by the energy of extraterrestrial collisions.

Planetary scientists hypothesize that icy comets delivered significant amounts of water to Earth. This water accumulated as pools of liquid and as clouds of gaseous water vapour in a newly forming atmosphere. Much of Earth's water and its atmosphere was also derived from **volcanic outgassing** of water, nitrogen, and carbon, originally trapped in minerals but released by melting within Earth's interior. According to one school of thought, comets may have supplied the majority of Earth's water during the heavy bombardment phase of the Solar System. If this is true, the chances increase that organic matter, which is also found in comets, played an important part in the origin of life on Earth. However, one study suggests that most of Earth's water may not have come from comets. Researchers found that comet Hale-Bopp (visible from Earth in 1997) contains substantial amounts of "heavy water," which is rich in deuterium, an isotope of hydrogen.

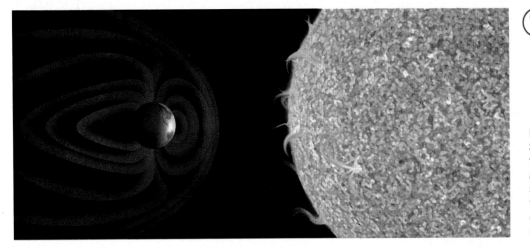

(?) Earth lost its early atmosphere to the solar wind. What is the source of Earth's present atmosphere?

FIGURE 2.15 Early in the history of the Solar System, the solar wind stripped the inner planets of their primitive atmospheres. In this image, the modern solar wind "blows" Earth's geomagnetic field into a streamlined shape with the blunt end facing into the solar wind and the tail extending downwind.

FIGURE 2.16 With our Moon rising in the background, a young, crater-scarred Earth sustains yet another impact. Comets streak over primitive Earth bringing water to fill impact basins. These impacts created the first shallow seas and added new gases to the young planet.

(?) A thin atmosphere is shown. Where did it come from? What is it made of?

If comet Hale-Bopp is typical in this respect, and if cometary collisions were a major source of terrestrial oceans, Earth's ocean water should be similarly rich in deuterium, but in fact it is not.

Heat stored in Earth's interior led to the creation of the first volcanoes. After it lost its hydrogen, helium, and other hydrogen-containing gases due to the Sun's radiation, Earth lacked an atmosphere. In time, delivery of icy gases by comets, along with volcanic outgassing of volatile molecules such as water, methane, ammonia, hydrogen, nitrogen, and carbon dioxide, produced Earth's second atmosphere. By 4.3 to 4.4 billion years ago, rain fell to form the first shallow oceans. Geologists hypothesize that the other terrestrial planets experienced a similar process in their early histories. The early Earth atmosphere contained very little oxygen, and it only started to build up to present levels when oxygen-producing organisms developed. Since then plant photosynthesis and other surface processes have also contributed to the chemical makeup of our atmosphere.

What about Our Moon?

Where did our Moon come from? There have been a number of hypotheses to describe how the Earth-Moon system formed. The *fission hypothesis* was that the centrifugal force associated with Earth's spin caused a bulge of material to separate from Earth in the area of the equator. Unfortunately, this hypothesis requires that Earth rotate once every 2.5 hours in order to develop the necessary force. The

capture hypothesis proposed that Earth's gravity captured a passing planetesimal, which then became the Moon. However, the chemistry of Moon rocks is similar to that of Earth rocks, suggesting that both developed either from a common source or at least at the same location in the planetary nebula. In the *double planet hypothesis,* Earth and the Moon were formed concurrently from a local cloud of gas and dust. However, this hypothesis fails to account for several aspects of our Moon's history, such as the unusual tilt of its axis, the melting of its surface rocks, and the fact that it is less than half as dense as Earth.

Currently, the *impact hypothesis* is the most widely accepted account of the Moon's formation (**FIGURE 2.17**). It states that at some point during planetesimal accretion, Earth suffered a massive collision with a huge object the size of Mars. The collision destroyed the object and part of the already accreted Earth, and it is thought that this material formed a ring of debris around the planet, which rapidly coalesced to form the Moon.

FIGURE 2.17 The impact hypothesis suggests that Earth suffered a massive collision that led to the formation of our Moon. This artist's re-creation shows Earth and a Mars-size object, each peppered by hundreds of smaller impacts, colliding with one another in the early Solar System.

(?) **Expand Your Thinking**—Several hypotheses have been forwarded to explain the origin of the Moon. How is this an example of critical thinking?

LET'S REVIEW "GEOLOGY IN OUR LIVES"

Now that you've finished the chapter, "Geology in Our Lives" will have taken on new meaning for you. Let us review it: Earth, the Sun, and other objects in the Solar System originated at the same time from the same source and have evolved in varying ways since then. The Solar System began with the collapse and condensation of a planetary nebula. Understanding Earth's place in the cosmos provides us with perspective and appreciation for the scale of nature. Exploring the Solar System is one of humanity's grand accomplishments. Someday our planetary neighbours may provide us with useful resources.

As far as we know, Earth is the only habitable planet in the universe. Certainly it is the only habitable planet in the Solar System. You now know the context of Earth's origin and why it is habitable (and why other planets are not). The stage is set for us to return to Earth and investigate its earliest history, how and why it became layered, and the processes of plate tectonics that govern Earth's evolution through time. These are the subjects of the next chapter, "Plate Tectonics."

STUDY GUIDE

2-1 Earth's origin is described by the solar nebula hypothesis.

- The **Solar System** consists of our Sun, eight classical planets (in order from the Sun: Mercury, Venus, Earth, Mars, Jupiter, Saturn, Uranus, and Neptune); five dwarf planets (Pluto, Ceres, Haumea, Makemake, and Eris); small Solar System bodies (including asteroids, comets, and objects in the Kuiper Belt, scattered disc, and Oort Cloud); 162 moons orbiting the classical planets; countless particles; and interplanetary space. The Sun is a star located at the centre of the Solar System; all other components orbit around it.

- The origin of the Solar System is described by the **solar nebula hypothesis**. This hypothesis holds that the components of the Solar System were all formed together approximately 4.6 billion years ago and have evolved in different ways since then.

2-2 The Sun is a star that releases energy and builds elements through nuclear fusion.

- Stars burn by means of **nuclear fusion**, which creates new elements. When stars use up their hydrogen fuel, they become brighter. A star the size of our Sun will become a red giant, engulf its inner planets, and eventually burn out. A field of mineral particles and gas, known as a planetary nebula, is produced when a star explodes at the end of its "lifetime." The planetary nebula provides the raw material for building new planets, such as Earth, as well as new stars, such as the Sun.

- The solar core is the site of nuclear fusion. There hydrogen is converted into helium; the energy created through this process is carried to the surface of the Sun by convection, where it is released as light and heat. Our Sun has been active for approximately 4.6 billion years and has enough fuel to last another 4 to 5 billion years.

2-3 Terrestrial planets are small and rocky, with thin atmospheres.

- In many ways, Earth is unlike the other planets. Earth's hot interior provides the energy to drive plate tectonics, yet its surface is cool enough to support life. Earth's position relative to the Sun provides enough warmth to maintain liquid water, yet we still experience freezing seasons. Earth's winds are relatively calm, the atmosphere is thin and transparent so it is easily penetrated by light, and daily extremes in temperature are comparatively modest. All these conditions promote the existence of life on Earth.

- The planet Mercury lies closest to the Sun and is dominated by unrelenting and intense heat during the day. During the night, because it lacks an atmosphere to trap that heat, temperatures plummet to extreme cold.

- Venus is radically different from Earth. Carbon dioxide, which is very efficient at trapping heat, makes up 96.5 percent of its atmosphere. As a result, temperatures on the surface reach 477°C, making Venus even hotter than Mercury. Also, Venus does not rotate in the same direction that it travels around the Sun, unlike all the other planets.

- Mars is the most Earthlike of the other planets. It has dry river channels that indicate that water once flowed across its surface.

2-4 Gas giants are massive planets with thick atmospheres.

- Jupiter is the first of the **gas giants**. It is composed mainly of hydrogen and helium, with small amounts of methane, ammonia, water vapour, and other compounds.

- Saturn is oblate at the poles, a condition that is caused by an outward push in the equatorial region due to a high rate of rotation. The atmosphere consists primarily of hydrogen, with lesser amounts of helium and methane.

- Uranus is the only planet that does not rotate perpendicular (or nearly so) to the ecliptic. That is, Uranus lies on its side—or tipped over—and its north pole points toward the Sun for half a Uranian year.
- The inner two-thirds of Neptune is likely composed of a mixture of molten rock, water, liquid ammonia, and methane. The outer third is a mixture of heated gases composed of hydrogen, helium, water, and methane.

2-5 Objects in the Solar System include the dwarf planets, comets, and asteroids.

- Pluto and its moon, Charon, are similar in size. This duo is more of a double dwarf-planet system than a planet and moon system. Pluto and Charon, Haumea, and Makemake are objects in the Kuiper Belt.
- Eris is an object in the scattered disc region, which overlaps with the Kuiper Belt but extends farther into space and is not confined to the plane of the ecliptic.

- Ceres is the fifth dwarf planet. It is a large asteroid located in the asteroid belt.
- Comets are "dirty snowballs" of frozen volatiles that are locked in gravitational orbit around the Sun.
- The asteroid belt is a zone of rocky debris between Mars and the giant gaseous planet Jupiter.

2-6 Earth's interior accumulated heat during the planet's early history.

- The earliest phase of Earth's history, the "Hadean Eon," was characterized by extraterrestrial impacts and heating of the planet's interior. Earth gained a new atmosphere as a result of **volcanic outgassing** together with the delivery of gases and water by ice-covered comets.

KEY TERMS

extrapolation (p. 31)
gas giants (p. 29)
Milky Way Galaxy (p. 28)
nuclear fusion (p. 30)

planetesimal accretion (p. 29)
solar nebula hypothesis (p. 28)
Solar System (p. 28)
solar wind (p. 32)

stellar nucleosynthesis (p. 30)
terrestrial planets (p. 29)
volcanic outgassing (p. 40)

ASSESSING YOUR KNOWLEDGE

Please answer these questions before coming to class. Identify the best answer.

1. The classical planets in order from the Sun are
   a. Mercury, Venus, Mars, Earth, Jupiter, Saturn, Neptune, Uranus.
   b. Jupiter, Venus, Mars, Earth, Mercury, Saturn, Neptune, Uranus.
   c. Mars, Venus, Earth, Jupiter, Saturn, Mercury Neptune, Uranus.
   d. Venus, Mars, Mercury, Earth, Jupiter, Saturn, Neptune, Uranus.
   e. None of the above.

2. The energy that is destined to become sunlight may take _____ years to travel from the Sun's core to its surface.
   a. 10
   b. 100
   c. 100,000
   d. 1,000,000
   e. 1,000

3. Nuclear fusion
   a. is the source of the Sun's energy.
   b. occurs when the nucleus of an atom fissions and releases energy.
   c. radiates throughout the universe.
   d. is the reason that Jupiter has no solid surface.
   e. has not yet occurred in the Sun.

4. Mercury, Venus, and Mars are different from Earth because
   a. they are closer to the Sun.
   b. Earth has volcanoes and they do not.
   c. Earth currently has liquid water and they do not.
   d. Earth is the only planet with ice this close to the Sun.
   e. they are all in retrograde orbit.

5. The largest storm in the Solar System is found on which planet?
   a. Uranus
   b. Earth
   c. Mars
   d. Jupiter
   e. Pluto

6. The fastest winds in the Solar System are found on which planet?
   a. Neptune
   b. Jupiter
   c. Venus
   d. Earth
   e. Mars

7. The basic structure of the Solar System is described as
   a. the ecliptic.
   b. the Oort Cloud.
   c. inner terrestrial and outer gaseous planets.
   d. the asteroid filter.
   e. rotating nuclear fission.

8. Why do the outer planets and their moons consist mostly of ice and gas while the inner planets are made up mostly of rock and metal?
   a. The solar wind stripped the inner planets of volatile compounds.
   b. The outer gas giants had greater volcanism, which produced large quantities of gases.
   c. Gravity sucked the gases from the inner planets into the Sun.
   d. Solar heat is so limited in the outer portion of the Solar System that solids turn into gas.
   e. Far from the Sun, the outer gas giants are made of ice.

9. The dwarf planets are
   a. Mercury, Earth, and Mars.
   b. Ceres, Pluto, Haumea, and Mercury.
   c. Eris, Ceres, Pluto, Haumea, and Makemake.
   d. There are no dwarf planets, only moons.
   e. None of the above

10. What is planetesimal accretion?
    a. The Kuiper Belt collapsed into the core region.
    b. Collisions of bits of ice, gas, and dust grew into planetesimals, planetary embryos, and eventually planets.
    c. Jupiter, with its huge mass, broke into pieces that eventually became the major planets.
    d. The solar wind tore the young planets into smaller pieces called "planetesimals," and these later grew together to form the present planets.
    e. The solar wind kept the planets in their positions.

11. Which of the following is the name of a hypothesis explaining the origin of the Solar System?
    a. planetesimal collision
    b. nebular expansion
    c. solar nebula
    d. nuclear fusion
    e. solar objects

12. The major gases in the Solar System include
    a. ice, argon, methane, and carbon.
    b. water, carbon dioxide, ammonia, helium, hydrogen, and carbon monoxide.
    c. lithium, beryllium, carbon, hydrogen, and carbon monoxide.
    d. ammonia, oxygen, helium, hydrogen, carbon monoxide, and water.
    e. water, hydrogen sulphide, ammonia, helium, lithium, and carbon monoxide.

13. Comets are made of
    a. molten rock.
    b. ice and mineral grains.
    c. gas and ice.
    d. rock and a thin atmosphere of argon.
    e. mostly potassium and oxygen.

14. Extraterrestrial impacts
    a. probably occurred in two waves.
    b. may have delivered water to Earth and an early atmosphere.
    c. may have originated at the Oort Cloud and Kuiper Belt regions.
    d. produced the scars on the Moon's surface.
    e. All of the above

15. The source of Earth's heat is a combination of
    a. extraterrestrial impacts, gravitational energy, and radioactivity.
    b. nuclear fusion, volcanism, and compression.
    c. compression, volcanism, and solar wind.
    d. solar wind, radioactivity, and gravitational energy.
    e. None of the above.

FURTHER RESEARCH

1. If you were in charge of a scientific mission to Mars, what goals for the mission would you establish?

2. Do you think plate tectonics exists on other planets? How would you be able to tell from here on Earth?

3. In this chapter, we stated that Jupiter's Great Red Spot has been in existence for hundreds of years. How would we know this?

4. When you look at any of the gas giants, what are you seeing: the land or the top of the clouds?

5. When you look at the terrestrial planets, what are you seeing: the land or the top of the clouds?

6. What is the difference in general chemistry between the gas giants and the terrestrial planets?

ONLINE RESOURCES

Explore more about planetary geology on these websites:

NASA images of the planets and other objects:
photojournal.jpl.nasa.gov/index.html

Images and discussion of the Solar System:
www.solarviews.com/eng/homepage.htm

Canadian Space Agency
www.asc-csa.gc.ca/index.html

Additional animations, videos, and other online resources are available at this book's companion website:
www.wiley.com/go/fletchercanada
This companion website also has additional information about WileyPLUS *and other Wiley teaching and learning resources.*

46

CHAPTER 3
PLATE TECTONICS

Chapter Contents and Learning Objectives (LO)

3-1 Earth has three major layers: core, mantle, and crust.
LO 3-1 *Describe Earth history during the Hadean Eon.*

3-2 The core, mantle, and crust have distinct chemical and physical features.
LO 3-2 *List Earth's internal layers and describe each.*

3-3 Lithospheric plates carry continents and oceans.
LO 3-3 *Describe the origin and recycling of oceanic crust.*

3-4 Paleomagnetism confirms the seafloor-spreading hypothesis.
LO 3-4 *Identify the evidence that the polarity of Earth's geomagnetic field has reversed in the past.*

3-5 Plates have divergent, convergent, and transform boundaries.
LO 3-5 *List the three types of plate boundaries.*

3-6 Oceanic crust subducts at convergent boundaries.
LO 3-6 *Describe the processes occurring at ocean–continent and ocean–ocean convergent boundaries.*

3-7 Orogenesis occurs at convergent boundaries.
LO 3-7 *Describe the origin of the Himalayan Mountains.*

3-8 Transform boundaries connect two spreading centres.
LO 3-8 *Describe transform faults.*

3-9 Earthquakes tend to occur at plate boundaries.
LO 3-9 *Describe where earthquakes tend to occur and why.*

3-10 Plate movement powers the rock cycle.
LO 3-10 *Describe the rock cycle and its relationship to plate tectonics.*

GEOLOGY IN OUR LIVES

The theory of plate tectonics describes Earth's surface as being organized into massive slabs of lithosphere called *plates* in which continents and ocean basins are embedded. Plates move across Earth's surface at about the same speed, on average, that your fingernails grow. At the edges of the plates, where they collide, separate, and slide past one another, rock is recycled, mountains are built, new crust forms, and numerous geologic hazards and geologic resources develop. The history of Earth's surface is largely interpreted by re-creating the past position of plates. The unifying theory of plate tectonics has advanced our understanding of geologic hazards and resources.

(?) Thingvellir National Park, Iceland, lies on the boundary of two plates that are pulling away from each other. What geologic hazards are likely to occur here?

3-1 Earth Has Three Major Layers: Core, Mantle, and Crust

LO 3-1 *Describe Earth history during the Hadean Eon.*

Recall from Chapter 2 that during the Hadean Eon, Earth's interior accumulated and stored heat produced by several processes: impacts of extraterrestrial objects, decay of radioactive elements within the planet, gravitational energy associated with compression of the rocky sphere, and friction generated by flowing liquid iron (Fe). As heat built up, a deep magma ocean formed on the surface; volcanic outgassing released compounds that were formerly bound in the rocks; and the chemical layering of Earth's interior was set in motion.

The **iron catastrophe**—perhaps the most significant single event in Earth history—occurred when the planet's temperature passed the melting point of iron (1,538°C). Based on laboratory experiments, calculations, and evidence of Earth's chemistry, scientists hypothesize that it was this event, known informally as the "iron catastrophe," that resulted in the internal layers that characterize Earth today.

The iron catastrophe began when molten iron, one of Earth's most abundant elements, flowed toward the planet's interior under the pull of gravity. Compounds of lighter elements (e.g., silicon [Si] and oxygen [O] and other light elements that were also molten) were displaced toward the surface. Iron (and nickel [Ni]) accumulated in Earth's deep interior, and less dense compounds accumulated near the surface. Ultimately this process, called **chemical differentiation**, created Earth's internal structure.

As the first great tide of molten iron moved slowly through Earth, frictional heating caused by the sinking of metal droplets and the heat produced as a by-product of their conversion of potential energy to kinetic energy raised the temperature another 2,000°C (**FIGURE 3.1**). This had a dramatic effect, causing the surface to develop a deep ocean of molten rock. Later, as Earth cooled, this ocean solidified to form a solid *primordial crust.*

© Don Dixon

FIGURE 3.1 During the Hadean Eon, Earth's surface developed a magma ocean that extended to a depth of several kilometres.

(?) How did the iron catastrophe change the distribution of compounds in Earth's interior?

Scientists date this molten period from two observations: (1) The oldest meteorites and lunar rocks are about 4.4 to 4.6 billion years old, and (2) the oldest known Earth rocks are about 3.8 to 4.1 billion years old (although some zircon grains from Australia are 4.4 billion years old). The difference in these two ages, roughly half a billion years, may represent the molten phase.

As Earth cooled and the rate of extraterrestrial bombardment waned, the surface formed a solid volcanic crust. The period of geologic time that followed the Hadean Eon, beginning 3.8 billion years ago, is called the **Archean Eon**. This was an important time in Earth's history because the planet could now support life.

Earth's Interior

During the iron catastrophe, about one-third of the primitive planet's mass sank toward the centre. It was through this process that Earth was transformed from a *homogeneous* body, with roughly the same kind of material at all depths, into a *heterogeneous,* or layered, body. It developed a dense iron **core** having a solid inner layer and a liquid outer layer; a brittle outermost rind, the **crust**, composed of less dense compounds with *lower melting points;* and between them, a solid **mantle** of intermediate density. The mantle is Earth's thickest layer and is composed of solid rock that is capable of flow.

Compare the abundance of elements in the crust and in Earth as a whole (**FIGURE 3.2**). Because most of the iron sank to the core, iron drops to fourth place in the crust. Conversely, silicon, aluminum (Al), calcium (Ca), potassium (K), and sodium (Na) are far more abundant in the crust than in the planet as a whole. The reason for this difference is that the elements that are more abundant in the crust form lightweight chemical compounds that tend to be more easily melted. Materials such as these melted early during the period of chemical differentiation, rose to the surface through *convection* (whereby hot materials rise) in the mantle, and accumulated to form the crust.

Much of the way Earth works is based on the structure of its interior. Using *inductive reasoning* (see Chapter 1) from observations of how seismic waves from earthquakes behave as they pass through Earth's interior, along with laboratory reconstructions of pressure and temperature conditions, scientists construct hypotheses describing conditions inside the planet. Earth seems to be composed of four zones, although research is revealing a more complex interior (**FIGURE 3.3**). At the centre is the **inner core**, a rock body made up of solid metal alloy consisting mostly of iron and nickel (two dense elements) that is very hot (about 5,000°C). Normally at this temperature, the rock would be molten, but deep inside the planet, high pressure prevents it from melting. The inner core therefore is solid, perhaps even crystalline.

The inner core is surrounded by the **outer core**, which is also extremely hot, but because it is under less pressure, the rock there is melted. The outer core is also rich in iron and nickel, probably mixed by convection. The mantle is massive (averaging 2,900 km thick) and

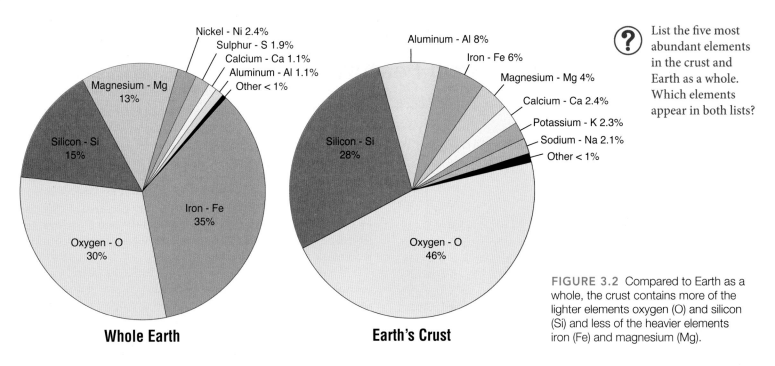

Whole Earth

Earth's Crust

(?) List the five most abundant elements in the crust and Earth as a whole. Which elements appear in both lists?

FIGURE 3.2 Compared to Earth as a whole, the crust contains more of the lighter elements oxygen (O) and silicon (Si) and less of the heavier elements iron (Fe) and magnesium (Mg).

consists of hot solid rock that can flow. How can solid rock flow? Geologists hypothesize that hot rock rises through the mantle very slowly, perhaps only a few centimetres in hundreds of years. This motion is thought to be due to the fact that rocks nearer the base of the mantle are hotter relative to those in the upper mantle, causing them to expand and rise, and then cool when they reach the upper mantle where they contract and eventually sink.

Above the mantle is the solid outer layer of rock that we live on, Earth's crust. Relative to the planet's radius, the crust is thinner than an eggshell. The crust is the outer portion of the **lithosphere**, which consists of the rigid upper mantle (lower lithosphere) and the crust (upper lithosphere). Below the lithosphere lies the *asthenosphere,* a weak and ductile layer in the upper mantle at depths between 150 km and 300 km but perhaps extending as deep as 700 km.

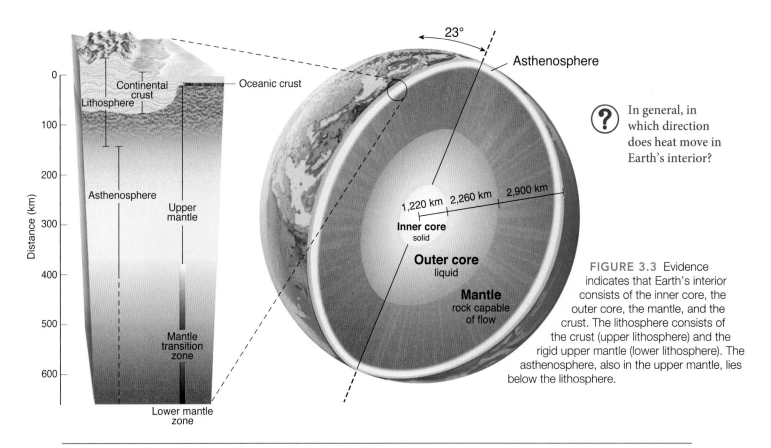

(?) In general, in which direction does heat move in Earth's interior?

FIGURE 3.3 Evidence indicates that Earth's interior consists of the inner core, the outer core, the mantle, and the crust. The lithosphere consists of the crust (upper lithosphere) and the rigid upper mantle (lower lithosphere). The asthenosphere, also in the upper mantle, lies below the lithosphere.

(?) **Expand Your Thinking**—If the crust is composed of lighter compounds than the mantle, is it appropriate to refer to it as "floating" on the mantle?

3-2 The Core, Mantle, and Crust Have Distinct Chemical and Physical Features

LO *3-2 List Earth's internal layers and describe each.*

Earth's interior consists of the core, mantle, and crust; each has distinct chemical and physical features.

The Core

The inner part of Earth is the core, which is about 2,900 km below the surface. The core is a dense ball estimated to be composed of 90 percent iron and about 10 percent nickel, oxygen, and sulphur. The inner core—Earth's centre—is solid and about 1,220 km thick, while the outer core is about 2,260 km thick. Pressures in the inner core are so great that it cannot melt, even though its temperature may exceed 5,000°C. The outer core, which experiences less pressure, is molten. Research indicates that because Earth rotates, the motion of the liquid outer core around the solid, mainly iron inner core creates the planet's **geomagnetic field**, a magnetic field that surrounds Earth (**FIGURE 3.4**). Some scientists hypothesize that the core may actually spin faster than the planet as a whole.

Elementary physics tells us that when an electric current moves around a magnet, it intensifies the magnetic force. Iron is a naturally magnetic metal that, as noted earlier, makes up most of Earth's core. Electric currents in the liquid outer core surround the solid, mainly iron inner core and enhance a magnetic field that spreads throughout the planet. Although it cannot be seen, this magnetic field emerges through the surface at the poles and surrounds the planet.

Since 1845, when the German mathematician *Carl Friedrich Gauss* made the first measurements of Earth's geomagnetic field,

the field's intensity has weakened by about 10 percent. This weakening suggests that the field is undergoing one of the variations in its strength that are known to have occurred many times in Earth's history. However, some scientists suggest that it presages a *reversal in magnetic polarity,* another process that has occurred at irregular intervals in Earth's history, for largely unknown reasons but with potentially disastrous effects. The impact on our electricity-dependent technology could be devastating.

The magnetic field fluctuates, but thankfully never vanishes because it acts as a shield against solar radiation, which is very damaging to living tissue. The sunburn you feel after spending an hour in direct sunlight is a reminder of the power of solar radiation. Without the geomagnetic field, living tissue on Earth's surface would be fried to a crisp. Humans (and, in fact, all life on Earth) exist today only because a relatively strong magnetic field protects us from the Sun's high-energy radiation. It does this by preventing high-energy particles from entering Earth's atmosphere. Thus, in a powerful way, life was made possible by the iron catastrophe.

The Mantle

The layer above the core is the mantle. Its upper surface is about 5 km to 10 km below the **oceanic crust** and about 20 km to 80 km below the **continental crust** (Earth's crust can be divided into an oceanic type and a continental type). The mantle accounts for nearly 80 percent of Earth's total volume. A unique property of the mantle is its *plasticity*—even though the rock is solid, high temperature and pressure enable it to flow extremely slowly.

Compared to the core, the mantle contains more silicon and oxygen, while compared to the crust, it contains more iron and magnesium, as well as larger amounts of calcium and aluminum. Near the bottom of the mantle, the temperature is about 4,000°C, but at the top, it is about 870°C. Hence, heat flows upward through the mantle. Like gravity, heat is an important agent that influences our planet's structure. Heat causes objects to expand. To envision this process in the mantle, think of a lava lamp. A light bulb in the base of the lamp heats wax, causing it to expand (become less dense) and rise to the surface. Cool wax at the top is denser and descends until it reaches the bottom, is heated again, and rises once again. The mantle convection hypothesis is based on observations of the seismic waves that travel outward from earthquakes. These energy waves travel not only along Earth's surface but also through its interior. By observing how these waves change speed as they pass through rising and descending columns, or *plumes,* of mantle rock, geophysicists calculate the shape of those convection columns and create three-dimensional models (**FIGURE 3.5**). These models reveal that Earth's interior is more complex than the simple layered model shown in Figure 3.3.

The Crust

As noted earlier, the rigid upper mantle forms the lower portion of the lithosphere. The upper lithosphere is the solid outer shell that we

Courtesy of NASA Mashall Space Flight Center (NASA-MSFC)

FIGURE 3.4 The geomagnetic field emanates from the core, surrounds Earth as a protective shield, and shelters sensitive living tissue from damaging solar radiation.

(?) Why is the geomagnetic field not symmetrical around Earth in this image?

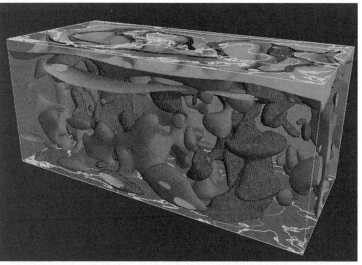

Adam Dziewonski

FIGURE 3.5 Earth's interior as viewed from the north. This complicated model of the mantle is based on records of the behaviour of seismic waves passing through warm (red) and cool (blue) rock. The structure of the planet is more complex than a simple sequence of uniform layers.

(?) How will the red areas and blue areas change over time?

live on, Earth's crust. The crust is composed of brittle rock that fractures easily. It varies in thickness and chemical composition, depending on whether it is oceanic or continental. Oceanic and continental crusts are formed by entirely different geologic processes.

- Oceanic crust is formed at places where iron- and magnesium-enriched magma emerges from the mantle, cools, and crystallizes to create new sea floor. The rock is dense and relatively lacking in lighter compounds compared to continental crust.
- Continental crust is generally composed of rock that is relatively enriched with lighter elements (e.g., silicon, aluminum, sodium, potassium) and depleted with respect to iron and magnesium. Continental crust essentially "floats" on the denser mantle.

Most of the heat moving upward through the mantle comes out through the sea floor. However, typically under the continents, cold rock sinks slowly toward the base of the mantle, where it may recycle. Earth's interior is thought to experience convection at very slow rates measured in millions of years (**FIGURE 3.6**). Check out the "Critical Thinking" exercise "Mantle Plumes" to gain a better understanding of how mantle convection might work.

CRITICAL THINKING

MANTLE PLUMES

Fill a clear, tall glass container 4/5 with water and 1/5 with vegetable oil. Drop a pinch of salt in the top. The salt will form an irregular ball and sink to the bottom of the glass. Then watch as plumes of vegetable oil begin to rise through the water toward the surface. Add more salt so that multiple plumes develop at once.

1. What is the composition of the plumes? Why do they rise through the water?
2. What fundamental physical property causes the plumes?
3. What is the shape of the plumes? Draw them. Geologists hypothesize that plumes of hot rock many kilometres wide move upward through Earth's interior and come to a stop against the underside of the lithosphere, causing it to arch upward and crack in a zone that can be hundreds of kilometres across.

Your experiment may simulate how rock migrates in the mantle (**FIGURE 3.6**).

(?) What is the nature of convection in the mantle?

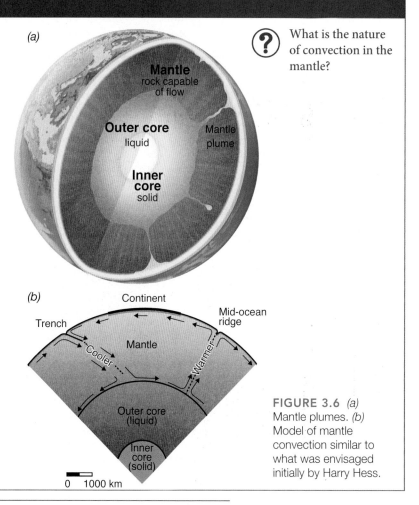

FIGURE 3.6 (a) Mantle plumes. (b) Model of mantle convection similar to what was envisaged initially by Harry Hess.

(?) **Expand Your Thinking**—Imagine there was no iron catastrophe. How would Earth be different today?

3-3 Lithospheric Plates Carry Continents and Oceans

LO 3-3 *Describe the origin and recycling of oceanic crust.*

Today, much of our understanding of Earth's geology is explained by the remarkable theory of plate tectonics. This theory has revolutionized and unified the science of geology over the last few decades. Scientists generally agree that plate tectonics is possible only because Earth's interior is a source of heat that rises to the surface and because the mantle experiences convection (as Hess originally hypothesized). They hypothesize that this convection contributes to the movement of lithospheric plates, thus keeping Earth's surface in constant motion.

Lithospheric Plates

We live on the surface of a planet that is actively shaped by plate tectonics. Mountain ranges, ocean basins, continents, chains of oceanic islands, and other features of the surface all owe their origin to, and can be explained in part by, the movement of massive lithospheric plates. Right now you are riding on a lithospheric plate. Which plate are you on? In what direction are you moving? How many different plates have you visited in your life? Before you finish reading this chapter, you will have the answers to these questions.

Tectonic theory describes Earth's lithosphere, which includes the relatively cold and brittle crust and rigid upper mantle, as being broken up into plates that move and interact with one another

(FIGURE 3.7). These plates move in response to forces in the mantle, and as a result, **plate boundaries**, where plates converge and grind against one another, are locations of great geologic change.

Plates come in a wide range of sizes; they may be hundreds or thousands of kilometres across. Because plates are portions of the lithosphere, a single plate may carry both continental and oceanic crust. Hence, plate movement is responsible for moving continents, thereby solving Alfred Wegener's problem of how to move the continents and build Pangaea. Twelve major plates have been identified. These are known as the African, Arabian, Eurasian, Antarctic, Indian–Australian (some geologists consider the Indian and Australian Plates to be separate), Pacific, Philippine, South American, North American, Cocos, Nazca, and Caribbean Plates. There are several minor plates as well, such as the Juan de Fuca, Bismarck, and Caroline Plates.

In places where two plates run into each other, mountains, volcanoes, and deep-sea trenches may form and earthquakes may occur. In places where two plates pull apart, long, deep *rift valleys* form, such as along the tops of high *mid-ocean ridges* on the sea floor. Because plates move, the continents and oceans riding on them shift over time. For instance, the Atlantic Ocean exists today because the North American, European, African, and South American Plates are moving away from one another, opening up the ocean between them.

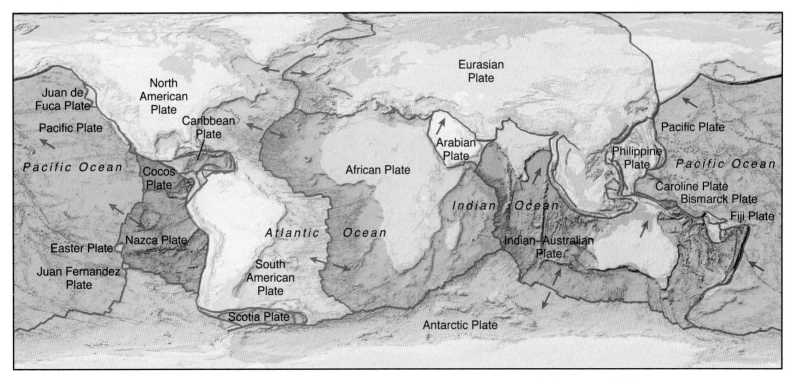

FIGURE 3.7 There are 12 major plates and several smaller ones. Plates are not static; they move several centimetres per year, on average about as fast as your fingernails grow, carrying continents and ocean basins across the planet's surface.

(?) Using the directions of plate movement as your guide, circle places where heat can most effectively escape from Earth's interior to the surface.

Subduction Zones and Spreading Centres

A significant aspect of many plates is that they slide or dive into the mantle along one edge and are renewed with magma welling up from the mantle along another edge. Places where plates are recycled into the mantle—*subduction zones*—are typically characterized by *deep-sea trenches* where the ocean above the downward-curving surface of a plate is extremely deep. At these locations, a denser plate dives, or *subducts,* beneath a less dense plate. For instance, if the edge of one plate is characterized by continental crust (relatively enriched in silicon dioxide [SiO_2]) and is pushed into another plate whose edge is characterized by oceanic crust (which is denser because it is relatively enriched in iron and magnesium), the oceanic crust will subduct beneath the continental crust because it is denser (**FIGURE 3.8**).

While subduction is occurring along one edge of a plate, new lithosphere may be added to another edge of the same plate at a *spreading centre.* Spreading centres are places where magma wells up from beneath the lithosphere and solidifies, adding fresh rock to the edge of the plate. Some geologists hypothesize that spreading centres are located above upward-moving convection currents in the mantle. Others disagree, proposing instead that spreading centres are complex cracks in the mantle that react to stress in the crust and, by lowering pressure, cause melting in the upper mantle. We also know that at a spreading centre, the lithosphere arcs up, developing a wide bulge on Earth's surface at a mid-ocean ridge. But is it pushed up from below or swollen from the heat escaping from Earth's interior? These and many other questions regarding how plate tectonics actually "works" are widely debated as part of the scientific process.

We do know that a spreading centre is characterized by fresh crust, delivered from the mantle and composed of rock enriched in iron and magnesium and depleted in silicon and oxygen. An opening, or *rift zone,* develops at the surface and is filled with this young rock. Rifting of a continent leads to the formation of a low valley that may evolve into a narrow seaway (like the Red Sea) or a continental rift valley (like the East African Rift Valley). In time (after tens of millions of years), rifted lithosphere may widen into an ocean basin with a mid-ocean ridge, the birthplace of young oceanic crust.

Between their birth at spreading centres and their recycling at subduction zones, plates move like a conveyor belt on a one-way journey, taking tens of millions of years to get from one zone to the other. The formation and movement of lithospheric plates has shaped our continents and ocean basins. To match your understanding of world geography with your new understanding of plate tectonics, try the "Critical Thinking" exercise "Global Tectonics" on the next page.

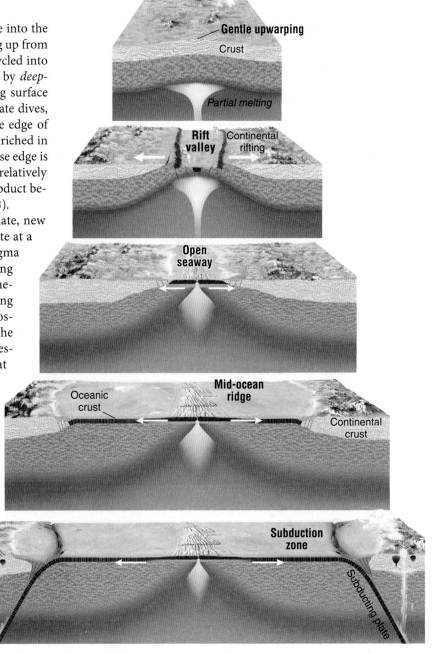

FIGURE 3.8 Lithosphere is produced at spreading centres and carried to subduction zones, where it is recycled into the mantle.

 On Earth's surface, what would be an early sign that a rift is developing?

 Expand Your Thinking—What is the significance of rock density in the movement of lithospheric plates?

CRITICAL THINKING

GLOBAL TECTONICS

The map (**FIGURE 3.9**) shows the topography of the continents and sea floor. Please work with a partner and answer the following. (You will need a marker.)

1. Draw the plate boundaries on the map, and label each plate. Include arrows indicating the direction in which each plate is moving.
2. Circle three subduction zones and three spreading centres. On another sheet of paper, write a description of the topography at each.
3. There is a third type of plate boundary other than a spreading centre and a subduction zone. Can you hypothesize the type of motion that would occur at a third type of plate boundary? What evidence would you look for?
4. What plate do you live on? In what direction is it moving? Describe the processes acting along its boundaries, the types of boundaries it has, and where the boundaries are located.
5. If you had to substantiate the theory of plate tectonics, what evidence (direct observations) would you be able to draw from this map?
6. Heat from Earth's interior and plate boundaries are closely related. But in what way? Imagine you made a map of the temperature of the crust (including the sea floor) using a satellite. What would that map look like? That is, describe the pattern of crust temperature across Earth's surface. What criteria would you use to assess the relationship between crustal temperature and movement of lithospheric plates?

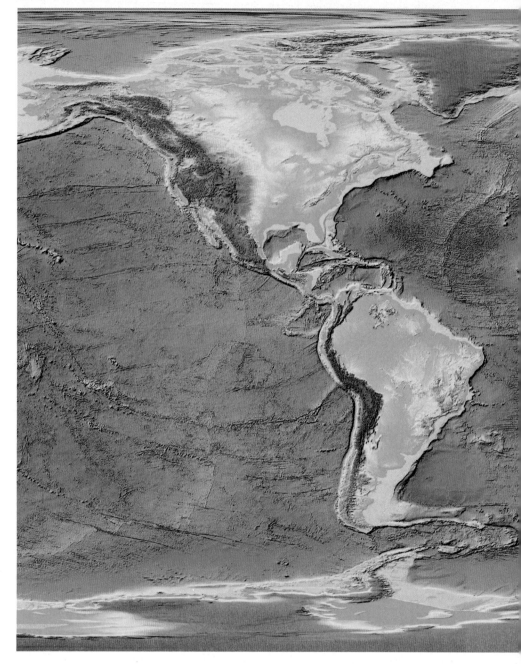

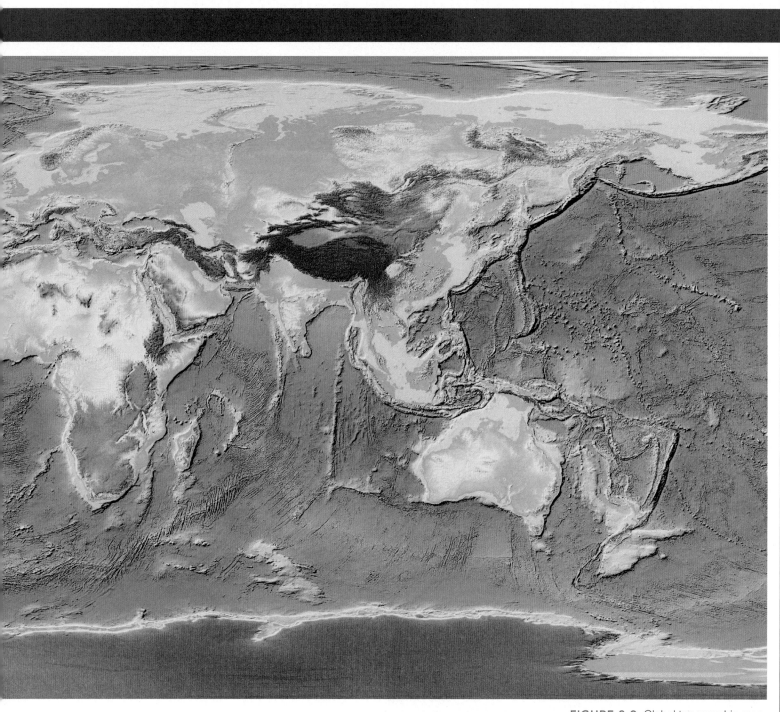

FIGURE 3.9 Global topographic map.

3-4 Paleomagnetism Confirms the Seafloor-Spreading Hypothesis

LO 3-4 *Identify the evidence that the polarity of Earth's geomagnetic field has reversed in the past.*

A magnetic field is *polarized,* meaning that it is characterized by two equal but opposite states, or *poles.* Earth's magnetic poles are named the *north magnetic pole* and the south *magnetic pole.* They emerge from the planet's surface close to, but not exactly, where its rotational (geographic) axis is located (FIGURE 3.10). Notably, Earth's magnetic poles apparently have switched at various times in the past. That is, a compass needle that today points to the north magnetic pole would in past times of reversed polarity point to the south magnetic pole. How do we know this? Because of **paleomagnetism** ("paleo" means "old").

Paleomagnetism

Iron-rich minerals are sensitive to the character of Earth's magnetic field and leave geologic evidence of its polarity in the crust, referred to as *paleomagnetism.* This paleomagnetism is locked into magnetic minerals as they crystallize in molten rock and cool through their *Curie temperature,* the temperature below which they incorporate into their structure the orientation of Earth's magnetic field. Likewise, iron-containing sediment particles settling through water physically rotate and align themselves to the geomagnetic field like compass needles. When these particles are transformed into hard rock, their magnetic orientation is permanently recorded. The orientation of ancient geomagnetic fields preserved in rocks can be measured by instruments called *magnetometers.*

An example of where paleomagnetism is recorded is in cooling magma at spreading centres. Iron-rich minerals crystallizing in newly erupted oceanic crust acquire Earth's prevailing magnetic polarity. When new crust cools, this magnetic "signature" is preserved. Geophysicists' measurements of these minerals reveal characteristics of the geomagnetic field at the time when the rock formed.

Much to the amazement of geologists, paleomagnetic studies conducted in the 1950s indicated that the polarity of the geomagnetic field has reversed numerous times in the past. Most scientists believe that these changes in polarity result from random, chaotic instabilities in the way the liquid iron moves in the outer core. In other words, every once in a while (at intervals ranging from hundreds of thousands to millions of years), Earth's magnetic field changes because (presumably) the movement of liquid iron in the core changes. Exactly why it changes, or exactly when it will change again, is unknown. What is known, however, is that the magnetic poles have reversed hundreds of times throughout Earth history—and likely will again.

Geologists studying this phenomenon recognize that rocks can be divided into two groups based on their magnetic properties. One group has *normal polarity,* characterized by magnetic minerals with the same polarity as today's magnetic field. The other group has *reversed polarity.* The age of a group of rocks showing a particular pattern of reversals can be determined by comparing it to a paleomagnetic reference sequence whose age has been determined by other methods of dating using geochemical techniques (a topic we cover in Chapter 13) (FIGURE 3.11).

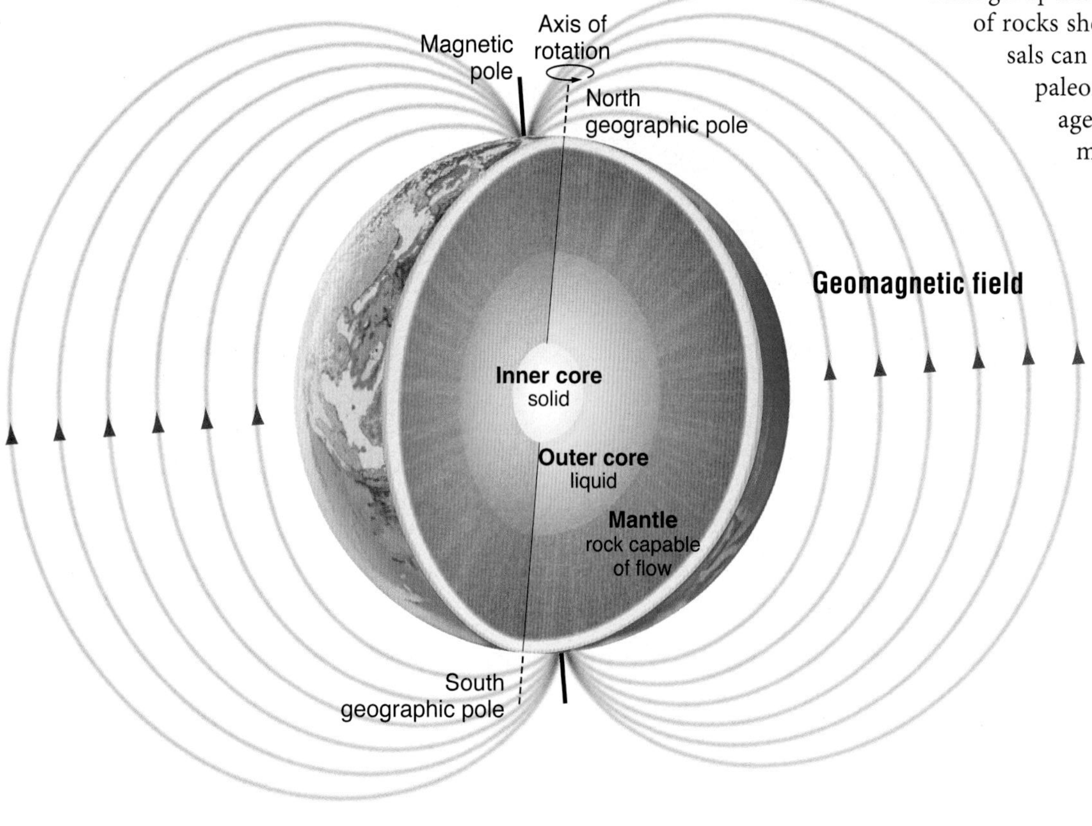

FIGURE 3.10 Earth's magnetic field is generated by convection in the liquid outer core.

? How do rocks record past changes in Earth's geomagnetic field?

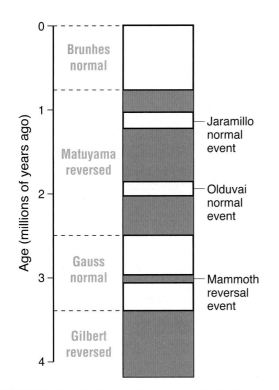

FIGURE 3.11 The history of global polarity for the past 4 million years consists of major *polarity reversals* that last for long periods and minor *polarity events* that are short-lived.

(?) What kind of plate boundaries produce a record of seafloor magnetism?

Seafloor Magnetism

Paleomagnetism is a valuable tool for understanding Earth's history, especially at **divergent plate boundaries** (where two or more plates pull away from each other). Because new rock, in which iron-rich minerals are abundant, is formed at oceanic spreading centres, the sea floor records changes in polarity through time. As it moves away from the spreading centre, the new rock carries with it a distinct magnetic signature. The moving oceanic crust thus preserves a record of switching polarity (**FIGURE 3.12**).

Oceanic crust formed during a period of normal polarity preserves a record of that polarity. When global polarity switches, crust forming at that time will record the change. Hence, a nearly symmetrical pattern of magnetic polarity is found on either side of an oceanic ridge as the sea floor continues to spread over time. Mapping the paleomagnetism of the sea floor adjacent to spreading centres is an important test of seafloor spreading that has confirmed the basic elements of the theory.

Geologists make maps of seafloor magnetic anomalies by colour coding the normal periods (usually black) versus the periods

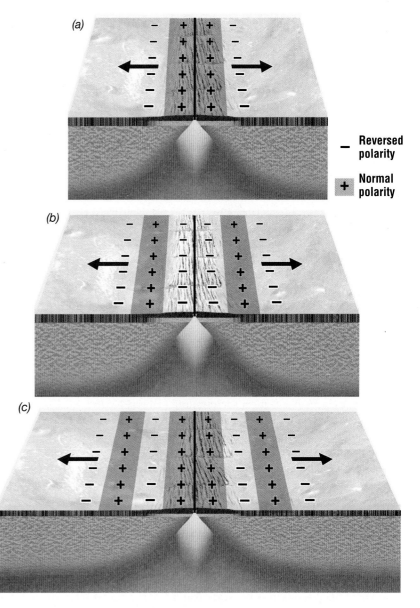

	Reversed polarity
−	
+	Normal polarity

FIGURE 3.12 New oceanic crust is formed continuously at the crest of an oceanic ridge. *(a)* A spreading ridge during a period of normal magnetism; *(b)* followed by a period of reversed magnetism; and *(c)* followed by a period of normal magnetism.

(?) Why is seafloor magnetism considered strong evidence of plate tectonics?

of reversal (white). In the map, this colour coding depicts a series of stripes on the sea floor, referred to as *magnetic striping*. Magnetic striping of oceanic crust usually produces nearly symmetrical patterns with an oceanic ridge in the centre. This process provides strong support for two basic elements of plate tectonic theory: (1) that new lithosphere is created from rock emerging from an upwelling source in the mantle, and (2) that the crust migrates away from these plate boundaries.

(?) **Expand Your Thinking**—Based on your reading of the first three chapters, describe the role of critical thinking in the development of the theory of plate tectonics.

3-5 Plates Have Divergent, Convergent, and Transform Boundaries

LO 3-5 *List the three types of plate boundaries.*

Since it was first proposed, the concept of seafloor spreading has developed from a hypothesis into the modern theory of plate tectonics, which is a unifying theory that provides an explanation of the major features at and beneath Earth's surface, past and present. For instance, it is used by exploration geologists to assist them in their hunt for oil, coal, and natural gas deposits. It successfully predicts why valuable minerals are found in some areas and not in others. The metals and fossil fuels you use are cheaper and easier to find because of the predictive power of tectonic theory. The theory also successfully predicts the location of important geologic hazards, such as volcanoes and earthquakes. Earthquakes and volcanic eruptions do not strike randomly but occur most often (but not always) along the boundaries where plates meet.

The Ring of Fire

One such area is the circum-Pacific **Ring of Fire**, where the Pacific Plate and other minor oceanic plates such as the Juan de Fuca, Nazca, and Cocos Plates meet surrounding plates (**FIGURE 3.13**). Bordered by active plate boundaries, the Ring of Fire is the most seismically and volcanically active zone in the world. People living in the Pacific region are, generally speaking, at greater risk of experiencing geologic hazards than are people living anywhere else in the world.

Plate Boundaries

To understand the dramatic processes that occur at plate boundaries, it helps to use **reasoning by analogy**, another tool of scientific thinking. Picture an orange. Peel the orange so that the skin is in, say, six pieces. Then fit the pieces back together on the orange. Now try sliding one of the pieces across the surface and observe how it interacts with the others.

If you do this with care, you will see three types of interaction: pieces *spreading apart* (divergent boundaries), pieces *sliding past one another* (**transform boundaries**), and pieces *pushing together* (**convergent boundaries**). If two pieces push together hard enough, one will subduct beneath the other or the two will crumple together. These are the only types of interaction that are possible when solid pieces are forced to move around a sphere. It is one of these motions that is shifting the plate beneath you. Because Earth is nearly a sphere, the same processes you observed with the orange peel segments apply to lithospheric plates. As plates move across Earth's surface, three types of boundaries are formed. These are defined by their relative motion (**FIGURE 3.14**), as follows:

1. Divergent boundaries—new lithosphere forms as plates pull away from each other.
2. Convergent boundaries—lithosphere subducts as one plate dives beneath another, or two plates collide head-on without either one subducting.
3. Transform boundaries—lithosphere is neither formed nor recycled; plates simply grind past each other.

Divergent Boundaries

Divergent boundaries occur in areas where two plates are moving away from one another due to seafloor spreading. Seafloor spreading creates new oceanic crust from iron-rich magma that wells up out of the mantle. How does this happen?

Rifting (tearing open) of the lithosphere at a divergent boundary exposes rock in the mantle to a dramatic decrease in pressure as it rises under the rift zone. The decreased pressure causes the rock to melt. The sudden release of pressure causing melting is similar to suddenly opening a can of soda causing carbon dioxide to bubble forth. The carbon dioxide is converted from its dissolved state into gas bubbles,

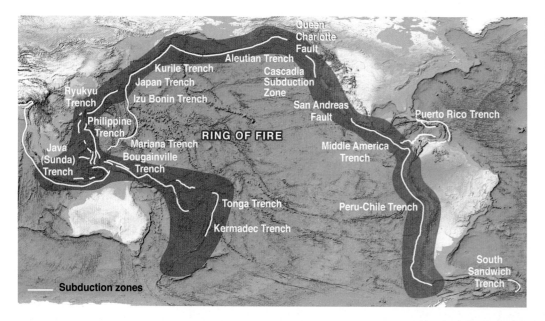

— Subduction zones

FIGURE 3.13 The Pacific Ocean is surrounded by plate boundaries where volcanoes, earthquakes, and tsunamis have the highest probability of occurrence.

(?) Describe the role of plate tectonics in forming the Ring of Fire.

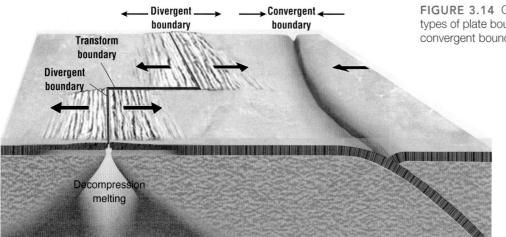

FIGURE 3.14 Geologists recognize three types of plate boundaries: divergent boundaries, convergent boundaries, and transform boundaries.

 Describe why "decompression melting" is more significant at divergent plate boundaries than at convergent plate boundaries.

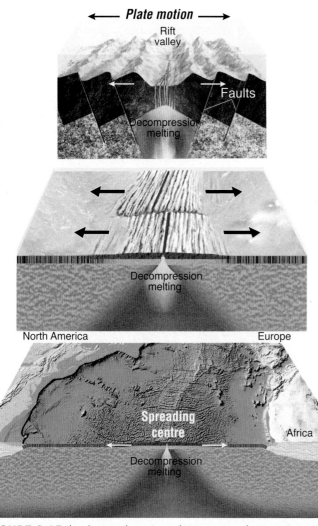

FIGURE 3.15 In places where two plates are moving apart, a central rift valley forms that is characterized by earthquakes and submarine volcanoes.

(?) Predict what geologists in a research submersible would discover if they dove into an active submarine rift valley.

and rock in the mantle is converted from solid to liquid (melted). *Decompression melting* creates magma, which fills the opening in the lithosphere made by rifting. This rock becomes new lithosphere forming high-density, iron-rich oceanic crust (**FIGURE 3.15**).

As newly formed sea floor is transported away from the rift zone on a moving plate, it is replaced by mantle upwelling from below. In this way, the sea floor acts like a conveyor belt carrying new lithosphere away from the rift zone. Earlier we discussed how this process can open up space between continents and lead to the formation of a narrow shallow sea. Rifting in the Red Sea is one such example; it is widening as Saudi Arabia moves away from Africa (see Figure 3.9).

Seafloor spreading leads to the formation of new crust that, compared to continental crust, is relatively enriched in iron and magnesium and depleted in silica (SiO_2) (because it reflects the chemistry of the mantle). As two plates continue to move apart, the rock in the sea floor grows older as its distance from the rift zone increases. As it ages, it cools and becomes denser and is buried under marine sediments that are deposited on the sea floor.

The crust is brittle, meaning that it has a tendency to fracture. Hence, during rifting, the edges of the plates typically break into blocks that slide downward along parallel fracture surfaces. Earthquakes occur when the crust breaks and the rift valley widens. In Figures 3.8 and 3.15, you can see that this breakage occurs both at the edges of the continent where the rifting originated and along the boundary between the oceanic plates in the active rift zone. The motion of opening and the downward settling of the blocks form a central rift valley marking the plate boundary at the spreading centre. These blocks tilt slightly outward, away from the centre of the valley. Magma migrates upward into the rift valley along fracture surfaces among the blocks. Submarine volcanic action is common in the rift zone.

Divergent boundaries on land form rift valleys that are typically 30 km to 50 km wide. The East African Rift Valley in Kenya and Ethiopia and the Thingvellir rift in Iceland (shown in the photo at the beginning of this chapter) are examples. Rift valleys on the sea floor tend to be much narrower, and they run along the tops of mid-ocean ridges. To learn more about seafloor spreading, go to the "Critical Thinking" box "Opening the North Atlantic Basin" and work through the exercise of measuring the history of the Atlantic Ocean Basin as it opened up, following the breakup of Pangaea.

(?) **Expand Your Thinking**—How would you test the hypothesis that there are three types of plate boundaries?

CRITICAL THINKING

OPENING THE NORTH ATLANTIC BASIN

You will need a ruler, coloured pencils, and a simple calculator for this exercise.

FIGURE 3.16 is a map of the sea floor between Africa and North America. These two land masses were last joined when they were part of the supercontinent known as Pangaea. The line labelled "0" is the Mid-Atlantic Ridge, where new sea floor is being formed. The age of the sea floor in millions of years before the present is marked on either side of the ridge. The distance between point A on North America and point B on Africa is 4,550 km.

Select a point on the sea floor and record its age. Measure the distance it has moved from the oceanic ridge where it formed. Use the scale on the map to measure the distance with a ruler.

1. Age of sea floor: _____ million years (my).
2. Distance to the Mid-Atlantic Ridge: _____ km.
3. Calculate the *half-rate* of seafloor spreading—that is, the velocity at which this rock has spread away from the ridge (distance/time = speed): _____ km/my.
4. What is this rate in centimetres per year? _____ cm/y.
5. Calculate the total rate of ocean widening (2 × half-rate = ocean widening rate): _____ cm/y.

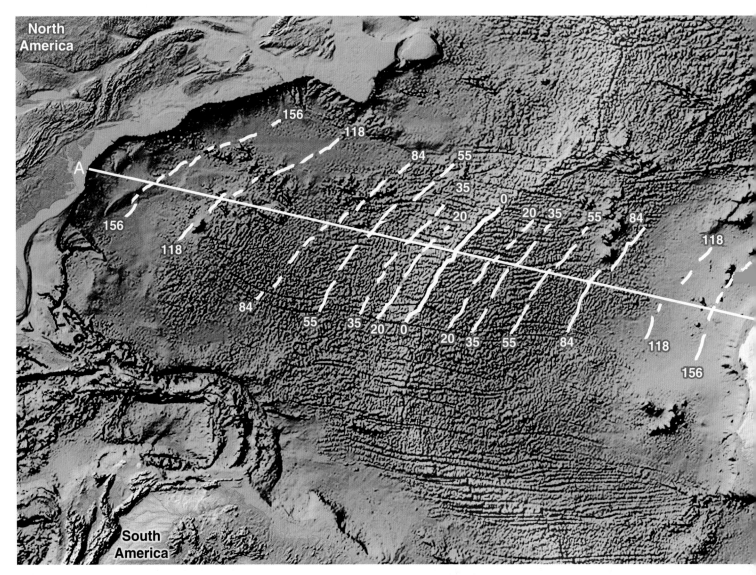

FIGURE 3.16 History of the North Atlantic Ocean.

6. What is it in kilometres per million years? _____ km/my.
7. What is the age of the North Atlantic Basin? _____ my.
8. Name the geologic period during which the North Atlantic Basin began to open up. (Use the geologic time scale in FIGURE 3.17.) _____.
9. How much has the distance between North America and Africa increased since you were born? _____.
10. How much does the distance increase during the average lifetime of a Canadian (~81 years)? _____.

11. An important phenomenon has been observed about the rate of seafloor spreading: It has not been constant through geologic time.
    a. How does this affect your calculations in the above questions?
    b. How might you correct for this observation in order to increase the accuracy of your answers?
12. Using the simplified paleomagnetic time scale in Figure 3.17, draw on the map with coloured pencils to indicate the approximate location of normal- and reversed-polarity sections of the sea floor. That is, make a map of magnetic striping.

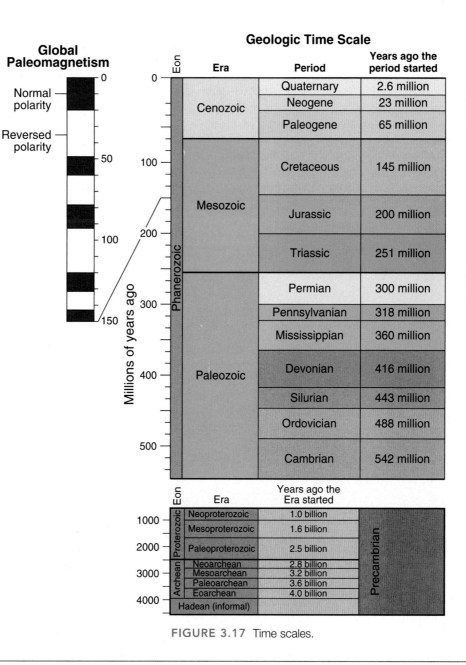

FIGURE 3.17 Time scales.

3-6 Oceanic Crust Subducts at Convergent Boundaries

LO 3-6 *Describe the processes occurring at ocean–continent and ocean–ocean convergent boundaries.*

The type of rock found at two converging plates is an important factor controlling what happens at the boundary between them. There are three types of convergent boundaries, each named for the type of crust present: ocean–continent, ocean–ocean, and continent–continent.

Ocean–Continent Convergence

FIGURE 3.18 illustrates an **ocean–continent convergent boundary**. Here, a plate composed of oceanic crust has collided with one composed of continental crust. Because oceanic crust is relatively high in iron and magnesium and is dense and thin, it subducts, or goes under, the continental plate. The continental plate, built of rock that is relatively enriched in low-density compounds, "floats" on the denser mantle and naturally overrides the denser oceanic plate.

The geology at an ocean–continent convergent zone is quite complicated. Thick layers of marine sediment, consisting of clay particles washed off the continents and microskeletal debris from plankton living in the ocean, have collected on the oceanic plate since it first formed at the spreading centre. In general, the older a plate is, the thicker the layer of sediment covering it. (Sediment thickness can reach hundreds of metres or more.) As an oceanic plate subducts,

some of this sediment is scraped off and collects, or "accretes," as a series of angular rock slabs, forming a wedge along the front of the overriding plate. This process is similar to what happens when snow accumulates in front of a snowplow. The wedge is known as an **accretionary prism** because of its three-dimensional shape. Continued accretion pushes the prism into a ridge called a *fore-arc ridge*. On the landward side of this ridge, a depression (known as a *fore-arc basin*) can develop; such a basin naturally will collect sand and mud coming off the continent. These features, the ridge and the sediment-filled basin, form the *continental shelf* that marks the front edge of the continental lithosphere. Some of the sediment is subducted along with the oceanic crust.

Our knowledge of the geologic processes occurring at convergence zones is complicated by the difficulty of studying these large-scale features. At some locations, it appears that ocean–continent convergence leads to accretion, as just described. At other locations, researchers have proposed that tectonic *erosion* actually scrapes rock off the underside of the overriding plate. Rather than building a continental shelf by forming an accretionary prism, tectonic erosion removes rock from the underside and leading edge of the overriding plate. Geologists hypothesize that thick layers of sediment on a subducting plate tend to produce accretion, whereas thin layers of sediment lead to erosion.

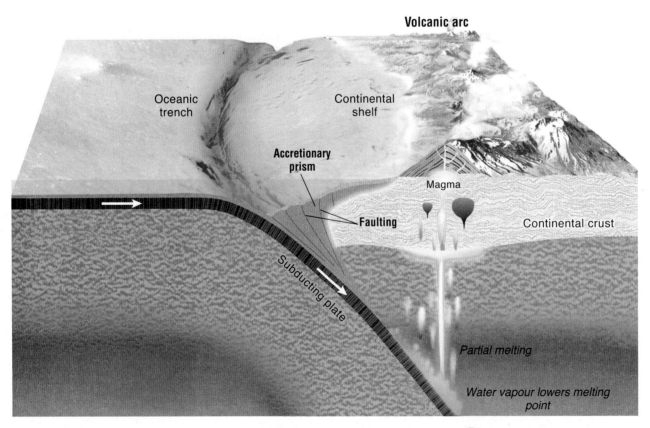

Volcanic arc

Oceanic trench

Continental shelf

Accretionary prism

Faulting

Magma

Continental crust

Subducting plate

Partial melting

Water vapour lowers melting point

FIGURE 3.18 At many sites, ocean–continent convergence is characterized by the growth of an accretionary prism and a line of volcanoes called a *volcanic arc*.

 What is the source of rock in an accretionary prism?

Releasing Water

As an oceanic plate subducts into the mantle, it experiences increases in temperature and pressure. These increases result in the release of large amounts of water vapour and other volatile fluids and gases trapped within the marine sediments and oceanic crust. Released water vapour invades upward into the hot rocks of the upper mantle and causes widespread melting; this process is similar to the way in which salt causes ice to melt prematurely—it lowers the melting point of the ice. Contrary to what we might think, the presence of water or water vapour in hot rocks will lower the melting point, causing it to melt more easily. Hence, as this water rises into the mantle, the rocks there melt, forming magma. An additional minor source of magma may come from melting of the subducting oceanic crust.

Magma generated above a subducting plate migrates upward into the overlying plate, taking advantage of fractures and zones of weakness in the rock, and partially melts the crust to form yet more magma. Eventually it may erupt onto the surface as a volcano, or crystallize within the crust as a *pluton*. An entire line of highly explosive volcanoes, called a **volcanic arc**, will grow along the edge of the continental plate on the landward side of the subduction zone. This volcanic range marks the presence of the convergent boundary.

Partial melting produces magma that is rich in silica (silicon and oxygen, and other low-density elements) compared to the parent rock. This means that the chemistry of the rock in the volcanic arc differs from that of the rock in the upper mantle—in relative terms, the volcanoes are enriched in silicon and oxygen and depleted in iron and magnesium compared to the upper mantle. This knowledge will prove valuable in Chapter 6, where we discuss why some volcanoes are violently explosive while others are not.

Oceanic Trenches

The deepest sea floor in the world is the *Challenger Deep*, located at the southern end of the Mariana Trench where the Pacific Plate subducts beneath the Philippine Plate at a steep angle. An *oceanic trench* is a deep, curved valley thousands of kilometres long and 8 km to 11 km deep that cuts into the ocean floor. Trenches are caused by subduction of one plate under another. They are curved because it is not possible to form a straight depression in Earth's curved surface. (Try pushing in the surface of a Ping-Pong ball with your thumb—the edge of the depression is curved.)

Ocean–Ocean Convergence

An **ocean–ocean convergent boundary** (FIGURE 3.19) is formed where two plates composed of oceanic crust collide. Usually the older and therefore, denser of the two plates subducts below the other. As in ocean–continent convergence, water vapour released from the rock and sediment of the subducting plate causes partial melting in the mantle above it. As the resulting magma migrates into the overlying oceanic crust, it causes partial melting and the eventual eruption of a line of volcanoes on the sea floor. In time, these volcanoes build a chain of volcanic islands known as an **island arc**.

Ocean–ocean boundaries differ from ocean–continent boundaries in that they typically do not exhibit well-developed accretionary prisms or fore-arc basins. Some geologists think that ocean–ocean boundaries are erosional and that, over time, the overriding plate is worn back (by tectonic erosion) toward the island arc.

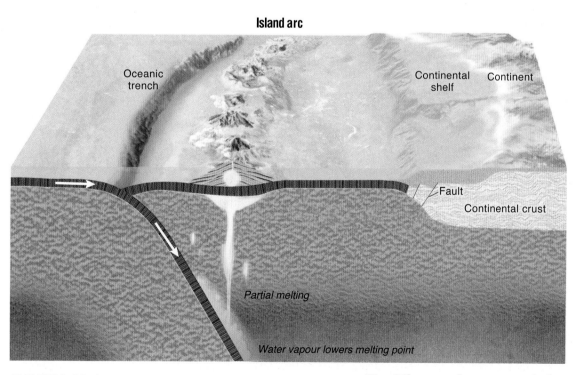

Island arc

Oceanic trench

Continental shelf

Continent

Fault

Continental crust

Partial melting

Water vapour lowers melting point

FIGURE 3.19 An ocean–ocean convergent boundary is characterized by a line of volcanic islands called an island arc.

(?) When two plates composed of oceanic crust converge, which one is likely to subduct?

(?) **Expand Your Thinking**—What features or parts of a volcanic arc and an island arc are similar, and what parts are dissimilar?

3-7 Orogenesis Occurs at Convergent Boundaries

LO *3-7 Describe the origin of the Himalayan Mountains.*

Mountain building in any form (e.g., volcanism, block faulting, crustal thickening) is referred to as **orogenesis** (Greek for "mountain" and "origin"). Mountain building resulting from the collision of two or more lithospheric plates is known as *collisional orogenesis*, characterized by thickening and deformation of crust at the collision site as well as by folding, breaking, and crustal uplift. The most spectacular type of collisional orogenesis occurs at the third type of convergent boundary, the **continent–continent convergent boundary**. It should be noted that another type of collisional orogenesis occurs where oceanic plateaus, island arcs, and micro-continents are carried in on the subducting oceanic plate and are accreted onto the leading edge of the overriding continental plate. A classic example of this is the Canadian Cordillera, which we will look at in more detail in Chapter 15.

Many plates consist of both continental and oceanic crust, and the entire plate moves toward any site where the oceanic portion is being subducted. If the subduction zone is characterized by ocean–continent convergence, and subduction of oceanic crust continues, any continent riding on the subducting plate will be drawn toward the overriding plate. The evolving margin of the continent on the overriding plate (known as an **active margin**), with its volcanic arc and accretionary prism, will collide with the tectonically quiet margin (known as a **passive margin**) of the continent riding on the subducting plate. When the ocean finally closes as the last of the oceanic crust is subducted and recycled into the mantle, a major collision zone will be formed because the continental crust is composed of relatively buoyant rock that resists subduction. As a result, when two plates experience continent–continent convergence, both undergo compression that leaves their edges broken, contorted, and deformed (**FIGURE 3.20**). In addition, the fore-arc ridge, the fore-arc basin, the accretionary prism, portions of the oceanic crust, and the volcanic arc are all squeezed between the two colliding continents. This process

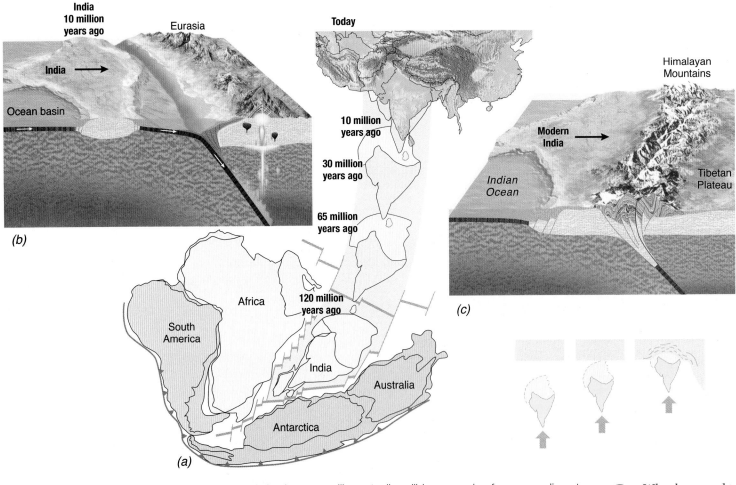

FIGURE 3.20 Two continents separated by a subduction zone will eventually collide, squeezing fore-arc sediments and rocks. The Himalayan Mountains, the tallest system of mountains in the world, are forming at a continent–continent convergent boundary resulting from the collision of the Indian–Australian and Eurasian Plates. *(a)* After separating from the southern supercontinent Gondwana approximately 120 million years ago, India migrated toward the Eurasian Plate. *(b)* As India approached, the ocean between India and Eurasia narrowed and the sea floor was characterized by a deep-sea trench. *(c)* Eventually India collided with Eurasia and produced the Himalayan Mountains and the Tibetan Plateau.

(?) What happened to the oceanic crust between India and Eurasia?

builds mountains in a dramatic fashion that results in the creation of a high range of mountain peaks with deep continental "roots." Geologists refer to upward movement of rock as *crustal uplift.* The resulting mountain range is embedded within a new, larger continent composed of two formerly separate continents. The new land mass sits higher in the mantle because of the combined buoyancy of the two continents. In this way, continents may grow larger over time. Continent–continent convergence is responsible for the formation of the Himalayan, Ural, and Appalachian Mountain Systems, as well as many others.

The Himalayan Mountains are a striking example of collisional orogenesis resulting from continent–continent convergence. They are the result of collision between the Indian–Australian Plate and the southern edge of the Eurasian Plate. Over 100 million years ago, India broke away from the huge southern continent, known as *Gondwana,* which was composed of South America, Africa, Madagascar, Antarctica, India, Sri Lanka, and Australia. Gondwana itself was assembled during late Precambrian and Cambrian time and later participated in the formation of Pangaea. The breakup of Pangaea, which initially rifted near the equator, formed a large northern continent called *Laurasia,* and the southern land mass, Gondwana, re-emerged in the Southern Hemisphere.

The Himalayan Mountains were formed as a result of orogenesis caused by continent–continent convergence. It is the highest system of mountains in the world, containing Mount Everest, the tallest mountain. In behind the Himalayan Mountains, this convergence has also resulted in the uplifting of a very large tract of land, the Tibetan Plateau, which is approximately 25,000 km^2 with an average elevation of 4.5 km. The mountain-building process continues, as India is still pushing its way to the north and the Himalayas are deforming upward—Mount Everest grows by over 6 mm and shifts northeast by about 44.5 mm each year. Deadly earthquakes in Pakistan and India are evidence of active tectonics in this region.

To review what you have learned about plate boundaries, test your understanding by completing the exercise in the "Critical Thinking" box "Tectonic Features" below.

CRITICAL THINKING

TECTONIC FEATURES

1. Label plate tectonic features in the cross-section below (**FIGURE 3.21**). Label as many features as you can—there are at least 10 different ones—and draw arrows indicating the direction of movement of the crust.
2. Label the Earth layers in the cross-section.
3. Identify the general chemistry and physical state of each layer, and indicate important processes and features that occur within it.
4. Two rift zones are shown. What are the differences and similarities between them?
5. Draw this scene after a period of time has passed. Identify an area of the world that looks like your drawing.

FIGURE 3.21 Major components of the plate tectonic model.

Expand Your Thinking—So far in this chapter, you have learned how two different types of mountain ranges are built: volcanic mountains (arcs—two types) and "collision" mountain ranges. If you were asked to determine the origin of a mountain range about which you had no prior information, what observations would you collect to build your hypothesis? How would you test your hypothesis? Identify some observations that would point you in the direction of one origin versus another.

3-8 Transform Boundaries Connect Two Spreading Centres

LO *3-8 Describe transform faults.*

Transform boundaries are characterized by side-to-side plate movement. That is, they occur where two plates slide past each other rather than colliding or separating. You modelled transform boundaries in your experiment with the orange peel.

This type of motion is called *shearing.* Typically, transform boundaries produce little direct collision or separation between plates and tend to be marked by linear valleys, offset ridges, long lakes that have been filled by groundwater, and stream valleys that zigzag across the plate boundary (**FIGURE 3.22**).

Transform boundaries are important for several reasons: (1) they complete our understanding of plate tectonics, (2) they explain significant features on Earth's surface, and (3) they are characterized by frequent earthquakes. Canadian geophysicist *J. Tuzo Wilson*, who first proposed the hotspot hypothesis, also proposed that transform boundaries connect two spreading centres (or, less commonly, two subduction zones or a subduction zone and spreading centre). Let us take a closer look at two common types of transform boundaries that connect two spreading centres.

Probably the most famous transform boundary in the world is the *San Andreas Fault, a transform fault* (**FIGURE 3.23**). A **fault** is a place where the crust is broken and the broken edges are offset relative to each other (either vertically or horizontally or both). The San Andreas Fault is a transform boundary between the North American Plate to the east and the Pacific Plate to the west. It should be noted that the terms "transform boundary" and "transform fault" are synonymous.

The Pacific Plate grinds northward along the edge of the North American Plate at a rate of about 6 cm/y. The City of Los Angeles, largely sitting on the Pacific Plate, is slowly

(?) What evidence would you look for to indicate the direction of displacement associated with movement along the San Andreas Fault?

(?) What has happened to the stream in this setting?

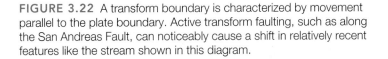

FIGURE 3.22 A transform boundary is characterized by movement parallel to the plate boundary. Active transform faulting, such as along the San Andreas Fault, can noticeably cause a shift in relatively recent features like the stream shown in this diagram.

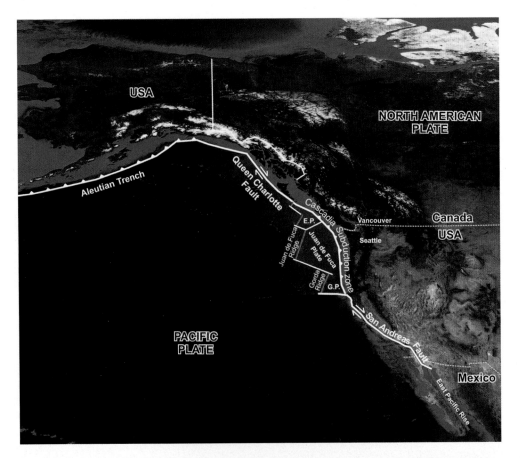

FIGURE 3.23 The San Andreas and Queen Charlotte Faults are transform boundaries between the North American Plate (right side) and the Pacific Plate (left side). Abbreviations: E.P.= Explorer Plate; G.P.= Gorda Plate.

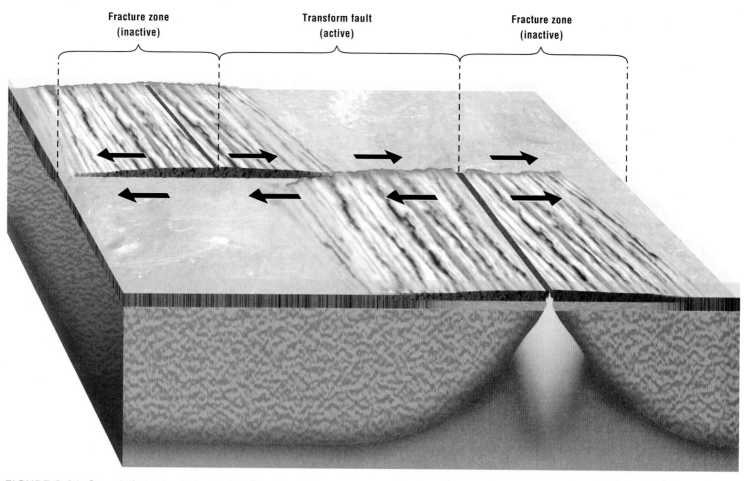

FIGURE 3.24 Oceanic fracture zones consist of transform faults, which are a type of active plate boundary, and fracture zones, which are inactive.

 Why are fracture zones inactive?

moving toward the City of San Francisco, which sits largely on the North American Plate. At the present rate of movement, the two will be joined in about 10 million years (plus or minus a couple of suburbs).

But what about Wilson's prediction? The San Andreas Fault connects two spreading centres. To the north, offshore of Washington and Oregon, is the *Gorda Ridge,* where a spreading centre separates the Gorda Plate from the Pacific Plate, and to the south is the *East Pacific Rise* in the Gulf of California, where a spreading centre separates the Pacific Plate from the North American Plate. Thus, the San Andreas Fault is a transform boundary that connects two spreading centres, just as Wilson predicted.

A common second type of transform boundary is an **oceanic fracture zone** consisting of inactive *(fracture zone)* and active *(transform fault)* portions. These are ocean-floor valleys that extend horizontally away from spreading centres (**FIGURE 3.24**). Some fracture zones are hundreds to thousands of kilometres long and as much as 6 km to 7 km deep. Examples include the Clarion, Molokai, and Pioneer fracture zones in the east Pacific between California and Hawaii. Only small portions of oceanic fracture zones, the transform fault portion, are active at any given time.

You may be curious about why transform faults develop. The answer, however, is not clear. Reasoning by analogy, the process that results in the formation of a fracture zone may be similar to what

happens when you tear a sheet of paper. If you look closely at one side of the tear, you will notice dozens of micro-tears at angles (often right angles) to the main direction of the torn sheet. Each of these micro-tears is an expression of the tendency of fractured objects (the torn paper in this case) to splinter. The torn paper is analogous to a spreading centre, with oceanic transform faults beginning as splinters of the main rift and having since extended.

Examine Figure 3.24. Notice that a single transform fault has split a spreading centre; as Wilson put it, "the transform boundary connects two spreading centres." Along the portion of the fracture between the two spreading centres, opposite sides of the boundary move in opposite directions, creating a shear zone in the transform fault portion. But where the fracture is not located between spreading centres, the two sides of the boundary are locked together and moving in the same direction. This is the inactive fracture zone.

Something else to notice is the age of the crust on either side of a fracture zone. Because under normal circumstances new sea floor is made at approximately the same rate on either side of a spreading centre, the sea floor at equal distances from an oceanic ridge should be similar in age. But because an oceanic fracture offsets a spreading centre, the sea floor on either side of a fracture may differ in age. On the side closer to where it was made, the sea floor is younger. On the side, farther from where it was made, it is older.

Expand Your Thinking—Using Figure 3.24, write the words "older," "younger," and "same" to indicate the age relationships of the sea floor on both sides of the transform fault and fracture zones.

3-9 Earthquakes Tend to Occur at Plate Boundaries

LO *3-9 Describe where earthquakes tend to occur and why.*

How do the three boundaries defined by crust type (ocean–ocean, ocean–continent, continent–continent) and the three plate motion settings (divergent, convergent, transform) relate to one another? TABLE 3.1 summarizes the geologic processes we have discussed and gives examples of the types of crust and styles of plate motion found in various tectonic settings.

Risky Business

In general, plate boundaries are geologically hazardous places. A notable example is the Pacific Ring of Fire, mentioned earlier. This region is hazardous because of the geologic processes occurring at plate boundaries. The forces associated with plate separation in rift valleys commonly lead to earthquakes (Chapter 12) and volcanic eruptions (Chapter 6). Transform boundaries, where one plate shears past another, are characterized by earthquakes that can be very destructive. For example, earthquakes caused by shearing along the Queen Charlotte Fault (Figure 3.23) just off the west coast of British Columbia are among Canada's most frequent and destructive. Convergent boundaries, the site of plate subduction and orogenesis, are subject to numerous earthquakes as well as explosive volcanism.

The most hazardous earthquakes occur when a section of a transform boundary or a subduction zone becomes "locked." That is, local plate motion comes to a halt because two segments of plate cannot slide past each other, even though overall plate movement continues. As the pressure created by plate movement builds, the potential for a large-scale earthquake grows as well. Finally, the locked segment can no longer withstand the accumulated stress and the boundary breaks free with a sudden jolt, causing an earthquake (FIGURE 3.25). In the case of a locked subduction zone, these are called *megathrust quakes.*

This was probably the situation that led to the massive Andaman-Sumatra quake on December 26, 2004, which caused a tsunami that killed more than 230,000 people.

Earthquakes cause the ground to shake because seismic energy released by the breakage of rock passes through the crust. The result can be disastrous as houses, bridges, and roadways fracture and topple as a result of the shaking (FIGURE 3.26). Earthquakes also loosen unstable hillsides and cause landslides and rock avalanches.

 Perhaps you are among the millions of people living in the beautiful regions created by plate interactions. If so, you are also at relatively high risk of experiencing geologic hazards such as earthquakes. For instance, the west coast, which includes large urban centres such as Vancouver, B.C., and Seattle, Wash., is susceptible to megathrust earthquakes induced by the Cascadia Subduction Zone (Figure 3.23). On January 26, 1700, a magnitude 8.7–9.2 megathrust earthquake struck the Pacific Northwest with such force it generated waves that travelled across the Pacific Ocean, causing a tsunami to hit Japan.

Mountain ranges, volcanic slopes, shorelines—these and other environments located on plate boundaries can be hazardous because of the very forces that formed them. In the course of human history, millions of people have been killed by earthquakes, volcanic eruptions, tsunamis, landslides, and other hazards produced at plate boundaries.

Occasionally major eruptions or earthquakes kill large numbers of people. In 1883, for example, an eruption of Krakatau volcano in Indonesia, along with the resulting tsunami, killed 37,000 people. An earthquake in China in 1557 killed an estimated 830,000 people living in hillside caves that collapsed. In 1983, a

TABLE 3.1 Geologic Processes at Plate Boundaries			
	Type of Plate Motion		
Type of Crust	**Divergent**	**Convergent**	**Transform**
Ocean–Ocean	Oceanic ridge, rift valley, spreading centre	Island arc, tectonic erosion	Oceanic fracture zone, transform fault
Example	Mid-Atlantic Ridge	Indonesian Islands	Molokai oceanic fracture zone
Ocean–Continent	Not common	Volcanic arc, subduction zone, oceanic trench	Transform fault zones
Example		Cascade Mountains, Andes Mountains	Queen Charlotte Fault
Continent–Continent	Continental rift valley	Collision zone, orogenesis, crustal uplift	Transform fault zone
Example	East African Rift	Himalaya Mountain System	Portions of San Andreas Fault

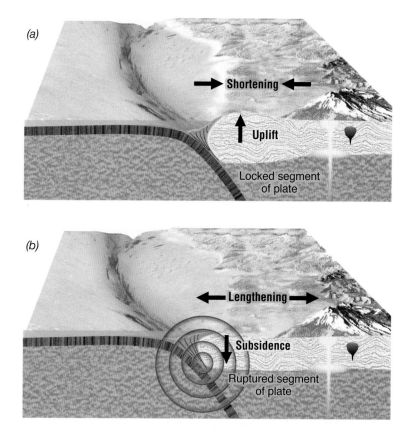

FIGURE 3.25 When a locked subduction zone ruptures, it creates a megathrust quake that can be very damaging. (a) As strain accumulates, the crust shortens and uplift occurs above the subduction zone. (b) When the plates unlock, the crust extends, the land subsides, and a megathrust earthquake occurs.

? What hazards and environmental changes does the coastal zone on the overriding plate experience during and following a megathrust earthquake?

mudflow on Nevada del Ruiz volcano in Colombia, the result of a volcanic eruption, killed 25,000 people. In 1976, an earthquake in Tangshan, China, killed 250,000 people. In December 2003, 45,000 people died as a result of a major earthquake in Iran, and in China, over 80,000 people died in 2008 from a massive earthquake. The two worst disasters in recent history happened only a few years apart. The Andaman-Sumatra tsunami in 2004 destroyed hundreds of communities on the shores of the Indian Ocean and killed about 230,000 people, and the 2010 Haiti earthquake in the Caribbean Sea killed up to 316,000 people. All these events had one thing in common—they all occurred on plate boundaries. Clearly, plate boundaries are dangerous places.

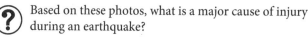

FIGURE 3.26 (a) A major earthquake hit the Caribbean nation of Haiti on January 12, 2010. It is estimated that up to 316,000 people died and over 1 million were left homeless. (b) In 2008, a strong earthquake hit China and killed over 80,000 people, caused 300,000 injuries, and left 5 million homeless.

? Based on these photos, what is a major cause of injury during an earthquake?

Where Do Earthquakes Occur?

Do you live near a seismic hazard zone? We know that earthquakes are generated by broken crust moving suddenly along a fault surface. We also know that the theory of plate tectonics predicts active faulting at plate boundaries. Hence, we may *deduce* that earthquakes are associated with plate boundaries (**FIGURE 3.27**), where crust is known to break and move suddenly. In large part, this is an accurate prediction. But even if you do not live near a plate boundary, you may be at risk. Damaging earthquakes sometimes occur in areas that are not located at plate boundaries. In Canada, these areas include the Charlevoix and Lower St. Lawrence Seismic Zone along the St. Lawrence River; northeastern Ontario in the region near Kapuskasing, Timmins, and Kirkland Lake; the southern Great Lakes region including Toronto and Hamilton; and even Nunavut, to name a few.

About 81 percent of the world's largest earthquakes occur in the *Circum-Pacific Seismic Belt*. This belt, which includes the rim of the Pacific Plate and nearby plates, corresponds to the Pacific Ring of Fire and extends from Chile, north along the west coast of North America,

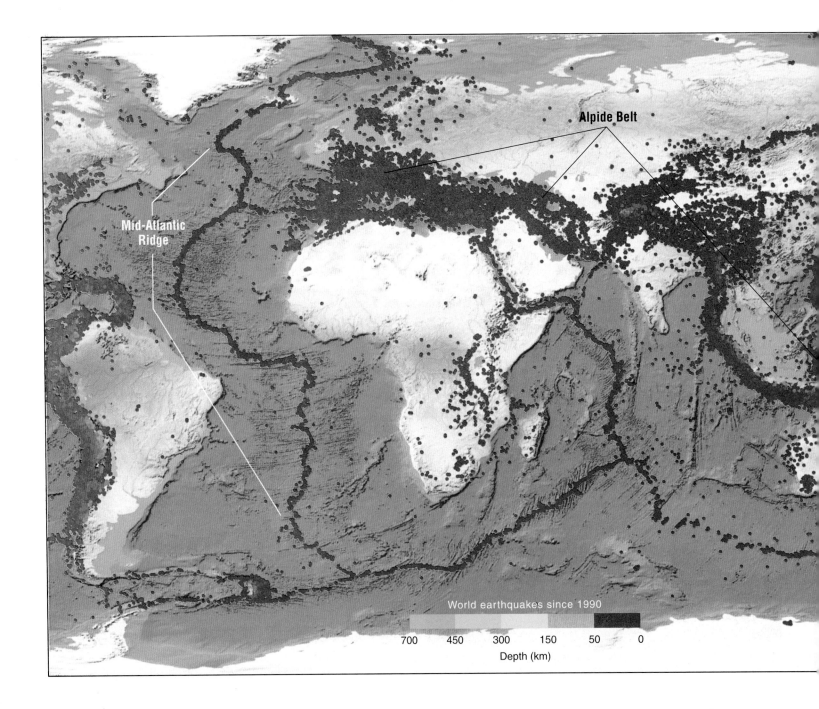

Alpide Belt

Mid-Atlantic Ridge

World earthquakes since 1990

700   450   300   150   50   0

Depth (km)

through the Aleutian Islands to Japan, and south through the Philippine Islands, the island groups of the southwest Pacific, and New Zealand. Eight of the 10 largest earthquakes since 1900 have occurred along this seismic belt.

The second most important earthquake belt, the *Alpide Belt,* extends from Java to Sumatra through the Himalayas and the Mediterranean and out into the Atlantic. This belt accounts for about 17 percent of the world's largest earthquakes, including some of the most destructive, such as the Andaman-Sumatra quake in 2004, the Iran shock in 2003, and tremors in Turkey in 1999 that killed more than 17,000 people. This belt is marked by largely convergent and transform boundaries between numerous microplates moving from the south against the Eurasian Plate to the north.

The third prominent earthquake belt follows the submerged *Mid-Atlantic Ridge.* Quakes in this belt have not been particularly destructive, as they generally occur far from population centres. Although earthquakes in prominent seismic zones are expected, damaging shocks can and do occur outside these areas.

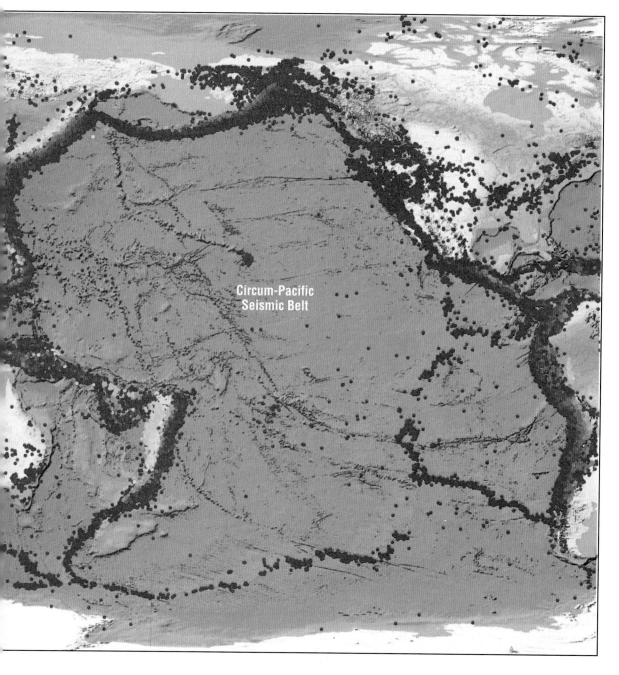

Circum-Pacific
Seismic Belt

 Based on the alignment of earthquakes, can you locate and name the 12 major lithospheric plates on the map in Figure 3.27?

FIGURE 3.27 Earthquakes tend to be aligned with plate boundaries. The Circum-Pacific Seismic Belt, the Alpide Belt, and the Mid-Atlantic Ridge are the most seismically active regions.

 Expand Your Thinking—Why do oceanic fracture zones typically experience shallow earthquakes?

3-10 Plate Movement Powers the Rock Cycle

LO 3-10 *Describe the rock cycle and its relationship to plate tectonics.*

Scientists are not sure what causes plates to move. Four possible mechanisms have been hypothesized: Plates may be (1) pushed away from spreading centres, (2) pulled down into trenches, (3) dragged along by the friction created by mantle convection on their undersides, or (4) driven by gravity to slide down the slope from the ridge to the trench. Conventional wisdom holds that all four mechanisms contribute to plate movement, although researchers are undecided on the proportion of total movement contributed by each mechanism (FIGURE 3.28).

Ridge push is the force applied to plates at spreading centres. As plates separate, new, hot magma is extruded between them from the upper mantle. The high temperature in the rift lowers the density of the new rock, causing it to float higher in the mantle; this creates a "push" effect from below. Gravity draws the young sea floor downhill and away from the rift zone, widening the gap and thus allowing hot magma to well up and push again.

Slab pull is another potential driving mechanism. Subduction zones typically occur far from spreading centres, and, therefore, the plates have had plenty of time (tens of millions of years) to cool. The lower temperature and changes in the mineral composition of this older sea floor make the plate denser than the hot mantle below it.

Hence the leading edge of the plate, or "slab," sinks under its own weight. As the slab is drawn under, it pulls the rest of the plate behind it. Suggesting that slab pull is an important process is the observation that plate movement is faster where the age (and thus thickness and density) of subducted crust is greater.

Several models of plate tectonics support the mantle convection hypothesis originally proposed by Harry Hess (see Figure 3.6). In the *plate drag* model, the lithosphere rides passively on currents in the upper mantle. In one model of plate drag, the upper mantle is separate from the lower mantle, with little or no transfer of rock between them. Small, local convection cells in the upper mantle occur beneath the interiors of plates and create frictional drag on the underside thereby reinforcing plate movement. Plate drag thus is the result of mantle convection currents dragging plates away from ridges. However, many geologists are not convinced that passive drag can generate the energy needed to move massive plates like the North American Plate.

The *ridge slide* mechanism relies on the gravitational slope between the crest of an oceanic ridge and the base of a deep-sea trench to cause a plate to slide downhill. Such action might lead to ripping of the plate (rifting) and pressure-release melting in the upper mantle.

The Rock Cycle

James Hutton (1727–1797), an 18th-century gentleman farmer and one of the founders of modern geology, originated the concept of the *rock cycle*. This concept illustrates the interrelationships among igneous, sedimentary, and metamorphic rocks. Today, we know that tectonic processes drive important aspects of the rock cycle.

The modern rock cycle concept proposes that the mantle, the crust, the atmosphere, the biosphere, and the hydrosphere can be envisioned as a giant recycling machine in which the elements that make up rocks are neither created nor destroyed but rather are redistributed and transformed from one natural form to another over time.

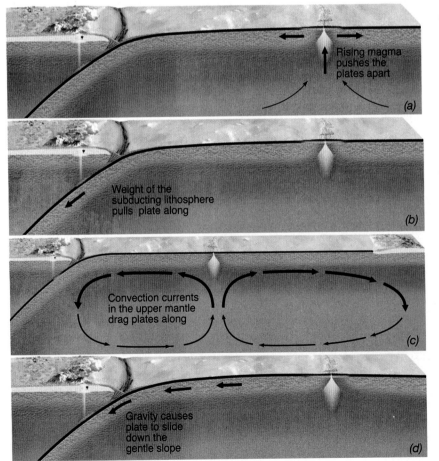

(a) Rising magma pushes the plates apart

(b) Weight of the subducting lithosphere pulls plate along

(c) Convection currents in the upper mantle drag plates along

(d) Gravity causes plate to slide down the gentle slope

(**?**) Each of these models of plate movement is a separate hypothesis. How would you test each one?

FIGURE 3.28 Researchers hypothesize four mechanisms of plate movement: *(a)* the ridge push model; *(b)* the slab pull model; *(c)* the plate drag model; *(d)* the ridge slide model.

The rock cycle is a continuous process and can begin at any step in the cycle. Heat from the Sun produces chemical reactions between living organisms, liquids, gases, and solid materials on Earth's surface, causing *weathering*. (Weathering is covered in Chapter 7.) Weathering reactions destroy minerals in sedimentary, igneous, and metamorphic rocks that are exposed to the atmosphere. New minerals form as a result of weathering processes.

Mechanical, chemical, and biological weathering produce sediment that consists of new minerals as well as the broken remains of older weathered rock. Sediments may be eroded (moved) and eventually deposited in the ocean or some other depositional environment. (Sedimentary processes are covered in Chapter 8.) As sediments accumulate, buried particles are turned into solid rock through compaction and cementation.

Sedimentary rock, when exposed to high temperatures and pressures at plate margins or deep in the crust, will metamorphose. If metamorphic rock is exposed to higher temperatures originating within Earth (i.e., geothermal heat), it will melt and turn into magma. Magma that becomes part of oceanic or continental crust will cool and form igneous rock. In time, continental crust composed of igneous rock will be uplifted by orogenesis and exposed to weathering—whereupon it turns to sediment and enters a new round of the rock cycle.

The rock cycle can be defined and linked to the tectonic framework (**FIGURE 3.29**). As we saw earlier, tectonics leads to the formation of new igneous rock (oceanic crust) at a spreading centre. As ocean crust moves slowly toward subduction, it collects thick layers of marine sediment consisting of fine particles from the land and skeletal remains of plankton that live in the ocean. Eventually the crust subducts at a convergent margin. The heat and pressure of subduction releases water and other volatile compounds from the crust as the plate descends into the mantle.

Sediment may subduct with the plate or collect as an accretionary ridge and prism above the subduction zone. Both subducted and accreting sediments are metamorphosed by pressure at the convergent margin. Water escaping from the subducting plate causes partial melting in the overlying mantle. The resulting magma is high in silicon and oxygen and other easily melted components. It intrudes into the overlying crust where it generates more magma through partial melting of the crust. Eventually magma extrudes onto Earth's surface and forms a volcanic arc or island arc composed of igneous rock.

Both oceanic crust and the sediments on it experience metamorphism in the subduction zone due to the extreme pressures and temperatures related to plate convergence. When subduction eventually leads to orogenesis, the level of metamorphism increases due to the enormous pressure generated by colliding continents causing the crust to thicken by folding and faulting—in fact, large segments of continental crust may metamorphose. Eventually these rocks will weather and the sediments will return to the sea.

Crust in the mountainous arcs experiences weathering and breaks down into particles of sediment. This sediment is eroded by wind and running water and collects on oceanic crust, forming sedimentary rock. This rock is eventually subducted (or joins the accretionary prism and ridge), thus completing the rock cycle.

The rock cycle is Earth's great recycling engine, perpetually renewing and destroying rock and ultimately providing humans with critically important natural resources. Simultaneously, it exposes us to dangerous natural hazards, such as earthquakes, volcanism, and landslides.

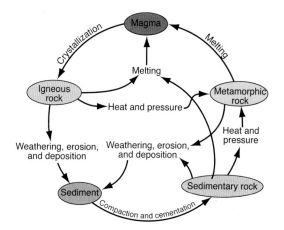

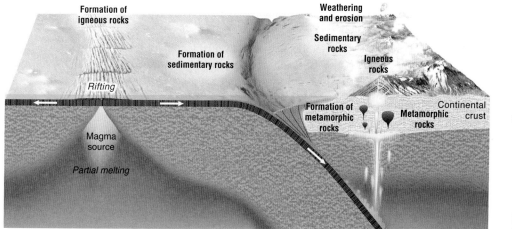

Identify each step of the rock cycle in the model of plate tectonics.

FIGURE 3.29 The rock cycle is closely tied to plate tectonics.

Expand Your Thinking—How would the rock cycle change if the rate of heat flow from the mantle doubled?

LET'S REVIEW "GEOLOGY IN OUR LIVES"

Now that you have finished the chapter, "Geology in Our Lives" will have taken on new meaning for you. Let us review it: The theory of plate tectonics describes Earth's surface as being organized into massive slabs of lithosphere called *plates* in which continents and ocean basins are embedded. Plates move across Earth's surface at about the same speed, on average, that your fingernails grow. At the edges of plates, where they collide, separate, and slide past one another, rock is recycled, mountains are built, new crust forms, and numerous geologic hazards and geologic resources develop. The history of Earth's surface is largely interpreted by re-creating the past position of plates.

The unifying theory of plate tectonics has advanced our understanding of geologic hazards and resources.

You have learned that Earth is not an unchanging ball of rock in orbit around the Sun. It is hoped that this knowledge will stay with you throughout your life and that you will continue to be aware that plate tectonics perpetually renews the crust and drives the rock cycle, that it is a major cause of hazardous earthquakes and volcanoes, and that these processes have been shaping Earth's history for many millions of years.

STUDY GUIDE

3-1 Earth has three major layers: core, mantle, and crust.

- When Earth's temperature reached the melting point of iron, the **iron catastrophe** occurred.

- The iron catastrophe led to the differentiation of Earth into layers. The **inner core** and **outer core** form the core complex, a region that is rich in iron and nickel. The **mantle** is rich in heavy elements compared to the **crust** but contains more silicon and oxygen than the **core**. The uppermost layer is the lithosphere, which consists of the crust and a portion of the rigid upper mantle. The crust is low in iron and magnesium and high in silicon and oxygen relative to Earth as a whole.

3-2 The core, mantle, and crust have distinct chemical and physical features.

- Earth's core is over 2,900 km below the surface. The inner core is thought to be composed of a solid iron-nickel alloy and the outer core a turbulent iron-rich fluid. The core complex generates a **geomagnetic field** that emanates from the planet's surface near the poles.

- Mantle rock has plasticity (the ability to flow). Because the lower mantle is hotter than the upper mantle, convection occurs, in which warm rock rises and cold rock descends.

- Earth's crust consists of continental and oceanic types. Together these form the uppermost portion of a thicker layer called the lithosphere. **Oceanic crust** is 5 km to 10 km thick; **continental crust** is 20 km to 80 km thick or more. Oceanic crust is relatively rich in iron and magnesium, while continental crust is rich in low-density compounds.

- The portion of the lithosphere underlying the crust belongs to the upper mantle. The density of continental rocks allows the crust to float on the mantle. The lower lithosphere rests on the asthenosphere.

3-3 Lithospheric plates carry continents and oceans.

- Tectonic theory describes Earth's lithosphere as being broken up into plates that move and interact with one another. These plates move in response to forces in the mantle. As a result, **plate boundaries**, where plates bump into and grind against one another, are locations of great geologic change.

- Twelve major plates have been identified. These are known as the African, Arabian, Eurasian, Antarctic, Indian–Australian (in some cases the Indian and Australian Plates are separate), Pacific, Philippine, South American, North American, Cocos, Nazca, and Caribbean Plates. There are several minor plates as well.

- Places where plates are recycled into the mantle, called subduction zones, are characterized by deep-sea trenches, where the ocean above the downward-curving surface of a plate is extremely deep. At these locations, a denser plate dives, or subducts, beneath a less dense plate.

3-4 Paleomagnetism confirms the seafloor-spreading hypothesis.

- As magnetic minerals crystallize in molten rock, they incorporate into their structure the orientation of Earth's magnetic field. Likewise, iron-containing sediment particles settling through water physically rotate and align themselves to the geomagnetic field like compass needles. When these particles are transformed into hard rock, their magnetic orientation is permanently recorded. This is called **paleomagnetism**.

- Paleomagnetism is a valuable tool for understanding Earth history, especially at **divergent plate boundaries** (where two or more plates pull away from each other).

- Geologists studying this phenomenon recognize that rocks can be divided into two groups based on their magnetic properties. One group has normal polarity, characterized by magnetic minerals with the same polarity as today's magnetic field. The other group has reversed polarity.

3-5 Plates have divergent, convergent, and transform boundaries.

- Bordered by active plate boundaries, the **Ring of Fire** is the most seismically and volcanically active zone in the world. People living in the Pacific region are, generally speaking, at greater risk of experiencing geologic hazards than are people living elsewhere in the world.

- **Divergent boundaries**—new lithosphere forms as plates pull away from each other.
- **Convergent boundaries**—lithosphere subducts as one plate dives beneath another or two plates collide head-on without either one subducting.
- **Transform boundaries**—lithosphere is neither formed nor recycled; plates simply grind past each other.

3-6 Oceanic crust subducts at convergent boundaries.

- An **ocean–continent convergent boundary** occurs where a plate composed of oceanic crust has collided with one composed of continental crust. Because oceanic crust is relatively high in iron and magnesium and is dense and thin, it subducts, or goes under, the continental plate.
- As a plate subducts, some of this sediment is scraped off and collects, or "accretes," as a series of angular rock slabs, forming a wedge along the front of the overriding plate. This process is similar to what happens when snow accumulates in front of a snowplow. The wedge is known as an **accretionary prism** because of its three-dimensional shape.
- An **ocean–ocean convergent boundary** is formed where two plates composed of oceanic crust collide. Usually the older and (therefore) denser of the two plates subducts below the other. These are sites of island arc formation.

3-7 Orogenesis occurs at convergent boundaries.

- The third type of convergent boundary occurs when an ocean basin closes and two plates composed of continental crust collide. Because neither of the two plates will subduct, a **continent–continent convergent boundary** builds mountains in a spectacular fashion.
- Continent–continent convergence is always preceded by ocean–continent convergence. As subduction of oceanic crust continues, any continent riding on the subducting plate will be drawn toward the overriding plate. The evolving margin of the continent on the overriding plate (known as an **active margin**), with its volcanic arc and accretionary prism, will collide with the tectonically quiet margin (known as a **passive margin**) of the continent riding on the subducting plate.
- Mountain building resulting from this kind of continent–continent collision is a type of **orogenesis**. The Himalayan Mountains are an example of orogenesis. They are the result of collision between the Indian Plate and the southern edge of the Eurasian Plate.

3-8 Transform boundaries connect two spreading centres.

- Transform boundaries are characterized by side-to-side plate movement. That is, they occur where two plates slide past each other, rather than colliding or separating. This type of motion is called *shearing*.
- Geophysicist J. Tuzo Wilson, who first proposed the hotspot hypothesis, also proposed that transform boundaries connect two spreading centres (or, less commonly, two subduction zones).

- Probably the most famous transform boundary in the world is the San Andreas Fault, a transform fault. A **fault** is a place where the crust is broken and the broken edges are offset relative to each other (either vertically or horizontally). The San Andreas Fault is a transform boundary between the North American Plate to the east and the Pacific Plate to the west.
- A common type of transform boundary is an **oceanic fracture zone** consisting of inactive (fracture zone) and active (transform fault) portions. These are ocean-floor valleys that extend horizontally away from spreading centres.

3-9 Earthquakes tend to occur at plate boundaries.

- The forces associated with plate separation in rift valleys lead to earthquakes and volcanic eruptions. Transform boundaries are characterized by earthquakes that can be very destructive. Convergent boundaries are subject to numerous earthquakes as well as explosive volcanism.
- The most hazardous earthquakes occur when a section of a transform boundary or a subduction zone becomes "locked." That is, local plate motion comes to a halt because two segments of plate cannot slide past each other even though overall plate movement continues. As the pressure created by plate movement builds, the potential for a large-scale earthquake grows as well. In the case of a locked subduction zone, these are called megathrust quakes.
- Earthquakes cause the ground to shake because seismic energy released by the breakage of rock passes through the crust. The result can be disastrous as houses, bridges, and roadways fracture and topple as a result of the shaking.
- About 81 percent of the world's largest earthquakes occur in the Circum-Pacific Seismic Belt. This belt, which includes the rim of the Pacific Plate and nearby plates, corresponds to the Pacific Ring of Fire. The second most important earthquake belt, the Alpide Belt, extends from Java to Sumatra through the Himalayas and the Mediterranean and out into the Atlantic. The third prominent earthquake belt follows the submerged Mid-Atlantic Ridge. Quakes in this belt have not been particularly destructive, as they generally occur far from population centres.

3-10 Plate movement powers the rock cycle.

- Scientists are not sure what causes plates to move. Four possible mechanisms have been hypothesized: Plates may be (1) pushed away from spreading centres, (2) pulled down into trenches, (3) dragged along by the friction created by mantle convection on their undersides, or (4) driven by gravity to slide down the slope from the ridge to the trench. Conventional wisdom holds that all four mechanisms contribute to plate movement, although researchers are undecided on the proportion of total movement contributed by each mechanism.
- James Hutton (1727–1797), an 18th-century gentleman farmer and one of the founders of modern geology, originated the concept of the rock cycle. The modern rock cycle concept proposes that the mantle, the crust, the atmosphere, the biosphere, and the hydrosphere can be envisioned as a giant recycling machine in which the elements that make up rocks are neither created nor destroyed but rather are redistributed and transformed from one natural form to another over time.

KEY TERMS

accretionary prism (p. 62)
active margin (p. 64)
Archean Eon (p. 48)
chemical differentiation (p. 48)
continent–continent convergent
 boundary (p. 64)
continental crust (p. 50)
convergent boundaries (p. 58)
core (p. 48)
crust (p. 48)
divergent plate boundaries (p. 57)

fault (p. 66)
geomagnetic field (p. 50)
inner core (p. 48)
iron catastrophe (p. 48)
island arc (p. 63)
lithosphere (p. 49)
mantle (p. 48)
ocean–continent convergent
 boundary (p. 62)
oceanic crust (p. 50)
oceanic fracture zone (p. 66)

ocean–ocean convergent boundary
 (p. 63)
orogenesis (p. 64)
outer core (p. 48)
paleomagnetism (p. 56)
passive margin (p. 64)
plate boundaries (p. 52)
reasoning by analogy (p. 58)
Ring of Fire (p. 58)
transform boundaries (p. 58)
volcanic arc (p. 63)

ASSESSING YOUR KNOWLEDGE

Please answer these questions before coming to class. Identify the best answer.

1. During the Hadean Eon, which of the following is thought to have occurred?
   a. growth of the modern seas
   b. formation of modern continents
   c. the iron catastrophe
   d. origin of life on Earth
   e. All of the above.

2. How does the chemical differentiation of Earth today reflect the influence of the iron catastrophe?
   a. There is more iron in the core than in the crust.
   b. The lower lithosphere stores most of Earth's iron.
   c. Much of Earth's iron has escaped as a result of extraterrestrial impacts.
   d. Iron is rare in Earth.
   e. None of the above.

3. What are the principal differences between the average chemistry of the crust and the average chemistry of Earth as a whole?
   a. The crust is relatively rich in less dense compounds and relatively depleted in iron.
   b. The crust is relatively rich in magnesium and relatively depleted in oxygen.
   c. Earth as a whole has a greater abundance of silicon than the crust does.
   d. The crust contains a greater abundance of heavier elements than Earth as a whole does.
   e. None of the above.

4. How is Earth organized?
   a. Earth has an inner and outer core, a mantle, and a crust.
   b. Earth has an inner mantle and an outer lithosphere, with a liquid inner core.
   c. Earth's crust rests atop the liquid mantle and the solid outer core.
   d. The inner core is solid, the mantle is solid, and the crust is solid under the continents and liquid under the oceans.
   e. None of the above.

5. Subduction occurs
   a. when one plate crashes into another.
   b. when a lithospheric plate is recycled into Earth's interior.
   c. when a plate enters the inner core.
   d. when a continent is recycled beneath an overriding plate.
   e. during orogenesis.

6. Oceanic crust
   a. is formed by asteroid impact.
   b. is enriched in iron and magnesium compared to continental crust.
   c. forms from sea salt.
   d. is made of metamorphic rock.
   e. None of the above.

7. Magnetic reversals are caused by
   a. lunar gravitational effects.
   b. changes in the rate at which Earth orbits the Sun.
   c. impacts of extraterrestrial objects.
   d. unknown causes.
   e. faster subduction rates across Earth.

8. Evidence that the polarity of Earth's geomagnetic field has reversed in the past is found
   a. in magnetic striping in volcanic arcs.
   b. in magnetic reversals recorded by iron minerals in oceanic crust.
   c. in accretionary prisms.
   d. where magma develops above a subducting slab.
   e. All of the above.

9. Three plate boundaries, defined by relative motion, are
   a. converging, diverging, and lateral.
   b. convergent, divergent, and transform.
   c. strike slip, hotspot, and spreading centre.
   d. spreading centre, transform, and divergent.
   e. All of the above.

10. The three types of convergent plate boundaries are
    a. convergent, divergent, and volcanic.
    b. ocean–ocean, ocean–continent, and continent–continent.
    c. subducting, divergent, and shearing.
    d. igneous, sedimentary, and metamorphic.
    e. None of the above.

11. At ocean–ocean convergent boundaries,
    a. younger, less-dense crust tends to subduct.
    b. island arcs tend to subduct.
    c. transform faults typically will develop.
    d. there are rarely earthquakes.
    e. None of the above.

12. The Himalayan Mountains are an example of
    a. extraterrestrial impact.
    b. continent–ocean convergence.
    c. a subduction zone.
    d. continent–continent convergence.
    e. All of the above.

13. At the San Andreas transform fault,
    a. lithosphere is subducted as one plate dives below another.
    b. new lithosphere is formed as two plates pull away from each other.
    c. pressure-release melting recycles old crust.
    d. the Pacific Plate moves to the north relative to the North American Plate.
    e. All of the above.

14. Earthquakes occur most often at
    a. divergent plate boundaries.
    b. ocean–ocean convergent plate boundaries.
    c. ocean–continent plate boundaries.
    d. transform boundaries.
    e. All of the above.

15. The rock cycle is a concept that
    a. has no relationship to plate tectonics.
    b. is not a well-accepted hypothesis.
    c. describes the recycling of rock.
    d. was first described only two decades ago.
    e. All of the above.

FURTHER RESEARCH

1. Search the web and find out when Earth's polarity last reversed.

2. What are some of the geologic characteristics of plate boundaries?

3. Why do oceanic ridges have such high elevation in comparison to the surrounding sea floor?

4. What would Earth be like if plate tectonics did not exist?

5. How do geologic processes on an active margin differ from those on a passive margin (see Section 3-7)?

6. How does the chemistry of lava at an island arc differ from that of lava at a volcanic arc?

7. Search the web for a story about a recent earthquake or volcanic eruption. Did it occur at a plate boundary? Name the plates involved and identify the type of boundary at which the event occurred. From this information, describe the most likely cause of the event.

ONLINE RESOURCES

Explore more about geology on the following websites:

On the Cutting Edge—Plate Tectonic Movement Visualizations:
serc.carleton.edu/NAGTWorkshops/geophysics/visualizations/PTMovements.html

USGS Story of plate tectonics:
pubs.usgs.gov/gip/dynamic/dynamic.html

Natural Resources Canada—Earthquake Processes:
www.nrcan.gc.ca/earth-sciences/energy-mineral/geology/geodynamics/earthquakeprocesses/8845

Natural Resources Canada—Earthquakes Canada:
www.earthquakescanada.nrcan.gc.ca/index-eng.php

USGS Discussion of Earth's interior:
pubs.usgs.gov/gip/interior

USGS Discussion of rock families:
geomaps.wr.usgs.gov/parks/rxmin

Paleomap Project of Christopher Scotese:
www.scotese.com/Default.htm

Additional animations, videos, and other online resources are available at this book's companion website:
www.wiley.com/go/fletchercanada
This companion website also has additional information about WileyPLUS and other Wiley teaching and learning resources.

CHAPTER 4
MINERALS

Chapter Contents and Learning Objectives (LO)

4-1 Minerals are solid crystalline compounds with a definite (but variable) chemical composition.
LO 4-1 *Define the term "mineral."*

4-2 A rock is a solid aggregate of minerals.
LO 4-2 *List five useful minerals, their composition, and how they are used.*

4-3 Geologists use physical properties to help identify minerals.
LO 4-3 *List and describe the common physical properties used in mineral identification.*

4-4 Atoms are the smallest components of nature with the properties of a given substance.
LO 4-4 *Describe the structure of an atom.*

4-5 Minerals are compounds of atoms bonded together.
LO 4-5 *Describe the three common types of bonds.*

4-6 Oxygen and silicon are the two most abundant elements in the crust.
LO 4-6 *Describe the formation of silicate structures.*

4-7 Metallic cations join with silicate structures to form neutral compounds.
LO 4-7 *State why single substitution and double substitution occur during crystallization.*

4-8 There are seven common rock-forming minerals.
LO 4-8 *List the seven common rock-forming minerals.*

4-9 Most minerals fall into seven major classes.
LO 4-9 *Describe each of the seven major classes of minerals.*

GEOLOGY IN OUR LIVES

Minerals are a critical geologic resource. Their products are used every day in manufacturing, agriculture, electronics, household items, and even the clothes you wear. Hence, it is vital that we ensure the sustainability of mineral resources.

Minerals are solid crystalline compounds with a definite (but variable) chemical composition that are found in the crust and mantle.

 This photo shows enormous crystals of gypsum discovered in a cave in Mexico. Is gypsum used in the making of any everyday product?

4-1 Minerals Are Solid Crystalline Compounds with a Definite (but Variable) Chemical Composition

LO 4-1 *Define the term "mineral."*

Your alarm rings in the morning and you switch on the light. After washing your face, brushing your teeth, and getting dressed, you turn on the radio and eat breakfast. Your day has hardly begun, and already you have used dozens of minerals or mineral-based products. Nearly everything you have done so far, and will do during the rest of the day, will involve the use of minerals. *Minerals are solid chemical compounds that we use every day.*

 Minerals are consumed in manufacturing the products we use, growing the food we eat, and building our homes, roads, and office buildings. This great demand for minerals places a huge burden on mining companies to find mineral reserves. Mining companies seek minerals in Earth's crust using remote sensing, the chemistry of rock samples, and an understanding of plate tectonics to assess the location and size of potential mineral deposits. Once an economical mineral deposit has been located, geological and mining engineers determine how it can be extracted safely and profitably without irreparably harming natural ecosystems. Balancing the need for mineral resources and the need to conserve natural environments is a constant challenge for the mineral industry.

Are we in danger of running out of minerals? That is unlikely. However, as we exhaust the more easily mined deposits, the cost of some mineral resources is likely to rise. Read the "Earth Citizenship" box "Will We Run Out of Minerals?" to learn more about this important question.

Of the thousands of known minerals, only about two dozen are common in crustal rocks. In this chapter, we learn about the most common minerals in rocks, how and where they form, and their importance to humans.

What Is a Mineral?

Perhaps none of nature's phenomena are as beautiful and diverse as the 4,000-plus minerals in Earth's crust. Minerals occur in every hue of the rainbow and in widely varied and fantastic shapes. Some minerals are harder than steel; others are so soft that you can scratch them with your fingernail. Minerals sometimes occur as tiny specks of dust but, when viewed through a microscope, are found to be composed of intricate crystals. All minerals can bend light, whereas some taste salty or sour, catch fire, dissolve in water, change colour, and even change shape on their own.

A **mineral** *is a naturally occurring, inorganic, crystalline solid with a definite chemical composition, which can vary within a restricted range.* This definition is quite a mouthful, but each of its four parts is important. Let us take a closer look at each one.

1. *Naturally occurring.* A mineral occurs in nature; it is not made by humans. Natural substances, like quartz, salt, gold, and gemstones, are minerals. Other substances, such as Formica, plastics, glass, and manufactured gems such as cubic zirconia, may appear similar to minerals but do not occur naturally and therefore are not minerals.

2. *Inorganic.* A mineral is inorganic, meaning that it cannot contain compounds composed of organic carbon. Organic carbon, the type of carbon found in all living things, bonds with hydrogen to create compounds that can be represented by multiples of the simple chemical compound *formaldehyde*, CH_2O (e.g., *glucose*, $C_6H_{12}O_6$). All five kingdoms of living organisms (Animals, Plants, Monera, Fungi, and Protista) are composed of organic carbon. Inorganic carbon is formed when carbon bonds with elements other than hydrogen; for instance, *calcium carbonate*, $CaCO_3$, is the mineral *calcite*. Geological processes can also form minerals such as *diamond* and *graphite*, which are made completely of inorganic carbon. Thus, coal cannot be a mineral because it is composed of plant remains made of organic carbon. But the mother-of-pearl lining of a shell can be a mineral because it is made of calcite.

3. *Crystalline solid.* The atoms in minerals are arranged in an orderly fashion to create a **crystalline structure**. Liquid (like water) or gaseous (like carbon dioxide) substances are not minerals. A mineral must have an internal gridwork of atoms that forms an orderly, repeating, three-dimensional pattern. Thus, when a mineral breaks, the shape of its pieces is predictable. For example, the mineral *halite*, NaCl (table salt, **FIGURE 4.1**) contains atoms of sodium (Na) and chloride (Cl) arranged in the shape of a cube. If you drop a large block of salt or smash it with a hammer, it breaks into many cubes (and rectangles), large and small.

4. *Chemical composition.* A mineral has a definite, but sometimes variable, chemical composition. Some minerals have a definite composition, such as *quartz* (SiO_2), whereas others can have a restricted range of composition, such as *olivine*. The mineral *olivine* consists of iron and magnesium atoms in various quantities, plus silicon and oxygen. The formula for olivine is definite: $(Mg, Fe)_2SiO_4$, or two parts magnesium (Mg) or iron (Fe) atoms, one part silicon atoms, and four parts oxygen atoms. Because magnesium and iron atoms are similar in size and have identical electrical charges, they are interchangeable in the crystalline structure of olivine. Thus, they can be present in any combination—all iron and no magnesium or vice versa or any ratio in between. The result can be an iron-rich olivine (tends to be brownish), a magnesium-rich olivine (tends to be a brilliant green), or something in between. Hence the chemical composition is "definite, but sometimes variable" in a mineral.

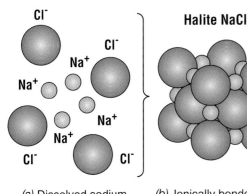
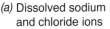
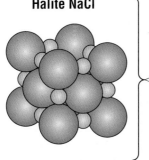

Courtesy of Chip Fletcher

(a) Dissolved sodium and chloride ions

(b) Ionically bonded sodium and chloride ions

(c) Crystal of halite (NaCl) composed of many bonded sodium and chloride ions

(d) Aggregate of many intergrown halite crystals

FIGURE 4.1 Minerals are crystalline; that is, the atoms within them are arranged in an orderly fashion. Halite (common table salt) consists of (a) sodium (Na) and chloride (Cl) atoms, which can be dissolved in water, and (b) arranged in the shape of a cube, (c) such that large groups of NaCl, (d) make the mineral in its natural form.

(?) Why does halite conform to the definition of a mineral?

EARTH CITIZENSHIP

WILL WE RUN OUT OF MINERALS?

Economically important minerals are vital in our day-to-day lives, and there is an ever-increasing demand for them fuelled by global population growth and economic development, especially in the developing economies of China, India, and Brazil. **FIGURE 4.2** shows the general trend of increasing demand for metals, as a country grows in population and "wealth," whatever the definition of the latter may be! In this case, the trends are for the United States, but every developed and developing country would show the same broad trend. Geologists know that minerals are a good example of a non-renewable resource. So does this suggest that eventually we will run out of minerals? Several factors need to be taken into account:

1. Increasing and improving recycling of metals will reduce the need to mine them.

2. New technologies have made it possible to find and mine low-quality mineral deposits that were previously unprofitable.

3. Technological developments in mineral processing and manufacturing allow metals to be extracted and used more efficiently.

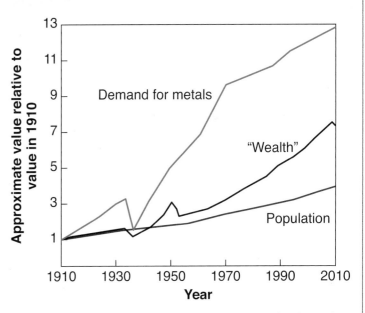

FIGURE 4.2 Typical trends showing the increase in the demand for metals, derived from minerals, as population and the economy of a country grows. Adapted from www.Masterresource.org by Dr. Indur Goklany, policy analyst for the United States government.

(?) **Expand Your Thinking**—What environment has dissolved Na and Cl? What process concentrates these elements and promotes their tendency to bond?

4-2 A Rock Is a Solid Aggregate of Minerals

LO 4-2 *List five useful minerals, their composition, and how they are used.*

Because minerals are the basic components of rocks, they are an essential part of the process through which the solid parts of Earth form, namely the crust and the mantle (**FIGURE 4.3**). Geologists define a rock as a solid aggregate of minerals or, less commonly, as a mass of naturally formed glass. Granite, for example, is an aggregate of several minerals—typically quartz, plagioclase feldspar, orthoclase feldspar, biotite, and others (**FIGURE 4.4**). Thus, to understand rocks, we must start with their basic components—minerals. The study of minerals is known as *mineralogy*, and geologists who study minerals are *mineralogists*.

Humans have exploited minerals, especially *gemstones* and *precious metals* (gold, silver, and others), since prehistoric times. The modern economy is heavily dependent on metals, which are extracted from minerals. Iron is an example. We extract iron from the mineral hematite, which is mined from the crust. Hematite is a compound of iron (Fe) and oxygen (O). Like all minerals, it can be described by the proportion of its chemical elements; every piece of hematite is composed of two parts iron and three parts oxygen, expressed chemically as Fe_2O_3. We have developed processes that can separate iron from hematite, as we need pure iron to make steel by combining it with carbon (C) and other elements, such as chromium (Cr) and silicon (Si).

Other minerals (**TABLE 4.1**) contain valuable metals such as copper, tin, zinc, lead, and titanium. But many minerals besides those that contain metals are useful (**FIGURE 4.5**); for instance, gypsum is used for making drywall and wallboard. Gypsum consists of the elements calcium (Ca), sulphur (S), and oxygen (O), which are bound to water molecules (H_2O) in the formula $CaSO_4 \cdot 2H_2O$.

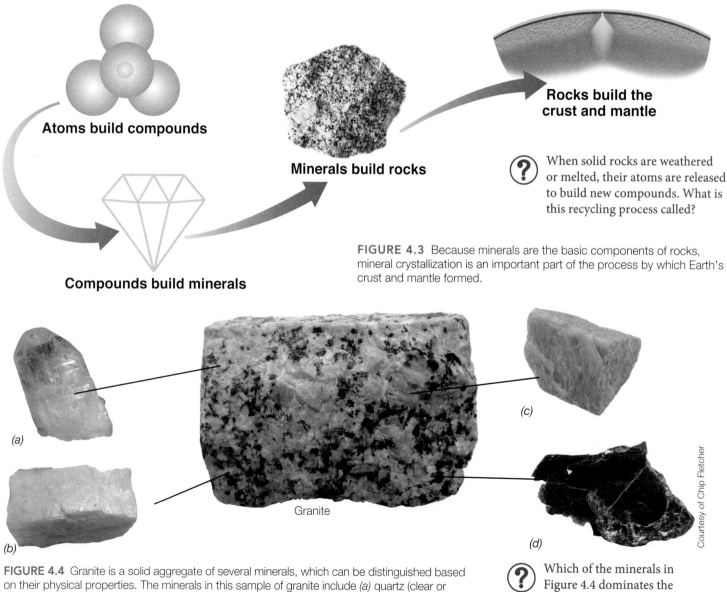

Atoms build compounds

Compounds build minerals

Minerals build rocks

Rocks build the crust and mantle

(?) When solid rocks are weathered or melted, their atoms are released to build new compounds. What is this recycling process called?

FIGURE 4.3 Because minerals are the basic components of rocks, mineral crystallization is an important part of the process by which Earth's crust and mantle formed.

Courtesy of Chip Fletcher

(a)

(b)

(c)

(d)

Granite

Courtesy of Chip Fletcher

FIGURE 4.4 Granite is a solid aggregate of several minerals, which can be distinguished based on their physical properties. The minerals in this sample of granite include (a) quartz (clear or white), (b) plagioclase feldspar (grey or white), (c) potassium feldspar (pink), and (d) biotite (black).

(?) Which of the minerals in Figure 4.4 dominates the colour of the granite sample?

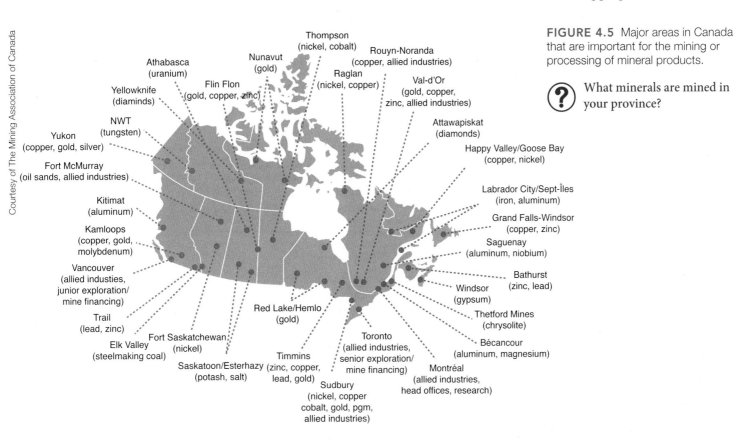

FIGURE 4.5 Major areas in Canada that are important for the mining or processing of mineral products.

? What minerals are mined in your province?

Courtesy of The Mining Association of Canada

TABLE 4.1 Common Minerals and Their Uses

Mineral	Composition	Uses
Chalcopyrite	Copper–iron–sulphur mineral; $CuFeS_2$	Mined for copper
Feldspar	Large mineral family; aluminum–silicon–oxygen composition; $x(Al,Si)_3O_8$, where x = various elements, such as sodium, calcium, and potassium	Ceramics and porcelain
Fluorite	Calcium–fluorine mineral; CaF_2	Mined for fluorine (its most important ore); used in steel making
Galena	Lead–sulphur mineral; PbS	Mined for lead
Graphite	Pure carbon; C	Pencil "lead" (replacing the actual lead metal once used in pencils); dry lubricant
Gypsum	Hydrous calcium–sulphur mineral; $CaSO_4 \cdot 2H_2O$	Drywall, plaster of Paris
Halite	Sodium–chloride mineral; NaCl	Table salt, road salt
Hematite	Iron–oxygen mineral; Fe_2O_3	Mined for iron (to make steel)
Magnetite	Iron–oxygen mineral; Fe_3O_4	Mined for iron
Pyrite	Iron–sulphur mineral; FeS_2	Mined for sulphur and iron
Quartz	Silicon–oxygen mineral; SiO_2	In pure form, used for making glass
Sphalerite	Zinc–iron–sulphur mineral; (Zn, Fe)S	Mined for zinc
Talc	Magnesium–silicon–oxygen–hydrogen mineral; $Mg_3Si_4O_{10}(OH)_2$	Used in ceramics, paint, talcum powder, plastics, and lubricants
Rutile	Titanium–oxygen mineral; TiO_2	Mined for titanium, a valuable metal used in aerospace and other industries
Calcite	Calcium–carbon–oxygen–mineral; $CaCO_3$	Toothpaste, cement, drywall, sheetrock

? **Expand Your Thinking**—Does a handful of sand grains qualify as a rock? Explain your thinking.

4-3 Geologists Use Physical Properties to Help Identify Minerals

LO *4-3 List and describe the common physical properties used in mineral identification.*

The discovery of gold in 1896 along the Klondike River, near Dawson City, Yukon, sparked a gold rush that attracted thousands of hopeful prospectors to the area. To be successful, these eager amateurs had to be able to distinguish between real gold and "fool's gold" (pyrite). The two minerals do look alike; both are golden and shiny. However, simple physical tests quickly betray the difference. If you bite pure gold, it will show tooth marks; if you bite pyrite, you will break a tooth. (Don't try it!) When a knife blade is dragged over real gold, it leaves a gouge; a knife makes no impression on pyrite. When rubbed on broken crockery, real gold leaves a bright golden streak; pyrite leaves a dirty black streak.

Distinguishing between minerals is usually not so easy. Geologists must learn to identify hundreds of minerals in the field so that they do not have to send samples to a lab for testing—a time-consuming, expensive procedure. Fortunately, many minerals have physical properties that aid in their identification. Each mineral's unique chemistry and crystalline structure at the atomic level determine its physical properties — for example, *cleavage, fracture, hardness, lustre, and crystal form* (TABLE 4.2). Essentially, the field geologist uses critical thinking to observe a mineral's physical properties and develop a hypothesis about it. If it is important to know the exact identity of a sample, the sample may be tested in a lab using more sophisticated methods.

Cleavage and Fracture

"Cleavage" is the tendency of a mineral to break, leaving a fairly smooth, flat surface. "Fracture" is the term used when breakage is irregular; that is, breakage does not occur along a cleavage plane. *Conchoidal* fracture exhibits seashell-shape curved surfaces like those of broken glass. A *fibrous* fracture looks like splintered wood.

Cleavage occurs in crystals that have specific planes of weakness. These planes, which are produced by weak chemical bonding, are inherent in the crystal's structure. Cleavage is reproducible, which means that a crystal can be cleaved along the same parallel plane many times (FIGURE 4.6). *Any sample of a particular type of crystal will have the same cleavage.* Cleavage planes are identifiable because they are usually repeated within the interior of a specimen, producing incipient breakage surfaces parallel to an external cleavage surface.

Hardness

How can you measure the hardness of a mineral? "Hardness" is the ease with which a mineral can be scratched. It is determined by scratching one mineral with another or by scratching an object of known hardness with a mineral sample (FIGURE 4.7). Two centuries ago, the German-born mineralogist *Friedrich Mohs* proposed the **Mohs hardness scale**, based on the relative hardness of different minerals. The scale uses numerical values from 1 to 10 as a relative measure of hardness based on comparison with specific minerals. Talc, the softest mineral, is assigned a hardness value of 1. Diamond, the hardest mineral, is assigned a value of 10. All other minerals fall somewhere between these two extremes.

Lustre

"Lustre" refers to the way in which a mineral's surface reflects light. It is related not to colour or shape but to the interaction between light and electrons in the crystal structure, the intensity of reflection of light at the surface, and how far light can penetrate the crystal. The terms used to describe lustre are descriptive rather than objective: *metallic, vitreous* (glassy), *silky, resinous, pearly,* and *earthy.*

Colour

Colour is the first thing you notice when viewing a mineral. Colour is one of the qualities that attract people to gemstones. Generally speaking, however, colour is a poor property to use in identifying minerals

TABLE 4.2 Physical Properties of Minerals

Property	Definition	Testing Method
Cleavage	Breakage along planes of weakness	Examine sample for planar breakage surfaces in one or more specific directions.
Fracture	Breakage, not along cleavage plane	Examine sample for either irregular or conchoidal breakage surfaces.
Hardness	Resistance to scratching or abrasion	Use materials of known hardness to determine hardness of sample.
Lustre	Character of reflected light	Does sample appear metallic or non-metallic?
Crystal Form	Geometric shape formed by the growth of crystal faces	Describe geometric shape: cubic, hexagonal, etc. Not commonly seen in most samples. Be careful to distinguish the faces of crystals from cleavage planes.
Reaction to HCl	Chemical interaction of weak hydrochloric acid (HCl) and calcium carbonate ($CaCO_3$)	Place a drop of HCl on sample and watch for a reaction (bubbles).
Streak	Colour of the mineral when the crystal is powdered	Rub sample on porcelain to determine colour of streak.

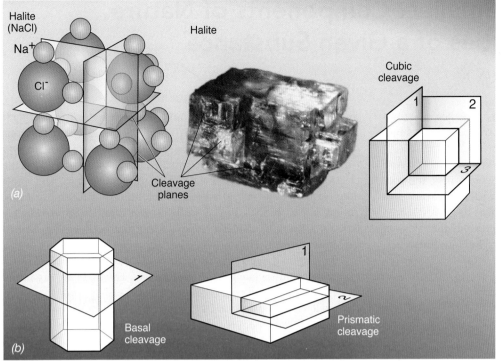

Halite
(NaCl)

Na+

Cl-

Halite

Cleavage
planes

(a)

Cubic
cleavage

Basal
cleavage

Prismatic
cleavage

(b)

Photo courtesy of The Mineralogical Association of Canada

FIGURE 4.6 *(a)* Halite breaks along three planes of cleavage; it has cubic cleavage. *(b)* Also shown are examples of basal cleavage (cleavage parallel to the base of a mineral) and prismatic cleavage (cleavage forming a six-sided prism).

(?) When a mineral cleaves, which is breaking, atoms or the bond between atoms? Is it easy to break an atom?

because many minerals exhibit multiple colours. Moreover, the colours of some minerals are identical to those of other minerals.

There is a scientific explanation for why minerals' colours are often misleading. Technically, the colour of a mineral is caused by the absorption, or lack of absorption, of various wavelengths of light (red is longest, purple is shortest). When pure white light (light containing all wavelengths of visible light) enters a crystal, some wavelengths are absorbed while others are reflected. The chemistry and structure of the mineral determine which of those wavelengths are reflected or absorbed. Often the presence of tiny amounts of an impurity—that is, some other element—will affect which wavelengths are reflected and, thus, the mineral's colour. For example, quartz can occur in a variety of colours, such as milky, rosy, smoky, and amethyst because of the presence of impurities. In short, colour is an unreliable basis for identifying a mineral.

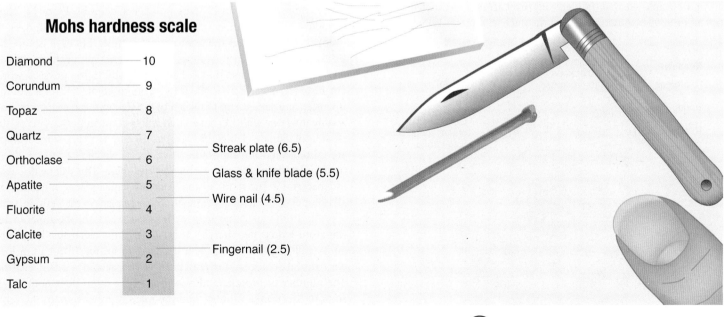

Mohs hardness scale

Diamond	10
Corundum	9
Topaz	8
Quartz	7
Orthoclase	6
Apatite	5
Fluorite	4
Calcite	3
Gypsum	2
Talc	1

Streak plate (6.5)

Glass & knife blade (5.5)

Wire nail (4.5)

Fingernail (2.5)

FIGURE 4.7 To test hardness, scratch one mineral across the surface of another. A mineral that leaves a scratch on another is harder.

(?) What is the hardness of plastic?

 Expand Your Thinking—How is the use of physical properties to identify a mineral an example of critical thinking?

4-4 Atoms Are the Smallest Components of Nature with the Properties of a Given Substance

LO 4-4 *Describe the structure of an atom.*

Minerals are composed of **atoms** bonded together in various configurations. Atoms are the smallest components in nature that have the properties of a given substance. For example, in a chemical reaction, every oxygen atom behaves in the same way. And every atom in a sample of copper has the properties of copper. It is possible to break atoms into smaller, subatomic particles (electrons, protons, and neutrons), but when we do, the atom loses its properties and no longer exists. Thus, the atom is the fundamental building block of minerals, the component responsible for a mineral's unique properties.

The Structure of Atoms

Every atom in the known universe is a tiny structural unit consisting of *electrons*, *protons*, and (usually) *neutrons* (**FIGURE 4.8**). An atom's centre, or *nucleus*, is composed of protons (large, heavy, and having a positive electrical charge, $+$) and neutrons (large, heavy, and having no electrical charge). A normal hydrogen (H) atom does not have any neutrons, but all other atoms do. Electrons surround the nucleus. They are tiny and light, and have a negative electrical charge ($-$).

Each element is distinctive because its atoms have a unique number of protons. Because opposite electrical charges are attracted to each other, the negatively charged electrons in an atom form a cloud around its nucleus, with its positively charged protons. For example, hydrogen is distinctive because it has only 1 proton and 1 electron (and no neutrons), making it the lightest element. Iron is distinctive because it has 26 electrons, 26 protons, and 26 neutrons.

The subatomic world is fascinating. If we were to look inside an atom, we would see that the electrons orbit around the nucleus but do not enter it. We would notice that the nucleus remains intact because the neutrons help "glue" the protons within it together, preventing it from flying apart in most cases. We would also notice that electrons orbit relatively far from the nucleus. For example, if we were to draw a hydrogen atom to scale with the nucleus the diameter of a pencil, the electron's orbit would be about 0.5 km from the nucleus. The whole atom would be the size of a baseball stadium—mostly empty space. Atoms are incredibly tiny. One hydrogen atom is approximately 0.00000005 mm in diameter. It would take almost 20 million hydrogen atoms to form a line as long as a hyphen (-).

Protons and neutrons are particles resembling tiny marbles. But electrons are more like light or radio waves. Electrons are organized in negatively charged energy levels (called shells or *orbitals*) around the nucleus. In most atoms, the innermost shell holds only 2 electrons, and each subsequent shell holds a maximum of 8 electrons.

Each element is defined by the number of protons in its nucleus: This is its **atomic number**. An atom whose nucleus contains a different number of protons is a different element. For example, carbon atoms have 6 protons (atomic no. 6), nitrogen atoms have 7 (atomic no. 7), and oxygen atoms have 8 (atomic no. 8). The number of neutrons plus protons in the nucleus is the atom's **mass number**; carbon, for instance, normally has 6 protons and 6 neutrons in its nucleus, for a mass number of 12.

Isotopes and Ions

As just noted, carbon atoms normally contain 6 protons and 6 neutrons; some, however, contain 7 or 8 neutrons. Hence, carbon always has an atomic number of 6, but its mass number may be 12, 13, or 14. These variations in mass number create **isotopes** of carbon (**FIGURE 4.9**). All isotopes of carbon have the same atomic number, but each isotope has a different mass number. Isotopes of an element typically have different physical properties and thus behave differently in the physical world. For instance, a water molecule (H_2O) composed of ^{16}O (mass number is written as a superscript to the left of the element) can evaporate more readily than one with ^{18}O, a fact that will become important in our discussion of climate change in Chapters 16 and 17. Also, some isotopes are unstable and undergo radioactive decay, which is a process that is used to determine the age of minerals and rocks. This will be discussed in Chapter 13. The term "isotope" means "equal place," meaning that all isotopes of an element belong in the same place in the periodic table.

An atom can either gain or lose electrons and become an **ion** (an atom with a net charge). If an atom gains electrons, it acquires an overall negative charge and is called an **anion**. If the atom loses electrons, it acquires an overall positive charge and is called a **cation**. The number of protons and neutrons in the atom does not change, so both the atomic number and the mass number remain the same. Generally (but not always) it is the electrons in the outermost shell, called *valence electrons*, that are involved in making ions.

Why do atoms gain and lose electrons? An atom does so *to achieve a stable electron configuration by filling its outer shell with a maximum number of electrons—that is, 8—or, alternatively, by losing its outer shell to reveal a filled inner shell.* Atoms that gain or lose only a few

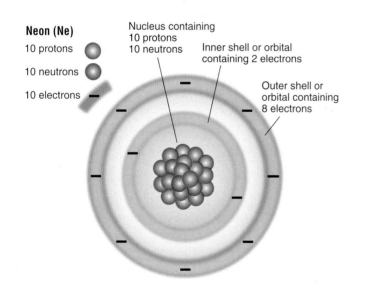

Neon (Ne)
10 protons
10 neutrons
10 electrons

Nucleus containing
10 protons
10 neutrons

Inner shell or orbital
containing 2 electrons

Outer shell or
orbital containing
8 electrons

FIGURE 4.8 Atoms consist of a nucleus constructed of protons and neutrons surrounded by electrons. The example shown is an atom of neon. Hydrogen is the only atom without a neutron in its nucleus.

(?) What makes one element different from another?

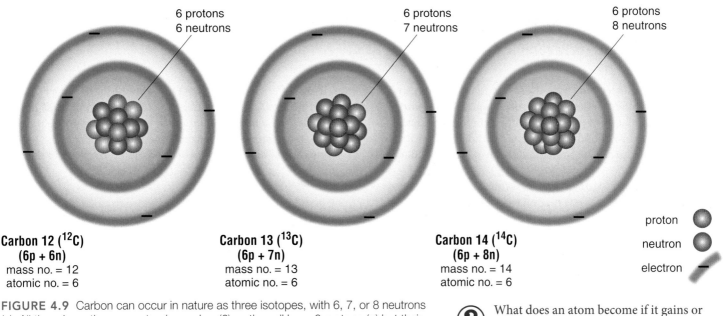

Carbon 12 (^{12}C)
(6p + 6n)
mass no. = 12
atomic no. = 6

Carbon 13 (^{13}C)
(6p + 7n)
mass no. = 13
atomic no. = 6

Carbon 14 (^{14}C)
(6p + 8n)
mass no. = 14
atomic no. = 6

6 protons
6 neutrons

6 protons
7 neutrons

6 protons
8 neutrons

proton
neutron
electron

FIGURE 4.9 Carbon can occur in nature as three isotopes, with 6, 7, or 8 neutrons (n). All three have the same atomic number (6) as they all have 6 protons (p) but their mass numbers vary: 12, 13, or 14.

(?) What does an atom become if it gains or loses electrons?

electrons to become stable do so readily and become ions. Those that give up their outer electrons to reveal a filled inner shell have a positive charge (cations). Those that acquire new electrons to fill their outer shell have a negative charge (anions). For instance, aluminum (Al) has 3 electrons in its outer shell. It readily loses these (often to other atoms seeking extra electrons) to reveal a filled inner shell. That is why most aluminum atoms are cations with a +3 charge (written as Al^{3+}).

Do atoms really give up electrons easily? They certainly do. That's why you get a shock when you scuff your feet on a carpet and then touch a metal doorknob (**FIGURE 4.10**). Friction between your shoe

and the rug transfers electrons from your body to the rug, giving your body a temporary positive charge. When you reach for the doorknob (or any metal surface), you feel a shock as electrons leap off the metal to your body. This happens because many common metals (such as aluminum) have only a few electrons in their outer shell, which they easily lose. When you approach with a positive charge, 3 electrons from every aluminum atom on the surface of the doorknob leap onto your fingertip, causing a shock. The metal atoms become ionized (cations), you return to a neutral state, and the shock you feel is the exchange of electrons in what is known as *static electricity*.

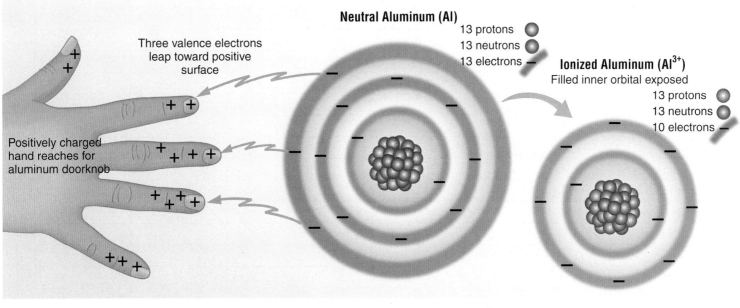

Neutral Aluminum (Al)
13 protons
13 neutrons
13 electrons

Three valence electrons leap toward positive surface

Positively charged hand reaches for aluminum doorknob

Ionized Aluminum (Al^{3+})
Filled inner orbital exposed
13 protons
13 neutrons
10 electrons

FIGURE 4.10 Aluminum atoms in a doorknob will readily give up the 3 electrons in their outer shells to a positively charged hand, creating static electricity.

(?) Why do electrons readily leave the aluminum?

(?) **Expand Your Thinking**—The tendency to fill the outer orbital with 8 electrons has been called the "octet rule." Why would this be called a "rule"?

4-5 Minerals Are Compounds of Atoms Bonded Together

LO 4-5 *Describe the three common types of bonds.*

Atoms of most elements have the ability to bond with certain atoms of other elements to form **compounds**. Hydrogen and oxygen atoms bond to form the compound water. Sodium and chlorine atoms bond to form halite (common salt). Most of Earth's 4,000-plus minerals are compounds formed by the bonding of two or more elements.

Why do atoms bond together? *They do so to achieve a stable electron configuration.* As we have seen, every atom has a tendency to fill its outermost shell with the maximum number it can hold (2 for the first shell, 8 for the others). A small group of elements have 8 electrons in their outer shells: helium, neon, argon, krypton, xenon, and radon. These are all known as *inert gases* because they do not tend to form compounds.

TABLE 4.3 shows the pattern of electrons in the first 20 elements (atomic numbers 1–20).

As we saw in the case of static electricity, atoms can exchange electrons quite easily. This tendency drives the chemical bonding process. A bond between chemical units (atoms, molecules, ions) is formed by the attraction of atoms to one another through the sharing or exchange of electrons. To put it more simply, atoms may share or exchange electrons to attain a full complement of 8 electrons in their outer shells. The tendency of atoms to exchange electrons, although the ease with which this happens can be quite variable, causes atoms to form one of three common types of bonds: ionic, covalent, or metallic.

Ionic Bonds

Think of ordinary salt, a mineral that is mined from thick layers underground and extracted from sea water to de-ice roads and season food. Salt is the mineral halite, a compound that is formed when atoms of sodium (a metal) and chlorine (a poisonous gas) are attracted to one another and form ionic bonds. In ionic bonding, 2 atoms exchange an electron. The atom receiving the electron becomes an anion, while the atom giving up the electron becomes a cation. The two oppositely charged ions are attracted to each other and form a bond. Thus, *ionic bonding* happens because cations and anions are attracted to one another. In the case of halite, the result of ionic bonding is the compound sodium chloride, which has no net charge (FIGURE 4.11a). The compound has properties

TABLE 4.3 Electron Patterns of the First 20 Elements (Shaded elements are inert gases)

Element	Symbol	Atomic Number	Number of Electrons in Each Shell			
			First (2 is stable)	Second (8 is stable)	Third (8 is stable)	Fourth (8 is stable)
Hydrogen	H	1	1			
Helium	He	2	2			
Lithium	Li	3	2	1		
Beryllium	Be	4	2	2		
Boron	B	5	2	3		
Carbon	C	6	2	4		
Nitrogen	N	7	2	5		
Oxygen	O	8	2	6		
Fluorine	F	9	2	7		
Neon	Ne	10	2	8		
Sodium	Na	11	2	8	1	
Magnesium	Mg	12	2	8	2	
Aluminum	Al	13	2	8	3	
Silicon	Si	14	2	8	4	
Phosphorus	P	15	2	8	5	
Sulphur	S	16	2	8	6	
Chlorine	Cl	17	2	8	7	
Argon	Ar	18	2	8	8	
Potassium	K	19	2	8	8	1
Calcium	Ca	20	2	8	8	2

that are vastly different from those of its components (it is tasty, it is non-metallic, it is not a dangerous gas, and it shatters easily).

Covalent Bonds

Now think of ordinary water, the liquid that makes up about 60 percent of our bodies, covers 71 percent of Earth's surface, and is essential to life. Water is a compound that is formed when 1 atom of hydrogen (a light explosive gas) and 2 atoms of oxygen (a heavier explosive gas) are joined by covalent bonds (Figure 4.11b). In *covalent bonding*, elements "share" electrons in order to fill their outermost shells. The hydrogen atom and the oxygen atom each donate 1 electron to form a chemical bond. The 2 electrons that form the bond are shared by both atoms, resulting in a single covalent bond.

Think of covalent bonding in terms of two blocks of wood joined with two nails. The pieces of wood are the atoms, and the nails are the electrons that the wood blocks share to form the covalent bond. Each piece of wood shares a portion of the nails.

Metallic Bonds

A third type of bond is a *metallic bond*. Native gold is a good example, as the gold atoms are joined by metallic bonds. Metallic bonding is similar to covalent bonding because both types involve the sharing of electrons. However, in metallic bonds, all the extra valence electrons of a large group of atoms (millions or billions of them) move easily from one atom to another, and thus help to bind the atoms together. All the atoms share this "sea of electrons."

Elements that Tend to Become Ions

Many elements tend to become ions. Those that have 1 or 2 electrons in their outermost shell have an unstable configuration, and the electrons are easily lost to form a cation. These include the important crustal elements sodium (Na^+), potassium (K^+), calcium (Ca^{2+}), and magnesium (Mg^{2+}), all of which have a tendency to give up their outermost electrons and become cations. Aluminum (Al^{3+}), the most abundant metal in Earth's crust, has 3 valence electrons and also readily loses them to become a cation.

Some elements have a tendency to gain electrons in their outer shells during the formation of covalent bonds. For instance, fluorine (F^-), which forms a toxic gas, and oxygen (O^{2-}), Earth's most abundant crustal element, tend to gain electrons in their outer shells and become anions. All elements with a similar valence state will seek to gain 1 or more electrons in order to fill their outer shell.

A semi-metal, silicon (Si^{4+}), is the second-most abundant element in Earth's crust. Silicon commonly bonds covalently with 4 oxygen atoms to build the largest class of minerals in the crust: the **silicates** (discussed in Sections 4-6 and 4-9).

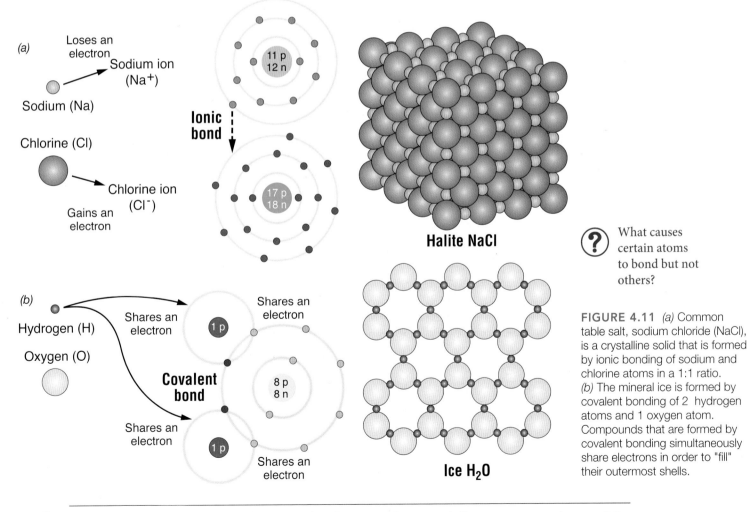

Halite NaCl

Ice H₂O

What causes certain atoms to bond but not others?

FIGURE 4.11 (a) Common table salt, sodium chloride (NaCl), is a crystalline solid that is formed by ionic bonding of sodium and chlorine atoms in a 1:1 ratio. (b) The mineral ice is formed by covalent bonding of 2 hydrogen atoms and 1 oxygen atom. Compounds that are formed by covalent bonding simultaneously share electrons in order to "fill" their outermost shells.

Expand Your Thinking—Why do elements with 8 electrons in their outer shells occur as a gas and not a solid?

4-6 Oxygen and Silicon Are the Two Most Abundant Elements in the Crust

LO 4-6 *Describe the formation of silicate structures.*

Plate tectonics plays an important role in mineral formation. Magma rising from the mantle and the melted crust cools as it penetrates solid rock. Cooling allows **igneous minerals** to crystallize. Igneous crystallization occurs in three important tectonic settings (**FIGURE 4.12**): subduction zones, spreading centres, and magma associated with hotspots.

Of course, other types of minerals do not have igneous origins. **Sedimentary minerals** (Chapters 7 and 8) crystallize from groundwater and help to cement the grains in the sediment together. They include calcite ($CaCO_3$), hematite (Fe_2O_3), and quartz (SiO_2). Some minerals tend to crystallize in the absence of oxygen; for instance, pyrite (FeS_2) will develop in iron- and sulphur-rich environments (such as *anoxic* mud). **Metamorphic minerals** (Chapter 9) also crystallize in particular environments. Where conditions in the crust lead to high heat and pressure, a rock may undergo *recrystallization* (crystallization of new minerals using elements of an old mineral without melting) to form metamorphic minerals.

Crystallization

Crystallization is the process through which atoms or compounds that are in a liquid state (e.g., in magma or dissolved in water) are arranged into an orderly solid state (mineral). This process occurs in two steps: nucleation and crystal growth.

Nucleation is the initial grouping of a few atoms that starts the process of crystal growth. A major barrier to nucleation is heat energy, which keeps magma in a liquid state in which atoms are vibrating too much with kinetic energy to bond together. This "energy barrier" can be circumvented if a solid is present in the liquid. The solid acts as a *nucleation seed*. It can be a particle of unmelted rock, another crystal, or even the solid wall of a magma chamber. Crystal growth will occur sooner on a nucleation seed than if no seed is present in the liquid.

Crystals grow as atoms from magma are deposited on the surface of a seed. Atoms arrange themselves to achieve an electrically neutral crystalline compound. After many millions of atoms have bonded together, the result is an orderly crystal—a mineral.

Rapidly growing crystals typically are smaller and less developed than slowly growing crystals. During laboratory experiments, mineralogists have observed that if crystals begin to grow at multiple nucleation sites, the resulting rock will contain many small crystals. Conversely, if there are fewer nucleation sites, the resulting rock will contain fewer and larger crystals.

The Eight Most Abundant Elements in the Crust

The eight most common elements in the crust, in order of average abundance, are shown in **TABLE 4.4**. Also shown are their typical ionized states. Although iron accounts for only 6 percent of the crust, it actually constitutes over a third of the entire planet, as it is much more abundant in the mantle and the core.

Silicates

Silicon and oxygen dominate the composition of Earth's crust. Oxygen, a large, bulky atom, has 6 electrons in its outermost shell, 2 short of the desired 8. It is typically found in an ionized state as O^{2-}. Silicon is the second-most abundant element by weight. It is a small atom with 4 electrons in its outermost shell, usually found as Si^{4+}. Together, silicon and oxygen readily form a covalent bond, creating a *silicate compound*.

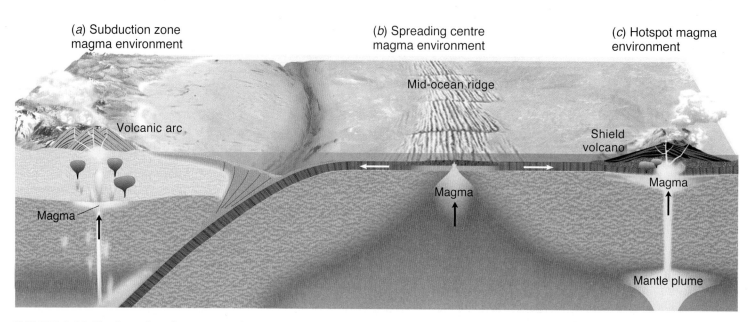

(a) Subduction zone magma environment

(b) Spreading centre magma environment

(c) Hotspot magma environment

Mid-ocean ridge

Volcanic arc

Shield volcano

Magma

Magma

Magma

Mantle plume

FIGURE 4.12 The formation of magmas and the ultimate igneous crystallization of these magmas occur in three tectonic settings: *(a)* subduction zones, *(b)* spreading centres, and *(c)* mid-plate hotspots.

? **What types of plate boundaries can you identify in Figure 4.12?**

Minerals made of silicate compounds are the most abundant naturally occurring inorganic substances in Earth's crust. In these compounds, 4 oxygen atoms surround a single silicon atom. Each oxygen atom covalently shares 1 electron with the silicon atom, jointly filling the outermost shells of both atoms.

Silicate compounds have a geometric shape known as a *tetrahedron* (a shape with "four faces"). A tetrahedron is a pyramid with 4 large oxygen atoms packed around a much smaller silicon atom (**FIGURE 4.13a**). The silicate compound has an overall charge of (-4), and can be described by the following chemical formula $(SiO_4)^{4-}$. This silicate compound is the building block of all silicate minerals.

Each oxygen atom in a silicate tetrahedron shares 1 electron with the silicon atom, but it still needs 1 more electron to achieve a filled outer shell. Since there is abundant silicon in most magma, it often finds a second silicon atom to bond with and thus forms a second tetrahedron. This process results in linked pairs of tetrahedra, created by the sharing of oxygen atoms. Two isolated tetrahedra ($[SiO_4]^{4-} + [SiO_4]^{4-}$) have a combined charge of -8. A pair of linked tetrahedra ($[Si_2O_7]^{6-}$) sharing a single oxygen atom has a charge of -6.

TABLE 4.4 The Eight Most Common Elements in the Crust

Element	Average Abundance	Typical Ionization
Oxygen	46%	O^{2-}
Silicon	28%	Si^{4+}
Aluminum	8%	Al^{3+}
Iron	6%	Fe^{2+} or Fe^{3+}
Magnesium	4%	Mg^{2+}
Calcium	2.4%	Ca^{2+}
Potassium	2.3%	K^+
Sodium	2.1%	Na^+

The first step in magma crystallization is the organization of these silicate tetrahedra to form a crystal of a new mineral.

By the time crystallization is completed, a mineral has become a neutral compound; that is, it does not have an electrical charge. Hence, before a silicate mineral can crystallize, all negative charges associated with silicate tetrahedra must be balanced with cations. How does this happen? Referring again to Table 4.4, the most abundant cations are aluminum (Al^{3+}), iron ($Fe^{2+, 3+}$), magnesium (Mg^{2+}), calcium (Ca^{2+}), potassium (K^+), and sodium (Na^+). Most crustal magma contains large quantities of silicon and oxygen, and, therefore, many negatively charged tetrahedra are formed as it cools. There is also an abundance of positively charged metallic cations and, therefore, tremendous potential for forming many different silicate minerals.

Tetrahedron linkage does not stop at just one pair. Because an oxygen atom achieves stability by bonding with 2 silicon atoms, complex structures of many linked tetrahedra are formed as magma cools. The crystalline structure of silicate minerals ranges from independent tetrahedra that depend entirely on cations of other elements to link them, to single chains, double chains, and sheets of tetrahedra. There are even three-dimensional "framework silicates" in which all the oxygen atoms are shared by adjacent tetrahedra (Figure 4.13*b*, *c*, *d*, *e*, and *f*). These are all variations of **silicate structures**.

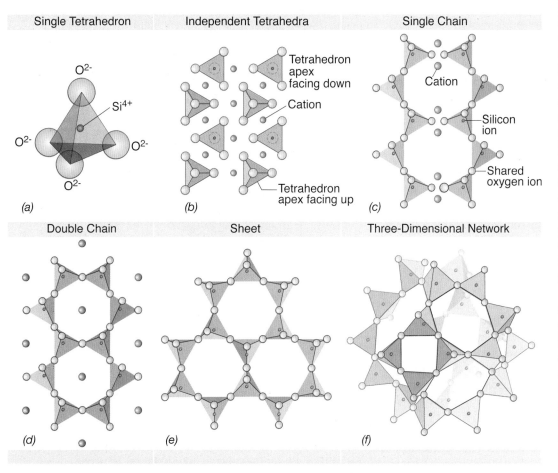

FIGURE 4.13 The silicate structures: *(a)* single tetrahedron; *(b)* independent tetrahedra bonded with metallic cations; *(c)* single chain structure; *(d)* double chain; *(e)* sheet silicate; *(f)* three-dimensional network silicate structure.

 What type of bond joins silicate tetrahedra to make silicate structures?

Expand Your Thinking—Common "rust" is actually a mineral composed of iron oxide. Is this mineral igneous, sedimentary, or metamorphic?

4-7 Metallic Cations Join with Silicate Structures to Form Neutral Compounds

LO 4-7 *State why single substitution and double substitution occur during crystallization.*

Alone, silicate compounds, including single tetrahedra, chains, sheets, and framework silicates, are not neutral. They carry a negative charge that must be neutralized through bonding with one or more cations. The most abundant cations in magma of average composition are the metallic cations we are familiar with: Al^{3+}, $Fe^{2+, 3+}$, Mg^{2+}, Ca^{2+}, K^+, and Na^+. The metallic cations differ in size and electrical charge. These differences are significant because they control which cations bond with which silicate structure and thus determine the resulting mineral structure and chemistry. Certain cations can substitute for one another because they have a similar size and charge. This cation *substitution* makes possible the formation of a wide variety of minerals. The cations that most often substitute for each other are Na^+ and Ca^{2+}, Al^{3+} and Si^{4+}, and Fe^{2+} and Mg^{2+}.

Cation Substitution

To illustrate the process of substitution, we return to the silicate mineral olivine (**FIGURE 4.14**). Olivine has an independent tetrahedron silicate structure (see Figure 4.13*b*). Hence, we can write a portion of its chemical formula as $(SiO_4)^{4-}$. These tetrahedra are linked to one another by iron cations with a charge of $+2$ in a ratio of 2 cations for each tetrahedron: $Fe_2^{2+}(SiO_4)^{4-}$; the total charge is then 0.

But that is not the end of the story. Magnesium also has a charge of $+2$, and it is similar in size to Fe^{2+}. During crystallization, Mg^{2+} may replace Fe^{2+} (or vice versa) because the total charge of the compound remains at 0. In this way, cation substitution produces the "variable chemical composition" referred to in the definition of a mineral discussed at the beginning of the chapter. The chemical composition of the resulting mineral, olivine $(Fe^{2+}, Mg^{2+})_2(SiO_4)^{4-}$, may range along a continuum from an iron-dominated form (known as fayalite, $[Fe_2^{2+}(SiO_4)^{4-}]$) to a magnesium-dominated form (known as forsterite $[Mg_2^{2+}(SiO_4)^{4-}]$).

Single substitution explains the interaction of silicon and oxygen with iron and magnesium. What about the remaining cations? Imagine that two cations of the same size but with different charges are available for the same space between tetrahedra. If one cation gets in, the resulting compound will be neutral, and a mineral will have formed. But if the other cation gets in, the compound will not be neutral. Since minerals must be neutral, one of the silicon or oxygen ions usually experiences a substitution that results in a neutral compound. This is called *double substitution*: The first cation substitution is followed by a second one that re-establishes the neutrality of the compound involving Si or O.

An example of double substitution can be seen in the case of plagioclase feldspar. Feldspar, a framework silicate (see Figure 4.13*f*), incorporates calcium, sodium, and aluminum cations within its structure through double substitution. Calcium and sodium ions are nearly the same size, so they freely substitute for each other during crystallization. However, since they have different charges, a second substitution must occur to restore neutrality. To do this, an aluminum ion will substitute for a silicon ion, since they, too, are similar in size.

The double substitution of cations in the mineral plagioclase feldspar occurs in the following manner. Initially the compound

$Ca^{2+}Al_2^{3+}Si_2^{4+}O_8^{2-}$ is formed in cooling magma. This compound is neutral, with the -16 charge of the 8 oxygens (O_8^{2-}) balanced by the $+2$ of the calcium (Ca^{2+}), the $+6$ of the 2 aluminums (Al^{3+}), and the $+8$ of the 2 silicon ions (Si_2^{4+}). Sodium (Na^{1+}) ions are able to substitute for the calcium ions because they are nearly the same size. However, this upsets the charge balance. (Check this: Is $Na^{1+}Al_2^{3+}Si_2^{4+}O_8^{2-}$ balanced?) To restore electrical neutrality, a second substitution takes place, with another silicon replacing one of the aluminum ions. The result— $Na^{1+}Al^{3+}Si_3^{4+}O_8^{2-}$ —is a neutral mineral.

Olivine and plagioclase feldspar are only two of many types of silicate minerals that are formed in the crust. **FIGURE 4.15** presents a guide to the chemistry and crystalline structure of some of the more important rock-forming minerals discussed in the next section. Through the formation of silicate tetrahedra and cation substitution, magma eventually crystallizes into a solid mass containing many

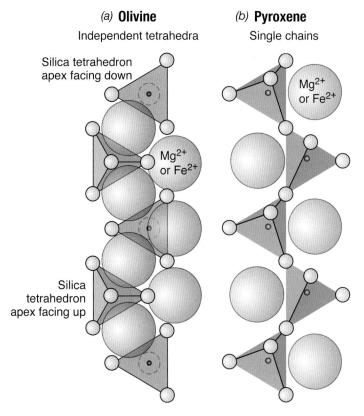

(a) Olivine
Independent tetrahedra

(b) Pyroxene
Single chains

Silica tetrahedron apex facing down

Mg^{2+} or Fe^{2+}

Mg^{2+} or Fe^{2+}

Silica tetrahedron apex facing up

FIGURE 4.14 (a) The mineral olivine, $(Mg, Fe)_2SiO_4$, has an independent silicate structure wherein separate tetrahedra are linked by iron (Fe^{2+}) and magnesium (Mg^{2+}) cations. (b) The mineral pyroxene, $(Mg, Fe)SiO_3$, has a single-chain silicate structure wherein iron and magnesium cations are bonded to a silicate chain.

 Would pyroxene be the same mineral if the tetrahedra were not linked into a chain but were instead isolated tetrahedra joined by Mg^{2+} or Fe^{2+}? Write the new chemical formula if this were the case.

Example Mineral and Formula	Cleavage	Silicate Structure	Example
Olivine $(Mg, Fe)_2 SiO_4$	None	Independent tetrahedra 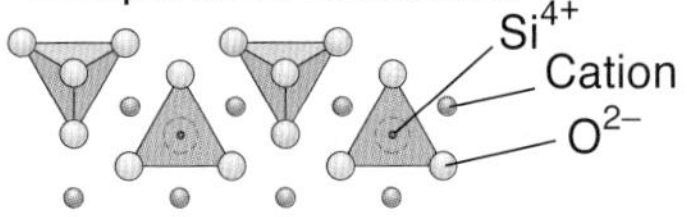 Si^{4+} Cation O^{2-}	 Olivine
Pyroxene group (Augite) $(Mg, Fe)SiO_3$	Two planes at right angles	Single chain 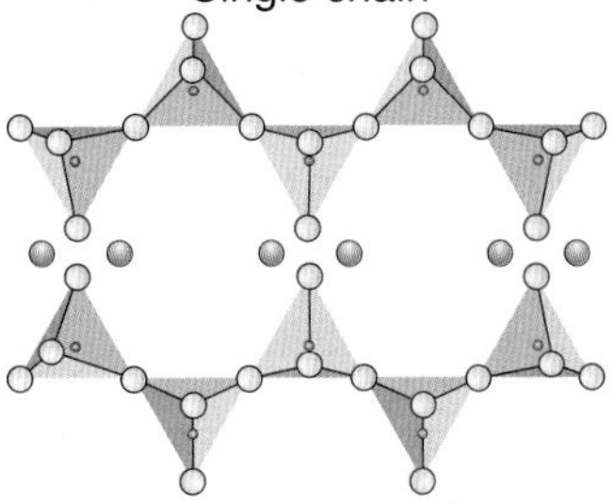	 Augite
Amphibole group (Hornblende) $Ca_2(Fe, Mg)_5Si_8O_{22}(OH)_2$	Two planes at 60° and 120°	Double chain 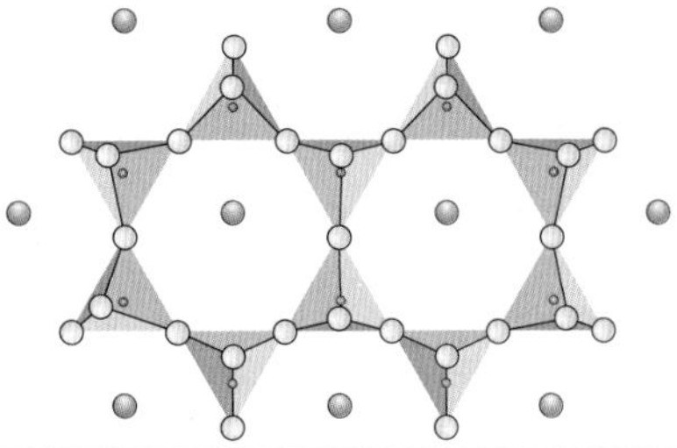	 Hornblende
Micas (variety: Biotite) $K(Mg, Fe)_3AlSi_3O_{10}(F, OH)_2$ (variety: Muscovite) $KAl_2(AlSi_3O_{10})(OH)_2$	One plane	Sheet 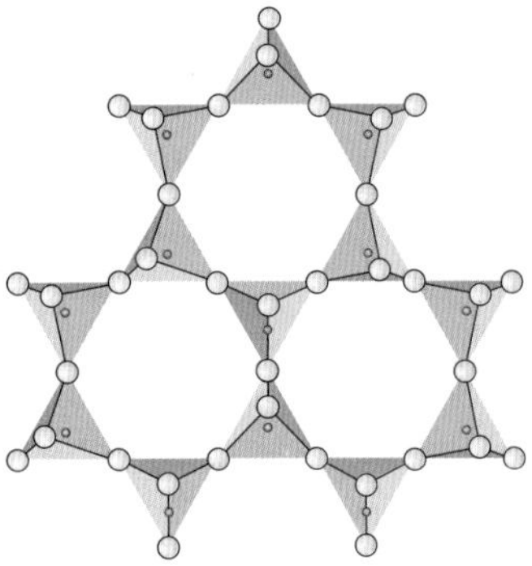	 Biotite 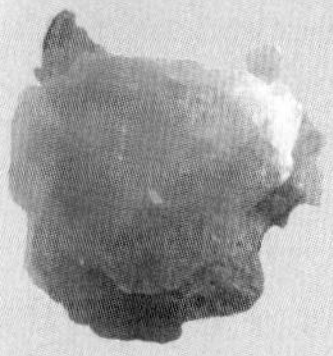 Muscovite
Feldspars Potassium feldspar (Orthoclase) $KAlSi_3O_8$ Plagioclase feldspar $(Ca, Na)(Al, Si) Al Si_2 O_8$	Two planes at 90°	Three-dimensional network 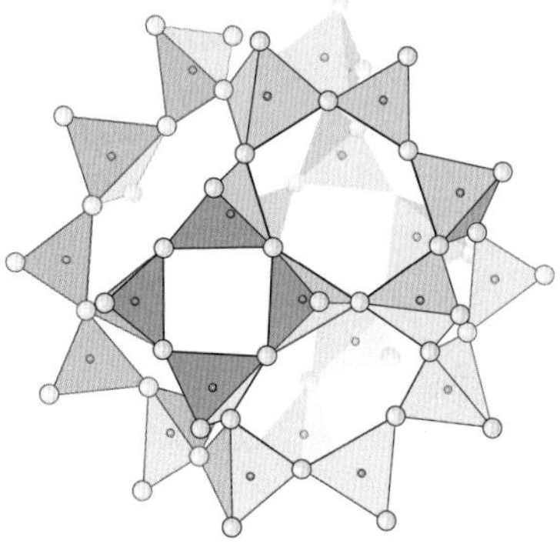	 Potassium feldspar  Quartz
Quartz SiO_2	None		

Photos courtesy of Chip Fletcher

FIGURE 4.15 Common silicate minerals and their structures.

types of minerals. In Chapter 5, we look more closely at igneous mineral formation and discover that, as magma cools, different minerals crystallize at different temperatures. For now, though, let's complete our study of minerals with an introduction to seven important rock-forming minerals, along with a short survey of the seven major classes of minerals found in Earth's crust.

(?) Expand Your Thinking—Why are both the size *and* charge of metallic cations significant in contributing to mineral diversity?

4-8 There Are Seven Common Rock-Forming Minerals

LO 4-8 *List the seven common rock-forming minerals.*

Because of their great chemical diversity, minerals display a wide variety of physical properties. Some minerals, such as galena (lead ore), are very dense; others, such as corundum, are very hard. Still others are light and soft; gypsum, for example, can be scratched with a fingernail. Individual minerals can also have variations—for example, fluorite can be found in green, blue, pink, clear, and other varieties. What allows minerals to display such a range of properties?

What Is a Mineral?

Let us return to the definition of a mineral: "a naturally occurring, inorganic, crystalline solid with a definite, but sometimes variable, chemical composition." The crystalline structure of a given mineral cannot vary; every sample of a specific mineral will exhibit the same geometric arrangement of atoms. However, the "definite, but sometimes variable" part of the definition indicates that a mineral's chemical composition can vary within limits. Thus, a mineral may exhibit some variation in its chemistry (and its colour), provided that its fundamental components are always present. For instance, as noted earlier, the proportions of iron and magnesium in olivine differ from one sample to another. Some variations are so common that they are given names.

Minerals that have a fixed crystalline structure and whose chemical composition varies are given a *group name*. Within a group, mineral *species* possess a particular and restricted chemical composition. Within a species, a common irregularity in crystallization or a chemical impurity that is often observed may produce a distinct mineral *variety*. Identifying the mineral's group, species, and variety accommodates chemical variation in a mineral. Consider the example of feldspar.

Group: Feldspar. All members have a three-dimensional silicate framework crystalline structure (Si_3O_8), and all have two cleavage planes forming nearly 90° angles.

Species: Plagioclase feldspar (**FIGURE 4.16a**) contains sodium and calcium ions that freely substitute for one another during crystallization, as described in Section 4-7.

Courtesy of Chip Fletcher

(a)   *(b)*

FIGURE 4.16 Two feldspar species: *(a)* plagioclase feldspar, and *(b)* orthoclase feldspar.

(?) What physical properties would you use to identify plagioclase feldspar and orthoclase feldspar in the field?

Varieties: Plagioclase ranges from sodium-rich albite ($NaAlSi_3O_8$) to calcium-rich anorthite ($CaAl_2Si_2O_8$). Between these two extremes are varieties with intermediate abundances of sodium and calcium. Two examples are andesine ($Na[70\%-50\%]Ca[30\%-50\%]$ $(Al, Si)AlSi_2O_8$) and labradorite ($Ca[50\%-70\%]Na[50\%-30\%](Al, Si)AlSi_2O_8$).

Species: Orthoclase feldspar (Figure 4.16*b*) is an example of an alkali feldspar and contains potassium ions ($KAlSi_3O_8$).

Common Rock-Forming Minerals

There are seven important rock-forming minerals or mineral groups: the olivine group, the pyroxene group, the amphibole group, the mica group, the feldspar group, quartz, and calcite (**FIGURE 4.17**).

1. **Olivine group** [$(Fe, Mg)_2SiO_4$]: Olivines are silicates that crystallize in cooling magmas and are rich in iron or magnesium. Olivine is named for its olive-green colour. It is an important component of iron/magnesium-rich volcanic rocks.

2. **Pyroxene group** [$(Fe, Mg, Al, Ca, Na)SiO_3$]: Pyroxene, an important single-chain silicate, is a common rock-forming mineral. It crystallizes at a high temperature in the absence of water. If water were present, a double-chained amphibole would most likely form instead.

3. **Amphibole group** [$(Ca, Na, Fe, Mg, Al)Si_8O_{22}OH_2$]: Hornblende, the most common species of amphibole, is found in many rocks. Hornblende is actually the name given to a series of minerals that are rather difficult to distinguish based on ordinary physical properties alone. The iron, magnesium, and aluminum ions can freely substitute for one another and form separate varieties of hornblende, of which there are many (magnesiohornblende, ferrohornblende, etc.).

4. **Mica group** [$K(Fe, Mg)_3AlSi_3O_{10}(F, OH)_2$]: Biotite, a member of the mica group, is present in many rocks and forms under a wide variety of conditions. Typically ranging in colour from black to brown, it has a layered structure composed of sheets of iron–magnesium–aluminum silicate weakly bonded together by layers of potassium ions. Muscovite, the other common member of the mica group, is lighter in colour, because it does not contain any iron or magnesium (See Figure 4.15, and note that muscovite is still electrically neutral.)

5. **Feldspar group** [$(Na, Ca)(Al, Si)AlSi_2O_8$]: As described earlier, the feldspar group contains both plagioclase and orthoclase feldspar species. The feldspars are the most common mineral group in Earth's crust. Because they crystallize under a wide range of pressures and temperatures, they occur in many rocks.

6. **Quartz** (SiO_2): Quartz is the second-most common silicate mineral, is found in nearly every geologic environment, and is a component of almost every type of rock. Quartz is also very diverse in terms of varieties, colours, and forms.

Photos courtesy of Chip Fletcher

FIGURE 4.17 Seven common rock-forming minerals: (a) olivine is the light olive-green mineral in this ultramafic rock, dunite, (b) block prismatic crystal of the pyroxene, augite, (c) prismatic crystal of the amphibole, hornblende, (d) biotite crystals from a coarse-grained pegmatite, (e) blocky pink orthoclase feldspar, (f) blocky white crystal of albite, a plagioclase feldspar, (g) crystal of quartz showing hexagonal prism and pyramid faces, and (h) optically transparent crystal of calcite showing perfect rhombohedral cleavage in three directions.

(?) Which of these minerals are silicates?

7. **Calcite** ($CaCO_3$): Calcite gets its name from *calx*, a Latin word meaning "lime." It is one of the most common minerals on Earth, comprising about four percent of the crust by weight. It is formed in many different geologic environments but is especially common as marine sediment because several types of marine organisms, such as coral and certain types of very abundant plankton, secrete exoskeletons of calcite. Calcite is used in the steel, glass, chemical, and optical industries, and as ornamental stone.

 Expand Your Thinking—Study the colour and chemistry of the seven common minerals. There is a strong relationship between colour and chemistry. What is it?

4-9 Most Minerals Fall into Seven Major Classes

LO 4-9 *Describe each of the seven major classes of minerals.*

No chapter on mineralogy would be complete without introducing the seven major classes of minerals (**FIGURE 4.18**), defined by the important anion or anion group in the mineral. **TABLE 4.5** presents the seven classes and some of their economic uses.

These classes will remain little more than abstractions unless you can see where each occurs in Earth's crust. We have therefore included an illustration of the environments in which each class of minerals form. Please study **FIGURE 4.19** in the "Critical Thinking" box "Mineral Environments" as you read the following summary of the seven mineral classes.

1. *Silicates* are compounds of silicon, oxygen, and other elements. They are the largest, most complex mineral group. About 30 percent of all minerals are silicates, and up to 90 percent of the crust is composed of silicates. This is not surprising, because oxygen and silicon are the two most abundant elements in the crust. Important silicates include quartz, feldspar, olivine, amphibole, and biotite. Silicates form in cooling magma, certain sedimentary environments, and metamorphic rocks. They also crystallize in chemically enriched **hydrothermal fluids** that form in the crust when groundwater comes into contact with hot rock.

2. *Native elements* include over 100 known minerals, mostly metals. The native elements are often pure masses of a single element, such as gold, silver, lead, copper, and sulphur. They accumulate in several environments, but they are especially concentrated in *vein deposits* (mineral deposits that fill pre-existing cracks in rock) formed by hydrothermal fluids (discussed more fully in Chapter 5).

3. *Oxides* are compounds of metallic cations (iron, copper, aluminum, and others) bonded to oxygen atoms. Many ores mined and refined to extract pure metals for commercial use are oxides. Examples include hematite (iron oxide), corundum (aluminum oxide), cuprite (copper oxide), and ice (hydrogen oxide, a mineral only at temperatures below 0°C). Oxides are present everywhere in the crust because of the prevalence of oxygen. They are formed in surface environments where iron, manganese, and other metals and native elements bond with oxygen from the atmosphere and elsewhere.

4. *Sulphides* are compounds of metallic cations (lead, iron, zinc, etc.) bonded with sulphur (an anion, S^{2-}), such as the mineral pyrite (FeS_2). As with oxides, commercially valuable metals constitute a high proportion of this group and are mined as valuable ores for refinement. Important sulphides include pyrite (iron sulphide), galena (lead sulphide), sphalerite (zinc sulphide), cinnabar (mercury sulphide), and chalcopyrite (copper-iron sulphide). Sulphides tend to form as vein deposits from hydrothermal fluids.

5. *Sulphates* are compounds of metallic cations bonded with the sulphate anion (SO_4^{2-}). They include anhydrite (calcium sulphate), which is used to make plaster, and barite (barium sulphate), important in the drilling industry. Sulphates form in surface environments.

(e)   *(f)*   *(g)*

Courtesy of Chip Fletcher

FIGURE 4.18 The seven major mineral classes and one example from each class: *(a)* silicates (orthoclase feldspar), *(b)* native elements (copper), *(c)* oxides (hematite), *(d)* sulphides (pyrite), *(e)* sulphates (gypsum), *(f)* halides (fluorite), and *(g)* carbonates (malachite).

(?) These classes are defined by chemical composition. Describe another criterion by which you could define major classes of minerals.

TABLE 4.5 Seven Major Mineral Classes

Mineral Group	Example	Chemical Formula	Common Economic Uses
Silicates	Quartz	SiO_2 (silicon dioxide)	Silicon source for electronics, abrasives
	Olivine	$(Mg,Fe)_2SiO_4$ (magnesium–iron silicate)	Abrasives
	Orthoclase	$KAlSi_3O_8$ (potassium–aluminum silicate)	Glass, porcelain
	Plagioclase	$(Na,Ca)AlSi_3O_8$ (sodium–calcium–aluminum silicate)	Ceramics, building stone
Native Elements	Copper	Cu	Electrical wiring
	Gold	Au	Trade, jewellery, electronics
	Sulphur	S	Fertilizers, explosives
	Graphite	C (carbon)	Pencil "lead"; lubricant
Oxides	Ice	H_2O (hydrogen oxide)	Crystalline form of water
	Hematite	Fe_2O_3 (ferric oxide)	Ore of iron, pigment
	Corundum	Al_2O_3 (aluminum oxide)	Jewellery (sapphire), abrasive
	Magnetite	Fe_3O_4 (ferrous oxide)	Ore of iron
Sulphides	Galena	PbS (lead sulphide)	Ore of lead
	Pyrite	FeS_2 (iron sulphide)	Sulphuric acid production
	Chalcopyrite	$CuFeS_2$ (copper–iron sulphide)	Ore of copper
	Sphalerite	ZnS (zinc sulphide)	Ore of zinc
Sulphates	Barite	$BaSO_4$ (barium sulphate)	Drilling mud
	Gypsum	$CaSO_4 \cdot 2H_2O$ (The dot represents water embedded within the crystalline structure as the water molecule.)	Plaster
	Anhydrite	$CaSO_4$ (calcium sulphate)	Plaster
Halides	Fluorite	CaF_2 (calcium fluoride)	Used in steel making
	Halite	$NaCl$ (sodium chloride)	Common salt
Carbonates	Calcite	$CaCO_3$ (calcium carbonate)	Portland cement, lime
	Dolomite	$CaMg(CO_3)_2$ (calcium–magnesium carbonate)	Cement, lime
	Malachite	$Cu_2CO_3(OH)_2$ (copper carbonate)	Jewellery

6. *Halides* are compounds whose anions are halogens, a group of highly reactive gases including fluorine, chlorine, bromine, iodine, and astatine. Halogens usually have a charge of -1 when chemically bonded; hence they tend to react with many other elements. Common halides include fluorite (calcium fluoride), halite (sodium chloride), and sylvite (potassium chloride).

7. *Carbonates* are compounds of cations bonded with the carbonate ion (CO_3^{2G}). Important carbonates include calcite (calcium carbonate) and dolomite (calcium–magnesium carbonate). Carbonates are formed in calcium-saturated waters such as oceans and other complex fluids (such as hydrothermal fluids).

Expand Your Thinking—Describe where an example of each of the seven major mineral classes may precipitate on Earth's surface as a sedimentary mineral (not an igneous or metamorphic mineral).

? CRITICAL THINKING

MINERAL ENVIRONMENTS

Figure 4.19 illustrates some of the geologic environments in which the seven major mineral groups occur:

- Silicates commonly form in cooling magma.
- Native elements tend to concentrate in hydrothermal veins or surface deposits.
- Oxides can be found in ordinary igneous rocks or surface environments.
- Sulphides form in oxygen-depleted environments where metallic cations bond with sulphur.
- Sulphates form in the presence of oxygen and sulphur with metallic cations.
- Halides form in places where evaporation of sea water or other chemical fluids leads to crystallization of dissolved compounds.
- Carbonates form as skeletal components of marine organisms such as plankton (deep sea) or corals (shallow sea), or where water evaporates.

Please work with a partner and answer the following:

1. In Figure 4.19, the minerals in the photo are named. Identify the mineral class for each mineral.
2. What compound forms the basic structure of the silicate minerals? Draw it.
3. Name an environment in which sulphates are found.
4. Give three examples of marine organisms that produce carbonate minerals.
5. What element must be present for oxide minerals to form? Give an example of an oxide mineral that you encounter in your everyday life.
6. Why are native elements economically valuable? Give examples of three valuable native elements and how you use them.
7. What are the differences between the environmental conditions that lead to formation of sulphides and those that lead to formation of sulphates?
8. Describe an environment in which halides form. Give an example of a halide mineral that is economically important.
9. List five economically important minerals that have been used in objects you encountered today.

Halite

Gypsum

Calcite

Calcite: © Editorial Image, LLC/Alamy; gypsum: © B Christopher/Alamy; halite: Photo courtesy of The Mineralogical Association of Canada; galena: © Gary Cook/Alamy; ice crystals: © Stocksearch/Alamy; olivine: Jacana/Science Source; granite: Science Photo Library/Science Source.

FIGURE 4.19 Examples of the environments in which the seven common mineral groups form.

LET'S REVIEW "GEOLOGY IN OUR LIVES"

Now that you have finished the chapter, "Geology in Our Lives" will have taken on new meaning for you. Let us review it: Minerals are a critical geologic resource used every day in manufacturing, agriculture, electronics, household items, and even the clothes you wear. Hence, it is vital to ensure the sustainability of mineral resources. Minerals are solid crystalline compounds with a definite (but variable) chemical composition that are found in the crust and mantle.

From the mineral nutrients ingested in our food to the metals that make cars, computers, and buildings, minerals sustain us, our cities and farms, and the world at large. Keep this important idea with you, and remember that minerals are not an endless resource. Use them sparingly, and recycle whenever possible.

STUDY GUIDE

4-1 Minerals are solid crystalline compounds with a definite (but variable) chemical composition.

- Minerals make possible the manufacture of almost every product sold today and are essential to maintaining our way of life.
- A mineral is a naturally occurring, inorganic, crystalline solid with a definite, but sometimes variable, chemical composition.

4-2 A rock is a solid aggregate of minerals.

- A rock is a solid aggregate of minerals. For example, granite is an aggregate of several minerals—typically quartz, plagioclase feldspar, orthoclase feldspar, biotite, and others.

4-3 Geologists use physical properties to identify minerals.

- Many minerals have several physical properties that can be used to identify them. Each mineral's unique chemistry and **crystalline structure** give it distinctive physical qualities, such as hardness, cleavage, fracture, lustre, colour, and streak.

4-4 Atoms are the smallest components of nature with the properties of a given substance.

- **Atoms** are the smallest components in nature that have the properties of a given substance. Each element is defined by its **atomic number**, which is the number of protons in its nucleus. The number of neutrons and protons in the nucleus is the element's **mass number**. Variations in mass number create **isotopes**. A variation in electrical charge forms an **ion**. A negative charge forms an **anion**; a positive charge forms a **cation**.
- Most atoms seek to achieve a stable electron configuration by bonding with another atom.

4-5 Minerals are compounds of atoms bonded together.

- Atoms of most elements have the ability to bond with atoms of other elements in order to achieve a stable electron configuration. The result is the formation of **compounds**.

- The tendency of atoms to seek to attain the full complement of electrons in their outermost shells causes them to form one of three common types of bonds: ionic, covalent, or metallic.

4-6 Oxygen and silicon are the two most abundant elements in the crust.

- Cooling magma allows igneous minerals to crystallize in three important tectonic settings: subduction zones, spreading centres, and hotspots. Most magma consists of the eight major elements in the crust: oxygen (O^{2-}), silicon (Si^{4+}), aluminum (Al^{3+}), iron ($Fe^{2+, 3+}$), magnesium (Mg^{2+}), calcium (Ca^{2+}), potassium (K^+), and sodium (Na^+) as well as dozens of trace elements.
- **Crystallization** is the process in which atoms or compounds that are in a liquid state (magma) are arranged into an orderly solid state. This process involves **nucleation** followed by crystal growth. Many nucleation sites produce smaller crystals that are less well developed; few nucleation sites produce larger crystals.
- Silicon and oxygen readily form a covalent bond to create a **silicate** compound. Four oxygen atoms surround a single silicon atom, forming the silicate compound $(SiO_4)^{4-}$. The four-sided geometric shape of silicate crystals is known as a tetrahedron.
- When the process of crystallization is complete, a mineral compound is neutral. All negative charges associated with silicate tetrahedrons are balanced by the positive charges of cations.

4-7 Metallic cations join with silicate structures to form neutral compounds.

- In magma, cations of similar size and charge jockey for position within the silicate network in a process called cation substitution. This process leads to the formation of a wide variety of minerals. The cation pairs that most often substitute for one another are Na^+/Ca^{2+}, Al^{3+}/Si^{4+}, and Fe^{2+}/Mg^{2+}. Olivine forms by single substitution and feldspar forms by double substitution.

4-8 There are seven common rock-forming minerals.

- Minerals whose composition varies over a particular range are given a group name. Within the group, each mineral species has a specific composition. Within species, a common irregularity in crystallization or a chemical impurity produces a distinct mineral variety.

- There are seven important rock-forming minerals: the olivine group, the pyroxene group, the amphibole group, the mica group, the feldspar group, quartz, and calcite.

4-9 Most minerals fall into seven major classes.

- There are seven major mineral classes based on their composition: silicates, native elements, oxides, sulphides, sulphates, halides, and carbonates.

KEY TERMS

amphibole group (p. 94)
anion (p. 86)
atomic number (p. 86)
atoms (p. 86)
calcite (p. 95)
cation (p. 86)
compounds (p. 88)
crystalline structure (p. 80)
crystallization (p. 90)

feldspar group (p. 94)
hydrothermal fluids (p. 96)
igneous minerals (p. 90)
ion (p. 86)
isotopes (p. 86)
mass number (p. 86)
metamorphic minerals (p. 90)
mica group (p. 94)
mineral (p. 80)

Mohs hardness scale (p. 84)
nucleation (p. 90)
olivine group (p. 94)
pyroxene group (p. 94)
quartz (p. 94)
sedimentary minerals (p. 90)
silicate structures (p. 91)
silicates (p. 89)

ASSESSING YOUR KNOWLEDGE

Please answer these questions before coming to class. Identify the best answer.

1. Which of the following is part of the definition of a mineral?
    a. liquid
    b. electrically charged
    c. inorganic
    d. synthetic
    e. None of the above.

2. Many minerals are useful in everyday life. Some examples include
    a. feldspar and quartz.
    b. clay and gypsum.
    c. graphite and chalcopyrite.
    d. copper and titanium.
    e. All of the above.

3. To quickly identify a mineral sample, geologists use
    a. physical size.
    b. colour.
    c. physical properties.
    d. laboratory analysis.
    e. None of the above.

4. "Fool's gold" is
    a. hematite.
    b. calcite.
    c. pyrite.
    d. native gold.
    e. None of the above.

5. One of the isotopes of the element carbon (atomic number 6) has a mass number of 13. How many neutrons does this isotope have in its nucleus?

    a. 5
    b. 6
    c. 7
    d. 14
    e. None of the above.

6. What are formed when sodium ions and chlorine ions combine to produce NaCl?
    a. ionic bonds
    b. covalent bonds
    c. organic structures
    d. isotopes
    e. native elements

7. What property causes the mineral biotite to break into flat sheets?
    a. its density
    b. its electrical charge
    c. its crystalline structure
    d. its hardness
    e. None of the above.

8. Silicates are constructed by
    a. carbon and hydrogen.
    b. iron and oxygen.
    c. silica and feldspar.
    d. silicon and oxygen.
    e. None of the above.

9. Single substitution occurs during crystallization because
    a. neutral compounds attract ions.
    b. the number of leftover ions must be balanced.

c. a charged compound is formed.

d. ions of similar size can substitute for one another.

e. forming dense compounds requires single substitution.

10. The two most abundant elements in the crust form
    a. oxides.
    b. sulphates.
    c. silicates.
    d. carbonates.
    e. halides.

11. The important rock-forming minerals include
    a. feldspars, biotite, and garnet.
    b. calcite, feldspars, biotite, and amphiboles.
    c. amphiboles, feldspars, quartz, and rutile.
    d. rutile, amphibole, calcite, and garnet.
    e. quartz, feldspar, granite, and basalt.

12. The silica compound takes the shape of
    a. a rectangle.
    b. a tetrahedron.
    c. a polygon.
    d. a polymer.
    e. magma.

13. Plagioclase feldspar is
    a. a mineral group.
    b. a mineral species.
    c. a mineral variety.
    d. a type of quartz.
    e. All of the above.

14. Which of the following best describes the difference between sulphates and sulphides?
    a. Sulphates include nitrogen; sulphides do not.
    b. Sulphides are metals bonded with sulphur; sulphates are metals bonded with the sulphate anion.
    c. Sulphates are metals bonded with inorganic carbon; sulphides are metals bonded with water molecules.
    d. Sulphates are formed only in igneous rocks; sulphides are formed in all types of rock.
    e. All of the above.

FURTHER RESEARCH

1. Look around you. Can you identify 10 substances made from minerals? (See Tables 4.1 and 4.5.)

2. Define "mineral" and explain the meaning of each component of the definition.

3. Referring to Table 4.3, draw the oxygen atom. Be sure to include the nucleus and its components as well as the electrons in their orbitals.

4. Why are cations needed for minerals to form? Refer to the charge of a silicate tetrahedron in your answer.

5. Go to the web and find photos of minerals from each of the seven mineral groups. Identify the minerals and describe their chemistry, their crystalline structure, and the environment in which they form.

6. What controls the order in which minerals crystallize from magma?

7. What part of Earth produces the magma that forms minerals with high iron and magnesium content? Where does the magma that forms silicon- and oxygen-enriched minerals come from?

ONLINE RESOURCES

Explore more about geology on the following websites:

Mineralogical Association of Canada website with lots of links:
www.mineralogicalassociation.ca

Classroom activities and review of key concepts:
csmres.jmu.edu/geollab/Fichter/Minerals/index.html

Additional animations, videos, and other online resources are available at this book's companion website:
www.wiley.com/go/fletchercanada
This companion website also has additional information about WileyPLUS and other Wiley teaching and learning resources.

CHAPTER 5
IGNEOUS ROCK

Chapter Contents and Learning Objectives (LO)

5-1 Igneous rock forms when molten, or partially molten, rock solidifies.
LO 5-1 *Describe igneous rock.*

5-2 Igneous rock forms through a process of crystallization and magma differentiation.
LO 5-2 *Describe the processes that lead to magma differentiation.*

5-3 Bowen's reaction series describes the crystallization of magma.
LO 5-3 *Describe the process of mineral crystallization in a magma.*

5-4 The texture of igneous rock records its crystallization history.
LO 5-4 *Identify, in detail, the information revealed by igneous texture.*

5-5 Igneous rocks are named on the basis of their texture and composition.
LO 5-5 *Identify how igneous rock colour relates to its chemical composition.*

5-6 The seven common types of igneous rock in more detail.
LO 5-6 *List the seven most common types of igneous rock and their composition and texture.*

5-7 All rocks on Earth have evolved from the first igneous rocks.
LO 5-7 *Describe the process of igneous evolution and the role of plate tectonics.*

5-8 Basalt forms at spreading centres, hotspots, and subduction zones.
LO 5-8 *Describe the environments where basalt accumulates and why it accumulates there.*

5-9 Igneous intrusions occur in a variety of sizes and shapes.
LO 5-9 *Describe the characteristics of the different types of intrusions.*

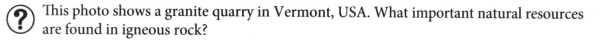

GEOLOGY IN OUR LIVES

Melting recycles Earth materials, and igneous rock forms when molten rock (magma) cools and crystallizes. Igneous rocks were the first rocks on Earth. All other rocks have evolved from igneous rocks that formed over 4 billion years ago. Igneous rocks provide critical building materials and mineral resources, and make up Earth's interior.

(?) This photo shows a granite quarry in Vermont, USA. What important natural resources are found in igneous rock?

5-1 Igneous Rock Forms When Molten, or Partially Molten, Rock Solidifies

LO 5-1 *Describe igneous rock.*

At this moment, **igneous rock** (*ignis* is Latin for "fire") lies beneath your feet. Whether visible at the surface (**FIGURE 5.1**) or deeply buried, igneous rock accounts for 80 percent of the material in Earth's crust. But although igneous rock is everywhere, it forms only in certain tectonic settings. These include spreading centres, where liquid **magma** rises from the upper mantle and cools to form the igneous sea floor; subduction zones, where hot fluids emerge from subducting oceanic crust and cause melting in the overlying upper mantle; and mid-plate settings, where a line of volcanoes grows from a magma source at a hotspot.

What Is Igneous Rock?

Igneous rock starts out as liquid magma created by **partial melting** either in the mantle or in the crust. As magma migrates toward the surface, it may undergo **differentiation**, and then cool and crystallize to form solid rock, adding to the mass of the crust. Magma that reaches the surface may form a volcano built up of accumulated **lava** and **pyroclastic debris** (erupted bits of rock, such as ash; see Chapter 6, "Volcanoes").

You may already be familiar with some types of igneous rock (**FIGURE 5.2**). *Granite* is a handsome building stone used to make floors, countertops, and gravestones. It forms the core of the Coast Mountains in British Columbia and the St. Elias Mountains, including Mount Logan, the highest point in Canada, in Yukon. Granite is also found in many places in the Canadian Precambrian Shield, and forms the spectacular cliffs in Yosemite National Park in California. *Basalt* builds chains of volcanic islands across the Pacific, Indian, and Atlantic Oceans and forms the bedrock of every ocean, covering millions of square kilometres. The Hawaiian Islands are probably the most famous volcanic islands, and many people visit them to see igneous rocks actually forming.

Carsten Peter/National Geographic/Getty Images, Inc.

FIGURE 5.1 When volcanoes erupt, ash, various gases, and magma are often discharged. Magma, whether within the crust or on it, solidifies to become igneous rock

 How would an igneous rock composed of ash look different from one composed of lava?

(a)

(b)

Courtesy of Chip Fletcher

FIGURE 5.2 *(a)* Granite and *(b)* basalt are two common igneous rocks built of silicate minerals. On average, granite tends to contain higher amounts of silica and lower amounts of iron and magnesium compared with basalt.

 Name two basic differences in the appearance of these two rocks.

Why Study Igneous Rocks?

Igneous rocks were the first to form as the young Earth's molten surface cooled and crystallized billions of years ago. Since that time, those original rocks have weathered and eroded, metamorphosed, melted, and crystallized many times over. This sequence of changes has led to the formation of the myriad sedimentary, metamorphic, and igneous rocks that make up our planet today. Igneous rock, thus, was Earth's first geologic generation, and all other rocks have evolved from it.

Although geologists have learned much about igneous rock, questions persist about how magma crystallizes, what conditions lead to the concentration of certain valuable minerals, and exactly how igneous processes have created continental crust. Although important clues to conditions deep in magma chambers have been revealed by laboratory experiments and field observations, questions concerning the origins of igneous rocks and their valuable mineral deposits are subjects of continuing geologic research and critical thinking.

Igneous rocks can be subdivided into two fundamental types: extrusive and intrusive. **Extrusive** igneous rock is lava and other volcanic products that have been extruded onto Earth's surface and then crystallize in contact with air or water. **Intrusive** igneous rock is

magma that has crystallized within the crust or mantle. As intrusive rock can form at great depths within the crust or mantle it is often referred to as **plutonic** (for Pluto, the Roman god of the underworld). The youngest igneous rocks in Canada are found in British Columbia, and they are related to the active plate tectonic boundary to the west. The Cascade Range, that extends from California to southern British Columbia, includes large and active **volcanoes**, such as Mount Rainier, Mount Hood, Mount St. Helens, and Mount Shasta. Mount Baker, Mount Garibaldi, and Mount Meager are very close to Vancouver and the Fraser Valley, although these volcanoes have not erupted recently. However, it is very important to understand how these volcanoes formed and how they might erupt in order to understand the hazards associated with them. This will be discussed in more detail in Chapter 6.

It is also important to understand how igneous rocks form because many concentrations of important metals are associated with them. For example, most of the world's chromium, nickel, and platinum, and most of the world's deposits of copper, gold, and rare earth elements are linked to igneous processes. This will be highlighted further in Chapter 10, "Geologic Resources."

 Expand Your Thinking—Igneous rock comes from magma. What plate tectonic environments produce magma?

5-2 Igneous Rock Forms through a Process of Crystallization and Magma Differentiation

LO *5-2 Describe the processes that lead to magma differentiation.*

Partial Melting

How does rock melt to create magma? Imagine mixing pennies, bits of candle wax, and water in a bowl, freezing the mixture, and then thawing it at room temperature. When the mass thaws, the ice melts first while the wax and copper remain solid. If you put the remaining mixture in the oven, the wax will melt but the pennies will not. Why? *Because the three substances have different melting points.* Rocks are more complicated, but the principle is the same. The different chemical compounds in the rock—that is, different minerals—all have different melting points. Thus, in order to generate a magma, the source rock, which might be in the mantle or the crust, must have reached a high enough temperature at a particular depth (or pressure) that minerals start to melt. The minerals, and the amounts of them, that melt will determine the composition of the magma that is produced. This process is called partial melting, and is how all magmas are generated. Volcanoes are the surface expression of this process, and as they are not distributed equally over Earth's surface there must be something special about the locations deep in the mantle or crust that allows melting to take place. This will be discussed further in sections 5-7, 5-8, and 5-9.

The mantle is the source of many magmas, especially those that eventually form basalt rock. Rock in the mantle is very hot, but it remains solid (although it is capable of flow) because of the high pressures there. However, *convection* is thought to cause some "mixing" in the mantle, as hot rock rises and cooler rock subsides. As hot rock rises to shallower depths, it experiences lower pressure, and this leads to **decompression melting**. Melting occurs because, as pressure is reduced, the rock's melting point is reduced as well. Decompression melting takes place under the crust where the upper mantle is exposed to decreases in pressure as a result of *crustal rifting, volcanic eruption, or mantle convection.* Magmas can form at temperatures of 600°C to 1,300°C, depending on the melting point of the minerals in the source rock. Higher temperatures are needed to melt the mantle than the crust, but once enough magma accumulates, it tends to rise, as it is less dense than the surrounding solid rock and intrudes into the crust (**FIGURE 5.3**). Some of it cools and crystallizes, forming (intrusive) igneous rock.

Magma Differentiation

Magma is a hot soup composed of Earth's major elements (Si, O, Ca, Al, Fe, Na, Mg, K) plus other elements in lesser quantities. As magma migrates into and through the crust, its composition changes. This process is called **magma differentiation**, and it explains how a single "parent" magma body can produce dozens of different types of igneous rocks (**FIGURE 5.4**). Magma differentiation can include four processes: *crystal settling, magma migration, magma assimilation,* and *magma mixing.*

Crystallization and Crystal Settling

As you might imagine, as magma cools, it experiences the reverse of partial melting. That is, during cooling, different compounds solidify and become minerals at different temperatures, with some portion of the magma remaining fluid until the very end. This process of *crystallization* was described in Chapter 4, "Minerals." The important point here is that the chemical composition of magma changes as elements are removed from it through crystallization. As the magma cools or moves to places where pressures are lower, the minerals formed differ in their chemical composition because the magma itself has a different chemical composition.

 How would Earth's surface be different if rock in the mantle did not experience decompression melting?

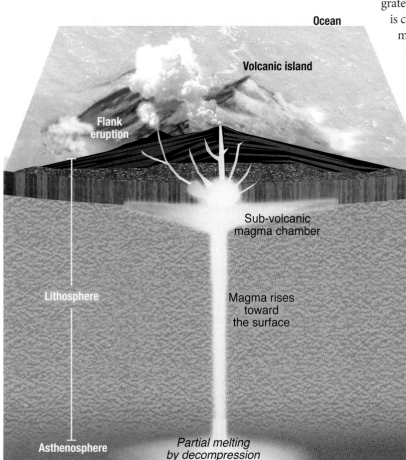

FIGURE 5.3 Magma may form when rock in the mantle experiences a decrease in pressure at shallow depths and undergoes decompression melting.

Magma crystallization probably takes thousands of years. Depending on the size of the magma body and its rate of cooling, roughly a million years may be needed to fully solidify a large accumulation. The magma may either crystallize within the crust or erupt at the surface and then crystallize quite rapidly. The characteristics of the resulting igneous rock—its chemistry, mineralogy, and crystal size—are thus partly determined by the depth at which crystallization occurs.

Usually, iron-, magnesium-, and calcium-based minerals crystallize first (at the highest temperatures). As cooling continues, therefore, the magma becomes relatively depleted in the elements that preferentially go into the higher temperature minerals. One way in which this depletion occurs is through crystal settling, in which the early-forming minerals, which are denser than the surrounding liquid, settle out of the magma body. As the minerals settle they may react with the still molten magma, but those that survive may produce a zone of highly concentrated minerals in the magma called a *cumulate*. Cumulates often contain economically valuable ore deposits, such as concentrations of chromium (see Chapter 10, "Geologic Resources").

Magma Migration

As we have seen, magma may migrate within the crust. The cumulate then becomes isolated from the magma and forms a separate body of igneous rock with a distinctive mineral composition. The separated magma continues to crystallize at its new location without some of its original components (i.e., those in the early-crystallizing minerals), forming new types of igneous rock. If magma erupts onto Earth's surface, still other types of minerals and rocks will be produced because crystallization will occur much more rapidly in the cool atmosphere or in the cold ocean if it erupts on the sea floor.

Magma Assimilation

As hot magma rises, the heat from the magma may be enough to melt the surrounding rocks in the crust or blocks of the wall rock that fall into the magma. The magma may also react chemically with the surrounding rocks. These processes will result in the addition (or "assimilation") of elements into the original magma, thereby changing its composition.

Magma Mixing

Sometimes magma encounters, and becomes mixed with, other bodies of magma. Alternatively, a body of magma may be separated into two or more masses. Each mass will form its own minerals as it cools, with iron-, magnesium-, and calcium-based minerals crystallizing first.

In each of these cases, magma passes through predictable *stages of crystallization,* each of which produces specific minerals. These stages establish the crystallization environment for the specific sets of minerals that make up each type of igneous rock. Clearly, since any body of magma can have a complex history, a wide variety of igneous rocks can form from a single magma body.

During the crystallization of most igneous rocks, oxygen and silicon form silicate tetrahedra bonded with metallic cations, producing minerals. In Chapter 4, we examined how silicates crystallize, but we did not look at the sequence in which particular minerals are formed or at the resulting types of igneous rock. We do this in the next section.

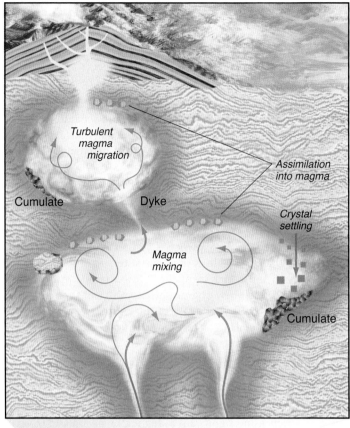

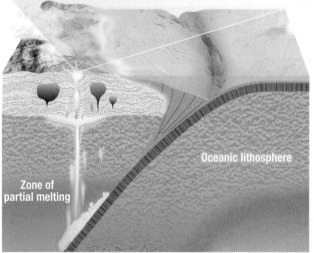

FIGURE 5.4 Magma differentiation can include four processes: crystal settling, magma migration, magma assimilation, and magma mixing. All these processes can change the composition of magma, producing many kinds of igneous rock.

 How would the magma ready to erupt from the volcano differ from the magma just entering the lower magma chamber?

 Expand Your Thinking—Magma is said to "evolve" by differentiation. What does this mean?

5-3 Bowen's Reaction Series Describes the Crystallization of Magma

LO 5-3 *Describe the process of mineral crystallization in a magma.*

As magma cools, minerals crystallize and separate from the liquid magma in a predictable sequence that depends on the chemistry and temperature of the magma body. Crystallization is essentially the change of state in which the random distribution of compounds in a liquid align themselves in the rigid crystal structures of solids or minerals.

Each mineral starts to form at different temperatures, and so as magma cools, the crystallization process forms different solid minerals in a predictable sequence. The first minerals to crystallize are those that are stable at the highest temperatures, followed progressively by those that are stable at lower temperatures. With the formation of each new crystal, the chemistry of the remaining magma changes to reflect the loss of the elements that went into the solidifying minerals.

In general, as crystallization progresses, magmas first lose their magnesium, iron, and calcium. These elements bind into early-forming minerals such as olivine, pyroxene, and calcium-rich plagioclase feldspar, which are stable in a solid state at high temperatures. Because these minerals have relatively low silica content, the magma experiences no significant reduction in silicon and oxygen. The magma thus evolves to a new chemical state that is relatively enriched in silica. This process is described in **Bowen's reaction series**, shown in **FIGURE 5.5**.

Bowen's Reaction Series

In the 1920s, mineralogist *N. L. Bowen* conducted experiments on simple mixtures of elements to determine the likely sequence in which common silicate minerals crystallize from magma. He developed an idealized progression, now known as Bowen's reaction series, which is still widely accepted as a general model of magma crystallization.

Bowen determined that certain minerals form at specific temperatures as magma cools. Minerals at the top of his series crystallize early in the cooling process, while the magma is still very hot. These minerals have the highest crystallization temperatures, are mostly dark in colour, and generally contain the most magnesium, iron, and calcium, when considering the common rock-forming silicate minerals. They also have the highest *specific gravity* (i.e., they are dense and feel noticeably heavy). We describe igneous rocks composed of these minerals, and others like them, as **mafic** (from *magnesium* + *ferric*) in composition. (**Ultramafic** rocks have even higher amounts of magnesium and iron.) The term "mafic" is used for any silicate mineral, magma, or rock that contains large amounts of magnesium, iron, and calcium. Common rock-forming mafic minerals include olivine, pyroxene, amphibole, and biotite mica. Calcium-rich plagioclase feldspars are also quite common in mafic rocks.

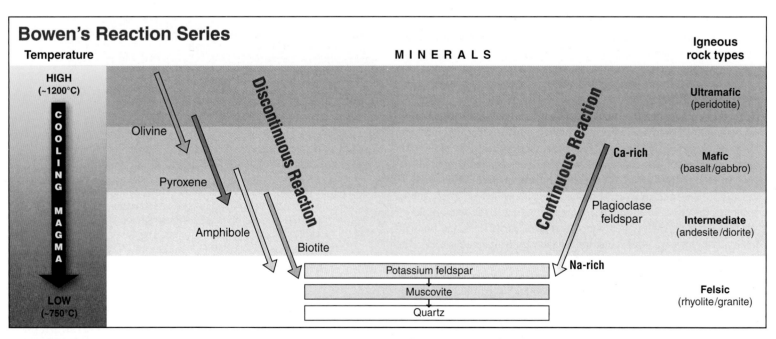

FIGURE 5.5 Mineralogist N. L. Bowen performed experiments and described two crystallization pathways leading to the formation of common silicate minerals in magma of average composition. The horizontal shaded bands highlight the mineral compositions of various types of rocks.

? Describe why the chemical composition of magma changes as the temperature decreases.

Minerals that form at the bottom of the series crystallize late in the cooling process, when the magma has cooled significantly. They have the lowest crystallization temperatures; are light in colour and relatively enriched in sodium, potassium, and silica; and have a relatively low specific gravity (i.e., they are less dense than mafic minerals). We describe igneous rocks composed of these elements as **felsic** (from *feldspar + silica*) in composition. Common felsic minerals include quartz, muscovite mica, the orthoclase feldspars, and the plagioclase feldspars. Note that sodium-rich plagioclase feldspars are more common in felsic rocks, whereas the calcium-rich plagioclase feldspars are more common in mafic rocks.

Igneous rocks whose composition is **intermediate** between that of mafic rocks and felsic rocks are composed of minerals that are somewhat abundant in iron, magnesium, calcium, potassium, sodium, and oxygen. Note in Figure 5.5 that there are two pathways for mineral crystallization: discontinuous, on the left, and continuous, on the right. This reflects Bowen's observation that crystallization at higher temperatures in mafic and intermediate magmas tends to separate into two simultaneous crystallization pathways (known as branches). The *discontinuous branch* describes the formation of the mafic minerals olivine, pyroxene, amphibole, and biotite mica. It is "discontinuous" because crystallized minerals formed at higher temperatures react chemically with the surrounding magma and re-crystallize at a lower temperature to create a different mineral with a different crystalline structure.

The *continuous branch* of crystallization describes the evolution of a common type of feldspar—plagioclase feldspar, a framework silicate—as it changes from calcium rich to sodium rich. The transition is smooth, as it involves single and double substitutions of cations without changes in the crystalline structure.

At lower temperatures, the two pathways merge and the minerals common to felsic rocks—potassium feldspar, muscovite mica, and quartz—form. TABLE 5.1 lists the most common igneous rock-forming minerals.

TABLE 5.1 Common Igneous Rock-Forming Minerals

Felsic Minerals (lighter in colour)		Intermediate Minerals		Mafic Minerals (darker in colour)	
Name	**Composition**	**Name**	**Composition**	**Name**	**Composition**
Plagioclase Feldspar	$NaAlSi_3O_8$ (albite)	Plagioclase Feldspar	$(Ca, Na)(Al, Si)AlSi_2O_8$	Plagioclase Feldspar	$CaAl_2Si_2O_8$ (anorthite)
Orthoclase Feldspar	$KAlSi_3O_8$	Amphibole Group	(Mg, Fe, Ca, Na) $Si_8O_{22}(OH)_2$	Olivine	$(Mg, Fe)_2SiO_4$
Quartz	SiO_2	Biotite Mica	$K(Mg,Fe)3AlSi_3O10(F,OH)_2$	Pyroxene	(Mg, Fe, Ca, Na, Al) SiO_3
Muscovite Mica	$KAl_3Si_3O_{10}(OH)_2$				

Framework Silicate	Double Chain Silicate	Sheet Silicate	Isolated Tetrahedra	Single Chain Silicate

Images: Marii Miller, Chip Fletcher

Expand Your Thinking—Bowen's reaction series is a *model* that predicts mineral crystallization in magma of average composition. Explain how this is an example of critical thinking.

5-4 The Texture of Igneous Rock Records Its Crystallization History

LO 5-4 *Identify, in detail, the information revealed by igneous texture.*

As we have seen, igneous rock forms as a result of magma crystallization. As minerals crystallize, magma becomes more viscous (resists flow) until the entire mass solidifies. The magma thus becomes a mass of interlocked crystals like the granite sample shown in FIGURE 5.6.

Texture and Intrusive/Extrusive Rocks

The texture of an igneous rock is a very important observation that can be used to interpret the cooling history, cooling rate, and the importance of volatiles. For example, the size of its minerals is an obvious observation that can provide an indication of how quickly the magma crystallized. Igneous rocks that cool beneath the surface for thousands of years may develop large crystals and a coarse texture. In essence, crystallization is slower and crystals tend to be larger than crystals in a rapidly cooled rock. Additional factors come into play in determining crystal growth but, in general, the rate of cooling exerts an important control on crystal size. Rocks that have larger crystals are generally considered to be intrusive. Magma that erupts at Earth's surface develops small crystals (or none at all) because it cools rapidly—within minutes to months—allowing far less time for crystallization. These rocks are generally considered to be extrusive or volcanic.

Based on this knowledge of the way crystals form, geologists can *infer* (or develop a hypothesis about) the crystallization history of magma from the texture of the resulting rock—that is, whether it is coarse- or fine-grained (FIGURE 5.7). "Coarse-grained" and "fine-grained" can vary widely, however, ranging from coarse-grained rocks with fist-size or even larger crystals to fine-grained rocks of volcanic glass that cooled so quickly that they contain no mineral crystals at all.

Common Igneous Textures

The best way to understand textures is to see them, so please refer to TABLE 5.2 and Figure 5.7 while reading the text.

Coarse-grained (intrusive) rocks are *phaneritic* (from the Greek word *phaneros,* meaning "visible")—the mineral grains within them are large enough to be seen with unaided eyes. Phaneritic rocks have large crystals because crystallization proceeded slowly and the resultant mineral grains are relatively large. Granite and gabbro are examples of phaneritic igneous rocks.

Fine-grained rocks are *aphanitic* (from *aphanes:* "hidden, invisible")—the mineral grains within them are too small to be seen with the unaided eye. These rocks are extrusive; basalt is an example. The finest rock texture is *glassy;* volcanic glass (obsidian) contains no crystals at all and, hence, no minerals. This can form when some magmas are extruded into water or air resulting in rapid quenching, which does not allow time for crystals to form and grow.

A rock with two distinct textures—mixed large and small grains—is *porphyritic.* The larger crystals are *phenocrysts,* and the finer ones are referred to as the *matrix* or *groundmass.* Rocks with porphyritic texture have undergone two separate stages of cooling: an intrusive stage, in which phenocrysts formed, and a later stage, usually at a higher level in the crust or after eruption, in which the matrix grains crystallize more rapidly.

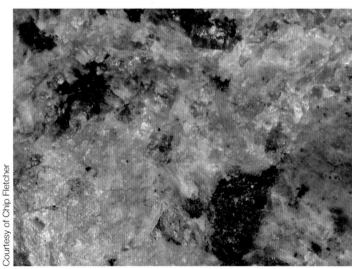
(a)

(b)

FIGURE 5.6 *(a)* Close-up of the surface of granite showing the interlocking nature of mineral grains; the crystals of the different minerals are each a few millimetres in size. *(b)* Photomicrograph showing the interlocking nature of minerals in an igneous rock. This is a photograph taken of a thin section of granite using a microscope. Each mineral exhibits different colours because of the way light interacts with the structure of that mineral. For scale, the base of the photomicrograph is 6 cm.

 What minerals can you identify in Figure 5.6a?

Aphanitic texture

Porphyritic texture

Pyroclastic texture

Phaneritic texture

Dirk Wiersma/Science Source

Courtesy of Chip Fletcher

FIGURE 5.7 Igneous rock texture reflects crystallization history.

 How does porphyritic texture form?

An igneous rock that is full of bubble-like holes is *vesicular*. The holes, or vesicles, were made by expanding gas bubbles trying to escape from the magma. This process occurs during volcanic eruptions because of the rapid expansion of gas as magma moves toward Earth's surface during an eruption. Basalt can have vesicles, but vesicles are more likely to be found in *pumice* and *scoria*. In fact, pumice contains so many vesicles that it can actually float on water because of its low density.

Finally, rocks that solidify when they are expelled into the air during violent volcanic eruptions are *pyroclastic* or *fragmental*.

Pyroclastic volcanic rocks may consist of numerous glassy fragments and shards that have been "welded" together by the heat of the eruption. They often feel grainy, like sandpaper, because shards of volcanic glass are embedded in the rock. The fragments may also consist of angular blocks broken off the walls of a volcanic vent, or large blobs of magma that crystallize as it flies through the air, forming a "bomb." Fine-grained pyroclastic-material is called *ash*, which eventually turns into a rock called a *tuff*.

TABLE 5.2 Common Igneous Textures

Texture	Definition	Example
Phaneritic	Minerals large enough to see with the unaided eye	Granite
Aphanitic	Minerals too small to see	Rhyolite
Glassy	No obvious minerals	Obsidian
Porphyritic	Two distinct mineral sizes	Porphyritic Basalt
Vesicular	Many holes or pits in rock surface caused by the trapping of escaping gas	Vesicular Basalt
Pyroclastic	Fused, glassy fragments, ash, and blocks of rock from explosive volcanic eruption	Tuff

? **Expand Your Thinking**—Why would volcanically erupted rocks be glassy?

5-5 Igneous Rocks Are Named on the Basis of Their Texture and Composition

LO 5-5 *Identify how igneous rock colour relates to its chemical composition.*

A close look at the size of mineral grains reveals a rock's texture. Similarly, a close look at a rock's colour reveals its general chemical composition.

The Colour of Igneous Rock

Minerals at the top of Bowen's reaction series, such as pyroxene and amphibole, tend to be dark-coloured and mafic to ultramafic in composition. Minerals at the bottom of Bowen's reaction series, such as sodium plagioclase feldspar, orthoclase feldspar, and quartz, tend to be light-coloured and felsic in composition. Geologists use these colour trends, along with texture, to identify and name igneous rocks.

The chemical content of an igneous rock determines its composition. However, analyzing rock chemistry requires expensive equipment and extensive training. In addition to conducting such analyses, geologists also use observations of physical properties to determine the composition of igneous rocks, just as we used physical properties to identify minerals (Chapter 4).

Colour is used to visually estimate the proportion of light minerals to dark ones in a rock sample. Light colours (white, light grey, tan, and pink) most often indicate a felsic composition. Felsic rocks are relatively high in silica, potassium, and sodium. Dark colours (black, dark grey, and dark brown) most often indicate a mafic composition. Mafic rocks are relatively low in silica, potassium, and sodium and relatively high in iron, magnesium, and calcium. Rocks with intermediate compositions are often grey or consist of equal parts of dark and light minerals. Geologists use a colour index to estimate the chemical composition of igneous rocks (**FIGURE 5.8**).

The colour index is not always an accurate indicator of a rock's composition. There are exceptions. Obsidian, for example, is volcanic glass that has cooled quickly after eruption. Obsidian is felsic in composition, yet it is commonly dark brown to black in colour because it contains trace amounts of iron. Dunite has an ultramafic composition (it is composed entirely of olivine, which has a high iron and magnesium content), yet it is apple green to yellowish green in colour.

In summary, while the colour index can be used to approximate the composition of many igneous rocks, judgments based on it should be viewed as estimates. Rocks with a felsic composition can, and usually do, contain dark-coloured minerals, and mafic rocks can contain light-coloured minerals.

Up to this point, we have discussed igneous mineralogy (Bowen's reaction series), texture (phaneritic, aphanitic, porphyritic, etc.), and the composition of igneous rocks (ultramafic, mafic, intermediate, and felsic). When we consider these characteristics together, we can use them to name the *most common types of igneous rock* (described in the next section).

Classifying the Most Common Types of Igneous Rock

Geologists classify igneous rocks based on a combination of the colour/texture, mineral/texture, or chemical composition/texture properties of the rock. These three classification systems are shown in **FIGURE 5.9**. Although at first glance this chart looks complex, it actually pulls together our knowledge about the minerals, texture, and composition of igneous rocks to make it easier to name them. And while it might

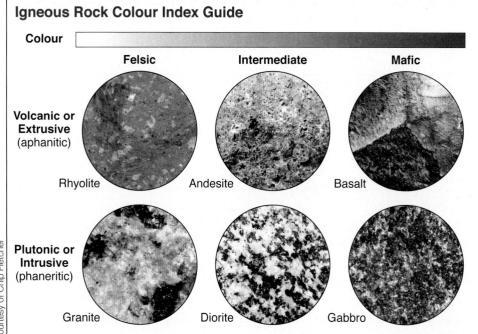

Igneous Rock Colour Index Guide

Colour

Felsic — Intermediate — Mafic

Volcanic or Extrusive (aphanitic)
Rhyolite — Andesite — Basalt

Plutonic or Intrusive (phaneritic)
Granite — Diorite — Gabbro

Courtesy of Chip Fletcher

(?) Use this information to describe the composition of the two rocks in Figure 5.2.

FIGURE 5.8 Igneous rock colour. Samples are arranged vertically by texture (top to bottom—aphanitic to phaneritic). Notice that the colour or shading of the rock generally corresponds to its composition—darker colours to mafic and lighter colours to felsic.

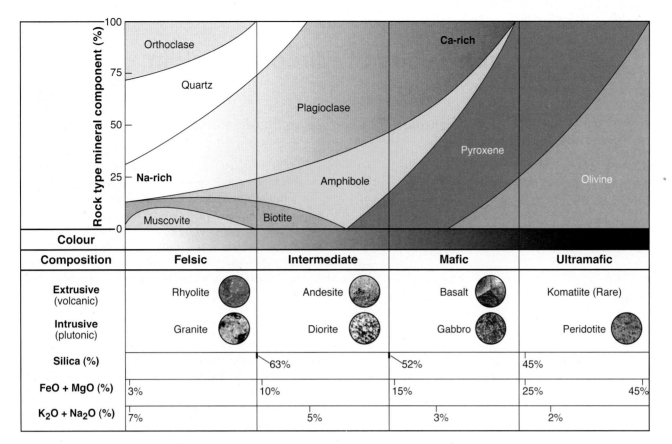

Composition	Felsic	Intermediate	Mafic	Ultramafic
Extrusive (volcanic)	Rhyolite	Andesite	Basalt	Komatiite (Rare)
Intrusive (plutonic)	Granite	Diorite	Gabbro	Peridotite
Silica (%)		63%	52%	45%
FeO + MgO (%)	3%	10%	15%	25% 45%
K₂O + Na₂O (%)	7%	5%	3%	2%

FIGURE 5.9 Classification of igneous rocks using mineralogy (upper portion of chart), colour, and chemistry (lower portion). See text for explanation. The internationally recognized divisions between felsic, intermediate, mafic, and ultramafic rocks is based on the amount of silica. The most silica-rich felsic rocks may contain up to 80% silica, whereas ultramafic rocks may contain as little as 30 percent silica.

? What would you call a fine-grained igneous rock that does not contain quartz?

not seem so at first, naming is important. It is how geologists around the world communicate about the rocks they find and study. By using the same naming system, for example, geologists everywhere can identify a particular sample of light-coloured, felsic, phaneritic rock of intrusive origin as granite.

Look at Figure 5.9 more closely and examine the top portion. It identifies the eight common igneous rock-forming minerals in Bowen's reaction series: orthoclase feldspar, quartz, plagioclase feldspar, muscovite mica, biotite mica, amphibole, pyroxene, and olivine. These minerals are found in the most common igneous rocks: *rhyolite, granite, andesite, diorite, basalt, gabbro, komatiite,* and *peridotite* (shown in the lower portion of the figure and described in detail in the next section).

The first column on the left shows the mineral percentage, colour, composition, and texture of each type of rock. Column 2 lists the felsic igneous rocks (rhyolite and granite) with their mineral content, colour, and chemistry. The lower portion shows the chemistry of the felsic rocks: approximately 6 percent to 7 percent potassium and sodium (K₂O + Na₂O), 3 percent to 9 percent iron and magnesium (FeO + MgO), and about 63 percent to 80 percent silica (SiO₂). Above the chemistry are the names of the intrusive rock (granite) and the extrusive rock (rhyolite) in the felsic category. Most of these rocks are light in colour.

The top portion of the chart shows the minerals in the rock and their abundance in typical samples. In the case of granite and rhyolite you would expect to find quartz, sodium-rich plagioclase feldspar, a potassium-rich feldspar like orthoclase, biotite mica, and maybe amphibole and muscovite mica. Column 3 lists the intermediate igneous rocks (andesite and diorite) and their characteristics; column 4 lists the mafic rocks (basalt and gabbro); and column 5 lists an ultramafic rock (peridotite). The extrusive equivalent of peridotite, the most common rock in the mantle, is komatiite, which is very rare.

This chart tells us three important things:

1. Intrusive and extrusive igneous rocks with the same composition contain the same minerals even though they differ in colour and texture.

2. Silica becomes less abundant as iron and magnesium become more abundant.

3. The abundance of potassium and sodium decreases as silica content decreases.

Using Figures 5.8 and 5.9, you can define the mineralogy, chemistry, texture, and colour of the most common igneous rocks. In the next section, we explore each of the major types of igneous rock in more detail.

? **Expand Your Thinking**—Describe the probable chemistry and mineralogy of magma that has experienced significant differentiation. What igneous rock will result?

5-6 The Seven Common Types of Igneous Rock in More Detail

LO *5-6 List the seven most common types of igneous rock and their composition and texture.*

The seven most common types of igneous rock are granite, rhyolite, diorite, andesite, gabbro, basalt, and peridotite. The extrusive equivalent of peridotite is komatiite, but is rare in comparison. By becoming familiar with them, we can better understand the *igneous rock system*—the major environments in which igneous rocks are formed.

Granite and Rhyolite

Granite is a coarse-grained, felsic intrusive rock that may be white, grey, pink, or reddish (**FIGURE 5.10a**). It is a rock of the continental crust, most commonly found in mountainous areas. The average granite contains coarse grains of quartz, potassium feldspar (orthoclase), and sodium-rich plagioclase feldspar as well as other common minerals, including the mica group (silvery muscovite and black biotite) and small amounts of amphibole.

Granites are the most abundant intrusive rocks of mountain belts and the bedrock of continents. They solidify in great magma chambers and form large intrusive structures called *batholiths* that may extend over hundreds of kilometres and are often associated with diorite and gabbro. Granite is an active intruder of continental crust, frequently also forming smaller intrusive bodies, called **stocks**, which cut across other rocks in the crust.

Rhyolite is a fine-grained extrusive rock of felsic composition, the volcanic equivalent of granite (**FIGURE 5.10b**). Granite and rhyolite are identical in terms of mineral content. Because one is intrusive and the other is volcanic, however, they have very different textures and hence have been given different names.

Rhyolite is usually light grey to pink and can be formed in either of two ways. It may be erupted and form small dome-like bodies of rhyolitic lava, which ultimately crystallize. Rhyolite lava is very viscous and does not flow very far from the vent. Often, these rhyolite lava domes plug the vent, and so sometimes, in spectacular fashion, rhyolite rock is made when a very explosive eruption hurls hot magma and fragments into the atmosphere. Gravity returns these pyroclasts to Earth's surface, depositing them in layers. If the pyroclasts are hot enough, they may weld together to form a rhyolite tuff (composed of pyroclasts and fragments less than 2 mm across). A *rhyolite breccia* (composed of pyroclasts and fragments more than 2 mm across) is also formed by pyroclasts welding together. We will learn more about these volcanic rocks in Chapter 6.

Diorite and Andesite

Diorite is a coarse-grained, intermediate intrusive rock composed of sodium- and calcium-bearing plagioclase feldspar (light-coloured) and amphibole (dark-coloured) (**FIGURE 5.11a**). These minerals give diorite its characteristic salt-and-pepper appearance. Small amounts of quartz and biotite mica may also be present. Diorite is the intermediate, intrusive member of the igneous rock family.

Andesite is the fine-grained volcanic equivalent of diorite (**FIGURE 5.11b**). It is formed by the eruption of intermediate or felsic magma. Because such eruptions often begin explosively, deposits of pyroclast layers are common on and around a volcano. The explosive phase is followed by a flow of lava that cools, hardens, and protects the underlying layer of pyroclasts from erosion. The volcanoes resulting from a series of such eruptions are layered (*stratified*) and are therefore called *stratovolcanoes*. Andesite is an important and frequent component of stratovolcanoes.

Gabbro and Basalt

Gabbro is a coarse-grained, mafic intrusive rock composed mostly of light-coloured calcium-rich plagioclase feldspar, dark pyroxene, and dark or green olivine (**FIGURE 5.12a**). The dark-coloured minerals

(a)

(b)

FIGURE 5.10 Granite *(a)*, an intrusive felsic igneous rock, is an important component of continental crust. Rhyolite *(b)* is its extrusive equivalent, identical chemically but fine-grained due to its volcanic origin.

(?) How is the appearance of these rocks related to their history?

Courtesy of Chip Fletcher

(a)

(b)

Courtesy of Chip Fletcher

Diorite and andesite are found in volcanic arcs. Explain the process whereby ultramafic rock in the mantle evolves to become intermediate in composition.

FIGURE 5.11 Diorite *(a)* is an intrusive igneous rock with an intermediate composition. It has a salt-and-pepper appearance. Andesite *(b)* has the same composition as diorite but is finer grained due to its volcanic origin.

(a)

(b)

Courtesy of Chip Fletcher

Describe where gabbro and basalt are found in greatest abundance and explain why mafic rock is predicted at that site.

FIGURE 5.12 Gabbro *(a)* is a mafic intrusive rock composed of olivine, pyroxene, amphibole, and Ca-plagioclase. Basalt *(b)* is a mafic volcanic rock of the same composition.

give gabbro a dark green to black colour that clearly indicates its mafic composition. At mid-ocean ridges, basaltic magma rises from the upper mantle and forms the intrusive component of oceanic crust. This makes gabbro among the most abundant forms of intrusive igneous rock on the planet and a very important component of oceanic crust.

Basalt is the fine-grained, mafic, extrusive equivalent of gabbro. It is usually black and forms when mafic magma cools at Earth's surface. FIGURE 5.12b shows a typical sample. Basalt magma forms by partial melting of mantle peridotite in a variety of tectonic settings (see section 5-8) so it is very abundant on planet Earth. For example, basalt magma erupts on the ocean floor into the rift at mid-ocean ridges during the process of seafloor spreading. The resulting basalt is the single most important component of oceanic crust.

Chains of high volcanic islands that develop at hotspots, such as the Hawaiian islands and Tahiti, are composed mostly of basalt that erupts from volcanoes on the sea floor and eventually builds up to create islands above the ocean's surface. For instance, the Big Island of Hawaii is composed of five huge volcanoes that have the classic, gentle slopes and rounded, shield-like shape found in volcanoes composed of basalt, which are known as *shield volcanoes*.

Peridotite and Komatiite

Peridotite is the only ultramafic igneous rock that occurs on Earth's surface in any appreciable quantity (FIGURE 5.13). It is composed of the ferromagnesian silicate minerals olivine and pyroxene. In some cases, it may represent cumulates that formed by the process of crystal settling (described earlier), as it contains the minerals that would

crystallize first, at the highest temperature. Because the magma that forms ultramafic rocks is so uniquely differentiated and composed of such dense compounds, it is very rarely found in a volcanic form. However, the extrusive equivalent, komatiite does exist, although it is usually found only in regions of the continental crust that are Precambrian in age. There are no volcanoes erupting ultramafic magmas at present, and geologists think that the Earth's mantle is cooler now than in the Precambrian so the temperatures necessary to generate these magmas are difficult to attain. Many geologists consider peridotite to represent the average composition of the upper mantle, making it one of the most abundant rocks on the planet, even though it is extremely rare on the surface.

Courtesy of Chip Fletcher

Explain why ultramafic rock is rare on Earth's surface.

FIGURE 5.13 Many geologists think that peridotite, although rarely seen in the crust, may represent the average composition of the upper mantle. This sample consists largely of green olivine crystals and black pyroxenes.

Expand Your Thinking—What might happen if mafic magma mixed with felsic magma? What minerals and rocks would result and in what order?

5-7 All Rocks on Earth Have Evolved from the First Igneous Rocks

LO 5-7 *Describe the process of igneous evolution and the role of plate tectonics.*

Igneous rocks were the first rocks on Earth. All other rocks have evolved from the igneous mafic/ultramafic parent rocks that formed over 4 billion years ago. Today, as a result of igneous evolution, igneous rocks are found in a profusion of types and tectonic environments. To deepen your understanding of the environments in which igneous rocks form, see the "Critical Thinking" box "Igneous Environments."

All Igneous Rocks Result from Magma Differentiation

Through the process of **igneous evolution**, the original, relatively uniform composition of primitive, molten Earth gave rise not only to igneous rocks but also, through the rock cycle, to sedimentary and metamorphic rocks. The realization that igneous rocks evolve over time is one of the most important ideas in modern geology because it explains the sequence of steps that have produced the great diversity of rocks that exist on Earth today.

Plate tectonic theory tells us that lithosphere is recycled into the mantle at subduction zones. Partial melting of the upper mantle above a subducting slab plus partial melting of the crust due, in some cases, to heat provided by a magma intrusion produce chemically distinct magmas, and ultimately igneous rocks that are relatively high in silica when compared to the rock that melts. This process, along with magma differentiation, has caused the original composition of Earth's crust to evolve into more chemically complex types of rocks.

The igneous rock system has evolved in a series of steps. In each step, the partial melting of the parent igneous rock results in the formation of two portions whose compositions differ from that of the parent rock (**FIGURE 5.14**). These portions are:

1. A relatively more felsic magma which is higher in silica than the parent. This is because the minerals in the parent rock with the lower melting points obviously start to melt first.

2. A more mafic fraction, or "residue," which is the solid mineral assemblage left behind in the parent rock after partial melting is complete, and the resulting magma (portion 1) has separated. It is high in minerals enriched in iron, magnesium, and calcium, such as olivine and calcium-rich plagioclase, and which would have needed higher temperatures to start to melt.

Igneous evolution occurs through partial melting of the crust or upper mantle, or through crystallization and other differentiation processes (see Section 5-2).

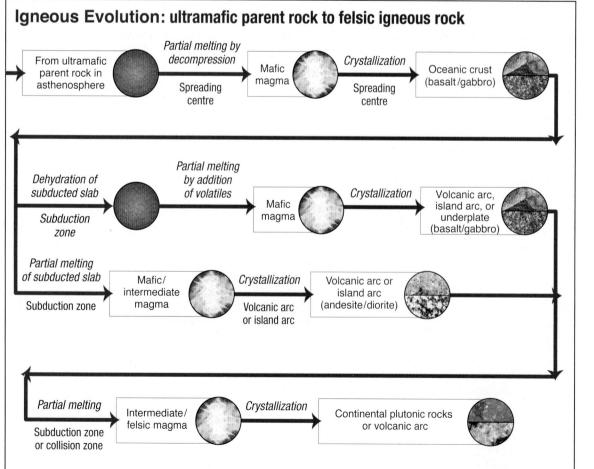

Igneous Evolution: ultramafic parent rock to felsic igneous rock

From ultramafic parent rock in asthenosphere → *Partial melting by decompression* / Spreading centre → Mafic magma → *Crystallization* / Spreading centre → Oceanic crust (basalt/gabbro)

Dehydration of subducted slab / Subduction zone → *Partial melting by addition of volatiles* → Mafic magma → *Crystallization* → Volcanic arc, island arc, or underplate (basalt/gabbro)

Partial melting of subducted slab / Subduction zone → Mafic/intermediate magma → *Crystallization* / Volcanic arc or island arc → Volcanic arc or island arc (andesite/diorite)

Partial melting / Subduction zone or collision zone → Intermediate/felsic magma → *Crystallization* → Continental plutonic rocks or volcanic arc

FIGURE 5.14 Igneous evolution is driven by magma differentiation, including crystallization and partial melting, which segregate parent material into chemically distinct products.

 At what step in this sequence would continental crust begin to develop?

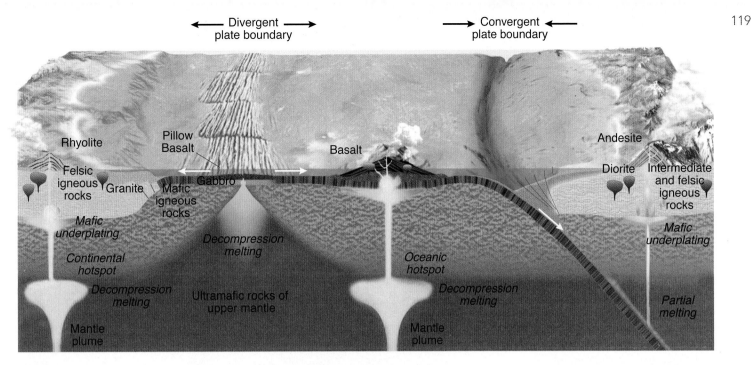

FIGURE 5.15 Igneous rock is a ubiquitous component of Earth's crust because it evolves as a product of tectonic processes. The composition of the magmas, and ultimately the igneous rocks, that form the oceanic and continental crust depend on the composition of the source rock, the processes that lead to partial melting, and the extent of differentiation. Mafic magmas generated in the mantle may also be trapped at the base of the crust (underplated), because they are denser than the continental crust. However, the heat added to the base of continental crust may cause it to melt producing felsic magmas, such as at the continental hotspot (left side of the diagram).

(?) Identify where partial melting is taking place in this illustration and describe the original magma type and the fractionated portions.

For example, during partial melting (say at a spreading centre where decompression melting takes place), ultramafic parent rock (peridotite) rises from the mantle and partially melts, producing two fractions: (1) basalt or gabbro magma that forms new oceanic crust and (2) ultramafic residue left behind in the upper mantle. Later, when the same basalt and gabbro oceanic crust is recycled during subduction, hot water and other volatiles escaping from the slab lower the melting point of the overlying upper mantle, leading to partial melting. In some situations, the ocean crust being subducted may undergo partial melting. The resulting magmas are mafic to intermediate in composition and intrude into the crust. The presence of these magmas may provide enough heat that partial melting of the crust may occur, and because the crust is higher in silica than the mantle, the resulting magmas will be intermediate to felsic in composition.

If time and conditions allow, magma that has intruded into continental crust can continue to crystallize, and intermediate magma produced above a subduction zone can itself differentiate into felsic magma (granite), leaving behind a crystal residue that is more mafic than the intermediate rock. This process of differentiation is also referred to by geologists as *fractionation*, meaning that chemically distinct fractions of the parent magma are formed. Fractionation continues until everything that can be fractionated from the original magma body has been removed. The process is complete when rocks have formed. However, these rocks will not remain solid forever. They will enter the rock cycle and be weathered, metamorphosed, or even remelted.

Magma Evolution Is Related to Tectonics

Fractionation occurs mainly at divergent and convergent plate boundaries because both settings have buoyant magma that rises

and crystallizes (**FIGURE 5.15**). At divergent plate boundaries (spreading centres), ultramafic parent rock from the upper mantle melts, and this melt fractionates to form mafic rocks—intrusive gabbro and extrusive basalt. At convergent plate boundaries, partial melting of upper mantle rock above subducting oceanic lithosphere generates mafic melts, which may form basalt and gabbro. If the subducting oceanic lithosphere (composed of basalt and gabbro) undergoes melting, or if gabbros that formed by the crystallization of mafic magma that was *underplated* at the base of the crust undergo melting, then intermediate magmas, such as diorite and andesite, may be generated. However, these may eventually fractionate to create felsic magmas, and thus granite and rhyolite.

Igneous fractionation is an extremely important process: It is responsible for the formation of volcanic and island arcs, the sea floors, the continents, land masses at hotspots, and other distinct crust types. One implication of this process is that Earth originally had no continents and that the total size of the continents has increased over geologic time as a result of igneous evolution into felsic products. If one imagines an Earth with no continents, it is easy to appreciate the importance of igneous evolution to all of Earth's systems.

A final outcome of igneous evolution is that different types of igneous rock occur in different places on Earth, and the distribution of rock types is related to plate tectonic processes and Earth's history. In general, continents are made of felsic igneous rocks (granite), oceanic crust is made of mafic igneous rocks (basalt and gabbro), and volcanic and island arcs are made of intermediate igneous rocks (diorite and andesite).

(?) **Expand Your Thinking**—Explain why the liquid fraction is more felsic than the parent after magma differentiates.

CRITICAL THINKING

IGNEOUS ENVIRONMENTS

Please work with a partner and answer the following questions using FIGURE 5.16.

1. Add labels to Figure 5.16 to identify all igneous processes and environments and plate tectonic processes and environments. On a separate sheet of paper carefully describe the igneous environments.

2. Summarize the melting processes that are responsible for the magmas in each environment you identified.

3. Identify locations in the diagram where each of the seven most common types of igneous rock are likely to be found.

4. Describe the texture and composition of each type of rock and why these characteristics occur in that environment. That is, what igneous processes in each environment produce the unique texture and composition of each type of rock?

5. Identify the magma source for each type of rock. How is the magma source responsible for the rock composition?

6. If you were asked to advise the premier of a province with a volcanic arc, what suggestions could you provide for reducing the province's vulnerability to volcanic hazards?

7. A shield volcano at a hotspot is on a moving plate. What is its long-term fate? Describe in detail.

8. Imagine a volcano made of rhyolite above a hotspot. What can you infer about the magma source and the geologic setting?

Stratovolcano

Volcanic arc

Mid-ocean ridg

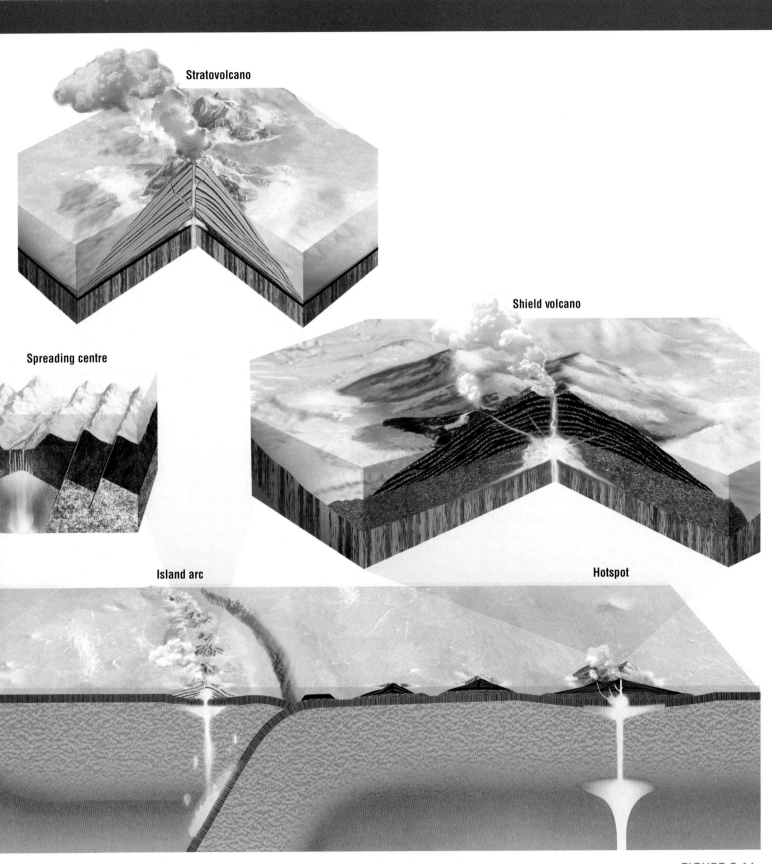

Stratovolcano

Shield volcano

Spreading centre

Island arc

Hotspot

FIGURE 5.16

5-8 Basalt Forms at Spreading Centres, Hotspots, and Subduction Zones

LO 5-8 *Describe the environments where basalt accumulates and why it accumulates there.*

Magma actively intrudes into the crust throughout the world, forming various kinds of igneous rock. These processes are not random. Tectonic activity creates the specific geologic conditions necessary for intrusive and volcanic processes to occur. The formation of igneous rock at spreading centres and hotspots are two such places where these processes occur.

Spreading Centres

Magma intrudes into the crust at spreading centres where two plates diverge to form a rift. As the plates spread apart, hot rock from the asthenosphere flows upward and fills voids in the fractured lithosphere. At the beginning of this process, mantle rock is solid and ultramafic in composition, probably composed of peridotite. Upward movement adds thermal buoyancy to the ridge area. In other words, the hot, less-dense rock rises under the cold crust, pushing upward and causing it to bulge up. This is why spreading centres in ocean basins are characterized by ridges rising 1,000 m or more above the surrounding sea floor (**FIGURE 5.17**).

As ultramafic rock from the upper mantle rises to shallow levels, it decompresses and the minerals with lower melting temperatures melt to produce mafic magma. Magma that rises beneath a spreading centre forms the intrusive igneous rock gabbro. Magma that rises through fissures in the crust to the surface forms its volcanic equivalent, basalt. Basalt eruptions are typically *fissure eruptions*, where lava emerges from cracks in the oceanic ridge. The many pathways that basalt takes in its upward journey to the sea floor form a dense network of dikes. They are referred to as *sheeted dikes* because they form vertical columns and curtains of intrusive rock in thick sequences, like cards in a deck stood on end. Basalt erupting onto the sea floor is quickly quenched, or "frozen," by cold sea water and generates submarine deposits of *talus* (in this case, talus is broken glass and ash of basaltic composition) and bulbous rocks called **pillow lava** (**FIGURE 5.18**).

The high heat flow at mid-ocean ridges produces numerous **hydrothermal vents** in the sea floor. These vents occur when cold sea water seeps downward through cracks in the crust and meets the hot intrusive rock below. The water warms as it nears hot rock or a magma body and becomes chemically enriched with dissolved metallic cations that it leaches from the surrounding crust. When the water reaches a critical temperature and turns into steam, it is forced back toward the surface and erupts as a hot chemical spring. Some of these springs are **black smokers**, so named for their release of dark, billowing clouds of metal-rich sulphide particles. Massive, economically valuable *metallic sulphide deposits* crystallize out of the hot water emerging from the vent.

As this process continues over millions of years, it builds new oceanic lithosphere consisting of intrusive gabbros in the lower part, sheeted vertical dikes, and basaltic lavas (of glassy talus and pillows) in the upper portions. Interlayered within the basaltic lava are mineral-rich metallic sulphide deposits associated with fields of black smokers. The interaction between cold sea water and the mafic rocks of the oceanic lithosphere that results in the black smokers, also results in alteration of the original minerals in the basalt and gabbro (pyroxene, calcium-rich plagioclase feldspar, and maybe olivine). The result is that new minerals containing water in their crystal structure, like chlorite, epidote, and serpentine (called *hydrated minerals*) may form. This is very important for the processes that ultimately take place when this oceanic lithosphere is subducted at a subduction zone (more about that later).

Mantle Plumes (Hotspots)

Several exceptionally active sites of plutonism and volcanism are found at hotspots located far from plate boundaries. One hypothesis suggests that massive plumes of anomalously hot mantle rock underlie

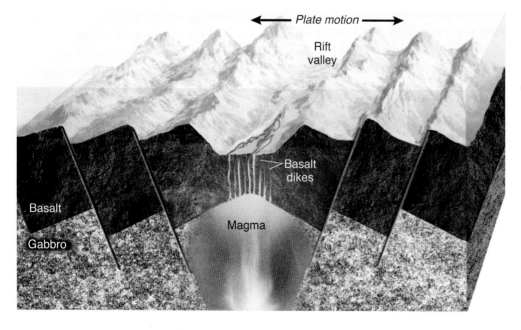

(?) Why is the upper crust composed of basalt and the lower crust composed of gabbro?

FIGURE 5.17 New sea floor is created at divergent boundaries where peridotite from the upper mantle rises and partially melts by the process of decompression melting to produce mafic magma. This is the source of the gabbro at the base of the crust and the basalt dikes that make up the shallow sea floor.

FIGURE 5.18 Pillow lava forms when hot lava meets cold sea water. The surface of the lava chills in the water (making glass) while the interior remains hot and pushes forward, stretching the glass layer into a round, bulbous shape. The larger pillows in this photo are about a metre in diameter.

(?) Why are the surfaces of pillow lavas made of glass?

active hotspots. These mantle plumes (**FIGURE 5.19**) appear to originate in the lower mantle and rise (presumably slowly) because they are less dense (hotter) than the surrounding rock. As an example, at the core-mantle boundary, the heat that is transmitted from the outer core into the mantle may vary from location to location, and so some mantle rock may be slightly hotter than nearby mantle rock. The hotter rock will be less dense and so will rise—a mantle plume has been born.

Research suggests that a mantle plume rises as a plastically deforming mass with a bulbous head fed by a long, narrow tail so that the overall shape resembles that of a tadpole. As the head encounters the base of the lithosphere, it spreads outward into a mushroom shape. As a mantle plume rises, it experiences a decrease in pressure that can lead to melting in its upper part. Decompression melting of these hot rocks can generate huge volumes of basalt magma that feed plutons at shallow depths in the crust. At the surface, massive **large igneous provinces** composed of *flood basalts* form above mantle hotspots in either continental or oceanic settings. (These are discussed more fully in Chapter 6, "Volcanoes.") Large igneous provinces contain huge volumes of basalt released over millions of years and may create immense plateaus covering thousands of square kilometres. The Columbia River Plateau in Idaho, Washington, and Oregon is one example. In Canada, one of the largest igneous events occurred about 1.27 billion years ago, and was related to a mantle plume under what is now the northern coast of the Northwest Territories. The Coppermine volcanics formed at

that time, as well as an enormous swarm of dikes, called the Mackenzie dikes, which radiate away from the centre of the plume and are found intruding about 3 million square kilometres of the Canadian Shield.

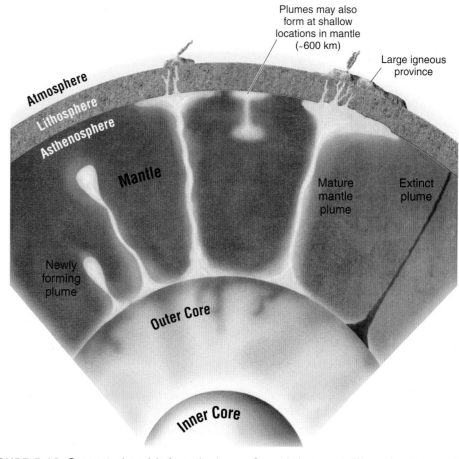

FIGURE 5.19 Conceptual model of mantle plumes. Several plumes, at different locations and in various stages of formation, may be active at the same time.

(?) **Expand Your Thinking**—Would you expect magma at a spreading centre to have the same composition as magma at a hotspot? Why or why not? What type of magma will feed a volcano where a hotspot intrudes into continental crust?

We still have much to learn about hotspots. Although most geologists accept the hotspot concept, the number of hotspots worldwide and exactly how they are formed are still subjects of debate and controversy.

Subduction Zones

The most volcanically active area on Earth is the "Ring of Fire," which surrounds the Pacific Ocean and includes the active volcanoes in South and North America, Japan, the Philippines, Indonesia, and New Zealand. Most of these are generated by the processes at convergent plate margins.

As a descending plate bends downward during subduction, it creates a long, arcing depression in the sea floor called an *oceanic trench*. These trenches are the deepest topographic features on Earth's surface. Because continental crust consists of rocks of a relatively low density, it is too buoyant to subduct. A plate will subduct only if it is composed of oceanic crust. Deposited on this basalt crust is *marine sediment* that contains large volumes of sea water within its countless pores. As well, the mafic igneous rocks that are being subducted have been altered by sea water, as part of the process that generated the black smokers described in the last section, so they are full of hydrated minerals.

As a plate subducts, it progressively encounters higher temperatures and pressures. These force water out of the descending slab and into the mantle lying above the subducting crust. The presence of this *water lowers the melting temperature of the mantle rock*, causing it to melt (**FIGURE 5.20**). The resulting magma is enriched in silica (thus, it is more felsic) compared to the ultramafic rock of the upper mantle; hence, it varies from basalt to andesite in composition. As the magma rises into the overriding plate, differentiation produces intrusive bodies of varying composition that feed a belt of volcanoes aligned parallel to the trench.

If the overriding plate is composed of continental crust, the resulting chain of volcanoes is a *continental volcanic arc*. Examples include the Cascade Range in southwestern British Columbia and the U.S. Pacific Northwest and the Andes Range in South America. If the overriding plate is composed of oceanic crust, the resulting chain is an *island arc*. The Aleutian Islands chain in the North Pacific is an example of an island arc.

Lava at subduction zones is typically basaltic at oceanic island arcs and andesitic at continental volcanic arcs. This difference occurs because basaltic magma rising through a continental plate to form a volcanic arc is initially surrounded by felsic rocks and may become more felsic (turning into andesite) as a result of differentiation processes, such as assimilation. Basaltic magma rising into an overriding plate of basalt at an ocean–ocean boundary experiences less chemical change because the surrounding rocks are similar to it in composition.

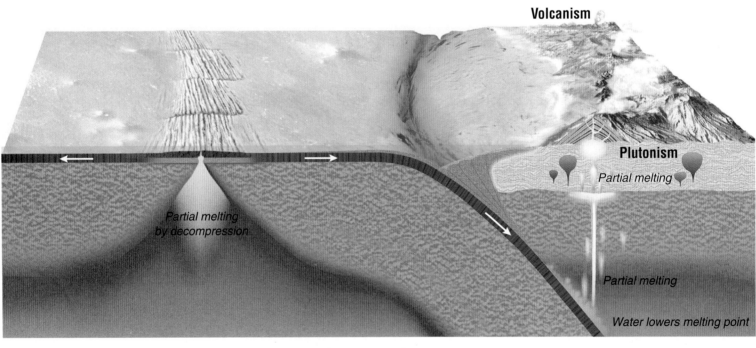

FIGURE 5.20 Subduction zones and spreading centres are areas in which magma is formed as a result of partial melting, plutonism, and volcanism.

 Describe the role of water in producing magma.

 Explain why island arcs tend to be composed of basalt and continental volcanic arcs tend to be composed of andesite.

5-9 Igneous Intrusions Occur in a Variety of Sizes and Shapes

LO 5-9 *Describe the characteristics of the different types of intrusions.*

Igneous intrusions, or **plutons**, occur in a variety of sizes and shapes, from small bodies measured in metres to massive units that occupy hundreds of cubic kilometres and form the cores of mountain ranges. Some plutons may become exposed at the surface as a result of erosion of the overlying rocks. This can happen during prolonged weathering that removes the overlying rock layers, a process that is enhanced by tectonic forces causing the pluton itself to slowly shove the crust upward. This leads to accelerated erosion because as the crust lifts to higher elevations, the gradient of the land increases, providing running water and gravity with more erosive energy.

When magma intrudes into cold crustal rocks, several things happen. First, the magma creates space for itself by wedging open the overlying rock layers, breaking off large blocks of rock. These rocks can be melted by the heat provided by the magma and the elements within the rocks added to the magma. If these blocks of crust or mantle rocks are not completely melted, they might be preserved in the final igneous rock. They are then called *xenoliths* (from the ancient Greek word *xenos*, meaning "foreign"). Second, heat from the magma may melt the walls of the intruded crustal rock. This melting adds new elements to the magma (assimilation). Third, heat from the magma bakes the surrounding crust for distances that can range from metres to kilometres and alters its mineral composition and texture (a process called "contact metamorphism," which will be discussed in Chapter 9).

Geologists often refer to the rock that is intruded by the magma as **country rock**. The processes by which magma intrudes into solid country rock is highly debated, although the final shape of the intrusive body may be related to these processes and the pathway taken by the magma. Geologists identify these bodies by their shape and orientation, as summarized in **FIGURE 5.21**.

The massive intrusion at the bottom of the figure is a batholith (note the xenoliths incorporated into the magma). Intrusive bodies smaller than batholiths (less than about 100 km²) are *stocks*. Magma forms a laccolith when it pushes up rock layers into a dome and then cools in that shape. Study the dike in the figure. It consists of magma that followed fissures and cracks in the overlying crust, generally vertically (or nearly so), and hardened into igneous rock. Note that the dike cuts across *strata* in the country rock. As the magma encountered different conditions, it intruded horizontally, parallel to those strata, forming a *sill* (like a windowsill).

Batholiths, stocks, and dikes are "discordant" intrusions because they cut across the layers of country rock into which they intrude. Sills and laccoliths are "concordant" intrusions because their boundaries are essentially parallel to those of the layers of country rock into which they intrude. It is important to remember that all intrusive rocks are younger than the country rock, even if they are parallel to the layers in the country rock. Magma that pushes to the surface feeds a volcanic eruption through a volcanic pipe. Volcanologists often refer to the "plumbing" of a volcano because magma can flow through complex systems of pipes and fissures to feed surface eruptions.

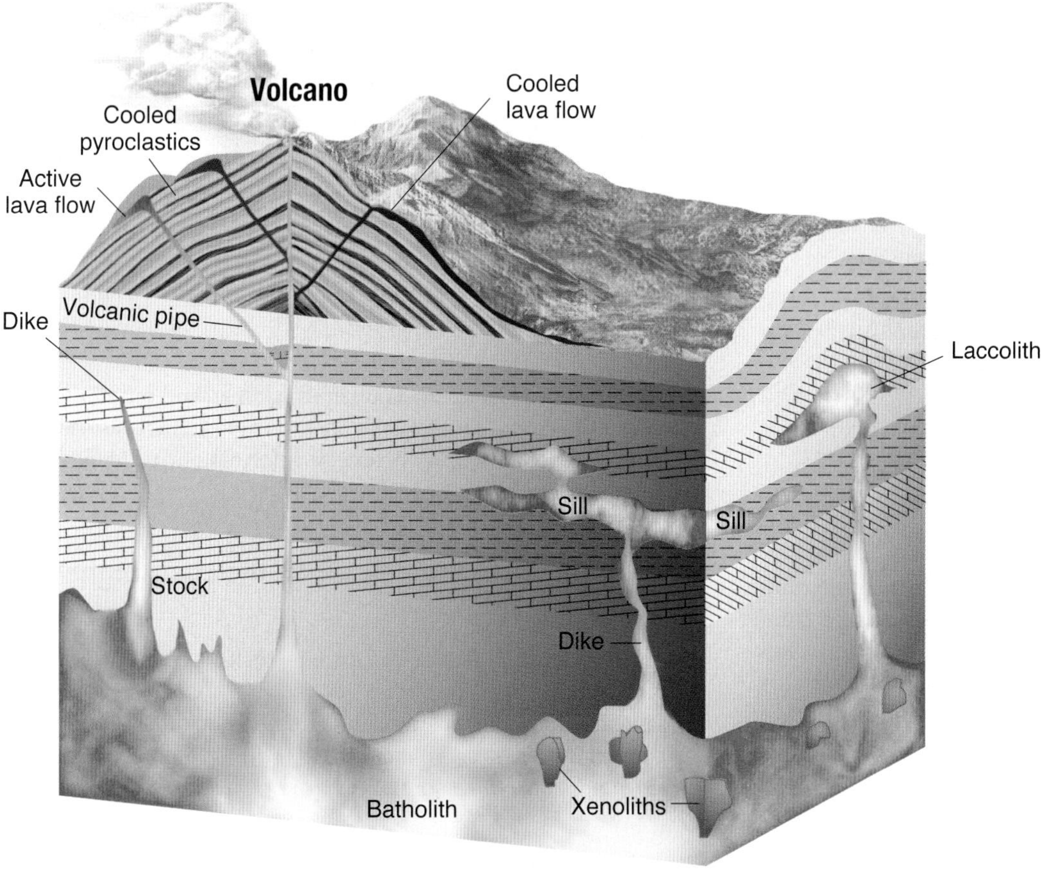

FIGURE 5.21 Plutons can take various shapes, depending on the volume of the magma, the force of the intrusion, and the structure and strength of the surrounding crust.

(?) Describe the influence that xenoliths had on the evolution of magma composition.

(?) **Expand Your Thinking**—Do you agree or disagree that over time the crust is growing? Do you think that more crust is added as plutons or as volcanic rocks? Why?

LET'S REVIEW "GEOLOGY IN OUR LIVES"

Now that you have finished the chapter, "Geology in Our Lives" will have taken on new meaning for you. Let us review it: Melting recycles Earth materials and igneous rock forms when molten rock (magma) cools and crystallizes. Igneous rocks were the first rocks on Earth. All other rocks have evolved from igneous rocks that formed over 4 billion years ago. Igneous rocks provide critical building materials and mineral resources, and make up Earth's interior.

Several important concepts emerge from our understanding of igneous rock formation: (1) The rock cycle renews and recycles rocks, sediments, and other materials of the crust; (2) because igneous rock is the most abundant rock on the planet, melting is an important and widespread process in the rock cycle; and (3) melting occurs in various tectonic environments, including spreading centres, subduction zones, and hotspots. As a result, igneous rock occurs in a wide variety of types and is formed in diverse environments.

Despite this diversity, all igneous rock originates from a single type of parent rock, the peridotite of the upper mantle. Through partial melting, this parent rock produces magma that is basaltic in composition, which in turn evolves into other types of magma. In the very earliest phases of Earth's history, igneous rock constituted 100 percent of the planet. Since then all rock—igneous, sedimentary, and metamorphic—has been produced by igneous evolution. All rock is related to, and derived from, the first molten rock on the ancient planet. Why? Because melting recycles Earth materials.

STUDY GUIDE

5-1 Igneous rock forms when molten, or partially molten, rock solidifies.

- Four-fifths of Earth's crust is composed of **igneous rock**. Igneous rocks result from melting in the upper mantle and crust.

- Volcanic igneous rock consists of **lava** and other volcanic products that are extruded at the surface of the crust or at very shallow levels below the surface, forming **extrusive** igneous rock. **Intrusive** igneous rock forms when magma crystallizes within the crust without being exposed to the cool temperatures of the atmosphere.

5-2 Igneous rock forms through a process of crystallization and magma differentiation.

- Rifting of the crust lowers the pressure in the upper mantle, and the resulting **decompression melting** causes magma to form, which migrates into the crust. The magma and crust interact as the magma crystallizes.

- Rocks undergo **partial melting**, which means that silica-rich compounds in the rock melt before other compounds.

- **Magma differentiation** is caused by magma mixing, crystal settling, magma assimilation, and magma migration. Through these processes, magmas of differing composition evolve from the single parent magma. In most cases, differentiation produces magma that is richer in silica than the parent magma.

5-3 Bowen's reaction series describes the crystallization of magma.

- **Bowen's reaction series** describes the order in which the minerals in a magma body crystallize. In magma of average composition, the crystallization process begins with **mafic** minerals (those that are high in iron, magnesium, and calcium), continues with minerals of **intermediate** composition, and ends with **felsic** (silica-rich) minerals.

- There are two branches of crystallization: (1) discontinuous crystallization occurs when minerals formed at higher temperatures react chemically with the surrounding magma and recrystallize to form a mineral with a new crystalline structure; (2) continuous crystallization involves progressive changes in plagioclase feldspar as it evolves from calcium-rich to sodium-rich composition.

5-4 The texture of igneous rock records its crystallization history.

- The texture of an igneous rock depends on the size of the crystals within it, which is influenced by the rate at which the rock cooled. Igneous rocks that cool slowly tend to contain large mineral grains and have a coarser texture (called phaneritic texture) than those that cool rapidly. These rocks are intrusive or plutonic. Igneous rocks that cool rapidly on Earth's surface usually contain smaller mineral grains and have a finer texture (called aphanitic texture). These rocks are extrusive or volcanic. Igneous rocks may also be glassy, pyroclastic, vesicular, or porphyritic.

- Minerals at the top of Bowen's reaction series (such as pyroxene and amphibole) tend to be dark in colour and mafic to **ultramafic** in composition because they are rich in iron and magnesium.

- Minerals at the bottom of Bowen's reaction series (such as sodium-rich plagioclase, potassium feldspar, and quartz) tend to be light in colour and felsic in composition. Geologists use these colour trends, along with texture information, to identify and name igneous rocks.

5-5 Igneous rocks are named on the basis of their texture and composition.

- The classification system for igneous rocks tells us three important things: (1) intrusive and extrusive igneous rocks with the same chemical composition contain the same minerals; (2) silica content decreases as iron and magnesium content increases; and (3) potassium and sodium contents decrease as silica content decreases.

5-6 There are seven common types of igneous rock.

- The seven common types of igneous rock are rhyolite, granite, andesite, diorite, basalt, gabbro, and peridotite.

5-7 All rocks on Earth have evolved from the first igneous rocks.

- Igneous rocks were the first rocks on Earth. All other rocks have evolved from parent igneous rocks that were formed over 4 billion years ago. **Igneous evolution** has produced the great diversity of rock on Earth today.

- The igneous rock system evolves in a series of steps. In each step, the partial melting of the parent igneous rock results in the formation of two fractions, each with a different composition from that of the parent. The two fractions are (1) a felsic fraction of compounds with lower melting points, typically high in silica relative to the parent—this is the magma produced by the partial melting, and (2) a mafic fraction of solid minerals (i.e., rock that is high in iron, magnesium, and calcium minerals relative to the parent rock—this consists of the minerals that do not melt during partial melting.

5-8 Basalt forms at spreading centres, hotspots, and subduction zones.

- Tectonic activity creates the geologic settings and conditions necessary for intrusive and extrusive processes to occur.

- Magma intrudes into the crust at spreading centres where two plates diverge. Oceanic crust consists of several layers of igneous rock. Fed by magma generated by decompression melting in the peridotite upper mantle, gabbro crystallizes at the base of the crust. Above the gabbro are sheeted dykes of basalt, and above these are layers of pillow basalt, glassy talus, and metallic sulphide deposits.

- Several active sites of plutonism and volcanism are located at hotspots far from plate boundaries. Scientists hypothesize that a massive plume of anomalously hot mantle rock underlies active hotspots. Plumes rise slowly through the mantle due to their positive buoyancy relative to the surrounding rock.

- Volcanism and plutonism occur at convergent plate margins (as well as in other settings).

5-9 Igneous intrusions occur in a variety of sizes and shapes.

- Plutons occur in a variety of sizes and shapes, from small bodies measured in metres to massive units of hundreds of cubic kilometres, forming the cores of mountain ranges.

KEY TERMS

black smokers (p. 122)
Bowen's reaction series (p. 110)
country rock (p. 125)
decompression melting (p. 106)
differentiation (p. 106)
extrusive (p. 107)
felsic (p. 111)
hydrothermal vents (p. 122)
igneous evolution (p. 118)

igneous rock (p. 108)
intermediate (p. 111)
intrusive (p. 107)
large igneous provinces (p. 123)
lava (p. 106)
mafic (p. 110)
magma (p. 106)
magma differentiation (p. 108)
partial melting (p. 106)

pillow lava (p. 122)
plutonic (p. 107)
plutons (p. 125)
pyroclastic debris (p. 106)
stocks (p. 116)
ultramafic (p. 110)
volcanoes (p. 107)

ASSESSING YOUR KNOWLEDGE

Please answer these questions before coming to class. Identify the best answer.

1. What is igneous rock?
   a. rock produced by melting
   b. rock composed of sediments
   c. rock derived from pressure
   d. rock that mixes the mantle and crust
   e. None of the above.

2. Most melting in the mantle is a result of
   a. high-pressure melting.
   b. decompression melting.
   c. sudden increases in temperature.
   d. turbulent mantle plumes.
   e. none of the above.

3. In most cases, magma differentiation produces magma with higher _____ content than the parent magma.
   a. iron
   b. silica
   c. calcium
   d. mineral
   e. None of the above.

4. Magma that is cooling undergoes
   a. crystallization.
   b. recrystallization.
   c. partial melting.
   d. refractionation.
   e. erosion.

5. Bowen's reaction series describes
   a. the sequence in which minerals melt in rapidly heating magma.
   b. the sequence in which plutons form in migrating magma.
   c. the sequence in which rocks form in average continental crust.
   d. the sequence in which minerals crystallize in cooling magma.
   e. None of the above.

6. The order of mineral crystallization is typically
   a. felsic, mafic, intermediate, ultramafic.
   b. felsic, intermediate, mafic, ultramafic.
   c. ultramafic, mafic, intermediate, felsic.
   d. mafic, ultramafic, felsic, intermediate.
   e. All of the above.

7. Fill in the blanks with the appropriate texture terms.
   a. two sizes of crystals:_____
   b. fused, glassy shards:_____
   c. small crystals:_____
   d. many small openings produced by escaping gas:_____

8. *Mafic* means_____; *felsic* means_____.
   a. high in iron, magnesium, and calcium; high in silicon and oxygen
   b. high in calcium and magnesium; high in silicon, oxygen, and iron
   c. high in iron and oxygen; high in silicon, calcium, and magnesium
   d. high in silicon, oxygen, and calcium; high in iron and magnesium
   e. volcanic; plutonic

9. The composition of dark igneous rock is likely to be
   a. felsic.
   b. mafic.
   c. rhyolitic.
   d. plutonic.
   e. None of the above.

10. List five common types of igneous rock.
    a. _____
    b. _____

    c. _____
    d. _____
    e. _____

11. Which of the following best describes igneous evolution?
    a. All rocks evolved as a result of hotspots.
    b. All rocks evolved as a result of spreading-centre volcanism.
    c. All rocks evolved as a result of differentiation of early igneous rocks.
    d. All rocks are a result of meteorite impacts.
    e. None of the above.

12. Which of the following statements are correct? (There can be more than one correct answer.)
    a. Granite is formed at spreading centres.
    b. Andesite is formed at subduction zones.
    c. Basalt is formed at hotspots.
    d. Gabbro is formed at spreading centres.
    e. None of the above.

13. Continental volcanic arcs are typically composed of
    a. granite and gabbro.
    b. gabbro and peridotite.
    c. rhyolite and andesite.
    d. andesite and diorite.
    e. All of the above.

14. Plutons are
    a. magma bodies within the deep crust.
    b. intrusive igneous rocks in the lower mantle.
    c. magma bodies produced by volcanism.
    d. igneous rocks produced by fissure eruptions.
    e. made by contact metamorphism.

15. Plate tectonics is important to igneous evolution because
    a. plate tectonics formed the first igneous rocks billions of years ago.
    b. melting does not occur at plate boundaries.
    c. mantle plumes occur only at spreading centres.
    d. plate tectonics provides for many igneous environments.
    e. plate tectonics does not allow for partial melting.

FURTHER RESEARCH

1. What tectonic environment leads to the production of granite?

2. What tectonic environment leads to the production of basalt?

3. What is pillow basalt, and how does it form?

4. Use a map of the world to identify several locations where you would expect to find hydrothermal vents.

ONLINE RESOURCES

Explore more about igneous rocks on the following websites:

Explore stories and photos about igneous rocks at these two websites:
www.usgs.gov/science/science.php?term=572
geology.com/rocks/igneous-rocks.shtml

Additional animations, videos, and other online resources are available at this book's companion website:
www.wiley.com/go/fletchercanada

This companion website also has additional information about WileyPLUS and other Wiley teaching and learning resource

CHAPTER 6
VOLCANOES

Chapter Contents and Learning Objectives (LO)

6-1 A volcano is any landform that releases lava, gas, or ash or has done so in the past.

LO *6-1 Define the term "volcano" and describe why geologists study volcanoes.*

6-2 There are three common types of magma: basaltic, andesitic, and rhyolitic.

LO *6-2 Compare and contrast the three common types of magma.*

6-3 Explosive eruptions are fuelled by violent releases of volcanic gas.

LO *6-3 Describe volcanic gases and the role they play in explosive versus effusive eruptions.*

6-4 Pyroclastic debris is produced by explosive eruptions.

LO *6-4 Describe the formation of pyroclastic debris.*

6-5 Volcanoes can be classified into six major types based on their shape, size, and origin.

LO *6-5 Compare and contrast central vent volcanoes with large-scale volcanic terrains.*

6-6 Shield volcanoes are a type of central vent volcano.

LO *6-6 Identify what gives a shield volcano its distinctive shape.*

6-7 Stratovolcanoes and rhyolite caldera complexes are central vent volcanoes.

LO *6-7 Compare and contrast the magma composition of stratovolcanoes versus rhyolite caldera complexes.*

6-8 Large-scale volcanic terrains lack a central vent.

LO *6-8 Identify the main characteristics of large-scale volcanic terrains and give examples.*

6-9 Most volcanoes are associated with spreading centre volcanism, arc volcanism, or intraplate volcanism.

LO *6-9 Describe the role of plate tectonics in spreading centre volcanism, arc volcanism, and intraplate volcanism.*

6-10 Volcanic hazards threaten human communities.

LO *6-10 List and define several types of volcanic hazards.*

GEOLOGY IN OUR LIVES

A volcano, such as this one on Hawaii erupting lava and pyroclastic material, is any landform that releases lava, gas, or ash or has done so in the past. Volcanoes are a critical component of the rock cycle. They rejuvenate the crust, renew soil, impact global climate, and contribute to gases in the atmosphere. However, eruptions may consist of violently explosive columns of gas and ash that penetrate high into the atmosphere, lava flows that burn and bury whatever they encounter, and avalanching pyroclastic flows of scalding fumes and glass shards that move extremely rapidly and catastrophically down volcanic slopes. To avoid volcanic hazards and utilize volcanic resources, it is important to understand how volcanoes work.

(?) We learned in Chapter 5 ("Igneous Rock") that magma (and lava) can be felsic, intermediate, or mafic. How does magma chemistry influence the nature of volcanic eruptions?

6-1 A Volcano Is Any Landform that Releases Lava, Gas, or Ash or Has Done So in the Past

LO 6-1 *Define the term "volcano" and describe why geologists study volcanoes.*

Few geologic processes are more familiar to the average person than a volcanic eruption. We frequently see erupting volcanoes in movies, on TV, and in newspapers and magazines. But despite this familiarity, the average person does not know how volcanoes work, why they erupt, or why they are located where they are. We will learn the answers to these questions in this chapter. We will also learn that an important clue to a volcano's origin and eruptive processes is found in its shape and size.

Not many spectacles in nature are as awe-inspiring as an erupting volcano. Because of this, volcanoes have long figured in mythology worldwide. In European culture, for example, people once believed that Vulcan, the Roman god of fire and metalworking, had a subterranean forge whose chimney was the tiny island of Vulcano in the Mediterranean Sea north of Sicily. They thought Vulcano's hot lava fragments and dark ash clouds came from Vulcan's furnace as he beat out thunderbolts for Jupiter, the king of the gods, and weapons for Mars, the god of war. Our word "volcano" is derived from the name of this island. Now we understand that Vulcano and its sister islands Stromboli and Lipari are fed by partial melting in the upper mantle as the Nubian Plate, which is the part of the African Plate to the west of the East African Rift Valley, subducts beneath the southern margin of the Eurasian Plate, creating an island arc of active volcanoes.

In Hawaiian culture, people attribute eruptions to the wrathful Pele (*PEL-lay*), the goddess of volcanoes. Hawaii is home to the world's longest continuously erupting volcano, Kilauea. Since 1952, Kilauea has erupted 34 times, and since January 1983, the eruption has continued without interruption. If you visit Kilauea in Hawaii Volcanoes National Park on the Big Island of Hawaii, a UNESCO World Heritage Site, you will see a barren and bleak landscape made up of several square kilometres of black crunchy rock. Some of this rock may be only weeks old and will break into glassy fragments underfoot. The air may smell of sulphur, and clouds of steam will boil up from cracks in the ground. The lavas of Kilauea flow like syrup. Although they are very hot, if you are careful you can approach small rivulets of molten rock. If you have the appropriate protection from the heat and are wearing gloves, it is possible to sample the lava and watch as it chills to glass (**FIGURE 6.1**).

Volcanoes are indeed spectacular and full of mystery. They are natural phenomena that are continually being explored and studied by **volcanologists**.

What Is a Volcano?

Volcanoes come in many shapes and sizes. A **volcano** is any landform that releases lava, gas, or ash or has done so in the past. Hot materials escape from an opening called a **vent** or fissure. Often a cone of volcanic rock builds up around the vent. The cone is generally composed of accumulated erupted materials, such as ash, pumice, and lava. The cone can be a few metres high or an entire mountain. The term "volcano" can refer both to the vent (opening) and to the cone.

In this chapter, we examine three basic types of **central vent volcanoes** (classic volcanoes built around a central vent) and three varieties of **large-scale volcanic**

© Roger Ressmeyer/Corbis

 What kind of lava is Kilauea Volcano made of?

FIGURE 6.1 The heat is intense, but the lava at Kilauea Volcano in Hawaii can be cautiously approached and sampled.

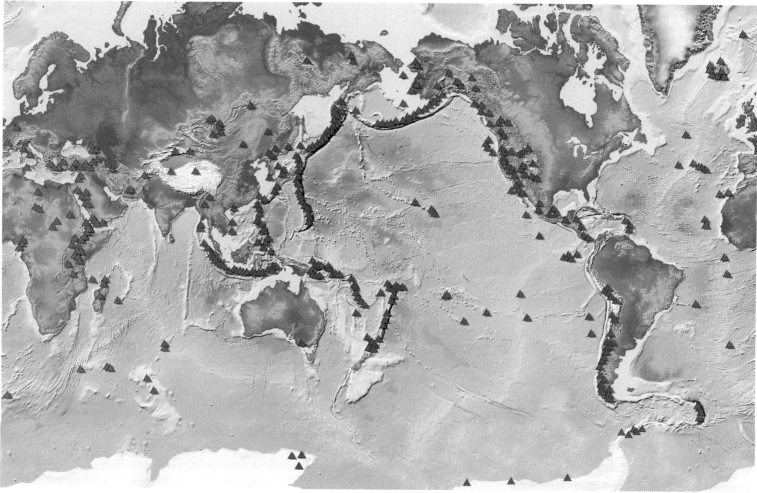

FIGURE 6.2 Global map of active volcanoes.

 Do the locations of the world's active volcanoes show any recognizable pattern?

U.S. Geological Survey, Dept. of the Interior

terrains formed by volcanic action. We will identify two sorts of lava (shiny, fluid *pahoehoe* and dark, jagged *aa*) and discuss other volcanic products. By the end of the chapter, you will understand that some volcanoes have **explosive eruptions** and rip open entire mountainsides, while others have **effusive eruptions** that pour out fluid lava, adding to the land surface with relative calm. These different behaviours can be explained by the chemistry of the magma and its origin, as revealed through our understanding of plate tectonics.

Why Study Volcanoes?

The biggest human-interest issue related to volcanoes is safety. In many places, Earth's surface—both above and below sea level—is of volcanic origin, so it is little wonder that volcanoes pose such a hazard to life and are so widely studied. Scientists estimate that today the total population at risk from volcanoes is at least 500 million. Since 1700, volcanic activity has killed more than 260,000 people, destroyed entire cities and forests, and severely disrupted local economies for months and even years. The 1991 eruption of Mount Pinatubo in the Philippines caused atmospheric cooling of about 0.6°C that temporarily offset a decade-long global warming trend. It is important, therefore, that

scientists continue to improve their understanding of volcanoes and how they work. *Millions of people are vulnerable to the effects of dangerous eruptions.*

The subject of this chapter is how volcanoes form, how they behave, and why they are located where they are. Volcanoes attract attention because they are such dynamic and awe-inspiring displays of power. Right now, volcanoes are active on every continent—even Antarctica—and on the floor of every major ocean (**FIGURE 6.2**).

Of more than 1,500 volcanoes considered active on Earth's land surface, an average of 10 are erupting each day. More than a dozen active volcanic islands dot the oceans. Many eruptions are small; however, even small eruptions can be powerful and occasionally cause great damage. It is thus vital to understand and monitor active volcanoes, which are those that have erupted at least once in the last 10,000 years, even though they are dormant. Dormant volcanoes are in the right geological setting and thus may erupt in the future.

Volcanoes occur on Earth, other planets, and moons. For example, Venus is highly volcanic, Mars used to be, and one of Jupiter's four big moons, Io, is the most volcanically active body in the Solar System.

 Expand Your Thinking—What tectonic environments promote volcanism?

6-2 There Are Three Common Types of Magma: Basaltic, Andesitic, and Rhyolitic

LO 6-2 *Compare and contrast the three common types of magma.*

What are the products of volcanism? Lava is one, but there are others, of which **pyroclastic debris** (*pyroclastic means "fiery pieces"*) and volcanic gas are the most important. In this section, we review the main types of magma that feed volcanoes; other volcanic products are discussed in later sections.

Magma Chemistry

Lava, a type of magma, is fluid rock that comes from a vent or fissure. It is also the name of the solid rock formed when the lava cools. In Chapter 5 ("Igneous Rock"), we described three types of extrusive rock, and these are also the main types of magma: **rhyolitic** (felsic in composition), **andesitic** (intermediate), and **basaltic** (mafic).

In studying volcanoes, researchers have found that three characteristics most influence the behaviour of magma—and, hence, volcanic processes and the shape and size of a volcano. These are (1) the proportion of silica in the magma, (2) the magma's temperature, and (3) the amount of dissolved gas the magma contains.

High-silica magma tends to be rich in dissolved gases. It is viscous (thick and "sticky") because it contains many tiny chains of silicate tetrahedrons. Hence, gas does not escape easily and tends to accumulate within the magma. High-silica magma does not flow easily, because it has high viscosity, and therefore will build steep slopes, resulting in tall volcanoes.

Low-silica magma has fewer sturdy silica bonds and therefore is characterized by low viscosity and lower levels of dissolved gas. It is "runny" and fluid. You cannot build a steep slope with runny material—for example, when poured, pancake batter quickly assumes the low, flat profile of a pancake. Low-silica magma tends to build long, low-lying volcanoes with gentle slopes. The characteristics of the three magma types are summarized in **TABLE 6.1**.

Basaltic Magma

Most magma is basaltic in composition. (The oceanic crust is basalt, and it represents 71 percent of Earth's surface.) Basaltic magma is

FIGURE 6.3 Basaltic lava flows easily because of its low viscosity.

(?) What aspect of lava chemistry influences its viscosity?

mafic—compared to felsic and intermediate rocks, it is high in magnesium and iron and low in silica. Basaltic magma is hotter and more fluid than andesitic and rhyolitic magma. Because it contains fewer silica chains, gases that come out of solution can rise unimpeded through the fluid magma, and there is no significant buildup of gas pressure. The outcome is a relatively "gentle," effusive eruption. Because basalt flows easily, some basalt eruptions release large volumes of lava. A good example of this is the long, sinuous rivers of lava issuing from Kilauea that flow for many kilometres across the island of Hawaii (**FIGURE 6.3**).

Basaltic lava is formed by partial melting of ultramafic or mafic sources. Since most mantle sources are ultramafic in composition, basaltic lava is commonly found at oceanic hotspots, mid-ocean ridges,

Magma Type	Composition	Silica Content and Viscosity	Gas Content	Explosivity	Lava Temperature	Examples of Volcanoes
Basaltic	Mafic	Least, ~50% (thin, runny = low viscosity)	0.5%–2%	Least	Hottest ~1,100°C to 1,200°C	Mid-ocean ridges; plateau basalts like the Columbia Plateau; Hawaiian Islands
Andesitic	Intermediate	Intermediate, ~60%	3%–4%	Intermediate	Cooler ~900°C to 1,000°C	Mount St. Helens, Mount Rainier
Rhyolitic	Felsic	Greatest, >70% (thick, stiff = high viscosity)	4%–6%	Greatest	Coolest ~700°C to 800°C	Yellowstone Volcano

TABLE 6.1 Magma Types

(a)

(b)

FIGURE 6.4 (a) "Aa" is a Hawaiian term for basaltic lava that has a rough, fragmented surface composed of broken lava blocks called clinkers. (b) "Pahoehoe" is a Hawaiian term for basaltic lava that has a smooth, shiny, and ropy surface.

(?) In which tectonic environments are you likely to find aa and pahoehoe?

and island arcs where ocean crust is subducted below ocean crust (because there is no continental crust, which tends to be relatively high in silica). Basaltic lava is even found at many ocean–continent convergent volcanic arcs as partial melting of the continental crust is not always a significant source of magma.

These fluid lava flows can be subdivided into two types based primarily on the nature of the lava's surface texture: aa (*ah-ah*) lava, which has a rough, fragmented surface and a jagged appearance; and pahoehoe (*pah-hoy-hoy*) lava, which has a smooth surface that is shiny and ropy in appearance (**FIGURE 6.4**). Pahoehoe lava is less viscous, and may turn into an aa lava as it cools and becomes more viscous.

Andesitic Magma

Andesitic magma has an intermediate composition. Compared to basaltic magma, it contains more silica and less iron and magnesium (making it lighter in colour). Andesite has more silica bonds, making it less fluid than basalt, with higher gas content.

When magma migrates toward the surface, it experiences a decrease in pressure. This frees gas from its dissolved state, forming bubbles in the ascending molten rock. The released gas tries to escape, but in magma with a high silica content, it is trapped by the formation of silicate chains during cooling. Gas pressure builds up, eventually generating explosive eruptions that blow out great volumes of pyroclastic debris as magma is converted to ash. If andesitic magma has had a chance to depressurize before erupting—that is, if conditions have made it possible for gases to escape slowly while the magma rises—it can erupt non-violently. However, because of its high gas content and the fact that it does not flow as easily as basaltic magma, volcanoes containing andesitic magma are more likely to erupt violently.

Whereas basaltic magma feeds aa and pahoehoe lava flows, andesitic magma feeds a different type of flow that consists of smooth-sided, angular blocks that are less porous than those in aa lava flows. Andesitic magma is viscous enough to form immobile plugs, called lava domes, that block vents and prevent lava extrusion. Because they block a volcanic vent and prevent gases from escaping, *lava domes* are often forcefully expelled in extremely violent eruptions.

Andesitic magma commonly erupts from high, steep-sided volcanoes known as *composite volcanoes* (also known as **stratovolcanoes**) that are formed above convergent plate margins. The lava typically emerges in small-volume flows that advance only short distances (hundreds of metres to a kilometre or so) down the flanks of a volcano. These flows tend to form stiff ridges that rise above the surrounding landscape, a result of their high viscosity.

Rhyolitic Magma

Rhyolitic magma has a felsic composition. Its higher silica content makes it more viscous and slower moving than other types of magma (**FIGURE 6.5**). Because of the high viscosity and gas content of rhyolitic magma, eruptions of volcanoes composed of this type are usually violently explosive and generate high volumes of pyroclastic debris.

However, such eruptions deplete dissolved gases in the magma source. The degassed magma can then rise to the surface and extrude less violently. Thus, after an initial eruption phase marked by enormous and catastrophic explosions, viscous rhyolitic magma generally oozes from a volcano's central vent to form symmetrical lava domes. Rhyolitic lava flows, like those produced by andesitic volcanoes, produce ridges of viscous lava that rise as much as 10 m above the surrounding landscape.

FIGURE 6.5 At the Laguna del Maule caldera in Chile, the brown weathering rocks along the far shore of the lake are blocky rhyolite lava flows.

(?) Describe the eruption style of rhyolitic magma.

(?) **Expand Your Thinking**—Explain the tectonic processes responsible for each of the three primary magma types.

6-3 Explosive Eruptions Are Fuelled by Violent Releases of Volcanic Gas

LO 6-3 *Describe volcanic gases and the role they play in explosive versus effusive eruptions.*

Knowledge about the dissolved gases contained in magma is key to understanding why volcanoes erupt. Magma bodies rise in the crust until they reach a point of neutral buoyancy. As gas expands and magma moves closer to the surface, lessening pressure and gaseous expansion produce an eruption. The interaction among the viscosity, temperature, and gas content of the magma determines whether an eruption will be explosive or effusive.

Explosive versus Effusive Eruptions

Most magma is stored in the mantle or crust prior to eruption. During this period, it typically reaches, or comes close to, saturation by a number of gases, particularly water vapour. Most explosive eruptions are fuelled by the explosive release of these gases at shallow levels in the volcano's "plumbing system." The potential for any magma to erupt explosively depends on how dissolvable the gases are and the ability of the magma to retain them during its ascent to the surface. If gases can escape passively, a lava eruption (effusion) will ensue. If gases cannot escape passively (for example, if the magma has a high silica content), an explosive eruption will ensue (**FIGURE 6.6**).

An explosive eruption occurs for two reasons: (1) The solubility (dissolvability) of water vapour and other gases typically is several times greater in silica-rich rhyolitic magma than in basaltic magma (typically 4–6 percent versus 1 percent), and (2) the high viscosity of rhyolitic magma inhibits the rise and escape of gas bubbles. (The buoyant rise rate of a bubble is 106 times faster in basalt than in rhyolite.)

FIGURE 6.6 *(a)* Explosive eruption of pyroclastic debris, Mount St. Helens (Washington, U.S.A.), 1980. *(b)* Effusive eruption of Piton de la Fournaise Volcano on Reunion Island, Indian Ocean, 2007.

? Why do explosive eruptions occur?

Where does volcanic gas come from? Gases are dissolved in molten rock within the crust or upper mantle because of high pressure. Pressure decreases as magma rises toward the surface, and the gases begin to form tiny bubbles. With continued bubble formation, the volume of magma increases, making it less dense and causing it to rise faster. Near the surface, as bubbles grow in number and size, the volume of gas may actually exceed the volume of molten rock. This creates "magma foam." The large number of rapidly expanding gas bubbles in the foam fragments the magma and produces **tephra** (airborne pyroclasts) when erupted. It is this gas fragmentation, and not just the explosive energy of the eruption, that leads to tephra formation.

The increase in magma volume due to gas fragmentation is truly remarkable. For example, if 1 m³ of 900°C rhyolitic magma containing 5 percent dissolved water were suddenly brought to the surface, it would expand over 600 times in size, occupying 670 m³ at atmospheric pressure. Large amounts of gas in the magma can lead to massive eruption columns that spew tephra high into the atmosphere.

Gas Chemistry

Gases spread from an erupting vent primarily in the form of acid aerosols (tiny acid droplets). These compounds attach themselves to tephra particles and microscopic salt particles or other dust in the atmosphere and travel in air currents. Typically, the most abundant volcanic gas is water vapour (H_2O), followed by carbon dioxide (CO_2) and sulphur dioxide (SO_2). Volcanoes also release smaller amounts of hydrogen sulphide (H_2S), hydrogen (H_2), carbon monoxide (CO), hydrogen chloride (HCl), hydrogen fluoride (HF), and helium (He). The gas content measured at three different volcanoes is illustrated in **TABLE 6.2**.

The gases released from a volcano can be as deadly as hot, fiery lava. Volcanic gases (principally sulphur dioxide, carbon dioxide, and hydrogen fluoride) can threaten the health of humans and animals, kill or damage crops, and interfere with air traffic and property. (See the "Geology in Our Lives" box "Bringing Europe to a Standstill.") Sulphur dioxide produces acid rain and air pollution downwind from a volcano, and large explosive eruptions inject massive volumes of sulphur aerosols into the upper atmosphere. These can lower the air temperature by blocking radiation from the Sun. They also damage Earth's ozone layer by forming new molecules with ozone gas.

The Hawaii city of Kona, located downwind of Kilauea Volcano, used to have high lead concentrations in its water because sulphuric rain dissolves the lead used in the plumbing of buildings. This was an important health concern because lead damages the nervous system.

(a)

US Geological Survey/Science Photo Library

(b)

Sylvain Grandadam/Stone/Getty Images, Inc.

GEOLOGY IN OUR LIVES

BRINGING EUROPE TO A STANDSTILL

On April 14, 2010, British civil aviation authorities ordered the country's airspace closed because a cloud of ash drifting from the erupting Eyjafjallajökull (Icelandic for "island-mountain glacier") volcano in Iceland had made it too dangerous to allow air traffic (**FIGURE 6.7**). Flying through volcanic ash can abrade the cockpit windshield, making it impossible for the pilots to see out; damage communication and navigation instruments on the outside of the aircraft; and worst of all, coat the inside of jet engines with concrete-like solidified ash, causing the engine to shut down.

Within 48 hours, the ash cloud had spread across northern Europe. Three hundred and thirteen airports in England, France, Germany, and other European nations closed for about one week causing the cancellation of over 100,000 flights. This affected airline schedules around the world and resulted in some 10 million stranded air travellers at a cost estimated at $3.3 billion. As a consequence of this event, caused by what volcanologists describe as a rather modest eruption, European authorities are considering bringing all European air traffic under the control of a single agency, a move designed to increase flexibility in response to future disruptive episodes.

Scientists familiar with Iceland's volcano system fear that additional eruptions could activate a nearby volcano named Katla. Katla has been dormant for decades, but it could reawaken with the activity at Eyjafjallajökull, and the two of them are known to

have erupted together in the past. Katla is a larger volcano, and it is buried under the glacier Myrdalsjökull, which is half a kilometre thick. The large amount of ice atop Katla is a recipe for ash clouds of greater magnitude than those seen from Eyjafjallajökull. Presumably, they would be capable of once again bringing Europe and airports around the world to a standstill.

FIGURE 6.7 The April 14, 2010 eruption of Eyjafjallajökull in Iceland brought air travel throughout Europe to a standstill for nearly a week. The cancellation of over 100,000 flights stranded some 10 million passengers around the world.

AP/Wide World Photos

The effects are particularly devastating when young children ingest lead, as it can cause learning difficulties and, in extreme cases, brain damage. Kona has now removed lead from its plumbing system.

Carbon dioxide is especially hazardous because it is heavier than air and flows into low-lying areas, killing by asphyxiation.

Late one night in 1986, carbon dioxide seeped out of Lake Nyos, a supposedly dormant, water-filled volcanic crater in Cameroon, Africa. The gas spread into the surrounding valleys, killing more than 1,700 people as they slept, as well as thousands of cattle, birds, and other animals.

TABLE 6.2 Percent Volcanic Gas Content at Three Volcanoes

Gas	Kilauea, Hawaii (basaltic magma, 1,170°C, hotspot, shield volcano)	Erta Ale, Ethiopia (basaltic magma, 1,130°C, divergent margin, shield volcano)	Momotombo, Nicaragua (andesitic magma, 820°C, convergent margin, stratovolcano)
H_2O	37.1	77.2	97.1
CO_2	48.9	11.3	1.44
SO_2	11.8	8.34	0.50
H_2	0.49	1.39	0.70
CO	1.51	0.44	0.01
H_2S	0.04	0.68	0.23
HCl	0.08	0.42	2.89
HF	—	—	0.26

Source: R. B. Symonds, W. I. Rose, G. Bluth, and T. M. Gerlach, "Volcanic Gas Studies: Methods, Results, and Applications," in M. R. Carroll and J. R. Holloway, eds., *Volatiles in Magmas, Mineralogical Society of America Reviews in Mineralogy*, 30 (1994): 1–66.

 Expand Your Thinking—Describe how lava chemistry influences the style of volcanic eruptions.

6-4 Pyroclastic Debris Is Produced by Explosive Eruptions

LO 6-4 *Describe the formation of pyroclastic debris.*

Although lava flows are better known, it is pyroclastic debris that makes up the largest volume of volcanic products on land. Most pyroclastic eruptions are explosive and associated with andesitic or rhyolitic magma. However, basaltic magma may form pyroclastic debris if the eruption is gas rich. Within a volcano, the rapid formation of gas bubbles tends to fragment magma into particles. Explosive eruptions are characterized by the violent expulsion of these fragments in the form of pyroclastic debris. These eruptions also contain high amounts of gas that, on reaching Earth's surface, expands to many hundreds of times its original volume.

Tephra

Explosive eruptions are typically accompanied by an **eruption column**—a massive, high-velocity, billowing cloud of gas, molten rock, and solid particles—that is blasted into the air with tremendous force (**FIGURE 6.8**). *Tephra* is the general term used by volcanologists to characterize airborne pyroclastic debris produced by an eruption column. Tephra typically includes volcanic *ash* (with particles less than 2.5 mm across), larger pyroclastic fragments called *lapilli* (particles 2.5 to 63 mm across), *blocks* (more than 63 mm across) that are ejected in a solid state, and *bombs* (also more than 63 mm across) that are ejected in a semi-molten or plastic condition. Together, these materials are known as pyroclastic debris (**FIGURE 6.9**).

In a major explosive eruption, most of the pyroclastic debris consists of lapilli and ash. Because bombs are ejected in a semi-molten form, they often take on a rounded, aerodynamic shape as they fly through the air. When pyroclastic debris accumulates on the ground and solidifies, it forms a rock called **tuff**. Tuff is made up of the various bits of volcanic debris that have been ejected; it usually consists of particles of various sizes contained in an ash matrix.

Pyroclastic Flows

It is common for the base of an eruption column to collapse and form searing **pyroclastic flows** that are driven down the slopes of the volcano by the force of gravity (**FIGURE 6.10**). A pyroclastic flow is a high-speed avalanche of hot ash, rock fragments, and gas that can reach temperatures of up to 800°C and move at speeds of up to 200 km/h to 250 km/h. A pyroclastic flow is capable of overcoming, knocking down, burying, and burning anything in its path. When these materials lose their energy and come to rest as thick beds of ash and lapilli, they eventually solidify into a welded mass of glass shards. The resulting rock is referred to as a **welded tuff**.

Pyroclastic flows move at incredible speeds and constitute a lethal hazard for communities living on the slopes or at the base of explosive volcanoes. In 1902, a pyroclastic flow raced down the side of Mount Pelée on the Caribbean island of Martinique. Despite warnings of an impending eruption, the mayor of a nearby town had convinced the residents not to leave because an election was scheduled within the next few days. Twenty-eight thousand people died in a matter of minutes.

One way in which volcanologists gauge the potential danger of a volcano is by examining rock in the vicinity for the occurrence of welded tuffs. Past pyroclastic flows will have produced a stratigraphic record consisting of sequences of welded tuffs whose thickness and frequency offer a guideline for assessing the probability of future flows, their likely direction, and their probable size.

Another important, and curious, product of explosive eruptions is *pumice*. As is well known, pumice is a volcanic rock that floats. Lumps of pumice resemble sponges because they contain a network of gas bubbles captured within the fragile volcanic glass and minerals composing the rock. It is created when gases rapidly escape from magma, producing a froth or foam. The foam quickly cools and solidifies to form pumice. Typically, pumice is silica-rich, whereas the lower silica, basaltic equivalent is called *scoria*.

Klaus Nigge/National Geographic/Getty Images

(?) What type of rock is formed by the accumulation of pyroclastic debris?

FIGURE 6.8 An eruption column is composed of pyroclastic debris, gas, and lava.

D. E. Wieprecht/Volcano Hazards Program/USGS

(a)

Courtesy of Dr. J. Alean/Stromboli

(b)

Photo Researchers, Inc. / Science Photo Library

(c)

J. P. Lockwood/Volcano Hazards Program/USGS

(d)

Courtesy Ted Eckmann, UC Santa Barbara
Department of Geography

(e)

© Scientifica/Visuals Unlimited/Corbis

(f)

FIGURE 6.9 Pyroclastic debris may consist of (a) ash, (b) lapilli, (c) pumice, (d) volcanic bombs, (e) volcanic block, and (f) welded tuff, a rock consisting of ash-size glass shards.

 What types of magma tend to produce pyroclastic debris?

FIGURE 6.10 In 1984 the explosive eruption of Mayon Volcano in the Philippines generated several large pyroclastic flows.

 What is the origin of pyroclastic flows?

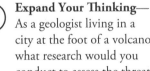

 Expand Your Thinking— As a geologist living in a city at the foot of a volcano, what research would you conduct to assess the threat from pyroclastic flows?

Bullit Marquez/AP/Wide World Photos

6-5 Volcanoes Can Be Classified into Six Major Types Based on Their Shape, Size, and Origin

LO *6-5 Compare and contrast central vent volcanoes with large-scale volcanic terrains.*

Volcanoes can be classified by their shape, size, and origin. In this section, we define two broad classes of volcanoes: central vent volcanoes and large-scale volcanic terrains (**FIGURE 6.11**).

Central vent volcanoes tend to build a volcanic landform from a more or less central vent. These volcanoes most often have a *summit crater* but may also produce *flank eruptions* (eruptions from the side of the volcano), *fissure eruptions* that originate from an elongated fracture on the side of a volcano (see the "Critical Thinking" box, "The Laki Fissure Eruption," at the end of this section), and other types of eruptions. Central vent volcanoes are the most widespread and best-known type—seen in the familiar public image of a volcano as a cone-shape mountain.

Large-scale volcanic terrains lack a central vent and are formed by eruptive products coming from a network of sources. They also generally (but not always) constitute massive features that have re-shaped the land (or sea floor) over an area of hundreds of thousands of square kilometres or more. Large-scale volcanic terrains are important because they are globally significant, account for gaps in our understanding of volcano origins that are not explained by the other types, and conform to the definition of volcanoes that we adopted at the beginning of the chapter (*A volcano is any landform that releases lava, gas, or ash or has done so in the past*).

While there are examples of volcanoes with features of both categories just described, as well as volcanoes that do not fit into either category, this classification system provides a sense of the diversity of scale and origin found among the world's volcanoes. Within each category we find at least three distinct varieties:

1. Central vent volcanoes
   a. Shield volcanoes
   b. Stratovolcanoes (or composite cones)
   c. Rhyolite caldera complexes

2. Large-scale volcanic terrains
   a. Monogenetic fields
   b. Large igneous provinces
   c. Mid-ocean ridges

The names of these six volcanic types may appear complex and technical, but you will find that they are easy to remember once you understand the concepts behind the words (**TABLE 6.3**).

TABLE 6.3 Types of Volcanoes

Type	Shape	Magma Type	Tectonic Setting	Example
Central Vent Volcanoes				
1. Shield volcano	Large volume, gentle, low-angle slopes	Basaltic, low-silica	Mid-plate setting (most often) or variable setting	Mauna Loa Volcano (Hawaii, U.S.A.)
2. Composite volcano or stratovolcano	Tall, with steep slopes, often irregular outline from past explosions and rugged dome areas	Andesitic, silica-rich magma at subduction zone	Convergent boundary (usually)	Mount Pinatubo (Philippines), Mount St. Helens (Washington, U.S.A.)
3. Rhyolite caldera complex	Low-relief system of collapsed calderas and many small vents	Rhyolitic, silica-rich magma (including melted crustal rock)	Convergent boundary or isolated mid-plate	Yellowstone National Park (Wyoming, U.S.A.)
Large-Scale Volcanic Terrains				
4. Monogenetic field	Low-relief system of vents, cones, occasional stratovolcano	Basaltic, low-silica	Convergent boundary, or isolated mid-plate	Ukinrek Maars (Alaska, U.S.A.)
5. Large igneous province	High plateau, massive volume (>100,000 km³), many layers of lava, no single distinct mountain	Basaltic, low-silica	Variable setting, often mid-plate or continental margin	Columbia Plateau, (Washington, U.S.A.)
6. Mid-ocean ridge	Broad slopes on long, linear ridge with central rift valley	Basaltic, low-silica	Divergent boundary, spreading centre along mid-ocean ridge	Segment of the East Pacific Rise

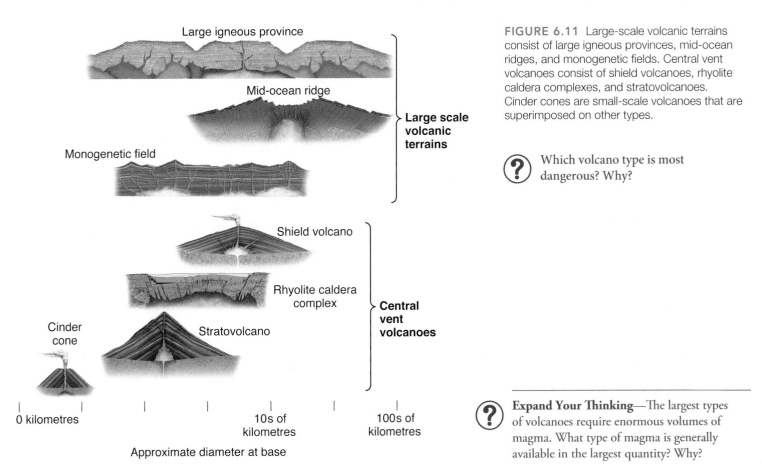

FIGURE 6.11 Large-scale volcanic terrains consist of large igneous provinces, mid-ocean ridges, and monogenetic fields. Central vent volcanoes consist of shield volcanoes, rhyolite caldera complexes, and stratovolcanoes. Cinder cones are small-scale volcanoes that are superimposed on other types.

(?) Which volcano type is most dangerous? Why?

(?) **Expand Your Thinking**—The largest types of volcanoes require enormous volumes of magma. What type of magma is generally available in the largest quantity? Why?

(?) ## CRITICAL THINKING

THE LAKI FISSURE ERUPTION

On June 8, 1783, the Laki fissure zone in Iceland began to erupt in what was to become the largest basalt eruption in recorded history. The eruption (**FIGURE 6.12**) lasted eight months, during which time 14.73 km³ of basaltic lava and some tephra were erupted, covering 565 km² of land. Haze from the eruption was reported from Iceland to Syria.

In Iceland, volcanic outgassing led to the loss of most of the island's livestock (due to eating grass contaminated by fluorine), crop failure (due to acid rain), and the death of one-quarter of the human residents (due to famine). Lava erupted from fissures located 45 km from the coast and flowed toward the ocean at speeds averaging 0.4 km/h.

As a graduate student working with volcanologist Professor Andre Morrell of the University of Marseille, you are studying the history of the Laki fissure eruption. Your first task is to graph the history of lava flow to the coast and determine the distance covered during every six-hour period.

1. Design a data table to show the distance travelled by the lava over a five-day period. Calculate the distance covered every six hours.
2. Plot the data on a graph with time on the *x*-axis and distance on the *y*-axis.

FIGURE 6.12 The Laki eruption site is now quiet, but in 1783 it erupted in long lines of lava fountains that discharged massive volumes of lava, forming a small basalt plateau.

3. How long did it take the lava to reach the coast?
4. How many kilometres did the lava travel in three days?
5. If you lived in a village only 10 km from the eruption site, how much time after the first sign of eruption would you have to evacuate? Imagine that you owned a farm with animals and crops. What would you be able to save, and what would most likely be lost to the eruption?

6-6 Shield Volcanoes Are a Type of Central Vent Volcano

LO *6-6 Identify what gives a shield volcano its distinctive shape.*

Shield volcanoes have a distinct summit crater. They are named for their fanciful resemblance to a warrior's shield laid on the ground (**FIGURE 6.13**) and are noted for their enormous size and broad, gently sloping profiles. Various stages of shield building may be characterized by flank eruptions, fissure eruptions, and the construction of secondary cones on the flanks of the volcano. Made of fluid basaltic lava, most shield volcanoes are built of thousands of thin basaltic layers that accumulate over time.

The key to understanding a shield volcano's shape is the chemistry of its magma. Shield volcanoes are fed by a low-silica, low-gas magma source in the upper mantle. Eruptions release repeated flows of pahoehoe and aa that create the volcano's layered structure. Shield volcanoes also tend to have a relatively constant magma supply rate, which keeps the magma "plumbing system" hot and allows the lava to maintain a high temperature, remaining fluid longer and hence flowing farther. The chemical composition of magma in a shield volcano probably has not changed significantly since its formation in the mantle.

Some of the largest volcanoes in the world are shield volcanoes. The most famous examples are found in the Hawaiian Islands, a linear chain of huge shield volcanoes that includes Kilauea and Mauna Loa. Both are among Earth's most active volcanoes, and Mauna Loa is the largest. Mauna Loa's true size is not obvious at first. It projects over 4 km above sea level, but its base on the sea floor lies beneath more than 4.5 km of water. Added together, Mauna Loa's total height is over 8.5 km above the sea floor, making it the most massive mountain on Earth.

Lava pours out of a shield volcano from vents at the summit or along its flanks. It is common for eruptions on shield volcanoes to include high lava fountains that form **cinder cones** at the vent (**FIGURE 6.14**). This style of eruption is spectacular but not explosive. Cinder cones, which some scientists consider to be a separate category of volcano, are built up by the ash, lapilli, blocks, and bombs of congealed lava ejected from a single vent. High gas content in the magma source tends to drive the fiery eruption. When the gas is blown violently into the air, it accumulates in the form of small fragments that solidify and fall as cinders around the vent, forming a nearly circular cone (or oval, if the wind is more persistent in one direction). Most cinder cones have a bowl-shape summit crater.

Numerous in western North America and other volcanic environments of the world, cinder cones are frequently found as secondary, short-lived eruption sites in association with a longer-lived central vent volcano. They are also found in association with large-scale volcanic terrains and thus are associated with both classes of volcanoes. It is uncommon to find a cinder cone more than 300 m high. There are actually numerous examples of cinder cones in British Columbia, including those within Wells Gray Provincial Park, and a beautiful example is the Eve Cone (Figure 6.14) which is one of a number that formed as part of the Mount Edziza volcanic complex. One of the most recent eruptions in Canada was in the 18th century, and led to the formation of the Tseax Cone. It is thought that over 2,000 Nisga'a people died during the eruption as a result of asphyxiation by volcanic gases. The "Geology in Action" box in this section describes some of the examples of Canadian volcanoes.

Shield volcanoes are approximately 90 percent lava and 10 percent pyroclastic deposits. Explosions do occur on shield volcanoes, often in the form of a **phreatomagmatic eruption** caused by the rapid expansion of steam when magma comes in contact with groundwater. The occurrence of these eruptions is unpredictable, and they can be highly dangerous to life and property in their immediate vicinity.

Shield volcanoes may form wherever a basaltic magma source erupts on the land surface or ocean floor. They commonly occur in mid-plate settings (oceanic hotspots). However, they can also be found in convergent settings, such as subduction-related volcanic arcs and island arcs. Mid-plate shield volcanoes form in places where

FIGURE 6.13 Shield volcanoes, such as Mauna Loa on the island of Hawaii, are built up by eruptions of fluid basaltic lava.

Courtesy of Patricia MacQueen, Simon Fraser University

(?) Do you agree that it is highly unlikely for rhyolitic magma to build a shield volcano? Why?

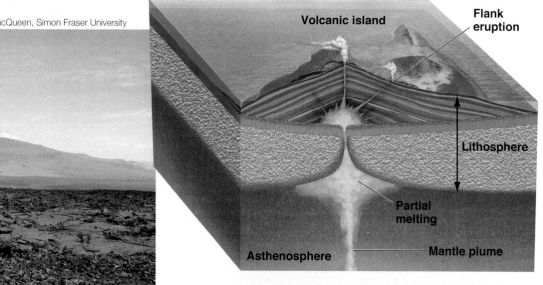

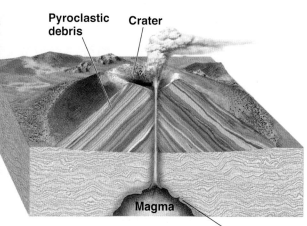

FIGURE 6.14 Cinder cones are small, steep cones with wide craters when compared to their overall size. The photograph is of the Eve Cone, which is part of the Mount Edziza volcanic complex in British Columbia.

a persistent source in the mantle continuously feeds basaltic lava to the surface, where it erupts and builds a volcano. Most often, this occurs in oceanic lithosphere, presumably because the crust is thin and easily breached, and the magma source is low in silica. As a plate moves over a hotspot, a volcano is born and grows, eventually dying as it moves off the hotspot; then a new one forms. In this way, a chain of volcanoes can develop. Many examples are found in the Pacific Ocean Basin, with the Hawaiian Islands being the best known.

(?) Expand Your Thinking—What is "cinder"? What conditions lead to the production of cinder?

🌧 GEOLOGY IN ACTION

CANADA HAS VOLCANOES AS WELL

The western margin of North America forms part of the "Ring of Fire" around the Pacific Ocean (Figure 6.2). Along the margin are a variety of different types of volcanoes, including a surprisingly large number in British Columbia and southern Yukon (**FIGURE 6.15**).

The Cascades Volcanic Belt in the United States, and its extension into southwestern British Columbia, the Garibaldi Volcanic Belt, is related to the subduction of the Juan de Fuca Plate below the North American Plate. Explosive volcanic eruptions are common in this volcanic environment. A spectacular example was the eruption of Mount St. Helen's in 1980, which ejected 1 km³ of ash that spread over an area of 57,000 km², killed 57 people, and lowered the mountain peak by 400 vertical metres. The youngest explosive eruption in Canada occurred 2,350 years ago when Mount Meager produced a cloud of volcanic ash that spread as far east as Alberta and a number of destructive pyroclastic flows.

The youngest volcanic eruptions are further north in British Columbia and are related to rifting of the crust as the Pacific Plate moves northward along the Queen Charlotte Fault. The eruption at Tseax occurred in about 1750, and it is thought that about 2,000 of the Nisga'a people that lived in the area died of carbon dioxide asphyxiation. The basalt lava flows also dammed the Tseax River.

Yellowstone National Park, one of the most popular tourist destinations in North America, sits on a potential *supervolcano*. If it were to erupt as it has in the past, it would cause the largest natural disaster in the history of humankind. The last time it erupted, 640,000 years ago (Lave Creek eruption), it ejected over 1,000 times the ash and lava of Mount St. Helens, and the ash deposits have been found as far south as the Gulf of Mexico. An earlier eruption, the "Huckleberry Ridge eruption" 2.1 million years ago, was more than 2,400 times the volume of the Mount St. Helens eruption. The Yellowstone Volcanoes consists of *rhyolite caldera complexes*, fed by a hotspot located under the felsic rock of the North American continent.

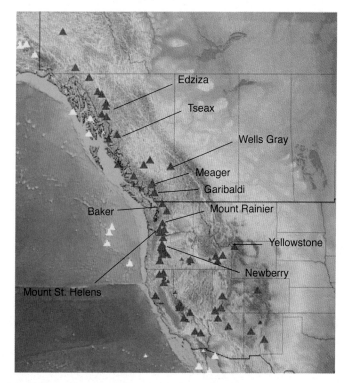

FIGURE 6.15 Volcanoes in western Canada and the United States. The volcanoes named are referred in the text in this chapter. (Source: Based on data from the Global Volcanism Program of the Smithsonian Institution, and the United States Geological Survey.)

6-7 Stratovolcanoes and Rhyolite Caldera Complexes Are Central Vent Volcanoes

LO 6-7 *Compare and contrast the magma composition of stratovolcanoes versus rhyolite caldera complexes.*

Stratovolcanoes account for the largest proportion (about 60 percent) of Earth's individual volcanoes, and the great majority of volcanic arcs and island arcs are composed of stratovolcanoes. As the name implies, stratovolcanoes are composed of alternating layers of stratified lava flows and pyroclastic deposits (FIGURE 6.16). There are dozens of famous examples, including Mount St. Helens (Washington State), Mount Pinatubo (Philippines), Mount Fuji (Japan), and Soufrière Hills (West Indies). Mount Garibaldi and Mount Meager in southwestern British Columbia and the impressive Mount Baker just over the border in Washington State are other examples. They form part of the same volcanic belt as Mount St. Helens (Figure 6.15).

Stratovolcanoes have steep slopes and are often explosively eruptive. They may be composed of several types of lava. Typically they build over many years of eruptions that produce layers of silica-rich lava and ash. Because these volcanoes are usually fed by intermediate-composition magma (originating at subduction zones), the magma source is gas rich and viscous. Internal gas pressure builds behind the sticky magma, eventually leading to explosive release producing pyroclastic debris, followed by lava flows down the sides of the volcano.

Magma Genesis

When a plate is recycled into the mantle by subduction, the increased heat causes water and other volatile substances to escape from the subducting oceanic crust. Because water tends to lower the melting point of most rocks, the result is partial melting in the overlying mantle. The magma (typically water-rich basaltic magma) migrates toward the surface, initiating partial melting in the silica-enriched crust and thereby generating molten rock that is also silica enriched. If the overriding plate is oceanic, the resulting magma could be either basaltic or andesitic. If the overriding plate is continental (typically composed of granite or a metamorphic rock called *gneiss*), the resulting magma is usually andesitic, but under certain conditions it can be rhyolitic.

As we learned in Section 5-2, during partial melting, silica-rich compounds melt early and mafic compounds melt later; hence, the thicker the overlying crust, the greater the potential for magma–crust interaction leading to enhanced production of silica-rich magma. For this reason, ocean–continent subduction sites typically have silica-rich eruptions whereas ocean–ocean sites often do not. The resulting magma feeds the growth of a line of volcanoes on the overriding plate: a *volcanic arc* if the overriding plate is continental; an *island arc* if the overriding plate is oceanic.

Stratovolcanoes contrast sharply with shield volcanoes. (See the "Critical Thinking" box "The Geology of Shield Volcanoes and Stratovolcanoes"). Whereas shield volcanoes are usually characterized by effusive eruptions of fluid basaltic lava, stratovolcanoes are just the opposite: They typically erupt with massively explosive force because the magma is gas rich and too viscous for volcanic gases to escape easily. This allows tremendous internal pressure to build up. Strong silica chains form in the andesitic magma, making it much less fluid, so gas cannot readily escape.

The same geologic conditions that form high-silica magmas also produce a high level of dissolved gas in the magma. Trapped gas creates explosive conditions by adding to the pressure buildup

(?) How does the chemistry of magma determine the shape and eruption style of stratovolcanoes?

FIGURE 6.16 *(a)* Mount Fuji, Japan, a classic stratovolcano; and *(b)* simplified view of the internal structure of a stratovolcano.

© kumakuma1216/iStockphoto.com

Andesite lava flow

Pyroclastic flow

Layers of lava and pyroclastic debris

(a)

(b)

that accompanies the ejection of highly viscous lavas. As a result, lava cannot easily flow from the vent, and the volcano may develop a plug of cooled lava—a lava dome.

When the pressure becomes high enough, the lava dome is explosively ejected. In the worst case, a **Plinian eruption** may occur. Plinian eruptions are large volcanic explosions that send thick, dark columns of pyroclasts and gas high into the stratosphere. Such eruptions are named for the Roman scholar Pliny the Younger, who carefully described the disastrous eruption of Mount Vesuvius in AD 79. This eruption formed a huge column of ash that rose high into the sky. Many thousands of people evacuated areas around the volcano; nevertheless, approximately 2,000 people were killed and the cities of Pompeii and Herculaneum were destroyed.

Stratovolcanoes cause far more human tragedy than do volcanoes of any other type. One reason is simply that stratovolcanoes are more numerous than other types of volcanoes, so more people live on their flanks. Other reasons reflect the inherent characteristics of a steep-sided cone built from highly viscous magma interlayered with pyroclastic deposits:

- Stratovolcanoes are steep piles of ash and lava. Both their steepness and their composition sharply increase the likelihood of slope failure (landslides, avalanches, mudflows).
- Stratovolcanoes have a strong tendency to erupt in mighty explosions that disrupt communities and threaten human life across broad areas.

Rhyolite Caldera Complexes

Rhyolite caldera complexes originate from magma that is so gas rich and viscous that it almost always produces catastrophic explosions. In fact, the explosions are so violent that the volcano tends to collapse on itself and become nothing more than a series of large depressions ("caldera complexes") in the ground. Rhyolite caldera complexes are the most explosive of Earth's volcanoes but often do not even look like volcanoes. Their tendency to erupt explosively is due to the high silica and gas content of their rhyolitic magma.

The collapsed depressions are **calderas** (Spanish for "cauldron") that may be several kilometres wide, indicating that huge magma chambers (10 km or more in width) are associated with the eruptions. In fact, layers of ash (either ash falls or pyroclastic flows) often extend over thousands of square kilometres in all directions from these calderas. A good example is the Yellowstone Volcano, which spread ash over several surrounding states when it last erupted 640,000 years ago.

A caldera is a large, usually circular depression that is formed when magma is withdrawn (usually by eruption) from a shallow underground magma reservoir (**FIGURE 6.17**). The rapid depletion of large volumes of magma removes structural support for the overlying roof of the reservoir, allowing it to collapse and thus create the caldera. Calderas are different from **craters**, which are smaller, circular depressions created primarily by the explosive excavation of rock during eruptions.

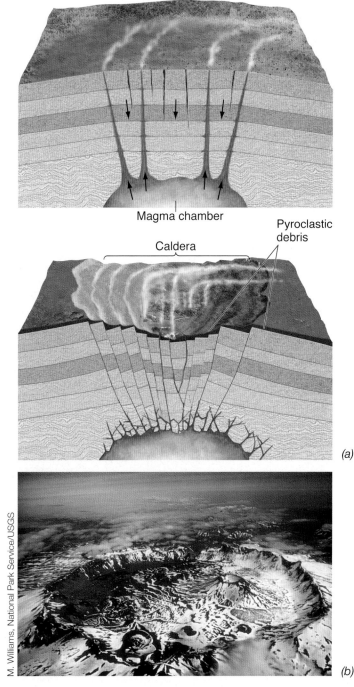

Magma chamber

Caldera

Pyroclastic debris

(a)

M. Williams, National Park Service/USGS

(b)

FIGURE 6.17 Caldera complex: (a) cross-sections during the formation of a caldera, (upper) before formation and (lower) after formation; and (b) Aniakchak Caldera, Alaska, which was formed during an enormous explosive eruption about 3,450 years ago that expelled more than 50 km³ of magma. The caldera is 10 km in diameter and ~1 km deep. Later eruptions formed domes, cinder cones, and explosion pits on the floor of the caldera.

 How does the chemistry of magma determine the shape and eruption style of a rhyolite caldera complex?

Expand Your Thinking—Someday it is likely that the Atlantic sea floor will begin to subduct beneath the eastern seaboard of North America. Describe the volcanism that will result.

? CRITICAL THINKING

THE GEOLOGY OF SHIELD VOLCANOES AND STRATOVOLCANOES

Please work with a partner to answer the following questions, based on **FIGURE 6.18**.

1. Add labels to Figure 6.18 that identify the major features associated with shield volcanoes and stratovolcanoes. Be sure to identify magma composition at various settings, volcanic products, features that make up the structure of the two volcanoes, and plate tectonics aspects.

2. What happens to a mid-plate shield volcano as it moves away from a hotspot?

3. Summarize the processes acting on a volcanic arc. How do weathering, erosion, explosive volcanism, and effusive volcanism combine to shape the appearance of the arc?

4. Describe the processes that create magma sources for the shield volcano and stratovolcano. How are they different? How are they similar?

5. Explain the differences and similarities in the internal structure of the two types of volcanoes.

6. Describe the geologic hazards associated with shield volcanoes and compare them to the hazards associated with stratovolcanoes. If you were a premier of a province or mayor of a community, how would you mitigate the impact of these hazards on nearby communities?

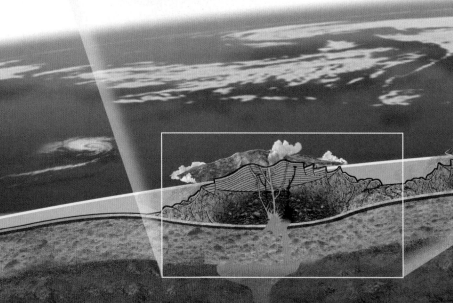

FIGURE 6.18

7. You are asked to study a new volcano. What criteria would you use to assess the danger it poses to local communities?
8. What might happen if basaltic magma mixed with rhyolitic magma and erupted? Describe the resulting volcano and the volcanic processes.

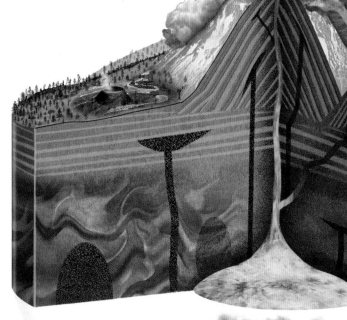

6-8 Large-Scale Volcanic Terrains Lack a Central Vent

LO *6-8 Identify the main characteristics of large-scale volcanic terrains and give examples.*

Monogenetic fields are collections ("fields") of vents and flows, sometimes numbering in the hundreds or thousands (FIGURE 6.19). They do not look like classic volcanoes. Although there may be a single magma source for the whole field (making them "monogenetic"), it sends magma to the surface at a low rate, in a diffuse manner, and through multiple vents. Studies indicate that in many cases, each vent erupts only once. Monogenetic fields grow laterally rather than vertically and form fields instead of mountains. A monogenetic field can be thought of as taking all the separate eruptions and flows of a single large volcano and spreading the pieces across a broad landscape as separate features.

Monogenetic fields are collections of cinder cones and/or *maar* vents and the lava flows and pyroclastic deposits associated with them. A maar is a low-relief, broad volcanic crater formed by shallow explosive eruptions. The explosions are usually caused when groundwater is invaded by magma, causing a phreatomagmatic eruption. Often a maar later fills with water to form a lake. Sometimes a stratovolcano is at the centre of the field, as in the San Francisco Volcanic Field in Arizona, which has produced more than 600 volcanoes in its short 6-million-year history. In this case, both volcanic categories appear in the same geographic area. A Canadian example is the Wells Gray–Clearwater Volcanic Field in British Columbia.

Large Igneous Provinces

Large igneous provinces (**LIPs**) form at locations where massive quantities of basaltic lava emanate from large magma chambers and pour onto the surface from systems of long fissures (ground cracks) instead of from central vents. An immense outwelling of lava floods the surrounding countryside or sea floor over a period of millions of years, forming broad plateaus hundreds of metres high. In fact, there are places where so much lava has issued from the ground that the lava is referred to as **flood basalt**. Layer after layer of this very fluid, silica-depleted basaltic lava forms high plateaus under the sea and on continents that extend over an area of at least 100,000 km². These lava flows are often visible in the walls of eroded river valleys, measuring a kilometre or more in total thickness (FIGURE 6.20). Clearly, such a phenomenon requires a massive magma source. Lava plateaus composed of flood basalts can be seen in Iceland, Washington and Oregon, India, Siberia, and elsewhere. The massive Ontong Java Plateau on the floor of the Pacific Ocean is built up of flood basalts.

LIPs have been formed, some in the oceans and some on land, at various times in the past. Notable examples are the Siberian Traps in Russia and the Deccan Traps in northwestern India. ("Trap" is a Sanskrit word meaning "step," referring to the steplike topography produced by the stacked layers of lava.) The most famous example of a LIP in North America is the Columbia Plateau, which covers most of southeastern Washington State and northeast Oregon and extends to the Pacific. Active 6 to 17 million years ago, the Columbia Plateau was created by the first eruption of the same hotspot that today feeds the Yellowstone Volcano.

Geologists hypothesize that large igneous provinces are caused by the arrival of an upward-moving mantle plume into the lithosphere (as described in Section 5-8). It is thought that mantle plumes are enriched in lighter elements and hotter than the surrounding

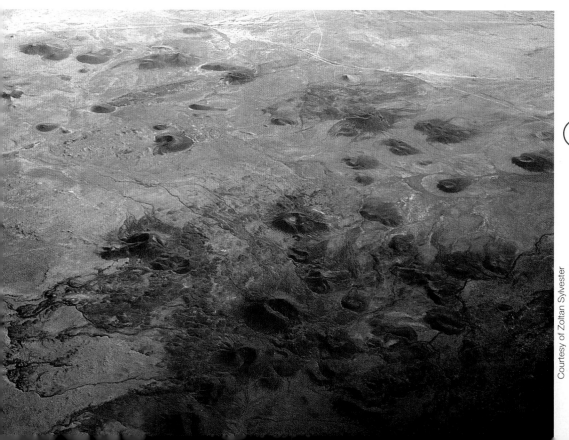

Courtesy of Zoltan Sylvester

FIGURE 6.19 The San Francisco Volcanic Field in Arizona has produced more than 600 volcanoes during its 6-million-year history. This image shows over two dozen separate volcanoes in one portion of the field.

? Monogenetic fields consist of dozens to hundreds of separate volcanoes. Are these each fed by a separate magma source or a single magma source? What can you infer from this about the tectonic setting?

FIGURE 6.20 A stack of several Columbia River basalt lava flows in the Columbia River Gorge, Washington State. Each flow is 15 m to 100 m thick.

 Why is it unlikely for lava plateaus to form above a rhyolitic magma source?

rock. As they rise and the pressure they encounter decreases, magma is generated by partial melting at the top of the plume. As the plume, which already contains magma, reaches the base of the lithosphere, it mushrooms outward, leading to the generation of more magma as the pressure on the plume decreases even further. This feeds magma that erupts into the crust to form huge basalt flows. Magma appears to be produced in greatest abundance during the first few million years after a mantle plume reaches the surface; therefore, flood basalts form quickly in geologic terms. If the plume maintains its position and magma is produced at a consistent rate, it could constitute a hotspot (though this is not always the case). Mantle plumes are thought to have a deep origin, perhaps at the core-mantle boundary for the larger ones and at a depth of about 600 km for the smaller ones.

Mid-Ocean Ridges

Mid-ocean ridges develop at locations where new oceanic crust is continually being formed by magma rising from the mantle (FIGURE 6.21). Because this young crust is still warm and is being pushed up from below, it sits higher than the surrounding sea floor, forming a long, broad ridge that can rise over 2 km above the adjacent ocean bottom. Eruptions along a central rift valley in the centre of the ridge tend to be quiet affairs composed of fluid, silica-depleted basalt. A mid-ocean ridge runs along the floor of every major ocean, tending to run down the middle except in the Pacific, where it runs along the south and southeast side of the ocean basin. These ridges are interconnected, forming a global underwater mountain chain approximately 60,000 km in length.

FIGURE 6.21 This photo shows the first direct observation of submarine volcanism, similar to what would be occurring along a spreading centre. The photo was obtained by a remotely operated unmanned robot called "Jason" controlled from a research ship (December 2009). It shows a volcano in the southwestern Pacific Ocean, north of Tonga. Glassy basaltic talus collects around a vent that is spewing sulphur-rich gases into the water. The water depth is approximately 1,200 m.

 Why is most submarine volcanism fed by basaltic magma?

 Expand Your Thinking—Why is it likely that all large-scale volcanic terrains would be composed of basaltic lava?

6-9 Most Volcanoes Are Associated with Spreading Centre Volcanism, Arc Volcanism, or Intraplate Volcanism

LO *6-9 Describe the role of plate tectonics in spreading centre volcanism, arc volcanism, and intraplate volcanism.*

Most volcanic activity is linked to plate tectonic processes, and many of the world's active above-sea volcanoes are located near convergent plate boundaries, particularly around the Pacific Basin's "Ring of Fire." However, much more volcanism—producing about three-quarters of all the lava erupted on Earth—takes place unseen beneath the waters of the world's oceans, mostly along spreading centres, such as the East Pacific Rise and the Mid-Atlantic Ridge. Some volcanoes erupt in mid-plate settings. These are probably tied to plumes in the mantle rather than to plate interaction in the lithosphere.

There are three tectonic settings known to foster volcanoes (**FIGURE 6.22**):

1. *Spreading centre volcanism*, characterized by fluid basaltic magma that occurs at divergent plate margins forming mid-ocean ridges.

2. *Arc volcanism*, characterized by explosive rhyolitic, andesitic, and basaltic magma that occurs at convergent margins.

3. *Intraplate volcanism* (also known as mid-plate volcanism), characterized by shield volcanoes, rhyolite caldera complexes, and monogenetic fields.

Spreading Centre Volcanism

Spreading centre volcanism occurs at divergent plate boundaries. Spreading centre magmas are basaltic, and the resulting erupted products reflect that chemistry. Because of their relatively low silica content, the lavas are characterized by low viscosity and low dissolved gas content. These magmas are least likely to produce violent eruptions.

Spreading centre magmas originate in partially molten chambers within the upper mantle. When the magma feeding a spreading centre moves toward the surface, it forms an intrusive *dike*. Dikes are magma-filled cracks in the crust that feed eruptions on the sea floor. Geologists hypothesize that a typical ridge eruption leaves behind a dike from tens of centimetres to a few metres in width extending between the magma chamber and the eruptive fissure at the surface.

Often a molten layer of liquid rock forms on the surface of a magma source. This molten rock is periodically tapped by vertical fractures, which provide conduits for the rapid rise of magma to the surface. After volcanism has occurred and magma crystallizes within these fractures, a system of *sheeted dikes* is created that extends upward from the source of the magma. These dikes are an important component of the rocks that make up the new oceanic crust that forms at spreading centres. The layering of ocean crust is discussed further in Chapter 22, "Coastal Geology."

Lavas extruding from a spreading centre fissure chill quickly in contact with cold sea water and form pillow basalts and deposits of glassy *talus* (broken rock). This process of magma ascent and eruption occurs at different times at different places on the oceanic ridge system. The rise of heat from the asthenosphere adds thermal buoyancy to the ridge area. This is the reason that ridges rise high

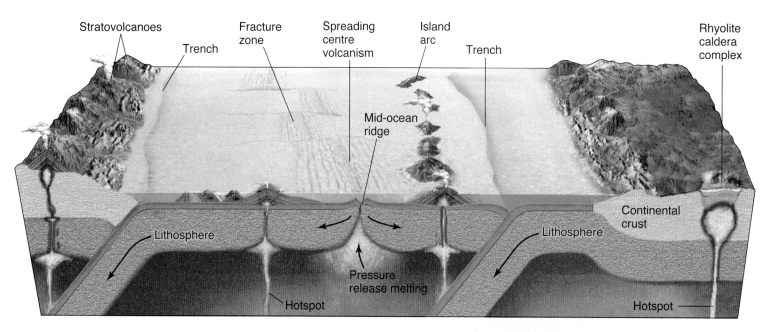

FIGURE 6.22 Most global volcanism is found at spreading centres, convergent arcs, or within plate interiors.

(?) Identify the magma chemistry directly beneath each of the active volcanoes in Figure 6.22.

above the surrounding sea floor. Eruptions produced in this manner are typically quiet fissure eruptions, and scientists in research submersibles have been able to study these eruptions. Eruptions may sometimes produce pillow basalt interbedded with hydrothermal metallic sulphide mineral deposits formed from "black smokers."

The base of this newly formed sea floor is composed of plutons made of gabbro originating from the same magma source. The resulting crust consists predominantly of a lower layer of gabbro, a layer of sheeted dikes, a layer of basalt pillow deposits, and a surface coating of marine sediments that thickens as time passes and the sea floor moves farther from the rift zone.

Arc Volcanism

Arc volcanism occurs at locations where two plates converge. One plate containing oceanic crust descends beneath the adjacent plate, which may be composed of either continental or oceanic crust. If the overriding plate is oceanic, an island arc will be created. If the overriding plate is continental, a volcanic arc will develop (**FIGURE 6.23**). The Cascade Range, which extends from southwestern British Columbia into the Pacific Northwest of the United States, is an example of a volcanic arc; the islands of Indonesia are an example of an island arc. The majority of volcanoes produced by arc volcanism are typically violent stratovolcanoes, although shield volcanoes may form from a basaltic magma source.

The crust portion of a subducting slab contains a significant amount of water in hydrated minerals (minerals that contain water molecules) and marine sediments. As the subducting slab encounters high temperatures and pressures, the water is released into the overlying mantle. *Water has the effect of lowering the melting temperature of rock*, thus causing it to melt more easily than it would if no water were present. The magma produced by this mechanism typically varies from basaltic to andesitic in composition and is gas rich. Plutonic bodies of mafic or intermediate composition may evolve through crystallization and produce felsic magmas. As the silica content increases, the violence of the resulting volcano increases, as does the tendency to create pyroclastic products.

If rhyolitic magma is produced through differentiation, the chances of a highly explosive eruption are great. This is most likely to happen with magma that has intruded and partially melted continental crust, producing a high-silica liquid. The high-silica content of continental crust can enhance the differentiation process that produces rhyolitic magma, leading to the formation of some rhyolite caldera complexes.

Intraplate Settings

Intraplate settings such as hotspots can be exceptionally active sites of volcanism. The exact relationship between large igneous provinces and hotspots is still somewhat unclear. However, if you follow some of the longer hotspot tracks from their current eruptions toward their oldest volcanoes, you end up at a LIP. This is true of the Yellowstone hotspot and the Columbia Plateau, for example. For this reason, it is

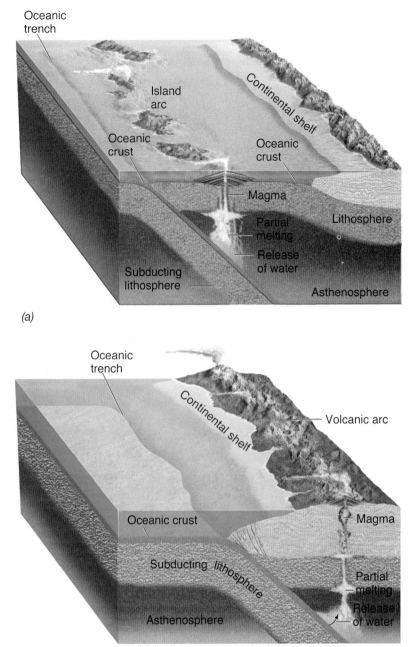

FIGURE 6.23 (a) Island arc volcanism occurs when a plate composed of oceanic crust subducts beneath another plate that is also composed of oceanic crust. (b) Volcanic arc volcanism occurs when oceanic crust subducts beneath continental crust.

(?) How will magma chemistry differ between an island arc and a volcanic arc?

thought that some LIPs are produced above mantle hotspots when the plume first arrives.

Intraplate volcanism is also characterized by monogenetic fields and, in some cases, by rhyolite caldera complexes. Indeed, intraplate volcanism is marked by a diversity of volcanic types and processes and is the subject of ongoing research by volcanologists.

FIGURE 6.24 illustrates the three tectonic settings we have just discussed.

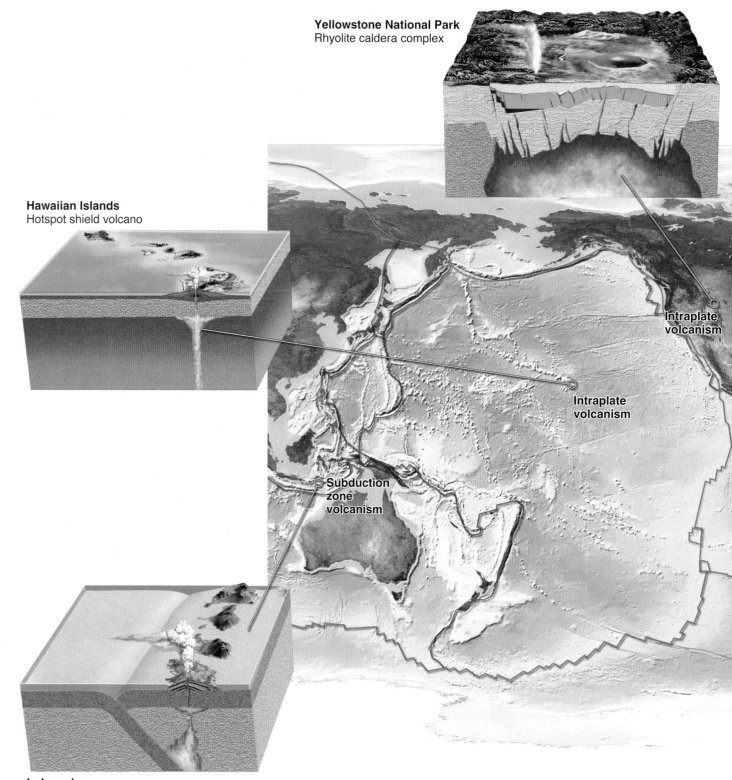

Yellowstone National Park
Rhyolite caldera complex

Hawaiian Islands
Hotspot shield volcano

Intraplate volcanism

Intraplate volcanism

Subduction zone volcanism

Indonesia
Island arc stratovolcano

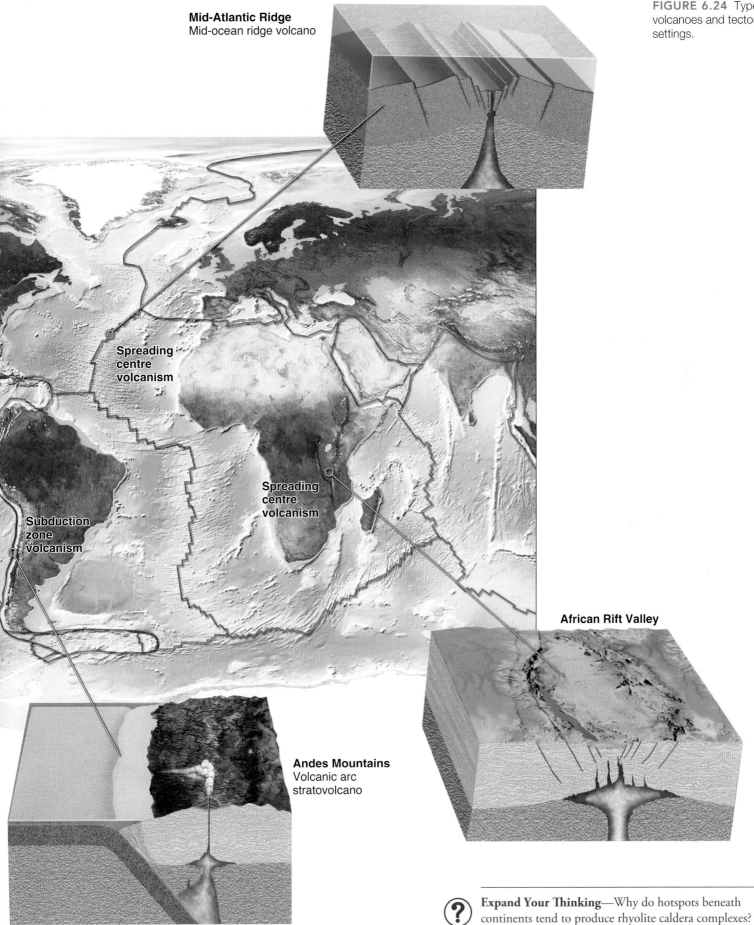

Mid-Atlantic Ridge
Mid-ocean ridge volcano

FIGURE 6.24 Types of volcanoes and tectonic settings.

Spreading centre volcanism

Spreading centre volcanism

Subduction zone volcanism

African Rift Valley

Andes Mountains
Volcanic arc stratovolcano

Expand Your Thinking—Why do hotspots beneath continents tend to produce rhyolite caldera complexes?

6-10 Volcanic Hazards Threaten Human Communities

LO 6-10 *List and define several types of volcanic hazards.*

We know that volcanoes are dangerous. They are dangerous for the obvious reasons that they explode and because their lava and pyroclastic flows burn, bury, and destroy anything in their path. But there are other kinds of volcanic hazards that can endanger life and threaten property (**FIGURE 6.25**).

FIGURE 6.25 Geologists measure a ground crack and monitor gas chemistry at the base of an active volcano.

Courtesy of Dr. Glyn Williams-Jones, Simon Fraser University

(?) What kind of gases are emitted by volcanoes?

Volcanic Hazards

Volcanoes tend to be seismically active. That is, they are places where rapid movement of magma through the ground breaks rock and causes earthquakes that can damage buildings, cause landslides, and open cracks and chasms in the ground. The ground will tilt, swell, and fracture as magma migrates beneath the surface from one location to another. Volcanologists place networks of tiltmeters on active volcanoes and use GPS measurements to track the changes in the surface of the volcano, which might indicate that magma is moving and a magma chamber may be growing, both of which may indicate that an eruption is imminent.

Many arc volcanoes exhibit catastrophic eruptive styles (because of their viscous, gas-rich andesitic and rhyolitic magmas), ejecting hot rock particles, ash, noxious gases, and needle-thin shards of glass. Even volcanoes at mid-ocean ridges, hotspots, and ocean–ocean convergent margins, which are characterized by more fluid, basaltic magma, can erupt explosively when they encounter groundwater, causing phreatomagmatic eruptions. Also, because so much of the material in massive shield volcanoes is rapidly chilled lava (i.e., glass), the mechanical strength of these volcanoes is poor; they fracture easily and frequently, producing large landslides.

As an eruption starts, gas-charged magma moves through the conduit system toward the surface. As it moves, the confining pressure decreases, allowing dissolved gases to be released in much the same way

StockTrek/Photodisc/Getty Images, Inc.

FIGURE 6.26 The buoyant plume of gas and ash formed by a Plinian eruption column can collapse around its outer edges and produce an avalanching, gas-charged cloud of ash, a pyroclastic flow that rolls down the slopes of the volcano.

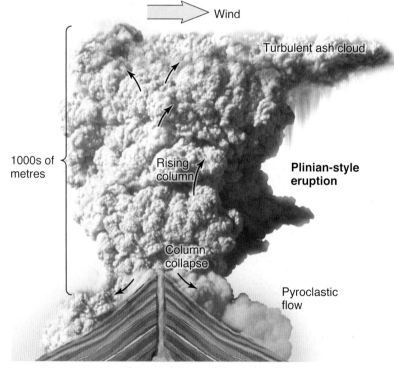

Wind

Turbulent ash cloud

1000s of metres

Rising column

Plinian-style eruption

Column collapse

Pyroclastic flow

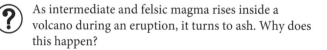
Stratovolcano

(?) As intermediate and felsic magma rises inside a volcano during an eruption, it turns to ash. Why does this happen?

FIGURE 6.27 Lahars from the slopes of Mayon Volcano in the Philippines buried residents and farms in Abay Province on December 2, 2006.

 What volcanic product mixes with water to make a lahar?

that the carbon dioxide in a soda drink is released when its container is opened. Where these conditions occur in basaltic magma, the eruption is characterized by fire fountains hundreds of metres high (similar to the spray from a shaken soda drink). When these conditions occur in the silica-enriched magmas that characterize arc volcanoes, the resulting eruption will be a cataclysmic Plinian-style eruption.

As mentioned earlier, in a Plinian eruption, the outer edge of the column of erupted gas and ash can collapse downward to create a moving avalanche of hot, gas-charged ash and glass (**FIGURE 6.26**). The resulting pyroclastic flow rolls with the speed of an express train down the slopes of the volcano. Anything in the way of this turbulent flow will be instantly burned and buried in thick layers of hot ash. Plinian eruptions are among the most devastating and deadly events in nature.

Heavy rain, tremors due to the movement of magma, or over-steepening of a slope due to internal magma intrusions can trigger slope failure. A common deadly hazard on stratovolcanoes is a *lahar* (an Indonesian word denoting any type of flow but used by volcanologists to describe ash flows on volcanoes). Lahars are rapidly flowing mixtures of rock, mud, and water on the slopes of a volcano (**FIGURE 6.27**). They can form in a variety of ways—because of rapid melting of snow and ice, intense rainfall on loose volcanic deposits, or the breaking through of a lake dammed by volcanic deposits, or as a consequence of avalanches. Lahars are deadly events that bury everything in their path.

Volcanoes and Humans

 As mentioned at the beginning of the chapter, scientists estimate that the total number of people at risk from volcanoes is at least 500 million. Volcanic activity can cause thousands of deaths, destroy towns, farmlands, and forests, and severely disrupt local economies for periods of months to years.

Clearly, volcanic eruptions are among Earth's most dramatic and violent agents of change. Not only can powerful explosive eruptions drastically alter land and water for tens of kilometres around a volcano, but tiny volcanic particles erupted into the stratosphere can temporarily change our planet's climate.

Eruptions often force people living nearby to abandon their land and homes, usually forever. Those living farther away are likely to avoid complete destruction, but their communities, crops, industries, transportation systems, and electrical grids may be damaged by explosively ejected tephra, lahars, earthquakes, pyroclastic flows, and landslides. Despite our improved ability to identify hazardous areas and warn of impending eruptions, increasing numbers of people are exposed to volcanic hazards. Scientists face a formidable challenge in providing reliable and timely warnings of eruptions to so many people at risk.

Volcanic Risk

As we have seen, the highest gas content is found in silica-rich magma. Higher gas content is associated with explosive eruptions, allowing us to create a ranking of volcanic explosivity (**FIGURE 6.28**). High-silica volcanoes (stratovolcanoes, rhyolite caldera complexes) are the most dangerous. But keep in mind that every volcano is dangerous: Long-quiet volcanoes have been known to release killer gases that seep across the countryside at night; passive shield volcanoes can suddenly explode when magma encounters groundwater; and lava moving downhill is an unstoppable force. Appreciate volcanoes for their might and majesty, respect them for their immense power and unpredictability, and avoid them because they are dangerous.

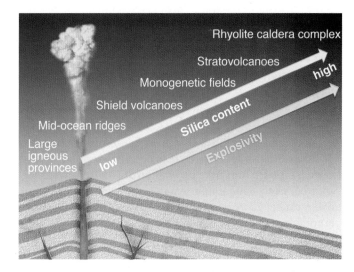

FIGURE 6.28 The higher the silica content of its magma source, the more likely a volcano is to erupt explosively.

 Why is silica content a good guide to the explosivity of a volcano?

 Expand Your Thinking—Consider the various types of volcanoes in Canada, particularly in British Columbia. Do an Internet search to find out about some of the different locations and volcanoes. Should residents of British Columbia be worried about explosive volcanic eruptions? Why or why not?

LET'S REVIEW "GEOLOGY IN OUR LIVES"

Now that you have finished the chapter, "Geology in Our Lives" will have taken on new meaning for you. Let us review it: A volcano is any landform that releases lava, gas, or ash or has done so in the past. Volcanoes are a critical component of the rock cycle. They rejuvenate the crust, renew soil, impact global climate, and contribute to gases in the atmosphere. However, eruptions may consist of violently explosive columns of gas and ash that penetrate high into the atmosphere, lava flows that burn and bury whatever they encounter, and avalanching pyroclastic flows of scalding fumes and glass shards that move extremely rapidly and catastrophically down volcanic slopes. To avoid volcanic hazards and utilize volcanic resources, it is important to understand how volcanoes work.

In general, three types of magma are found in and around the world's volcanoes: basaltic, andesitic, and rhyolitic. The variation in silica content among these is an important factor that governs magma gas content. High-silica magma is viscous and tends to prevent gases from escaping; low-silica magma is less viscous and allows gases to escape to the atmosphere, preventing the buildup of pressure. Gas content and therefore pressure determine whether an eruption will be effusive or explosive.

We identified two main categories of volcanoes: central vent volcanoes and large-scale volcanic terrains. Shield volcanoes, stratovolcanoes, and rhyolite caldera complexes are characterized by increasing amounts of silica and therefore increasing amounts of explosivity, with caldera complexes being the source of the largest explosions in Earth's history. Monogenetic fields, large igneous provinces, and mid-ocean ridges are large-scale volcanic terrains. Most volcanoes are associated with spreading centre volcanism, arc volcanism, or intraplate volcanism.

STUDY GUIDE

6-1 A volcano is any landform that releases lava, gas, or ash or has done so in the past.

- **Volcanoes** occur throughout Earth's oceans and continents. Although they are magnificent and awesome, these natural phenomena exhibit distinctive behaviours for reasons that we can understand through application of scientific thinking.

- It is important to improve our understanding of the behaviour of volcanoes in order to protect people who live near volcanoes. In addition, volcanoes provide information about other important natural processes, such as plate tectonics, climate change due to volcanism, and earthquakes related to the movement of magma.

- Since 1700, volcanic activity has killed more than 260,000 people and destroyed entire cities. Even though geologists have greatly improved abilities to identify hazardous areas and warn of impending eruptions, increasing numbers of people face certain danger. Today the population at risk from volcanoes is approximately 500 million. As a result, scientists face a formidable challenge in providing reliable and timely warnings of eruptions.

6-2 There are three common types of magma: basaltic, andesitic, and rhyolitic.

- A magma's silica content (SiO_2) exerts a fundamental control on the behaviour of **lava** through silicate tetrahedron bonding. Silica content is often related to gas content, and the presence of gases in a magma source can result in massive eruptions and produce pyroclastic debris.

- Three types of magma are commonly associated with volcanoes: **basaltic**, **andesitic**, and **rhyolitic**. Basaltic magma normally erupts without explosion. The lava is very fluid and flows out of the volcano with ease. Andesitic magma tends to contain more gas and is viscous (resistant to flow). It is often expelled in an **eruption column** as part of an **explosive eruption**. Rhyolitic magma is so viscous and gas rich that it usually creates massively explosive eruptions that devastate the nearby countryside.

6-3 Explosive eruptions are fuelled by violent releases of volcanic gas.

- The gases expelled in an explosive eruption include carbon dioxide, water vapour, sulphur dioxide, and others.

- Magma bodies rise in the crust until they reach a point of neutral buoyancy. As gas expands and magma moves closer to the surface, the lowering of pressure and expansion of gases drive eruptions. The interaction among the viscosity, temperature, and gas content of magma determines whether an eruption will be explosive or **effusive** (characterized by relatively fluid lava flow).

6-4 Pyroclastic debris is produced by explosive eruptions.

- **Pyroclastic debris** is another volcanic product that is usually associated with explosive eruptions or with fire fountains that develop in gas-rich magma that does not discharge explosively. It consists of ash, lapilli, blocks, and bombs, all of which are called **tephra** if they are expelled into the atmosphere. These may cool and solidify on the ground, forming a rock called **tuff**. A **pyroclastic flow** can occur when the base of an eruption column collapses and rushes down the slopes of the volcano like an avalanche and burns and buries everything in its path in hot ash and gases.

6-5 Volcanoes can be classified into six major types based on their shape, size, and origin.

- Geologists distinguish between two general classes of volcanoes: **central vent volcanoes** and **large-scale volcanic terrains**. Within each class, they have identified three types of volcanoes. Each type has a distinctive shape and size (morphology), behaviour, and chemistry.

6-6 Shield volcanoes are a type of central vent volcano.

- **Shield volcanoes** are fed by low-silica, low-gas magmas originating in the mantle. These produce basaltic lava in the form of pahoehoe and aa flows that are fluid and rarely explosive. Because of their low viscosity, basaltic lavas cannot hold a steep slope, and the resulting volcano morphology is broad and gentle. Occasionally, explosive **phreatomagmatic eruptions** occur as a result of sudden contact with groundwater.

6-7 Stratovolcanoes and rhyolite caldera complexes are central vent volcanoes.

- **Stratovolcanoes**, or composite volcanoes, consist of alternating andesitic lava flows and layers of explosively ejected pyroclastic deposits. The chemistry of the magma source is intermediate, making the lava viscous and difficult to erupt. The result is explosive eruptions due to the buildup of gases within the magma.
- **Rhyolite caldera complexes** are the most explosive of Earth's volcanoes, typically leaving only gaping holes in the crust as a record of their behaviour. High-silica, high-gas magmas lead to massive explosions followed by collapse of the volcanic edifice, producing an inverse volcano.

6-8 Large-scale volcanic terrains lack a central vent.

- Poorly understood by geologists, **monogenetic fields** are collections of hundreds of separate **vents**, lava flows, and cinder cones that are thought to be fed by a single magma source.

- **Large igneous provinces** are composed of **flood basalts** fed by massive mantle plumes. These are characterized by especially fluid basaltic lavas discharged over time, leading to the creation of large plateaus of basalt.
- Mid-ocean ridges develop at spreading centres and are characterized by basalt flows and the creation of young crust.

6-9 Most volcanoes are associated with spreading centre volcanism, arc volcanism, or intraplate volcanism.

- Plate tectonics can help explain many aspects of volcanoes. In general, volcanoes are found in three types of plate settings. Spreading centre volcanism is characterized by fluid basaltic magma that occurs at divergent plate margins; arc volcanism is characterized by explosive rhyolitic, andesitic, and basaltic magma that occurs at convergent margins; and intraplate volcanism is characterized by shield volcanoes, rhyolite caldera complexes, and monogenetic fields.
- Arc volcanism occurs in places where two plates converge. One plate, containing oceanic crust, descends beneath the adjacent plate, which can be composed of either continental or oceanic crust. If the overriding plate is oceanic, an island arc will develop. If the overriding plate is continental, a volcanic arc will develop.
- Geologists hypothesize that some hotspots are fed by a mantle plume with a source near the core-mantle boundary.
- Because both convergent settings and intraplate hotspots are characterized by partial melting, there is ample opportunity to produce high-silica magma and explosive volcanism.

6-10 Volcanic hazards threaten human communities.

- Volcanoes are extremely hazardous. Pyroclastic flows, lahars, ash falls, volcanic bombs, and massive lethal violent explosions are all sources of danger to humans and their possessions.

KEY TERMS

andesitic magma (p. 134)
basaltic magma (p. 134)
calderas (p. 145)
central vent volcanoes (p. 132)
cinder cones (p. 142)
craters (p. 145)
effusive eruptions (p. 133)
eruption columns (p. 138)
explosive eruption (p. 133)
flood basalt (p. 148)

large igneous provinces (LIPs) (p. 148)
large-scale volcanic terrains (p. 132)
lava (p. 134)
monogenetic fields (p. 148)
phreatomagmatic eruption (p. 142)
Plinian eruption (p. 145)
pyroclastic debris (p. 134)
pyroclastic flows (p. 138)
rhyolite caldera complexes (p.145)
rhyolitic magma (p.134)

shield volcanoes (p. 142)
stratovolcanoes (p. 135)
tephra (p. 136)
tuff (p. 138)
vent (p. 132)
volcano (p. 132)
volcanologists (p. 132)
welded tuff (p. 138)

ASSESSING YOUR KNOWLEDGE

Please answer these questions before coming to class. Identify the best answer.

1. What is the best definition of a volcano?
    a. Any landform that releases lava, gas, or ash.
    b. A large mountain that spews lava, gas, or ash.
    c. Any feature on Earth that emits gas and lava or has done so in the past.
    d. A hole in the ground from which lava escapes.
    e. A layer of ash on sloping ground.

2. Why are hotspots characterized by both explosive and nonexplosive volcanism?
    a. Hotspots always occur near sea water.
    b. Hotspots are never explosive.
    c. Hotspots have high temperatures.
    d. Hotspots may have different types of magma sources.
    e. Hotspots are only explosive if they have a basaltic magma source.

3. Look at the rhyolitic lava flow in Figure 6.5. Why does this flow stand so high above the surrounding land?
    a. The surrounding land was forced down by the weight of the flows.
    b. The flows are composed of low-viscosity lava.
    c. The surrounding land was low, and the flows filled in valleys.
    d. The flows are composed of high-viscosity lava.
    e. It is actually many flows on top of each other.

4. What are the three products of volcanism?
    a. gas, water, and dust
    b. rock, lava, and intrusions
    c. volcanic gas, lava, and pyroclastic debris
    d. pyroclastic debris, fluid lava, and hardened lava
    e. volcanoes, craters, and vents

5. The viscosity of magma increases when
    a. high silica content builds tetrahedral chains.
    b. low silica content dissolves tetrahedral chains.
    c. high gas content breaks down tetrahedral chains.
    d. low gas content breaks down tetrahedral chains.
    e. the number of tetrahedra decreases.

6. A Plinian-style eruption is characterized by
    a. phreatomagmatic explosions.
    b. fissure eruptions and high volumes of lava.
    c. a massive eruption column of gas and pyroclastic debris.
    d. a lack of pyroclastic debris.
    e. lava effusion.

7. True or false?
    a. _____Shield volcanoes always occur at convergent margins.
    b. _____The Mid-Atlantic Ridge is a monogenetic field.
    c. _____Few volcanoes actually erupt lava; most erupt granite.
    d. _____A large igneous province may mark the point at which a mantle plume first discharged lava onto the crust.
    e. _____Rhyolite caldera complexes most commonly form in places where continental crust is partially melted, producing a high-silica magma source.

8. A lava dome
    a. is formed by a large pile of cinders in the shape of a cone.
    b. acts like a plug in a vent and hence is usually explosively expelled.
    c. is found only in shield volcanoes over hotspots.
    d. is typically composed of low-silica basalt.
    e. All of the above.

9. The Yellowstone Volcano
    a. is associated with a hotspot under continental crust.
    b. is an example of a rhyolite caldera complex.
    c. is the site of the largest volcanic eruption in known geologic history.
    d. still has an active magma chamber below ground.
    e. All of the above.

10. Name three tectonic settings in which volcanism occurs and the type of volcanism found in each.

Settings	Type
1._____	_____
2._____	_____
3._____	_____

FURTHER RESEARCH

Go online and research one of the volcanoes listed here and then answer the questions that follow. Your instructor may require you to write a report as part of your course.

- Vesuvius, Italy
- Kilauea, Hawaii
- Etna, Sicily
- Mount St. Helens, Washington
- Krakatau, Indonesia
- Tambora, Indonesia
- Novarupta and Amak, Alaska
- Unzen, Japan
- Rotorua, New Zealand
- Yellowstone, Wyoming
- Juan de Fuca Ridge
- East Pacific Rise

1. What is the name of the volcano or eruption that you selected?

2. Describe the volcanic processes occurring in your volcano, and classify the volcano using the classification system presented in this chapter.

3. Is the volcano currently active? If not, when did it last erupt? What is its eruption history?

4. What is its eruption style: explosive, effusive, or some other type? Please elaborate.

5. What type of magma is present? How does its chemistry influence the volcanic processes occurring in this volcano?

6. What is the tectonic setting?

7. Is the volcano near a human settlement? How is this significant?

8. Describe unique and important features of the activity or eruption.

9. Find additional information online about your volcano.

10. Turn in your report, word-processed (not handwritten) with a title page, illustrations, reference sources, and other research information. Your instructor will tell you the proper format.

ONLINE RESOURCES

Explore more about volcanoes on the following websites:

The NASA Earth Observatory site contains up-to-date information on the latest eruptions:
earthobservatory.nasa.gov/NaturalHazards

Smithsonian Institution Museum of Natural History Glocal Volcanism Program, and check out their fabulous Google Earth map showing the location of all volcanoes in the world:
www.volcano.si.edu/index.cfm

The USGS Cascades Volcano Observatory site is full of activity updates, event reports, and an excellent photo glossary:
volcanoes.usgs.gov/observatories/cvo/

Descriptions of Canadian volcanic areas:
vulcan.wr.usgs.gov/Volcanoes/Canada/description_canadian_volcanics.html
www.volcanodiscovery.com/canada.html

The USGS Hawaii Volcano Observatory site contains excellent information about the Hawaiian hotspot, shield volcanoes, and the ongoing eruption of Kilauea Volcano, the most active volcano on the planet:
hvo.wr.usgs.gov

Additional animations, videos, and other online resources are available at this book's companion website:
www.wiley.com/go/fletchercanada
This companion website also has additional information about WileyPLUS and other Wiley teaching and learning resources.

CHAPTER 7
WEATHERING

Chapter Contents and Learning Objectives (LO)

7-1 Weathering includes physical, chemical, and biological processes.
LO 7-1 *Compare and contrast the three types of weathering.*

7-2 Physical weathering causes fragmentation of rock.
LO 7-2 *Describe several types of physical weathering.*

7-3 Hydrolysis, oxidation, and dissolution are chemical weathering processes.
LO 7-3 *Describe the role of water in chemical weathering.*

7-4 Biological weathering involves both chemical and physical processes; sedimentary products result from all three types of weathering.
LO 7-4 *List four common weathering products, and describe how they are produced.*

7-5 Rocks and minerals can be ranked by their vulnerability to weathering.
LO 7-5 *State the concept behind predicting mineral and rock stability on Earth's surface.*

7-6 The effects of weathering can produce climate change.
LO 7-6 *Describe the relationship between climate and weathering.*

7-7 Weathering produces soil.
LO 7-7 *Compare and contrast the weathering processes in humid tropical and arctic climates.*

7-8 Soil, spheroidal weathering, and natural arches are products of weathering.
LO 7-8 *List and define the typical soil profile layers.*

7-9 Soil erosion is a significant problem.
LO 7-9 *Describe the various ways in which sediments are eroded.*

7-10 There are 10 orders in the Canadian soil classification system.
LO 7-10 *Describe the dominant soil orders in polar soils, temperate soils, desert soils, and tropical soils.*

GEOLOGY IN OUR LIVES

Weathering is the physical, chemical, and biological degradation of the crust. Weathering produces soil, which yields much of our food and the ores of critical metals, such as aluminum and iron. Almost everything humans need can be traced back to weathering: food, clothing, paper, timber, medicines, shade, oxygen, and much more. But erosion caused by human activities destroys soil and other weathering products faster than they can be replaced, and deforestation destroys soil so that natural ecosystems can rarely recover on their own. Understanding weathering enables people to conserve geologic resources and protect natural environments.

? The Alberta Badlands are the product of weathering. What weathering processes contributed to the development of these remarkable rock formations?

7-1 Weathering Includes Physical, Chemical, and Biological Processes

LO 7-1 *Compare and contrast the three types of weathering.*

In this chapter, we describe **weathering,** a set of physical, chemical, and biological processes that break down rocks and minerals in the crust to create sediment, new minerals, soil, and dissolved ions and compounds (**FIGURE 7.1**). Sediment produced by weathering consists of broken pieces of rock, mineral grains released from the crust, and new minerals that crystallize during the weathering process or later from the dissolved compounds produced by weathering.

Weathering produces sediment, and **erosion** transports it across the landscape. Streams, rivers, waves, gravity, and wind are responsible for the erosion that carries sediments to their ultimate resting places, known as *environments of deposition* (which we will study in Chapter 8, "Sedimentary Rock"). It is in depositional environments, such as the ocean, that sediments are deposited and buried over time. Eventually, the sediments are *lithified* (compacted and cemented) into sedimentary rocks and enter the rock cycle.

 An important result of weathering is the formation of **soil.** Soil is unconsolidated mineral and organic material forming the uppermost layer on Earth's surface. It serves as a natural medium for the growth of land plants; hence, it is an essential resource. Almost everything we need can be traced back to soil: food, clothing, paper, timber, medicines, shade, oxygen, and much more. But eroded soil caused by human activities is a major pollutant of rivers and streams. It takes many centuries to form a few centimetres of soil, but only minutes to erode it away. Fortunately, with careful management, sustainable farming practices, and an understanding of weathering and erosion processes, soil can be conserved for future generations.

Physical weathering occurs when rock is fragmented by physical processes that do not change its chemical composition. It involves the mechanical breakdown of minerals and rocks by a variety of processes. If you wanted to "weather" a piece of paper using mechanical force, you would tear it into pieces by physically ripping it apart.

Biological weathering occurs when rock disintegrates due to the chemical and/or physical activity of a living organism. The types of organisms that cause weathering range from bacteria to plants and animals. A piece of paper would be biologically "weathered" by a mouse nibbling at it and by mould growing on it, causing the paper to decay.

Chemical weathering is the chemical decomposition of minerals in rock. This process may result in the formation of any or all of the following:

1. new **sedimentary minerals** (minerals formed by weathering),

2. compounds that are dissolved in water, and

3. gases that escape to the atmosphere, are dissolved in water, or trapped in cavities in soil and sedimentary deposits.

If you wanted to chemically "weather" a piece of paper, you could burn it, producing gases and a residue of ash.

Chemical, physical, and biological weathering work together to break down the crust. Chemical weathering is generally the most effective of the three processes in that it attacks all surfaces that are exposed to gases and fluids. Physical and biological weathering cause rock to fragment into particles, thus increasing the surface area that is vulnerable to chemical weathering (**FIGURE 7.2**). The effectiveness of chemical weathering is greatly enhanced by mechanical and biological processes.

In Figure 7.2, the effect of increasing the amount of exposed surface area is illustrated on the large block on the left, which has a surface area of 6 m^2. When it is divided into eight equal cubes, the surface area doubles, to 12 m^2; when it is divided into 64 equal cubes, the surface area doubles again, to 24 m^2. Continuing this process for 10 sequential halvings produces over 1 billion cubes, each 1 mm on a side (the size of a grain of sand). The combined surface area of these cubes is 1,000 times greater than that of the original cube. With over 1,000 times more surface area to act on, chemical weathering becomes 1,000 times more effective at destroying the rock.

Is the Moon's surface weathered? See the "Geology in Action" box titled "Space Weathering" to find out.

(?) Describe the weathering evident in this photograph.

© Craig Aurness/Corbis

FIGURE 7.1 Physical, chemical, and biological weathering cause rocks and minerals to decompose.

(?) Expand Your Thinking—Nearly everything gets weathered when left exposed to the elements (including your skin, which you'll begin to notice when you get older). Describe a common example of weathering that you find outside your building.

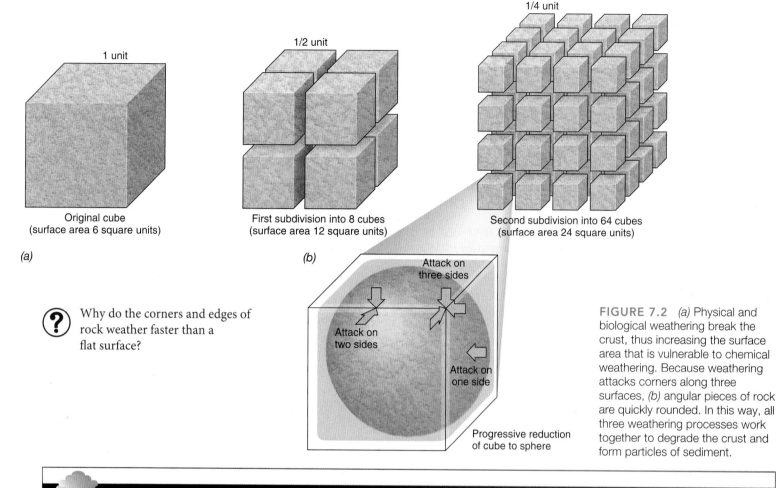

Original cube
(surface area 6 square units)

First subdivision into 8 cubes
(surface area 12 square units)

Second subdivision into 64 cubes
(surface area 24 square units)

(a)

(b)

Attack on three sides

Attack on two sides

Attack on one side

Progressive reduction of cube to sphere

(?) Why do the corners and edges of rock weather faster than a flat surface?

FIGURE 7.2 *(a)* Physical and biological weathering break the crust, thus increasing the surface area that is vulnerable to chemical weathering. Because weathering attacks corners along three surfaces, *(b)* angular pieces of rock are quickly rounded. In this way, all three weathering processes work together to degrade the crust and form particles of sediment.

GEOLOGY IN ACTION

SPACE WEATHERING

On Earth's surface, weathering of exposed rocks is caused by the action of water, natural gases, and life. Where there is little atmosphere, such as on the Moon or asteroids, a different kind of weathering, called *space weathering*, occurs. *Dr. Bruce Hapke*, a professor at the University of Pittsburgh, was the first scientist to correctly identify the process of space weathering.

The Moon's surface is bombarded by meteorites of all sizes, ranging from objects the size of mountains to tiny particles smaller than a grain of sand (called *micrometeorites*). These hit the Moon at speeds of over 16 km/s. In addition to making craters, the largest of which can be seen through telescopes, high-velocity impacts break up exposed rock into a fine powder referred to as *regolith*. (Regolith is the blanket of soil and loose rock fragments that overlies the bedrock; it is found on Earth as well as the Moon.) The entire lunar surface is covered with regolith, which can be seen in pictures and videos taken by astronauts who have landed on the Moon (**FIGURE 7.3**).

However, there is more to space weathering than the mere smashing of rocks. Material that is exposed to space darkens over time. If a Moon rock is ground into a fine powder similar to regolith, the powder will be considerably darker than the original rock. This darkening is caused by thin layers of vaporized rock that coat the surfaces of grains of regolith. Freshly exposed rock surfaces on Earth also tend to darken with time, although they do so by chemical weathering, not space weathering.

FIGURE 7.3
Space weathering has produced the layer of weathered debris called regolith that covers the Moon's surface.

© Eugene Cernin/Corbis

On the Moon, rocks can be vaporized by two processes: meteorite impacts and radiation from the Sun. Enough heat is generated by high-speed impacts to melt and evaporate some particles of regolith. When the solar wind hits the lunar surface at nearly 500 km/s, it knocks atoms off the rock surface in a process known as *sputtering*. Both evaporated and sputtered material coat surrounding grains of regolith, making them darker.

All lunar rocks contain iron, often combined with oxygen. However, the processes of evaporation, sputtering, and redeposition of the rock vapour cause some oxygen atoms to be lost, leaving the iron behind. The leftover iron forms tiny grains of metal less than one-millionth of a centimetre across. These tiny grains of metallic iron are very efficient absorbers of light. Hence, any object coated with these grains turns dark. If it were not for these fine iron particles, the Moon would be about three times brighter than it is.

7-2 Physical Weathering Causes Fragmentation of Rock

LO 7-2 *Describe several types of physical weathering.*

Physical weathering takes many different forms. Keep in mind that physical weathering plays the crucial role of increasing the surface area of rocks, making them more vulnerable to chemical decomposition. It also changes the appearance of Earth's surface as a result of several processes: pressure release, abrasion, freeze-thaw, hydraulic action, salt-crystal growth, and others.

Pressure Release

Rock is brittle, and it breaks when overlying pressure is released by tectonic processes or by erosion. This *pressure release* leads to the growth of fractures, known as **joints.** Joints are openings or "partings" in rock where no lateral displacement has occurred. If lateral displacement has occurred, the fracture is termed a *shear fracture or fault.* (See Chapter 11, "Mountain Building," where we discuss the process of faulting.) Sets of joints composed of repeating parallel fractures may occur, exposing large areas of crust to weathering (**FIGURE 7.4**).

Sheeted joints develop when rock is slowly uplifted by tectonic forces or by the removal of overlying layers by erosion. As the weight of overlying rock is released, the crust expands and fractures into flat horizontal slabs. *Exfoliation* occurs when these slabs shift and uncover the underlying rock. Unicorn Horn, a peak capping a smooth 200 m cliff in the Adamant Range of the Selkirk Mountains of British Columbia is part of a large exfoliated dioritic batholith (Adamant pluton) that was exposed by glacial erosion. The removal of the overlying weight by glacial scouring, as well as the melting of the glacier itself, resulted in sheeted joints and exfoliation (**FIGURE 7.5**).

FIGURE 7.4 Sets of joints composed of repeating parallel fractures expose the crust to weathering.

(?) How do joints affect the rate of rock weathering and why?

Abrasion

Abrasion is an important source of physical weathering. Abrasion occurs when sedimentary particles collide, leading to mechanical wearing or grinding on their surfaces. This happens when small particles of rock carried by wind, water, or ice collide with larger rocks. Blowing wind and running water are usually laden with suspended particles that abrade any surface they encounter. (This process is often referred to as "sandblasting.") Abrasion creates sand

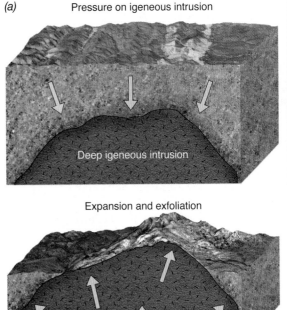

(a) Pressure on igneous intrusion

Deep igneous intrusion

Expansion and exfoliation

Exhumation and erosion

FIGURE 7.5 When overlying weight is removed, *(a)* rock expands and forms sheeted joints. *(b)* Exfoliation shapes the landscape in the Adamant Range in British Columbia.

(b)

Courtesy of Dr. Dan Gibson, Simon Fraser University

 The presence of exfoliation is evidence of pressure release. Suppose you observed exfoliation; use inductive reasoning to build a hypothesis that expresses a general principle about pressure release and its effect on rocks at Earth's surface.

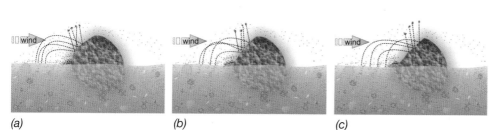

(a)   (b)   (c)

(d)

Mark A. Wilson (Dept. of Geology, College of Wooster)

FIGURE 7.6 Ventifacts are formed when wind-blown particles abrade rock *(a–c)*, creating a flat surface that faces in the direction of the prevailing wind *(d)*.

(?) Draw an arrow on the photograph indicating the wind direction.

and silt, and because abrasion acts quickly, over geologic time, it can cause significant and dramatic alteration of the crust. Rocks with unusual shapes and "fluted" or flat faces may have been abraded by windblown sediment (**FIGURE 7.6**). Such rocks are called **ventifacts.**

Freeze-Thaw

Ice wedging (**FIGURE 7.7**) occurs when water flows into a joint and freezes. Because water increases in volume by 9 percent when it turns into ice, the growth of ice crystals forces the joint to split open. This process plays a major role in weathering the crust in temperate, arctic, and alpine regions. Ice wedging is most effective at –5°C. It, along with gravity, is responsible for the formation of **talus**, slopes of fallen rock that collect at the base of cliffs and steep hillsides.

Hydraulic Action

On rocky shorelines, the powerful force of breaking waves forces water into cracks and fractures in the rock. Air trapped at the bottom of these openings is compressed against the rock and weakens it. As the wave recedes, the pressurized air is released explosively, cracking the rock and expelling fragments of the rock face. This process widens the crack so that more air is trapped by the next wave, leading to a greater explosive force. This hydraulic action damages cliffs and other types of coastal rock outcrops.

Growth of Salt Crystals

When saline water seeps into fractured crust and then evaporates, it may cause the growth of salt crystals that break the rock. Crystal growth usually accompanies evaporation, so for it to occur, heat must

accumulate in the rock during the day. The salt crystals expand as they are heated and exert pressure on the surrounding rock. The most effective salts are compounds of calcium chloride, magnesium sulphate, and sodium carbonate, which are all dissolved in sea water. Some of these salts can expand by three times or more. Salt-crystal growth is common in places where persistent salt-laden winds blow against sea cliffs and other types of rocky shorelines.

Questionable Factors in Physical Weathering

A number of other physical weathering phenomena are subjects of ongoing debate. For instance, weathering by expansion and contraction due to daily changes in rock temperature, known as *insolation weathering* (insolation refers to solar radiation), has been one of the most intensely disputed topics in research on rock weathering. It was formerly thought by researchers that the continual heating and cooling of a rock—for example, on the floor of a desert—would cause the rock to first expand and then contract. This repeated swelling and shrinking would lead to cracking and breakage of the outer surface. However, repeated laboratory experiments have failed to show this process at work, and scientists therefore question the role of insolation weathering as an important process.

Alternate wetting and drying of rocks, known as *slaking,* is also thought to play a role in weathering. Slaking occurs when successive layers of water molecules accumulate between the mineral grains of a rock. The increasing thickness of the water creates tensional stress, which pulls the rock grains apart. Laboratory research has shown that slaking, in combination with naturally dissolved sodium sulphate found in soil moisture, can disintegrate samples of igneous and metamorphic rock in only 20 cycles of wetting and drying.

FIGURE 7.7 Ice wedging on the steep side of a cliff causes blocks to fall and collect at the cliff's base. The resulting deposit of ice-wedged boulders is called a talus slope.

(?) What will this place look like eventually as frost wedging continues to weather the cliff face over time? What if crustal uplift occurs?

(a)   (b)

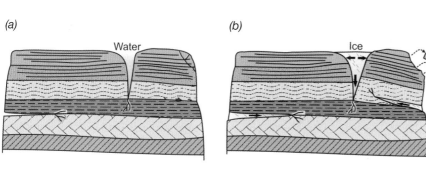

Courtesy of Dr. James MacEachern

(?) **Expand Your Thinking**—Washing clothes makes them look weathered. Is this physical, biological, or chemical weathering? Why?

7-3 Hydrolysis, Oxidation, and Dissolution Are Chemical Weathering Processes

LO 7-3 *Describe the role of water in chemical weathering.*

Most chemical weathering is the result of water interacting with minerals in a rock. Water can be a particularly effective agent of decomposition because of the nature of the water molecule.

The Water Molecule

A water molecule forms when two atoms of hydrogen bond covalently with one atom of oxygen (**FIGURE 7.8**). The bonding involves the sharing of electrons by the hydrogen and oxygen atoms; however, in water, the sharing is not equal. The oxygen atom attracts the shared electrons more strongly than the hydrogen does, and the result is a slightly charged molecule (known as a *polarized molecule*). The two hydrogen atoms tend to be bonded strongly to one side of the oxygen atom (across an arc of ~104°), leaving the other side of the molecule open and able to form more bonds.

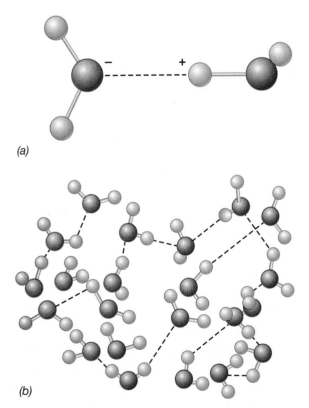

(a)

(b)

FIGURE 7.8 Water molecules *(a)* are polarized because hydrogen (white) bonds to one side of the oxygen (red) while the other side is exposed. *(b)* Water molecules bond to one another by sharing a hydrogen atom. A hydrogen atom of one molecule will bond to the open side of another molecule. When a number of hydrogen bonds act in unison, they create a strong combined effect, forming liquid water.

(?) Does liquid water have a crystalline structure?

Close to the hydrogen atoms, there is a partial positive charge; elsewhere, across the face of the oxygen atom, there is a partial negative charge. The negatively charged region tends to attract cations (positively charged ions), especially another hydrogen cation. This property also allows water to bond with ions in other molecules, such as metallic cations in most minerals.

By bonding with, and removing cations from, a solid mineral surface, water is highly effective at dissolving many substances, earning its title "the universal solvent." Water therefore plays a role in most chemical weathering reactions.

Hydrolysis

Hydrolysis is the most common chemical weathering process because it causes the decomposition of silicate minerals, the most common inorganic substances in Earth's crust. Many minerals are decomposed at least partially by hydrolysis (**FIGURE 7.9**). In this reaction, ions in a mineral react with hydrogen (H^+) and hydroxyl (OH^-) ions in water. The hydrogen ions replace some of the cations in the mineral, thereby changing the mineral's composition. In addition, other ions and compounds may be dissolved from the mineral and carried away in the water.

Hydrolysis is not very effective in pure water, but when some acid (dissolved H^+) is added, it becomes an important agent of weathering. Acidic rainwater forms naturally when carbon dioxide in the atmosphere (or the ground) dissolves in water to produce carbonic acid. Carbonic acid is formed by this reaction: CO_2 (gas in the atmosphere or ground) + H_2O (water vapour in the atmosphere or ground) = H_2CO_3 (carbonic acid; usually precipitates as moisture).

Hydrolysis breaks down feldspar, the most abundant mineral in Earth's crust, and creates **clay**, the most abundant sediment. Hydrolysis is the primary method by which some clay minerals, the primary minerals in the sedimentary rocks *shale* and *mudstone*, are formed. Dissolved ions released during the hydrolysis of feldspar may combine to form other sedimentary minerals. The following equation is a hydrolysis reaction involving a common feldspar, orthoclase:

$$2KAlSi_3O_8 + 2H_2CO_3 + 9H_2O \rightleftharpoons Al_2Si_2O_5(OH)_4$$
$$+ 4H_4SiO_4 + 2K^+ + 2HCO_3^-$$

orthoclase + carbonic acid + water [yield] clay (kaolinite) + silicic acid in solution + potassium and bicarbonate ions in solution

In this reaction, the igneous mineral orthoclase decomposes to produce a clay mineral, *kaolinite* (also known as china clay). The physical properties of kaolinite are quite different from those of orthoclase (**FIGURE 7.10**). Unweathered orthoclase is a relatively hard mineral with two directions of cleavage while kaolinite is soft, white, and microscopic.

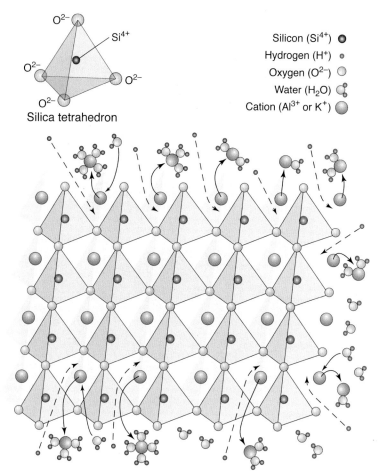

Silicon (Si⁴⁺) — Si^{4+}
Hydrogen (H⁺) — H^+
Oxygen (O²⁻) — O^{2-}
Water (H₂O) — H_2O
Cation (Al³⁺ or K⁺) — Al^{3+} or K^+

Silica tetrahedron

FIGURE 7.9 Because of the polar nature of the water molecule, it will easily bond with cations on a solid mineral surface. The solid mineral is dissolved when the ions are removed by joining with water molecules. In some cases, the cation is replaced with a hydrogen atom.

(?) Why is water such an effective weathering agent?

Oxidation

Oxidation is another important chemical weathering process. Oxidation involves the loss of an electron from a cation in a crystal to free oxygen in the environment. In the process, the oxygen and the cation bond to form a new mineral that is a member of the *oxide mineral class*. Readily oxidized elements include iron, sulphur, and chromium.

Iron is the most commonly oxidized cation. For example, Fe^{2+} (ferrous iron) or Fe^{3+} (ferric iron), when combined with gaseous oxygen (O_2), yields the compound Fe_2O_3, *hematite*, a sedimentary mineral that is an important iron ore. A common example of oxidation is the formation of rusty brown oxides of iron on the surface of iron-containing rocks and minerals (**FIGURE 7.11**). For example, in the olivine variety *fayalite*, iron released by hydrolysis undergoes oxidation to ferric oxide (hematite). The next equations show hydrolysis followed by oxidation.

$$Fe_2SiO_4 + 4CO_2 + 4H_2O \rightleftharpoons 2Fe^{2+} + 4HCO_3^- + H_4SiO_4$$

fayalite + carbon dioxide + water [yield] iron and bicarbonate ions in solution + silicic acid in solution

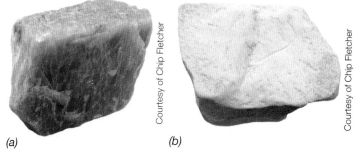

(a)   (b)

FIGURE 7.10 Hydrolysis of orthoclase feldspar *(a)* produces the sedimentary mineral kaolinite *(b)*.

$$2Fe^{2+} + 4HCO_3^- + 1/2\,O_2 + 2H_2O \rightleftharpoons Fe_2O_3 + 4H_2CO_3$$

iron + bicarbonate in solution + gaseous oxygen + water [yield] ferric oxide mineral (hematite) + carbonic acid

Oxidation is accelerated by wet conditions and high temperatures. During the oxidation process, the volume of the mineral structure may increase, usually making the mineral softer and weaker and rendering it more vulnerable to other types of weathering.

Dissolution

Dissolution is a chemical weathering reaction in which carbonic acid dissolves the mineral *calcite,* usually found in limestone (a common sedimentary rock). Dissolution of calcite is similar to hydrolysis except that all the products are dissolved; there is no solid residue unless later precipitation (of the dissolved calcite) occurs at another location (e.g., forming *travertine* in a cave stalagmite). Bicarbonate (HCO_3^-), a product of dissolution, is a major part of the

FIGURE 7.11 Oxidation of iron often leads to a brown and orange coloration in rocks and soil.

(?) What is the common name for iron oxide?

dissolved chemical load of most streams. The next equation shows the dissolution of calcite.

$$CaCO_3 + H_2CO_3 \rightleftharpoons Ca^{2+} + 2HCO_3^-$$

calcite + carbonic acid [yield] dissolved calcium + dissolved bicarbonate

Dissolution of limestone on a large scale can result in a unique kind of landscape called **karst topography.** Karst is formed when crust composed of limestone bedrock experiences widespread dissolution. This occurs when carbonic acid, in the form of groundwater, percolates along joints and bedding planes. As the rock dissolves, large underground caverns are created. This process is called *karstification.* Over time, the caverns grow, coalesce, and undermine an area so that the roof eventually collapses and forms a depression called a **sinkhole** (**FIGURE 7.12**). The destructive effect of carbonic acid is also threatening the vitality of Earth's oceans. Learn more about this worrisome problem in the "Earth Citizenship" box, titled "Ocean Acidification".

 Expand Your Thinking—Why does rinsing your hands with water make them feel clean? Describe the chemical reactions taking place.

 What might happen to a building sitting on the roof of a cavern in a karst landscape?

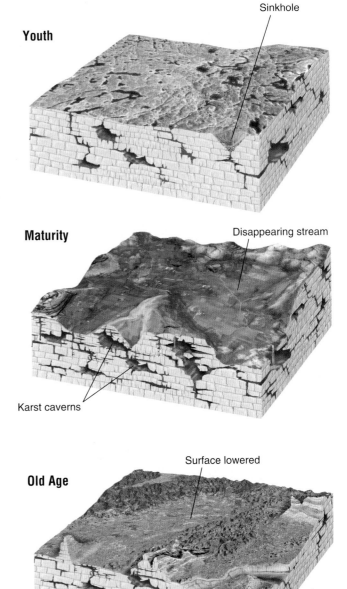

FIGURE 7.12 Limestone experiencing dissolution will develop karst topography. Karst topography progresses through stages labelled "youth," "maturity," and "old age." When acidic groundwater dissolves limestone, valleys, large caverns, sinkholes, and underground streams are created. As karst valleys widen and merge, the crust gradually weathers until eventually the ground surface is lowered and a new level forms.

EARTH CITIZENSHIP

OCEAN ACIDIFICATION

Acid Bath

The same process involved in dissolution, namely the reaction of CO_2 and H_2O to produce H_2CO_3 (carbonic acid), is responsible for *ocean acidification*, a growing danger from global warming that is gaining worldwide attention. Humans release CO_2 to the atmosphere—lots of it by burning coal and oil, making cement, and clearing forests to make farmland. This CO_2 traps heat and leads to global warming. (See Chapter 16, "Global Warming," for more details on the problems of global warming.) However, the ocean absorbs approximately one-third of our CO_2 emissions, leading to another set of problems related to lowering the *pH of the water.*

The pH scale measures how acidic or *basic* a substance is. Acidic and basic are two extremes that describe chemicals, just like hot and cold are two extremes that describe temperature. Mixing acids and bases can cancel out their extreme effects, much like mixing hot and cold water can moderate water temperature. The pH ranges from 0 to 14; a pH of 7 is *neutral*, a pH less than 7 is acidic, and a pH greater than 7 is basic. Each whole number decrease of pH value is 10 times more acidic than a pH of 5 and 100 times (10 times 10) more acidic than a pH of 6.

Earth Citizenship: Ocean Acidificatation

Because the ocean has absorbed CO_2 since the beginning of the Industrial Revolution, the average pH of ocean surface waters has decreased approximately 0.1 unit (from about 8.2 to 8.1), making them more acidic (**FIGURE 7.13**). Research indicates an additional drop of 0.2 to 0.3 by the end of the century, the greatest change in hundreds of thousands of years.

Impacts

Lowering sea water pH with CO_2 affects biological processes, such as photosynthesis, nutrient acquisition, growth, reproduction, and individual survival depending on the amount of acidification. Studies show decreases in shell and skeletal growth in a range of marine organisms, including reef-building corals, commercially important mollusks such as oysters and mussels, and several types of plankton at the base of marine food webs. These members of the food chain are essentially being "weathered" by the very ocean water in which they have evolved to thrive.

Data suggest that there will be ecological winners and losers, leading to shifts in the composition and function of many marine ecosystems. Such changes could threaten coral reefs, fisheries, protected species, and other natural resources.

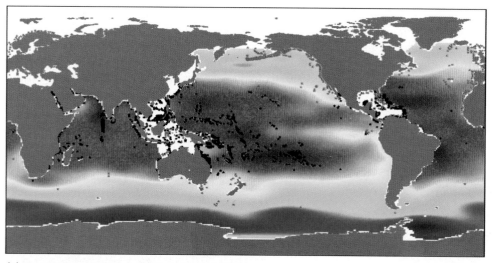

High $CaCO_3$ saturation
(promotes shell growth)

(a) Relative $CaCO_3$ saturation prior to industrialization.

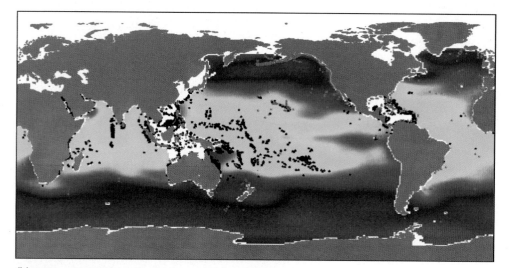

Low $CaCO_3$ saturation
(corosive to shells)

(b) Relative $CaCO_3$ saturation by the end of the century.

FIGURE 7.13 Coral reefs (black and pink symbols) require the right levels of light and temperature and the presence of $CaCO_3$ in sea water. Dissolved CO_2 reduces the $CaCO_3$ saturation of sea water and affects the ability of all organisms to precipitate shells and other skeletal materials. The red/yellow = high $CaCO_3$ saturation (480 percent to 330 percent) and the blue/purple = low $CaCO_3$ saturation (180 percent to 50 percent [corrosive]). As oceans acidify, the ability of organisms such as coral and certain types of plankton to precipitate shells and other skeletal materials decreases. By 2100, surface waters south of 60° S and portions of the North Pacific will become undersaturated (purple [corrosive]) with respect to $CaCO_3$.

7-4 Biological Weathering Involves Both Chemical and Physical Processes; Sedimentary Products Result from All Three Types of Weathering

LO 7-4 *List four common weathering products, and describe how they are produced.*

Biological weathering is the product of organisms causing weathering that is either chemical or physical in character, or a combination of the two. The types of organisms that can cause weathering range from bacteria in soil and rock to plants to animals. Four of the more important biological weathering processes are:

1. *Movement and mixing of materials.* Burrowing organisms (FIGURE 7.14) cause soil particles to turn over, move to new locations, and change depth. This movement can introduce the materials to new weathering processes found at distinct depths under the ground and expose new particle surfaces to attack by chemical weathering.

2. *Simple breaking of particles.* Rocks can be fractured as a result of burrowing by animals or the pressure from growing roots, known as **root wedging** (FIGURE 7.15).

3. *Production of carbon dioxide by animal respiration or organic decay.* Carbon dioxide increases the acidity of water, which then attacks and dissolves minerals and other compounds in rocks. When carbon dioxide combines with water, the resulting carbonic acid (H_2CO_3) is an effective chemical weathering agent causing hydrolysis, dissolution, and oxidation.

4. *Changes in the moisture content of soils.* Organisms influence the moisture content of soils and thus enhance weathering. Shade from leaves and stems, the presence of root masses, and high levels of organic material in soil all increase the amount of water in the soil. This higher moisture content, in turn, enhances physical and chemical weathering processes.

FIGURE 7.14 Burrowing organisms produce carbon dioxide, overturn and mix soil particles, increase porosity, and increase soil moisture content.

FIGURE 7.15 Root wedging.

 Why does "overturning" a soil particle increase its weathering susceptibility? What role does increasing soil moisture play?

How would root wedging and rock joints work together to enhance weathering?

Products of Weathering

Weathering occurs because conditions at Earth's surface are different from the high-temperature and high-pressure conditions prevailing within the crust, where most igneous and metamorphic rocks and minerals form. The products of weathering are new minerals that are in equilibrium with surface conditions (and therefore tend to resist weathering themselves).

Chemical weathering yields *weathering products* that are significant components of the rock cycle and the sedimentary family of rocks. Some of the weathering products are new sedimentary minerals that result from crystallization, such as **sedimentary quartz,** various types of clay, calcite, and hematite. Other weathering products include dissolved compounds and some gases.

Sedimentary quartz, hematite, and calcite are important natural cements that precipitate from groundwater and bind sedimentary particles to form solid sedimentary rock. Clays are sheet silicates that trap water between layers. They have many economic uses in manufacturing, drilling, construction, and paper production. As a significant component of soils, clays are important in crop production and natural ecosystems.

Chemical weathering also produces dissolved ions and compounds, such as dissolved silica, cations, and bicarbonate, which are washed into the oceans and used by marine organisms to build shells (**TABLE 7.1**).

TABLE 7.1 Some Chemical Weathering Processes and Products

Process	Product	Dissolved Material	Example of Weathering Product
Hydrolysis of quartz	Sedimentary quartz (chert, agate, etc.)	Silica SiO_2	Chert
Hydrolysis of orthoclase feldspar	Kaolinite clay	K^+, Silica SiO_2	Kaolinite
Oxidation of olivine	Hematite	$4H_2CO_3$	Hematite
Dissolution of calcite	Dissolved calcite, travertine (product of calcium carbonate precipitation)	$CaCO_3$	Travertine

Images courtesy of Chip Fletcher

Expand Your Thinking—Biological weathering introduces organic material to regolith. When organic material decays, CO_2 is produced. Do you agree that this can greatly increase the rate of chemical weathering? Explain why or why not.

7-5 Rocks and Minerals Can Be Ranked by Their Vulnerability to Weathering

LO 7-5 *State the concept behind predicting mineral and rock stability on Earth's surface.*

It has been estimated that 95 percent of the crust is composed of igneous or metamorphic rock. Even sedimentary rocks are often composed of igneous and metamorphic mineral grains. Having been formed under conditions of high temperature and pressure, these materials are unstable in the low temperature and pressure of Earth's surface. Therefore, they tend to be decomposed by weathering (**FIGURE 7.16**).

Rock and Mineral Stability

Not all minerals decompose at the same rate. Various natural factors control how quickly any mineral will weather. *Climate* and *mineral chemistry* are the primary factors that regulate the rate at which various rocks and minerals are weathered. Climate governs weathering and soil formation directly, through precipitation and temperature, and indirectly, by influencing the kinds of plants, animals, and bacteria found in a given region, which cause biological weathering and influence the chemistry of groundwater. For instance, warm, wet climates promote rapid chemical weathering, whereas cool dry climates promote physical weathering.

A mineral's chemistry determines its vulnerability to specific weathering processes and the degree to which a rock is out of equilibrium with the chemistry of the immediate environment. In fact, mineral chemistry is so important in controlling weathering that it is possible to rank the vulnerability of minerals to weathering based on their chemistry (**TABLE 7.2**).

Table 7.2 ranks some common minerals in order of their *stability,* or resistance to weathering. This ranking generally reflects the fact that many minerals formed at high temperatures and pressures (i.e., mafic and ultramafic minerals), which are the first to crystallize in a cooling magma chamber, are among the least stable on Earth's surface, where the environment differs most from their environment of crystallization. The last minerals to crystallize (i.e., felsic minerals) are among the most stable on Earth's surface, where the environment is most similar to their environment of crystallization. Sedimentary minerals are the most stable of all because they are formed on Earth's surface as a result of weathering processes. These sedimentary minerals include oxides of silicon (SiO_2 [quartz], which comes in many forms, one of which is *chert),* aluminum (Al_2O_3, known as alumina; the main component of the aluminum ore *bauxite),* and iron (Fe_2O_3, known as rust; iron oxide is the mineral hematite). You may not consider rust as stable since it is easy to remove. But rust and other types of oxides will typically not weather. They simply grow with time.

For the silicate minerals, the order of their stability in Earth's surface environments is the reverse order of Bowen's reaction series. That is, the most stable silicates from a weathering perspective are the last to crystallize in a cooling magma chamber, while the least stable silicates are the first to crystallize—the ranking of mineral stability is sometimes referred to as the Goldich stability series.

Based on their mineral content, rocks also have predictable susceptibility to weathering (Table 7.2). Although mineralogy strongly influences the rate of a rock's decay, multiple weathering processes also affect it. Most rocks are vulnerable to some form of decomposition by physical, biological, and chemical weathering processes.

Courtesy of Dan Gibson, Simon Fraser University

(?) What weathering processes are responsible for shaping these cliffs and ridges of diorite? List the minerals that are common to diorite and explain how each is likely to weather.

FIGURE 7.16 Diorite, common in some mountain ranges, is not stable on Earth's surface and tends to weather quickly.

TABLE 7.2 Mineral and Rock Stability on Earth's Surface

Mineral Stability		Stability on Earth's Surface	Rock Stability	
Igneous and Metamorphic Minerals	Sedimentary Minerals		Igneous and Metamorphic Rocks	Sedimentary Rocks
Olivine	Halite	Least Stable	Basalt	Rock salt
Pyroxene	Calcite	↓	Granite	Limestone
Ca-Plagioclase feldspar	Hematite		Marble	Rock gypsum
Biotite	Kaolinite		Gneiss	Siltstone
Orthoclase feldspar	Bauxite	Most Stable	Schist	Shale
Quartz	Chert		Quartzite	Quartz sandstone

Clay Minerals

You are undoubtedly familiar with clay, but you probably do not realize that *clay* is the name of an important group of minerals. Most of us think of clay as mud or small particles of soil. Clay particles are indeed small. They are actually sedimentary minerals that crystallize on Earth's surface. The most common clays are *kaolinite, montmorillonite,* and *illite.*

Clays are *phyllosilicates,* meaning that their crystalline structure consists of layers of silica tetrahedra organized in sheets. Typically, the silica sheets are sandwiched between layers of cations (**FIGURE 7.17**). The sheets are weakly bonded to these layers, and water molecules and other neutral atoms or molecules can be trapped between the sheets. Water and other molecules trapped in the silicate structure cause clay to swell, which has the effect of making it soft to the touch. An example is the clay mineral *talc,* which is used in talcum powder.

Although clays tend to be soft, they are remarkably resilient in surface environments. Members of this mineral group are the last to break down chemically at Earth's surface and hence are abundant in soils and sedimentary rocks. When mixed with water, clays become easily deformable and can be moulded and shaped in ways that most people associate with "potter's clay."

Many everyday objects contain clay in some form. Clays have many uses—for example, in ceramics and as filler for paint, rubber, and plastic. The largest quantities of clay are used by the paper industry, where kaolinite is used to produce the glossy paper found in most magazines. Montmorillonite is used in facial powder and as filler for paints and rubber. Clay is even used in your toothpaste to give it consistency.

(a)

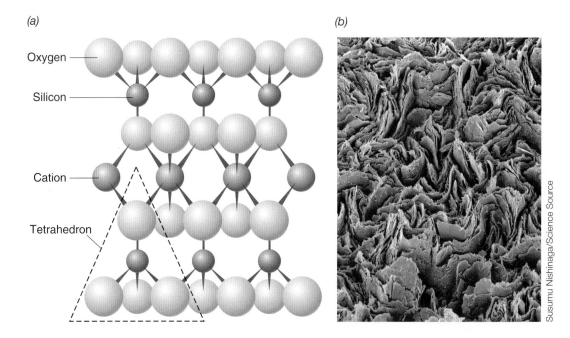

(b)

Oxygen
Silicon
Cation
Tetrahedron

Susumu Nishinaga/Science Source

(?) How is the silicate structure of clay related to its physical characteristics?

FIGURE 7.17 *(a)* Clays are a group of minerals built of layers of silica tetrahedra. Different layers are connected by cations. *(b)* Clay consists of small particles shaped like platelets. Clay has properties of plasticity, compressibility, swelling, and shrinkage that make it useful for industrial purposes.

 Expand Your Thinking—Imagine that climate change causes a formerly arid region with limestone crust to experience greater rainfall over time. How will the weathering processes change? Should people living in the area be concerned? Why?

7-6 The Effects of Weathering Can Produce Climate Change

LO 7-6 *Describe the relationship between climate and weathering.*

Weathering influences the chemistry of the atmosphere, oceans, and rocks. One reason Earth is habitable is that weathering governs the amount of carbon dioxide (CO_2) in the atmosphere, and carbon dioxide influences the average temperature of the air.

Carbon Dioxide

For example, although Venus is closer to the Sun than Earth is, thick clouds of sulphuric acid reflect much of the incoming solar radiation so that, at its surface, Venus receives only half as much solar heat as Earth. Yet the average surface temperature on Venus is 460°C. The temperature is so high because Venus's atmosphere is 96 percent carbon dioxide, a gas that traps heat, analogous to glass panels trapping heat in a greenhouse. In fact, carbon dioxide is known as a **greenhouse gas.** Earth's atmosphere is only 0.2 percent carbon dioxide, a perfect level for keeping the air warm but not too warm. In fact, if the air contained no carbon dioxide, its average temperature would be 31°C cooler, making life as we know it impossible.

How does weathering control this important gas? Carbon dioxide is a major component of the **carbon cycle,** the production, storage, and movement of carbon on Earth. (We discuss the carbon cycle in more detail in Chapter 16.) Over geologic time, carbon dioxide enters Earth's atmosphere through *volcanic outgassing,* the activity of hot springs and geothermal vents, and the oxidation of organic carbon in sediments. Volcanic outgassing alone is sufficient to contribute all the carbon dioxide in our atmosphere in only 4,000 years. This input must be balanced by withdrawal of carbon dioxide from the atmosphere. Otherwise, the amount of carbon dioxide present would have a runaway greenhouse effect similar to that prevailing on Venus. One of the ways in which carbon dioxide is removed from the atmosphere is through chemical weathering of silicate minerals.

As we learned earlier, carbon dioxide in the atmosphere easily combines with water during hydrolysis to form carbonic acid (H_2CO_3). The reaction of H_2CO_3 with continental rocks (represented by $CaSiO_3$) produces $CaCO_3 + SiO_2$ and H_2O. Streams carry these products to the ocean where plankton use them to build shells, which fall to the sea floor when they die. One way these biogenic sediments re-enter the rock cycle is when the lithospheric plate they ride on is subducted and recycled. Magma created by the subduction process migrates toward the crust and eventually can lead to volcanic outgassing of carbon dioxide through a volcanic arc (**FIGURE 7.18**). Through this series of processes, carbon migrates through a complete turn of the carbon cycle.

Over geologic time, the rate of carbon dioxide removal by weathering must approximately equal the rate of carbon dioxide production by volcanoes. One can think of the balance maintained by chemical weathering and volcanic outgassing as a thermostat governing Earth's climate. However, this long-term equilibrium is characterized by a highly variable and dynamic climate in the short term. Ice ages, warm periods, and extreme shifts from so-called hothouse climates to ice-house climates are prevalent throughout Earth history.

The Uplift Weathering Hypothesis

Carbon dioxide is a greenhouse gas that traps heat and plays a significant role in regulating global climate. Volcanic outgassing is an important long-term source of carbon dioxide in the atmosphere, and weathering is one way to *lower* carbon dioxide abundance. The most important

(?) Is the carbon cycle related to the rock cycle? Explain your answer.

FIGURE 7.18 Weathering of rock and volcanic outgassing regulate the amount of carbon dioxide in the atmosphere.

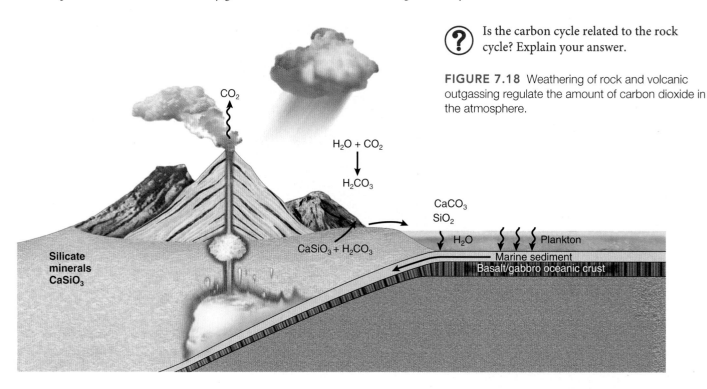

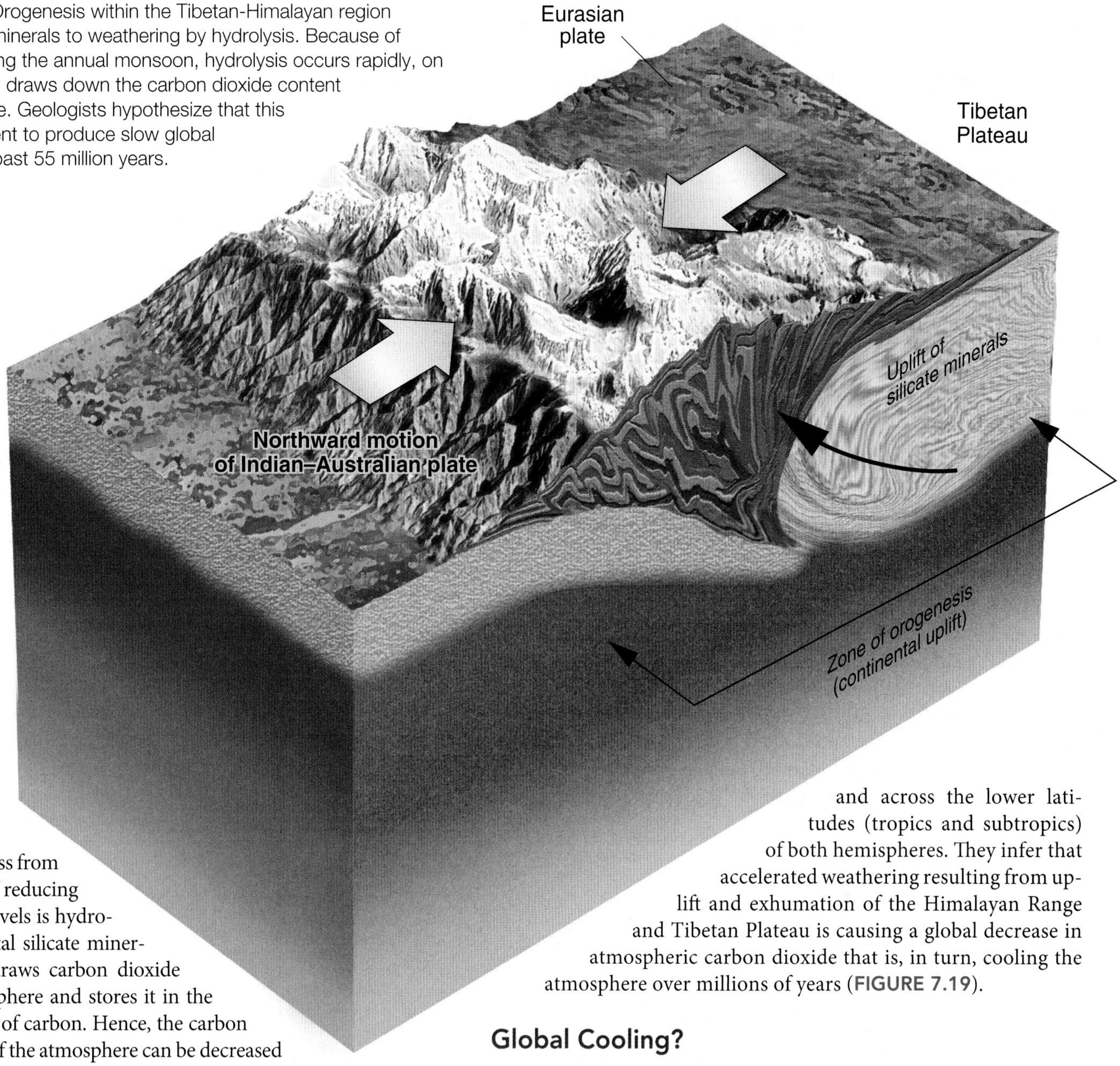

FIGURE 7.19 Orogenesis within the Tibetan-Himalayan region exposes silicate minerals to weathering by hydrolysis. Because of intense rains during the annual monsoon, hydrolysis occurs rapidly, on a large scale, and draws down the carbon dioxide content of the atmosphere. Geologists hypothesize that this process is sufficient to produce slow global cooling over the past 55 million years.

Eurasian plate

Tibetan Plateau

Uplift of silicate minerals

Northward motion of Indian–Australian plate

Zone of orogenesis (continental uplift)

weathering process from the standpoint of reducing carbon dioxide levels is hydrolysis of continental silicate minerals. Hydrolysis draws carbon dioxide out of the atmosphere and stores it in the crust in the form of carbon. Hence, the carbon dioxide content of the atmosphere can be decreased by exposing new crust to weathering.

Orogenesis, the building of mountain ranges by collision of tectonic plates, exposes large areas of fresh silicate rock to hydrolysis, leading to withdrawal of carbon dioxide from the atmosphere. This idea is described in the *uplift weathering hypothesis,* which proposes that hydrolysis is an active agent in regulating global climate change. The hypothesis asserts that the global rate of chemical weathering is dependent on the availability of fresh rock that can be attacked by hydrolysis. In this case, tectonic uplift and exhumation of mountains causes accelerated weathering, leading to drawdown of carbon dioxide and shifts in global climate (cooling).

Scientists have discovered that, starting about 55 million years ago in the Paleogene Period, Earth's atmosphere began a long but steady cooling that continues today and is most evident at the poles and across the lower latitudes (tropics and subtropics) of both hemispheres. They infer that accelerated weathering resulting from uplift and exhumation of the Himalayan Range and Tibetan Plateau is causing a global decrease in atmospheric carbon dioxide that is, in turn, cooling the atmosphere over millions of years (**FIGURE 7.19**).

Global Cooling?

You may find it confusing that global cooling is occurring when the media have had so much to say about global warming. The explanation lies in the time scales of the two trends. Over tens of millions of years, uplift of the Himalayas has cooled the atmosphere, playing a role in promoting several ice ages over the past million years. But more recently—over the past few centuries—pollution of the atmosphere with carbon dioxide and other greenhouse gases from the burning of *fossil fuels* is inferred to have caused the atmosphere to warm. Thus, the uplift weathering hypothesis describes a process that operates over millions of years, while the modern global warming hypothesis describes a process that has been operating for only a few centuries. These two very different processes, both involving carbon dioxide, influence the global climate on vastly different time scales.

Expand Your Thinking—What evidence would you look for to support the hypothesis that global cooling has been taking place over the past 55 million years?

7-7 Weathering Produces Soil

LO 7-7 *Compare and contrast the weathering processes in humid tropical and arctic climates.*

Soil is a critical resource that provides food for the world's population. In fact, almost everything we need can be traced back to soil: food, clothing, paper, timber, medicines, shade, oxygen, and more. Soil serves as a bridge between living things and the inanimate world. The dozen or so elements that are essential for sustaining animal and plant life come largely from soil: iron, magnesium, calcium, sodium, potassium, and others.

Plants and micro-organisms use elements and compounds taken from soil and assimilate them into the food chain, making them available to other plants and animals. Ultimately, the reason your food sustains you day in and day out is that it is imbued with minerals and nutrients that come from soil and, ultimately, from weathering of rock. But soil is much more than a rock-weathering product. It is an ecosystem with a complex network of organisms that interact among themselves as well as with the gas, fluid, and mineral constituents.

Components of Soil

Soil is a more complex substance than most people realize. It is composed of more than just mineral particles or dirt (**FIGURE 7.20**). A true soil is the product of a living environment. It consists of air, water, mineral particles, and organic material. The formation of soil is influenced by biological processes, the nature of the parent rock, climate, topography, and time. But the most important of these factors is *climate*. Climate is weather averaged over a long period, as expressed in the saying "Climate is what you expect, and weather is what you get."

Soil Formation Is Controlled by Climate

Climate is a critical factor not only in weathering but in soil formation as well (**FIGURE 7.21**). In regions where both temperature and rainfall are low (such as in cooler regions or at high elevations), mechanical weathering in the form of ice wedging and wind abrasion is the most important weathering process. Conversely, chemical weathering (hydrolysis, dissolution, and oxidation) is most important in regions where temperature and rainfall are both high, such as the humid tropics.

Hot, Arid Climates

Arid environments, such as deserts, allow little growth of vegetation and provide too little water to permit much chemical weathering. Consequently, these areas develop a special type of soil. Salts accumulate at the surface due to evaporation, and erosion, frost, abrasion, and slaking break down the rocky surface into sand or gravel. Low moisture also means that fine particles are easily blown away while large particles remain behind and form a tightly packed layer known as *desert pavement*. Because of the negligible rainfall, there is little biological activity, and organic matter is scarce. The result is a dry place full of sand, minerals formed by evaporation (such as halite and calcite), and desert pavement.

Cold Climates

Arctic and alpine environments can also be dry because, in these regions, water has turned to snow and ice and is unavailable for chemical weathering. Moreover, the biological activity of plants and

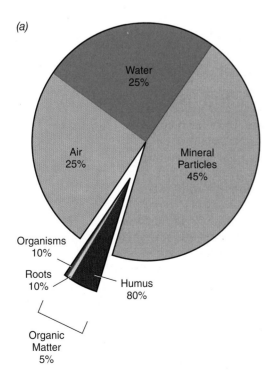

(a)

(b)

FIGURE 7.20 Average soil contains four basic components: mineral particles, water, air, and organic matter. Organic matter can be subdivided further into humus, roots, and living organisms.

(?) Describe the source of minerals that are found in soil.

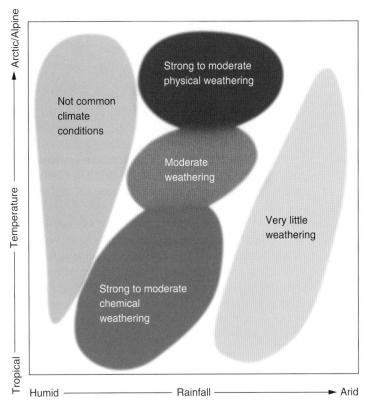

FIGURE 7.21 Temperature and rainfall (climate) control the stability of minerals at Earth's surface. Mineral stability, in turn, governs the rate of soil formation. The intensity of weathering differs across the world because of variations in climate and the geology of the crust.

Would feldspar survive longer in the cold arctic or the humid tropics? Consider how it is affected by weathering.

micro-organisms proceeds slowly in cold climates. Because organic material decays very slowly, it can accrue and form wetlands known as *bogs* that contain thick accumulations of peat known as *muskeg*. Mechanical breakdown (by ice wedging) is the major weathering process affecting the crust in cold climates. However, in the summer, weathering is accelerated by the presence of fluid water, organisms, and the long daylight of the high latitudes.

Hot, Humid Climates

In humid environments such as a tropical rainforest that receives extensive rainfall, the groundwater reaches almost to the surface for most of the year. Deep soils cannot develop, and most minerals and nutrients are stored in the living vegetation of the forest and in a rich, deep layer of leaf litter on the forest floor. Vegetation can grow only if nutrients in this litter or the forest ecosystem become available for new growth—for example, if a tree sheds a leaf that then decomposes or if an animal dies and its body decomposes. Although tropical rainforest ecosystems look very rich, they are fragile and depend on a tightly balanced and rapid exchange of organic material and nutrients from the forest floor into living organisms. Insects, tropical air temperature, and the heavy rainfall play an important role in accelerating the decomposition process.

Humid tropical climates lead to intense chemical weathering that produces soils largely composed of **insoluble residues,** or mineral products of weathering. These residues include thick soil crusts of iron oxide, known as **laterite,** and aluminum oxide, known as **bauxite** (FIGURE 7.22). In forming bauxite, an important ore of aluminum, the metallic cations usually are removed from a parent rock composed of orthoclase feldspar.

(a)

(b)

FIGURE 7.22 Bauxite (a) and laterite (b) are both reddish soils that are important sources of aluminum and iron ore, respectively.

Expand Your Thinking—Some plants such as spinach are a good source of iron for our diet. Where does this iron come from, and why is it available?

7-8 Soil, Spheroidal Weathering, and Natural Arches Are Products of Weathering

LO 7-8 *List and define the typical soil profile layers.*

Humid mid-latitude climates with seasonal freezing allow dead vegetation to accumulate in the soil. This occurs because microbes and insects have only the warm season during which to work on recycling the organic material, and this is not enough time to decompose it all. Hence, the soil surface becomes rich in partially decayed organic plant debris, known as the **humus layer.**

As rain soaks through the humus layer, the water picks up dissolved ions and compounds and removes them by percolation, a process called *leaching.* Leached materials may recombine to form clay minerals, thus creating a set of layers, or *soil horizons.* These horizons are collectively known as the **soil profile** (**FIGURE 7.23**). The thickness of the soil profile depends on the climate and the nature of the parent rock as well as the length of time over which it has formed. Because of the variability of natural settings, as well as disturbance by humans, not all the soil horizons shown in the figure may be present at any given location.

The Soil Profile

The uppermost horizon in the soil profile is the *organic (O) horizon,* or humus layer. It is composed of vegetation debris, leaf litter, and other organic material lying on the surface. This layer is not present in cultivated fields.

Below the humus layer is the *A horizon,* or *topsoil.* Usually it is darker than lower layers, loose and crumbly and containing varying amounts of organic matter. In cultivated fields, the plowed layer is topsoil. It is generally the most productive layer of the soil profile.

The *B horizon,* or *subsoil,* is usually lighter in colour, dense, and low in organic content. Most of the material leached from the A horizon ends up in this zone. In some arid climates, a *K horizon* may develop below the B horizon.

Still deeper is the *C horizon.* It is a transitional area between the soil and the parent rock of the crust. Partially disintegrated parent rock and mineral particles may be found in this horizon. At some point, the C horizon will end and the final *R horizon, bedrock,* will begin. Bedrock is solid sedimentary, igneous, or metamorphic rock beneath the soil.

Climate and the composition of the parent rock control the development of the soil profile. Forested regions with moderate rainfall tend to develop strong O and A horizons with well-developed B and C horizons. Grasslands have a restricted O horizon and a wide topsoil or A horizon.

Spheroidal Weathering

Beneath the soil profile the bedrock often undergoes **spheroidal weathering** (**FIGURE 7.24**). This is a form of exfoliation that occurs on rock characterized by intersecting sets of joints such as shown in Figure 7.4. Spheroidal weathering occurs when rectangular blocks outlined by two or more perpendicular sets of joints are attacked by chemical weathering. Chemical reactions can effectively weather the

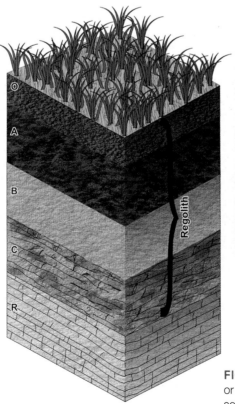

O horizon (humus)
Organic horizon developed mainly from mosses, rushes, and woody materials.

A horizon (topsoil)
Mineral horizon, mixed with some humus. Dark and organic, this is usually the most fertile part of the soil profile.

B horizon (subsoil)
Mineral horizon enriched with clay, organic matter, and iron and aluminum oxides or by in situ weathering.

C horizon
Mineral horizon of partially altered bedrock that represents the transition zone between bedrock and soil.

R horizon (unweathered bedrock)
Consolidated bedrock layer too hard to break by hand (>3 on Mohs' scale) or to dig with a spade.

 What are the primary factors that govern soil development?

© Daniel J. Pennock, University of Saskatchewan

FIGURE 7.23 A typical soil profile consists of a series of layers, or horizons. The O horizon is formed only if a well-developed source of organic matter is present. The topsoil, or A horizon, is most fully developed in grasslands and is very thin in arid regions.

FIGURE 7.24 Rectangular blocks outlined by joints are attacked by chemical weathering. Because corners and edges are weathered most rapidly, the block takes on a spheroid shape.

 When weathering finally makes a sphere, will the bottom of the sphere weather as fast as the upright surfaces? Explain your answer.

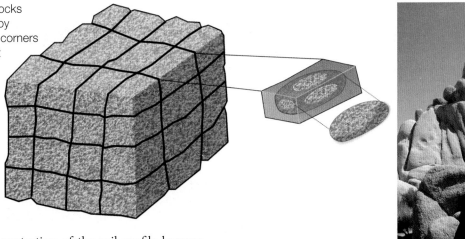

Marli Miller

bedrock surface beneath the protection of the soil profile because water percolating through the soil is chemically enhanced, it does not readily evaporate, and geothermal heat is present to stimulate chemical reactions (such as hydrolysis). Spheroidal weathering also occurs on the surface.

Because corners and edges are formed by two, three, or more intersecting surfaces that are simultaneously attacked by weathering, sharp angles quickly become smooth, rounded surfaces. Some researchers think this process may be enhanced by insolation weathering. The rock disintegrates along *dilation* (expansion) *fractures* that conform to the surface topography. When a block has been weathered so that it is spherical, its entire surface is weathered evenly and no further change in shape occurs (except that it grows smaller).

Natural Arches

Natural arches (**FIGURE 7.25**) are formed when sheeted joints undergo weathering to produce narrow walls of rock called *fins*. Ice wedging causes the exposed rock on each side of a fin to gradually fall away. Eventually the fin is cut through to produce a natural arch. Continued weathering causes the hole in the arch to grow until the arch becomes unstable and collapses.

Arches National Park in Utah has the greatest concentration of natural arches in the world. Its more than 200 arches range in size from 1 m to over 89 m wide. So many arches are present in this area because of the particular conditions there: a well-developed parallel pattern of joints, low amounts of precipitation, and a large fluctuation in daily temperatures, with the result that water freezes and thaws regularly. Hence, ice wedging is the dominant weathering mechanism. Exposed joints are continually weathered into fins that in turn produce a landscape of arches.

To improve your understanding of weathering and soil production, work on the critical thinking exercise in the "Critical Thinking" box titled "Rock Weathering."

 Expand Your Thinking—Why is it important to understand the soil profile?

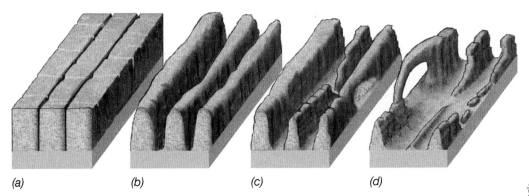

(a)   (b)   (c)   (d)

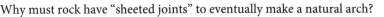

FIGURE 7.25 Sheeted joints (a) weather into a set of fins (b). Exfoliation arches (arches that do not penetrate to the other side) develop in the sides of fins as a result of ice wedging (c). Eventually exfoliation arches break through the fin, producing a window. The window grows into an arch that (d) eventually widens until it is unstable and collapses.

 Why must rock have "sheeted joints" to eventually make a natural arch?

© Bret Edge/Age Fotostock

? CRITICAL THINKING

ROCK WEATHERING

Study **FIGURE 7.26** and note the locations of the lettered sites.

Site A: High elevation; strong winter to summer temperature changes; steep slopes; little vegetation; freezing conditions in winter; granite bedrock.

Site B: Hilly; forested; drained by streams; abundant rainfall; strong seasonal temperature changes; basalt bedrock.

Site C: Gentle slopes; grasslands; mild winters; hot summers; abundant rainfall; occasional drought lasting for several years; limestone bedrock.

Site D: Flat, parched land; little vegetation; occasional intense rainfall; normally arid; intense drought; hot, dry conditions all year; limestone.

Site E: Abundant forest; hot, humid conditions; intense daily rainfall throughout the year; warm climate; granite bedrock.

1. Fill in the following chart describing the conditions at each site.
2. Rank the sites by order from most stable to least stable.
3. Describe the criteria you used for your ranking.
4. Predict the sediment composition produced at each site.
5. Rank the relative significance of physical, chemical, and biological weathering at each site and explain your reasoning.

	General Climate Description	Most Important Climate Factors	Minerals Vulnerable to Weathering	Weathering Process(es)	Weathering Product(s)	Soil Profile Layers	Dominant Erosion Process(es)
Site A							
Site B							
Site C							
Site D							
Site E							

FIGURE 7.26

7-9 Soil Erosion Is a Significant Problem

LO 7-9 *Describe the various ways in which sediments are eroded.*

Active movement of particles by running water is called **fluvial erosion.** (The term "fluvial" refers to water running in a *channel,* the subject of Chapter 19, "Surface Water.") Other forms of erosion include *mass wasting* (studied in Chapter 18, "Mass Wasting'), the movement of sediment by gravity—as occurs, for example, in a landslide or mudflow—and aeolian erosion (Chapter 21, "Deserts and Wind"), in which sediment is moved by wind. (*Aeolus* was the ancient Greek god of wind.)

Erosion by Water

Fluvial erosion (**FIGURE 7.27**) occurs when particles of sediment are removed by flowing water in a channel. Water carries sediment in one of two ways: *suspended load* or *bed load.*

Suspension is a process in which sediments are kept in a perpetual state of movement in the water without touching the floor of the channel. Light (small) grains of sediment are held suspended in the water column by turbulence in streams and rivers. *Bed load* refers to sediments that roll or slide along the channel floor or bounce up temporarily into the water column. *Saltation* is a type of bed load transport where sediments move by continually bouncing into the water column. Together, suspended load and bed load move sediments to a final environment of deposition, where they come to rest and may eventually become sedimentary rock.

Most streams transport the largest component of their sediment load in suspension. This is certainly true if the source of sediment yields large quantities of clay and silt. Small particles are easily kept in suspension by fast-flowing, turbulent waters. Suspended load is visible as a cloud of sediment in the water.

Soil erosion takes place in four stages:

1. The impact of falling raindrops detaches clay, silt, and sand grains.

2. *Sheet erosion* removes grains in places where conspicuous water channels are lacking. Sheet erosion can be a serious source of soil loss on steep slopes but also operates effectively on very gentle slopes.

3. *Rill erosion* removes soil through many small, conspicuous channels, where runoff is concentrated.

4. *Gully erosion* (**FIGURE 7.28**) consists of water cutting downward into the soil and forming deep channels. Gullies are a result of confined flow within the banks of a channel. They form in plow furrows, animal trails, and vehicle ruts and between rows of crop plants. In contrast to rills, gullies cannot be obliterated by ordinary tillage. Deep gullies grow rapidly and are the cause of large quantities of lost sediment.

Erosion by Wind

Aeolian, or wind, erosion occurs when heavy winds (**FIGURE 7.29**) strip sediments and soil from the land. Wind erosion is a problem in many parts of the world. It is worse in arid and semi-arid regions, where there is little vegetation to hold the sediment.

During the 1930s, a prolonged dry spell in the Canadian Prairies culminated in dust storms and soil destruction in disastrous proportions. The resulting "dust bowl" inflicted great hardship on the

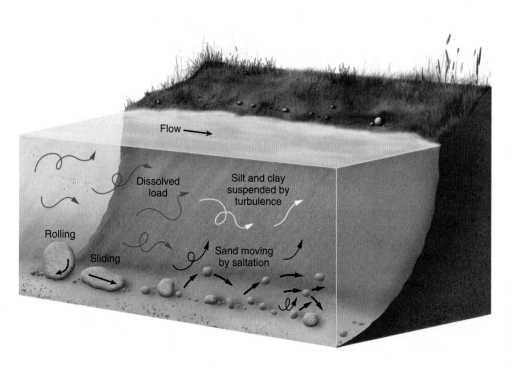

(?) Which type of sediment transport moves the greatest volume of sediment over the course of an average year: suspended load or bed load? Explain your reasoning.

FIGURE 7.27 Sediments are moved downstream in bed load or suspended load. Saltation is a type of bed load transport in which sediments bounce into the water column for short periods.

people and the land. Nearly 80 years after the dust bowl, wind erosion continues to threaten the sustainability of natural soil resources. Wind erosion is still a significant source of concern for agricultural land throughout the Canadian Prairies. Many scientists predict that with increasing global warming, dust bowl conditions will return.

Soil Erosion

 Use of land for agriculture, forestry, and transportation has accelerated soil erosion and has affected all agricultural regions of Canada. Soil erosion is both a national and global problem. Each year in Canada about 36 million hectares of croplands (~4 percent total land area of Canada) are eroded by wind and water. On average, aeolian erosion is responsible for about 40 percent of this loss.

The impacts of soil erosion can significantly reduce the effectiveness of croplands because it reduces the lighter, less dense constituents, such as organic matter, clays, and silts. These are the fertile components of the topsoil that help to stabilize the soil, provide nutrients for plants (e.g., crops), and help retain, store, and recycle water. We must continue to minimize soil erosion if we want to maintain sustainable agricultural production. In Canada, erosion is most widespread in the Prairie provinces, but it is also a serious problem on deforested slopes, along river valleys, and many other areas. Soil erosion is often caused by poor land management.

FIGURE 7.28 Gully erosion marks the start of true confined (channelized) flow. Soil loss due to gully erosion is a global problem resulting from poor land conservation practices.

Robert Harding Picture Library Ltd/Alamy

 Why is gully erosion an important problem for farmers to be aware of?

Jerry Clement/Prairies to Mountains Photography

FIGURE 7.29 Wind erosion, such as this dust storm in southern Alberta, can cause significant damage to croplands.

 How might poor land management lead to wind erosion? How would you protect your soil if you were a farmer in a windy area?

 Expand Your Thinking—It has been said that soil erosion has brought down civilizations. What is the reasoning behind this?

7-10 There Are 10 Orders in the Canadian Soil Classification System

LO **7-10** *Describe the dominant soil orders in polar soils, temperate soils, desert soils, and tropical soils.*

It is critically important for governments to conserve and manage their soil resources for the economic, agricultural, and environmental needs of their citizens. This need has led to the science of **soil classification.** Soils are classified by Agriculture and Agri-Food Canada (AAFC) into 10 dominant soil orders (**TABLE 7.3**). Soils are classified into these categories based on several factors, including mineral content, texture, organic composition, and moisture content.

Canadian Soil Orders

Approximately 85 percent of Canada is covered by soil, of which the following six types account for approximately 95 percent of Canadian soils (**FIGURE 7.30**):

- *Cryosols* (permafrost), which dominate in northern Canada
- *Podzols,* which form in relatively wet boreal forested regions
- *Luvisols,* which are derived from loamy till (glacial till) in boreal forested regions
- *Brunisols,* which are immature soils that have not developed mature soil horizons in relatively dry boreal forested regions
- *Chernozems,* which form in grasslands that serve as valuable agriculture lands in the Interior Plains
- *Organic* soils, which are the major soils of peatlands (e.g., bogs, swamps, fens)

FIGURE 7.31 shows a map depicting the distribution of dominant soil orders across Canada. If you study the map, you will note several regional patterns.

- By far the largest area covered by soil consists of cryosols found in the colder permafrost regions of northern Canada, which includes Yukon, the Northwest Territories, and Nunavut, as well as smaller regions within northernmost British Columbia, Alberta, Manitoba, Ontario, and Quebec.

TABLE 7.3 The 10 Soil Orders of Canada

Order	Description	Approximate Soil %
Brunisolic	Poorly developed soils that have not developed beyond the B horizon, but are not so poorly developed as to be considered regosolic	10%
Chernozemic	Grassland soils in the southern Interior Plains with the A horizon darkened by high organic matter	5%
Cryosolic	Soils of northern cold climates with permafrost within 1 m of the surface	47%
Gleysolic	Mottle-coloured, water-saturated soils depleted of oxygen	2%
Luvisolic	Soils with textural contrast between the A and B horizons, where clay has been moved by water from A and deposited in B	10%
Organic	Soils composed dominantly of organic matter (>30% by weight) in the upper 0.5 m; often called peat, bog, or fen soils	5%
Podzolic	Sandy, acidic forest soils, typical of cool, wet climates	18%
Regosolic	Incipient soils with little or no development of recognizable horizons, especially the B horizon	1%
Solonetzic	Same ecozone as chernozemic soils; characterized by a hard B horizon when dry, which swells and becomes sticky when wet	1%
Vertisolic	Clay-rich soils that shrink and swell markedly with drying and wetting, respectively	<1%

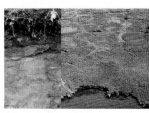
(a)

(b)

(c)

(d)

(e)

(f)

(?) In which regions of the country would each of these be common? What criteria will you use for your predictions and why?

FIGURE 7.30 The six most common soil types in Canada: *(a)* cryosols, *(b)* podzols, *(c)* luvisols, *(d)* brunisols, *(e)* chernozems, and *(f)* organic soils.

FIGURE 7.31 Canada soil map. Maps of soil orders assist governments in managing their precious soil resources. They allow scientists to understand the relationships between soil development and the influence of climate and local geology.

Why is it important to manage soil effectively?

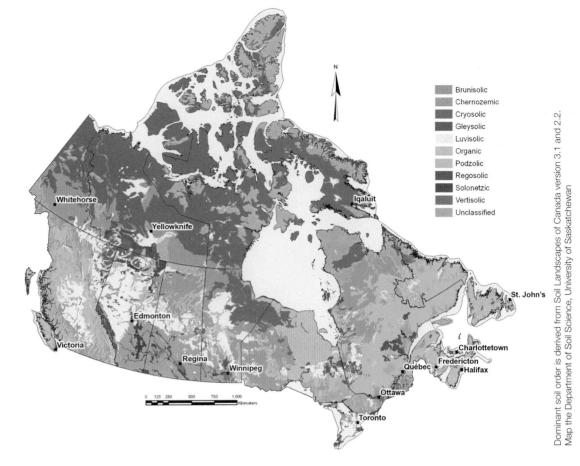

Dominant soil order is derived from Soil Landscapes of Canada version 3.1 and 2.2. Map the Department of Soil Science, University of Saskatchewan

- A large northwest-southeast trending band of brunisols can be tracked from Whitehorse in Yukon, into the Northwest Territories, and onto the Canadian Shield within northern Saskatchewan and Manitoba and western Ontario; it then swings up into northern Quebec.
- Because of the temperate, moist climate and abundance of coniferous forests, large parts of British Columbia, Quebec, and the Atlantic provinces, including Labrador, are dominated by podzols.
- The southern Interior Plains, including Alberta, Saskatchewan, and Manitoba are dominated by chernozemic soils, which provide fertile farmlands.
- Organic soils are the dominant wetland soils in the forested regions that track around the southern shores of Hudson Bay, which include northern portions of Manitoba, Ontario, and western Quebec.
- Much of the northern half of Alberta, and parts of central British Columbia, Saskatchewan, and Manitoba consist of luvisolic soils, which tend to develop in the forested boreal regions of the Canada.

General Climate Conditions

This complex system of naming soils is more understandable when described using familiar climate zones. Within North America, the dominant soil orders can be classified into four general groups based on their affinity for certain climate conditions: polar soils, temperate soils, desert soils, and tropical soils.

Polar soils are dominated by mechanical weathering and water drainage. Because chemical weathering is much slower and weaker in the cold, clay production by feldspar hydrolysis is relatively unimportant. As a result, the clay-rich B horizon is not well developed in arid polar soils. These soils are usually classified as brunisols because they lack well-developed horizons. In humid polar settings, tundra vegetation may grow on permafrost that never thaws. These conditions prevent water drainage so that in the summer, these soils become rich wetlands. Permafrost soils (which are perpetually frozen) are classified as cryosols.

Temperate soils are characterized by high organic content and well-developed soil horizons. Podsols with an organic-rich A horizon and an iron-rich B horizon form in evergreen forests. Prairies and grasslands are characterized by chernozems with deep, rich A horizons.

Desert soils form in dry climates with high rates of evaporation. Because this reduces leaching, calcium carbonate minerals (calcite) tend to collect. They accumulate below the B horizon and form a distinctive sub-horizon composed of calcrete.

Tropical soils develop where there is high rainfall and warm temperatures that encourage rapid chemical reactions. Vertisols form with high clay content causing the soil to alternately swell and shrink in wet and dry seasons. Elsewhere, where rainfall is more intensive throughout the year (i.e., rainforests), laterite and bauxite develop, which are infertile due to relentless leaching of essential nutrients out of the soil profile.

 Expand Your Thinking—Predict what soil order is dominant where you live. Explain your reasoning.

LET'S REVIEW "GEOLOGY IN OUR LIVES"

Now that you have finished the chapter, "Geology in Our Lives" will have taken on new meaning for you. Let us review it: Weathering is the physical, chemical, and biological degradation of the crust. Weathering produces soil, which yields much of our food and the ores of critical metals, such as aluminum and iron. Almost everything humans need can be traced back to weathering: food, clothing, paper, timber, medicines, shade, oxygen, and much more. But erosion caused by human activities destroys soil and other weathering products faster than they can be replaced, and deforestation destroys soil so that natural ecosystems can rarely recover on their own. Understanding weathering enables people to conserve geologic resources and protect natural environments.

We have learned that soil resources are lost at unsustainable rates to wind and water erosion when we are not attentive to the need to manage these processes. Turning land from a wild state with endemic vegetation into an agricultural state requires careful husbandry of fragile soil resources. The rate of soil production is scaled to geologic time spans: centuries, millennia, and eons. The rate of soil loss due to poor land-use practices is scaled to human time spans: seasons, years, and decades. Without careful mitigation, the mismatch between rates of soil production and soil loss cannot last for long. Eventually soil resources will dwindle and disappear, to the detriment of future generations. These impacts are governed by the climate—the same climate that governs rock weathering.

STUDY GUIDE

7-1 Weathering includes physical, chemical, and biological processes.

- **Weathering** is a series of physical, biological, and chemical processes that modify rocks, minerals, and sediments in the crust. Although weathering decays the crust, it also leads to the growth of new minerals through crystallization.

- **Chemical weathering** can attack only surfaces that are exposed to gases and fluids.

- **Physical weathering** and **biological weathering** cause rock to fracture and fragment, thus increasing the surface area that is vulnerable to chemical weathering.

7-2 Physical weathering causes fragmentation of rock.

- Physical weathering takes many different forms. **Joints** are openings or partings in rock where no lateral displacement has occurred. Sets of joints, composed of repeating parallel fractures, expose large areas of crust to weathering processes. As the weight of overlying rock is released, the crust expands and fractures into flat, horizontal slabs of sheeted joints. Exfoliation occurs when these slabs move and uncover the underlying rock.

- Abrasion occurs when a force causes two rocks to collide, leading to mechanical wearing or grinding of their surfaces.

- Growth of ice crystals, known as **ice wedging,** occurs when liquid water flows into joints and freezes, forcing the joints to split open further.

- Alternate wetting and drying of rocks, known as slaking, can be an important factor in weathering.

7-3 Hydrolysis, oxidation, and dissolution are chemical weathering processes.

- A water molecule forms when two atoms of hydrogen bond with an oxygen atom. Hydrogen atoms tend to bond strongly to one side of the oxygen atom, leaving the other side of the molecule open. Close to the hydrogen atoms is a partial positive charge; elsewhere on the molecule is a partial negative charge. This region of the molecule attracts cations.

- Chemical weathering involves several common chemical reactions, including **hydrolysis, oxidation, dissolution,** and others. Hydrolysis is the chemical reaction of a compound with acidic water, usually resulting in the formation of one or more new compounds. Oxidation involves the loss of an electron from a cation in a crystal and its use by free oxygen in the environment. In the process, the oxygen and the cation bond to form a new mineral that is a member of the oxide mineral family. Dissolution is a chemical weathering process in which carbonic acid dissolves the mineral calcite, usually found in the sedimentary rock limestone.

- **Karst topography** occurs when limestone bedrock undergoes widespread dissolution as carbonic acid percolates along joints and bedding planes. As the rock dissolves, large underground caverns are created.

7-4 Biological weathering involves both chemical and physical processes; sedimentary products result from all three types of weathering.

- Biological weathering is the product of organisms causing weathering that is either chemical or physical in nature, or a combination of the two. It includes breaking and movement of particles, exposure to carbon dioxide, and changes in the amount of moisture present.

- **Sedimentary quartz,** hematite, and calcite are important weathering products that act as natural cements. They precipitate from groundwater, bind sedimentary particles together, and thus make solid sedimentary rock. Clay minerals, chert, hematite, and travertine are important products of weathering. Chemical weathering also produces dissolved ions and compounds, such as dissolved silica and dissolved calcite, which are washed into the oceans and used by marine organisms to build shells.

7-5 Rocks and minerals can be ranked by their vulnerability to weathering.

- Weathering occurs because conditions at Earth's surface are different from those prevailing in the environments where most igneous and metamorphic rocks and minerals form.
- For the silicate minerals, the order of stability in Earth's surface environments is the reverse order of Bowen's reaction series. The most stable are the last to crystallize; the least stable are the first to crystallize.

7-6 The effects of weathering can produce climate change.

- Weathering is a geologic system that works with tectonic processes to change global climate, alter the chemistry of the atmosphere, and create soil that is critical to sustaining life.
- The uplift weathering hypothesis asserts that the global rate of chemical weathering is dependent on the availability of fresh rock to be attacked by hydrolysis. The carbon dioxide content of the atmosphere is decreased when new crust is exposed to weathering by tectonic uplift during orogenesis. Orogenesis exposes large areas of fresh silicate-rich rock to hydrolysis, leading to withdrawal of carbon dioxide from the atmosphere.

7-7 Weathering produces soil.

- Soil consists of air, water, mineral particles, and organic material. The formation of soil is influenced by biological processes, the nature of the parent rock, climate, topography, and time.
- Climate is a critical factor in soil formation. In regions where temperature and rainfall are both low (such as in cooler regions or at high elevations), mechanical weathering is the most important weathering process. Conversely, chemical weathering is most important in regions where temperature and rainfall are both high, such as the humid tropics.
- Humid tropical climates lead to intense chemical weathering that produces soils composed largely of **insoluble residues.** These soils include thick soil crusts of iron oxides, known as **laterite,** and aluminum oxides, known as **bauxite.**
- Humid mid-latitude climates with seasonal freezing allow vegetation debris to accumulate in the soil. The soil surface becomes extremely rich in organic plant debris and is known as the **humus layer.**

7-8 Soil, spheroidal weathering, and natural arches are products of weathering.

- As rain percolates through the humus layer, dissolved ions in the percolating water may not be removed but instead may recombine to form clay minerals that make up a set of layers or soil horizons. These horizons are collectively known as the **soil profile.** The thickness of the soil profile varies, depending on the climate and bedrock.
- **Spheroidal weathering** occurs when rectangular blocks outlined by two perpendicular sets of joints are attacked by chemical weathering. Edges and corners are rapidly worn down by chemical weathering to create curved surfaces.
- Natural arches form when joints undergo weathering to produce narrow walls of rock called *fins.* Ice wedging causes the exposed rock on each side of a fin to gradually fall away. Eventually the fin is cut through to produce a natural arch. Continued weathering causes the hole in the arch to grow larger until the arch becomes unstable and collapses.

7-9 Soil erosion is a significant problem.

- Fluvial erosion occurs when particles of sediment are removed by flowing water in a channel. Water carries sediment in one of two ways: suspended load or bed load. Other forms of active erosion include mass wasting, in which sediment is moved by gravity, as in a landslide or mudflow, and aeolian erosion, in which sediment is moved by wind.
- Soil erosion is a national and global problem. In Canada, about 36 million hectares of cropland are eroded by wind and water at rates that are greater than the rate of soil replenishment required for sustainable agricultural production in unfertilized soil. On average, aeolian erosion is responsible for about 40 percent of this loss.

7-10 There are 10 orders in the Canadian soil classification system.

- It is critically important for governments to conserve and manage their soil resources for the economic, agricultural, and environmental needs of their citizens. This need has led to the science of **soil classification.** Soils are classified by Agriculture and Agri-Food Canada (AAFC) into 10 dominant soil orders: brunisols, chernozems, cryosols, gleysols, luvisols, organic soils, podzols, regosols, solonetz, and vertisols.

KEY TERMS

bauxite (p. 177)
biological weathering (p. 162)
carbon cycle (p. 174)
chemical weathering (p. 162)
clay (p. 166)
dissolution (p. 167)
erosion (p. 162)
fluvial erosion (p. 182)
greenhouse gas (p. 174)
humus layer (p. 178)

hydrolysis (p. 166)
ice wedging (p. 165)
insoluble residues (p. 177)
joints (p. 164)
karst topography (p. 168)
laterite (p. 177)
oxidation (p. 167)
physical weathering (p. 162)
root wedging (p. 170)
sedimentary minerals (p. 162)

sedimentary quartz (p. 171)
sinkhole (p. 168)
soil (p. 162)
soil classification (p. 184)
soil profile (p. 178)
spheroidal weathering (p. 178)
talus (p. 165)
ventifacts (p. 169)
weathering (p. 162)

ASSESSING YOUR KNOWLEDGE

Please answer these questions before coming to class. Identify the best answer.

1. Weathering consists of
   a. erosion, tectonics, and uplift.
   b. chemical, biological, and physical degradation.
   c. crust age, chemistry, and sedimentary minerals.
   d. sedimentary quartz, hematite, and alumina.
   e. None of the above.

2. Spheroidal weathering is caused by
   a. sand abrasion in running water.
   b. crystal growth in cold climates.
   c. chemical weathering of angular rocks.
   d. a combination of slaking and mass wasting.
   e. None of the above.

3. The chemical interaction of oxygen with other substances is known as
   a. dissolution.
   b. hydrolysis.
   c. saturation.
   d. oxidation.
   e. None of the above.

4. The most important form of chemical weathering of silicate minerals is
   a. crystal growth.
   b. slaking.
   c. hydrolysis.
   d. dissolution.
   e. frost wedging.

5. Insoluble residues are
   a. minerals produced by weathering.
   b. dissolved compounds resulting from chemical weathering.
   c. soils that are rich in organics.
   d. typically dissolved in hydraulic acid.
   e. All of the above.

6. The tendency of silicates to weather on Earth's surface is predicted by
   a. mineral texture.
   b. rock colour and environment of deposition.
   c. Bowen's reaction series.
   d. tectonic setting.
   e. their roundness.

7. Which of the following statements about carbon dioxide is true?
   a. It is an important gas that regulates Earth's climate and influences groundwater chemistry.
   b. The amount present in the atmosphere is affected by the rate of crustal weathering.
   c. Its decrease has caused net global cooling over recent geologic history.
   d. It is a greenhouse gas.
   e. All of the above.

8. The variables that most affect the weathering process are rock composition and
   a. topography.
   b. surface area.
   c. living things.
   d. climate.
   e. None of the above.

9. Which of the following statements about soil erosion is true?
   a. It is a form of pollution that affects biological communities.
   b. It is a major problem affecting million of hectares of croplands.
   c. It threatens to impact food production.
   d. It takes centuries to make soil and only minutes to erode it.
   e. All of the above.

10. Soil profiles in temperate climates are characterized by
    a. a rich humus layer.
    b. calcrete horizons.
    c. a thin, stony A horizon.
    d. aluminum and iron oxides in the upper layer.
    e. None of the above.

11. Karst topography is the result of
    a. soil erosion.
    b. biological weathering of silicate rock.
    c. chemical weathering of carbonate rock.
    d. spheroidal weathering.
    e. All of the above.

12. The major agricultural lands in Canada are based on which soil?
    a. gleysols
    b. solonetz
    c. podzols
    d. chernozems
    e. cryosols

13. Aluminum ore comes from
    a. ice wedging.
    b. hot, arid climates.
    c. humid tropical settings.
    d. physical weathering.
    e. All of the above.

14. Iron oxide and aluminum oxide soils are
    a. insoluble residues.
    b. regosols.
    c. vertisols.
    d. typically in polar soils.
    e. None of the above.

15. The uplift weathering hypothesis
    a. explains global cooling.
    b. explains global warming.
    c. explains agricultural development.
    d. causes orogenesis.
    e. None of the above.

FURTHER RESEARCH

1. Why is water an effective agent in chemical weathering?

2. How does the weathering of silicate minerals influence Earth's climate?

3. What are joints, and how are they formed? Why are they important in weathering?

4. What factors determine the stability of a mineral on Earth's surface?

5. Draw the rock cycle. Indicate where weathering fits in the cycle, and create a special section that describes some of the details of weathering.

6. Draw an average soil profile for the area where you live.

7. Search the web for an explanation of which Canadian agency is responsible for soil classification. Study its mission.

ONLINE RESOURCES

Explore more about weathering on the following websites:

Soil Landscapes of Canada:
http://sis.agr.gc.ca/cansis/nsdb/slc/intro.html

Virtual Soil Science:
http://soilweb.landfood.ubc.ca/promo

Soils of Canada:
http://www.soilsofcanada.ca/orders/luvisolic/index.php

Agriculture and Agri-Food Canada: "Canadian System of Soil Classification, 3rd edition":
http://sis.agr.gc.ca/cansis/taxa/cssc3/intro.html

U.S. National Resources Conservation Service:
www.nrcs.u sda.gov

U.S. National Soil Erosion Research Lab:
www.ars.usda.gov/main/site_main.htm?modecode=36-02-15-00

Weathering–Wikipedia, The Free Encyclopedia:
http://en.wikipedia.org/w/index.php?title=Weathering&old id=553611019

Additional animations, videos, and other online resources are available at this book's companion website:
www.wiley.com/go/fletchercanada
This companion website also has additional information about WileyPLUS *and other Wiley teaching and learning resources.*

CHAPTER 8
SEDIMENTARY ROCK

Chapter Contents and Learning Objectives (LO)

8-1 Sedimentary rock is formed from the weathered and eroded remains of the crust.

LO 8-1 *Describe the reasons that geologists study sedimentary rocks.*

8-2 There are three common types of sediment: clastic, chemical, and biogenic.

LO 8-2 *Compare and contrast the three types of sediment.*

8-3 Sediments change as they are transported across Earth's surface.

LO 8-3 *List the ways that sediments change as they are transported across Earth's surface.*

8-4 Clastic grains combine with chemical and biogenic sediments.

LO 8-4 *Describe how the composition of sediments changes between the weathering site and the formation of sedimentary rock.*

8-5 Sediment becomes rock during the sedimentary cycle.

LO 8-5 *Describe the sedimentary cycle.*

8-6 There are eight major types of clastic sedimentary rock.

LO 8-6 *List the eight major types of clastic sedimentary rock, and give a brief description of each.*

8-7 Some sedimentary rocks are formed by chemical and biological processes.

LO 8-7 *List the major types of chemical sedimentary rock and biogenic sedimentary rock.*

8-8 Sedimentary rocks preserve evidence of past environments.

LO 8-8 *Describe how a geologist uses inductive reasoning to interpret Earth's history using sedimentary rocks.*

8-9 Primary sedimentary structures record environmental processes.

LO 8-9 *List four common sedimentary structures, and explain their significance.*

GEOLOGY IN OUR LIVES

Much of our knowledge of Earth's history is derived from the study of sedimentary rocks. Sedimentary rock is formed from the weathered and eroded remains of the crust. It contains fossils and other evidence of the history of life, changes in the environment, climate shifts, plate tectonic movement, and many other details of how the planet has developed and changed over time. Sedimentary rock is the source of fossil fuels such as coal, petroleum, natural gas, and oil shale. We get valuable metals from sedimentary accumulations such as placer deposits of gold and platinum, and the ores of iron and aluminum. Construction materials including building stone, sand and gravel to make road beds, and cement for the building industry all come from sedimentary rocks. Perhaps of greatest significance, most of our communities obtain fresh water for drinking, irrigation, and manufacturing from groundwater found in the pore spaces of sedimentary rocks.

(?) Many of the rock layers you can see in this photograph, which is a view from the south rim of the Grand Canyon, are composed of sediments that accumulated on the sea floor in a shallow ocean. What evidence would reveal to a geologist that a rock formed in a shallow marine environment?

8-1 Sedimentary Rock Is Formed from the Weathered and Eroded Remains of the Crust

LO 8-1 *Describe the reasons that geologists study sedimentary rocks.*

Most of Earth's surface is covered with layers of loose **sediment** (particles of mineral, broken rock, and organic debris; **FIGURE 8.1a**), and over 75 percent of the land surface is covered with **sedimentary rock** (rock composed of sediment; **FIGURE 8.1b**). For billions of years, weathered particles of sediment have been eroded from the land—removed from mountains and hills through the action of gravity, wind, and running water. These little bits of rock and mineral grains are washed downstream, where they eventually come to rest on the bottoms of rivers, lakes, and, ultimately, oceans. Over time, they accumulate as layer upon layer of eroded sediment that is deposited in environments across the planet's surface. As they are buried, the overlying pressure builds, and these layers *consolidate* and *compact*. The grains are *cemented* by dissolved compounds in groundwater and in due course become solid sedimentary rock.

Sedimentary rock and loose sediments can be "read" by geologists as records of earlier environmental conditions. Rock composed of gravel may reveal the former presence of a stream or glacier; rock composed of **sand** tells of a past beach or sand dune; **mud** reveals ancient deltas, marshes, or deep lakes. The greatest collection of sedimentary rock comes from the sea floor, where fossil plankton collects to form *biogenic ooze* and **clay** accumulates to form *abyssal clay*. In each of these cases, a geologist interprets the sediments and buried fossils to learn about former surface conditions on Earth. This chapter explains that sedimentary rock records the history of surface environments and life on Earth.

Sedimentary rock is formed from the weathered and eroded remains of igneous, metamorphic, and other sedimentary rocks as well as from organic sources. Many sedimentary rocks also contain both direct and indirect evidence of living organisms, such as *skeletal parts* or impressions of plants and animals (leaf imprints, footprints, etc.). Some minerals in sedimentary rocks crystallize from dissolved compounds in groundwater, and help to cement the grains together, the most common cements being calcite and quartz. Some minerals result

© Hans Strand/Corbis   Thomas Kitchen & Victoria Hurst/All Canada Photos/Corbishil Degginger/Alamy

(a)   *(b)*

FIGURE 8.1 *(a)* Loose sediment is found on Earth's surface in many types of environments. Here sand forms a dune in the Namib Desert, Namibia. *(b)* Sediment collects in a sedimentary environment where it may consolidate and become hard sedimentary rock made of individual particles cemented and/or compacted together. These sandstones in Dinosaur Provincial Park, Alberta, form from sands that were deposited in river channels and estuaries.

Based on your knowledge of minerals that are resistant to weathering, what is the likely mineralogy of these sediments?

from chemical reactions between the crust and gases in the atmosphere (such as oxygen and carbon dioxide), leading to the formation of, for example, **iron oxides** (such as *hematite*). In fact, many of the minerals in sedimentary rocks are formed by the weathering processes we studied in Chapter 7 ("Weathering"). Other sedimentary minerals are produced by living organisms that secrete a mineral coating, such as corals that produce calcite. Sedimentary rocks composed of **silt** and clay harden when fine particles become consolidated and water is expelled. Although sedimentary rocks can be formed through several different processes, they generally take on characteristics that reflect specific conditions on Earth's surface. Hence, geologists study them for clues to the history of Earth's surface environments.

Why Study Sedimentary Rock?

Our society depends on many resources that come from sedimentary rock (**FIGURE 8.2**). These resources include *fossil fuels* that are critical to industry and transportation: petroleum, coal, oil shale, and natural gas. Forty percent of global energy needs are met by petroleum, 34 percent by coal, and another 27 percent by natural gas. In Canada, there is less dependence on coal, although the fossil fuels together still provide about 80 percent of our energy needs. Other sedimentary resources include accumulations of valuable metals, sand and gravel for construction, and the stored groundwater that we drink.

Rocks and sediments can be thought of as pages in the book of Earth's history. Fossilized skeletal fragments of long-extinct plants and animals (many of them microscopic plankton and pollen) are evidence of the shifting climates and changing environments of our restless planet. Without sedimentary rocks, not only would we be unable to heat our homes or run our machines, but we would know little about what Earth was like during the billions of years that have preceded our short tenure.

© Lowell Georgia/Corbis

FIGURE 8.2 The fossil fuels that power modern society come entirely from sedimentary rock. The photograph is of an offshore drilling platform, approximately 300 km southeast of Halifax, Nova Scotia, that is exploring for and extracting oil from sedimentary rocks on continental shelves in deep water.

(?) List the various roles that fossil fuels play in your life.

(?) Expand Your Thinking—Use the texture and composition of sedimentary rock as described in Section 8-1 to propose a classification system for sedimentary rock. What will you name your various erock types?

8-2 There Are Three Common Types of Sediment: Clastic, Chemical, and Biogenic

LO *8-2 Compare and contrast the three types of sediment.*

Sedimentologists—geologists who study sediments—have identified three types of sediment based on their composition: **clastic sediments,** which are broken pieces of rock deposited by water, wind, ice, or some other physical process; **chemical sediments,** which are produced by inorganic (non-biological) precipitation of dissolved compounds (e.g., through evaporation); and **biogenic sediments,** which are produced by organic (biological) precipitation of compounds or consist of the remains of living organisms. We explore these three types of sediment in this section.

Clastic Sediments

Clastic sediments are solid and often broken pieces of pre-existing rocks or minerals that have been physically transported from one place to another (**FIGURE 8.3**). Since most of the crust is composed of silicate minerals, the majority of clastic sediments are composed of silicate minerals. But this generalization may or may not be true from place to place.

The silicate minerals that are most likely to survive the rigours of weathering are those that are stable on Earth's surface and are formed late in Bowen's reaction series. Unstable mafic minerals usually are not important constituents of most clastic sediments because they are vulnerable to chemical deterioration and do not readily survive the weathering process. These include ferromagnesian silicate minerals such as olivine and amphibole, and the calcium-rich members of the plagioclase feldspar series. As we learned in Chapter 7, many igneous and metamorphic minerals formed at high temperatures or pressures are not stable under the low temperatures and pressures typical of sedimentary environments on Earth's surface; they rapidly break down and are not commonly found in sedimentary rocks.

Stable minerals found in clastic sediments include quartz, orthoclase feldspar, muscovite, clay minerals, and some valuable *heavy minerals* (dense minerals such as garnet, zircon, chromite, magnetite, tourmaline, and others). Also common in clastic sediments are eroded fragments of crustal rocks, known as *lithic fragments* (such as bits of shale, basalt, limestone, and others).

Chemical Sediments

Chemical sediments form as a result of chemical reactions that are not related to the actions of living organisms. Chemical sediments form from water that is saturated with dissolved cations and anions. Crystallization occurs when these ions develop covalent and ionic bonds and thus form chemical compounds, producing solid minerals such as calcite, quartz, and halite. The solid crystals then precipitate from the water to create sediment. This inorganic precipitation can occur in groundwater, in the ocean, in hot springs that flow onto Earth's surface, and even in freshwater lakes.

The most common chemical sediments form as a result of *evaporation* in a lake or ocean (**FIGURE 8.4**). During the process of evaporation,

 What is the mineral composition of evaporites?

FIGURE 8.4 Evaporite deposits, such as these salts in the desiccated bed of a desert lake, consist of chemical sediments.

 Would you agree that sand composition may reflect climate? Explain.

FIGURE 8.3 Clastic sediment, in this case sand-sized, consists of the stable minerals quartz and orthoclase feldspar, as well as clay minerals, micas, rock fragments (called *lithic* fragments), and others.

Courtesy of Chip Fletcher

Franck Guiziou/Hemis/Corbis

(a)

(b)

(c)

Courtesy of Chip Fletcher

David Nunuk/Science Photo Library

(d)

FIGURE 8.5 Sandstones (and other clastic sedimentary rocks) contain inorganically precipitated cements—a type of chemical sediment. The colour of the rock reflects the minerals that form the cement. *(a)* Calcium carbonate cemented sandstone. *(b)* Silica cemented sandstone. *(c)* Iron oxide cemented sandstone. *(d)* Concretions eroded from sedimentary rocks in Red Rock Coulee, Alberta. The largest are about 2 m across.

 What are the mineral names of these three types of cements?

water molecules change from liquid to gas, but dissolved compounds containing calcium, silica, iron oxide, sodium, and chlorine (and others) are left behind. As a result, any remaining water becomes enriched in these compounds, and they begin to crystallize. Minerals known as **evaporites** form in the briny solution and settle to the bottom as a layer of chemical sediments. The minerals halite, calcite, and gypsum, among others, are all inorganically precipitated when water evaporates.

 What chemical compound(s) are most common in marine biogenic sediments?

Courtesy of Chip Fletcher

FIGURE 8.6 Biogenic sediments are composed of skeletal fragments, usually of marine organisms, such as plankton or reef-dwelling corals and algae.

Inorganic precipitation also occurs in groundwater flowing through accumulations of porous sediments. A precipitated compound will grow around individual grains and fill empty spaces between grains, often cementing adjacent grains together. *Common mineral cements* formed in this way are calcite, quartz, goethite, and hematite. In certain cases, cement starts to precipitate preferentially around a nucleus, such as a fossil, and continued addition of cement forms concretions. These can vary considerably in size (**FIGURE 8.5**).

Most chemical sediments are deposited near the locations where they form rather than being transported from afar. That is, they precipitate at the same location where they ultimately will be transformed into sedimentary rock. The dissolved compounds of which the sediments are composed, however, typically have travelled long distances from weathering sites located high in adjacent watersheds or discharged from hydrothermal vents at spreading centres.

Biogenic Sediments

Biogenic sediments are composed of the fossil remains of plants and animals (**FIGURE 8.6**). Biogenic sediments include skeletal fragments of reef-dwelling organisms such as corals and shells (usually made of calcite or its polymorph, aragonite), highly organic plant remains that form coal (composed of carbon with various impurities, such as sulphur), and biogenic ooze found on the deep-sea floor, which is composed of microscopic skeletal fragments of marine plankton (made of organically precipitated calcium carbonate or silica).

 Expand Your Thinking—What is similar about biogenic and chemical sediments that makes them fundamentally different from clastic sediments?

8-3 Sediments Change as They Are Transported across Earth's Surface

LO 8-3 *List the ways that sediments change as they are transported across Earth's surface.*

From the moment they are freed by weathering and erosion to travel across Earth's surface, sediments are imprinted with the physical and chemical "signatures" of the various environments they encounter.

Particle Size Reflects Environmental Energy

Flows of air, water, and ice, along with gravity, are responsible for most of the work of moving sediment from its origin in weathered crust toward the sedimentary basins in which it finally comes to rest. These sedimentary basins are also known as **environments of deposition.** Strong currents move more and larger grains; weak currents move only smaller grains of clay and silt. Geologists describe this relative strength of natural processes as the *environmental energy*. Gravity drives this transportation system, and its effect is relentless. Grain by grain, ion by ion, through the ages, mountains are worn down and delivered to the sea.

To describe sedimentary particles, geologists use common, yet precise, terms denoting the size of grains. Hence, the familiar terms "gravel," "sand," "silt," and "clay" each denote a particular range in the sizes of sedimentary particles (**FIGURE 8.7**).

Gravel includes any particle that is more than 2 mm across in its largest dimension. In decreasing size order, gravel can consist of boulders (larger than 256 mm), cobbles (64 to 256 mm), pebbles (4 to 64 mm), and granules (2 to 4 mm). Sand refers to particles between

? In what environment might you find each of these grain sizes deposited? Give one example for each.

FIGURE 8.7 *(a)* Gravel includes all grains with a large dimension greater than 2 mm. *(b)* Sand is between 0.06 mm and 2 mm. *(c)* Silt is smaller than 0.06 mm but larger than 0.004 mm. *(d)* Clay is 0.004 mm or finer.

0.06 mm and 2 mm across. Geologists often subdivide sand into coarse, medium, and fine sand based on specific size limits. In practical terms, fine sand is about the smallest grain that can be easily seen with the naked eye. Silt is sediment that is finer than sand, and clay is the finest sediment of all. Mud is a combination of silt and clay.

As environmental energy increases, the speed and turbulence of air and water currents increase. More and larger sedimentary particles roll and drag in the currents and are suspended above the bottom. In high-energy environments, such as streams flowing at rates of over 50 cm/s and ocean beaches with strong waves, sedimentary deposits tend to collect gravel and sand. In environments with moderate energy levels, such as quiet streams (flowing at rates of 20 cm/s to 50 cm/s), lake beaches, many rivers, and lagoons, sedimentary deposits tend to contain more sand and mud. In low-energy environments with currents flowing at rates of less than 20 cm/s, such as the deep-sea floor, deep-lake floors, wetlands, and parts of estuaries and deltas, sedimentary deposits contain no gravel and little sand and are dominated by mud.

Sedimentary deposits are governed not only by environmental energy but also by the abundance of sediment. That is, if most of the available sediment is gravel, the deposit will reflect this fact. The same is true for other types of sediments, with the result that *the final deposit reflects the combined influences of environmental energy and sediment availability.*

As grains of sediment are transported, they undergo changes related to the energy of the environment. Two processes in particular occur in sediments: **sorting,** in which grains are separated by density and size, and **abrasion,** in which grains become smaller and more rounded the longer they are transported.

Sorting

Sorting is the tendency for environmental energy to separate grains according to size (**FIGURE 8.8, TABLE 8.1**). A *well-sorted deposit* is one in which the grains are nearly all the same size. A *poorly sorted deposit* is one in which the grains vary widely in size; such a deposit is often a mixture of mud, sand, and gravel. Sandy dunes are an example of a well-sorted deposit. Sedimentary deposits resulting from a glacier are usually poorly sorted because the ice typically carries grains of all sizes, from clay to boulders, simultaneously.

Water has a tendency to sort particles according to their mass, which is determined by their size and composition. The size of particles with the same composition will determine when moving currents have enough energy to erode and transport them and when they will be deposited. The latter is when the currents have too little energy to move them. Fine sand tends to erode earlier than large grains. Clay and silt tend to form compact layers and are not easily eroded. In streams, lakes, and the ocean, sands tend to collect with other sands and gravels with other gravels. This sorting occurs because as the energy of moving water waxes and wanes, grains of equal mass (determined by density and size) tend to erode together and be deposited together.

Wind is probably the most effective sorting mechanism for finer sediments. Windblown sand forms fields of dunes, and

Why do windblown deposits tend to be well sorted?

FIGURE 8.8 *(a)* If all grains in a sedimentary deposit are nearly the same size, then it is well sorted. *(b)* If sediment contains pieces of gravel as well as sand and mud, such as in this stream channel, it is poorly sorted.

windblown silt forms huge accumulations called *loess*. Sorted gravel accumulations called *pavements* form when wind removes finer grains of sand and silt. Ice (in glaciers) is the poorest sorting mechanism, transporting and depositing sediments of all sizes with equal ease.

Abrasion

As sediment is transported, individual grains are subject to abrasion because they strike or rub against other particles. Minerals such as feldspar break easily along cleavage planes. However, for minerals without cleavage, such as quartz, the process of abrasion produces *rounding*; that is, it rounds off the sharp edges or corners of grains. A well-rounded grain probably has travelled a great distance from its original source; an angular grain has not travelled as far. Boulders tend to be rounded more quickly than sand grains because they strike with greater force.

Sedimentary deposits tend to become smaller grained and better sorted with greater distance from their source area (**FIGURE 8.9**). For example, sediments in mountainous watersheds are poorly sorted, with angular clasts and an abundance of large-diameter particles. In contrast, the sediments far downstream tend to be well rounded, finer grained, and better sorted. This change in sediment texture is due to the length and duration of transport and the sorting and abrasion that take place. Different environments reflect different degrees of abrasion (and sorting), depending on the energy and the type and abundance of sediment.

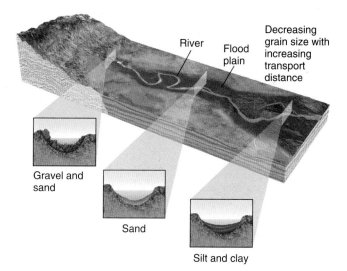

FIGURE 8.9 Sediment that is transported far from its source tends to be finer grained, more rounded, and better sorted than sediment that is deposited close to its source.

Explain why grain size decreases in sediments far from their source.

Expand Your Thinking—Where do eroded grains eventually come to rest?

TABLE 8.1 Features of Sedimentary Environments

Sedimentary Environment	Transportation Process	Sediment Size	Sorting	Abrasion (degree of rounding)
Wind-deposited dune	Wind	Fine to medium sand	Very strong	Strong
Ocean beach	Waves, currents, wind	Fine sand to gravel	Strong	Very strong
Wetland	Weakly circulated water, wind	Clay to silt	Strong	Weak
Stream channel	Running water	Silt to gravel	Moderate	Very strong
Glacier	Moving ice	Clay to large gravel (boulders)	Weak	Weak
Steep hillside	Mass wasting	Clay to large gravel (boulders)	Weak	Moderate

8-4 Clastic Grains Combine with Chemical and Biogenic Sediments

LO **8-4** *Describe how the composition of sediments changes between the weathering site and the formation of sedimentary rock.*

Sorting and abrasion change the character of clastic sediments as they travel from mountains to basins. Dissolved components also move from weathering sites, where they originate, into basins (such as the ocean), where they are deposited. Dissolved ions and compounds are deposited through either organic or inorganic precipitation. Both processes produce a solid that joins the sedimentary system as either chemical or biogenic sediment.

Among porous sediments in sedimentary basins, dissolved silica, calcium carbonate, and iron oxide may precipitate inorganically as cement that binds grains to one another. (Refer to Figure 8.5.) Inorganic precipitation is also important at locations where evaporation raises the concentration of dissolved ions in the water. For example, dissolved sodium and chlorine may be concentrated enough that ultimately sodium chloride, called halite (salt), precipitates. In the ocean, marine organisms combine dissolved calcium ions with dissolved bicarbonate ions to form a wide variety of solid calcium carbonate shells. Some types of plankton, such as diatoms, use dissolved silica to build their shells.

An important aspect of these organic and inorganic reactions is that they regulate the chemistry of ocean water. Oceans are fed by the discharge of streams and rivers around the world. Hydrothermal vents, the hot water springs that develop where there is submarine volcanism, also discharge into the world's oceans. In the course of geologic time, the input of these sources is balanced by evaporation, which removes water from the oceans and temporarily stores it in the atmosphere. The same streams and hydrothermal vents are also the source of the dissolved chemicals that make sea water salty. Evaporation does not remove these chemicals. Unless there was some mechanism to regulate them, they would perpetually build up in sea water, making the ocean saltier and saltier. There is such a mechanism: Marine organisms extract chemicals from sea water to make their shells and skeletons. These shells and skeletons, in turn, fall to the sea floor

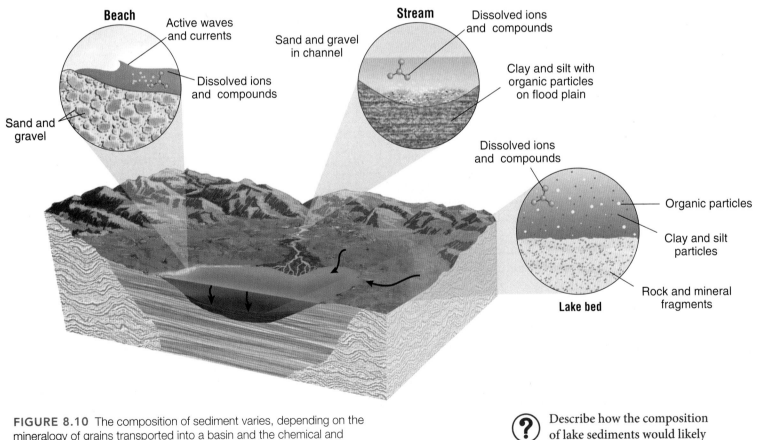

FIGURE 8.10 The composition of sediment varies, depending on the mineralogy of grains transported into a basin and the chemical and biogenic sediments produced within the basin.

(?) Describe how the composition of lake sediments would likely differ from stream sediments.

as biogenic sediment when the organisms die. Thus, the chemistry of sea water is regulated as chemicals are removed from the water by biogenic precipitation and stored in sedimentary rocks composed of fossil skeletal debris. As a result, seawater chemistry is regulated by a combination of biologic, volcanic, and weathering processes and remains fairly stable.

The Changing Composition of Sediments

As sediments accumulate in sedimentary basins, they change not only in texture but also in composition. The composition of sedimentary deposits evolves over time as a result of two processes:

1. Grains are subjected to continuous chemical weathering as they are transported: This weathering tends to remove unstable grains (olivine, pyroxene, calcium-rich plagioclase, and amphibole) and increase the relative abundance of more stable grains (quartz, clays, muscovite, sodium-rich plagioclase, and orthoclase).

2. New types of sediment are introduced, such as biogenic sediment (fossils and organic debris) produced within the sedimentary basin (FIGURE 8.10), local weathering products washed into the basin (soil, mineral assemblages with some unstable components, oxides, etc.), and locally precipitated chemical sediments.

Dissolved compounds mix with sediments in the basin and add cements or other types of chemical sediments that influence the overall composition of a deposit. The "Critical Thinking" box titled "Sediment Deposition" describes a simple experiment showing how sediments of various sizes tend to accumulate in layers.

Organic sediments are most likely to originate *within* sedimentary basins. These sediments include vegetation preserved from decay in wetlands, which turns into peat (and eventually coal), and the remains of algae, bacteria, and other types of microscopic organisms, which may become petroleum. (We discuss these fuels further in Chapter 10, "Geologic Resources.") These materials mix with clastic sediments or accumulate in concentrations of organic sediment.

CRITICAL THINKING

SEDIMENT DEPOSITION

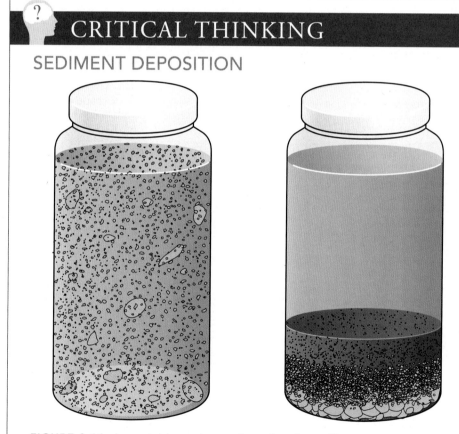

Here is a simple experiment to demonstrate how sediments accumulate and why sedimentary rocks tend to form in layers.

1. Obtain approximately $\frac{1}{4}$ L (about 1 cup) of sediment consisting of a mixture of mud, sand, and gravel. Place it in a tall, narrow jar made of clear glass (FIGURE 8.11).
2. Add water to the jar until it is nearly full. Place a lid or cover on it so that it is sealed.
3. Shake the jar thoroughly so that the sediment and water are completely mixed.
4. Quickly turn the jar upright and set it on a flat surface.

Draw a diagram of the final accumulation of sediments and answer the following questions:

1. How does the final accumulation differ from the sediment before it was mixed with water?
2. What process caused this difference? How and where would the same process occur in nature?
3. What feature of sedimentary rocks have you recreated? Find a picture of this feature in the book and interpret what it means.

FIGURE 8.11 Jar containing water, mud, sand, and gravel just after shaking, and after sediments have accumulated.

 Expand Your Thinking—Several factors control the salinity of sea water. Write a hypothesis that predicts the processes governing ocean salinity. Describe several ways to test your hypothesis.

8-5 Sediment Becomes Rock during the Sedimentary Cycle

LO *8-5 Describe the sedimentary cycle.*

Because there are many types of sediment, there are many types of sedimentary rock. Most sedimentary rock consists of some or all of four basic components (**FIGURE 8.12**): grains, cement, matrix, and pore spaces. Grains are particles of sediment (clasts, chemical grains, or biogenic grains); cement binds the grains together; the matrix is fine sediment (mud or other fine-grained material) that fills in the space around grains and cement; and pore spaces are empty spaces within the rock. **Porosity** is a measure of the amount of empty space contained in a rock.

When sediments accumulate and are buried, they eventually turn into solid rock. This process, called **lithification,** takes place in two ways:

1. *Cementation.* Particles are bound together by natural cements that are inorganically precipitated from dissolved compounds in groundwater.

2. *Compaction.* Particles are compacted and consolidated by pressure from the weight of overlying layers of sediment.

Most sedimentary rocks *lithify* as a result of a combination of these two processes. As sediment is buried in the crust, it is subjected to heat and pressure, which produce physical and chemical changes. On average, temperature rises by 30°C for every kilometre of burial in the crust, and pressure increases by the equivalent of 1 atmosphere (the pressure of the atmosphere pressing down on Earth's surface at sea level) for every 4.4 m of depth. At a depth of 4 km, the temperature can be 120°C and the pressure can exceed 4,000 atmospheres. These high temperatures and pressures cause cementation and compaction, which bind sediments and form sedimentary rock.

Compaction happens when sediment is compressed during burial (**FIGURE 8.13**). Clay and silt typically experience more compaction than sand and gravel do. Because mud is usually deposited in water, the pore spaces are filled with liquid. Compaction involves slow squeezing, with the result that this water is expelled and usually travels upward to locations where the compaction pressure is lower.

The resulting *dewatered* sediment is denser because it has lower porosity—that is, less pore space between grains. Mud particles (clay and silt size) may include clay minerals, which are shaped like flat dinner plates. They tend to align when compacted, forming thin layers of sediment called *laminations.* Particles adhere to one another with weak, grain-to-grain electrostatic forces. As sands become compacted, they are reorganized into a closely packed stack of grains. Mud decreases in volume by as much as 50 percent to 80 percent during compaction, whereas sand may decrease in volume by 10 percent to 20 percent.

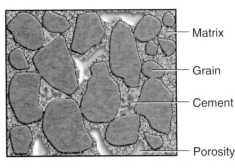

FIGURE 8.12 Sedimentary rock is composed of four constituents: grains, cement, matrix, and pore spaces.

(?) On what basis would you assign names to sedimentary rocks?

Cementation occurs when dissolved ions in groundwater inorganically precipitate to fill pore space (usually in sands), thus binding grains together. The most common cements come from dissolved compounds that originate within the surrounding sediments. Hence, silica (SiO_2) often cements quartz and feldspar grains; iron oxides (FeO or Fe_2O_3) often cement iron-rich grains; and calcium carbonate ($CaCO_3$) often cements calcite grains.

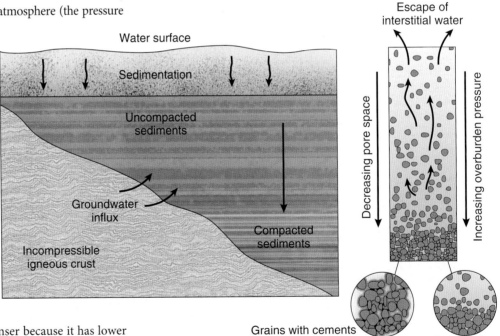

FIGURE 8.13 Compaction tends to drive water out of sediment, and cementation fills the empty space between grains with secondary minerals that precipitate from groundwater.

(?) How are sand and gravel grains lithified? How are mud particles lithified?

Sedimentary Rocks within the Rock Cycle

The rock cycle perpetually supplies sediment from the weathering and erosion of tectonically uplifted crust (**FIGURE 8.14**). Each stage of the sedimentary part of the rock cycle is characterized by an important set of processes that influence the resulting sedimentary rock. These processes include tectonic uplift, weathering, erosion and transportation, deposition, burial, and lithification, followed by renewed uplift, which starts the cycle over again.

Most tectonic uplift is associated with plate boundary convergence. When crust is uplifted at a convergent plate margin, orogenesis or volcanism produces mountains. Mountainous slopes are immediately attacked by chemical, physical, and biological weathering, producing solid clasts and dissolved ions and compounds. Various physical processes of erosion (gravity, running water, wind, and glacial ice) remove these products of weathering. Eroded sediments move across Earth's surface until they come to rest in an environment of deposition. As they move, the texture and composition of the sediments may be influenced by the chemical and physical characteristics of the surface environments they encounter.

Sediments eventually come to rest, accumulate, and form layers in response to varying rates of delivery. Sustained deposition through time requires a sedimentary basin where tectonics continuously lowers the crust so that it does not fill up. Buried sediment is subjected to compaction due to the pressure imposed by overlying sediments and to cementation by dissolved ions in groundwater. As a result of these processes, it becomes sedimentary rock.

During later tectonic uplift, these sedimentary rocks, as well as igneous and metamorphic rocks, are exposed to weathering. The sedimentary part of the rock cycle then starts again, and this part of the cycle helps to answer some important questions:

- Why hasn't Earth's surface eroded down to sea level long ago? *Answer:* New mountains perpetually arise from orogenesis, arc volcanism occurring at convergent plate boundaries, and other types of mountain building. (Mountain building is discussed in Chapter 11.)
- What is the relationship between plate tectonics and Earth surface environments? *Answer:* Mountains are a prolific source of sediment because they are exposed to intense weathering, and the resulting sediment becomes an integral component of Earth's surface, in many cases influencing the character of environments and life in them.
- Why is sedimentary rock such an effective recorder of Earth's history? *Answer:* The texture and composition of sedimentary rocks may be influenced by environmental conditions. They contain fossils and other constituents that are the result of geologic and biologic processes.

The sedimentary cycle is powered by solar energy, which drives weathering; geothermal energy (and gravity), which drives plate motion; and gravity, which drives sediment erosion and transportation from mountains to sedimentary basins. It is estimated that some grains of highly resistant quartz have participated in as many as eight cycles of mountain building and erosion, repeatedly eroding from one sedimentary rock and eventually becoming part of another.

(?) Why does uplift subject the crust to renewed weathering and erosion?

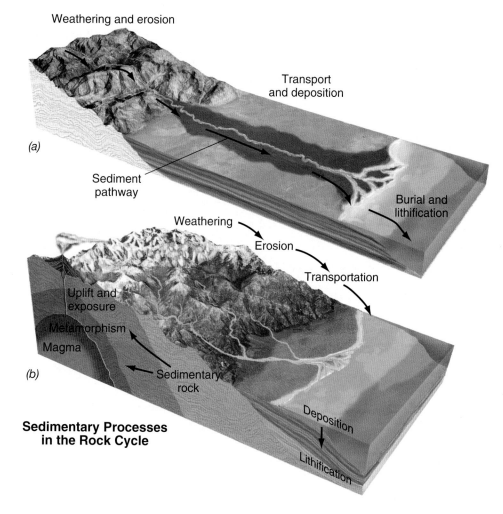

FIGURE 8.14 *(a)* Weathering forms sediment, which is moved by wind, water, ice, and gravity into environments of deposition. *(b)* There they lithify and become solid sedimentary rock. Eventually they may be tectonically uplifted and subjected to a renewed cycle of weathering and erosion.

(?) **Expand Your Thinking**—Describe the relationship between the sedimentary cycle and plate tectonics.

8-6 There Are Eight Major Types of Clastic Sedimentary Rock

LO 8-6 *List the eight major types of clastic sedimentary rock, and give a brief description of each.*

Sedimentary rocks are assigned names based on the lithology of their components—that is, their grains and cements. A rock's *lithology* is a description of its composition (mineralogy) and texture (the size, shape, and arrangement of the sediments).

Clastic Sedimentary Rock

Lithified clastic sediments produce clastic sedimentary rock (**TABLE 8.2**). Rocks in this category are distinguished by the size of their particles (Figure 8.7 and **FIGURE 8.15**) as well as their mineral content.

Conglomerate (**FIGURE 8.16a**) is composed of gravel particles that have been rounded by abrasion in stream or beach environments. Rounding of large, angular clasts indicates that they have been transported some distance from the weathered crust where they originated.

Breccia (**FIGURE 8.16b**) is similar to conglomerate, except that the gravel particles have very angular corners and edges. These indicate that the clasts have not been transported far from their place of origin.

Sandstone is commonly found in three varieties: *quartz sandstone* (**FIGURE 8.16c**), *arkose* (**FIGURE 8.16d**), and *lithic sandstone* (**FIGURE 8.16e**). Quartz sandstone is composed of quartz grains with minor amounts of other types of sediments, usually cemented by silica cement or iron oxide. Arkose is a sandstone that is dominated by quartz grains, but also contains more than 25 percent feldspar grains. It often has a matrix of clay and iron oxide, and is pink or red in colour. Arkose is usually formed by rapid weathering of granite. Lithic sandstone is a dark grey or brown sandstone consisting of poorly sorted angular quartz along with lithic fragments and feldspar grains.

Siltstone (**FIGURE 8.16f**) is composed of mostly silt particles, and *claystone* (**FIGURE 8.16g**) is composed of clay particles. Mudstone consists of a mixture of clay- and silt-size particles. *Shale* (**FIGURE 8.16h**) is a sedimentary rock composed predominantly of clay particles, but usually silt is also present. It has a laminated structure as a result of the preferred orientation of sediment grains (e.g., clay minerals) during deposition, which is sometimes accentuated by compaction.

Reading the Sands of Time

Sandstone can reveal stories about its history. The sand found around active volcanoes consists of volcanic rock fragments, volcanic glass, and other igneous minerals (**FIGURE 8.17**). Sands in river valleys or on beaches close to nearby mountain ranges may consist of quartz (the most durable common mineral), some feldspar (also durable, but more easily weathered to clay), and other clastic minerals weathered from plutonic igneous rocks in the mountains. In areas where there is no abundant source of land-based sediment, sand may be composed of shell fragments, coral, and marine skeletal fragments.

The region where crust weathers and erodes to produce sediment is known as the *source area*. By analyzing sand grains in sediment (**TABLE 8.3**), geologists can infer the type of rock present at the source

TABLE 8.2 Clastic Sedimentary Rocks (listed on the basis of grain size)

Rock	Texture	Composition
Conglomerate	Rounded gravel	Clastic, lithic fragments
Breccia	Angular gravel	Clastic, lithic fragments
Quartz sandstone	Sand	Quartz
Arkose	Sand	Quartz, feldspar
Lithic sandstone	Sand	Quartz, feldspar, lithic fragments
Siltstone	Silt, massive	Silicate minerals
Claystone	Mud, massive	Silicate minerals and clay minerals
Shale	Mud, laminated	Silicate minerals and clay minerals

(?) What criteria are used to determine the texture of clastic sedimentary rock?

FIGURE 8.15 Clastic sedimentary rocks are named for the size of their clasts.

Sediment

Gravel — Sand — Silt — Clay and silt

Sedimentary rock

Conglomerate — Sandstone — Siltstone — Shale

FIGURE 8.16 The eight major clastic sedimentary rocks: *(a)* conglomerate, *(b)* breccia, *(c)* quartz sandstone, *(d)* arkose, *(e)* lithic sandstone, *(f)* siltstone, *(g)* claystone, and *(h)* shale.

TABLE 8.3 Inferring Geologic History from the Composition of Sand

Sand Type	Transport Energy and Distance	Weathering Intensity	Probable Source Rock
Quartz	High energy and distance	High	Granite or metamorphic rocks
Feldspar and quartz	High energy, moderate distance	Moderate to low	Granite or gneiss
Carbonate	Any energy, low distance	Low	Reef or marine organisms
Heavy minerals	High energy, any distance	Moderate to low	Igneous or metamorphic

area, the climate, the relative rate of weathering, tectonic activity, processes acting on the sand within the environment of deposition, and the length of time the sand has resided in the depositional environment.

Quartz grains come from source rocks that are rich in quartz, such as granite, gneiss (a metamorphic rock), or other sandstones containing quartz. Feldspar sand in arkose comes from weathering of granite or gneiss. Rock fragments in lithic sandstone come from fine-grained source rocks, such as shale, slate, basalt, rhyolite, andesite, chert, and various metamorphic rocks.

Using information about the composition of sand, geologists can infer the climate in the source area. Weathering in humid climates will quickly create clay, iron oxides, and aluminum oxides from unstable host rocks containing feldspar and minerals rich in iron and aluminum. The presence of feldspar in sandstone indicates either that the climate was arid and chemical weathering was not intense or that erosion rates were rapid. (If tectonic uplift were fast, the resulting steep slopes would produce rapid erosion.) If quartz is dominant, the climate was probably humid, destroying all feldspars and other easily weathered minerals.

FIGURE 8.17 The composition of the sand grains depends on their source. The sand on this beach in Iceland is black, because the grains are derived from the erosion of basalt volcanic rock.

 Expand Your Thinking—What can the eight major types of clastic sedimentary rock tell you about their source areas? Interpret each.

8-7 Some Sedimentary Rocks Are Formed by Chemical and Biogenic Processes

LO **8-7** *List the major types of chemical sedimentary rock and biogenic sedimentary rock.*

For the most part, chemical sedimentary rocks are identified based on their mineral composition; however, in some cases, rock *texture* is the basis for identification (**TABLE 8.4**).

Rock salt, which is an evaporite, is composed of the mineral halite (a member of the halide mineral group; recall Chapter 4, "Minerals") (**FIGURE 8.18a**). It is typically formed by the evaporation of salty water or sea water that contains dissolved sodium and chlorine ions. *Rock gypsum* is composed of the mineral gypsum ($CaSO_4 \cdot 2H_2O$; a member of the sulphate mineral group) (**FIGURE 8.18b**) and, like rock salt, usually results from the evaporation of sea water.

Limestone is composed of the mineral calcite (from the carbonate mineral group). Although most limestone is biogenic, *travertine* is a type of inorganic limestone (**FIGURE 8.18c**). Travertine forms wherever water that is saturated with $CaCO_3$ evaporates. Under such conditions, layers of inorganically precipitated calcium carbonate accumulate, forming terraces, towers, pools, and other odd shapes. Travertine also forms in caves where groundwater enriched with $CaCO_3$ drips from the ceiling. In time, structures called *stalactites* develop and hang down from the ceiling. Water that drips to the floor may form *stalagmites,* which gradually grow upward toward the ceiling. When stalagmites and stalactites meet and merge, they form a *column.*

TABLE 8.4 Chemical Sedimentary Rocks

Rock	Texture	Composition
Chert	Microcrystalline	Microcrystalline silica
Dolostone	Crystalline	Dolomite
Rock gypsum	Crystalline	Gypsum
Rock salt	Crystalline	Halite
Travertine	Microcrystalline	Calcite from saturated fluids

Dolostone (**FIGURE 8.18d**) is similar to limestone but is composed mostly of the mineral *dolomite* [$CaMg(CO_3)_2$; a carbonate]. It forms when dissolved magnesium in groundwater substitutes for some of the calcium in the original limestone; it can also form as a result of direct precipitation of dolomite. Most dolostone is a result of chemical alteration of limestone by magnesium-rich groundwater.

Chert (**FIGURE 8.18e**) is a sedimentary rock composed of microcrystalline quartz that comes from inorganic precipitation in

(a)

(b)

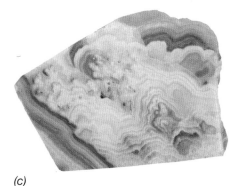

(c)

(d)

(e)

Courtesy of Chip Fletcher

(?) What is the primary criterion for recognizing chemical sedimentary rocks?

FIGURE 8.18 Chemical sedimentary rocks: *(a)* rock salt, *(b)* rock gypsum, *(c)* travertine, *(d)* dolostone, and *(e)* chert.

groundwater. A biogenic form of chert originates in organic beds of marine silica-rich fossils of two types of plankton: radiolarians and diatoms. Black or grey chert is called *flint*, and red chert is called *jasper*.

Biogenic Sedimentary Rock

Biogenic sedimentary rocks (**TABLE 8.5**) are also identified based on their mineral composition and texture.

Coquina (**FIGURE 8.19a**) is a form of limestone composed entirely of broken and worn shell *fragments*. The pieces usually are weakly cemented together, especially in recent deposits. Coquina consists of mechanically transported fossil shell debris with little or no matrix, loosely cemented so that the rock looks very porous. Typically the shells are those of clams and snails from a beach environment.

Skeletal limestone (**FIGURE 8.19b**) is composed of fossil fragments of carbonate organisms. Some of these are visible fragments, but many limestones are composed of microscopic skeletal fragments. Skeletal limestone may also consist of large pieces of reef organisms such as coral or some other type of shelled fossil. However, most limestone accumulates in marine environments from biogenic ooze. Biogenic ooze is calcium carbonate in the form of skeletal fragments of microscopic plankton. *Micrite* is "lime mud," dense, dull-looking sediment composed of clay-size crystals of calcite. Overall, carbonate rock of all types accounts for about 15 percent of Earth's sedimentary crust.

Chalk (**FIGURE 8.19c**) is a biogenic limestone composed of the microscopic skeletons of coccolithophorids, single-celled plankton, often called "coccoliths." Coccolith skeletons are continuously settling toward the bottoms of oceans around the world. But deep water is so cold and acidic that the skeletons dissolve.

TABLE 8.5 Biogenic Sedimentary Rocks		
Rock	**Texture**	**Composition**
Chalk	Clay or mud	Skeletal coccolitho-phorids
Coal	Massive, blocky	Concentrated carbon
Coquina	Sand or gravel	Shell fragments
Skeletal Limestone	Visible or micro-scopic skeletal fragments	Calcite

Therefore, chalk accumulates only at shallower depths, which today means either along the crests of mid-ocean ridges or on shallow continental shelves.

Coal (**FIGURE 8.19d**) is an organic sedimentary rock that is the result of physical and chemical alteration of *peat* by processes involving bacterial decay, compaction, heat, and time. Peat deposits typically form in wetlands where plant debris accumulates at a rate that exceeds the rate of bacterial decay. The rate of decay is low because the oxygen in organic-rich water is used up by the decay process. For peat to become coal, it must be buried by sediment. Burial causes compaction, which squeezes out water. Continued burial and the addition of heat and time cause complex hydrocarbon compounds in the deposit to break down and change in a variety of ways. Through this complex process, peat eventually is transformed into coal.

Courtesy of Chip Fletcher

(a)   (b)   (c)   (d)

FIGURE 8.19 Biogenic sedimentary rocks: *(a)* coquina, *(b)* skeletal limestone, *(c)* chalk, and *(d)* coal.

 Each biogenic rock tells a story about the history of life on Earth. Imagine that the four rocks in Figure 8.19 were the same age (say, 10 million years old) and found within 100 km of each other. Describe a logical geographic pattern for where they originally formed. What would they signify about past environments? Build a testable hypothesis.

? Expand Your Thinking—What is the dominant mineral in biogenic and chemical sedimentary rock? Where does it come from, and why is it so abundant?

8-8 Sedimentary Rocks Preserve Evidence of Past Environments

LO *8-8 Describe how a geologist uses inductive reasoning to interpret Earth's history using sedimentary rocks.*

Every sedimentary rock bears the imprint of the natural forces that shaped it. By examining layers of sedimentary rock that accumulate over time, geologists decipher Earth's history as if reading the pages of an encyclopedia.

Using *inductive reasoning,* for instance, geologists infer that at locations where sandstones are found, there was once a river or beach; places where marine skeletal limestones are found today were formerly part of the sea floor; organic sediments tell us where life has been active in the past; poorly sorted but weakly abraded sediments might reveal a former glacier. Seashells can be found in the high mountaintops of the Himalayas; ancient coral reefs dot the Rocky Mountains; and evidence of massive river deltas far from any modern coastline has been found in the Appalachian Mountains.

Some of these sedimentary deposits contain valuable ores and fuel. Others hold fossils of long-extinct species (**FIGURE 8.20**). Throughout the world, sedimentary rocks present the most complete record of past Earth surface environments.

Continental Environments

Glaciers (**FIGURE 8.21a**) are characterized by a cold climate, moving ice that mechanically grinds the crust and plucks and abrades clasts ranging in size from clay to boulders the size of houses, and voluminous discharges of sediment-laden water from the front of the ice (where it is perpetually melting). The sediment, which is usually poorly sorted, contains particles of many different sizes and an abundance of silt and clay.

Deserts (**FIGURE 8.21b**) form in arid environments with little vegetation. In such environments, wind is especially effective in shaping the land surface. Windblown sand is abraded and well sorted, and may be formed into dunes. The action of wind may also erode and remove sand, leaving behind gravel particles that form a desert pavement.

Alluvial fans (**FIGURE 8.21c**) often develop in arid environments where rainfall is infrequent but torrential. When it does rain, massive discharges of sediment emerge from steep valleys that drain mountainous areas and form fan-shaped deposits of mud, sand, and gravel on the adjacent flatter areas.

Streams (**FIGURE 8.21d**) are characterized by channels that may carry sediment of all sizes. Sedimentary deposits within the channels are typically coarser grained than those on any adjoining flood plain.

Wetlands (**FIGURE 8.21e**) are characterized by standing water in which most of the sediment is organic, produced by abundant vegetation. Biogenic sediments such as peat tend to accumulate, as do clays and silts. Wetlands form where the water table intersects the land surface.

Lakes (**FIGURE 8.21f**) are highly diverse settings that vary greatly in size and may be fresh or saline, shallow or deep. These environments accumulate clastic, chemical, and biogenic sediments. Some lakes dry up each year, producing chemical sediment through evaporation. Other lakes are deep and large and are characterized by high-energy physical processes, such as frequent storms, that promote coarse clastic sediments.

 What might have killed this fish 48 million years ago? What evidence would you look for to support your hypothesis?

© Visuals Unlimited/Corbis

FIGURE 8.20 Sedimentary rocks contain fossils that provide important clues to past Earth environments. This rock is from a former lake in the Green River Formation in Wyoming. We know the formation (a "formation" is a distinct rock unit) was a former lake because of the fine sediments (silt) and the types of fossils we find there. This fossil shows a larger fish eating a smaller fish that apparently died in mid-meal.

FIGURE 8.21 Continental environments of deposition: *(a)* glacier, *(b)* desert and dune, *(c)* alluvial fan, *(d)* stream, *(e)* wetland, and *(f)* lake.

(?) What is the typical sediment type found in each of these environments: clastic, chemical, biogenic, or some combination?

Coastal Environments

Deltas (**FIGURE 8.22a**) contain deposits of clay, silt, and sand that form an irregular fan shape at locations where a stream flows into a standing body of water, such as a lake or the ocean. Deltas form where large amounts of sediment accumulate and are usually associated with major rivers, such as the Fraser, the Mississippi, and the Nile.

Beaches (**FIGURE 8.22b**) are often exposed to high-energy waves that deposit grains of material of many types and size. Many beaches, however, consist predominantly of coarse sand-size grains of quartz. They tend to erode or grow seaward according to the season and are frequently pounded by storms that cause substantial erosion, and may transport the sand and redeposit it somewhere else along the shoreline.

Barrier islands (**FIGURE 8.22c**) are composed of coarse sand and are separated from the mainland by a barrier lagoon. Chains of barrier islands form as a result of the accumulation of sand driven by waves and currents. They are separated by tidal inlets that control the flow of sea water into the lagoon.

Lagoons (**FIGURE 8.22d**) contain mud and fine sand with a large biogenic component. They are protected from the ocean by barrier

islands or reefs, yet they experience the effect of tidal currents through inlets in these barriers.

Tidal wetlands (**FIGURE 8.22e**) contain silt and clay with large amounts of organic sediment and skeletal fragments of marine organisms, such as plankton. They support abundant growth of salt-tolerant plants that trap suspended mud that is carried onto the wetland surface by high tides during the course of the day.

Marine Environments

Reefs (**FIGURE 8.23a**) are massive biological constructions of calcium carbonate that extend along the edges of continental shelves or volcanic islands. They are built by corals and other invertebrates, and various types of algae in warm, sunlit seas in the tropics between latitudes 30°N and 30°S of the equator.

Continental shelves (**FIGURE 8.23b**) are the flooded edges of continents. They accumulate clastic, biogenic, and chemical sediments over long periods of geologic time. A shelf can be shallow (less than 200 m deep), flat, and up to hundreds of kilometres wide. Mapping the ocean floor is very difficult. But using the techniques described in the "Geology in Our Lives" box titled "Mapping the Sea Floor Using

(a)

(c)

(d)

(b)

(e)

FIGURE 8.22 Coastal environments of deposition: *(a)* delta; *(b)* beach; *(c)* barrier island; *(d)* lagoon; and *(e)* tidal wetland.

? What is the typical sediment type found in each of the environments in Figure 8.22: clastic, chemical, biogenic, or a combination?

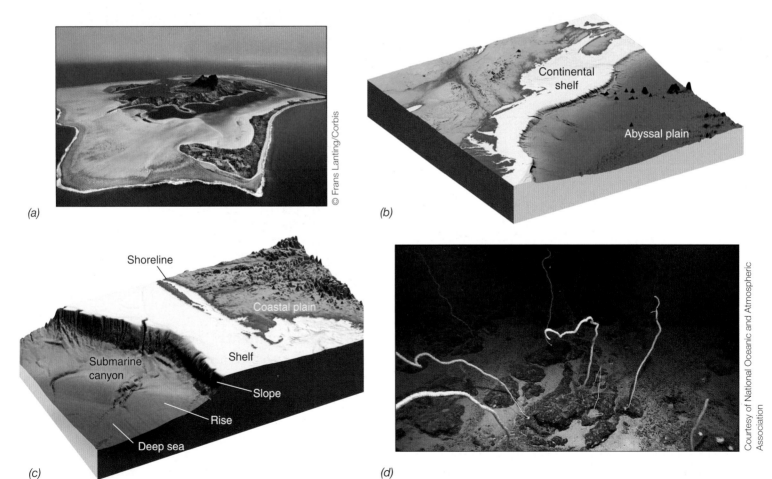

(a)

(b)

Continental shelf

Abyssal plain

(c)

Shoreline

Coastal plain

Shelf

Submarine canyon

Slope

Rise

Deep sea

(d)

FIGURE 8.23 Marine environments of deposition: *(a)* reef; *(b)* continental shelves; *(c)* continental margin; and *(d)* deep-sea environments.

 What is the typical sediment type found in each of the environments in Figure 8.23: clastic, chemical, biogenic, or a combination?

Satellite Radar," geologists are able to decipher the topography of continental shelves and the deep sea.

Continental margins (**FIGURE 8.23c**) extend from the landward edge of continental shelves (the shoreline), described above, to the seaward edge of the continental rise. Margins thus include the *continental slope,* which descends steeply, and the *continental rise,* which has a gentle slope. The rise is composed of biogenic marine sediments and clastic sediments delivered to the area by *turbidity currents,* undersea avalanches of sediment that originate on the slope.

Deep-sea environments (**FIGURE 8.23d**) include the deep floors of all the oceans and consist of the oceanic trenches, mid-ocean ridges, and *abyssal plains,* the flattest parts of the ocean floors. They may contain fine-grained clastic material derived from the continental margins and fine-grained biogenic ooze derived from the remains of plankton. The crests of the mid-ocean ridges, where new oceanic crust is formed, have only thin sedimentary coatings because of their young age.

Please review your understanding of these environments by reading the "Critical Thinking" box titled "Environments of Deposition."

GEOLOGY IN OUR LIVES

MAPPING THE SEA FLOOR USING SATELLITE RADAR

You do not feel it, but as you move from place to place, you experience changes in gravity created by variations in the density of the rock beneath your feet. Regions with denser rock exert a stronger gravitational pull. We do not feel these changes because our bodies are not sensitive enough. But water instantly responds to variations in gravity by accumulating in places where gravitational force is high. On the deep-sea floor, high and low topographic features cause variations in the gravity field because they have greater or lesser mass. The surface of the ocean forms bumps, ridges, and depressions that mimic these features, even though they may be over 5 km below the surface. These bumps and dips in the ocean surface are caused by minute variations in Earth's gravitational field resulting from variations in seafloor topography.

A typical undersea volcano is 2,000 m tall and has a radius of about 20 km. The surface bump it creates cannot be seen with the naked eye because its sides slope very gently, but it can be measured using satellite technology. Mid-ocean ridges, deep-sea trenches, undersea volcanoes, oceanic fracture zones, and many other types of seafloor and tectonic features are "reflected" in the topography of the ocean surface (**FIGURE 8.24**).

These tiny bumps and dips on the ocean's surface can be measured using very accurate radar mounted on a satellite. In 1985, the U.S. Navy launched GEOSAT (Geodectic Satellite) to map the ocean surface at a horizontal resolution of 10 km to 15 km and a vertical resolution of 0.03 m. GEOSAT orbits on a path that takes it nearly over the North and South Poles 14.3 times per day. In this orbit, the satellite maps the sea surface at a speed of about 7 km/s, taking about 1.5 years to make a map of Earth's entire surface.

Two very precise distance measurements must be made in order to map the topography of the ocean surface with such a high degree of accuracy. First, a global network of stations constantly tracks the location of the satellite; second, the satellite measures its height above the ocean surface using the two-way travel time of pulses of microwave radar. A sharp radar pulse emitted from the satellite measures a spot on the ocean surface 1 km to 5 km in diameter. The "footprint" of the pulse must be large enough to average out the local irregularities in the surface due to ocean waves. The satellite emits 1,000 pulses per second to improve the accuracy of the measurement. The radar data are corrected for variations in the composition of the atmosphere (e.g., moisture content) that may cause changes in the radar pulse. The sea surface height is also corrected for tidal fluctuations.

This application of physics provides detailed information about seafloor topography, the features of which can be explained by the theory of plate tectonics.

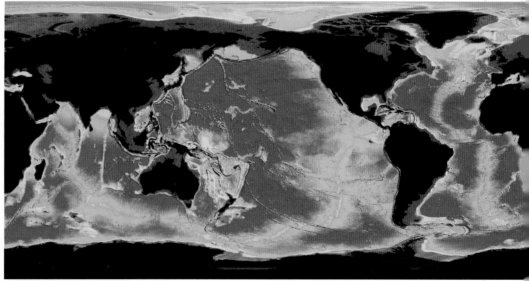

FIGURE 8.24 Most seafloor topography is caused by the creation or movement of lithospheric plates.

 Expand Your Thinking—Why do geologists study the record of past environments in sedimentary rocks? Identify some important reasons why you would study a sedimentary formation in detail.

? CRITICAL THINKING

ENVIRONMENTS OF DEPOSITION

Clastic sediments are exposed to a wide range of natural conditions as they travel from their source to the environments of deposition where they come to rest and are joined by chemical and biogenic sediments. Along the way, they are abraded, sorted, rounded, temporarily buried, and subjected to chemical weathering. FIGURE 8.25 illustrates some of the many environments in the sedimentary cycle from the mountains to the sea floor.

    Please work with a partner and answer the following questions.

1. What specific characteristics could you use to distinguish between marine and non-marine sedimentary deposits?
2. Imagine that you were asked to sample the sediments in a stream channel at three locations: (a) the upper portion of the stream, (b) the middle portion, and (c) the lower portion. Assuming that granite and lithic sandstone are the source of weathering for your sediments, describe the composition and texture of each sample at each location.
3. Choose four environments of deposition in Figure 8.25. For each, describe the texture and composition of the sediments you would be likely to find there.
4. Name the sedimentary rock that would form from the sediments in each of your four environments.
5. Drawing on your knowledge of natural resources, name three environments of deposition, the rock that will form in each, and the natural resource that could be extracted from that rock.
6. List three high-energy environments and three low-energy environments in Figure 8.25.
7. Consider the continental shelf, the streams, the desert, and the glacier. If the sediment source for each were granitic mountains, predict the texture and composition of sediments at each environment.

FIGURE 8.25 Environments of deposition where sediments collect and lithify.

8-9 Primary Sedimentary Structures Record Environmental Processes

LO **8-9** *List four common sedimentary structures, and explain their significance.*

Depending on the depositional environment, sedimentary rocks record certain features reflecting the physical processes that act on the sediments. These features are called **primary sedimentary structures.** They include constructions, such as ripple marks and mud cracks, which form in a sedimentary environment and are preserved in the solid rock after it lithifies. Some sedimentary structures are created by water or wind as it moves grains of sediment. Others form after sediments have been deposited. Primary sedimentary structures provide information about the environmental conditions under which sediment is deposited: Some form in quiet water under low-energy conditions, others in moving water or high-energy conditions.

Many primary sedimentary structures are originally created as *bedforms,* features that develop in the *active* environment as a result of wind or water shaping loose sediments. These bedforms include dunes, ripples, and sand waves.

The most basic sedimentary structure is *stratification,* the layering that occurs as sediments are deposited over time (FIGURE 8.26). You experimented with stratification earlier in the chapter in the "Critical Thinking" box titled "Sediment Deposition." Stratification is the most obvious feature of sedimentary rocks because the strata, or layers, are visible due to differences in the colour or texture of the various layers.

Because grains tend to settle through water or air under the influence of gravity, most sedimentary environments produce horizontal strata, with older sediment on the bottom and younger sediment on the top. However, when sediments accumulate on the steep slope of a sand dune, the front of a delta, or in certain kinds of stream channels, the beds may form at an angle to the horizontal. Such beds are called *cross beds* (FIGURE 8.27), and they provide a clue to the physical processes operating at the time when the sediments accumulated. Usually cross beds indicate that blowing wind or running water caused the deposition of these sediments.

Graded beds form when deposition occurs in water containing sediment consisting of grains of many sizes. As water velocity decreases, the larger, heavier grains fall through water more rapidly than smaller grains, so the resulting bed has large particles at the bottom and smaller particles at the top (FIGURE 8.28). You may have

FIGURE 8.26 Sedimentary rocks usually display layers of deposited sediments called *strata*. Stratification occurs because of changes in the grain size, sorting, colour, or matrix from one layer to the next. These sedimentary rock layers are beautifully preserved in the Horseshoe Canyon area, near Drumheller in Alberta.

© Barrett & MacKay/All Canada Photos/Corbis

FIGURE 8.27 Cross bedding is apparent in this sandstone composed of fossil sand dunes in Zion National Park, Utah. The dipping beds indicate the former face of a dune migrating from left to right.

Bernhard Edmaier/Science Source

 Do strata reflect changes in environment?

 Which way did the wind blow in this ancient desert?

observed the formation of graded bedding if you used sandy soil in the experiment described in the "Critical Thinking" box titled "Sediment Deposition."

A *ripple mark* is an important sedimentary structure that forms when running water or blowing wind shapes loose sediment into ripples. These ripples are commonly preserved during lithification (**FIGURE 8.29a and b**) and provide important indicators of past environments. For example, symmetrical ripples preserved in a sandstone may suggest that the original sandy sediment was deposited in a tidal environment, in which there was regular back-and-forth movement of water forming and shaping the ripples.

Perhaps you have seen the cracks that develop in wet mud that lies drying in the sun. As the mud dries, it contracts in all directions and forms polygonal features known as *mud cracks* (**FIGURE 8.29c and d**). These can be preserved in sedimentary rock, showing that the same process occurred in past environments.

© Mike Grandmaison/All Canada Photos/Corbis

FIGURE 8.28 This accumulation of stream sediment at Hopewell Rocks on the Bay of Fundy, Nova Scotia, displays graded beds with large pebbles at the base of channels with smaller sand grains towards the top.

(?) Why do graded beds form?

(?) **Expand Your Thinking**—Ripple marks in streams and rivers tend to be asymmetrical: One slope is short and steep (facing downstream) and the other slope is long and gentle (facing upstream). This is because the shape of the ripples is controlled by the direction that water is flowing, and the direction that the grains are being transported. However on the shallow sea floor, ripples are symmetrical; both slopes are the same. Why is this?

iStock.com/Mlenny

(a)

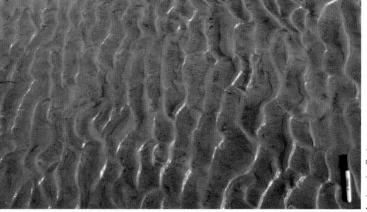
Andrew La Croix

(b)

© Theo Allofs/Corbis

(c)

Marli Miller

(d)

(?) What kind of sediment typically forms mud cracks?

FIGURE 8.29 Bedforms: *(a)* modern ripple marks; *(b)* ancient ripple marks in sandstone; *(c)* modern mud cracks form by desiccation; *(d)* ancient mud cracks lithified in sedimentary rock.

LET'S REVIEW "GEOLOGY IN OUR LIVES"

Now that you have finished the chapter, "Geology in Our Lives" will have taken on new meaning for you. Let us review it: Much of our knowledge of Earth's history is derived from the study of sedimentary rocks. Sedimentary rock is formed from the weathered and eroded remains of the crust. It contains fossils and other evidence of the history of life, changes in the environment, climate shifts, plate tectonic movement, and many other details of how the planet has developed and changed over time. Sedimentary rock is the source of fossil fuels such as coal, petroleum, natural gas, and oil shale. We get valuable metals from sedimentary accumulations such as placer deposits of gold and platinum, and the ores of iron and aluminum. Construction materials including building stone, sand and gravel to make road beds, and cement for the building industry all come from sedimentary rocks. Perhaps of greatest significance, most of our communities obtain fresh water for drinking, irrigation, and manufacturing from groundwater found in the pore spaces of sedimentary rocks.

Sediments can provide valuable information. For instance, if you live in an earthquake-prone region, sediments near your home may reveal how often damaging earthquakes have occurred in the past. If you live near the ocean, nearby sediments may record the pattern and frequency of hurricanes and tsunamis. Sediments can reveal how often a river floods, a mountain produces landslides, or a volcano erupts. By "reading" the sedimentary record, geologists help make our world safer, find critical natural resources, and learn about human origins. Keep this important idea in mind when we learn about geologic time (Chapter 13) and look more closely at Earth's history (Chapter 14).

STUDY GUIDE

8-1 Sedimentary rock is formed from the weathered and eroded remains of the crust.

- Most of Earth's surface is covered with layers of loose **sediment,** and more than 75 percent of the land surface is **sedimentary rock.**

- From the moment they are freed by erosion to travel across Earth's surface, sediments may be imprinted with the physical and chemical characteristics of the various environments they encounter. Geologists interpret these "signatures" and are able to infer geologic history from them.

- Sedimentary rock provides natural energy and mineral resources that are critically important to modern industry, technology, and society.

- Sedimentary rock forms from weathered and eroded remains of igneous, metamorphic, and other sedimentary rock. Many sedimentary rocks contain direct and indirect evidence of living organisms, such as skeletal parts or impressions of plants and animals (leaf imprints, footprints, etc.).

8-2 There are three common types of sediment: clastic, chemical, and biogenic.

- There are three types of sediment, defined by their composition:

  1. **Clastic sediments** are broken pieces of crust deposited by water, wind, ice, or other physical process. Minerals that are commonly found in clastic sediments include quartz, orthoclase feldspar, muscovite, clay minerals, and some heavy minerals. Also common in clastic sediments are eroded fragments of rocks, known as lithic fragments.

  2. **Chemical sediments** are produced by inorganic precipitation of dissolved compounds: calcite, gypsum, halite, microcrystalline quartz, iron oxide, and others.

  3. **Biogenic sediments** are produced by living organisms and organically precipitated compounds such as calcite and chert.

8-3 Sediments change as they are transported across Earth's surface.

- Clastic sediments are solid components of pre-existing rock that have been transported from one location to another. Since most rock is composed of silicate minerals, most clastic sediment consists of silicate minerals.

- As sedimentary grains are transported, they undergo two kinds of changes related to the energy of the environment—the force of the wind or water current carrying them:

  1. They experience **sorting**, in which they are separated based on their density.

  2. They experience **abrasion**, and the longer they are transported, the smaller and more spherical they become.

8-4 Clastic grains combine with chemical and biogenic sediments.

- Dissolved components move from weathering sites, where they are created, into sedimentary basins, where they are deposited. Dissolved ions and compounds may be deposited through either organic or inorganic precipitation. Both processes result in a crystalline solid that joins the sedimentary cycle as either chemical or biogenic sediment.

- Sediments that accumulate within sedimentary basins display changes in composition. Unstable grains (olivine, pyroxene, calcium-rich plagioclase, amphibole, and others) become less abundant while stable grains (quartz, clays, muscovite, sodium-rich plagioclase, orthoclase) become more abundant. Biogenic sediments accumulate, and chemical sediments may become more abundant.

8-5 Sediment becomes rock during the sedimentary cycle.

- Sedimentary rock forms through a process called **lithification** in two ways:

  1. Larger particles, such as sand and gravel, are bound together by natural cements that are inorganically precipitated from dissolved compounds in groundwater.

  2. Smaller particles, such as silt and clay, are compacted and consolidated by pressure created by the weight of overlying layers of sediment.

- The sedimentary cycle is a component of the rock cycle that supplies sediment created by the weathering and erosion of tectonically uplifted crust. The cycle consists of tectonic uplift, weathering, erosion and transportation, deposition, burial, and lithification.

8-6 There are eight major types of clastic sedimentary rock.

- Sedimentary rocks are identified based on their lithology (the composition and texture of the particles and cements within them). Specific combinations of texture and composition are characteristic of particular types of sedimentary rock. The composition of clastic sedimentary rock can be interpreted in terms of the history of the sediments: transport energy and distance, weathering intensity, and probable source rock. These provide abundant information about Earth's history.

- Clastic rocks are distinguished on the basis of grain size as well as composition. They consist of conglomerate, breccia, quartz sandstone, arkose, lithic sandstone, siltstone, claystone, and shale.

8-7 Some sedimentary rocks are formed by chemical and biological processes.

- Chemical sedimentary rocks are distinguished mostly based on their mineral composition, but in some cases, texture is the critical feature. The rocks in this category are rock salt, rock gypsum, travertine, dolostone, and chert.

- Biogenic sedimentary rocks are also distinguished based on their mineral composition and texture. The rocks in this category are limestone, coquina, chalk, and coal.

8-8 Sedimentary rocks preserve evidence of past environments.

- Most sedimentary rocks reflect the environment in which they formed and provide a record of past conditions on Earth's surface.

- A number of sedimentary environments are found on continents, along coasts, and in the oceans. Continental environments include glaciers, deserts and dunes, alluvial fans, streams, wetlands, and lakes. Coastal environments include deltas, beaches, barrier islands, lagoons, and tidal wetlands. Marine environments include reefs, continental shelves, continental margins, and deep-sea environments.

8-9 Primary sedimentary structures record environmental processes.

- **Primary sedimentary structures** are physical features that form in a sedimentary environment and are preserved in the solid rock after it has lithified. They provide important clues for geologists analyzing past surface conditions. Sedimentary structures include stratification, cross beds, graded beds, ripple marks, and mud cracks.

KEY TERMS

abrasion (p. 196)
biogenic sediments (p. 194)
chemical sediments (p. 194)
clastic sediments (p. 194)
clay (p. 192)
environments of deposition (p. 196)

evaporites (p. 195)
iron oxides (p. 193)
lithification (p. 200)
mud (p. 192)
porosity (p. 200)
primary sedimentary structures (p. 215)

sand (p. 192)
sediment (p. 192)
sedimentary rock (p. 192)
silt (p. 193)
sorting (p. 196)

ASSESSING YOUR KNOWLEDGE

Please answer these questions before coming to class. Identify the best answer.

1. Geologists study sedimentary rocks because
   a. they provide a record of Earth's history.
   b. they are sources of fossil fuels.
   c. they may contain important mineral resources.
   d. they may contain fossils, providing a history of life, including human evolution.
   e. All of the above.

2. Sediments produced by the action of living organisms are called
   a. chemical sediments.
   b. physical sediments.
   c. clastic sediments.
   d. biogenic sediments.
   e. None of the above.

3. Well-sorted and well-rounded sand grains indicate that sediment
   a. came from a nearby source area.
   b. was deposited at the location where it was found.
   c. travelled from a distant source area.
   d. has not been influenced by weathering.
   e. None of the above.

4. "Lithification" refers to
   a. the set of natural processes that turn sediment into rock.
   b. the processes of erosion and tectonic uplift.
   c. the effects of chemical weathering.
   d. All of the above.

5. After being created by weathering, sediments
   a. experience more weathering.
   b. combine with chemical sediments.
   c. combine with biogenic sediments.
   d. experience sorting and abrasion.
   e. All of the above.

6. The "sedimentary cycle" refers to
   a. the continual erosion of sediments from mountainsides.
   b. the process of recycling sediments.
   c. the formation of rock through compaction of sediments.
   d. the formation of rock through chemical precipitation of sediments.
   e. None of the above.

7. Which of the following statements is correct?
   a. Clastic sedimentary rocks include arkose and lithic sandstone.
   b. Biogenic sedimentary rocks include coquina and coal.
   c. Chemical sedimentary rocks include travertine and rock salt.
   d. Clastic sedimentary rock includes conglomerate and shale.
   e. All of the above.

8. Chemical sedimentary rocks include
   a. lithic sandstone.
   b. coal.
   c. breccia.
   d. chert.
   e. None of the above.

9. Biogenic sedimentary rocks are formed by
   a. evaporation.
   b. inorganic precipitation.
   c. glaciers.
   d. floods.
   e. None of the above.

10. Rock fragments are known as
    a. lithic fragments.
    b. bioclastic sediments.
    c. evaporites.
    d. natural cements.
    e. None of the above.

11. Particle sizes are described using the following terms
    a. sand, gravel, lithic fragments, and natural cements.
    b. gravel, sand, silt, and clay.
    c. conglomerate, sandstone, arkose, and shale.
    d. abraded, sorted, rounded, and spherical.
    e. None of the above.

12. Continental environments include
    a. beaches, continental margins, rivers, and lakes.
    b. beaches, rivers, lakes, and glaciers.
    c. beaches, rivers, deltas, and barrier islands.
    d. rivers, lakes, wetlands, and glaciers.
    e. None of the above.

13. Organisms play a significant role in the origin of _____ sedimentary rock.
    a. clastic
    b. biogenic
    c. chemical
    d. lithologic
    e. None of the above.

14. Sedimentary rocks are classified by
    a. mineralogy and fossils.
    b. cementation and compaction.
    c. environment of precipitation and environment of deposition.
    d. composition and texture.
    e. All of the above.

15. Primary sedimentary structures are
    a. physical features of a rock related to the environment of deposition.
    b. physical features of a rock related to the process of cementation.
    c. chemical features of a rock produced by the motion of water and wind.
    d. sediment forms produced by biogenic processes.
    e. None of the above.

FURTHER RESEARCH

1. What types of sedimentary environments are found in your area? What types of sediments accumulate there?

2. What role does calcium carbonate play in the rock cycle?

3. What types of sedimentary rock would you expect to find on the Moon? On Mars?

4. Choose a sedimentary environment and describe the primary sedimentary structures you would expect to find preserved in the resulting rocks.

5. Why do sediments tend to move from mountainous regions toward the sea?

6. As unconsolidated sediments are compacted, what happens to the water within the pore spaces?

7. How can quartz move through the sedimentary cycle several times in the course of geologic history?

ONLINE RESOURCES

Explore more about sedimentary rocks on the following websites:

U.S. Geological Survey—sedimentary rocks:
http://geomaps.wr.usgs.gov/parks/rxmin/rock2.html

NASA Earth Observatory—satellite images of modern environments:
http://earthobservatory.nasa.gov

Explore the geological map of Canada and find out where there are sedimentary rocks, as well as the other rock types:
www.nrcan.gc.ca/earth-sciences/products-services/mapping-product/geoscape/canada/6175

Chris Hadfield, the Canadian commander of the NASA Space Station sent back spectacular images of the Earth, including many sedimentary environments:
https://twitter.com/Cmdr_Hadfield/media/grid

Additional animations, videos, and other online resources are available at this book's companion website:
www.wiley.com/go/fletchercanada

This companion website also has additional information about WileyPLUS and other Wiley teaching and learning resources.

CHAPTER 9
METAMORPHIC ROCK

Chapter Contents and Learning Objectives (LO)

9-1 Metamorphic rocks are composed of sedimentary, igneous, or metamorphic minerals that have recrystallized.

LO 9-1 *Describe the process of metamorphism.*

9-2 Changes in heat and pressure can cause metamorphism.

LO 9-2 *Compare and contrast regional and contact metamorphism.*

9-3 Chemically active fluids transport heat and promote recrystallization.

LO 9-3 *Describe the role of water in metamorphism.*

9-4 Rocks evolve through a sequence of metamorphic grades.

LO 9-4 *Describe the metamorphic grades and how they are identified.*

9-5 Foliated texture is produced by directed stress related to regional metamorphism.

LO 9-5 *Compare and contrast the four principal types of foliated texture.*

9-6 Nonfoliated rocks may develop during regional or contact metamorphism.

LO 9-6 *List the common nonfoliated rocks, and describe the products of contact metamorphism.*

9-7 The relationship between mineral assemblage and metamorphic grade is expressed by metamorphic facies.

LO 9-7 *Define the term "metamorphic facies."*

9-8 Metamorphism is linked to plate tectonics.

LO 9-8 *Describe how metamorphic facies are related to plate tectonics.*

GEOLOGY IN OUR LIVES

All rocks can be exposed to rising temperature and pressure conditions where they are intruded by magma, buried deep in the crust, or involved in processes at convergent margins. Increases in pressure and heat, usually in the presence of chemically active fluids, recrystallize igneous, sedimentary, and metamorphic minerals to make new metamorphic rock. Conditions that lead to metamorphism are linked to plate tectonics or deep burial within the crust or both. Metamorphic rocks provide important building materials and mineral resources.

(?) Metamorphic rocks, such as these found in the northern Monashee Mountains of southern B.C., can form when the crust is exposed to rising pressure and temperature conditions. What processes can raise the pressure and temperature of the crust?

9-1 Metamorphic Rocks Are Composed of Sedimentary, Igneous, or Metamorphic Minerals that Have Recrystallized

LO 9-1 *Describe the process of metamorphism.*

Deep within the crust, pressures can reach a crushing 10,000 atmospheres (10,000 times the pressure of the atmosphere on Earth at sea level) and temperatures can be over 600°C. What happens when rocks are subjected to such high pressures and temperatures? How do they change and are the changes predictable? What factors influence the evolution of rock under these extreme conditions? This chapter answers these and additional questions regarding rock **metamorphism**.

Geologists use the term *metamorphism* to describe changes in mineralogy, texture, and/or composition that occur in rocks affected by changes in temperature, pressure (also called **stress**), and/or chemically active fluids. In the world of rocks, metamorphism occurs predominantly by a phenomenon called *solid-state reaction* or a solid-state change. During metamorphism, minerals that *recrystallize* and remain solid are referred to as *metamorphic minerals*, if part, or all, of the rock melts, that part is classified as igneous.

What Is Metamorphic Rock?

Metamorphic rocks are sedimentary, igneous, or metamorphic rocks that take on a new texture and mineralogy in order to maintain equilibrium under changing conditions within the crust. That is, minerals within a rock change in order to be stable under new conditions.

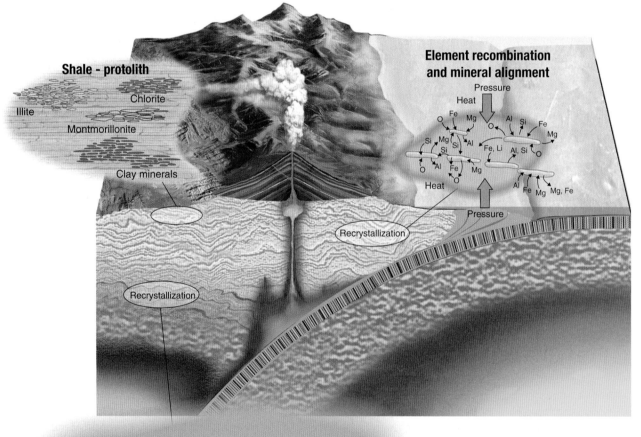

 Do clay minerals melt before recrystallizing?

FIGURE 9.1 During metamorphism, minerals recrystallize. Here, shale composed of clay minerals made of silicon, oxygen, aluminum, iron, magnesium, sodium, and other elements experiences metamorphism due to increasing pressure and temperature as the crust thickens at a convergent plate margin. As conditions within the crust become more intense, elements in clay recombine to form new, more stable minerals. This can eventually result in a metamorphic rock called *gneiss* that consists of an assemblage of metamorphic minerals aligned in dark and light bands.

In most cases, this occurs through **recrystallization** of minerals in the original (or "parent") rock (**FIGURE 9.1**). Recrystallization is the growth of new mineral grains in a rock by recycling old minerals. Rarely does recrystallization involve only one mineral. Rather, a metamorphic rock is formed because a new *assemblage* of minerals results from recrystallization.

Metamorphic rocks are a major source of building material. Slate and marble are commonly used as finishing stone in buildings and for statuary. (See the "Geology in Our Lives" box titled "Master of Marble.") Metamorphism of ultramafic rock (magnesium- and iron-rich igneous rocks) can produce chrysotile, the principal source of *asbestos,* which is an important component of fire-retardant materials, and *talc* (from metamorphism of dolostone), a very soft silicate mineral that is widely used as filler in paints, rubber, paper, asphalt, and cosmetics. Metamorphism also has the effect of bringing together certain elements, such as metals, into dense accumulations that form valuable ores, such as the gold deposits in the Abitibi region of Quebec, Canada.

Like the other types of rock we have studied, metamorphic rocks are named based on their texture and mineral composition. If a metamorphic rock is heavily squeezed by tectonic forces, newly crystallizing minerals will align perpendicular to the maximum stress (as discussed in Section 9-2), thus creating textures known as **foliation** (planar structures) and **lineation** (linear structures). Different types of foliation and lineation give rise to different types of rock (as we will see later in the chapter). Because of tectonic forces that pull, squeeze, and shear the crust and often accompany metamorphism, these rocks typically acquire new texture as well as new mineralogy. The new minerals align as they form, layers develop, colour bands of minerals appear, and crystals may stretch and even rotate. Features of the parent rock, such as vesicles, fossils, and primary sedimentary structures (cross-bedding, stratification, etc.), usually disappear. The resulting metamorphic rock is a product of the composition of its parent rock, referred to as the *protolith*, and the conditions the protolith has experienced. Hence, it offers a window into the environment of the deep crust.

GEOLOGY IN OUR LIVES

MASTER OF MARBLE

Because of its beauty and softness, *marble* (made of metamorphosed limestone or dolomite) has been the medium for the world's greatest statuary. If ever an artist was able to breathe warm life into cold stone, it was the famous Italian painter, poet, architect, and sculptor Michelangelo di Lodovico Buonarroti Simoni (1475–1564).

Michelangelo visited the 1,500-year-old quarry at Carrara in Tuscany, Italy (**FIGURE 9.2**), several times to search for ideal blocks of marble for his masterpieces. He took advantage of several important aspects of the stone to produce stunning sculptures, including his two most famous: *David* and the *Pietà*. The marble at Carrara is famous for having a uniform and homogeneous crystal pattern that can be predictably shaped. It is soft and resistant to shattering; it has a degree of transparency that lets light penetrate and scatter just below the stone surface, giving the appearance of life; and it forms in a range of shades.

Michelangelo's approach to sculpture was to liberate the human body encased in the cold marble. He would draw his figure on the face of a block and, working from all sides, hammer away the stone to free his works from the depths of rock where they resided, as the human soul is encased within the physical body.

The *Pietà* (Italian for "mercy") is considered by many to be Michelangelo's masterpiece. It is known for the soft flowing lines of the robes and the realistic depiction of the dead body of Jesus in the arms of his mother, Mary, after the crucifixion. The statue, carved from a solid block of Carrara marble that Michelangelo spent months looking for, is in the Vatican in Rome. The figures in the *Pietà* are out of proportion, owing to the difficulty of depicting a fully grown man cradled full length in a woman's lap. If Christ were

© Araldo de Luca/Corbis

Mauro Fermariello/Photo Researchers, Inc.

(a)   *(b)*

FIGURE 9.2 *(a)* The marble quarry at Carrara, Italy, has been mined for building stone and sculpture since the time of ancient Rome. *(b)* Michelangelo's most famous sculpture is the *Pietà*, which depicts the Virgin Mary holding the body of Christ after the crucifixion.

carved at human scale, the Virgin, standing, would be nearly 5 m tall. However, owing to Michelangelo's skill, much of the Virgin's size is concealed within her drapery, and the figures look peaceful and natural. Michelangelo's interpretation of this moment in Christian history was different from those previously created by other artists. He decided to create a youthful, serene, and celestial Virgin Mary instead of a brokenhearted and somewhat older woman.

(?) **Expand Your Thinking**—How do metamorphosed rocks become exposed at the surface where we can study them?

9-2 Changes in Heat and Pressure Can Cause Metamorphism

LO *9-2 Compare and contrast regional and contact metamorphism.*

Metamorphism is not an isolated event. Rather, it is the cumulative effect of all the continuous changes that a rock has undergone. A metamorphic rock is the product of shifting pressure, temperature, and chemical conditions that cause changes in the mineralogy and texture of a protolith. Hence, the temperature, pressure, and composition of the protolith are primary factors in determining the mineral assemblage and texture of the resulting metamorphic rock.

The Influence of Heat

Geothermal heat is conducted upward through Earth's interior, eventually radiating out to space. Heat increases with depth in the crust along what is known as the *geothermal gradient.* The average geothermal gradient is an increase of about 30°C per kilometre of depth in the crust, but the actual increase varies in different tectonic settings. In volcanically active areas like island arcs, the temperature may increase by about 30°C to 50°C per kilometre. In areas such as ocean trenches, where cold oceanic crust is driven into the interior, the temperature may increase by as little as 5°C to 10°C per kilometre (**FIGURE 9.3**).

When rocks are buried, typically due to tectonic activity (e.g., at a convergent margin), they immediately experience geothermal heat. As burial continues and temperatures rise, *solid-state reactions* begin to take place. Temperatures between 150°C and 200°C (at depths of approximately 6 km to 7 km) cause clay minerals, the most common sediment, to become unstable. They recrystallize to become the metamorphic mineral *chlorite,* followed by *muscovite.* These minerals often mark the early stages of metamorphism.

A second source of heat is plutonic intrusions. Magma that intrudes into the crust bakes the solid rock around it. Intruded crust is generically referred to as *country rock.* It experiences a type of metamorphism called **contact metamorphism,** or *thermal metamorphism,*

because conditions are related to high heat levels in the absence of greatly elevated pressure (**FIGURE 9.4**). Contact metamorphism is a result of the high geothermal gradient surrounding a pluton. This type of metamorphism usually is restricted to relatively shallow depths (< 10 km), where there is a large contrast in temperature between the intruding magma and the cool surrounding country rock.

The area of rock affected by contact metamorphism is usually a baked region, known as an *aureole.* The metamorphic aureole, also called a *contact aureole,* consists of zones of recrystallized minerals approximately parallel to the edges of the pluton. Multiple zones usually develop around an intrusion, each characterized by a distinct mineral assemblage reflecting the temperature gradient from the high-temperature magma to the unaltered country rock.

When a crystalline structure is heated, ions within it begin to vibrate. Great heat leads to greater vibration, and when the bonds forming compounds are eventually broken, the rock melts. However, before melting occurs, some atoms vibrate free and migrate to new locations within the mineral structure or in the rock. These atoms are prime candidates for recrystallization.

As this process continues, new atomic structures (new minerals) form that are stable at the higher temperature. Overall, the composition of the resulting metamorphic rock is very similar to that of the protolith because the parent atoms have been recycled; however, the minerals within it are different. In addition, the composition of the rock may change when chemically active fluids are introduced.

The Influence of Pressure

Within the crust, the pressure caused by the overlying layers of rock is called *lithostatic stress;* it acts on rock simultaneously and equally in all directions. Lithostatic stress increases with depth in the crust and

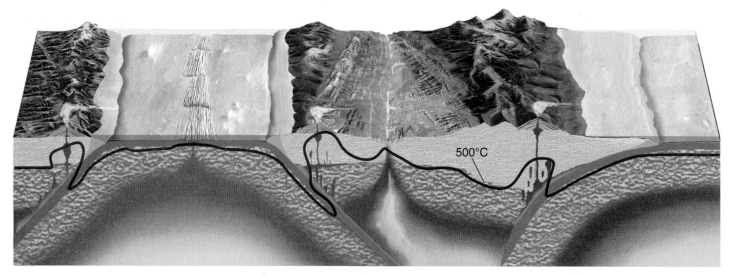

FIGURE 9.3 Heat within Earth's crust varies among tectonic settings. At spreading centres and volcanic arcs, the crust is warm near the surface. At subduction zones, cold sea floor descends into the mantle. The purple line represents the 500°C isotherm (a line connecting points with equal temperature).

(?) Describe the processes that cause the 500°C isotherm to vary its location in the crust.

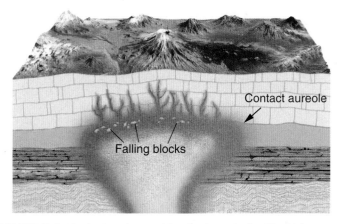

FIGURE 9.4 Contact metamorphism is characterized by high-temperature conditions surrounding plutonic intrusions, which results in a contact aureole of metamorphic rock.

(?) How might an intrusion affect groundwater in the vicinity?

plays an important role in metamorphism, closing the pore spaces in rocks, causing compaction, and, at great depths, causing minerals to recrystallize to form more compact atomic structures.

Directed stress differs from lithostatic stress in that pressure acting on the crust is greatest in one direction. This causes folding, bending, breaking, or even overturning of rock strata. Directed stress causes rocks to be shortened in one direction (the direction of the maximum principal stress) and elongated or lengthened in the direction perpendicular to the least principal stress.

Directed stress at convergent margins is created by the collision of two plates such that pressure is greatest in one direction (i.e., the direction of the maximum principal stress). Metamorphic minerals growing under directed stress crystallize with their long axis extending in the direction perpendicular to the maximum principal stress. This creates foliation and lineation because minerals are lined up in an easily visible manner. (Foliation is similar to the alignment of cards in a deck; lineation is similar to the alignment of a bunch of pencils held in a hand.) Under directed stress, platy minerals (which have the shape of dinner plates), such as members of the mica family, and elongate minerals (which are especially long in one direction, like a pencil), such as members of the amphibole family, can display lineation (**FIGURE 9.5**). These minerals lengthen into the low-stress area (i.e., direction of least principal stress) and hence become elongated in a direction perpendicular to the direction of maximum principal stress.

When pressure and temperature combine to alter rock over large areas, the result is **regional metamorphism.** This kind of metamorphism, which can extend over thousands of square kilometres, is not necessarily accompanied by plutonism. Most regionally metamorphosed rocks occur in places where the crust has been deformed during orogenesis or by burial deep in the crust. Two colliding plates (a convergent margin) create an extensive zone of high pressure (produced by compression, or "squeezing") characterized by directed stress in the crust and usually producing foliation and/or lineation. Thus, regionally metamorphosed rocks occur in the cores of orogenic mountain belts formed at continent–continent convergent margins and on the leading edges of continents and oceanic plates overriding subduction zones.

Regional metamorphism results from a general increase of temperature and pressure over a large area. Due to the great depths at which it occurs, high pressure seals the pore spaces in the rock, preventing water from circulating. Hence, groundwater may play only a limited role in transporting heat and dissolved ions involved in the recrystallization process. In terms of volume, the process of regional metamorphism is far more significant than contact metamorphism. Rocks produced in this way are distinguished by their foliated texture and mineral assemblage.

Courtesy of Chip Fletcher

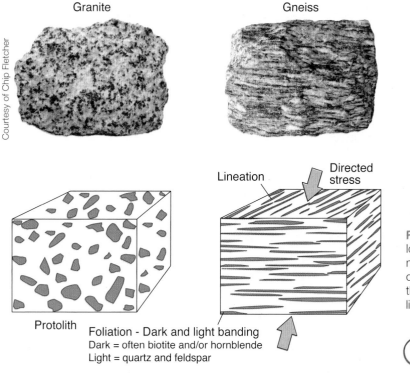

Granite   Gneiss

(?) **Expand Your Thinking**—Groundwater can transport heat and dissolved compounds. Explain how groundwater might influence contact metamorphism.

FIGURE 9.5 Minerals crystallize toward the direction of lowest stress, which is perpendicular to the direction of maximum principal stress. Directed stress is strongest in one direction and produces mineral layering that is perpendicular to the direction of highest stress. The gneiss shows foliation and lineation, the granite protolith does not.

 (?) What was the direction of maximum principal stress?

9-3 Chemically Active Fluids Transport Heat and Promote Recrystallization

LO *9-3 Describe the role of water in metamorphism.*

There is water in the crust. This *groundwater* (Chapter 20, "Groundwater") plays a role in determining the texture and composition of many types of sedimentary, igneous, and metamorphic rocks. Groundwater that comes into contact with intruding magma or leaks out of marine sediment in subducting oceanic crust usually contains a rich medley of dissolved compounds and metallic cations. Such chemically enhanced waters are known as *chemically active fluids.*

Chemically active fluids are especially important in recrystallizing the country rock surrounding a magma chamber. These fluids transport heat and influence the chemistry and mineralogy of newly forming metamorphic rocks. Chemically active fluids come from groundwater if metamorphism is occurring in shallow crust (< 2 km in depth). They are also produced when water (usually steam) escapes from magma; when sea water gets buried in the crust with a subducting oceanic plate; or when water-rich minerals such as micas or amphiboles are dehydrated by heat in the crust, and the escaping water circulates through the surrounding rock. Water is found throughout the crust and it plays an important role in nearly all metamorphic reactions. It lowers the melting point of many minerals, thus promoting early partial melting of felsic compounds and the production of granitic magmas.

Chemically active fluids increase mineral recrystallization by promoting the dissolution and migration of ions. During metamorphism, as temperature and pressure increase, water becomes more chemically reactive and dissolves ions and compounds more readily. For instance, when two mineral grains are squeezed together by directed stress, any water that is present will greatly accelerate the dissolution of ions from those grains. Because the water flows from areas of high stress to areas of low stress, those ions will be carried away. Dissolved ions may bond again in a nearby area of lower stress. As a result, as described earlier, minerals tend to lengthen into the low-stress area, growing in a direction perpendicular to the direction of maximum principal stress.

Certain types of chemically active fluids, known as *hydrothermal fluids,* are produced by the interaction of groundwater and magma in the final stages of cooling. Hydrothermal fluids are hot and contain dissolved ions and compounds that react with country rock, causing crystallization of valuable *metals,* such as lead, silver, platinum, and gold.

Hot water that is enriched in dissolved minerals will gain access to cracks and fractures in stressed rocks and line these openings with **hydrothermal vein fillings** of newly crystallized minerals (FIGURE 9.6). The dissolved compounds will greatly modify any solid-state reaction that is taking place. Chemical alteration in the presence of hydrothermal fluids is known as *metasomatism.*

Hydrothermal fluids causing metasomatism alter the baked zones of contact metamorphic rocks so that they contain many new types of metamorphic minerals that would not form if only dry country rock was metamorphosed. Hydrothermally precipitated minerals characterize the metamorphic rocks that develop around contact metamorphic zones. Hydrothermal fluids are particularly important in metasomatism of limestone and dolomite to produce metamorphic rocks called *skarns,* which can host valuable ore deposits.

The important industrial mineral asbestos may be found as a hydrothermal vein deposit. Asbestos is widely used as a fire retardant, but certain types of asbestos have been identified as a health hazard. Read the "Geology in Our Lives" box titled "Asbestos—A Real Danger or an Overreaction?" to learn more about this controversial problem.

 Expand Your Thinking—Where do fluids in the crust come from?

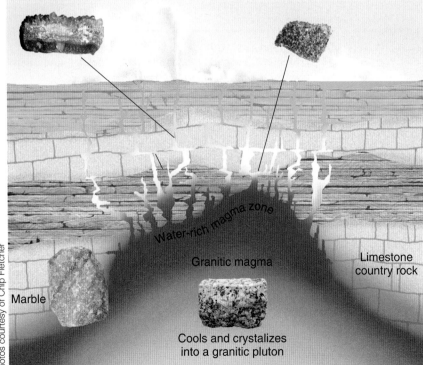

Photos courtesy of Chip Fletcher

Water-rich magma zone

Granitic magma

Limestone country rock

Marble

Cools and crystalizes into a granitic pluton

FIGURE 9.6 Water plays an important role in nearly all metamorphic reactions. It lowers the melting point of many minerals, thus promoting early partial melting of felsic compounds and the production of granitic magmas, of which granite is the end product upon cooling and crystallizing. Water also transports dissolved compounds into cracks and fractures in overlying rocks and produces hydrothermal vein deposits that are rich in valuable minerals. Shown at the top of the diagram are veins of quartz (white) and amethyst (purple), a type of quartz.

 How did minerals get into the fractures above the magma chamber?

GEOLOGY IN OUR LIVES

ASBESTOS—A REAL DANGER OR AN OVERREACTION?

To people around the world, the word "asbestos" elicits an instant reaction of fear. Once commonly used in everyday items such as toothpaste and as insulation in our homes, asbestos, a term applied to various types of fibrous minerals from the amphibole and serpentine groups (FIGURE 9.7), is now viewed as a cause of two principal types of cancer: cancer of the lung tissue itself and mesothelioma, a cancer of the thin membrane that surrounds the lungs and other internal organs.

Asbestos fibres do not readily melt or ignite. Hence, they are useful for making fireproof clothing and insulation and are mixed into plaster and paint to make fire-retardant building materials. There are six commercial varieties of asbestos: five made from a member of the igneous mineral amphibole (called "brown" and "blue" asbestos) and one made from the metamorphic mineral chrysotile (called "white" asbestos).

The asbestos panic began in 1986 when the U.S. Environmental Protection Agency implemented the Asbestos Hazard Emergency Response Act requiring all schools to be inspected for the presence of asbestos. The fear that asbestos causes cancer comes from studies of mine workers in South Africa and western Australia who were exposed to high levels of blue and brown asbestos. The workers showed unusually high rates of mesothelioma, in some cases after less than a year of exposure. Further studies have documented that when the thin, rodlike fibres of brown and blue asbestos are inhaled into the lungs, they are neither broken down nor exhaled, but can remain there indefinitely, serving as possible nucleation sites for severe and deadly lung disease.

However, white asbestos (chrysotile) is the most widely used industrial variety in Canada, the United States, and elsewhere. Studies of workers in chrysotile mines in Canada and Italy show that health risks are minimal to nonexistent. Mortality rates from mesothelioma and lung cancer among these workers differ very little from those in the general population. Chrysotile fibres are curly and less rodlike than those of blue and brown asbestos. They tend to break down if inhaled and are more easily expelled from the lungs.

Even though most of the asbestos used in Canada and the United States is relatively harmless chrysotile, asbestos has been removed from schools and other buildings at a cost of $50 to $100 billion. Asbestos removal has depreciated the value of commercial buildings by $1 trillion and actually has increased, not decreased, the risk of injury or death, both from increased exposure to asbestos dusts during the dirty "ripouts" and from accidents experienced by workers removing the asbestos.

Does asbestos present a risk to students across the nation? Data indicate that the level of asbestos in school buildings is equivalent to the level found outdoors produced by natural sources in the crust. Nearly all the asbestos found in schools is chrysotile, which is relatively harmless in low concentrations. The risk from asbestos is far less than other common risks, such as school sports (100 times more dangerous), aircraft accidents (60 times more dangerous), and car accidents (320 times more deadly).

These data indicate that the asbestos panic was unwarranted and has placed an expensive burden on government agencies that feel compelled to remove asbestos from public buildings even though risks from everyday hazards are far greater.

FIGURE 9.7 (a) The mineral asbestos is found in at least six common forms. White asbestos (chrysotile) is the most frequent source of asbestos fibres used in schools and public buildings. Chrysotile fibres are not as hazardous as those from brown or blue forms of asbestos and have been proven to pose little to no health risk to exposed persons. (b) Nonetheless, worldwide, expensive asbestos removal programs continue to cost taxpayers billions of dollars each year.

Courtesy of Chip Fletcher

(a)

© Ted Spiegel/Corbis

(b)

9-4 Rocks Evolve through a Sequence of Metamorphic Grades

LO 9-4 *Describe the metamorphic grades and how they are identified.*

Different locations in the crust experience different levels of heating and stress, some intense, some relatively mild. As a result, rock may experience different *grades* of metamorphism, depending on the kind of recrystallization that is stimulated by these conditions. The concept of gradational change from **low-grade metamorphism** to **high-grade metamorphism** reflects the sequence of mineral and texture changes that occur under different conditions (**FIGURE 9.8**).

The changes that occur during metamorphism are recorded in the form of textures and mineral assemblages. High-grade metamorphic rock is greatly altered from its original form and often has a completely different mineralogy than the protolith. Of course, the mineralogy of the protolith, the pressure and temperature conditions, and the presence of any chemically active fluids will govern the mineralogy of the metamorphic product. Low-grade metamorphic rock, while experiencing less severe conditions, nonetheless displays a unique texture and mineralogy.

Index Minerals

The change from a protolith to a metamorphic product is characterized by evolving mineral assemblages. As we saw earlier, the assemblage of minerals within a rock evolves in response to changing temperature and pressure conditions. Hence, geologists reason that specific mineral assemblages serve as geothermometers and geobarometers of conditions within the crust. Moreover, certain minerals identify specific metamorphic conditions. This is one way in which critical thinking is used to interpret Earth's history as it is recorded in metamorphic rocks. Geologists can build a simple hypothesis predicting conditions in the crust based on the presence of specific metamorphic minerals.

Minerals that serve as such "guideposts" are known as **index minerals.** Different types of protoliths evolve different types of index minerals. Shale, the most common sedimentary rock, develops these index minerals (from lowest to highest metamorphic grade): *chlorite, muscovite, biotite, garnet, kyanite,* and *sillimanite* (**FIGURE 9.9**). These minerals serve as indicators of the degree of metamorphism the shale underwent. For instance, chlorite tends to develop under relatively low pressure and low temperature conditions. Its presence in a rock

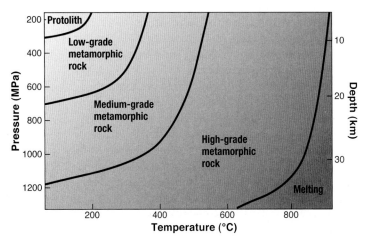

FIGURE 9.8 As rock is buried deeper in the crust, the temperature and pressure increase. Higher temperatures and pressures lead to metamorphism.

(?) Why do texture and mineral assemblage change with metamorphic grade?

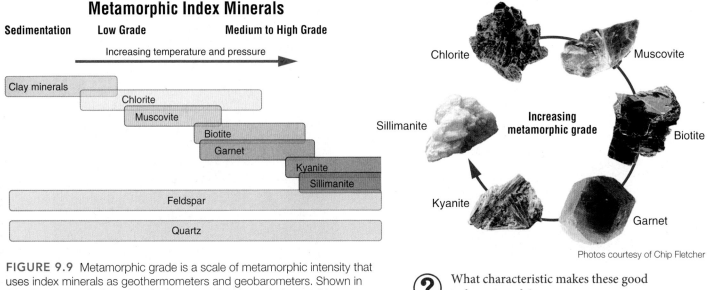

Metamorphic Index Minerals

Photos courtesy of Chip Fletcher

FIGURE 9.9 Metamorphic grade is a scale of metamorphic intensity that uses index minerals as geothermometers and geobarometers. Shown in this diagram are index minerals that are typical for a shale protolith.

(?) What characteristic makes these good index minerals?

is an indicator of low-grade metamorphism. Alternatively, sillimanite tends to form in high temperature and pressure conditions, and its presence in a rock is an indicator of high-grade metamorphism.

Quartz and feldspar also appear in metamorphic products. However, because they tend to crystallize in low-grade as well as high-grade conditions, they are ubiquitous. They do not indicate a specific metamorphic condition and are not considered index minerals.

Texture

Rocks exposed to directed stress (usually related to regional metamorphism) may develop foliation and/or lineation. Again, we will use shale as our example. Like the mineral assemblage, foliation in shale evolves with increasing metamorphic intensity. Foliated rocks have crystals oriented in near-parallel layers. Shale evolves through four common foliated textures, in order of increasing metamorphic grade: *slate, phyllite, schist,* and *gneiss* (**FIGURE 9.10**). Of these, those with large-grained textures (schist and gneiss) typify medium- to high-grade metamorphism; those with fine-grained textures (slate and phyllite) generally indicate low- to medium-grade metamorphism.

Contact metamorphism, and in some cases regional metamorphism, may produce nonfoliated products. Shale exposed to high temperature (such as during contact metamorphism) evolves into a nonfoliated rock called *hornfels*. Another example of a nonfoliated rock is *anthracite* coal, the metamorphic product of bituminous coal. Other nonfoliated rocks include *marble, quartzite,* and *metaconglomerate.* For the most part, they are identified on the basis of their mineral assemblage; we discuss them later in the chapter.

TABLE 9.1 summarizes the common protoliths and their foliated and nonfoliated metamorphic products. These rocks and products are discussed in the next two sections.

TABLE 9.1 Protoliths and Metamorphic Products

Protolith	Low to Medium Grade	High Grade
Foliated Product		
Fine-grained sedimentary (e.g., shale)	Slate, phyllite, schist	Gneiss* [*paragneiss*]
Igneous rock (e.g., basalt)	Greenschist, blueschist	Amphibolite, granulite, eclogite Gneiss* [*orthogneiss*]
Nonfoliated Product		
Conglomerate	Metaconglomerate	[Gneiss (foliated)]
Sandstone	Quartzite	Quartzite
Shale, mudstone, siltstone	Hornfels	Hornfels
Bituminous coal	Anthracite	Anthracite
Igneous rock (e.g., basalt)	Hornfels	Hornfels
Limestone, dolomite	Marble	Marble

*A gneiss that originated as a sedimentary rock is called a *"paragneiss"*; a gneiss that originated as an igneous rock is called an *"orthogneiss."*

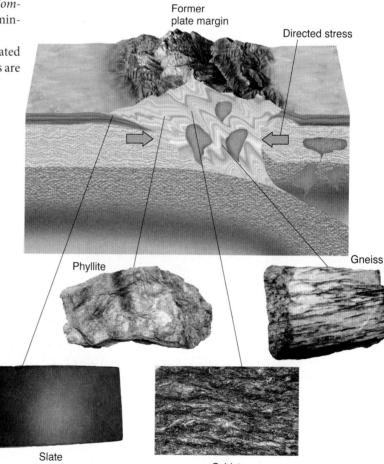

 In general, what happens to the foliation of regionally metamorphosed rock as the grade increases?

FIGURE 9.10 Regional metamorphism of shale can produce a sequence of foliated rocks whose textures evolve with increasing pressure.

Photos courtesy of Chip Fletcher, Dr. Dan Gibson, SFU

 Expand Your Thinking—Is a mineral that crystallizes in low-grade, medium-grade, and high-grade conditions a good index mineral? Why or why not?

9-5 Foliated Texture Is Produced by Directed Stress Related to Regional Metamorphism

LO *9-5 Compare and contrast the four principal types of foliated texture.*

Just as igneous and sedimentary rocks are classified by their texture and mineralogy, so are metamorphic rocks. As mentioned earlier, metamorphic textures fall into two categories: foliated and nonfoliated. Foliated texture (TABLE 9.2, FIGURE 9.11) is produced by directed stress related to regional metamorphism, whereas **nonfoliated texture** is generally found in metamorphic rocks that are undergoing contact metamorphism or in rocks that, when viewed with the unaided eye, show no evidence of foliated texture.

Slate

Slaty texture is caused by the parallel orientation of microscopic grains. This texture is characteristic of slate, which has a tendency to separate along parallel planes. Slaty texture results in a property known as **rock cleavage**, also called *slaty cleavage* (FIGURE 9.12).

Rock cleavage should not be confused with cleavage in a mineral, which is related to internal atomic structure. Rock cleavage develops because new minerals, usually platy minerals such as chlorite, muscovite, and biotite, grow during metamorphism and align in a direction perpendicular to the dominant direction of stress. Hence, rock cleavage describes a relationship between minerals, whereas mineral cleavage describes the arrangement of atoms within a mineral.

Geologists use a compass to measure the orientation and direction of rock cleavage, which is typically caused by directed stress generated by plate convergence. This information is used to infer the

TABLE 9.2 Foliated Metamorphic Rocks

		Types of Foliated Texture		
	Slate	**Phyllite**	**Schist**	**Gneiss**
Grain Size	Fine grained; minerals not visible to the unaided eye	Fine grained; most minerals not visible to the unaided eye	Medium to coarse grained; minerals visible	Medium to coarse grained; minerals visible
Minerals	Clay minerals, chlorite, muscovite	Chlorite, muscovite	Muscovite, biotite, garnet, kyanite, and others	Feldspars, quartz, muscovite, biotite, and other ferromagnesian minerals
Appearance	Dense	Satiny lustre	Shiny lustre	Banded

(a)

(b)

(c)

(d)

(e)

Photos courtesy of Chip Fletcher

FIGURE 9.11 In foliated rocks, the texture evolves from a protolith, such as *(a)* shale, to *(b)* slate, *(c)* phyllite, *(d)* schist, and *(e)* gneiss as directed stress and coarsening of metamorphic minerals increases.

 Why does directed stress tend to produce foliated texture?

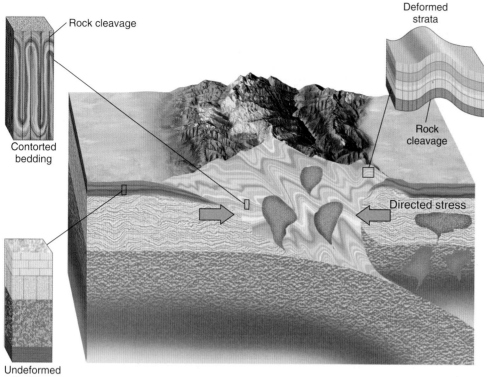

FIGURE 9.12 The growth of chlorite during low-grade metamorphism leads to rock cleavage. The study of rock cleavage allows geologists to re-create past tectonic conditions and former plate boundaries.

(?) Would contact metamorphism produce rock cleavage? Explain your answer.

past location of plate collision zones. Slaty cleavage has been used to re-establish the former position and movement of continents and in reconstructing the supercontinent Pangaea.

Phyllite

Phyllitic texture refers to the parallel arrangement of platy minerals, usually micas (such as chlorite and muscovite) that are barely visible to the naked eye. The mineral arrangement produces a lustre that is satiny, silky, and/or wavy. Abundant mica grains give rock specimens a silky sheen with phyllitic texture, representing low- to medium-grade metamorphism.

Schist

Schist is metamorphosed phyllite. *Schistose texture* refers to foliation resulting from near-parallel to parallel orientation of platy minerals like chlorite, muscovite, and biotite. Quartz and amphibole are also common in schist. The average size of minerals is generally smaller in schist than in gneiss. The main difference, however, is that foliation is more distinct in schist due to recrystallization and enlargement of platy minerals such as chlorite and muscovite.

Samples of schist often contain large crystals of minerals, such as garnet and tourmaline, called *porphyroblasts*, which are significantly

bigger than the surrounding matrix minerals. These form as increasingly intense metamorphic conditions transform several different minerals into a new and distinct metamorphic mineral. There are a variety of schists, named for their dominant minerals. Examples include *garnet-mica schist, greenschist* (dominated by chlorite), *blueschist* (dominated by the mineral glaucophane), and *biotite schist* (dominated by biotite). Each is formed under different pressure and temperature conditions.

Gneiss

Gneissic texture refers to coarse foliation in which minerals are segregated in discontinuous bands ranging from 1 mm to several centimetres in thickness. Each band is dominated by one or two minerals and hence is characterized by distinctive dark or light colouring (**FIGURE 9.13**), giving the rock a striped appearance. Light bands commonly contain quartz and feldspar; dark bands commonly contain hornblende and biotite.

Gneiss is a high-grade metamorphic rock that can evolve from a number of different protoliths, the most common being granite, diorite, and schist. Intense conditions cause solid-state migration of atoms, leading to the formation of mineral bands oriented perpendicular to the direction of maximum principal stress. As gneiss approaches its melting point, the mineral bands bend and become contorted. Of course, if the temperature rises too high, the rock will melt and return to the igneous phase of the rock cycle.

FIGURE 9.13 The famous Acasta gneiss, located along the Acasta River in the Northwest Territories, is one of the oldest known rock on Earth, dated at 4.03 billion years old.

Reproduced with the permission of Natural Resources Canada 2013, courtesy of the Geological Survey of Canada (Photo 2007-178 by Marc St-Onge).

 Expand Your Thinking—Explain the relationship between recrystallization, rock cleavage, directed stress, and plate motion. State your answer in the form of a testable hypothesis.

9-6 Nonfoliated Rocks May Develop during Regional or Contact Metamorphism

LO *9-6 List the common nonfoliated rocks, and describe the products of contact metamorphism.*

Metamorphic rocks in which there is no visible orientation of mineral crystals have a nonfoliated texture (TABLE 9.3). Nonfoliated rocks (FIGURE 9.14) may develop during regional or contact metamorphism, and thus nonfoliated texture does not always signal an absence of directed stress. This is because some minerals growing under directed stress do not develop foliation or lineation. For instance, when limestone or dolomite metamorphoses to become marble, calcite crystals enlarge but may not become foliated even under directed stress. When quartz sandstone metamorphoses to become quartzite, the quartz grains may not appear foliated unless viewed through a microscope.

Marble and quartzite usually contain uniformly sized or *equigranular* grains of a single mineral (calcite, dolomite, quartz). Metamorphosed conglomerates may retain the original texture of the protolith, including the outlines and colours of larger grains, such as pebbles. However, because metamorphism has caused recrystallization of the matrix, the metamorphosed conglomerate is known as a *metaconglomerate*. In cases where metamorphism has deformed the shape of the large grains, the rock can be referred to as a *stretched pebble conglomerate*.

Quartzite and metaconglomerate are distinguished from their sedimentary protoliths by the fact that they break *across* the quartz grains, not around them. Marble has a crystalline appearance, and its mineral grains

TABLE 9.3 Nonfoliated Metamorphic Rocks

	Marble	**Quartzite**	**Metaconglomerate**	**Hornfels**	**Anthracite Coal**
Grain Size	Medium to coarse grained; minerals visible	Medium to coarse grained; minerals visible	Medium to coarse grained; minerals visible	Fine to coarse grained	Fine grained; minerals not visible
Minerals	Calcite Dolomite	Quartz	Any conglomerate	Most varieties of metamorphic minerals	Carbon-rich material
Appearance	Hardness of 3 for calcite, 3.5 for dolomite; calcite fizzes readily with dilute hydrochloric acid; dolomite must be powdered before it will fizz	Hardness of 7; breaks across grains	Breaks across grains and around them	Dense; often dark coloured	Black, shiny; conchoidal fracture

(a)

(b)

(c)

(d)

(e)

Explain what factors lead to foliated texture and what factors lead to nonfoliated texture.

FIGURE 9.14 Nonfoliated metamorphic rocks: *(a)* marble, *(b)* quartzite, *(c)* metaconglomerate, *(d)* hornfels, and *(e)* anthracite.

Photos courtesy of Chip Fletcher

are larger than those of its limestone or dolomite protolith. In many cases, calcite crystals become larger during metamorphism. Marble and quartzite are characterized by the interlocking nature of their crystals.

Hornfels, a nonfoliated rock formed by metamorphism of shale and other types of rocks, has a nondescript appearance. It is usually a medium to dark shade of grey, lacks foliation, and contains few recognizable minerals that may include larger *porphyroblasts*. Anthracite, often overlooked in discussions of metamorphism, is a nonfoliated rock produced by metamorphism of bituminous coal.

Zoned Mineral Recrystallization

When a hot pluton, or magma body, intrudes into the crust, it can produce zones of minerals formed by the combined effects of the added heat, chemical interaction of the magma and country rock, and chemically active fluids. The composition of the country rock plays an important role in determining the mineralogy of the final metamorphic product, which may be economically important (**FIGURE 9.15**).

In dry limestone country rock, calcite and dolomite are stable at temperatures above 700°C. Therefore, contact metamorphism of a pure limestone or dolostone may yield no more than a coarsening (enlarging) of grain size, producing marble. However, such country rock often contains other materials besides calcite or dolomite, such as silica in the form of chert or feldspar sand grains, and calcite-dolomite-quartz mixtures are also common. When they undergo metamorphism, clay minerals in the country rock will add aluminum to the mix. And when water is present, mineral reactions are enhanced and dissolved ionic compounds are transported throughout the contact zone. In any of these cases, if the protolith is chemically complex, the resulting metamorphic product can be quite different from the parent. For instance, contact between a silicate-rich pluton and impure limestone country rock can result in the formation of the mineral *wollastonite* and the production of carbon dioxide gas:

$$CaCO_3 + SiO_2 \rightarrow CaSiO_3 + CO_2$$
calcite + silica → wollastonite + carbon dioxide

This reaction occurs at temperatures of about 500°C. Thus, when wollastonite is present in a rock, it can be interpreted as a record of the interaction between a silica intrusion and limestone crust (**FIGURE 9.16**). Wollastonite, along with garnet and diopside, typically comprise the innermost zone of the metamorphic aureole, and the minerals calcite, serpentinite, chlorite, dolomite, and olivine form assemblages that occur as parallel outer zones. Marble is often found at the outermost edges of the aureole within the intruded limestone.

Intrusions into silicate sedimentary rocks, such as shale and sandstone, also produce zoned aureoles. The contact metamorphic hornfels, containing higher-grade minerals such as garnet and mica, occurs in the immediate vicinity where the intruding magma meets shale. Farther away, the quartz, clay, and carbonate components of shale and sandstone may be converted into biotite, chlorite, and calcite as well as other minerals. In the cases we have described involving limestone, shale, and sandstone, the contact aureole zone surrounding a large pluton may be hundreds of metres wide.

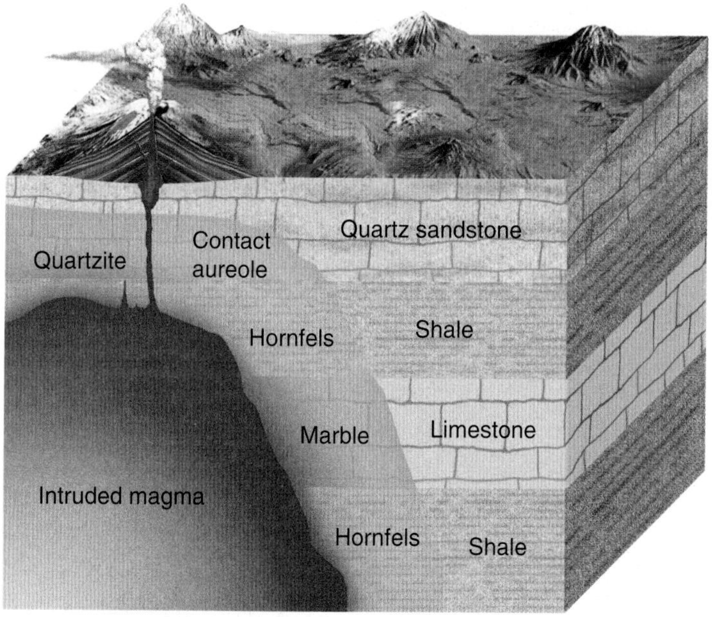

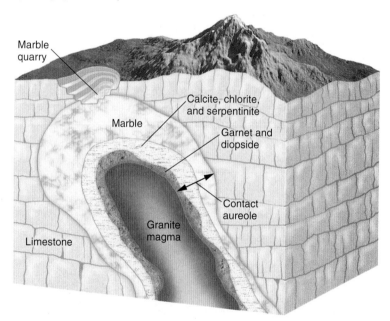

FIGURE 9.15 In general, contact metamorphism produces nonfoliated rocks.

FIGURE 9.16 Impure limestone is altered by contact metamorphism. For example, the minerals garnet and diopside may characterize the baked zone closest to the magma. Calcite, chlorite, and serpentinite minerals may develop in the next zone. An outer zone consists of marble, which is metamorphosed calcite.

(?) What factors contribute to the growth of chemically diverse minerals in contact metamorphism?

(?) Marble is an economically important rock. How is it used?

(?) **Expand Your Thinking**—At a subduction zone, intrusion (high temperature) occurs with convergence (high pressure). Describe how these conditions combine to produce metamorphic products from the following: limestone, shale, quartz sandstone, and granite.

9-7 The Relationship between Mineral Assemblage and Metamorphic Grade Is Expressed by Metamorphic Facies

LO 9-7 *Define the term "metamorphic facies."*

Depending on the composition of a protolith, the mineral assemblage of a metamorphic product reflects the temperature and pressure conditions prevailing during metamorphism. The relationship between mineral assemblage and metamorphic grade is expressed by the concept of **metamorphic facies**: a set of metamorphic mineral assemblages repeatedly found together and indicating certain pressure and temperature conditions.

Metamorphic Facies

A metamorphic *facies* (Latin for "face") is a mineral assemblage that reflects specific pressure and temperature conditions in the crust. A portion of crust that is exposed to metamorphic conditions, such as a convergent plate margin, will develop a range of facies, each of which represents a stage of low- to high-grade conditions. Hence, a *sequence* of metamorphic facies might be observed in a metamorphosed region reflecting the range of low-grade to high-grade conditions that have occurred.

The nature of those facies will depend on the *geothermal gradient*, the rate at which pressure and temperature increase with depth in the crust. A geothermal gradient characterized by a very rapid increase in temperature but relatively little change in pressure might be present around an igneous intrusion and would result in a specific texture and assemblage of minerals. Another geothermal gradient—for instance, one reflecting a rapid increase in pressure and relatively little change in temperature—would result in a very different texture and mineral assemblage and, therefore, a different metamorphic facies (**FIGURE 9.17**).

The concept of metamorphic facies was developed by a Finnish geologist, *Pentti Eskola*, in 1915. In the course of a study of basaltic rocks, Eskola identified the various pressure (P) and temperature (T) conditions that basalt may experience in the course of metamorphism on its way to melting. Figure 9.17 illustrates Eskola's concepts as they have been modified to account for the effects of plate tectonics. Although Eskola's work was largely with basalts, today's facies names apply to many types of mafic, ultramafic, and sedimentary protoliths.

Note in Figure 9.17 that the least pressure and lowest temperature are represented within the upper few kilometres of the crust in the upper left corner of the diagram. This region of P-T conditions is named the *zeolite facies* after a mineral group found in low-grade metamorphosed oceanic crust. As seafloor basalt undergoes early metamorphism associated with shallow burial in the crust, it produces zeolites, which are characterized by interlocking tetrahedrons of silicate and aluminum oxide. Zeolites are distinctive for their ability to lose and absorb water without damage to their crystal structures. Remember that the zeolite facies, like all facies, is a conceptual zone and may consist of other types of minerals, depending on the composition of the protolith.

Metamorphic Pathways

With zeolite as the point of origin, five "generalized" pathways of metamorphism are diagrammed (Figure 9.17). It should be noted that during any given metamorphic episode, there can be deviations from these pathways. For example, a rock at a convergent plate boundary

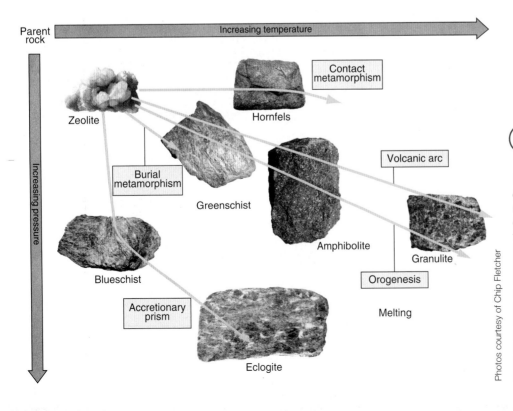

(?) What is the origin of zeolites?

FIGURE 9.17 Five basic pathways of metamorphism are contact metamorphism, metamorphism typical of volcanic arcs, metamorphism typical of collisional mountain belts (orogenesis), burial metamorphism (low-grade metamorphism caused by burial in sedimentary basins), and metamorphism typical of accretionary prisms at subduction zones. A rock that is undergoing metamorphic change along any of these pathways will pass through phases, or facies, characterized by crystallization of certain sets of minerals.

Photos courtesy of Chip Fletcher

may begin following the contact metamorphism pathway, but then deviate to higher pressures as it undergoes orogenesis.

The uppermost pathway shows a typical geothermal gradient associated with contact metamorphism. Temperature increases rapidly toward the plutonic body, but pressure remains relatively stable. Igneous and sedimentary rocks in contact with a plutonic body metamorphose to hornfels and establish the *hornfels facies* (FIGURE 9.18). Depending on the composition of the protolith, the mineral assemblage may consist of chlorite, plagioclase, quartz, garnet, staurolite, pyroxene, hornblende, and other minerals. The remaining pathways fall under the broad category of regional metamorphism.

The second pathway shows the geothermal gradient associated with plutonic-volcanic arc complexes at plate boundaries. Metamorphism progresses from the zeolite facies into the *greenschist facies* consisting of chlorite, serpentine, quartz, biotite, plagioclase, and other minerals, depending on the protolith. As metamorphism continues, the rock passes through the *amphibolite facies* and into the *granulite facies.* A zone of partial melting is encountered in the amphibolite facies, where minerals with low melting points will begin to form magma. This partial melting is enhanced when water is present to lower the melting point of minerals in the evolving rock. This type of partial melting is an important process for producing magmas associated with igneous activity in subduction zones.

The third metamorphic pathway is associated with collisional mountain building during the process of orogenesis. The orogenesis

pathway passes through the same low- and medium-grade facies as does volcanic arc metamorphism but enters each facies at a higher pressure.

The fourth pathway is characteristic of burial in the crust under the influence of the geothermal gradient. Burial metamorphism typically occurs within sedimentary basins >10 km deep. The rocks are metamorphosed due to increased pressure and temperature related to progressive burial within a basin, and generally only reach zeolite to greenschist facies.

A fifth metamorphic pathway is encountered at plate convergent zones that build accretionary prisms. This pathway is characteristic of locations where oceanic sediments on subducting crust mix with sediments from continental crust to build accretionary prisms. The resulting pile of sediment experiences a rapid rise in pressure, especially if the sediments are taken partly down into the subduction zone, and a relatively slow increase in temperature. Rocks progress from the zeolite facies to the *blueschist facies* and finally to the eclogite facies. Blueschists are encountered in the *forearc area* of a subduction zone, the region between the deep-sea trench and the volcanic arc. Depending on the exact composition of the protolith, blueschists consist of the minerals glaucophane, sodium-rich amphibole (which has a bluish hue), chlorite, epidote, quartz, and kyanite.

Read the "Critical Thinking" box titled "Metamorphism and Tectonics" to apply your understanding of plate tectonics to the concept of metamorphic facies.

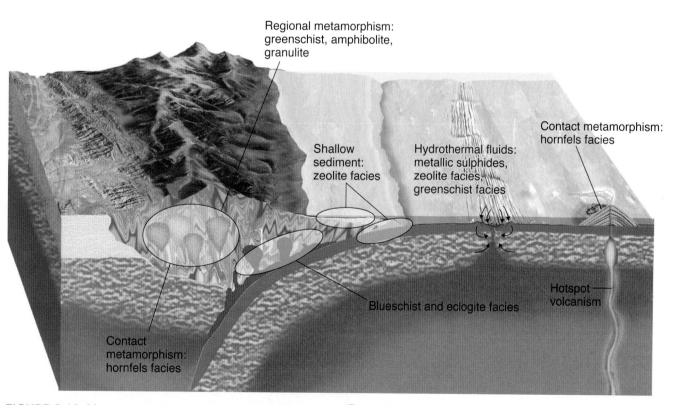

FIGURE 9.18 Metamorphic facies are typically related to tectonic environments or deep burial in continental crust.

(?) What environmental conditions are found with deep burial and how are they similar or different relative to a convergent margin?

(?) **Expand Your Thinking**—Metamorphic facies can be a difficult concept. Explain it in the simplest terms you can think of.

? CRITICAL THINKING

METAMORPHISM AND TECTONICS

Metamorphic environments are closely tied to plate tectonics (**FIGURE 9.19**). For example, nearly all types of metamorphism can occur at subduction zones. Regional and contact metamorphism occurs in the temperature and pressure conditions that develop at convergent plate boundaries. Divergent plate boundaries, at spreading centres, are locations where zeolite and greenschist facies develop in the presence of hydrothermal circulation. The pressure and temperature conditions at convergent and divergent boundaries produce the majority of the metamorphic rocks in the crust.

Please work with a partner and answer the following questions:

1. Examine Figure 9.19 and identify where contact metamorphism is likely to occur.
2. Identify where regional metamorphism is likely to occur.
3. Identify where you would find each of the following: marble, hornfels, phyllite, metaconglomerate, gneiss. Explain why you picked these sites.
4. What pressure and temperature conditions and protoliths determine the location of the major metamorphic facies (zeolite, hornfels, blueschist, greenschist, amphibolite, eclogite, and granulite)? Indicate the location of each.

5. Describe the pressure and temperature conditions that prevail where burial metamorphism occurs.
6. What role do chemically active fluids play in the growth of metamorphic rocks at convergent and divergent boundaries? Where do these fluids come from?
7. Is the accretionary wedge a site of high temperature or high pressure or both? What is the origin of the accretionary wedge, and hence, what is its composition? Is it likely to have chemically active fluids?
8. Describe how chemically active fluids will differ from one site to another in Figure 9.19. Explain why they differ.
9. For each facies you have identified, describe its protolith, the T/P conditions in which it evolved, the role of chemically active fluids, and its present mineralogy.

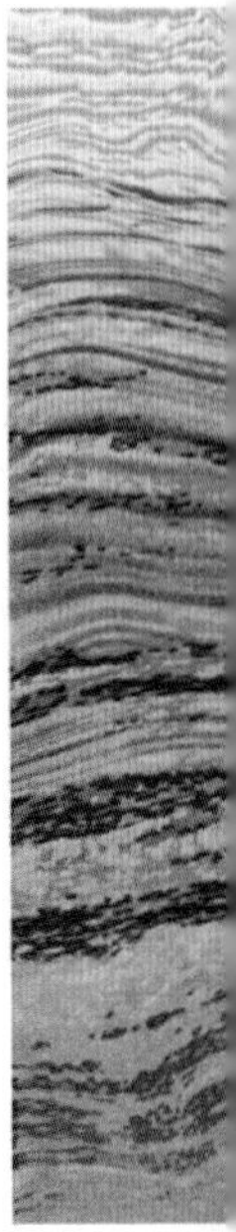

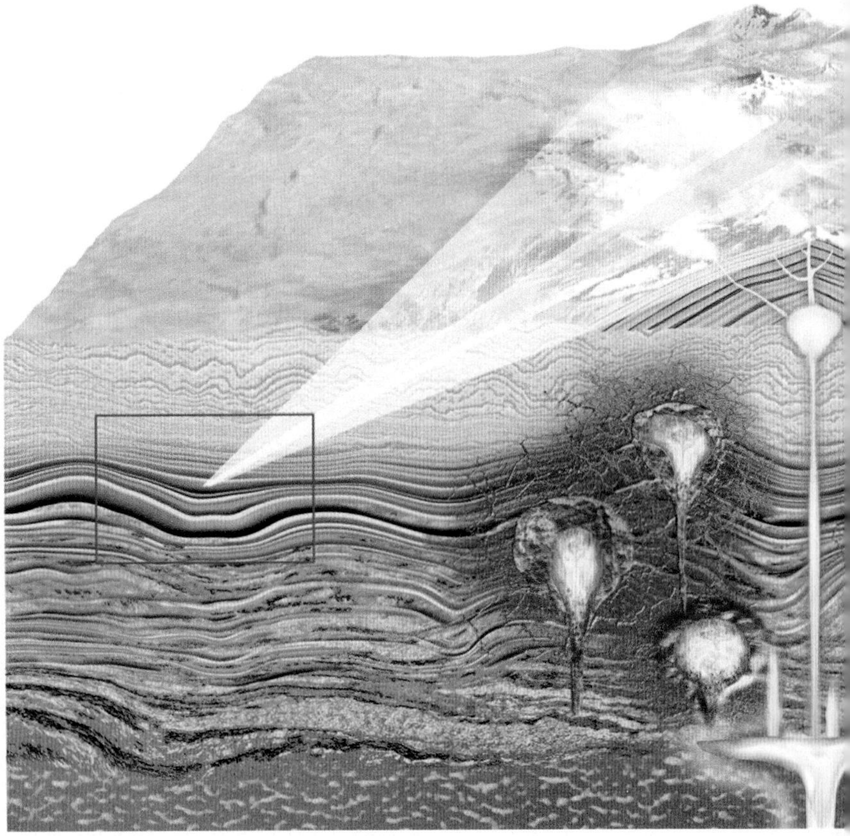

FIGURE 9.19 Metamorphic environments

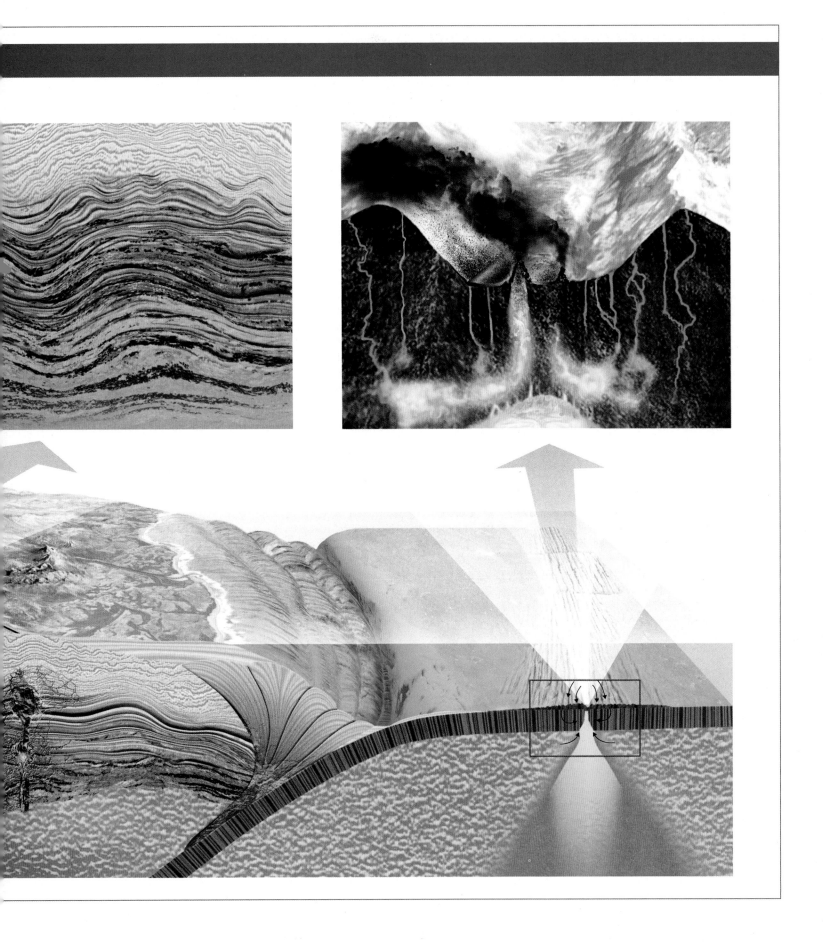

9-8 Metamorphism Is Linked to Plate Tectonics

LO 9-8 *Describe how metamorphic facies are related to plate tectonics.*

Throughout our discussion of metamorphic rocks, we have connected specific tectonic processes to the pressure and temperature conditions that lead to metamorphism. Metamorphism and plate tectonics are intimately linked (**FIGURE 9.20**).

Divergent Margins

At spreading centres, hydrothermal fluids interact with mafic and ultramafic oceanic crust. Hydrothermal fluids lead to the deposition of valuable metallic sulphide deposits (Chapter 5, "Igneous Rock"), which accumulate in fractures and fissures among the young basalts in the mid-ocean ridge. These deposits contain minerals from the native elements and sulphide classes. Where they are found on continents (at former convergent margins where ancient sea floor is preserved), they are mined for economically valuable metals (Chapter 10, "Geologic Resources"). In most cases, these sites are characterized by low pressure and moderately elevated temperatures. The most significant metamorphic process occurring at mid-ocean ridges is the heat-driven circulation of hydrothermal fluids leading to mineral reaction and recrystallization and the development of zeolite and greenschist facies.

Convergent Margins

As oceanic crust moves away from a divergent margin, it is buried by marine sediment. Burial continues as both basaltic and sedimentary protoliths move directly toward active subduction. During subduction, the crust is exposed to increasing directed stress, with an accompanying rise in temperature. Slate, phyllite, and schist textures develop during plate convergence as blueschist and eclogite facies evolve in the forearc region of greatest directed stress.

Large-scale regional metamorphism takes place in the core of the overriding plate. Greenschist, amphibolite, and granulite metamorphic facies develop across the broad geographic band between the subduction zone and the interior region of the overriding plate. In this region, increases in pressure due to plate convergence are accompanied by temperature increases associated with partial melting of deeply buried continental lithosphere and descending oceanic crust.

Hotspots

Igneous intrusions are sites of contact metamorphism at convergent boundaries and hotspots. The composition of the protolith (intruded

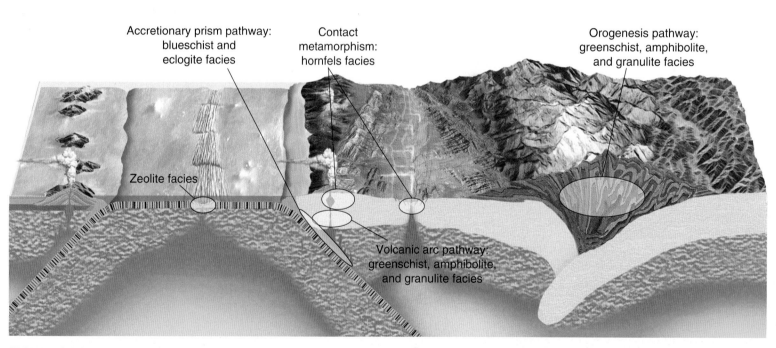

FIGURE 9.20 Metamorphism develops at divergent and convergent boundaries, as well as mid-plate settings.

(?) In which tectonic environment does pressure play the greatest role? In which tectonic environment does heat play the greatest role?

crust), the presence of chemically active fluids, and the level of thermal metamorphism occurring in contact aureoles determine the mineral assemblage that will develop at such sites. Contact metamorphism in oceanic crust typically forms pyroxene and hornblende-rich hornfels, while intrusion into sedimentary country rock results in plagioclase, muscovite, quartz, and calcite-rich hornfels or marble. Intruded limestone results in the formation of marble.

Deep Burial

The stable interior of continental lithosphere, far from plate boundaries, is characterized by deep burial metamorphism where temperatures exceeding 600°C produce eclogite and amphibolite facies. If enough water is present, partial melting of high-grade schists and gneiss occur. This magma forms granitic intrusions that add to the overall increase in regional temperature.

Basalt, the most abundant form of crust, offers a real-world example of tying tectonic locations and metamorphic processes together under the facies concept.

Metamorphism of Basalt Crust

Basalt is the most common igneous rock (TABLE 9.4). It is rich in the ferromagnesian silicates olivine, pyroxene, and calcium-rich plagioclase feldspar. These minerals are *anhydrous;* that is, they do not contain water. This means that when water is present, they will readily use it to form new minerals. For example, when chemically active fluids enter the metamorphic process, new assemblages of minerals develop from the ferromagnesium silicates present in basalt. We saw already that this leads to zeolite and greenschist facies at the spreading centre.

Water is present at convergent boundaries because it is trapped in the pores and bonded to the minerals in basalt and marine sedimentary rocks. In some cases, water is trapped in the atomic structure of minerals within the crust. Riding within an oceanic plate as it moves toward a subduction zone, these minerals readily release their trapped water in the earliest stages of metamorphism during subduction.

Hot water is released from the subducting plate. This water lowers the melting point of rocks that it encounters in the upper mantle and crust in the overriding plate. Magma formation in the wedge of mantle above the subducting plate plays an important role in raising the regional temperature of the crust as well as producing plutonic bodies that intrude into overlying rock, resulting in contact metamorphism.

In seafloor basalt that is undergoing regional metamorphism at a subduction zone, water typically is present in abundance, making possible the crystallization of new assemblages of *hydrous* (water-rich) zeolites. As basalt undergoes metamorphism in the presence of water, it evolves into low-grade greenschist that contains a mineral assemblage consisting of chlorite, epidote, plagioclase, and calcite. Greenschist is equivalent in metamorphic grade to slate and phyllite, and has distinctive foliation and a greenish hue produced by the green minerals chlorite and epidote.

As pressure and temperature increase during the later stages of subduction, chlorite is replaced by amphibole. The resulting rock, amphibolite, is dark and coarse grained, lacking distinct foliation. Chlorite and other platy micas have been replaced by dark minerals. As pressure and temperature conditions increase, amphibole is replaced by pyroxene, and granulite, the high-grade member of the basalt sequence, forms. Granulite consists largely of pyroxene, plagioclase, and garnet. Like amphibolite, granulite is weakly foliated. Continued increase in heat and pressure eventually leads to partial melting.

TABLE 9.4 Metamorphic Products of Basalt

Increasing temperature and pressure ⟹

Protolith Basalt	Not Metamorphosed	Low Grade	Medium Grade	High Grade
Metamorphic Products	Basalt + H_2O	Zeolite, greenschist	Amphibolite	Granulite
Foliation	None	Distinct schistosity	Indistinct; occasional due to parallel amphibole crystals	Indistinct due to absence of micas
Size of Grains	Visible with magnifier	Visible with magnifier	Obvious by eye	Large and obvious
Minerals	Olivine, pyroxene, plagioclase	Chlorite, epidote, calcite, plagioclase	Amphibole, plagioclase, quartz	Pyroxene, plagioclase, garnet

 Expand Your Thinking—Besides plate boundaries, describe where metamorphism occurs and the source and role of chemically active fluids there.

LET'S REVIEW "GEOLOGY IN OUR LIVES"

Now that you have finished the chapter, "Geology in Our Lives" will have taken on new meaning for you. Let us review it: All rocks can be exposed to rising temperature and pressure conditions where they are intruded by magma, buried deep in the crust, or involved in processes at convergent margins. Increases in pressure and heat, usually in the presence of chemically active fluids, recrystallize igneous, sedimentary, and metamorphic minerals to make new metamorphic rock. Conditions that lead to metamorphism are linked to specific tectonic processes occurring at plate boundaries or deep burial within the crust or both. Metamorphic rocks provide important building materials and mineral resources.

Metamorphic rocks contain a variety of different mineral assemblages. Exactly which minerals make up a metamorphic rock depends primarily on two factors: (1) the mineralogy and chemical composition of the parent rock and any fluids present, and (2) the temperature and pressure prevailing during metamorphism (metamorphic grade). Metamorphism is a good example of how natural forces interact with natural materials to make predictable products. The same assemblage of protolith minerals will result in different metamorphic products, depending on the specific heat and temperature pathways they follow. Add fluids to the mix, and the variety of resulting products becomes even greater.

Yes, pressure and heat metamorphose rocks in the crust. But the real lesson is that geologic processes leave behind a trail of information that enables us to use critical thinking to discover how Earth works.

With the completion of Chapter 9, we come to the end of our survey of the three large families of rocks composing the solid shell of Earth: igneous, sedimentary, and metamorphic.

STUDY GUIDE

9-1 Metamorphic rocks are composed of sedimentary, igneous, or metamorphic minerals that have recrystallized.

- **Metamorphic rocks** are sedimentary, igneous, or metamorphic rocks that have taken on a new texture and mineralogy in order to maintain equilibrium under changing conditions within the crust. This occurs as a result of solid-state **recrystallization** of minerals in the original ("parent") rock. Recrystallization produces a new assemblage of minerals by recycling existing minerals.

- Metamorphic rocks represent a distinct phase in the life of the crust and are a key element of the rock cycle.

9-2 Changes in heat and pressure can cause metamorphism.

- Rocks that are undergoing **metamorphism** alter their form and mineralogy in order to achieve a more stable configuration when experiencing increased temperature, increased pressure, and chemically active fluids.

- Heat increases with depth in the crust along what is known as the geothermal gradient. The average geothermal gradient is an increase of about 30°C to 50°C per kilometre of depth in the crust, but the actual increase varies in different tectonic settings. At temperatures between 150°C and 200°C (at a depth of ~8 km), clay minerals (the most ubiquitous sediment) become unstable and recrystallize to become the metamorphic minerals chlorite and muscovite, marking the onset of early metamorphism.

- Lithostatic **stress** increases with depth in the crust and plays an important role in metamorphism, closing the pore spaces in rocks, causing compaction, and, at great depths, causing minerals to recrystallize to form more compact atomic structures. Lithostatic stress deforms the rock equally in all directions. **Directed stress** differs from lithostatic stress in that pressure acting on the crust is greatest in one direction. This causes folding, bending, breaking, or even overturning of rock strata.

9-3 Chemically active fluids transport heat and promote recrystallization.

- Chemically active fluids play a significant role in metamorphism. They transport heat and have a profound influence on the chemistry and mineralogy of metamorphic rocks. Important fluids may come from groundwater if metamorphism is occurring in shallow crust (less than ~2 km deep); be produced by dewatering of magma; come from sea water; or be produced by dehydration of water-rich minerals, such as micas or amphiboles.

- Hydrothermal fluids come from the interaction of groundwater and magma in the final stages of cooling. They are hot and contain dissolved ions and compounds that react with country rock, causing crystallization of valuable metals, such as lead, silver, platinum, and gold.

- Chemical alteration in the presence of hydrothermal fluids is known as metasomatism.

9-4 Rocks evolve through a sequence of metamorphic grades.

- Rock may undergo different grades of metamorphism, depending on the conditions in which metamorphism takes place. The concept of gradational change, from **low-grade metamorphism** to **high-grade metamorphism,** reflects the nature of the mineral and textural changes that occur.

- Shifting mineral assemblages characterize the change from protolith to metamorphic product. **Index minerals** that track these changes include (from lowest to highest metamorphic grade) chlorite, muscovite, biotite, garnet, kyanite, and sillimanite. When identified in rock samples, these minerals serve as indicators of the grade of metamorphism experienced by the rock. For instance, the presence of chlorite indicates low-grade conditions while the presence of sillimanite indicates high-grade conditions.

- Shale evolves through four common foliated textures, in order of increasing metamorphic grade: slate, phyllite, schist, and gneiss. Of these, those with large-grained textures (schist and gneiss) typify medium- to high-grade metamorphism; those with fine-grained textures (slate and phyllite) generally indicate low- to medium-grade metamorphism.

9-5 Foliated texture is produced by directed stress related to regional metamorphism.

- Within the crust, directed stress is created by the collision of two plates or by simple burial in such a way that pressure is greatest in one direction. Rocks formed under directed stress usually exhibit a texture known as **foliation** created by aligned platy minerals, and a **lineation** formed by the preferred alignment of elongate minerals.
- **Regional metamorphism** occurs when elevated pressures and temperatures are created at convergent tectonic margins. Two colliding plates create an extensive zone of elevated pressure characterized by directed stress in the deep crust. Directed stress typically causes the formation of foliation in the products of regional metamorphism.

9-6 Nonfoliated rocks may develop during regional or contact metamorphism.

- **Contact metamorphism** takes place when a rock is subjected to high temperature because of intrusion by an igneous plutonic body. In general, contact metamorphism is not accompanied by significant changes in pressure.

- Fine-grained rocks with a **nonfoliated texture** include hornfels and anthracite coal. The large-grained nonfoliated rocks are marble, quartzite, amphibolite, and metaconglomerate. These rocks are, for the most part, identified based on their mineral assemblages.
- The result of contact metamorphism of shale is hornfels. When limestone is changed by contact metamorphism, the resulting rock is marble.
- Contact between a silicate-rich pluton and impure limestone country rock results in the formation of the mineral wollastonite and the production of carbon dioxide gas.

9-7 The relationship between mineral assemblage and metamorphic grade is expressed by metamorphic facies.

- Relationships between mineral assemblage and metamorphic grade are expressed in the concept of **metamorphic facies,** which states that, depending on the composition of the original rock, the mineral assemblage in the metamorphic rock reflects the temperature and pressure conditions prevailing during metamorphism.
- The major facies are zeolite, hornfels, greenschist, amphibolite, granulite, blueschist, and eclogite.

9-8 Metamorphism is linked to plate tectonics.

- The pressure and temperature conditions that lead to metamorphism are linked to specific tectonic processes occurring at plate boundaries. Metamorphism and plate tectonics thus are intimately linked.

KEY TERMS

contact metamorphism (p. 222)
directed stress (p. 223)
foliation (p. 221)
high-grade metamorphism (p. 226)
hydrothermal vein fillings (p. 224)
index minerals (p. 226)

lineation (p. 221)
low-grade metamorphism (p. 226)
metamorphic facies (p. 232)
metamorphic rocks (p. 220)
metamorphism (p. 220)
nonfoliated texture (p. 228)

recrystallization (p. 221)
regional metamorphism (p. 223)
rock cleavage (p. 228)
stress (p. 220)

ASSESSING YOUR KNOWLEDGE

Please answer these questions before coming to class. Identify the best answer.

1. Metamorphic rocks are often formed by increased
   a. pressure and cementation.
   b. heat and melting.
   c. pressure and heat.
   d. cooling and solidification.
   e. None of the above.

2. Metamorphism occurs when
   a. minerals partially melt and quickly recrystallize.
   b. recrystallization occurs in the solid state.
   c. loose sediments grow new crystals that cement grains together.
   d. igneous minerals have solidified.
   e. None of the above.

3. What type of metamorphism is local in extent and results from the rise in temperature in country rock surrounding an igneous intrusion?
   a. regional
   b. contact
   c. burial
   d. metasomatism
   e. plutonism

4. Chemically active fluids promote
   a. regional stress.
   b. formation of oceanic intrusions.
   c. stable conditions deep in the crust.
   d. dissolution and migration of ions.
   e. None of the above.

5. The metamorphic index minerals are
   a. kaolinite, garnet, quartz, chlorite, biotite, and schist.
   b. chlorite, garnet, sillimanite, hornfels, schist, and muscovite.
   c. slate, phyllite, schist, chlorite, greenschist, and gneiss.
   d. chlorite, muscovite, biotite, garnet, kyanite, and sillimanite.
   e. gneiss, slate, chlorite, and quartz.

6. Foliated metamorphic rocks, in order of increasing metamorphic grade, are
   a. clay, chlorite, muscovite, biotite, garnet, and sillimanite.
   b. marble, quartzite, mylonite, and gneiss.
   c. slate, phyllite, schist, and gneiss.
   d. shale, slate, quartzite, marble, and schist.
   e. gneiss, slate, schist, chlorite, and phyllite.

7. Marble is related to limestone in the same way that
   a. basalt is related to andesite.
   b. slate is related to shale.
   c. gravel is related to breccia.
   d. gneiss is related to marble.
   e. sandstone is related to basalt.

8. Which of the following statements about foliated rocks is correct?
   a. They reflect the influence of directed stress in the crust.
   b. They are usually formed within intruded country rock.
   c. They are the product of metasomatism.
   d. They rarely develop at convergent margins.
   e. None of the above.

9. The eclogite facies
   a. reflects low temperature and low pressure conditions.
   b. develops only from carbonate sedimentary protoliths.

   c. is formed in shallow conditions along the orogenesis pathway.
   d. reflects conditions on the accretionary prism and burial metamorphism pathways.
   e. None of the above.

10. Which of the following usually develops when magma intrudes into shale?
    a. hornfels
    b. zeolite
    c. schist
    d. marble
    e. subduction

11. Which of the following tectonic processes is (are) most important to metamorphism?
    a. plate rotation
    b. sediment accumulation and erosion
    c. subduction and plate convergence
    d. paleomagnetic wandering
    e. Plate tectonics is not related to metamorphism.

12. Common contact metamorphic rocks include
    a. zeolite, hornfels, and shale.
    b. slate, gneiss, and marble.
    c. quartzite, marble, and hornfels.
    d. basalt, granulite, and blueschist.
    e. None of the above.

13. Regional metamorphosis of shale occurs in the following sequence:
    a. zeolite, gneiss, slate
    b. slate, phyllite, schist, gneiss
    c. gneiss, marble, schist, hornfels
    d. greenschist, slate, hornfels, basalt
    e. None of the above.

14. Metamorphism of basalt makes this sequence:
    a. zeolite, greenschist, amphibolite, granulite
    b. marble, hornfels, schist, zeolite
    c. zeolite, marble, hornfels, granulite
    d. granulite, blueschist, hornfels, slate
    e. None of the above

15. List the metamorphic rocks produced by contact metamorphism of the following protoliths:
    a. Basalt: _____
    b. Shale: _____
    c. Limestone: _____
    d. Quartz sandstone: _____

FURTHER RESEARCH

1. Is metamorphism possible without plate tectonics?

2. Why does most marble not display visible foliation?

3. A convergent plate margin can produce both regional and contact metamorphism. Explain how this is possible.

4. Why does rock cleavage develop?

5. How do conditions produced by regional metamorphism differ from conditions produced by contact metamorphism?

6. What are the index minerals, and how do geologists interpret them?

7. Describe the concept of metamorphic facies. What are the five pathways of metamorphism?

8. How does metamorphism in mid-continent settings differ from metamorphism at convergent margins?

ONLINE RESOURCES

Explore more about sedimentary rocks on the following websites:

The Canadian Encyclopedia—Metamorphic Rock
www.thecanadianencyclopedia.com/articles/metamorphic-rock

Wikipedia open source online encyclopedia entry for metamorphic rocks:
http://en.wikipedia.org/w/index.php?title=Metamorphic_rock&oldid=553934110

University of Oxford site that compiles links on metamorphism:
www.earth.ox.ac.uk/~davewa/metpet.html

Additional animations, videos, and other online resources are available at this book's companion website:
www.wiley.com/go/fletchercanada

This companion website also has additional information about WileyPLUS and other Wiley teaching and learning resources.

242

CHAPTER 10
GEOLOGIC RESOURCES

Chapter Contents and Learning Objectives (LO)

10-1 The crust contains metals, building stone, minerals, and sources of energy.

LO 10-1 *Compare and contrast mineral resources and energy resources.*

10-2 Mineral resources include non-metallic and metallic types.

LO 10-2 *Describe in detail non-metallic and metallic mineral resources.*

10-3 Ores are formed by several processes.

LO 10-3 *Describe the four general categories in which mineral deposits can form.*

10-4 Fossil fuels, principally oil, provide most of the energy that powers society.

LO 10-4 *Describe the forms of energy used in modern society.*

10-5 Oil is composed of carbon that is derived from buried plankton.

LO 10-5 *Describe the origin of oil and how it accumulates in the crust.*

10-6 About 77 percent of the world's oil has already been discovered.

LO 10-6 *Describe the status of the world's oil supply.*

10-7 Coal is a fossil fuel that is found in stratified sedimentary deposits.

LO 10-7 *Describe the origin of coal.*

10-8 Nuclear power plants provide about 17 percent of the world's electricity.

LO 10-8 *Identify and describe the special feature of uranium-235 that makes it useful as an energy source.*

10-9 Renewable energy accounts for more than 20 percent of Canada's energy sources.

LO 10-9 *Identify the ways energy can be provided from renewable sources.*

GEOLOGY IN OUR LIVES

Nearly every substance you use in your life has either been mined from the crust or manufactured using energy from fossil fuels. All humans are dependent on mineral resources and fossil fuels, yet these geologic resources are non-renewable on the human time scale. Hence, use of non-renewable resources must be tempered with awareness that they may become unaffordable if they are not replenished, used in a sustainable manner, or replaced with renewable alternatives. Knowing the facts about geologic resources can help in your decision making as a consumer and as a caretaker of Earth.

(?) At this moment, identify at least five things around you produced with geologic resources.

10-1 The Crust Contains Metals, Building Stone, Minerals, and Sources of Energy

LO 10-1 *Compare and contrast mineral resources and energy resources.*

Our day-to-day world is filled with materials made from rocks and minerals. Industries and entire economies are powered by fossil fuels taken from the crust. Rocks contain valuable metals, building stone, precious minerals, and hydrocarbon compounds that can be burned to produce energy. Without geologic resources, civilization would have barely progressed beyond the days of the "caveman."

The use of geologic resources began millions of years ago, when *Homo habilis* ("handy man") broke open a flint nodule to make a cutting tool (**FIGURE 10.1**). Our Stone Age ancestors discovered that when they used stones creatively, their quality of life improved. For over 2 million years, human culture has been closely linked to geologic resources.

By 10,000 years ago, humans had learned to make pottery and quarry rock for shelter, heralding the dawn of the Neolithic, or New Stone Age. The Copper Age was born when humans learned to extract copper from the mineral malachite by heating it over a fire. Someone found that mixing tin with copper would strengthen it, and

the Bronze Age began. The Iron Age followed and eventually was replaced by the Fossil Fuel Age. Today, it is sometimes claimed that we live in the Silicon Age or the Technology Age. But this does not mean that we no longer rely on rocks and minerals; even today, the primary material used to make silicon computer chips is quartz from grains of beach sand. Throughout our history, our existence and prosperity as a species has been defined by our ability to find and effectively use **geologic resources**.

What Are Geologic Resources?

Consider all the materials you encounter and use in your daily life: metals, water, glass, plastics, food, fabrics, chemicals, and the things that we make from them—your clothing, beauty aids, car, the components of your cellphone and computer, your home. How many of these items have their origin in the crust? How many are manufactured using energy resources? The answer is all of them. Every element used to make these items comes from rocks mined out of Earth's crust. Rocks, minerals, water, soil, and sources of energy that are useful to humans are termed *geologic resources.*

Geologic resources are non-living materials that are mined to maintain our system of industry and quality of life. The water you use is taken from the ground (Chapter 20, "Groundwater") or from streams (Chapter 19, "Surface Water"). Soil is the product of weathering the crust (Chapter 7, "Weathering"). In this chapter, we review rocks, minerals, and fuels that play a critical role in modern society. These geologic resources include two broad categories: **mineral resources** and **energy resources**. The majority of geologic resources are **non-renewable**. That is, the total amount available on Earth is reduced every time they are used, and what is used is not replenished within a human generation (25–30 years). Geologic processes like the rock cycle naturally renew mineral and energy resources, but the *rate* of geologic renewal is vastly exceeded by the rate of human consumption. Hence, our use of non-renewable resources must be tempered with awareness that they may be depleted if we do not replenish them or use them in a sustainable manner.

Mineral resources are useful rock and mineral materials and range from antimony to zinc, asbestos to zircon (**TABLE 10.1**). Some of these resources are directly mined from the crust and have value as building stone, gravel, and sand. Other mineral resources are not useful in their natural state and must be processed because the commodity they hold, such as emeralds, asbestos, or native gold, must be obtained through complex and expensive technologies. These include *ore bodies* that hold metals that can be profitably mined, such as copper or lead, whose processing can generate *toxic waste*.

Energy resources include both non-renewable (oil, coal, natural gas, and uranium) and renewable resources (solar power, wind power, bioenergy, geothermal energy, fuel cells, hydroelectric power, and ocean energy). At present, society meets most of its energy needs with non-renewable resources. These account for approximately

Philippe Plailly/Science Source

FIGURE 10.1 Our early ancestor *Homo habilis* first used hand tools to improve the quality of life. This marked the beginning of a long relationship between humans and the prized geologic resources available in Earth's crust.

(?) What materials were easily available from the crust for ancient communities?

80 percent of the world's total energy consumption and 95 percent of its commercial energy consumption. But because these sources are non-renewable, a growing **renewable energy** industry is emerging. In coming years, as the cost of oil continues to rise (because fewer new reserves are being found), renewable forms of energy, such as solar, hydroelectric, and geothermal energy, will increasingly become part of our economy.

Why Study Geologic Resources?

One hundred years ago, the world's population totalled 1.7 billion. Today, it is nearly 7 billion, more than a threefold increase. One-quarter of a billion people are added every 3.2 years. This population explosion strains natural resources and draws on buried wealth in Earth's crust at an accelerating rate. *In the next 10 years, we will consume more energy and mineral resources than we have throughout all of human history.* Modern industry now acknowledges that geologic resources have limits and that their use has environmental consequences. Important questions have arisen: How soon will oil supplies be exhausted? Are there cleaner technologies for burning coal? Can recycling metals reduce the need for mining? Can renewable energy replace our dependence on fossil fuels? Metals may become unaffordable in the future; are there substitutes?

We study geologic resources because they are fundamentally important to our way of life, to the structure of our society and civilization, and even to world peace. As the world's population expands, the demand for geologic resources increases. Heavy demand places stress on supplies of those resources—so much so that supplies of some critical resources, such as oil, will be increasingly limited, possibly to the point of exhaustion, within your lifetime if our current rate of consumption continues.

TABLE 10.1 Ten Common Metal Ores			
Metal	**Ore Minerals**	**Major Sources**	**Common Uses**
Aluminum	Bauxite [$Al(OH)_3$]	Australia, Guinea, Jamaica	Most abundant metal in crust; used in packaging (31%), transportation (22%), building (19%)
Cobalt	Cobaltite ($CoAsS$)	Zaire, Zambia, China, Cuba	Used in alloys for jet engines, chemicals, magnets, carbides for cutting tools Japan and China use 50% of world production.
Copper	Native copper (Cu), chalcopyrite ($CuFeS_2$), bornite (Cu_5FeS_4), chalcocite (Cu_2S)	Chile, United States, Canada (British Columbia), Peru	Electric wiring, plumbing, heating, construction, chemical machinery, alloys (brass, bronze)
Iron	Hematite (Fe_2O_3)	Australia, Brazil, China, Canada (Labrador and Quebec)	Most important industrial metal Used in steel, auto parts, medicine, biochemical research, paints, plastics, cosmetics, paper
Lead	Galena (PbS)	China, Australia, United States (Colorado)	Used in batteries, electronic applications, TV tubes and glass, X-ray and gamma radiation shielding, ammunition
Molybdenum	Molybdenite (MoS_2)	Canada (British Columbia), Chile, United States (Colorado)	Used in steel (47% of all uses) to make automotive parts, construction equipment, tools, cast iron, lubricants
Platinum group metals (PGM)	Native platinum (Pt)	South Africa	Among the scarcest of metallic elements; used as a catalyst for industrial plant emissions, films
Titanium and titanium dioxide	Ilmenite ($FeTiO_3$), rutile (TiO_2)	Western and central United States, United Kingdom, China, Japan	Metal in jet engines, airframes, and space and missile applications; used as a white pigment in toothpastes, pills, inks, plastics, and paints
Uranium	Uraninite (UO_2)	Canada (Saskatchewan), Australia	Used to produce more than 15% of Canada's energy; also used in medicine, radiometric dating, powering nuclear submarines, and other defence purposes
Zinc	Sphalerite [(Zn,Fe)S]	China, Australia, Peru, United States (Tennessee, Missouri)	Coating on steel, chemicals, rubber, galvanizing iron, automotive parts, dry cell batteries, pennies, medicine

 Expand Your Thinking—How would a world without iron, aluminum, and copper be different?

10-2 Mineral Resources Include Non-metallic and Metallic Types

LO *10-2 Describe in detail non-metallic and metallic mineral resources.*

Mineral resources fall into two categories: *non-metallic* (e.g., building stone, sand, and gravel) and *metallic* (TABLE 10.2).

Non-metallic Mineral Resources

Non-metallic mineral resources (FIGURE 10.2) include sand for making glass, rock salt for de-icing roads, gypsum for plaster and wallboard, phosphate minerals for fertilizer, and clays for ceramics, bricks, and other uses. Per person, we use 9 tonnes of (non-fuel) mineral resources in North America per year. Of this, 94 percent is non-metallic minerals—mostly in the form of construction materials.

There are three groups of non-metallic minerals:

1. *Industrial minerals,* used for industrial purposes other than agriculture or construction activities (e.g., the technology industry, steel making, automobile and goods manufacturing)

2. *Agriculture minerals,* including nitrate (produced by bacteria), potassium, and phosphate (mined from sedimentary deposits), used as fertilizer in food production

3. *Construction minerals,* including *aggregates* (crushed stone, sand, gravel—largely used for building roads); cement (made from limestone and shale); clay and gypsum

With the world population nearing the 7 billion mark and likely to reach 9 billion by 2040, food production and the need for fertilizers and agricultural chemicals will continue to expand.

Metallic Mineral Resources

Five *abundant metals* are used in industry: aluminum, iron (Figure 10.2), magnesium, manganese, and titanium. These constitute over 0.1 percent of Earth's crust. *Scarce metals,* the term for all others (including copper, lead, zinc, gold, silver, etc.), constitute less than 0.1 percent.

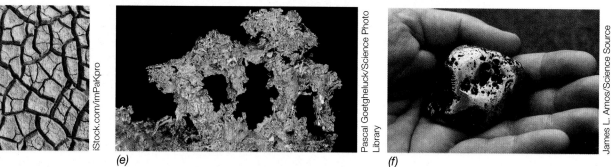

(a)   Martin Bond/Photo Researchers, Inc.   (b)   Courtesy of Chip Fletcher   (c)   © photogen/Alamy

(d)   iStock.com/imPaKpro   (e)   Pascal Goetgheluck/Science Photo Library   (f)   James L. Amos/Science Source

(g)   Courtesy of Chip Fletcher

FIGURE 10.2 Common mineral resources: *(a)* limestone quarry; *(b)* construction aggregate; *(c)* industrial sand processing quarry; *(d)* hematite, the principal ore of iron; *(e)* sedimentary deposit of clay that is mined for industrial uses; *(f)* native gold; and *(g)* platinum from a sedimentary deposit.

? Which of these resources are mined near where you live?

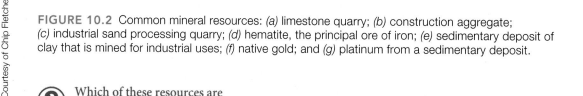

TABLE 10.2 Mineral Resources

Non-metallic Mineral Resources	Metallic Mineral Resources
Industrial Minerals Corundum, diamond, feldspar, fluorite, garnet, pumice, quartz *Agricultural Minerals* Potash, nitrate, peat, phosphate, salt, sulphur	*Abundant Metals* Aluminum (Al), iron (Fe), magnesium (Mg), manganese (Mn), titanium (Ti)
Construction Minerals Building stone, clay, gravel, gypsum, limestone, sand, shale	*Scarce Metals* Bismuth (Bi), chromium (Cr), cobalt (Co), copper (Cu), gold (Au), lead (Pb), mercury (Hg), molybdenum (Mb), nickel (Ni), platinum (Pt), silver (Ag), tin (Sn), tungsten (W), uranium (U), zinc (Zn)

TABLE 10.3 Common Ore Types and Metallic Content

Ore Type	Typical Metals
Carbonates and silicates	Beryllium, iron, lithium, magnesium, manganese, nickel
Native metals	Copper, gold, iron, osmium (Os), platinum, silver
Sulphides	Antimony (Sb), arsenic (As), bismuth, cobalt, copper, lead, mercury, molybdenum, nickel, silver, zinc
Oxides and hydroxides	Aluminum, chromium, iron, manganese, niobium (Nb), tantalum (Ta), tin, titanium, tungsten, uranium

Approximately 40 metals are available through mining and **smelting** (heating or melting a mineral to obtain the metal it contains). The minerals that are smelted or processed to yield the metal of interest are called the **ore minerals**, of which there are about 100. They tend to form in concentrated masses called *mineral deposits* mixed with valueless minerals, called *gangue*. The more concentrated an ore mineral deposit is—that is, the less gangue it contains—the greater its value, but to be profitably mined, a deposit must meet a minimum standard of size and concentration. A profitable assemblage of minerals plus gangue is an **ore**.

Ore minerals of abundant metals are found in many rocks, while ore minerals of scarce metals are found only in locations where a special geologic process has enriched the ore mineral to a profitable concentration.

Common Ore Minerals

Native metals, sulphides, oxides and hydroxides, and silicates and carbonates (**TABLE 10.3**) are used as ore minerals. Of primary importance in smelting an ore mineral is the amount of energy needed to break the chemical bonds and release the desired metal. Typically, less energy is needed to smelt sulphide, oxide, or hydroxide minerals (to obtain iron, aluminum, copper, and other ores) than is required to smelt silicate minerals. Hence, few silicate minerals serve as ore minerals. Sometimes ore is extracted from silicate and carbonate minerals if a metal is not found in a more desirable ore mineral or if it is formed in a desirable mineral but not in large enough deposits.

Some metals, such as copper, tin, and iron, may be found in several types of ore minerals.

Native Metals

Only two metals, gold and platinum, are commonly found in their native state, and in both cases, the native metals are themselves the primary ore minerals. Silver, copper, iron, and osmium occur less often in their native state, and only a few such occurrences are large enough to be profitable as ore deposits.

Sulphides

Sulphides are the largest group of ore minerals. They consist of one or more metals combined with sulphur. For example, chalcocite (Cu_2S) is an important copper ore that has been profitably mined for centuries. It has a high copper content (nearly 80 percent) and is easily separated from sulphur. Copper, lead, zinc, nickel, molybdenum, silver, arsenic, cobalt, and mercury all readily form sulphide ore minerals.

Oxides and Hydroxides

Oxides and hydroxides are a large and diverse group of ore minerals that are formed in a variety of igneous, metamorphic, and sedimentary environments. Hematite is an oxide that is an important iron ore. Because of the strength of the bonds between oxygen and metals, oxides are resistant to further chemical attack and generally form strong, hard minerals.

(?) **Expand Your Thinking**—Give some examples of non-metallic minerals and how they are used in your daily life.

10-3 Ores Are Formed by Several Processes

LO 10-3 *Describe the four general categories in which mineral deposits can form.*

Mineral deposits form in several ways (TABLE 10.4). They fall into four general categories: *magmatic segregation deposits, hydrothermal deposits, residual mineral deposits,* and *sedimentary mineral deposits.* We review a few examples of each of these ore-forming processes to get an idea of how mineral accumulations develop.

Magmatic Segregation Mineral Deposits

A magmatic segregation deposit forms during magma crystallization, usually in **cumulates** and **pegmatites**. Magmatic cumulates form as a result of *density segregation* and *immiscible melts.* A cumulate is a layer of dense minerals forming a concentrated deposit on the floor of a magma chamber. This happens when early-forming crystals either settle because they are denser than the remaining magma, or are carried by currents to the floor of a magma chamber. Magma begins life as a homogeneous liquid, but magmatic segregation through crystallization produces cumulates of varying composition (FIGURE 10.3). One example, the Bushveld Igneous Complex of South Africa, consists of layers that include concentrated deposits of chromite (iron-magnesium-chromium oxide) and magnetite (iron oxide).

A different kind of magmatic segregation involves immiscible melts that form cumulates. Cooling magma can precipitate droplets of a second magma with a different composition. Like oil and water, the two do not mix. (They are "immiscible.") When magma is saturated with the melt of a particular mineral, solid precipitation of that mineral occurs. If saturation is reached at a temperature higher than the melting point of a mineral, a drop of liquid forms instead of a solid mineral grain. One of the most important examples is thought to be the formation of sulphur-rich immiscible melts, which can then form heavy liquid layers in magma in a similar manner to that in which a cumulate is formed. As the sulphur-rich melt cools, a deposit of copper, nickel, and platinum metals in a gangue of iron sulphide minerals may form. These cumulate and immiscible melt-related mineral deposits typically form from ultramafic or mafic magmas. Pegmatites form in certain kinds of magma, especially granitic magma, that contain dissolved water. As this "wet" magma cools, the first minerals to crystallize (such as feldspar) tend to exclude water from their chemical makeup, leaving a water-rich residue. Some elements, such as lithium, beryllium, and niobium, are concentrated in this residue. If the residual magma migrates (say, into a fracture), it may form bodies of igneous rock enriched in rare elements, called *rare-metal pegmatites.* These rocks can be very coarse grained, with individual crystals of mica, feldspar, and beryl over 1 m long.

Hydrothermal Mineral Deposits

Hydrothermal deposits form when hot, salty water dissolves and transports metallic ions that become ore minerals. Water may be released by cooling magma or expelled during metasomatism; or sea water or groundwater may percolate through fractures and porous rocks into the crust. The simplest hydrothermal deposit is a *vein,* which forms when hydrothermal solutions precipitate compounds and native metals along the walls of a crack. Often veins occur where igneous plutons provide a heat source. Hydrothermal solutions that come into contact with a pluton are thermally energized and charged with dissolved compounds; they will migrate through crust that is fractured by the intrusion. Crystallization occurs as the fluid reacts chemically with rocks lining a fracture, or if conditions change along the fracture, such as a decrease in temperature. The result is that economic minerals may precipitate.

Another hydrothermal deposit is **porphyry copper**. Typically found in association with intrusive rock containing *disseminated minerals,* these deposits occur in fractured country rock containing a network of tiny, closely spaced veins of quartz. Each vein carries grains of the copper ore minerals, such as chalcopyrite and bornite, or the molybdenum ore molybdenite. Some porphyry deposits also contain significant amounts of gold. The volume of mineralized rock is huge and can amount to billions of tonnes of ore. Though averaging only 0.5 percent to 1.5 percent copper, mined tonnages of porphyry are so large that more than 50 percent of all copper comes from porphyry copper deposits.

TABLE 10.4 Formation of Ore Minerals		
Category	**Ore Type**	**Typical Metals**
Magmatic segregation deposits	Cumulates Immiscible melts Pegmatites	Chromite, ilmenite, magnetite Iron sulphides with copper, nickel, platinum Beryllium, lithium, niobium
Hydrothermal deposits	Vein Porphyry copper deposit Volcanogenic massive sulphide Sedimentary exhalative Unconformity-type uranium	Gold, native metals, silver Copper, molybdenum, gold Copper, gold, lead, silver, zinc Zinc, lead Uranium
Residual mineral deposits	Residual soils Secondary enrichment	Aluminum, iron, nickel Gold, lead, silver, zinc
Sedimentary mineral deposits	Banded iron formation Evaporite Placer deposit	Iron Potash, gypsum, halite, borax Gold, platinum, zircon, others

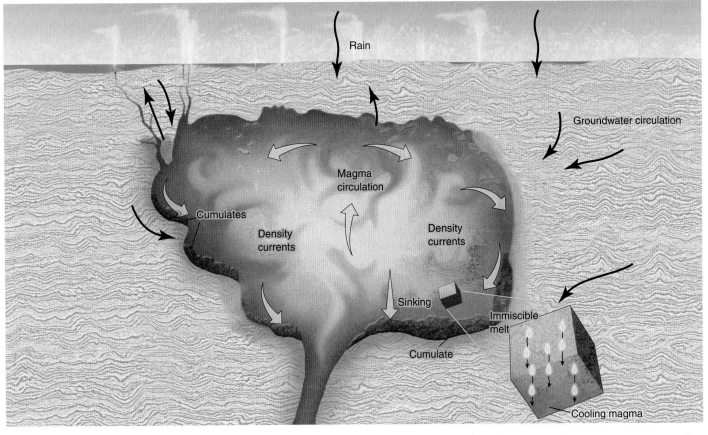

FIGURE 10.3 Magmatic cumulates in magma chambers filled with mafic and ultramafic magmas form through two processes. Density segregation forms layers of dense minerals by settling of grains, in-place crystallization, and magmatic currents sweeping grains from the walls and roof within a pluton. Immiscible melts may develop if a compound reaches its saturation state before it passes its crystallization temperature. The result is a dense liquid that eventually becomes a concentrated mineral deposit.

(?) Explain why density is the key factor segregating these minerals.

Volcanogenic massive sulphide (VMS) deposits form at volcanoes on the sea floor (in spreading centre, hotspot, or subduction zone environments) (**FIGURE 10.4**). Sea water penetrates the crust and becomes a hydrothermal fluid that dissolves and concentrates scarce metals. A hydrothermal vent develops where gushing, mineralized water precipitates an ore deposit onto the adjacent sea floor. VMS deposits are among the richest metal-yielding ores in the world, with as much as 20 percent of a deposit yielding copper, lead, zinc, gold, and silver. These types of deposits can be seen forming today on the sea floor. The well-known "black smokers" are locations where the hydrothermal fluid is pumped to the surface, and the reaction between the hot hydrothermal fluid and the cold sea water leads to the formation of sulphide minerals (the "black" in the "smoke").

Other important hydrothermal deposits include sedimentary exhalative deposits, which form from fluids that have interacted with sedimentary rocks and then move along faults to the sea floor (where they are "exhaled" into the sea water). The fluids are cooled, and interact with sea water that is depleted in oxygen, leading to the precipitation of sphalerite and galena, the main ores of zinc and lead. The Sullivan Mine at Kimberley (B.C.), now closed, is an example of this type of deposit. There are also deposits of this type around the town of Faro and in the MacMillan Pass area in the Yukon.

(?) How does a VMS deposit end up on a continent where it can be mined?

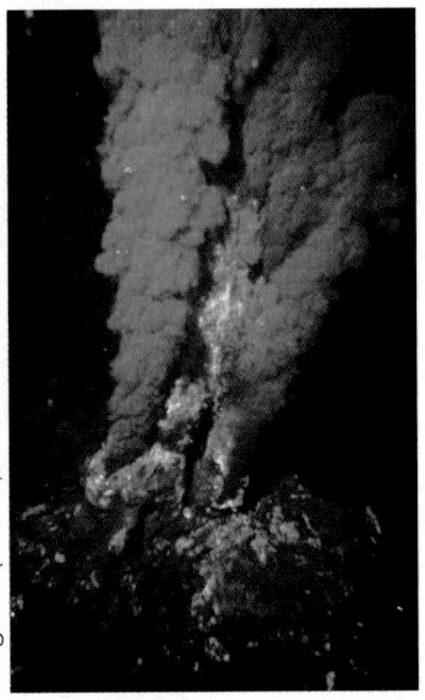

FIGURE 10.4 *(a)* Wherever volcanic action is found on the sea floor, sea water can penetrate the crust and come into contact with hot rocks and magma. The resulting hydrothermal fluids react with the crust and dissolve and concentrate scarce metals. When the resulting solution forms a hot spring (hydrothermal vent) on the sea floor, the gushing, mineralized water quickly cools and deposits its dissolved load. *(b)* Deposits formed in this way are known as volcanogenic massive sulphide (VMS) deposits.

OAR/National Undersea Research Program (NURP); NOAA

(a)

(b)

Woods Hole Oceanographic Institution

The world's richest uranium deposits are found in the Athabasca Sedimentary Basin of northern Saskatchewan, the best example being the McArthur River deposit. They are called unconformity-type uranium deposits, and are located at the place where younger faults crosscut the unconformity between older metamorphic rocks and younger sandstones. The main uranium ore mineral is uraninite, and it forms when the hydrothermal fluid carrying uranium interacts with rocks that contain a reductant. Uranium is found in two main valence states: U^{6+} can be carried easily in fluids, whereas U^{4+} is difficult to transport. Thus, any process that converts U^{6+} to U^{4+} will lead to the formation of uraninite.

Residual Mineral Deposits

Residual mineral deposits consist of metallic ore minerals that have been concentrated by chemical weathering in heavily weathered soils (FIGURE 10.5). Chemical weathering carries away compounds and ions dissolved in water. This process, called *leaching*, creates concentrated deposits of residual elements in weathered soils. These deposits may include aluminum, iron, manganese, and other metals that form oxides.

(a)

(b)

FIGURE 10.5 Intense weathering leaches ions and compounds out of bedrock and creates a concentrated mineral deposit in the residual soil. *(a)* Bauxite is the world's primary source of aluminum. It is a residual mineral deposit produced by intense weathering of aluminum-bearing rocks in tropical climates. *(b)* This bauxite operation is in Jamaica.

 How do residual soils form?

The most important ore of aluminum, **bauxite**, is a residual mineral deposit characterized by intense weathering in tropical climates. Heavily leached soils form bauxites that develop either on rocks that are low in iron or where iron has been dissolved by leaching. Bauxites currently developing in tropical regions in Australia, Brazil, and elsewhere all contain gibbsite $(Al[OH]_3)$ as the primary ore mineral.

Ions leached by the weathering process may be redeposited in nearby environments. This process can produce highly enriched ores, a process known as *secondary enrichment* (FIGURE 10.6). Secondary enrichment is especially important in two circumstances: (1) when gold-bearing rocks are leached and the dissolved components migrate in groundwater to form an enrichment layer of nuggets in soil, and (2) where leached sulphide minerals produce sulphuric acid that percolates dissolved metals into groundwater and forms an enriched layer (usually copper ore). Secondary enrichment of porphyry copper deposits is an important factor in the development of copper ores in the southwestern United States, Mexico, Peru, Canada, and Chile. Lead, zinc, and silver are also subject to secondary enrichment when rock containing them is heavily weathered.

Sedimentary Mineral Deposits

Sedimentary processes may form sedimentary mineral deposits, including ores of iron, which is by far the most important metal from an economic and technical point of view. Sedimentary iron deposits are among the world's great mineral treasures. The most important iron ores are **banded iron formations (BIFs)**, so called because of their fine

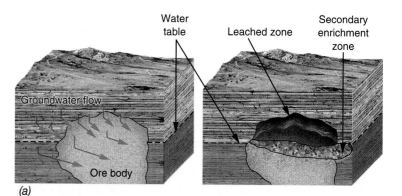

(a)

(b)

FIGURE 10.6 *(a)* When groundwater percolates through an ore body, it may leach out metallic cations and redeposit them at the water table in an enriched zone above the original ore body. This is called secondary enrichment. *(b)* Secondary enrichment of porphyry coppers is an important factor in the development of copper ores in Canada, the United States, Mexico, Peru, and Chile. Shown is native copper, although spectacular green malachite crystals, and blue azurite crystal may also form in this environment.

Why is the water table the location of enrichment for these ore bodies?

FIGURE 10.7 The most important ore of iron, banded iron formations, formed during a restricted period of time in the Precambrian, during which the composition of the oceans and atmosphere were changing from oxygen-poor to oxygen-rich.

Courtesy of Chip Fletcher

(?) What is the source of iron in BIFs?

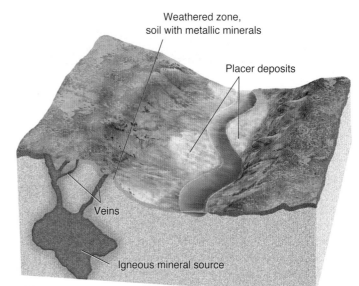

(a)

layering of cherty silica and iron, generally in the form of hematite, magnetite, or siderite (FIGURE 10.7).

BIFs are formed in ancient sedimentary basins. They are characterized by alternating bands of silica and iron-rich layers only a few centimetres thick but with extensive continuity—some layers are over 100 km long. It is also remarkable that most BIFs were formed only between 2.5 and 1.8 billion years ago, which geologists hypothesize to be the time during which the atmosphere was changing from being oxygen-poor to oxygen-rich. The chemistry of the oceans is linked to the composition of the atmosphere and thus was also very different from the chemistry of today's oceans. Iron is soluble in oxygen-poor water, but is virtually insoluble when oxygen is added to the water, and will precipitate as iron-rich minerals. It is thought that the iron and silica in these sedimentary basins may have been added to the water from hydrothermal fluids as well. Another type of sedimentary deposit is the *evaporite,* discussed in Chapter 8 ("Sedimentary Rock"). Evaporites form as a result of the evaporation of lake or sea water. As water evaporates, the remaining liquid becomes saturated in any dissolved substance it contains. Typically, deposits of halite (table salt), sylvite (potash; from which potassium is extracted for fertilizers), gypsum (used in wallboard), and borax (soap) develop under these conditions.

Placer deposits have a sedimentary origin. Such mineral deposits are found in stream channels draining watersheds that contain metallic ores (FIGURE 10.8). The stream may erode the rocks that contain the ore minerals, and the mineral grains are transported downstream. The deposits form when the energy level of the water decreases, at which time the higher-density grains are deposited to form a concentrated deposit. The less-dense grains are transported further down the stream. To be concentrated effectively, placer minerals must not only have a high density (greater than about 3.3 g/cm³), they must be mechanically durable and chemically stable in order to resist dissolution or reaction with surface water.

Ore minerals with suitable properties for forming placers are the oxides cassiterite (tin), chromite (chromium), ilmenite and rutile (titanium), and magnetite (iron), and the silicate zircon (zirconium). In addition, native gold and platinum have been mined from placers, as have several gemstone minerals, including diamond, ruby, and sapphire.

Try the "Critical Thinking" exercise "Mineral Deposits" to test your understanding of mineral deposits.

(b)

© Bettmann/CORBIS

(c)

Parks Canada Collection/Klondike National Historic Sites

FIGURE 10.8 *(a)* Minerals and native metals that are eroded out of ore rocks will collect in the channel of adjacent streams as a sedimentary placer deposit. Flowing water reworks these sediments and removes less dense grains, thus leaving a concentrated deposit of heavy minerals that can be mined profitably. *(b)* Miners uses sluice boxes to separate gold from one of the placer mines active during the Klondike Gold Rush in the Yukon near Dawson City. *(c)* One of the largest dredges used to excavate the gravels that contained gold in the Bonanza Creek valley, near Dawson City. It is now on display at the Dredge No. 4 National Historic Site.

(?) What well-known native element in western Canada is found in placer deposits?

(?) **Expand Your Thinking**—Water plays a role in nearly all these mineral deposits. Explain the various roles played by water. You may want to make a table to organize the details.

? CRITICAL THINKING

MINERAL DEPOSITS

Mineral resources can occur in many environments, but are often closely tied to igneous processes (FIGURE 10.9). In the vicinity of a crystallizing granitic plutonic body, interactions with groundwater and production of water-rich magma in the late stages of crystallization lead to the development of hydrothermal vein deposits and pegmatites. Within magma chambers containing mafic or ultramafic magma, magmatic segregation leads to cumulates formed by density segregation or by the separation of immiscible melts. In intruded crust, porphyry coppers and vein-hosted gold deposits may develop. The metal-rich ores produced by these processes can be weathered to form residual mineral deposits and secondary enrichment layers. Evaporites and placer deposits also form on the weathered surface of the crust.

Please work with a partner and answer the following questions using Figure 10.9.

1. Identify regions where magma interaction with country rock has formed mineral ore deposits. Identify the likely composition of ores there and the metals that can be smelted from those ores.
2. Where do residual mineral deposits and sedimentary mineral deposits form, and what economic minerals are found there?
3. Describe the magma segregation processes at work within a pluton that result in mineral ore deposits.
4. What role do chemically active fluids play in the growth of mineral ore deposits near magma intrusions? Where do these fluids come from?
5. Identify the tectonic conditions that may be responsible for the overall environment depicted in Figure 10.9.
6. Make a table that summarizes all the mineral resources in Figure 10.9. Include all the relevant details that you know. How will you organize your table? How will you group various resources?

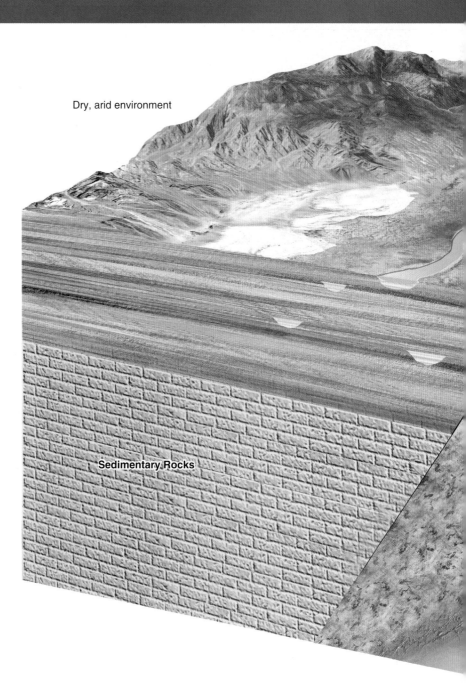

Dry, arid environment

Sedimentary Rocks

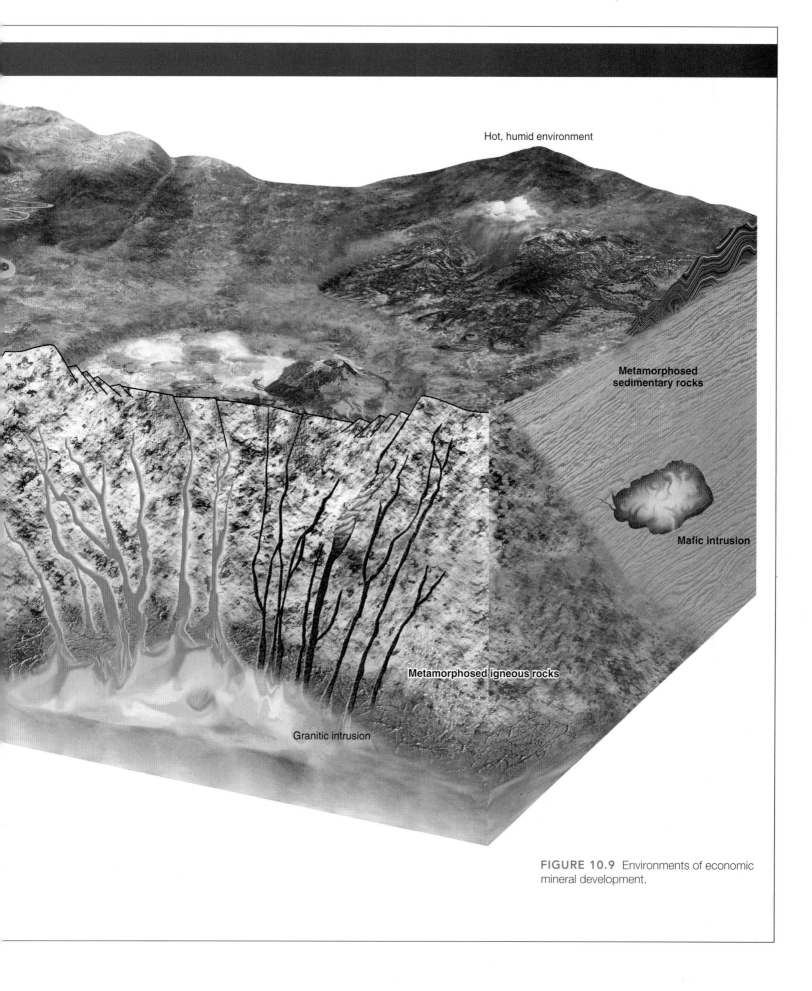

Hot, humid environment

Metamorphosed
sedimentary rocks

Mafic intrusion

Metamorphosed igneous rocks

Granitic intrusion

FIGURE 10.9 Environments of economic mineral development.

10-4 Fossil Fuels, Principally Oil, Provide Most of the Energy that Powers Society

LO 10-4 *Describe the forms of energy used in modern society.*

Energy is the vital force powering business, manufacturing, and the transportation of goods and services to the world. The supply of and demand for energy play key roles in national security and a nation's economy. Hence, it is not surprising that Canada and the United States spend nearly $700 billion annually on energy consumption. Energy on Earth's surface takes two main forms: (1) **solar energy**, which originates with nuclear reactions taking place within the Sun's core and is delivered to our planet as electromagnetic radiation, and (2) **geothermal energy**, which originates within Earth's interior through the radioactive decay of atoms, also a nuclear reaction.

As solar energy interacts with Earth's surface—including the atmosphere, the oceans, the land, and living organisms—it can assume other forms. These forms include *wind energy, hydroelectric energy, biomass energy* (produced by burning wood or other organic products), and **fossil fuels**, (concentrated forms of burnable carbon contained in rock). It can also exist simply as *direct solar energy* (sunlight that generates heat).

Fossil fuels are enriched with carbon from the remains of organisms that lived millions of years ago. These fuels include coal, oil, and natural gas. They provide 75 percent of all the energy consumed in Canada: over one-fifth of the electricity and almost all transportation fuels. Moreover, our reliance on fossil fuel, especially oil, to power the economy will increase in the next two decades, even with aggressive development of other energy sources. But a major debate is under way among geologists, energy economists, and government officials over whether supplies of affordable oil will be exhausted in a few years or a few decades. See the "Earth Citizenship" box titled "Peak Oil?" for a discussion of this issue.

Geothermal energy is heat that radiates from Earth's interior and through the surface, where it enters the atmosphere. Geothermal heat is produced by the radioactive decay of unstable nuclei of atoms in the core, mantle, and crust. It collects in Earth's interior because rock has excellent insulation properties. This heat can be tapped for commercial purposes by accessing hot groundwater circulating near igneous intrusions. For instance, Reykjavik (the capital of Iceland) derives 99 percent of its energy from geothermal sources and is known as a clean, self-sufficient city.

It is also possible to tap heat from nuclear decay by making concentrated *fuel rods* of radioactive materials, such as uranium. This is how nuclear power plants are fuelled. Like the heat produced by geothermal energy, the heat produced by nuclear decay is used to turn water into steam that can drive a *turbine*. A turbine is a machine that generates electricity by turning a shaft with running water, steam, wind, or some other type of energy.

Four **non-renewable energy resources** provide the majority of our energy needs: oil, coal, natural gas, and uranium (**FIGURE 10.10**). These energy resources are the focus of much geologic research, and it is important to understand the geologic background of these deposits and their use. We start with fossil fuels, the backbone of the world energy economy.

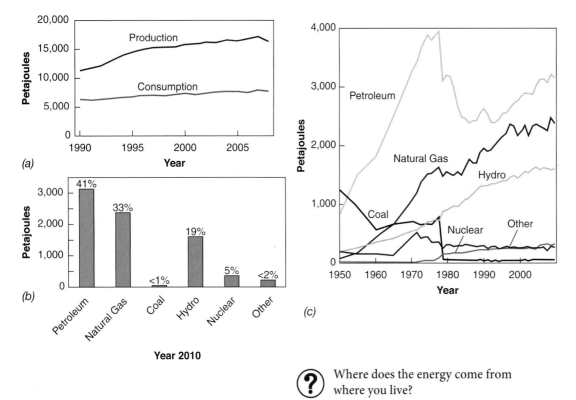

FIGURE 10.10 Energy use in Canada: *(a)* Canada's domestic energy production has consistently outpaced its consumption, which has been very good for its economy as most of the surplus is sold to the United States, which has run a significant energy deficit since the 1970s. *(b)* Currently, 41 percent of Canada's energy needs are met by petroleum, 33 percent by natural gas, 19 percent by hydroelectricity, 5 percent by nuclear power, < 2 percent by other means (e.g., biomass, steam, photovoltaic, wind), and slightly less than 1 percent by coal. *(c)* Over the past half-century, most of the growth in Canadian energy consumption has been supplied by petroleum, natural gas, and hydroelectric power. Source: Data compiled from Statistics Canada and the Canadian Association of Petroleum Producers (CAPP).

(?) Where does the energy come from where you live?

EARTH CITIZENSHIP

PEAK OIL?

In recent years, the price for a litre of gasoline has risen from approximately $0.75 in 2005 to over $1.40 in 2012. This is the most rapid rise in gas prices in history, and everyone from the average consumer to economic experts are wondering why.

According to Statistics Canada, Canada consumes about 1.8 million barrels of oil products per day, and about 42 percent is used as gasoline for motors of various types (including your car). Each barrel of oil contains 159 L. So, in Canada, about 120 million litres of gasoline are consumed every day. The price of gasoline fluctuates from day to day, week to week, and month to month in response to many factors, but principally in response to changes in production and demand. When demand is greater than production (oil scarcity), prices go up. When production is greater than demand (oil excess), prices usually go down. Production is controlled by the countries that pump and process oil, such as Saudi Arabia, Nigeria, Venezuela, Iraq, Canada, and several others. Since 2005, producers have limited oil production to about 86 million barrels per day despite rapidly rising demand from developing countries, such as China, India, and others. This growing demand and limited production lies at the root of the rising gasoline prices today.

However, looking at the longer-term picture of oil availability, the situation is more complicated. You might not realize it, but the world is in the process of using up the last easily available oil. It is a stark fact that Earth holds a finite supply of oil (**FIGURE 10.11**).

The ability to continue producing crude oil from fields around the world eventually will peak and then dwindle. Some analysts estimate that the current cap on production (86 million barrels a day) is a sign that we have already reached that point—known as *peak oil*.

This peak of oil production will mark the end of affordable oil (and perhaps already has). We will not recognize exactly when the peak is reached until after the fact, when we can look back and identify when oil production reached its maximum and then fell. Peak oil is likely to mark a turning point where the world changes from relying on an increasing supply of cheap oil to relying on a dwindling supply of expensive oil. It is likely to be a time of economic consequences: gasoline shortages, price spikes, lifestyle disruption. Energy companies are already squeezing oil from "unconventional" sources, such as tar sands, oil shale, and "heavy crude" that was formerly too expensive to extract. There will also be a push to pump oil fields that previously were deemed too environmentally damaging to drill. (This push has already begun in offshore oil fields and the Arctic National Wildlife Refuge in Alaska.)

But humans are ingenious. Pushed by expensive oil prices, society will increasingly rely on sustainable forms of energy, such as biofuels, greater use of mass transportation, and vehicles running on electricity generated by alternative energy sources. There are no easy options, and all will take time to explore. With every visit to the gas pump, after all, the end of cheap oil draws closer.

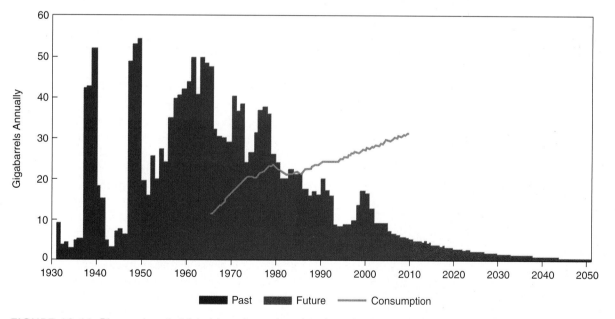

FIGURE 10.11 Discoveries of oil (blue) have been dropping since the 1960s. The red curve indicates predicted future discoveries. The orange line indicates annual consumption of oil. The trend of these data (running in opposite directions) indicates that eventually we will run out of affordable oil.

Expand Your Thinking—How are you planning to manage the problem of increasingly expensive gasoline in your daily life?

10-5 Oil Is Composed of Carbon that Is Derived from Buried Plankton

LO 10-5 *Describe the origin of oil and how it accumulates in the crust.*

The world is addicted to **petroleum**, commonly known as oil. Altogether, the world now consumes a staggering *1,000 barrels of oil per second.* That means that by the time you finish reading this page, nearly one-quarter million barrels of oil will have entered the air as a mixture of carbon dioxide, carbon monoxide, nitrous oxide, sulphur dioxide, unburned hydrocarbons, and particulate matter known as *black soot.* These pollutants enter the atmosphere and are major contributors to urban smog, human respiratory disease, and human-induced global warming. Ironically, after oil was first discovered in the middle of the nineteenth century, it was quickly adopted by industrialized nations because it was cheaper and cleaner than coal, the primary source of energy at the time. Now, with the price of oil skyrocketing, many predict a return to reliance on coal power in the future. Others want to see clean energy sources, such as wind and solar power, replace dirty fossil fuels.

The word "petroleum" means "rock oil," from the Latin words *petra,* "stone," and *oleum,* "oil." Places where natural gas and oil escape from Earth's surface, called *seeps,* have been known from early times (**FIGURE 10.12**). The Sumerians, Assyrians, and Babylonians used crude oil collected from large seeps on the Euphrates River in what is now Iraq. The Egyptians used oil as a wound dressing, liniment, and laxative. Spanish explorers discovered oil pools in present-day Cuba, Mexico, Bolivia, and Peru. Native Americans in New York and Pennsylvania used oil for medicinal purposes.

In the 19th century, the industrial revolution was under way. The need to lubricate moving machinery, as well as illuminate dark factories, spurred the search for new ways to obtain oil. Energy to drive machinery, formerly supplied by wood and coal, was supplanted with energy supplied by oil. Liquid petroleum was more concentrated and easier to transport than any fuel previously available.

These conditions set the stage for drilling the first oil well, which was done by James M. Williams in what is now called Oil Springs, Ontario, near the city of Sarnia. In 1857, Williams struck oil at a depth of 20 m while digging a water well for his plant where he intended to process asphalt and kerosene removed from the gum beds near Black Creek. This was nearly two years before the first oil well was drilled in Titusville, Pennsylvania, on August 27, 1859, by Edwin L. Drake. Some historians credit Drake with the first oil well because his was drilled into a bedrock reservoir whereas Williams's well was located above the bedrock. Regardless, with the spread of Drake's drilling techniques, inexpensive oil flowed from the crust and was processed at already existing coal-oil refineries (where kerosene was produced from coal). By the end of the 19th century, oil wells had been drilled across southern Ontario all the way to Alberta, in 14 U.S. states and across Europe and east Asia.

Where Does Oil Come From?

Petroleum comes from plankton. Oil and natural gas are the products of compressing and heating carbon-rich deposits of single-celled plants, such as *diatoms* and blue-green algae, and single-celled animals, such as *foraminifera,* over geologic time. The plankton once lived in environments of marine, brackish, or fresh water. After they died, they were buried rapidly, which prevented decay by oxidation and enabled later conversion into petroleum.

Anaerobic micro-organisms in silt and clay convert buried organic matter into *kerogen* and *methane* (natural gas). Kerogen is a waxy, dark-coloured product of chemically altered plant and animal debris. In time, as kerogen is buried deeper and subjected to higher temperatures and pressures, it experiences thermal degradation and *cracking* (a process in which heavy organic molecules are broken up into lighter ones). Depending on the amount and type of organic matter, cracking occurs at depths between 760 m and 4,880 m and temperatures between 65°C and 150°C, an environment termed the *oil window.* The resulting compounds, or kerogen products, share a basic molecular structure of carbon and hydrogen atoms linked in a ring in which atoms of sulphur, oxygen, and nitrogen are found. With continued cracking in the oil window, these products evolve into thousands of different organic compounds, all built of hydrogen and carbon atoms. These constitute a class of compounds known as *hydrocarbons,* of which the most important is oil.

© Linda Robshaw/Alamy

FIGURE 10.12 Oil seeps, where petroleum naturally flows out of the ground, are found in many countries and provided early oil hunters with their first clues about where to drill.

(?) Why did oil quickly replace coal as the primary source of energy after it was discovered?

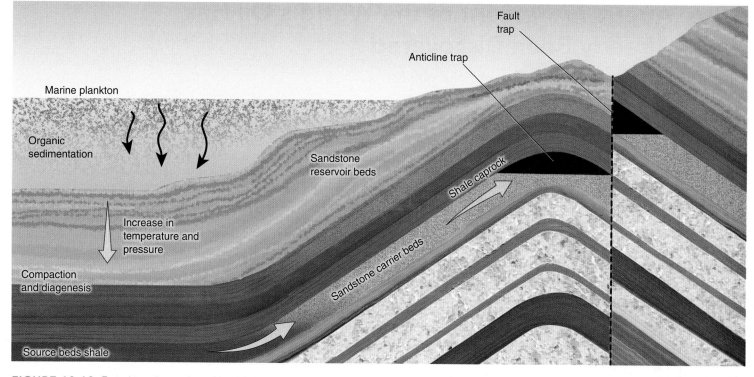

FIGURE 10.13 Petroleum is produced by the burial and chemical alteration of plankton. As the organic material is subjected to increased temperature and pressure, it enters the oil window and is converted into kerogen—concentrated carbon—and eventually into petroleum. Liquid petroleum migrates through carrier beds until it encounters a trap or escapes on the surface as a seep.

(?) What causes oil to accumulate in a reservoir?

Source Beds and Reservoir Beds

Hydrocarbons are expelled from their *source bed* (the place where they form), usually a deposit of organic shale, by pressure related to compaction of the sediments. Oil migrates through coarse-grained *carrier beds,* usually consisting of sandstone or carbonates (FIGURE 10.13), until it is trapped in a *reservoir bed* (also usually sandstone or carbonate) and forms a *reservoir.* In some cases, however, it may migrate to the surface and become a seep.

Most buried sediments are saturated with water, and therefore oil migration takes place in an environment of groundwater. Because oil is lighter than water, in time it floats to the top and accumulates in the highest portion of a reservoir bed. The **porosity** and **permeability** of carrier and reservoir beds are important factors in the migration and accumulation of oil. Porosity is the percentage of open (pore) space in a rock; permeability is the degree to which a fluid is able to flow through a rock. Permeability depends on the number of interconnected pore spaces present. A rock can have high porosity and low permeability (an example is shale, in which pore spaces are not connected), or moderate porosity and high permeability (as in some sandstones, in which pore spaces are connected). Most petroleum is found in *clastic reservoirs* (composed of sandstone and siltstone), where the oil resides in pores between sediment grains. Next in abundance are *carbonate reservoirs* (composed of limestone and dolomite), where the oil resides in cavities formed by dissolution

of the rock during weathering as well as between sediment grains. Reservoir rocks usually have a porosity ranging from 5 percent to 30 percent, but not all pore spaces are occupied by oil; a certain amount of groundwater cannot be displaced and is always present.

Oil Traps

Oil collects in a *trap,* a closed layer of porous and permeable reservoir rock that is sealed by an impermeable *cap rock* usually composed of shale or evaporite layers. The trap may be any shape as long as it is a closed geologic "container" preventing oil leakage.

Traps form in several ways. Many are formed by tectonic processes at convergent margins that produce folded or faulted rock units. These are known as *structural traps.* The most common structural traps are *anticlines*—"upfolds" of strata formed by compression in the crust (see Figure 10.13). The porous reservoir rock contains oil, and the impermeable cap rock prevents the oil from leaking out. The oil migrates to the arch in the anticline because it is a zone of low pressure. About 80 percent of the world's oil is found in anticlinal traps. Another kind of structural trap is the *fault trap,* in which strata are displaced by a fracture that forms a barrier to oil migration. The barrier may be the fault itself or an impermeable bed, acting as cap rock that is brought into contact with a carrier bed by the fault. Faults and folds often combine to produce traps, each forming part of the enclosed container.

(?) Expand Your Thinking—What physical characteristic must carrier beds have to be effective pathways for petroleum migration?

10-6 About 77 Percent of the World's Oil Has Already Been Discovered

LO 10-6 *Describe the status of the world's oil supply.*

Two overriding principles apply to world oil supplies: (1) Most oil is found in a few large fields, and most fields are small; and (2) as exploration progresses, the average size of discoveries decreases, as does the amount of oil found by exploratory drilling. In any region, the large fields are usually discovered first.

Although geological limitations ultimately govern the amount of oil available to the world, the price of oil is controlled by the nations that produce it. These nations have banded together to form alliances with the common goal of keeping oil prices stable in order to secure steady income for the members. In other words, these nations control the availability of oil to the world, and therefore the price, to ensure a profit. The largest alliance is the *Organization of the Petroleum Exporting Countries,* or OPEC. OPEC includes 13 nations in the Middle East, South America, Africa, and Southeast Asia. Together this group controls approximately two-thirds of the world's oil reserves and over 40 percent of the oil production.

A total of about 50,000 oil fields have been discovered worldwide; however, over 90 percent of them have insignificant global impact. The largest oil fields are the "supergiants," holding 5 billion or more barrels of recoverable oil, and the "world-class giants," containing 500 million to 5 billion barrels. Approximately 280 world-class giant fields have been discovered. These, plus the supergiants, hold 80 percent of the world's known recoverable oil. Less than 5 percent of the known fields contain roughly 95 percent of the world's known oil.

Fewer than 40 supergiant fields are known worldwide, yet originally (before being pumped) these contained about half of all the oil that had been discovered up to that time. The Arabian-Iranian Sedimentary Basin in the Middle East contains two-thirds of these supergiant fields (**FIGURE 10.14**). The remaining supergiants are found in the United States (2), Russia (2), Mexico (2), Libya (1), Algeria (1), Venezuela (1), and China (1).

Although the United States consumes roughly 19.5 million barrels of oil a day, mostly to power its 200 million automobiles, it produces only about 5 million barrels a day, enough to cover a little more than 25 percent of its needs. The other roughly 75 percent, about 14.5 million barrels a day, has to be imported from other countries. Canada provides the majority of the petroleum needs of the United States—a little more than 19 percent. About 18 percent of U.S. oil imports come from Persian Gulf countries (e.g., Saudi Arabia, Bahrain, Iraq, and others). Mexico (10 percent), Venezuela (9 percent), Nigeria (7.7 percent), and several other international sources provide the remainder of the United States' oil needs. Meanwhile, the global demand for oil is growing, with India, China, and South Korea in particular fuelling recent sharp increases in demand. It is inevitable that there will be greater competition among nations for a diminishing world oil supply.

The Status of the World Oil Supply

The first 200 billion barrels of oil were produced in the 109 years from 1859 to 1968. Now the same amount is produced in less than a decade, at a rate of 22 billion barrels a year. Figure 10.14a shows the world's oil supply. Estimates of total undiscovered oil resources, made by exploration geologists who have researched geologic deposits around the world, range from 275 billion to 1.469 trillion barrels. The world's total oil supply (past, present, and future) amounts to about 2.39 trillion barrels. Of this, 77 percent has already been discovered and 30 percent has already been produced and used. If this estimate proves to be reasonably accurate, current rates of oil production could be sustained until about the middle of this century. At that point, a shortage of oil will force a production decline and major changes in the lifestyles of both oil-producing and oil-consuming nations. However, significant increases in oil usage will shorten the time to peak oil.

Over 50 percent of world oil supplies are located in the Middle East. With 175.2 billion barrels of proven reserves, Canada places third globally, behind Saudi Arabia and Venezuela. However, even with Canada's rich endowment of oil, North America is still a distant second to the Middle East, and has already produced almost half of its total oil. Eastern Europe, because of large deposits in Russia, is also well endowed with oil, but western Europe is not. Most of its oil is in the North Sea, which makes it difficult and expensive to access. Africa, Asia, and South America have moderate amounts of oil. Large undiscovered oil resources are believed to exist in North America, eastern Europe, and the Middle East.

Natural Gas

Natural gas, also known as **methane**, is a colourless, odourless, clean-burning fossil fuel. It is easy to transport and a convenient fuel for heating, cooling, lighting, and cooking; operating home appliances; generating electricity; and many industrial purposes.

The origin of most natural gas is similar to that of oil; in fact, natural gas and oil may form simultaneously. Natural gas originates when organic material is buried and compressed within the crust. Plankton that rapidly accumulates in shallow seas is one source of natural gas. Buried vegetation that has not oxidized is another source. Over millions of years, the pressure and heat of burial converts this carbon-rich material into methane and other light hydrocarbons. In petroleum source beds at low temperature (in shallower deposits), more oil is produced than natural gas. At higher temperatures, however, more natural gas is created than oil. That is why natural gas is usually associated with oil in deposits that are relatively deeply buried in Earth's crust (2,500 m to 3,200 m

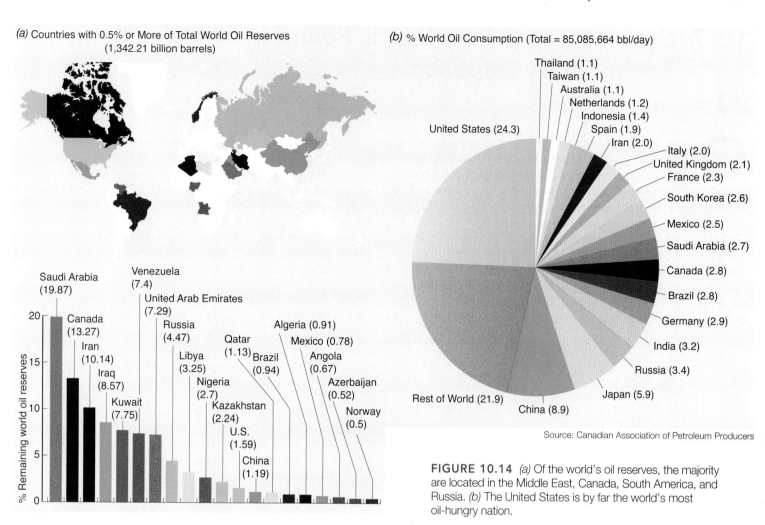

(a) Countries with 0.5% or More of Total World Oil Reserves
(1,342.21 billion barrels)

(b) % World Oil Consumption (Total = 85,085,664 bbl/day)

Source: Canadian Association of Petroleum Producers

FIGURE 10.14 (a) Of the world's oil reserves, the majority are located in the Middle East, Canada, South America, and Russia. (b) The United States is by far the world's most oil-hungry nation.

or more). Deeper deposits, very far underground, usually contain primarily natural gas.

Like oil, natural gas migrates through porous and fractured carrier beds and accumulates in reservoir beds. Because it is lighter, it separates from oil and forms a surface layer within the reservoir. It may also form in coal deposits, where it is found dispersed throughout the pores and fractures of a coal bed.

Natural gas can also be formed through the conversion of surface organic matter by micro-organisms, called *methanogens,* which chemically break down organic matter to produce methane. These micro-organisms are commonly found in areas near the surface that lack oxygen; they also live in the intestines of most animals, including humans. An example of *biogenic methane* is landfill gas. Decomposition of waste materials in a landfill produces natural gas that can be harvested and added to the energy supply.

According to the Canadian Association of Petroleum Producers, Canada has natural gas reserves of about 700 to 1,300 trillion cubic feet (Tcf), or 12 to 22 percent of world reserves, giving Canada one of the largest endowments of natural gas in the world. The National Energy Board of Canada and Statistics Canada place Canada as the world's third-largest producer and exporter of natural gas at approximately 6 Tcf per year. U.S. (> 20 Tcf), and Canadian (2.9 Tcf) consumption totals more than 23 Tcf per year. The majority of U.S. natural gas imports come from Canada, mainly the western provinces of Alberta, British Columbia, and Saskatchewan. Overall, Canada depends on natural gas for about one-third of its total primary energy requirements.

In Canada, the production of energy based on natural gas (33 percent) alone is second only to that based on petroleum, with hydroelectric energy ranking third.

(?) Expand Your Thinking—Several events have occurred in the past few years that are characteristic of "peak oil." Name a couple.

10-7 Coal Is a Fossil Fuel that Is Found in Stratified Sedimentary Deposits

LO **10-7** *Describe the origin of coal.*

On average, Canada uses about 4.7 kg of coal per person every day to generate 18 percent of the nation's electricity using coal-burning power plants. Canada has 8.7 billion tonnes (b.t.) of proven coal resources, of which 6.6 billion tonnes are considered recoverable. Canada potentially also has another 193 billion tonnes of unproven coal resources. Approximately 97 percent of Canada's coal resources are found in western Canada (**FIGURE 10.15b**). Canada's current estimate of recoverable coal alone (6.6 b.t.) is enough to last at least 112 years at Canada's current rate of coal consumption. This would be a total energy value equal to more than 34 billion barrels of crude oil—approximately equivalent to 20 percent of Canada's proven petroleum reserves.

Worldwide, coal is the most abundant and most widely distributed fossil fuel (**FIGURE 10.15a**). The amount of recoverable global coal reserves is estimated at about 1,081 billion tonnes. At current rates of production, the reserves could last for another two centuries. The United States, with 26 percent, and Russia, with 19 percent, account for nearly half of global coal reserves. China (12 percent), Australia (8 percent), Germany (7 percent), South Africa (5 percent), Poland (2 percent), and Canada (1 percent) also have significant amounts of the world's recoverable coal.

What Is the Origin of Coal?

Coal is the product of ancient vegetation that has been altered over long periods of time. Dense layers of wet, matted vegetation growing in warm, humid wetlands are converted by micro-organisms into a compressed and concentrated deposit known as *peat,* which eventually becomes coal. Of course, much of the plant matter that accumulates on Earth's surface never turns into peat or coal because it is removed by fire or decomposed by organic decay. As a result, geologists infer that the vast coal deposits found in ancient rocks must represent periods during which several favourable biological and physical processes occurred at the same time: broad tracts of low-lying, freshwater wetlands; prolific growth of vegetation in a moderate climate; and rapid burial and compaction of that vegetation by clastic sediments.

Coal is classified into three grades: *lignite* (brown coal), **bituminous coal** (soft coal), and **anthracite** (hard coal). These grades are formed in a continuous sequence from sedimentary (lignite, bituminous coal) to metamorphic rock (anthracite). When peat is exposed to heat and pressure due to burial, it is compressed to about 20 percent of its original thickness and becomes lignite. As lignite is further compressed and heated to 100°C to 200°C, it is transformed into bituminous coal. This process drives out water and volatile gases, concentrating the carbon. Longer exposure to high temperature further drives out gases and stimulates solid-state changes that produce anthracite. Anthracites are compressed to

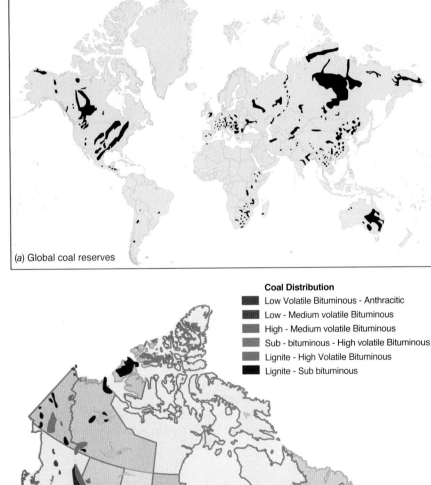

(a) Global coal reserves

Coal Distribution
- Low Volatile Bituminous - Anthracitic
- Low - Medium volatile Bituminous
- High - Medium volatile Bituminous
- Sub - bituminous - High volatile Bituminous
- Lignite - High Volatile Bituminous
- Lignite - Sub bituminous

 Where would you find the modern equivalent of a coal wetland today?

FIGURE 10.15 *(a)* Canada, the United States, and Russia account for over half of the global coal reserves. *(b)* Distribution of coal resources within Canada.

5 percent to 10 percent of their original thickness and contain less than 10 percent water and volatile gases.

Direct evidence of coal's origin comes from two sources: (1) All grades of coal may contain distinct plant remains, and (2) rock layers above, below, and adjacent to **coal seams** contain plant fossils in the form of impressions, thin films of carbonized leaves and stems, and larger vegetation, such as roots, branches, and trunks. When examined microscopically, cell walls, spores, and other minute plant structures can also be recognized.

Although coal is abundant and is a major energy source, there are problems associated with burning coal; see the "Earth Citizenship" box "Problems with Coal Use" below.

EARTH CITIZENSHIP

PROBLEMS WITH COAL USE

When burned, coal produces many compounds, some of which can cause cancer. Burning coal is a leading cause of smog, acid rain, and air toxicity.

Annually, a typical coal power plant generates these products:

- 3.4 million tonnes of carbon dioxide, the primary cause of global warming. This is equivalent to the amount of carbon dioxide contained in 161 million trees.
- 9,000 tonnes of sulphur dioxide, which causes acid rain that damages forests, lakes, and buildings, and forms small airborne particles that can cause respiratory ailments.
- 450 tonnes of airborne particles called fly ash, which causes bronchitis, asthma, and atmospheric haze.
- 10,200 tonnes of nitrogen oxide, equal to the amount emitted by half a million cars. Nitrogen oxide leads to the formation of ozone (smog) that inflames the lungs.
- 650 tonnes of carbon monoxide, a source of headaches and stress for people with heart disease.
- 200 tonnes of hydrocarbons, volatile organic compounds that form ozone.
- 77 kg of mercury, of which just 1/70 of a teaspoon deposited in a 10-ha lake can make the fish unsafe to eat.
- 102 kg of arsenic, which causes cancer in 1 out of 100 people who drink water containing 50 parts per billion.
- 51 kg of lead, 2 kg of cadmium, other toxic metals, and traces of radioactive uranium.

Children are particularly susceptible to air pollution because they breathe more air per kilogram of body weight than adults do, and their lungs are still developing. Researchers at the Harvard School of Public Health found that people living within 48 km of coal-powered utility smokestacks had a three to four times greater chance of premature death than those living farther away.

Industrial devices that "scrub" coal combustion can trap some of these pollutants. The efficiency and environmental compatibility

Photo courtesy of Vivian Stockman/www.ohvec.org. Flyover courtesy of Southwings.org.

FIGURE 10.16 Surface mining causes massive environmental damage. In West Virginia, 1,600 km of streams have been buried under mine waste and 120,000 ha of forests destroyed due to surface coal mining.

of coal-burning plants are improving as they increasingly turn to innovative combustion methods and coal liquefication processes.

Another problem associated with coal is the environmental damage caused by mining. About 90 percent of Canadian coal and 60 percent of U.S. coal is stripped from surface mines; the rest comes from underground mines. Surface mining dramatically alters the landscape. Coal companies often remove entire mountaintops to expose the coal below, and wastes generally are dumped into valleys and streams (FIGURE 10.16). Acid rock drainage from waste rock damages watersheds and groundwater. In West Virginia, more than 120,000 ha of hardwood forests (almost a quarter the size of Prince Edward Island) and 1,600 km of streams have been destroyed by strip mines and mountaintop mining.

In sum, although the United States and Canada have abundant coal reserves, the environmental and health costs are large and not easily mitigated.

Expand Your Thinking—Many authorities point to the abundance of coal as a solution to energy shortages related to declining oil reserves. Describe the benefits and drawbacks of this to someone who does not understand energy challenges.

10-8 Nuclear Power Plants Provide about 17 Percent of the World's Electricity

LO 10-8 *Identify and describe the special feature of uranium-235 that makes it useful as an energy source.*

In France, about 80 percent of all the electricity is generated by nuclear power. Nuclear power provides 5 percent of the energy consumed in Canada, and 15 percent of the electricity generated. There are more than 400 nuclear power plants around the world, supplying 17 percent of global energy, with 5 plants located in eastern Canada (Ontario = 3;

Quebec = 1; New Brunswick = 1) that house 22 nuclear power reactors. Canada is world-renowned for making one of the most efficient and safe nuclear reactors, the CANDU reactor (**Can**ada **D**euterium **U**ranium). Deuterium (atomic mass of 2) is the heavier of the two stable isotopes of hydrogen found in nature, but accounts for only 0.0156 percent of

---

EARTH CITIZENSHIP

THE PROBLEM WITH POWER

Where will we turn for power over the next few decades? The debate has never been hotter. Canada has plenty of coal, enough to last into the next century if we so choose. But mining is environmentally destructive and dangerous; in 1958 an explosion in a Nova Scotia mine killed 74 men and trapped another 100. Coal is also unhealthy; if you live near a coal plant you have a three to four times greater chance of premature death.

The power behind planes, buses, and automobiles is oil; 95 percent of all transportation is fuelled by oil. No one loves to drive as much as North Americans. We drive everywhere, even if it is only around the corner for a litre of ice cream. But burning oil pollutes the air and, along with burning coal and natural gas, is the major source of heat-trapping carbon dioxide that causes global warming. Like coal mining, drilling for oil can be dangerous and environmentally damaging. In April 2010, an oil rig in the Gulf of Mexico caught fire and sank, killing 11 workers and unleashing a spill of millions of barrels of crude oil (**FIGURE 10.17**).

There are also problems with nuclear power. The 1986 Chernobyl accident released more than 100 times as much radiation as the atomic bombs dropped on Nagasaki and Hiroshima in World War II. An estimated 4,000 people will ultimately lose their lives as a result of the radiation sickness the Chernobyl accident caused. Fearing a similar catastrophe, the world was gravely concerned following the Tōhoku earthquake and tsunami on March 11, 2011, which severely damaged the Fukushima nuclear power plant. Thankfully, the radiation was largely contained and there were no deaths associated with radiation exposure. In addition to the obvious difficulties of preventing nuclear accidents and the safe disposal of radioactive waste, other problems with nuclear power include: (1) Building a single power plant costs many billions of dollars and can take over a decade, and (2) terrorists around the world are desperately trying to get their hands on fissionable materials to build nuclear weapons. Nuclear reactors are the easiest and the most prolific source of raw materials to build these weapons.

Proponents of renewable energy like to point out that wind and solar power do not have these difficulties, but in truth, they have

FIGURE 10.17 The sinking of the drilling rig Deepwater Horizon in April 2010 released millions of barrels of crude oil. Wildlife, marine ecosystems, tourism, the fishing industry, and the economy of the Gulf of Mexico will be damaged for years to come.

Courtesy of Jeff Schmaltz, MODIS Rapid Response Team, NASA

their own thorny issues. Energy demands are so great and renewable energy so undeveloped, that it is unlikely to become a major energy provider any time soon. For instance, cars run on gasoline; converting them to electricity will require an astounding shift in American industry. And where will electric cars "refuel"? A national system of electricity vendors needs to be built, essentially replacing the corner gas stations. Also, the batteries for electric cars have their own inherent environmental problems in their construction, use, and eventual disposal. Although there is no shortage of wind and sunlight in parts of the country, it is difficult to convert that energy into usable form and transport it to consumers. Renewable energy is intermittent, and storing energy (essentially building giant batteries) will be enormously expensive. There are also environmental issues because solar panels and wind farms require large amounts of land.

How will the problem with power ultimately be resolved? In the end, it will be the economic marketplace that will decide. People will vote on new energy with their wallets, and their choice will be influenced by lifestyle and personal ethics. The future of energy will be the source that establishes the most acceptable combination of cost, safety, environmental preservation, and convenience.

 Expand Your Thinking—Why is induced fission considered by many an important long-term source of energy? What are the problems associated with its use?

naturally occurring hydrogen, whereas protium (atomic mass of 1) accounts for 99.98 percent. One of the innovations of the CANDU is in its use of "heavy water" enriched with deuterium (99.75 percent of the hydrogen atoms). In part, because of this innovation, the CANDU does not require "enriched" uranium (see below), and can even use the spent fuel from other nuclear reactors.

Uranium

Uranium, the fuel used by nuclear power plants, is a fairly common element, one that was incorporated into the planet during its formation. Various isotopes of uranium are radioactive. The rate at which they decay (or change into another element) is described by their *half-life*. Half-life is the time it takes for half of a certain amount of a radioactive isotope to decay into a daughter isotope. (This process is explained in depth in Chapter 13, "Geologic Time.") Uranium-238 (^{238}U) has an extremely long half-life (4.5 billion years), meaning that its rate of radioactive decay is slower than those of isotopes with shorter half-lives; therefore, it is still present in rocks in fairly large quantities. Uranium-238 accounts for 99 percent of the uranium on Earth. Uranium-235 makes up about 0.7 percent of the remaining uranium found naturally; uranium-234, formed by the decay of uranium-238, is even rarer.

Uranium-235 has an interesting property that makes it useful for nuclear power production. It is one of the few materials that can undergo **induced fission** (in other words, it can be caused to decay radioactively). If a free neutron runs into a uranium-235 nucleus, the nucleus will absorb the neutron, become unstable, and immediately undergo fission—that is, split. Fission produces two lighter atoms and throws off two or three new neutrons. The two new atoms emit heat as they settle into their new states. The new neutrons may cause fission in nearby uranium-235 nuclei in a process known as a *continuous reaction*. Three things make induced fission especially useful as a power source: (1) The probability of a uranium-235 atom capturing a neutron as it passes is fairly high; (2) the process of capturing the neutron and splitting happens very quickly; and (3) an incredible amount of energy (in the form of heat) is released when a single atom splits.

This process is such an effective source of energy that only 0.4 kg of highly enriched uranium (an amount the size of a baseball) can power a nuclear submarine or aircraft carrier to the same extent as about 3.8 million litres of gasoline (which would require a container the size of a five-storey building). In order for these properties of uranium-235 to work, however, a sample of uranium must be "enriched" so that it contains 2 percent to 3 percent or more uranium-235. "Light water" nuclear reactors use enriched uranium formed into bundles of *fuel rods* that are immersed in regular water, which acts as a moderator that absorbs neutrons and a coolant (otherwise the bundles would overheat and melt). However, the light water is very good at absorbing neutrons so enriched uranium is needed. Without enriched uranium, the fuel would not reach what is termed "criticality," a situation where a steady rate of fission over time is achieved. In the CANDU reactor, natural uranium with only 0.72 percent uranium-235 can be used because deuterium-rich heavy water has a significantly reduced ability to absorb neutrons. Thus, much less uranium-235 is required to reach steady state fission. In both types of reactors, neutron-absorbing *control rods* (made of either cadmium, boron, stainless steel, or cobalt) are inserted into each bundle to help control the rate of the nuclear reaction. When an operator wants the uranium core to produce more heat, the control rods are raised out of the bundle. To create less heat and slow the continuous reaction, the control rods are lowered into the uranium bundle. The rods also can be lowered completely into the uranium bundle to shut the reactor down in the case of an accident or to replace the rods.

The uranium bundle acts as a high-energy source of heat. It boils water and turns it into steam. The steam drives a turbine, which spins a generator to produce power (**FIGURE 10.18**). Well-constructed nuclear power plants have an important advantage when it comes to electrical power generation: They are extremely clean when compared with coal-fired plants. A coal-fired power plant actually releases more radioactivity into the atmosphere than does a properly functioning nuclear power plant.

(a) Courtesy of Ontario Power Generation

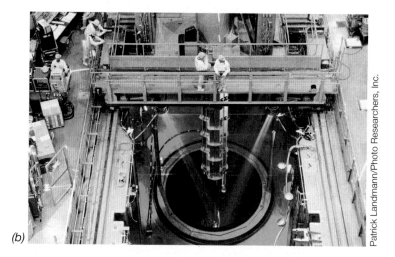
(b) Patrick Landmann/Photo Researchers, Inc.

FIGURE 10.18 *(a)* Like other types of power plants, this Canadian-built CANDU reactor at the Pickering Nuclear Generating Station on the north shore of Lake Ontario uses heat to produce steam. The steam is used to generate electricity by turning the blades of a turbine. The source of heat in a nuclear plant is fission of the nucleus of uranium-235. *(b)* The reactor core houses control rods to control the rate of nuclear reaction.

? Describe the special properties of uranium-235 that make it a useful power source.

10-9 Renewable Energy Accounts for More than 20 Percent of Canada's Energy Sources

LO *10-9 Identify the ways energy can be provided from renewable sources.*

Before 1973, most industrial nations enjoyed almost unlimited access to fossil fuel supplies. Coal and oil together provided over 90 percent of Canadian and U.S. energy needs and were thought to be plentiful despite rising demand. But during the following decade, North America endured two severe oil shortages arising from political tensions in the Middle East, where, as we have seen, a large share of the world's oil supply is located. This drove up gasoline prices, sent the economy into a tailspin, and prompted lawmakers and government agencies to develop new energy policies.

With important help from the government, the modern renewable energy industry was born from this turmoil. Renewable energy from inexhaustible sources such as the Sun, water, and wind offered a tempting answer to the nation's energy problems (**FIGURE 10.19**). Renewable energy sources include:

- *Hydroelectric power,* which is generated by flowing water. Hydro is by far the largest source of renewable energy in Canada, accounting for at least 63 percent of its electricity production, making Canada second in the world for hydroelectricity generation, with China leading the way. Canada's total hydroelectric capacity is more than 72,000 megawatts. Hydroelectricity will remain a dominant source of electricity supply in Canada (about 59 percent; the rest is exported to the United States) for the foreseeable future. It has the advantages of being a flexible, low-cost source of non-CO_2 emitting base-load electricity, which contributes to maintaining competitive and stable electricity prices. Hydro power is flexible in the sense that the output from hydro generating stations can be adjusted quickly with variations in demand. Hydro power can contribute to maintaining price stability because it is not subject to the volatility of fuel costs.

- *Solar power,* which relies on the Sun's energy to provide heat, light, hot water, electricity, and even cooling for homes, businesses, and industry. Solar technologies include absorptive materials that convert sunlight into electricity (photovoltaic solar cells), mirrors that concentrate solar power, sunlight used to heat and cool buildings, and solar heating of hot water and living space.

- *Wind power,* which is used to generate electricity, charge batteries, pump water, and grind grain. Large wind turbines in *wind farms* produce electricity that can be sold to customers. Canada has over 5,400 megawatts of wind-generating capacity.

- *Bioenergy* technologies, which burn crops grown specifically as an energy source to make electricity, fuel, heat, chemicals, and other materials. In Canada, bioenergy accounts for 4 percent to 6 percent of energy production and ranks as a crucial renewable energy source.

- *Geothermal energy,* which is harnessed from the heat produced inside Earth to provide energy that we can use and sell, such as electricity. In Canada, most geothermal resources are concentrated in the west, but are still vastly underutilized.

- *Geoexchange energy,* which uses the near-constant temperature of shallow (about 6 m) soil, rock, or surface water to regulate the indoor climate in buildings; often referred to as a "ground source heat pump" or "geothermal heat pump." However, the term "geothermal heat pump" should be avoided in order to distinguish it from the medium- to high-temperature geothermal technology. These ground source heat pumps are an excellent energy alternative as they can be used almost anywhere.

(a) (b) (c)

? Why are hydroelectricity, solar, and wind energy "renewable"?

FIGURE 10.19 Fields of (a) wind farms and (b) solar collectors have advanced technology, but the cost of these energy sources remains higher than that of fossil fuels. (c) Robert-Bourassa hydroelectric power station, eighth largest in the world, showing the spillway, often called the "giant's staircase." By far the most cost-effective and reliable source of renewable energy in Canada is hydroelectricity. Until the cost of other forms of renewable energy decreases or fossil fuel becomes more expensive, this will remain the most effective renewable energy competitor in the energy market.

Arctic Photo/All Canada Photos

- *Hydrogen fuel cells,* which work like batteries but do not run down or need to be recharged. As long as fuel (hydrogen) is supplied, they will produce electricity and heat. Fuel cells can be used to power vehicles or provide electricity and heat to buildings.
- *Ocean energy,* which draws on the energy of ocean waves and tides or on the heat stored in the ocean, can be used for many applications, including electricity generation.

Despite decades of research and development, renewable energy still plays a minor role in the energy market. It accounts for 24 percent of energy production in Canada and only 7 percent in the United States (FIGURE 10.20). These energy sources are not widely used because they cost more than fossil fuel energy, sometimes two to four times as much as traditional fuel.

Federal programs in Canada and the United States attempt to stimulate the market for renewable energy with tax breaks, energy mandates, and grants that make renewable energy more competitive. Without these policies, renewable energy's share of the energy market plummets to 1 percent. But such programs are controversial and do not fully achieve the desired goal of widespread renewable energy use.

Proponents of renewable energy urge our governments to maintain or increase its commitment to energy alternatives. They say that fossil fuels spew pollutants into the air, causing health and environmental problems and adding to global warming. They point out that renewable energy reduces our dependence on foreign oil and therefore improves national security. But opponents say that oil and coal are still the cheapest sources of energy. They argue that government efforts lead to higher energy prices and suppress economic growth. They also argue that renewable energy cannot provide enough power to meet most needs. If the marketplace is left alone to set energy costs, they say, consumers will reap the benefits of lower prices and an improved economy.

This debate is likely to increase in the near future as oil costs escalate and renewable energy offers the promise of a solution.

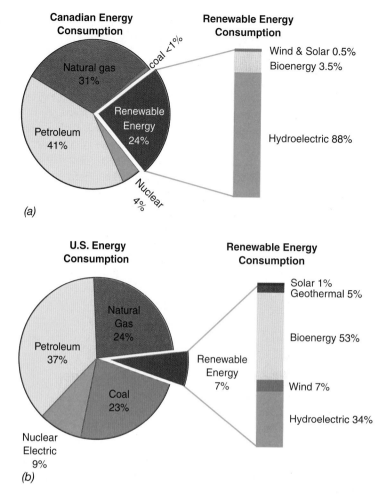

(a)

(b)

FIGURE 10.20 *(a)* Renewable energy in Canadian accounts for at least 24 percent of our overall energy consumption (based on 2008 statistics). *(b)* Renewable energy accounts for only 7 percent of overall energy sources in the United States.

(?) How does renewable energy in Canada compare to the United States, and why are there such big differences?

 GEOLOGY IN OUR LIVES

THE UBIQUITOUS TURBINE

If you take a physics course, you will learn that by spinning a coil of copper wire between the two poles of a magnet you can make an electric current. Physicist *Michael Faraday* discovered this principle in the 1800s. The machine that performs this work is known as a turbine, and it is employed by the power industry in making the electricity that is such an integral part of our lives. Approximately 80 percent of all electricity used by the world today is generated by steam turbines, wherein the coil is turned by steam pushing fan blades inside the turbine (FIGURE 10.21). Making steam requires boiling water, and the heat for this can come from burning coal, oil, natural gas, or vegetation; geothermal heat; nuclear fission—in other words, nearly every energy source we have discussed in this chapter. Even wind, waves, tide currents, and falling water are used to turn a coil inside a magnet to create an electrical current.

FIGURE 10.21 Steam turbines consist of fan blades turned by steam to generate electricity.

(?) **Expand Your Thinking**—Why is it desirable to increase the role of renewable energy in the world?

LET'S REVIEW "GEOLOGY IN OUR LIVES"

Now that you have finished the chapter, "Geology in Our Lives" will have taken on new meaning for you. Let us review it: Nearly every substance you use in your life has either been mined from the crust or manufactured using energy from fossil fuels. All humans are dependent on mineral resources and fossil fuels, yet these geologic resources are non-renewable on the human time scale. Hence, use of non-renewable resources must be tempered with awareness that they may become unaffordable if they are not replenished, used in a sustainable manner, or replaced with renewable alternatives. Knowing the facts about geologic resources can help in your decision making as a consumer and as a caretaker of Earth.

Throughout your day, you rely on geologic resources. Every substance used to make metals, plastics, glass, ceramics, electronics, and everything else in your home, classroom, and place of work comes from rocks mined out of Earth's crust or is manufactured with fossil fuels. But as these resources grow harder to find, they will become more expensive until they are unaffordable. For this reason, it is important to recycle mineral resources and develop alternatives to those that are in short supply. Similarly, as supplies of affordable oil diminish, it is increasingly important to use alternative energy sources that are renewable.

STUDY GUIDE

10-1 The crust contains metals, building stone, minerals, and sources of energy.

- **Geologic resources** are non-living materials that are mined to maintain our economy and quality of life. These include two broad categories: **mineral resources** and **energy resources**. The majority of geologic resources are **non-renewable**. We study geologic resources because they are fundamentally important to our way of life and the structure of our society and civilization.

- Society meets most of its energy needs with four non-renewable resources: oil, coal, natural gas, and uranium. These account for 80 percent of total energy consumption and 95 percent of commercial energy consumption.

10-2 Mineral resources include non-metallic and metallic types.

- Mineral resources include all useful rocks and minerals. They fall into two groups: non-metallic resources and metals. There are three groups of non-metallic resources: industrial, agricultural, and construction.

- Approximately 40 metals are produced by mining and **smelting** minerals. Minerals that are smelted more readily than others are **ore minerals**, of which there are about 100. A profitable assemblage of minerals plus gangue (valueless minerals) is an **ore**.

10-3 Ores are formed by several processes.

- Mineral deposits fall into four categories: magmatic segregation deposits, hydrothermal deposits, residual mineral deposits, and sedimentary mineral deposits.

10-4 Fossil fuels, principally oil, provide most of the energy that powers society.

- Energy on Earth's surface arrives in two fundamental forms: **solar energy**, which originates with nuclear reactions in the Sun's core, and **geothermal energy** originating in Earth's interior through the radioactive decay of atoms.

- Four **non-renewable energy resources** provide the majority of our energy needs: coal, oil, natural gas, and uranium. **Fossil fuels**—coal, oil, and natural gas—are enriched with carbon, the remains of organic matter millions of years old. They provide about 75 percent of all the energy consumed in Canada, representing virtually all transportation fuels. Chief among the fossil fuels is oil, or **petroleum**.

10-5 Oil is composed of carbon that is derived from buried plankton.

- Most oil is derived from single-celled plants and animals such as plankton that live in marine, brackish, or fresh water. When they die, they decay through oxidation unless they are rapidly buried in clay and silt.

- Hydrocarbons form in source beds, migrate through carrier beds, and collect in reservoir beds. Reservoir beds collect petroleum because they act as a trap. The various types of traps include structural traps (anticlines and faults) and stratigraphic traps. Petroleum does not migrate out of a trap because of the presence of a cap rock. Petroleum moves and collects in the crust based on the **porosity** and **permeability** of the rocks it encounters.

10-6 About 77 percent of the world's oil has already been discovered.

- Two principles apply to world oil supplies: (1) Most petroleum is found in a few large fields, and most fields are small; and (2) as exploration progresses, the average size of discoveries decreases. About 50,000 oil fields have been discovered.

- The first 200 billion barrels of world oil were produced in the 109 years from 1859 to 1968. Now the same amount is produced in less than a decade, at a rate of 22 billion barrels a year. The total world oil supply (past, present, and future) amounts to about 2.39 trillion barrels. Of this, 77 percent has already been discovered and 30 percent has already been produced and used.

- Natural gas, also known as **methane**, is a colourless, odourless, clean-burning fossil fuel.

10-7 Coal is a fossil fuel that is found in stratified sedimentary deposits.

- Coal is carbon-rich rock that is found in stratified sedimentary deposits. It is one of the most important fuels, used primarily in energy production by utility companies. On average, Canadians consume about 4.7 kg of coal per person per day, and 18 percent of the nation's electricity is generated by coal-burning power plants. Canada has 6.6 billion tonnes of accessible coal, an amount that could meet Canada's demand at current levels for at least 112 years.

- There are problems associated with the use of coal. When coal is burned, compounds are produced that cause smog, acid rain, and air toxicity. Although Canada and the United States have abundant coal reserves, the environmental and health costs are large and not easily mitigated.

- Coal is formed by the chemical alteration of dense layers of matted vegetation growing in warm, humid wetlands. Micro-organisms convert organic material into peat. Coal is classified into three grades: lignite (brown coal), **bituminous coal** (soft coal), and **anthracite** (hard coal). These grades form in a continuous metamorphic sequence.

10-8 Nuclear power plants provide about 17 percent of the world's electricity.

- In France, about 80 percent of all electricity is generated by nuclear power plants. In Canada, nuclear power supplies only 5 percent of total energy consumption and is fourth in importance after oil, natural gas, and hydroelectricity. There are more than 400 nuclear power plants around the world, with only 5 in Canada. Nevertheless, Canada is world-renowned for developing one of the most efficient and safe nuclear reactors, the CANDU reactor.

- Uranium-235 undergoes **induced fission** when its nucleus absorbs a neutron, produces two lighter atoms, and throws off two or three new neutrons. The two new atoms emit heat, and the neutrons cause fission in nearby uranium-235 nuclei.

10-9 Renewable energy accounts for more than 20 percent of Canada's energy sources.

- Renewable energy includes solar power, wind power, bioenergy, geothermal energy, hydrogen fuel cells, hydroelectric power, and ocean energy. Despite Canada's laudable accomplishment of producing more than 20 percent of its energy needs via renewable resources, and decades of research and development, worldwide, renewable energy plays a minor role in the energy market.

KEY TERMS

anthracite (p. 260)
banded iron formations (BIFs) (p. 250)
bauxite (p. 250)
bituminous coal (p. 260)
coal seams (p. 261)
cumulates (p. 248)
energy resources (p. 244)
fossil fuels (p. 254)
geologic resources (p. 244)

geothermal energy (p. 254)
induced fission (p. 263)
methane (p. 258)
mineral resources (p. 244)
non-renewable (p. 244)
non-renewable energy resources (p. 254)
ore (p. 247)
ore minerals (p. 247)
pegmatites (p. 248)

permeability (p. 257)
petroleum (p. 256)
placer deposits (p. 251)
porosity (p. 257)
porphyry copper (p. 248)
renewable energy (p. 245)
smelting (p. 247)
solar energy (p. 254)

ASSESSING YOUR KNOWLEDGE

Please answer these questions before coming to class. Identify the best answer.

1. Energy resources and mineral resources are
   a. types of geologic resources.
   b. produced by smelting.
   c. the result of residual weathering.
   d. generally produced during the Precambrian.
   e. None of the above.

2. Metallic ores segregate from magma in which of the following processes?
   a. immiscible melting
   b. evaporation
   c. secondary enrichment
   d. residual weathering
   e. All of the above.

3. Examples of a non-metallic ores are
   a. iron, lead, and copper.
   b. gypsum, calcite, and clay.
   c. sulphides and oxides.
   d. cassiterite and uraninite.
   e. methane and coal.

4. Sedimentary mineral deposits include
   a. secondary deposits.
   b. immiscible liquids.
   c. porphyry coppers.
   d. placers.
   e. petroleum.

5. The four general categories in which mineral deposits can form are
   a. magmatic segregation deposits, hydrothermal deposits, residual mineral deposits, and sedimentary mineral deposits.
   b. anticlinal traps, carrier beds, source beds, and reservoir beds.
   c. lignite, peat, bituminous coal, and anthracite.
   d. pegmatites, vein deposits, residual soils, and placer deposits.
   e. None of the above.

6. Which of the following is a deposit of hematite that can be mined profitably?
   a. gangue
   b. secondary deposit
   c. enriched layer
   d. banded iron formation
   e. coal seams

7. Sulphides are common ores of
   a. gravel, sand, and clay.
   b. beryllium, lithium, and calcite.
   c. aluminum, uranium, and iron.
   d. copper, lead, and molybdenum.
   e. None of the above.

8. The most widely used commercial source of energy is
   a. fossil fuels.
   b. nuclear power.
   c. renewable energy sources.
   d. solar energy.
   e. None of the above.

9. Which of the following is *not* true?
   a. Most renewable energy sources are more expensive than fossil fuels.
   b. Burning coal produces hazardous air pollution.
   c. Citizens are concerned about the safety records of nuclear power plants.
   d. Global oil costs are likely to remain low until the end of this century.
   e. All of the above.

10. The oil window is the result of
    a. extensive lignite compaction.
    b. oxidized groundwater flow into reducing zones.
    c. deep burial of fossil plankton beds.
    d. residual soil formation.
    e. banded iron formations.

11. Which of the following statements about oil is true?
    a. It is widely abundant and likely to last for at least two centuries.
    b. It is radioactive, creating problems of long-term storage.
    c. It will probably become unaffordable when peak production levels are passed.
    d. It is a source of groundwater contamination that is of great concern.
    e. None of the above.

12. Which of the following has the highest carbon content?
    a. weathered soils
    b. bituminous coal
    c. anthracite
    d. gangue
    e. residual mineral deposits

13. Which of the following is an example of a fossil fuel?
    a. uranium-235
    b. hydroelectric power
    c. solar power
    d. immiscible liquids
    e. methane

14. Uranium-235 is a useful source of energy because of its tendency to undergo
    a. nuclear fusion.
    b. spontaneous combustion.
    c. fission with no waste.
    d. induced fission.
    e. None of the above.

15. An important challenge for the world economy in the future will be
    a. developing clean sources of energy.
    b. finding sustainable sources of energy.
    c. developing alternatives to mining expensive metals.
    d. sustaining economic prosperity after "peak oil."
    e. All of the above.

FURTHER RESEARCH

1. Can fossil fuels be found on other planets? Explain your answer.

2. Considering all the sources of energy and their various problems, what do you think will be the dominant source of energy in Canada when you are 60 years old? What problems must be overcome for that to occur? How will it affect your lifestyle?

3. Do some research and determine the source of electricity where you live. Trace the electricity all the way to the geologic source that is used to generate it.

4. Look around the room. Identify two metallic minerals and two construction minerals that may have been used in its construction. What were the possible ores from which those minerals are derived?

5. Name the primary ore of aluminum, and describe the way it is formed.

6. Where does most iron come from?

7. A body of magma intrudes into sedimentary crust. Describe the various ways in which the presence of the pluton will result in the formation of ores.

8. Describe the problems associated with using coal as an energy source.

ONLINE RESOURCES

Explore more about geologic resources on the following websites:

Statistics Canada, "Energy Statistics Handbook":
www5.statcan.gc.ca/bsolc/olc-cel/olc-cel?catno=57-601-XIE&lang=eng

National Energy Board of Canada:
www.neb-one.gc.ca/clf-nsi/rcmmn/hm-eng.html

Canadian Nuclear Association:
www.cna.ca

Natural Resources Canada—Energy Sector:
www.nrcan.gc.ca/energy/home

Natural Resources Canada—Minerals and Mining Sector:
www.nrcan.gc.ca/minerals-metals

Canadian Association of Petroleum Producers:
www.capp.ca/Pages/default.aspx

The Mining Association of Canada:
www.mining.ca/site/index.php/en

USGS International Minerals Statistics and Information:
minerals.usgs.gov/minerals/pubs/country

Association for the Study of Peak Oil and Gas:
www.peakoil.net

The U.S. Nuclear Energy Institute:
www.nei.org

The National Renewable Energy Laboratory:
www.nrel.gov

Additional animations, videos, and other online resources are available at this book's companion website:
www.wiley.com/go/fletchercanada

This companion website also has additional information about WileyPLUS and other Wiley teaching and learning resources.

CHAPTER 11
MOUNTAIN BUILDING

Chapter Contents and Learning Objectives (LO)

11-1 Rocks in the crust are bent, stretched, and broken.
LO 11-1 *Define the types of stress that are present in the crust.*

11-2 Strain takes place in three stages: elastic deformation, ductile deformation, and fracture.
LO 11-2 *Define the three stages of strain.*

11-3 Strain in the crust produces joints, faults, and folds.
LO 11-3 *Compare and contrast joints, faults, and folds.*

11-4 Dip-slip and strike-slip faults are the most common types of faults.
LO 11-4 *Compare and contrast dip-slip faults and strike-slip faults.*

11-5 Rock folds are the result of ductile deformation.
LO 11-5 *Describe simple folds and complex folds.*

11-6 Outcrop patterns reveal the structure of the crust.
LO 11-6 *Describe the use of strike-and-dip symbols in making a geological map.*

11-7 Mountain building may be caused by volcanism, faulting, and folding.
LO 11-7 *Compare and contrast volcanic mountains, fault-block mountains, and fold-and-thrust mountains.*

11-8 Volcanic mountains are formed by volcanic products, not by deformation.
LO 11-8 *Describe the geology of the Cascade Mountains.*

11-9 Crustal extension formed the Basin and Range province.
LO 11-9 *Describe the origin of the Basin and Range province.*

11-10 Fold-and-thrust belts produce some of the highest and most structurally complex mountain belts.
LO 11-10 *Describe the origin of fold-and-thrust belts.*

GEOLOGY IN OUR LIVES

Stress (pressure) in the crust causes rock to fold and break. This produces earthquakes and landslides, builds mountains, and challenges our ability to find and extract geologic resources. Past and present plate boundaries are regions where stress tends to be the greatest and the crust has the most complicated structure. Understanding how stress affects the crust improves our access to oil and coal, groundwater, metallic ores, and other geologic resources that we depend on. Studying the structure of the crust advances our understanding of geologic hazards and it enhances our knowledge of Earth's history.

 What kind of plate boundary produces mountains like these found in the southern Canadian Rockies within Mount Assiniboine Provincial Park?

11-1 Rocks in the Crust Are Bent, Stretched, and Broken

LO 11-1 *Define the types of stress that are present in the crust.*

Solid rock of the crust may experience bending, pulling, and fracturing. Geologists refer to this process as **deformation** (FIGURE 11.1). The forces that cause deformation are **stresses**.

We first explored stress in Chapter 9 ("Metamorphic Rock"), where we defined *lithostatic stress* as uniform pressure acting on a rock simultaneously and equally from all directions (Section 9-2). In many cases, rock may experience unequal stress due to tectonic forces. This *directed stress* is force acting on a rock to deform it in a particular direction. There are three basic kinds: **tensional stress** (stretching), **compressional stress** (squeezing), and **shear stress** (side-to-side movement). These stresses generally originate at plate boundaries and lead to faulting and folding of the crust—and, hence, to orogenesis (mountain building). This chapter explores these subjects in more detail.

Rocks Are Deformed by Stress

Stress is defined as force applied over an area. All of the crust, oceanic and continental, is exposed to a complex set of stresses often related to plate tectonics. Hence, the crust at plate boundaries is usually highly deformed because plate interactions generate extreme stresses. Some of the force at the boundaries can radiate throughout a plate and even be present at mid-plate settings. Former plate boundaries, such as a convergent, continent–continent boundary that is now far from the edge of a plate, still may experience stress related to their tectonic history. In addition to stresses related to plate interaction, lithostatic stress, associated with the confining weight of rock (*confining stress*), is present throughout the crust. Confining stress increases as depth increases (FIGURE 11.2a).

Types of Differential Stress

Tensional stress (also called "extension" or "pull-apart" stress) is a force that tends to pull rock apart. That is, *tension lengthens or stretches the crust* (FIGURE 11.2b). Crustal lengthening occurs when rock layers are bent—for example, when they are pushed up from below and have to stretch across the greater length. The region at the top of a fold (or the top of a dome) experiences stretching as a layer changes shape to accommodate the greater length needed to extend around the circumference of a bend.

Crustal extension is found at divergent boundaries where two plates diverge or pull apart. The plate boundary is pushed upward by magma intrusion and by buoyancy associated with high heat flow through the crust. As new crust moves outward from the rift valley, it is pulled by gravity down the slopes of an oceanic ridge. The combination of all these processes generates tensional stress within the crust that results in fractures and thinning.

Courtesy of Dr. Dan Gibson, Simon Fraser University

(?) Describe the directions of stress that caused the folding and faulting in Figure 11.1.

FIGURE 11.1 Stress in the crust leads to rock deformation. In this photo, taken near Stewart, British Columbia, we see brown sedimentary rocks sitting on grey volcanic rocks. These rocks have been folded (bent) and faulted (broken), causing them to be displaced relative to each other.

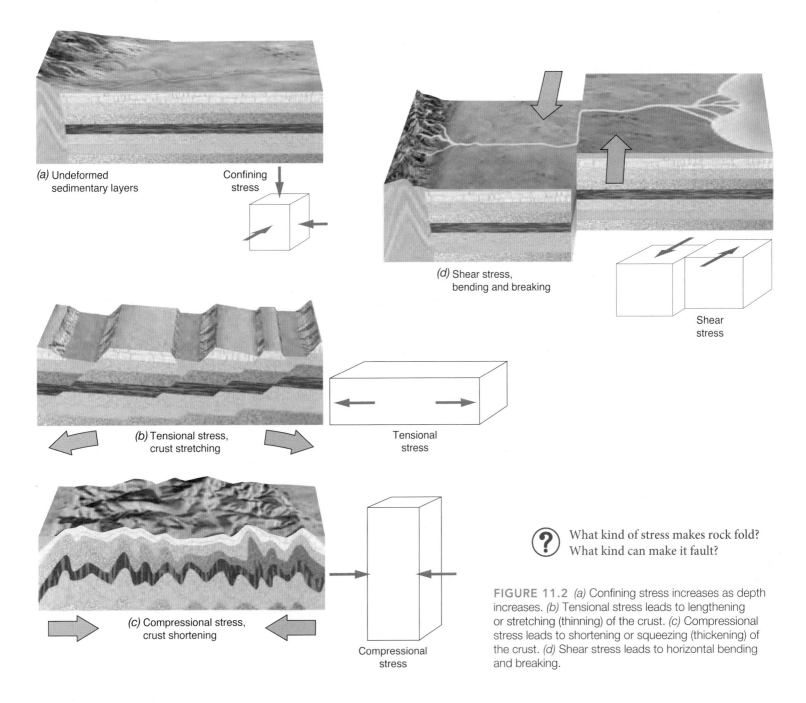

(a) Undeformed sedimentary layers

Confining stress

(b) Tensional stress, crust stretching

Tensional stress

(c) Compressional stress, crust shortening

Compressional stress

(d) Shear stress, bending and breaking

Shear stress

(?) What kind of stress makes rock fold? What kind can make it fault?

FIGURE 11.2 (a) Confining stress increases as depth increases. (b) Tensional stress leads to lengthening or stretching (thinning) of the crust. (c) Compressional stress leads to shortening or squeezing (thickening) of the crust. (d) Shear stress leads to horizontal bending and breaking.

Compressional stress is a force that squeezes or shortens rock. Crustal shortening (FIGURE 11.2c) is common at convergent plate boundaries where folding, crushing, and fracturing of rock layers can occur. Compression thickens crust by forcing rock into a smaller space. Remember, however, that even though compression causes rock to fold, layers in the region of greatest curvature of a fold undergo tension and thinning as they extend around the bend. Hence, both compressional and tensional stresses are at work on the crust as it is shortened.

Shear stress (FIGURE 11.2d) is force that causes slippage, or "translation" (migration), of rock layers past one another. Crustal slippage produces rock faulting, a displacement or offset across a fracture zone. Shear stress is characteristic of *transform plate boundaries* that move in opposite directions sideways. As two plates grind past each other, shear stresses at the boundary fracture and crush the crust, creating broad fault zones like the San Andreas fault zone in California and the Queen Charlotte fault off the west coast of British Columbia.

 Expand Your Thinking—Stress deforms rock. How might rock react to stress that is applied very quickly compared to stress that is applied very slowly?

11-2 Strain Takes Place in Three Stages: Elastic Deformation, Ductile Deformation, and Fracture

LO 11-2 *Define the three stages of strain.*

When rocks are deformed, they are said to undergo **strain.** *Strain is the change in shape and/or volume of a rock caused by stress.* When rocks undergo strain, they pass through three successive stages of deformation: elastic deformation, ductile deformation, and fracture.

Elastic Deformation

Elastic deformation is fully reversible; that is, the deformation is not permanent. Most rocks are elastically deformed when they first experience stress, which causes the atomic bonds within and between minerals to stretch, flex, and bend. When the stress is removed, the bonds snap back to their original form and the rock regains its original shape. For instance, as a glacier grows on bedrock at the beginning of an ice age, the rock slowly yields to the growing weight of the ice over thousands of years and elastically deforms downward. When climate changes and the ice age ends, the crust elastically rebounds as the glacier retreats and the ice melts. (This is called *glacioisostatic rebound.*)

In most types of material, elastic behaviour occurs under small amounts of stress, usually when the stress is first applied and before it has a chance to build up to significant levels. But as stress increases and the atomic bonds are further deformed, rocks (and other materials) approach their *elastic limit:* the point beyond which deformation becomes permanent and is not reversible.

A spring holding a weight is a good example of elastic behaviour. As stress is applied to a spring by adding more and more weight, the increase in strain is displayed by the stretching of the spring. The spring displays elastic strain as long as it fully recovers its shape when the weight is removed. However, as more weight is added, eventually the elastic limit of the spring will be exceeded, and when the weight is removed, the spring will not fully recover its original shape.

The British scientist *Robert Hooke* (1635–1703) studied various materials under states of stress. He found that a graph or plot of stress (units of pressure) versus strain (units of deformation) yields a straight line as long as the elastic limit is not exceeded. Although this relationship is more easily demonstrated using a spring, Hooke's law holds true for rocks as well (**FIGURE 11.3**).

Ductile Deformation

Ductile deformation, also known as *plastic deformation,* occurs when a material accumulates strain that is not reversible. Ductile deformation takes the form of a permanent change in the size and/or shape of a rock that has passed its elastic limit. As further stress is applied, the deformation (strain) becomes increasingly pronounced. Most materials will stop exhibiting ductile deformation at some point and begin to fracture. Rocks exhibit ductility when they are folded (**FIGURE 11.4**).

Fracture

Fracture occurs when a material that is accumulating strain, or deforming, finally breaks. This stage is termed *brittle deformation.* Of course, like ductile strain, brittle deformation is permanent. Once a fracture has formed, further stress opens the fracture more but does not cause any additional change in the volume and/or shape of the rock. Hence, at the point of fracture, strain no longer grows as stress is applied; the fracture simply experiences increased displacement.

FIGURE 11.5 plots the rate of strain accumulation as stress is applied to a rock sample in a laboratory; note that the rock first responds elastically, showing linear strain accumulation (consistent with Hooke's law). When the rock passes its elastic limit, it experiences ductile deformation. Finally, as stress builds, the rock fractures. In the crust, many such fractures develop into **faults:** zones or planes of breakage across which layers of rock are displaced relative to one another.

(?) Give another example of elastic deformation.

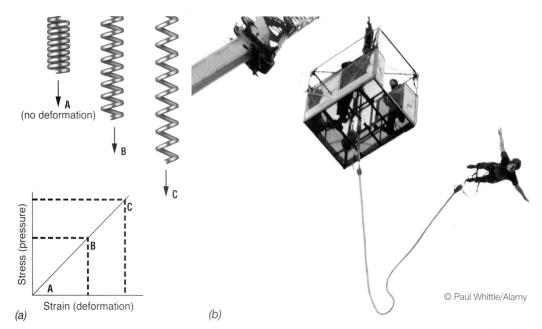

(a) A (no deformation) — B — C

Stress (pressure) vs. Strain (deformation), points A, B, C

© Paul Whittle/Alamy

(b)

FIGURE 11.3 Hooke's law. In elastic behaviour, strain is proportional to stress. *(a)* Hooke's law predicts that as weights are added to a spring, the stress (weight) and strain (change in length) can be plotted as a straight line. *(b)* Hooke's law allows you to calculate the progress made by a bungee jumper on an elastic line.

FIGURE 11.4 This metamorphosed gneiss in the Canadian Cordillera displays ductile strain by creating many small folds in response to compressional stress.

 Describe an example of ductile deformation in some common materials.

Factors that Govern Strain

The type of strain experienced by a rock depends on several factors: temperature, confining pressure, strain rate, rock composition, and water. At high temperatures, rock exhibits greater *ductility* because molecules, and their bonds, stretch and move more readily at high temperatures than at lower temperatures. Cold rock tends to be brittle and fractures more easily. Confining pressure has the same effect. At higher confining pressure—that is, under greater lithostatic stress—rock tends to be more ductile because the pressure of the surrounding environment hinders the formation of fractures.

The rate at which strain accumulates can also influence the behaviour of rocks. When stress is applied rapidly, as with a sudden hammer blow, rocks and other solids tend to fracture more readily. When stress is applied slowly and strain accumulates slowly, atomic and molecular bonds have a chance to accommodate the stress by deforming.

The composition of a rock influences the type of strain it undergoes. Some minerals, such as quartz, olivine, and feldspars, are very brittle. Others, such as micas, clays, calcite, and gypsum, are more ductile. This difference is due largely to the types of atomic bonds holding the minerals together. As a result, the mineralogy of a rock is an important factor influencing the type of strain it undergoes.

Last, the presence or absence of water can influence how a rock is deformed. It is thought that water changes and weakens bonds and may form a film around mineral grains. These properties allow atomic and molecular bonds to bend, break, and reform more readily, encouraging wet rocks to behave in a ductile manner, while dry rocks tend to be brittle.

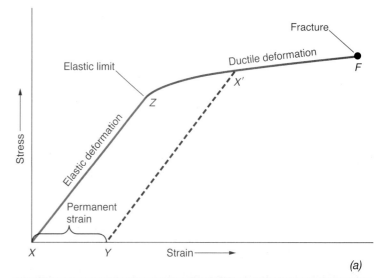

(a)

Courtesy of Dr. Dan Gibson, Simon Fraser University

(b)

FIGURE 11.5 *(a)* A typical sample of rock tested in a laboratory will exhibit a straight-line relationship between stress and strain (X to Z) when deforming elastically, as predicted by Hooke's law. At the elastic limit (Z), the rock experiences ductile deformation until the stress is removed at X'. At this point, the rock returns to an unstressed state along line X'Y but retains some permanent deformation (strain amount XY). If the rock continues to experience stress past point X, it will eventually fracture (point F). *(b)* A fault is formed when rock fracture leads to displacement parallel to the fracture surface.

 In Figure 11.5b, does the fault reveal crustal shortening or crustal extension? What type of stress caused this?

Expand Your Thinking—With your finger, push down on your forearm and then remove your finger. What kind of strain did your arm exhibit?

11-3 Strain in the Crust Produces Joints, Faults, and Folds

LO 11-3 *Compare and contrast joints, faults, and folds.*

Stress applied to the crust will cause rocks to deform. In most cases, rocks undergo ductile strain (folds) or fracture (joints and faults). **Joints** are fractures in the crust that display no net displacement between the two sides parallel to the fracture surface. Many joints form when confining stress is removed and rock expands. Other joints are a result of stress spread across a broad region within the crust, as when brittle rock attempts to conform to the curvature of Earth's surface.

Faults form when brittle rocks fracture and a distinct offset occurs along the plane of breakage. In some cases, the offset is minor (centimetres to metres) and the displacement is easily measured. In other cases, the displacement is large (hundreds to thousands of metres) and it is difficult to obtain an accurate measure of the amount of displacement. When strain accumulates slowly and prevailing conditions, such as high temperature and pressure or the presence of water, allow rocks to experience ductile deformation, they may develop into one or many **folds** (FIGURE 11.6a).

Joints

Joints are common in the crust and can develop in all types of rock. Since most rock at Earth's surface is relatively cold, it is brittle and tends to fracture when stressed or when stress is removed and the rock expands. As with any brittle material, rock joints will be formed at any point of weakness, such as around a fossil, between minerals, or along bedding planes.

Joints can develop under a number of conditions. Igneous rock shrinks as it cools and develops sets of intersecting joints called *columnar jointing* (FIGURE 11.6b), which reflect the decrease in the rock's volume. When a large weight is removed from the crust, such as when a glacier melts, the confining pressure is reduced. The crust expands upward and develops fractures that follow flaws in the rock. Joints form along the edges of cliffs or at **outcrops** (places where layers of rock are exposed), because at such locations the crust is not

David Parker/Science Source

(a)

Courtesy of Dr. Glyn Williams-Jones, Simon Fraser University

(b)

Courtesy of Dr. Dan Gibson, Simon Fraser University

(c)

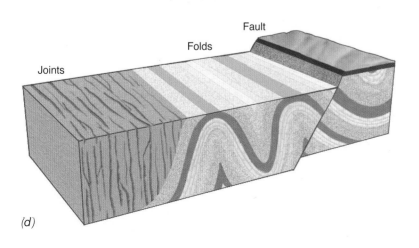

(d)

(?) How does a geologist determine if a fracture is a joint or a fault?

FIGURE 11.6 *(a)* Folds, *(b)* joints, and *(c)* faults are common geologic structures. *(d)* At times joints, folds, and faults may occur in the same vicinity.

confined and tends to expand outward. Joints can also develop as a result of tectonic stresses that travel far from plate boundaries and cause fracturing at locations hundreds of kilometres away.

Joints play an important role in enhancing other geologic processes. Jointed rocks provide easy routes for groundwater that moves along fracture planes in otherwise impermeable crust. Chemically active fluids that emanate from cooling magma often travel along fractures in the country rock and deposit hydrothermal veins of ore minerals (Chapter 10, "Geologic Resources"). Joints are attacked by acidic fluids and subjected to enhanced dissolution, hydrolysis, and other types of chemical weathering (Chapter 7, "Weathering"). Freeze-thaw processes lever joints open to form natural arches and cause cliff collapse, which forms talus.

Strike and Dip

Geologists seek to understand the formation of stresses within the crust in order to gain a better understanding of tectonic processes and their history. As a first step, they create maps of geologic structures such as folds, joints, and faults. *Maps are two-dimensional representations of three-dimensional surfaces oriented to Earth's coordinate system* (north, south, east, and west). On a proper map, north is always at the top. To show crustal structure on a map, geologists have devised a simple symbol to display the spatial orientation of rock layers that make up folds, faults, and joints. This symbol has two parts: the **strike** and the **dip.**

Imagine the surface of a tilted rock layer or a fault fracture as a two-dimensional plane. You could draw a line on that plane that is perfectly horizontal. Geologists call that horizontal line the *strike.* A line drawn perpendicular to the strike is known as the *dip.* (**FIGURE 11.7**).

Strike and *dip* refer to the orientation of a geologic feature. The strike of a planar body of rock, or stratum, is a line representing the intersection of that feature with the horizontal. (Remember your geometry class—a line is formed at the intersection of two planes.) On a map, this is represented with a short straight line oriented to the compass direction of the strike. Strike is given as a *compass bearing* (e.g., N30°E) or as a single three-digit number representing the *azimuth* (e.g., 030°). The dip gives the angle of maximum inclination below the horizontal of a geologic feature. It can be thought of as the direction in which water would flow if rain fell on the fault surface or rock layer. On a map, the dip is a shorter perpendicular line off one side of the strike line. Often the angle of inclination is written next to the dip line (e.g., 40°). *The strike is always measured perpendicular to the dip.* A little later in the chapter, we discuss the strike-and-dip symbol again and explain how it is used to interpret various types of folds and faults.

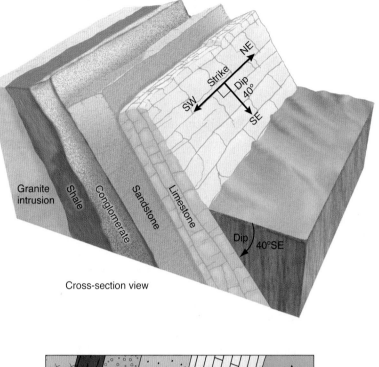

Cross-section view

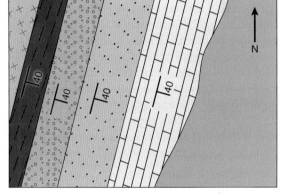

Map view

FIGURE 11.7 *(a)* The line of strike (denoted by a compass bearing or azimuth) marks the orientation of the trend of a geologic plane such as a fault surface or (as shown) a rock layer. The line of dip (always oriented 90° to the strike) is also marked in proper compass orientation, along with the angle of inclination of the geologic plane. *(b)* Strike-and-dip symbols are used on a map to represent the spatial orientation of geologic features such as rock strata, faults, joints, and others. Because the top of a map always indicates north, the orientation of the strike-and-dip symbol represents the real-life orientation of the geologic feature.

(?) What type of instruments does a geologist need to measure the strike and dip of a plane?

 Expand Your Thinking—Describe the stress and strain history of Figure 11.6d. It is not simple and involves at least two stages.

11-4 Dip-Slip and Strike-Slip Faults Are the Most Common Types of Faults

LO 11-4 *Compare and contrast dip-slip faults and strike-slip faults.*

Two types of faults are common: dip-slip faults and strike-slip faults.

Dip-Slip Faults

Dip-slip faults are formed at locations where a fault plane is inclined, or at an angle to the horizon, and one block of crust is displaced up or down along the plane relative to another. *In dip-slip faults, the displacement occurs up or down parallel to the dip of the fault surface;* hence the term "dip-slip." The main types of dip-slip faults are **normal faults** and **reverse faults**; in addition, a special type of reverse fault is known as a **thrust fault.** For any inclined fault, the block that lies above or "hangs" above the fault plane is the *hanging wall block* and the block below it is the *footwall block* (**FIGURE 11.8**).

These terms originated with miners. They found faulted rock easy to excavate, and fault surfaces often were filled with economic minerals precipitated from fluids that migrate along the fault plane. Miners named the ceiling of their tunnels the "hanging wall block" and the floor of a mine shaft the "footwall block." Normal faults and reverse faults are defined by the relative displacement of the hanging wall block to the footwall block: in normal faults the hanging wall block moves down relative to the footwall block; in a reverse fault, the hanging wall block moves up.

FIGURE 11.9 displays the major types of faults. *Normal faults develop in brittle rocks that are exposed to tensional stress.* In the case of a normal fault, an inclined fracture develops as the crust is pulled apart and the hanging wall block slips downward along the inclined fault plane relative to the footwall block. Such faults are formed at locations where crustal stretching or lengthening is occurring. Rift valleys, for example, are places where crustal extension occurs and normal faults are common.

Reverse faults (**FIGURE 11.9b**) *are caused by compressional stress;* the hanging wall block is displaced upward along the fault plane relative to the footwall block. A thrust fault is a special type of a reverse fault in which the dip of the inclined fault plane is less than 45°, more commonly around 30°. Thrust faults develop when stresses are distributed over a wide geographic region. As a result, thrust faults can have considerable displacement, measuring hundreds of kilometres, and may result in the formation of major topographic features composed of rock layers in which older strata overlie younger ones.

Strike-Slip Faults

Strike-slip faults occur at locations where the motion on a fault plane is horizontal (**FIGURE 11.9d and e**); that is, two blocks of crust travel (slip) parallel to the strike of a fault rather than along the dip. *Strike-slip faults are caused by shear stress acting parallel to the surface of the crust.* They take two forms, depending on the relative displacement across the fault. To an observer standing on one side of the fault and looking across to the far side, where a block is relatively displaced, if the block on the other side has moved to the left, the fault is a *left-lateral strike-slip fault.* If the block has moved to the right, the fault is a *right-lateral strike-slip fault.*

Strike-slip faults include some of the world's most dangerous fault systems, such as the North Anatolian fault in Turkey, the San Andreas fault in California, and the Enriquillo-Plantain Garden fault in Haiti. All three are renowned for causing devastating earthquakes that occur suddenly when there is movement along the strike of the fault (**FIGURE 11.11**).

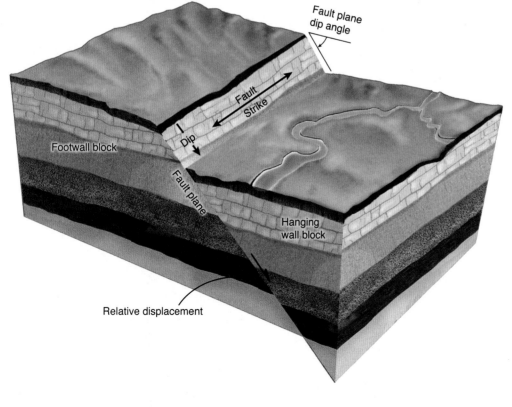

Fault plane dip angle

Fault Strike

Dip

Footwall block

Fault plane

Hanging wall block

Relative displacement

(?) What type of fault is shown in Figure 11.8?

FIGURE 11.8 The footwall block lies below the fault plane, and the hanging wall block lies above it. The strike of a fault plane is perpendicular to the dip of the fault plane.

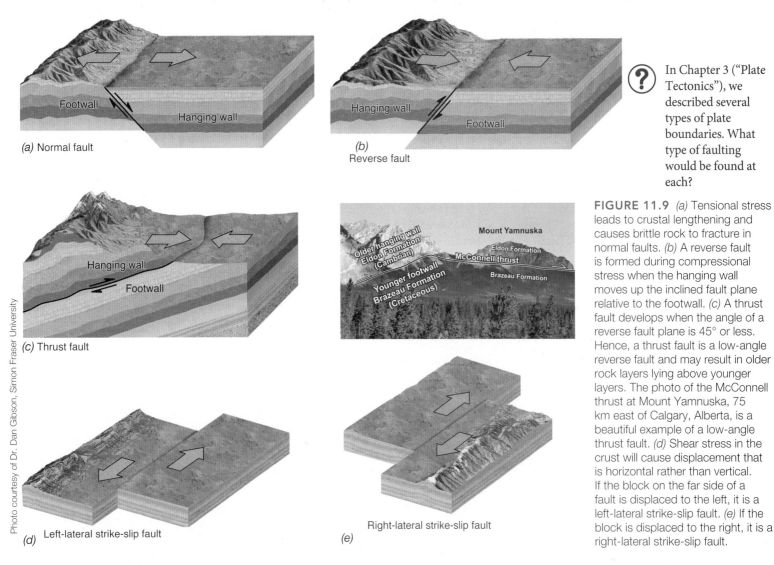

(a) Normal fault

(b) Reverse fault

(c) Thrust fault

(d) Left-lateral strike-slip fault

(e) Right-lateral strike-slip fault

Photo courtesy of Dr. Dan Gibson, Simon Fraser University

(?) In Chapter 3 ("Plate Tectonics"), we described several types of plate boundaries. What type of faulting would be found at each?

FIGURE 11.9 (a) Tensional stress leads to crustal lengthening and causes brittle rock to fracture in normal faults. (b) A reverse fault is formed during compressional stress when the hanging wall moves up the inclined fault plane relative to the footwall. (c) A thrust fault develops when the angle of a reverse fault plane is 45° or less. Hence, a thrust fault is a low-angle reverse fault and may result in older rock layers lying above younger layers. The photo of the McConnell thrust at Mount Yamnuska, 75 km east of Calgary, Alberta, is a beautiful example of a low-angle thrust fault. (d) Shear stress in the crust will cause displacement that is horizontal rather than vertical. If the block on the far side of a fault is displaced to the left, it is a left-lateral strike-slip fault. (e) If the block is displaced to the right, it is a right-lateral strike-slip fault.

(?) CRITICAL THINKING

INFERRING THE HISTORY OF BROKEN ROCK

New Zealand straddles the boundary of the Australian and Pacific plates. **FIGURE 11.10** shows one of many outcrops there that reveal the structure of the crust. Study the photo and answer the following questions.

1. What type of stress caused this fault?
2. What is your evidence for identifying the movement of the two sides?
3. Describe the geological history of this place; that is, what happened first, second, and so on?
4. From your analysis, infer a plausible tectonic history for the crust here. Describe this history.
5. What additional observations would you wish to collect to test your hypothesis?

© iStock.com/mikeuk

FIGURE 11.10 Fault within a New Zealand cliff face.

(?) Draw a line along the fault in Figure 11.10 and indicate with arrows the relative displacement of the two sides.

In Turkey, right-lateral motion along the North Anatolian fault system generated a 45-second earthquake at 3:01 a.m. on August 17, 1999. The epicentre was southeast of Izmit, an industrial city near Istanbul. The earthquake was felt over a large area, as far east as Ankara, about 320 km away. Unofficial estimates place the death toll between 30,000 and 40,000. Most of the deaths and injuries were caused by collapsed commercial and residential buildings, typically four to eight stories high. At 5:00 p.m. on October 17, 1989, the San Andreas fault suddenly awoke in northern California. The quake was responsible for 62 deaths, 3,757 injuries, and over $6 billion in damage. More than 18,000 homes and 2,600 businesses were damaged, and about 3,000 people were left homeless. On January 12, 2010, the Enriquillo-Plantain Garden fault in Haiti broke open after 250 years of quietly building stress. The rupture was roughly 65 km long with an average slip of 1.8 m. By January 24, the devastated nation had endured 52 aftershocks, each of sufficient strength to cause extensive damage. After all three of these quakes, teams of geologists rushed to the scene to assist in rescue operations and study the evidence of fault movement in hopes of reducing the damage caused by future quakes.

In China, 9 m of reverse faulting on the Longmenshan fault generated enormous damage in Sichuan Province; see the "Geology in Action" box titled "The Sichuan Quake of 2008."

Evidence of Faulting

Finding a fault can be difficult. Evidence of faulting can be obvious or subtle, depending on the extent of fault movement, the amount of time that has passed since the last movement, and the nature of the ground surface where fault movement last occurred. Weathering and sediment deposition at the surface tend to obscure the features formed when rocks rupture and slide past each other. Geologists look for several types of features that provide evidence of faulting:

- *Slickensides:* polished-looking surfaces that indicate the fault plane, which often contain *slickenlines,* consisting of parallel striations (i.e., scratches), grooves, and ridges (usually in the form of mineral precipitates) that result from frictional sliding of one block past another (**FIGURE 11.12**); the orientation of the slickenlines indicates the direction of the most recent movement.
- *Fault breccia:* loose, coarse fragments of rock formed in the fault zone by grinding and crushing due to fault movement. It occurs on either or both sides of the fault plane or may form a zone that obscures the fault plane entirely.
- *Fault gouge:* very fine-grained, pulverized rock formed by grinding and crushing along the fault plane.

(a)

(b)

(c)

? Building collapse is a major hazard during an earthquake. Propose a hypothesis that relates ground motion to how a building fails. How would you test your hypothesis?

FIGURE 11.11 *(a)* A collapsed seven-storey building in Turkey in 1999. *(b)* Buildings destroyed by the San Andreas quake of 1989. *(c)* Devastation in Port-au-Prince, Haiti, where more than 200,000 people died in January 2010.

FIGURE 11.12 Slickenlines on a slickenside surface.

FIGURE 11.13 This fault scarp is formed by normal faulting along the Wasatch Mountain Range in Utah. The crust in the lower portion of the photo has dropped down relative to the crust in the upper portion. Arrows point to the fault scarp at the base of the much larger hill.

(?) What type of stress caused the faulting in Figure 11.13?

Evidence may also include the joining of two geologic rock units that would otherwise not be in contact, such as two sedimentary rock formations of different ages, or intrusive rock in sharp contact with country rock lacking a hornfels or baked zone in between. Geologists also examine topographic maps and aerial photographs for linear features on the surface, such as *fault scarps* (a cliff or escarpment formed by a fault) (FIGURE 11.13).

GEOLOGY IN ACTION

THE SICHUAN QUAKE OF 2008

On Monday, May 12, 2008, the most damaging earthquake to hit China since 1975 violently shook the ground for about two or three minutes, according to witnesses (**FIGURE 11.14**). The earthquake was generated by reverse faulting along the Longmenshan fault located in south-central China in the province of Sichuan. Measurements indicate that the hanging wall shifted approximately 9 m along the fault plane during the quake.

According to Chinese officials, the quake killed 87,000 people and injured another 400,000. Sadly, these figures include 158 earthquake relief workers killed in landslides as they tried to repair roads. The earthquake left at least 4.8 million people without housing, although the number could be as high as 11 million.

The seismicity of central and eastern Asia is caused by the northward movement of the Indian-Australian plate at a rate of 5 cm/year as it collides with Eurasia. This results in the uplift of the Himalayan mountain belt and Tibetan Plateau and associated earthquake activity. China frequently suffers large and deadly earthquakes as a result of the convergence taking place between the two plates.

FIGURE 11.14 Many people were trapped in collapsed buildings following the Sichuan earthquake.

(?) **Expand Your Thinking**—Imagine that the wall directly in front of you is lined up perfectly with the compass so that the left side points due west and the right side points due east. Now in your mind, tilt the wall away from you 45°. Imagine filling the room halfway with water. Where the water level hits the tilted wall creates a line of strike. Draw the strike and dip of the wall on a sheet of paper that is a proper map.

11-5 Rock Folds Are the Result of Ductile Deformation

LO 11-5 *Describe simple folds and complex folds.*

Rocks that have exceeded their elastic limit and behave in a ductile manner tend to bend or contort; the resulting geologic structures are called *folds*. Most folds are the result of compressional stresses acting over long periods. Because the *rate of strain accumulation* is low, rocks that might normally experience brittle or elastic deformation can instead undergo ductile (or *plastic)* deformation and produce a number of different types of folds. Folds are most visible in rocks composed of layers.

For ductile deformation of rock to occur, a number of conditions must be met:

- The rock must have the ability to deform under pressure and heat.
- The higher the temperature of the rock, the more plastic it becomes.
- The level of pressure must not exceed the internal strength of the rock. If it does, fracturing will occur.
- Deformation must take place slowly.

Folds are described by their geometry and orientation (**FIGURE 11.15**). The sides of a fold are called *limbs*. The limbs intersect at the tightest part of the fold, called the *hinge or hinge point*, through which runs the *hinge line*[1] of the fold. The hinge line represents the line connecting all points of maximum curvature of a folded surface or layer. If the hinge line is not horizontal, the fold is called a *plunging fold*. An imaginary plane that includes the hinge line and divides the fold as symmetrically as possible is called the *axial plane* of the fold.

Simple Folds

The simplest types of folds are symmetrical and upright. That is, their limbs dip away from their axes at the same angle and their axial plane is vertical. Two types of simple folds are the **anticline** (defined as a convex upward arch that has the oldest rocks in the centre of the fold) and the **syncline** (defined as a concave upward arch that has the youngest rocks in the centre of the fold) (**FIGURE 11.16**). An upright, symmetrical anticline is a fold in which rock layers have been bent to form an arch. The axial plane is vertical, and the two limbs of the anticline dip away from the centre of the arch at the same angle. The inverse of an anticline is an upright, symmetrical syncline. A syncline is a concave upward bend in of the rock layers. Note that if the age of the folded rocks is unknown, then the convex upward arch is simply referred to as an **antiform**, and the concave upward arch as a synform.

Another type of simple fold is the **monocline,** in which there is a slight bend in parallel layers of rock. For example, monoclines often develop at locations where horizontal strata are bent upward by a buried fault that does not break the surface, so that the one of the limbs is inclined and the other limb remains horizontal. The axial plane of the fold is tilted at an angle to the vertical. Notice that the occurrence of a fault is a sign of brittle rock behaviour while the occurrence of folded layers is a sign of ductile behaviour.

Complex Folds

As stresses in the crust increase in magnitude and complexity, the resulting folds become more complicated (**FIGURE 11.17**). Asymmetrical anticlines and synclines can develop, as well as overturned structures in which one limb is tilted beyond the vertical. *Plunging anticlines* and *plunging synclines* have hinge lines that are not horizontal and that intersect the surface. Groups of anticlines and synclines, known as *S-* and *Z-folds,* which, as the name implies, describes the shape of the folds, formed when one fold limb became overturned while the other remained upright (**Figure 11.17d**). Other types of complex folds include *domes,* which are circular or slightly elongated upward displacements of rock layers, and *basins,* which are circular or slightly elongated downward displacements of rock.

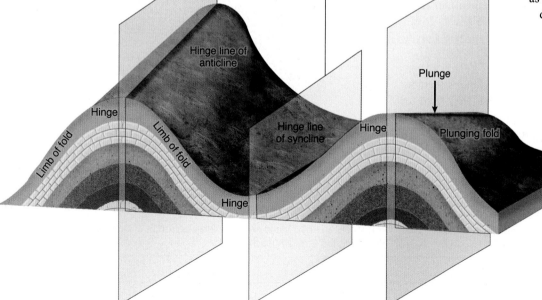

 What type of stress is responsible for the illustrated structures?

FIGURE 11.15 Folds are described using terms that refer to aspects of their geometry.

[1]Note, this can also be referred to as the *fold axis,* but only if the fold is cylindrical in geometry with a relatively straight hinge line. If the hinge line is bent or wavy, it cannot be called a fold axis.

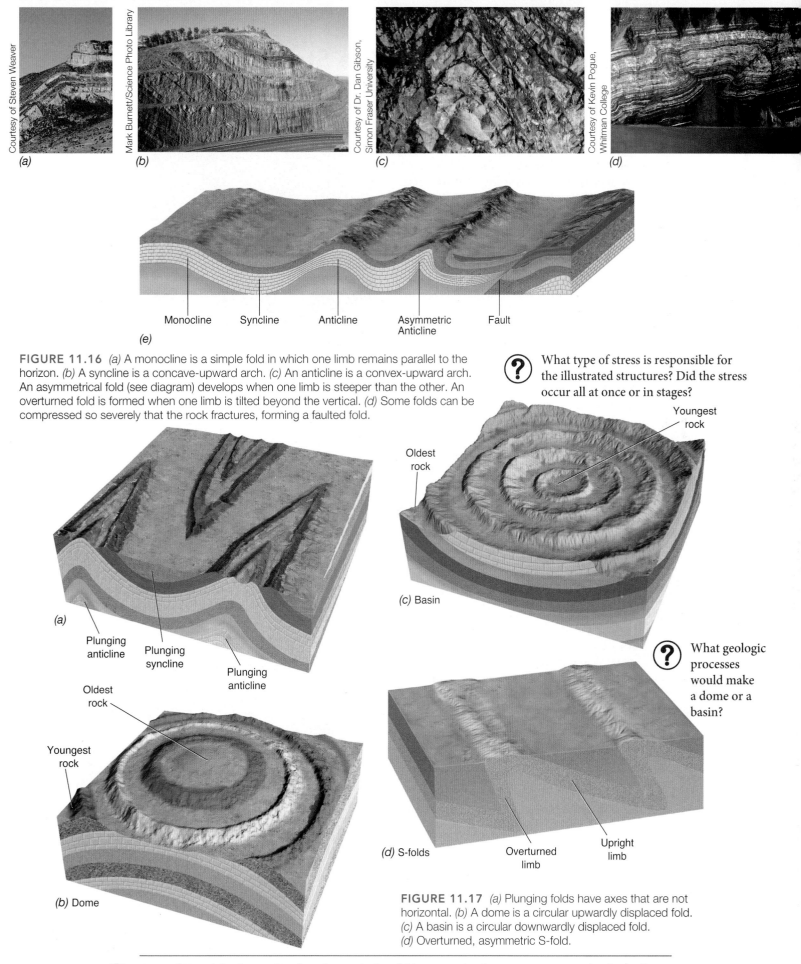

(a) *Courtesy of Steven Weaver*

(b) *Mark Burnett/Science Photo Library*

(c) *Courtesy of Dr. Dan Gibson, Simon Fraser University*

(d) *Courtesy of Kevin Pogue, Whitman College*

Monocline Syncline Anticline Asymmetric Anticline Fault

(e)

FIGURE 11.16 *(a)* A monocline is a simple fold in which one limb remains parallel to the horizon. *(b)* A syncline is a concave-upward arch. *(c)* An anticline is a convex-upward arch. An asymmetrical fold (see diagram) develops when one limb is steeper than the other. An overturned fold is formed when one limb is tilted beyond the vertical. *(d)* Some folds can be compressed so severely that the rock fractures, forming a faulted fold.

? What type of stress is responsible for the illustrated structures? Did the stress occur all at once or in stages?

Youngest rock

Oldest rock

(c) Basin

(a)

Plunging anticline

Plunging syncline

Plunging anticline

Oldest rock

Youngest rock

(b) Dome

? What geologic processes would make a dome or a basin?

(d) S-folds

Overturned limb

Upright limb

FIGURE 11.17 *(a)* Plunging folds have axes that are not horizontal. *(b)* A dome is a circular upwardly displaced fold. *(c)* A basin is a circular downwardly displaced fold. *(d)* Overturned, asymmetric S-fold.

? **Expand Your Thinking**—You have been asked to fold a granite kitchen countertop into the shape of an anticline. You have any technology you need at your disposal. How will you achieve the task?

11-6 Outcrop Patterns Reveal the Structure of the Crust

LO 11-6 *Describe the use of strike-and-dip symbols in making a geological map.*

It is one thing to discuss folds while observing their full three-dimensional geometry on a sheet of paper. It is quite another to piece together the nature of folded crust "in the field" without knowing what lies hidden underground. This is the challenge faced by *structural geologists.* **Structural geology** is concerned with determining the geometry of the crust and inferring from it the nature of the stresses it has undergone. In order to do this, geologists must determine the nature of folds and faults in rocks that usually are below ground and therefore not in full view. Structural geologists analyze outcrops to determine the spatial orientation of rock layers and the organization (or "structure") of the crust below ground. From this information, they can infer the relative age of the layers and the stresses the rocks have experienced.

Because gravity acts uniformly on sedimentary particles settling through water, we know that sedimentary rocks typically form as horizontal layers. (This is known as the *principle of original horizontality* and is discussed more fully in Chapter 13, "Geologic Time"). When these previously flat-lying strata are tilted, they provide a record of their deformation history. Geologists measure the angle and orientation of tilting in order to determine the direction and source of the deforming stresses. That is, geologists determine the strike (compass bearing) and dip (inclination) of the rock.

As we discussed briefly in Section 11-3, geologists do this by representing a layer of tilted rock as a geometric plane. The orientation of that plane in space can be defined using the strike and dip. The strike is a horizontal line on a dipping plane described relative to north or by a compass azimuth (0° to 360°), and dip is an angle to the horizon that varies from 0° (horizontal) to 90° (vertical). As mentioned, the dip can also be imagined as the direction and angle in which water would flow if rain were to fall on the rock layer.

As we saw earlier, the strike is measured perpendicular to the dip. The strike can be envisioned as the line formed by the intersection of the plane of the dipping rock layer with a horizontal plane. One way to determine a horizontal plane is to imagine flooding the outcrop with water so that an imaginary water level (a horizontal plane) intersects the surface of the tilted rock. The water's edge, formed where the two planes intersect, is the strike, and it is measured using a compass. Strike

(?) What types of folds are shown in Figure 11.18c?

(a)

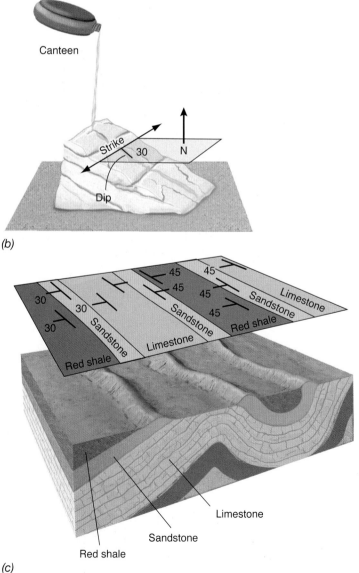

(b)

(c)

FIGURE 11.18 *(a)* These geology students are measuring the orientation of rock layers at an outcrop. *(b)* Strike is the compass bearing of a plane representing a rock layer at an outcrop. In this example, if north is toward the back of the outcrop (note north arrow), the strike is approximately northeast-southwest and the dip is 30° to the southeast. Dip is the direction in which water will flow across the outcrop. *(c)* A geological map shows the lithology, age, and strike and dip of rock units.

and dip are represented by a symbol that can be placed on a **geological map**. Geological maps show the topography of an area, the lithology and age of rock units exposed at the surface or shallowly buried under loose sediment and soil, and their strike and dip (**FIGURE 11.18**).

Geological maps display the location of colour-coded rock units and plot the measurements of strike and dip at outcrops. In Figure 11.18a, geology students interpret the dip of rock layers to determine the geologic structure of the crust. This information is used to make a map where strike and dip symbols are plotted on outcrop locations (Figure 11.18c). In this figure, adjacent dip indicators point away from one another to depict an anticline. Adjacent dip symbols pointing toward each other indicate a syncline. Notice also how the rock units crop out in repeating patterns. Such patterns develop when erosion lowers Earth's surface by removing overlying rock and sediment so that portions of the structures are gone, and other portions, such as the limbs, are exposed to view.

Certain other aspects of outcroppings are useful for determining the structure of the crust. The pattern of outcropping may reveal the orientation of the underlying structure. Parallel strips of outcrops can indicate folds with horizontal axes. That is, the folds do not plunge. An outcrop pattern of Vs can indicate plunging folds (**FIGURE 11.19**). A V that opens in the direction of plunge is a syncline, while a V that closes in the direction of plunge is an anticline.

If the relative age of rock layers is known, this information can be used to interpret the history of the crust. If an outcrop pattern reveals younger rock layers (deposited first) between older rocks (deposited later), a syncline is present. If the pattern consists of older rock layers between younger layers, an anticline is present. Likewise, in the case of a V outcrop pattern for a plunging fold, older rocks will be found in the middle of the V if an anticline is present and younger rocks in the middle of the V if a syncline is present.

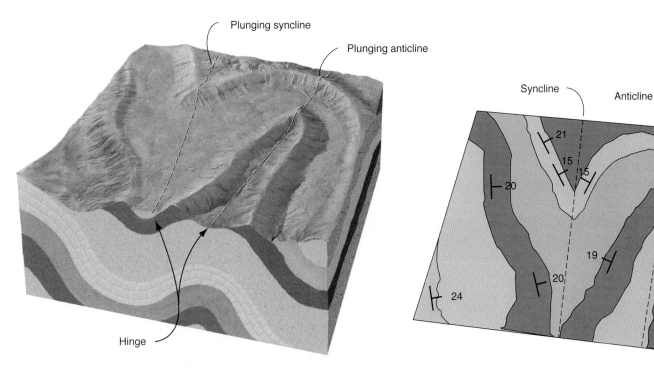

FIGURE 11.19 Erosion exposes internal rock layers forming outcrop patterns that reveal the orientation of underlying structures. These patterns can be mapped, and the strike and dip of the rocks can be used to reveal the structure of the underlying folds.

 Imagine you are mapping a dome; what would the strike-and-dip pattern be?

Expand Your Thinking—What characteristics of maps make them such important and effective tools in structural geology?

11-7 Mountain Building May Be Caused by Volcanism, Faulting, and Folding

LO 11-7 *Compare and contrast volcanic mountains, fault-block mountains, and fold-and-thrust mountains.*

A mountain is an area of land that rises abruptly up from the surrounding region. Generally, a mountain is considered to project at least 300 m above the surrounding land. A *mountain range* consists of several closely spaced mountains with a common origin. **Mountain belts** are made up of several mountain ranges, usually with related histories. Familiar mountain belts include the *North American Cordillera* (including the Rockies and Sierra Nevada Ranges), the *Himalayas,* the *Alps,* and the *Appalachians.* These are all examples of mountain belts composed of numerous mountain ranges.

Mountain-building processes fall into three general categories, producing three types of mountain belts: (1) **volcanic mountains** usually form at volcanic arcs above subduction zones, at divergent boundaries, or at hotspots; (2) **fault-block mountains** are characterized by tensional stress, crustal thinning, and normal faulting; and (3) **fold-and-thrust mountains** develop where tectonic plates collide along convergent margins.

Volcanic Mountains

Volcanic mountains are the result of rising magma that breaks through Earth's surface and builds a volcano (Chapter 6, "Volcanoes"). Volcanic mountains are usually related to hotspots (Hawaii), volcanic arcs at convergent tectonic boundaries (Mount Meager, British Columbia), or spreading-centre volcanism at divergent boundaries. The volcanic mountains of Hawaii contain both the tallest mountain on

Earth when measured from its base to its summit (the Mauna Kea shield volcano; 10,203 m from the Pacific sea floor to the highest point) and the most massive mountain on the planet (the Mauna Loa shield volcano; 40,000 km³ in total volume) (**FIGURE 11.20**).

Fault-Block Mountains

Fault-block mountains are the product of tensional stresses that pull apart the crust in a horizontal direction, breaking it into a number of separate blocks. As each block moves vertically, it compensates for the tensional forces by undergoing normal faulting that produces sequences of up-thrown ranges and down-thrown valleys (**FIGURE 11.21**).

Fold-and-Thrust Mountains

Although volcanic and fault-block mountains are common, the largest mountain belts are formed by tectonic forces that deform and raise the crust. These tectonic belts are known as **orogenic belts**, which involve mountain building by folding and thrusting, a process referred to as **orogeny**. Fold-and-thrust mountains may occur as a single range (the Ural Mountain Range in Eurasia) or as a belt of several mountain ranges (the Rocky Mountains). In general, fold-and-thrust mountains require tectonic convergence (plate collision), the only force great enough to generate the enormous stresses that produce this scale of rock deformation and uplift.

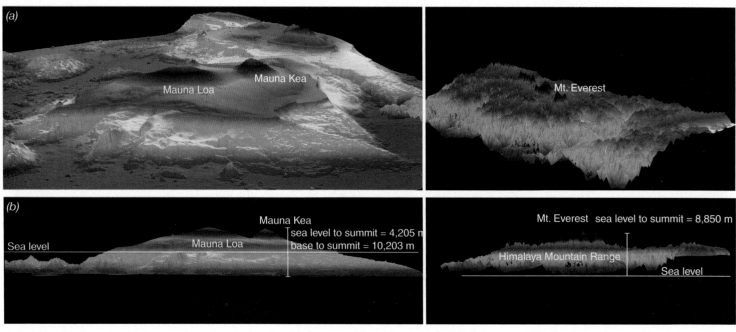

FIGURE 11.20 *(a)* The Mauna Loa shield volcano is the most massive mountain on the planet with a total volume of 40,000 km³. The 3-D mass of Mount Everest is shown to the right for comparison. *(b)* Mauna Loa's neighbor, Mauna Kea, is the tallest mountain in the world, with a total height of 10,203 m measured from the sea floor at its base to its summit, making it taller overall than Mount Everest (8,850 m) whose base is considered to be sea level where it rises up from the Ganges Plain to the south.

(?) What is the source of magma in Hawaiian volcanoes?

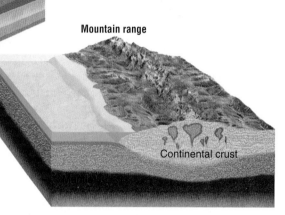

FIGURE 11.21 Fault-block mountains form in areas where the crust is stretched. Stretching leads to normal faulting, which produces blocks of crust that are displaced downward, making valleys. Blocks that stand higher, or are displaced upward, create mountain ranges.

(?) What type of stress produces fault-block mountains?

Newly formed fold-and-thrust mountains are subject to weathering and erosion, which ultimately reduce their mass because sediments are transported to other locations by wind and running water. The crust compensates for this loss of mass by floating higher in the mantle (geologists use the term *rebound*), bringing deeper levels in the crust to relatively higher levels, a process known as **isostasy** (from the Greek words *iso,* "equal," and *stasis,* "standing"). Because of isostasy, the crust continually rebounds, bringing new rock that was originally deeper in Earth's crust to Earth's surface, where it is worn down by weathering and erosion; this process is referred to as *exhumation*. This can eventually wear down most of topography of a mountain belt, leaving a low-relief *peneplain* in its place.

Test your understanding of mountain belts with the "Critical Thinking" box titled "Mountain Building."

Isostasy

Like an iceberg that must have 90 percent of its mass below the ocean surface in order to support 10 percent of its mass above, mountains usually require deep roots,[2] which add to the overall buoyancy of continental crust. These deep roots are necessary to support the mass of lofty mountain belts.

According to the principle of isostasy, mountain belts typically stand high above the surface because they have thick, buoyant roots of felsic crust that extend into the mantle. Isostasy also means that the vertical position of the crust must perpetually change in order to adjust to changes in mass, a process termed *isostatic adjustment.* For instance, constant erosion of rock from mountains into the sea constitutes a shift of mass, and a deeper level of a mountain belt will rise closer to the surface to compensate for it (**FIGURE 11.22**). Hence, as erosion wears mountains away, the deeper levels rise in response to the loss of mass and the roots shorten until an equilibrium is achieved approaching the average elevation of continents (about 840 m). Eventually all that remains of the mountain belt is the exposed core of highly deformed and metamorphosed rock.

 Expand Your Thinking—Describe a common example of isostasy that you encounter in your life.

[2]Deep roots are not always required. A hot, buoyant mantle near a convergent margin can support high-standing mountain belts without deep roots, which is the case for the Canadian Cordillera.

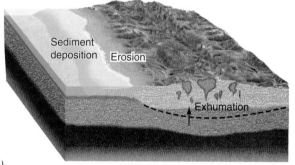

(a)

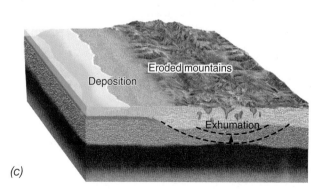

(b)

(c)

FIGURE 11.22 As erosion wears away the high elevation of a mountain belt, the loss of mass is compensated by vertical adjustment of the crust and shortening of the root.

(?) Which is more dense, felsic crust or ultramafic mantle? How is this fact relevant to the process of isostasy?

? CRITICAL THINKING

MOUNTAIN BUILDING

There are three common types of mountain-building processes (FIGURE 11.23). Volcanic mountains form at locations where magma intrudes into the crust: at mid-plate hotspots, divergent plate boundaries, or convergent plate boundaries. Fault-block mountains form by extension and thinning of the crust. Fold-and-thrust belts develop where continental plate convergence leads to large-scale reverse and thrust faulting caused by compressional stress.

Please work with a partner and answer the following questions.

1. Which of the three mountain-building environments is most likely to develop folding? Why?
2. How do basic tectonic processes differ between volcanic and fold-and-thrust mountain belts?
3. Why are fault-block mountains characterized by normal faulting and fold-and-thrust mountains by reverse and thrust faulting? Describe how this difference is related to the type of stress acting at each setting.
4. Describe physical differences between all three types of mountains. Which of the three mountain-building processes produces the highest mountains? The most extensive mountain belt? Describe other differences.
5. Refer to a map of the world. List the names of two examples of each of the three types of mountains.
6. Label the tectonic- and mountain-building processes in Figure 11.23.
7. Based on topography only, what criteria would you use to identify each of the three types of mountains?
8. Do you agree that there should be a sequence of volcanic rocks in the fold-and-thrust belt? Explain your reasoning.
9. Explain why it is unlikely that a fault-block system and a fold-and-thrust system would develop simultaneously in the same place.

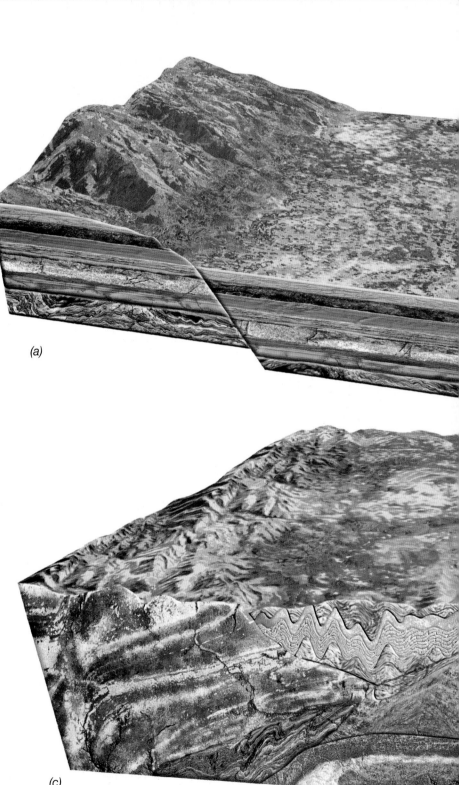

(a)

(c)

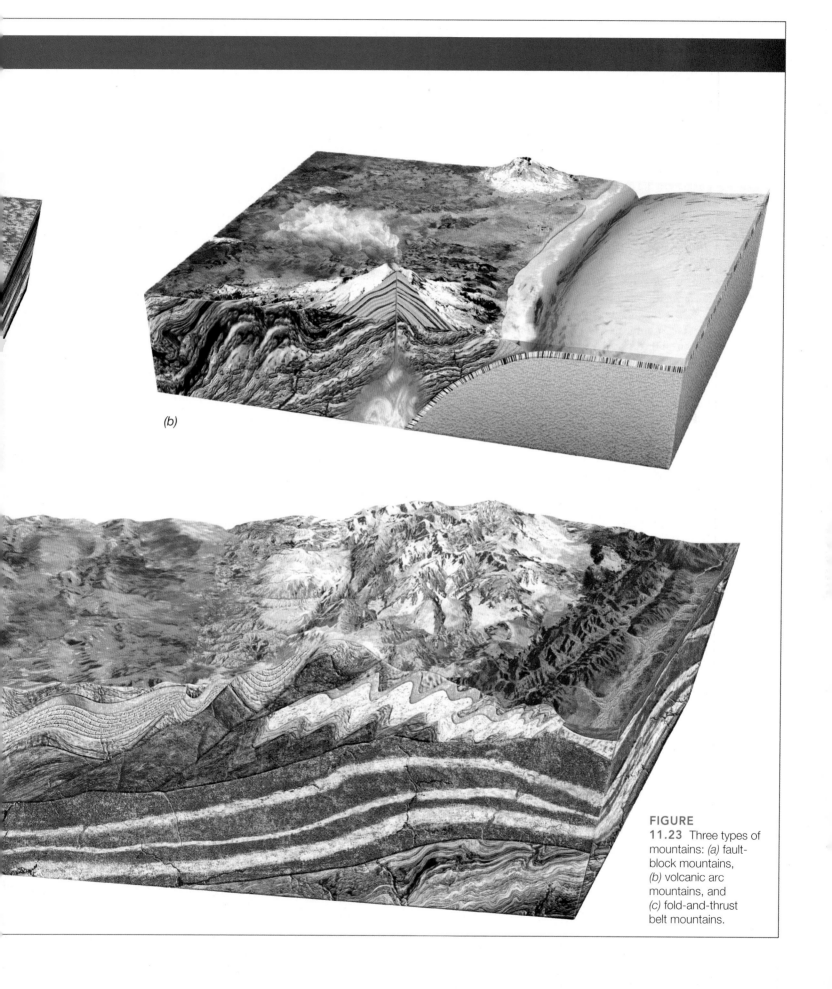

(b)

FIGURE
11.23 Three types of mountains: (a) fault-block mountains, (b) volcanic arc mountains, and (c) fold-and-thrust belt mountains.

11-8 Volcanic Mountains Are Formed by Volcanic Products, Not by Deformation

LO *11-8 Describe the geology of the Cascade Mountains.*

Volcanic mountains are formed not by deformational processes but by volcanic products on Earth's surface. Since we reviewed volcanic processes thoroughly in Chapter 6, here we will discuss the history of the most active volcanic range in the conterminous Canada and United States: the Coast and *Cascade Mountains* in southwestern British Columbia, and the Cascade Range of Washington, Oregon, and northern California. Herein, we'll simply refer to them as the Coast-Cascade Ranges.

The Coast-Cascade Ranges consist of several active volcanoes as well as many apparently dormant volcanic centres that were active during late Neogene and Quaternary Periods. This volcanic arc developed above a subduction zone reaching from southern British Columbia to northern California (see the "Geology in Action" box titled "Strain Buildup at the Cascadia Subduction Zone"). It is formed by the subduction of the Explorer, Juan de Fuca, and Gorda plates (fragments of the former Farallon plate) beneath the western edge of the North American plate (**FIGURE 11.24**).

Unlike the situation in typical subduction zones, no trench is present along the coast of the Coast-Cascade Ranges. Instead, *accreted terranes* and the accretionary wedge have been uplifted to form a series of coastal ranges and mountain peaks. Accreted terranes are blocks of crust with relatively low density (andesite or granite). They are former oceanic and continental slivers like Japan, and former island arcs like Indonesia that have converged onto the North American plate. They are covered in more detail in Chapter 15 ("The Geology of Canada"). As the lithospheric plate to which they belong subducts, they are scraped off and accrete onto the leading edge of the overriding plate. Each terrane has a unique genesis and geologic history; well over 20 such terranes have been identified.

Inland from these coastal mountains are a series of valleys: Fraser, Puget, Willamette, and Shasta. Some of these valleys were former forearc basins that were uplifted (as part of the accreted-terrane process) during the growth of the western margin of the continent as it has moved to the west. Major cities are located in these valleys: Vancouver, Seattle, and Portland.

The Coast-Cascade Ranges include more than a dozen large volcanoes. Although they share some general characteristics, each has unique geologic traits. There are two regions of current volcanic activity. Volcanoes are most active in Washington and northern Oregon, and a second region of eruptions is found in northern California. In contrast, central and southern Oregon is calm, as is southern British Columbia. The locations of the calm zones correspond to the positions of fracture zones that offset the Explorer, Juan de Fuca, and Gorda oceanic ridges.

(?) What will happen in the Coast-Cascade Ranges when the Explorer, Juan de Fuca, and Gorda plates are completely subducted?

FIGURE 11.24 *(a)* The volcanic arc of the Pacific Northwest is built of magma generated by partial melting of the Explorer, Juan de Fuca, and Gorda plates. Red arrows indicate relative plate motions. *(b)* Fifteen principal volcanic centres are found within the Coast-Cascade Ranges of the Pacific Northwest. *(c)* The Coast and Cascade Ranges are situated within the Cascade volcanic arc.

(a)
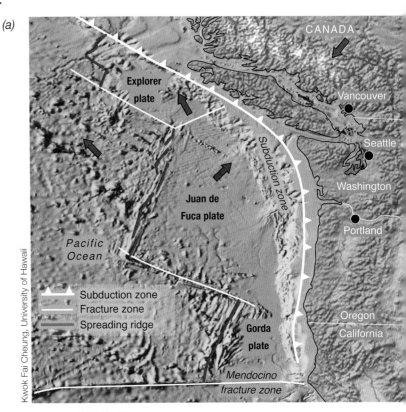
Kwok Fai Cheung, University of Hawaii

(b)
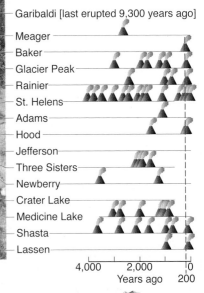

(c)
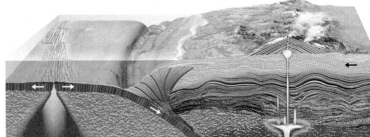

The magma of the Cascade volcanic arc is primarily generated by partial melting in the upper sediment-laden portion of the subducting oceanic plates. The typical lava of Cascade eruptions is andesite, which is characteristic of subduction zones around the world. Other types of lava that are present range in composition from basalt to rhyolite; they are products of various combinations of partial melting and igneous evolution. Each major Cascade volcano has a distinct "signature" in terms of its lava composition.

Eruptions of basalt tend to be relatively calm events dominated by lava flows. Exceptions occur where such magmas encounter substantial groundwater (under snow cover or glaciers), which may create steam-driven *phreatomagmatic explosions.* Andesite and rhyolite magmas are more viscous and tend to develop higher gas pressures. Such magmas may explode violently in catastrophic eruptions that release immense volumes of tephra, ash, and pumice. Mount Meager, Glacier Peak, Mazama (Crater Lake), and Mount St. Helens (Figure 11.24b) have experienced this type of eruption.

Recent eruptions, particularly that of Mount St. Helens in 1980, have provided important clues for better understanding of ancient volcanism in the Cascade system. One result of the 1980 eruption was recognition of the importance of landslides in the development of volcanic landscapes. A huge section on the north side of Mount St. Helens slid away and created chaotic topography many miles from the volcano itself. Accompanying the eruption were pyroclastic flows and mudflows that swept across the landscape. Similar events occurred at Mount Meager, Mount Shasta, and other Cascade volcanoes in prehistoric times.

(?) **Expand Your Thinking**—Explain why different Cascade volcanoes would have different magma types.

GEOLOGY IN ACTION

STRAIN BUILDUP AT THE CASCADIA SUBDUCTION ZONE

Subduction zones are dominated by three types of stress. Compressional stress develops at locations where two plates collide, with one (the denser of the two) sliding below the other. This sliding is called *interplate slip.* Tensional stress also develops. The broad region of bending where one plate curves below another is characterized by extension. Shear stress is high at the surface of the subducting plate where it slides below the overriding plate.

Because the subduction process may not always be smooth, it is possible for two plates to become "locked." That is, interplate slip ceases as the top of the subducting plate and the bottom or front of the overriding plate refuses to move. In these cases, strain accumulates as pressure builds around the locked portion of the subduction zone until it unlocks.

Geologists hypothesize that a portion of the Cascadia subduction zone is locked. This zone is the area where the Juan de Fuca plate subducts beneath the North American plate along the coast of southern British Columbia, Washington, and Oregon. Because compressional stress is active at the subduction zone and interplate slip has ceased, strain is accumulating in the form of crustal shortening and vertical uplift of the ground surface (**FIGURE 11.25**). The land above the locked region is deforming elastically upward to an extent that can be measured using a network of global positioning system (GPS).

Because the deformation is still in an elastic phase, Hooke's law predicts that it is fully reversible. This means that when the locked region once again experiences interplate slip, the land surface will elastically recover, probably rapidly, producing a large and dangerous earthquake. In fact, scientists have determined from a combination of geologic evidence and Japanese diaries that the last time the locked crust was released, in 1700, it rebounded so rapidly that it caused a massive earthquake and a *tsunami* (a seismic sea wave) that flooded not only the Cascadia coast but the coast of Japan as well (**FIGURE 11.26**). The same type of event is expected to occur again, but exactly when is unknown.

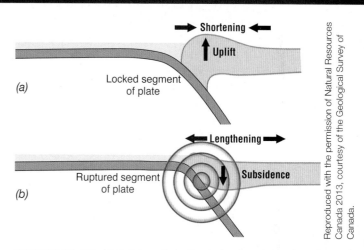

FIGURE 11.25 (a) Buildup of elastic strain along locked portions of the Cascadia subduction zone causes the crust to shorten and the land surface to rise. (b) When interplate slip occurs, the lock ruptures, the crust extends, and the surface subsides. This event may happen rapidly, causing a damaging earthquake and producing a tsunami. Source: Natural Resources Canada.

Reproduced with the permission of Natural Resources Canada 2013, courtesy of the Geological Survey of Canada.

FIGURE 11.26 Researchers have determined that giant earthquakes have occurred along the Cascadia subduction zone. The most recent, on January 26, 1700, caused a tsunami that flooded portions of the North American coast and crossed the Pacific Ocean, causing damage to coastal Japan. This is a computer model of the 1700 tsunami as it crossed the Pacific.

11-9 Crustal Extension Formed the Basin and Range Province

LO 11-9 *Describe the origin of the Basin and Range province.*

The *Basin and Range province* of the western United States and Mexico (**FIGURE 11.27**) experiences crustal extension related to plate movement. It is characterized by unique topography that reflects normal faulting of the crust. Blocks of faulted crust displaced downward form linear valleys, known as *grabens*. These valleys are separated by linear mountain ranges of high-standing blocks, known as *horsts*. In the U.S. portion of the Basin and Range province, horsts form approximately 400 short, linear mountain ranges separated by arid valleys. This topography is typical of crust that is being pushed upward from below, causing regional extension.

Sedimentology

Within many of the graben valleys are lakes that are filled with water in the winter and spring but evaporate in summer and fall. Broad sedimentary deposits of precipitated minerals result from the evaporation, and in the wet season, these deposits are buried by clays, silts, and layers of algae that bloom in the wet conditions. Because of continued faulting, the valleys drop down over time, and the layered deposits become very thick. This is the sort of environment that produces *oil shale*, a fuel source that is potentially valuable but environmentally destructive to extract (**FIGURE 11.28**).

The entire Basin and Range province has an arid climate because of the *rain shadow* created by the high Sierra Nevadas to the west, which block and collect moisture coming off the Pacific Ocean.

However, in winter and spring, short, intense bursts of rainfall sweep tonnes of mud, sand, and gravel onto the valley floors. Over hundreds of thousands of years, the valleys fill with sediment as the ranges are worn down. Slowly the valley floors bury the slopes of the mountains in their own sediment until a new round of faulting rejuvenates the landscape and raises the fault-block mountains once again.

Tectonics

The *Sierra Nevadas* of California mark the western border of the Basin and Range province, and another province, called the *Colorado Plateau*, marks the eastern border. Nearly the entire southwestern portion of the United States, including all three of these regions, has been uplifted, and the crust extended (or stretched), over the last 20 million years by plate tectonics. This tectonic activity has led to a widespread pattern of normal faulting and the formation of fault-block mountain ranges.

Regional uplift of the western United States is the result of the western movement of the North American plate and its interaction with plates to the west. Approximately 30 million years ago, the central California coastline was an active subduction zone where the now-extinct *Farallon plate* subducted beneath the North American plate (**FIGURE 11.29**). Compressional stress associated with plate convergence caused crustal thickening east of the subduction site (western North America). This led to uplift of the Colorado Plateau

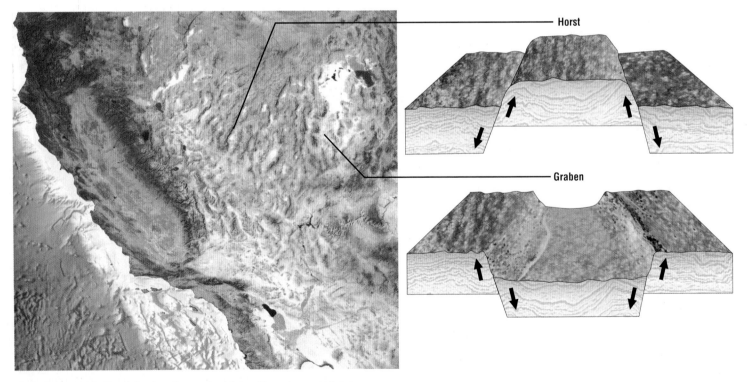

FIGURE 11.27 Crustal extension under Idaho, Nevada, and Utah has formed the Basin and Range province. The province is characterized by normal faulting that creates numerous horsts and grabens.

 What is the cause of crustal extension in the Basin and Range province?

FIGURE 11.28 The United States has the world's largest oil shale deposits, and these could potentially be an important fuel source. However, surface mining of oil shale, similar to the excavation of oil sands in Fort McMurray, Alberta (shown here), is environmentally destructive.

(?) Why are oil shale deposits so thick in the Basin and Range province?

and downward erosion of the Colorado River and other stream systems in the region, which remained at their original elevation while the land around them rose. The Grand Canyon formed in this way.

About 20 million years ago, subduction ended in the area of Los Angeles as remaining fragments of the Farallon plate became separate plates in their own right: the Explorer, Juan de Fuca, and Gorda plates to the north and the Cocos plate to the south. (These plates still exist and are subducting under different portions of the North American plate.) The former subduction zone was replaced by the present-day strike-slip faulting of the San Andreas transform boundary.

Over the next 20 million years, the transform boundary extended to the north and south, slowly replacing the subduction process along most of the California coast and adjoining north-south areas. Although some aspects of this transition are poorly understood, it is thought that the former Farallon

(?) What is the source of heat that extends the crust in the Basin and Range area?

FIGURE 11.29 (a) Approximately 30 million years ago, the western boundary of the North American plate experienced crustal thickening and compression related to subduction of the Farallon plate. (b) Subduction of the rift zone or spreading centre that formerly marked the boundary of the Farallon and Pacific plates created what is termed a *slab window*, which produced a region of decompression melting and high heat flow under the North American continent. This lifted the continental crust, generating tension and producing the Basin and Range province as well as some volcanic action.

plate remained intact after subducting, perhaps at a very shallow angle, beneath the North American plate. When subduction ceased, the last portion of the Farallon plate under the continent became detached and may have scraped along the underside of the North American plate, which continued to migrate to the west. This process may continue today.

When the subducted Farallon plate became detached, an opening (i.e., *slab window*) was created behind it in the upper mantle under the southwestern United States. This was the rift zone marking the former boundary of the Pacific and Farallon plates. This region was characterized by decompression melting, high heat flow, and plutonism, raising the crust and creating a tensional stress environment. As upwelling magma pushed into the base of the crust, it arched upward and stretched and thinned, creating a broad rift that became the Basin and Range province of Nevada, Utah, and parts of Idaho, New Mexico, Arizona, and Mexico. This tensional environment created the hundreds of fault-bounded horsts and grabens that characterize the rugged modern topography.

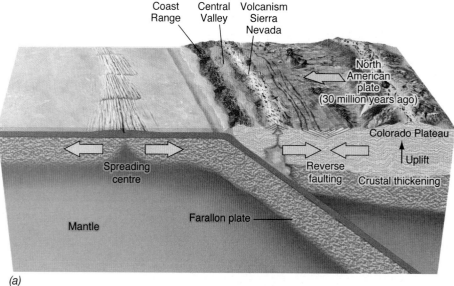

(a)

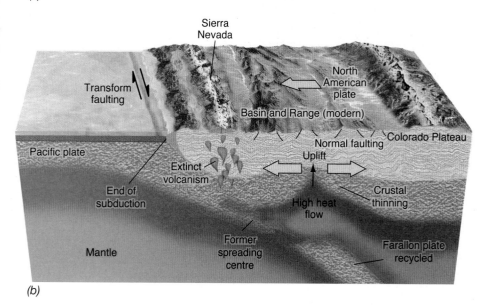

(b)

(?) **Expand Your Thinking**—A subducted plate under Nevada makes the Basin and Range province, but under Oregon, it makes the Cascades Range. Why are there different geologic products from the same (subduction) process?

11-10 Fold-and-Thrust Belts Produce Some of the Highest and Most Structurally Complex Mountain Belts

LO 11-10 *Describe the origin of fold-and-thrust belts.*

Large compressional stresses are generated in the crust at convergent margins, especially where continents collide. The Himalayan Mountains (the highest on Earth) were formed when the Indian–Australian plate collided with the Eurasian plate—a process that is still underway. (We discussed the origin of the Himalayas in Section 3-7.) See the "Geology in Action" box titled "Raising Everest" for more on this subject.

The Appalachian Belt

The Appalachian mountain belt extends from Alabama through Maine in the United States and continues across southern Quebec and the Maritime provinces of Canada all the way to the eastern tip of Newfoundland. The Appalachian mountain belt can be subdivided into seven major lithotectonic zones (**FIGURE 11.30a**), meaning the rocks within each zone were formed or deposited within a similar tectonic environment within a defined period of time. These lithotectonic zones can be grouped geographically as belonging to the southern or northern Appalachians based on their position relative to the New York promontory. They can also be grouped based on their interpreted paleogeographic birth place: those terranes that were formed on or near the ancient North American continent are referred to as *Laurentian*; the *Piedmont* and *Dunnage* zones contain terranes formed within and around the Iapetus Ocean, precursor to the Atlantic; and *Carolinia, Avalonia, Ganderia,* and *Meguma* zones that were originally a part of the Gondwana supercontinent.

The modern landscape along the east coast of North America is greatly influenced by the tectonic history that formed the Appalachian mountain belt. In the United States, geologists divide the southern Appalachians into five main *physiographic provinces*[3] (**FIGURE 11.30b**):

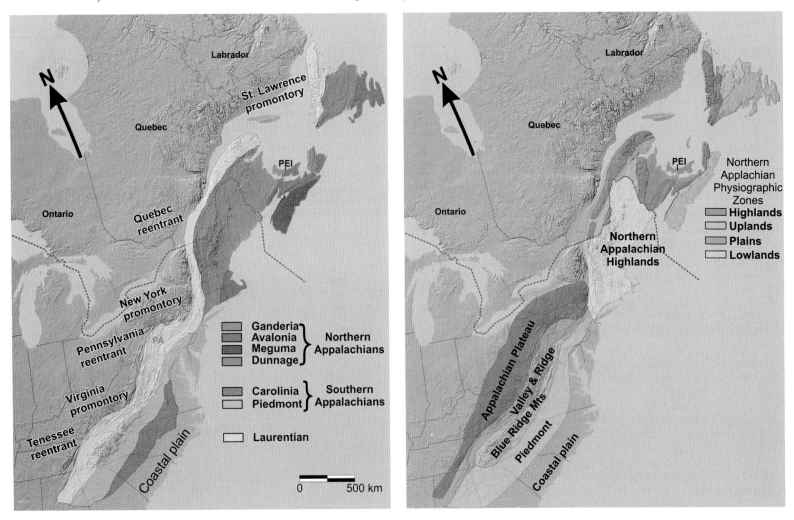

(a)   *(b)*

FIGURE 11.30 *(a)* Major lithotectonic zones of the Appalachian mountain belt. *(b)* The main physiographic provinces (United States) and zones (Canada) within the Appalachian mountain belt.

 What name do geologists give the continental fragments and island arcs that collided with Africa and North America?

[3]These are regions with similar geologic structures associated with landforms and topography that differ significantly from adjacent regions.

1. the *Coastal Plain,* extending from the seaward edge of the continental shelf to the foothills of the Appalachians;

2. the *Piedmont* (French for "foothills"), reaching from Alabama to New York and forming the *easternmost* portion of the mountainous topography;

3. the *Blue Ridge,* a narrow ridge composed of Precambrian rocks separating the Piedmont from the Valley and Ridge province to the west;

4. the *Valley and Ridge,* consisting of Paleozoic sedimentary rocks that have been thrusted and folded into large anticlines and synclines; and

5. the *Appalachian Plateau,* which lies above gently folded and tilted Paleozoic sedimentary strata.

In Canada, the physiography of the northern Appalachians can be divided into a series of highlands, uplands, plains, and lowlands (Figure 11.30b), and can be broadly categorized into the following physiographic zones:

1. Newfoundland, which includes the *Atlantic Uplands* on the east side, the *Central Lowlands,* and the rugged *Newfoundland Highlands* on the west side;

2. Nova Scotia, which includes the *Nova Scotia Highlands* across the north, and the *Nova Scotia Uplands* and *Annapolis Lowlands* to the south;

3. New Brunswick and Prince Edward Island, which includes the *New Brunswick Highlands* on the west, south, and east sides of New Brunswick, forming a horseshoe enclosure of the intervening *Maritime Plain,* comprising central New Brunswick and all of PEI; and

4. Quebec, which consists of the *Chaleur Uplands,* bounded to the west by the rugged *Quebec Highlands,* which includes the Notre Dame, Chic-Choc, and Sutton Mountains.

The story of how the Appalachians formed began nearly 1,100 million years ago during the Proterozoic Eon when convergent tectonics created an early supercontinent named *Rodinia* (similar in origin to Pangaea). The Adirondack Mountains of New York date from this time. By the start of the Phanerozoic Eon, Rodinia began a process of rifting as the continents slowly separated. Over time, the early Appalachians eroded, and sediments were deposited in a widening rift basin—the predecessor of the Atlantic, known as the *Iapetus Ocean.* By the Ordovician Period, the Iapetus basin, which had been widening for hundreds of millions of years, began to close as Africa, North America, and

Europe moved together. This gave rise to the first of several orogenies that would build the Appalachian fold-and-thrust belt.

The tectonic history of the modern Appalachians (**FIGURE 11.31**) is divided into three main phases of mountain building, or orogenies.

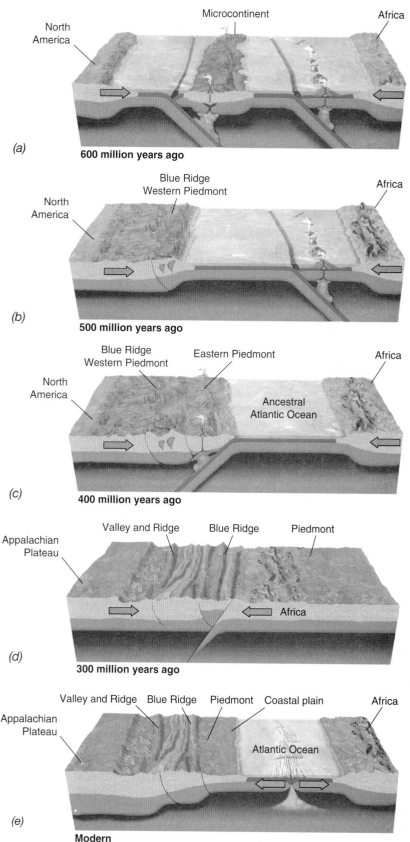

FIGURE 11.31 In the Proterozoic Eon, the Iapetus Ocean opened during rifting of Rodinia (an early supercontinent). During the latest part of the Proterozoic Eon and into the Paleozoic era, the Iapetus began to close, *(a)* squeezing a series of continental fragments and island arcs between the North American and African continents. Each collision was a mountain-building event: *(b)* the Taconic Orogeny, *(c)* the Acadian Orogeny, and *(d)* the Alleghenian Orogeny. *(e)* Mountain building ended when Pangaea rifted and the modern Atlantic Ocean was born.

Each orogeny begins with accumulation of thick marine sediments and volcanic deposits, continues with deformation and uplift of mountains, and culminates with the calming of tectonic conditions. A particular result of the crustal uplift associated with each orogeny was production of sediment in great volumes due to rapid erosion of the steep highlands. Thus, each phase is marked by the buildup of a so-called *clastic wedge* filling shallow seas on the continental (west) side of the Appalachians.

- The *Taconic Orogeny* was the first important tectonic activity in the Appalachians and took place during the Ordovician Period. Thrusting and folding occurred mainly in the northern portion of the mountain belt.

- The *Acadian Orogeny* was the major orogeny of the northern Appalachians. It occurred in the Devonian Period and was centred in New England and the Maritime provinces.
- The *Alleghenian Orogeny* was the major orogeny responsible for forming the southern Appalachians. It ended during the Pennsylvanian Period. A clastic sedimentary wedge spread over western Pennsylvania, West Virginia, Kentucky, and Tennessee.

All three orogenies are interpreted as collisions that occurred during the closing of the Iapetus Ocean between North America, Europe, and Africa as part of the formation of Pangaea. Seafloor spreading, subduction-zone volcanism, and clastic wedges were all part of a long, drawn-out sequence of tectonic events.

GEOLOGY IN ACTION

RAISING EVEREST

Long known as the highest peak on the planet, Mount Everest (**FIGURE 11.32**) is called Chomolungma ("goddess mother of the world") in Tibet and Sagarmatha ("goddess of the sky") in Nepal.

The Indian mathematician and surveyor *Radhanath Sikdar* was the first to identify Mount Everest as the world's tallest peak. In 1852, using trigonometric calculations, he measured the elevation as 29,002 ft. (8,840 m) above sea level. (His actual measurement was 29,000 ft. exactly, but it was felt that this number would be viewed as a rounded adjustment, so an arbitrary 2 ft. was added to it.) For this accomplishment, some believe that the peak should be named after Sikdar, not Sir George Everest, who was the surveyor general of India in the early nineteenth century and is not known to have actually laid eyes on the great peak. For more than a century, this measurement remained the mountain's officially accepted height. But in 1954, the Indian surveyor B. L. Gulatee calculated a new height by averaging 12 measurements from around the mountain. His calculations adjusted the height upward by 8 m to 8,848 m.

On May 5, 1999, a team of climbers armed with a state-of-the-art global positioning system (GPS) reached the summit of Everest. Climbers Pete Athans and Bill Crouse were supported by five Sherpas (Nepalese guides), who carried the GPS equipment and an extra supply of oxygen. The team climbed through the night so that they would reach the top and make their measurements during the warmest part of the day. Based on their calculations, Athans and

FIGURE 11.32 Mount Everest.

Jodi Cobb/National Geographic Creative

Crouse revised the height upward by 2 m, to 8,850 m. Since then, there has been no significant change in this altitude. However, the inexorable Indian plate seems to be shifting Everest's horizontal position steadily and slightly to the northeast at a rate of about 6 cm a year.

 Expand Your Thinking—What does the term "fold-and-thrust belt" describe?

LET'S REVIEW "GEOLOGY IN OUR LIVES"

Now that you have finished the chapter, "Geology in Our Lives" will have taken on new meaning for you. Let us review it: Stress (pressure) in the crust causes rock to fold and break. This produces earthquakes and landslides, builds mountains, and challenges our ability to find and extract geologic resources. Past and present plate boundaries are regions where stress tends to be the greatest and the crust has the most complicated structure. Understanding how stress affects the crust improves our access to oil and coal, groundwater, metallic ores, and other geologic resources that we depend on. Studying the structure of the crust advances our understanding of geologic hazards and it enhances our knowledge of Earth's history.

Stress produces strain in rock. Strain is exhibited by deformation, which may initially be elastic (fully reversible). However, if the elastic limit is exceeded, permanent deformation results and rocks exhibit ductile strain. Eventually fracture may occur—the usual outcome—resulting in sudden, damaging earthquakes. The greatest strain develops at plate boundaries, and these are regions of mountain building.

The next chapter, Chapter 12, "Earthquakes," looks into the source and impacts of earthquakes, a cause of severe damage and loss of life over the centuries. Every third or fourth day, on average, an earthquake occurs somewhere in the world that damages buildings, roads, and human communities. The energy released by earthquakes is a direct result of strain.

STUDY GUIDE

11-1 Rocks in the crust are bent, stretched, and broken.

- Rocks may be bent, pulled, and fractured. This process is called **deformation**, and the forces that cause deformation are called **stresses**. Three types of directed stress act on rock and cause it to deform: **tensional stress**, **compressional stress**, and **shear stress**. In general, these stresses originate at plate boundaries.

11-2 Strain takes place in three stages: elastic deformation, ductile deformation, and fracture.

- **Strain** is the change in shape and/or volume of a rock caused by stress. When rocks undergo strain, they pass through three successive stages of deformation: elastic deformation, ductile deformation, and **fracture.**

- The type of strain experienced by a rock depends on temperature, confining pressure, rate of strain, presence of water, and composition of the rock. When the rock passes its elastic limit, it experiences ductile deformation. As stress builds, the rock fractures. In the crust, many such fractures develop into **faults**: zones or planes of breakage across which layers of rock are displaced relative to one another.

11-3 Strain in the crust produces joints, faults, and folds.

- Stress causes deformation in the form of ductile strain (folds) or fracture (joints and faults). **Joints** are fractures that display no net displacement parallel to the crack. Faults are formed when brittle rocks fracture, and there is a distinct offset parallel to the plane of breakage. Rocks are most likely to **fold** when strain accumulates slowly, there is high temperature and pressure, the rock is composed of minerals that tend to be ductile, and fluids are present.

- Researchers create maps of geologic structures in order to better understand the crustal architecture, stress history, and geologic evolution of a region. To show crustal structure on a map, geologists use a simple symbol to display the spatial orientation of the layers of rock that make up folds, faults, and joints. This symbol has two parts, called the **strike** and the **dip**. The strike of a body of rock is a line representing the intersection of that inclined feature with the plane of the horizon. The dip gives the angle below the horizontal of a geologic feature. It can be thought of as the direction in which water would flow if rain fell on the fault surface or rock layer.

11-4 Dip-slip and strike-slip faults are the most common types of faults.

- **Dip-slip faults** may be either **normal faults** or **reverse faults**; a special type of reverse fault is known as a **thrust fault**. Normal faults develop in brittle rocks that are exposed to tensional stress. Reverse faults are caused by compressional stress; the hanging wall (which sits above the fault plane) is displaced upward along the fault plane relative to the footwall (which sits below the fault plane). A thrust fault is a special case of a reverse fault in which the dip of the inclined fault plane is less than 45°.

- **Strike-slip faults** occur where the relative motion on a typically steep to vertical fault plane is horizontal. There are left-lateral strike-slip faults and right-lateral strike-slip faults.

11-5 Rock folds are the result of ductile deformation.

- Folds are the result of compressional stresses acting over long periods. The simplest folds are the **anticline**, the **syncline**, and the **monocline**.

- Complex folds include asymmetrical and overturned anticlines and synclines, plunging anticlines and synclines, groups of anticlines and synclines, domes, basins, and contorted folds.

11-6 Outcrop patterns reveal the structure of the crust.

- To unravel the geometry and history of the crust, a structural geologist analyzes **outcrops.** The goal is to determine the spatial orientation of rock layers and, from this, the relative age of the layers and the history of deforming stresses experienced by the rock.
- A layer of tilted rock can be represented with a plane. The orientation of that plane in space can be defined using strike-and-dip notations. The dip is described by a compass direction (0° to 360°) and an angle to the horizon that varies from 0° (horizontal) to 90° (vertical).
- A **geological map** shows the topography of an area, the lithology and age of rock units exposed at the surface or shallowly buried under loose sediment and soil, and the rock units' strike and dip.
- Patterns of rock reveal the orientation of crustal structure. For folded rocks, parallel strips of outcrops usually indicate folds with horizontal axes. An outcrop pattern of Vs indicates plunging folds. A V that opens in the direction of the plunge is a syncline, while one that closes in the direction of the plunge is an anticline.

11-7 Mountain building may be caused by volcanism, faulting, and folding.

- The three common types of mountains are **volcanic mountains, fault-block mountains,** and **fold-and-thrust mountains.**
- When land rises during mountain building, it is immediately attacked by weathering and erosion. As the mass of the mountain is reduced, its root compensates by rising higher in the mantle. This process is called **isostasy.** Due to isostasy, mountain belts stand high above the surface because they have thick roots of felsic rock that extend deep into the underlying mantle. Also, the vertical position of the crust must change continuously in order to adjust to changes in mass; this is referred to as isostatic adjustment. For instance, constant erosion of rock from mountains into the sea constitutes a shift of mass that is compensated for by a slight rise in the mountain belt, although its actual height may stay the same or even be lower.

11-8 Volcanic mountains are formed by volcanic products, not by deformation.

- Volcanic mountains are formed not by deformational processes but by volcanic products on Earth's surface.
- The Cascade Mountains of British Columbia, Washington, Oregon, and California consist of several active volcanoes as well as many volcanic centres that were active during the Neogene and Quaternary Periods. This volcanic arc developed above a subduction zone that stretches from southern British Columbia to northern California and was formed by the recycling of the Juan de Fuca and Gorda plates beneath the western edge of the North American plate.

11-9 Crustal extension formed the Basin and Range province.

- Fault-block mountain building results from tensional stress that stretches areas of continental crust. Mountains formed in this way are found throughout the southwestern United States in the Basin and Range province. As crust is stretched, it develops normal faults that may build horst and graben structures.

11-10 Fold-and-thrust belts produce some of the highest and most structurally complex mountain belts.

- Large compressional stresses are generated in the crust at convergent margins, where tectonic plates collide. Rocks located in the collision zone between two tectonic plates are folded, faulted, and thrust faulted as part of the process of crustal thickening. This process pushes peaks upward and builds deep roots to form fold-and-thrust mountains. The Appalachian mountain belt was formed by this same fold-and-thrust process.

KEY TERMS

anticline (p. 282)
antiform (p. 282)
compressional stress (p. 272)
deformation (p. 272)
dip (p. 277)
dip-slip faults (p. 278)
fault-block mountains (p. 286)
faults (p. 274)
fold-and-thrust mountains (p. 286)
folds (p. 276)
fracture (p. 274)

geological map (p. 285)
isostasy (p. 287)
joints (p. 276)
monocline (p. 282)
mountain belts (p. 286)
normal faults (p. 278)
orogenic belts (p. 287)
orogeny (p. 287)
outcrops (p. 276)
reverse faults (p. 278)
shear stress (p. 272)

strain (p. 274)
stresses (p. 272)
strike (p. 277)
strike-slip faults (p. 278)
structural geology (p. 284)
syncline (p. 282)
tensional stress (p. 272)
thrust fault (p. 278)
volcanic mountains (p. 286)

ASSESSING YOUR KNOWLEDGE

Please answer these questions before coming to class. Identify the best answer.

1. The three types of stresses are
   a. anticline, syncline, and monocline.
   b. folds, faults, and fractures.
   c. compressional, tensional, and shearing.
   d. elastic, ductile, and brittle.
   e. volcanic, fold and thrust, fault-block.

2. Strain is
   a. a hotspot.
   b. only found where rock has been partially melted, usually in a subduction zone.
   c. a condition of being sheared.
   d. a change in shape and/or volume of a rock caused by stress.
   e. None of the above.

3. The three types of strain are
   a. elastic, ductile, and brittle.
   b. fracture, folding, and faulting.
   c. bending, breaking, and stretching.
   d. compressional, tensional, and folding.
   e. convergent, divergent, and transform.

4. A reverse fault is formed when
   a. the footwall moves upward relative to the hanging wall block.
   b. the fault plane shifts along the line of strike.
   c. the hanging wall is displaced upward along the fault plane.
   d. tensional stress causes brittle strain.
   e. All of the above.

5. An anticline is the product of
   a. tensional stress.
   b. compressional stress.
   c. shear stress.
   d. elastic stress.
   e. Anticlines are not formed by stress.

6. How do dip-slip and strike-slip faults differ from one another?
   a. They describe different directions of movement along a fault plane.
   b. Each is a different type of fold.
   c. Only one of them involves ductile behaviour.
   d. They are not different—they are the same thing.
   e. Both are the result of ductile strain.

7. A V outcrop pattern may indicate
   a. an upright fold.
   b. a dip-slip fault.
   c. shear stress.
   d. a plunging fold.
   e. that no folding has occurred.

8. Why does the formation of a fold require ductile behaviour?
   a. Ductile behaviour is necessary because a rock must plastically deform around the bend.
   b. Ductile behaviour is necessary because formation involves a reverse fault.
   c. It does not require ductile behaviour.
   d. A fold only forms under shear stress.
   e. A fold is a type of dip-slip fault.

9. What are strike and dip?
   a. descriptions of partial melting
   b. the result of volcanic mountain building
   c. symbols on a map indicating the orientation of a feature
   d. the angle that a fault is folded after elastic behaviour
   e. None of the above.

10. What types of faults are formed by tensional, compressional, and shear stresses, respectively?
    a. reverse fault, strike-slip fault, normal fault
    b. normal fault, strike-slip fault, reverse fault
    c. strike-slip fault, normal fault, normal fault
    d. normal fault, reverse fault, strike-slip fault
    e. All of the above.

11. Which of the following are equivalent to stress and strain?
    a. force and deformation
    b. uplift and erosion
    c. breaking and bending
    d. action and dissolving
    e. pushing and pulling

12. Volcanic mountains are
    a. found only at plate boundaries.
    b. found at plate boundaries and hotspots.
    c. formed by dip-slip faulting.
    d. found only at continent–continent convergent boundaries.
    e. None of the above.

13. The Cascade Mountains
    a. formed in the Precambrian Eon.
    b. formed at a divergent plate boundary.
    c. are caused by fault-block movement.
    d. are a volcanic arc.
    e. All of the above.

14. Fault-block mountain systems are
    a. caused by volcanic activity.
    b. related to transform plate boundaries.
    c. created by compressional stress.
    d. the product of tensional stress.
    e. There are no fault-block mountains.

15. What is a fold-and-thrust belt?
    a. another name for a V outcrop pattern
    b. a mountain belt at a spreading centre
    c. a mountain belt at a continent–continent convergent margin
    d. the result of crustal thinning by tensional stress
    e. a mountain belt resulting from transform faulting

FURTHER RESEARCH

Try the following exercise in interpreting a geological map.

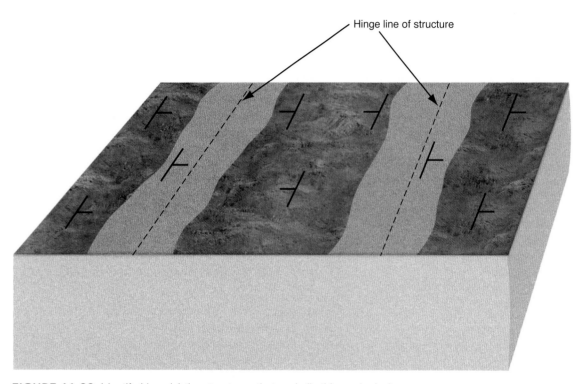

FIGURE 11.33 Identify (draw in) the structures that underlie this geological map.

Cindy Hanson is a second-year student majoring in education. She is taking a course in physical geology to fulfill her science core requirement. She knows that after she graduates and finds a job as an elementary-school teacher, one of the subjects she will be teaching is earth science. So she is working hard to learn as much geology as she can.

Cindy's geology teacher has asked her to interpret a geological map and draw the types of structures found within the crust. Please examine **FIGURE 11.33** and help Hanson answer the following questions.

1. Fill in the rock layers in the empty sides.

2. Label the structures in the crust.

3. What is the approximate angle of dip of the rock layers on the left side of the cross-section?

4. Draw arrows indicating the stresses acting on the crust. What type of stress is this?

5. Describe a tectonic setting that might produce this type of deformation in the crust.

ONLINE RESOURCES

Explore more about mountain building on the following websites:

The "Structural Geology" entry in the online free web encyclopedia Wikipedia:
http://en.wikipedia.org/w/index.php?title=Structural_
geology&oldid=539552000

The Wikipedia page on the "Appalachian Mountains":
http://en.wikipedia.org/w/index.php?title=Appalachian_
Mountains&oldid=554449612

The Wikipedia entry on "Geology of the Himalaya":
http://en.wikipedia.org/w/index.php?title=Geology_of_the_
Himalaya&oldid=552701499

NASA's Earth Observatory, for images and discussion of Earth surface topography:
http://earthobservatory.nasa.gov

Additional animations, videos, and other online resources are available at this book's companion website:
www.wiley.com/go/fletchercanada

This companion website also has additional information about WileyPLUS and other Wiley teaching and learning resources.

302

CHAPTER 12
EARTHQUAKES

Chapter Contents and Learning Objectives (LO)

12-1 An earthquake is a sudden shaking of the crust.
> **LO 12-1** *Describe why the risk from earthquakes has increased in recent decades.*

12-2 There are several types of earthquake hazards.
> **LO 12-2** *List and describe earthquake hazards.*

12-3 The elastic rebound theory explains the origin of earthquakes.
> **LO 12-3** *Define the elastic rebound theory.*

12-4 Most earthquakes occur at plate boundaries, but intraplate seismicity is also common.
> **LO 12-4** *Describe the relationship of earthquakes to plate boundaries.*

12-5 Divergent, convergent, and transform boundaries are the sites of frequent earthquake activity.
> **LO 12-5** *Compare and contrast seismicity at divergent, convergent, and transform plate boundaries.*

12-6 Earthquakes produce four kinds of seismic waves.
> **LO 12-6** *List and describe the four kinds of seismic waves.*

12-7 Seismometers are instruments that measure and locate earthquakes.
> **LO 12-7** *Describe seismometers and how they are used to locate the epicentre of an earthquake.*

12-8 Earthquake magnitude is expressed as a whole number and a decimal fraction.
> **LO 12-8** *Compare and contrast the ways to characterize earthquakes: moment magnitude, Richter magnitude, and intensity.*

12-9 Seismology is the study of seismic waves in order to improve understanding of Earth's interior.
> **LO 12-9** *Describe how seismic wave characteristics result in P-wave and S-wave shadow zones.*

12-10 Seismic data confirm the existence of discontinuities in Earth's interior.
> **LO 12-10** *List the major discontinuities in Earth's interior and their depths.*

12-11 Seismic tomography uses seismic data to make cross-sections of Earth's interior.
> **LO 12-12** *Describe seismic tomography and what it reveals about Earth's interior.*

GEOLOGY IN OUR LIVES

An earthquake is a sudden shaking of the crust, usually caused by the rupture of a fault. Earthquakes produce seismic waves that scientists can use to improve our understanding of Earth's interior. Earthquakes are also dangerous geologic hazards that are responsible for enormous damage to many communities, such as that shown in the opening photo of Beichuan, China, damaged by the 2008 Sichuan earthquake. More than 380 major cities lie on or near unstable regions of Earth's crust, and, because of swelling urban populations, geologists fear that a devastating earthquake capable of killing 1 million people could occur this century. But crowded inner cities are not the only concern—global growth in human population in the past 100 years means that people now live in areas that were previously remote and thought to be too dangerous for communities. Based on knowledge of seismic risk, designing and constructing buildings that can withstand shaking and establishing effective earthquake response and disaster assistance networks have become important elements of managing these hazards in our communities.

(?) Where are earthquakes most likely to occur?

12-1 An Earthquake Is a Sudden Shaking of the Crust

LO 12-1 *Describe why the risk from earthquakes has increased in recent decades.*

A trembling sensation forces you out of your sleep. Groggy, you realize that a mild but persistent vibration is running through your bed. Creaks and groans emanate from the walls, and you hear the window shaking. A low rumbling fills the air, sounding like an endless truck driving past your house. Two thoughts enter your mind at once: *The shaking won't stop and it's getting stronger.* Now fully awake, you suddenly grasp the situation as you hear the sharp sound of breaking dishes coming from the kitchen. Panic sets in as you rush from the room, yelling to other occupants of the house, "It's an earthquake!"

What Causes an Earthquake?

An **earthquake** is a sudden shaking of the crust. This shaking is typically caused by the abrupt release of strain that has slowly accumulated in rock. Recall from Chapter 11 ("Mountain Building") that *strain* is the change in shape and/or volume of a rock caused by *stress* (pressure). When stress is applied slowly and continuously over time, most rocks deform elastically. That is, they rebound to their original shape if the stress is removed or when the rock fractures. The same is true along a fault that may be "stuck"; the elastic deformation accumulates until the fault becomes "unstuck" **(FIGURE 12.1)**. When rock breaks or a fault abruptly ruptures, it releases strain in a form of energy called **seismic waves,** which radiate through Earth's interior and along its surface. Seismic waves cause the shaking of an earthquake.

Most earthquakes are associated with the rupture of a fault. But they may also result from volcanic action, landslides, or any other rapid release of energy that shakes the ground. Strain accumulation (elastic deformation) is persistent at the edges of lithospheric plates. Hence, plate boundaries are among the most dangerous places on Earth because of the high frequency of earthquakes. But *intraplate earthquakes* also occur and are capable of causing serious damage and loss of life.

Today earthquakes endanger more people than ever before. The world's population has more than doubled in the past half century. Half of the world's people live in densely developed urban centres. More than 380 major cities with growing urban populations lie on or near unstable regions of Earth's crust, and seismologists fear that a devastating earthquake capable of killing 1 million people could occur this century **(FIGURE 12.2)**.

In any given year, more than 3 million earthquakes occur throughout the world. However, most of these cause no damage because they are relatively small. On average, Canada experiences approximately 4,000 earthquakes per year, half of which happen in or offshore of British Columbia, Canada's earthquake hotspot. Most of these earthquakes go unfelt and are only detected by very sensitive equipment. As the population grows and urban centres swell, the infrastructure on which modern life depends (high-rises, bridges, pipelines, communication towers, energy grids, emergency services, etc.) expands

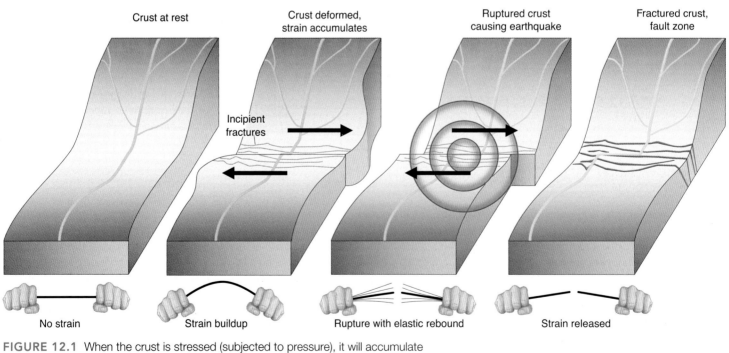

| Crust at rest | Crust deformed, strain accumulates | Ruptured crust causing earthquake | Fractured crust, fault zone |

Incipient fractures

No strain — Strain buildup — Rupture with elastic rebound — Strain released

FIGURE 12.1 When the crust is stressed (subjected to pressure), it will accumulate strain (deformation). When strain accumulates slowly, rocks can behave elastically. They spring back to their original shape when the stress is removed. This can happen when they break or when a fault ruptures. Bending and breaking a stick is the same process. The strain is released in the form of energy called seismic waves, which pass through Earth's interior and along its surface.

(?) What type of plate boundary will produce the conditions shown in Figure 12.1?

© Roger Ressmeyer/Corbis

(?) How do buildings react when they are shaken?

FIGURE 12.2 Seismic waves cause shaking, which can damage buildings. This house was largely destroyed by the Loma Prieta earthquake that struck San Francisco in 1989.

into previously undeveloped regions. With more people living across more of Earth's surface, earthquakes now pose a greater hazard to our lives and communities than ever before. A century ago, even a large earthquake could go unnoticed if it occurred in a remote area. But today there are few remote areas, and a small earthquake may be felt by thousands of people, a large one by millions. The 2008 *Sichuan earthquake* caused over 80,000 deaths and destroyed over 4 million homes

in a region that just two decades ago was largely open countryside. Because of China's rapid economic expansion, the area is now heavily populated, and unfortunately, buildings have not been constructed to withstand shaking, and emergency services were unprepared for a major seismic event.

By studying the nature of the destruction caused by past quakes (**TABLE 12.1**), we have learned how to build strong-yet-flexible buildings that can withstand strong shaking. Earthquakes can be deadly and frequent events: the 2003 Bam quake in Iran killed over 26,000 people; the 2004 Sumatra quake in Indonesia (which spawned a tsunami) killed 230,000; the 2005 Kashmir earthquake killed at least 86,000; the 2008 Sichuan earthquake killed 87,000; the 2010 Haiti earthquake killed more than 200,000; and the 2010 Chile earthquake killed hundreds of people. These recent events demonstrate that the risks associated with earthquakes are still high, even in our modern world.

How frequent are major quakes? Since 1900, 114 quakes have each caused the deaths of more than 1,000 people. That is an average of more than one quake per year. Clearly, research on the distribution and causes of earthquakes is critically important to increasing the safety of our communities.

TABLE 12.1 Historical Quakes Causing at Least 50,000 Deaths

Year (Date)	Deaths	Location (Country, Area)	Comments
856 (December 22)	200,000	Iran, Damghan	Few details available
893 (March 23)	150,000	Iran, Ardabil	Few details available
1138 (August 9)	230,000	Syria, Aleppo	Few details available
1268 (no specific date)	60,000	Asia Minor, Silicia	Few details available
1290 (September 27)	100,000	China, Chihli	Few details available
1556 (January 23)	830,000	China, Shaanxi	Total building collapse in some towns, large ground fractures
1667 (November)	80,000	Caucasia, Shemakha	Few details available
1693 (January 11)	60,000	Italy, Sicily	Few details available
1727 (November 18)	77,000	Iran, Tabriz	Few details available
1755 (November 1)	70,000	Portugal, Lisbon	Caused deadly tsunami
1783 (February 4)	50,000	Italy, Calabria	Few details available
1908 (December 28)	72,000	Italy, Messina	Caused deadly tsunami
1920 (December 16)	200,000	China, Haiyuan	Major landslides
1923 (September 1)	143,000	Japan, Kwanto	Great Tokyo fire
1948 (October 5)	110,000	Soviet Union (Turkmenistan)	Widespread building collapse
1970 (May 31)	70,000	Peru, Chimbote	Great rockslide, floods
1976 (July 27)	255,000	China, Tangshan	655,000 deaths probable
1990 (June 20)	50,000	Western Iran	Landslides
2004 (December 26)	230,000	Sumatra	Caused deadly tsunami
2005 (October 8)	86,000	Kashmir, India, and Pakistan	3.3 million homeless
2008 (May 12)	87,000	China, Sichuan	4.8 million homes destroyed
2010 (January 13)	222,521	Haiti, Port-au-Prince	Primarily building collapse

(?) **Expand Your Thinking**—What causes the shaking experienced in an earthquake?

12-2 There Are Several Types of Earthquake Hazards

LO 12-2 *List and describe earthquake hazards.*

Scientists cannot predict when an earthquake will happen. But it is possible to define areas that are especially vulnerable to *earthquake hazards* (dangerous events caused by earthquakes). The principal earthquake hazards are (1) ground shaking (leading to building collapse), (2) landslides, (3) fire, and (4) tsunamis. Each hazard is discussed in this section.

Ground Shaking

Ground shaking caused by earthquakes actually poses little direct danger to a person; an earthquake cannot shake you to death. However,

shaking produces effects that can cause damage and loss of life. One of these is *sediment liquefaction.* Liquefaction occurs when shaking causes groundwater to rise to the surface and mix with sandy or muddy soil. The sediment loses strength, and any buildings or roads on the sediment will be destroyed as they sink into the wet, soupy quicksand (or quickclay). The sediment may firm up again after the earthquake as the groundwater drains to its natural location deeper in the crust. Shaking causes vibration at the base of a building, a dynamic process that many buildings are not designed to withstand. If you place strain onto the structure and then let it snap back, it will sway back and forth with an amplitude that decreases with time. The swaying will happen

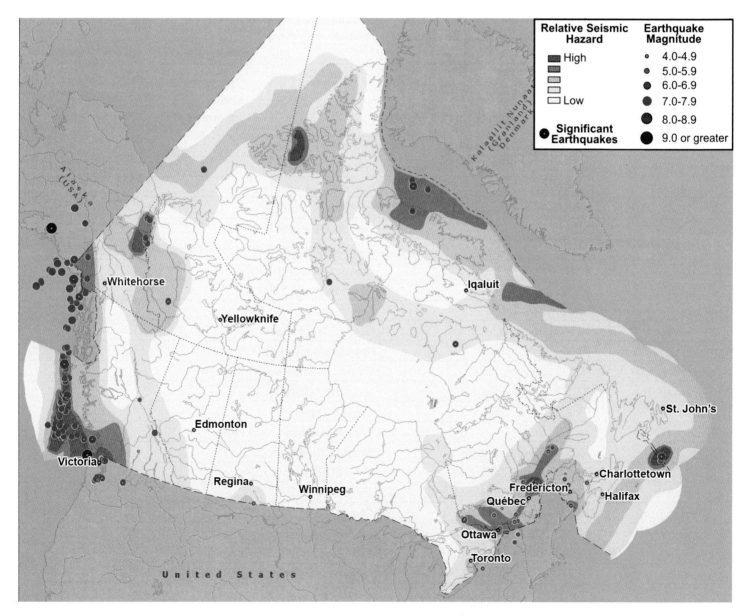

FIGURE 12.3a Natural Resources Canada uses data on past earthquakes to create a seismic hazard map for all regions of Canada.

with a particular frequency called the *natural frequency*. If the ground shakes with the same frequency as a building's natural frequency, the amplitude of sway will get larger such that the shaking is in *resonance* with the building's natural frequency. Seismic shaking that matches a building's natural frequency produces the most strain on the building and can cause it to collapse.

Natural Resources Canada and the U.S. Geological Survey (USGS) maintain earthquake research programs that define probable levels of building shaking throughout the country. These agencies produce maps of **seismic hazards** that are used by engineers and city planners in designing and constructing buildings, bridges, highways, and utilities that are better able to withstand earthquakes (**FIGURE 12.3**).

Landslides

Ground shaking can cause steep hillsides to collapse. This process, known as *mass wasting*, is discussed in Chapter 18 ("Mass Wasting"). Mass wasting may occur for many reasons that do not include earthquakes, such as *undercutting* the base of a steep hill (e.g., by lateral river erosion or highway construction) and heavy rainfall that saturates and lubricates sediments or rock layers; even lightning strikes have been known to trigger a mass wasting event. But mass wasting triggered by ground shaking during an earthquake is very common (**FIGURE 12.4**).

Unfortunately, tens of thousands of communities around the world are located at the base of steep hills and mountains in earthquake-prone areas. This is an extremely dangerous situation.

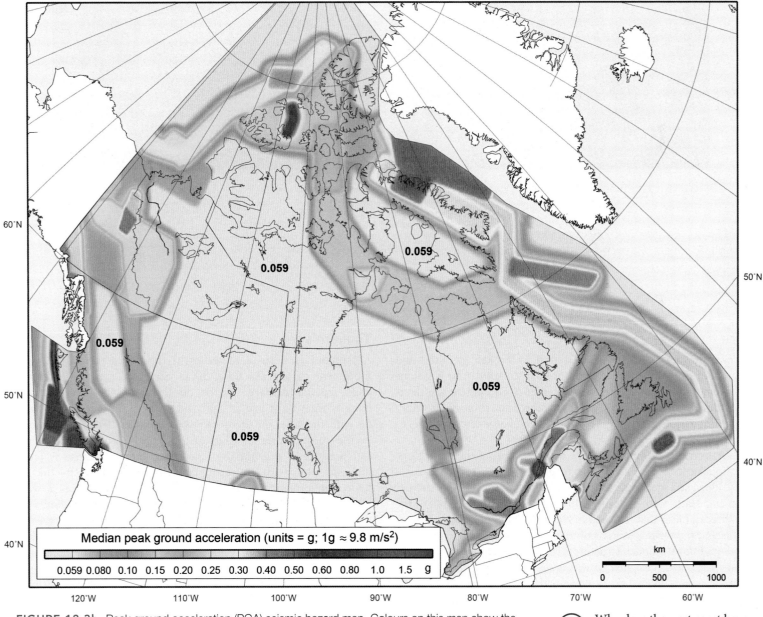

FIGURE 12.3b Peak ground acceleration (PGA) seismic hazard map. Colours on this map show the relative level of risk based on horizontal shaking that has a 1-in-50 (2 percent) chance of occurring in a 50-year period. Shaking is expressed as g (g is the acceleration of a falling object due to gravity). Areas in hot colours are at greatest risk for seismic shaking that can cause damage. This information is useful for purposes such as developing safer building codes. See the "Online Resources" section at the end of this chapter for links with more information.

Why does the west coast have a high seismic shaking rating?

Image by W. Thompson, 2005

FIGURE 12.4 These landslides were triggered by the 2005 Kashmir earthquake.

(?) Although seismic shaking produced these landslides, undercutting had made the cliff unstable. What caused the undercutting?

Fire

Fires following an earthquake, started by broken gas lines and power lines or tipped-over wood or coal stoves, can be a serious problem. Often water lines and emergency equipment are damaged during an earthquake, and roads are made impassable, making it impossible for firefighters and other emergency personnel to do their work.

A horrible example of a fire occurred during the 1995 Kobe earthquake in Japan. According to official statistics, 5,472 people were killed and more than 400,000 were injured. Most of the destruction and deaths were not caused directly by the earthquake but by the hundreds of fires that followed it. Newer buildings had been built to withstand the seismic stress, but many of the city's older residences were traditional wooden structures. They collapsed easily under their heavy tile roofs (designed to withstand hurricanes), thus trapping their occupants. Even if rescue crews had been able to reach them, the water supply was not functional. Authorities were unable to control the fires that raged through neighbourhoods for days after the quake (**FIGURE 12.5a**)

Tsunamis

A **tsunami** is a series of waves in the ocean caused by rapid movement of the sea floor. If an earthquake generates a tsunami, coastal damage can be deadly and widespread. A tsunami is what most people call a tidal wave, but it has nothing to do with the tides on the ocean. Any movement of the sea floor, such as an undersea landslide or a submarine volcanic explosion, can produce a tsunami. But the most common cause is faulting associated with an earthquake that violently shifts the sea floor and displaces the overlying water column, causing a wave (**FIGURE 12.6**). A tsunami is not like the more common *wind wave* that we are used to seeing on the ocean's surface. Tsunamis are imperceptible in the open ocean (except to instruments), but when they run aground they may flood across the shoreline for many minutes and cause extensive inundation. If the wave trough between wave crests happens to hit the shoreline first, there will be *drawback,* causing the shoreline to recede dramatically. This is a clear warning sign a tsunami surge is imminent and evacuation of the area should take place immediately.

(a) *(b)*

FIGURE 12.5 *(a)* Ruptured power and gas lines caused hundreds of fires in Kobe, Japan, following the 1995 earthquake. *(b)* Tsunami wave overtops seawall and inundates a street in Miyako City, Japan, following the 2011 Tōhoku earthquake.

? Why was it difficult for emergency crews to suppress the fires?

The period of time between successive wave crests can be anywhere from five minutes to an hour or more, depending on how far away the waves originated. The time from the front of the wave crest to the back can be many minutes. Hence, when one wave hits the shore it floods the land for many minutes until the back of the wave arrives. The famous Tōhoku earthquake in 2011, the largest ever to hit the island of Japan, caused a deadly tsunami that flooded lands over 40 m above sea level, travelled up to 10 km inland, caused three nuclear power plants to melt down, and destroyed many coastal cities and villages (**FIGURE 12.5b**).

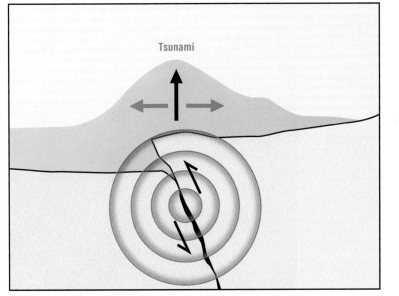

FIGURE 12.6 A tsunami is caused by sudden movement of the sea floor. Faulting associated with an offshore earthquake is a common cause of tsunamis.

? What is the major difference between a 2 m wind wave and a 2 m tsunami? Why is one more dangerous?

? **Expand Your Thinking**—How would you design a city that could quickly recover from an earthquake?

12-3 The Elastic Rebound Theory Explains the Origin of Earthquakes

LO 12-3 *Define the elastic rebound theory.*

Earthquakes are usually associated with slippage along a fault, most often (but not always) associated with a plate boundary (**FIGURE 12.7**). The point where slippage first occurs is called the earthquake **focus,** and the **epicentre** is the geographic spot on Earth's surface that is located directly above the focus. When an earthquake occurs, energy is released from the focus and travels in all directions in the form of **surface waves** and **body waves.** Surface waves travel along Earth's surface like the ripples that travel across a pond. Body waves travel into and through Earth's interior. When they emerge on the other side of the planet, they can be detected by instruments known as **seismometers.**

Although earthquakes are reported in the media only when they are particularly damaging, they are an everyday occurrence on our planet. According to the U.S. Geological Survey, more than 3 million earthquakes occur every year. That's about 8,000 a day, or 1 every 11 seconds. The vast majority of quakes are extremely weak and go unnoticed except by the sensitive seismometers designed to measure them. It is the big quakes, such as the 2010 Haiti earthquake, that occur in highly populated areas that get our attention and usually cause horrifying damage.

Elastic Rebound Theory

The **elastic rebound theory** describes the mechanism by which earthquakes are generated. The theory was formulated in the aftermath of the devastating 1906 San Francisco earthquake. After the quake, geologists discovered a fault that could be followed along the ground in a more or less straight line for 435 km. The crust on one side of the fault had slipped laterally relative to the crust on the other side by up to 7 m; the average amount of slippage was 4.7 m. The fault and slippage drew the curiosity of scientists, especially since, prior to that time, no one had been able to explain what happens within the crust to cause earthquakes. Previously, it had been assumed that stresses causing earthquakes must be located close to the earthquakes themselves; the concept of plate tectonics was unknown.

After studying the fault line created by the 1906 quake, *Harry Fielding Reid* of Johns Hopkins University postulated that earthquakes are caused by distant forces. He suggested that these forces cause a gradual buildup of *elastic energy* in the crust over tens to thousands of years, slowly accumulating strain (deformation) and distorting the crust in much the same way that a stick stores elastic energy by bending before it breaks. (Refer back to Figure 12.1.)

Eventually a fault (**FIGURE 12.8**) or a pre-existing weakness in the rock is no longer able to accommodate the buildup of more strain, and it ruptures. Slippage of rock at the focus produces stress that travels along the fault, causing displacement of both sides of the fault plane until the strain has been largely released. At the time of the earthquake, strained rock elastically snaps back into position, although it experiences net displacement (breakage) in one direction. Vibrations of the rock as it returns to its original shape cause seismic waves. Reid described this process as "elastic rebound." It can be compared to gradually pulling a rubber band until it is stretched as far as possible; the strain is released when the band breaks and the two pieces snap back to their original shape.

(?) If you were to study this fault plane closely, what evidence would you find that movement had occurred between the two sides of the fault?

FIGURE 12.8 Strain stored in the crust is suddenly released when rocks fracture or slip along a fault. As the rock elastically returns to its original shape, it generates vibrations that become seismic waves travelling through Earth's interior and along Earth's surface.

© Robert J. Varga, Pomona College

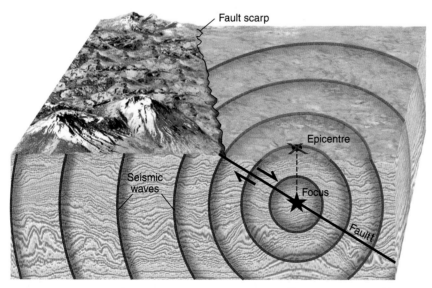

FIGURE 12.7 Earthquakes are usually associated with slippage along a fault.

(?) What type of fault is shown in Figure 12.7?

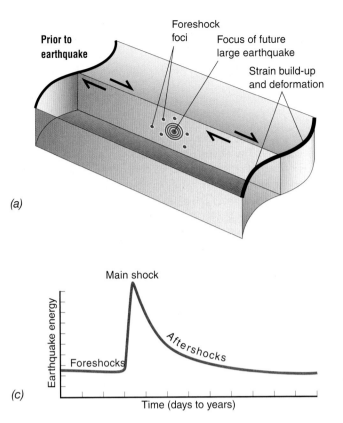

(a)

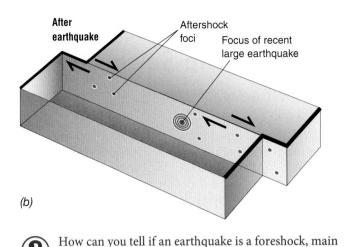

(b)

(c)

FIGURE 12.9 (a) Foreshocks tend to cluster around the future focus of a main shock. (b) Aftershocks may be spread out along the rupture surface of the main shock as well as at nearby faults. (c) The main shock is the largest among a cluster of earthquakes.

(?) How can you tell if an earthquake is a foreshock, main shock, or aftershock?

Foreshocks and Aftershocks

Part of living in a region that is prone to earthquakes is learning to live with **foreshocks** and **aftershocks.** As strain accumulating in the crust approaches the elastic limit of the rock, earthquakes tend to occur in *clusters.* The largest quake in a cluster is the **main shock;** those occurring before it are foreshocks, and those occurring after it are aftershocks. Foreshocks tend to occur in the immediate vicinity of the focus of the main shock, whereas aftershocks may be spread out along the rupture surface (the fault plane) as well as at other, nearby faults (**FIGURE 12.9**).

You can learn about the largest fault rupture ever recorded by reading the "Geology in Action" box titled "The Sumatra Earthquake Moved the Entire Planet."

GEOLOGY IN ACTION

THE SUMATRA EARTHQUAKE MOVED THE ENTIRE PLANET

The Sumatra earthquake of 2004 set many records: it was the most powerful quake in more than 40 years; it triggered a devastating tsunami that killed more people than any natural catastrophe in modern history; it lasted an unbelievable 500 to 600 seconds (most quakes last only a few seconds); and it shook the ground everywhere on Earth's surface, causing Earth to wobble on its axis by up to 2.5 cm. Weeks later, the planet was still trembling. The quake was produced by the longest fault rupture ever observed: 1,255 km. That is the distance from Victoria to Saskatoon or Toronto to Halifax. The fault shifted continuously for 10 minutes, also a record.

The earthquake and resulting tsunami (**FIGURE 12.10**), which swept across the Indian Ocean and was recorded on instruments around the world, killed approximately 230,000 people in 11 countries and left over 1 million people homeless. Ground movement of as much as 1 cm occurred everywhere on Earth's surface, although the movement was too small to be felt in most areas.

(a)

(b)

FIGURE 12.10 Banda Aceh, Indonesia, (a) before and (b) after flooding by the tsunami created by the 2004 Sumatra earthquake. The tsunami reached the shores of Indonesia, Sri Lanka, India, Thailand, and other countries with waves up to 30 m high. It caused serious damage and deaths as far away as Port Elizabeth, South Africa, 8,000 km from the source.

(?) **Expand Your Thinking**—What sequence of events characterizes the buildup of elastic energy in the crust?

12-4 Most Earthquakes Occur at Plate Boundaries, but Intraplate Seismicity Is Also Common

LO *12-4 Describe the relationship of earthquakes to plate boundaries.*

In 1969, researchers *Muawia Barazangi* of Cornell University and *James Dorman* of Columbia University published a map showing the locations of earthquakes occurring between 1961 and 1967. This map has become famous as one of the primary pieces of evidence in support of the theory of plate tectonics. The earthquake epicentres depicted in the Barazangi/Dorman map clearly lined up along long narrow zones that define the boundaries of lithospheric plates. The interiors of the plates themselves are, for the most part, not prone to the high numbers of quakes that occur at plate boundaries, and parts of plate interiors are earthquake free, or *aseismic* (FIGURE 12.11). But don't be mistaken, plate interiors can also experience seismic activity.

Intraplate Earthquakes

Although the great majority of earthquakes occur at plate boundaries, Figure 12.11 reveals that some quakes occur within plate interiors far from modern plate margins. Many of these *intraplate earthquakes* probably are linked to stresses transmitted into plate interiors that nonetheless originate at the plate edge. Often these quakes have shallow foci and are related to the reactivation of old faults that are normally quiet. These faults may be left over from a period of tectonic action that has since ceased, or they may mark the establishment of a new region of deformation.

Intraplate volcanism, such as at a hotspot, is another source of intraplate seismicity. The region around Yellowstone National Park, Wyoming, is famous for its frequent earthquakes caused primarily by the movement of magma (or gases) within the crust beneath the park. Visitors who have never felt a quake before often are thrilled to experience their first while viewing the park's hot springs and geysers. The Big Island of Hawaii, location of Kilauea volcano at Volcano National Park, one of the world's most active volcanoes, is the most seismically active location in the United States, despite the fact that it is thousands of kilometres from the nearest plate boundary. Hawaiian quakes are the result of magma moving through the crust and breaking rock as it intrudes, or when the crust ruptures as a result of having to bear the great weight of a shield volcano.

Earlier in the chapter, we mentioned one of the worst intraplate seismic events to occur in the modern era. This quake occurred on May 12, 2008, in the province of Sichuan in south-central China. It has been estimated that 87,000 lives were lost and damage exceeded

(?) Deep earthquakes occur in areas where oceanic crust is being actively subducted. About 90 percent of all earthquakes occur at depths less than 100 km. Why do subduction zones have deep earthquakes?

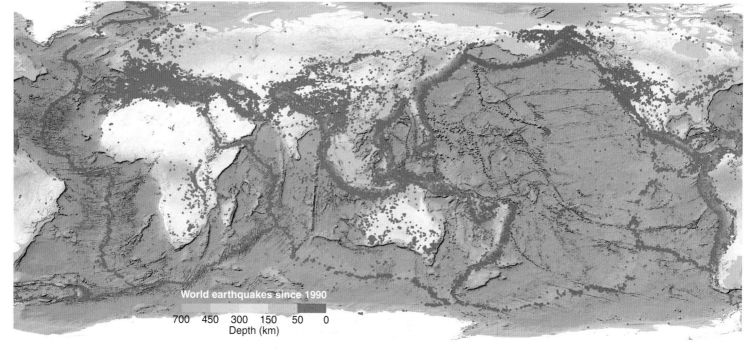

World earthquakes since 1990

700 450 300 150 50   0
Depth (km)

FIGURE 12.11 This map depicts the global distribution of foci for moderate to large earthquakes since 1990. The foci are colour-coded by depth.

$20 billion. The event is probably related to the ongoing collision of the Indian-Australian plate with the Eurasian plate and the resulting movement along the Longmenshan thrust fault at a depth of 20 km. This seismicity is the latest in a long history of damaging earthquakes extending through the Himalayan fold-and-thrust belt, across the Tibetan Plateau, and northward into central and northern China.

Modern plate boundaries cannot satisfactorily explain the 1925 magnitude (M) 6.2 earthquake within the Charlevoix seismic zone (CSZ), Quebec, or the 1811 and 1812 earthquakes at New Madrid, Missouri. However, there may be a past tectonic explanation for these quakes. The CSZ and New Madrid earthquakes may have occurred along "failed rifts" known as the *St. Lawrence paleo-rift faults* in Quebec and the *Reelfoot rift* in Missouri. A "failed rift" is a weakened zone of normal faults (horsts and grabens) in the crust. In this case, they were left behind from a separation of continents that took place 700 to 550 million years ago during the initial rifting and opening of the Iapetus Ocean at the end of the Proterozoic Eon. The CSZ also happens to be located above the *Charlevoix impact structure,* which likely contributes to weaknesses in the Precambrian crust that facilitate the seismicity in the Charlevoix region. The CSZ averages > 250 earthquakes per year, while the New Madrid fault area averages > 100 earthquakes per year, but very few of the earthquakes are actually felt because most seismic events in these regions are minor. At times, though, these faults cause larger earthquakes. Within the CSZ there have been at least five earthquakes that exceeded M 6 since 1663. Geologists who study the New Madrid earthquakes estimate that every 70 to 90 years, there is a moderate to strong earthquake and every 250 to 500 years a major earthquake that can cause severe damage occurs.

Earthquake Prediction

How can plate tectonics help in earthquake prediction? Earthquakes occur in four settings: (1) intraplate regions, (2) divergent boundaries, (3) convergent boundaries, and (4) transform boundaries. (Types 2 to 4 are discussed in the next section.) Thus, it is possible to identify regions of Earth's surface where large earthquakes can be expected to occur. We know that each year about 140 earthquakes of moderate strength or greater will occur in these settings, which together account for about 10 percent of Earth's surface.

Although it is possible to predict *where* quakes are likely to happen on a worldwide basis, it is not possible to determine precisely *when* earthquakes will occur. The cyclical processes that drive plate tectonics—strain buildup followed by an earthquake followed by more strain buildup—have been going on for hundreds of millions of years; as a result, plates have moved, on average, several centimetres per year. But at any instant in geologic time—for example, the year 2013—we do not know exactly what point in the cycle of strain buildup and strain release has been reached by any given portion of crust. Only by monitoring the stress and strain in small areas in great detail can scientists hope to predict when renewed seismic activity is likely to take place. Even then, the prediction will be a statement of statistical probability that lacks specific information about timing, location, or the amount of energy released.

In summary, plate tectonics is a powerful but broad tool for predicting earthquakes. It tells us *where* 90 percent of Earth's major earthquakes are likely to occur, but it cannot tell us much about *when* they will occur.

Learn more about quakes in the "Geology in Our Lives" box titled "Major Canadian and U.S. Earthquakes in the Past 70 Years."

GEOLOGY IN OUR LIVES

MAJOR CANADIAN AND U.S. EARTHQUAKES IN THE PAST 70 YEARS

- **October 27, 2012**. The second largest Canadian earthquake ever recorded by a seismometer, magnitude 7.8, struck Haida Gwaii (formerly the Queen Charlotte Islands), British Columbia. Fortunately, there were no casualties or significant structural damage reported in any of the nearby communities.
- **December 12, 1985**. The surprisingly large magnitude 6.9 Nahanni earthquake was recorded in the Nahanni region of the Mackenzie Mountains in the Northwest Territories.
- **February 9, 1971**. In California's San Fernando Valley, a strong quake left 65 people dead.
- **June 24, 1970**. A magnitude 7.4 earthquake along the Queen Charlotte fault rocked Haida Gwaii.
- **March 28, 1964**. Known as the Good Friday Earthquake, the second largest quake in history struck near Prince William Sound, Alaska, and killed 128 people.
- **August 22, 1949**. The magnitude 8.1 Haida Gwaii earthquake along the Queen Charlotte fault was Canada's largest since 1700. The shaking was so severe on Haida Gwaii that cows were knocked off their feet.
- **June 23, 1946**. Canada's largest recorded onshore earthquake (M 7.3) had an epicentre in central Vancouver Island, just west of the communities of Courtenay and Campbell River, causing two deaths (one due to a capsized boat, the other due to a heart attack).

Expand Your Thinking—Provide a testable hypothesis explaining intraplate earthquakes.

12-5 Divergent, Convergent, and Transform Boundaries Are the Sites of Frequent Earthquake Activity

LO 12-5 *Compare and contrast seismicity at divergent, convergent, and transform plate boundaries.*

Plate boundaries are characterized by frequent earthquakes. In this section, we discuss divergent boundaries, convergent boundaries, and transform boundaries in more detail.

Divergent Seismicity

The narrow earthquake belt extending along the world's mid-ocean ridges provides direct support for the concept of seafloor spreading. Because the lithosphere is thin and weak at divergent boundaries and large amounts of strain cannot accumulate, seismic activity tends to be of low to moderate energy and to occur at shallow depths. While major earthquakes are rare, moderate quakes can be extremely damaging. Volcanic activity along the axis of the ridges is associated with this type of seismicity. Earthquakes at divergent boundaries are related to normal faulting in which blocks of crust along the rift valley walls slide inward in the overall tensional environment (**FIGURE 12.12**). Strike-slip faulting is also a source of seismicity at spreading centres, when crustal slippage occurs along the transform boundaries found on most mid-ocean ridge systems.

Convergent Seismicity

Earth's largest and most active seismicity occurs at convergent boundaries, where one plate subducts beneath another in a zone of accumulating compressive stress. These regions of plate interaction are characterized by progressively deeper earthquakes along a line proceeding from the trench across the overriding plate (**FIGURE 12.13**). In most cases, the quakes' foci are located within or along the surface of the subducting plate. These foci define a steeply dipping plane called the **Wadati-Benioff zone**, or WBZ, which corresponds to faulting associated with the subducting oceanic slab. The epicentres of deep-focus quakes are located far from the surface expression of the plate boundary and illustrate how far below the overriding plate a subducted slab extends as it is recycled. These epicentres appear to be intraplate earthquakes, but in reality they are tied to the subduction of oceanic lithosphere deep below the surface and are part of the process of tectonic convergence.

Transform Seismicity

Tectonic boundaries may be characterized by strike-slip faulting, in which two plates slip past each other at a transform margin. At transform boundaries, earthquakes are shallow, running only as deep as 25 km. Three excellent, but dangerous, examples of transform seismicity are found at the *Queen Charlotte fault* along British Columbia's west coast, *the San Andreas fault* in California, and the *North Anatolian fault* in Turkey. All are right-lateral strike-slip faults (**FIGURE 12.14**).

Both the Queen Charlotte and San Andreas faults represent transform plate boundaries where the Pacific plate moves northwestward relative to the North American plate. The movement along these faults causes some of the biggest earthquakes in North America. The Queen Charlotte and San Andreas faults are actually the "master" faults of a complex fracture network that cuts its way along the west coast of British Columbia and through western California, respectively. The Queen Charlotte fault system is more than 850 km long, extending from just north of Vancouver Island all the way up the west side of the Alaskan panhandle. The entire San Andreas system is more than 1,200 km long and extends to depths of about 16 km. The fault zone is a complex corridor of crushed rock ranging from a few hundred metres to 2 km wide. Many smaller faults, the product of regional stresses emanating from the master fault, branch out from and rejoin the zone. During the 1906 San Francisco earthquake, roads, fences, and rows of trees that crossed the fault were offset by several metres. In each case, the ground west of the fault moved northward relative to the ground east of the fault.

Turkey has a long history of large seismic events that occur as a series of progressive adjacent earthquakes. Most recently, starting in 1939, the North Anatolian fault produced a sequence of major earthquakes. The eleventh, in 1999, killed over 17,000 and left approximately half a million people homeless. Researchers had estimated that there was a 12 percent chance of this earthquake occurring in the 30 years between 1996 and 2026. On March 8, 2010, the North Anatolian fault shook again killing at least 51 people and destroying traditional stone and mud-brick homes as well as minarets in at least six villages. The next day over 100 aftershocks were felt.

In the Queen Charlotte, San Andreas, and North Anatolian faults, the sudden faulting that initiates a great earthquake

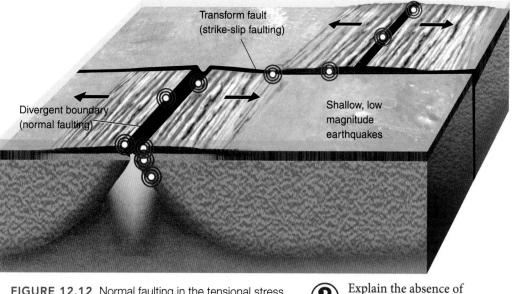

FIGURE 12.12 Normal faulting in the tensional stress environment of spreading centres causes shallow, low-energy earthquakes. Strike-slip faulting at transform faults also produces shallow earthquakes.

Transform fault (strike-slip faulting)

Divergent boundary (normal faulting)

Shallow, low magnitude earthquakes

 Explain the absence of deep-focus earthquakes at divergent and transform boundaries.

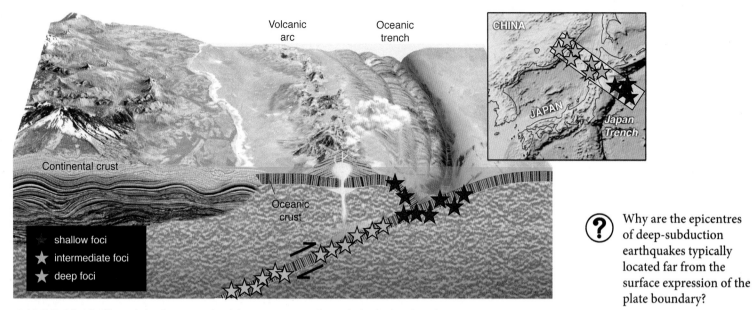

FIGURE 12.13 The subducting oceanic slab at convergent boundaries is the site of numerous earthquakes that extend from shallow to deep regions under the overriding plate.

(?) Why are the epicentres of deep-subduction earthquakes typically located far from the surface expression of the plate boundary?

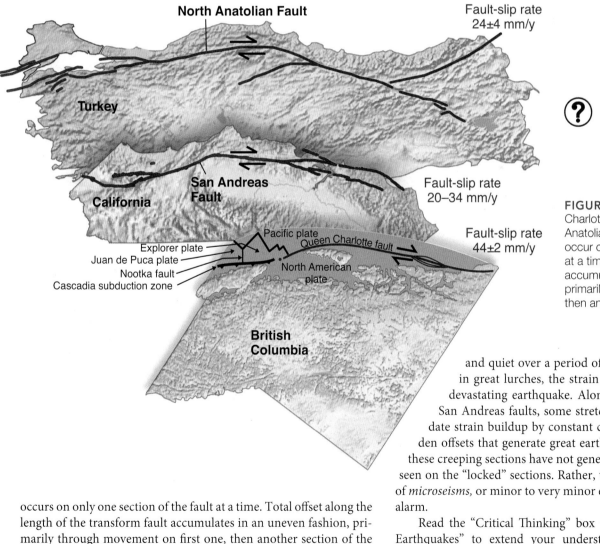

(?) What type of faulting typically occurs at the Queen Charlotte, San Andreas, and Anatolian faults?

FIGURE 12.14 At the Queen Charlotte, San Andreas, and North Anatolian faults, great earthquakes occur on only one section of the fault at a time. Offset along the fault plane accumulates in an uneven fashion, primarily by movement on first one, then another, section of the fault.

occurs on only one section of the fault at a time. Total offset along the length of the transform fault accumulates in an uneven fashion, primarily through movement on first one, then another section of the fault. The sections that produce great earthquakes remain "locked" and quiet over a period of years as strain builds; then, in great lurches, the strain is released in the form of a devastating earthquake. Along the Queen Charlotte and San Andreas faults, some stretches of the faults accommodate strain buildup by constant creeping rather than by sudden offsets that generate great earthquakes. In historical times, these creeping sections have not generated earthquakes of the size seen on the "locked" sections. Rather, they inch forward in a series of *microseisms*, or minor to very minor earthquakes that do not raise alarm.

Read the "Critical Thinking" box titled "Tectonic Settings for Earthquakes" to extend your understanding of the relationship between seismicity and plate tectonics.

(?) **Expand Your Thinking**—Describe the stress and strain environment at divergent, convergent, and transform boundaries.

CRITICAL THINKING

TECTONIC SETTINGS FOR EARTHQUAKES

Earthquakes tend to occur at plate boundaries and some intraplate settings (**FIGURE 12.15**).

Please work with a partner on the following problems.

1. Construct a table describing the characteristics of earthquake foci at divergent, convergent, transform, and intraplate settings. Describe each category in terms of depth of foci (shallow, intermediate, deep), frequency of occurrence (often, infrequent), tendency to produce major quakes (more likely, less likely), and likelihood of being heavily developed by human communities (more likely, less likely).
2. Describe the nature of the prevailing stress at each of the four settings.
3. At each setting, does strain typically accumulate over long periods? Explain.
4. What is the nature of strain release—localized or regional?
5. Describe the nature of rock breakage at each setting. That is, what type of faulting is likely to occur in an earthquake at that location?
6. The concepts of stress and strain provide an overarching framework for synthesizing a general hypothesis of earthquake occurrence. Write a general hypothesis describing earthquake occurrence and describe appropriate tests for it.
7. For each illustration in Figure 12.15, label the types of stresses, strains, faults, and other seismic, tectonic, and structural features.

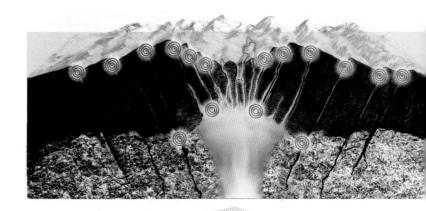
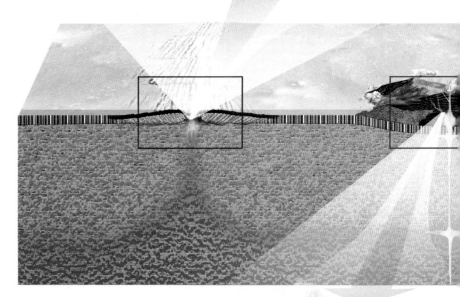
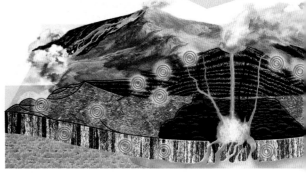

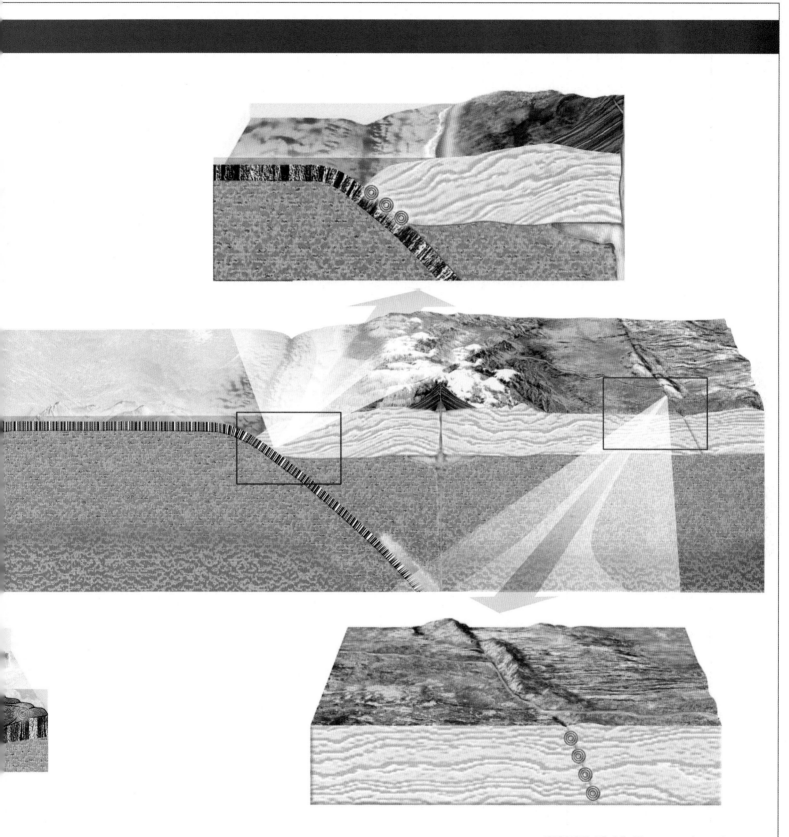

FIGURE 12.15 Plate tectonic settings. Earthquakes tend to occur at plate margins and some intraplate regions.

12-6 Earthquakes Produce Four Kinds of Seismic Waves

LO 12-6 *List and describe the four kinds of seismic waves.*

Earthquakes release energy in the form of seismic waves that travel outward from the focus through the crust and into Earth's interior (**FIGURE 12.16**). There are four kinds of seismic waves, and they all move in different ways (**TABLE 12.2**).

Seismic Waves

Earthquakes radiate seismic energy in the form of two types of *surface waves:* **Love waves** and **Rayleigh waves;** and two types of *body waves:* primary compressional waves, or **P waves,** and secondary shear waves, or **S waves.** Body waves travel through Earth's inner layers, but most (not all) of the energy in surface waves moves along the planet's surface.

The seismic P wave is the fastest kind of wave. It moves through solid rock as well as fluids, such as water or Earth's liquid layer (the outer core). A P wave compresses and *dilates* (stretches) the material through which it moves. Hence, any material that resists changes in volume (such as solids and liquids) will transmit a P wave rather than absorb it. A P wave pushes and pulls on the rock or liquid it moves through, just as sound waves push and pull on air (**FIGURE 12.17a**). Have you ever heard a big clap of thunder and noticed the windows rattle at the same time? The windows rattled because the sound waves were pushing and pulling on the window glass in much the same ways that P waves push and pull on rock. Sometimes animals can hear the P waves of an earthquake. Usually humans only feel the bump and rattle of these waves.

The S wave travels slower than a P wave and can move only through solid materials, such as rock. The S wave moves up and down or side to side. Hence, any material that is resistant to changes in shape will transmit an S wave. Because liquids do not resist changes in shape, they will absorb S waves rather than transmit them. For this reason, S waves travel through the solid layers of Earth (crust and mantle) but not the liquid layer (outer core). As we will discover later, the different properties of S and P waves (and large-energy surface waves) allow scientists to use seismic waves to explore the structure of Earth's interior.

Seismic surface waves are the slowest waves and therefore arrive at distant seismometer stations after the P and S waves generated by a quake. They may cause a rolling motion like that of swells at sea or a very strong side-to-side shaking similar to that caused by S waves. Generally surface waves cause the strongest vibrations and worst damage.

The fastest surface wave is the Love wave, named for *A.E.H. Love*, a British mathematician who worked out the mathematical model for this kind of wave in 1911 (**FIGURE 12.17B**). It is usually the first surface wave to be recorded by a seismograph. Love waves move the ground from side to side, causing widespread damage to buildings, roads, and other solid objects. The second kind of surface wave is the Rayleigh wave, named for *John William Strutt, Lord Rayleigh*, who made the mathematical calculations that predicted the existence of this kind of wave in 1885. A Rayleigh wave rolls along the ground just as a wave rolls across a lake or ocean. Because it rolls, it moves the ground up and down and forward and backward in the direction that the wave is moving. Most of the shaking felt from an earthquake is due to the Rayleigh wave, which can be much larger than the other waves.

Measuring Waves

A *seismogram* is made by a seismometer; it is a record of the arrival and size of body waves and surface waves. When an earthquake occurs, the first wave recorded on a seismogram is a P wave.

? How do these four types of waves differ from one another?

TABLE 12.2 Properties of Seismic Waves

Type	Relative Speed	Passes Through/Along	Type of Motion
Body waves			
P wave	Fastest	Solid and liquid	Pushes and pulls
S wave	Second fastest	Solid only	Up and down or side to side
Surface waves			
Love wave	Third fastest	Earth's surface	Side to side
Rayleigh wave	Slowest	Earth's surface	Up and down, forward and back

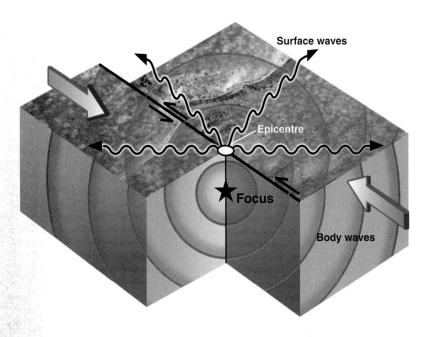

FIGURE 12.16 Earthquakes radiate seismic energy in the form of surface waves and body waves. There are two types of body waves: P (primary) waves and S (secondary) waves; and two types of surface waves: Love waves and Rayleigh waves.

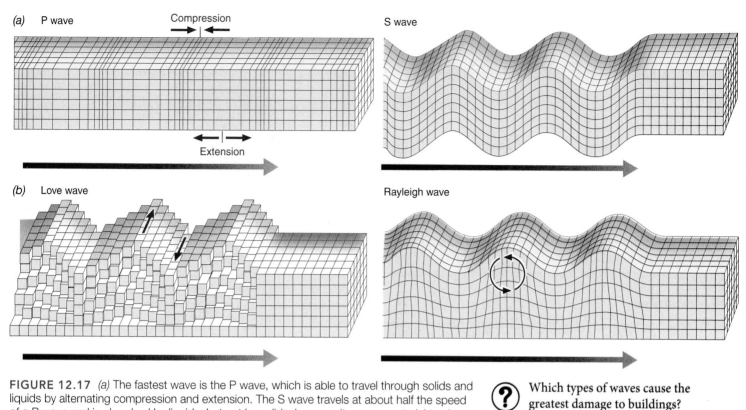

FIGURE 12.17 (a) The fastest wave is the P wave, which is able to travel through solids and liquids by alternating compression and extension. The S wave travels at about half the speed of a P wave and is absorbed by liquids, but not by solids, because it causes material to change shape but not volume. (b) Love waves cause the ground to shift laterally, from side to side. Rayleigh waves cause the ground to roll up and down and shift forward and backward.

(?) **Which types of waves cause the greatest damage to buildings?**

The next wave is the S wave, followed by the surface waves (FIGURE 12.18). These waves arrive at different times because of their different speeds. For instance, the velocity of P waves travelling through granite is about 6 km/s, while S waves travel at about 3.5 km/s. In most conditions encountered in the solid crust, P waves travel about 1.7 times faster than S waves. Surface waves travel at about 90 percent of the speed of S waves.

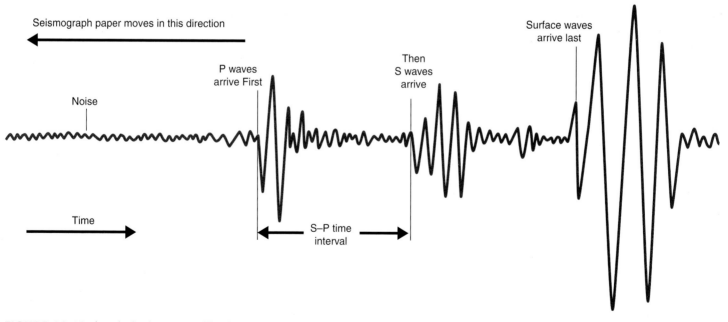

FIGURE 12.18 A typical seismogram. The first wave to arrive at a seismometer is a P wave, followed by the S wave, and then by the surface waves. On some seismograms, the two surface waves are indistinguishable.

(?) Why is the P wave typically the first wave to be recorded by a seismogram?

(?) **Expand Your Thinking**—Explain why *resisting* wave energy is the key to transmitting it.

12-7 Seismometers Are Instruments that Measure and Locate Earthquakes

LO 12-7 *Describe seismometers and how they are used to locate the epicentre of an earthquake.*

Because earthquakes cause such terrible destruction, people have been trying for centuries to measure them and understand how they work. Today geophysicists are able to use instruments that measure and record the vibrations produced by earthquakes. Seismometers can detect, record, and measure the arrival of body waves (P and S waves) and surface waves (Love and Rayleigh waves). Using these measurements, scientists can determine the location of an earthquake's epicentre, the depth of its focus, time of occurrence, and type of faulting. They can also estimate how much energy the earthquake released.

Seismometers

Modern seismometers use highly precise electronic signals to measure seismic energy. Taking advantage of inertia (the tendency of a body to maintain its state of rest or uniform motion unless acted upon by an external force), an electric current keeps a free-hanging weight moving along with the rest of the seismometer (which is attached to the ground) as shaking occurs. Variations in the amount of current needed to keep the weight synchronized to the motions of the ground

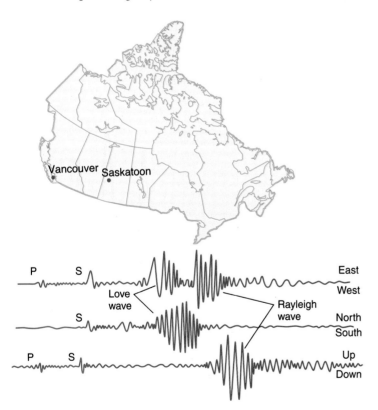

FIGURE 12.19 Seismometer stations measure vibrations in three directions. If an earthquake in Vancouver is measured by an instrument in Saskatoon, seismic waves approaching from the west will register as P waves on the east-west and up-down sensors, as Rayleigh waves on the east-west and up-down sensors, as S waves on all three sensors, and as Love waves on the east-west and north-south sensors.

(?) Why do seismometers measure vibrations in three directions?

provide a sensitive record of seismic energy. Seismometers are so precise that they can even record the vibrations caused by breaking waves and changing tide levels on shorelines many kilometres away.

Since most seismic stations measure three components of vibration (east-west, north-south, and up-down), different directions of motion registered by a seismometer result in records of different kinds of seismic waves. For instance, imagine that a seismometer in Saskatoon records the energy emitted by an earthquake in Vancouver (**FIGURE 12.19**). The seismic waves approaching from the west will register as P waves on the east-west and up-down sensors, as Rayleigh waves on the east-west and up-down sensors, as S waves on all three sensors, and as Love waves on the east-west and north-south sensors.

Finding an Epicentre

Seismologists use networks of seismometer stations to determine the location and estimated depth of an earthquake's focus, exact time of the earthquake, type of faulting, and amount of energy released. These calculations are based on the arrival times of P and S waves. Because these waves travel at different velocities, the spacing between their arrival times at a seismometer station is proportional to the distance from the station to an epicentre (**FIGURE 12.20**). Hence, it is possible for a single station to calculate the distance to an earthquake, but not the direction in which the waves are travelling.

To locate an epicentre, at least three seismic stations are needed. Each station calculates the distance to the epicentre. On a map, the location of each station is surrounded by a circle with a radius equal to the distance to the epicentre. The point at which the three circles intersect marks the location of the epicentre. This process is called *triangulation*.

Earthquake Intensity

There are many ways to measure how much energy an earthquake releases. We now have seismometers that can help us quantify the amount of energy released. (This topic is discussed in the next section.) However, before the widespread use of seismometers, we could only say that an earthquake was "big" or "small" based on the amount of shaking and damage that occurred.

Earthquake *intensity* is used as a qualitative measure of earthquake energy. Intensity is a measure of the physical effects of shaking and the amount of damage caused by an earthquake. For a single earthquake, surrounding areas may have experienced different amounts of damage. Hence, the intensity of an earthquake can differ at different locations. Intensity in a given area depends not only on the energy of the earthquake but also on the area's geology and its distance from the epicentre.

The intensity scale currently in use is the *modified Mercalli (MM) intensity scale,* named after *Giuseppe Mercalli,* an Italian scientist who first described the scale in 1902. The MM scale was developed in 1931 by the American seismologists *Harry Wood* and *Frank Neumann.* Their ranking system is composed of 12 levels of intensity ranging from imperceptible shaking to catastrophic destruction.

The levels are designated by Roman numerals (I–XII). The lower numbers of the intensity scale deal with the manner in which people feel an earthquake. The higher numbers of the scale are based on

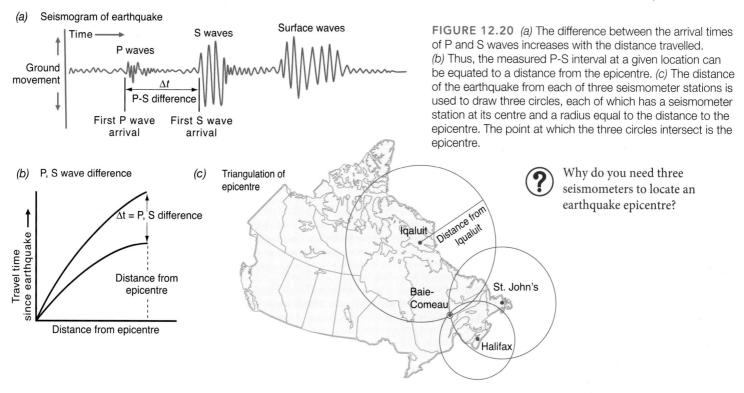

(a) Seismogram of earthquake

(b) P, S wave difference

(c) Triangulation of epicentre

FIGURE 12.20 *(a)* The difference between the arrival times of P and S waves increases with the distance travelled. *(b)* Thus, the measured P-S interval at a given location can be equated to a distance from the epicentre. *(c)* The distance of the earthquake from each of three seismometer stations is used to draw three circles, each of which has a seismometer station at its centre and a radius equal to the distance to the epicentre. The point at which the three circles intersect is the epicentre.

? Why do you need three seismometers to locate an earthquake epicentre?

observed structural damage (**TABLE 12.3**). The MM scale does not have a mathematical basis; rather, it is an arbitrary ranking based on observed effects. Observed effects include descriptions such as people awakening, movement of furniture, damage to chimneys, and total destruction of buildings.

The MM scale is most meaningful to non-scientists because it refers to effects actually experienced in a particular location. To develop maps of earthquake intensity, the Geological Survey of Canada asks persons in a region that has recently experienced an earthquake to fill out a questionnaire about the effects they experienced or witnessed. (These are called *felt reports*.) The information is used to assign intensity values based on firsthand reports. In general, the maximum intensity occurs near the quake's epicentre.

	TABLE 12.3 The Modified Mercalli Intensity Scale
I.	Not felt except by a very few under especially favourable conditions.
II.	Felt only by a few persons at rest, especially on upper floors of buildings.
III.	Felt quite noticeably by persons indoors, especially on upper floors of buildings. Many people do not recognize it as an earthquake. Standing motor cars may rock slightly; vibrations similar to the passing of a truck. Duration estimated.
IV.	Felt indoors by many, outdoors by few during the day. At night, some awakened. Dishes, windows, doors disturbed; walls make cracking sound. Sensation like heavy truck striking building. Standing motor cars rocked noticeably.
V.	Felt by nearly everyone; many awakened. Some dishes, windows broken. Unstable objects overturned. Pendulum clocks may stop.
VI.	Felt by all; many frightened. Some heavy furniture moved; a few instances of fallen plaster. Damage slight.
VII.	Damage negligible in buildings of good design and construction; slight to moderate in well-built ordinary structures; considerable damage in poorly built or badly designed structures; some chimneys broken.
VIII.	Damage slight in specially designed structures; considerable damage in ordinary substantial buildings with partial collapse. Damage great in poorly built structures. Fall of chimneys, factory stacks, columns, monuments, walls. Heavy furniture overturned.
IX.	Damage considerable in specially designed structures; well-designed frame structures thrown out of plumb. Damage great in substantial buildings, with partial collapse. Buildings shifted off foundations.
X.	Some well-built wooden structures destroyed; most masonry and frame structures destroyed with foundations. Rails bent.
XI.	Few, if any, (masonry) structures remain standing. Bridges destroyed. Rails bent greatly.
XII.	Damage total. Lines of sight and level are distorted. Objects thrown into the air.

 ? Expand Your Thinking—An earthquake causes public concern. Why is it important to quickly locate the epicentre?

12-8 Earthquake Magnitude Is Expressed as a Whole Number and a Decimal Fraction

LO **10-2** *Compare and contrast the ways to characterize earthquakes: moment magnitude, Richter magnitude, and intensity.*

Most authorities measure and report the energy released by an earthquake by describing its *magnitude*. Magnitude is described using the **Richter scale,** which measures the largest squiggle on a seismogram. Because this number describes the size of an *earthquake source,* it is the same no matter where you are or what the shaking feels like.

Richter Magnitude

The *intensity* of an earthquake depends on local factors, such as types of buildings and the geologic setting. Because these factors differ from one location to another, the MM scale cannot be used to make objective comparisons between earthquakes in different places—for example, India and New York. The Richter magnitude scale solves this problem by offering a single mathematical value for comparing the size of all earthquakes, and it has been accepted for use around the world. *Charles F. Richter* of the California Institute of Technology proposed the scale in 1935. It calculates earthquake magnitude from the *logarithm of the maximum seismic wave amplitude* recorded by seismometers (**FIGURE 12.21**).

Richter knew that astronomers used a scale to describe the brightness of stars. He reasoned that a similar scale could be useful for studying earthquakes. Like the brightness of stars, the magnitude of earthquakes varies over a wide range. In order to compress the scale to a more manageable range, Richter decided to use the logarithm of the seismic wave amplitude. To use seismic wave amplitude accurately, instruments everywhere must be calibrated to the same degree of sensitivity so that their records can be compared. Adjustments are made to account for differences in distance between various seismometers and an earthquake epicentre as well as the type of displacement that takes place along a fault plane, both of which can influence the nature of the seismogram.

Earthquake magnitude is expressed as a whole number and a decimal fraction (e.g., a light quake may rank 4.8 and a major one may rank 7.1). Since the Richter scale is logarithmic, each whole-number increase in magnitude represents a *tenfold increase* in the amplitude of a measured wave. That is, 1 unit of Richter magnitude describes a factor of 10 difference in ground vibration. The vibration caused by an earthquake of magnitude 4 is 10 times larger than an earthquake of magnitude 3, and a magnitude 7 quake is 100 times greater than a magnitude 5 quake. In terms of energy released by an earthquake, one whole-number increase on the scale is equivalent to about 33 times more energy than the amount associated with the preceding whole-number value.

A *micro-earthquake* typically has a magnitude of about 2.0 or less. It is not commonly felt by people and usually is recorded only on local seismometers. Sensitive seismometers all over the world are capable of recording events of about 4.5 or greater; there are several thousand such shocks annually (**TABLE 12.4**). On average, one *great earthquake* (with a magnitude of 8.0 or greater) occurs somewhere in the world each year. On February 27, 2010, a great earthquake of Richter magnitude 8.8, the seventh strongest earthquake in recorded history, occurred in the South American nation of Chile. The earthquake was so strong that it shifted Earth's axis of rotation by about 8 cm.

The Richter scale has no upper limit, nor is it used to express damage. A moderate quake in a densely populated area that causes many deaths and considerable damage may be of the same magnitude as a quake in a remote area that does nothing more than shake a forest.

 Why is the Richter magnitude preferable to the modified Mercalli scale for comparing the sizes of two earthquakes in different locations?

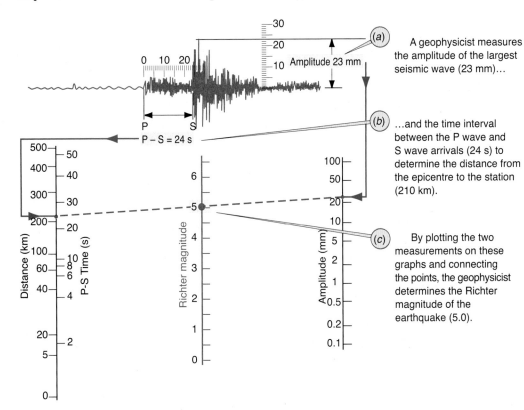

(a) A geophysicist measures the amplitude of the largest seismic wave (23 mm)...

(b) ...and the time interval between the P wave and S wave arrivals (24 s) to determine the distance from the epicentre to the station (210 km).

(c) By plotting the two measurements on these graphs and connecting the points, the geophysicist determines the Richter magnitude of the earthquake (5.0).

FIGURE 12.21 To assign a Richter magnitude to an earthquake, (a) a geophysicist measures the amplitude of the largest seismic wave on a seismogram. (b) The P-S wave interval is measured to determine the distance to the epicentre. Then (c) a line is used to connect these two measurements on a graph that identifies the Richter magnitude of the quake.

TABLE 12.4 Frequency of Occurrence of Earthquakes Since 1900

Descriptor	Richter Magnitude	Average Number Annually
Great	8 and higher	1
Major	7–7.9	18
Strong	6–6.9	120
Moderate	5–5.9	800
Light	4–4.9	6,200 (estimated)
Minor	3–3.9	49,000 (estimated)
Very Minor	< 3.0	1,000 per day

Earthquake Moment Magnitude

Another scale, called the *moment magnitude* scale, has been devised for more precise study of great earthquakes. Moment magnitude is a measure of the amount of energy released by an earthquake. The energy can be estimated based on the amount of rupture that a fault experiences. The moment magnitude tends to spread out the larger earthquakes along a more refined scale of measurement so that they

are not all grouped together, as they are at the top of the Richter scale (**FIGURE 12.22**). Scientists generally prefer moment magnitude over the Richter scale because it is more precise.

The moment magnitude describes something physical about an earthquake: It describes the area of the fault that ruptured. It is calculated in part by multiplying the area of the fault's rupture surface by the distance over which the crust is displaced along the fault. Hence, the most accurate calculation of moment magnitude must be based on field measurements of the rupture surface after the event has taken place. But because many earthquakes originate from faults buried underground (known as *blind faults*), scientists have devised a way to estimate moment magnitude using the recordings on a seismogram.

Large earthquakes occur far less frequently than small ones. As Figure 12.22 shows, each year approximately 1 million earthquakes occur with a moment magnitude greater than 2. The frequency of more powerful quakes decreases by an order of magnitude for every increase in the magnitude of the event. For instance, each year approximately 100,000 earthquakes exceed magnitude 3; 13,000 quakes exceed magnitude 4; and 1,000 exceed magnitude 5. This fact suggests that events larger than magnitude 8 should occur an average of once per year. But in fact they are rarer than that, occurring every three to five years. The very largest recorded quakes, the 9.2 event in Alaska (1964) and the 9.5 event in Chile (1960), appear to be exceptionally uncommon events. The "Critical Thinking" box titled "Earthquake Analysis" offers an opportunity to use what you have learned to analyze a hypothetical quake.

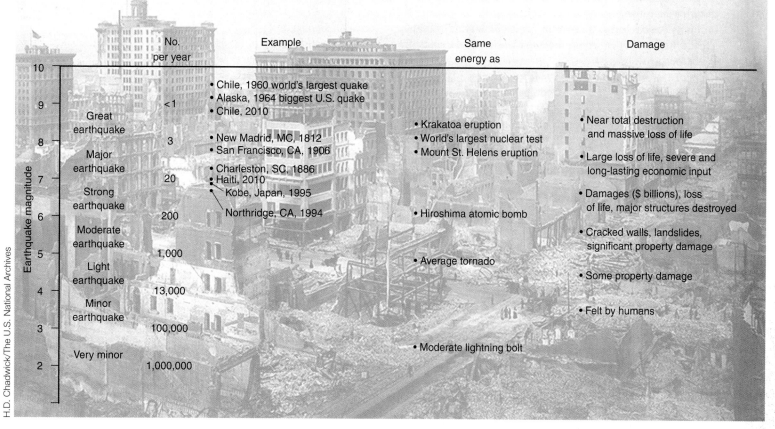

FIGURE 12.22 This chart shows a comparison of earthquakes in terms of magnitude and familiar events.

(?) Some earthquakes last longer than others. What does this tell you about what is happening on the rupturing fault?

(?) **Expand Your Thinking**—What method do the media typically use to characterize an earthquake? Why?

? CRITICAL THINKING

EARTHQUAKE ANALYSIS

You will need the following tools for this exercise: pencil, ruler, and a compass to draw circles.

An earthquake has just occurred in British Columbia. To analyze the earthquake please follow the directions below. FIGURE 12.23a shows an example of how to measure the S-P interval on a seismogram. Figure 12.23b is a chart to calculate the distance from the epicentre using the S-P interval.

1. Use the seismogram in Figure 12.23c from Victoria, B.C.; Figure 12.23d from Merritt, B.C.; and Figure 12.23e from Bella Coola, B.C., to complete the following table.

Station	S-P Interval	Distance from Epicentre
Victoria, B.C.	(s)	(km)
Merritt, B.C.	(s)	(km)
Bella Coola, B.C.	(s)	(km)

2. Use the map in Figure 12.23f to draw a circle around each seismograph station. The radius of each circle should be equal to the distance from the epicentre. Use the map scale to measure the radius of the circle. Find a map of British Columbia and identify the location of the nearest town to the earthquake epicentre. Write this below:

   Epicentre: _____.

3. Two measurements are needed to calculate the magnitude of an earthquake: the S-P interval and the maximum amplitude of the seismic waves. Although only one amplitude measurement is necessary to estimate the magnitude of an earthquake, confidence in the result goes up if measurements from several seismograph stations are used. On each seismogram measure the maximum amplitude of the S wave (use the vertical axis and measure from 0 in one direction). Fill in the following table.

Station	Distance from Epicentre	S-Wave Amplitude
Victoria, B.C.	(km)	(mm)
Merritt, B.C.	(km)	(mm)
Bella Coola, B.C.	(km)	(mm)

4. Use Figure 12.23g to calculate the earthquake magnitude. Plot the station distance and S-wave amplitude from the above table. A line drawn between each pair of points provides an estimated magnitude. Write the magnitude for each station below:

   Earthquake magnitude at:

   Victoria, B.C. _____

   Merritt, B.C. _____

   Bella Coola, B.C. _____

5. What is the average magnitude of the earthquake?
6. Write a one-paragraph press release describing your methodology and conclusions.
7. Using the location of the epicentre, look up the earthquake history of this real event on the web. What is the name of this earthquake and when did it occur?
8. If assigned by your instructor, write a short paper recounting the history of this event.
9. Based on the epicentre and what you have read, build a hypothesis that describes the cause of the earthquake. How would you test this hypothesis? Extrapolate the implications of this earthquake with regard to the seismic hazard in the surrounding region.

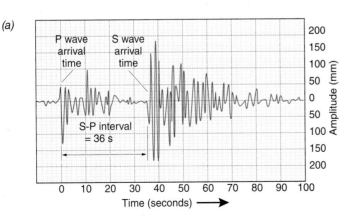

(a)

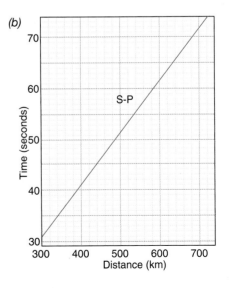

(b)

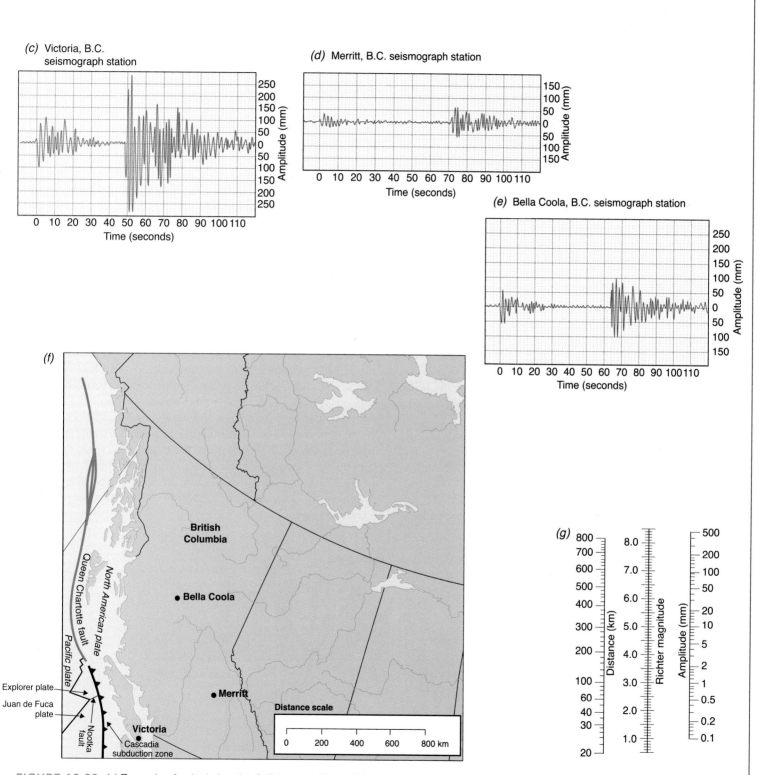

FIGURE 12.23 *(a)* Example of calculating the S-P interval. The S-P interval is 36 seconds as counted on the vertical lines, each of which is equal to 2 seconds; *(b)* relates the S-P interval to distance. Notice 36 seconds is roughly equal to 350 km. Seismograms from *(c)* Victoria, *(d)* Merritt, and *(e)* Bella Coola show the arrival of waves from one earthquake. *(f)* Map of British Columbia. *(g)* Nomogram used to calculate the earthquake magnitude.

12-9 Seismology Is the Study of Seismic Waves in Order to Improve Understanding of Earth's Interior

LO 12-9 *Describe how seismic wave characteristics result in P-wave and S-wave shadow zones.*

One of the great discoveries in the science of geophysics was the realization that seismic waves from earthquakes can be used to improve our understanding of Earth's interior.

An important property of seismic waves is that their velocity changes when they encounter differences in rock layers due to changes in pressure, temperature, or composition. In addition, they *refract* (change direction) and *reflect* (bounce off) when passing across the interface between two layers in Earth's interior. Because of this property, differences in the nature of solid and fluid layers within Earth's interior, as well as differences between solid layers, are recorded in the behaviour of seismic waves. For this reason, hundreds of seismometer stations around the world create a network that is able to record detailed aspects of Earth's interior. With every new large earthquake, we gain new data about Earth's internal layers.

The P-Wave Shadow Zone

According to Snell's law, **wave refraction** is caused by changes in wave velocity due to differences in the density of the materials through which a wave passes. Seismic waves have a higher velocity in denser rock and a lower velocity in less dense rock. If a seismic wave passes into rock in which its velocity increases, it is refracted upward relative to its original path. A wave passing into rock in which its velocity decreases is refracted downward. **FIGURE 12.24** illustrates this natural phenomenon.

Within Earth's interior, seismic wave velocities generally increase with depth because rock density generally increases with depth and waves travel faster in denser material. As a result, waves are continually refracted along curved paths (called *ray paths*) that arc gently back toward Earth's surface (**FIGURE 12.25a**). However, if there is a sudden change in rock density, the wave velocity will change suddenly;

such an interface is called a **discontinuity.** Discontinuities found at well-mapped depths in the interior of the planet help us mark boundaries between various layers, such as core, mantle, and crust.

P waves change the *volume* of material that they encounter. Hence, any material that resists changing volume, including most liquids and solids, will transmit P waves rather than absorb them. As P waves generated by an earthquake travel through Earth's interior, they refract as they cross a seismic discontinuity at about 2,900 km depth, which marks the boundary between the mantle and the outer core. P waves refract again when they encounter the inner core and when they exit the inner and outer cores. These changes in the path of P waves leave a region of Earth's surface, on the far side of the earthquake source, that does not receive any P waves. The result is a *P-wave shadow zone* on the far side of the core that is created by wave refraction (**FIGURE 12.25b**).

The S-Wave Shadow Zone

S waves change the shape of materials; therefore, liquids, which do not resist changes in shape, absorb S waves rather than transmit them. As a result, S waves that encounter the liquid outer core are absorbed, creating an *S-wave shadow zone* on the far side of the core (**FIGURE 12.25c**). Any seismometer located such that the outer core lies between it and an earthquake focus will not receive any S waves.

Together, wave refraction and the P- and S-wave shadow zones reveal important details about Earth's interior that would not otherwise be available. The locations of the P- and S-wave shadow zones depend on where the earthquake occurs as well as the shape and size of the mantle and core. Because there are many earthquakes in many locations, seismologists have been able to use this information to map Earth's interior.

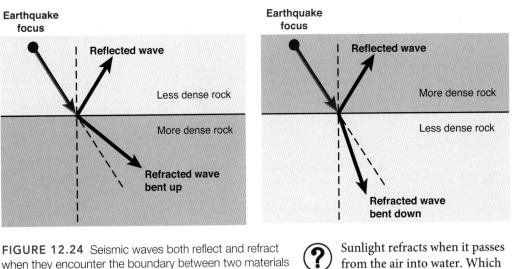

FIGURE 12.24 Seismic waves both reflect and refract when they encounter the boundary between two materials of differing density.

(?) Sunlight refracts when it passes from the air into water. Which way does a light ray refract: upward or downward?

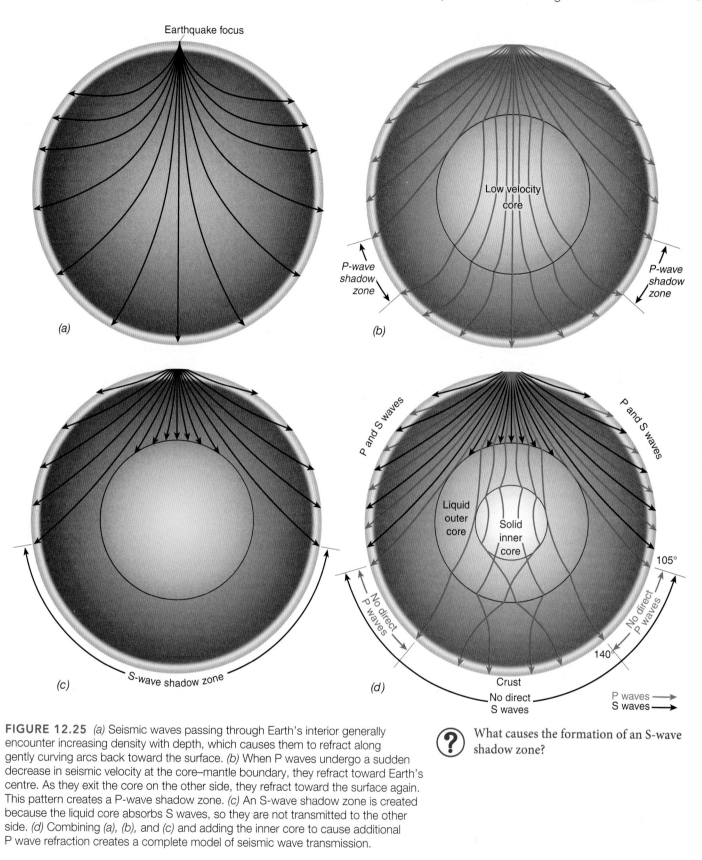

FIGURE 12.25 *(a)* Seismic waves passing through Earth's interior generally encounter increasing density with depth, which causes them to refract along gently curving arcs back toward the surface. *(b)* When P waves undergo a sudden decrease in seismic velocity at the core–mantle boundary, they refract toward Earth's centre. As they exit the core on the other side, they refract toward the surface again. This pattern creates a P-wave shadow zone. *(c)* An S-wave shadow zone is created because the liquid core absorbs S waves, so they are not transmitted to the other side. *(d)* Combining *(a)*, *(b)*, and *(c)* and adding the inner core to cause additional P wave refraction creates a complete model of seismic wave transmission.

 What causes the formation of an S-wave shadow zone?

 Expand Your Thinking—If Earth had no core, how would we know?

12-10 Seismic Data Confirm the Existence of Discontinuities in Earth's Interior

LO 12-10 *List the major discontinuities in Earth's interior and their depths.*

Seismologists have accumulated data that confirm the existence of several pervasive discontinuities marking distinct layers in Earth's interior. These layers are the result of variations in the chemistry and physical properties of rock lying beneath the crust.

The Moho

The first major discontinuity marks the border between the crust and the mantle. This boundary was identified in the early 20th century by a Croatian scientist named *Andrija Mohorovicic*. Studying seismic records, Mohorovicic noticed the arrival of a pair of P and S waves, followed by another, slower pair of P and S waves. The slower pair, he reasoned, had travelled more or less directly through the crust, and the faster set had probably refracted at a boundary in Earth's interior and travelled at some depth in a higher-velocity zone. Using critical thinking, he inferred that beneath the crust was a layer of denser rock (thus permitting higher seismic velocity) that must have refracted and sped along the first-arriving set of waves. Based on this inference, Mohorovicic hypothesized the existence of a distinct boundary that separates the crust from some lower-lying portion of Earth's interior that has a different rock composition.

The boundary between the crust and the mantle, its existence now confirmed many times since it was first described by Mohorovicic, is named the **Mohorovicic discontinuity** or *Moho* for short. The depth of the Moho is known to vary from approximately 8 km beneath ocean basins to 20 km to 70 km beneath the continents. As we saw in Chapter 11 ("Mountain Building"), thicker portions of continental crust are typically found beneath mountain belts in the form of deep roots needed to sustain the mass of the crust. By comparing laboratory measurements of the velocity with which seismic waves travel through various kinds of rock, it is possible to infer the composition of the crust. Oceanic crust is generally composed of basalt, with the lower portion consisting of gabbro, while continental crust varies from granite near the top to gabbro near the Moho. These findings are consistent with other evidence, such as drilling, mapping of rock outcrops, and seismic studies based on artificial sound sources.

The Mantle

Beneath the Moho lies the mantle. Laboratory tests of rocks that are rich in dense minerals such as olivine, pyroxene, and garnet show seismic velocities of about 8 km/s, while rocks of granite, gabbro, and basalt have velocities of 6 km/s to 7 km/s. These measurements are consistent with P-wave velocities that increase abruptly at the Moho from about 6 km/s to 7 km/s to 8 km/s to 12 km/s. From this evidence, seismologists infer that the mantle is composed of dense, ultramafic rock such as *peridotite* (Chapter 5, "Igneous Rock"). The few actual samples of mantle rock that are available—mantle fragments found in magmas and diamond-rich *kimberlite pipes* of mantle rock that intrude the crust—confirm the composition of the mantle inferred from seismic studies.

Within the upper mantle, at a depth of about 100 km, there is a significant, though subtle, decrease in both P- and S-wave velocities that marks the base of the lithosphere and the top of the *asthenosphere*. Recall from Chapter 3 ("Plate Tectonics") that the lithosphere is composed of both the crust and the upper part of the mantle. The lithosphere is brittle solid rock broken into plates that sit on a *low-velocity zone* (LVZ) marking the asthenosphere. Lithospheric plates appear to float and move around on top of the more ductile asthenosphere. The seismic velocity of P waves increases below the Moho until they reach a depth of about 100 km, beyond which it decreases to below 8 km/s. This second discontinuity is thought to be the boundary between the base of the lithosphere and the top of the asthenosphere, marked by the LVZ (FIGURE 12.26).

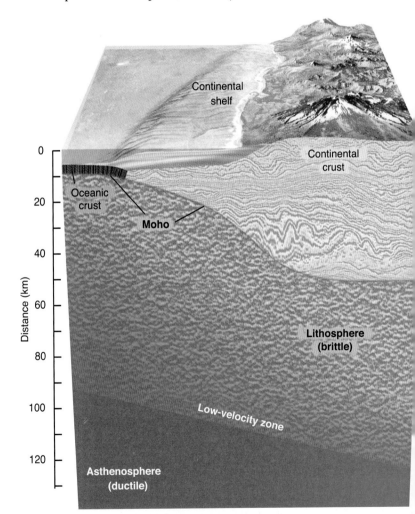

FIGURE 12.26 Below the Moho lies the lower lithosphere, which is a brittle solid, similar to the crust but denser, that represents the lower portion of plates. The lithosphere forms a boundary with the asthenosphere at the low-velocity zone (LVZ), which exhibits ductile deformation, perhaps related to partial melting.

(?) Why is the thickness of oceanic crust different from that of continental crust?

Much has yet to be learned about the LVZ, but it is thought to extend to a depth of approximately 350 km and to be composed of rock that is closer to its melting point than the rock above or below it. This makes rock in the LVZ less rigid and more ductile than the layers above and below. One hypothesis for the origin of the LVZ is based on the fact that minerals of the mantle are susceptible to melting over a range of temperatures. Some amount of melting might have produced a liquid component that could act as a lubricant. If this were the case, it would increase the ductile behaviour of the LVZ. However, because S waves, which do not travel through liquid, are successfully transmitted by the LVZ, we know that relatively little melting must occur.

At a depth of 400 km, there is another abrupt, though small, increase in the velocities of both S and P waves that marks the lower boundary of the upper mantle and the top of the *mantle transition zone* (FIGURE 12.27). This boundary, known as the *400-km discontinuity,* is thought to be the result of a denser reorganization of the atoms in the mineral olivine, the most abundant mineral in the mantle. Laboratory experiments show that when samples of olivine are subjected to pressure and temperature conditions equivalent to those at a depth of 400 km, the atoms recrystallize into a form known as the *spinel* mineral family, which is about 10 percent denser than olivine. The increase in P- and S-wave velocities at the 400-km discontinuity is probably the result of the olivine–spinel transition and the resultant increase in density.

Another abrupt increase in seismic wave velocities occurs at a depth of 660 km, marking the base of the mantle transition zone and the top of the lower mantle. It is uncertain whether this discontinuity, known as the *660-km discontinuity,* is the result of a mineral change in the mantle or a change in chemistry, or both. One hypothesis describes a switch from pyroxene in mantle rocks to a mineral known as *perovskite,* which crystallizes at high pressure. Pyroxene, a silicate mineral, has four oxygen atoms packed around one silicon atom (recall Chapter 4, "Minerals"). Perovskite possesses a denser structure, with six oxygen atoms packed on one silicon atom. Laboratory experiments suggest that minerals in the spinel family, supposedly formed at the 400-km discontinuity, also reorganize themselves to form perovskite at pressures found at a depth of 660 km.

The Core

At a depth of 2,900 km, seismic P-wave velocities suddenly decrease and S-wave velocities go to zero. This is the top of the outer core. Seismologists infer that this layer is liquid, since S waves are not transmitted. A sudden increase in P-wave velocities at a depth of approximately 4,800 km indicates that the inner core is solid, composed mostly of iron and small amounts of nickel and sulphur.

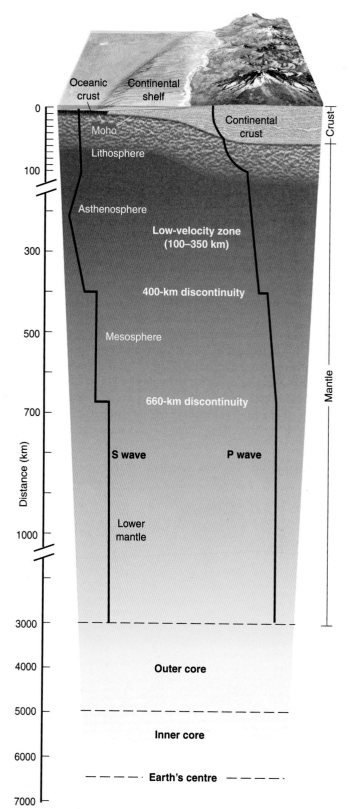

FIGURE 12.27 Changes in the seismic velocity of P and S waves mark discontinuities at 100 km (the low-velocity zone), 400 km (the base of the upper mantle), 660 km (the top of the lower mantle), 2,900 km (the top of the core), and 4,800 km (the inner core boundary).

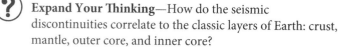

Expand Your Thinking—How do the seismic discontinuities correlate to the classic layers of Earth: crust, mantle, outer core, and inner core?

Why would the mineralogy of the mantle change from olivine to denser spinel and then to even denser perovskite?

12-11 Seismic Tomography Uses Seismic Data to Make Cross-Sections of Earth's Interior

LO 12-11 *Describe seismic tomography and what it reveals about Earth's interior.*

In recent decades, the P and S waves from hundreds of earthquakes have been tracked at hundreds of seismometer stations, producing a global database of seismic velocities at all levels within Earth. The science of **seismic tomography** utilizes these data to make cross-sectional slices through Earth in order to examine the nature of the planet's interior. *Tomography* means literally "drawing slices"—in this case, slices of Earth. The method is borrowed from the medical computerized axial tomography technique (CAT scan) that is widely used in hospitals. CAT scans use small differences in X-rays that sweep the body to map variations in tissue and organs in order to construct cross-sections and three-dimensional images of organs.

Variations in the velocity of seismic waves depend on the temperature, pressure, and chemical composition of Earth's interior. However, if the physical parameters that determine seismic velocities varied only with depth, our planet would be a very dull place, basically consisting of little more than parallel layers. No mountains would be formed; there would be no active volcanism; and there would be no earthquakes. However, as revealed in Figure 3.5, seismic tomography has shown us that Earth's interior is much more compelling. It is *lateral variations* in temperature, pressure, and composition (and hence

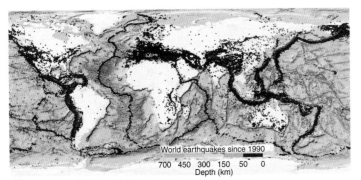

(a) Global earthquake epicentres

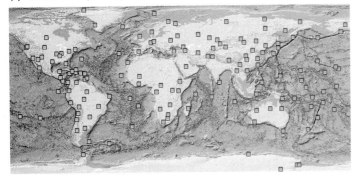

(b) Global seismograph stations

FIGURE 12.28 The uneven distribution of earthquakes *(a)* largely along plate margins and seismometer stations *(b)* largely in the Northern Hemisphere and continental settings means that global tomographic data are rich in some areas and sparse in others.

(?) How would you solve the problem of having no seismometers in oceanic areas?

in seismic wave velocities) that create hot and cold regions in Earth's interior. These regions reflect various phenomena discussed in earlier chapters, such as mantle plumes, continental roots, subducted slabs, and other complexities inside the planet. Seismic tomography allows seismologists to map these features by determining the size, shape, and location of three-dimensional variations in wave speed.

One of the major problems in developing accurate three-dimensional images of Earth's interior stems from the uneven distribution of earthquakes and seismometers around the globe (**FIGURE 12.28**). The majority of earthquakes occur along plate margins. Hence, large sections of crust provide no source data for P- and S-wave velocities because they have not experienced large earthquakes in the modern era of instrumentation.

Another problem is that seismometer stations are located predominantly in continental settings in the Northern Hemisphere. Hence, data from the Southern Hemisphere and oceanic locations are underrepresented. As a result, although seismic data are globally abundant, they are not necessarily always available in critical areas needed to define details of Earth's interior. Researchers are still experimenting with all types of seismic waves—Rayleigh waves, Love waves, P waves, and S waves—to determine which type offers the greatest ability to form accurate images and thus improve our understanding of Earth's interior.

Tomographic mapping of Earth's interior has revealed new information about the organization of the mantle and core. Because of differences in the density of rock within the mantle, the various discontinuities discussed in Section 12-10 possess rugged topography, some of which may be related to subducted slabs of oceanic crust that have fallen into the mantle. **FIGURE 12.29** shows the P-wave tomography of the mantle. Areas in blue depict cold, dense bodies of rock where P-wave velocities are high, while areas in yellow and red reveal regions of hot, less dense rock, where P-wave velocities are low. This image displays a fallen slab (also called a *stagnant slab*) of the subducted Pacific plate, lying on the 660-km discontinuity under Japan and the Sea of Japan. (Researchers think that stagnant slabs are largely inactive former plates located in the mantle.) Below this is an extensive high-velocity area in the lower mantle that is interpreted to be a "slab graveyard" where other cold slabs of subducted oceanic crust have fallen and collected on the core–mantle boundary. Also shown are two mantle plumes: the Pacific plume and the African plume.

As shown in Figure 12.29, tomographic mapping at the latitude of California (~38°N) reveals features that are explained by plate tectonic theory. Cold regions below oceanic trenches appear to be descending slabs of oceanic lithosphere. Hot regions below mid-ocean ridges suggest the presence of thermally buoyant plumes in the mantle supplying new rock to the lithosphere. Slab graveyards are identified where cold oceanic crust has not been fully recycled into the mantle. **FIGURE 12.30a** uses variations in S-wave velocity to construct a tomographic cross-section of the mantle below the North American plate (NAM). The cross-section reveals several features that are consistent with plate tectonic theory. At the surface, the cold, dense rock of the North American plate is revealed in dark blue. Beneath NAM,

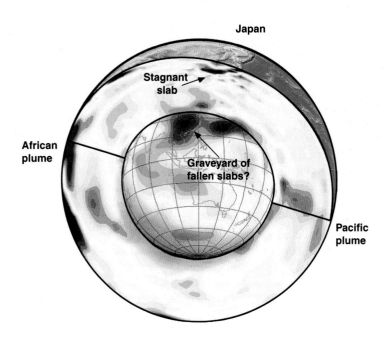

FIGURE 12.29 Earth's mantle mapped using the velocity of seismic P waves. Red areas consist of hot, low-density rock and blue areas consist of cold, high-density rock. This cross-section reveals two mantle plumes: the Pacific plume and the African plume. Also displayed is a stagnant slab of the Pacific plate where it has been subducted beneath Japan (the Honshu arc) and lies on the 660-km discontinuity. Other stagnant slabs have fallen to the deep mantle and collected as a "slab graveyard" at the core–mantle boundary.

(?) Why does the mantle have low-density and high-density areas?

the figure shows remnants of the Farallon plate (FAR) as a strong greenish-blue feature. In the western region, FAR is bluish and has stagnated on the 660-km discontinuity. In the central and eastern regions, FAR is greenish-blue and has penetrated the mantle below the 660-km discontinuity and extends to the central and lower mantle. **FIGURE 12.30b** is an interpretation of how FAR, NAM, and the

Pacific plate (PAC) have interacted over the past 80 million years. Westward movement of NAM eventually overran the spreading centre at the divergent boundary of PAC and FAR. FAR subducted below North America between 80 and 40 million years ago, but by 20 million years ago, it had detached from the crust, and portions of it lie as a stagnant slab on the 660-km discontinuity.

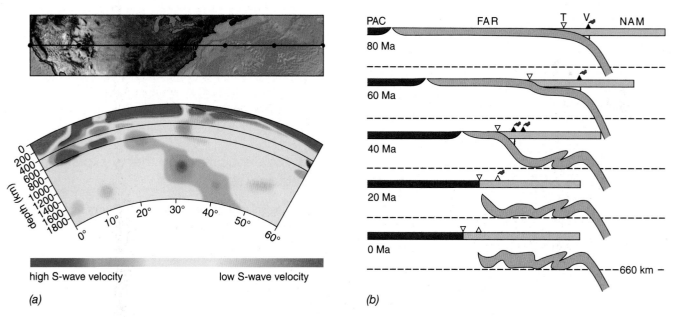

(a)

(b)

FIGURE 12.30 *(a)* Tomographic cross-section of the mantle spanning the North American plate at the latitude of California (~38°N). S-wave velocities are colour-coded. Each dot on the map corresponds to a 10° shift in longitude shown under the cross-section. *(b)* Interpretation of how the Farallon plate (FAR), North American plate (NAM), and the Pacific plate (PAC) have interacted over the past 80 million years. T = trench and V = volcano. (*Source:* Reprinted with permission from Christian Schmid, et al., "Fate of the Cenozoic Farallon Slab from a Comparison of Kinematic Thermal Modeling with Tomographic Images," *Earth and Planetary Science Letters*, November 20, 2002. © Elsevier.)

(?) What sorts of features do you see in Earth's interior in Figure 12.30a? Explain the processes responsible for these features.

 Expand Your Thinking—How do you reconcile the model of mantle convection that we discussed in Chapter 3, "Plate Tectonics" (see Figure 3.6), with major discontinuities consistent with a layered interior?

LET'S REVIEW "GEOLOGY IN OUR LIVES"

Now that you have finished the chapter, "Geology in Our Lives" will have taken on new meaning for you. Let us review it: An earthquake is a sudden shaking of the crust usually caused by the rupture of a fault. Earthquakes produce seismic waves that scientists can use to improve our understanding of Earth's interior. Earthquakes are also dangerous geologic hazards that are responsible for enormous damage to many communities. More than 380 major cities lie on or near unstable regions of Earth's crust, and, because of swelling urban populations, geologists fear that a devastating earthquake capable of killing 1 million people could occur this century. But crowded inner cities are not the only concern—global growth in human population in the past 100 years means that people now live in areas that were previously remote and thought to be too dangerous for communities. Based on knowledge of seismic risk, designing and constructing buildings that can withstand shaking and establishing effective earthquake response and disaster assistance networks have become important elements of managing these hazards in our communities.

Earthquakes can cause horrible devastation and loss of life. Ground shaking, hillside failure, tsunamis, and fires all result from seismic events. Over history, more than 20 earthquake events have each caused the loss of more than 50,000 human lives. An earthquake is a sudden shaking of the crust typically caused by the abrupt release of strain that has slowly accumulated in rock. Seismic energy, released by a quake, consists of surface waves and body waves that travel through Earth's interior and along the surface. Earthquakes occur at faults, and many faults are related to the dynamic interaction of plate margins. But quakes also occur in plate interiors away from the margins. Seismic waves are measured by seismometers, hundreds of which, located throughout the world, monitor and record the passage of seismic waves created by earthquakes occurring every day at an average of one every 11 seconds.

STUDY GUIDE

12-1 An earthquake is a sudden shaking of the crust.

- An **earthquake** is a sudden shaking of the crust. It is typically caused by the abrupt release of strain that has slowly accumulated in rock. Most rocks are elastic, so when stress is applied slowly and continuously over time, rocks will deform elastically. That is, they will rebound back to their original shape if the stress is removed or when the rock fractures.

- Most earthquakes are associated with the rupture of a fault. When rock breaks or a fault abruptly ruptures, it releases strain in a form of energy called **seismic waves,** which radiate through Earth's interior and along its surface. Seismic waves cause the shaking of an earthquake.

- More than 380 major cities lie on or near unstable regions of Earth's crust, and seismologists fear that a devastating earthquake capable of killing 1 million people could occur this century. With greater infrastructure located across more of Earth's surface, our investments are increasingly vulnerable to earthquake damage.

12-2 There are several types of earthquake hazards.

- *Ground shaking* caused by earthquakes can produce *sediment liquefaction, mass wasting,* and building collapse. Liquefaction occurs when groundwater rises to the surface and causes sediment to lose strength, leading to the collapse of buildings and roads. Any hillside that is overlain by weak or fractured bedrock, thick soil, or layers of unconsolidated sediment can experience mass wasting when exposed to ground shaking. Buildings that have a natural frequency that resonates with the shaking of an earthquake will tend to collapse.

- The National Research Council of Canada and the U.S. Geological Survey produce maps of **seismic hazards** that are used by engineers and city planners in designing and constructing structures to withstand shaking.

- Fires following an earthquake can be a serious problem. Often water lines and emergency equipment are damaged during an earthquake, making it impossible for firefighters and other emergency personnel to do their work.

- Earthquakes may generate a **tsunami,** a series of waves in the ocean created when the sea floor is rapidly displaced. Tsunamis can be several metres high when they hit a shoreline and can do enormous damage.

12-3 The elastic rebound theory explains the origin of earthquakes.

- Geologists use the **elastic rebound theory** to explain why earthquakes occur. Distant forces cause a gradual buildup of elastic energy in the crust. Rocks accumulate strain, much as a stick stores elastic energy by bending. Eventually, a fault or a pre-existing weakness in the rock fails by fracturing.

- The earthquake **focus** is the location where rocks first fail. Slippage occurring at the focus is a source of stress that travels along the fault and causes displacement until the built-up strain is largely released. At the time of the earthquake, strained rock elastically snaps releasing the pent-up stress, causing vibrations of the rock that are referred to as *seismic waves.*

- As strain accumulating in the crust approaches the elastic limit of rocks, earthquakes tend to occur in clusters. The largest of a cluster is the **main shock;** anything occurring before it is a **foreshock,** and anything occurring after it is an **aftershock.**

12-4 Most earthquakes occur at plate boundaries, but intraplate seismicity is also common.

- Earthquakes are not randomly spread across the globe. For the most part, they occur at locations where plate boundaries meet and brittle rocks of the crust fracture as plate edges separate, collide, and shear across each other.

- Intraplate earthquakes are often related to stresses transmitted into plate interiors that originate at the plate edge. Many of these are caused by the reactivation of old rift zones or faults that are normally quiet. Such quakes tend to have shallow foci and are left over from former tectonic action that has ceased, or they may mark the establishment of a new region of deformation. Intraplate seismicity may also be related to magma movement associated with hotspots in plate interiors, such as at Yellowstone or in Hawaii.

- Earthquakes occur in four kinds of plate settings: (1) intraplate regions, (2) divergent boundaries, (3) convergent boundaries, and (4) transform boundaries.

12-5 Divergent, convergent, and transform boundaries are the sites of frequent earthquake activity.

- At divergent boundaries, seismic activity tends to occur at shallow depths. The lithosphere at a divergent boundary is thin and weak; hence large amounts of strain do not accumulate to cause major earthquakes. However, significant damaging quakes can occur.

- The largest and most active seismicity occurs at convergent boundaries, where one plate subducts beneath another in a zone of actively accumulating compressive stress. These regions of plate interaction are characterized by progressively deeper earthquakes along a line proceeding from the trench toward the overriding plate.

- Earthquake foci at convergent margins define a steeply dipping plane called the **Wadati-Benioff zone,** or WBZ, that corresponds to faulting between the subducting oceanic slab and the upper mantle of the overriding plate.

- At transform boundaries, earthquakes are shallow, running only as deep as 25 km. Three dangerous examples of transform seismicity are found at the Queen Charlotte fault in British Columbia, the San Andreas fault in California, and the North Anatolian fault in Turkey. All are right-lateral strike-slip faults.

12-6 Earthquakes produce four kinds of seismic waves.

- Earthquakes radiate seismic energy in the form of two types of *surface waves* and two types of *body waves*. Surface waves consist of **Love waves** and **Rayleigh waves.** Body waves consist of primary waves, or **P waves,** and secondary waves, or **S waves.** Because liquids do not resist changes in shape, they absorb S waves rather than transmit them.

12-7 Seismometers are instruments that measure and locate earthquakes.

- A **seismometer** is an instrument that records seismic waves. A record of seismic waves is called a seismogram. One seismometer station will measure three types of ground motion: north-south, east-west, and up-down. Seismograms allow scientists to estimate the distance, direction, size, and type of faulting of an earthquake.

- Earthquake intensity is measured by the modified Mercalli (MM) intensity scale, which ranks the level of damage caused by the quake on a scale from I to XII.

12-8 Earthquake magnitude is expressed as a whole number and a decimal fraction.

- The magnitude of a quake is measured using the **Richter scale,** which measures the largest wave amplitude on a seismogram. Earthquake magnitude is calculated from the logarithm of the maximum seismic wave amplitude recorded by seismometers.

- The moment magnitude is calculated in part by multiplying the area of the fault's rupture surface by the distance the crust has been displaced along a fault. Hence, the most accurate calculation of moment magnitude is based on field measurements of the rupture surface.

12-9 Seismology is the study of seismic waves in order to improve understanding of Earth's interior.

- Seismic **wave refraction** is caused by changes in wave velocity due to differences in density in Earth's interior. If a wave passes into rock of higher seismic velocity, the wave will be refracted upward relative to its original path. A wave passing into rock of lower seismic velocity will be refracted downward. Wave refraction is used to identify a **discontinuity.**

- P-wave and S-wave shadow zones on the far side of the core are created by wave refraction and absorption due to changes in the material through which they pass.

12-10 Seismic data confirm the existence of discontinuities in Earth's interior.

- Seismologists have confirmed the existence of several pervasive discontinuities marking distinct layers in Earth's interior. The discontinuity between the crust and the mantle is named the **Mohorovicic discontinuity,** or Moho, at a depth of approximately 8 km beneath ocean basins to 20 km to 70 km beneath continents. The base of the lithosphere and the top of the asthenosphere is marked by a discontinuity called the *low-velocity zone* (LVZ) at an average depth of about 100 km.

- At a depth of 400 km, an abrupt, though small, increase in the velocities of both S and P waves marks the lower boundary of the upper mantle and the top of the *mantle transition zone*. Another abrupt increase in seismic wave velocities occurs at a depth of 660 km, marking the base of the mantle transition zone and the top of the lower mantle. At a depth of 2,900 km, seismic P-wave velocities suddenly decrease and S-wave velocities go to zero. This is the top of the liquid outer core. A sudden increase in P-wave velocities at a depth of approximately 4,800 km indicates that the inner core is solid, composed mostly of iron and small amounts of nickel and sulphur.

12-11 Seismic tomography uses seismic data to make cross-sections of Earth's interior.

- **Seismic tomography** uses changes in P- and S-wave velocity to make cross-sectional slices in order to examine the nature of rock in Earth's interior. Tomography reveals significant differences in rock density within the mantle, which researchers interpret as mantle plumes, stagnant slabs of subducted crust, and slab graveyards on the core–mantle boundary.

KEY TERMS

aftershocks (p. 311)
body waves (p. 310)
discontinuity (p. 326)
earthquake (p. 304)
elastic rebound theory (p. 310)
epicentre (p. 310)
focus (p. 310)
foreshocks (p. 311)

Love waves (p. 318)
main shock (p. 311)
Mohorovicic discontinuity (p. 328)
P waves (p. 318)
Rayleigh waves (p. 318)
Richter scale (p. 322)
S waves (p. 318)
seismic hazards (p. 307)

seismic tomography (p. 330)
seismic waves (p. 304)
seismometers (p. 310)
surface waves (p. 310)
tsunami (p. 308)
Wadati-Benioff zone (p. 314)
wave refraction (p. 326)

ASSESSING YOUR KNOWLEDGE

Please answer these questions before coming to class. Identify the best answer.

1. Earthquake risk has increased in recent decades because
   a. there are more earthquakes.
   b. earthquakes have grown larger.
   c. the human population has expanded.
   d. plate margins have become more dangerous.
   e. All of the above.

2. In regions lacking modern engineering standards, the greatest hazard during an earthquake is
   a. building collapse.
   b. ground fractures.
   c. violent shaking.
   d. groundwater withdrawal.
   e. None of the above.

3. Earthquake hazards include
   a. shaking, fires, landslides, and tsunamis.
   b. tsunamis and liquefaction.
   c. lack of communication and lack of food and water due to damaged public services.
   d. lack of emergency help, building collapse, and mass wasting.
   e. All of the above.

4. What is an earthquake?
   a. a sonic boom
   b. any rock fracture
   c. sudden rapid shaking of Earth's crust
   d. any breakage of rock
   e. None of the above.

5. The elastic rebound theory
   a. describes how seismic waves pass through Earth's interior.
   b. describes how folding in the crust dampens earthquake magnitude.
   c. is a theory describing the behaviour of faulting crust.
   d. was proven wrong in the middle of the 20th century.
   e. was first defined by Archimedes.

6. Earthquakes at divergent margins tend to be
   a. deep and strong.
   b. produced by normal faulting.
   c. on the surface of a subducting slab.
   d. related to compression.
   e. caused by anticlines.

7. Examples of Canadian intraplate seismicity include
   a. Vancouver.
   b. Baie Comeau.
   c. Halifax.
   d. Whitehorse.
   e. St. John's.

8. Seismicity at divergent, convergent, and transform plate boundaries is caused by the following stresses (in correct order):
   a. compressive, shear, and subductive
   b. transverse, subductive, and decompressive
   c. tensional, compressive, and shearing
   d. normal, reverse, and plunging
   e. None of the above.

9. Which type of seismic wave causes changes in the volume of material?
   a. P wave
   b. S wave
   c. Rayleigh wave
   d. Love wave
   e. None of the above.

10. Finding an epicentre requires
    a. firsthand accounts.
    b. seismic tomography.
    c. wave refraction.
    d. triangulation.
    e. calculating the intensity.

11. The first seismic wave recorded on a seismogram is the
    a. Rayleigh wave.
    b. S wave.
    c. P wave.
    d. Love wave.
    e. All body waves arrive at the same time.

12. Earthquake magnitude is measured by
    a. the Richter scale.
    b. the modified Mercalli scale.
    c. the epicentre.
    d. triangulation.
    e. the level of damage.

13. Seismic shadow zones are the result of
    a. wave refraction at discontinuities and absorption in the outer core.
    b. wave reflection and interference in the mantle.
    c. seismic amplitude causing discontinuities in the crust.
    d. a lack of large earthquakes.
    e. a lack of seismometers in the Southern Hemisphere.

14. Seismic discontinuities are found at the following depths:
    a. 8 km to 20–70 km, 100 km, 400 km, 660 km, 4,800 km, and 2,900 km
    b. 20 km to 50 km, 70 km, 3,500 km, and 2,900 km
    c. 20 km to 50 km, but only at subduction zones
    d. 8 km, 100 km, 1,000 km, 2,000 km, 3,000 km, and 4,000 km
    e. None of the above.

15. Seismic tomography reveals that Earth's interior is
    a. like a layer cake with four layers.
    b. like a layer cake but interrupted by plumes.
    c. highly complex with hot regions and cool regions.
    d. too complex to understand.
    e. made entirely of olivine to the inner core.

FURTHER RESEARCH

1. Are there likely to be earthquakes on other planets? Why or why not?

2. How do seismic waves destroy buildings?

3. What can we predict about future earthquakes?

4. How do earthquakes start?

5. What is meant by elastic rebound?

6. What factors contribute to the amount of destruction caused by an earthquake?

ONLINE RESOURCES

Explore more about geologic resources on the following websites:

Learn all about earthquakes, Canadian earthquake history and hazard maps, and earthquake research in Canada:
www.earthquakescanada.nrcan.gc.ca

Learn about tips for earthquake preparedness provided by Emergency Management British Columbia:
http://embc.gov.bc.ca/em/hazard_preparedness/earthquake_
information.html

Investigate research by the U.S. Geological Survey at the USGS Earthquake Hazards Program:
http://earthquake.usgs.gov

Learn about quake hazards in Washington and Oregon at the Pacific Northwest Seismograph Network:
www.pnsn.org

The U.S. Federal Emergency Management Agency (FEMA) has an earthquake hazard reduction page:
http://www.fema.gov/national-earthquake-hazards-reduction-
programs

Additional animations, videos, and other online resources are available at this book's companion website:
www.wiley.com/go/fletchercanada

This companion website also has additional information about WileyPLUS and other Wiley teaching and learning resources.

CHAPTER 13
GEOLOGIC TIME

Chapter Contents and Learning Objectives (LO)

13-1 Geology is the "science of time."
LO 13-1 *Compare and contrast relative dating and isotopic dating.*

13-2 Earth history is a sequence of geologic events.
LO 13-2 *Describe early attempts to determine Earth's age.*

13-3 Seven stratigraphic principles are used in relative dating.
LO 13-3 *List and describe the seven stratigraphic principles.*

13-4 Relative dating determines the order of geologic events.
LO 13-4 *Describe the use of stratigraphic principles to construct a history of geologic events.*

13-5 Isotopic dating uses radioactive decay to estimate the age of geologic samples.
LO 13-5 *Define the concept of "half-life."*

13-6 Geologists select an appropriate radioisotope when dating a sample.
LO 13-6 *Compare and contrast the primary radioisotopes and the cosmogenic radioisotopes.*

13-7 Accurate dating requires understanding various sources of uncertainty.
LO 13-7 *Describe some potential sources of uncertainty in isotopic dating.*

13-8 Potassium-argon and carbon provide important isotopic clocks.
LO 13-8 *Describe why radiocarbon dating is useful for dating relatively recent organic material.*

13-9 Earth's age is measured using several independent observations.
LO 13-9 *Describe the evidence that Earth's age is about 4.6 billion years.*

◆ GEOLOGY IN OUR LIVES

Knowing the history of rock layers in the crust is critical to predicting the location of geologic resources and discerning the events that make up Earth's past. Geologists analyze geologic time using two methods that are based on critical thinking: (1) relative dating, which determines the *order* of geologic events, and (2) isotopic dating, which estimates the *age* of a geologic event. These methods have revealed that Earth has a long and fascinating history and that it is approximately 4.6 billion years old. It is necessary to integrate relative and isotopic dating to unravel the history of the crust in order to find resources, such as groundwater, metals, coal, and petroleum, that our communities depend on every day.

? What geologic events are represented in this photo showing the petroglyphs within Sego Canyon, Utah?

13-1 Geology Is the "Science of Time"

LO *13-1 Compare and contrast relative dating and isotopic dating.*

Every field of science is known for one major characteristic: Physics is the science of motion, chemistry is concerned with the composition of matter, and biology seeks to understand life. Geology is known as the "science of time."

Time

The 18th-century naturalist *James Hutton* is often called the father of geologic time. He is credited with the description of Earth as so old, and time so vast, that *"We find no vestige of a beginning, no prospect of an end."* Since then the integrated work of geologists, physicists, and chemists has indeed found a vestige of the beginning and established that Earth formed approximately 4.6 billion years ago (bya[1]), an estimate that is based on several lines of evidence. The evidence includes techniques for determining the age of Moon rocks and meteorites, the abundance of lead (Pb) isotopes created by radioactive decay in Earth rocks, and the direct measurement of very old rocks and minerals in continental crust (**FIGURE 13.1**). We examine each of these later in this chapter.

(a)

© Don Francis, Department of Earth & Planetary Sciences, McGill University

(b)

Northwest Territories Geoscience Office

FIGURE 13.1 Canada lays claim to the oldest Earth rocks ever found. However, it is still hotly debated whether *(a)* the Nuvvuagittuq greenstone belt, located along the eastern shore of Hudson Bay in northern Quebec, or *(b)* the Acasta gneiss, about 300 km north of Yellowknife, N.W.T., holds that title. Both are thought to have formed more than 4 billion years ago.

 These rocks have been involved in several episodes of orogeny (mountain building). Which family of rock is it likely to be in: sedimentary, igneous, or metamorphic?

Geologists are trained to analyze and interpret evidence gleaned from the vast archive of Earth's rocks. Sedimentary rocks, in addition to recording environmental change, also provide a history of changing life on Earth. Igneous rocks document magma production, volcanism, oceanic crust, and important aspects of plate tectonics. Metamorphic rocks preserve evidence of past fold-and-thrust belts, continental collisions and crustal thickening, and the pressure and temperature conditions of deep burial within the crust. In a very real sense, geologists are Earth historians. Hidden among the minerals and strata of the crust is the writing of the ages (**FIGURE 13.2**). We will examine the critical thinking skills geologists use to interpret this evidence and learn the history of our planet as it is written in Earth's *rock record* (the record of Earth history preserved in the crust).

Dating

Imagine that your history professor gives you an assignment on the War of 1812. You recall reading a magazine article sometime last winter about the American invasions of Upper and Lower Canada—probably in January. Pulling out a stack of magazines from the past year, you dig down, encountering issues from March and February. January must be next. How did you know that the January issue would be at the bottom of the pile? As you progressed down through the pile, you knew the older issues would be under the more recent issues—so you worked backward through time. You applied the geologic *principle of superposition*, perhaps the most powerful dating tool in our critical thinking toolbox.

This simple scenario illustrates the critical thinking that geologists use to interpret the history of the crust. Two types of time analysis are used: **relative dating** and **isotopic dating**. Relative dating determines the order of formation (like the magazines—which came first, which came second, etc.) among layers of rock based on their relationship to one another. Isotopic dating uses naturally occurring **radioactive decay** to estimate the age of geologic samples. (Radioactive decay is explained in Section 13-5.) Together, relative and isotopic dating methods can be combined to build a potent understanding of the history of Earth through time.

Michael Quinn for the National Park Service

FIGURE 13.2 The Grand Canyon reveals a thick sequence of igneous, metamorphic, and sedimentary rocks that preserve a history of changing geologic environments. These rocks have been dated using a combination of relative and isotopic dating. The oldest among them formed over 1 billion years ago, recording Earth events before the advent of multicellular life.

 In what part of the Grand Canyon will you find the oldest rocks?

[1]Also referred to as "Ga" (giga-annum), which stands for billion years before present.

Why Study Geologic Time?

The study of geologic time has direct implications in our lives. The same tools that are used to understand the first organisms that populated early oceans and walked on ancient continents are applied to finding coal and oil, valuable metals, groundwater, and other geologic resources. As we learned in Chapter 10 ("Geologic Resources"), petroleum is created by the exposure of highly organic layers of sediment to specific pressure and temperature conditions (the "oil window"). Sediments that are "cooked" in the crust too long may hold only methane, and those that are not buried long enough will not have been converted into petroleum.

In some areas, sedimentary rock layers of a specific age are known to hold oil. Hence, the ability to determine the age of potentially oil-bearing rocks simplifies the process of oil exploration—resulting in a direct payoff at the pump. The same is true of coal. Earth's history includes two great periods of coal formation: the Carboniferous and the Permian. Geologists searching for new coal seams explore rocks from these periods to increase their probability of success. We can apply the same deductive reasoning in searching for economic minerals, construction materials, and other types of geologic resources.

Above and beyond the economic incentives for studying geologic time, human curiosity drives us to answer questions about our origins, our history, and our place in Earth's sequence of natural events. It is humbling to realize that the entire span of human history represents less than 1 percent of the length of time that dinosaurs walked the land. Or, as Mark Twain once observed, "If the Eiffel Tower were now representing the world's age, the skin of paint on the pinnacle-knob at its summit would represent man's share of that age; and anybody would perceive that that skin was what the tower was built for." Twain was observing the lack of logic in assuming that all of geologic history took place in preparation for the arrival of our species. Clearly, Twain was a man who grasped the full magnitude of geologic time.

A Walk through Time

Imagine that Earth's entire history is contained in the last 24 hours (1,440 minutes) (FIGURE 13.3). Recorded human history began only 2.5 seconds ago. The first evidence of our species, *Homo sapiens,* appeared in the last 40 seconds. The dinosaurs disappeared from Earth 20 minutes ago, and the first primitive land plants appeared only 2 hours ago. Amazingly, the first evidence of multicellular life occurred in rocks only 4 hours ago. Oxygen accumulated in Earth's atmosphere about 8 hours ago, and the great continental land masses had mostly reached their current size 13 hours ago. Life itself, the hallmark of our venerable planet, first appeared in its tenuous and fragile form as a tiny single-celled creature approximately 18 hours ago.

The remarkable thing about Earth's history is that the events many believe to be ancient—for example, dinosaurs, primitive plants, and huge marine reptiles—are in reality the recent cousins of modern life forms, when compared to life's truly ancient ancestors. Why, 20 hours of our "day" had already passed before the first multicellular life forms evolved. When viewed from the colossal expanse of Earth's history, our species has been present for the mere blink of an eye.

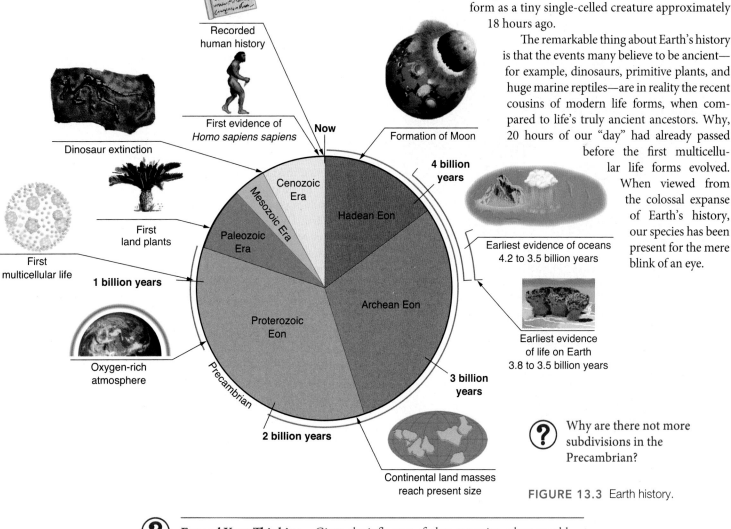

Recorded human history

First evidence of *Homo sapiens sapiens*

Dinosaur extinction

First land plants

First multicellular life

1 billion years

Oxygen-rich atmosphere

Cenozoic Era

Mesozoic Era

Paleozoic Era

Precambrian

Proterozoic Eon

2 billion years

Now

Formation of Moon

4 billion years

Hadean Eon

Earliest evidence of oceans 4.2 to 3.5 billion years

Archean Eon

Earliest evidence of life on Earth 3.8 to 3.5 billion years

3 billion years

Continental land masses reach present size

FIGURE 13.3 Earth history.

(?) Why are there not more subdivisions in the Precambrian?

(?) **Expand Your Thinking**—Given the influence of plate tectonics, where would you search for Hadean and early Archean rocks? Why?

13-2 Earth History Is a Sequence of Geologic Events

LO 13-2 *Describe early attempts to determine Earth's age.*

Not so long ago, most people's ideas about Earth's age were determined by dogma, faith, and religious belief rather than critical thinking. Through much of the 18th and 19th centuries, most people in Western cultures believed that the world had been brought into existence exactly 4,004 years before the birth of Christ.

Early Notions about Earth's Age

Many biblical scholars provided interpretations of Earth's age by detailed analyses of the Bible. James Ussher, the Bishop of Armagh, Ireland (FIGURE 13.4), proposed the date of Creation with exact precision. According to his analysis of the Bible, Earth came into existence at nightfall, Sunday evening, October 23, 4004 BC. Ussher arrived at this date by using a combination of astronomical cycles, historical accounts, and the chronology reported in the Biblical book of Genesis. He placed the beginning of Creation just 4,000 years before the birth of Christ; the last 4 in 4004 was occasioned by the date of Christ's birth, believed to be 4 BC. Ussher wrote about his findings between 1650 and 1658, and his ideas became popular as the result of being cited in the margin of the Great (1701) Edition of the English Bible, remaining there without explanation for 200 years. In 1900, on publication of a new edition, Cambridge University Press removed all reference to Ussher's date.

FIGURE 13.4 Based on a careful analysis of the Bible, Archbishop James Ussher concluded that God had created Earth on Sunday, October 23, 4004 BC.

Ussher's age for Earth is an example of precision. What is the difference between precision and accuracy?

Deep Time

Ussher's concepts were widely accepted, in part because they posed no threat to the social order of the time. They were comfortable ideas that followed the logic of other theologians and did not upset the authority of church and state. However, not everyone believed Ussher's proposed age for Earth. Some people were beginning to consider the idea of "deep time."

The discovery of deep time, the idea that the age of the universe can be measured in billions of years, goes back at least to the speculations of the German philosopher Immanuel Kant (1724–1804), the French mathematician and astronomer Pierre Simon Laplace (1749–1827), and the Scottish naturalist James Hutton (1726–1797). Kant and Laplace were the originators of the solar nebula hypothesis for the origin of the Solar System (Chapter 2, "Solar System"). Obviously, such an origin implies enormous amounts of time. Hutton proposed immensely long time spans to explain how observable rates of sediment erosion and deposition, and ongoing volcanic activity, could have created great valleys, thick sediment sequences, mountain ranges, and other features of Earth's surface.

Not all such speculators thought in terms of billions of years. Other examples of critical thinkers included Leonardo da Vinci (1452–1519), who calculated rates of sedimentation in the Po River of Italy and concluded that it had taken some 200,000 years to form nearby rock deposits. In 1760, the Frenchman Georges-Louis Leclerc de Buffon (1707–1788) estimated Earth's age to be 75,000 years by calculating the time it would have taken for it to cool from the molten state. In 1831, Charles Lyell (1797–1875) arrived at an age of 240 million years based on changes recorded in fossils found in rock in the English countryside. In 1901, John Joly (1857–1933) calculated the rate of salt delivery from rivers to the ocean and estimated the time needed to produce sea water. His answer was 90 to 120 million years.

Earth's age has long been a subject of debate among scientists and between Biblical creationists and scientists. However, for scientists, that debate has been resolved: Calculations based on the phenomenon of radioactivity have shown conclusively that Earth formed approximately 4.6 bya. Until the beginning of the 20th century, scientists had only vague ideas of Earth's age. It was generally believed that Earth was essentially unchanged since its formation. Most people thought that oceans were constant, continents remained unaltered, and any given hill or valley had stood, with little modification, since the beginning of time. Until improvements in understanding radioactivity provided by Marie and Pierre Curie along with Henri Becquerel in 1898, there was no reliable method for determining the numerical age of any rock, fossil, geologic feature, or geologic event.

In the absence of any absolute age calculations, 18th- and 19th-century naturalists such as William Smith, Charles Lyell, James Hutton, and others developed the science of relative dating. Their investigations and methods convinced them that Earth was much older than most people imagined. We still apply the principles of relative dating developed by these pioneering geologists.

(a) Earth Observatory, NASA

(b) David Davis/Photo Researchers

(c) Ken M. Johns/Science Source

(d) G.R. Roberts/Science Source

(e) François Gohier/Science Source

? In each photo describe the geologic process at work.

FIGURE 13.5 Five fundamental geologic events can be ordered using relative dating. *(a)* Sediment deposition leads to the formation of sedimentary rock. *(b)* Erosion is slow but effective at removing portions of the rock record. *(c)* Igneous intrusion creates dikes and other types of intrusive bodies. *(d)* Faulting is revealed when two sides of a fracture are displaced relative to each other. *(e)* Deformation is produced by stress in the crust.

Five Fundamental Geologic Events

Relative dating is a system of *reasoning* that is used to determine the chronological sequence or order of a series of **geologic events** (notable occurrences of common geologic processes). Relative dating does not provide a precise age; it only indicates that a particular geologic event occurred before or after another. It was not until a few decades ago, when numerical dating techniques based on radioactivity had significantly improved, that sequences of events could be dated in terms of "years ago."

Relative dating is based on the identification of five types of geologic events:

1. **Deposition** of sediments and the formation of sedimentary rock strata in environments of deposition (**FIGURE 13.5a**).

2. **Erosion** of the crust such that a gap in time (known as an **unconformity**[2]) is preserved in the geologic record of past events (**FIGURE 13.5b**). The duration and extent of the gap depends on the duration and extent of the erosion. Erosion can affect any rock that is exposed to the weathering processes of the atmosphere and ocean.

3. **Intrusion** of magma through the crust (**FIGURE 13.5c**) results in intrusive igneous rocks if the magma cools and crystallizes within the crust. Intrusive rock typically cuts across layers of older country rock.

4. **Faulting** of the crust. A fault is a break or fracture where rock layers on one side of the break are displaced relative to rock layers on the other side. A fault may occur as an entire zone, as in the San Andreas fault zone of California, which is over a kilometre wide in places and consists of several main faults and many small secondary faults. Or a fault may be a single fracture that runs through the crust. The distinguishing feature of a fault is that it displays relative *movement* of the rocks on either side (**FIGURE 13.5d**).

5. **Deformation** and **uplift** of rock layers so that they become tilted, folded, or even turned upside down. Deformation typically occurs because of tectonic activity due to plate movement that can disrupt the crust at plate boundaries and within plate interiors (**FIGURE 13.5e**). This process *uplifts* the crust, subjecting it to weathering, erosion, and the formation of unconformities.

 Expand Your Thinking—What is the fundamental difference between scientific thinking about a problem and religious thinking about a problem?

---

[2] An unconformity can also represent an extended period of non-deposition.

13-3 Seven Stratigraphic Principles Are Used in Relative Dating

LO 13-3 *List and describe the seven stratigraphic principles.*

Recognizing and identifying one or more of the five geologic events (deposition, erosion, intrusion, faulting, and deformation) in an outcropping of rock or a geologic cross-section of the crust is the first step in relative dating. This is not always easy. The presence of sedimentary rock indicates that deposition has obviously taken place. Erosion can be hard to recognize, but if a layer of rock appears to be cut into and filled by the overlying layer, erosion likely has occurred. Occasionally, the presence of features such as a surface cutting down through stratigraphy, pieces of the underlying layers incorporated in the overlying layer, or a layer of gravel may indicate that erosion has occurred. Erosion also is recognizable when there is a significant difference in the ages of two adjacent rock units. The presence of an igneous dike or pluton cutting across stratigraphy indicates intrusion. Faulting and deformation are indicated by broken and displaced layers and folded layers, respectively.

To properly interpret a complex section of crust in which all or many of these events have played a role, geologists have established a set of rules to guide their thinking. These rules, called **stratigraphic principles**, are basic, relatively simplistic, but important concepts that provide the foundation for relative dating. There are seven stratigraphic principles:

1. The *principle of superposition.* In a normal sequence of sedimentary rocks (also referred to as *strata*), the oldest layer is at the bottom and the youngest layer is at the top (FIGURE 13.6a). This is sometimes referred to as "layer cake geology"—the bottom layers of a cake are placed down first, with subsequent layers placed on top. The layer on the bottom would be the oldest, and the layers on the top the youngest. An exception to this principle occurs when the layers have been disturbed and overturned by folding due to tectonic forces.

2. The *principle of original horizontality.* Sedimentary rocks are deposited in layers parallel to Earth's surface. There are some exceptions to this rule, but they are uncommon and recognizable. When rocks are found in a non-horizontal configuration, we can conclude that some geologic event has tilted them (FIGURE 13.6b). For instance, when two continents collide at a convergent margin, originally flat layers of sedimentary rock are bent and broken so that they are tilted. This is the signal that the geologic event of deformation has taken place.

3. The *principle of original lateral continuity.* Sedimentary beds originally are laterally continuous within their environment of deposition. Hence, similar rock units at different locations may, in fact, be the same even though they are not now connected. Faulting, severe folding, and erosion may have separated the originally continuous beds. For instance, when a stream erodes the crust, it may cut downward through

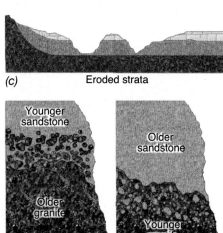

(a) Youngest / Oldest

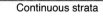
Continuous strata

(c) Eroded strata

(e) Younger sandstone / Older sandstone / Older granite / Younger granite

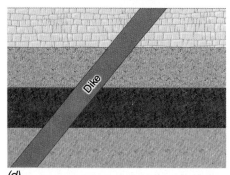

Horizontal strata

(b) Deformed strata

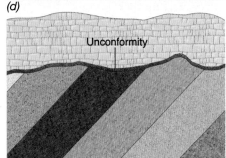

Dike

(d)

Unconformity

(f)

(?) In each example, can you identify the oldest geologic *event*?

FIGURE 13.6 Illustration of six stratigraphic principles. *(a)* Principle of superposition: Among undeformed layers of sedimentary rock, the lower layers are older than the upper layers. *(b)* Principle of original horizontality: Most sedimentary rocks are originally deposited as horizontal strata. Tilted and folded strata are a sign of deformation. *(c)* Principle of original lateral continuity: Faulting, severe folding, and erosion may separate originally continuous strata. *(d)* Principle of cross-cutting relationships: Geologic events (intrusion, erosion, faulting, and deformation) that cut across pre-existing rocks are younger than the rocks. *(e)* Principle of inclusions: Clasts of existing rock that are incorporated into a sedimentary layer or intrusion come from rocks that are older than the sedimentary layer or intrusion into which they have been incorporated. *(f)* Principle of unconformities: Unconformities develop when erosion or non-deposition interrupts the continuity of the geologic record.

layers of rock and form a deep, wide canyon. Despite the distance between similar rock outcrops on either side of the canyon, the principle of lateral continuity tells us that they once formed a continuous layer of rock from the same environment (**FIGURE 13.6c**).

4. The *principle of cross-cutting relationships.* Rocks or other geologic features (e.g., erosion or intrusions) that cut across pre-existing rocks are younger than the rocks that they cut across. **FIGURE 13.6d** illustrates this concept well; layers of sedimentary rock are intruded by a basalt dike. The principle of cross-cutting relationships tells us that the basalt is younger than the sedimentary beds because the dike cuts across them.

5. The *principle of inclusions.* Any part of an existing rock that is incorporated into a sedimentary layer or igneous intrusion is older than the sedimentary layer or intrusion into which it has been incorporated (**FIGURE 13.6e**). For instance, a granite batholith that intrudes sandstone crust may have pieces of sandstone incorporated within the cooled granite. This may be because blocks of sandstone fell off the walls of the magma chamber into the molten granite. A geologist would interpret this as a sign that the granite is younger than the sandstone inclusions.

6. The *principle of unconformities.* Unconformities develop in a geologic cross-section when erosion or non-deposition (environments where no sediments have been deposited for extended periods of geologic time) interrupts the continuity of the geologic record (**FIGURE 13.6f**). Their position in a relative dating sequence can sometimes be determined by the principle of cross-cutting relationships. The presence of an unconformity indicates that a portion of the geologic record is missing, as if pages had been torn out of a book. Without fossils or numerical dating to assign an exact age to the rocks on either side of the unconformity, it is difficult to know how much of the rock record has been removed by erosion.

7. The *principle of fossil succession.* Plants and animals change through time; this change is termed **evolution**. (There is more on evolution in Chapter 14, "Earth's History.") Rock layers record the changes because the fossils of plants and animals that lived at one time will not be found in rocks that were formed at a different time, as long as the fossils have not been weathered out of older rocks and incorporated into younger strata. The typical succession of fossils serves as a relative dating tool. That is, the assemblage of fossils occurring in a rock can serve to identify and date that rock. This principle is incredibly useful and warrants more attention.

The principle of fossil succession is an especially powerful geologic tool. It can be illustrated as follows (**FIGURE 13.7**): If we begin at the present and examine increasingly older layers of rock, we will come to a level where no fossils of humans are present. As we step farther back in time, each step brings us to levels where no fossils of flowering plants are present, the next level has no birds, the following level has no mammals, then no reptiles, no four-footed vertebrates, no land plants, no fish, no shells, and, finally, we reach ancient rocks where there are no plants or animals whatsoever. Each of these steps takes us back in time, and the presence of each fossil group indicates a single, specific interval of geologic time.

(?) **Expand Your Thinking**—Explain how fossils help geologists establish the relative ages of different rock layers or outcrops.

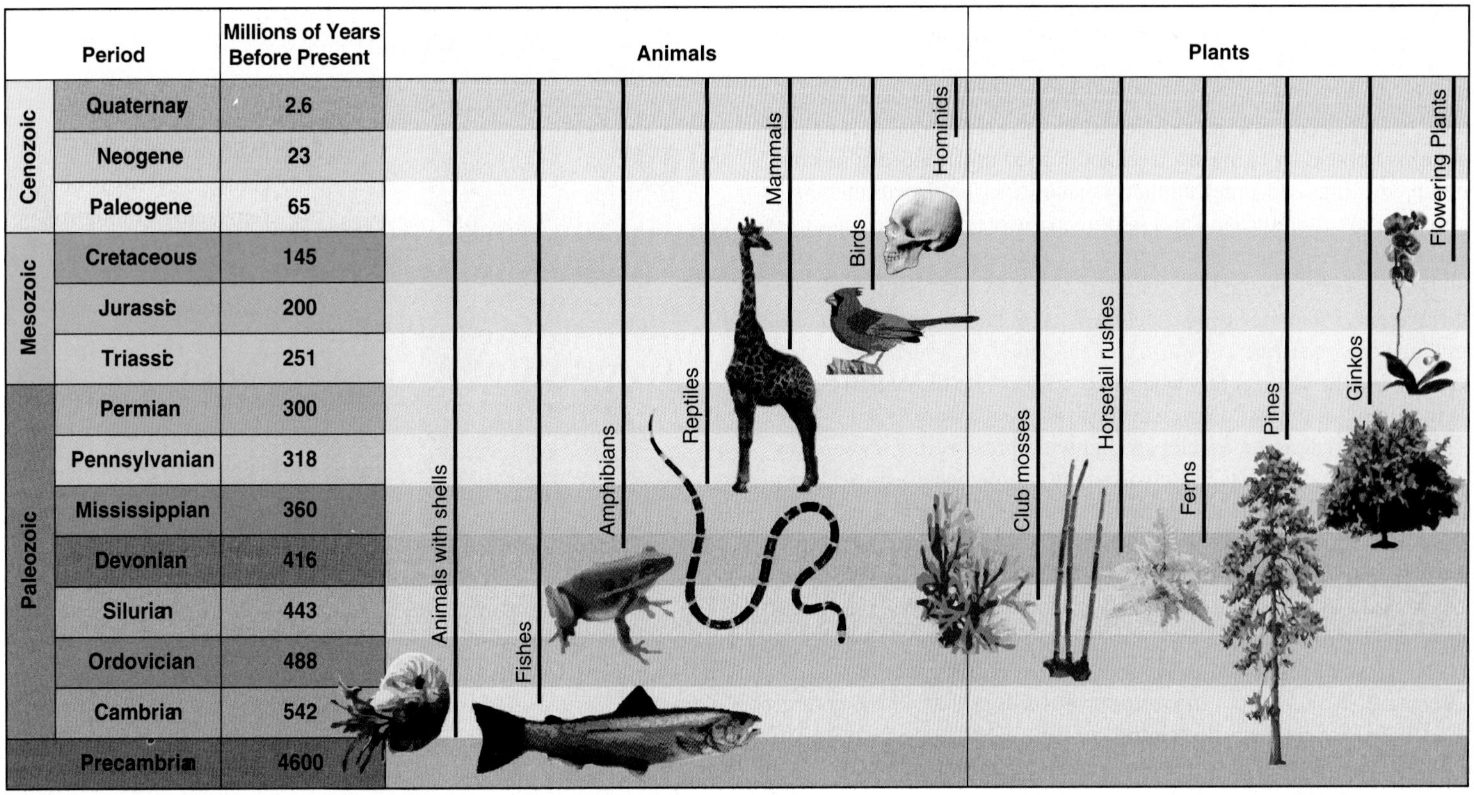

	Period	Millions of Years Before Present	Animals	Plants
Cenozoic	Quaternay	2.6		
Cenozoic	Neogene	23		
Cenozoic	Paleogene	65		
Mesozoic	Cretaceous	145		
Mesozoic	Jurassic	200		
Mesozoic	Triassic	251		
Paleozoic	Permian	300		
Paleozoic	Pennsylvanian	318		
Paleozoic	Mississippian	360		
Paleozoic	Devonian	416		
Paleozoic	Silurian	443		
Paleozoic	Ordovician	488		
Paleozoic	Cambrian	542		
	Precambrian	4600		

(Animal labels: Animals with shells, Fishes, Amphibians, Reptiles, Mammals, Birds, Hominids. Plant labels: Club mosses, Horsetail rushes, Ferns, Pines, Ginkos, Flowering Plants)

FIGURE 13.7 Sedimentary strata may contain groups of fossils that are unique to that period of Earth's history.

 Explain the relationship between evolution and the principle of fossil succession.

13-4 Relative Dating Determines the Order of Geologic Events

LO 13-4 *Describe the use of stratigraphic principles to construct a history of geologic events.*

Imagine that you have studied the geology near your school for several years and have constructed a geologic cross-section that depicts the layers and organization of rocks in the crust. You find various types of sedimentary rocks, an unconformity, a basalt dike and sill, and evidence of uplift and deformation. By applying stratigraphic principles, you can use this information to create a model of the history of the crust. Such a model is shown in **FIGURE 13.8**. It depicts the step-by-step sequence of geologic events that led to the formation of the crust as it is today.

An Example of Relative Dating

Figure 13.8a begins the history of a section of the crust by illustrating the deposition of sedimentary layers A to D during a period of rising sea level. The principle of superposition tells us that these layers were deposited in order from A (first) to D (last). Geologists have ascertained that sea levels rise and fall through time as a result of changes in climate—a warmer climate melts glaciers and warms the oceans, resulting in higher sea levels, while a colder climate expands icecaps, drawing water from the ocean to make ice and thus lowering sea level. Changes in the vertical position of a continent also occur; a continent may rise because of plate convergence and may fall because of plate rifting. The relative difference in elevation between a continent and sea level is referred to as *continental freeboard*. When warm climate (high sea level) and a low continent (low freeboard) occur at the same time, shallow seas can flood broad tracts of land. This occurred during the Ordovician and Cretaceous Periods when shallow seas flooded the Canadian and U.S. midwest and deposited thick layers of sandstone, limestone, and shale (discussed in Chapter 15, "The Geology of Canada").

In the hypothetical case depicted in Figure 13.8a, a watershed environment (an energetic stream depositing gravel that makes conglomerate) is flooded by rising sea level and buried under a beach (a sandy beach, forming sandstone), a muddy shallow marine environment (silts and clays swept off the land and settling to the sea floor, forming shale), and finally a fully marine environment (calcareous plankton collecting on the sea floor, forming limestone). Just such an event happened when the *Western Interior Seaway* developed across North America 90 to 65 million years ago (mya[3]). A warm climate caused sea level to rise and advance from the south across what is now Texas into the central portion of the United States and all the way north across Canada (**FIGURE 13.9**). This warm, shallow sea developed after widespread rifting and breakup of Pangaea, an event that also involved the continental freeboard lowering across North America. During this time, the seaway was formed by the downward flexing of the crust in front of the eastward propagating North American Cordillera, an orogen so large and thick that it caused a downward warping of the lithosphere in front of it to the east (discussed in Chapter 15).

In Figure 13.8b, an igneous sill (E) intrudes between layers C and D. The principle of cross-cutting relationships tells us that sill E is younger than the strata into which it intrudes. Figure 13.8c reveals the intrusion of dike F. Again, the principle of cross-cutting relationships tells us that dike F must be younger than the rocks (layers A to E) into which it intrudes.

[3] Also referred to as "Ma" (mega-annum), which stands for million years before present.

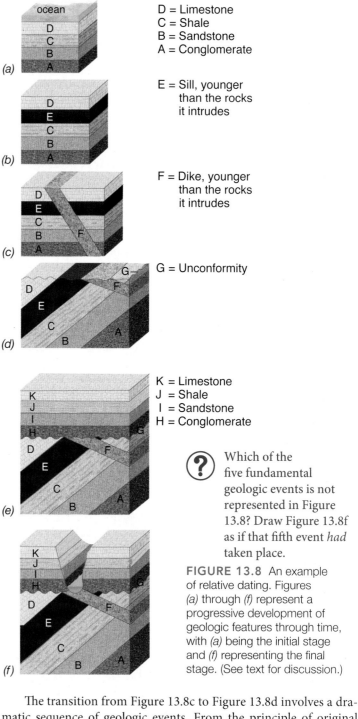

D = Limestone
C = Shale
B = Sandstone
A = Conglomerate

E = Sill, younger than the rocks it intrudes

F = Dike, younger than the rocks it intrudes

G = Unconformity

K = Limestone
J = Shale
I = Sandstone
H = Conglomerate

? Which of the five fundamental geologic events is not represented in Figure 13.8? Draw Figure 13.8f as if that fifth event *had* taken place.

FIGURE 13.8 An example of relative dating. Figures *(a)* through *(f)* represent a progressive development of geologic features through time, with *(a)* being the initial stage and *(f)* representing the final stage. (See text for discussion.)

The transition from Figure 13.8c to Figure 13.8d involves a dramatic sequence of geologic events. From the principle of original horizontality, we can infer that tectonic forces (perhaps plate convergence) have caused the deformation (tilting) of all the layers that have been formed so far (layers A to F). These strata have been tilted at a high angle so that weathering and erosion have levelled the upturned strata and formed a flat surface environment. This is an unconformity (labelled G) on which sedimentary strata H, I, and J are

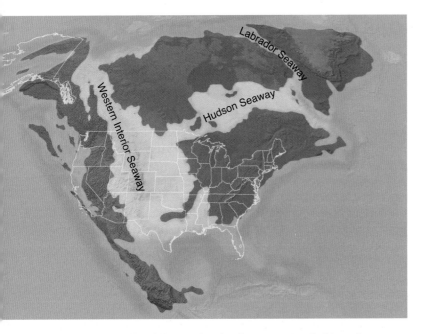

FIGURE 13.9 The Western Interior Seaway extended from the Gulf of Mexico to the northern reaches of Canada about 90 to 65 million years ago. There is no modern analog to this shallow marine basin, which was about 1,000 km wide and 800 m to 900 m deep.

(?) What types of sediment would preserve a record of the Western Interior Seaway?

deposited during another episode of rising sea level in Figure 13.8e. In Figure 13.8f, newly deposited layers H to J are eroded at the modern surface by a stream channel.

In constructing this geologic history, we identified four types of geologic events (deposition, erosion, deformation, and intrusion). In addition, we employed the stratigraphic principles of superposition, cross-cutting relationships, original horizontality, and unconformities to unravel the details of crust formation. In summary, layers A, B, C, and D were laid down horizontally, with A deposited first and D last. The intrusion of an igneous sill (labelled E) occurred next, followed by intrusion of a dike (F). A tectonic event occurred, probably an orogeny resulting from continental collision at a convergent margin, tilting and lifting the strata. This led to erosion of the land surface, forming an unconformity (G). Layers H to J were then deposited during a rise in sea level, H first and J last. These layers currently are being eroded by a modern stream cutting down through the rock layers and forming a canyon.

The First Geologic Map

William "Strata" Smith made the world's first geologic map by applying these very same principles. He was born in Oxfordshire, England, on March 23, 1769. As a young man trained in surveying, he was employed to supervise the digging of canals in southern England. At the time, canals were dug by hand. It was important to know the local geology when planning the route of a new canal so that diggers could have access to the softest rock. Smith observed that the fossils found in a section of sedimentary rock always appeared in a certain order from the bottom to the top. This order of appearance could also be seen in other rock sections, even those on the other side of England. From these observations, he was able to predict the organization of the crust based on fossil evidence.

Why is it significant that fossils appear in a predictable sequence? Three factors are important here: (1) Fossils represent the remains of once-living organisms; (2) many fossils are the remains of extinct organisms; that is, they belong to species that are no longer living anywhere on Earth; and (3) extinct fossilized organisms offer unique timepieces for identifying and dating the strata in which they are found. Different fossils are found in rocks of different ages because life on Earth has changed through time. Hence, when we find the same kinds of fossils in rocks from different places, we know that the rocks are the same age. By the early 1800s, Smith had proposed the principle of fossil succession and developed England's first geologic map. Subsequently, thousands of geological maps have been produced building on the principles established by William Smith (**FIGURE 13.10**).

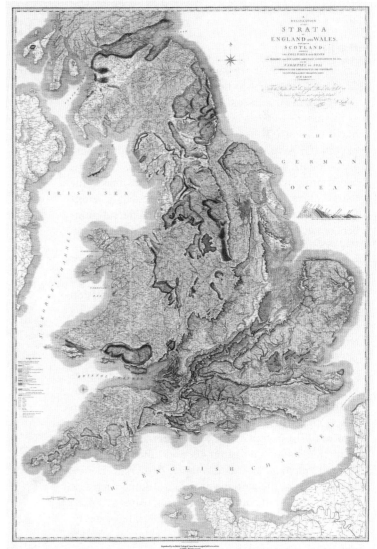

FIGURE 13.10 The first geologic map. Each colour represents strata of a different age.

(?) **Expand Your Thinking**—Describe a testable scientific hypothesis for why the Western Interior Seaway came into existence.

13-5 Isotopic Dating Uses Radioactive Decay to Estimate the Age of Geologic Samples

LO 13-5 *Define the concept of "half life."*

Numerical geologic time identifies the age of natural materials in "years before present." Because the most common tool used in determining numerical geologic time employs the natural phenomenon of radioactive decay of *isotopes*, this type of age determination is called isotopic dating. Radioactive decay is a process wherein the unstable nucleus of an atom spontaneously emits radiation and a *subatomic particle*. This results in an atom of one type (called the *parent isotope*) transforming to an atom of another type (called the *daughter isotope*). During the first half of the 20th century, much of modern physics was devoted to exploring why this change occurs.

Radioactivity

Let us review what we know about atoms. You may recall from Chapter 4 ("Minerals") that the nucleus of each chemical element is composed of a specific number of protons, known as the *atomic number,* which is how the element is defined. Carbon (C) has 6 protons, nitrogen (N) has 7, oxygen (O) has 8, and so on. The nucleus of an element also contains a variable number of neutrons. The number of protons plus the number of neutrons makes the *mass number*. Atoms of the same element that have different mass numbers (because they have different numbers of neutrons) are known as *isotopes* of that element. For example, carbon with 6 protons may have 6, 7, or 8 neutrons yielding the isotopes carbon-12, carbon-13, and carbon-14. However, carbon-14 is not stable, and its nucleus will radioactively decay by releasing radiation and a beta particle (an electron). When a neutron loses a beta particle, it becomes a proton and, in the process, the atomic number changes from 6 (carbon) to 7 (nitrogen), producing nitrogen-14.

Within most igneous, sedimentary, and metamorphic minerals are radioactive atoms called **radioisotopes**. Examples of radioisotopes include uranium (U) isotopes such as uranium-238 and uranium-235, thorium (Th) isotopes such as thorium-232, the potassium (K) isotope potassium-40, and others. *Each type of radioisotope decays at a unique, fixed rate.* Hence, as time passes, the amount of parent isotope decreases and the amount of daughter isotope increases at a fixed rate within a mineral. If you know the rate of decay, the changing ratio of parents to daughters is a direct reflection of how much time has elapsed since the isotopes became trapped in the mineral. Researchers, utilizing sensitive laboratory instruments to measure the abundance of parent and daughter isotopes in a mineral, use this ratio and the rate of decay to estimate the time since the mineral crystallized and thereby the age of the geologic event it represents (such as intrusion or metamorphism). Using radioisotopes to date Earth materials is a powerful technique, and it has been employed to reveal many important events in Earth history.

How does decay occur? There are three types of decay: *alpha decay, beta decay,* and *electron capture* (**FIGURE 13.11**). In alpha decay, the nucleus of an atom emits a single alpha particle consisting of two protons and 2 neutrons. Alpha decay results in the formation of a daughter isotope with 2 fewer protons than the parent isotope; the atomic number decreases by 2 and the mass number decreases by 4. For instance, the isotope uranium-234 (useful for dating fossil coral) experiences alpha decay and produces the daughter isotope thorium-230. The atomic number of uranium is 92 and that of thorium is 90 (2 fewer protons). See the "Geology in Our Lives" box titled "Alpha Decay, Silent Killer" for a description of the environmental hazard associated with alpha decay.

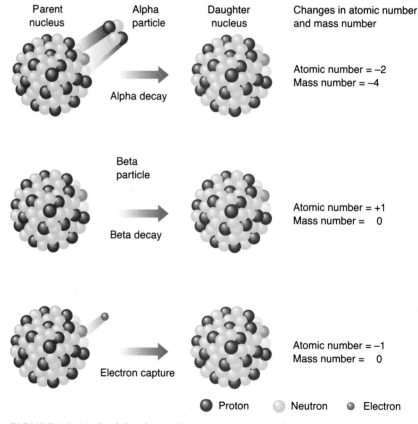

FIGURE 13.11 In alpha decay, the atomic number of the parent isotope decreases by 2 and the mass number decreases by 4. In beta decay, the atomic number increases by 1 and the mass number is unchanged. In electron capture decay, the atomic number decreases by 1 and the mass number remains unchanged.

(?) What aspect of radioactive decay allows it to be useful for dating geological materials? Explain your answer.

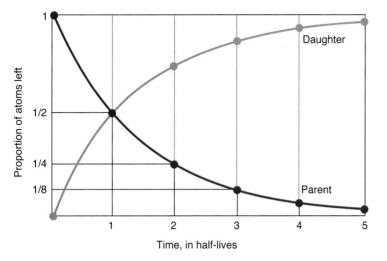

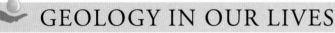

FIGURE 13.12 With each half-life, the amount of parent radioisotope in a sample decreases by half and the amount of daughter radioisotope increases by half.

(?) The amount of an isotope is 1/32 of its original amount. How many half-lives have passed?

During beta decay, the nucleus emits a negative beta particle that changes one of the neutrons into a proton. In this case, the parent isotope gains 1 proton and loses 1 neutron. Hence, the mass number does not change but the atomic number increases. For instance, thorium-234 experiences beta decay and becomes the daughter isotope

protactinium-234. Notice that protactinium-234 has the same mass as thorium-234, but the element protactinium (atomic number 91) is 1 proton heavier than thorium (atomic number 90).

During electron capture, an atom loses 1 proton (it becomes a neutron) and is transformed into a lighter element. However, the mass number does not change because the total number of protons and neutrons remains the same. For example, the radioactive isotope potassium-40 decays through electron capture into argon-40. The potassium isotope loses 1 proton and becomes argon, but the mass number stays the same.

Half-life

Because it is impossible to determine exactly when a single radioisotope will decay, geologists express the probability of decay with the statistical **half-life**. *The half-life is the probable time required for half of the radioisotopes in a sample to decay.* Half-life is a value that has been measured for each type of radioisotope. As time passes, the amount of parent isotope in a sample decreases by half during each half-life. Therefore, after one half-life, the sample will contain half the number of atoms that were originally present. After two half-lives, one-fourth of the original number of atoms will be present. After three half-lives, one-eighth of the original amount will remain; after four half-lives, one-sixteenth will remain; and so forth (**FIGURE 13.12**).

Half-life is a statistical description that says *there is a very high probability that half of the atoms of a radioisotope in a sample will decay in a defined period of time.* Each radioactive isotope has been studied carefully in order to make a reasonably accurate estimate of its half-life. "Geology In Our Lives, Alpha Decay, Silent Killer" discusses how radioisotopes can impact our everyday lives.

GEOLOGY IN OUR LIVES

ALPHA DECAY, SILENT KILLER

Radon-222 (^{222}Rn) is a radioactive gas that can pose serious health risks. According to Health Canada, about 7 percent of homes in Canada have radon levels above the Canadian guideline (200 becquerels/m^3). Radon-222 undergoes alpha decay and has a half-life of only 3.8 days; hence, if you live or work in a building infiltrated by radon-222, it is impossible to avoid inhaling this radioisotope. Alpha particles cause damage to sensitive lung tissue. Breathing radon-laden air over a lifetime can substantially increase your risk of developing lung cancer.

Radon-222 gas is produced by the decay of uranium-238 in the rock, soil, and groundwater beneath our communities (**FIGURE 13.13**). The gas cannot be seen, smelled, or tasted, and as it filters up through the soil or outgases from our drinking water, it becomes trapped in the buildings where we live and work and can build up to dangerous levels.

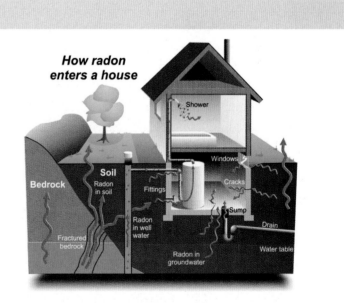

FIGURE 13.13 Radon may enter a building through several pathways.

(?) **Expand Your Thinking**—If radioactive isotopes had variable rates of decay, would isotopic dating still be possible? Why or why not?

13-6 Geologists Select an Appropriate Radioisotope when Dating a Sample

LO 13-6 *Compare and contrast the primary radioisotopes and the cosmogenic radioisotopes.*

Not every type of radioisotope (TABLE 13.1) is appropriate to use for dating a geologic sample. There is no point in looking for, say, potassium-40 in a rock that is known to contain little potassium. Another important consideration is that only certain radioisotopes will accurately characterize the time frame of the sample being dated. That is, *young samples should be dated with short-lived radioisotopes and old samples with long-lived radioisotopes.* If a very young event is being dated, such as a lava flow approximately 500 to 1,000 years old, it is not appropriate to pick a radioisotope with a half-life of a few billion years—there simply will not be enough daughter isotope to be measured accurately because not enough time has passed. Likewise, if a sample to be dated is several hundred million years old, such as an ancient granite pluton, it is not appropriate to try to date it using a radioisotope with a half-life of only a few thousand years. There will not be enough parent radioisotope left to be measured accurately because too much time will have passed and most of the parent radioisotope will have decayed.

Isotopic Clocks

Scientists use an atom-measuring instrument called a *mass spectrometer* to measure the abundance of isotopes in a sample. Information about the ratio of parent radioisotopes to daughter isotopes in a sample is used to calculate an age based on the amount of time necessary for a given ratio to develop at the specified decay rate. Well over 40 isotopic dating techniques are in use around the world, each based on a different radioisotope. In general, useful isotopes fall into two categories: *primary radioisotopes* and *cosmogenic radioisotopes.*

Earth inherited the primary radioisotopes when it accreted during nebular condensation. (See Chapter 2, "Solar System.") These radioisotopes have been held in rocks for the entire 4.6 billion years of Earth's history. They are released as a rock melts, only to be incorporated into new rock when mineral crystallization occurs. The primary radioisotopes typically are used to date rocks from continental and oceanic crust.

Cosmogenic radioisotopes are formed by the interaction of subatomic particles (such as neutrons) produced by the Sun with nuclei of atoms in the atmosphere or on the surface of mineral grains in the crust that are exposed to the atmosphere. Because these isotopes have relatively short half-lives, they would not be present on Earth today if they were not continuously created by cosmic rays. The cosmogenic radioisotopes typically are used to date young samples involved in physical, chemical, and biological processes occurring in surface environments.

Try the exercise titled "The Age of Geologic Events" in the next "Critical Thinking" box to apply the principles of isotopic dating. See if you can calculate the age of three types of rock using the ratio of parents to daughters and the half-life. Once you are comfortable with the application of isotopic dating, try working out the geologic history in the "Relative and Isotopic Dating" box by applying both relative and isotopic dating principles.

TABLE 13.1 Common Radioisotopes and Half-Lives Used in Dating Samples

Radioactive (Parent)	Product (Daughter)	Half-Life (Years)	Radioactive (Parent)	Product (Daughter)	Half-Life (Years)
Samarium-147	Neodymium-143	106 billion	Uranium-235	Lead-207	0.7 billion
Rubidium-87	Strontium-87	48.8 billion	Beryllium-10	Boron-10	1.52 million
Thorium-232	Lead-208	14 billion	Chlorine-36	Argon-36	300,000
Uranium-238	Lead-206	4.5 billion	Uranium-234	Thorium-230	248,000
Potassium-40	Argon-40	1.25 billion	Thorium-230	Radium-226	75,400
			Carbon-14	Nitrogen-14	5,730

 Expand Your Thinking—Using the half-life of each radioisotope pair in Table 13.1, describe the type of geologic event or geologic material that could be dated.

CRITICAL THINKING

THE AGE OF GEOLOGIC EVENTS

FIGURE 13.14 is a cross-section of the crust consisting of sedimentary, igneous, and metamorphic rocks. You can determine the age of basalt, granite, and marble samples using uranium-235. TABLE 13.2 provides the results of laboratory isotopic analysis of the samples.

The ratio of the daughter atoms to the parent atoms is proportional to the age. This proportion, combined with the half-life, permits calculating the age in years with this formula: P_i percent $= P_t/(P_t + D_t)$.

P_i percent is the percent of initial parent atoms remaining in the system; P_t is the number of parent atoms today; D_t is the number of daughter atoms today. When you divide the number of parent atoms today by the number of parent atoms originally ($P_t + D_t$), you get a ratio, such as 1:2 (= 50 percent) or 1:4 (= 25 percent) or some other value. This ratio is the percent of parent atoms remaining in the system today (P_i percent), and it is directly related to the number of half-lives that have passed since the mineral sample crystallized. For example, if a single half-life has passed, 50 percent of the parent atoms will remain in a mineral sample. If two half-lives have passed, 25 percent of the parent atoms will remain. If you know the number of half-lives, you can calculate the age of the rock. Using the half-life for uranium-235 (700 million years), these results would give an age of 700 million years and 1.4 billion years, respectively.

Please answer the following questions.

1. What are the ages (in years) of the basalt, granite, and marble samples?
2. What can you say about the ages of the sandstone and shale units?
3. Identify the geologic events that built this section of crust and their proper sequence. Apply a combination of relative dating and isotopic dating to derive the geologic history.
4. In your geologic history, are there any events whose sequence is not clear? What additional information would help to clarify the correct sequence of events?
5. What is the very last event?
6. Describe the igneous processes that could produce basalt and granite in the same place.
7. Describe the sedimentary environments that might have produced this sequence of rocks.
8. What stratigraphic principles did you use in constructing your geologic history?

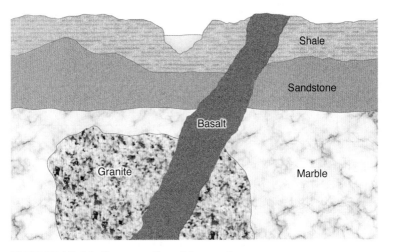

FIGURE 13.14 Cross-section of the crust.

TABLE 13.2 Results of Isotopic Analysis

Rock Unit	Number of Parent Atoms	Number of Daughter Atoms
Basalt	3,726	3,726
Granite	1,250	8,750
Marble	785	24,335

CRITICAL THINKING

RELATIVE AND ISOTOPIC DATING

FIGURE 13.15 shows a section of crust with a complex geologic history. Sedimentary, metamorphic, and igneous rocks are all present, as well as faulting, intrusions, and unconformities. Your job is to construct a complete geologic history of this region, using a combination of relative and isotopic dating. The rock types are defined using symbols that are explained in the legend.

TABLE 13.3 provides the half-life and the measured ratio of daughter to parent isotopes in a number of samples. The actual material analyzed that is used to constrain the age of each sample is included in brackets. The numerical ages you calculate can be assigned to geologic events, and they provide a framework for estimating the timing of other events that are recognized using relative dating. We will assume that the following relationship holds true for all the samples in this problem:

$$\text{Age} = \text{measured ratio} ? 1/`$$
$$` = 0.69/\text{half-life}$$

Once you have calculated the ages of samples A to H construct a history of geologic events and assign them to their proper geologic periods in historical order (youngest at the top, oldest at the bottom of your list). You will have to estimate the age of some events, but that is the nature of critical thinking. Be sure you have a reason for each of your estimates. There may be some geologic events that are not obvious but must have occurred based on the available evidence. Write a complete geologic history of this crust.

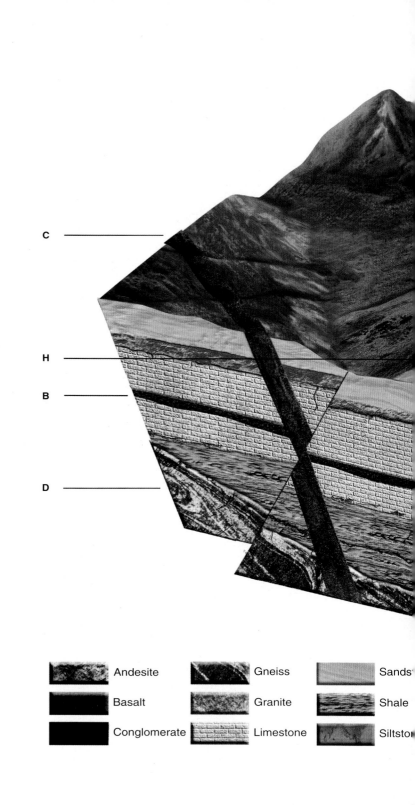

TABLE 13.3 Results of Isotopic Analysis

Sample	Isotope System	Half-life	Measured Ratio
A	Uranium-238/Lead-206 (mineral = zircon)	4.5×10^9	0.090
B	No values obtained	—	—
C	Potassium/Argon (mineral = hornblende)	1.3×10^9	0.006
D	Rubidium/Strontium (whole rock)	4.7×10^{10}	0.050
E	Potassium/Argon (mineral = biotite)	1.3×10^9	0.097
F	Uranium-235/Lead-207 (mineral = zircon)	7.1×10^8	0.046
G	Carbon-14 (wood fragment)	5.7×10^3	1.870
H	Lead-210 (stream sediment)	2×10	2.070

Andesite   Gneiss   Sands

Basalt   Granite   Shale

Conglomerate   Limestone   Siltsto

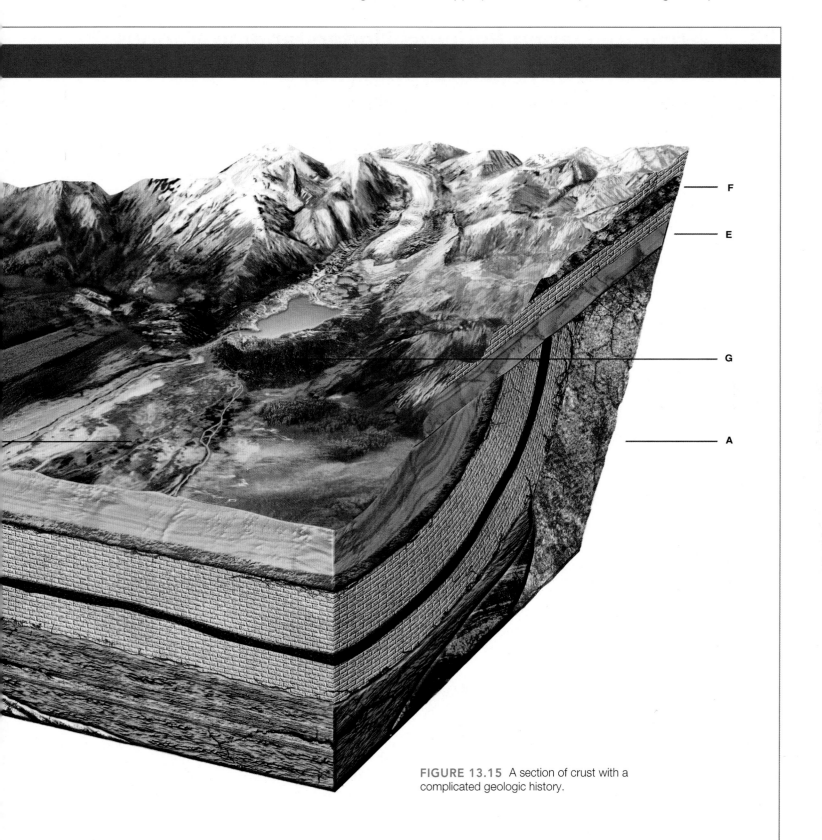

FIGURE 13.15 A section of crust with a complicated geologic history.

13-7 Accurate Dating Requires Understanding Various Sources of Uncertainty

LO 13-7 *Describe some potential sources of uncertainty in isotopic dating.*

Igneous rocks are good candidates for isotopic dating. The ages of minerals within an igneous rock document the time when the mineral crystallized from a magma source. However, accurate dating requires understanding sources of uncertainty.

Sources of Uncertainty

When molten rock cools and crystallizes, atoms are no longer free to move about. Radioisotopes of the right size and charge will be incorporated into the crystalline structure. Often daughter isotopes will not be included because they are the wrong size, but this is not always the case. Hence, a newly formed mineral may be free of daughter isotopes, which simplifies the process of determining its age. Any daughter atoms resulting from radioactive decay *after* crystallization will be locked in place within the crystalline structure and can be counted.

A geologist who takes a sample of rock and brings it to a laboratory for analysis and dating has several decisions to make:

- What is the relationship between the sample and the event it represents?
- What is the approximate age of the sample?
- What radioisotope is appropriate for both the type of material being dated and the approximate age of the sample?
- What are the potential sources of uncertainty in calculating the sample's age, and how can the age be tested?

To determine an accurate age, the geologist assumes that the sample being dated has remained a *closed system* throughout its life. This means that neither parent nor daughter isotopes have left or entered the sample since its formation. If the sample has not remained a closed system, metamorphism or alteration by groundwater probably has occurred. If parent isotopes have left the sample, the calculated age will be too old. If daughter isotopes have left, the calculated age will be too young.

Metamorphic conditions of high heat and/or pressure, or the presence of groundwater (especially hydrothermally active groundwater) containing parent or daughter isotopes, can lead to *open-system behaviour,* in which parent and daughter isotopes migrate in and out of a sample (**FIGURE 13.16**). However, if daughter isotopes are completely driven from a sample during metamorphism, any calculated age would be useful for determining how much time has passed since metamorphism occurred—a potentially valid scientific goal.

Determining the age of a sample is a two-step process. First the ratio of parent atoms to daughter atoms of a selected element in a sample is calculated. Then the half-life is used to calculate the time it took to produce that ratio of parent atoms to daughter atoms. In actual practice, in some cases, it is reasonable to assume that no daughter isotope was present at the time of crystallization. But in other cases, the initial amount of the daughter product must be determined. Usually the geologist will analyze several samples from one rock body. The differing amounts of parent and daughter isotopes found in these analyses can be used to estimate how much daughter isotope was present originally.

Now we will explore some specific radioisotopes that geologists use to determine the absolute age of a rock or fossil.

The Uranium Decay Chain

All isotopes of the element uranium are at least mildly radioactive. Uranium-238, which occurs naturally in most types of granite and soil in varying degrees, decays with a half-life of 4.5 billion years (**FIGURE 13.17**) and produces the stable daughter isotope lead-206. However, before lead-206 is produced, 14 separate steps are involved, each leading to the production and subsequent decay of a radioactive daughter isotope. These steps are referred to as a **decay chain**.

 How could hydrothermally active fluids influence the isotopic age of a rock sample?

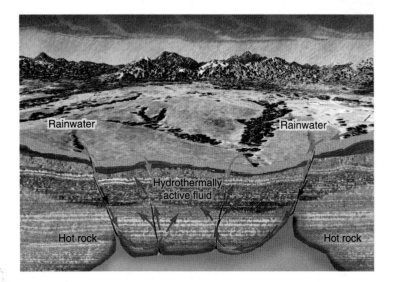

FIGURE 13.16 Open-system behaviour can be caused by metamorphism, interaction with groundwater, and alteration by hydrothermally active fluids. A sample that has experienced open-system behaviour will not provide an accurate isotopic age of its original crystallization. It will instead reflect some combination of that age and the degree to which open-system behaviour altered the parent-to-daughter ratio.

Figure 13.17 shows the decay chain of uranium-238 and the type of decay and half-life of each daughter isotope. The chain of decay products is formed as a result of the production of radioactive daughter isotopes. One daughter radioisotope decays to form another radioisotope, which in turn decays further until the decay process finally ends with an isotope that is stable. In this case, that stable isotope is lead-206. Some of the intermediate isotopes have very short half-lives; examples include protactinium-234, with a half-life of slightly more than one minute, and polonium-214, which decays in a few ten-thousandths of a second. The uranium decay chain is an example of how the process of radioactive decay can lead to the production of many daughter isotopes, each with its own distinctive half-life.

The use of uranium as an isotopic clock was the first method developed for dating very old rocks. As mentioned, half-life of uranium-238 is 4.5 billion years. Uranium-235 has a half-life of 700 million years, and thorium-232 has a half-life of 14 billion years. Each decays, in steps, to form other elements, including polonium and finally lead. The different parent isotopes yield different daughter isotopes of lead: uranium-238 yields lead-206; uranium-235 yields lead-207; thorium-232 yields lead-208. To determine a sample's age, the amounts of these lead isotopes in the sample are measured and compared with the amounts of the parent isotopes remaining in the sample.

Uranium-lead dating is carried out on only a limited set of minerals: namely, zircon, quartz, and apatite. Zircon is a particularly valuable mineral for this purpose because when it crystallizes in magma, it incorporates uranium into its crystal structure but does not incorporate significant amounts of lead. Therefore, the proportion of lead it contains is produced by the radioactive decay of uranium, which provides a powerful isotopic clock. Zircon is also extremely resistant to chemical and physical weathering, and once it has crystallized, it is very difficult to add or remove lead or uranium from it. For these reasons, zircon crystals survive for a long time in geologic materials, and they act as a closed system that does not easily metamorphose. Because of its resistance to metamorphism and its longevity, zircon has become an important mineral for dating ancient rocks. Zircon grains, formed over 4 billion years ago, have been found in sandstones in Australia (see Figure 13.21); zircon has been used to date other extremely old rocks (4.28 billion years in northeast Canada; refer to Figure 13.1). Zircon dating is also used to determine the age of granite batholiths that form huge intrusive complexes on several continents.

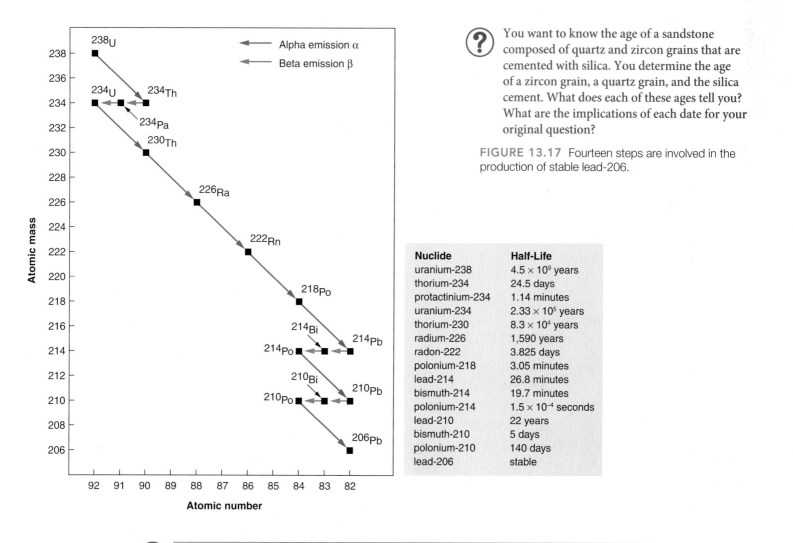

(?) You want to know the age of a sandstone composed of quartz and zircon grains that are cemented with silica. You determine the age of a zircon grain, a quartz grain, and the silica cement. What does each of these ages tell you? What are the implications of each date for your original question?

FIGURE 13.17 Fourteen steps are involved in the production of stable lead-206.

Nuclide	Half-Life
uranium-238	4.5×10^9 years
thorium-234	24.5 days
protactinium-234	1.14 minutes
uranium-234	2.33×10^5 years
thorium-230	8.3×10^4 years
radium-226	1,590 years
radon-222	3.825 days
polonium-218	3.05 minutes
lead-214	26.8 minutes
bismuth-214	19.7 minutes
polonium-214	1.5×10^{-4} seconds
lead-210	22 years
bismuth-210	5 days
polonium-210	140 days
lead-206	stable

(?) **Expand Your Thinking**—What is a closed system and why is it important for accurate dating?

13-8 Potassium-Argon and Carbon Provide Important Isotopic Clocks

LO 13-8 *Describe why radiocarbon dating is useful for dating relatively recent organic material.*

Certain radioisotopes have been particularly useful in deciphering Earth's history: potassium-argon (Ar) and radiocarbon are two examples.

Potassium-Argon

Potassium is abundant in the crust and has been used to date igneous events in Earth's history in thousands of locations. One isotope, potassium-40, is radioactive and decays to form two daughter products, calcium-40 and argon-40, through two different methods of decay. The ratio of these two daughter products to each other is constant: 11.2 percent of the parent becomes argon-40 and 88.8 percent becomes calcium-40. Argon is an inert, noble gas. When rock melts, argon tends to escape. Once the rock solidifies, it traps new argon produced by radioactive decay. This process creates a potassium-argon "clock." The clock is reset to zero by melting followed by crystallization.

You might think that a geologist simply needs to measure the amounts of potassium-40 and argon-40 in a rock to establish a date.

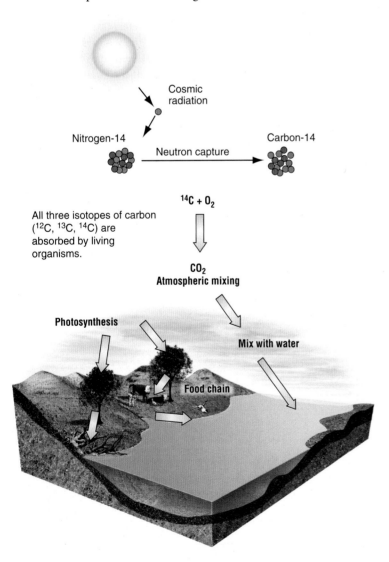

All three isotopes of carbon (^{12}C, ^{13}C, ^{14}C) are absorbed by living organisms.

However, in many cases, a small amount of argon remains in a rock when it hardens. This argon usually is trapped in tiny air bubbles that enter from the atmosphere. (One percent of the air we breathe is argon.) Any extra argon from air bubbles must be taken into account if it is significant relative to the amount of daughter argon produced by radioactive decay. This would most likely be the case in young rocks that have not had time to produce much daughter argon or in rocks that are low in parent potassium.

To be able to date the rock, the geologist must have a way to determine how much *original* atmospheric argon was in the rock when it crystallized. It is possible to do this because atmospheric argon has a couple of other isotopes, the most abundant of which is argon-36. We know that the ratio of argon-40 to argon-36 in air is 295. Thus, if we measure argon-36 and argon-40, we can calculate and subtract the excess atmospheric argon-40 to obtain an accurate age.

Radiocarbon

Willard F. Libby (1908–1980) and a team of scientists at the University of Chicago developed the *radiocarbon dating method* in the 1940s. It has since become the most powerful method of dating relatively young geologic and archaeological materials and has greatly advanced our understanding of human chronology, climate change, and environmental processes that have occurred over the last 40,000 years. (See the "Geology in Action" box titled "Radiocarbon Dating the First Map of North America.") For his efforts, Libby won the 1960 Nobel Prize in chemistry.

The radioisotope carbon-14 (^{14}C) is formed in the atmosphere when high-energy particles from the Sun (mostly protons and neutrons) interact with nitrogen in the air (**FIGURE 13.18**). When a neutron hits the nucleus of nitrogen, it drives out a proton. This changes the nitrogen into an isotope of carbon with a mass number of 14. The carbon isotope is unstable and will undergo beta decay to become nitrogen-14. The half-life for this reaction is 5,730 years.

Newly formed radiocarbon combines with oxygen to form radioactive carbon dioxide (^{14}CO$_2$). In today's atmosphere, one atom of carbon-14 exists for every trillion atoms of the most abundant isotope of the element carbon, carbon-12. Plants absorb radiocarbon as they take up carbon dioxide during photosynthesis. The radiocarbon is then passed from plants to animals, and eventually into soil, through the food chain. Although the amount of radiocarbon in plant and animal cells is perpetually decreasing through radioactive decay, eating and breathing renew it as long as the organism is alive.

When an organism dies, the amount of radiocarbon in its tissue decreases consistently according to the half-life clock. By measuring the residual carbon-14 in organic samples (bones, plant fragments, soil

(?) How does carbon dioxide containing radioactive carbon enter the food chain?

FIGURE 13.18 Carbon-14 is formed in Earth's atmosphere, where it combines with oxygen to form radioactive CO$_2$. Radiocarbon will eventually enter the food chain and provide a means of dating fossil organic materials.

samples, wood, coral, shells, organic-rich mud, etc.), it is possible to calculate the time elapsed since the plant or animal tissue was formed, provided that they have not been contaminated by younger carbon (from bacteria or carbon-rich acid [carbonic acid] in the soil or groundwater) or older carbon (such as calcium carbonate from rocks).

The amount of radiocarbon produced in Earth's atmosphere fluctuates through time because solar radiation varies in intensity, the geomagnetic field fluctuates, and the amount of carbon dioxide (CO_2) in the atmosphere changes from year to year. This means that the amount of carbon-14 entering the food chain is not consistent from one year to another. To take this variability into account, the geologist must calibrate a calculated radiocarbon age to an actual calendar year. Counting annual tree rings and dating the wood from each ring has produced a calibration curve that can be used for this purpose. Since certain, long-lived trees grow a new ring each year (principally U.S. bristlecone pine and German and Irish oak), the amount of carbon-14 in each tree ring provides an exact record of radiocarbon changes through time. In addition to tree rings, carbon-14 measurements of annually banded corals and annually layered organic-rich mud in lakes and bogs have extended the calibration to about 45,000 years—approximately the entire time span of the radiocarbon clock (**FIGURE 13.19**).

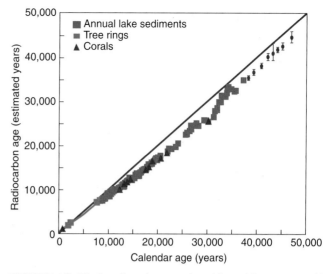

FIGURE 13.19 A radiocarbon age (*y*-axis) must be converted into true calendar years (*x*-axis) to be accurate. The blue symbols are from annual layers of organic lake mud; the green data come from annual tree rings; and the red symbols are from annually banded corals. If all the samples could be plotted on the straight line, carbon-14 years would exactly equal calendar years. But they do not, hence the need for a calibration system to correct radiocarbon years to "true" calendar years.

 Expand Your Thinking—Why must the calibration record have *annual* resolution?

 Why is it necessary to calibrate a radiocarbon date with calendar years obtained by other means?

GEOLOGY IN ACTION

RADIOCARBON DATING THE FIRST MAP OF NORTH AMERICA

Scholars believe that the Vinland Map (**FIGURE 13.20**) is a 15th-century map depicting Viking exploration of North America half a century before the arrival of Columbus's expedition. If genuine, the Vinland Map is one of the great documents of Western civilization; if fake, it is an amazing forgery.

According to one team of scientists, the kind of ink used on the map was made only in the 20th century. According to another, the parchment dates from the mid-1400s. Doubters argue that the parchment's age is irrelevant, while others, who believe that the analysis of the ink is flawed, have introduced evidence that many medieval documents used similar ink.

The Vinland Map shows Europe, the Mediterranean, northern Africa, Asia, and Greenland, all of which were known to travellers of the time. In the northwest Atlantic Ocean, however, it also shows the "Island of Vinland," which represents a part of present-day North America. Text on the map reads, in part: *"By God's will, after a long voyage from the island of Greenland to the south toward the most distant remaining parts of the western ocean sea, sailing southward amidst the ice, the companions Bjarni and Leif Eiriksson discovered a new land, extremely fertile and even having vines,… which island they named Vinland."*

In 1995, scientists from the U.S. Department of Energy, the University of Arizona, and the Smithsonian Center for Materials Research and Education radiocarbon dated the parchment on

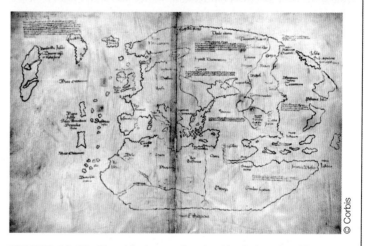

FIGURE 13.20 The oldest map showing North America. The map shows Europe (including Scandinavia), northern Africa, and Asia. A previously unmapped portion of North America—the region around Labrador, Newfoundland, and Baffin Island—is displayed in the upper left corner of the map.

which the map was drawn. The scientists were allowed to trim a 3-inch-long sliver off the bottom edge of the map for analysis. Using the National Science Foundation–University of Arizona's accelerator mass spectrometer, the scientists determined that the parchment dates to 1434 AD, plus or minus 11 years. The unusually high precision of the date was possible because it fell within a very favourable region of the carbon-14 calibration curve.

13-9 Earth's Age Is Measured Using Several Independent Observations

LO 13-9 *Describe the evidence that Earth's age is about 4.6 billion years.*

The commonly accepted age of Earth is 4.5 to 4.6 billion years. At first, this may not appear to be a particularly useful piece of information. But the intellectual process of conceiving where to find geologic records of Earth's age, the technological accomplishment of analyzing those records, and efforts to find independent data that test and refine initial findings are examples of the power of critical thinking and the scientific process.

Our understanding of Earth's age comes from multiple lines of scientific support, including (1) the age of primordial crust, (2) the age of Moon rocks, (3) the age of meteorites, (4) the abundance of lead isotopes, and (5) the consistency of these independent observations.

Primordial Crust

Despite widespread attempts, efforts to identify crust from the time of Earth's origin have failed. There are two reasons for this: (1) The oldest rocks were probably destroyed in the molten phase of Earth's early history (see Chapter 3, "Plate Tectonics"), and (2) processes of the rock cycle—namely, erosion and plate tectonics—have recycled the crust and destroyed direct evidence of Earth's early history.

The oldest whole rock of terrestrial origin (located in northeast Canada) formed 4.28 billion years ago. (See Figure 13.1.) However, an older particle from Australia, a zircon grain sampled in ancient sedimentary layers, has been dated as 4.4 billion years old (**FIGURE 13.21**). Rocks this old are rare, however, and most old crust, found in the ancient cores of continents, dates from about 3.5 billion years ago.

Moon Rocks

Although the Moon's exact origin still is not settled, study of lunar samples helps us understand Earth's geology because the two bodies are probably closely related. Since tectonics never operated in the Moon, and wind and water are absent, lunar rocks have not been subjected to nearly as much recycling. Lunar geology is complicated, however, by the effects of meteoric bombardment and by the small number of samples available for geologists to work with.

The oldest Moon rocks are from *lunar highlands* composed of old crust. The rocks were formed when the early crust was partially or entirely molten. They are mostly basalt in composition, but the *Apollo* missions have brought back only a few samples. **TABLE 13.4** shows that these rocks formed between 4.3 and 4.6 billion years ago.

If the Solar System truly has a common origin such that the Moon, Earth, and the other planetary bodies have a common origin and have since evolved, as proposed by the solar nebula hypothesis, the age of the oldest Moon rocks should reflect the age of Earth. The data in Table 13.4 are consistent with the oldest ages of Earth rocks that have been influenced by rock cycle processes that probably destroyed the earliest components of the crust.

Meteorites

The majority of the 70 or so well-dated meteorites have ages of 4.4 to 4.7 billion years (**TABLE 13.5**). These meteorites, which are fragments

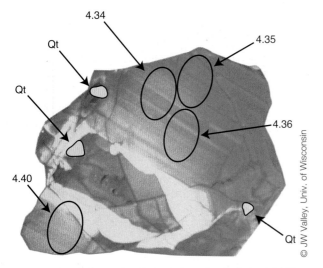

FIGURE 13.21 A tiny grain of zircon from Australia is 4.4 billion years old, the oldest date yet determined for a rock from Earth. Circles identify other dated samples in billions of years before present, and Qt indicates quartz grains.

© JW Valley, Univ. of Wisconsin

? Why is it unlikely to find an Earth rock that dates from the time of Earth's origin?

TABLE 13.4 Oldest Moon Rocks

Mission	Dating Technique	Half-Life	Age (billions of years)
Apollo 17	Rubidium (Rb)-Strontium (Sr)	48.8 billion	4.55 ± 0.1
Apollo 17	Rb-Sr	48.8 billion	4.60 ± 0.1
Apollo 17	Samarium (Sm)-Neodymium (Nd)	106 billion	4.34 ± 0.05
Apollo 16	$^{40}Ar/^{39}Ar$	1.25 billion	4.47 ± 0.1

TABLE 13.5 Ages of Some Meteorites

Meteorite Name	Dating Technique	Age (billions of years)
Allende fragment	Ar-Ar	4.52 ± 0.02
Juvinas fragment	Sm-Nd	4.56 ± 0.08
Angra dos Reis fragment	Sm-Nd	4.55 ± 0.04
Mundrabrilla fragment	Ar-Ar	4.57 ± 0.06
Various ordinary chondrites	Various methods	4.6 to 4.69 ± 0.14

FIGURE 13.22 Meteorites are among the most primitive materials in the Solar System.

(?) What is the primary source of meteorites?

of asteroids and represent some of the most primitive material in the Solar System, have been dated by several independent isotopic dating methods using primary radioisotopes. Meteorites are, in a sense, ideal for age studies since there is very little chance of their composition being altered after their formation. Although meteorite ages have no direct bearing on Earth's age, they say something about the age of the Solar System and, therefore, the age of the planets.

Geologists assume that meteorites (**FIGURE 13.22**) and Moon rocks were not subjected to the extensive alteration that Earth rocks have experienced. Therefore, their ages indicate when they were formed. Because major objects within the Solar System are thought to have been formed at the same time, and since evolved, Earth must be the same age as meteorites and a bit older than the Moon—that is, about 4.55 to 4.6 billion years old. Recall that the impact hypothesis for the origin of the Moon suggests that it is somewhat younger than Earth and was formed soon after the origin of the Solar System. This history is also consistent with the absolute ages of these materials.

Lead Isotopes

A fourth line of evidence for Earth's age comes from the abundance of lead isotopes in Earth rocks (**FIGURE 13.23**). This method yields results that are consistent with both meteorite and lunar ages. Natural lead is a mixture of four stable isotopes; three of these isotopes (lead-206, lead-207, and lead-208) result from the decay of radioisotopes of thorium and uranium. The fourth, lead-204, is not the result of radioactive decay. This means that all the lead-204 on Earth has been around since the planet's formation.

Based on extensive sampling of Earth's crust, geologists have determined the present-day abundances of the four lead isotopes relative to one another and to the parent isotopes (uranium-238, uranium-235, and thorium-232) that produced three of them. Because the original amount of lead on the planet cannot be measured, scientists use meteorites to determine Earth's original lead content.

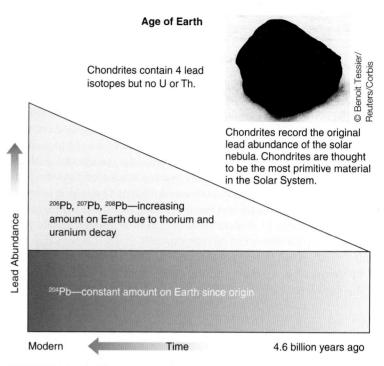

Age of Earth

Chondrites contain 4 lead isotopes but no U or Th.

Chondrites record the original lead abundance of the solar nebula. Chondrites are thought to be the most primitive material in the Solar System.

^{206}Pb, ^{207}Pb, ^{208}Pb—increasing amount on Earth due to thorium and uranium decay

Lead Abundance

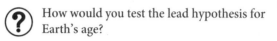

^{204}Pb—constant amount on Earth since origin

Modern   Time   4.6 billion years ago

FIGURE 13.23 The accumulation of lead-206, lead-207, and lead-208 provides a measure of elapsed time since Earth was formed.

(?) How would you test the lead hypothesis for Earth's age?

Some primitive meteorites (called **chondrites**) contain the four lead isotopes but no uranium or thorium parents. Scientists infer that the lead in these meteorites is not the product of radioactive decay and therefore has not changed since they were formed. Scientists believe that the lead content of chondrites is a reasonable approximation of Earth's original lead content, the so-called *primordial lead*.

By comparing the amounts of the four lead isotopes in primordial lead to their current amounts on Earth, scientists can determine how much lead has been added by radioactive decay since Earth was formed. Knowing the half-life of each parent, it is possible to calculate how long it took to create modern differences between the amounts of present-day and primordial lead for each of the three isotopes produced by radioactive decay. These calculations yield an age of about 4.6 billion years for Earth, which is consistent with the ages independently determined from meteorites and lunar rocks.

Consistency

It is noteworthy that lunar and meteorite ages display remarkable consistency. These results have been produced by different researchers using different isotopic dating techniques from different samples. In general, the oldest meteorite ages tend to be older than the oldest lunar ages, a fact that fits with our understanding of the formation of the Moon as occurring sometime after Earth was formed. It is also noteworthy that the calculation of radiogenic lead accumulation in Earth's crust corresponds perfectly to the age of Moon rocks and meteorites. Taken together, these four lines of evidence provide strong justification for the conclusion that Earth formed 4.55 to 4.6 billion years ago.

(?) **Expand Your Thinking**—What geologic environment would you search to look for the oldest rock on Earth?

LET'S REVIEW "GEOLOGY IN OUR LIVES"

Now that you've finished the chapter, "Geology in Our Lives" will have taken on new meaning for you. Let us review it: Knowing the history of rock layers in the crust is critical to predicting the location of geologic resources and discerning the events that make up Earth's past. Geologists analyze geologic time using two methods that are based on critical thinking: (1) relative dating, which determines the *order* of geologic events, and (2) isotopic dating, which estimates the *age* of a geologic event. These methods have revealed that Earth has a long and fascinating history and that it is approximately 4.6 billion years old. It is necessary to integrate relative and isotopic dating to unravel the history of the crust in order to find resources such as groundwater, metals, coal, petroleum, and others that our communities depend on every day.

Geology is the science of time. We have learned that time can be measured in several ways. The study of relative time is based on the application of basic principles of stratigraphy that identify the step-by-step sequence of geologic events leading to the structure of the crust. Early naturalists applied these principles to interpreting the origin of rock outcrops and recognized that geologic events recorded immense periods of time. They were the first to propose that Earth was hundreds of millions of years old.

With improved understanding of the physics of radioactivity came the realization that minerals hold "isotopic clocks" that are useful for isotopic dating of geologic events. In a breathtaking example of critical thinking, isotopic dating and relative dating combined to address the fundamental question of Earth's age, which is now estimated to be 4.5 to 4.6 billion years. On the basis of these technical advances, geologists have built and refined the geologic time scale, which tells the story of changes that characterize the planet's past and the diverse life that populates its surface. That is the story of the next chapter.

STUDY GUIDE

13-1 Geology is the "science of time."

- Every field of science is known for one great characteristic; geology is known as the science of time. Geologists are trained to interpret vast time spans in the details of Earth's rocks.

- Geologists apply two types of time analysis: **relative dating** and **isotopic dating**. Relative dating determines the order of formation (which came first, which came next, etc.) among layers of rock based on their interrelationships. Isotopic dating uses radioisotopes within minerals to calculate their chronological age.

- The study of geologic time is used to find fossil fuels, precious metals, and other geologic resources. In addition, human curiosity demands answers to questions about our origin, our history, and our place in the sequence of natural events.

13-2 Earth history is a sequence of geologic events.

- Early estimates of Earth's age were based on religious dogma. Scientific analysis of the problem of Earth's age led to estimates of hundreds of millions of years.

- Major geologic events fall into five main classes: **intrusion**, **erosion**, **deposition**, **faulting**, and rock **deformation**.

13-3 Seven stratigraphic principles are used in relative dating.

- **Stratigraphic principles** are used to guide scientific thinking when unravelling the history of crust formation.

- There are seven stratigraphic principles: the principle of superposition, the principle of original horizontality, the principle of original lateral continuity, the principle of cross-cutting relationships, the principle of inclusions, the principle of unconformities, and the principle of fossil succession.

13-4 Relative dating determines the order of geologic events.

- A block of crust contains a complicated system of rock layers. Geologists can interpret these by applying stratigraphic principles to determine the sequence of geologic events that produced the crust.

13-5 Isotopic dating uses radioactive decay to estimate the age of geologic samples.

- **Radioactive decay** is a process wherein the unstable nucleus of an atom spontaneously emits radiation and a subatomic particle. This results in an atom of one type (called the parent isotope) transforming to an atom of another type (called the daughter isotope). The rate of radioactive decay is fixed for a given radioisotope. It produces daughter isotopes that can be counted, and it is possible to estimate the original amounts of parent isotopes.

- There are three main types of radioactive decay: alpha decay, beta decay, and electron capture. Alpha decay emits a single alpha particle composed of 2 protons and 2 neutrons and makes a daughter isotope with 2 fewer protons. Beta decay occurs when the nucleus emits a negative beta particle that changes one of the neutrons into a proton. The parent isotope gains 1 proton at the expense of 1 neutron. In electron capture, an electron is drawn into the nucleus, where it combines with a proton to form a neutron.

- Because the rate of decay is fixed, the passage of time can be tracked by the reduction in parent isotopes and the increase in daughter isotopes. Since it is impossible to determine exactly when a single atom will decay, geologists have defined the statistical **half-life** of decay. The half-life is the time it takes for half the radioactive atoms in any sample to undergo radioactive decay.

13-6 Geologists select an appropriate radioisotope when dating a sample.

- Not every type of **radioisotope** can be used to date every type of rock. A geologist must choose among the many dating techniques available and select a radioisotope that is appropriate and useful for the material to be dated. Young materials should be dated with short-lived radioisotopes and old materials with long-lived ones. In general, useful isotopes fall into two categories: primary radioisotopes and cosmogenic radioisotopes.

13-7 Accurate dating requires understanding various sources of uncertainty.

- To date a rock accurately, the geologist must determine the answers to several questions: What is the relationship between the mineral sample and the event it represents? What is the approximate age of the sample? What radioisotope is appropriate for both the type of material being dated and the approximate age of the sample? What are the sources of potential error in a calculated age, and how can the age be tested?

- To determine an accurate age, the geologist assumes that the mineral being dated has remained a closed system throughout its life. This means that neither parent nor daughter isotopes have left or entered the mineral since its formation. When this is not the case, it is usually as a result of metamorphism.

- All isotopes of the element uranium are at least mildly radioactive. Uranium-238, which occurs naturally in most types of granite and soil in varying degrees, decays with a half-life of 4.5 billion years and produces the stable daughter isotope lead-206. This process consists of 14 steps, each leading to the production and subsequent decay of a radioactive daughter isotope. These steps are referred to as a **decay chain**.

13-8 Potassium-argon and carbon provide important isotopic clocks.

- The potassium-argon decay pair is an effective tool for dating igneous rocks, and radiocarbon is an effective tool for dating recent organic materials in the time frame of the last 40,000 years.

13-9 Earth's age is measured using several independent observations.

- The scientifically measured age of Earth is 4.55 to 4.6 billion years.
- Estimates of Earth's age are derived from several lines of evidence: (1) the age of primordial crust, (2) the age of Moon rocks, (3) the age of meteorites, (4) the abundance of lead isotopes, and (5) the consistency of these separate lines of evidence.

KEY TERMS

chondrites (p. 359)
decay chain (p. 354)
deformation (p. 341)
deposition (p. 341)
erosion (p. 341)
evolution (p. 343)

faulting (p. 341)
geologic events (p. 341)
half-life (p. 349)
intrusion (p. 341)
isotopic dating (p. 338)
radioactive decay (p. 338)

radioisotopes (p. 348)
relative dating (p. 338)
stratigraphic principles (p. 342)
unconformity (p. 341)
uplift (p. 341)

ASSESSING YOUR KNOWLEDGE

Please answer these questions before coming to class. Identify the best answer.

1. Relative dating is the process of
   a. calculating the age of a rock sample.
   b. determining how old a mineral is.
   c. determining the sequence of events in a period of geologic history.
   d. calculating when a mineral was renewed by metamorphism.
   e. determining the sources of uncertainty in a date.

2. Isotopic dating is the process of
   a. estimating the age of a sample using radioisotopes.
   b. documenting the unique fossil assemblage in a rock.
   c. determining the geologic events that formed a rock.
   d. determining the sequence of geologic events in the crust.
   e. assessing the rate of sediment accumulation in the ocean.

3. Early critical thinkers estimated Earth's age using
   a. calculations of the rate of delivery of salt to the sea.

   b. estimates of sediment accumulation over time.
   c. calculations of the time needed for Earth to cool.
   d. assessments of time needed to allow for evolution as recorded in rocks.
   e. All of the above.

4. The five fundamental geologic events described by stratigraphic principles are
   a. dating, eroding, depositing, removing, and folding.
   b. deformation, unconformities, superposition, deposition, and faulting.
   c. deposition, erosion, deformation, faulting, and intrusion.
   d. intrusion, deposition, erosion, tectonism, and unconformities.
   e. None of the above.

5. The principle of superposition states that
   a. the lowest layers in an undeformed sequence of strata are the oldest.
   b. the highest layers in an undeformed sequence of strata are the youngest.
   c. a rock layer lying above another must be the younger of the two.
   d. older rock units typically are found at the base of a sequence of rocks.
   e. All of the above.

6. When an intrusion invades the crust, its relative position in a sequence of geologic events is determined using the principle of
   a. unconformities.
   b. superposition.
   c. cross-cutting relationships.
   d. original lateral continuity.
   e. None of the above.

7. William "Strata" Smith is known for
   a. his use of the principle of fossil succession.
   b. using isotopic dating to determine the geology of England.
   c. developing modern map-making techniques.
   d. identifying England's earliest life-forms.
   e. None of the above.

8. Which of the following statements explains how the principle of cross-cutting relationships is related to the principle of unconformities?
   a. Unconformities are made by intrusions that cut across strata.
   b. Unconformities cut across pre-existing strata.
   c. These two principles are not related.
   d. Unconformities are younger than the strata that lie above them.
   e. Unconformities are not used in geologic time.

9. To date a very old rock, a geologist should use an isotope that
   a. is very old.
   b. is very young.
   c. has a long half-life.
   d. has a short half-life.
   e. has been contaminated by groundwater.

10. A radioisotope with a short half-life should be used to date
    a. a very old geologic event.
    b. a very young geologic event.

c. a sample that has not had a closed system.
d. the age of Moon rocks.
e. meteorites.

11. Typically
    a. primary radioisotopes have shorter half-lives than cosmogenic radioisotopes.
    b. primary radioisotopes have longer half-lives than cosmogenic radioisotopes.
    c. primary radioisotopes have the same half-lives as cosmogenic radioisotopes.
    d. primary radioisotopes come from the decay of cosmogenic radioisotopes.
    e. primary radioisotopes cannot be compared to cosmogenic radioisotopes.

12. One problem with isotopic dating is
    a. open-system behaviour.
    b. dangerous radioactivity.
    c. minerals contaminated by sand.
    d. dating a mineral with many types of radioactivity.
    e. There are typically very few problems with isotopic dating.

13. Which method would a geologist use to date a fossil bone fragment of a mammoth?
    a. potassium-argon
    b. rubidium-strontium
    c. carbon-14
    d. There is no way to date bone.
    e. None of the above.

14. Moon rocks are
    a. about the same age as the continents.
    b. evidence that Earth and the Moon are unrelated.
    c. evidence that Earth's age is approximately 4.6 billion years.
    d. not datable because they are radioactive.
    e. not allowed to be dated.

15. Which of the following statements is true?
    a. Moon rocks, meteorites, and continents are all the same age.
    b. The oldest Earth rock is the age of Earth.
    c. Earth, the Moon, and meteorites all formed at the same time.
    d. It is unlikely that any rocks are left on Earth from its origin.
    e. Meteorites come from the Moon.

FURTHER RESEARCH

1. The half-life of uranium-234 decaying to thorium-230 is 248,000 years. What is the age of a coral sample that has a uranium-234 to thorium-230 ratio of 1:3? What if the ratio were 3:1?

2. Why does isotopic dating rely on an unchanging rate of decay to be effective?

3. How can deformation interfere with use of the principle of superposition?

4. What are the major sources of uncertainty in isotopic dating?

ONLINE RESOURCES

Explore more about geologic resources on the following websites:

PBS—"Evolution—A Journey into Where We're from and Where We're Going":
www.pbs.org/wgbh/evolution/index.html

International Commission on Stratigraphy geologic time scale:
www.stratigraphy.org/index.php/ics-chart-timescale

Geological Society of America geologic time scale:
www.geosociety.org/science/timescale

American Scientific Affiliation—"Radiometric Dating—A Christian Perspective":
www.asa3.org/ASA/resources/Wiens.html

U.S. Geologic Survey—"Geologic Time":
http://pubs.usgs.gov/gip/geotime/contents.html

The TalkOrigins Archive—"Radiometric Dating and the Geological Time Scale":
www.talkorigins.org/faqs/dating.html

Additional animations, videos, and other online resources are available at this book's companion website:
www.wiley.com/go/fletchercanada

This companion website also has additional information about WileyPLUS and other Wiley teaching and learning resources.

364

CHAPTER 14
EARTH'S HISTORY

Chapter Contents and Learning Objectives (LO)

14-1 Earth's history has been unveiled by scientists applying the tools of critical thinking.
LO 14-1 *Describe the forebears of geology and their contributions.*

14-2 Fossils preserve a record of past life.
LO 14-2 *Describe the process of fossilization.*

14-3 There are several lines of evidence for evolution.
LO 14-3 *Describe the geological evidence for evolution.*

14-4 Molecular biology provides evidence of evolution.
LO 14-4 *Describe the process of evolution.*

14-5 Mass extinctions influence the evolution of life.
LO 14-5 *List the major extinctions and when they occurred.*

14-6 The geologic time scale is the "calendar" of events in Earth's history.
LO 14-6 *Describe the geologic time scale and how it is organized.*

14-7 The Archean and Proterozoic Eons lasted from 4.0 billion to 542 million years ago.
LO 14-7 *Compare and contrast the geologic processes that occurred in the Archean and Proterozoic Eons.*

14-8 In the Paleozoic Era, complex life emerged and the continents reorganized.
LO 14-8 *List the periods of the Paleozoic Era and a notable characteristic of each.*

14-9 In the Mesozoic Era, biological diversity increased and continents reorganized.
LO 14-9 *Compare and contrast the three periods of the Mesozoic Era.*

14-10 Modern mammals, including humans, arose in the Cenozoic Era.
LO 14-10 *Describe the major events of the Cenozoic Era.*

GEOLOGY IN OUR LIVES

The study of Earth's history provides humans with a sense of place and perspective in the natural world. It reveals that time is deep and great change has occurred in both geological environments and biological communities. The fossil record provides strong evidence of the evolution of life. Tens of thousands of fossils define animal and plant lineages that fully document evolutionary changes. These fossils prove that evolution is responsible for Earth's vast diversity of life forms. There are practical aspects to the study of Earth's history. Evolution guides understanding and treatment of medical problems. Geologists use fossils to identify the rocks that carry natural resources, such as oil, coal, and many others. Fossils tell us that significant environmental change leads to mass extinctions; that humans and other four-limbed animals share a common ancestor; that chimpanzees are more closely related to humans than they are to gorillas; and that, ultimately, all living organisms are related in a great web of life. In the end, this story tells us that if we wish to live sustainably on this planet, we absolutely must see beyond the horizon defined by our short lives.

(?) The photograph is of *Triceratops*, a dinosaur that lived during the latest Cretaceous. How do we know when it lived, and why do we think it was herbivorous?

14-1 Earth's History Has Been Unveiled by Scientists Applying the Tools of Critical Thinking

LO 14-1 *Describe the forebears of geology and their contributions.*

Use of the dating tools we studied in Chapter 13 ("Geologic Time") has led to fundamental advances in our knowledge of Earth's history—a field known as **historical geology.** Many of these advances came through the efforts of pioneering scientists who, a century and a half ago, walked the land, studied natural materials, and formulated many of the principles that define the science of geology.

Two significant contributions of these forebears have become hallmarks of science: the **theory of evolution** (changes in the inherited traits of a biological population from one generation to the next) and the recognition that Earth is profoundly old, a characteristic that scientist and author Stephen J. Gould called *deep time.* These two great contributions are presented in the "Geology in Our Lives" discussion at the start of this chapter: (1) Earth's history is characterized by deep time and great change, and (2) the fossil record provides strong evidence of the evolution of life. This chapter takes you on a walk through time to examine the major tectonic and biologic events of the Precambrian and Phanerozoic Eons (**FIGURE 14.1**).

The science of geology owes its origin to several visionary thinkers (**FIGURE 14.2**) who viewed Earth as a complex system with a history characterized by perpetual change across great spans of time. They realized that to fully understand that history, they needed analytical tools based on critical thinking, and they set forth to create those tools.

Nicholas Steno

Nicholas Steno (1638–1686), considered the father of geology and stratigraphy, was a Danish priest and anatomist who became interested in "figured stones" (fossils), some of which looked familiar while others did not. He was the first to suggest that fossils that looked like living organisms had in fact once been alive. Steno also proposed the principles of original horizontality, superposition, and lateral continuity that we studied in Chapter 13.

James Hutton

James Hutton (1726–1797), a Scottish physician, farmer, and geologist, was a staunch proponent of the idea that geologic processes alter Earth's surface. The common view in his time, called *Neptunism,* was that all rocks had been formed by precipitation or sedimentation from a single great ocean (the Biblical flood) and remained essentially unchanged from that time. Hutton proposed that geologic time was indefinitely long and that Earth was self-renewing: As mountains eroded away, new ones were uplifted; as the sea covered some lands, it receded from others. This, of course, is the basis of the *rock cycle* (Chapter 1, "An Introduction to Geology"). Hutton is famous for his claim that Earth has "no vestige of a beginning—no prospect of an end."

Perhaps Hutton's most lasting contribution is the concept of **uniformitarianism,** which states that Earth's history is best explained by observations of modern processes. In other words, geologic principles have been uniform over time. Sir Archibald Geikie (1835–1924) famously summed up uniformitarianism in stating, "the present is the key to the past."

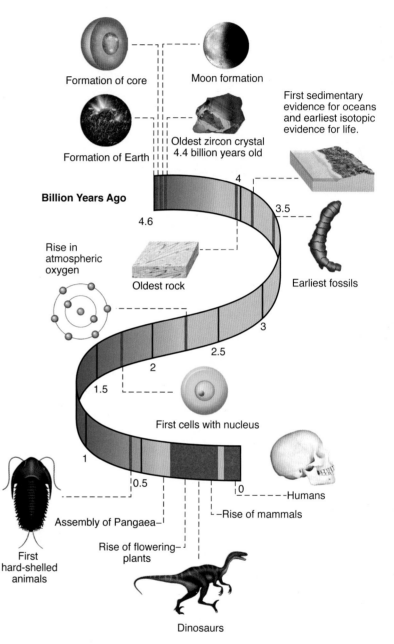

FIGURE 14.1 Earth's history is characterized by the evolution of life from simple early forms to highly diverse modern ecosystems. This history includes tectonic events that have built mountain systems, split and assembled continents, and formed oceans.

(?) It is interesting to compare the Earth's history to a period of one year. If the Earth formed on January 1, and today is December 31, when did a) the oldest rock form, b) the first cells with a nucleus develop, c) the first hard-shelled animals appear, and d) humans first appear?

George Bernard/Science Source
Bob Thomas/Popperfoto/Getty Images, Inc.

(a)   *(b)*   *(c)*   *(d)*

FIGURE 14.2 *(a)* Nicholas Steno, *(b)* James Hutton, *(c)* Charles Lyell, *(d)* Charles Darwin.

(?) Describe the fundamental advance in critical thinking that each of these early scientists contributed.

Charles Lyell

British lawyer and naturalist **Charles Lyell** (1797–1875) is often cited as the father of modern geology. His five-volume *Principles of Geology* became an indispensable reference for every 19th-century geologist. Lyell explained the principle of cross-cutting relationships and the principle of inclusions. He was best known for his ability to explain the meanings of discoveries by other geologists. For instance, he expounded at great length on Hutton's concept of uniformitarianism and brought the idea into full flower among the scientists of his time. As one of his colleagues said, "We collect the data and Lyell teaches us the meaning of them."

Charles Darwin

British biologist **Charles Darwin** (1809–1882) is credited with the theory of evolution, the scientific theory that accounts for changes seen in the fossil record. Today Darwin's theory is a cornerstone of scientific thought, especially in the fields of geology and biology.

In 1831, Darwin secured an unpaid position as naturalist aboard the HMS *Beagle,* a British ship that circumnavigated the globe. He eventually published his observations in *A Naturalist's Voyage on the Beagle* (1839). His theory of evolution, described in *On the Origin of Species by Means of Natural Selection* (1859) and *The Descent of Man* (1871), stated that all living things develop from a very few simple forms over time through **natural selection.** Natural selection is the process by which favourable traits that are *heritable* (passed on to offspring) become more common in successive generations of a population of reproducing organisms (**FIGURE 14.3**). Organisms that are best adapted to their environment (i.e., have variations that are most favourable for their survival) tend to survive and reproduce, transmitting their genetic characteristics to the next generation. Those organisms that are not as well suited to their environment die out. In this way, populations change over time, a phenomenon that is predicted by the theory of evolution. Natural selection is how evolution takes place. These changes are recorded in the geologic record of fossils, reflecting the principle of fossil succession.

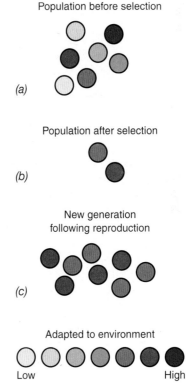

Population before selection

(a)

Population after selection

(b)

New generation
following reproduction

(c)

Adapted to environment

Low   High

FIGURE 14.3 Schematic representation of natural selection. *(a)* A population of reproducing organisms. *(b)* Population after natural selection takes place. *(c)* New generation of organisms following reproduction.

(?) Disease is one mechanism of natural selection. Name some others.

Expand Your Thinking—Natural selection is at work all around us. Describe an example of natural selection that you have encountered.

14-2 Fossils Preserve a Record of Past Life

LO 14-2 *Describe the process of fossilization.*

Fossils are the remains of animals and plants, or traces of their presence, that have been preserved in Earth's crust (FIGURE 14.4). *Fossilization* is the process that turns a once-living organism into a fossil. There are many fossils to be found, but they represent only a tiny fraction of all the animals and plants that ever lived.

Without fossils, our knowledge of Earth's history and the history of life on Earth would be very limited. Fossils provide details about how individual species come into existence and how they lived and interacted. In addition, because of the principle of fossil succession, they can tell us the age and history of sedimentary beds. Fossils provide a record of how life changed through time and offer strong evidence that supports and refines Darwin's theory of evolution. Evolution is the most significant natural process influencing Earth's biological communities through the long years of geologic history.

How Are Fossils Made?

When an animal or plant dies, it is usually completely destroyed. Another animal may eat it, or it may decay. But sometimes the remains of an animal are buried before they can be destroyed, and if the conditions are just right, the remains will be preserved in rock as fossils. Usually only the *hard parts* of an animal, such as teeth, bones, and shells, become fossilized. (Think about your own body. The parts that are most likely to become fossilized include teeth, hard bones, and nails. The same is true of other animals.) In some very rare cases, scientists have found fossils of bird feathers and dinosaur skin (FIGURE 14.5).

Almost all fossils are preserved in sedimentary rocks because the high temperature of magma, which ultimately crystallizes to form igneous rock, and the pressure of metamorphism tend to destroy fossils. However, fossilization does not always occur as simple burial. Imagine a cardboard coffee cup carelessly thrown into a mud puddle. There are four ways to fossilize the cup (FIGURE 14.6):

- formation of a *cast* when sediment fills the inside of the cup before the cardboard of the cup is dissolved
- formation of an *internal mould* when sediment or later minerals fill what used to be the cup and preserve an impression of the inner part of the cup

Leonello Calvetti/Science Photo Library

FIGURE 14.4 An artist's impression of what a *Tyrannosaurus rex,* the most famous carnivorous dinosaur, might have looked like based on evidence provided by fossils. *T. rex* lived during the late Cretaceous, and although not the largest dinosaur, it was still impressive as it attained lengths of about 12 metres, stood about 4 metres high, and weighed about 7 tonnes. However, there is still discussion as to whether T. rex was a predator or a scavenger, although most likely it was both. The most complete skeleton discovered in Canada, now named Scotty, was near Eastend, Saskatchewan, in 1991, and is now on display at the T. rex Discovery Centre in the same town.

(?) When a *Tyrannosaurus rex* died, which parts of its body would be the most likely to be preserved in the fossil record?

© -/epa/Corbis

FIGURE 14.5 In 1989, researchers uncovered a fossilized dinosaur nest in Patagonia, South America. They found the preserved skin of embryonic dinosaurs within the eggs that are late Cretaceous in age.

(?) What special conditions are needed to preserve soft body parts as fossils?

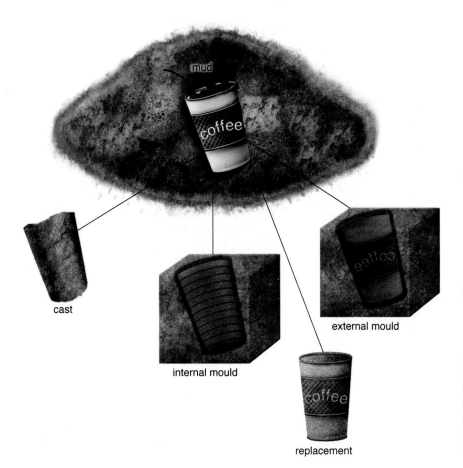

cast

internal mould

external mould

replacement

(?) Imagine you make a footprint in sand that is later filled in with mud. What kind of fossil have you made?

FIGURE 14.6 A coffee cup can be preserved as a cast (a hardened sample of the mud that fills the cup), as an internal mould (the impression of the inside of the cup), as an external mould (an impression of the outside of the cup), or by replacement (a mineral is substituted for the cardboard by precipitation from groundwater).

- formation of an *external mould* when the cup is pressed into the sediment to form an impression of the outside of the cup before it dissolves
- *replacement* when the cup is slowly replaced by new minerals

Another fossilization process is called *permineralization*, in which the spaces within an organism are filled with new minerals. This process can result in the preservation of the internal structure of an organism. Remember that a mould is a fossil in reverse: Ridges on the original organism become grooves, knobs become depressions, and cavities become bumps (**FIGURE 14.7**).

Replacement occurs if the hard parts, and sometimes the soft parts, are replaced by a hard mineral. This happens when chemically active groundwater dissolves the original material of a body part and replaces it with a hard mineral of equal volume and shape. Replacement can be a very delicate and precise mode of fossilization as it occurs atom by atom. Replacement can preserve an exact replica of the fossil, such as a shell replaced by calcium carbonate or a tree limb replaced by silica.

L. K. Broman/Science Source

FIGURE 14.7 Brachiopod fossils from the Paleozoic Era are preserved in dolomite. These fossils are about two centimetres across.

(?) Which of these are moulds and which are casts?

(?) **Expand Your Thinking**—The fact that (usually) only hard parts of an organism are preserved as a record of its existence tends to limit our understanding of past ecosystems. Why is this so?

14-3 There Are Several Lines of Evidence for Evolution

LO 14-3 *Describe the geological evidence for evolution.*

The early proponents of Darwin's theory had a difficult time convincing some of their colleagues that evolution was at work in the plant and animal kingdoms. Because evolution can take generations to manifest its changes, evolutionary processes are not easy to observe firsthand among communities of living organisms. Fossils provide a solution to this problem.

Phylogeny

A famous example of fossil evidence of evolution is the evolutionary lineage, or **phylogeny,** of the modern horse (**FIGURE 14.8**). Nearly 35 now-extinct ancestral species make up the family tree of the modern horse. Over time, a set of changes occurred among these ancestors that were passed from one generation to the next and produced traits characterizing the horse family today: the growth of a strong, grass-chewing set of teeth; an increase in size, strength, and speed; the development of a single toe (the hoof) from forebears originally with four toes; lengthening of the jaws to raise the eyes away from the mouth; and an increase in brain size.

The pattern of horse evolution is typical of many phylogenies. The horse phylogeny is a complex tree with numerous "side branches," some leading to extinct species and others leading to species closely related to modern *Equus*. This branched family tree is the result of random genomic variations and natural selection in a changing climate. Evidence suggests that horses evolved at the same time that forests were giving way to grasslands across the continents. Consistent with Darwin's theory, changes in the environment exerted natural selection on early populations of horse ancestors so that individuals with characteristics favourable to survival in open grassy plains were more likely to endure and reproduce.

Homologous Structures

Another clue to evolution is seen in **FIGURE 14.9**. **Homologous structures** are similar characteristics of organisms that are due to their shared ancestry. These structures are found in such diverse organisms as birds, swimming mammals, four-legged animals, humans, and insect-eating reptiles. This pattern suggests that these animals evolved from a common ancestor and that survival pressure (natural selection) has preferentially selected individuals possessing the specific traits and functions served by each limb. In four-limbed vertebrates, for example, limb bones may vary in size and shape, but all contain the same number and position of specific bones. This phenomenon indicates that totally different families of organisms share a common ancestor. If this were not the case, it is difficult to imagine how specific and fundamentally unique selective pressures could have converged on a single basic blueprint for limb design.

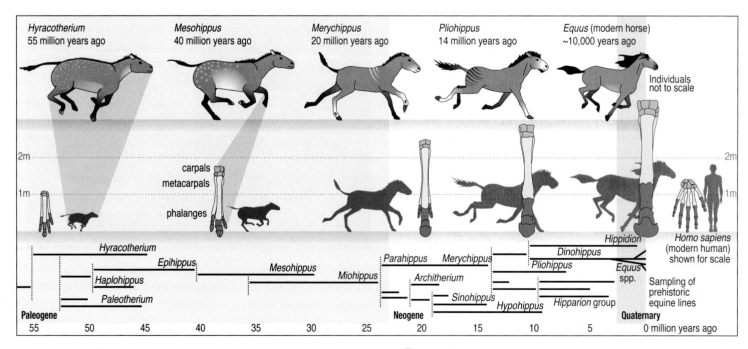

FIGURE 14.8 The phylogeny of the horse family shows consistent changes in the lower leg area, the development of molars designed for tough grasses, and an increase in body size and strength.

(?) Apply inductive reasoning (drawing general conclusions from specific observations) and use the changing characteristics of horse ancestors to describe the natural processes exerting selection pressure on the population.

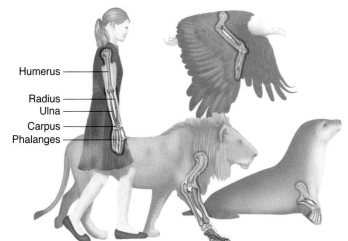

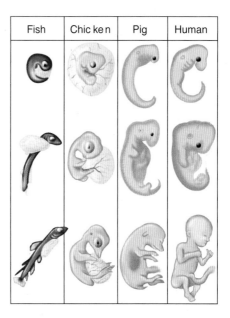

FIGURE 14.9 The limbs of various animals are modified for different functions. Notice that individual bones have been modified in different ways to accomplish different types of tasks (grasping, walking, flying, and paddling). All limbs possess the same basic set of bones in the same order.

 How does evolution theory account for the fact that all four-limbed animals have the same basic set of bones?

FIGURE 14.11 Many animals display remarkably similar features during their development as embryos. Humans, for instance, have a tail and primitive gill slits.

Vestigial Structures

In another important line of evidence for evolution, biologists identify **vestigial structures** in certain organisms as having resulted from evolution under selective pressure. Vestigial structures are the relics of body parts that were used by ancestral forms but are now non-essential (**FIGURE 14.10**). Natural selection does not exert enough pressure to completely eliminate vestigial parts once they have been replaced with newer functions. As a result, organisms tend to retain some aspect of the original structure. Humans possess more than 100 vestigial structures, including ear muscles, wisdom teeth, the appendix, the coccyx (tail vertebrae), hair, ridges on the upper lip, nipples on males, finger- and toenails, and others.

Embryology

Evidence of evolution is also present in studies of *embryology*. In their early stages of development, embryos of various fish, birds, and mammals display strikingly similar characteristics. Human embryos have tails (which eventually grow into the base of the spine) and gill slits (which become the Eustachian tube), as do other life forms in their early stages. It is thought that all these animals have a distant common ancestor from which they inherited a set of genes that control early embryologic development (**FIGURE 14.11**). As embryo development progresses, other genes assume control and produce an individual of a particular species.

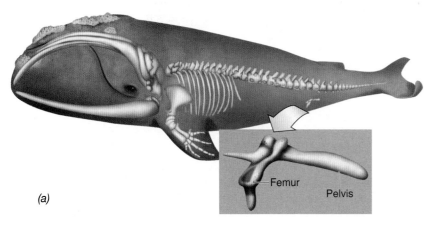

(a)

External structure

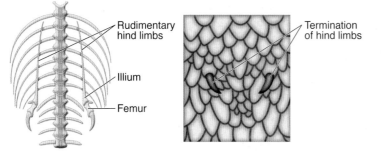

(b)

FIGURE 14.10 *(a)* Whales possess a vestigial pelvis and femur originally designed for walking. Proof that modern whales have evolved from walking ancestors was found in 1994 in the form of a fossil whale that had flippers for front legs and long hind limbs with elongate toes for webbed feet. *(b)* Boa constrictors also have vestiges of legs.

 The ridges on our upper lips are vestigial. What was their original form and function?

 Expand Your Thinking—Do you find these lines of evidence for evolution personally compelling? What questions do you have?

14-4 Molecular Biology Provides Evidence of Evolution

LO 14-4 *Describe the process of evolution.*

One of the ways in which evolution is thought to proceed occurs when a species becomes separated into geographically isolated populations that experience different environmental conditions. For instance, two groups of the same species may become separated and isolated from one another in two mountain valleys. Unique characteristics may arise among the separate populations, eventually causing them to become distinct and form new species. These differences are the product of **genetic mutation** (random changes to genetic material—RNA or DNA) and **genetic variation** (differences in inherited traits) among individuals in the population. Favourable variations are naturally selected through environmental pressure (they enhance the ability to survive in the environment) or community pressure (they enhance the ability to survive in a community of individuals). Variations that burden the organism with disadvantages will be *selected against,* making individuals with these variations less likely to survive and reproduce.

Controlled breeding is a modern form of isolation practised by horse, dog, and cat breeders (and others). In controlled breeding, humans allow specific individual animals with certain traits to breed, increasing the opportunity for these traits to collect in the next generation. Controlled breeding is responsible for the over 400 types of dogs that have been bred into existence by humans. These include breeds ranging from the tiny Chihuahua to the massive Great Dane. In fact, the entire family of dogs all evolved from a few wolves that were domesticated by humans in Asia less than 15,000 years ago.

Molecular Biology

The field of molecular biology provides evidence of evolution using molecules of DNA, RNA, and protein that are present in every organism. If one organism has evolved from another, their DNA sequence will be similar, and if two organisms are distantly related, their DNA will be different. For example, children with the same parents are closely related and therefore have very similar DNA sequences. A comparison of the DNA of chimpanzees with that of gorillas and humans reveals that chimpanzees share as much as 96 percent of their genes with humans (more than with gorillas). This finding indicates that humans and chimpanzees are more closely related to each other than to gorillas.

Molecular comparisons such as this have allowed biologists to build a *relationship tree* of the evolution of life (**FIGURE 14.12**). Molecular comparisons permit researchers to draw conclusions about organisms whose common ancestors lived such a long time ago that little obvious similarity is apparent in their appearance today. In the past three decades, biologists have identified three primary lineages of living organisms: (1) the *eukaryote* branch, which developed into animals, plants, fungi, and protists (including protozoa and most algae); (2) the relatively little-known *archaea* branch of organisms that had previously been found only in extreme environments, such as hot springs, but are now thought

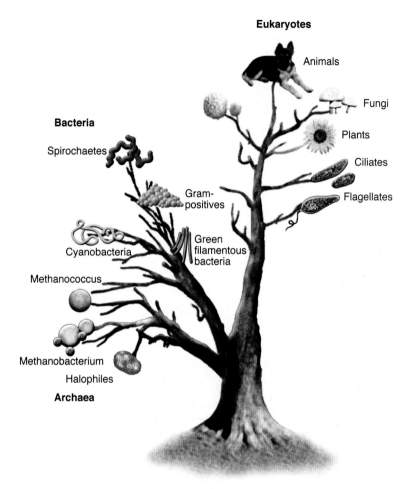

FIGURE 14.12 This simple portrayal of the relationship tree of life depicts the three major lineages of organisms (bacteria, eukaryotes, and archaea) that have been identified in the field of molecular biology.

(?) How do genetic mutation and genetic variation lead to evolution and, ultimately, produce the relationship tree of life?

to occupy most ecological niches on Earth; and (3) the *bacteria* branch, a large group of single-celled micro-organisms, some of which serve beneficial functions in animals and humans while others cause infections and disease.

AIDS Evolves

Why has it not been possible to conquer the common flu or AIDS? These diseases (and others) are caused by *viruses* that infect human victims. To fight viruses, we use *vaccines,* specific chemicals that trigger the production of antibodies. *Antibodies* are soldier cells that kill a specific type of virus before it can do much damage. *Viruses are living organisms that evolve* like all organisms. AIDS is caused by human immunodeficiency virus, or HIV (**FIGURE 14.13**). Because the HIV evolves, it is difficult to treat effectively.

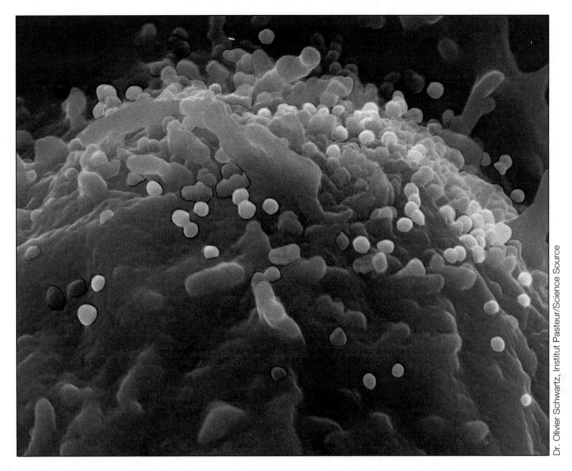

FIGURE 14.13 HIV, a living organism within the human body, evolves new strains that are resistant to drug treatment. It enters the cell, hijacks the cell's machinery to make more copies of the virus, and the new virus particles (yellow) then burst from the membrane of the cell (blue), killing it.

 Explain how evolution allows a virus to escape treatment by medicine.

Because viruses evolve, antibodies designed to attack one form of a virus will be useless against a new form of that virus. This is why vaccines have not succeeded in eliminating the common flu or AIDS, for example. To be effective, vaccines against particular strains of a flu virus must be constantly updated, a process that is carried out each year in preparation for "flu season."

In the case of AIDS, as the HIV virus multiplies within a patient, some of the offspring will have genetic mutations and variations that are drug resistant (purely by accident). These variations can change those parts of the virus that antibodies are designed to recognize. All it takes is a very small change—one that otherwise is completely irrelevant to the structure or function of the virus—for the antibody to become ineffective. Because of the huge number of viral cells that inhabit an infected person and the rapid turnover of the viral population, which can occur in a few hours or days, the statistics of mutation become overwhelming. That is, evolution can completely change the viral "strain" in a matter of months. Although a patient might be able to fight off the first strain, the virus can evolve faster than scientists can develop medicines to fight it. In scientific parlance, the virus "escapes" drug therapy by evolving.

Researchers are concerned with drug resistance not only in HIV but also in other viruses and bacteria. In recent decades, physicians have over-prescribed antibiotics and antiviral agents, inadvertently strengthening germs because new strains have evolved that are resistant to treatment. But patients have to shoulder some of the blame for the current situation. Too often, they demand antibiotics for colds or flu, which are not affected by the drugs. Also, patients frequently stop taking antibiotics before they have completed the full dose. This leaves a small amount of bacteria that are strong enough to resist the drugs in the patient's system—bacteria that may be passed on to others. Researchers note that the mechanisms of evolution need to be better understood in order to remedy this problem.

 Expand Your Thinking—Imagine that the theory of evolution did not exist. How might medical researchers account for the ability of viruses and bacteria to escape treatment?

14-5 Mass Extinctions Influence the Evolution of Life

LO 14-5 *List the major extinctions and when they occurred.*

One of the more startling revelations of historical geology is that **mass extinctions**—events in which large numbers of species permanently die out within a very short period—have occurred several times during Earth's history. These events are more than curious anomalies. They exert an important influence on the direction of animal and plant evolution. Consider this: At the time of their extinction 65 million years ago (mya), dinosaurs had been the ruling class of animals on land for nearly 200 million years. (Modern humans are estimated to have been on Earth for only 200,000 years.) Had dinosaurs not gone extinct, and thus allowed mammals to spread and adapt to new habitats, humans and many of the animals we share the planet with would probably not exist.

Why is this so? Mass extinctions wipe out hundreds, even thousands, of dominant species (**FIGURE 14.14**). Once those species are absent, ecological niches become available for other species to live in, grow, and reproduce. The dying-out of the dinosaurs eliminated the most important predators of mammals, which at the time were mainly small shrew- and mouse-like creatures. As a result of this mass extinction, mammals flourished. Environments and resources not previously available to them were now ripe for the picking. Mammals experienced accelerated evolution and blossomed in diversity and distribution around the globe.

Although the actual number of mass extinction events is not clearly known, there are five that appear to have been particularly significant. These five extinctions (**FIGURE 14.15**) occurred at the following times:

1. Ordovician-Silurian extinction: 450–440 mya

2. Late Devonian extinction: 375–360 mya

3. Permian-Triassic extinction: 251 mya

4. Triassic-Jurassic extinction: 200 mya

5. Cretaceous-Paleogene extinction: 65 mya

The Ordovician-Silurian extinction occurred at a time when life had just recently emerged out of the oceans and onto the continents. According to the fossil record, 25 percent of marine families and 60 percent of marine genera were lost. The Late Devonian extinction, about 375 to 360 mya, killed 20 percent of marine families and 55 percent of marine genera.

The Permian-Triassic extinction was the worst: 95 percent of all marine life on Earth died off, and 70 percent of land organisms became extinct. In the only known mass extinction of insects, it is estimated that about 55 percent of all families and about 85 percent of all genera disappeared. The Triassic-Jurassic extinction is thought to have been caused by massive floods of lava erupting from the central Atlantic sea floor. The volcanism might have caused deadly global warming and the demise of 20 percent of all families and 50 percent of all genera. The dinosaurs had little competition on land after this extinction event.

The Cretaceous-Paleogene extinction (often referred to as the Cretaceous-Tertiary extinction)—the one responsible for the demise of the dinosaurs—occurred about 65 mya. It was likely caused by the impact of the huge asteroid (several kilometres wide) that created the *Chicxulub Crater,* which is now hidden below the surface of the Yucatan Peninsula and beneath the Gulf of Mexico (**FIGURE 14.16**). The collision would have generated a cloud of gas and dust that blocked the Sun for years and changed global environmental conditions. Also, thin layers of sediment laid down at that time are often enriched with the element iridium, which is extremely rare in Earth's crust but

Interfoto/Alamy

 What is the difference between families, genera, and species?

FIGURE 14.14 *Dimetrodon* was a synapsid (a mammal-like reptile) that was an apex predator during the Permian Period. *Dimetrodon* was not a dinosaur, and actually became extinct before the first dinosaurs existed. Reptiles tend to gulp their food down without chewing, whereas synapsids like *Dimetrodon* developed teeth that sheared meat into smaller pieces. These eventually gave rise to the various kinds of teeth present in modern mammals.

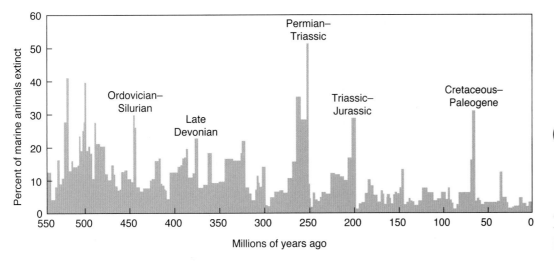

? Why are mass extinctions important in the course of evolution?

FIGURE 14.15 Five major mass extinctions have occurred during the last 500 million years of Earth's history.

much more abundant in meteorites. The extinction killed 16 percent of marine families, 47 percent of marine genera, and 18 percent of land vertebrate families, including the dinosaurs.

Not all mass extinctions occur for the same reasons. Various hypotheses have been forwarded for each one, and these are subjects of intense scientific debate. In general, extinctions are thought to result from drastic global changes that follow catastrophic events, such as asteroid or comet impacts or massive volcanic eruptions. The latter may generate enormous volumes of volcanic dust and gases, like sulphur dioxide, which can dissolve in water vapour in the atmosphere to generate acid rain. The dust and aerosols may have reduced the amount of sunlight reaching Earth's surface, and resulted in cooling of the climate. Scientists disagree as to whether such effects are great enough to cause worldwide and long-term cooling that would ultimately lead to the formation of new glaciers and ice sheets. If this had happened, water would have become locked up as ice in these massive glaciers, lowering the level of shallow seas and eliminating or changing marine habitats across the planet. Other theories attempting to explain extinction events include nearby supernova explosions that bathed Earth in radioactivity and destroyed the ozone layer, which protects life forms from ultraviolet radiation emitted by the Sun.

Whatever their cause, mass extinctions have occurred on Earth and have played a crucial role in determining the character of the flora and fauna that are present on the planet today.

Richard Bizley/Science Source

? On a global scale, how would a large extraterrestrial impact disturb the human community?

FIGURE 14.16 The impact of an extraterrestrial body striking Earth's surface can release billions of tons of ash and gas into the atmosphere, leading to global environmental changes that can cause mass extinctions. The body that created the Chicxulub Crater at the end of the Cretaceous is thought to have been about 15 km across, significantly smaller than the one in this artist's rendition.

? **Expand Your Thinking**—It has been said that today humans are driving the greatest mass extinction in geologic history. How would you test this hypothesis, and how might you use the results of your study to help in ecosystem management?

14-6 The Geologic Time Scale Is the "Calendar" of Events in Earth's History

LO 14-6 *Describe the geologic time scale and how it is organized.*

In Chapter 1, we learned about the **geologic time scale,** the "calendar" of events in Earth's history (**FIGURE 14.17**). The geologic time scale subdivides all time since the beginning of Earth's history (~4.6 billion years ago [bya]) into named units of variable length. Just as you use days, weeks, months, and years, the time scale uses **epochs** (tens of thousands to millions of years), **periods** (millions to tens of millions of years), **eras** (tens of millions to hundreds of millions of years), and **eons** (hundreds of millions to billions of years). An important difference, however, is that geologic units are not fixed at a certain length; they vary in duration because they are determined by events, not by lengths of time.

The geologic time scale is based on *stratigraphy,* the correlation and classification of sedimentary strata. Fossils found in strata provide the chief means of establishing a time scale. Because living things evolve over time, they change in appearance. The *principle of fossil succession* tells us that particular kinds of organisms existed only at particular times in Earth's history and therefore are found only in particular parts of the geologic record. In other words, evolution does not reproduce an organism once it has gone extinct. By correlating strata in which certain types of fossils are found, researchers can reconstruct the geologic history of various regions (and of Earth as a whole).

The geologic time scale consists of a succession of names that represent various intervals of time. The scale includes the ages, in years before present, marking the boundaries of the intervals. The names of many time units have been in use since the19th century and come from locations where rocks of that age are common or were first described. The ages of all the time units were established during the 20th century as new techniques became available to measure the absolute age of rocks and minerals, and are still being revised as more detailed information becomes available.

Time boundaries are based mostly on notable events in Earth's history as indicated by fossils. For instance, the appearance or disappearance of certain fossils may mark the end of one time unit and the beginning of another; hence, as mentioned earlier, extinctions play an important role in defining geologic time units. A special characteristic of the time scale is that the first 4 billion years of Earth's history (the Precambrian) are not divided in the same amount of detail as the last 542 million years. The latter marks the beginning of the Paleozoic Era, placed where the first complex life appears in the fossil record.

Hadean Time

As we learned in Chapter 13, Earth's age has been determined by applying the tools of radiometric dating to Moon rocks and meteorites as well as to Earth rocks. Few rocks on Earth remain from Hadean time because the processes of the rock cycle have largely destroyed them. Nonetheless, it is desirable to infer the nature of Earth's surface during that time.

In Chapter 2 ("Solar System"), we studied the origin of the Solar System and the processes associated with planetesimal accretion. Scientists deduce that Earth was initially a cool body of rock in space. But heat-producing processes, such as extraterrestrial bombardment, internal radioactive decay, and consolidation of the planet by gravity, led to the formation of a magma ocean as Earth reached a molten state. When the planet cooled, its exterior solidified as a primitive crust characterized by widespread volcanism. The oldest rock yet identified in the world is a 4.28 billion-year-old basaltic rock in northern Quebec, which was subsequently metamorphosed; it may be a piece of this primitive crust.

However, tiny crystals of zircon found in the Jack Hills conglomerate of Western Australia challenge the accepted thinking. Zircon is a common mineral that is especially resistant to weathering and is often found as small grains in sandstones and conglomerates. Zircon also incorporates uranium into its crystal structure, and over time, the isotopes of uranium undergo radioactive decay to isotopes of lead, which means that zircon is a great geochronometer. A zircon from the Jack Hills sample has been dated to 4.4 bya—a time when Earth was supposedly molten. Contained within this mineral are isotopes of oxygen (^{16}O and ^{18}O) that allow scientists to estimate the temperatures of processes leading to the formation of magmas and rocks. Geochemists measure the ratio of ^{18}O (a rare isotope representing about 0.2 percent of all oxygen on Earth) to ^{16}O (the common oxygen isotope, which accounts for about 99.8 percent of all oxygen). The proportions of ^{18}O and ^{16}O in a crystal depend on the temperature at which it was formed. The $^{18}O/^{16}O$ ratio in Earth's mantle—about 5:3—is well known, and geologists expected that Jack Hills zircons would reflect the presumed molten nature of Earth's surface, with isotope ratios similar to those found in the mantle. But when scientists completed their analysis of the crystals, they discovered that their predictions had been wrong; the isotopic ratios ranged up to 7:4.

At first researchers were confused by these results. What could these high oxygen isotope ratios mean? In younger rocks, the answer would be obvious because such samples are common. Rocks at low temperatures on Earth's surface can acquire a high oxygen isotope ratio if they interact chemically with rain or ocean water. Those high ^{18}O rocks, if buried and melted, form magma that retains the high ^{18}O value, which is passed on to zircons during crystallization. Thus, liquid water and low temperatures are required for the formation of zircons and magmas with high ^{18}O content. The presence of high oxygen isotope ratios in the Jack Hills zircons implies that liquid water must have existed on Earth's surface at least 400 million years earlier than the oldest known sedimentary rocks at Isua, Greenland. If this is correct, entire oceans may have existed, making Earth's early climate more like a sauna than a Hadean fireball. Research in this area continues, and the true nature of Earth's surface during Hadean time remains unresolved.

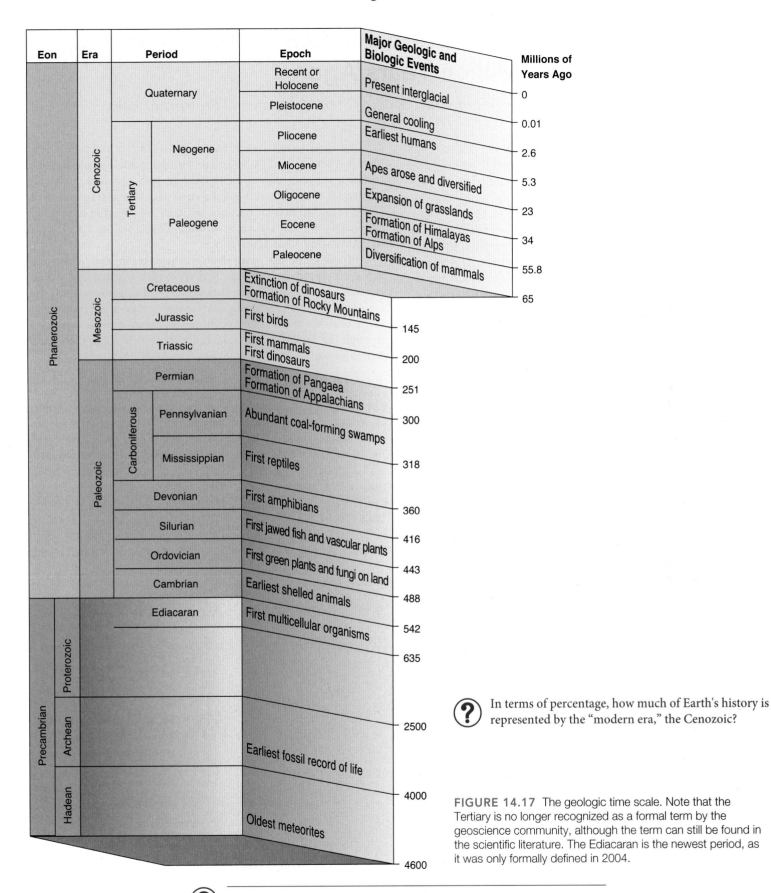

In terms of percentage, how much of Earth's history is represented by the "modern era," the Cenozoic?

FIGURE 14.17 The geologic time scale. Note that the Tertiary is no longer recognized as a formal term by the geoscience community, although the term can still be found in the scientific literature. The Ediacaran is the newest period, as it was only formally defined in 2004.

Expand Your Thinking—What epoch, period, era, and eon do we live in?

14-7 The Archean and Proterozoic Eons Lasted from 4.0 Billion to 542 Million Years Ago

LO 14-7 *Compare and contrast geologic processes occurring in the Archean and Proterozoic Eons.*

Following the Hadean Eon, Earth entered the *Archean Eon* (4.0 to 2.5 bya), followed by the *Proterozoic Eon* (2.5 bya to 542 mya).

Archean Eon

Archean rocks are peculiar and interesting because of their great age and the unique geologic processes that formed them. The surface of the Moon consists mainly of Archean rock. Interestingly, some ultramafic igneous Archean rocks on Earth are similar to some of the Moon rocks collected by astronauts. This similarity suggests that similar geologic processes occurred on the Moon and Earth during their earliest histories.

During the Archean, the atmosphere was composed of noxious, unbreathable gases. Methane, ammonia, carbon dioxide, and water vapour were the major components. As explained in Section 2-6, these were delivered by comets and asteroids, but volcanic outgassing from Earth's hot interior also played an important role. Some researchers infer that these gases provided the basis for building the organic molecules required for life to emerge from a lifeless environment. Other researchers favour the presence of organic molecules found on meteorites as the source of early life's building blocks.

It is estimated that only about 30 percent of the continental crust that exists now was formed in the Archean. However, studies have shown that modern continents are built around cores of this extremely ancient rock, called **cratons,** that formed during the late Archean and Proterozoic Eons (**FIGURE 14.18**).

The fossil record reveals that abundant life first appeared during the Archean Eon. These earliest fossils are microscopic bacteria, and their remains occur as strings of *blue-green cyano-bacteria* cells. Eventually, they formed domelike structures by trapping sediments between fine strands of bacteria (**FIGURE 14.19**). The mounds or domes they formed are called **stromatolites** from the Greek terms *stroma* (stratum) and *lithos* (rock).

Stromatolites, like other photosynthetic organisms, release oxygen as a by-product of metabolism. As oxygen released by these organisms accumulated in the atmosphere, it displaced the methane, ammonia, and carbon dioxide that were abundant at the time. Eventually, with the spread of other successful photosynthetic organisms through the shallow sunlit seas, our oxygen-rich atmosphere came into existence.

The Proterozoic Eon

The Proterozoic Eon saw the continued development and growth of the continents (a process that took about 1 billion years), the emergence of a new atmosphere that was richer in oxygen, and the building of new mountain ranges. Although these mountains were long ago worn down by erosion, the heavily metamorphosed rocks that formed their base can still be found.

During the Proterozoic, the first known supercontinent, *Rodinia* (**FIGURE 14.20**), formed around 1 bya and then began to break up due to plate divergence around 800 mya. Several pre-existing continents converged to form Rodinia during a series of orogenic events,

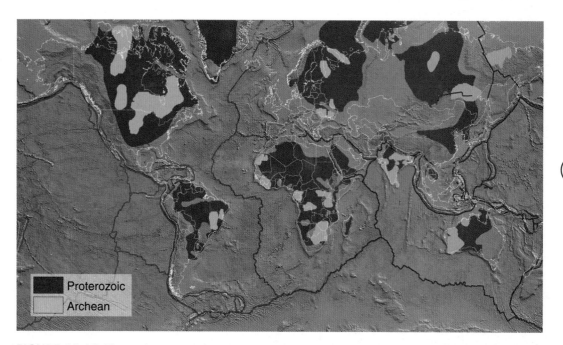

Proterozoic
Archean

FIGURE 14.18 The ancient cores of continents were formed in the late Archean and Proterozoic Eons. These are called cratons, and today they are found in the interior of most of the major land masses on Earth. Younger sedimentary rocks cover some of these older cratons.

? Describe tectonic processes that could build the remaining portions of the continents.

FIGURE 14.19 (a) Modern stromatolites, Shark Bay, Australia. (b) Fossil stromatolites. Stromatolites are formed by layers of bacteria that trap and bind sedimentary particles.

? What role did stromatolites play in changing the chemistry of the Proterozoic oceans and atmosphere?

Georgette Douwma/Science Source

(a)

(b)

Sinclair Stammers/Photo Researchers

the final being an event known as the *Grenville Orogeny.* (Recall from Chapter 11 ["Mountain Building"] that an *orogeny* is a tectonic episode characterized by orogenesis, or mountain building.) Among these pre-existing continental fragments were the ancient cratons of Laurentia (North America), South America, Australia, Africa, Siberia, the Baltics of Europe, South China, and Antarctica.

Diversification of life was a hallmark of the Proterozoic Eon. Life progressed from single-celled organisms to the first multicellular, ocean-dwelling plants and animals. Photosynthetic plants flourishing in the oceans produced oxygen gas, which changed the chemistry of air and sea water. As dissolved oxygen accumulated in the oceans, it combined with dissolved iron to form hematite (iron oxide), which settled onto the sea floor. Evidence for this change in seawater chemistry is the appearance of *banded iron formations* composed of layers of hematite (**FIGURE 14.21**) deposited in Proterozoic oceans.

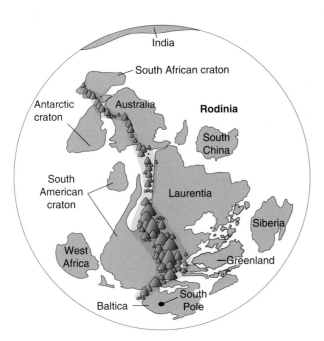

FIGURE 14.20 During the Proterozoic Eon, between approximately 1.3 bya and 800 mya, continents were clustered as one large mass, called Rodinia, in the Southern Hemisphere. A large mountain belt developed as the continents came together. This mountain-building event is called the Grenville Orogeny.

? What tectonic process builds supercontinents, and what process breaks them up?

Blue Gum Pictures/Alamy

FIGURE 14.21 Oxidized iron accumulated on the sea floor in the Proterozoic as microscopic marine plants produced Earth's first abundant atmospheric oxygen gas. This photograph shows a good example of a banded iron formation from Karijini National Park, Western Australia.

? **Expand Your Thinking**—Why does the Moon's surface still consist of Archean rock?

14-8 In the Paleozoic Era, Complex Life Emerged and the Continents Reorganized

LO 14-8 *List the periods of the Paleozoic Era and a notable characteristic of each.*

The *Phanerozoic Eon* (542 mya to present) is recorded in much more detail by rocks and fossils than are earlier eons. Major mountain-building events are evident; there is widespread evidence of the past positions and movement of plates; and the detailed fossil record can be used to reconstruct the history of macro-organisms (FIGURE 14.22).

Cambrian Period

Complex forms of life with shells and hard external body parts appeared in abundance about 542 mya, marking the beginning of the *Cambrian Period* (542 mya to 488 mya) of the **Paleozoic Era.** The Cambrian was a time of rapid and unprecedented evolutionary diversification. Of special note is a Cambrian rock unit in western Canada known as the *Burgess Shale.* Designated a World Heritage Site because of its special significance, the Burgess Shale records the "Cambrian explosion"—the sudden

appearance of a highly diverse animal assemblage (FIGURE 14.23) after billions of years during which single-celled organisms dominated.

The Ordovician Period

During the *Ordovician Period* (488 mya to 443 mya), the area north of the tropics was almost entirely ocean, and most of the world's land was collected into the southern supercontinent known as *Gondwana.* Throughout the Ordovician, Gondwana shifted toward the South Pole. Jawless fish with thick bony armoured plates appeared, and corals and bivalves came into existence. Shallow seas covered much of North America and then receded, leaving thick limestone units.

The Silurian Period

During the *Silurian Period* (443 mya to 416 mya), global warming contributed to a substantial rise in sea levels; broad tracts of coral

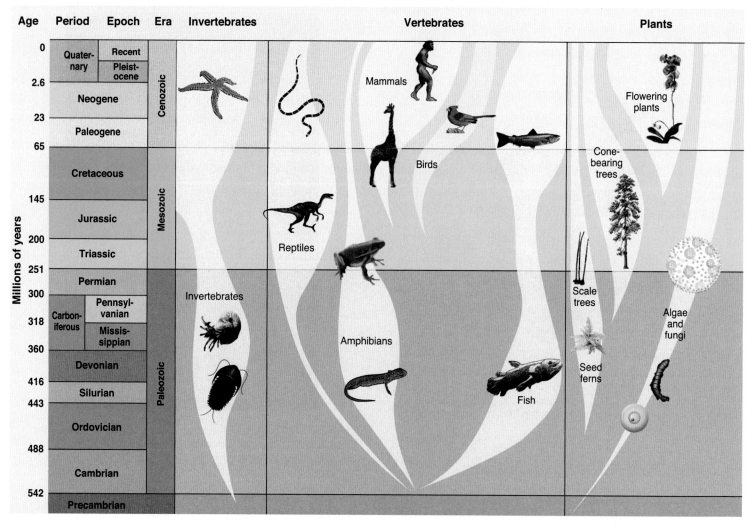

FIGURE 14.22 Following the Precambrian Eon, the Paleozoic Era witnessed an explosion of life forms. Marine and continental life diversified during the Paleozoic Era.

(?) What important change in the fossil record marks the transition from the Precambrian Eon to the Phanerozoic Eon? Why is it notable?

(a) Alan Sirulnikoff/Science Photo Library (b) © Kevin Schafer/ Alamy (c) L. Newman & A. Flowers/Science Photo Library (d) Publiphoto/Science Source

FIGURE 14.23 *(a)* The Walcott Quarry in the Burgess Shale, from which many famous fossil specimens have been obtained. *(b)* This is thought to be a claw from the predatory *Anomalocaris canadensis*, which also had a circular mouth ringed with fangs. The spines on the claw are about 1 cm long. *(c)* A photograph of a rock surface with a fossil of a worm-like organism, called *Ottoia*. *(d)* An artist's rendition of examples of multicelled organisms preserved in the Burgess Shale, which is estimated to be about 505 million years old.

(?) Why is the Burgess Shale such an important rock unit?

reefs made their first appearance; and insects, sharks, and bony fish (both marine and freshwater) appeared. This is the first period in which there is strong evidence of life on land, including fossils of organisms related to centipedes and leafless land plants.

The Devonian Period

The *Devonian Period* (416 mya to 360 mya) is known as the Age of Fish because abundant forms of armoured fish, lungfish, and sharks are found preserved in rock layers from that time. The global ozone layer formed (composed of the oxygen molecule O_3, which blocks harmful solar radiation), making possible the spread of air-breathing organisms, such as spiders and mites, on the land. Fossils of the first transitional forms of vertebrates emerging from the water onto the land (ancestors of amphibians) are found in rocks dating from the end of the Devonian (**FIGURE 14.24**).

The Carboniferous Period

The *Carboniferous Period* extended from 360 mya to 300 mya and is named for the fact that layers of coal are a distinctive part of the sequences of sedimentary rocks. Two major land masses existed in this period: Euramerica (North America, Greenland, northern Europe, and Scandinavia) to the north of the equator, and Gondwana (South America, Africa, peninsular India, Australia, and Antarctica) to the south. The land was covered by vast areas of forest, and amphibians were the most important land vertebrate. In North America, and particularly in the United States, the Carboniferous Period is divided into the limestone-rich *Mississippian Period* (360 mya to 318 mya) and the coal-rich *Pennsylvanian Period* (318 mya to 300 mya).

The Permian Period

The supercontinent Pangaea formed during the *Permian Period* (300 mya to 251 mya). It reached nearly from pole to pole and was surrounded by an immense ocean. Several notable biological events occurred: Insects evolved into modern forms, including dragonflies and beetles; amphibians declined; and reptiles underwent a spectacular development of carnivorous and herbivorous, and terrestrial and aquatic forms. Ferns and conifers spread in the cooler air.

(a) (b)

Science Source

Ted Daeschler/National Science Foundation, Courtesy of VIREO

FIGURE 14.24 *Tiktaalik roseae* was a transitional organism between fish (with fins) and amphibians (with legs); *(a)* artist's depiction and *(b)* fossil form.

(?) **Expand Your Thinking**—Develop a hypothesis that explains the development of shells and external hard parts among animals. How would you test your hypothesis?

14-9 In the Mesozoic Era, Biological Diversity Increased and Continents Reorganized

LO 14-9 *Compare and contrast the three periods of the Mesozoic Era.*

The **Mesozoic Era** (251 mya to 65 mya), composed of the *Triassic, Jurassic,* and *Cretaceous Periods*, was characterized by the breakup of Pangaea and the formation of the Atlantic Ocean in the space created by the separation of Africa and Europe, and North and South America. The Atlantic Ocean was formed by divergence and seafloor spreading along the Mid-Atlantic Ridge, which was born at this time. As North America moved westward, the leading (western) edge of the North American Plate collided with island arcs, continental fragments, and oceanic crust, building the cordilleran thrust-and-fold belt and producing vigorous arc volcanism. This process continues today at the Cascadia Subduction Zone, which stretches from offshore northern Vancouver Island to northern California.

New groups of invertebrates appeared, including new forms of many established organisms such as marine corals, molluscs, and echinoderms. Land plants flourished, and modern types of seed-bearing trees and flowering plants developed. Dinosaurs, mammals, and birds made their appearance, and the dinosaurs became dominant. The end of the Mesozoic was marked by the Cretaceous-Paleogene extinction, a major episode of extinction that is widely thought to have been caused by a meteorite impact. This event defines the boundary between the Cretaceous and Paleogene Periods.

The Triassic Period

The Triassic (251 mya to 200 mya), unlike previous periods, witnessed few significant tectonic events. Ammonoids, now-extinct cephalopods with a chambered shell, evolved into diverse forms. Reptiles experienced an explosion of diversity that produced the dinosaurs, and marine reptiles such as ichthyosaurs and plesiosaurs. The first mammals evolved from mammal-like reptiles. Pangaea covered nearly one-quarter of Earth's surface but slowly began to break apart when continental rifting became widespread toward the end of the Triassic. The general climate was warm, becoming semi-arid to arid in continental areas.

The Jurassic Period

Named for the Jura Mountains on the border between France and Switzerland, where rocks of this age were first studied, the Jurassic Period (200 mya to 145 mya) became a household word with the success of the *Jurassic Park* movies. Outside of Hollywood, the Jurassic remains important to us today, both because of its wealth of fossils and because of its economic importance. The oil fields of the North Sea, for example, are Jurassic in age.

The Atlantic Ocean began to appear as North America separated from Africa and South America (**FIGURE 14.25**). Plate subduction along western North America caused the crust to fold, creating mountains in the western part of the continent. Reptiles adapted to the sea, air, and land. Dinosaurs were the dominant animal form on land. The first bird, *Archaeopteryx*, evolved at this time from the therapod dinosaurs. The earliest modern amphibians, succeeded by the first frogs, toads, and salamanders, also evolved during the Jurassic.

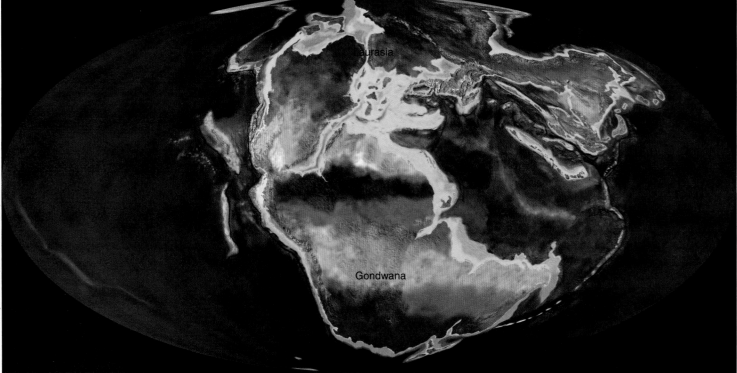

Ronald C. Blakey, Colorado Plateau Geosystems, Inc.

FIGURE 14.25 In the Early Jurassic, approximately 190 mya, the Atlantic Ocean was born as the supercontinent, Pangea, separated into Laurasia, consisting of North America, Europe and Asia, and Gondwana, consisting of Africa, and South America.

 What process is driving North America to the west?

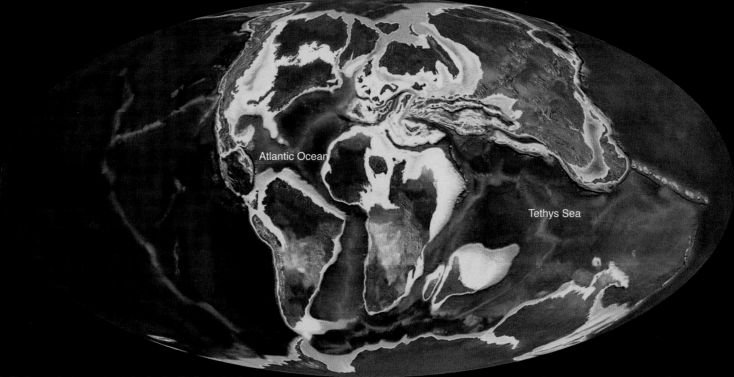

FIGURE 14.26 During the Cretaceous Period, approximately 90 mya, the continents took on their current configurations but were still moving to modern locations, largely due to continued widening of the Atlantic Ocean. However, sea level was higher than today, and approximately 30 percent of the present-day land surface was flooded. India, Australia, and much of the Middle East and southern Asia were still being formed.

 Why was sea level much higher in the Cretaceous than it is today?

Mammals continued to evolve, but were relegated to a minor role in the ecosystem, and mainly consisted of small, shrewlike animals. Forests of conifers and ginkgos became widespread. The oceans and lands thronged with life forms, most of which no longer exist.

The Cretaceous Period

During the Cretaceous Period (145 mya to 65 mya), the continents were shaped much as they are today. South America separated from Africa to form the South Atlantic, the North Atlantic widened, and the *Tethys Sea* wrapped the globe at the equator. The westward motion of North America raised the ancestral Rocky Mountains and the Sierra Nevada Mountains. Sea level continued to rise, ultimately submerging 30 percent of Earth's current land surface (FIGURE 14.26).

Climates were generally warm, and the poles were free of ice. Dinosaurs were the dominant vertebrate life form on Earth, and their range extended throughout every continent. Gastropods, corals, and sea urchins flourished. Early flowering plants evolved, including modern trees, as did modern insects. At the end of the Cretaceous, a mass extinction wiped out five major reptilian groups: dinosaurs, pterosaurs, ichthyosaurs, plesiosaurs (FIGURE 14.27), and mosasaurs. At the same time, many ammonites, corals, and other invertebrates became extinct. Mammals survived and flourished in the newly opened environments. The mass extinction at the end of the Cretaceous thus made it possible for groups that had previously played secondary roles to come to the forefront.

Study the "Critical Thinking" box titled "Reconstructing Earth's History" to increase your understanding of Earth's history.

FIGURE 14.27 At the end of the Cretaceous, most marine reptiles, such as this plesiosaur, became extinct.

 Expand Your Thinking—How might life today be different if the Cretaceous-Paleogene extinction had not happened?

CRITICAL THINKING

RECONSTRUCTING EARTH'S HISTORY

Please work with a partner and answer the following questions.

1. Using the global reconstructions in **FIGURE 14.28** as a guide, describe the changes in continents that led to their modern configuration.
2. Speculate and propose a hypothesis that describes a general trend that life forms followed during the course of evolution from the Archean to the late Quaternary.
3. How could the construction of a supercontinent such as Pangaea contribute to the occurrence of a major extinction?
4. What general characteristics of life mark the Proterozoic and Phanerozoic Eons?
5. How old is the Atlantic Ocean? How was it formed? What evidence would you look for to support your description?
6. Summarize general trends in Earth's history in terms of the atmosphere, living organisms, and the crust.
7. What evidence would you look for to test the hypothesis that a supercontinent (Pangaea) existed about 255 mya?
8. Construct an alternative hypothesis for the extinction of the dinosaurs. Identify the evidence you would need to support it.

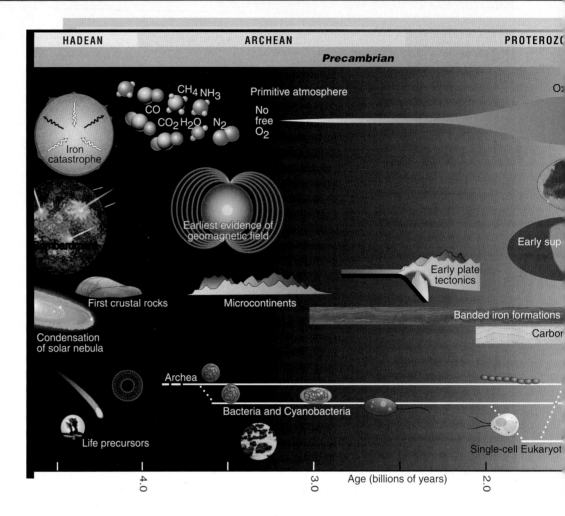

HADEAN   ARCHEAN   PROTEROZO

Precambrian

CH_4 NH_3 Primitive atmosphere   O:
CO
CO_2 H_2O N_2 No
free
O_2
Iron
catastrophe

Earliest evidence of
geomagnetic field

Early sup

Early plate
tectonics

First crustal rocks   Microcontinents

Banded iron formations

Carbo

Condensation
of solar nebula

Archea

Single-cell Eukaryot

Bacteria and Cyanobacteria

Life precursors

4.0   3.0   Age (billions of years)   2.0

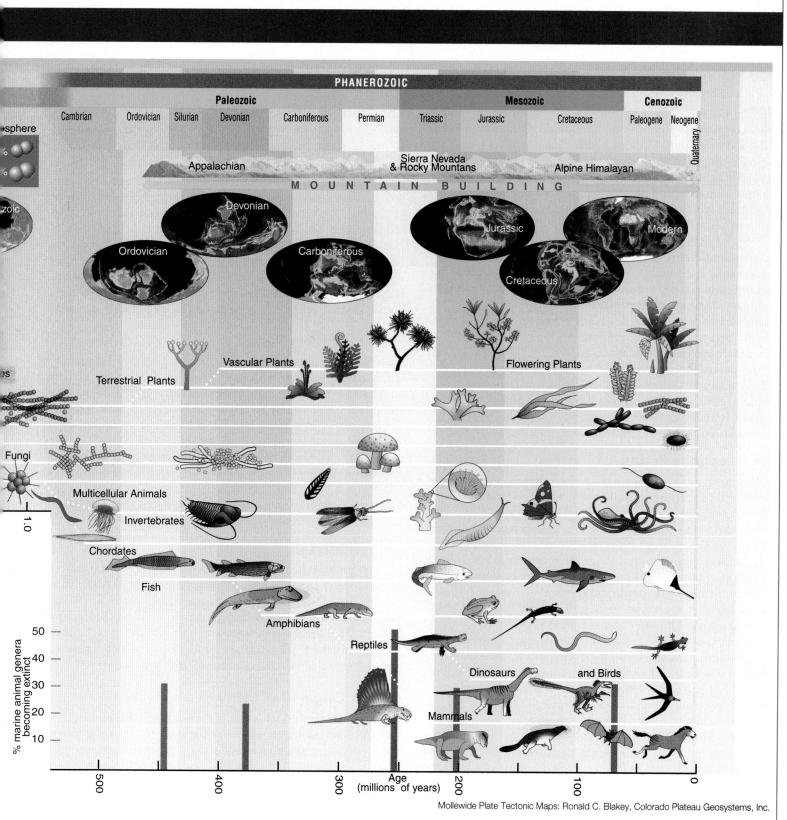

Mollewide Plate Tectonic Maps: Ronald C. Blakey, Colorado Plateau Geosystems, Inc.

FIGURE 14.28

14-10 Modern Mammals, Including Humans, Arose in the Cenozoic Era

LO 14-10 *Describe the major events of the Cenozoic Era.*

The **Cenozoic Era** (65 mya to the present) is relatively short compared to previous eras. The Atlantic Ocean continued to widen and push North and South America to the west as the mid-ocean ridge system produced new sea floor. This process activated additional processes, including subduction, mountain building, earthquakes, and igneous activity along the western coast of North and South America, all related to the westward movement of the two continents (discussed further in Chapter 15, "The Geology of Canada").

The end of the Cretaceous brought the end of many previously successful and diverse groups of organisms that dominated marine and terrestrial environments. Niches in the food chain that previously had been dominated by Cretaceous life forms opened up. New access to scarce resources made it possible for groups that previously had played secondary roles, such as birds and mammals, to move to the forefront and expand their diversity.

Modern groups of invertebrates appeared in the oceans, including the most common molluscs, echinoderms, and crustaceans. Fish became abundant, flowering plants diversified, and the first grasses appeared. Birds and then mammals underwent major increases in diversity. Humans appeared approximately 4 mya, but "modern" humans appeared only about 200,000 years ago.

The Paleogene Period

The *Paleogene Period* (65 mya to 23 mya) consists of the Paleocene, Eocene, and Oligocene Epochs. The period follows immediately after the Cretaceous extinction, which spelled the end of the dinosaurs and several other animal groups. Lasting 42 million years, the Paleogene is notable for the evolution and diversification experienced by mammals, which evolved from relatively small, simple forms into a group of highly diverse animals. Mammals developed large, dominant forms specialized to marine, terrestrial, and airborne environments. The Paleogene was also a period in which birds experienced considerable change and evolved into approximately their modern forms.

After the extinction of dinosaurs, mammals radiated into newly opened ecological niches. Cetaceans (baleen whales, toothed whales, dolphins) evolved from terrestrial meat-eating hoofed animals, as indicated by the presence of vestigial hips and femurs in modern whales and the discovery of intermediate fossil forms of whales. The first primates appeared in the early Paleogene as one of the first groups of mammals having a placenta. Monkeys and the first true apes appeared more than 25 mya. Early horses, goats, pigs, deer, rodents, and carnivores also emerged. Plate tectonics and volcanic activity formed the Rocky Mountains in western North America, and collisions between the Indian and Asian Plates raised the Alpine-Himalayan Mountain System. Antarctica and Australia separated and drifted apart, and Greenland split from North America. The development of the glaciers on Antarctica started as it drifted to the area of the South Pole.

Climates were subtropical and moist throughout North America and Europe. Continents converged to form the Middle East and active volcanism characterized Central America. Spreading grasslands replaced forests over large areas on several continents, setting the stage for the evolution of modern species of horses.

The Neogene Period

The *Neogene Period* extends from 23 mya to the beginning of the *Quaternary Period* approximately 2.6 mya. The Neogene consists of the Miocene and Pliocene Epochs. The climate cooled over the Neogene Period, generating glaciers in the Northern Hemisphere, but the exact causes are not well understood. One idea, the *uplift weathering hypothesis* (discussed in Chapter 7, "Weathering"), suggests that a general decrease in atmospheric carbon dioxide occurred due to amplified weathering of the newly exposed rock of the Himalayan fold-and-thrust belt. The last opening between the Atlantic and Pacific Oceans closed with the growth of new volcanic landforms in Central America. Closing the Pacific-Atlantic connection through Central America may have influenced global heat transport in oceanic currents, enhancing the cooling effect already underway.

The Quaternary Period

The *Quaternary Period* (2.6 mya to present) is notable because it is characterized by a series of **ice ages** and *interglacials* resulting from the advance and retreat of continental glaciers on every major land mass (except Australia). During the last 500,000 years or so, there have been five or more separate glacial periods, occurring approximately every 100,000 years, a pattern that continues today (FIGURE 14.29). This pattern is related to variations in Earth's orbit around the Sun, producing changes in the amount and location of solar radiation reaching the planet's surface (discussed in Chapters 16, "Global Warming," and 17 "Glaciers and Paleoclimatology"). During ice ages, as much as 30 percent of Earth's surface is covered by glaciers, which distinctly alter the landscape. Lowered sea levels expose land bridges between Siberia and North America and among islands in southwestern Asia.

One probable human ancestor, *Homo habilis*, evolved and diversified into modern forms during the Quaternary Period. Mammalian evolution included species of woolly mammoth, woolly rhinoceros, muskox, moose, reindeer, elephant, mastodon, bison, and ground sloth. In the Americas, large mammals such as horses, camels, mammoths, mastodons, sabre-toothed cats, and ground sloths were extinct by the end of the final ice advance 20,000 years ago.

Human Origins

Humans are members of a family known as the **Hominidae.** The hominid fossil record is fragmentary, but enough is known to formulate hypotheses outlining the evolutionary history of humans. Many hominid species have been identified (FIGURE 14.30), but how they relate to one another, which species are the direct ancestors of modern humans *(Homo sapiens sapiens),* and how and when human ancestors and apes diverged are subjects of active research.

The record of evolution from early *Australopithecines* (an early genus that is ancestral to humans) to recent humans has certain characteristics: increasing brain and body size, increasing use of and sophistication in tools, decreasing tooth size, and decreasing skeletal robustness or overall strength. Currently, the oldest candidate for a human ancestor is *Sahelanthropus tchadensis* (nicknamed "Toumai"). A nearly complete skull (Figure 14.30a) dating to 6 to 7 million years ago was found in the

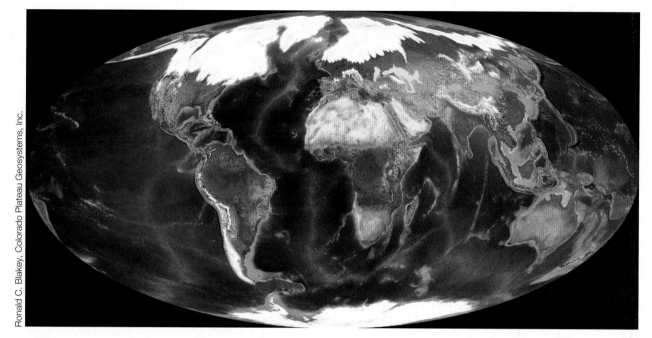

FIGURE 14.29 Since late in the Quaternary Period, Earth has experienced climate shifts from ice ages to interglacials approximately every 100,000 years. The most recent ice age culminated 20,000 to 25,000 years ago with as much as 30 percent of Earth's land surface covered by glaciers.

(?) How is Earth different during an ice age?

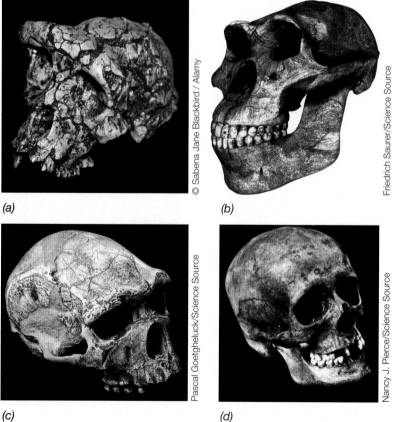

(a)

(b)

(c)

(d)

FIGURE 14.30 Four examples of probable human ancestors: *(a) Sahelanthropus tchadensis,* 6 to 7 million years old; *(b) Australopithecus afarensis,* 3 to 4 million years old; *(c) Homo erectus,* 300,000 to 1.8 million years old; *(d) Homo sapiens sapiens,* modern human.

African nation of Chad. Toumai has primitive features, such as a small brain (~350 cm^3) and pronounced brow ridges. These, plus its age, suggest that it is close to the common ancestor of humans and chimpanzees. However, until additional fossils from the same period are found, *paleontologists* (geologists who study past life) will not be able to determine whether Toumai represents an early ape or an early human.

Australopithecus afarensis lived between 3.9 mya and 3.0 mya (Figure 14.30b). *A. afarensis* had an apelike face and a low forehead, a bony ridge over the eyes, a flat nose, and poorly developed chin. The skull (375–550 cm^3) is similar to that of a chimpanzee, except for the more humanlike teeth and the shape of the jaw, which combines the rectangular shape of apes and the parabolic shape of humans. The pelvis and leg bones of *A. afarensis* resemble those of modern humans more closely than do earlier forms, leaving no doubt that these individuals walked upright on two feet.

Homo erectus (1.8 mya to 300,000 years ago) had a protruding jaw, large molars, no chin, thick brow ridges, and a long low skull (Figure 14.30c). Brain size varied between 750 and 1,225 cm^3. Fossil skeletons indicate that *H. erectus* may have been more efficient at walking than are modern humans, whose skeletons have adapted to allow for the birth of larger-brained infants. *H. erectus* was wide ranging; fossils have been found in Africa, Asia, and Europe. There is evidence that *H. erectus* probably used fire and made stone tools.

Modern humans are named *Homo sapiens sapiens.* Our species first appeared about 200,000 years ago (Figure 14.30d). The average brain size of modern humans is about 1,350 cm^3; the forehead rises sharply; eyebrow ridges are small or absent; and the chin is prominent. About 40,000 years ago, humans began making more sophisticated tools, using a variety of materials, such as bone and antler, to produce implements for engraving, sculpting, and making clothing. Artwork, in the form of beads, carvings, clay figurines, musical instruments, and cave paintings, appeared over the next 20,000 years.

(?) What looks different about the skull of *Homo sapiens sapiens* compared to the skulls of our ancestors?

(?) **Expand Your Thinking**—What major changes characterize the evolution of humans through time?

LET'S REVIEW "GEOLOGY IN OUR LIVES"

Now that you have finished the chapter, "Geology in Our Lives" will have taken on new meaning for you. Let us review it: The study of Earth's history provides humans with a sense of place and perspective in the natural world. It reveals that time is deep and great change has occurred in both geological environments and biological communities. The fossil record provides strong evidence of the evolution of life. Tens of thousands of fossils define animal and plant lineages that fully document evolutionary changes. These fossils prove that evolution is responsible for Earth's vast diversity of life forms. There are practical aspects to the study of Earth's history. Evolution guides understanding and treatment of medical problems. Geologists use fossils to identify the rocks that carry natural resources, such as oil, coal, and many others. Fossils tell us that significant environmental change leads to mass extinctions; that humans and other four-limbed animals share a common ancestor; that chimpanzees are more closely related to humans than they are to gorillas; and that, ultimately, all living organisms are related in a great web of life. In the end, this story tells us that if we wish to live sustainably on this planet, we absolutely must see beyond the horizon defined by our short lives.

As recently as 100 years ago, European cultures thought Earth was a young, essentially unchanging ball of rock. However, as the ideas of early geologists such as Steno, Hutton, Lyell, Darwin, and others became more widely accepted, people came to understand that Earth is a restless and evolving system with a long history characterized by perpetual change. The history of Earth can be described as the chemical and physical reorganization of the planet's interior, continents, oceans, and atmosphere, and the evolution of life.

The Precambrian, a massive expanse of time consisting of the first 88 percent of Earth's history, was a period of slow oxygenation of the atmosphere and oceans, the formation of an organized crust consisting of continents and ocean basins, and the early development of life. Since then, evolution has led to a perpetual increase in the diversity and complexity of life, punctuated by occasional, but significant, mass extinctions that, in some cases, profoundly affected the course of evolution.

Geologic time is organized in the geologic time scale, the calendar of Earth's history being subdivided into a number of eons, eras, and periods. Each of these witnessed special and distinctive episodes of evolving life, mountain building, and plate tectonic reorganization that shaped the character of our modern world.

In Chapter 15, we look more closely at the pattern of geologic change in order to improve our understanding of the geologic history of Canada.

STUDY GUIDE

14-1 Earth's history has been unveiled by scientists applying the tools of critical thinking.

- Earth's history is characterized by deep time and great change. The fossil record provides strong evidence of the evolution of life, the dynamic history of the crust, the interconnectedness of all living things, and the changing environmental conditions that led to mass extinctions on several occasions.

- The science of geology owes its origin to visionary men who saw that Earth was a complex system with a history characterized by perpetual change. **Nicholas Steno** was the first to suggest that fossils that looked like living organisms had in fact once been living organisms. Steno also developed the principles of original horizontality, superposition, and lateral continuity. **James Hutton** proposed that geologic time was indefinitely long and Earth was like a self-renewing machine: As mountains eroded away, new ones were uplifted; as the sea covered some lands, it receded from others. **Charles Lyell,** often cited as the father of modern geology, was able to craft sensible hypotheses from the observations of scientists of his day. Finally, **Charles Darwin** is credited with the **theory of evolution,** which states that all living things develop from a very few simple forms over time through **natural selection.** Natural selection is the process by which favourable traits that are heritable (passed on to offspring) become more common in successive generations of a population of reproducing organisms.

14-2 Fossils preserve a record of past life.

- **Fossils** are the remains of animals and plants, or the record of their presence, preserved in the rocks of Earth. Fossilization is the process that turns a once-living thing into a fossil. There are many fossils to be found, but only a tiny portion of all the animals and plants that ever lived have been fossilized.

- Usually only the hard parts of an animal, such as teeth, bones, and shells, become fossilized. These may be preserved in four ways: replacement, formation of an internal mould, formation of an external mould, or formation of a cast.

14-3 There are several lines of evidence for evolution.

- Fossils provide evidence of evolution. For instance, nearly 35 now-extinct ancestral species make up the family tree, or **phylogeny,** of the modern horse. Consistent with Darwin's theory, natural changes in the environment through time exerted selective pressure on early communities of horse ancestors so that populations with certain characteristics survived at higher rates.

- The bone structure of such diverse organisms as birds, swimming mammals, four-legged animals, humans, and insect-eating reptiles is evidence of evolution. This pattern of **homologous structures** suggests that these animals have evolved from a common ancestor and that natural selection results in the survival of individuals with the specific traits and functions served by each limb.

- **Vestigial structures** are body parts that were used by ancestral forms but are now non-essential. Selective pressure for completely eliminating vestigial parts is weak once they have been replaced with newer forms, so they tend to be retained in organisms for a long time.

- Evidence of evolution is also present in studies of embryology. In their early stages of development, embryos of various fish, birds, and mammals display strikingly similar characteristics.

14-4 Molecular biology provides evidence of evolution.

- Unique characteristics may arise among separate populations of the same species causing them to become new species. These differences are the product of **genetic mutation** and **genetic variation** that are naturally selected through environmental pressure or community pressure.

- Molecular biology documents the relationship of living organisms to their ancient ancestors. Three branches on the tree of living relationships describe all life forms: eukaryotes, archaea, and bacteria. Modern medical treatments are based on an understanding that various parasitic microbes, viruses, and bacteria experience evolution and thereby may escape treatment.

14-5 Mass extinctions influence the evolution of life.

- The history of life is characterized by five major **mass extinctions:** the Ordovician-Silurian extinction (450–440 mya), Late Devonian extinction (375–360 mya), Permian-Triassic extinction (251 mya), Triassic-Jurassic extinction (200 mya), and the Cretaceous-Paleogene extinction (65 mya). These events led to the extinction of large percentages of all living species, and they exerted an important influence on the direction of animal and plant evolution for the species that survived.

- Mass extinctions do not all occur for the same reasons. Various hypotheses have been forwarded for each one, and they are subjects of intense debate. In general, extinctions are thought to result from drastic global changes that follow catastrophic events, such as meteorite or comet impacts or massive volcanic eruptions.

14-6 The geologic time scale is the "calendar" of events in Earth's history.

- The **geologic time scale** subdivides all time since the beginning of Earth's history (4.6 bya) into named units of variable length: **epochs, periods, eras,** and **eons.**

- Fossils found in strata provide the chief means of establishing the geologic time scale. For instance, the appearance or disappearance of certain fossils may mark the end of one time unit and the beginning of another; hence, extinctions play an important role in defining geologic time units. The principle of fossil succession tells us that particular kinds of organisms are characteristic of particular times in Earth's history and particular parts of the geologic record. By correlating strata in which certain types of fossils are found, scientists can reconstruct the geologic history of various regions and of Earth as a whole.

- Hadean time corresponds to the period between Earth's formation 4.6 bya to the start of the Archean Eon 4.0 bya. Few Hadean rocks survive, but based on theoretical arguments, the period was characterized by an initially cool Earth that gradually was heated past the melting point of iron and developed a molten surface. This event was followed by cooling and the establishment of the first hard crust and very early oceans.

14-7 The Archean and Proterozoic Eons lasted from 4.0 billion to 542 million years ago.

- The atmosphere during the Archean Eon was composed of unbreathable gases: methane, ammonia, carbon dioxide, and water vapour. These came from extraterrestrial impacts by comets and asteroids as well as volcanic outgassing from the planet's hot interior.

- The Proterozoic Eon saw the development and growth of the continents (a process that took about 1 billion years), the emergence of a new atmosphere rich in oxygen, and the building of new mountain ranges.

14-8 In the Paleozoic Era, complex life emerged and the continents reorganized.

- The Phanerozoic Eon is recorded in much more detail by rocks and fossils than earlier eons. It contains information for reconstructing events in shorter time intervals consisting of the **Paleozoic, Mesozoic,** and **Cenozoic Eras,** each of which consist of several periods of variable length.

- The Cambrian Period is the time when most of the major groups of animals first appear in the fossil record. In particular, complex forms of life with shells and hard external body parts, such as marine invertebrates and fish, appeared in abundance. The Silurian Period witnessed the first coral reefs, the evolution of fish, the first evidence of life on land including relatives of spiders and centipedes, and the earliest fossils of vascular plants. The Devonian Period is known as the Age of Fish because various forms of armoured fish, lung-fish, and sharks are preserved in rock layers from the time. During this period, the global ozone layer formed, allowing the spread of air-breathing spiders and mites on land. The Carboniferous Period witnessed the convergence of major continental areas and the flourishing of insects and reptiles. In the Permian Period, the supercontinent Pangaea formed, surrounded by an immense world ocean.

14-9 In the Mesozoic Era, biological diversity increased and continents reorganized.

- The Mesozoic Era is characterized by the breakup of Pangaea and the formation of the Atlantic Ocean.

- The Triassic Period had few significant geologic events. Reptiles experienced an explosion of diversity, producing the dinosaurs and marine reptiles, such as ichthyosaurs and plesiosaurs. The first mammals evolved from mammal-like reptiles. In the Jurassic Period, reptiles adapted to the sea, the air, and the land. Dinosaurs were the dominant animal form. Mammals played a minor role in the ecosystem as small, shrewlike animals. Conifers and gingkos became widespread. During the Cretaceous Period, the continents were shaped much as they are today. The Tethys Sea wrapped the globe at the equator, and the westward motion of North America raised ancestral Rocky Mountains and the Sierra Nevada Mountains.

14-10 Modern mammals, including humans, arose in the Cenozoic Era.

- The Cenozoic is relatively short compared to previous eras. The Atlantic Ocean continued to widen and push North America to the west as the mid-ocean ridge system produced new sea floor. After the extinction of dinosaurs, mammals radiated into newly opened ecological niches in the Paleogene Period. The climate cooled over the Neogene Period, but the exact causes are not well understood. In the Neogene, the last opening between the Atlantic and Pacific Oceans closed with volcanism in Central America uniting North and South America.

- During the Quaternary Period, Earth entered a series of **ice ages.** The fossil record of human evolution is characterized by

increasing brain size and body size, increasing use of and sophistication in tools, decreasing tooth size, and decreasing skeletal robustness or overall strength.

KEY TERMS

Cenozoic Era (p. 386)
Charles Darwin (p. 367)
Charles Lyell (p. 367)
cratons (p. 378)
eons (p. 376)
epochs (p. 376)
eras (p. 376)
fossils (p. 368)
genetic mutation (p. 372)

genetic variation (p. 372)
geologic time scale (p. 376)
historical geology (p. 366)
Hominidae (p. 387)
homologous structures (p. 370)
ice ages (p. 387)
James Hutton (p. 366)
mass extinctions (p. 374)
Mesozoic Era (p. 382)

natural selection (p. 367)
Nicholas Steno (p. 366)
Paleozoic Era (p. 380)
periods (p. 376)
phylogeny (p. 370)
stromatolites (p. 378)
theory of evolution (p. 366)
uniformitarianism (p. 366)
vestigial structures (p. 370)

ASSESSING YOUR KNOWLEDGE

Please answer these questions before coming to class. Identify the best answer.

1. Nicholas Steno was the originator of
   a. igneous geology.
   b. stratigraphy.
   c. metamorphism.
   d. planetary geology.
   e. radiometric dating.

2. The term "fossil" refers to
   a. any evidence of evolution.
   b. any evidence of changes in species over time.
   c. any trace of soft parts of organisms.
   d. any evidence of past life on Earth.
   e. None of the above.

3. Evidence for evolution includes
   a. vestigial structures, homologous structures, and phylogeny.
   b. phylogeny, intrusion, and atmospheric oxidation.
   c. vestigial structures, population growth, and global climate change.
   d. phylogeny and fossil fragments remaining after mass extinctions.
   e. written accounts.

4. Natural selection is
   a. the tendency of populations with favourable variations to survive.
   b. the tendency of weaker individuals to produce more offspring.
   c. the tendency of a species to improve through time through random mating.
   d. the tendency of weak species to dominate over stronger individuals.
   e. the tendency of life to develop machines.

5. Evolution is
   a. change in the inherited traits of a population from one generation to the next.
   b. change in physical traits due to individual effort.
   c. genetic variation that is not passed on to future generations.
   d. the emergence of stronger individuals as dominant in a community.
   e. divine intervention.

6. Mass extinctions are important because they
   a. reduce competition and allow rapid evolution of surviving species.
   b. generally cause the death of carnivorous species.
   c. stop evolution because so many species remain.
   d. lead to the formation of large land masses.
   e. closed the Tethys Seaway.

7. The longest episode in Earth's history is the
   a. Cenozoic.
   b. Paleozoic.
   c. Proterozoic.
   d. Quaternary.
   e. Holocene.

8. The "Age of Fish" was the
   a. Precambrian.
   b. Silurian.
   c. Mesozoic.
   d. Devonian.
   e. Quaternary.

9. What characteristics of life on Earth mark the Archean Eon?
   a. Life was highly diverse.
   b. Life was confined to land alone.
   c. Life was characterized by increasing diversity among plants and insects.
   d. Life was characterized by simple forms, such as cyanobacteria.
   e. Mammals increased in diversity.

10. A major development at the end of the Precambrian was the
    a. evolution of birds on land.
    b. emergence of the first complex life forms.
    c. cooling of the crust.
    d. appearance of whales and other marine mammals.
    e. appearance of the first water on Earth.

11. Cratons are
    a. regions of excessive volcanic activity.
    b. Precambrian rock at the core of most continents.
    c. systems of mountain ranges formed by tectonic convergence.
    d. heavily metamorphosed roots of young divergent zones.
    e. island arcs and hotspots.

12. The phylogeny of the horse shows
    a. decreasing size, strength, and speed over time.
    b. lengthening of the forelegs to adapt to life in forests.
    c. increasing size, strength, and speed to adapt to life in grasslands.
    d. random changes in response to the breakup of Pangaea.
    e. smaller and more stealthy body arrangement.

13. During the Cenozoic,
    a. mammals went extinct.
    b. dinosaurs were dominant.
    c. the first plants and animals developed in the ocean.
    d. We are not sure what happened in the Cenozoic since it is the oldest era.
    e. None of the above.

14. What is uniformitarianism?
    a. The past is a key to the present.
    b. Understanding the future requires understanding the present.
    c. It is the study of fossils in order to predict evolution.
    d. Earth's history is best explained by observations of modern processes.
    e. It is a description of evolution.

15. When did human ancestors emerge?
    a. the Paleozoic
    b. the Mesozoic
    c. the Neogene
    d. the Paleogene
    e. the Archean

FURTHER RESEARCH

1. What is the name of the rock unit that contains the earliest multicellular living community?

2. What was the Tethys Sea?

3. Which continents made up Pangaea?

4. How does natural selection cause evolution?

5. What characteristics of the human immunodeficiency virus allow it to evolve quickly?

6. What is the dominant class of animals in the Cenozoic?

7. How can plate tectonics influence evolution?

8. Geologists recognize that "most of the geologic time scale is blank." What does this mean, and why would this be?

ONLINE RESOURCES

Explore more about Earth's history on the following websites:

"Geologic Time," a U.S. Geological Survey online publication that provides a thorough introductory discussion of many aspects of geologic time:
http://pubs.usgs.gov/gip/geotime

"Geologic Time Scale": Wikipedia online encyclopedia provides a thorough discussion of the geologic time scale:
http://en.wikipedia.org/wiki/Geologic_timescale

"Geologice Time—The Story of a Changing Earth": Smithsonian National Museum of Natural History online tour of Earth's history at an educational and easy-to-navigate site:
http://paleobiology.si.edu/geotime/index.htm

"The Burgess Shale": The Royal Ontario Museum's superb online resource for more information on the Burgess Shale:
http://burgess-shale.rom.on.ca/

Additional animations, videos, and other online resources are available at this book's companion website:
www.wiley.com/go/fletchercanada

This companion website also has additional information about WileyPLUS and other Wiley teaching and learning resources.

Sarah Leen/National Geographic Creative

© Thomas Kitchin & Victoria Hurst /All Canada Photos/Corbis   © Radius Images/Corbis   © Mike Grandmaison/Corbis

CHAPTER 15
THE GEOLOGY OF CANADA

Chapter Contents and Learning Objectives (LO)

15-1 Canada can be divided into six major geologic provinces.

LO 15-1 *Describe the basis for identifying a geologic province.*

15-2 The Canadian Shield contains the oldest rocks in the world.

LO 15-2 *Describe how the Canadian Shield formed.*

15-3 The continental platform consists of exposed flat-lying Phanerozoic rocks.

LO 15-3 *Describe the origin of sedimentary rocks in the continental platform.*

15-4 The Appalachian Orogen has been evolving since the end of the Precambrian.

LO 15-4 *Describe the role of orogenies during the Appalachian Orogen.*

15-5 The Innuitian Orogen is Canada's northernmost mountain belt.

LO 15-5 *Describe the two major orogenic events of the Innuitian Orogen*

15-6 The Cordilleran Orogen is Canada's youngest mountain belt.

LO 15-6 *Describe the development of the Cordilleran Orogen and the role of terrane accretion through time.*

15-7 Continental shelves and slopes form the margins of Canada.

LO 15-7 *Describe the origin of the continental shelf along the Atlantic coast.*

GEOLOGY IN OUR LIVES

The North American continent consists of two major geologic features: the craton and the surrounding orogenic belts. The craton is the stable core of the continent, consisting of the Precambrian shield and the Phanerozoic platform. The orogenic belts formed during the Phanerozoic when island arcs, oceanic rocks, and fragments of other continents accreted to the ancient continental core during plate convergence. This accretion resulted in the formation of mountain ranges. All Canadians depend on geology. Geology influences our water quality and food supplies and provides our energy require-

ments and manufacturing materials. Exported minerals, metals, and hydrocarbons bring vast amounts of money into the country. Understanding our geology is a shared effort of the federal, provincial, and territorial geological surveys, along with university and industry researchers across the country. These people and their institutions serve a critically important function of providing the geologic framework needed for the discovery and management of natural resources, identification of geologic hazards, and evaluation of long-term environmental changes.

(?) The geology of Canada can be organized into six major geologic provinces. What geologic province do you live in? What characterizes its geology?

15-1 Canada Can Be Divided into Six Major Geologic Provinces

LO 15-1 *Describe the basis for identifying a geologic province.*

How and when were the Appalachian Mountains and Canadian Cordillera formed? Why is the central region of Canada so flat, and how did its fertile soil develop? Why is British Columbia prone to earthquakes? You can answer these and other questions by learning about the geology of Canada. The geologic character of our land is a product of the ongoing forces of the rock cycle: melting, crystallization of magma, metamorphism, weathering, erosion, deposition, tectonic activity, and the interaction of these processes over the long history of our planet. Understanding the geologic features of our country is critically important to discovering, exploiting, and managing natural resources; dealing with geologic hazards; and protecting the environment.

The North American Plate and Its Boundaries

Canada is situated largely within the northern part of the North American Plate, which includes the North American continent, Greenland, and the western halves of Iceland and the North Atlantic Ocean floor (**FIGURE 15.1**). The eastern boundary of the plate is the spreading *Mid-Atlantic Ridge*, and the eastern margin of our continent is the wide, within-plate passive margin between the Maritime provinces,

Newfoundland, and the Atlantic Ocean floor. The northern boundary of the North American Plate is formed where the Mid-Atlantic Ridge extends east of Greenland up into the Arctic Ocean and continues past the North Pole into northeastern Siberia. The narrower western plate boundary bordering on the Pacific Ocean is an active plate margin. From northern California to a point northwest of Vancouver Island the relatively small oceanic Juan de Fuca Plate disappears beneath the North American Plate along the *Cascadia Subduction Zone*. The spreading Juan de Fuca Ridge separates the Juan de Fuca Plate from the enormous Pacific Plate and the triple junction, where the North American, Juan de Fuca, and Pacific Plates come together northwest of Vancouver Island. North of this, the Pacific Plate slides past the North American Plate along the offshore Queen Charlotte transform fault to eventually subduct beneath southern Alaska.

Fundamental Geologic Divisions of Canada

The North American continent consists of two major geologic components: the craton and orogenic belts (**FIGURE 15.2**). The **craton** (the word comes from the Greek word for "strength") is the stable core of

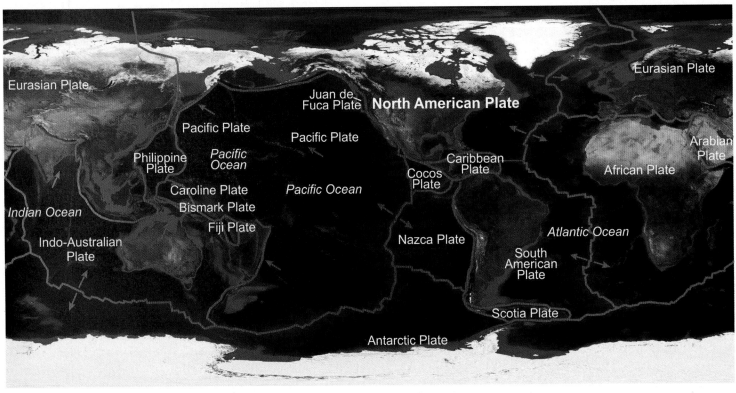

FIGURE 15.1 Satellite image showing the distribution of Earth's tectonic plates. Note the scale of the North American Plate which extends from the spreading ridge in the Atlantic Ocean to the boundary with the Pacific Plate along the eastern side of the Pacific Ocean.

? Why does the eastern boundary of the North American plate extend all the way to the Mid-Atlantic Ridge?

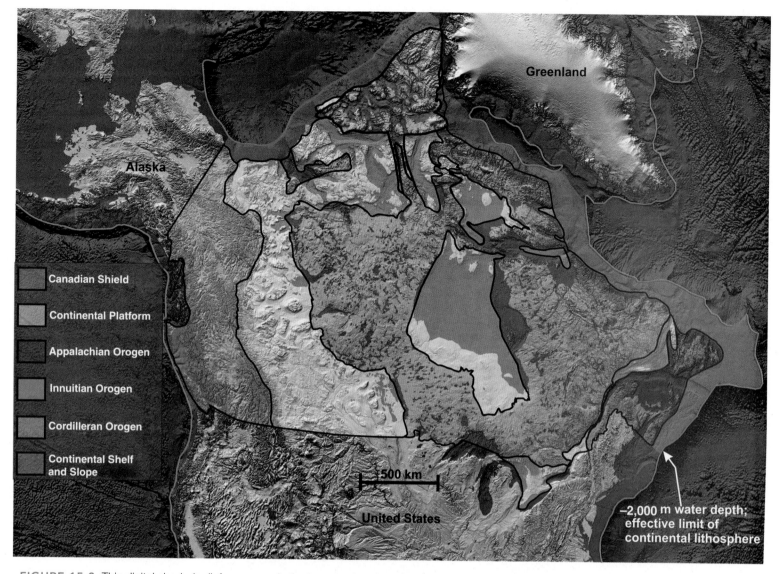

FIGURE 15.2 This digital shaded relief map reveals the topography of the land. Superimposed on the map are the six geologic provinces of Canada. The map shows that North America is characterized by two major geologic features: the craton which largely underlies the Canadian Shield and Continental Platform, and the flanking Appalachian, Innuitian, and Cordilleran orogens.

 What are the major geologic resources and geologic hazards in your province?

the continent. It consists of the **Canadian Shield**, where Precambrian rocks are exposed at the surface and which has been tectonically stable for nearly 1 billion years. The region where the Precambrian rocks are buried beneath flat-lying Phanerozoic sedimentary rocks is the **continental platform**. The orogenic belts are regions formed during the Phanerozoic when island arcs, ocean floor rocks, and fragments of other continents accreted to the craton during plate convergence. This process resulted in the formation of the mountainous regions around the craton (recall Chapter 11, "Mountain Building").

The complex geology of Canada can be subdivided into a number of separate **geologic provinces**, which are regions with distinctive geologic histories. These provinces are identified on the basis of two

criteria: (1) geologic structure and history, and (2) their landscapes or *physiography* (e.g., mountains versus plains).

Why Study Geologic Provinces?

All Canadians rely on water, soil, minerals, building material (e.g., stone and concrete), energy, and many other natural resources within our borders. The realization that geologic information was needed in a modern society to guide discovery, exploitation, and management of Canada's natural resources led to the formation of the Geological Survey of Canada in 1842. When Canada was surveyed during the 19th century, great differences were recognized, not only in landforms but in the underlying rocks as well. This led to the identification of the geologic provinces, each with its own distinctive geologic history, specific resources, particular hazards, and topographic "signatures" (Figure 15.2).

The Geological Map of Canada

Basic geologic information is essential for a modern society. Since 1842, mapping by the Geological Survey of Canada (GSC) has provided information on the basic geologic framework of Canada, with the newest version of the *Geological Map of Canada* published in 1996 (FIGURE 15.3). Mapping by the federal government continues today, in collaboration with other government agencies, the provincial geological surveys, and universities.

What is a geological map? Maps show features on Earth's surface at a reduced scale. Most geological maps depict the location, age, and lithology of rocks in the uppermost crust. Traditionally, much of the information was provided by **field geology**, which gathered data directly from the rocks exposed at Earth's surface or those known to exist at very shallow depths under the soil and sediment of the surface. Today, we have at our disposal many technologies that provide a more in-depth understanding of the geologic attributes mapped at Earth's surface. These include using geophysical techniques such as aeromagnetic and gravity surveys from planes, *radar* and *lidar* data acquired from satellites, and seismic data from natural and human-made shaking of the ground. These data can provide critical information for

FIGURE 15.3 The *Geological Map of Canada* with the outlines of the geologic provinces.

? What geologic province do you live in? Do you live on the craton or in one of the flanking orogens?

interpreting subsurface geology and geology in regions difficult to explore on the ground. They are used extensively by oil and mining companies to find, and better understand, their deposits.

Do you see patterns in the map in Figure 15.3 that reflect the geologic provinces shown in Figure 15.1? These patterns are related to the structure of the crust resulting from the orogenies that built much of the North American continent. These patterns are mainly related to two kinds of processes: (1) formation of supercontinents, and (2) accretion of **terranes**, which consist of fragments of continental crust, island arcs, or oceanic rocks. In this chapter, we review each geologic province in Canada. Although there may be time overlap in the evolution of some geologic provinces, they are presented in a general chronological order (**FIGURE 15.4**), starting with the oldest one, the Canadian Shield, and ending with a discussion of the youngest additions to Canada's land mass, the continental shelves and slopes.

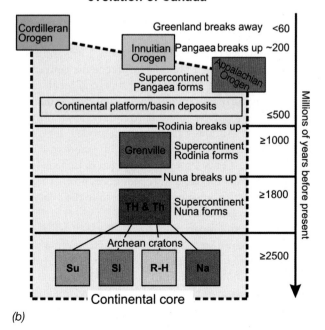

- - - - Covered limits of orogen

· · · · Covered limits of continental platform/basin deposits

−2,000 m water depth; ~limit of continental lithosphere

500 km

J.W.H. Monger

(a)

Timeline for geological evolution of Canada

(b)

How many supercontinents were involved with the geologic evolution of Canada?

FIGURE 15.4 (a) Major geologic components of Canada and (b) a timeline for their geologic evolution. The five Archean cratons of the Canadian Shield are shown in shades of pink (Su = Superior; Sl = Slave; R-H = Rae and Hearn; Na = Nain), and the major Proterozoic orogens that weld them together are shown in red (NQ = New Quebec; TH = Trans-Hudson; Th = Thelon; To = Torngat; W = Wopmay).

Expand Your Thinking—What methods would you use to make a geological map of Canada?

15-2 The Canadian Shield Contains the Oldest Rocks in the World

LO 15-2 *Describe how the Canadian Shield formed.*

The Canadian Shield is by far the largest of Canada's geologic provinces, occupying about half of Canada's land area (Figure 15.3), and is the largest area of continuously exposed Precambrian rocks on Earth. It consists of five Archean cratons containing rocks that are about 4 to 2.5 billion years old (Figure 15.4), including the oldest known rocks on Earth. The cratons were welded together by plate convergence during the Proterozoic; the welds or *sutures* are marked by the linear orogens of that age.

Most of the Shield is low lying at less than 500 m in elevation and part of it is submerged beneath the waters of Hudson Bay and around the eastern Arctic Islands. The land surface was shaped during its burial beneath the continental ice sheets during the last ice ages and features innumerable lakes connected by drainage systems (FIGURE 15.5). The rivers on the Shield discharge into Hudson Bay in the northeast, the Beaufort Sea in the northwest, and the St. Lawrence River in the southeast. The ancient rocks exposed within the Shield can be traced beneath younger rock formations by using deep drilling and geophysical methods. Mountains are present only in the northeastern part of the Shield that flanks the Labrador Sea, Davis Strait, and Baffin Bay which separates Canada from Greenland. It is thought that this strait, or rift, developed at the same time as the northern part of the North Atlantic but did not develop into an ocean basin. These mountains are called "rift-shoulder" mountains and rise to as much as 2,500 m.

Formation of the Oldest Rocks

As planet Earth differentiated from the cloud of cosmic dust orbiting the Sun to form a planet, the amalgamated dust particles melted and eventually became separated into the mantle and the core (Chapter 2, "Solar System"). The earliest crust began to form in the Hadean and early Archean over 3,600 million years ago by partial melting of the

mantle and segregation of elements into new magma, which eventually crystallized to form lower density igneous rocks in the first crust. Very little of this earliest crust remains intact on Earth's surface. However, between 3,600 and 2,500 million years ago, extensive regions of continental crust formed. These regions make up the bulk of the Archean cratons, the remains of which make up about 70 percent of the Canadian Shield by area. These regions consist mainly of granitic rock and gneiss with lesser amounts of metamorphosed volcanic and sedimentary rocks. The areas of Archean crust form the Slave, Rae, Hearne, Superior, and Nain cratons. Of these, the Superior craton that underlies much of Ontario and Quebec is by far the biggest and is about the same size as the present Indian subcontinent.

The Canadian Shield is home to the oldest rocks on Earth. In 1989, granitic gneisses, called the Acasta gneisses, were discovered in the western part of the Slave craton. Zircons in these rocks, which had crystallized from the original granitic magma, were dated using the radioactive decay of uranium to lead and yielded ages of just over 4 billion years. At the time, this was the oldest rock on Earth. However, in 2008 a graduate student at McGill University sampled metamorphosed volcanic rocks along the Quebec coast of Hudson Bay. Using the radioactive decay of samarium to neodymium, he discovered that the rocks could be as old as 4.28 billion years. Although the amount of Hadean rock in the Canadian Shield is small, a detailed study of them can provide important information about the processes that formed the first crust on Earth.

The Proterozoic Orogens

By the Early Proterozoic, between about 2,500 and 1,800 million years ago, it appears that processes similar to present-day plate tectonics had developed. During this time, most of the Archean cratons were separated from one another by rocks that formed on ocean floors or in island arcs. A major change occurred between about 1,950 and 1,800 million years ago, when Archean continental fragments converged, collided, and amalgamated to form a single supercontinent called **Nuna**. The intervening early Proterozoic rocks, squeezed, crunched, and metamorphosed between the colliding Archean cratons, formed the Proterozoic orogens. Although they are long eroded, they once were sites of great mountain belts comparable to the modern Himalayas and Alps. The biggest early Proterozoic orogen in Canada is the *Trans-Hudson*, which spans Saskatchewan, Manitoba, northern Quebec, and Baffin Island. Other smaller orogens include the *Wopmay*, *Thelon-Taltson*, *New-Quebec*, and *Torngat*.

Nuna subsequently split apart into pieces, some of which can now be found within other present-day continents. However, a single large fragment made of Archean crustal fragments welded together by Proterozoic orogens underlies most of Canada. About 1,100 million years ago, the fragments of Nuna converged and recombined in a different paleogeographic configuration to form a new supercontinent called **Rodinia**. The process created the Grenville Orogen, which in Canada extends from southern Labrador, across southern Quebec north of the St. Lawrence River, and into southeastern Ontario. It

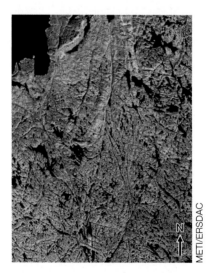

FIGURE 15.5 The Canadian Shield consists of Precambrian igneous and metamorphic rocks that have been exposed by erosion over millions of years. The scarred and gouged nature of the crust (filled by thousands of lakes) shown in this satellite image of Nunavut reflects this erosional history, but particularly the erosion caused by ice during the Pleistocene. The drainage patterns may be controlled by structures, such as faults, in the shield rocks.

METI/ERSDAC

(?) What is the preferred directions for the linear features shown in Figure 15.5? Propose a reason for this orientation.

continues southwestward in the subsurface right across our continent to northwestern Mexico.

In the following 500 million years or so, erosion wore down the ancient Shield rocks to a gently undulating surface. In the Late Proterozoic, about 700 million years ago, evidence from the northern and western margins of Canada shows that the Rodinian supercontinent was starting to break up. By latest Proterozoic–earliest Cambrian time, 550 to 540 million years ago, a discrete new continent called **Laurentia** separated from the other fragments of Rodinia. Laurentia's boundaries roughly approximated those of the present North American continent, except that it also contained Greenland and lacked the surrounding Phanerozoic orogens. The deposits of the continental platform were laid down on Laurentia and the *Appalachian, Innuitian* and *Cordilleran Orogens* formed at different times along its margins.

A Wealth of Mineral Resources

At the end of the 19th and the beginning of the 20th centuries, the resource-rich Canadian Shield was transformed from a largely undeveloped hinterland into a region teeming with mining activity and exploration. This coincided with the development of new, more effective techniques for processing ores containing metals such as gold, copper, and nickel and improved technology for producing hydroelectric power used by the mines and refineries. This meant that the Shield could meet the demand for mineral resources in the growing industrial sectors of central Canada, setting the stage for Canada to become largely autonomous and a major player in the world economy.

Today, the Canadian Shield is recognized as having one of the richest endowments of mineral resources in the world and is Canada's leading source of precious metals such as gold (e.g., Abitibi, Quebec, and Hemlo Mine, Ontario) and base metals that include nickel (e.g., Sudbury, Ontario), copper and zinc (e.g., Flin Flon, Manitoba),

and iron ore (e.g., Labrador and Quebec). The Shield has also become a world supplier of diamonds (e.g., Northwest Territories) (FIGURE 15.6).

Glaciation

The ancient rocks of the Canadian Shield, some of which had been buried under Phanerozoic sediments, are now exposed, partly because of normal and slow erosion over millions of years starting in the Precambrian and partly because continental glaciers eroded the overlying rock and soil during the ice ages of the last 500,000 years. Five major periods of glaciation during this period of time have been identified, with the most recent culminating 20,000 to 25,000 years ago.

The Canadian Shield was the site of several ice accumulation centres during the last ice age, which eventually combined into a continent-size glacier, in some places up to 4 km thick (FIGURE 15.7). The front edge of the ice crept south into the United States, excavating the massive basins that would become the Great Lakes. The ice stripped soil and sedimentary rock off the crust to expose the Precambrian basement. It left behind thick deposits of sand and gravel to mark its advance and its subsequent retreat over the last 10,000 years as the ice age came to an end, marking the start of Earth's current *interglacial period* (Chapter 17, "Glaciers and Paleoclimatology").

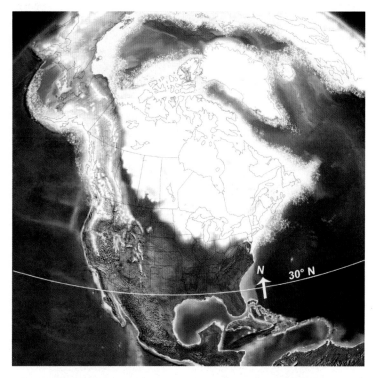

FIGURE 15.7 At the culmination of the last ice age most of Canada was covered by ice up to 4 km thick. The ice carved the bedrock and left deep basins that filled with glacial meltwater when the ice sheets receded. These basins became the Great Lakes. Many of the surface features, lakes, and drainage systems in the Canadian Shield, such as those in Figure 15.5, formed during the last glacial advance.

© Jason Pineau

FIGURE 15.6 Ekati diamond mine, Northwest Territories

 Why are the pits around the Ekati mine so circular in shape?

 Why is the Precambrian crust of the Canadian Shield not buried by younger rocks?

Expand Your Thinking—The oldest rocks in the Archean cratons in the Canadian Shield may have formed before plate tectonics as we know it today was operating. What processes may have been important in the Hadean and early Archean?

15-3 The Continental Platform Consists of Exposed Flat-Lying Phanerozoic Rocks

LO 15-3 *Describe the origin of sedimentary rocks in the continental platform.*

By about 500 million years ago, all the Precambrian continents on Earth had come together to form the supercontinent Rodinia and built the mountain ranges of the Grenville Orogen. Erosion was gradually wearing down its surface to a region of low relief. Rodinia began to break up about 700 million years ago, and by the Late Proterozoic–Early Cambrian, about 600 to 530 million years ago, a continent-sized fragment called Laurentia had broken free and was surrounded by oceans (**FIGURE 15.8**). Laurentia was the ancestor of the North American continent and nearly the same size, although it included Greenland and lacked the surrounding orogenic belts. Most of Laurentia forms the present craton.

The Precambrian rocks exposed in the Canadian Shield are buried beneath the Phanerozoic sedimentary rocks of the continental platform. Because the craton is strong, at over 200 km thick, it protected the platform deposits from deformation by the enormous tectonic forces that created the surrounding orogenic belts. Bedding in these deposits is at most gently tilted or broadly warped.

Topography

The platform deposits probably once covered much of the continental interior but erosion over millions of years and during the last ice ages has removed much of the cover in Canada, leaving deposits in three regions. The largest region is the Interior Plains of western Canada, where the rocks form the *Western Canada Sedimentary Basin*. The

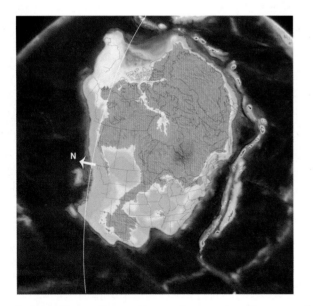

FIGURE 15.8 The continent of Laurentia surrounded by oceans in Cambrian time. The white line shows the estimated position of the equator at that time, and so the location and orientation of Laurentia on the globe was very different than it is now.

 What do you think the climate was like for Laurentia in Cambrian time?

region continues northward into the Arctic islands, and southward into the United States where it is continuous across the continent between the Appalachian and Cordilleran Orogens. Deposits of the *St. Lawrence Lowland* in southern Ontario are the northern tip of those in the United States. A third, isolated area is around southern Hudson Bay (Figure 15.4).

Erosion of the nearly flat-lying platform deposits resulted mostly in flat and gently undulating land surfaces that in places are covered with glacial deposits from the ice ages. Most bedrock exposures are along water courses and where escarpments, such as the Niagara Escarpment of southern Ontario, are formed by gently tilted rocks that are resistant to erosion. In parts of the western Interior Plains, erosion of Late Mesozoic sandstone and mudstone creates *badlands*, especially in southern Alberta and Saskatchewan. Much information on the older platform strata comes from the many exploratory boreholes and seismic reflection surveys made in the search for hydrocarbons and evaporite deposits.

The platform rises from sea level near the St. Lawrence River, Hudson Bay, and the Arctic Ocean margins, through a few hundred metres in southern Ontario, to over 1,000 m near the mountain front in western Canada. It is crossed by rivers, most of which in the west drain into Hudson Bay, although the big Mackenzie River in northwestern Canada flows into the Arctic Ocean. The region where platform strata are thinned by erosion and disappear against the Canadian Shield contains Canada's largest lakes. From northwest to southeast, these are Great Bear Lake, Great Slave Lake, Lake Winnipeg, Lake Superior, Lake Huron, and Lake Ontario.

Bedrock Geology

Sedimentary rocks of the continental platform range in age from late Proterozoic to Cenozoic, a time span of about 600 million years. This period includes incorporation of Laurentia into a new supercontinent called **Pangaea** about 300 million years ago and Pangaea's breakup about 100 million years later. Sedimentation on the platform was controlled by four main factors. First, during much of the Paleozoic and Mesozoic, the surface of the platform was close to sea level, so that worldwide rise and fall of sea water meant that, at times, the platform was covered by shallow seas in which sedimentary rocks were deposited. At other times, the surface was exposed and subjected to erosion (**FIGURE 15.9**).

Second, slow and gentle vertical movements within the craton formed basins that are hundreds of kilometres in diameter, within which the strata thicken toward the middle of the basin. The basins are largely separated by wide swells and arches, which are sometimes called highs, across which strata become thinner or are absent (**FIGURE 15.10**). The basins and arches evidently developed in response to very slow vertical movements, probably controlled by processes in the subcontinental mantle and structures in the Precambrian basement. Third, climate controlled the types of sediments deposited. During much of the Paleozoic, Laurentia lay near

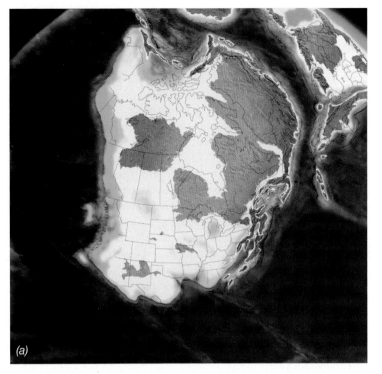

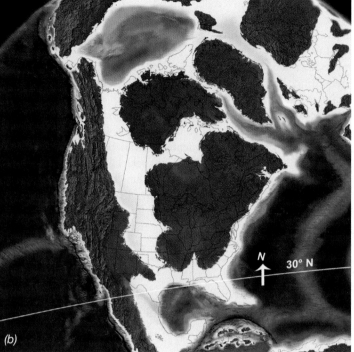

Ronald Blakey, Colorado Plateau Geosystems, Inc.

FIGURE 15.9 *(a)* The Ordovician Period in the continental interior was characterized by deposition of marine sediment on the broad, stable platform under a shallow sea. *(b)* Much later, during the Cretaceous Period, marine waters again swept across the continental interior but were restricted largely to the western portion within the *Western Interior Seaway* that flanked the emerging Cordillera.

(?) Describe a modern-day example of the geologic setting of the continental platform during the Ordovician or Cretaceous Period.

(?) Would you expect the sequence of sedimentary rocks in each of the Phanerozoic basins to be identical? Why, or why not?

FIGURE 15.10 Arches, highs, and major Phanerozoic basins of Canada, and their location relative to the cratons of the Canadian Shield, shown in shades of pink, and the major orogens. There are similar basins in the central part of the United States. For example, in addition to underlying much of Alberta, southern Saskatchewan, and Manitoba, the Williston and Western Canada Sedimentary Basins also extend into eastern Montana and North and South Dakota. If you compare this figure with Figure 15.4 you will notice that some of the sedimentary rocks in the Phanerozoic basins are affected by the Innuitian Orogen and the Cordilleran Orogen.

the tropics and was covered at times by warm shallow seas. Within these, carbonates were deposited in open waters and evaporites in bays and restricted basins. Toward the end of Paleozoic and through most of the Mesozoic time, the continent drifted slowly northward into cooler climates that favoured clastic rather than carbonate sedimentation. Fourth, the margins of the continental platform were depressed at times by the load of rock thrust onto them from the surrounding orogens. This created elongate *foreland basins* that received floods of detritus eroded from the new mountains.

The thickness of platform deposits varies, but averages about 1 to 3 km, far thinner than age-equivalent rocks in the surrounding orogens. Strata in some basins are about 4 km thick, and as much as 6 km thick in those basins near the front of the Cordillera. They are thinner on the arches, and feather out toward the Canadian Shield as a result of erosion. The succession contains at least five major regional disconformities—unconformities with parallel bedding below and above them, which were caused by the rise and fall of sea level. These disconformities divide the stratigraphic succession into *sequences*. In the Paleozoic, sequences generally start with clastic rocks followed by carbonates. Attempts have been made to link sequences to global changes of sea level caused by closing and opening of oceans and other orogenic events.

A Closer Look at the Continental Platform

During the Paleozoic, warm shallow seas advanced and retreated across the continental platform. There was widespread carbonate deposition in Cambrian to Early Ordovician time (about 510 to 470 million years ago), but after this the seas retreated, only to return in the Late Ordovician (about 460 million years ago) and cover much of the continental platform. Sea level fell again at the end of the Ordovician (about 440 million years ago) only to rise again by the Middle Silurian

(about 425 million years ago) when carbonate laid down in southern Ontario formed the Niagara Escarpment, best known for underpinning Niagara Falls (**FIGURE 15.11**).

In Late Silurian to Middle Devonian time (about 416 to 400 million years ago), slow vertical movements caused arches to rise and basins to subside. Above a widespread disconformity, younger Devonian strata (390 million years or more old) include carbonate reefs separated by basins containing shale, silt, and limestone. The reefs are major hydrocarbon reservoirs, and strata of this age are well known because thousands of boreholes have been drilled across the Interior Plains. In Saskatchewan, restricted circulation led to evaporation of sea water and the accumulation of valuable potash deposits. At the end of the Devonian and in the Early Carboniferous (about 360 million years ago), a small influx of clastic sedimentation into the Western Canada Sedimentary Basin may reflect tectonic events in the Innuitian and Cordilleran Orogens. By Late Carboniferous–Permian time (about 300 million years ago), fine clastic rock and sandy limestone prevail. Marine incursions onto the platform were less widespread in Late Paleozoic and Early Mesozoic, and Triassic and Jurassic strata are preserved only below the westernmost and southernmost Interior Plains. They contain sandstone, siltstone, and mudstone, some carbonate, and widespread evaporites.

By the Early Jurassic, the craton in western Canada was north of about the latitude of southernmost Ontario (about latitude 40°N), climates were cooler, and carbonate deposition minimal. By the Late Jurassic, about 150 million years ago, clastic detritus was being shed eastward from the rising Cordillera. The western margin of the platform was depressed under the load of thrust sheets advancing from the Cordillera, and in the Early Cretaceous, seas began to invade the platform again. Eventually the vast moat called the *Western Interior Seaway* extended from the Arctic Ocean to

 What present and past geologic processes led to the formation of Niagara Falls?

FIGURE 15.11 Horseshoe Falls, the Canadian portion of Niagara Falls, pour over the Niagara Escarpment composed of Middle Silurian carbonate.

© iStock.com/Orchidpoet

GEOLOGY IN OUR LIVES

THE FERTILE PRAIRIES

The Canadian prairies (**FIGURE 15.12**) are within the heart of the continental platform. These fertile grasslands are a product of sediments associated with the last ice age. They consist for the most part of glacial sediments deposited on horizontally bedded Paleozoic limestone and shale and Mesozoic sandstones and mudstones. The largest area of the prairies consists of plains of glacial sediment consisting of gravel, sand, silt, and clay. These sediments are 10 m to 30 m thick, and cover the underlying rock surface for thousands of square kilometres except where stream erosion has cut down to it.

The prairies have an extraordinarily even surface. The surface originated largely as rock waste that was transported by the creeping ice sheets, blown by cold winds sweeping off the ice, or carried by waters from the melting ice. Although some crystalline rocks from the Shield are found as boulders and cobbles, much of the glacial sediment has been crushed and ground to a fine texture appropriate to forming rich soils.

Russ Heinl/All Canada Photos, Getty Images, Inc.

FIGURE 15.12 The Canadian prairies are composed of glacial sediments, which give them their flat surface and high agricultural productivity.

The black, fertile soils of the prairies are a natural consequence of this history. During the period of glaciation, no vegetation was present to remove the minerals essential to plant growth, which is the case in the soils of normally weathered and dissected surfaces. Prairie soil (predominantly *chernozemic*) has been thoroughly mixed by glacial processes, and the prairies are continuously fertile for kilometres on end. The true prairies were once covered with a rich growth of natural grass and annual flowering plants, but today they are covered with farms reaping the benefits of this glacial heritage.

the Gulf of Mexico and covered the western continental platform (Figure 15.9b). Within this foreland basin, sand and mud eroded from the emerging mountains was deposited. By the Early Cenozoic, about 60 million years ago, thrust faulting had ceased, and rivers carried sand, silt, mud, and gravel across the platform.

Despite mostly unspectacular surface geology, the continental platform is economically the most important geologic province in Canada. It is the dominant agricultural region in the country. The well-known tar sands of northern Alberta, currently being mined, were fed by oil and gas resources that are widespread in the subsurface. Potash in Saskatchewan is a major export for use as fertilizer, and salt and gypsum are extracted in Ontario.

Beyond the Platform—Canada's Three Phanerozoic Orogens

The North American craton and platform is surrounded by three orogens, or mountain belts, that formed during the Phanerozoic. All three orogens have features in common, although they differ considerably in their evolution. In all three, rocks closest to the continental platform were deposited on continental margins created some 600 million years ago when Laurentia broke free of the supercontinent Pangaea. Some rocks in the orogens originated within or across the ocean basins that surrounded Laurentia. Carried by plate convergence, they eventually accreted to the former continental margins and raised mountains. Unlike the continental platform deposits protected by a strong Precambrian foundation, rocks in the orogens were deformed, intruded, and metamorphosed to create the Appalachian Orogen along the southeastern side of the continental platform, the Innuitian Orogen on the northern side, and the Cordilleran Orogen on the western side.

(?) Expand Your Thinking— Why does changing sea level produce a variety of sedimentary rocks?

15-4 The Appalachian Orogen Has Been Evolving Since the End of the Precambrian

LO 15-4 *Describe the role of orogenies that created the Appalachian Orogen.*

The **Appalachian Orogen** in Newfoundland, the Maritimes, and Quebec is up to 1,500 km long and 600 km wide. It continues southwestward for another 1,700 km as far as Alabama (**FIGURE 15.13**). Other parts of the same orogen were separated when Pangaea broke up and the North Atlantic Ocean opened, and today are found in western Europe and northwestern Africa.

Topography

The Appalachian Orogen in Canada today is but a shadow of its former glory in Late Paleozoic time, having been worn down by 250 million years of erosion that culminated in intense glacial scouring during the ice ages. Today, much of the region in Canada is less than 500 m in elevation, with the highest part over 1,200 m along the south shore of the St. Lawrence River in Quebec. Many of the best rock exposures are found along the shorelines of peninsulas, deep inlets, and islands.

Bedrock

The evolution of the Appalachian Orogen has been painstakingly unravelled by geologists working for the last 170 years. Appalachian evolution started in very Late Proterozoic time, some 600 million years ago, when Laurentia broke free from the supercontinent Rodinia and became surrounded by oceans. On the other side of the ocean from what is today eastern North America lay the great continent of *Gondwana*, composed of South America, Africa, India, and Australia, and the continent of Baltica, which included Scandinavia and parts of western Russia. Within the ocean were volcanic arcs and continental margin fragments. These arcs and fragments, now embedded in the Appalachian Orogen, are called *terranes*. They are typically identified as fault-bounded bodies each with its own geologic record that originated in different regions called *paleogeographic realms* (**FIGURE 15.13**).

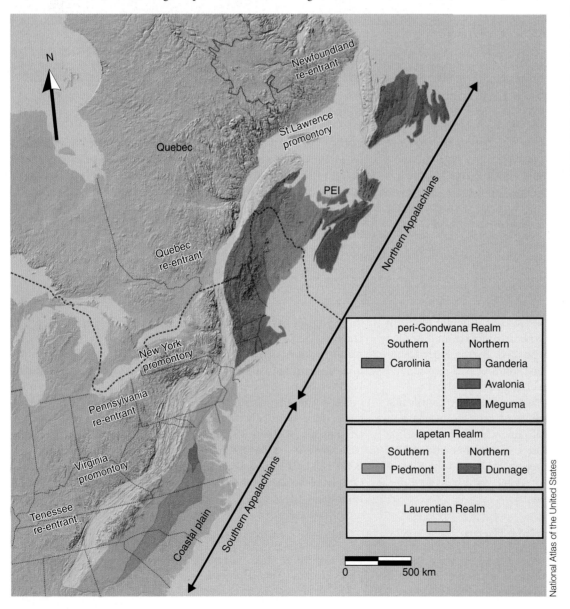

? How do the northern Appalachians differ from the southern Appalachians?

FIGURE 15.13 The Appalachian orogenic belt within eastern Canada and the United States showing the major lithotectonic units and paleogeographic realms.

National Atlas of the United States

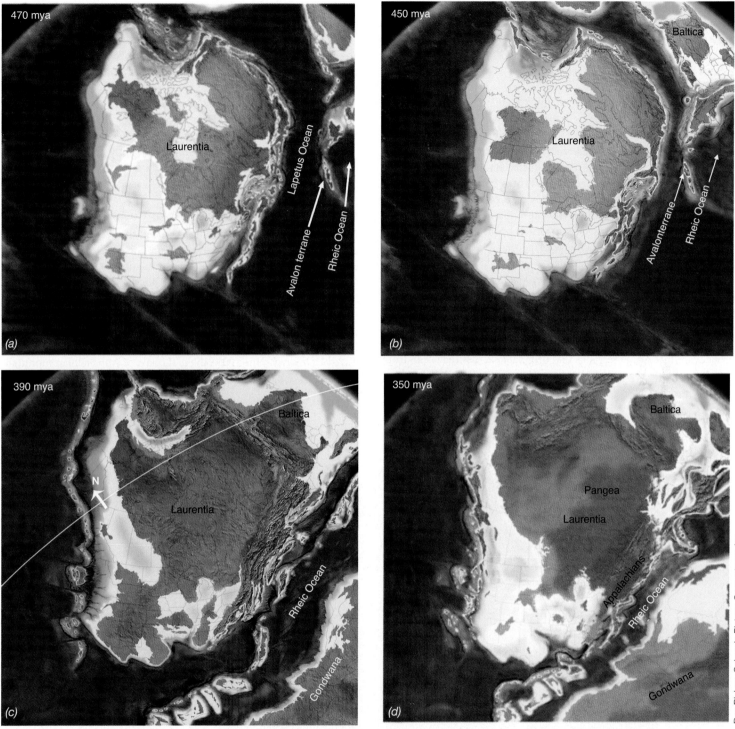

FIGURE 15.14 Evolution of the Appalachian Orogen. From 460 to 390 million years ago (mya) terranes were sequentially accreted to the Laurentian continental margin: In *(a)* and *(b)*, during the Taconic Orogeny the Iapetus Ocean is closing, causing accretion of terranes within this ocean. *(c)* By 390 mya, the Acadian Orogeny is underway as the Rheic Ocean is closing. White line represents the equator at that time. *(d)* By 350 mya, the last gaps of the Rheic Ocean are closing, leading to the continent–continent collision of Laurentia, Gondwana, and Baltica during the Appalachian Orogeny. This resulted in the formation of the supercontinent Pangaea.

? Would the Appalachian Orogen shown in Figure 15.14 produce fold-and-thrust mountains, fault-block mountains, or volcanic mountains? Explain your answer.

The terranes sequentially accreted to the Laurentian continental margin between the Middle Ordovician and Middle Devonian, 460 to 390 million years ago. By the Late Carboniferous, some 350 million years ago, Laurentia, Gondwana, and Baltica had collided to form the supercontinent Pangaea (**FIGURE 15.14**). Along the continent–continent collision zone, a great mountain belt was raised.

The part of this belt found today in North America is the Appalachian Orogen; other parts underlie Europe and northwest Africa. Pangaea started to break up about 250 million years ago and by 170 million years ago, the North Atlantic Ocean started to widen, ripping the great mountain belt apart and separating the Appalachians from their old world counterparts.

The Early Paleozoic rocks in the Canadian Appalachians are arranged in orogen-parallel belts of rocks that originated in three paleogeographic realms (Figure 15.13). First, on the northeastern side of the orogen, in Quebec, western New Brunswick, and northwestern Newfoundland, is the *Laurentian Realm*. It includes fragments of the nearby Canadian Shield, very Late Proterozoic rift-related clastic rocks, and Cambrian to Middle Ordovician carbonates and deeper water clastic rocks deposited on the Laurentian passive margin. This margin flanked the Iapetus Ocean, which opened at the end of the Proterozoic, nearly 600 million years ago, and closed by the Early Silurian, about 435 million years ago.

Second, within the oceanic *Iapetan Realm* are rocks preserved in the central part of the orogen in eastern New Brunswick, Cape Breton, and central Newfoundland. There, scraps of Early Paleozoic ocean floor, volcanic arcs, and arc-related basins have been grouped as *Dunnage terrane*. Third, and preserved in more southeastern parts of the orogen, are three large terranes that originated across the ocean from the Laurentian margin in the *peri-Gondwanan Realm*, meaning near

the former continent called Gondwana. The *Gander terrane* in New Brunswick and east-central Newfoundland consists of Late Proterozoic sedimentary and metamorphic rocks overlain by Middle Cambrian–Early Ordovician sandstone and shale. *Avalon terrane* in eastern and northern Nova Scotia and western Newfoundland is composed of relatively undeformed Late Precambrian arc-related rocks, about 600 million years old, and overlying sedimentary rocks that contain Cambrian trilobite faunas very different from those of the same age along the Laurentian margin. The *Meguma terrane* forms the peninsula of southeastern Nova Scotia and comprises mainly Cambrian and Early Ordovician sandstone and shale formed on a continental margin and is overlain by mid-Ordovician tillite.

Appalachian Orogenies in Canada

Appalachian mountain building occurred in three main stages—the older ones coincided with terrane accretion, and the final one with continent–continent collision. Parts of the floor of the Iapetus Ocean were converging and generating arcs above subduction zones in Late Cambrian to Middle Ordovician time, between about 500 and 460 million years ago. By the Middle Ordovician, 460 million years ago, rocks of the continental margin and parts of Iapetus were being thrust faulted and folded toward the continent during what is called the *Taconic Orogeny*. This marks the first phase of Appalachian mountain building in Canada (**FIGURE 15.15**).

FIGURE 15.15 The Taconic Orogeny was responsible for developing these beautiful folds within the Humber Arm allochthon of Newfoundland. Of note is the person in the red jacket, Dr. Hank Williams, a titan of Appalachian geology.

Explain how you would apply radiometric dating to determine when rocks in Humber Arm were folded.

In Late Ordovician–Silurian time, about 440 million years ago, rocks of the Gander terrane were accreted, deformed, and metamorphosed. This event was shortly followed by accretion of the Avalon terrane in the Early Silurian, 435 million years ago, which marked the closure of the Iapetus Ocean. A second ocean, called the Rheic Ocean opened between the Avalon terrane and Gondwana about 470 million years ago at about the same time as the Meguma terrane separated from the margin of Gondwana. The Meguma terrane travelled across the Rheic Ocean to collide with the Avalon terrane by the Middle Devonian, about 390 million years ago. This occurred during the earlier part of the *Acadian Orogeny*, which was the second phase of Appalachian mountain building. During and following accretion, great volumes of Silurian and Devonian granitic rock intruded into the accreted terranes. The terminal stage of mountain building, the *Appalachian Orogeny*, occurred when the remaining part of the Rheic Ocean closed in the Early Carboniferous, some 350 million years ago, when Gondwana and Laurentia collided to form the supercontinent Pangaea. Along the continent collision zone a great intra-continental mountain belt was raised that was torn apart when the North Atlantic Ocean opened (**FIGURE 15.16**). Today, the region contains the passive continent–ocean boundary that lies within the North American Plate.

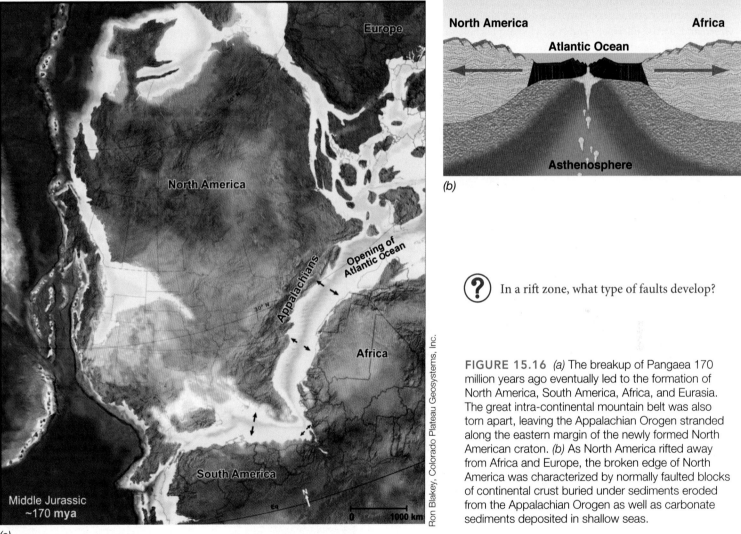

(?) In a rift zone, what type of faults develop?

FIGURE 15.16 *(a)* The breakup of Pangaea 170 million years ago eventually led to the formation of North America, South America, Africa, and Eurasia. The great intra-continental mountain belt was also torn apart, leaving the Appalachian Orogen stranded along the eastern margin of the newly formed North American craton. *(b)* As North America rifted away from Africa and Europe, the broken edge of North America was characterized by normally faulted blocks of continental crust buried under sediments eroded from the Appalachian Orogen as well as carbonate sediments deposited in shallow seas.

(?) **Expand Your Thinking**—If the Atlantic Ocean started to subduct beneath eastern North America at a rate of 4 cm per year, how long would it take for another "Appalachian-style" orogeny to occur?

15-5 The Innuitian Orogen Is Canada's Northernmost Mountain Belt

LO 15-5 *Describe the two major orogenic events of the Innuitian Orogen.*

The **Innuitian Orogen** is north of latitude 75° in northernmost Canada and distributed across the northern Arctic Islands (**FIGURE 15.17**). This remote region is known for difficult working conditions, so our understanding of the geology there is relatively new. In addition, the nature of the Arctic Ocean floor is the most poorly understood of any ocean basin and is currently the subject of much research.

The orogen extends southwestward from the northwestern tip of Greenland, across Ellesmere and Axel Heiberg Islands, to Melville and Prince Patrick Islands in the western Arctic, a distance of over 1,500 km.

The northwestern boundary of the orogen is the Arctic Ocean. On the east, it is bounded by the northern tip of Baffin Bay, which leads into the narrow strait between northwestern Greenland and Ellesmere Island. To the south is the Arctic platform, separated by the north-pointing finger of Boothia Peninsula, made of exposed Precambrian rocks of the Canadian Shield.

The Innuitian Landscape

The northern and eastern parts of the Innuitian Orogen feature rugged mountains that carry vast icecaps, with valley glaciers that descend to sea level and iceberg-choked fjords (**FIGURE 15.18**). The highest mountains on northern Ellesmere Island reach just over 2,600 m. Relief is much lower in western parts of the orogen, but what appears to be relatively featureless barren tundra at ground level can be seen from high-flying aircraft or satellites to contain giant folds.

Evolution of the Orogen

The Innuitian Orogen includes two major unrelated orogenies separated in time by a long period of relative tectonic quiet. The **Ellesmerian Orogeny** is of very Late Devonian to Early Carboniferous age, about 360 million years old. It was followed by a 200-million-year period of subsidence and extension in the **Sverdrup Basin**. During the **Eurekan Orogeny** of Early Cenozoic time, about 50 million years ago,

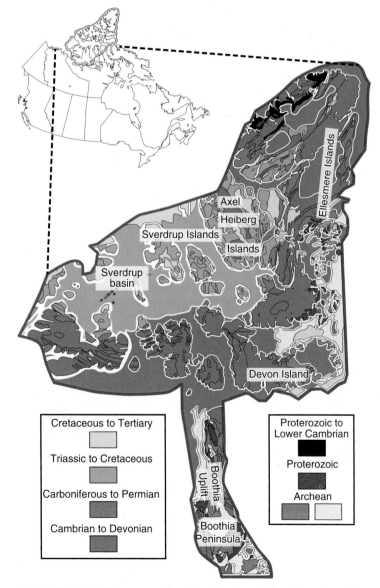

FIGURE 15.17 The remote Innuitian Orogen, outlined in red, within the northernmost Northwest Territories and Nunavut.

(?) Why do you think some of the oldest rocks within the Innuitian orogen are found in the Boothia uplift?

FIGURE 15.18 Axel Heiberg Island (foreground) and the mountains beyond consist mostly of Mesozoic strata of the Sverdrup Basin.

(?) Can you identify the tell-tale signs that suggest massive glaciers used to cover this landscape?

many older rocks and structures were further deformed. Unlike the relatively linear Appalachian and Cordilleran Orogens, the structural trends of the Innuitian Orogen are sinuous and oriented northeast–southwest in northern Ellesmere Island, north–south in eastern Ellesmere Island, and then bend sharply east–west on Devon Island, south of Ellesmere Island.

Beginning of the Orogen

Like the two other Phanerozoic orogens, the history of the Innuitian Orogen begins with the breakup of Rodinia, which started over 700 million years ago and ended in the Cambrian, about 530 million years ago, when Laurentia was an isolated continent. In the Early Paleozoic, the Arctic was a passive margin. Deposits of Cambrian to Silurian carbonate and local evaporite were laid down on the Arctic platform, which extended from northernmost Yukon to Greenland and was continuous to the south with the Interior Plains. To the northwest, from northernmost Greenland to northern Melville Island, Early Paleozoic deposits of shale and chert were laid down in deep water off the Laurentian continental margin.

Somewhere across the ocean basin that, in present coordinates, lay to the north of Laurentia, lurked a terrane of uncertain size known as *Pearya*, whose history is similar to that of older parts of the Appalachian Orogen and its extension in northwestern Europe. Pearya consists of metamorphic rocks intruded by Middle Proterozoic plutons about 1 billion years old, Late Proterozoic to Early Ordovician shelf rocks, and ocean floor and arc rocks of late Early Ordovician age. All were deformed, intruded, and metamorphosed and then unconformably overlain by a thick succession of sedimentary and arc-related volcanic rocks of Middle Ordovician to Late Silurian age, about 460 to 420 million years old.

The Ellesmerian Orogeny

Part of Pearya is preserved on northwestern Ellesmere Island and accreted to the Laurentian margin in the Late Silurian, about 420 million years ago. Accretion initiated a long episode of intermittent folding and thrusting mainly directed toward Laurentia. It involved both the deeper water sediments and continental shelf strata and culminated in the Ellesmerian Orogeny in the very Late Devonian–Early Carboniferous, about 360 million years ago. The deformed belt extends from northwestern Greenland to Melville and Prince Patrick Islands in the western Arctic.

Unlike the abundant magmatic rocks in the Appalachian and Cordilleran Orogens, few granitic and volcanic rocks involved in the Ellesmerian Orogeny are restricted to Pearya. Also unlike the other orogens, Canadian Shield rocks were activated at about the time Pearya was accreted. These rocks are exposed in the Boothia uplift, whose north–south orientation reflects the trend of Precambrian structures. The main uplift was in Late Silurian–Early Devonian time, when strata of the Arctic platform and continental shelf were eroded from the crest of the Boothia uplift and those on its flank were folded and faulted.

The Sverdrup Basin

During the Early Carboniferous, some 340 million years ago, the compressive forces that caused the Ellesmerian Orogeny relaxed. At first, sedimentation was mainly in rift basins containing lake and river deposits, but toward the end of the Carboniferous, about 300 million years ago, marine incursions became extensive. Great thicknesses of salt were deposited, and eventually deeper water black shale and muddy carbonate were laid down. By the Permian, about 270 million years ago, a single seaway called the Sverdrup Basin formed and persisted for the next 200 million years.

The Sverdrup Basin is a triangular-shaped embayment, bounded on its eastern and southern sides by rocks deformed during the Ellesmerian Orogeny (Figure 15.19). Today, the Arctic Ocean is found off the northwest side of the Basin, but during the Early Mesozoic, it was flanked by rocks now in eastern Siberia and northern Alaska. During the Mesozoic, about 9 km of sediment accumulated in the Sverdrup Basin, which lay within the continent and was connected at times to the seas that ebbed and flowed across the western continental platform. During the Early Mesozoic, deltas received detritus from big river systems feeding into the basin from the east. To the west were organic-rich marine muds. These are potentially important source rocks for hydrocarbons trapped in salt domes and anticlines squeezed up from underlying Late Paleozoic salt layers.

During and following the Middle Jurassic, about 170 million years ago, there was renewed rifting, and by the Early Cretaceous, about 130 million years ago, rifting, accompanied at times by extrusion of mafic volcanic rocks, led to the opening of the parts of the Arctic Ocean off northwestern Canada. Toward the end of the Cretaceous, about 70 million years ago, the northeastern Sverdrup Basin was uplifted during deformation that heralded the onset of the Eurekan Orogeny.

The Eurekan Orogeny

The latest phase in the evolution of the Innuitian Orogen was the Eurekan Orogeny. It is associated with events related to the separation of Greenland from Canada between 65 and 35 million years ago. As the Atlantic Ocean continued to open, the mid-Atalantic spreading ridge branched off to the northeast and began to rift Greenland away from North America, but it "failed" to continue spreading, and is referred to as a *failed rift*. The initial movement of Greenland was northeastward, along a transform fault that separates Ellesmere Island from northern Greenland. After 57 million years ago, Greenland changed direction and then acted as a giant battering ram that moved northwestward toward Ellesmere Island, causing folding and thrust faulting there and in eastern parts of the Sverdrup Basin. At the end of the Eocene, about 35 million years ago, rifting between Greenland and Canada ceased, so Greenland rejoined the North American Plate. Active spreading is now taking place right across the Arctic Ocean, past the North Pole and heading toward the Siberian continental platform.

The Labrador Sea

In the far northeastern Canadian Shield, along Canada's coast between Ellesmere Island and Labrador, and on the opposing western Greenland coast, are mountains, some as high as 2,500 m. The mountains were raised not by plate collision, but by extension associated with the Early Cenozoic (65–35 million years ago) failed rift that lies between Greenland and Canada. Arching accompanied initial seafloor spreading and left elevated areas along the sides of the rift that formed what are called *rift-shoulder mountains*.

 Expand Your Thinking—Why would a rift fail? Is this the fate of the East African rift?

15-6 The Cordilleran Orogen Is Canada's Youngest Mountain Belt

LO 15-6 *Describe the development of the Cordilleran Orogen and the role of terrane accretion through time.*

The mountains of western Canada belong to a geologically young segment of a vast, roughly 13,000-km-long mountain chain that forms the western part of the North and South American continents, and extends from the Arctic Ocean to the southern tip of South America. The mountains in South America are the Andes, and those in the north are called the North American Cordillera. "Cordillera" is a Spanish word used for systems of interconnected mountain ranges and intervening plateaus, valleys, and basins. The commonly used name "Rocky Mountains" applies only to the easternmost ranges of

the North American Cordillera from New Mexico to northern British Columbia.

The Cordilleran Landscape and the Underlying Rocks

Landscapes of the **Canadian Cordillera** are grouped into one of three physiographic systems (**FIGURE 15.19**). The mountainous Eastern System contains the Rocky Mountains, the Mackenzie and

(a)

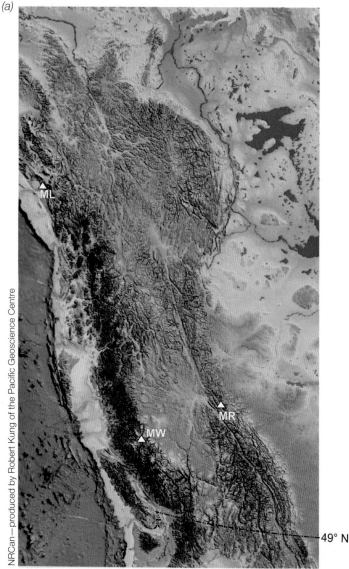

NRCan—produced by Robert Kung of the Pacific Geoscience Centre

(b)

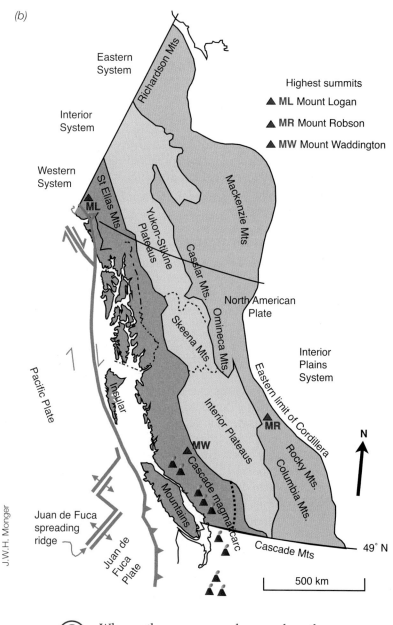

J.W.H. Monger

FIGURE 15.19 *(a)* False-colour digital elevation map highlighting the mountains belts within the Canadian Cordillera (darker reds and purples) flanked by the lower topography of the plateaus and Interior Plains (shades of pink, yellow, and green). *(b)* Major physiographic divisions, plates, and plate boundaries within the Canadian Cordillera.

? Why are there so many volcanoes along the western side of the Cordillera?

FIGURE 15.20 Canada's highest peak, Mount Logan, within the St. Elias Mountains of Yukon, stands at nearly 6,000 m.

(?) What processes are responsible for raising Mount Logan?

Richardson Mountains, and the Columbia, Omineca, and Cassiar Mountains farther west. The Interior System features the lower relief Interior and Yukon-Stikine Plateaus and Skeena Mountains. The mountainous Western System contains the Coast Mountains, the very high Saint Elias Mountains, and the Insular Mountains of the coastal islands. The highest summits are Mount Robson in the Rocky Mountains, Mount Waddington in the Coast Mountains, both about 4,000 m high, and Mount Logan (FIGURE 15.20) in the St. Elias Mountains, which, at nearly 6,000 m, is the highest peak in Canada.

Today, running water is the main agent of erosion, but during the ice ages almost the entire region, except for part of northern Yukon and the Northwest Territories, was covered by the Cordilleran Ice Sheets. These were as much as 2 km thick in central parts of the Cordillera. Only the higher peaks stood above the ice, and even those were sculpted by alpine glaciers. Lower mountains were rounded and glacial till was deposited in lowlands. Today, some of the largest icefields and glaciers on Earth outside of the polar regions are in the St. Elias Mountains and adjoining parts of Alaska. Many higher parts of the Coast Mountains contain glaciers, but these occur only locally in the Rocky Mountains.

The Eastern System shown in Figure 15.20 is mostly underlain by great thicknesses of sedimentary rock, whereas the Interior and Western Systems contain volcanic, sedimentary, and granitic rocks (FIGURE 15.21). A belt of metamorphic and granitic rock lies near the boundary between the Eastern and Interior Systems, in the Columbia, Omineca, and Cassiar Mountains. A second belt of mainly granitic rocks forms the Coast Mountains in the Western System and is one of the largest continuous masses of Phanerozoic granitic rock on Earth. Both belts contain intensely deformed rocks that in places, were buried as deep as 30 km and were subsequently uplifted and exposed by erosion during mountain building. The eastern limit of the Cordillera is where the rocks folded and thrust faulted during mountain building meet undeformed strata of the continental platform. The western boundary is the present active plate margin. West of Vancouver Island is the Cascadia Subduction Zone, and north of this, the vast Pacific Plate slides northward past the North American Plate along the Queen Charlotte transform fault.

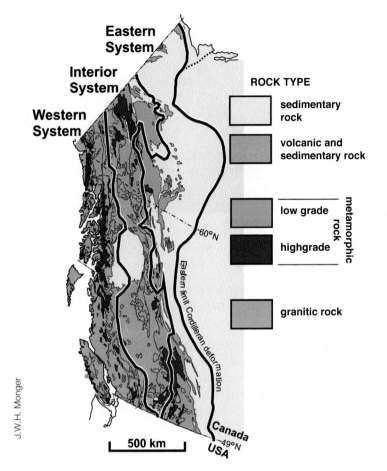

FIGURE 15.21 Simplified bedrock geology map of the Canadian Cordillera.

(?) Where is the greatest concentration of granitic rocks? Why do you think they are more concentrated in that region?

Ages, Tectonic Settings, and Evolution

In the eastern Cordillera, the dominantly sedimentary rocks range in age from Proterozoic to Cenozoic (about 1.7 billion years to 50 million years old) although most are less than 700 million years old. The oldest strata, about 1.7 to 1.2 billion years old, are made of great thicknesses of mainly shale and quartz-rich sandstone deposited in both deep and shallow water, lesser amounts of limestone, and very minor volcanic rocks.

Late Proterozoic deposits (700–600 million years old) are the oldest deposits that extend along the entire North American Cordillera and thus mark its inception. They were laid down in rift valleys that formed when the supercontinent Rodinia started to break up. They are composed of sandstone, shale, and minor volcanic rock and carbonate.

Paleozoic–Early Mesozoic deposits of mainly limestone and shale, and lesser amounts of sandstone (540 to 160 million years old) were mostly deposited in relatively shallow sea water in low latitudes on the edge of the newly formed continent of Laurentia. The strata can be traced westward from relatively thin formations deposited on the continental platform into the Rocky Mountains where the thickness is about nine times thicker than that of age-equivalent strata on the continental

J.W.H. Monger

FIGURE 15.22 Folded Mississippian Rundle Group stratigraphy in an anticline-syncline pair (outlined by black dashed line) within Mount Kidd, Alberta.

(?) If the view in the photo is to the north, what direction do you think the rocks were pushed to create the folds?

platform. It is these strata, thrust faulted and folded (**FIGURE 15.22**), that form the well-known scenery in the Rocky Mountains west of Calgary.

Late Mesozoic and Cenozoic deposits (less than 150 million years ago) in the Rocky Mountains are mainly sandstone, shale, and conglomerate, which reflect the northwest drift of the continent to higher, colder latitudes. Late Jurassic–Early Cretaceous deposits (about 150 to 100 million years old), preserved in thrust-faulted slices within the eastern Cordillera, host big deposits of coal. In Late Cretaceous and Early Cenozoic time, 100 to 60 million years ago, rocks in the eastern Rockies were folded and thrust faulted eastward up onto the eastern edge of the continental platform. This load of overthrust rocks depressed the crust and created a vast foreland basin in front of the emerging mountains to the west. Detritus eroding from the mountains was deposited in the foreland basin. In the Cretaceous, many dinosaurs lived in this region, some of which are on display in the Royal Tyrrell Museum, in Drumheller, Alberta, located in the badlands.

In the western two-thirds of the Canadian Cordillera, rocks of Devonian to Middle Mesozoic age (about 400 to 170 million years old) represent fragments of volcanic arcs, their accompanying *accretionary prisms* (Chapter 3, "Plate Tectonics"), and in Yukon and southeastern British Columbia, continental fragments torn off the edge of the craton. These fragments are called terranes, and like the terranes in the Appalachians, originated somewhere other than their present positions in the Canadian Cordillera. Note that terranes in the central Canadian Cordillera are probably fragments of offshore volcanic arcs (**FIGURE 15.23**) that persisted from the Devonian until the Middle Jurassic (about 390 to 170 million years ago). The granitic rocks that formed the roots of these arc volcanoes host major copper (**FIGURE 15.24**), gold, and molybdenum deposits in the western Cordillera.

Accompanying the volcanic arcs in the Early Mesozoic was an accretionary prism whose remnants contain exotic fossils that once lived in atolls far out in the ancestral Pacific, which at that time was a vast ocean called Panthalassa. Panthalassa occupied well over half Earth's surface area, as all continents were together in Pangaea, the most recent supercontinent, which existed about 300 to 200 million years ago (**FIGURE 15.25**).

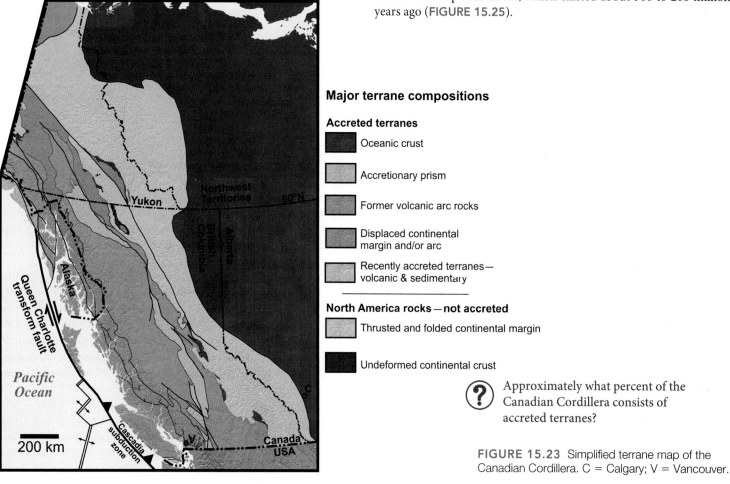

Major terrane compositions

Accreted terranes

◼ Oceanic crust

▦ Accretionary prism

▩ Former volcanic arc rocks

▨ Displaced continental margin and/or arc

▢ Recently accreted terranes— volcanic & sedimentary

North America rocks—not accreted

▢ Thrusted and folded continental margin

◼ Undeformed continental crust

(?) Approximately what percent of the Canadian Cordillera consists of accreted terranes?

FIGURE 15.23 Simplified terrane map of the Canadian Cordillera. C = Calgary; V = Vancouver.

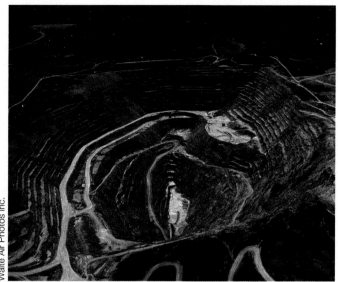

FIGURE 15.24 Highland Valley copper mine in south-central British Columbia. Minor amounts of gold and molybdenum are also produced there.

(?) What rock type do you think hosts the Highland Valley deposit?

About 200 million years ago Pangaea started to rift, and by 170 million years ago the North Atlantic began to open. As Pangaea broke up and North America drifted away from Africa and Europe, it plowed into the terranes along its western margin, accreted them, and initiated Cordilleran mountain building. Terranes now found in the Interior System overrode the outer continental margin about 180 million years ago, deeply burying it, and thickening the crust. Terranes now found in the Western System were accreted to those underlying the Interior System by the Cretaceous, 100 million years ago. This raised the ancestral Coast Mountains at the same time as the initiation of thrust faulting that carried rocks now in the Rocky and Mackenzie Mountains over the edge of the continental platform. Thus, between the Middle Jurassic and Late Cretaceous (from 180 to 90 million years ago), all major arc terranes accreted to the former continental margin, marking a major phase of Cordilleran mountain building. Subduction of oceanic lithosphere beneath the new continental margin continued to generate arc volcanoes on land above major granitic intrusions, the latest arc being the active Cascade Magmatic Arc, which extends from southwestern British Columbia to northern California, landward of the subducting Juan de Fuca Plate.

By the Late Cretaceous, the Canadian Cordillera was probably higher and narrower than it is today. It was eventually torn apart by large strike-slip faults that moved more westerly parts of the Cordillera northward. By Eocene time, some 50 million years ago, widespread normal faulting stretched and thinned the crust of the southern half of British Columbia.

Mountain building continues today, as shown, for example, by the prevalence of earthquakes and displacements in the westernmost North American Cordillera in particular. The adjoining Pacific Plate is moving northwestward, carrying great volumes of sediment eroded from the Cordillera and sliding past the coast of central British Columbia on the Queen Charlotte transform fault. In the past 5 million years as the Pacific Plate subducted beneath southern Alaska, the mass of sediment scraped off and thrust beneath the continental margin raised the highest mountains in North America: the St. Elias Mountains and the Alaska Range (Figure 15.21). In southwestern British Columbia, the small oceanic Juan de Fuca Plate intermittently sticks and slips as it subducts beneath the North American Plate, creating the potential for great earthquakes, the last of which occurred about 300 years ago.

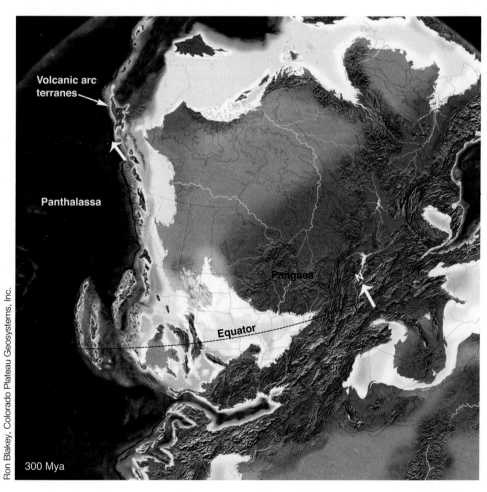

FIGURE 15.25 Possible Late Carboniferous position of volcanic arc terranes within the Panthalassic ocean, shown relative to Pangaea.

(?) Why are there so many volcanic arc terranes off the western margin of Pangaea in the Late Carboniferous?

(?) **Expand Your Thinking**—The Canadian Cordillera has a complex geologic history related largely to convergence. Describe the tectonic processes occurring along the active margin.

15-7 Continental Shelves and Slopes Form the Margins of Canada

LO *15-7 Describe the origin of the continental shelf along the Atlantic coast.*

Vast volumes of sediment are being eroded from North America, transported by rivers, and deposited around the margins of the continent, some on the continental shelves, some on the continental slopes, and some on the ocean floors. The shelves form submarine terraces as much as 700 km wide and up to 300 m deep. The continental slopes along the oceanward side of the shelves descend to the deep ocean floor at depths of about 2,000 m (Figure 15–4). The shelves on Canada's east and north sides developed along the margins of opening oceans and are wide, whereas the narrow western shelf is above an active plate boundary where the continental margin either overrides subducting oceanic lithosphere or is bounded by a transform fault along which oceanic lithosphere slides past the continent. The eastern and northern shelves are underlain by relatively undeformed Mesozoic and Cenozoic clastic sedimentary rocks that host vast reserves of oil and gas, whereas even young rocks underlying the western shelf are deformed.

The Atlantic Continental Shelf

Our continent, embedded within the North American Plate, is located far to the west of the plate's eastern edge at the spreading centre along the Mid-Atlantic Ridge. The eastern margin of the continent lies well within the plate and so is not subject to the violent forces found along many plate boundaries. It is a *passive margin* but it was not always that way, as described in Section 15.4 on the Appalachian Orogen.

The formation of the passive margin originated with the breakup of the supercontinent Pangaea in the Late Paleozoic to Early Mesozoic, about 280 to 230 million years ago. This process originated with a rift zone that separated Africa and Europe plus Asia from North and South America. As the new continents separated, a basin developed between them that gradually grew to become the Atlantic Ocean (Figure 15.16). The North Atlantic Ocean formed first, about 170 million years ago when North America separated from Africa, followed by the separation of Europe some 100 million years ago and the far northern Atlantic about 70 million years ago. The process continues today as the Mid-Atlantic Ridge provides new basaltic crust for the widening ocean floor.

As the North American Plate slowly moved away to the west, the thick continental crust along its eastern edge collapsed into a series of normally faulted blocks. Today, these blocks lie beneath the continental shelf and are deeply buried by sediments eroded from the Appalachian Highlands. This once-active divergent plate boundary has now become the passive margin on the eastern edge of westward-moving North America. In fact, the eastern margin of North America is widely recognized as a classic example of a passive continental margin.

The continental shelf of southeastern Canada consists of three main areas of sedimentary deposition: the Georges Bank, Scotian Shelf, and Grand Banks of Newfoundland (FIGURE 15.26). Below each of these areas are three main geologic units. The oldest, lower rift assemblage that formed in faulted basins, consists of clastic rocks, evaporites, and volcanics of Triassic and Early Jurassic age that lie on Precambrian and Palaeozoic rocks. This succession is overlain by a more widespread Jurassic and Cretaceous clastic-carbonate shelf assemblage deposited during the continental drifting phase, after the continents had separated and the continental crust of the margin subsided. Gradually, these sediments covered the faulted continental margin, burying it under sedimentary layers thousands of metres thick. During the Cenozoic, the supply of sediment was much reduced, leading to the deposition of thin layers of clastic sediments. All these rocks are capped by a layer of Pleistocene glacial deposits. The present submarine topography of the continental shelf and slope is related to the ice ages when lower sea levels allowed erosion by ice and rivers, and glacial sediments were deposited.

The Continental Shelf in the Arctic

The Arctic shelf developed as the Arctic Ocean basin opened in part by the extension of the Mid-Atlantic Ridge and in part possibly by the counterclockwise rotation of northern Alaska away from the western Canadian Arctic Islands, starting in the Early Cretaceous. The rocks that were deposited during the rifting phase were primarily sandstones and shales, derived from the North American continent, and are about 4 km thick. They are overlain by up to 10 km of Upper Cretaceous and Cenozoic clastic sedimentary rocks eroded from the emerging northern Cordillera, and channelled via the Mackenzie River and its vast delta in northwestern Canada (FIGURE 15.27). The Cenozoic rocks contain a vast storehouse of hydrocarbons, although exploration for hydrocarbons in this area is particularly challenging given the extreme climate, long periods of darkness, and sea ice.

The Pacific Shelf—an Active Continental Margin

The narrow Pacific shelf is an active continental margin, marked by numerous earthquakes and, in places, volcanoes, because it is located close to a plate boundary. West of Vancouver Island, the small, relatively buoyant oceanic Juan de Fuca Plate is descending beneath the continent along a northeast-dipping subduction zone, above which even young sediments are highly deformed. The buoyancy of the oceanic plate causes it to stick and build up strain. When that strain is suddenly released, large earthquakes may occur. Northwest of Vancouver Island, however, oceanic crust of the Pacific Plate slides northward past the continent along a transform fault, the Queen Charlotte Fault.

Expand Your Thinking—What are the most important differences between passive and active continental margins?

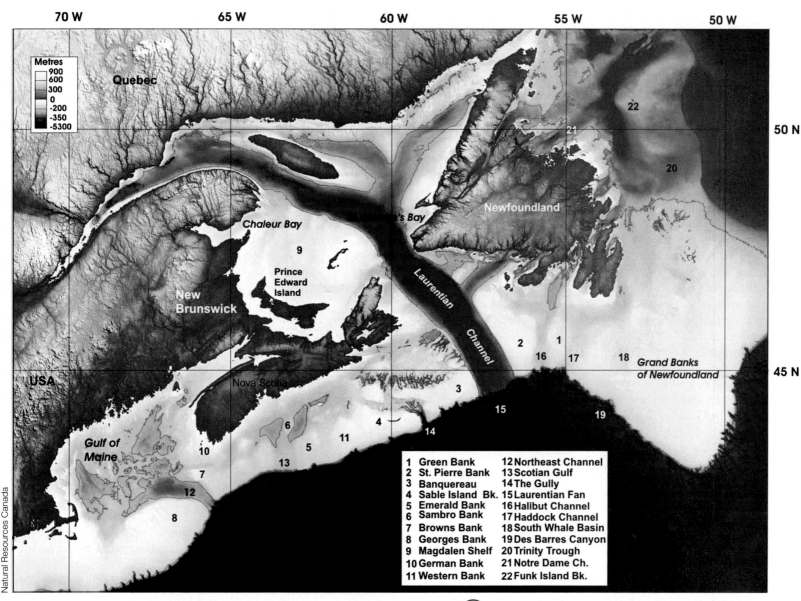

70 W   65 W   60 W   55 W   50 W

50 N

45 N

Metres
900
600
300
0
-200
-350
-5300

Quebec

Chaleur Bay

9

Prince
Edward
Island

New
Brunswick

Nova Scotia

Gulf of
Maine

10

7

12

8

13

5

11

4

14

6

3

15

19

2   1
16   17   18   Grand Banks
of Newfoundland

Newfoundland

Laurentian Channel

's Bay

22

21

20

USA

1 Green Bank	12 Northeast Channel
2 St. Pierre Bank	13 Scotian Gulf
3 Banquereau	14 The Gully
4 Sable Island Bk.	15 Laurentian Fan
5 Emerald Bank	16 Halibut Channel
6 Sambro Bank	17 Haddock Channel
7 Browns Bank	18 South Whale Basin
8 Georges Bank	19 Des Barres Canyon
9 Magdalen Shelf	20 Trinity Trough
10 German Bank	21 Notre Dame Ch.
11 Western Bank	22 Funk Island Bk.

Natural Resources Canada

FIGURE 15.26 The continental shelf off the southeastern coast of Canada showing the location of the Georges Bank, Scotian Shelf, and Grand Banks of Newfoundland.

 Rocks are submerged on the continental shelf. How do geologists know the shelf's structure and history?

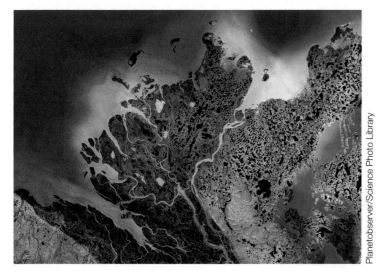

Planetobserver/Science Photo Library

FIGURE 15.27 Satellite image of the delta of the Mackenzie River. The river has been providing sediment to the continental shelf of the Arctic Ocean for millions of years.

 Do you think that the Mackenzie delta will look the same in the future?

LET'S REVIEW "GEOLOGY IN OUR LIVES"

Now that you have finished the chapter, "Geology in Our Lives" will have taken on new meaning for you. Let us review it: The North American continent consists of two major geologic features: the craton and the surrounding orogenic belts. The craton is the stable core of the continent, consisting of the Precambrian shield and the Phanerozoic platform. The orogenic belts formed during the Phanerozoic when island arcs, oceanic rocks, and fragments of other continents accreted to the ancient continental core during plate convergence. This accretion resulted in the formation of mountain ranges.

All Canadians depend on geology. Geology influences our water quality and food supplies and provides our energy requirements and manufacturing materials. Exported minerals, metals, and hydrocarbons bring vast amounts of money into the country. Understanding our geology is a shared effort of the federal, provincial, and territorial geological surveys, along with university and industry researchers across the country. These people and their institutions serve a critically important function of providing the geologic framework needed for the discovery and management of natural resources, identification of geologic hazards, and evaluation of long-term environmental changes.

The central craton, composed of heavily metamorphosed Archean and Proterozoic crust, consists of two parts: a shield and a platform. The Canadian Shield is the largest of the geologic provinces of Canada and the largest continuous area of Precambrian rocks on Earth. It also contains the oldest rocks on Earth. The continental platform is a stable, flat interior region covered with generally flat-lying beds of mainly Phanerozoic sedimentary rock.

There are three orogenic belts in Canada. The Appalachian Orogen, exposed in the Maritime provinces of Canada, was active mainly in the Paleozoic during the formation of Pangaea. In the far north of Canada, the Innuitian Orogen consists of rocks affected by two distinct mountain-building events, one in the Late Paleozoic and one in the Cenozoic. The third orogenic belt is the Cordilleran Orogen, consisting of the Eastern System, which includes the Rocky Mountains, the Interior System, and the Western System, which includes the mountains along the Pacific coast. These regions were formed as a result of the westward movement of the North American Plate following the breakup of Pangaea, which led to the accretion of several terranes onto the western side of North America during the Mesozoic. The western margin of Canada remains active today, whereas the eastern and northern continental margins of Canada are good examples of passive continental margins.

In the remaining chapters, we focus on modern geologic processes characterizing the major environments on Earth's surface. These chapters deal with global warming, glaciation, the hydrologic cycle, deserts, and the oceans.

STUDY GUIDE

15-1 Canada can be divided into six geologic provinces.

- Canada can be divided into six **geologic provinces** that are identified on the basis of their physiography and their geologic structure and history. They are the Canadian Shield, the **continental platform**, the Appalachian Orogen, the Innuitian Orogen, the Cordilleran Orogen, and the continental shelf and slope.

- The interior lands of Canada consist of the **craton**, which has the Precambrian rocks of the Canadian Shield at its core. These were periodically covered by shallow seas in which the Paleozoic and Mesozoic sedimentary rock of the continental platform were deposited. To the east, west, and north of the continental platform, the Appalachian, Innuitian, and Cordilleran Orogenies added new material to its edges, in a process known as continental accretion. These lands are composed of former island arcs, slivers of continental rock, and basaltic crust and marine sediments delivered by plates that converged on North America.

15-2 The Canadian Shield contains the oldest rocks in the world.

- The **Canadian Shield** is the largest of the geologic provinces and is the largest area of exposed Precambrian rocks on Earth. It contains rare rocks slightly older than 4 billion years.

- The shield contains five Archaen cratons welded together during Proterozoic collisions, forming a supercontinent called **Nuna.** A later collision, called the Grenville Orogen, led to the incorporation of the Shield into a new supercontinent called **Rodinia.** This was followed by the formation of the supercontinent called **Laurentia.** Finally, the Shield was fully developed, and this represents the stable core of present-day North America.

- During the Quaternary Period, the region was glaciated by the Laurentide Ice Sheet. Erosion by the ice sheets helped to expose the Canadian Shield.

15-3 The continental platform consists of exposed flat-lying Phanerozoic rocks.

- The continental platform deposits, which are typically flat-lying, undeformed sedimentary rocks, occur in three areas. The largest area is the Interior Plains of western Canada, where the rocks form the Western Canada Sedimentary Basin. Two smaller areas are the St. Lawrence Lowland, and the area around Hudson Bay. These deposits are largely undisturbed by mountain building and are composed of flat-lying or gently dipping sedimentary strata of Paleozoic and Mesozoic age. They have buried a portion of the craton known as the "stable platform."

- The geologic history of the platform deposits reflects the rise and fall of sea level, migration of the continent through tropical regions, and the addition of clastic detritus mostly eroded from flanking orogens. In the Early Paleozoic, warm, shallow seas dominated with the formation of fossil-rich carbonate rocks. In the Devonian, broad warping led to the development of basins bordered by carbonate reefs and filled by shale, silt, and sometimes evaporites. By the Jurassic, clastic sedimentary rocks became dominant, and during the Mesozoic, much of the detritus in the Interior Plains was derived by erosion from the mountains built during the Cordilleran Orogen.

15-4 The Appalachian Orogen has been evolving since the end of the Precambrian.

- The Appalachian Orogen formed a great mountain system extending along the eastern side of North America. In Canada, the rocks can be subdivided based on their distinctive geologic history. From northwest to southeast, they are the Laurentian Realm, the Dunnage terrane of the Iapetan Realm, and the Gander, Avalon, and Meguma terranes of the peri-Gondwana Realm.

- A series of orogenies occurring in the Paleozoic—the Taconic, Acadian, and Appalachian Orogenies—led to the formation of the Appalachian Highlands. All three orogenies are interpreted as collisions that occurred during the closing of the Iapetus Ocean between North America, Europe, and Africa. Various terranes accreted to the margin of North America during the early orogenies, prior to continent–continent collisions leading to the formation of **Pangaea**.

15-5 The Innuitian Orogen is Canada's northernmost mountain belt.

- The rocks of the Innuitian Orogen are preserved in rugged mountains and low-lying tundra in the northern Canadian Arctic islands of Ellesmere, Axel Heiberg, Devon, Melville, and Prince Patrick.

- The orogen consists of two mountain-building events separated by a period of sediment deposition. The accretion of the Pearya terrane with the northern margin of Laurentia culminated in the Ellesmerian Orogeny about 360 million years ago. The Sverdrup Basin then developed and accumulated sediments in rift basins, deltas, and deeper marine environments. The second mountain-building event, the Eurekan Orogeny, resulted in deformation of some of these sedimentary rocks and is thought to be linked to the formation of the failed rift between Greenland and Canada.

15-6 The Cordilleran Orogen is Canada's youngest mountain belt.

- The Canadian part of the Cordilleran Orogen consists of three parallel physiographic and tectonic systems. The easternmost is a mountainous system including the Rocky Mountains, and the Mackenzie and Richardson Mountains in the Yukon and Northwest Territories. The central part is the Interior System and has more subdued topography. The western part includes the rugged Coast Mountains and the St. Elias Mountains, including Mount Logan, the highest peak in Canada.

- The oldest rocks in the Cordillera are Late Proterozoic and formed during rifting as Rodinia broke apart. Thick sequences of Paleozoic and Early Mesozoic marine limestone and shale deposited on the new continental margin are preserved in the thrust sheets in the Rocky Mountains. The thrust sheets that developed in the Jurassic-to-Cretaceous time thickened the crust, causing a downward flexure to the east in front of the evolving mountain belt. This resulted in the formation of a foreland basin, which collected the detritus eroded from the mountains to the west.

- The interior and western physiographic systems of the Cordillera result from accretion of terranes consisting of volcanic arc rocks with their accompanying accretionary prisms and older continental fragments. Accretion of most terranes was initiated at the time the North Atlantic Ocean started to open in the Middle Jurassic, which led to westward movement of North America. Ultimately the Coast Mountains began to take shape about 100 million years ago at about the same time as the thrust faults developed that are now widely preserved in the Rocky Mountains to the east.

15-7 Continental shelves and slopes form the margins of Canada.

- Canada has passive continental margins along its Atlantic Ocean and Arctic Ocean coastlines, and an active continental margin on its west side. The Atlantic passive margin of the North American Plate originated with the breakup of the supercontinent Pangaea in the Late Paleozoic to Early Mesozoic, about 280 to 230 million years ago. Sediments were initially deposited in fault-bounded rift basins before being covered by Jurassic and Cretaceous marine sedimentary rocks thousands of metres in thickness.

- The Arctic continental shelf developed in part as Alaska rifted and rotated toward its present position in the Cretaceous. The early rift-basin sedimentary rocks are overlain by 10 km of Upper Cretaceous to Cenozoic clastic deposits related to the northward expansion of the Mackenzie Delta.

- The sedimentary rocks on the passive continental margins of Canada contain vast reserves of hydrocarbons.

KEY TERMS

Appalachian Orogen (p. 404)
Canadian Cordillera (p. 410)
Canadian Shield (p. 394)
continental platform (p. 395)
craton (p. 394)
Ellesmerian Orogeny (p. 408)

Eurekan Orogeny (p. 408)
field geology (p. 396)
geologic provinces (p. 395)
Innuitian Orogen (p. 408)
Laurentia (p. 399)
Nuna (p. 398)

Pangaea (p. 400)
Rodinia (p. 398)
Sverdrup Basin (p. 408)
terranes (p. 397)

ASSESSING YOUR KNOWLEDGE

Please answer these questions before coming to class. Identify the best answer.

1. What is a geologic province?
   a. a fault-block mountain system
   b. a change in crustal densities
   c. a condition of equilibrium
   d. a region of crust with similar physiography
   e. None of the above.

2. The rocks on the Atlantic continental shelf are
   a. highly deformed.
   b. composed largely of Paleozoic strata.
   c. composed largely of Mesozoic and Cenozoic strata.
   d. a result of the breakup of Rodinia.
   e. composed of Precambrian igneous and metamorphic rocks.

3. The regions of the Appalachian Orogen from northwest to southeast are
   a. Laurentia, Dunnage, Gander, Avalon, Meguma.
   b. Avalon, Gander, Dunnage, continental shelf.
   c. Coast Mountains, Interior System, Rocky Mountains.
   d. Laurentia, Continental Platform, Meguma, Dunnage.
   e. None of the above.

4. The craton of North America consists of the
   a. Shield and orogenic belts.
   b. orogenic belts and the continental shelf.
   c. Platform and shield.
   d. Glacial deposits and shield.
   e. Innutian Orogen and the Cordilleran Orogen.

5. Sedimentary rocks on the continental platform come from
   a. largely volcanic depositions.
   b. largely marine and coastal environments.
   c. largely deep sea environments.
   d. thick sequences of glacial sediments.
   e. None of the above.

6. The Cordilleran Orogen
   a. does not extend into Alaska.
   b. is composed of only two accreted terranes.
   c. reaches from Alaska to the tip of South America.
   d. was created by the formation of Pangaea.
   e. is Precambrian in age.

7. Canadian orogenies were largely
   a. Paleozoic in the east and north, and Mesozoic to Cenozoic in the west.
   b. related to destruction of the stable craton.
   c. the product of the formation of Eurasia.
   d. driven by northern migration of the stable craton.
   e. None of the above.

8. The Rocky Mountains were formed by
   a. continental accretion in the Early Paleozoic.
   b. the breakup of Rodinia.
   c. the formation of Pangaea.
   d. terrane accretion on the western margin of North America.
   e. the Taconic Orogeny.

9. The Innuitian Orogen consists of
   a. a core of Precambrian metamorphic rocks.
   b. sedimentary rocks affected by the Ellesmerian and Eurekan Orogenies.
   c. sedimentary rocks eroded from the Rocky Mountains.
   d. volcanic arc rocks.
   e. granitic intrusions.

10. The North American Plate is moving to the
    a. north.
    b. south.
    c. east.
    d. west.
    e. None of these.

11. What is an accreted terrane?
    a. rock accreted onto the edge of a plate by convergence
    b. sedimentary layers deposited in shallow interior seas
    c. faulted blocks produced by rifting
    d. new continental outlines produced by the breakup of Pangaea
    e. uplift caused by the Juan de Fuca Plate

12. The western margin of Canada is
    a. an active continental margin.
    b. a passive continental margin
    c. the product of continental accretion.
    d. uplifted by isostatic rebound.
    e. formed by plate movement over a hotspot.

13. The geologic structure of Canada consists of
    a. a central craton and three orogenic belts.
    b. a stable orogenic belt and a craton.
    c. accreted terranes forming a craton and stable orogenic belts to the north.
    d. a central orogenic belt and a hotspot.
    e. a central carbonate platform with four Precambrian fold-and-thrust belts.

14. What is the name of the oldest supercontinent, elements of which are preserved in the Canadian Shield?
    a. Pangaea
    b. Rodinia.
    c. Nuna.
    d. Meguma.
    e. Laurentia.

15. The thickest sequences of Phanerozoic sedimentary rocks can be found
    a. on the Continental Platform.
    b. on the active continental margin of western Canada.
    c. in the Avalon terrane of the Appalachians.
    d. on the Canadian Shield.
    e. in Yukon.

FURTHER RESEARCH

1. Does orogeny happen on other planets?

2. What factors control the flooding of the craton by marine environments?

3. British Columbia has several deep valleys separated by steep mountain ranges. What geologic processes led to this type of topography?

4. What is a *jokulhlaup*?

5. Why do continents rift apart?

ONLINE RESOURCES

Explore more about the geology of Canada and North America on the following websites:

Dr. Ron Blakey, a professor of Geology at Northern Arizona University, maintains a website showing global reconstructions of paleogeography through time. Some of these are used in this chapter. Explore the best global reconstructions at:
http://jan.ucc.nau.edu/~rcb7/RCB.html

The CBC series, Geologic Journey *visits geologic locations across the country:*
www.cbc.ca/geologic/field_guide/table_of_contents.html

The book Four billion years and counting: Canada's geological heritage *presents a more detailed overview of the geology of Canada:*
www.earthsciencescanada.com/4by

Natural Resources Canada provides extensive information about the geology of Canada, as well as maps, in the Atlas of Canada:
http://atlas.nrcan.gc.ca/site/english/maps/geology.html

The U.S. Geological Survey provides "A Tapestry of Time and Terrain":
http://tapestry.usgs.gov/na-info.html

Additional animations, videos, and other online resources are available at this book's companion website:
www.wiley.com/go/fletchercanada

This companion website also has additional information about WileyPLUS *and other Wiley teaching and learning resources.*

CHAPTER 16
GLOBAL WARMING

Chapter Contents and Learning Objectives (LO)

16-1 "Global change" refers to changes in environmental processes affecting the whole Earth.
LO 16-1 *Describe the cause of global warming.*

16-2 Heat circulation in the atmosphere and oceans maintains Earth's climate.
LO 16-2 *Describe how heat distribution on Earth depends on circulation in the oceans and atmosphere.*

16-3 The greenhouse effect is at the heart of Earth's climate system.
LO 16-3 *Explain what causes the amount of greenhouse gas in the atmosphere to change.*

16-4 The global carbon cycle describes how carbon moves through natural systems.
LO 16-4 *List the components of the global carbon cycle.*

16-5 Climate modelling improves our understanding of global change.
LO 16-5 *Describe the Sun's role in climate change.*

16-6 Human activities have increased the amount of carbon dioxide in the atmosphere.
LO 16-6 *Describe the human activities that contribute to greenhouse gas production.*

16-7 Earth's atmospheric temperature has risen by about 0.8°C in the past 100 years.
LO 16-7 *Identify some patterns and consequences of global warming.*

16-8 Global warming leads to ocean acidification and warming, melting of glaciers, changes in weather, and other impacts.
LO 16-8 *Describe how global warming is changing the ocean and the world's glaciers.*

16-9 Several international efforts are attempting to manage global warming.
LO 16-9 *Describe "carbon quotas" and "emission trading."*

GEOLOGY IN OUR LIVES

Global warming is changing Earth's climate, leading to rising sea levels, changes in weather, and ecosystem impacts. Global warming poses an extraordinary challenge to our economy and our environment. The weight of scientific evidence indicates that human activities play a significant role in global warming by releasing greenhouse gases into the atmosphere. Only by limiting greenhouse gas production can the worst future impacts of global warming be avoided. If we want to maintain a livable planet, we must reduce greenhouse gas production.

(?) Among scientists there is little controversy about the cause or existence of global warming. But within the public media there continues to be wide debate. Why is this?

16-1 "Global Change" Refers to Changes in Environmental Processes Affecting the Whole Earth

LO 16-1 *Describe the cause of global warming.*

The Earth is forever changing, through natural processes over geologic time or through our more recent activities (**FIGURE 16.1**). At present, it is estimated that on a typical day, we lose about 320 km² of rainforest to logging and agriculture in particular, over 60 km² of land to encroaching deserts, and potentially over 50 species to extinction. The world's human population increases by 200,000; 95 billion tonnes of carbon dioxide (CO_2) are added to the atmosphere from fossil fuels and cement production; and we burn an average of 85 million barrels of oil (13 billion litres). By the end of the day, Earth is a little hotter and its waters are more acidic.

Here are some more important facts about global environmental changes:

- The average atmospheric carbon dioxide concentration is about 400 parts per million (ppm) (June 2013), growing at an average annual rate of about 2 ppm over the last decade. This is the highest concentration in 15 million years, as measured in cores of glacier ice and seafloor sediments. Fifteen million years ago sea level was 23 m to 36 m higher and global temperatures were 2.8°C to 5.5°C warmer, which might mean that Earth has not equilibrated with the present increase in carbon dioxide.
- The National Oceanic and Atmospheric Administration in the United States produced a "State of the Climate" report in 2009 that indicated that since modern record keeping began, the period from 2000 to 2009 was the hottest decade on record. However, the next year, 2010, was the warmest on record globally, and 2012 was the warmest year in the United States on record.
- Except for a levelling off between the 1940s and 1970s, Earth's surface temperature has increased since 1880; in total, the average global temperature has increased by about 0.8°C.
- The world's population reached a record 7 billion people in late 2011. The United Nations predicts that global population could exceed 8 billion by 2025.

What do these startling facts have to do with geology? It is geologists who are experts in the natural processes that maintain the health of Earth's environment, and it is geologists to whom experts in other fields turn for factual guidance on how to manage natural resources so that they are not lost forever.

Global Warming

You have probably heard the term **global warming** both in school and through media such as newspapers, websites, and television. This term refers to an increase in the average temperature of Earth's surface (including the oceans). In fact, the decade 2000 to 2009 was the warmest recorded since 1880 when the instrumental record of global temperature began. The year 2010 was the hottest ever measured, breaking the record set in 2005. Several published reports have concluded that Earth's average surface temperature today is most likely the highest of the past 1,300 years. Researchers have also found that the rate of **sea-level rise,** a reflection of heat stored in the ocean and melting ice on land, has accelerated, more than doubling in recent decades. Scientific evidence indicates that natural influences cannot explain all the observed warming and human activities most likely play a significant role.

Climate and Weather

Weather is the short-term state of the atmosphere at a given location. It affects the well-being of humans, plants, and animals and the quality of our food and water supply. Weather is somewhat predictable because of our understanding of Earth's global climate patterns. *Climate* is the long-term average weather pattern in a particular region and is the result of interactions among land, ocean, atmosphere, ice, and the biosphere. Climate is described by many weather elements, such as temperature, precipitation, humidity, sunshine, and wind. Both climate and weather result from processes that accumulate and move heat within and between the atmosphere and the oceans.

Global climate change is any large-scale change in climate over time, whether natural or as a result of human activity. Changes in global climate are particularly important because they involve the natural balance of Earth's oceans, land surface, and atmosphere, leading to large-scale changes in ice and vegetation cover, sea level, weather patterns, and human welfare. Many experts believe that

© shaunl/Vetta Collection/iStockphoto.com

FIGURE 16.1 Natural environments sustain life on Earth. Humans, in turn, must employ critical thinking to minimize negative impacts on the environment.

 What negative impacts on the environment are you responsible for?

understanding global climate change, and taking action to manage it, is humanity's greatest challenge in the 21st century.

Global Change

The circulation of Earth's atmosphere and oceans links together the planet's living organisms and environments, from soil at the equator to ice at the poles. Even though Earth is 40,075 km in circumference and has a surface area of 509,600,000 km², the poles and tropics, deserts and forests, and continents and oceans are all *connected* by global processes. The term "global change" refers to changes in these processes on the scale of the whole Earth.

Our planet is dynamic and constantly changing, as it has changed throughout its 4.6-billion-year history. For most of Earth's history, those changes have been natural, and many of them have been enormous (such as the movement of continents and the evolution of life). The natural processes that cause global climate change include plate tectonics, volcanic eruptions, ocean circulation, cycles of solar output, extraterrestrial impacts, and variations in Earth's orbit. Global climate change is also caused by human activities.

On modern Earth, human activities have caused significant global changes in land use, air and water quality, and the abundance of natural resources, particularly in the last two centuries. There is scientific consensus that human activities are also altering Earth's climate, largely due to increasing levels of carbon dioxide and other **greenhouse gases** (atmospheric gases that trap heat and therefore warm the atmosphere) released by the burning of *fossil fuels,* and to reduced uptake of carbon dioxide resulting from deforestation. Studies indicate that the climate change observed during the 20th and early 21st centuries is due to a combination of changes in solar radiation, volcanic activity, and land use, and increases in atmospheric greenhouse gases. Of these, greenhouse gases are the dominant long-term influence.

Why Study Global Change?

The distribution of heat on our planet is important in every region and every environment on Earth (FIGURE 16.2). The total amount of heat and its variation across the planet surface drives global winds that circulate the atmosphere and control regional weather patterns, growing seasons, and living conditions. Earth is at the right distance from the Sun (about 148 million kilometres), with the right combination of gases in its atmosphere, and with water covering more than 70 percent of the planet's surface, to allow for the origin and evolution of life and the resources necessary to sustain life. So far as we know, no other planet in our Solar System has the thermal, physical, and chemical conditions that allow life to exist. These conditions are what make our blue planet unique and habitable. By studying global change, we gain the knowledge to sustain and enhance this livable condition rather than counteract it.

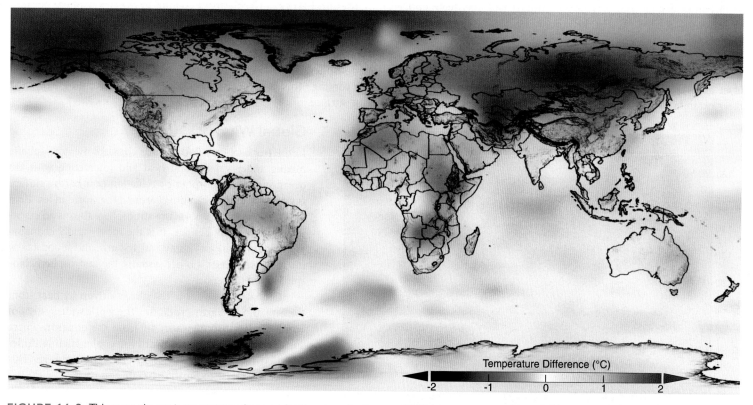

Temperature Difference (°C)

-2   -1   0   1   2

FIGURE 16.2 This map shows temperature changes for the decade January 2000 to December 2009, the warmest on record, relative to the 1951 to 1980 average. Warmer areas are in red, cooler are in blue. The largest temperature increases occurred in the Arctic and a portion of Antarctica. It will be interesting to see this map at the end of this decade. *Source:* NASA, www.nasa.gov/topics/earth/features/temp-analysis-2009.html

(?) What are some consequences of the Arctic and Antarctic warming faster than the rest of the world?

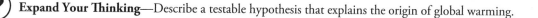

(?) Expand Your Thinking—Describe a testable hypothesis that explains the origin of global warming.

16-2 Heat Circulation in the Atmosphere and Oceans Maintains Earth's Climate

LO 16-2 *Describe how heat distribution on Earth depends on circulation in the oceans and atmosphere.*

Climate change is the product of changes in the accumulation and movement of heat in the oceans and atmosphere. To understand both natural and human influences on global climate, we must explore the physical processes that govern heat movement in the atmosphere and oceans.

The Atmosphere

The atmosphere is the envelope of gases that surrounds Earth and extends from the surface to an altitude of about 145 km, although the transition from the atmosphere to space is considered gradational (**FIGURE 16.3**). Around the world, the composition of the atmosphere is similar, but when looked at vertically, the atmosphere is not a uniform blanket of air. It can best be described as several layers, each with distinct properties, such as temperature and chemical composition. The red line in Figure 16.3 shows how atmospheric temperature changes with altitude.

In the layer nearest Earth, the *troposphere,* or "weather zone," the air becomes colder with increasing altitude; you may have noticed this if you have ever hiked in the mountains. This layer extends to an altitude at which the temperature is about –55°C, which is about 8 km in the polar regions and up to nearly 17 km above the equator. Above the troposphere is the *stratosphere,* where the protective ozone (O_3) layer absorbs much of the Sun's harmful ultraviolet radiation. This layer extends to an altitude of about 50 km; it becomes hotter with

increasing altitude because it absorbs ultraviolet radiation. The ozone layer is vital to the survival of plants and animals on Earth because it blocks the intense solar radiation that damages living tissue. Above the stratosphere is the *mesosphere,* which extends to an altitude of about 80 km. Like the troposphere, this layer grows cooler with increasing altitude. Finally, the highest layer is the *thermosphere,* which gradually merges with space. Temperatures *increase* with altitude in the thermosphere because it is heated by cosmic radiation from space. The upper part of the thermosphere is called the *ionosphere,* which is the part of the atmosphere that is ionized by solar radiation and may extend to an altitude of 1,000 km.

The boundaries between these layers are called "pauses." For example, the tropopause is the boundary between the troposphere and the stratosphere.

Very little vertical mixing of gases occurs in the atmosphere. This means that one layer can be warming while at the same time another is cooling. For example, global warming in the troposphere, the layer closest to Earth's surface, causes cooling in the stratosphere because as more heat is trapped in the lower atmosphere, less heat reaches the upper atmosphere. In fact, some parts of the upper atmosphere appear to be cooling at a rate of about 0.05°C per decade, even as the troposphere warms. To an observer in space, Earth would appear to be cooling. This is because greenhouse gases are trapping heat near the surface.

Global Winds

In the troposphere, global winds circulate the atmosphere, mixing water vapour and heat and interacting with the ocean surface. Atmospheric circulation near Earth's surface is vigorous, and air can travel around the world in less than a month (**FIGURE 16.4**). The atmosphere thus is an essential component of climate. It is the most rapidly changing and dynamic of Earth's physical systems, and it constantly interacts with Earth's other systems: the hydrosphere, biosphere, and lithosphere.

Global circulation is essentially driven by heat from the Sun and by Earth's rotation. The worldwide system of winds that transport warm air from the equator, where solar heating is greatest, toward the cooler high latitudes is called the *general circulation of the atmosphere.* This pattern gives rise to Earth's climate zones.

The general circulation consists of a number of *cells* (distinct regions of air flow), such as *Hadley cells* (shown in Figure 16.4). The impact of sunlight is strongest nearest the equator. Air that has been heated there rises and

 Why does Earth appear to be cooling to an observer in space?

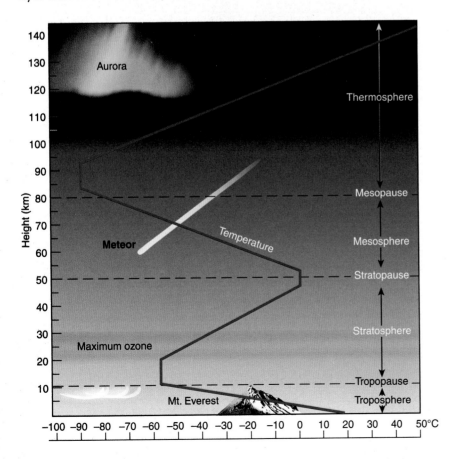

FIGURE 16.3 Vertical structure of Earth's atmosphere.

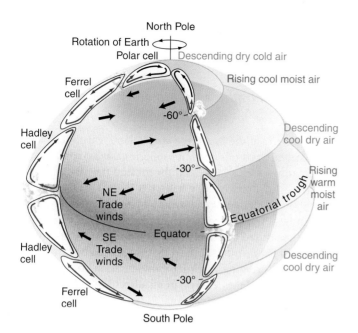

FIGURE 16.4 General circulation of the troposphere. The black arrows indicate the dominant wind direction at Earth's surface.

 How would climate be different if there were no general circulation of the atmosphere?

spreads out to the north and south, releasing water as it cools. After cooling, the air sinks back to Earth's surface within the subtropical climate zone between latitudes 25° and 40°. This cool, and now drier, descending air stabilizes the atmosphere at these sites and, as a result, many of the world's desert climates are found in the subtropical climate zone under these dry descending air masses (discussed more in Chapter 21, "Deserts and Wind"). Surface air from subtropical

regions returns toward the equator to replace the rising air, thus completing the cycle of air circulation within the Hadley cell. This flow of air, modified by the Coriolis effect caused by Earth's rotation, creates what are known as the **trade winds.** "Trade winds" were named several centuries ago by ships carrying trade goods around the world that found these winds reliable for planning their routes. North of the Hadley cell is the *Ferrel cell,* and north of this is the *Polar cell,* both of which operate in a fashion similar to the Hadley cell.

Global Water

The **hydrosphere** consists of all the liquid water on Earth—in the oceans, rivers, and lakes. Like the atmosphere, the oceans are characterized by a pattern of circulation that transports heat around the globe and affects Earth's climate (**FIGURE 16.5**). Global ocean **thermohaline circulation** is like a giant conveyor belt driven by the temperature ("thermo") and salinity ("haline") of ocean water.

Thermohaline circulation starts in the North Atlantic Ocean. When the warm surface water of the *Gulf Stream* (a current running north along the eastern coast of North America) reaches the cold polar North Atlantic, it cools and sinks. During its journey, Gulf Stream water has been evaporating (making it saltier) and cooling; therefore, it has become denser and sinks readily. When it reaches the deep ocean, it travels southward as a cold, deep, salty current called the *North Atlantic Deep Water.* In the Southern Hemisphere, it is joined by cold, salty water from Antarctica and enters the Indian Ocean. The current flows eastward past Australia into the Pacific Ocean, eventually rising in the North Pacific. There it becomes a warm surface current, called the *Equatorial Current,* and flows westward around Africa into the Atlantic. It may take up to 1,600 years for water to complete the entire global cycle. On its journey, the current transports heat, solid particles, and dissolved compounds (such as carbon dioxide) around the globe. The state of ocean circulation thus has a large impact on Earth's climate.

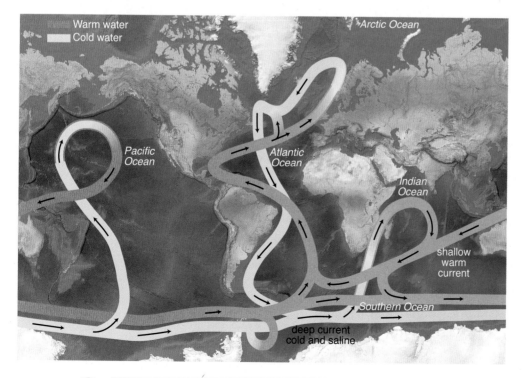

 Has global warming heated the entire ocean?

FIGURE 16.5 Thermohaline circulation carries heat around the globe and affects climate. The warm surface water in the North Atlantic Ocean is the Gulf Stream, which eventually cools and becomes saltier and denser forming the deep, southward-flowing North Atlantic Deep Water.

Expand Your Thinking—What is the impact of thermohaline circulation on the climate of northern Europe?

16-3 The Greenhouse Effect Is at the Heart of Earth's Climate System

LO *16-3 Explain what causes the amount of greenhouse gas in the atmosphere to change.*

The **greenhouse effect** is the atmosphere's natural ability to allow much of the incoming solar radiation to pass through to the surface of the Earth, but to store at least some of the heat radiated from Earth. The heat absorbed by certain gases (such as water vapour [H_2O], carbon dioxide [CO_2], and methane [CH_4]) maintains Earth's *surface temperature* at an average (and comfortable) 14°C. In contrast, as we learned in Chapter 2 ("Solar System"), a runaway greenhouse effect on Venus is responsible for raising the temperature on its surface to a scalding 477°C, which is one reason why life is unlikely to exist on that planet. Earth's greenhouse effect is more moderate and, hence, conducive to life, which depends on the presence of liquid water. Without the greenhouse effect, the average temperature of Earth's atmosphere would be –18°C, below the freezing point of water.

Among all the known planets, conditions conducive to life exist only on Earth. They exist because of Earth's *heat budget,* a complex balance of heat distribution and exchange between the atmosphere, hydrosphere, lithosphere, and biosphere (**FIGURE 16.6**).

The Heat Budget

The Sun's radiation is balanced at the top of the atmosphere so that the amount entering the atmosphere equals the amount leaving it. The total incoming solar energy is about 340 W/m² (watts of energy per square metre) at the top of the atmosphere. Part of this solar energy is absorbed by clouds and atmospheric gases and part is reflected by clouds, atmospheric gases, and the ground (i.e., Earth's land and water surfaces). Approximately half (170 W/m²) is absorbed by the ground. Some of the energy absorbed by the ground is re-radiated upward; some is transferred to the atmosphere as *sensible heat* (heat that can be measured by a thermometer); and some is transferred to the atmosphere as *latent heat* (heat that is released by processes such as evaporation, freezing, melting, condensation, or sublimation). The atmosphere radiates energy in all directions. When balance is achieved in the atmosphere, the total radiation leaving the top of the atmosphere equals the 340 W/m² received from the Sun. However, at present, it is thought that increasing levels of greenhouse gases are upsetting this balance.

Most of the light energy that penetrates Earth's atmosphere is *short-wave ultraviolet (UV) radiation*, which is mostly absorbed by the protective ozone layer in the stratosphere. Of the radiation that reaches the troposphere, more than half is reflected back into space. About 45 percent reaches Earth's surface and is absorbed by the oceans and land, then re-radiated back into the atmosphere in the form of *long-wave infrared (IR) radiation*. This long-wave IR radiation is absorbed by the greenhouse gases in the atmosphere. The heat

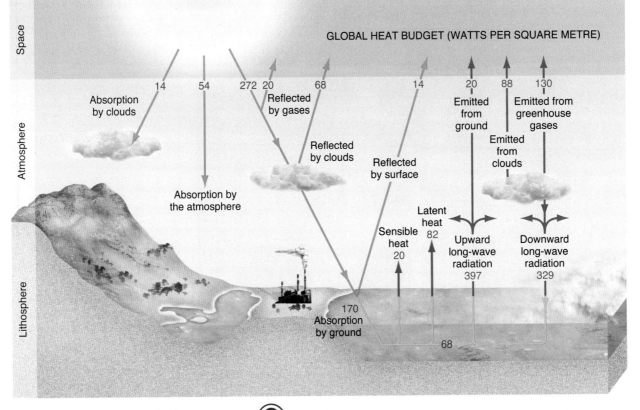

FIGURE 16.6 Earth's heat budget.

(?) Why types of radiation do the greenhouse gases trap?

absorbed by these gases warms the surface and the lower atmosphere, maintaining a surface temperature that sustains life.

Earth's atmosphere is composed mostly of oxygen and nitrogen, but neither of these gases absorbs infrared energy, so neither plays a role in warming Earth. Six principal greenhouse gases in Earth's atmosphere absorb long-wave radiation and keep the planet warm: water vapour, carbon dioxide, ozone, methane, fluorocarbons, and nitrous oxide (N_2O). Combined, these gases make up about 1 percent of the atmosphere. Because greenhouse gases are efficient at trapping long-wave IR radiation from Earth's surface, even their small percentage is enough to keep temperatures in the ideal range for liquid water (and life) to exist on Earth.

Carbon Dioxide (CO_2)

The amount of carbon dioxide in the atmosphere has varied significantly during Earth's history and began doing so long before modern humans inhabited the planet. Natural sources of carbon dioxide include volcanic outgassing, animal respiration, and decay of organic matter. Scientists can measure the history of atmospheric carbon dioxide concentrations by analyzing air bubbles trapped in ice cores (FIGURE 16.7) and employing other techniques using fossils (such as marine sediments composed of fossil plankton or the number of stomata observed on well-preserved fossil tree leaves). These methods have helped scientists understand long-term trends in carbon dioxide variability and global climate change.

Additional carbon dioxide is added to the atmosphere by human activities—in particular, the burning of fossil fuels (oil, natural gas, and coal), solid waste, and wood (such as during deforestation). These *anthropogenic emissions* are at the centre of research on

global warming. Although the sources and heat-trapping properties of greenhouse gases are undisputed, scientists are not sure how Earth's climate responds to increasing concentrations of the various gases. There is wide consensus among scientists around the world, particularly those involved with the Intergovernmental Panel on Climate Change (IPCC), that if anthropogenic emissions of carbon dioxide continue to rise, average global temperatures will increase, perhaps by as much as 6°C, by the end of this century. Most scientists also agree that anthropogenic emissions have been the dominant influence on climate change over the past 50 years, overwhelming natural causes. This is often called the enhanced greenhouse effect, as our activities are adding greenhouse gases to the atmosphere and thus increasing the atmosphere's ability to trap heat.

In the geologic past, climate changes occurred naturally (as we discuss more thoroughly in Chapter 17, "Glaciers and Paleoclimatology"). In the past half million years, carbon dioxide concentration remained between about 180 ppm during *glacial periods* and 280 ppm during *interglacial periods*, such as today (Figure 16.7). But carbon dioxide content has been much greater in other periods of Earth's history. Estimates of carbon dioxide content during the Phanerozoic Eon are based on the chemistry of fossilized soils, fossil plants, and fossil shells of plankton. These estimates indicate that concentrations as high as 1,000 to 4,000 ppm may have occurred for sustained periods; twice this level may even have been reached. The cause of such high levels is controversial: Episodes of extreme global volcanism, changes in land surface area due to plate tectonics, absence of polar ice, mountain building, and other mechanisms have all been suggested. However, it is clear that the level of only 180 ppm during glaciations is not far from the lowest that has ever occurred since the rise of macroscopic life in the last half-billion years.

In the 150 years since the **Industrial Revolution** in the mid-1800s, humans have altered Earth's environment through agricultural, urbanization, and industrial practices. The growth of the human population and activities such as deforestation and burning of

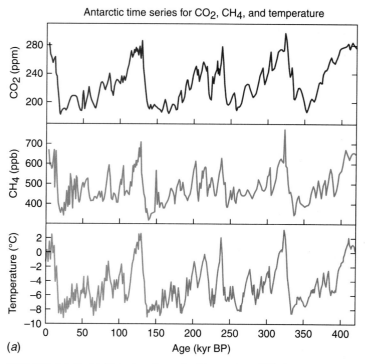

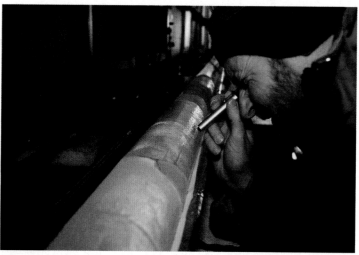

FIGURE 16.7 *(a)* Global carbon dioxide content (CO_2 in parts per million [ppm]), methane content (CH_4 in parts per billion [ppb]), and temperature (in °C) over the past 400,000 years have been measured using air bubbles trapped in ice in Antarctica. *(b)* Scientists drill ice cores on mountain glaciers, as well as in Greenland and Antarctica, to obtain evidence of past atmospheric composition.

What climate processes are reflected in the peaks and valleys of the ice core record?

fossil fuels have affected the mixture of gases in the atmosphere. The amount of carbon dioxide in the atmosphere prior to the Industrial Revolution was about 280 ppm. (We know this from ice cores.) Today the concentration of carbon dioxide of about 400 ppm, as measured at the Mauna Loa Observatory in Hawaii, is higher than at any other time in the past 15 million years.

Although carbon dioxide is not the most effective absorber of heat compared to other greenhouse gases, it is one of the most abundant, and once it is in the atmosphere, it stays for a long time (a century or more). Atmospheric carbon dioxide concentrations collected at the Mauna Loa Observatory every day since 1958 show seasonal oscillations superimposed on a long-term increase in carbon dioxide in the atmosphere (FIGURE 16.8). (This is the longest continuous record of this type of data anywhere in the world, the collection of which was initiated by Dr. Charles Keeling, a geochemist and oceanographer from the United States.) This increase is attributed to the activities of humans since the Industrial Revolution, primarily the burning of fossil fuels and deforestation. Of all the greenhouse gases released by human activities, carbon dioxide is the largest individual contributor to the enhanced greenhouse effect, accounting for about 60 percent. Unfortunately, the increase in global emissions of carbon dioxide from fossil fuels over the past 5 years was four times greater than the increase over the preceding 10 years. Rather than recognizing that the enhanced greenhouse effect is a potentially dangerous trend that can be curtailed by decreasing our emissions of heat-trapping gases, we are accelerating our carbon-burning activities.

Methane (CH₄)

Natural sources of methane include microbial and insect activity in wetlands and soils, wildfires, and the release of gases stored in ocean sediments. The current global atmospheric concentration of methane is 1.78 ppm, more than double what it was prior to the Industrial Revolution. Methane levels increased steadily in the 1980s, but the rate of increase slowed in the 1990s and was close to zero from 2000 to 2007. Researchers attribute this lull to a temporary decrease in emissions during the 1990s related to the decline of industry and farming when the former Soviet Union collapsed, along with a slowdown in wetland emissions during prolonged droughts. Scientists warn that with methane levels on the rise again, a more typical rate of increase will have a significant impact on climate.

Methane is responsible for about 20 percent of global warming, second only to carbon dioxide. About 60 percent of annual methane emissions come from anthropogenic sources. Human activities that release methane into the atmosphere include deforestation (burning logged tracts of forest), mining and burning fossil fuels, processing human wastes, raising livestock, and cultivating rice in paddies (agricultural wetlands). Methane is also trapped in ice, glaciers, seafloor sediment, and tundra; as melting occurs, the gas is released from these frozen sources to the atmosphere. Unlike carbon dioxide, methane is destroyed by reactions with other chemicals in the atmosphere and soil, so its atmospheric lifetime is only about 10 to 15 years.

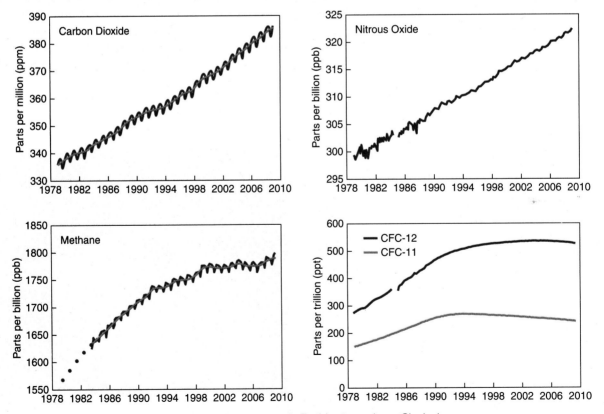

FIGURE 16.8 Concentrations of the most important greenhouse gases in Earth's atmosphere. Clockwise from upper left: Levels of carbon dioxide and nitrous oxide continue to climb. Levels of chlorofluorocarbons (CFCs) have declined since an international agreement was signed and implemented in 1987 to reduce their production and use. The concentration of methane slowed late in the 20th century due to droughts and a temporary decline in industrial emissions but has since returned to its previous pattern of steady increases. *Source:* Earth System Research Laboratory, U.S. National Oceanic and Atmospheric Administration, www.esrl. noaa.gov/gmd/aggi.

 What is the predominant cause of the growing abundance of greenhouse gases?

Nitrous Oxide (N₂O)

The main natural source of nitrous oxide is microbial activity in swamps, soil, the ocean surface, and rainforests. Human sources of this greenhouse gas include fertilizers, industrial production of nitric acid and nylon, and burning fossil fuels and solid waste. The current concentration of N_2O is 323 ppb (parts per billion). It has increased by 16 percent since the beginning of the Industrial Revolution and is responsible for 4 percent to 6 percent of global warming.

Ozone (O₃)

Ozone in the stratosphere absorbs and protects Earth from harmful UV radiation, but ozone in the troposphere is a pollutant. Because it absorbs heat in the lower atmosphere, it is considered a greenhouse gas. Natural sources of ozone include chemical reactions that occur between carbon monoxide (CO), hydrocarbons, and nitrous oxides, as well as lightning and wildfires. Human activities increase ozone concentrations indirectly by emitting pollutants that are "precursors" of ozone, including carbon monoxide, sulphur dioxide (SO_2), and hydrocarbons that result from the burning of biomass and fossil fuels. Tropospheric ozone is a strong absorber of heat, but it does not stay in the atmosphere for long, only a few weeks to a month or two. Nonetheless, its concentration is increasing at a rapid rate.

Fluorocarbons

A number of very powerful heat-absorbing greenhouse gases in the atmosphere do *not* occur naturally. They include chlorofluorocarbons (CFCs), hydrofluorocarbons (HFCs), perfluorocarbons (PFCs), and sulphur hexafluoride (SF_6), all of which are generated by industrial processes. Chlorofluorocarbons are used as coolants in air conditioning (the chemical known as Freon is a CFC), aerosol sprays, and the manufacture of plastics and Styrofoam. Chlorofluorocarbons did not exist on Earth before humans created them in the 1920s. They are very stable compounds, have long atmospheric lifetimes, and are now abundant enough to cause global changes in atmospheric chemistry, the biosphere, and climate.

Fluorocarbons contribute to warming by enhancing the greenhouse effect in the troposphere. Chlorofluorocarbons also chemically react with and destroy ozone in the stratosphere, creating the "ozone hole" over the Southern Hemisphere. Stratospheric ozone is important because it blocks 95 percent of the Sun's harmful UV rays from reaching Earth's surface. Loss of ozone allows more UV radiation to reach Earth's surface, where it is harmful to living organisms, influences photosynthesis, and contributes to global warming.

The good news is that many of the effects of CFCs are reversible. Thanks to the *Montreal Protocol,* signed by 27 nations in 1987, CFCs were recognized as dangerous pollutants, and their production and use was significantly reduced. Canada, one of the signers of the protocol, has been working to make sure that all systems that used CFCs are retrofitted to eliminate their use, and that all stocks are disposed of safely so that they are not released into the atmosphere. Chlorofluorocarbons already in the atmosphere have lifetimes of 75 to 150 years, so ozone depletion may continue for decades.

Water Vapour (H₂O)

Earth's climate is able to support life because of the greenhouse effect and the availability of water. Water vapour (a gas) is a key component of both of these processes. It is the most abundant greenhouse gas and is an important link between Earth's surface and its atmosphere. The concentration of water in the atmosphere is constantly changing, controlled by the balance between evaporation and precipitation. In fact, the average water molecule spends only about nine days in the air before precipitating back to Earth's surface.

Water vapour is a powerful natural greenhouse gas. The amount of water in the atmosphere is not directly affected by human activities but does respond to feedbacks related to human activities. A **climate feedback** is a secondary change that occurs within the climate system in response to a primary change, which can be either *positive* or *negative*. For example, as the temperature of Earth's surface and atmosphere increases, the atmosphere is able to "hold" more water vapour. That is, more evaporation occurs. The additional water vapour, acting as a greenhouse gas, absorbs more energy and causes further warming. This is a *positive feedback* in the sense that warming leads to more water vapour, which in turn leads to more warming, and so on. However, increased water vapour may cause increased cloud cover, which reflects more sunlight back to space. This condition would favour global cooling, a *negative feedback*. Which of these two opposite effects dominates is unknown.

Cooling Aerosols

Burning fossil fuels not only produces heat-trapping gases, it produces cooling *aerosols*. Aerosols are fine solid particles or liquid droplets suspended in the atmosphere that "scatter" (reflect) light. This behaviour increases Earth's albedo, the tendency to reflect sunlight and thus have a cooling effect. Most anthropogenic aerosols are sulphates (SO_4) that are released during the combustion of coal and petroleum. So much aerosol production accompanied industrial growth in the middle of the 20th century that global cooling occurred between the 1950s and the 1970s.

Volcanic eruptions can have the same effect. They blast huge clouds of particles and gases (including sulphur dioxide) into the atmosphere. In the stratosphere, sulphur dioxide converts to tiny persistent sulphuric acid (sulphate) particles that reflect sunlight. Particularly large eruptions can produce widespread cooling. For example, Mount Pinatubo in the Philippines erupted in June 1991 and cooled the planet nearly an entire degree, temporarily offsetting global warming.

Global Warming Potential

Molecule for molecule, some greenhouse gases are stronger than others. Each differs in its ability to absorb heat and in the length of time it remains in the atmosphere. The ability to absorb heat and warm the atmosphere is expressed by a gas's *global warming potential* (GWP), usually compared to carbon dioxide over some given time horizon. Methane absorbs 21 times more heat per molecule than carbon dioxide. Nitrous oxide absorbs 270 times more heat per molecule than carbon dioxide. Fluorocarbons are the most heat absorbent, with GWPs that are up to 30,000 times stronger than those of carbon dioxide. Understanding the GWPs of various gases is a useful way to describe the impact of human emissions and identify what changes can have the most positive impact.

 Expand Your Thinking—Identify two natural geologic processes that can change the greenhouse effect on Earth. Describe how they work.

16-4 The Global Carbon Cycle Describes How Carbon Moves through Natural Systems

LO 16-4 *List the components of the global carbon cycle.*

Natural movement of nutrients, various dissolved compounds, and certain elements through living and non-living systems are essential for life on Earth. Most of the chemical compounds that occur on our planet move between the atmosphere, hydrosphere, lithosphere, and biosphere and through *ecosystems* (communities of animals and plants and the environment they inhabit), which combine aspects of all these "spheres." Because of the many biologic, geologic, and chemical exchanges and reactions involved, this movement is known as *global biogeochemical cycling.* The key elements required for life move through biogeochemical cycles; they include oxygen, carbon, phosphorus, sulphur, and nitrogen. The rates at which elements and compounds move between places where they are temporarily stored *(reservoirs)* and where they are exchanged *(processes)* can be measured directly and modelled using computer programs.

One of the most important cycles that affects global climate is that of the element carbon (**FIGURE 16.9**). The **global carbon cycle** describes the many forms that carbon takes in the reservoirs and processes for carbon on Earth. These include:

- *rocks* in the lithosphere, such as limestone and carbon-rich black shale
- *hydrocarbon deposits* in the lithosphere, such as coal, oil, and natural gas
- *gases* in the atmosphere, such as carbon dioxide and methane
- *carbon dioxide* dissolved in water (oceans and fresh water)
- *organic matter* in the biosphere, such as the simple carbohydrate glucose ($C_6H_{12}O_6$), found in plants and animals

Most of the carbon on Earth is contained in the rocks of the lithosphere where it has been deposited slowly over millions of years in

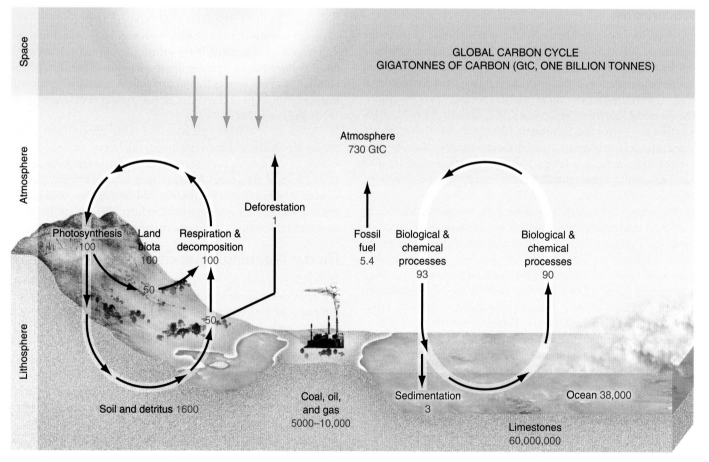

FIGURE 16.9 Carbon is cycled through Earth's atmosphere, hydrosphere, lithosphere, and biosphere. The values given here are global carbon reservoirs in gigatonnes (1 Gt C = 1,012 kg). Annual exchange and accumulation rates are in gigatons of carbon per year (Gt C/year), calculated over a decade. Two other geological processes that recycle carbon include the weathering of silicate rocks which uses CO_2 at a rate of approximately 0.2 Gt C/yr, and volcanic activity that releases CO_2 into the atmosphere at a rate of about 0.1 Gt C/yr. These are unimportant over short time scales, but very important on geological time scales.

(?) What is the largest reservoir of carbon?

the form of dead organisms (mostly plankton). Carbon is stored in the lithosphere in two forms: (1) *Oxidized carbon* is buried as carbonate (such as limestone, which is composed of calcium carbonate [$CaCO_3$]); (2) *reduced carbon* is buried as organic matter (such as dead plant and animal tissue).

Most of Earth's carbon is contained in limestone, which provides effective long-term storage for carbon that has been taken from the atmosphere and transferred to the lithosphere. The process of "storing" carbon in this way occurs in several steps. Gaseous carbon dioxide in the atmosphere is constantly entering the surface waters of the hydrosphere, where it dissolves into the *bicarbonate ion* (HCO_3^-) in this (simplified) chemical reaction:

$$2CO_2 + 2H_2O \longrightarrow 2HCO_3^- + 2H^+$$

The bicarbonate ion combines with dissolved calcium (Ca^{2+}) in sea water to form calcium carbonate ($CaCO_3$, calcite) in a reaction called *calcification*:

$$2HCO_3^- + Ca^{2+} \longrightarrow CaCO_3 + CO_2 + H_2O$$

In the first reaction, two molecules of CO_2 are taken from the atmosphere; in the second, only one molecule of CO_2 is released. This means that as limestone forms, some atmospheric carbon dioxide is trapped and buried in the most stable of forms: rock. Coral and other marine organisms, such as molluscs, some types of algae, and the plankton *foraminifera*, are excellent calcifiers. Most calcification occurs in the ocean, but some also occurs in fresh water. Have you ever seen stalagmites and stalactites in caves? They are made of limestone that was formed by the same chemical calcification reaction, but without the help of plants and animals (**FIGURE 16.10**).

The movement of carbon does not end there. Limestone may eventually be broken down by weathering, as you learned in Chapter 7 ("Weathering"). This process consumes atmospheric carbon dioxide in a chemical reaction that is essentially the reverse of calcification:

$$CaCO_3 + CO_2 + H_2O \longrightarrow Ca^{2+} + 2HCO_3^-$$

The weathering of silicate rocks (which account for most of the rocks in the lithosphere) also uses carbon dioxide, and in fact, this reaction provides the dissolved calcium that combines with bicarbonate ions in sea water to form limestone. (See below, although this is a simple reaction using a rare calcium-bearing silicate mineral, called wollastonite. If you substitute plagioclase feldspar, for example, the general reaction is similar although more complex.) Overall, it is the weathering of silicate rocks that controls the global CO_2 levels on geologic time scales:

$$CaSiO_3 + 2CO_2 + H_2O \longrightarrow Ca^{2+} + 2HCO_3^- + SiO_2$$

Cycling of carbon also occurs in the biosphere. Plants and some forms of bacteria can "use" inorganic carbon dioxide and convert it into organic carbon (such as carbohydrates and proteins), which is then consumed by all other forms of life, from zooplankton to humans, through the food chain. During *photosynthesis*, plants remove carbon dioxide from the atmosphere and convert it into organic carbon in plant tissues. Photosynthesis occurs on land in trees, grasses, and aquatic (freshwater) plants and

J. Debru/StockImage/Getty Images, Inc.

FIGURE 16.10 An example of calcification is the formation of stalagmites and stalactites in caves. Cave deposits are composed of travertine, a type of calcium carbonate that is formed by inorganic precipitation in fresh water.

 What weathering process breaks down limestone?

in phytoplankton, algae, and kelp in ocean surface waters that are penetrated by sunlight. The reaction requires sunlight and chlorophyll, and in its simplest form, it can be represented in this way:

$$6CO_2 + 6H_2O \longrightarrow C_6H_{12}O_6 + 6O_2$$
$$\text{carbon dioxide + water} \longrightarrow \text{organic matter + oxygen}$$

Because carbon dioxide is a greenhouse gas, vegetation plays an important role in global climate. Through the process of photosynthesis, plants remove *about 100 billion tonnes* of carbon dioxide from Earth's atmosphere each year. This is about 12.5 percent of the total amount of carbon in the atmosphere.

Some of the organic carbon created by plants during photosynthesis is consumed by animals and transferred through the food chain to higher forms of life. Eventually, the organic matter decays or is used in *respiration*, and the carbon is returned to the atmosphere as gaseous carbon dioxide. Respiration is the reverse of photosynthesis, and it occurs when animals consume organic material to produce the energy they need to live. These organisms (from bacteria to humans) breathe, die, and decay, all processes that convert organic carbon into carbon dioxide, which is released back into the atmosphere. The basic chemical reaction for respiration is:

$$C_6H_{12}O_6 + 6O_2 \longrightarrow 6CO_2 + 6H_2O$$
$$\text{organic matter + oxygen} \longrightarrow \text{carbon dioxide + water}$$

The cycling of carbon through photosynthesis and respiration is so rapid and efficient that all of the carbon dioxide in the atmosphere passes through the biosphere every four to five years.

 Expand Your Thinking—Explain why the carbon cycle is an important area of global warming research.

16-5 Climate Modelling Improves Our Understanding of Global Change

LO 16-5 *Describe the Sun's role in climate change.*

You are probably starting to realize that Earth's climate system is very complex, with cycles and feedbacks and reservoirs (some natural, some related to human activities) that all interact with one another over different lengths of time.

Climate Models

Researchers who study global change often use climate models, called **global circulation models (GCMs)**, which are computer-based

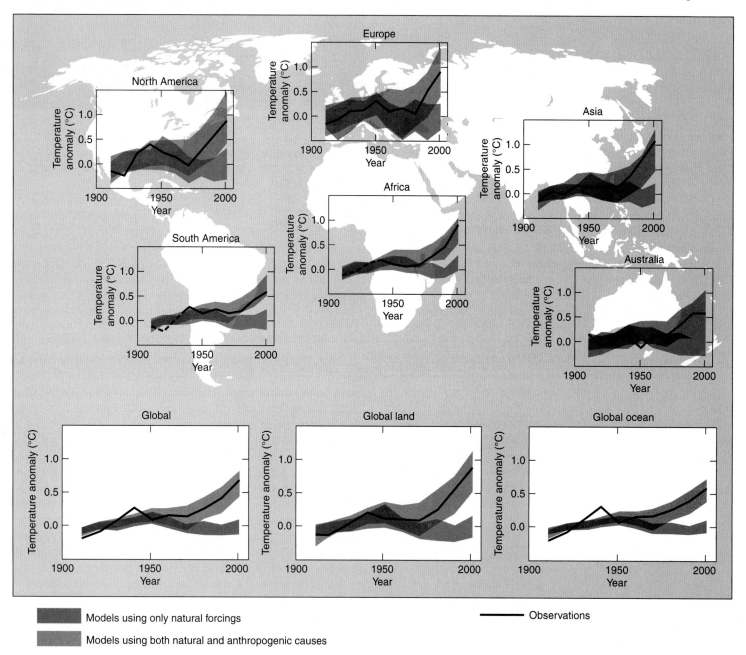

Models using only natural forcings

Models using both natural and anthropogenic causes

Observations

FIGURE 16.11 Simulating surface temperature using mathematical models (blue = natural factors only; red = natural and human factors combined) and comparing the results to measured changes (temperature = black line) can provide insight into causes of major temperature changes. The red band shows that human factors combined with natural factors best account for observed temperature changes. (*Source: Climate Change 2007: The Physical Science Basis. Working GroupIContribution to the Fourth Assessment Report of the Intergovernmental Panel on Climate Change*, Figure SPM.4. Cambridge University Press, Cambridge, UK, and New York, NY, USA.)

 What is the likely cause of the observed climate changes in Figure 16.11?

mathematical programs that simulate the physics (behaviour and interaction) of Earth's oceans, land, and atmosphere. Global circulation models consist of thousands of mathematical equations that are calculated by programs on supercomputers. The supercomputers receive input in the form of data on ocean currents and temperature, the concentration of carbon dioxide and other greenhouse gases in the atmosphere, the amount of sunlight, and the cover of vegetation, ice, and snow that affect the heating of Earth's surface.

Climate models are designed to simulate climate on a range of scales, from global to regional (hundreds of kilometres). But few GCMs regularly tackle climate changes at the local level (tens of kilometres). Most break up the atmosphere into 10 or 20 vertical levels between Earth's surface and outer space, where those cycles, feedbacks, and reservoirs are represented by complex calculations. Many models also include human population growth, economic behaviour, human health, and resource use. The model's output might include predictions of long-term precipitation patterns, or a map of the sea level 100 years from now, or an estimate of future global temperatures at a certain concentration of greenhouse gases.

GCMs can simulate surface temperatures and compare the results to observed changes. In **FIGURE 16.11**, 100 years of observed temperature changes are plotted as a black line. Two different model results are plotted in red and blue. Blue simulations were produced using only natural factors: solar variation and volcanic activity. They do not match the observed temperature changes very well. Red simulations were produced with a combination of natural *and* human factors, including anthropogenic emission of greenhouse gases and other products of industrial pollution. It is clear that the combination of human and natural factors provides the best match with measured temperatures, leading to the conclusion that human pollution by greenhouse gases is at least partially responsible for global warming.

The Sun's Role

Because energy from the Sun drives Earth's heat budget, changes in **insolation** (the amount of energy Earth receives from the Sun) have an effect on global temperature and climate. We observe this daily as temperatures fall after sunset and seasonally as Earth's distance from the Sun and the planet's tilt change temperatures and climate.

Although solar output may vary, on average, about the same amount of solar energy has reached our planet for millions of years. Scientists have measured this *solar constant* and found it to be approximately 1,360 W/m² (watts per square metre) of Earth's surface. In the decades since satellite measurements began in 1979, natural fluctuations of only 0.1 percent in the solar constant have been recorded. Greenhouse gases are about four times more effective at warming Earth's surface than this change in solar output, so most scientists agree that the effect of solar variability on global climate is small by comparison.

Although it does not appear that the Sun's output has varied enough to cause significant climate change, subtle variations in our orbital configuration have led to small but significant changes in the distribution of solar heating on Earth. For example, between 10,000 and 20,000 years ago, when Earth's climate shifted from an *ice age* to today's *interglacial*, global temperatures increased by several degrees and atmospheric carbon dioxide concentration rose from 180 ppm to 280 ppm (**FIGURE 16.12**).

In fact, as we saw in Figure 16.7, for the past several hundred thousand years, Earth's climate has shifted dramatically between glacial and interglacial states approximately every 100,000 years. This history is the result of cyclical changes in the intensity and contrast of the seasons, which in turn are due to changes in the geometry of Earth and the Sun. For instance, mild winters and cool summers tend to promote ice buildup from one year to the next, while hot summers tend to discourage it. Such patterns can lead to ice ages and interglacials.

Earth does not orbit the Sun in a perfect circle. It wobbles, tilts, has eccentricities, and other irregularities due to the changing gravitational influence of other planets. These cyclical changes in Earth–Sun orbital geometry do not affect the total amount of solar energy reaching Earth. Rather, they control the *location* and *timing* of solar energy around the planet, thus influencing whether seasons are particularly hot or cold (known as *seasonal intensity*) and contrasts between the seasons. In moving toward a glacial climate, they allow snowfall from one winter to survive through cooler summers and last until the following winter. This pattern leads to net accumulation of ice, a necessary condition for glacier formation. Other changes in orbital geometry lead to the end of ice ages and the occurrence of interglacials. We study this process in more detail in Chapter 17 ("Glaciers and Paleoclimatology").

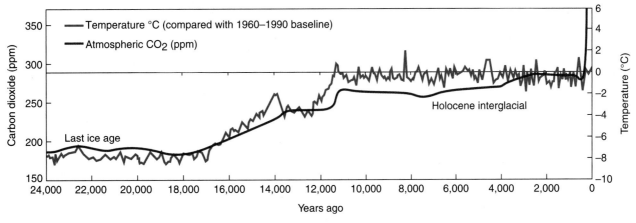

FIGURE 16.12 When Earth's climate shifted from the last ice age to the current interglacial, global temperatures increased and atmospheric carbon dioxide concentration rose from approximately 180 ppm to 280 ppm.

(?) What is the cause of the temperature change that occurred between approximately 20,000 and 10,000 years before the present?

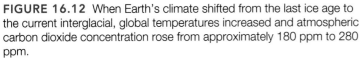

(?) **Expand Your Thinking**—Are GCMs effective for predicting future climate change in small geographic areas?

16-6 Human Activities Have Increased the Amount of Carbon Dioxide in the Atmosphere

LO 16-6 *Describe the human activities that contribute to greenhouse gas production.*

It is a common misconception to think that humans and their activities cannot have a serious impact on our planet. You might wonder how everyday activities like driving cars, heating and cooling homes, and harvesting natural resources can affect our environment on a global scale. It is important to recognize that human population growth and use of resources have an impact on our planet and on the processes that allow it to function as a healthy system.

Fossil Fuels

Remember all the carbon that exists in the form of organic matter in the biosphere? Much of that carbon was buried in sediments or sedimentary rocks and became part of the lithosphere for millions of years. Four trillion tonnes of carbon was stored in Earth's crust in this way. Over time, these organic carbon deposits were transformed into "fossil fuels"—coal, oil, and natural gas (recall Chapter 10, "Geologic Resources")—which humans extract and burn to produce energy (**FIGURE 16.13**). Burning fossil fuels and organic biomass are among the activities that humans perform to create usable forms of energy for electricity, industry, motorized vehicles, and building heating systems.

Coal forms from plant material that has been buried for millions of years. It is the most abundant of fossil fuels and supplies about 30 percent of global energy. At current levels of consumption, coal supplies could last for another 200 to 500 years. However, burning coal releases large amounts of carbon dioxide, soot, sulphur, and other pollutants into the atmosphere. Coal is considered one of the dirtiest forms of energy because of its detrimental effects on the environment and human health. In the future, our ability to use coal without contributing to global warming hinges on our ability to develop methods for preventing or trapping dirty emissions, or to capture and store in deep geological environments the carbon dioxide generated by coal burning.

Most oil and gas began in the ocean as the remains of *phytoplankton* (microscopic floating algae) that was deposited on the sea floor and deeply buried. At the elevated temperatures and pressures that prevail during long-term burial, the organic material was converted into oil and methane. These fossil fuels are a natural resource that humans extract from the crust by means of drilling and mining. Burning fossil fuel is similar to respiration but requires the addition of heat (denoted by the symbol Δ):

$$C_6H_{12}O_6 + 6O_2 + \Delta \longrightarrow 6CO_2 + 6H_2O$$
$$\text{organic matter} + \text{oxygen} + \text{heat} \longrightarrow \text{carbon dioxide} + \text{water}$$

Burning of fossil fuels speeds up the natural cycling of carbon between the lithosphere and the atmosphere and adds carbon dioxide, fine particles, and pollutants to the atmosphere at a much faster rate than would occur through normal tectonic and volcanic cycles. These energy resources are also non-renewable because fossil fuels take millions of years to form, and they exist in limited amounts within the crust. Think about how far you can travel in your car with 1 L of gasoline—it does not take very long to burn that litre, and then it is gone for good!

Respiration and the burning of fossil fuels are not the only sources of carbon dioxide. Volcanic eruptions and metamorphic processes introduce carbon dioxide into the atmosphere, as they have throughout Earth's history. Carbon dioxide is a natural component of Earth's atmosphere, so why does it matter that human activities add more? How do we know what carbon dioxide levels in the atmosphere were in the past? And why do we care?

Carbon Dioxide

Activities such as burning fossil fuels, deforestation, and various land-use practices account for about 12 percent of the total amount of carbon dioxide emitted to the atmosphere annually. About 75 percent of global human greenhouse gas emissions are produced by a relatively small number of countries, which include developed countries such as the United States, Canada, Japan, and the nations of European Union, and the highly populated rapidly developing countries such as China and India (**TABLE 16.1**).

Over the past 200 years, carbon dioxide levels in the atmosphere have risen by over 40 percent (from 280 ppm to 400 ppm), and they continue to rise by approximately 2.1 percent each year. Most scientists now believe that human activities are responsible for these observed increases in atmospheric carbon dioxide. Scientists

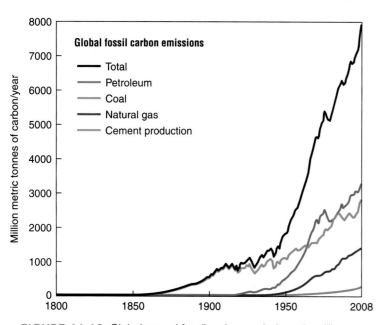

FIGURE 16.13 Global annual fossil carbon emissions, in million tonnes of carbon. The carbon dioxide releases from cement production result from the thermal decomposition of limestone into lime.

(?) What is the largest source of global fossil carbon emissions?

TABLE 16.1 Top 10 Greenhouse Gas Emitters

Country	Percent of Global Emissions	Tonnes per Person per Year	Percent of World Population
China	24.6	5.6	19.1
United States	16.4	16.0	4.5
India	6.2	1.5	17.1
Russia	5.0	10.7	2.0
Japan	3.4	8.1	1.8
Germany	2.3	8.4	1.2
South Korea	1.7	10.4	0.7
United Kingdom	1.5	7.2	0.9
Canada	1.5	13.5	0.5
Saudi Arabia	1.5	16.5	0.4

using GCMs predict that by the end of the 21st century, we could see carbon dioxide concentrations between 490 ppm and 1,260 ppm (75 percent to 350 percent higher than natural concentrations). The global warming associated with this increase in carbon dioxide levels will be greater and faster than the natural fluctuations seen in recent geologic history.

The global carbon cycle keeps carbon dioxide levels in the atmosphere and ocean fairly stable over the long term, but rapid, short-term additions of carbon dioxide to the atmosphere do affect global climate. Even seemingly small effects, such as an increase in global average temperatures by just a few degrees, can have a significant impact on climate, winds, ocean circulation, and regional weather patterns. In addition, because biogeochemical cycles are strongly interconnected, interacting with and affecting one another over very short periods (the time it takes to breathe) to very long time scales (millions of years), changes in the carbon cycle affect the cycles of other elements and compounds. Most of the threat to climate comes from the carbon dioxide released by the burning of fossil fuels, but human activities affect the global carbon cycle, and thus global climate, in other ways too.

Population Growth

Between 1930 and 2000, the number of humans on Earth tripled, from 2 billion to 6.1 billion (**FIGURE 16.14**). As a result of population growth and industrial development, the use of natural resources has increased by 1,000 percent in the past 70 years. Global population growth is expected to continue, with the world's population reaching 9 billion people by 2050 and 11 billion by 2100. This will put tremendous pressure on Earth's natural resources. Among the most important resources that humans depend on, and that result in the most environmental pollution and depletion of resources, is energy.

The developed nations account for 25 percent of the world's population but use more than 70 percent of the total energy produced. The United States consumes about 20 percent of the total to sustain its population of more than 300 million, which amounts to less than 5 percent of the world's population. In contrast, India's population of more than 1 billion (17 percent of world population) uses only 5 percent of global energy, although this proportion is increasing. One indicator of energy use collected by the United Nations is kilograms of oil equivalent per capita. The global average is about 1,800 kg/capita, and at present China uses that amount of energy. In contrast, Canada uses about 7,400 kg/capita.

Commercial energy is traded in the global marketplace in the form of oil, coal, gas, hydropower, and nuclear power. Oil supplies about 40 percent of the world's energy requirements, including 96 percent of the energy needed for transportation. There are approximately 700 million cars in the world today, and that number is expected to jump to more than 1.2 billion by 2025. The global demand for and consumption of oil and gasoline will likely double in the same period.

Coal supplies 30 percent of world energy needs and natural gas supplies 20 percent, meaning that fossil-fuel burning accounts for 90 percent of the global commercial energy trade. Recall from Chapter 10 that fossil fuels are non-renewable sources of energy; that there is a finite (limited) amount of these materials on Earth; and that significant cost and effort are involved in finding, extracting, and processing these materials. To date, the world's population has consumed over 875 billion barrels of oil. Experts estimate the global volume of recoverable oil at about 1,000 billion barrels. At the current rate of consumption, therefore, all the oil on Earth, which was formed over a period of millions of years, could be depleted in less than 40 years.

Natural environments, ecosystems, and habitats are also considered among Earth's resources because they sustain the complex web of life that supports humanity. Global changes are occurring in these natural resources as well. The World Resources Institute estimates that 50 percent of the world's coastal ecosystems are severely at risk. Average global temperature has increased by 0.8°C since the late 19th century and by about 0.5°C over the past 40 years, and scientists expect it to increase by an additional 1°C to 6°C during the next 100 years. This may not sound like much, but consider this: At the peak of the ice age about 21,000 years ago, global average temperature was only 5°C to 6°C colder than it is today. Under those conditions, glaciers covered most of North America and Europe, and woolly mammoths flourished. The warming of 5°C to 6°C from the last ice age to the present interglacial took approximately 5,000 to 8,000 years. Today humans are poised to achieve the same thing in only a century.

Changes in Land Use

The term "land use" refers to how humans use Earth's surface: urban, rural, agricultural, forest, and wilderness areas. On a global scale, carbon dioxide released by changes in land surface (deforestation to harvest timber and clear land for crops and pastures) represents about 18 percent of total annual carbon dioxide emissions. This source constitutes one-third of total emissions from developing countries and more than 60 percent of emissions from the least developed countries.

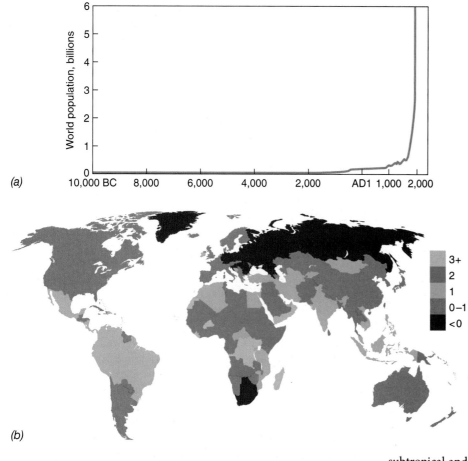

_(a)_

FIGURE 16.14 Human population growth. _(a)_ The rate of human population growth increased rapidly in the 20th century due to medical advances and increased agricultural productivity. _(b)_ Percentage annual population growth rates.

 Why is population growth relevant to the use of fossil fuels?

_(b)_

James P. Blair/National Geographic Creative

Deforestation is one of the most significant ways in which humans have caused global warming. It contributes to the enhanced greenhouse effect in two ways: (1) Burning trees releases carbon dioxide, and (2) dead trees no longer photosynthesize (store) carbon dioxide from the atmosphere. Removing forests also causes erosion of sediment, increases runoff during rainy seasons, and depletes the natural nutrients in soil. One-fifth of our planet's forests have been cleared, including 35 percent of all temperate forests and 25 percent of all subtropical and deciduous forests. It is important to realize that these are _averages_. In some parts of the world, 50 percent to 100 percent of the natural forest has been cleared (**FIGURE 16.15**). For example, between the Atlantic coast and the Mississippi River in the United States, wooded forests occupied 1.7 million kilometres in the 1700s. Today, only 1.0 million kilometres remain. In the past, 75 percent of Canada was forested. Even though the amount in Canada is much reduced, the boreal forest in northern Canada still covers about a third of the country and remains one of the largest areas of forest on the planet.

The impacts of poor land use are not limited to forest environments. Polluted runoff resulting from increased use of fertilizer and eroded sediment affect marine and freshwater ecosystems as well. Coral reefs below deforested watersheds around the world are stressed by sediment-laden water running off mountains that have been cleared of trees and vegetation. Salmon, trout, and other river fish depend on clean, clear streams in which to lay their eggs. Increasing global awareness of the impacts of changes in land use has brought forestry issues to the global political agendas of both the United Nations and the world's major developed nations. Awareness and education are the first steps toward global changes that will have positive effects both on humans and on the natural environment. In the next section, we explore additional issues related to global change.

What is the source of the global warming problem? Put yourself in the shoes of a decision maker with the "Critical Thinking" box titled "Where Do Greenhouse Gases Come From?"

FIGURE 16.15 The border between Haiti, on the left, and the Dominican Republic, on the right, highlights the difference between a forested area and one that has been stripped of vegetation. Deforestation leads to soil erosion, higher atmospheric carbon dioxide, and ecosystem destruction.

 Humans need wood. How would you manage a forest so that it provided timber but minimized negative impacts?

 Expand Your Thinking—What human activities have increased the amount of greenhouse gases in the atmosphere?

CRITICAL THINKING

WHERE DO GREENHOUSE GASES COME FROM?

Please work with a partner to answer the following questions.

1. Use **FIGURE 16.16** to fill in **TABLE 16.2** with percentages.
2. What are the major sources of each type of gas?
3. Based on these data, what do you think are the easiest first steps to take in reducing greenhouse gas emissions?
4. How would you go about taking these steps?
5. What are the potential consequences of taking these steps?
6. Who would be affected by your first steps? How? What questions would they have and what are your answers?
7. Do you need additional information to answer questions 3 to 6?

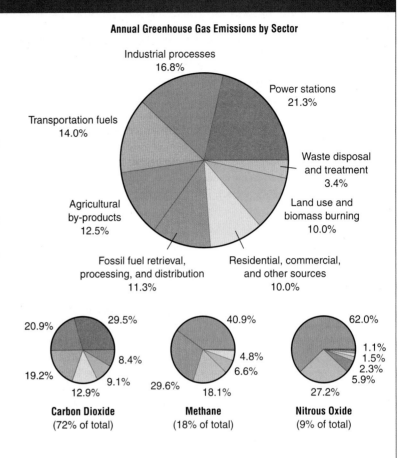

FIGURE 16.16 Relative percentages of anthropogenic greenhouse gases from eight categories of sources.

TABLE 16.2 Annual Sources of Greenhouse Gases by Percentage			
Gas Source	**Type of Greenhouse Gas**		
	Carbon Dioxide	**Methane**	**Nitrous Oxide**
Industrial processes			
Transportation fuels			
Power stations			
Waste disposal and treatment			
Land use and biomass burning			
Agricultural by-products			
Fossil fuel retrieval, processing, and distribution			
Residential, commercial, and other sources			

16-7 Earth's Atmospheric Temperature Has Risen by About 0.8°C in the Past 100 Years

LO 16-7 *Identify some patterns and consequences of global warming.*

According to published research, global surface temperatures have increased about 0.8°C (±0.18°C) since the late 19th century, and the linear trend for the past 50 years of an increase of 0.2°C (±0.03°C) per decade is nearly twice that for the past 100 years. The year 2012 was only a fraction of a degree cooler than 2010, the warmest year since record keeping began. January 2000 to December 2009 was the warmest decade on record, although it appears as though the decade we are in now will be warmer still. Scientific analysis of modern and past climates has shown that Earth is warmer now than at any time in the past 1,300 years (FIGURE 16.17).

The Melting Arctic

Warming has not been uniform around the globe (FIGURE 16.18). High latitudes are warming more and faster than low latitudes. The Arctic has warmed faster than any other region on Earth: by nearly 2.7°C in the last 30 years. The accelerated change in the polar regions has occurred partly because, as permafrost thaws, glaciers melt, and sea ice disappears, the area of white reflective surfaces decreases, so the atmosphere absorbs more of the Sun's energy. This is the kind of feedback that makes predicting future change so complicated.

Glaciers and sea ice all over the world are melting at the fastest rates ever measured. Alaska has 2,000 large glaciers, and it is estimated that 99 percent of them are "retreating," or melting. In Canada, glaciers and icefields are found in the mountains of Alberta, British Columbia, and Yukon, as well as in the eastern Arctic islands, particularly on Baffin Island. They cover an estimated area of 200,000 km², or about 2 percent of the area of Canada. Snow cover in the Northern Hemisphere has decreased by about 10 percent since 1966. Satellites have made continual observations of the extent of Arctic sea ice since 1978, recording a general decline of 10 percent per year throughout that period. On September 14, 2007, the extent of Arctic sea ice dropped below 4 million square kilometres, the lowest extent yet recorded in the satellite record and about 38 percent lower than the average (FIGURE 16.19). However, in 2012, the extent of Arctic sea ice dropped even further—to 3.4 million square kilometres. Geologic models predict that the warming trend in polar regions will continue, with these regions warming by 4°C to 7°C in the next 100 years. This warming and melting will affect the distribution of ice-dwelling animals, such as polar bears and penguins, and the feeding and migration patterns of many species of birds and mammals. In a practical sense, humans living in high latitudes will be affected not only by changes in food availability but also by changes in ice-dependent infrastructure, including oil rigs and ice roads. In contrast, coastal communities may become accessible by sea throughout more of the year. Polar warming may also have negative effects on the global circulation of heat through the atmosphere and oceans.

Rising Seas

Sea level has been rising at the rate of 1 mm to 2 mm per year for the past 100 years. This rate of increase may not seem significant, but it means that during the last 100 years, sea level has risen by about 15 cm worldwide. Recent studies have found that the global rate of sea-level rise has increased to over 3 mm/y, approximately twice the rate of the past century.

Global warming can cause sea level to rise in two ways: thermal expansion of ocean water and melting of icecaps and glaciers. At present, scientists believe that these two causes have an approximately equal influence on global sea-level rise. Further global warming of 1°C to 6°C, predicted to occur by 2100, probably will cause the rate of global sea-level rise to continue accelerating. Estimates of how high the sea level will rise by the end of the century vary, but studies suggest between about 0.75 m and 2.0 m.

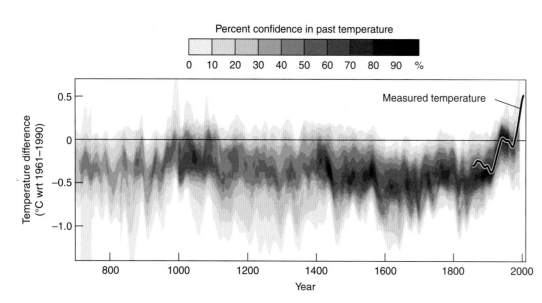

 Describe the patterns you see in this global temperature record.

FIGURE 16.17 Reconstructions of global temperature (using geologic materials such as tree rings and corals that provide a record of temperature) indicate that Earth is warmer now than at any time in the past 1,300 years. The black line is the record provided by modern reliable instruments; the level of brown shading represents probability of the temperature reconstruction. (*Source: Climate Change 2007: The Physical Science Basis. Working Group I Contribution to the Fourth Assessment Report of the Intergovernmental Panel on Climate Change,* Figure 6.10. Cambridge University Press, Cambridge, UK, and New York, NY, USA.)

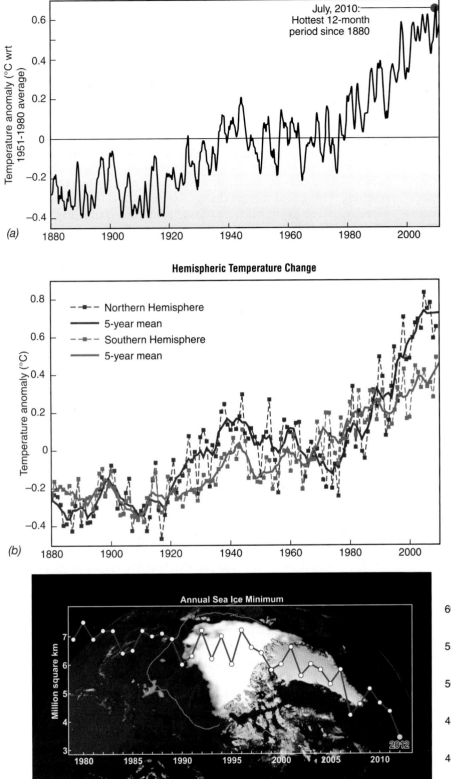

(a)

Hemispheric Temperature Change

(b)

Rising sea levels threaten coastal development and human populations, and have to be considered even in Canada (FIGURE 16.20). More than half of the world's population lives within 60 km of the ocean. For example, a 1 m rise in sea level could flood 17 percent of the area of Bangladesh, and submerge the Maldives in the Indian Ocean. The impacts of sea-level rise on coastal communities are financial as well as environmental and sociological. Millions of dollars are spent moving and protecting coastal infrastructure, such as roads and buildings. Home insurance in coastal communities is extremely expensive because the risks of property damage and loss are so high. Think of a place like New York City, which has almost 1,000 km of coastline. Sea-level rise will cause flooding of roads, subways, airports, and buildings. Flood control policies in coastal regions have thus become a very important topic in communities around the world.

(?) Where is the rate of warming the highest? What impacts will it have?

FIGURE 16.18 (a) July 2010 marked the end of the hottest 12-month period since record keeping began in 1880. Throughout the last five decades, global average temperatures have trended upward about 0.2°C per decade. Since 1880, the year that modern scientific instrumentation became available to monitor temperatures precisely, a clear warming trend is present, although there was a levelling off between the 1940s and 1970s. In total, average global temperatures have increased by about 0.8°C since 1880. *Source*: NASA. (b) This graph shows that the Northern Hemisphere is warming more rapidly than the Southern Hemisphere.

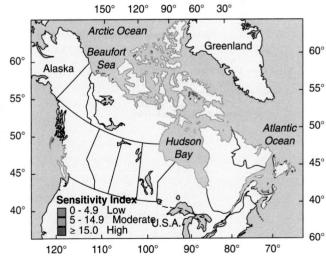

FIGURE 16.19 The history of Arctic sea ice.

(?) Describe the positive feedback to climate caused by melting sea ice.

Expand Your Thinking—What is the evidence for global warming?

FIGURE 16.20 Natural Resources Canada has evaluated how sensitive the coastline of Canada may be to a rise in sea level. As it turns out, very little of Canada's present coast will be submerged, and the most sensitive areas of Prince Edward Island, Nova Scotia, and New Brunswick may see higher rates of coastal erosion.

(?) Why is sea-level rise a problem in other parts of the world?

16-8 Global Warming Leads to Ocean Acidification and Warming, Melting of Glaciers, Changes in Weather, and Other Impacts

LO 16-8 *Describe how global warming is changing the ocean and the world's glaciers.*

Sea-surface temperature (SST) has increased by an average of 0.6°C in the past 100 years (**FIGURE 16.21**), and the acidity of the ocean surface has increased tenfold. Corals cannot tolerate severely warming waters, and the temperature stress causes a phenomenon known as *bleaching,* in which they expel the symbiotic algae that live in their tissues. In 1997 and 1998, coral bleaching was observed in almost all of the world's reefs in response to high SSTs. An estimated 16 percent of the world's corals died in that strong bleaching event, an unprecedented occurrence.

The oceans have absorbed about 40 percent of the carbon dioxide emitted by humans over the past two centuries. Increasing acidity, brought on by dissolved carbon dioxide that mixes with sea water to form *carbonic acid* (H_2CO_3), makes it difficult for calcifying organisms (corals, molluscs, many types of plankton) to secrete the calcium carbonate they need for their skeletal components. This **ocean acidification** is one of the consequences of carbon dioxide buildup that could have a great impact on the world's ocean ecology, which depends on the secretion of calcium carbonate by thousands of different species.

More than half the reefs in the Caribbean and Red Sea have gone from pristine to near extinction in the past century. The loss of healthy coral reefs affects all the species that dwell there (such as turtles, seals, molluscs, crabs, and fish), as well as the animals that depend on reef habitats as a food source (including seabirds, mammals, and humans). One-quarter of all sea animals spend time in coral reef environments during their life cycle. There are economic impacts as well. Tourism tied to coral reefs and commercial fisheries generate billions of dollars in revenue annually. Biodiversity, food supplies, and economics thus may all be affected by reef loss resulting from global climate change. However, reef loss is a complex issue. Reefs can suffer from coastal pollution, overfishing, and other types of human stresses. Exactly what roles warming temperatures, ocean acidity, and other anthropogenic impacts play in global reef health have yet to be fully defined by researchers, but they are all negative factors.

Droughts, Storms, and Severe Events

Even small changes in climate can result in large changes in the things that people depend on, including a clean water supply, the growth of food crops, and the position of the shoreline. Most scientists agree that the likely results of global warming include more frequent heat waves, droughts, extreme storm events, wildfires, changes in agriculture and vegetation, sea-level rise, and coastal erosion. Climate (and weather) will become more unstable and less predictable because the difference in temperature between the poles and the equator drives global winds and atmospheric circulation, which distribute heat and water (precipitation) around the world. Changes in the heat budget will lead to changes in wind and water circulation, leading to drought in some areas, while other areas will experience increases in the strength and frequency of storms and increased precipitation. Computer models and scientific experts suggest that these types of extreme events and weather anomalies will become more common in the future as global climate continues to change. Extreme events cause damage to property as well as threaten the health and well-being of communities. It has been estimated in 2012 that natural disasters, and in particular extreme weather events, cost insurance companies about $65 billion. The United States accounted for 90 percent of this loss, particularly as a result of Hurricane Sandy in October 2012. (**FIGURE 16.22**).

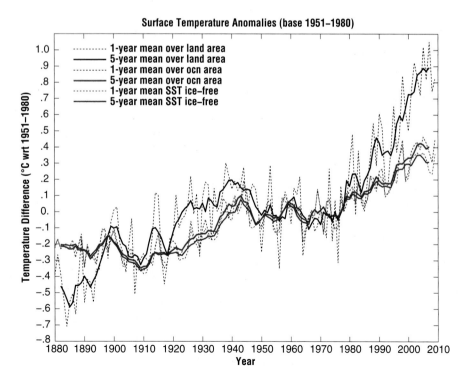

Surface Temperature Anomalies (base 1951–1980)

- ········· 1-year mean over land area
- —— 5-year mean over land area
- ········· 1-year mean over ocn area
- —— 5-year mean over ocn area
- ········· 1-year mean SST ice–free
- —— 5-year mean SST ice–free

Temperature Difference (°C wrt 1951–1980)

Year

 Why is ocean warming slower than land warming?

FIGURE 16.21 Annual and five-year average global temperature changes for the land (black), the global oceans (red), and the oceans that are ice-free (pink).

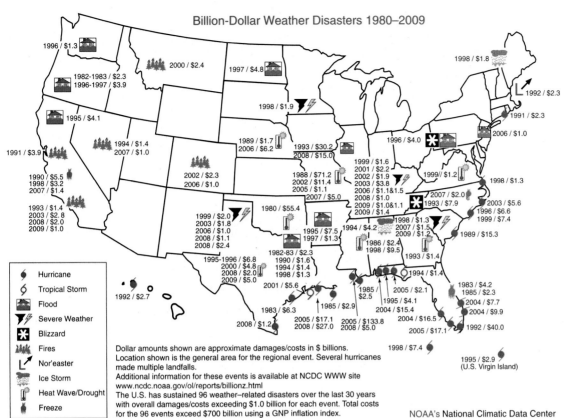

Billion-Dollar Weather Disasters 1980–2009

Hurricane
Tropical Storm
Flood
Severe Weather
Blizzard
Fires
Nor'easter
Ice Storm
Heat Wave/Drought
Freeze

Dollar amounts shown are approximate damages/costs in $ billions.
Location shown is the general area for the regional event. Several hurricanes
made multiple landfalls.
Additional information for these events is available at NCDC WWW site
www.ncdc.noaa.gov/ol/reports/billionz.html
The U.S. has sustained 96 weather–related disasters over the last 30 years
with overall damages/costs exceeding $1.0 billion for each event. Total costs
for the 96 events exceed $700 billion using a GNP inflation index.

NOAA's National Climatic Data Center

FIGURE 16.22 Map of billion-dollar U.S. weather disasters, 1980–2009. In October 2012, Hurricane Sandy caused significant amounts of damage, amounting to an estimated $68 billion in the coastal United States and eastern Canada. Environment Canada has a website that details significant weather disasters in Canada since 1996 (ec.gc.ca/meteo-weather, and then scroll to Climate and Historical Weather), although there have been much fewer billion dollar disasters.

 Why are there more billion dollar weather disasters in the United States than in Canada?

Melting Ice

A study has shown that the net loss of about 148 km³ of water per year from the Antarctic Ice Sheet outpaces the ice sheet's tendency to accumulate snow. That is, the Antarctic Ice Sheet is melting, not growing. Three broad ice-covered regions characterize Antarctica: the Antarctic peninsula, west Antarctica, and east Antarctica. In west Antarctica, the rate of ice loss has increased by 59 percent over the past decade, to about 132 billion tonnes a year, while the yearly loss along the peninsula has increased by 140 percent, to 60 billion tonnes. In east Antarctica, the ice is relatively stable, with interior snowfall approximately equalling melting by warm ocean currents along the coastal regions. Researchers have stated that there's no doubt Antarctica as a whole is losing ice yearly, and this loss is increasing each year. Studies also show that outlet glaciers feeding into the ocean from Antarctica and Greenland are accelerating and approximately doubled their discharge of ice between 2002 and 2008.

The total volume of land-based ice in the Arctic has been estimated at about 3.1 million cubic kilometres. If it were all to melt, this much ice corresponds to a sea-level rise of about 8 m. Most arctic glaciers and icecaps have been in decline since the early 1960s, with this trend speeding up in the past two decades. The extent of seasonal surface melt on the Greenland Ice Sheet has been observed by satellite since 1979; these observations show an increasing trend—*doubling over the past decade.* Recent years have marked record levels of summertime melting at higher altitudes on Greenland. Melting in areas above 2,000 m increased by 150 percent from the long-term average, with melting occurring on 25 to 30 more days than the average for the previous 19 years.

Impacts in the Biosphere

Signs and impacts of climate change in the biosphere include degradation and shrinkage of wildlife habitats, changes in the distribution of food sources for animals and people, disruption of the timing of animal and bird migration patterns, and **desertification.** Did you know that as much as one-third of the world is considered Desert? Many deserts are expanding as the global climate warms. The Sahara Desert in North Africa is larger than the United States and has expanded by more than 650,000 km² in the past 50 years. In the 1990s, the Gobi Desert in China expanded by more than 50,000 km², an area about the size of Nova Scotia. In many cases, global warming and poor land use by humans jointly cause desertification. Changes in the distribution of precipitation around the planet may be exacerbated by activities such as overfarming, not rotating crops, overgrazing, and deforestation. These activities play a role in stripping soils of moisture and nutrients, leading to desertification.

The ways in which humans use land and other natural resources affect the distribution and quality of plant and animal habitats. The area of undeveloped space for wildlife is continually declining under the pressure of growing human populations. Essential freshwater systems are affected by pollution, damming, and diversion of water for human use. Changes in habitat quality cause changes in the distribution of food sources and place wildlife populations under stress. For example, seabird populations are declining around the world, and researchers believe that the decline is due largely to changes in global atmospheric and ocean temperatures that affect the availability of food and the timing of migration cues. Similar declines have been noted in fish and marine mammal populations.

 Expand Your Thinking—Describe the impacts of global warming.

CRITICAL THINKING

GLOBAL WARMING

Working with a partner, please answer the following questions.

1. Study the significant climate events of 2009 in FIGURE 16.23. Make a list that summarizes the *types* of events that occurred.
2. Compare your list to the potential impacts of global warming described in this chapter. Discuss differences and similarities.
3. Discuss the longer-term impacts of sea-level rise (both positive and negative) and how a higher sea level of, say, a half metre, in the year 2075 will influence some of the events shown on this map.
4. "Anomalies" are events that are not normal. Identify three anomalies or events and discuss how they may change as warming continues over the next couple of decades.
5. Describe events that appear to be consistent with a warming world and those that do not.
6. Identify four events or anomalies and describe the impact of each if they become regular events due to global warming.
7. Considering these events and your reading of this chapter, construct a testable, alternative hypothesis that explains climate patterns of the past half century.

Alaska
Had its second warmest July, behind 2004, on record.

North America Snow Cover Extent
Third largest October snow cover extent on record.

Arctic Sea Ice
Third lowest extent on record in September, behind 2007 and 2008. September 1996 was the last year that had above-average sea ice extent.

Canada
A tornado claimed the lives of three people in Ontario. These were the first tornado-related fatalities in Canada since 1995 (Jul).

Hurricane Rick (Oct)
Maximum winds—285 km/hr. The second-most intense Eastern North Pacific hurricane on record, behind Linda of 1997, and the strongest hurricane to form in October since reliable records began.

USA
Record floods on the Red River in the northern Plains region (Mar). Wettest October since records began 115 years ago. Below average tornado season after record activity in 2008.

Hurricane Ida (Nov)
Maximum winds—165 km/hr. Produced deadly floods and landslides that claimed 192 lives and left over 14,000 people affected in El Salvador.

Mexico
Experienced severe to exceptional drought conditions (Sep).

Eastern North Pacific Hurricane Season
Near average activity: 20 storms, 8 hurricanes.

Guatemala
Drought decreased harvests by up to 50 percent. 400,000 families were affected (Sep).

Atlantic Hurricane Season
Below average activity: 9 storms, 3 hurricanes.

Hurricane Andres (Jun)
Maximum winds—130 km/hr. Andres was Eastern North Pacific's latest arrival of a named storm since reliable records began in 1970.

Colombia
Copious rainfall triggered widespread floods across western Colombia. Nearly 2,500 families were affected (Feb).

ENSO
La Niña conditions transitioned into a warm phase ENSO (El Niño) in June.

Brazil
Torrential downpours caused floods and mudslides, affecting over 186,000 residents. This region experienced its worst deluge in over 20 years (Apr).

Peru
Torrential rainfall fell over Peru's southwestern region of Puno, resulting in a deadly landslide (Mar).

Argentina
Heavy rainfall caused a devastating landslide in northern Argentina, destroying over 300 homes and affecting nearly 20,000 people (Feb).

Significant Climate Anomalies and Events in 2009

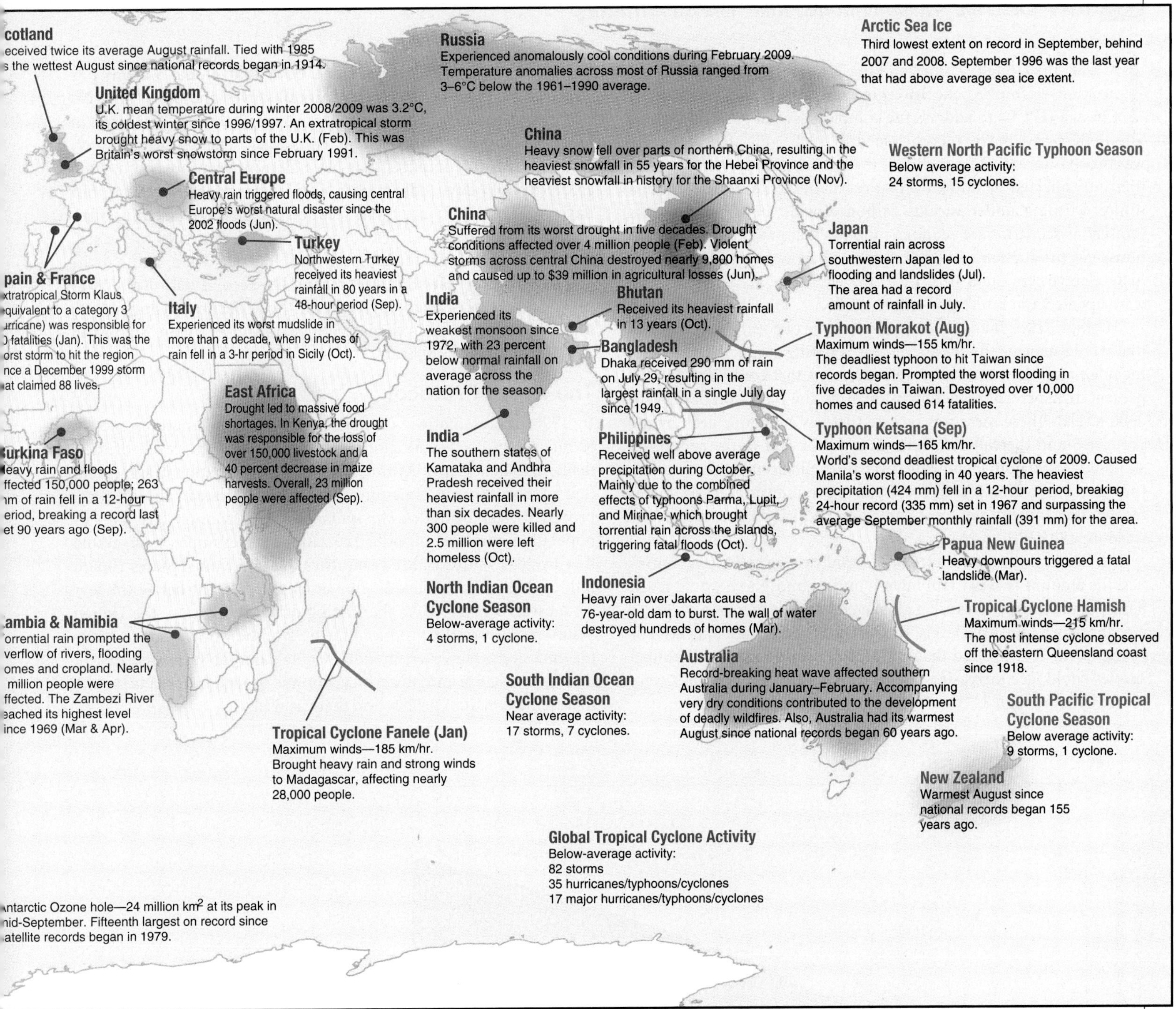

Scotland
Received twice its average August rainfall. Tied with 1985 as the wettest August since national records began in 1914.

United Kingdom
U.K. mean temperature during winter 2008/2009 was 3.2°C, its coldest winter since 1996/1997. An extratropical storm brought heavy snow to parts of the U.K. (Feb). This was Britain's worst snowstorm since February 1991.

Central Europe
Heavy rain triggered floods, causing central Europe's worst natural disaster since the 2002 floods (Jun).

Turkey
Northwestern Turkey received its heaviest rainfall in 80 years in a 48-hour period (Sep).

Spain & France
Extratropical Storm Klaus (equivalent to a category 3 hurricane) was responsible for 0 fatalities (Jan). This was the worst storm to hit the region since a December 1999 storm that claimed 88 lives.

Italy
Experienced its worst mudslide in more than a decade, when 9 inches of rain fell in a 3-hr period in Sicily (Oct).

East Africa
Drought led to massive food shortages. In Kenya, the drought was responsible for the loss of over 150,000 livestock and a 40 percent decrease in maize harvests. Overall, 23 million people were affected (Sep).

Burkina Faso
Heavy rain and floods affected 150,000 people; 263 mm of rain fell in a 12-hour period, breaking a record last set 90 years ago (Sep).

Zambia & Namibia
Torrential rain prompted the overflow of rivers, flooding homes and cropland. Nearly 1 million people were affected. The Zambezi River reached its highest level since 1969 (Mar & Apr).

Antarctic Ozone hole—24 million km² at its peak in mid-September. Fifteenth largest on record since satellite records began in 1979.

Russia
Experienced anomalously cool conditions during February 2009. Temperature anomalies across most of Russia ranged from 3–6°C below the 1961–1990 average.

China
Heavy snow fell over parts of northern China, resulting in the heaviest snowfall in 55 years for the Hebei Province and the heaviest snowfall in history for the Shaanxi Province (Nov).

China
Suffered from its worst drought in five decades. Drought conditions affected over 4 million people (Feb). Violent storms across central China destroyed nearly 9,800 homes and caused up to $39 million in agricultural losses (Jun).

India
Experienced its weakest monsoon since 1972, with 23 percent below normal rainfall on average across the nation for the season.

Bhutan
Received its heaviest rainfall in 13 years (Oct).

Bangladesh
Dhaka received 290 mm of rain on July 29, resulting in the largest rainfall in a single July day since 1949.

India
The southern states of Kamataka and Andhra Pradesh received their heaviest rainfall in more than six decades. Nearly 300 people were killed and 2.5 million were left homeless (Oct).

Philippines
Received well above average precipitation during October. Mainly due to the combined effects of typhoons Parma, Lupit, and Mirinae, which brought torrential rain across the islands, triggering fatal floods (Oct).

North Indian Ocean Cyclone Season
Below-average activity: 4 storms, 1 cyclone.

Indonesia
Heavy rain over Jakarta caused a 76-year-old dam to burst. The wall of water destroyed hundreds of homes (Mar).

Australia
Record-breaking heat wave affected southern Australia during January–February. Accompanying very dry conditions contributed to the development of deadly wildfires. Also, Australia had its warmest August since national records began 60 years ago.

South Indian Ocean Cyclone Season
Near average activity: 17 storms, 7 cyclones.

Tropical Cyclone Fanele (Jan)
Maximum winds—185 km/hr. Brought heavy rain and strong winds to Madagascar, affecting nearly 28,000 people.

Global Tropical Cyclone Activity
Below-average activity:
82 storms
35 hurricanes/typhoons/cyclones
17 major hurricanes/typhoons/cyclones

Arctic Sea Ice
Third lowest extent on record in September, behind 2007 and 2008. September 1996 was the last year that had above average sea ice extent.

Western North Pacific Typhoon Season
Below average activity: 24 storms, 15 cyclones.

Japan
Torrential rain across southwestern Japan led to flooding and landslides (Jul). The area had a record amount of rainfall in July.

Typhoon Morakot (Aug)
Maximum winds—155 km/hr. The deadliest typhoon to hit Taiwan since records began. Prompted the worst flooding in five decades in Taiwan. Destroyed over 10,000 homes and caused 614 fatalities.

Typhoon Ketsana (Sep)
Maximum winds—165 km/hr. World's second deadliest tropical cyclone of 2009. Caused Manila's worst flooding in 40 years. The heaviest precipitation (424 mm) fell in a 12-hour period, breaking 24-hour record (335 mm) set in 1967 and surpassing the average September monthly rainfall (391 mm) for the area.

Papua New Guinea
Heavy downpours triggered a fatal landslide (Mar).

Tropical Cyclone Hamish
Maximum winds—215 km/hr. The most intense cyclone observed off the eastern Queensland coast since 1918.

South Pacific Tropical Cyclone Season
Below average activity: 9 storms, 1 cyclone.

New Zealand
Warmest August since national records began 155 years ago.

FIGURE 16.23 Significant weather and climate events in 2009.

16-9 Several International Efforts Are Attempting to Manage Global Warming

LO 16-9 *Describe "carbon quotas" and "emission trading."*

In 1988, the United Nations and the World Meteorological Organization established the **Intergovernmental Panel on Climate Change (IPCC)** to address the complex issue of global warming. The IPCC's first major report, issued in 1990, indicated that there was broad international agreement that global climate was being significantly affected by human activities. In 2001, the IPCC projected future warming under various carbon dioxide *emissions scenarios* (emissions scenarios are economic models that project future greenhouse gas production). In 2007, the fourth assessment of global climate change was published. In it, the emissions scenarios of 2001 were updated and future projections made.

Using GCMs, the IPCC analyzed a range of government policies that would manage the release of heat-trapping gases. These policies included ambitious but achievable measures that could lead to a 50 percent to 85 percent reduction in emissions by 2050 (compared with 2000 levels). These measures require that gas emissions peak by the year 2015 and thereafter stabilize around the end of the century at 445 ppm to 490 ppm. Following this path could keep global temperature increases within 2°C to 2.4°C above pre-industrial levels, thereby avoiding some of the most damaging and irreversible impacts of warming (**FIGURE 16.24**).

If, however, the current trend of rapidly rising emissions continues, the amount of heat-trapping gas in the atmosphere would reach 855 ppm to 1,130 ppm, causing global temperature increases of 5°C to 6°C above pre-industrial levels, which would have severe impacts. For example, the IPCC found that up to 30 percent of plant and animal species could face increasing risk of extinction if temperatures were to rise by more than 2.5°C.

In 2007, the IPCC concluded both that global warming is "unequivocal" and that most of the observed increases in average global temperatures were "very likely" due to increases in greenhouse gases from human emissions. The report went on to define various impacts that have already been observed, such as these: Cold days, cold nights, and frosts are less frequent; hot days and hot nights are more frequent; heat waves are more frequent; the frequency of heavy precipitation events has increased; the incidence of extreme high sea level has increased; and Northern Hemisphere temperatures during the second half of the 20th century were *very likely* higher than those occurring during any other 50-year period in the last 500 years and *likely* the highest in at least the past 1,300 years.

The Kyoto Protocol

In 1997, representatives of Canada and 83 other nations met in Kyoto, Japan, to discuss global climate change. The result of their deliberations was the *Kyoto Protocol,* an international agreement that carbon dioxide emissions should be regulated. The protocol proposed establishing *carbon quotas* for each country based on its population and level of industrialization. For example, it suggested that by 2012, industrialized countries like the United States should reduce their carbon dioxide emissions to 7 percent below the levels measured in 1990. As the leading developed nation, the United States has historically produced the bulk of the world's carbon dioxide emissions. However, in 2007, China's output surpassed that of the United States, and powerful economic growth in India is rapidly increasing emission production there as well.

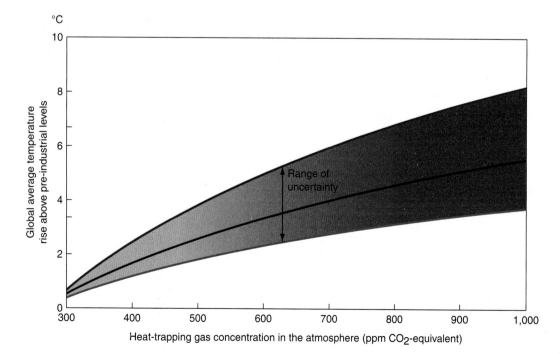

? Why does this projection of future temperature grow less certain with time?

FIGURE 16.24 As greenhouse gas emissions rise, the global average temperature also rises.

For the protocol to take effect, the countries that ratified it had to account for a total of at least 55 percent of global carbon dioxide emissions. Between 1997 and 2004, 122 nations ratified the Kyoto Protocol, representing 42.2 percent of global carbon emissions—not enough to bring the treaty into effect. The United States was not among the ratifying nations; if it had been, the percentage would have jumped to 78.3 percent. However, in November 2004, Russia ratified the treaty, thereby meeting the emissions percentage requirement and enacting the Kyoto Protocol on a truly global, international scale. By February 2005, 141 industrialized countries had signed the protocol, including all the nations of Europe. In 2012, 37 of the original countries that ratified the protocol agreed to an extension to 2020 to further reduce their greenhouse gas emissions. Canada did not commit to this extension, and became the first, and only, nation to withdraw from the protocol in December 2012. The United States continues to not make a formal commitment to reducing greenhouse gas emissions, whereas Australia and most European countries have done so and have set specific targets.

The Kyoto Protocol suggests that *emission trading* would allow more developed, wealthy countries to legally exceed their emissions quota by purchasing "credits" from other, less industrialized countries that would be unlikely to reach their allowable carbon quota. Emission trading focuses on carbon dioxide because it represents more than half of the warming attributed to greenhouse gases and because most of the emissions come from powered vehicles and agricultural activities.

How would emission trading work? Here is an example: Across the midwestern United States, seasonal wetlands known as prairie potholes, called sloughs in Canada, have been a nuisance to farmers for decades (**FIGURE 16.25**). Now they have the potential to become a new kind of crop—a greenhouse gas crop. Scientists from the U.S. Geological Survey discovered that these sloughs, when not farmed, are highly organic environments and can store 2.3 tonnes of carbon per 0.4 hectare per year. Multiplying that by 16 million hectares of sloughs acting as carbon "sinks" for a decade results in the removal of *360 million tonnes* of carbon from the atmosphere, equivalent to taking 4 million cars off the road. Imagine a coal-fired power plant in Saskatchewan paying a farmer in Saskatchewan to *not* plow and instead conserve his or her acres of sloughs, and you can understand how the carbon trade would work.

A power plant is a *source* of carbon dioxide emissions; the sloughs are a carbon *sink*. Bringing these together to match carbon, tonne for tonne, means that no net carbon would be added to the atmosphere, even though the power plant goes on generating energy for its customers and profiting from it. The farmer, too, benefits because he or she is paid for the "use" of the land without having to put in the time and expense of draining, filling, plowing, and farming the troublesome sloughs. Who would broker this deal? The power plant and the farmer would not even meet. They would simply buy and sell "carbon credits" in a market much like the stock exchange. In North America, the market was the Chicago Climate Exchange, established in 2003 by 14 companies, including Ford, DuPont, and Motorola. In 2010 it had 400 companies, institutions, and farming groups as member, before it was taken over by the Intercontinental Exchange, which has an office in Calgary. The potential market for

FIGURE 16.25 Sloughs in the Canadian Prairies, such as these east of Kenaston, Saskatchewan, can store carbon. These environments could remove 400 million tonnes of carbon from the atmosphere—equivalent to taking 4 million cars off the road—per decade.

 How can sloughs offset carbon dioxide production?

carbon dioxide is enormous. In the United States alone, estimates reach hundreds of billions of dollars.

The carbon trade could be global, because for some substances, such as carbon dioxide, it makes no difference where reductions are made. Because the troposphere is constantly mixing, moving, and changing, carbon emitted in North America could be absorbed anywhere in the world.

Supporters of global carbon trading say that it would allow realistic emissions reduction targets to be set in developed countries, which are the biggest carbon emitters (sources), without hurting the industrial economy. On the other side of the carbon trade (the sinks), developing or poor countries would benefit from the money invested by carbon-producing countries in reforestation and clean energy projects. Promoting energy efficiency during economic and industrial growth could help developing countries in another way as well: They could benefit from what the developed nations have learned in the two centuries since the Industrial Revolution.

Critics of carbon trading say that it is not a solution to greenhouse warming because it does not set specific goals for reducing carbon dioxide emissions. Most people who are familiar with the concept of carbon trading agree that it is only part of the solution and that cutting back carbon dioxide emissions should be a key component in managing the effects of human activities on global climate. The concept is also spurring businesses large and small to ask such questions as: "What are my company's options for reducing greenhouse gas emissions? Are there new business opportunities associated with addressing climate change?" Many believe that this is a form of global change from which Earth's citizens can benefit.

 Expand Your Thinking—Describe four ways to significantly reduce greenhouse gas emissions. What are the costs and benefits of each? You will have to do some independent research (on the web) to properly answer this.

LET'S REVIEW "GEOLOGY IN OUR LIVES"

Now that you have finished the chapter, "Geology in Our Lives" will have taken on new meaning for you. Let us review it: Global warming is changing Earth's climate, leading to rising sea levels, changes in weather, and ecosystem impacts. Global warming poses an extraordinary challenge to our economy and our environment. The weight of scientific evidence indicates that human activities play a significant role in global warming by releasing greenhouse gases into the atmosphere. Only by limiting greenhouse gas production can the worst future impacts of global warming be avoided. If we want to preserve our planet so that it continues to support the development of human civilization, we must reduce greenhouse gas production.

Earth's climate has changed over the past century, and there is new and stronger evidence that most of the warming observed in the past 50 years is attributable to human activities. Moreover, computer models are predicting that temperatures will continue to rise during the 21st century and beyond. This trend is described in the fourth report of the IPCC, which is based on the findings of an international team of climate scientists.

Meeting the challenge of global warming will require sustained effort over decades: on the part of governments, which must guide their cultures and societies as the effects of climate change unfold and as technological solutions are developed; on the part of industry, which must innovate, manufacture, and operate under new conditions in which climate change will drive many decisions; and on the part of the public, which must begin to pursue a more environmentally conscious path in purchases and lifestyles. In the next chapter we study the environments that are most sensitive to climate change—glaciers.

STUDY GUIDE

16-1 "Global change" refers to changes in environmental processes affecting the whole Earth.

- **Global warming** is the increase in the average temperature of Earth's surface (including the oceans). The decade 2000 to 2009 was the warmest recorded since 1880. However, the year 2010 was the hottest ever measured, and so the present decade may end up being the warmest on record.

- Climate is the long-term average pattern of weather of a particular region (say, over 30 years) and is the result of interactions among land, ocean, atmosphere, ice, and biosphere. It is the product of processes that accumulate and move heat within and between the atmosphere and ocean.

- Human activities have caused significant global changes in land use, air and water quality, and abundance of natural resources, particularly in the past two centuries. There is scientific consensus that human activities are also altering Earth's climate, largely due to increasing levels of carbon dioxide and other **greenhouse gases** released into the atmosphere by the burning of fossil fuels.

16-2 Heat circulation in the atmosphere and oceans maintains Earth's climate.

- Climate change is the product of changes in the accumulation and movement of heat in the ocean and atmosphere. To understand both natural and human influences on global climate, we must explore the physical processes governing the atmosphere and the ocean.

- Very little vertical mixing of gases occurs between the layers of the atmosphere. This means that one layer can be warming while another is cooling at the same time. The general circulation of air at the surface is broken up into a number of cells, of which the most common is the Hadley cell.

- **Thermohaline circulation** starts in the North Atlantic Ocean. When the warm surface water of the Gulf Stream reaches the cold North Atlantic, it cools and sinks. It travels southward as a cold, deep, salty current called the North Atlantic Deep Water. In the Southern Hemisphere, the current turns and flows northward past Australia into the Pacific Ocean. Eventually it rises in the North Pacific, becoming a warm surface current that flows westward around Africa. It may take about 1,600 years for water to complete the entire global cycle.

16-3 The greenhouse effect is at the heart of Earth's climate system.

- The **greenhouse effect** is the atmosphere's ability to store heat radiated from Earth. The heat absorbed by certain gases (such as water vapour [H_2O], carbon dioxide [CO_2]), and methane [CH_4]) maintains Earth's temperature at an average (and comfortable) 14°C. Without it, Earth's average temperature would be −18°C.

- Six principal greenhouse gases in Earth's atmosphere absorb long-wave radiation and keep Earth warm: water vapour, carbon dioxide, ozone (O_3), methane (CH_4), fluorocarbons, and nitrous oxide (N_2O). Combined, these gases make up about 1 percent of the atmosphere.

16-4 The global carbon cycle describes how carbon moves through natural systems.

- The **global carbon cycle** describes the forms (and exchange rates) that carbon takes in the reservoirs (storage points) and sources (points of origin) of carbon on Earth.

- Most of the carbon on Earth is contained in the rocks of the lithosphere; it was deposited slowly over millions of years. Carbon is stored in the lithosphere in two forms: Oxidized carbon is buried as carbonate (such as limestone, which is composed of calcium carbonate [$CaCO_3$]), and reduced carbon is buried in the form of organic matter (such as dead plant and animal tissue) and may be incorporated in shale and other organic-rich forms, such as fossil fuels.

- Carbon is also found in gases in the atmosphere, such as carbon dioxide and methane; carbon dioxide dissolved in the **hydrosphere** (oceans and fresh water); and organic matter in the biosphere (such as the simple carbohydrate glucose [$C_6H_{12}O_6$], in plants and animals).

16-5 Climate modelling improves our understanding of global change.

- Earth's climate system is very complex, with cycles and feedbacks and reservoirs that all interact with one another over different lengths of time. People who study global change use climate models, which are computer-based mathematical programs that simulate the behaviour and interaction of Earth's oceans and atmosphere.

- The concentration of carbon dioxide in the atmosphere has varied naturally by a factor of 10 in the last 600 million years, and today it is close to its minimum level, suggesting that we are in a relatively cool period in Phanerozoic history. Because of its role in warming Earth, carbon dioxide has affected global climate on long time scales. Significant global changes in climate, sea level, and atmospheric composition have been part of Earth's history for millions of years.

16-6 Human activities have increased the amount of carbon dioxide in the atmosphere.

- Four trillion tonnes of carbon is stored in the rocks of Earth's crust. The burning of fossil fuels (such as coal, oil, and gas) for energy and organic biomass (such as trees during **deforestation**) for energy, agricultural expansion or urban development are among human activities that release stored carbon into the atmosphere.

- In the last 200 years, carbon dioxide levels in the atmosphere have risen by 40 percent (from 280 ppm to over 400 ppm), and they continue to rise by approximately 2.1 percent each year.

- As a result of population growth and industrial development, the use of natural resources has increased by 1,000 percent in the last 70 years. Global population growth is expected to continue, reaching 9 billion people by 2050 and 11 billion by 2100. This population growth will put tremendous pressure on Earth's natural resources.

- Carbon dioxide released by changes in land use represents about 18 percent of total annual emissions. This source constitutes one-third of total emissions from developing countries and more than 60 percent of emissions from the least developed countries.

16-7 Earth's atmospheric temperature has risen by about 0.8°C in the past 100 years.

- According to published research, global surface temperatures have increased about 0.8°C (±0.18°C) since the late 19th century, and the linear trend for the past 50 years of 0.2°C (±0.03°C) per decade is nearly twice that of the past 100 years. The year 2012 was only a fraction of a degree cooler than 2010, the warmest year since record keeping began. January 2000 to December 2009 was the warmest decade on record, although it appears as though the decade we are in now will be warmer still. Scientific analysis of modern and past climates has shown that Earth is warmer now than at any time in the past 1,300 years.

- High latitudes are warming more and faster than low latitudes. The Arctic has warmed faster than any other region on Earth—by nearly 2.7°C in the last 30 years alone.

- Glaciers and sea ice all over the world are melting at the fastest rates ever measured. There are glaciers and icefields in Alberta, British Columbia, Yukon, Baffin Island, and Alaska and it is estimated that 99 percent of them are "retreating," or melting. The extent of snow cover in the Northern Hemisphere has decreased by about 10 percent since 1966. In 2007, the extent of Arctic sea ice dropped below 4 million square kilometres, the lowest extent yet recorded in the satellite record, although by 2012, the extent of Arctic sea ice had dropped even further—to 3.4 million square kilometres.

- Sea level has been rising at the rate of 1 mm to 3 mm per year for the past 100 years. This rate of increase may not seem significant, but it means that during the last 100 years, sea level has risen by up to 20 cm worldwide.

16-8 Global warming leads to ocean acidification and warming, melting of glaciers, changes in weather, and other impacts.

- Sea-surface temperature (SST) has increased by an average of 0.6°C in the last 100 years. **Ocean acidification** threatens marine organisms that rely on calcium carbonate.

- Signs and impacts of climate change in the biosphere include degradation and shrinkage of wildlife habitats, changes in the distribution of food sources for animals and people, disruption of the timing of migration patterns, and desertification.

16-9 Several international efforts are attempting to manage global warming.

- The **Intergovernmental Panel on Climate Change (IPCC)** was established in 1988 by the United Nations and the World Meteorological Organization to address the complex issue of global climate change, particularly global warming.

- The Kyoto Protocol is an international agreement stating that carbon dioxide emissions resulting from human activities are affecting Earth's climate and should be regulated over the long term.

KEY TERMS

climate feedback (p. 429)
deforestation (p. 436)
desertification (p. 441)
global carbon cycle (p. 430)
global circulation models (GCMs) (p. 432)
global climate change (p. 422)

global warming (p. 422)
greenhouse effect (p. 426)
greenhouse gases (p. 423)
hydrosphere (p. 425)
Industrial Revolution (p. 427)
insolation (p. 433)

Intergovernmental Panel on Climate Change
    (IPCC) (p. 444)
ocean acidification (p. 440)
sea-level rise (p. 422)
thermohaline circulation (p. 425)
trade winds (p. 425)

ASSESSING YOUR KNOWLEDGE

Please answer these questions before coming to class. Identify the best answer.

1. Earth's atmosphere has warmed by about
   a. 8°C in the last 100 years.
   b. 0.08°C in the last 1,000 years.
   c. 80°C in the last 100,000 years.
   d. 0.8°C in the last 100 years.
   e. The climate has not warmed.

2. The carbon dioxide content of the atmosphere has risen in recent decades. This is likely due to
   a. increased volcanic outgassing.
   b. decreased eccentricity and obliquity.
   c. increased burning of fossil fuels.
   d. the breakup of Pangaea.
   e. None of the above.

3. The country that emits the most greenhouse gases is
   a. China.
   b. Brazil.
   c. Russia.
   d. the United States.
   e. India.

4. Heat circulation in the atmosphere governs climate. It includes
   a. global circulation in the form of a number of "cells."
   b. thermohaline circulation, in which air takes tens of thousands of years to circle the globe.
   c. trade winds, which "trade" heat vertically from the troposphere into the ionosphere.
   d. oceanic upwelling, which releases heat into the atmosphere from warm deep waters.
   e. There is very little heat circulation in the atmosphere.

5. The greenhouse effect
   a. is the atmosphere's natural ability to store heat radiated from Earth.
   b. is governed in large part by the ability of ice to absorb heat.
   c. includes the process of short-wave radiation from the biosphere.
   d. does not include heat production by the Sun.
   e. All of the above.

6. The most important greenhouse gases include
   a. CO_2, CH_4, CFCs, O_3, N_2O, and H_2O.
   b. CO_4, FH_4, H_2O, CO_3, N_2O, and SO_2.

   c. CO_2, CO, NH_3, O_2, N_3O, and HO.
   d. HSO, CO_3, CHO_2, N_3O, and NH_4.
   e. Only CO_2 is important.

7. The atmosphere has several layers. These layers include the
   a. lithosphere, biosphere, and ionosphere.
   b. troposphere, stratosphere, and mesosphere.
   c. troposphere, ionosphere, and hydrosphere.
   d. biosphere, lithosphere, and atmosphere.
   e. All of the above.

8. Fluctuations in solar output in the 20th century are
   a. all that is needed to account for global warming.
   b. a relatively small component of modern climate change.
   c. not presently monitored.
   d. responsible for the rise in greenhouse gas production.
   e. Solar heating does not fluctuate.

9. Climate is
   a. the long-term average pattern of weather.
   b. the result of interactions among land, ocean, atmosphere, ice, and biosphere.
   c. the product of processes that accumulate and move heat within and between the atmosphere and ocean.
   d. changing due to global warming.
   e. All of the above.

10. The amount of greenhouse gas in the atmosphere is increasing due to
    a. increased volcanic activity.
    b. increased solar output.
    c. changes in oceanic circulation.
    d. decreased carbon cycling.
    e. human activities.

11. Which of the following statements about sea-level rise is true?
    a. It is not yet occurring but is likely to occur in the future.
    b. It is occurring now and is one of the major hazards associated with global warming.
    c. It is occurring now at the highest rate in the last half-million years.
    d. It is a major cause of global warming.
    e. It is not a major worry.

12. Global warming is changing the world's glaciers and oceans because of
    a. increased snowfall and ocean circulation.
    b. decreased melting and ocean cooling.
    c. increased melting and ocean acidification.
    d. decreased snowfall and greater thermohaline circulation.
    e. Glaciers and oceans are not changing.

13. Which of these reactions describes the calcification process?
    a. $2CO_2 + 2H_2O \longrightarrow 2HCO_3^- + 2H^+$
    b. $CaCO_3 + CO_2 + H_2O \longrightarrow Ca^{2+} + 2HCO_3^-$
    c. $2HCO_3^- + Ca^{2+} \longrightarrow CaCO_3 + CO_2 + H_2O$
    d. $H_2O + 2HCO_3^- \longrightarrow 2HCO_3^- + 2H^+$
    e. None of the above.

14. A carbon quota is
    a. the amount of carbon a single person is allowed to produce in one year.
    b. a limit of greenhouse gas emissions for each city in the world.
    c. a target greenhouse gas emission level for each country.
    d. the amount of greenhouse gas that will cause a certain level of damage.
    e. None of the above.

15. Some consequences of global warming include
    a. sea-level rise and coral bleaching.
    b. ocean acidification and glacier retreat.
    c. shifts in species ranges and competition.
    d. drought, storms, and rainfall changes.
    e. All of the above.

FURTHER RESEARCH

1. Describe how Earth's climate has changed over the following time scales by filling in the table.

Time Scale	Pattern of Temperature Change	Primary Causes of Change	Scientific Evidence of the Change	Resulting Environmental Changes
50 years				
1,500 years				
21,000 years				
125,000 years				
400,000 years				

2. How could climate change affect us in the future?

3. How could greenhouse gas emissions be reduced?

4. What human activities contribute to climate change?

5. How is climate change related to anomalous weather and climate events?

ONLINE RESOURCES

Learn more about global warming on the following websites.

The Center for Climate and Energy Solutions in the United States advocates an aggressive approach to mitigating climate change: www.c2es.org

These Canadian organizations have different approaches to climate change policies:
Pembina Institute: www.pembina.org
David Suzuki Foundation: www.davidsuzuki.org

The Intergovernmental Panel on Climate Change provides the latest research on climate change: www.ipcc.ch

The debate over climate change is summarized at: http://climatedebatedaily.com

NASA's Earth Observatory website has lots of images of Earth, and an overview of Earth's energy budget: http://earthobservatory.nasa.gov/Features/EnergyBalance

You can find out more about carbon dioxide and keep track of changes in concentrations in the atmosphere in real time at: http://co2now.org

Additional animations, videos, and other online resources are available at this book's companion website: www.wiley.com/go/fletchercanada

This companion website also has additional information about WileyPLUS and other Wiley teaching and learning resources.

CHAPTER 17
GLACIERS AND PALEOCLIMATOLOGY

Chapter Contents and Learning Objectives (LO)

17-1 A glacier is a river of ice.
  LO 17-1 *Describe glaciers and where they occur today.*

17-2 As ice moves, it erodes the underlying crust.
  LO 17-2 *List and define the various types of glaciers.*

17-3 Ice moves through the interior of a glacier as if on a one-way conveyor belt.
  LO 17-3 *Identify how ice moves inside a glacier and how this movement affects the front of the glacier.*

17-4 Glacial landforms are widespread and attest to past episodes of glaciation.
  LO 17-4 *Compare and contrast depositional and erosional features formed by glaciers.*

17-5 The majority of glaciers and other ice features are retreating due to global warming.
  LO 17-5 *Describe the response of various ice environments to global warming.*

17-6 The ratio of oxygen isotopes in glacial ice and deep-sea sediments is a proxy for global climate history.
  LO 17-6 *Describe how oxygen isotopes serve as a proxy for global climate history.*

17-7 Earth's recent history has been characterized by cycles of ice ages and interglacials.
  LO 17-7 *Define "marine isotopic stages" and describe what they indicate.*

17-8 During the last interglacial, climate was warmer and sea level was higher than at present.
  LO 17-8 *List characteristics of the last interglacial that are relevant to understanding the consequences of modern global warming.*

17-9 Glacial-interglacial cycles are controlled by the amount of solar radiation reaching Earth.
  LO 17-9 *Define the Earth–Sun orbital parameters.*

17-10 Together, orbital forcing and climate feedbacks produced paleoclimates.
  LO 17-10 *Describe climate feedbacks and give examples.*

GEOLOGY IN OUR LIVES

A glacier is a large, long-lasting river of ice that forms on land, undergoes internal deformation, erodes the crust, deposits sediment, and creates glacial landforms. Glaciers have shaped the mountains and nearby lands where they expanded during past climate changes. Earth's recent history has been characterized by cool climate periods, called ice ages, during which glaciers around the world expanded, and warm periods, called interglacials, during which glaciers retreated. By studying the history of climate changes, scientists learn how natural systems respond to shifts in temperature and ice volume. Today, most of the world's ice sheets, alpine glaciers, ice shelves, and sea ice are dwindling because we are in an interglacial period.

(?) The opening photo shows a glacier in Kluane National Park, Yukon. As you can see, glaciers are rivers of ice that sculpt the land surface by eroding the crust and depositing sediments as they move. Almost every mountain system in the world has evidence of past glaciation. What would evidence of past glaciation look like?

17-1 A Glacier Is a River of Ice

LO 17-1 *Describe glaciers and where they occur today.*

In 1837, a young Swiss zoologist named **Louis Agassiz** (1807–1873) used critical thinking to develop a radical new hypothesis. He suggested that northern Europe had once been covered by thick layers of ice, similar to those covering Greenland, during a period that he called the *ice age.* Having spent time walking among the many **glaciers** that fill the valleys of the Alps, Agassiz observed that these "rivers of ice" were in motion. He saw that their fronts advanced and retreated at various times and that they carried, locked within their icy grip, large amounts of bedrock, gravel, sand, and mud. He also observed glacial sediments deposited many kilometres from their source. Agassiz compared the *glacial striations* (linear gouges in bedrock made by rocks suspended in glacial ice) exposed by retreating glaciers to identical gouges found in bedrock far from any ice. In fact, he was able to explain many aspects of northern European (and later North American) sedimentary geology and *geomorphology* (the study of landforms) that until then had no good scientific explanation. For instance, he pointed out that the deep valleys and sharp peaks and ridges of the high Alps were the result of glacial erosion.

Today, Agassiz's *theory of the ice ages* is well established among scientists studying the **Quaternary Period**, the most recent period of geologic time (2.6 million years ago to the present). Having endured two decades of criticism from his peers before his hypothesis became generally accepted later in his career, Agassiz would be pleased to see how his original ideas now form the basis of scientific thinking about the origins of modern surface environments as well as global **paleoclimatology** (the study of past climates).

Glaciers

A glacier is a large, long-lasting river of ice that forms on land, undergoes internal deformation, and creates **glacial landforms** (**FIGURE 17.1**). Glaciers are made of compressed, recrystallized snow, usually carrying a large sediment load, which range in length and width from several hundred metres to hundreds of kilometres. Most glaciers are several thousand years old, but the glaciers covering Greenland and Antarctica have been in existence in one form or another for hundreds of thousands of years.

Glaciers develop on land at high elevations (mountains) and high latitudes (the Arctic and Antarctic) above the *snow line,* the elevation above which snow tends to accumulate from one year to the next rather than completely melting during the summer. Even mountains

FIGURE 17.1 Glaciers, such as those emanating from the Heiltskuk Icefield in the Coast Mountains of British Columbia, are large, long-lasting rivers of ice made of compressed, recrystallized snow.

(?) What are the dark stripes in the middle of the glaciers? Develop a hypothesis about how they form.

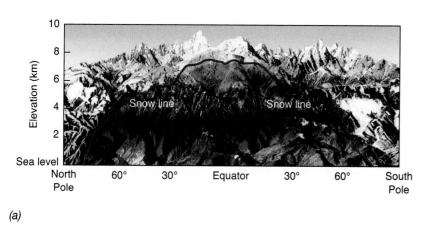

(a)

(b)

© Accent Alaska.com/Alamy

FIGURE 17.2 *(a)* In today's climate, glaciers form at elevations above 5,000 m between 0° and 30° latitude. At higher latitudes, they may form near sea level. *(b)* Permafrost is permanently frozen soil with a seasonal active layer that can support plant life.

 Why does permafrost form? What climate conditions are necessary?

in the tropics, such as the Andes Mountains of Peru, have a snow line, and mountainous regions in equatorial New Guinea and Africa have glaciers. But snow lines occur at lower elevations in temperate and Arctic regions than they do in the tropics.

In fact, the farther from the equator, the lower the snow line, until in the high Arctic, the ground at sea level is permanently frozen in what is known as **permafrost** (**FIGURE 17.2**). Permafrost is soil whose temperature remains at or below the freezing point of water for more than two years. It is found at high latitudes and on mountain slopes at high elevations in *periglacial environments* (environments dominated by ground ice and freeze/thaw processes, as in the tundra). The upper portion of permafrost is called the *active layer.* This layer is typically 0.6 m to 4 m thick, seasonally thaws and refreezes, and can support plant life for part of the year. In northern Siberia, the permafrost is over 1.4 km thick. Today, approximately 20 percent of Earth's land surface is covered by permafrost or glacial ice.

Above the snow line, snow accumulates from one year to the next in relatively large quantities. This is a fundamental requirement for a glacier to form, as it means that the snow cover can become so thick that the lower layers recrystallize and form *glacier ice.* This process may involve repeated freezing and thawing, forming a type of granular ice called *névé,* which, with continued compression, fuses to form *firn.* The weight of the overlying mass eventually causes firn to recrystallize as glacier ice (**FIGURE 17.3**).

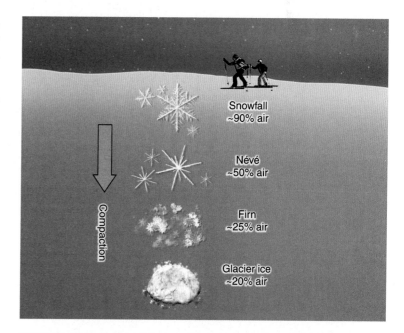

FIGURE 17.3 Snow will fuse under pressure to form névé and firn, which eventually recrystallizes as glacier ice.

 Why does snow turn to ice as it is buried?

Expand Your Thinking—Describe the theory of the ice ages. What is the evidence supporting the theory? Can you provide an alternative theory? How would you test your theory?

17-2 As Ice Moves, It Erodes the Underlying Crust

LO 17-2 *List and define the various types of glaciers.*

Ice behaves as a solid that is brittle; this is most obvious by the way it fractures, creating large fissures called *crevasses* as it moves over irregularities in the bedrock that forces the ice to bend and buckle. However, once glaciers reach thicknesses of about 40 m to 60 m or more their lower layers of ice experience *plastic flow* as weakly bonded layers of stacked water (H_2O) molecules begin to slide past each other (**FIGURE 17.4**).

Another way in which a glacier moves is by *basal sliding.* Basal sliding is dominant in glaciers whose temperature hovers around the melting point. It occurs when portions of the glacier move across the crust, which is lubricated by a layer of **meltwater**. Meltwater forms when increased pressure toward the base of a glacier lowers the melting point of ice; hence, the ice at the base melts more readily than the ice above it. Geothermal heat from Earth's interior also contributes to melting. The thicker a glacier becomes, the more efficient it is at trapping geothermal heat and the greater the role of this heat in creating meltwater.

As ice moves, it erodes the underlying bedrock. Most of this erosion occurs through two processes: plucking and abrasion (**FIGURE 17.5**). *Plucking* occurs when meltwater beneath the glacier penetrates fractures and crevices in bedrock and the resulting *ice wedging* (expansion of water when it freezes) breaks off pieces of rock. When the freezing water expands, it pries loose blocks of rock, which the moving ice picks up. Blocks of rock carried in the ice cause *abrasion* as they grind and pulverize the bedrock beneath a glacier, creating linear scars termed *glacial striations.* Pulverized rock, known as *rock flour,* is so abundant that the meltwater flowing away from a glacier is usually bluish-grey in colour due to suspended sediments. Through this erosion, sediments of all sizes become part of the glacier's load, and as a result, the glacial deposits are poorly sorted.

Types of Glaciers

Glaciers exist in a great diversity of forms and environments. However, there are two principal types: **alpine glaciers**, which are in mountainous areas, and **continental glaciers**, which cover larger areas and are not confined to the high elevations of mountains. Most of the concepts we discuss in this chapter apply equally to both types (**FIGURE 17.6**).

There are many types of alpine and continental glaciers. *Valley glaciers* are the smallest type of alpine glacier (see Figure 17.1). The ice in *temperate glaciers* is near its melting point throughout the year, from its surface to its base. Small variations in temperature—for example, from one season to the next—produce abundant water flowing from the front of the glacier, and basal slip is the dominant mode of ice movement. The ice in *polar glaciers,* in contrast, is always below freezing. Melting is rare, and ice loss occurs by *sublimation,* in which a solid (ice) turns into a gas (water vapour) without melting first. *Subpolar* glaciers have characteristics of both temperate and polar glaciers. They usually experience a seasonal zone of melting near the surface that generates some meltwater but little to no basal melt.

Larger *icecaps* can cover an entire mountain, a range of mountains, or even a volcano (provided it is not active). The edges of icecaps typically are characterized by tongues of ice called *outlet glaciers* that flow

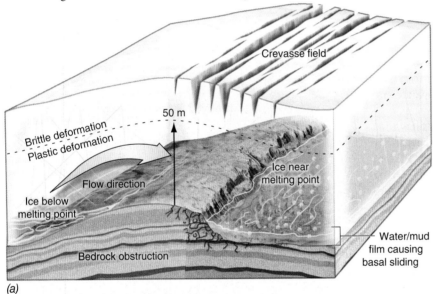

(a)

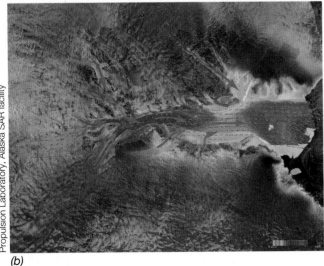

Canadian Space Agency/NASA/Ohio State University, Jet Propulsion Laboratory, Alaska SAR facility

(b)

 In the form of a testable hypothesis, describe how glacier movement occurs.

FIGURE 17.4 The temperature near the base of a glacier is an important factor in how it moves. *(a)* If the ice is near the melting point, it is more likely to move by basal sliding. If it is well below the melting point, the glacier is more likely to move by plastic flow. *(b)* Movement of the world's largest glacier, Lambert Glacier in East Antarctica, has been mapped by researchers. The smaller tributary glaciers flow at relatively low velocities (green, 100–300 m/y), but the downstream portion of the main Lambert Glacier flows at higher rates (blue to red, 400–800 m/y).

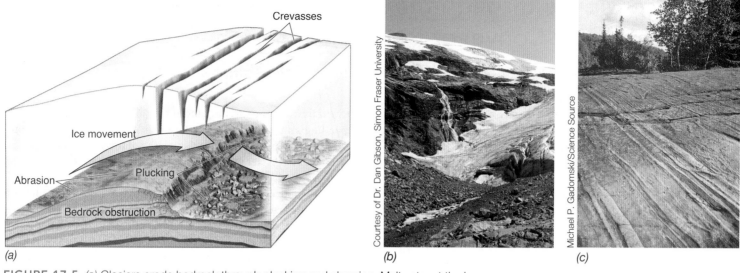

FIGURE 17.5 (a) Glaciers erode bedrock through plucking and abrasion. Meltwater at the base of the glacier enters fractures in the bedrock. When it freezes, ice wedging causes pieces of bedrock to break off and become incorporated into the flowing ice. Suspended blocks of rock and sediment abrade the bedrock, producing rock flour (b) and creating glacial striations (c).

? What glacial erosion process produces rock flour?

outward from the main ice mass, usually through a mountain pass. Outlet glaciers are often very large; the Lambert Glacier (Figure 17.4) is an outlet glacier of the East Antarctic Ice Sheet.

The largest glaciers are called *continental ice sheets*. They completely cover the landscape and are so massive that the topography of the land may not visibly influence them except near their edges. They are referred to as *ice sheets* because they have the geometry of a bed sheet: They are thin relative to their great surface area. The only continental ice sheets in existence today are on Greenland and Antarctica. These glaciers are so large that if all the ice on Greenland were melted, it would raise global sea level by about 7 m; the ice on Antarctica, if melted, would raise sea level by over 70 m.

? How do alpine glaciers differ from continental glaciers?

FIGURE 17.6 There are two principal types of glaciers: (a) alpine glaciers, which form in mountain systems, and (b) continental glaciers, such as the Greenland Ice Sheet, which are remnants of the last ice age, when they were much larger.

? **Expand Your Thinking**—How would you "prove" that glaciers cause erosion?

17-3 Ice Moves through the Interior of a Glacier as if on a One-Way Conveyor Belt

LO *17-3 Identify how ice moves inside a glacier and how this movement affects the front of the glacier.*

The interior of an active glacier is like a one-way conveyor belt, continuously delivering ice, rock, and sediments to the front, or *terminus*. This system feeds new ice from the **zone of accumulation**, which usually accounts for 60 percent to 70 percent of the glacier surface area, to the **zone of wastage**, where ice is lost not only by melting but also by sublimation and *calving,* in which large blocks of ice break off the front of the glacier. Between the zone of accumulation and the zone of wastage is the *equilibrium line,* marked by the elevation at which accumulation and wasting are approximately equal (**FIGURE 17.7**).

Although ice within a glacier continuously moves forward, the terminus will retreat if the rate of wastage exceeds the rate of ice delivery. The terminus will advance when the rate of delivery exceeds the rate of wastage. The rate of ice delivery to the terminus often is controlled by snow buildup in the zone of accumulation, but it also may respond to short-term accelerations and decelerations in basal sliding and plastic flow. Hence, in an active glacier, *ice is always moving*

forward through the interior, and the terminus advances or retreats, depending on the relative rates of ice delivery and wastage. This process is characterized as the **glacier mass balance**. Ice flow may slow and stop, but a glacier will not "flow" uphill or somehow retreat other than by wasting.

As ice moves forward within the body of a glacier, it delivers sediment toward the glacier's surface along fractures in the zone of wastage. Sediment is transported from the base of the glacier to the surface along a curved internal trajectory, aided by the constant wastage at the surface and a series of upward-curving fractures within the ice along which shearing action causes sediment to migrate upward. In this way, sediment is delivered from the interior of the glacier and accumulates on the surface as the ice wastes, thus forming a *lag deposit* of rocky blocks, gravel, sand, and mud that can be so thick and pervasive that it completely blankets the surface. For this reason, the surface of many glaciers in the zone of wastage is rich in sediment (**FIGURE 17.8**).

Vestiges of the Last Ice Age

Approximately 160,000 glaciers exist today, and glaciers can be found in approximately 47 countries and on every continent except Australia. Nevertheless, more than 94 percent of all the ice on Earth is locked up in the continental glaciers on Greenland and Antarctica.

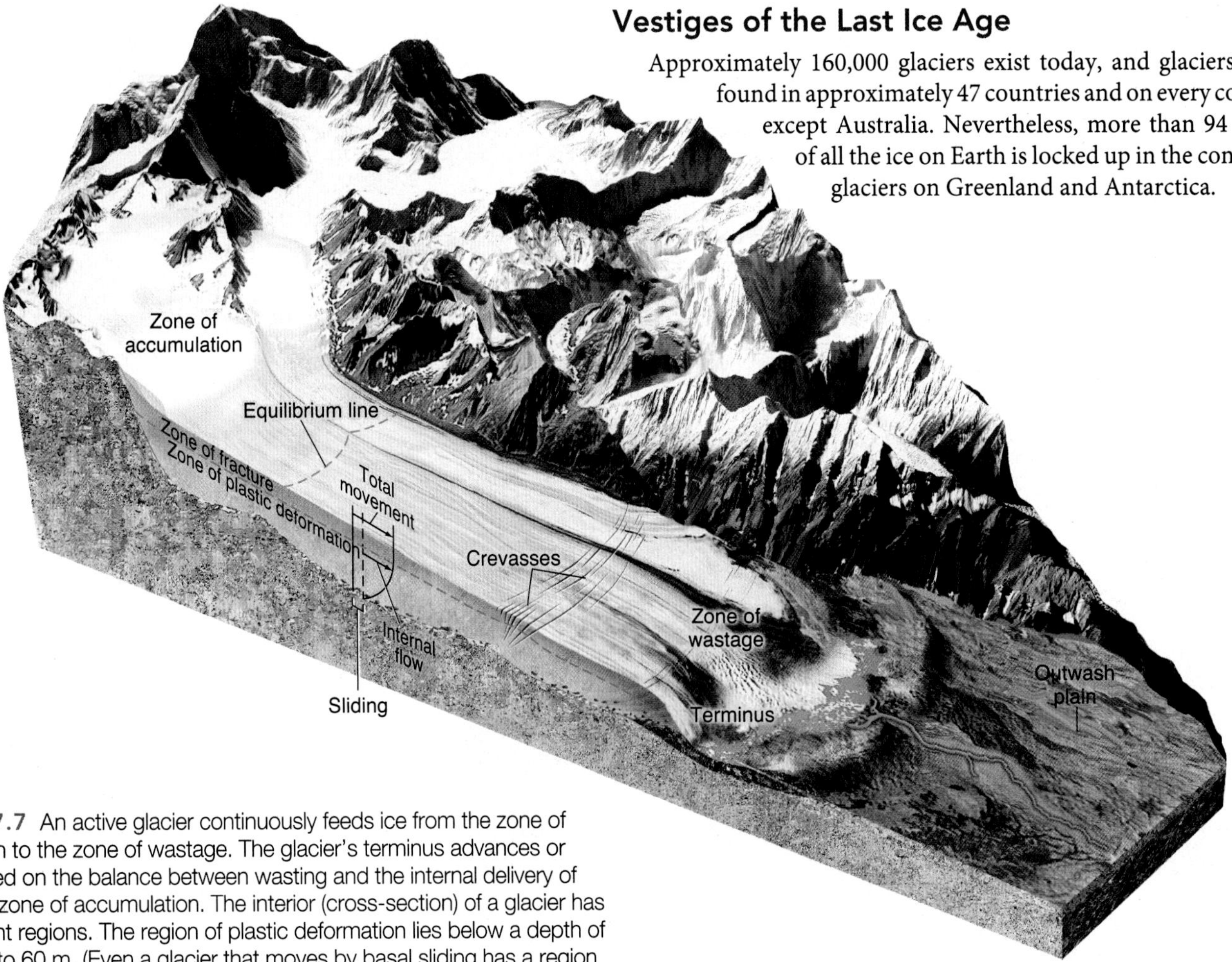

Zone of accumulation

Equilibrium line

Zone of fracture
Zone of plastic deformation

Total movement

Internal flow

Sliding

Crevasses

Zone of wastage

Terminus

Outwash plain

FIGURE 17.7 An active glacier continuously feeds ice from the zone of accumulation to the zone of wastage. The glacier's terminus advances or retreats based on the balance between wasting and the internal delivery of ice from the zone of accumulation. The interior (cross-section) of a glacier has two important regions. The region of plastic deformation lies below a depth of about 50 m to 60 m. (Even a glacier that moves by basal sliding has a region of plastic deformation.) Above this is the zone of fracture. The rate of ice movement is slowest at the base of the glacier due to friction on the bedrock.

(?) Describe ice movement inside a glacier that is retreating.

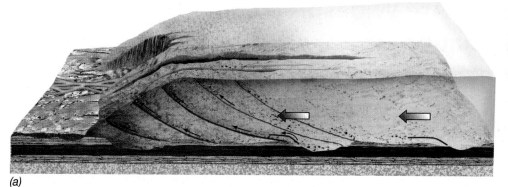

(a)

(b) © FLPA/Terry Whittaker/Age Fotostock America, Inc.

FIGURE 17.8 *(a)* Sediment is delivered to the zone of wastage along internal shear planes (i.e., curved black lines with arrows). *(b)* Poorly sorted sediment accumulates on the surface of a glacier.

 Describe the sources, pathways, and fate of sediments in an alpine glacier.

Greenland and Antarctica are vestiges of massive ice sheets that formed during the last ice age. The weight of these glaciers has depressed the underlying bedrock so that the interior land area of Greenland lies more than 300 m below sea level and portions of Antarctica lie 3 km below sea level. If you were to remove all the ice from these two places, the ocean would flow in and cover the majority of the land. Eventually the land would *glacioisostatically rebound* (a term, based on the concept of isostasy that we studied in Chapter 11 ["Mountain Building"], referring to upward recovery of the crust once ice has retreated), but this process would take several thousand years.

The ice on Greenland and Antarctica has two components: (1) thick inland ice sheets that rest on the crust, and (2) thinner floating ice shelves, ice streams, outlet glaciers, and glacier tongues that form at the edges of the sheets (**FIGURE 17.9**). As successive layers of snow build up in the interior regions where the main ice sheet develops, the layers beneath are gradually compressed into solid ice. Snow input is balanced by glacial outflow at the edges, so, unless the climate changes, the height of these ice sheets stays approximately constant through time.

The ice flowing away from the main accumulation area follows a pressure gradient toward the outer edges of the glacier. This process differs from that in alpine glaciers, where ice flow is generally downhill. In the case of ice sheets, flow is away from the area of highest compressive stress and toward

areas of lower stress along the coastline. There, ice either melts or is carried away as *icebergs,* which also eventually melt, thus returning the water to the ocean from whence it came. Outflow from accumulation sites is organized into drainage basins separated by *ice divides* (usually mountain ranges or stagnant ice) that concentrate the flow of ice into either narrow mountain-bounded outlet glaciers or fast-moving *ice streams* surrounded by slow-moving ice rather than by rock walls. Coastal settings where outlet glaciers are located are often deeply eroded into steep valleys known as *fjords* that are flooded by the ocean. In Antarctica, however, much of the outward-flowing ice has reached the coast and spread over the surface of the ocean to form floating *ice shelves* that are attached to ice on land. There are ice shelves along more than half of Antarctica's coast but very few in Greenland.

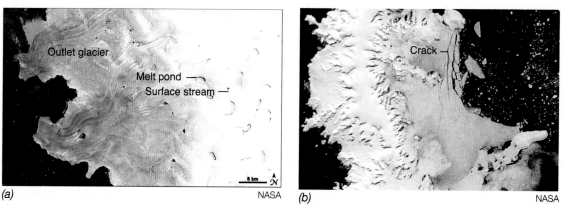

(a)   NASA   (b)   NASA

FIGURE 17.9 *(a)* Where the Greenland Ice Sheet flows into the ocean, it is organized into numerous outlet glaciers that break up into icebergs. *(b)* On the right side of the photo, outlet glaciers flow into the sea from the Antarctic Peninsula and feed floating ice shelves that are attached to their glacier sources. When these shelves break off, they may form massive icebergs.

 What drives ice flow in a continental glacier? In an alpine glacier? Does this difference affect sediment transport?

Expand Your Thinking—How would the line of equilibrium respond to a warming climate? To a cooling climate? Why?

17-4 Glacial Landforms Are Widespread and Attest to Past Episodes of Glaciation

LO 17-4 *Compare and contrast depositional and erosional features formed by glaciers.*

Over the past 500,000 years or so, Earth's history has been characterized by great swings in global climate, from extreme states of cold (ice ages) to warm periods called **interglacials.** Ice ages were characterized by the growth of massive continental ice sheets reaching across North America and northern Europe. At their maximum, these glaciers were over 4 km thick in places. Accompanying the spread of ice sheets was dramatic growth in alpine glaciers, many of which expanded into icecaps that covered large areas of mountainous territory and extended down valleys onto lands surrounding the mountains.

The formation of all this ice required a vast reservoir of water, and the oceans were the obvious source of that water. Evaporation of ocean water fed snow precipitation in the cold climate,

and glaciers expanded around the world. As a result, sea level fell by perhaps as much as 130 m, exposing shallow sea floor around the continents over a period of many thousands of years. In several places newly exposed sea floor connected adjacent lands that were previously separated by water. These "land bridges" allowed early communities of humans and animals to migrate to new lands. If you had been alive then, you could have walked on the newly exposed lands between Siberia and Alaska (this land bridge is presumed to have aided the peopling of the Americas), from France into England, from England to Ireland, and from Malaysia across Indonesia and on to Borneo (**FIGURE 17.10**). In many places, today's shore is tens or even hundreds of kilometres from where the shore existed during the ice ages.

We currently live in the latest interglacial, known as the *Holocene Epoch*, which began about 10,000 years ago. The last ice age began approximately 75,000 years ago and peaked between 20,000 and 30,000 years ago. For 50,000 years, the last ice age dominated climate. It was a time of great changes to the landscape as glaciers expanded and contracted leaving myriad *erosional* and *depositional* glacial landforms.

Erosional Glacial Features

During the last ice age, alpine glaciers carved the bedrock of mountainous regions and created numerous glacial landforms. Now that the ice age has ended and most glaciers have retreated, these exposed lands are being carved again by running water. *Glacial valleys*, or *glacial troughs*, that have been eroded by alpine glaciers are distinctly U shape in cross-section (**FIGURE 17.11**). Valleys shaped by stream erosion, in contrast, have a distinct V shape (as we will see in Chapter 19, "Surface Water"). In many cases, a mountain valley combines both shapes if it was once glaciated but is now being eroded by running water.

Mountainous regions contain many glacial landforms (**FIGURE 17.12**). A *cirque* is a bowl-shaped, amphitheatre-like depression formed by glacial plucking and abrasion at the head of a glacier. Cirques mark the birthplace of valley glaciers where conditions are favourable for perennial snow accumulation and the formation of glacial ice. In the Northern Hemisphere, many cirques are on the north-facing sides of mountains, where there is less direct sunlight than on south-facing slopes. Cirques may contain a small lake, called a *tarn*, and if there is a series of these lakes connected by a stream, these lakes are called

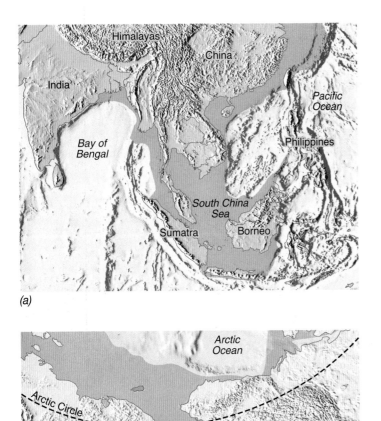

(a)

(b)

Describe changes in the water cycle that occur during an ice age.

FIGURE 17.10 During the last ice age sea level fell by approximately 130 m. Immense tracts of land were uncovered around *(a)* the island nation of Indonesia and *(b)* the Bering Strait, exposing important land bridges across which early humans migrated to new lands.

© Dennis Fandrich

FIGURE 17.11 The broad U shape of the Bow Valley north of Peyto Lake, Alberta, is a result of glacial erosion during the last ice age. During the current Holocene interglacial, the lower portion of the valley is being resculpted by running water. The final shape of the valley will be a combination of both types of erosional processes.

? What does the shape of a valley reveal about its history?

paternoster lakes. These high-elevation lakes, often set in idyllic alpine meadows, are known for their water clarity and beauty. Where three or more cirques carve out opposite sides of a mountain, the resulting sharp peak is a *horn.* Two adjacent valleys that are filled with glacial ice may carve a sharp ridgeline called an *arête.* These landscapes often include *hanging valleys* formed during the last ice age, when a glacier in a smaller tributary valley joined a larger "main" valley. The *tributary glacier* would not have an opportunity to erode its base to the floor of the main valley, so when glacial ice melted, the floor of the tributary valley was left "hanging." Hanging valleys are often marked by spectacular waterfalls plunging to the floor of the main valley.

Depositional Features

Glaciers transport sediment of all sizes (**FIGURE 17.13**), from huge boulders the size of buildings to tiny clay particles. They carry this material either on the surface of the ice or embedded within the body of the glacier. As a result, depositional features range from very poorly sorted to well sorted, and the nature of the sorting depends on whether the sediment was transported by water or ice (or a combination of the two), how long it was transported, and what happened to it after it was released from the ice by wasting. All sediment deposited as a result of glacial erosion is called *glacial drift.*

Glacial drift includes **glacial till,** unstratified and unsorted sediment that has been deposited by glaciers and, in large part, the melting of glaciers. Till consists of a chaotic mixture of angular rocks in a matrix of mud and sand produced by abrasion and plucking at the base of the glacier. *Glacial erratics* are blocks of bedrock with a foreign lithology that have been directly deposited by glaciers and the melting of glaciers. Some erratics are found hundreds of kilometres from their source. By mapping the distribution of erratics and using their lithology to trace them back to their source, geologists can determine the movement of ice that has long since disappeared.

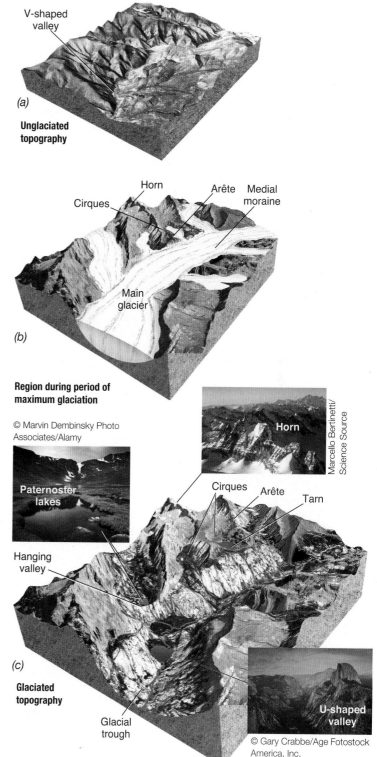

FIGURE 17.12 (a) Unglaciated topography is usually characterized by V-shaped stream valleys and low, rounded mountains. (b) During an ice age, valley glaciers carve and excavate the rock. (c) When ice melts, the resulting topography is characterized by dramatically steeper terrain, high relief, and sheer walls.

 State a hypothesis that predicts mountainous landforms carved by erosional processes.

One of the principal glacial landforms is the **glacial moraine** (FIGURE 17.14). A moraine is a deposit of till that marks a former position of the ice. There are several types of moraines. *Ground moraines* tend to be relatively horizontal deposits with a hummocky topography and are deposited beneath moving ice. An *end moraine* is a ridge of till marking the front edge of ice movement. *Terminal moraines* are ridges of till left behind at the front edge of a retreating glacier at the line of its farthest advance. There may be several end moraines marking various stopping points during the history of a glacier, but there is only one terminal moraine that marks its farthest advance. *Lateral moraines* are deposits of till marking the edges of a glacier against a valley wall; and a *medial moraine* is formed when two valley glaciers coalesce, and till from each mingles to create a boundary between the two. Medial moraines are seen as black streaks in systems of valley glaciers composed of two or more tributary ice streams (see Figures 17.6 and 17.14). All types of moraines except the ground moraine are usually found in a ridge composed of till. The shape and position of a moraine allow scientists to reconstruct the size, shape, and extent of the former glacier. Radiometric dating (primarily using ^{14}C) of material such as fossil wood or peat within a moraine also allows geologists to determine the timing of glacier advance and retreat.

Sediments within glacial drift can be picked up and moved by streams of meltwater that emanate from a glacier. The resulting deposit is known as *stratified drift*. One form of stratified drift is the *outwash plain* (FIGURE 17.15). Streams running from the end of a melting glacier are usually choked with sediment and form *braided streams,* which are channels with a large sediment load, forming numerous islands as well as sand and gravel bars (as discussed in Chapter 19, "Surface Water"). Much of the sediment is ultimately deposited as poorly sorted stratified layers in an outwash plain. These deposits often are referred to simply as **glacial outwash**. If meltwater streams erode into outwash deposits, the banks form river terraces called *outwash terraces*.

Depressions called *kettle holes* may form in outwash where a large block of ice has been buried by sediment and later melts. Kettle holes may eventually fill with water to become *kettle lakes* in which fine-grained sediment usually collects. Streams and lakes may even form on top of a stagnant glacier and accumulate stratified sediment on the ice. When the glacier melts, these deposits are set down on the ground surface and become hills known as *kames*. Sediment in kames is stratified, because the ice melts uniformly. Former stream deposits on the ice become *kame terraces*.

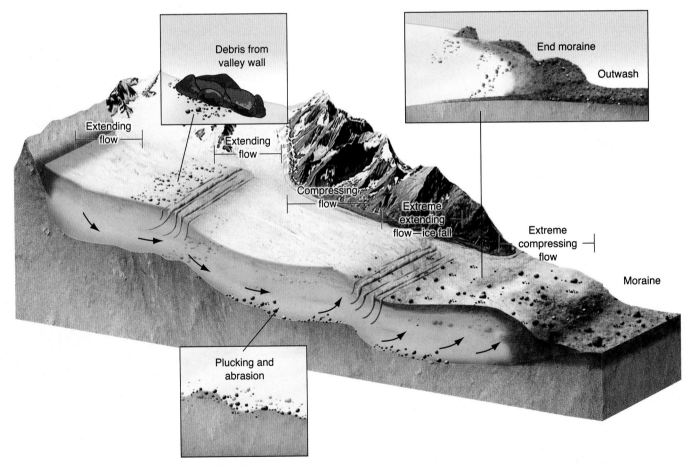

FIGURE 17.13 Glaciers transport sediment of all sizes. As the ice moves over uneven terrain, portions of the glacier compress and extend, causing changes in the surface topography of the glacier as well as differences in the internal movement of the ice. Glacial sediments are eroded from bedrock through plucking and abrasion. They may fall from surrounding valley walls onto the ice and be carried to the front of the glacier by the internal movement of ice into the zone of wastage. When the ice retreats, the resulting deposit is known as glacial drift.

(?) List and describe the ways that glaciers move and deposit sediment.

(a)   Courtesy of Dr. Dan Gibson, Simon Fraser University

(b)   Copyright © William W. Shilts

(c)   © Ed Darack/Science Faction/Corbis

FIGURE 17.14 *(a)* Glacial till is unstratified and unsorted sediment that has been deposited directly by wasting ice. *(b)* Moraines are deposits of till formerly in direct contact with ice. This photo of a glacier on Bylot Island, Nunavut, shows a terminal moraine and lateral moraine. *(c)* When valley glaciers coalesce, like those shown from the Kaskawulsh Glacier in the St. Elias range of Yukon Territory, their lateral and medial moraines come together to form strips of sediment (till) on the surface of the ice.

 What determines the composition and texture of sediment in all moraines?

It is common to find glaciated landforms composing *kame and kettle topography* where kettle lakes and kame deposits are spread across the land.

Eskers are long, sinuous ridges of sand and gravel deposited by streams that formerly ran under or within a glacier. As the ice melts, the channel deposits are lowered to the ground and form a ridge known as an esker. The land surface beneath a moving continental ice sheet can be moulded into smooth, elongated forms called *drumlins*.

Other types of glacial deposits include *glaciomarine drift,* which collects when glaciers reach the ocean or a lake. After entering the water, glaciers typically calve off large icebergs that float on the surface until they melt. Upon melting, the rock debris in the iceberg is deposited on the sea floor or lake bed as an unsorted layer of sediment. Sometimes single large rock fragments fall out on the floor of the water body; these are called *dropstones.* Fjords are narrow inlets along the seacoast that were once occupied by a valley glacier, called a *fjord glacier.*

 What is the difference between "till" and "outwash"?

FIGURE 17.15 Glacial landforms associated with sediment deposition.

Expand Your Thinking—How would you determine the age of the past advance and retreat of an alpine glacier? What would the significance of that history be?

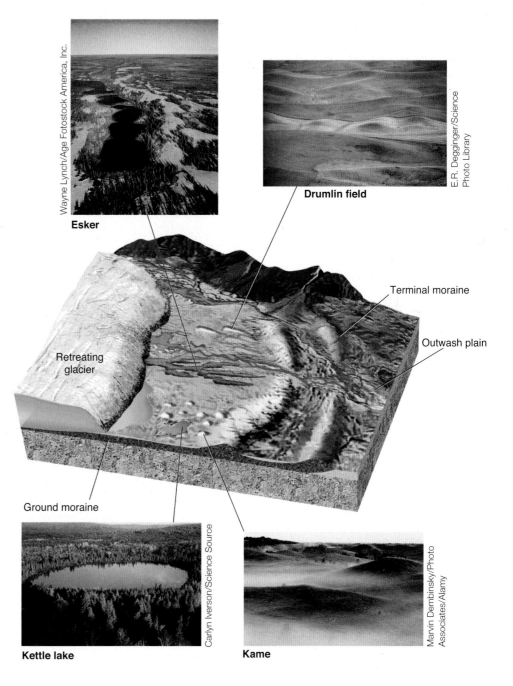

Esker

Drumlin field

Terminal moraine

Outwash plain

Retreating glacier

Ground moraine

Kettle lake

Kame

Wayne Lynch/Age Fotostock America, Inc.

E.R. Degginger/Science Photo Library

Carlyn Iverson/Science Source

Marvin Dembinsky/Photo Associates/Alamy

? CRITICAL THINKING

INTERPRETING GLACIAL LANDSCAPES

Please work with a partner and answer the following questions using **FIGURE 17.16**.

1. Label as many glacial landforms and features as you can.
2. Describe the past maximum extent of glaciation. What is your evidence?
3. Several glaciers are depicted in Figure 17.16. Indicate if any are advancing or receding.
4. What criteria did you use in answering question 3? What evidence did you use? What additional information, if any, do you need to develop an accurate answer?
5. Glacial environments are rich sources of sediment. Describe the sources of sediment and sediment characteristics in the various environments of Figure 17.16.
6. As a glaciologist, you are asked to study this region and report on the impacts of global warming. Describe the hypothesis you want to test, the data you will collect, and the methods you will use. Consider questions such as: How will monitoring play a role in your project? How will you establish a history of glacier trends? How will you project/predict future changes?
7. Assuming that global warming is causing accelerated glacial recession here, how will this impact communities that rely on these glaciers for water resources? What are the short-term and long-term impacts? How might communities prepare today to adapt to these impacts?

FIGURE 17.16

17-5 The Majority of Glaciers and Other Ice Features Are Retreating due to Global Warming

LO 17-5 *Describe the response of various ice environments to global warming.*

Because global warming threatens the stability of glaciers around the world, scientists use satellite monitoring to track changes in alpine glaciers, ice sheets, sea ice, and ice shelves. The data show that the majority of glaciers and other ice features are retreating due to global warming, and on Greenland and Antarctica, the rate of retreat has accelerated in the past decade.

Alpine Glaciers

Several research groups are monitoring the world's alpine glaciers, and the most recent data indicate that these glaciers are disappearing (FIGURE 17.17). Statistics indicate that in 2006–2007, the average rate of melting and thinning among alpine glaciers more than doubled: an average of 1.5 m of ice thinning occurred compared to earlier losses averaging half a metre per year. Among some key outlet glaciers on Antarctica, the rate of thinning is up to 9 m/y. This increased thinning continues a long trend in accelerated ice loss over the past 25 years and brings the total global average glacier thinning since 1980 to over 11.5 m of ice.

Arctic Sea Ice

In the summer of 2012, the area covered by Arctic sea ice shrank to a 33-year low based on satellite data that has been collected since 1979. The 2012 record is significantly below the minimum set in 2007 (see Figure 16.19). Ironically, the accelerating retreat of sea ice may be due to changes in climate brought on by the lack of sea ice itself.

When there is less sea ice in the summer, the Arctic Ocean receives more heat from the Sun. The warmer water makes it harder for the ice to recover in the winter; hence, there is a greater likelihood that sea ice will retreat during the summer. This process repeats itself year after year. Researchers call the switch from white snow and ice to black sea water an *albedo flip* (the term "albedo" describes the reflectivity of Earth's surface to sunlight). White (heat-reflecting) ice is being converted into black (heat-absorbing) water. The longer this process continues, the less likely the ice will recover.

Ice Shelves

Ice shelves are thick slabs of ice that are attached to coastlines and extend out over the ocean. In the natural course of events, ice shelves may break off and become large icebergs. Beginning in the mid-1990s, however, some ice shelves began exhibiting a new behaviour: rapid disintegration into small pieces, most likely as the result of warming temperatures (FIGURE 17.18). In the Antarctic summer of 2002, the 220-m-thick Larsen B ice shelf on the Antarctic Peninsula broke up. Within barely one month's time, it had disintegrated into about 3,250 km² of fragmented ice, an area greater than half of Prince Edward Island. An event of this magnitude had never been witnessed before. As a consequence of this breakage, glaciers in the adjacent mountains have surged forward, thinning at an accelerated rate because ice shelves no longer buttress where the glaciers enter the ocean. Since 2002, six other ice shelves have collapsed around the Antarctic Peninsula.

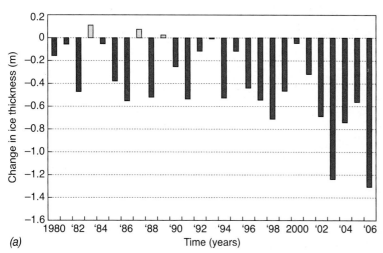

(a)

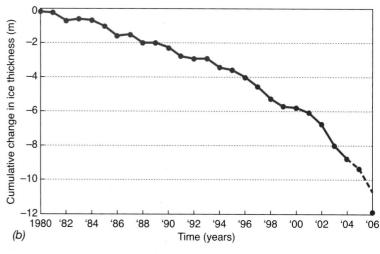

(b)

FIGURE 17.17 Global alpine glacier thinning. *(a)* The annual change in mean ice thickness has been accelerating since 1980. *(b)* The same data, plotted as cumulative ice thickness, show a total reduction in thickness of over 11.5 m.

 What evidence would you expect to find in alpine valleys that glaciers are retreating?

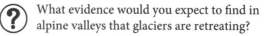 **Expand Your Thinking**—How might warming air cause glacier thinning?

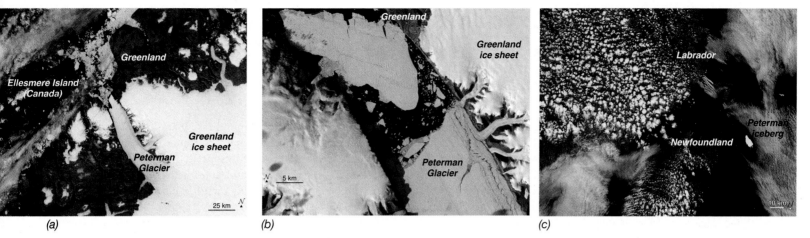

(a) (b) (c)

FIGURE 17.18 On August 5, 2010, satellite imagery revealed that an enormous 251 km² chunk of ice broke off the floating portion of Peterman Glacier that connects the Greenland Ice Sheet to the Arctic Ocean along the west coast of Greenland. This created a massive iceberg equivalent to about 10 percent of the volume of the Peterman Glacier. The iceberg eventually made its way to the coast of Newfoundland. *(a)* Intact Peterman glacier on July 28, 2010, prior to the *(b)* August 16, 2010, satellite image of the ice island that calved off the Peterman Glacier. *(c)* August 1, 2011: The massive iceberg, calved from Peterman Glacier the year before, makes its way to the coast of Newfoundland.

(?) Aside from increasing air temperature due to global warming, what other factors may have contributed to the collapse of Peterman Glacier?

GEOLOGY IN OUR LIVES

MELTING ICE SHEETS

The Greenland Ice Sheet

Since 1979, scientists have tracked the extent of summer melting of the Greenland Ice Sheet (**FIGURE 17.19a**). In 2007, the extent of melting broke the record set in 2005 by 10 percent, making it the largest season of melting ever recorded. Melting on portions of the glacier rose 150 percent above the long-term average, with melting occurring on 25 to 30 more days in 2007 than the average for the previous 19 years. In the past decade, the total mass deficit (the annual difference between snowfall and melting) tripled, and the amount of ice lost in 2008 was nearly three times the amount lost in 2007. In 2009, scientists announced that Greenland's ice was melting at a rate three times faster than it was only five years earlier.

The Antarctic Ice Sheet

The Antarctic Ice Sheet (**FIGURE 17.19b**) is divided into three regions: *West Antarctica*, the *Antarctic Peninsula*, and *East Antarctica*. In West Antarctica, ice loss has increased by 59 percent over the past decade, to about 132 billion tonnes per year, while the yearly loss along the Antarctic Peninsula has increased by 140 percent, to 60 billion tonnes, and 87 percent of the glaciers there are retreating. In East Antarctica, melting is largely confined to the seaward edge of the ice (where it is caused by warm ocean currents), but every year the melting encroaches farther inland and onto higher elevations. Despite expectations that snowfall in Antarctica would increase in the warmer atmosphere and thereby offset melting, there has been no statistically significant increase in snowfall, and recent data suggest that snowfall actually has decreased slightly. Overall, Antarctica is losing a total of 100 km³ of ice each year, and the rate of loss is accelerating.

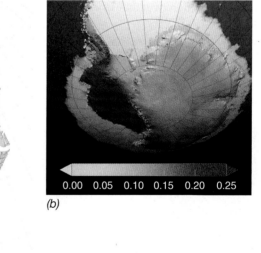

0.00 0.05 0.10 0.15 0.20 0.25

(b)

Melt day anomaly

−30 −15 0 15 30

(a)

FIGURE 17.19 *(a)* Melting on the Greenland Ice Sheet has tripled over the past decade. This map compares the number of days on which melting occurred in 2007 to the average annual number of melting days from 1988 to 2006. *(b)* Antarctica has warmed at a rate of about 0.12°C per decade since 1957, for a total average temperature rise of 0.5°C.

17-6 The Ratio of Oxygen Isotopes in Glacial Ice and Deep-Sea Sediments Is a Proxy for Global Climate History

LO *17-6 Describe how oxygen isotopes serve as a proxy for global climate history.*

As you learned in Chapter 16 ("Global Warming"), carbon dioxide levels in the atmosphere have undergone dramatic shifts in the past several hundred thousand years. But in that time, they never reached the high levels prevailing today. We know this because scientists can measure past carbon dioxide content of the atmosphere in samples obtained by drilling in continental ice sheets and alpine glaciers as well as in biogenic sediment composed of fossilized plankton on the sea floor.

In *ice cores* (FIGURE 17.20), the past carbon dioxide content of the atmosphere is measured directly from bubbles of air that were trapped during the formation of glacial ice. The longest ice cores (over 3 km in length) come from Antarctica. In this case, carbon dioxide is being used as a **climate proxy**—an indicator of climate in the past—because it is directly related to the heat-trapping ability of the atmosphere. But cores provide other measures of past climates as well. For instance, fossil snow also contains information about the temperature of the atmosphere and the amount of sunlight-blocking dust. Deep-sea cores can record changes in global ice volume and ocean chemistry.

Oxygen Isotopes and Global Ice Volume

Deep-sea sediment is composed of the microscopic shells of fossil plankton. The chemistry of these shells—for instance, tiny planktonic *foraminifera*—provides chemical clues to the climate prevailing when they formed. Cores of these sediments offer a record of climate history extending hundreds of thousands to millions of years back through time.

Foraminifera use dissolved compounds and ions in sea water to precipitate microscopic shells of calcite ($CaCO_3$). Both the $CaCO_3$ of foraminifera skeletons and a molecule of water (H_2O) in sea water contain oxygen (O). In nature, oxygen occurs most commonly as

the isotope ^{16}O, but it is also found as ^{17}O and ^{18}O. Water molecules composed of the heavier isotope ^{18}O do not evaporate as readily as those composed of the lighter isotope ^{16}O. Likewise, in atmospheric water vapour, heavier water molecules with ^{18}O tend to precipitate (as rain and snow) more readily than those composed of lighter ^{16}O (FIGURE 17.21).

Both evaporation and precipitation of oxygen isotopes occur in relation to temperature. $H_2^{18}O$ tends to be left behind when water vapour is formed during the evaporation of sea water, and it tends to be the first molecule to condense when rain and snow are forming. Hence, since most water vapour in the atmosphere is formed by evaporation from the tropical ocean, by the time it travels the long distance to the high latitudes and elevations where ice sheets and glaciers are located, it is relatively depleted in $H_2^{18}O$ and enriched in $H_2^{16}O$. This means that during an ice age, vast amounts of $H_2^{16}O$ are locked up in global ice sheets for thousands of years. At the same time, the oceans are relatively enriched in $H_2^{18}O$. Since the ratio of ^{18}O to ^{16}O in the shells of foraminifera mimics the ratio of these isotopes in sea water, the oxygen isotope content of these shells provides a record of global ice volume through time.

Oxygen Isotopes and Atmospheric Temperature

Oxygen isotopes in fossil foraminifera provide a record of global ice volume, and in ice cores, oxygen isotopes provide a record of changes in air temperature above a glacier. Because the atmosphere is so well mixed, the isotopic content of air above a glacier is indicative of the temperature of the atmosphere; hence, the isotopic content of snow is useful as a proxy for global atmospheric temperature. At the poles, as an air mass cools and water vapour condenses to snow, molecules of $H_2^{18}O$ condense more readily than do molecules of $H_2^{16}O$, depending on the temperature of the air. Typically, above a glacier, the

Marc Steinmetz/Aurora Photos Inc.　　　Carlos Muñoz-Yage/Science Source

Marc Steinmetz/
Aurora Photos Inc.　　Pasquale Sorrentino/Science Source　　Marc Steinmetz/Aurora
Photos Inc.

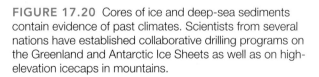

FIGURE 17.20 Cores of ice and deep-sea sediments contain evidence of past climates. Scientists from several nations have established collaborative drilling programs on the Greenland and Antarctic Ice Sheets as well as on high-elevation icecaps in mountains.

 What is the point of taking cores of old ice and marine sediments?

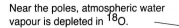

Near the poles, atmospheric water vapour is depleted in ^{18}O.

Heavy ^{18}O-rich water condenses over mid-latitudes.

Meltwater from glacial ice is depleted in ^{18}O.

Water slightly depleted in ^{18}O evaporates from warm subtropical waters.

FIGURE 17.21 The H$_2$^{18}O water molecule does not evaporate as readily as the H$_2$^{16}O molecule. Once in the atmosphere, water vapour composed of H$_2$^{18}O tends to condense and precipitate more readily in cooling air than a molecule of H$_2$^{16}O. Because most water vapour originates from the tropical ocean, by the time it travels to high latitudes and high elevations where glaciers form, it is enriched in H$_2$^{16}O relative to sea water. Hence, snow and ice are also enriched in the H$_2$^{16}O molecule.

(?) ^{16}O and ^{18}O behave differently in the water cycle of evaporation and precipitation. How does this reveal climate?

condensation falls out of a cloud as snow. This fractionation of oxygen isotopes, where ^{18}O is more readily incorporated versus ^{16}O, increases with decreasing global temperature; the colder the air temperature, the greater the proportion of ^{18}O versus ^{16}O that is incorporated into the snow. Thus, the oxygen isotopic content of snow (measured as the ratio of ^{18}O to ^{16}O) is a proxy for air temperature; hence, cores of glacial ice record variations in air temperature through time.

Because global ice volume and air temperature are related, the records of oxygen isotopes in foraminifera and glacial ice show similar patterns. These records provide researchers with two independent proxies for the history of global climate. Many researchers have tested and verified the history of global ice volume as inferred from oxygen isotope values in deep-sea cores and the history of temperature preserved in glacial cores from every corner of the planet. Past episodes of cooler temperature recorded in ice cores strongly correlate to periods of increased global ice volume recorded by the oxygen isotopes of marine sediments. Likewise, past episodes of warmer climate correlate well to periods of decreased ice volume.

An example of this record is shown in **FIGURE 17.22**. The variation in abundance of oxygen isotopes in ice (a proxy for atmospheric temperature) and in marine foraminifera (a proxy for global ice volume) do indeed display strong agreement and provide researchers with a global guide for interpreting past climate patterns and events.

(?) What makes oxygen isotopes such good recorders of global climate?

(?) **Expand Your Thinking**—Describe other types of climate proxies you can think of.

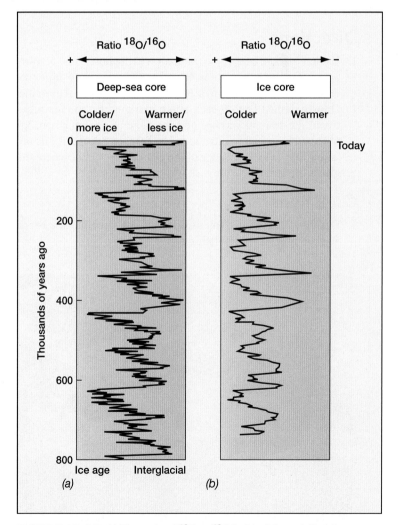

FIGURE 17.22 (a) The ratio of ^{18}O to ^{16}O in fossil foraminifera in deep-sea cores provides a proxy for global ice volume. (b) The ratio of ^{18}O to ^{16}O in cores of glacial ice documents changes in atmospheric temperature, confirming that decreased ice volume inferred from deep-sea cores correlates to times of warmer atmosphere, while increased ice volume recorded in deep-sea cores correlates to times of cooler atmosphere.

CRITICAL THINKING

MASS BALANCE OF A GLACIER

The U.S. Geological Survey (USGS) operates a long-term program to monitor glacier mass balance at certain key *benchmark glaciers* in the United States. "Glacier mass balance" refers to changes in ice volume that occur when glacier wasting and accumulation increase or decrease from one year to the next. *South Cascade Glacier* in the state of Washington is a long-term, high-quality mass balance monitoring site operated by the USGS.

FIGURE 17.23 shows five photos of South Cascade Glacier from 1928, 1958, 1979, 2003, and 2006. A comparison of these reveals much about the recent history of this glacier. FIGURE 17.24 shows monitoring data collected by USGS glaciologists. The data show trends in annual mass balance, cumulative mass balance, and air temperature over the past several decades. Please work with a partner to carry out the following exercise.

Directions

1. Using a marker, label these features on each photo: zone of wastage, zone of accumulation, lateral moraine, ground moraine, crevasse field, glacial drift, glacial outwash, recessional moraine.
2. Draw a dashed line on each photo marking the boundary between the zone of wastage and the zone of accumulation. Ignore patches of snow surrounded by bare ice. Label this line the "snow line." Along the edge of the glacier, draw a line where the glacier surface meets the rock wall on the left side of the valley.

Questions

1. Describe the evidence you used to delineate the zone of wastage and the zone of accumulation.

2. During the elapsed time, has the snow line generally remained stationary, moved to a lower elevation, or moved to a higher elevation? What does this indicate about the glacier?
3. Compare the photographs from 1928 and 1979. What is the evidence that the glacier has thinned in the zone of wastage?
4. Comparing the photographs from 1979 and 2006, what is the evidence that the glacier has thinned in the zone of wastage?
5. Describe the behaviour of the glacier terminus over the elapsed time. Describe the glacial processes that caused this change.
6. Using Figure 17.24, answer the following: What is the relationship between Figures 17.24a and 17.24b? Why is it that in Figure 17.24a, the lines do not show steep slopes, but in Figure 17.24b, the line does show a steep slope?

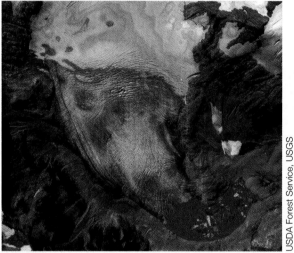

(b) 1958

(a) 1928

(c) 1979

FIGURE 17.23 South Cascade Glacier in *(a)* 1928, *(b)* 1958, *(c)* 1979, *(d)* 2003, and *(e)* 2006.

7. Compare the trends in Figure 17.24b with your interpretation of evidence in Figure 17.23. Do they agree or disagree? What do these two independent lines of evidence indicate about South Cascade Glacier? Is there any additional evidence that supports your conclusion?

8. State a hypothesis accounting for your observations of the glacier for the 1928–2006 period.

9. Describe a research project in which a glaciologist tests your hypothesis. Write a proposal asking for funding to conduct this research.

(d) 2003 · USDA Forest Service, USGS

(e) 2006 · USGS

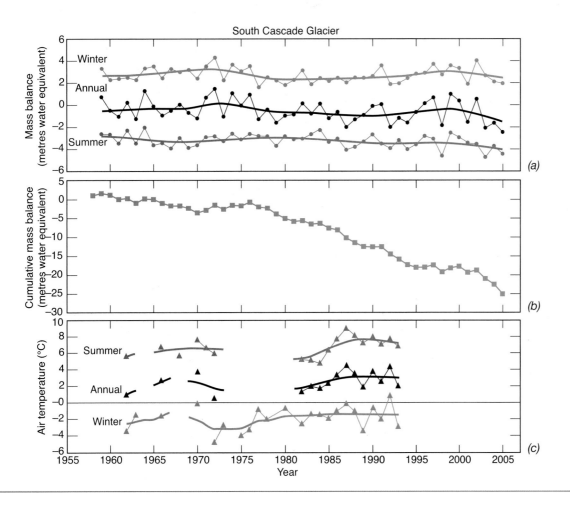

FIGURE 17.24 Observational measurements at South Cascade Glacier: (a) mass balance; (b) cumulative mass balance; and (c) air temperature.

17-7 Earth's Recent History Has Been Characterized by Cycles of Ice Ages and Interglacials

LO *17-7 Define "marine isotopic stages" and describe what they indicate.*

The microscopic plankton called foraminifera reveal an intriguing history of global ice volume that agrees well with ice core records from both Greenland and Antarctica (**FIGURE 17.25**). These natural archives show that global climate change is characterized by alternating warm episodes and ice ages occurring approximately every 100,000 years. To name these episodes, scientists have developed a numbering system referred to as **marine isotopic stages,** or MIS for short. Odd-numbered stages are warm and even-numbered stages are cool. We are living in the latest warm episode, known as MIS 1, and the last ice age is known as MIS 2. Figure 17.25 illustrates this naming system.

This history of cooling and warming has several important features:

1. Major glacial and interglacial cycles are repeated approximately every 100,000 years.

2. Numerous minor episodes of cooling (called *stadials;* MIS 4 is an example) and warming (called *interstadials;* MIS 3 is an example) are spaced throughout the entire record.

3. Global ice volume during the peak of the last interglacial, approximately 125,000 years ago, was lower than at present, and global climate was warmer.

4. Following the last interglacial, global climate deteriorated in a long, drawn-out cooling phase culminating in MIS 2, approximately 20,000 to 30,000 years ago, with a major glaciation.

5. The current interglacial has lasted approximately 10,000 years.

Interglacial Cycles

During the culmination of the last ice age (MIS 2), snow and ice accumulation built up massive continental glaciers in Europe, North America, Greenland, and Antarctica (**FIGURE 17.26**). In places snow and ice accumulations were 3 km to 4 km thick and as high as some modern mountain ranges. The ice sheet across eastern and central Canada, known as the *Laurentide Ice Sheet* (farther to the west, it was called the *Cordilleran Ice Sheet),* and the ice sheet across northern Europe, known as the *Fennoscandian Ice Sheet,* weighed enough to depress the crust as much as 600 m and deform the mantle beneath. Today, the crust under portions of northern Europe, Canada, and the United States is still rebounding upward now that the weight of the ice is gone.

The current interglacial, the Holocene Epoch, began about 10,000 years ago. But the Pleistocene Epoch (starting approximately 2.6 million years ago and ending at the onset of the Holocene Epoch) was a time of many glacial and interglacial episodes. Sedimentary deposits are fragmentary because each new ice age tends to erode and bury the evidence of those that came before. However, based on evidence from glacial deposits on land in North America and Europe, geologists have been able to document four glaciations during the Pleistocene.

The longer-preserved marine oxygen isotope record suggests that over the past 500,000 years, each glacial-interglacial cycle has lasted about 100,000 years; hence, there have been approximately five glaciations in this period. During the length of a typical 100,000-year cycle, climate gradually cools and ice slowly expands until it reaches peak cooling. At the peak, glaciers cover most of Canada southward

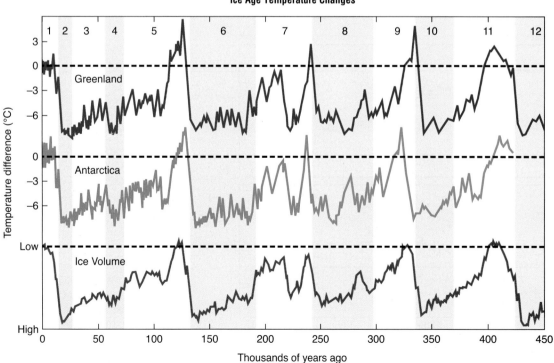

Ice Age Temperature Changes

(?) Explain the term "marine isotopic stage."

FIGURE 17.25 Ice cores from Greenland and Antarctica record atmospheric temperature, and marine cores record global ice volume. These sets of cores reveal similar histories of major warm-cold cycles occurring approximately every 100,000 years. Researchers have divided the record into marine isotopic stages that are numbered beginning with MIS 1, the current warm period. Odd-numbered stages are warm, and even-numbered stages are cool. MIS numbers appear along the top of the figure.

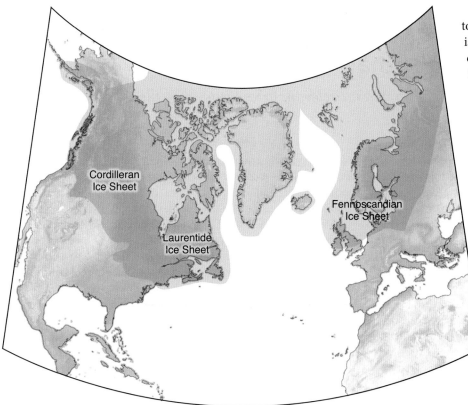

to the Great Lakes, Iceland, Scandinavia, and the British Isles. In the Southern Hemisphere, part of Chile is covered, and the ice of Antarctica covers part of what is now the Southern Ocean. In mountainous regions, the snow line lowers by 1,000 m in altitude from the warmest to the coldest periods of a cycle.

Once ice cover reaches a maximum during a glacial episode, within a couple of thousand years, global temperature rises again, marking the onset of the glacial retreat and the eventual return to the glaciers' minimum extent and volume (FIGURE 17.27). The last ice age culminated about 20,000 to 30,000 years ago, and by approximately 5,000 years ago, most of the ice had melted (except for remnants on Greenland and Antarctica). Since then, glaciers generally have retreated to their smallest extent, with the exception of short-term climate fluctuations such as the so-called *Little Ice Age,* a cool period that lasted from about the 16th century to the 19th century.

FIGURE 17.26 During the last ice age, continental ice sheets developed over much of North America and northern Europe.

? What evidence would geologists have likely used to reconstruct the position of past ice sheets?

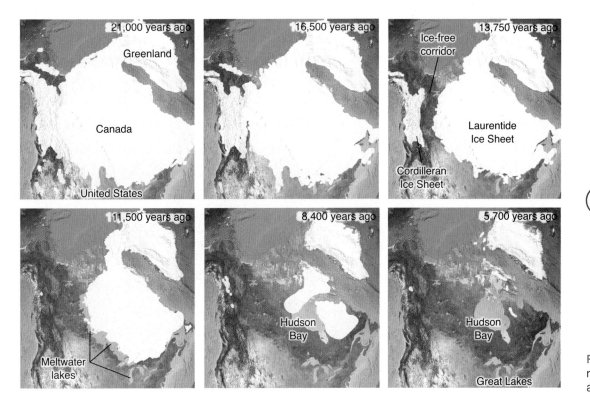

? Where in Canada would you expect to find glacial landforms? What would they look like? Where in Canada would you not expect to find glacial landforms?

FIGURE 17.27 The history of retreating ice in North America after the end of the last ice age.

? **Expand Your Thinking**—Describe the nature of global climate change over the past half million years. How would you test your description?

17-8 During the Last Interglacial, Climate Was Warmer and Sea Level Was Higher than at Present

LO 17-8 *List characteristics of the last interglacial that are relevant to understanding the consequences of modern global warming.*

The last interglacial period occurred between approximately 130,000 and 75,000 years ago. Climate during this 55,000-year period was not always warm. Rather, researchers have identified five major phases consisting of three interstadials and two stadials. These show up clearly in the ice core records as well as the deep-sea record. FIGURE 17.28 shows these phases, using the naming system for marine isotopic stages. The last interglacial is named MIS 5, and the stadials and interstadials are labelled MIS 5a to 5e. Of these, MIS 5e was the warmest and most like the Holocene Epoch.

Marine isotopic stage 5e is a good example of a warm period with characteristics similar to those of our current interglacial. Because it is also a relatively recent event, rocks and sediments that record climate conditions from that time have not been subjected to extensive erosion, metamorphism, or weathering. MIS 5e lasted approximately 12,000 years, from 130,000 to 118,000 years ago, and the average age of fossil corals from around the world that grew at that time is 123,000 years. For example, FIGURE 17.29 shows a fossil reef on the Hawaiian island of Oahu that illustrates another important feature of MIS 5e: Sea level was higher than present by as much as 4 to 6 m. Hence, researchers have concluded that climate was somewhat warmer and that melted ice contributed to the higher sea level. This conclusion is supported by deep cores of ice from both Greenland and Antarctica that preserve temperature records from MIS 5e. In Figure 17.25, you will notice that ice core records from Antarctica and Greenland and marine sediment all preserve a record for MIS 5e that is warmer, with lower ice volume, than the present-day climate.

MIS 5e has been cited as a possible analog for a future climate under continued global warming. Studies have shown that carbon dioxide concentrations in the atmosphere were relatively high (although not as high as they are today due to the current contribution of industrial greenhouse gases); temperatures were higher than at present; and sea level was as much as 4 m to 6 m higher. Scientists study MIS 5e to develop an accurate estimate of the duration of the last interglacial period and improve our understanding of its primary causes. Both of these goals are intended to provide a basis for improving *global circulation models* that are used to project climate change (see Chapter 16, "Global Warming").

MIS 5e as an Analog for This Century

Computer models of climate change during MIS 5e indicate that sea-level rise started with melting of the Greenland Ice Sheet, not the Antarctic Ice Sheet. Research also suggests that ice sheets across both the Arctic and Antarctic could melt more quickly than expected this

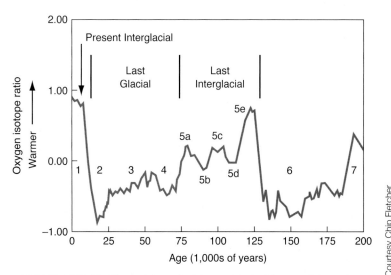

FIGURE 17.28 The last interglacial consisted of five stadials and interstadials, named MIS 5a to 5e. The last glacial consisted of two stadials, MIS 4 and MIS 2, as well as one interstadial, MIS 3. The current interglacial is MIS 1.

FIGURE 17.29 This rocky shoreline in Hawaii is composed of limestone formed by a fossil reef that grew under higher-than-present sea level during MIS 5e.

 Describe the general history of global climate over the last interglacial cycle.

 Explain how this fossil reef can be used to reconstruct the past position of sea level.

century because temperatures are likely to rise higher than they did during MIS 5e, especially in the polar regions. If these predictions are correct, by 2100, the Arctic could warm by 3°C to 5°C in summer. During MIS 5e, meltwater from Greenland and other Arctic sources raised sea level by as much as 4 m. However, since sea level actually rose 4 m to 6 m, researchers have concluded that Antarctic melting and thermal expansion of warm sea water must have been responsible for the remainder of the rise in sea level. This conclusion is supported by the discovery of marine plankton fossils and isotopes formed by solar radiation (possible only if ice retreat exposed Antarctic bedrock to sunlight) beneath the West Antarctic Ice Sheet, indicating that parts of the ice disappeared at some point over the last several hundred thousand years.

The rise in sea levels produced by Arctic warming and melting could have floated, and thus destabilized, ice shelves sitting on the shallow continental shelf of Antarctica. Ice shelves that float do not buttress their glaciers, which may accelerate the rate at which the glaciers flow into the sea, a condition scientists refer to as "glacial collapse." If such a process occurred today, it could lead to rapid sea-level rise and more glacial collapse—a positive feedback response. In the past few years, sea level has begun rising more rapidly; now it is rising at a rate of over 3 cm per decade. Recent studies have also found accelerated rates of glacial retreat along the margins of both the Greenland and Antarctica Ice Sheets.

During MIS 5e, the amount of global warming needed to initiate this melting was less than 3.5°C above modern summer temperatures, similar to the amount that is predicted to occur by mid-century if carbon dioxide levels continue to rise unchecked. The amount of Greenland Ice Sheet melting that produced higher sea levels is shown in **FIGURE 17.30**. According to this reconstruction, sea levels rose more than 1.6 m per century—a rate that would be catastrophic for coastal communities worldwide if it were to happen today. Computer modelling indicates several other features of MIS 5e that may portend global conditions by the end of the 21st century: Global carbon dioxide will rise by 1 percent per year (similar to the current rate of rise); 2100 will be significantly warmer than MIS 5e, so Greenland is already headed toward a state similar to that depicted in Figure 17.30; and the West Antarctic Ice Sheet will contribute significantly to global sea-level rise by 2100.

The later part of MIS 5e was characterized by precipitous changes in global climate. Although the period has been studied intensively, global climate during MIS 5a to 5d is poorly understood. Researchers speculate that temperatures during MIS 5b and 5d were significantly cooler than current temperatures, that global ice volume expanded, and that global sea level dropped perhaps by as much as 25 m below the current level. These were, in effect, mini ice ages that lasted a few thousand to 10,000 years each. MIS 5b was likely cooler than present, and ice volume was greater than at present. MIS 5a may have had sea level that was 1 m higher than present and is thus studied as a potential analog to our warming world.

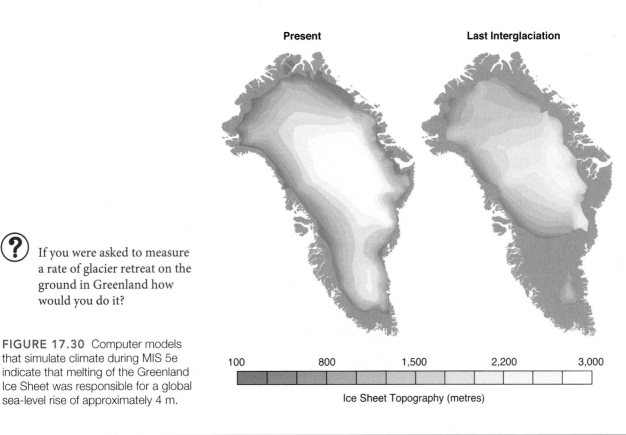

Present

Last Interglaciation

| 100 | 800 | 1,500 | 2,200 | 3,000 |

Ice Sheet Topography (metres)

 If you were asked to measure a rate of glacier retreat on the ground in Greenland how would you do it?

FIGURE 17.30 Computer models that simulate climate during MIS 5e indicate that melting of the Greenland Ice Sheet was responsible for a global sea-level rise of approximately 4 m.

Expand Your Thinking—What do climate researchers hope to learn through the study of MIS 5e?

17-9 Glacial-Interglacial Cycles Are Controlled by the Amount of Solar Radiation Reaching Earth

LO 17-9 *Define the Earth–Sun orbital parameters.*

Scientists are still uncertain about all the factors that drive variations in paleoclimate. However, there is agreement that regular and predictable differences in Earth's exposure to solar radiation over the past half million years must play an important role because they match the timing of climate swings.

Insolation and Axial Tilt

The past 500,000 years or so have been characterized by particularly intense glacial-interglacial cycles occurring approximately every 100,000 years as well as by stadial-interstadial episodes that have occurred more often. Fluctuations in the timing and location of *insolation* (solar radiation received at Earth's surface) caused by regular changes in the geometry of Earth's orbit around the Sun are the most likely cause of these climate instabilities. This solar variable was neatly described by the Serbian mathematician *Milutin Milankovitch* (1879–1958) in 1930.

To understand these **orbital parameters,** we first must appreciate the effect of Earth's tilted axis on the amount of insolation received through the year. Earth's axis is tilted an average of 23.5° from the vertical (**FIGURE 17.31**). As Earth orbits the Sun, this tilt means that during one part of the year, the Northern Hemisphere is tilted toward the Sun and receives greater insolation (summer), whereas six months later, it is tilted away from the Sun and receives less insolation (winter); the reverse applies to the Southern Hemisphere. These annual changes in insolation create the seasons.

Orbital Parameters

Three aspects of Earth's orbital geometry change in regular cycles under the influence of the combined gravity of Earth, the Moon, the Sun, and the other planets. These orbital parameters dictate the timing and location of insolation reaching Earth's surface and thus regulate climate (**FIGURE 17.32**).

Earth's orbit changes from a nearly perfect circle to more of an ellipse and back again in a 100,000-year cycle (and a 400,000-year cycle) known as *eccentricity*. Eccentricity affects the amount of insolation received at *aphelion* (the point in its orbit when Earth is farthest from the Sun) and at *perihelion* (the point in its orbit when Earth is closest to the Sun). The effect of eccentricity is to shift the seasonal contrast in the Northern and Southern Hemispheres. For example, when Earth's orbit is more elliptical, one hemisphere will have hot summers and cold winters; the other will have moderate summers and moderate winters. When the orbit is more circular, both hemispheres will have similar contrasts in seasonal temperature.

The second orbital parameter is called *obliquity*. The angle of Earth's axis of spin changes its tilt between 21.5° and 24.5° every 41,000 years. Changes in obliquity cause large changes in the seasonal distribution of sunlight at high latitudes and in the length of the winter dark period at the poles. This makes for cool winters and summers that have less variation in temperature between the two seasons, which is optimal for the growth of ice sheets. The change in obliquity has little effect on low latitudes.

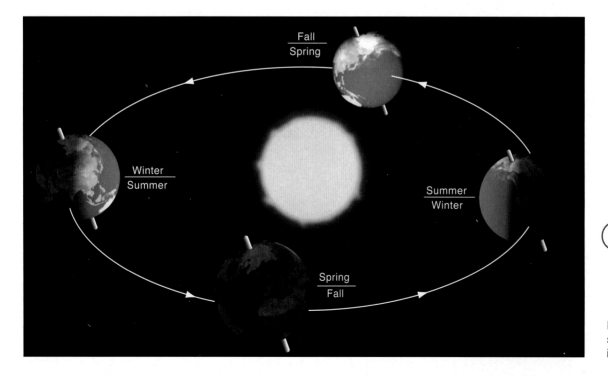

? How would life on Earth be different if the axis were not tilted?

FIGURE 17.31 Earth's seasons are the result of a tilt in the planet's axis.

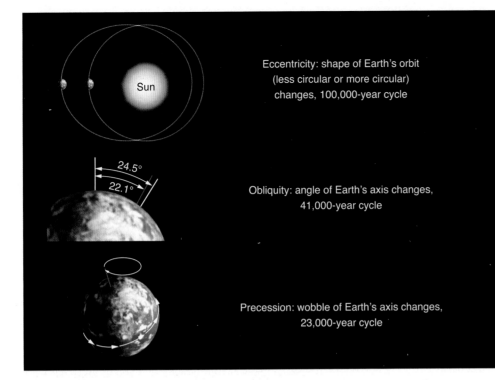

 What is the principal effect of changes in Earth's orbit?

FIGURE 17.32 The primary orbital parameters driving climate changes over the past half million years are eccentricity, obliquity, and precession.

Finally, Earth's axis of spin slowly wobbles. Like a spinning top running out of energy, the axis wobbles toward and away from the Sun over the span of 19,000 to 23,000 years. This is known as *precession*. Precession affects the timing of aphelion and perihelion, and has important implications for climate because it affects the seasonal balance of insolation. For example, when perihelion falls in January, winters in the Northern Hemisphere and summers in the Southern Hemisphere are slightly warmer than the corresponding seasons in the opposite hemispheres. The effects of precession on the amount of radiation reaching Earth are closely linked to the effects of obliquity (changes in tilt). The combined variation in these two factors causes radiation changes of up to 15 percent at high latitudes, greatly influencing the growth and melting of ice sheets.

Milankovitch theorized that ice ages occur when orbital variations cause lands in the region of 65° north latitude (the approximate latitude of central Canada and northern Europe) to receive less sunshine in the summer. This allows snow to build up from year to year to form glacial ice. Why is the Northern Hemisphere the crucial location for glacier formation? Two reasons: Most of the continents are located in the Northern Hemisphere, and glaciers form on land, not water.

Based on this reasoning and his calculations, Milankovitch predicted that the ice ages would peak every 100,000 and 41,000 years, with additional significant variations every 19,000 to 23,000 years.

FIGURE 17.33 plots variations in all three orbital parameters across the past 1 million years. Indeed, Milankovitch's predictions have been verified by the ocean sediment record of global ice volume. Although we are unable to determine the exact timing of ice ages and interglacials from the sediment or ice core records, they do indicate that ice ages and interglacials occur roughly every 100,000 years, and that stadials and interstadials are more frequent as approximated by his predictions of 41,000, and 19,000 to 23,000 years, respectively.

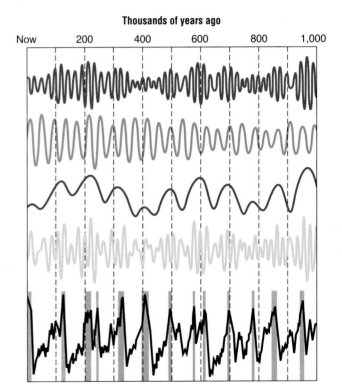

Thousands of years ago

FIGURE 17.33 This plot shows the relative solar forcing due to precession (red), obliquity (green), and eccentricity (blue). The cumulative solar forcing (insolation) at 65° north latitude caused by the combined influence of the three orbital parameters is plotted in yellow. Black shows a history of global ice volume with interglacials (vertical grey bars) having regular variations in climate approximately every 100,000 years and more often, just as Milankovitch predicted.

 Expand Your Thinking—Describe how plate tectonics and orbital parameters may combine to govern paleoclimate.

 Examine Figure 17.33. Does solar forcing account for all changes in ice volume?

17-10 Together, Orbital Forcing and Climate Feedbacks Produced Paleoclimates

LO *17-10 Describe climate feedbacks and give examples.*

If you compare the total solar forcing in Figure 17.33 (yellow line) to the paleoclimate record in marine sediments (black line), you will notice that the timing and magnitude of the two do not exactly match. For instance, the solar forcing that led to MIS 5e is considerably greater than the forcing that is producing the modern interglacial. Also, the drop in insolation at MIS 5d is greater than the low at the peak of the last ice age 20,000 to 30,000 years ago—yet the last ice age was much colder. What creates these disparities? The answer is that *Earth's climate is not driven solely by insolation.*

Earth's environmental system exercises positive and negative *climate feedbacks* that can enhance or suppress the timing and intensity of the Earth–Sun geometry. A climate feedback is a process taking place on Earth that amplifies (a positive feedback) or minimizes (a negative feedback) the effects of insolation. Climate feedbacks are responsible for the difference between orbital forcing and Earth's actual climate. Following are some examples of climate feedbacks.

Emerging from the Last Ice Age

When scientists tried to build computer models to simulate paleoclimate, they could not get them to reproduce past climate change unless they added changes in carbon dioxide levels to accompany the changes in insolation caused by orbital parameters. Although scientists are still trying to understand what causes natural changes in carbon dioxide levels, most believe that past climate changes were initiated by orbital forcing and then enhanced and continued by a rise in greenhouse gases.

Ice cores record past greenhouse gas levels as well as temperature; hence, they allow researchers to compare the history of the two. In the past, when the climate warmed, the change was accompanied by an increase in greenhouse gases, particularly carbon dioxide. However, increases in temperature *preceded* increases in carbon dioxide. This pattern is opposite to the current pattern, in which industrial greenhouse gases are *causing* increases in temperature. The difference is related to climate feedbacks.

One idea for relating paleoclimate to feedbacks was developed by scientists analyzing a core of marine sediments from the ocean floor near the Philippines. That area of the Pacific contains foraminifera that live in tropical surface water. When they die and settle to the bottom, they preserve a record of changing tropical air temperatures. But different types of foraminifera living on the deep sea floor are bathed in *bottom waters* (water that travels along the sea floor, not at the surface) fed from the Southern Ocean near Antarctica. These foraminifera record the temperature of those cold southern waters. The fossils of both types of foraminifera are deposited together on the sea floor. On radiocarbon dating both types of foraminifera, scientists found that water from the Antarctic region warmed before waters in the topics—as much as 1,000 to 1,300 years earlier. The explanation for this difference, they believe, is a positive climate feedback.

First, predictable variations in Earth's eccentricity and obliquity increased the insolation at high southern latitudes during spring in the Southern Hemisphere. That increase warmed the Southern Ocean. As a result, sea ice shrank back toward Antarctica, uncovering ocean waters that had been isolated from the atmosphere for millennia. As the Southern Ocean warmed, it was less able to hold dissolved carbon dioxide, and great quantities of carbon dioxide escaped into the atmosphere. (Warm water can hold less dissolved gas than cold water can.) The released gas proceeded to warm the whole world. This process was responsible for driving climate out of its glacial state and into an interglacial state. It explains how small temperature changes caused by orbital parameters led to a positive feedback in global carbon dioxide that warmed the world.

The Younger Dryas

When scientists first analyzed paleoclimate evidence in marine and glacial oxygen isotope records, they discovered that the Milankovitch theory predicted the occurrence of ice ages and interglacials with remarkable accuracy. But they also found something that required additional explanation: Some climate changes appeared to have occurred very rapidly. Because the Milankovitch theory tied climate change to slow and regular variations in Earth's orbit, scientists had assumed that climate variations would also be slow and regular. The discovery of rapid changes was a surprise. Here again, the answer lay in the climate feedback system.

Cores show that while it took thousands of years for Earth to emerge totally from the last ice age and warm to today's balmy climate, fully one-third to one-half of the warming—about 10°C at Greenland—occurred within mere decades, according to ice records in Greenland (**FIGURE 17.34**). At approximately 12,800 years ago, following the last ice age, temperatures in most of the Northern Hemisphere rapidly returned to near-glacial conditions and stayed there during a climate event called the *Younger Dryas* (named after the alpine flower *Dryas octopetala*). The cool episode lasted about 1,300 years, and by 11,500 years ago, temperatures had warmed again. Ice core records show that the recovery to warm conditions occurred with startling rapidity—less than a human generation. Changes of this magnitude would have a huge impact on modern human societies. There is an urgent need to understand and predict such abrupt climate events.

A look at marine sediments confirms that this pattern is present and must be a global characteristic of climate change. Hence, scientists *conclude* that although the general timing and pace of climate change is set by the orbital parameters, some feedback process must play an important role in its precise timing and magnitude. What might that process be? *Global thermohaline circulation* is thought to play a key role.

Today, warm water near the equator in the Atlantic Ocean is carried to the north on the *Gulf Stream,* which flows on the surface from southwest to northeast in the western Atlantic. The Gulf Stream releases heat into the atmosphere through evaporation, and the heat in turn moves across northern Europe and moderates the climate. As a

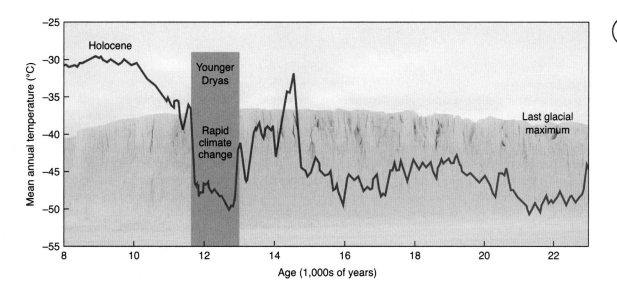

Why is understanding rapid climate change important?

FIGURE 17.34 The Younger Dryas climate event was a dramatic cooling that lasted approximately 1,300 years during the transition between glacial and interglacial states. The return to warm conditions was equally rapid, occurring in less than a human generation.

result, the Gulf Stream becomes cool and salty as it approaches Iceland and Greenland. This makes it very dense; consequently, it sinks deep into the North Atlantic (making a current called the *North Atlantic Deep Water)* before it can freeze. From there, it is pulled southward toward the equator. The Gulf Stream continuously replaces the sinking water, warming Europe and setting up the global oceanic conveyor belt we call the *thermohaline circulation.*

As we learned in Chapter 16, thermohaline circulation transports heat around the planet and hence plays an important role in global climatology. Acting as a conveyor belt carrying heat from the equator into the North Atlantic, it raises Arctic temperatures, discouraging the growth of ice sheets. However, influxes of fresh water from melting ice on the lands that surround the North Atlantic (such as Greenland) can slow or shut down the circulation by preventing the formation of deep water. Fresh water in the North Atlantic would decrease the Gulf Stream (and its role as a source of heat) and lead to cooling in the Northern Hemisphere, thereby regulating snowfall in the crucial region where ice sheets shrink and grow (65°N). Hence, a shutdown of the thermohaline circulation could play a role in a negative climate feedback pattern beginning with ice melting (warming) that ironically leads to glaciation (cooling).

The key to keeping the belt moving is the saltiness of the water. Saltier water increases in density and sinks. Many scientists believe that if too much fresh water entered the ocean—for example, from melting Arctic glaciers and sea ice—the surface water would freeze before it could become dense enough to sink toward the bottom. If the water in the north did not sink, the Gulf Stream eventually would stop moving warm water northward, leaving Northern Europe cold and dry within a single decade.

This hypothesis of rapid climate change is called the *conveyor belt hypothesis,* and the paleoclimate record found in ocean sediment cores appears to support it. Paleoclimate studies have shown that, in the past, when heat circulation in the North Atlantic Ocean slowed, the climate of northern Europe changed. Although the last ice age peaked about 20,000 to 30,000 years ago, the warming trend that followed it was interrupted by cold spells at 17,500 years ago and again at 12,800 years ago (the Younger Dryas). These cold spells happened just after melting ice had diluted the salty North Atlantic water, slowing the ocean conveyor belt.

Hence, we have seen two types of climate feedback:

1. *A positive climate feedback in the Antarctic that ended the last ice age.* Predictable variations in Earth's tilt and orbit caused warming, which triggered the withdrawal of sea ice in the Southern Ocean. This led to additional warming of ocean water, reducing its ability to hold dissolved carbon dioxide. The carbon dioxide escaped into the atmosphere and warmed the planet beyond the temperatures that would have been achieved by orbital parameters alone.

2. *A negative climate feedback late in the transitional phase between the last ice age and the modern interglacial.* Warming at approximately 12,800 years ago produced abundant fresh water in the North Atlantic that diluted the salty Gulf Stream. This put a temporary end to the global thermohaline circulation and triggered rapid cooling in the Northern Hemisphere. Later, after a period of cooling lasting approximately 1,300 years called the Younger Dryas, thermohaline circulation once again became a source of heat transport throughout the world's oceans. Apparently very rapidly, the renewed circulation triggered global warming that would not have been predicted by orbital parameters alone.

These climate feedbacks, and others that are still being discovered, work in parallel with orbital parameters to ultimately determine the nature of Earth's climate.

Expand Your Thinking—State a hypothesis predicting global climate that integrates plate tectonics, orbital parameters, and climate feedbacks. How would you test your hypothesis?

LET'S REVIEW "GEOLOGY IN OUR LIVES"

Now that you have finished the chapter, "Geology in Our Lives" will have taken on new meaning for you. Let us review it: A glacier is a large, long-lasting river of ice that forms on land, undergoes internal deformation, erodes the crust, deposits sediment, and creates glacial landforms. Glaciers have shaped the mountains and nearby lands where they expanded during past climate changes. Earth's recent history has been characterized by cool climate periods, called ice ages, during which glaciers around the world expanded, and warm periods, called interglacials, during which glaciers retreated. By studying the history of climate changes, scientists learn how natural systems respond to shifts in temperature and ice volume. Today, most of the world's ice sheets, alpine glaciers, ice shelves, and sea ice are dwindling because we are in an interglacial period that may be accentuated by human-induced global warming.

Past climate, called *paleoclimate,* led to the repeated expansion and contraction of ice sheets on the continents. These expansions and contractions sculpted the mountains, deposited sediments across the landscape, and produced much of the topography of northern lands that we see today. Changes in climate that produced past glaciations were caused by changes in the amount of sunlight reaching Earth (insolation), a result of shifting orbital parameters. But orbital parameters alone are not sufficient to cause the rapid and dramatic changes in climate revealed by ice cores and marine sediments. These changes are also influenced by climate feedbacks that enhance or diminish the effects of changing insolation. Indeed, today, most of the world's ice sheets, alpine glaciers, and ice shelves are melting due to a dramatic climate forcing process: human pollution of the atmosphere with greenhouse gases.

STUDY GUIDE

17-1 A glacier is a river of ice.

- **Louis Agassiz** proposed the "theory of the ice ages" and pointed out that the deep valleys and sharp peaks and ridges of the mountains of Europe and North America were once buried in ice just as Greenland is today.

- A **glacier** is a large, long-lasting river of ice that forms on land, undergoes internal deformation, and creates **glacial landforms.** Glaciers range in length and width from several hundred metres to hundreds of kilometres, depending on their type. Most glaciers are several thousand years old, but those covering Greenland and Antarctica have been in existence in one form or another for hundreds of thousands of years.

17-2 As ice moves, it erodes the underlying crust.

- Ice behaves as a solid that is brittle (i.e., it fractures) until it reaches thicknesses of 40 to 60 m or more. At these thicknesses, the lower layers of ice experience plastic flow. Basal sliding occurs when portions of the glacier move across the crust on a layer of **meltwater.**

- As ice moves, it causes erosion to the underlying bedrock through plucking and abrasion. Pulverized rock, known as rock flour, is composed of silt-size grains of ground-up rock. Sediments of all sizes become part of the glacier's load, and as a result, many glacial deposits are poorly sorted.

- Glaciers exist in a wide variety of forms and environments. There are two principal types of glaciers: **alpine glaciers,** which are in mountainous areas, and **continental glaciers,** which cover larger areas and are not confined to the high elevations of mountains.

17-3 Ice moves through the interior of a glacier as if on a one-way conveyor belt.

- The interior of an active glacier is like a one-way conveyor belt, continuously delivering ice, rock, and sediments to the front of the glacier. This system feeds new ice from the **zone of accumulation,** usually accounting for 60 percent to 70 percent of the glacier's surface area, to the **zone of wastage,** where ice is lost not only by melting but also by sublimation and calving (when large blocks of ice break off the front). Between the zone of accumulation and the zone of wastage is the equilibrium line (also known as the snow line), marked by the elevation at which accumulation and wasting are approximately equal.

- Approximately 160,000 glaciers exist today. They are present on every continent except Australia and in approximately 47 of the world's countries. Yet more than 94 percent of all the ice on Earth is locked up in the continental glaciers covering Greenland and Antarctica.

17-4 Glacial landforms are widespread and attest to past episodes of glaciation.

- Earth's history has been characterized by great swings in global climate, from extreme states of cold (ice ages) to warm periods called **interglacials.** Ice ages are characterized by the growth of massive continental ice sheets reaching across North America and northern Europe.

- During ice ages, sea level dropped by as much as 130 m, exposing lands between Siberia and Alaska, from Ireland to England to France, and from Malaysia to Indonesia to Borneo. In many places, the shoreline was tens to hundreds of kilometres seaward of its present location.

- Glacial valleys that have been eroded by alpine glaciers are U shape in cross-section. Valleys that have been shaped by stream erosion have a V shape. A mountain valley will have a combination of both shapes if it was once glaciated but is now being eroded by a stream channel.

- All sediment that has been deposited by glacial erosion is called glacial drift. Glacial drift includes **glacial till,** unstratified and unsorted sediment that is deposited directly by melting ice. Glacial erratics are blocks of bedrock deposited by melting ice that have a foreign lithology. A **glacial moraine** is a deposit of till that marks a former position of the ice. Ground moraines tend to be relatively horizontal deposits beneath moving ice. An end moraine is a ridge of till marking the front edge of ice movement. A terminal moraine is the ridge of till left behind at the line of a glacier's farthest advance when it retreats. Lateral moraines are deposits of till that mark the edges of the glacier against a valley wall. A medial moraine forms when two valley glaciers coalesce, and till from each of them combines to form a boundary between them. Other depositional features include **glacial outwash** plains, kames, eskers, kettle lakes, and drumlins.

17-5 The majority of glaciers and other ice features are retreating due to global warming.

- Global warming threatens the stability of glaciers around the world. Scientists are monitoring alpine glaciers, ice sheets, Arctic sea ice, and ice shelves. Data show that the majority of glaciers and other ice features are retreating due to global warming.

- In 2006–2007, the average rate of melting and thinning of alpine glaciers more than doubled. In 2007, the extent of melting on Greenland broke the record set in 2005 by 10 percent, making it the largest season of melting ever recorded. In West Antarctica, ice loss has increased by 59 percent over the past decade, to about 132 billion tonnes a year, while the yearly loss along the Antarctic Peninsula has increased by 140 percent, to 60 billion tonnes, and 87 percent of glaciers there are retreating. In East Antarctica, melting is largely confined to the seaward edge of the ice (where it is caused by warm ocean currents), but every year melting encroaches inland and onto higher elevations.

17-6 The ratio of oxygen isotopes in glacial ice and deep-sea sediments is a proxy for global climate history.

- Foraminifera in deep-sea sediment use dissolved compounds and ions to precipitate microscopic shells made of calcite ($CaCO_3$). The oxygen isotope content of these shells is used to track changes in global ice volume. Oxygen isotopes in ice cores provide a record of changing air temperatures. Because global ice volume and air temperature are related, the records of oxygen isotopes in foraminifera and glacial ice show similar patterns.

17-7 Earth's recent history has been characterized by cycles of ice ages and interglacials.

- Oxygen isotope records from marine sediments and ice cores show Earth's climate alternating between warm periods and ice ages. These are named with a numbering system called **marine isotopic stages,** or MIS. Odd-numbered stages are warm; even-numbered stages are cool.

- Major glacial and interglacial cycles are repeated approximately every 100,000 years. Minor episodes of cooling (stadials) and warming (interstadials) occur as well. Global ice volume during the last interglacial was lower than at present, and the climate was warmer. Following the last interglacial, the climate entered a cooling phase that culminated 20,000 to 30,000 years ago with a major glaciation. The current interglacial has lasted approximately 10,000 years.

17-8 During the last interglacial, climate was warmer and sea level was higher than at present.

- The last interglacial occurred 130,000 to 75,000 years ago and consisted of three interstadials and two stadials. The last interglacial (MIS 5) consisted of a sequence of stadials and interstadials labelled MIS 5a to 5e. MIS 5e is the best example of a warm period with characteristics similar to those of the current interglacial.

- Computer models of climate change during the last interglacial show that sea-level rise started with melting of the Greenland Ice Sheet, not the Antarctic Ice Sheet. Results suggest that ice sheets across the Arctic and Antarctic could melt more quickly than expected during this century because temperatures are likely to rise higher than they did during the last interglacial. By 2100, Arctic summers may be as warm as they were 130,000 years ago, when sea level was up to 6 m higher than it is today.

17-9 Glacial-interglacial cycles are controlled by the amount of solar radiation reaching Earth.

- Differences in Earth's exposure to solar radiation play an important role in paleoclimate because they match the timing of climate swings.

- The past 500,000 years had particularly intense glacial-interglacial cycles spaced approximately every 100,000 years, with stadial and interstadial episodes occurring at greater frequency. Fluctuations in insolation caused by changes in the geometry of Earth's solar orbit are the most likely cause of instabilities in Earth's climate.

- Solar variables were described by the Serbian mathematician Milutin Milankovitch. He calculated the timing of three components of Earth's solar orbit that contribute to changes in climate: a cyclical change in the shape of Earth's orbit known as eccentricity; a change in the angle of the axis of spin called obliquity; and a wobble in Earth's axis known as precession.

- Small variations in insolation related to eccentricity, obliquity, and precession influence the amount of sunlight each hemisphere receives. Ice ages occur when lands at 65°N (the latitude of central Canada and northern Europe) receive less sunshine in summer than usual.

- Milankovitch predicted that ice ages peak every 100,000 and 41,000 years, with additional significant variations every 19,000 to 23,000 years.

17-10 Together, orbital forcing and climate feedbacks produced paleoclimates.

- Paleoclimate events do not exactly match insolation because Earth's environmental system exercises positive and negative climate feedbacks that can enhance or suppress the timing and intensity of the Earth–Sun geometry. A climate feedback is a process on Earth that amplifies (a positive feedback) or minimizes (a negative feedback) the effect of the orbital parameters. Climate feedbacks are responsible for the difference between orbital forcing and Earth's actual climate.

- A positive climate feedback ended the ice age. Predictable variations in Earth's orbit and tilt increased the amount of sunlight hitting southern latitudes, causing Antarctic sea ice to retreat. The reduced ice cover resulted in warming of ocean water, which caused dissolved carbon dioxide to escape into the atmosphere. The released gas warmed the whole world.

- The Milankovitch theory ties climate change to slow and regular variations in Earth's orbit. Thermohaline circulation may contribute to rapid climate change. The conveyor belt carries heat from the equator toward the poles, where it raises Arctic temperatures, discouraging the growth of ice sheets. Influxes of fresh water slow or shut down the circulation by preventing bottom water formation. The cessation of the conveyor belt cools the Northern Hemisphere and regulates snowfall at 65°N.

KEY TERMS

alpine glaciers (p. 454)
climate proxy (p. 466)
continental glaciers (p. 454)
glacial landforms (p. 452)
glacial moraine (p. 460)
glacial outwash (p. 460)
glacial till (p. 459)

glacier mass balance (p. 456)
glaciers (p. 452)
interglacials (p. 458)
Louis Agassiz (p. 452)
marine isotopic stages (p. 470)
meltwater (p. 454)
orbital parameters (p. 474)

paleoclimatology (p. 452)
permafrost (p. 453)
Quaternary Period (p. 452)
zone of accumulation (p. 456)
zone of wastage (p. 456)

ASSESSING YOUR KNOWLEDGE

Please answer these questions before coming to class. Identify the best answer.

1. What is a glacier?
   a. a frozen river
   b. snow on a steep slope that is slowly sliding downhill
   c. an ice avalanche composed of compressed, recrystallized snow
   d. a large, long-lasting river of ice
   e. a type of ice in mountains that moves up and down hills

2. The smallest type of alpine glacier is a(n)
   a. temperate glacier.
   b. icecap.
   c. outlet glacier.
   d. polar glacier.
   e. valley glacier.

3. Ice within a glacier is
   a. perpetually moving forward toward the terminus.
   b. moving backward and forward, depending on the rate of supply and melting.
   c. mostly immobile.
   d. perpetually frozen to the bedrock beneath.
   e. Ice is not the primary component of a glacier.

4. A moraine is
   a. a deposit of till that was in contact with glacial ice.
   b. a deposit of shale that forms under glacial ice.
   c. stratified but unsorted conglomerate.
   d. a deposit of sand formed only when a glacier advances.
   e. a hump of old ice.

5. Which glacial landforms are left by a retreating continental ice sheet? (Circle all that are true.)
   a. drumlins, eskers, and terminal moraines
   b. ground moraines, kettle lakes, end moraines
   c. deposits of till, outwash plains, glacial erratics
   d. horns, arêtes, and tarns
   e. hanging valleys, U-shaped valleys

6. In response to global warming, alpine glaciers around the world are
   a. generally retreating.
   b. relatively stable.
   c. generally advancing.
   d. moving in random ways.
   e. None of the above.

7. Retreat of Arctic sea ice
   a. has no effect on global warming.
   b. causes sea level to fall due to melting sea ice.
   c. may accelerate snowfall on the Antarctic Ice Sheet.
   d. may accelerate growth of the Greenland Ice Sheet.
   e. will change albedo and be a positive feedback to warming.

8. Oxygen isotopes provide a climate proxy because
   a. when they freeze in sea ice, they can be detected with satellites.
   b. they are sensitive to temperature, evaporation, and precipitation.
   c. some of them are never detected.

    d. they offer a natural record of plankton abundance.

    e. None of the above.

9. Marine isotopic stages

    a. are not useful in studying paleoclimate.

    b. are expressed in double digits for warm periods and in single digits for cool periods.

    c. are used to name glacials and interglacials and stadials and interstadials.

    d. occur principally during ice ages, when glaciers expand.

    e. are used to name glacier advances and retreats.

10. During the interstadial 5e,

    a. sea level was lower and temperature was warmer.

    b. sea level was about the same and temperature was higher.

    c. sea level was higher by 4 m to 6 m and temperature was warmer.

    d. Greenland was larger and sea level was lower.

    e. None of the above.

11. Which of the following statements about orbital parameters are true? (Circle all that apply.)

    a. They govern the weather due to their relationship to the Moon.

    b. They include precession, obliquity, and axial tilt.

    c. They produce climate cycles with a 100,000-year periodicity.

    d. They correlate with observations of paleoclimate.

    e. They do not influence global climate.

12. The Younger Dryas

    a. was a short-term climatic warming at the end of the last ice age.

    b. occurred during the last interglacial.

    c. was a cool period that followed the end of the last ice age.

    d. is the scientific name for modern global warming.

    e. occurred during the last ice age.

13. Climate feedbacks

    a. may magnify or suppress the influence of orbital parameters.

    b. explain the timing of major climate cycles.

    c. are the results of the gravitational influence of the nearby planets.

    d. rarely influenced the nature of paleoclimate.

    e. are the processes that control Earth's exposure to insolation.

14. For a glacier to advance,

    a. snowfall must slow and then stop.

    b. retreat must equal advance.

    c. wastage must exceed accumulation.

    d. accumulation must exceed wastage.

    e. the internal conveyor belt must reverse direction.

15. Rapid climate changes

    a. are well predicted by the orbital parameters.

    b. have not been shown to be real phenomena.

    c. are caused by climate feedbacks, not by orbital parameters.

    d. are important only in the tropics.

    e. are typical of changes in insolation.

FURTHER RESEARCH

1. Use the Internet to find the glacier closest to your location. Is it advancing or retreating?

2. Do you live in a coastal community? Has your local newspaper or TV news station presented any stories about sea-level rise? If so, what do they say about it?

3. Find out what your region was like during the last ice age by doing research on the Internet, interviewing local geology professors, and/or obtaining the geological survey for your locality. Is there any evidence of processes that were active in your area during the last ice age?

4. How does the landscape of Canada reflect the influence of the Laurentide or Cordilleran Ice Sheet?

ONLINE RESOURCES

Explore more about glaciers and paleoclimatology on the following websites:

Natural Resources Canada—The Atlas of Canada: Scroll down to "Glaciers and Icefields":
http://atlas.gc.ca/site/english/maps/water.html

U.S. National Snow and Ice Data Center, "All about Glaciers":
http://nsidc.org/glaciers

U.S. Geological Survey "Benchmark Glaciers":
http://ak.water.usgs.gov/glaciology

U.S. Geological Survey, "Satellite Image Atlas of Glaciers of the World":
http://pubs.usgs.gov/fs/2005/3056/

NASA Earth Observatory, "Glaciers, Climate Change, and Sea-Level Rise":
http://earthobservatory.nasa.gov/IOTD/view.php?id=5668

U.S. National Climatic Data Center, "Paleoclimatology":
www.ncdc.noaa.gov/paleo/paleo.html

NASA Earth Observatory, "Paleoclimatology: Introduction":
http://earthobservatory.nasa.gov/Features/Paleoclimatology

Additional animations, videos, and other online resources are available at this book's companion website:
www.wiley.com/go/fletchercanada

This companion website also has additional information about WileyPLUS and other Wiley teaching and learning resources.

482

CHAPTER 18
MASS WASTING

Chapter Contents and Learning Objectives (LO)

18-1 Mass wasting is the movement of rock and soil down a slope due to the force of gravity.

LO **18-1** *Describe mass wasting.*

18-2 Creep, solifluction, and slumping are common types of mass wasting.

LO **18-2** *Define "creep," "solifluction," and "slumping."*

18-3 Fast-moving mass-wasting events tend to be the most dangerous.

LO **18-3** *Distinguish among mass-wasting processes that are flows, slides, and falls.*

18-4 Avalanches, lahars, and submarine landslides are special types of mass-wasting processes.

LO **18-4** *Compare and contrast avalanches, lahars, and submarine landslides.*

18-5 Several factors contribute to unstable slopes.

LO **18-5** *Describe the factors that contribute to mass wasting.*

18-6 Mass-wasting processes vary in speed and moisture content.

LO **18-6** *Describe how various types of mass-wasting processes differ in terms of their speed and moisture content.*

18-7 Human activities are often the cause of mass wasting.

LO **18-7** *List some ways in which humans cause mass wasting.*

18-8 Research improves knowledge of mass wasting and contributes to the development of mitigation practices.

LO **18-8** *Describe methods of mitigating mass-wasting hazards.*

GEOLOGY IN OUR LIVES

Mass wasting is the set of processes that move weathered rock, sediment, and soil down a slope due to the force of gravity. Mass wasting can be hazardous if humans (or buildings and roads) are in the way. Over the past century, the world's population has grown enormously, and many people now live in places that were formerly considered too dangerous or risky. As buildable land becomes scarce, communities are expanding into environments that are exposed to mass wasting and other geologic hazards, turning them into neighbourhoods. The best way to manage mass-wasting hazards is to know the characteristics of a hazardous slope and avoid building there. Doing this is called "hazard avoidance." Avoidance is the most effective and inexpensive approach to managing geologic hazards.

(?) What geologic factors might contribute to causing a landslide such as this one in Morro do Bumba, Brazil, that killed 267 people?

18-1 Mass Wasting Is the Movement of Rock and Soil Down a Slope Due to the Force of Gravity

LO 18-1 *Describe mass wasting.*

During the early morning hours on January 19, 2005, heavy rains destabilized a steep embankment in North Vancouver, British Columbia, that resulted in a fatal landslide, carrying with it two homes until it came to rest some 90 m downhill (**FIGURE 18.1**). One person died and others were severely injured. The total monetary cost of this event for North Vancouver has never been fully divulged (easily exceeding $10 million). Nevertheless, coupled with the human suffering, there is no doubt this tragic event carried an enormous cost. In hindsight, it may have been avoided if proactive measures had been undertaken, such as increasing the stability of the steep slopes and improving the drainage network in the area.

According to Natural Resources Canada, in a typical year, slope failure causes $200 million to $400 million in damage in Canada. Such failures have accounted for approximately 600 human fatalities in recorded Canadian history. In the United States, the numbers are even more staggering, with $1.5 billion in damage and 25 to 50 deaths per year. In many other countries, the loss is higher because of poor land-use practices and building codes that ignore geologic hazards. The process in which gravity pulls soil, debris, sediment, and broken rock (collectively known as **regolith**) down a hillside or cliff is called **mass wasting** (**FIGURE 18.2**).

Mountain building, weathering, and erosion are perpetually changing the slope of the land. As a result, gravity inevitably causes loose and unstable accumulations of rock and sediment to move downhill. Mass wasting actually occurs in several forms (discussed in the next section), from the instantaneous event that affected North Vancouver to longer and slower processes that shift soil and regolith to the base of slopes over the course of years, decades, and centuries. All of these events can be hazardous if humans (or buildings and roads) are in the way. Mass wasting is another example of a *geologic hazard* that is best dealt with by

FIGURE 18.1 A landslide in North Vancouver, January 19, 2005, swept 90 m down a steep embankment, destroying two homes and killing one person.

 What preventative measures would you recommend in order to avoid future landslides in this area?

FIGURE 18.2 Mass wasting is the movement of rock and soil down a slope due to the force of gravity.

 Imagine that human activity caused this landslide. Describe three human activities that could cause mass wasting like this.

being avoided. By learning to recognize the characteristics of an unstable slope and not building or engaging in other activities there, we avoid the threat posed by mass wasting. This approach is termed **avoidance.**

The Human Toll

 Over the past century, the world's population has grown enormously, and many people now live in places that were formerly considered too dangerous or risky. As buildable land becomes scarce, communities are expanding into hazardous environments, turning them into neighbourhoods. These environments include stream valleys that are prone to flooding, the slopes of active volcanoes, fault zones with earthquake hazards, shorelines where the ocean is a constant threat, and hillsides that are vulnerable to mass wasting (**FIGURE 18.3**).

Unfortunately, at the same time that humans move onto risky hillsides, we alter the landscape in ways that make slopes unstable and more susceptible to mass wasting. That is, we not only live in dangerous places, we alter them and unintentionally make them more dangerous. **TABLE 18.1** lists a few of the many recent landslide events that have resulted in loss of human life.

The Vaiont Dam

Although it happened over 50 years ago, the Vaiont Dam disaster still serves as a startling reminder of the tragic consequences of neglecting to consider **slope stability** in engineering projects. One of the world's highest (266 m), the Vaiont Dam spans the Vaiont River Valley in the Italian Alps. The purpose of the dam was to trap the Vaiont River in order to create a reservoir and generate hydroelectric power for neighbouring communities. However, shortly after it was constructed

FIGURE 18.3 As human populations expand into previously undeveloped areas, activities such as building this wilderness road across an unstable slope may unwittingly place people in the path of geologic hazards. In early 2009, 33 coffee pickers walking home from work were killed by this landslide in northern Guatemala.

(?) How could building this road have contributed to the landslide?

and the narrow valley behind the dam began to fill with water, large cracks were seen forming in the soil near the top of Monte Toc, the mountainside rising 1,000 m above the reservoir. Engineers grew concerned that landslides into the reservoir might reduce its capacity. As it turned out, that was the least of their worries.

When the Vaiont Dam was completed and the reservoir filled with water, groundwater in the surrounding valley and its slopes began to rise. This had the effect of decreasing the strength of rock on the valley walls as they became saturated and as water pressure separated internal fracture surfaces. As summer turned to autumn, heavy rains added to the weight of rock and soil perched on steep slopes above the reservoir valley. By early October, engineers monitoring the hillside measured shifts in the rock that were moving at the startling rate of 40 cm per day. Sensing danger, wildlife abandoned the area, and the hills fell silent. Alarmed, engineers opened the dam's floodgates in an effort to lower the lake level, but the water continued to rise. The accelerating downward movement of the mountainside was filling the reservoir.

On October 9, 1963, at about 10:15 p.m., a massive block of soil and rock 2 km long, 1.5 km wide, and several hundred metres thick accelerated down Mount Toc into the reservoir. There it created a titanic wave 100 m high that overtopped the dam and raced through the valley below. Without warning, it slammed into the town of Longarone, killing 2,600 people. The entire event lasted only six minutes.

With fatal consequences, the engineers designing the reservoir had neglected to account for the geology of the valley's slopes. A steeply dipping layer of highly fractured limestone rested on a bed of clay that became slick and unstable when wet. Over the centuries, river erosion had removed rock at the base of the hillside that previously buttressed the limestone, a process known as **undercutting**. Rainfall and the rising water table reduced the friction holding the limestone in place and caused the clay layer to lose strength. Although the landslide may have eventually occurred naturally, the reservoir accelerated the process. In short, this was the wrong place to build a dam.

Amazingly, the dam survived the giant wave and still stands; yet the destruction of towns down the valley and the loss of life and injury were disastrous. After the accident, the reservoir was closed and abandoned, but the Vaiont catastrophe was not forgotten. As word of the disaster spread worldwide and the causes were analyzed, more attention was focused on the science of **geological engineering,** the study of geological processes and the role they play in engineering projects and safety. The *natural processes* and *geological products* associated with mass wasting have since become the subject of classroom lessons in universities around the world, and engineers have learned to highlight the importance of geological analysis in designing major construction projects, such as dams, highways, and bridges. Study "The Vaiont Dam" exercise in the "Critical Thinking" box .

TABLE 18.1 Recent Landslide Events

Year	Location	Types	Deaths
1993	Papua New Guinea	Earthquake triggered landslide	37
1996	Washington State	Sea cliff collapse	9
1997	Australia	Landslide	18
2001	El Salvador	Earthquake triggered landslide	1,200
2002	Russia	Landslide	111
2005	California	Heavy rains triggered landslide	16
2005	North Vancouver, British Columbia	Heavy winter precipitation initiated a landslide	1
2006	Philippines	Heavy rains and an earthquake triggered a huge debris slide	1,126
2010	Uganda	Heavy rains triggered landslide	100 to 300
2010	Saint-Jude, Quebec	After a prolonged period of rain, groundwater liquefied Leda clay, causing a large translational slide	4
2010	Mount Meager, British Columbia	One of Canada's biggest landslides on record: a massive rock slide/debris flow with a volume of 48,500,000 m³	0
2012	Johnson's Landing, British Columbia	Heavy rains and snowmelt triggered a landslide in southeast British Columbia	4

(?) **Expand Your Thinking**—The landslide in North Vancouver appeared to strike without warning, or did it? Besides heavy rain, what other factors could have triggered the debris slide? What could North Vancouver do to avoid future landslides?

CRITICAL THINKING

THE VAIONT DAM

The Google Earth image in **FIGURE 18.4** shows the Vaiont River Valley in Italy and the landslide that occurred there. Please study the image and answer the following questions.

1. With a marker, outline the scar left by the landslide.
2. Draw an outline around the regolith that slid into the Vaiont River Valley.
3. Label the remaining portion of the reservoir, and as best you can, try to outline the edges of the total reservoir had the landslide not happened. Start from the top edges of the dam on the right and left, and try to contour around what you think was the reservoir (the inset photos may be helpful; you may also want to look at different perspectives using Google Earth).
4. The landslide caused a giant wave that jumped over the dam and killed 2,600 people in the town of Longarone. Show with arrows and labels the sequence of events, the path of the landslide, and the direction of the wave.
5. Imagine you are a geological engineer. What would you study and why to insure that a reservoir you are building is safe? Be detailed and specific in your answer.

Rolls Press/Popperfoto/Getty Images, Inc.

Before (January 1962)

Archivio Cameraphoto Epoche/Hulton Archive/Getty Images, Inc.

After (October 1963)

FIGURE 18.4 The Vaiont Dam was the highest in Europe when it was built in 1959. However, on October 9, 1963, a massive landslide catastrophically fell into the central portion of the Vaiont reservoir, causing a 100-m-high wave that overtopped the dam, killing 2,600 people in the town of Longarone located down the valley below the dam.

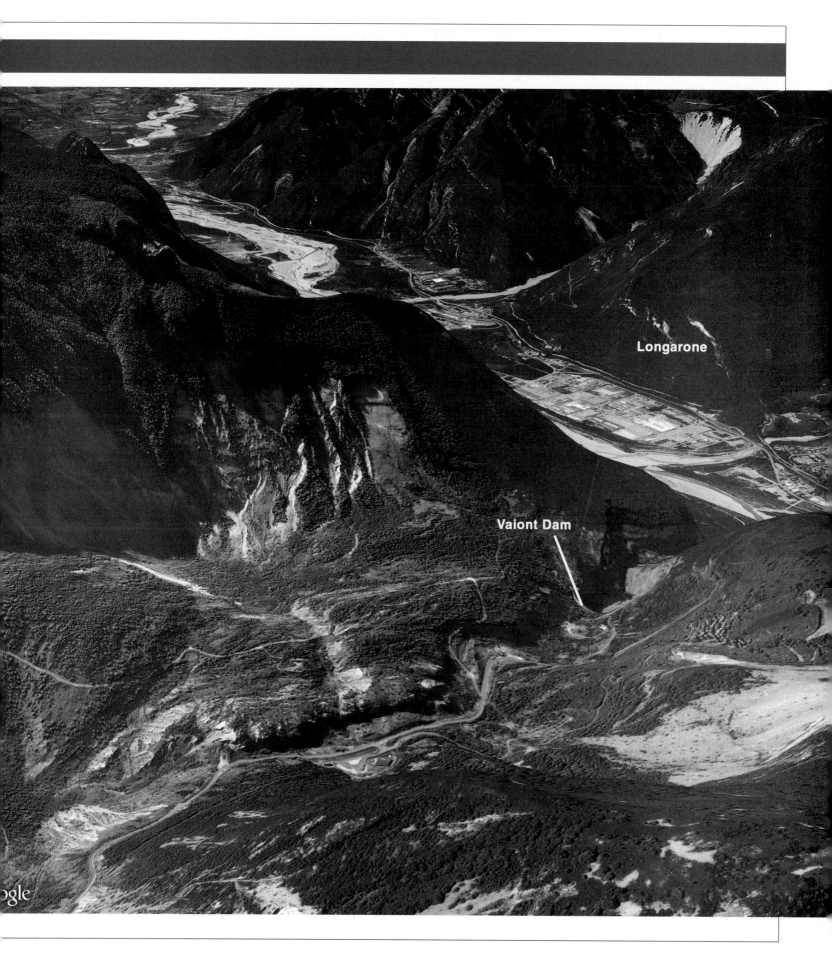

18-2 Creep, Solifluction, and Slumping Are Common Types of Mass Wasting

LO 18-2 *Define "creep," "solifluction," and "slumping."*

Mass wasting is the movement of soil, sand, regolith, and rock down a slope due to the force of gravity. Most regolith is the product of weathering. Hence, mass wasting is the first type of erosion to move these sediments from the place where they formed. Without mass wasting, valleys cut by streams would be narrow, steep-sided gorges with vertical walls. But the tendency for rock and regolith to become unstable and move to the base of slopes (where they are carried away by running water) leads to the widening and expansion that occurs in stream valleys (**FIGURE 18.5**). As we learned in Chapter 17 ("Glaciers and Paleoclimatology"), glacial erosion is another important process that sculpts valleys in mountain systems around the world. Most canyons and valleys have been formed by a combination of glacial erosion, stream erosion, and mass wasting.

Several forms of mass wasting occur. Although we tend to use the word "landslide" to describe any kind of slope movement, geologists distinguish among several types of mass-wasting processes. Knowing the characteristics, causes, and outcomes of each of these processes is the first step toward avoiding them.

FIGURE 18.5 Mass wasting creates broad valleys from steep gorges that have been cut by streams. The width of the Grand Canyon is a result of mass wasting.

 Erosion by the Colorado River has made the Grand Canyon deep. What type of mass wasting has made the canyon wide?

Creep

Soil **creep** is the slow, down-slope migration of soil under the influence of gravity. Creep occurs over periods ranging from months to centuries. Curved tree trunks, cracks in slopes, tilted power poles, and bent fences or walls are common indicators that creep is taking place (**FIGURE 18.6**). The mechanics of soil creep depend on the role of gravity and the rise and fall of the slope surface. Because water increases in volume by over 9 percent when it freezes, the surface of a slope will expand upward (a process known as frost heave) in winter and contract again in summer. Alternate wetting and drying of clay in the soil can have the same effect. (Clay may expand when wet.) These processes work in essentially the same way: Individual soil or rock particles are raised at right angles to the slope by swelling or expansion of the soil and settle vertically downward during compaction or contraction. The net result is slow downslope creep.

Solifluction

Solifluction is a type of soil creep that occurs in water-saturated regolith in cold climates. Most solifluction is seen in *permafrost* zones. Permafrost (recall Chapter 17) is ground that is permanently frozen but a thawed layer develops above it in the summer. During warm weather, the ground in permafrost areas begins to thaw from the surface downward. Much of the freshly melted water cannot be absorbed by or move through the permafrost layer, so it saturates the upper layer of vegetation and regolith. The saturated layer flows down the slightest of slopes as it slips over the frozen layer beneath it. The result is lobes (or wave forms) of slowly flowing soil, the distinctive sign of solifluction (**FIGURE 18.7**).

Slump

A **slump** occurs when regolith suddenly drops a short distance down a slope, usually as a cohesive block of earth that simultaneously slides and rotates along a *failure surface* shaped like the bowl of a spoon (**FIGURE 18.8**). As material shifts downward, one or more crescent-shape *headwall scarps* develop at the upslope end of the slide. The base of the slide is characterized by the *toe* of the slump.

In a typical slump, the upper surface of the rotating block remains relatively undisturbed. Slumps leave a curved scar or depression on the slope. They can be isolated or occur in large *slump complexes* covering thousands of square metres. Slumps are especially common in places where clay-rich sediments are exposed along a steep bank that has been undercut. They often occur as a result of human activities (such as deforestation) and are common along roads where hillsides have been oversteepened by undercutting at the base during construction. They are also common along riverbanks and coastlines, where erosion has undercut the slopes. Heavy rains and earthquakes can trigger slumps.

© Tom Bean/Alamy

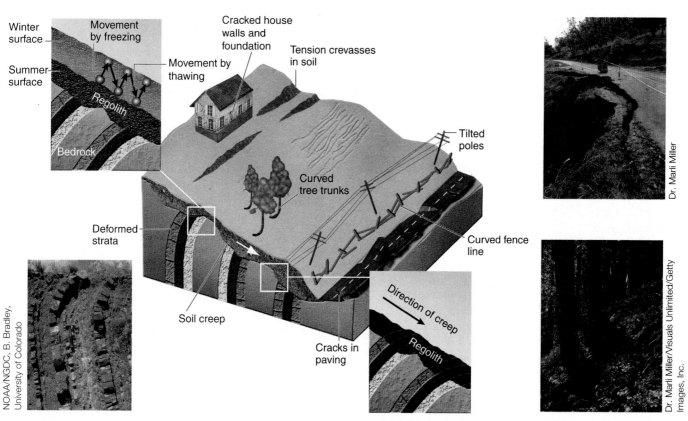

Winter surface

Summer surface

Movement by freezing

Regolith

Bedrock

Movement by thawing

Cracked house walls and foundation

Tension crevasses in soil

Tilted poles

Curved tree trunks

Curved fence line

Deformed strata

Soil creep

Cracks in paving

Direction of creep

Regolith

NOAA/NGDC, B. Bradley, University of Colorado

FIGURE 18.6 Soil creep is often marked by warped fence lines, soil ripples and cracks, curving of rock units, broken road surfaces and building walls, tilted poles and trees, and curved tree trunks.

Dr. Marli Miller

Dr. Marli Miller/Visuals Unlimited/Getty Images, Inc.

? What environmental conditions promote soil creep?

FIGURE 18.7 Important in polar regions and some high mountains, solifluction is a very slow type of slope failure. It occurs in areas of permafrost where saturated soil flows over the frozen permafrost layer beneath it.

Dr. Marli Miller

? Describe a home with design features that are resistant to damage by solifluction.

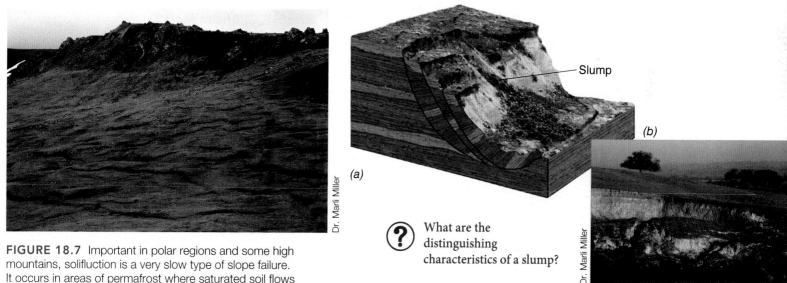

Slump

(a)

(b)

Dr. Marli Miller

? What are the distinguishing characteristics of a slump?

FIGURE 18.8 (a) A slump is defined by a curved failure surface and a rotating block (or series of blocks). (b) The tops of slump blocks remain undisturbed even though they have been displaced downhill by several metres. Note the headwall scarp.

? **Expand Your Thinking**—Describe a geologic environment where creep, slump, and solifluction are all taking place near each other. What triggers might be at work?

18-3 Fast-Moving Mass-Wasting Events Tend to Be the Most Dangerous

LO 18-3 *Distinguish among mass-wasting processes that are flows, slides, and falls.*

On February 17, 2006, after 10 days of heavy rains and a minor earthquake, a deadly **debris flow** on the Philippine island of Leyte caused catastrophic damage and loss of life (**FIGURE 18.9**). An entire elementary school filled with students and their teachers (1,126 people) was buried by the massive debris flow that surged out of the nearby hills and fanned out across the fields where the school was located.

Mudflows and Debris Flows

Mudflows and debris flows are rivers of rock, earth, and other debris saturated with water. Mudflows tend to have lower viscosity (i.e., they are more watery) than debris flows because they are composed principally of water-saturated mud. Debris flows carry clasts ranging in size from clay particles to boulders and often a large amount of woody debris as well. Both types of flows develop when water accumulates rapidly in the ground during heavy rainfall or rapid snowmelt, changing stable regolith into a flowing river of sediment, called a "slurry." They can flow rapidly, striking with little or no warning at high speeds.

The distinguishing feature of a debris or mudflow is the tendency for the material to exhibit *turbulent flow characteristics*. Although other types of mass-wasting processes are chaotic, they may not exhibit true flow characteristics similar to those of a fluid, whereas a debris or mudflow does exhibit these characteristics. Both types of flows typically consist of a well-mixed, turbulent mass of saturated regolith that literally *flows* rather than *slides* (**FIGURE 18.10**). Mudflows, because they have such low viscosity, can travel several tens of kilometres from their source, growing in size as they pick up trees, boulders, cars, and other materials (debris flows typically do not travel as far). Both mudflows and debris flows result from heavy rains in areas with an abundance of unconsolidated sediment, such as steep slopes that have been deforested. In these areas, it is common after a heavy rain for streams to turn into mud or debris flows as they pick up loose sediment and break free of their banks. As we learned in Chapter 6 ("Volcanoes"), flows of muddy ash called *lahars* can also result from volcanic eruptions that cause snow or ice on a volcano's slopes to melt or a crater lake to drain.

U.S. Marine Corps photo by Lance Cpl. Raymond D. Petersen III

FIGURE 18.9 This debris flow on the Philippine island of Leyte buried a school full of children and their teachers under 40 m of water-saturated soil, mud, trees, regolith, and other debris.

(?) Based on field evidence, how would you tell the difference between a debris flow and a mudflow?

Quick Clay

Quick clay is a special type of marine clay that can spontaneously change to a fluid state in a process called *liquefaction*. Large volumes of quick clay, called *Leda clay*, are found within the St. Lawrence Lowland in southern Quebec and Ontario. The clay was deposited within the St. Lawrence Valley in the Champlain Sea, a large marine embayment formed at the end of the Pleistocene during the mass melting of a glacial ice sheet that covered most of central and eastern Canada. In normal, stable conditions the clay can have very high water contents (up to 80 percent) and still support many times its weight (**FIGURE 18.11a**). The surface tension of water-coated flakes of clay holds the clay together as a cohesive mass. However, slight disturbances such as ground shaking due to small-magnitude earthquakes in the St. Lawrence Valley can cause liquefaction of the clay and result in devastating slides (**FIGURE 18.11b**).

Rock Slides and Debris Slides

Rock slides and **debris slides** result when a slope fails along a plane of weakness. A debris slide is characterized by unconsolidated rock, debris, and regolith that have moved downslope along a relatively shallow failure plane. A rock slide typically lacks a variety of debris; usually it consists of blocks of rock in a chaotic mass. The most catastrophic landslide event ever recorded in Canadian history involved a rock slide at Turtle Mountain, which buried part of the town of Frank, Alberta, under 82 million tonnes of rock in the middle of the night on April 29, 1903 (**FIGURE 18.12**). More than 90 people died. Contributing factors included dramatic changes in weather and an unstable failure surface, formed by bedrock riddled with joints and bedding planes that paralleled the steep slope.

Debris slides form steep, unvegetated scars in the upper region of the slide and irregular, hummocky (bumpy) deposits in the toe region. Debris slide scars are likely to "ravel"—that is, continue to pull apart into separate pieces—and remain unvegetated for many years. Such slides are most likely to occur on slopes with an incline of over 65 percent where regolith overlies shallow soil or bedrock. The slide surface is usually less than 5 m deep. The probability of sliding is low in places where bedrock is exposed, unless weak bedding planes and extensive bedrock joints and fractures parallel the slope.

Rock slides and debris slides are popularly known as *landslides*. The mass-wasting event at Vaiont Dam was a debris slide. Slides can develop on a bedding plane, a clay layer, a buried erosional surface, or a *joint* surface (**FIGURE 18.13**). Recall from earlier chapters that joints are fractures in the crust that result from the cooling or expansion of a rock layer.

Rockfalls and Debris Falls

Rockfalls occur when an accumulation of consolidated rock is dislodged and falls through the air, or free-falls, due to the force of gravity (**FIGURE 18.14**). **Debris falls** are similar, except that they usually involve a mixture of soil, regolith, vegetation, and rocks. A rockfall may consist of a single rock or a mass of rocks, and the falling rocks may dislodge others as they collide with the cliff face. Because this process involves free-falling through the air, rockfalls and debris falls are associated with steep cliffs.

FIGURE 18.10 (a) This Indonesian village was destroyed by a mudflow. (b) This debris flow damaged the town of Glenwood Springs, Colorado.

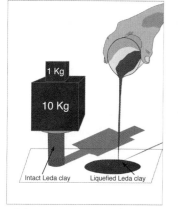

FIGURE 18.11 (a) A demonstration of the strength of intact Leda clay (left) versus the same clay with the same water content, which liquefies when disturbed. *Source*: Based on C.B. Crawford, "Quick Clays of Eastern Canada," *Engineering Geology, Vol 2* (1968) pp. 239–265. (b) Photo showing the 1 km × 0.5 km path of destruction caused by liquefaction of Leda clay near St. Jude, Quebec, on May 10, 2010. Four people died in one of the houses that collapsed within the crater.

FIGURE 18.12 Photo looking southwest at the Frank slide, which left a deep scar in the side of Turtle Mountain on April 29, 1903. It was the most catastrophic landslide in Canadian history. The run-out of the rock avalanche covered the eastern part of the town of Frank, Alberta. The new townsite (lower right corner) was established to the north after the slide.

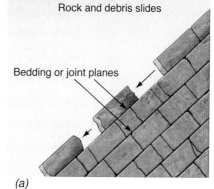

FIGURE 18.13 (a) Rock or debris slides tend to occur along a plane of weakness, such as a bedding surface, an unconformity, a clay layer, or a joint. (b) This debris/rock slide occurred in the spring of 2002 along McAuley Creek, British Columbia.

 What geologic clues would you look for to assess if your home was vulnerable to a debris flow or a debris slide?

Rockfalls and debris falls are identified by one primary feature. What is it?

FIGURE 18.14 (a) Rockfall. (b) When a single rock or a mass of debris free-falls through the air, it may dislodge others as it collides with the cliff face. (c) Debris fall.

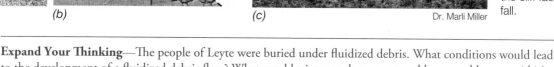

Expand Your Thinking—The people of Leyte were buried under fluidized debris. What conditions would lead to the development of a fluidized debris flow? What would trigger such an event and how would you avoid it?

18-4 Avalanches, Lahars, and Submarine Landslides Are Special Types of Mass-Wasting Processes

LO 18-4 *Compare and contrast avalanches, lahars, and submarine landslides.*

Certain special types of mass-wasting processes pose unique dangers in many areas of the world. Avalanches (a type of debris flow), lahars (a type of mudflow), and submarine landslides (a type of debris flow) occur under specific geologic conditions and may not be widespread, but in certain areas, they are prevalent and their impact may be catastrophic.

Avalanches

An **avalanche** is a fast-flowing, fluidized (by air, not water) mass of snow, ice, air, and occasionally some regolith that cascades down a mountainside due to the collapse of a snowfield (**FIGURE 18.15**). Large avalanches, moving over 10 million tonnes of snow, can travel at speeds exceeding 300 km/h.

Craig Ellis

FIGURE 18.15 Powder snow avalanches, such as this one in the northern Monashee Mountains of British Columbia, tend to be the largest type of avalanche.

(?) What can trigger an avalanche?

Snow accumulates in layers of varying strength due to differences in the weather conditions during the snowfall or deposition of newer snow on older snow that has undergone melting and recrystallization in the snowpack due to fluctuating temperatures. As snow builds up on a steep slope and increases in weight, a weak layer may fail because of vibration or some other cause. The overlying mass detaches from the slope, slides over the weak layer, and gains strength and size as it gathers more material on its downhill journey.

Loose snow avalanches are the most common type. They consist of newly fallen snow that has not had time to compact. These avalanches usually form in steep terrain on high slopes and originate at a single detachment point. The avalanche gradually widens on its way down the slope as more snow accumulates, forming a teardrop shape. *Slab avalanches,* consisting of a strong, stiff layer of snow known as a "slab," account for about 90 percent of avalanche-related fatalities. Slabs develop when the wind deposits snow on a *lee slope* (the mountainside not facing the wind) in compact masses. The avalanche starts with a crack that quickly grows and moves across the slab, in some cases releasing a solid mass of snow hundreds of metres long and several metres thick. Still another type of avalanche, a *powder snow avalanche*, develops as air mixes with the snow at the front of an avalanche, forming a powder cloud, a turbulent suspension of snow particles. The largest avalanches are usually of this type.

Lahars

We first discussed **lahars** (**FIGURE 18.16**) in Chapter 6 ("Volcanoes"), but since they cause such catastrophic damage, it is worthwhile to revisit them here. A lahar is a mudflow composed of pyroclastic ash and lapilli and water that flows down the slopes of a volcano. Lahars have the consistency of concrete—they are fluid when moving but become solid after coming to rest. Lahars can be created by snow and ice melted by a pyroclastic flow during a volcanic eruption or by a flood caused by a melting glacier, a flash flood in a stream, or a heavy rainfall. The key feature of a lahar is that it consists of mud originating in an existing volcanic ash deposit. Lahars often start in a stream channel, but as they gather mass and speed, they break out of the channel when their momentum prevents them from turning. This is when they become most dangerous to a local population.

Several volcanoes near population centres are considered particularly dangerous as potential sources of lahars. They include Mount Rainier in the United States, Mount Ruapehu in New Zealand, and Mount Galunggung in Indonesia. At the base of Mount Rainier, several hundred thousand people live in the Puyallup Valley near Seattle, where towns have been built on lahar deposits that are only about 500 years old. It is believed that lahars flow through the valley every 500 to 1,000 years; hence, these communities face considerable risk. Local authorities have set up a warning system and made evacuation plans to be put into effect at the first sign of a developing lahar.

FIGURE 18.16 A lahar is a mudflow composed of pyroclastic material and water that flows down the slopes of a volcano. This lahar originated on the slopes of Mount Pinatubo in the Philippines in 1991.

Richard P. Hoblitt/USGS

(?) How do geologists determine if an area is vulnerable to a lahar?

Submarine Landslides

Submarine landslides are a type of underwater mass-wasting that can remove large masses of rock and sediment from continental shelves and oceanic islands. Among the most awe-inspiring are the beautiful *palis* (Hawaiian for "cliffs") of the Hawaiian Islands. These sheer, vertical rock faces of ancient basalt lava flows are hundreds of metres high. Many palis drop directly into the sea (**FIGURE 18.17**).

The geologic history of these immense cliffs is a tale of mass wasting on a prodigious scale. The cliffs are formed by vast debris flows that take away whole pieces of an island in a single event. To understand this process, recall the discussion of mineralogy in Chapter 4 ("Minerals"). The Hawaiian Islands are composed of two types of mafic igneous rock: extrusive and intrusive. The extrusive rock (basalt) is lava that solidifies within a few minutes of reaching the atmosphere. With so little time in which to crystallize, the lava, though it consists of basalt, has the texture of glass and is very brittle because it lacks a crystalline structure.

Although no one was present when these palis were formed, it is hypothesized that the swelling of a volcano and deep-seated earthquakes associated with magma intrusion prior to an eruption occasionally cause a flank of the volcano literally to break off and slide into the sea. Such an occurrence is possible because the islands are essentially a pile of glassy debris. Today's palis are the eroded remnants of those fractures, which have since retreated and been reshaped by weathering due to waves, wind, and rain.

Several observations provide evidence in support of this hypothesis. Maps of the sea floor show topography that is consistent with a landslide source. Fields of chaotic debris can be traced back to the islands where some of the highest palis are located. In addition, the main shield volcanoes composing the islands have incomplete outlines. Rather than having the broad circular or oval shape typically found at the base of a shield volcano, each of the two shield volcanoes forming the island of Oahu shows only half its base—each forms a crescent rather than an oval. The same is true of the island of Molokai. Finally, marine scientists have identified the remnants of more than 25 giant landslides surrounding the Hawaiian Islands. These slides were some of the largest on Earth, and most took place within the past 4 million years.

(a) Koshtra Tolle/Getty Images, Inc. (b)

(?) What is the evidence for submarine landslides in Hawaii?

FIGURE 18.17 The Hawaiian palis *(a)* are the highest sea cliffs in the world. A large debris field *(b)* off the windward side of the islands of Oahu and Molokai is connected to each island at the location of two large sea cliffs.

 Expand Your Thinking—What features would you design into a community to make it less vulnerable to a lahar?

18-5 Several Factors Contribute to Unstable Slopes

LO 18-5 *Describe the factors that contribute to mass wasting.*

Mass wasting may occur in the form of phenomena ranging from slow, imperceptible creep to fast-moving landslides.

Gravity

Mass wasting is governed by *gravity*. A fundamental force of nature, gravity is the force of attraction between objects, and on Earth's surface, it causes things to move downhill. Gravity can be divided into two stress components for objects resting on sloping surfaces (FIGURE 18.18). One component, the *shear stress*, is parallel to the slope (g_s) and the other, the *normal stress*, is perpendicular to the slope (g_p). On steep slopes (more than 45°), the stress parallel to the slope is greatest and tends to pull objects downhill. On gentler slopes (less than 45°), the normal stress component perpendicular to the slope is greatest and tends to hold the object in place.

Because Earth's gravity is the primary force that controls the stability of materials on a slope, *friction* and *density* play important roles in determining when and how slope movement occurs. Friction, created by resistance to movement at the point of contact between grains of material, tends to be a stabilizing factor in slope deposits. But when frictional forces are temporarily or permanently disrupted, mass wasting may occur. If the amount of friction is low, materials are more likely to move. In contrast, mass wasting is less likely if the amount of friction is high.

Water

Friction can be disrupted in a number of ways. For instance, heavy rainfall can saturate rock and sediment on a slope and increase its weight so that it exceeds the force holding it in place. The same rainfall may lubricate the surface on which a deposit rests, also disrupting the friction that helps to hold it in place. Slight shaking from a nearby earthquake or desiccation in hot sunlight causing sediment to shrink and crack is another way in which friction on a slope may be disrupted.

The density of sedimentary deposits is reduced when they become saturated with water. That is, their overall density approaches that of water, and they may achieve a state of temporary buoyancy. It is at this moment that gravity may control the deposit and cause slope failure. *Water content is an important controlling factor* that influences the stability of earth materials on slopes. Water destroys the cohesion that exists when one grain of sediment is in contact with another. As the pore space between grains fills with water, water pressure increases. Pressure can build so that pore water literally lifts one grain off another in a process called *dilation* (expansion). Once grains become dilated, the entire assemblage of rock and regolith gains buoyancy and gravity causes the slope to fail (FIGURE 18.19).

Water plays other roles as well. As rock and sediment on a slope become saturated, they gain weight, and once motion occurs, that weight contributes to the *momentum* of the mass. Momentum is the tendency of a moving mass to continue moving. A large mass of saturated rock and sediment weighing millions of tonnes possesses an amount of momentum that can be hard to stop. Heavy, water-saturated lahars and mudflows have been known to travel dozens of kilometres at speeds of over 100 km/h, crossing ridges, filling valleys, and burying entire landscapes because the momentum of the mass carried it far beyond the original slope on which it started.

Water also plays a role in changing the physical properties of regolith: What was formerly stable becomes unstable. Dry clay is among the strongest sedimentary deposits because of the grain-to-grain cohesion that exists. But once it becomes wet, clay turns into a slick, cohesionless mass that not only possesses little internal strength but actually acts as a lubricant that enhances downslope movement of other material.

Oversteepening

Another factor that controls slope stability is **oversteepening.** Loose sediment tends to be stable on slopes less than the *angle of repose.*

FIGURE 18.18 On steep slopes, shear stress, which is the gravity component parallel to the slope (g_s) moves materials downhill. On gentler slopes, normal stress, which is the gravity component perpendicular to the slope (g_p) tends to hold material in place.

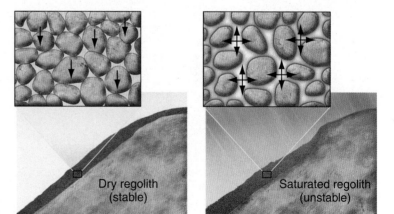

FIGURE 18.19 As the pore space between grains fills with water, water pressure lifts one grain off another in a process called "dilation." This can cause a slope to fail.

 What factors can reduce friction on a slope?

 Describe the forces at work in saturated regolith with increasing pore pressure.

The angle of repose is the slope beyond which an unconsolidated mass will spontaneously fail. Coarse-grained sediment tends to have a steeper angle of repose, while fine-grained sediment tends to have a lower angle of repose. In general, angles of repose vary between 25° and 40°, depending on the content of the sediment. Adding material to the top of a sedimentary accumulation or *undercutting* the slope at its base has the effect of increasing the angle of the slope to the point at which its stability is exceeded. Most often, this leads to slope failure and frequently to catastrophic mass wasting. Examples of undercutting a slope at the base are common where roadways are carved into valley walls or wave erosion on coastlines undercuts the base of a cliff (FIGURE 18.20).

Triggering Slope Failure

Water content and oversteepening control slope stability. They may create conditions in which regolith or rock is poised on the threshold of failing. To initiate a mass-wasting event, some kind of *trigger* typically is needed. The most common trigger of slope failure is *vibration*. Vibration, or ground shaking, can be caused by passing trucks or trains, an earthquake, a breaking wave, the rumbling of a volcano, or even thunder. Another common trigger is melting. Ice will hold unstable materials together, but with the warming that happens

seasonally or even daily, ice can melt and loosen its grip on failure-prone slope deposits. On a much larger scale, the melting of permafrost in the Canadian Arctic related to climate warming has caused large tracts of land to become unstable, leading to massive slumping, sinkholes, and landslides (FIGURE 18.21).

Ice wedging, the physical weathering process by which ice expansion breaks off pieces of rock, can lead to mass wasting (see Chapter 7, "Weathering," Figure 7.7). It may occur when direct morning sunlight warms a mountain slope and melts ice that was formed during the previous evening. This process of daily ice wedging followed by slope failure is responsible for much of the fallen rock that collects at the base of steep cliffs. The resulting apron of loose rock is known as **talus** or *scree* (FIGURE 18.22).

FIGURE 18.21 Large slumps along the coast of Herschel Island, Yukon, are caused by the thawing of the underlying permafrost.

Hugues Lantuit

FIGURE 18.20 Landslides along coastal cliffs occur when high waves batter the shoreline. A coastal slope is oversteepened when wave erosion removes material from its base. This is known as undercutting.

© Aerial Archives/Alamy

 Describe several processes, both human and natural, that produce oversteepening.

FIGURE 18.22 Talus lines the base of these cliffs in the Selkirk Mountains of British Columbia.

Courtesy of Dr. Dan Gibson, Simon Fraser University

 Use evidence in Figure 18.22 and your own reasoning to write a hypothesis explaining the origin of this pile of debris. How would you test your hypothesis?

Expand Your Thinking—Describe the role of water in mass wasting.

18-6 Mass-Wasting Processes Vary in Speed and Moisture Content

LO 18-6 *Describe how various types of mass-wasting processes differ in terms of their speed and moisture content.*

We can categorize mass-wasting processes by their relative speed (fast to slow) and moisture content (wet to dry) (**FIGURE 18.23**). The fastest processes are mudflows, debris flows, debris falls, and rockfalls. These move at rates ranging from 1 m/s to several tens of metres per second. Mudflows require that muddy sediments be saturated with water, whereas the formation of rockfalls can be a completely dry process. The speed of these types of failures makes them the most hazardous. They can overwhelm individuals and even entire communities in a matter of minutes, bury victims, and cause catastrophic damage. Mass-wasting processes that move at moderate rates include debris slides, rock slides, and slumps. The slowest forms include solifluction, which requires a high degree of water saturation, and soil creep, which is a largely dry process, although it can be caused by alternating wetting and drying or freezing and thawing.

Protection from these hazards requires awareness of their presence, mapping and monitoring of controlling factors and triggers, and, finally, direct engineering aimed at altering the environmental processes that lead to slope failure (see **FIGURE 18.24** in the Earth Citizenship box).

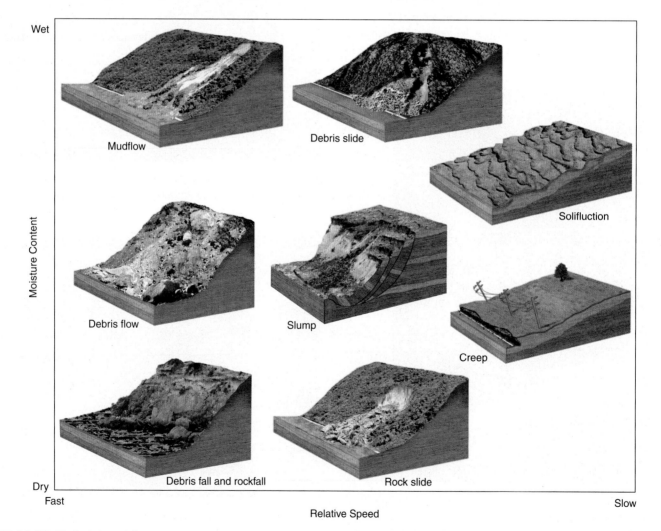

FIGURE 18.23 Typical slope failure processes can be categorized based on moisture content and relative speed.

(?) All these mass-wasting events require a trigger. Give some examples of a mass-wasting trigger, and include a discussion of factors (including human) that promote mass wasting.

(?) **Expand Your Thinking**—As a geological engineer responsible for analyzing potential mass-wasting hazards on a nearby hillside, why would you pay special attention to the speed and moisture content of potential slope processes? Describe the methods you would use to assess the hazard potential of a slope.

EL SALVADOR'S DEADLY LANDSLIDE

On January 13, 2001, a large earthquake off the coast of El Salvador triggered a debris slide of rock and soil that roared down on the city of Santa Tecla. In an instant, hundreds of homes and more than 1,200 people were buried. International search teams worked for weeks to rescue the living, provide care for the injured, and retrieve bodies (**FIGURE 18.24**).

Local residents had warned authorities for years that *deforestation*, hillside *undercutting* to make room for new homes, *soil* placed at the hilltop, and lack of proper geological engineering analysis would lead to a disaster if the hillside should fail in the next earthquake. "The developers kept digging further and further into the base of the mountainside despite everything we said, and the government let them do it," said Miguel Cordero, a survivor who lost his family.

Local groups of citizens originally initiated a lawsuit to halt developers. Although the court ruled that development must cease, the ruling was only temporary. Soon the ruling was overturned, and developers went on clearing trees and cutting away the base of the slope, eventually resulting in the deadly debris slide.

(a)

© AP/Wide World Photos

(b)

FIGURE 18.24 A large earthquake triggered the collapse of this oversteepened hillside. More than 1,200 people were buried in the debris slide at Santa Tecla, El Salvador, in 2001.

© AP/Wide World Photos

18-7 Human Activities Are Often the Cause of Mass Wasting

LO 18-7 *List some ways in which humans cause mass wasting.*

Knowing the relationship between local geology and mass-wasting processes can lead to better *land-use planning*.

Planning

Many geologists are involved in land-use planning. Their job is to assess the presence of geologic hazards and resources, plan appropriate land modifications that mitigate hazards and conserve resources, and recommend sustainable practices to ensure a safe and productive future for land use. One goal of good land-use planning is to reduce human vulnerability to geologic hazards, including slope failure. Examples of such planning include analyzing the geology of hillsides before building on them, not building on or below unstable slopes, and avoiding actions that may cause slope failure, such as undercutting the base of a hillside or removing vegetation that holds regolith in place. Thus, it is important when developing land to be familiar with the various types of mass-wasting processes, their underlying causes, factors that affect slope stability, and what humans can do to reduce vulnerability and risk from slope failure.

The human tendency to alter hillsides often leads to mass wasting. FIGURE 18.25 provides examples of how development on steep hillsides may create an unsafe situation.

Human Activities

In Figure 18.25, a large water tank is built high on a hillside in order to make use of the gravity force necessary to supply a community with water pressure. This is a common practice; most water delivery to homes and businesses relies on natural water pressure from an elevated storage tank. Either a tank is built on a nearby hillside or, where the land around a community has low relief, a high-standing tank is constructed on its own steel legs, often 30 m high or more. Water is mechanically pumped into the tank and allowed to naturally flow downhill to be used by homes and businesses.

On a hillside, the weight of a water tank may destabilize the slope and cause oversteepening. In addition, any leakage from the tank or delivery lines may saturate regolith on the slope or lubricate underlying planes of weakness. The results may not be apparent for years, and the mass wasting that eventually occurs could be sudden and catastrophic or take the form of slow creep that is barely perceptible (yet nonetheless damages the tank).

When houses or other buildings are built on slopes, it is common building practice to cut into the slope to create a level *bench*. Soil and gravel are then added to provide a foundation for features such as garages, pools, and even the main structure of the building. This practice can oversteepen the hillside because it cuts into the toe of the upslope regolith and rock. The addition of soil oversteepens the slope and adds weight to unstable portions of the hillside.

Another problem that develops is the result of in-ground waste disposal. In-ground waste disposal involves moving fluid waste from a building into a temporary holding tank that is buried nearby. The tank is perforated so that waste fluid seeps into the ground. The soil acts as a natural filter that cleans the waste water and provides a growth environment for microbes and various bacteria that consume the organic waste. These systems are commonly referred to as *septic tanks, cesspools,* and *leach fields.* In-ground waste disposal is frequently used in communities on hill slopes because of the difficulty of extending waste delivery pipes up steep slopes to carry waste materials to a centralized waste-water processing plant. As waste water from multiple homes on a hillside soaks into the ground, it can saturate regolith, increase the weight of the unconsolidated material, and further weaken and lubricate underlying zones of weak rock.

Driveways, streets, and other roadways carved into hillsides oversteepen the slope and undermine the base of regolith deposits and rock strata. Overall, construction on hillsides increases the vulnerability of buildings, both on the slope and below it, to hazardous slope failure events.

EARTH CITIZENSHIP

GLOBAL WARMING AND MASS WASTING

As global warming heats the air at Earth's surface, rates of evaporation and precipitation increase. Greater rates of evaporation put more water in motion through the water cycle. In many places, this has increased the intensity of rainfall. Because water is a significant factor in mass wasting, especially acting as a trigger in many cases, heavier rainstorms may increase the tendency for mass wasting in certain areas.

While it sounds counterintuitive, a warmer world produces both wetter and drier conditions. Even though total global precipitation increases, the regional and seasonal distribution of precipitation changes and more precipitation comes in heavier rains (which can cause flooding and mass wasting) rather than light events. According to Statistics Canada, in the past 50 years, averaged over the entire country, total precipitation in Canada has increased by about 17 percent. This has been especially noteworthy in the Arctic regions, where the percent increase of precipitation, more than 25 percent, has been most dramatic. This will increase the melting of the permafrost and quicken mass wasting and the erosion of the destabilized soil. Increased precipitation also poses a risk for other regions in Canada such as southern Manitoba and Saskatchewan. There, the rivers drain though the lacustrine deposits of glacial Lake Agassiz that cover a large area and have produced a drainage basin with very low relief, and consequently, a high susceptibility to flooding.

Flooding often occurs in the spring following snowmelt, especially if accompanied by ice jams in rivers and periods of heavy precipitation. Extended periods of heavy precipitation have been increasing over the past century, most notably in the past two to three decades. Regions such as the Pacific coast and the St. Lawrence Lowlands where these conditions combine with other types of triggers, such as earthquakes, may see increased vulnerability to mass wasting due to global warming.

FIGURE 18.25 Human development activities may reduce the stability of regolith and rock, leading to mass wasting.

 Identify the various ways that human activities reduce slope stability.

 Expand Your Thinking—Describe some mitigation steps that a community can take to reduce the threat of damage and injury due to mass wasting.

CRITICAL THINKING

MASS WASTING

Please work with a partner and answer the following questions using **FIGURE 18.26**.

1. Describe the difference between a mudflow and a debris flow. What conditions would lead to the formation of one rather than the other?

2. Describe the shapes of various types of mass movement. What natural conditions influence mass wasting? How are these conditions related to the *shape* of a mass movement?

3. Summarize the processes acting on slopes in your area. How do weathering, rock types, relief, climate, soil type, and various trigger mechanisms combine to influence mass wasting where you live?

4. Construct a table listing the mass-wasting processes shown in Figure 18.26 and the conditions that govern them. Use a marker to number the processes in the figure.

5. What are the hazards associated with mass wasting? How are human activities and structures vulnerable to mass wasting?

6. You are the mass-wasting expert employed by North Vancouver (see Figure 18.1). Describe the research program you have set up to improve understanding of mass wasting in nearby hills and valleys.

7. As the expert in question 6, you have to convince the municipal council each year to fund your program. Write one compelling paragraph explaining why they should continue to spend money on your program.

FIGURE 18.26 Forms of mass wasting.

18-8 Research Improves Knowledge of Mass Wasting and Contributes to the Development of Mitigation Practices

LO 18-8 *Describe methods of mitigating mass-wasting hazards.*

A number of commonsense practices are used to reduce the risk of mass wasting. The first step in mitigating any risk is to identify locations where the risk is high. To do this, the Earth Sciences Sector of Natural Resources Canada and Public Safety Canada provide information related to natural hazards and hazard mitigation on an ongoing basis and in response to emergency situations. They encourage research and promote public education about natural hazards, including mass wasting.

These government agencies provide an effective approach to understanding all types of mass wasting. First, they identify areas that are likely to have unstable soil and slopes. These are places that are located:

- where the surficial geology is particularly susceptible to instability (e.g., glacial deposits such as Leda clay).

- in areas of higher than average mean annual precipitation.

- close to major water bodies such as lakes, rivers, and the ocean.

- on existing landslides.

- on or at the base of slopes.

- in or at the base of minor drainage hollows.

- at the base or top of an old fill slope.

- at the base or top of a steep cut slope.

- on developed hillsides with leach-field septic systems.

Using these and other criteria, the Geological Survey of Canada has compiled a map that identifies areas of landslide susceptibility across Canada (**FIGURE 18.27**).

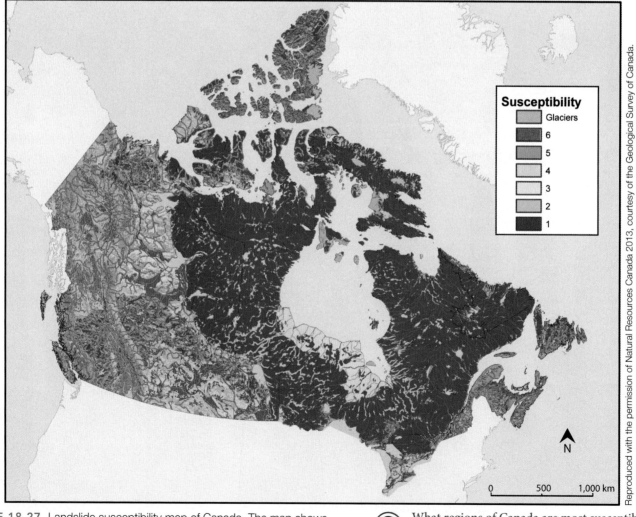

Susceptibility
- Glaciers
- 6
- 5
- 4
- 3
- 2
- 1

Reproduced with the permission of Natural Resources Canada 2013, courtesy of the Geological Survey of Canada.

0   500   1,000 km

N

FIGURE 18.27 Landslide susceptibility map of Canada. The map shows regional landslide susceptibility in hot (red) to cold (green) colours based on high (6) to low (1) susceptibility ranking, respectively.

(?) What regions of Canada are most susceptible to mass wasting? In general, what are the characteristics of these areas that promote mass wasting?

In Figure 18.27 regional landslide susceptibility is shown in hot (red) to cold (green) colours representing high (6) to low (1) susceptibility ranking, based on the criteria outlined above. The term "susceptibility" indicates whether a location is exposed to mass wasting hazards. For instance, if your home is located at the base of a steep cliff, it is in a location that likely has higher susceptibility and would be mapped as brown to red on the map. With maps like this one, it is possible to decrease vulnerability to mass-wasting events by avoiding construction in regions that are mapped as susceptible to landslide hazards.

Once a region has been identified as being at risk of mass wasting, local governments, communities, and individual homeowners can take several steps to reduce vulnerability. The first and most obvious step is avoidance. The best way to protect against mass-wasting hazards is to avoid them in the first place. This is accomplished by knowing the signs of a potentially hazardous slope and choosing not to build or develop in the area.

Avoidance is always the most effective strategy to reduce vulnerability to natural hazards of any type. Unfortunately, because of decades of past building practices, there are many sites where it is too late for avoidance, and **direct mitigation** must be used to reduce vulnerability to mass-wasting hazards. Direct mitigation practices include:

- maintaining and creating *vegetation cover* so that dense and deep root growth can help stabilize a slope by increasing soil cohesion and reducing water pressure by taking in water.

- *regrading* land so that oversteepening is reduced or removed by redistributing regolith and rock mass to less hazardous locations.

- *preventing undercutting* at the base of a hill by reinforcing it with *rip-rap* (e.g., loose boulders) or a retaining wall.

- *reducing water infiltration* so that hill slopes do not become saturated; this is accomplished by diverting drainage and/or installing perforated drainage pipes into the hillside.

- *stabilizing hillsides* using engineering measures designed to hold rock and regolith against a hill slope (**FIGURE 18.28**).

- *frequent monitoring and inspection*, either manually or remotely or both, especially in those areas deemed to be at risk.

Several engineering measures are used to protect roadways and buildings from the hazards of mass wasting. These measures include *retaining walls* to keep fallen rock debris off roads, diverting traffic under avalanche and *rockfall sheds*, erecting *rock nets* to prevent rock and debris falls, putting in *rock bolts and metal strapping* to stabilize cliff faces, and spraying rock faces with *shotcrete* (cement slurry) to stabilize weathering rock walls.

(a)
© Justin Kase z08z/Alamy

(b)
Dr. Marli Miller

(c)
Courtesy of Dr. Dan Gibson, Simon Fraser University

FIGURE 18.28 *(a)* Rock netting protects roadways from falling rock. *(b)* Shotcrete (sprayed cement) can be used on unstable slopes to reduce mass wasting. *(c)* Metal strapping with rock bolts, like that used in the wall rocks of the Revelstoke Dam in southeastern British Columbia, can be used to stabilize cliff and rock faces.

 Why would vegetation on a slope help to stabilize it?

Expand Your Thinking—What is the best way to reduce vulnerability to mass wasting? Why is it the best? What problems prevent this method from being used in *all* cases?

LET'S REVIEW "GEOLOGY IN OUR LIVES"

Now that you have finished the chapter, "Geology in Our Lives" will have taken on new meaning for you. Let us review it: Mass wasting is the set of processes that move weathered rock, sediment, and soil down a slope due to the force of gravity. Mass wasting can be hazardous if humans (or buildings and roads) are in the way. Over the past century, the world's population has grown enormously, and many people now live in places that were formerly considered too dangerous or risky. As buildable land becomes scarce, communities are expanding into environments that are exposed to mass wasting and other geologic hazards, turning them into neighbourhoods. The best way to manage mass-wasting hazards is to know the characteristics of a hazardous slope and avoid building there. Doing this is called "hazard avoidance." Avoidance is the most effective and inexpensive approach to managing geologic hazards.

When the gravitational force acting on a slope exceeds the slope's ability to resist it, slope failure occurs. Mass wasting may take the form of rapid downslope movement of rock and/or regolith or slower types of movement that may continue for decades. It is a major geologic hazard that is responsible for $200 million to $400 million in damage in a typical year, and has accounted for approximately 600 human fatalities in recorded Canadian history. Landslides cause damage to bridges, roadways, buildings, and public infrastructure. By knowing the signs of a potentially hazardous hill slope and choosing not to build or develop in the area, we can protect ourselves and our communities from unstable slopes.

In the next chapter, we study streams. Stream environments are similar to hillsides in that they pose hazards to human use and development, most often where we have failed to acknowledge the natural processes that we should avoid rather than ignore.

STUDY GUIDE

18-1 Mass wasting is the movement of rock and soil down a slope due to the force of gravity.

- According to Natural Resources Canada, in a typical year, landslides cause $200 million to $400 million in damage. They have accounted for approximately 600 human fatalities in recorded Canadian history. In many other countries, the toll is higher due to poor land-use practices and building codes that ignore geologic hazards.

- **Regolith** consists of soil, debris, sediment, and broken rock. **Mass wasting** is the movement of regolith down a slope due to the force of gravity. Mass wasting is an example of a geologic hazard that is best avoided by learning to recognize the signs of an unstable slope and not building or engaging in any sort of unsafe activity there. This policy is called **avoidance.**

- The world's population has grown tremendously, and people now live in places that formerly were considered too dangerous or risky. We not only live in dangerous places, we alter them and unintentionally make them more dangerous.

18-2 Creep, solifluction, and slumping are common types of mass wasting.

- Most canyons and valleys have been formed by a combination of glacial and stream erosion and mass wasting.

- Soil **creep** is the slow downslope migration of soil under the influence of gravity. Creep occurs over periods ranging from months to centuries.

- **Solifluction** is a type of soil creep that occurs in cold climates when regolith is saturated with water. Most solifluction is seen in permafrost zones. Solifluction produces distinctive lobes on hill slopes where the soil remains saturated with water for long periods but is able to slide on an underlying layer of ice.

- A **slump** occurs when regolith suddenly drops a short distance down a slope, usually as a cohesive block of earth that simultaneously slides and rotates along a failure surface shaped like the bowl of a spoon. As material shifts downward, one or more crescent-shape headwall scarps develop at the upslope end of the slide. The base of the slide is characterized by the toe of the slump.

18-3 Fast-moving mass-wasting events tend to be the most dangerous.

- A **mudflow** is a river of rock, earth, and other debris saturated with water. Mudflows tend to have low viscosity because they are very watery. **Debris flows** are also saturated with water and exhibit flow characteristics, but they carry clasts ranging in size from clay particles to boulders, and often carry a large amount of woody debris as well. Both types of flows develop when water rapidly accumulates in the ground during heavy rainfall or rapid snowmelt, changing stable regolith into a flowing river of sediment, called a "slurry."

- **Quick clay** is a special type of marine clay, known as *Leda clay*, that can spontaneously change to a fluid state in a process called *liquefaction*. In normal, stable conditions the clay can have a very high water content (up to 80 percent) and still support many times its weight. The surface tension of water-coated flakes of clay holds the clay together as a cohesive mass. However, slight disturbances such as ground shaking due to small magnitude earthquakes can cause liquefaction of the clay and result in devastating slides.

- **Rock slides** and **debris slides** result when a slope fails along a plane of weakness. A debris slide is characterized by unconsolidated rock, debris, and regolith that have moved downslope along a relatively shallow failure plane. A rock slide typically lacks much debris, usually consisting of blocks of rock in a chaotic mass.

- **Rockfalls** occur when an accumulation of rock is dislodged and falls through the air, or free-falls under the force of gravity. **Debris falls** are similar except that they usually involve a mixture of soil, regolith, vegetation, and rocks.

18-4 Avalanches, lahars, and submarine landslides are special types of mass-wasting processes.

- An **avalanche** is a fast-flowing, fluidized (by air, not water) mass of snow, ice, air, and occasionally some regolith that cascades down a mountainside due to the collapse of a snowfield. As snow builds up on a steep slope, a weak layer may fail due to the increasing weight or from a vibration or some other cause. The overlying mass detaches itself from the slope, slides along the weak layer, and gains strength and size as it gathers additional material on its downhill journey.

- A **lahar** is a mudflow composed of pyroclastic material and water that flows down the slopes of a volcano. Lahars have the consistency of concrete; they are fluid when moving but solidify after they have stopped flowing.

- Submarine landslides are vast debris flows that take away whole pieces off continental shelves or oceanic islands in a single event. It is hypothesized that the swelling of a volcano and deep-seated earthquakes associated with magma intrusion may cause a flank of a volcano literally to break off and slide into the sea. Such an event is possible because a volcanic island is essentially a pile of glassy talus.

18-5 Several factors contribute to unstable slopes.

- In all cases, mass wasting is governed by gravity. Gravity can be divided into two components for objects resting on sloping surfaces. One component is parallel to the slope (g_s) and one is perpendicular to the slope (g_p). On steep slopes, the force parallel to the slope will be greatest and will tend to pull objects downhill. On gentler slopes, the component perpendicular to the slope will be greatest.

- Friction tends to hold regolith in place, but it can be disrupted in a number of ways. Heavy rainfall can saturate rock and sediment on a slope and increase its weight so that it exceeds the force holding it in place. The same rainfall may lubricate the surface on which a deposit rests, also disrupting the friction that holds it in place. Slight shaking from a nearby earthquake or desiccation in hot sunlight, causing sediment to shrink and crack, may also disrupt friction on a slope.

- Water destroys the cohesion that exists when one grain of sediment is in contact with another. As the pore space between grains fills with water, water pressure increases. As a result, pore water literally lifts one grain off another in a process called "dilation."

- Water content and oversteepening affect slope stability. A trigger is needed to initiate a mass-wasting event. The most common trigger of slope failure is vibration.

18-6 Mass-wasting processes vary in speed and moisture content.

- We can categorize mass-wasting processes based on their relative speed (fast to slow) and moisture content (wet to dry). The fastest processes are mudflows, debris flows, debris falls, and rockfalls. Mass-wasting processes that move at moderate rates include debris slides, rock slides, and slumps. The slowest forms include solifluction, which requires a high degree of water saturation, and soil creep, a largely dry process, although it can be caused by alternating wetting and drying as well as freezing and thawing.

18-7 Human activities are often the cause of mass wasting.

- One goal of land-use planning is to reduce human vulnerability to slope failure. Examples include analyzing the geology of hillsides before building on them, not building on or below unstable slopes, and avoiding actions that may cause slope failure, such as undercutting the base of a hillside or removing vegetation that holds soil in place.

18-8 Research improves knowledge of mass wasting and contributes to the development of mitigation practices.

- Areas that are generally prone to landslides are found on existing landslides, on or at the base of slopes, in or at the base of minor drainage hollows, at the base or top of an old fill slope, at the base or top of a steep cut slope, and on developed hillsides where leach-field septic systems are used.

- The best way to protect ourselves against mass-wasting hazards is to avoid them in the first place. This is accomplished by knowing the signs of a potentially hazardous hill slope and choosing not to build or develop in the area.

KEY TERMS

avalanche (p. 492)
avoidance (p. 484)
creep (p. 488)
debris falls (p. 490)
debris flow (p. 490)
debris slides (p. 490)
direct mitigation (p. 484)

geological engineering (p. 485)
lahars (p. 492)
mass wasting (p. 484)
mudflows (p. 490)
oversteepening (p. 494)
quick clay (p. 490)
regolith (p. 484)

rockfalls (p. 490)
rock slides (p. 490)
slope stability (p. 484)
slump (p. 488)
solifluction (p. 488)
talus (p. 495)
undercutting (p. 485)

ASSESSING YOUR KNOWLEDGE

Please answer these questions before coming to class. Identify the best answer.

1. What is mass wasting?
   a. chemical and physical weathering of soil and rock
   b. movement of regolith down a slope due to gravity
   c. weathering of regolith by rainfall
   d. oversaturation of loose rock and soil by water
   e. flash flooding by a stream

2. Very slow movement of rock and soil down a hillside without the formation of a scarp is termed
   a. a slump.
   b. a debris slide.
   c. creep.
   d. earth flow.
   e. an avalanche.

3. The primary requirement for solifluction is
   a. a tropical environment.
   b. a steep hillside above a river channel.
   c. glaciation.
   d. permafrost.
   e. None of the above.

4. Flows, slides, and falls are characterized by (respectively)
   a. high water content, high shaking, and glaciation.
   b. high water content, lower water content, and permafrost.
   c. free-fall through the air and fluidization.
   d. turbulent flow, sliding on a plane of weakness, and free-falling.
   e. fluidization, turbulent flow, and sliding on a plane of weakness.

5. What is a dangerous type of mass wasting on active volcanoes?
   a. creep
   b. lahars
   c. rock slides
   d. solifluction
   e. rockfalls

6. Which of the following is most prone to slope failure?
   a. conglomerate
   b. dipping fractured limestone on a clay bed
   c. horizontally bedded quartz sandstone
   d. basalt dike
   e. marine sediment

7. Which of the following contributes to mass wasting?
   a. high water content
   b. vibration
   c. strata inclined downhill
   d. oversteepened slope
   e. All of the above.

8. What is the difference between a slump and a debris slide?
   a. The presence or absence of water determines if it is a slump or debris slide.
   b. Slumps occur only on permafrost.
   c. A slump consists of rotating but coherent blocks, and a debris slide is chaotic.

d. Slumps consist of solid rock; debris slides do not.
   e. All of the above.

9. A mass-wasting process that is fast and has high water content is
   a. solifluction.
   b. slumping.
   c. creep.
   d. a mudflow.
   e. a debris fall.

10. Talus is a result of
    a. a mudflow.
    b. a rockfall.
    c. earth flow.
    d. soil creep.
    e. None of these.

11. What triggers mass wasting?
    a. desiccation
    b. vibration
    c. saturation
    d. freeze-thaw
    e. All of the above.

12. When developing an area, it is best to _____ any mass-wasting hazard that may be present.
    a. mitigate
    b. avoid
    c. fix
    d. engineer
    e. oversteepen

13. Human causes of mass wasting include
    a. cutting the base of a hillside.
    b. overweighting the top of a slope.
    c. saturating a slope.
    d. removing vegetation.
    e. All of the above.

14. The potential for mass wasting can be reduced by
    a. reducing water infiltration.
    b. vegetating hillsides.
    c. regrading land.
    d. using rock netting.
    e. All of the above.

15. Imagine a steep stream valley whose long axis is oriented north–south and whose sedimentary rock layers are dipping to the east. On which side of the valley is a rock slide more likely to occur?
    a. southwest
    b. northwest
    c. west
    d. east
    e. None of the above.

FURTHER RESEARCH

1. What tectonic settings lead to frequent mass wasting?

2. Would mass wasting occur on other planets? Why or why not? What factors would govern the types of mass wasting occurring on other planets?

3. In your area, what factors trigger mass wasting? What types of mass-wasting events are most common?

4. What steps can people take to reduce vulnerability to mass wasting?

5. Is a snow avalanche a type of mass wasting? Why or why not?

ONLINE RESOURCES

Explore more about mass wasting on the following websites:

Wikipedia, the free online encyclopedia—"Mass wasting":
http://en.wikipedia.org/wiki/Mass_wasting

Natural Resources Canada, Earth Sciences Sector—"Natural Hazards":
www.nrcan.gc.ca/earth-sciences/natural-hazard/natural-hazard/3612

Public Safety Canada—"Natural Hazards of Canada":
www.publicsafety.gc.ca/cnt/mrgnc-mngmnt/ntrl-hzrds/index-eng.aspx

Landslide Susceptibility Map of Canada:
http://geogratis.gc.ca/api/en/nrcan-rncan/ess-sst/68f835fa-f143-5b92-ab87-e77eb22151ab.html

U.S. Geological Survey Landslide Hazards Program:
http://landslides.usgs.gov

Additional animations, videos, and other online resources are available at this book's companion website:
www.wiley.com/go/fletchercanada

This companion website also has additional information about WileyPLUS *and other Wiley teaching and learning resources.*

CHAPTER 19
SURFACE WATER

Chapter Contents and Learning Objectives (LO)

19-1 The hydrologic cycle moves water between the atmosphere, the ocean, and the crust.
LO 19-1 *List and describe the five processes in the hydrologic cycle.*

19-2 Runoff enters channels that join other channels to form a drainage system.
LO 19-2 *Describe the concept of drainage systems.*

19-3 Discharge is the amount of water passing a given point in a measured period of time.
LO 19-3 *Identify how discharge, flow, and channel characteristics vary between headwaters and base level.*

19-4 Running water erodes sediment.
LO 19-4 *Compare the various types of sediment transport (suspended load, bed load, and dissolved load) with the current velocities they need, as described in the Hjulstrom diagram.*

19-5 There are three types of stream channels: straight, meandering, and braided.
LO 19-5 *Describe how meandering channels change with time and what causes the formation of braided channels.*

19-6 Flooding is a natural process in streams.
LO 19-6 *Identify the causes of flooding and the hazards associated with building on a flood plain.*

19-7 Streams may develop a graded profile.
LO 19-7 *Define the concept of a graded stream.*

19-8 Fluvial processes adjust to changes in base level.
LO 19-8 *Identify the influence of base-level changes, such as dams and waterfalls, on fluvial processes.*

19-9 Fluvial sediment builds alluvial fans and deltas.
LO 19-9 *Describe an alluvial fan and the avulsion process in delta building.*

19-10 Water problems exist on a global scale.
LO 19-10 *Describe the freshwater problems that are increasing worldwide.*

GEOLOGY IN OUR LIVES

Water is naturally cleansed and renewed as it moves through the hydrologic cycle between the atmosphere, the ocean, and the crust. The hydrologic cycle includes condensation, precipitation, infiltration, runoff, and evapotranspiration. However, rates of water use by humans may outpace natural rates of water renewal. Water scarcity is increasing in dozens of nations, and polluted runoff transports pathogens and contaminants into natural ecosystems and human drinking water supplies. Another widespread problem is stream flooding that threatens communities built on flood plains. Flood plains are natural extensions of stream channels, and periodic flooding is a normal process. It is best to avoid building on flood plains because attempts to stop flooding usually result in environmental damage and actually worsen flooding problems in the next community downstream.

? How does water move through the natural world?

19-1 The Hydrologic Cycle Moves Water between the Atmosphere, the Ocean, and the Crust

LO 19-1 *List and describe the five processes in the hydrologic cycle.*

Would you believe that a dinosaur once may have ingested the very same molecules that were in your last drink of water? From the time water vapour condensed on the cooling Earth in the late Hadean Eon, the water environment has been in motion. Following the commands of gravity and the Sun's energy, water moves restlessly between the atmosphere, the ocean, and the crust in what is termed the **hydrologic cycle** (FIGURE 19.1).

 Water covers 71 percent of Earth's surface; it is the dominant agent governing environmental processes. Woven into water's geologic journey are the many ways in which humans rely on clean water. We use water for manufacturing, to irrigate crops, as a source of energy, for recreation, for washing, and, of course, for drinking. Although nature continually refreshes water resources, rates of human usage generally outpace natural rates of water renewal. For this reason, it is important that we carefully manage water as a valuable resource for our own use as well as for that of future generations. To do this, it is necessary to understand the hydrologic cycle and the nature of surface water (discussed in this chapter) and groundwater (discussed in Chapter 20, "Groundwater").

Five Processes in the Hydrologic Cycle

Each year the hydrologic cycle circulates nearly 577,000 km³ of water (about 500 quadrillion litres, or 500×10^{15} L). An important part of the cycle is the storage of water in natural reservoirs such as the ground, the ocean, lakes, and various forms of ice. After leaving the atmosphere as rain, snow, or some other type of condensation, water may run quickly to the sea in channels called **streams,** or may be held

in a lake for a hundred years, in a glacier for thousands of years, or in the ground for 10,000 years or more. Or it may evaporate immediately. Regardless of how long the water may be detained, it is eventually released to re-enter the hydrologic cycle.

Five natural processes keep water moving through the hydrologic cycle: *condensation, precipitation, infiltration, runoff,* and *evapotranspiration.* Condensation occurs when water changes from vapour (gas) to liquid. Water vapour moves mostly by convection. That is, warm, humid air rises and cool air descends. As warm water vapour rises into the atmosphere, it loses heat and cools. This causes the rate of condensation to exceed the rate of evaporation; thus, the vapour changes into liquid or ice.

You can see condensation in action as water drops form on the outside of a cold glass or can. This water obviously does not come from inside the glass or can; it comes from the warm air that touches the cool surface of the container: Water vapour in warm air condenses into liquid water on the outside of the cool container. The same process forms clouds. When warm water vapour in the air cools and condenses, it creates clouds of microscopic droplets held aloft by air currents. As condensation continues, the droplets grow in size until they are too large and heavy to stay in the air, and the result is rain.

Precipitation is water that falls from the atmosphere as rain or snow. Sometimes it evaporates before reaching the ground. More often, however, it reaches Earth's surface, adding to streams and lakes, or soaks into the soil to become *groundwater* (water temporarily stored in the crust).

Infiltration occurs when water seeps into the ground. The amount of water that infiltrates the soil varies with land slope, vegetation, soil

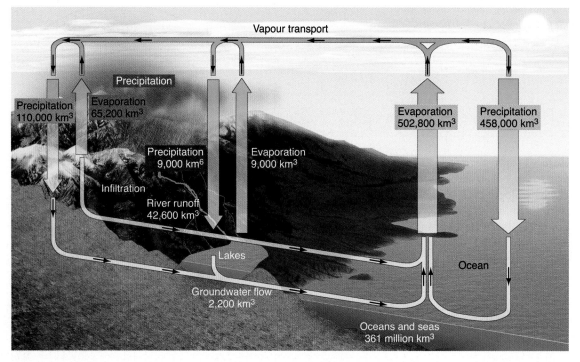

Vapour transport

Precipitation

Precipitation 110,000 km³

Evaporation 65,200 km³

Precipitation 9,000 km⁶

Evaporation 9,000 km³

Infiltration

River runoff 42,600 km³

Lakes

Groundwater flow 2,200 km³

Evaporation 502,800 km³

Precipitation 458,000 km³

Ocean

Oceans and seas 361 million km³

? What are the major processes in the hydrologic cycle?

FIGURE 19.1 Nearly 577,000 km³ of water circulates through the hydrologic cycle every year. The cycle consists of five major processes—condensation, precipitation, infiltration, runoff, and evapotranspiration—that keep water continuously moving through Earth's environments.

type, and rock type and with whether the soil is already saturated with water. In the ground, water is stored in the small spaces between the grains of soil and rock of the crust, so the amount of this open space determines the amount of water that can be stored.

Runoff is precipitation that reaches the surface but does not infiltrate the soil. Runoff also comes from melted snow and ice. During heavy precipitation, soils become saturated with water so that additional water cannot soak in and must flow on the surface. Because of gravity, surface water travels downhill. Hence, runoff drains downward into streams, lakes, and eventually the ocean. Normally less than 20 percent of rainfall runs off the surface; the remaining 80 percent soaks into the ground or evaporates. During times of abnormally high rainfall, however, runoff can approach 100 percent of the total rainfall.

Water vapour re-enters the atmosphere by evapotranspiration. Water moves into the atmosphere by evaporating from the ground and transpiring from plants. *Transpiration* is part of plant metabolism. It occurs when plants take in water through the roots and release it through the leaves, a process that can clean water by removing contaminants and pollution. *Evaporation* occurs when energy from the Sun heats water, causing water molecules to become so active that some of them rise into the atmosphere as water vapour.

Fresh Water for Human Use

Freshwater resources are vitally important to humans, yet the supply is very limited. The total volume of water on Earth is about 1.4 billion km^3; of this total, only 2.5 percent is fresh water (35 million km^3). Fully 68.9 percent of all fresh water is locked up in the form of ice and permanent snow cover in mountainous areas, the Antarctic, and Arctic regions. Another 30.8 percent is stored underground in the form of groundwater of various types, including *soil moisture,* shallow and deep aquifers, *wetlands,* and permafrost. Freshwater lakes and rivers contain only 0.3 percent of the world's fresh water.

According to the United Nations, the total usable freshwater supply for ecosystems and humans is only 200,000 km^3 of water, or less than 1 percent of all freshwater resources and only 0.01 percent of all the water on Earth. Read the "Earth Citizenship" box titled "Water Facts" to learn more about water use and abundance.

EARTH CITIZENSHIP

WATER FACTS

Water is part of our daily lives. In wealthy countries, it has become so familiar and is so readily available in every water fountain and sink that we hardly give it a second thought. The total usable freshwater supply for ecosystems and humans constitutes only 0.01 percent of all the water on Earth. The following list of water facts provides more information about this most common and most valuable of Earth's natural resources (FIGURE 19.2).

- About 80 percent of all evaporation is from oceans; 20 percent comes from surface water and plants.
- Water vapour is the third most abundant gas in the atmosphere (after nitrogen and oxygen).
- If all the water vapour in the atmosphere were to fall at once, Earth would be covered with only about 2.54 cm of water.
- About 75 percent of the human body consists of water.
- Showering and bathing is one of the largest domestic uses (27 percent) of water.
- In Canada, about 1.5 million cubic metres of water are used per day (including industrial and agricultural needs), which is second only to the United States among developed countries.
- A leaky faucet can waste 380 L a day.
- An average family of four uses 3,335 L of water per week just to flush the toilet.

- The average shower uses about 75 L of water; it uses 150 L in 10 minutes.
- You use about 19 L of water if you leave the water running while brushing your teeth.

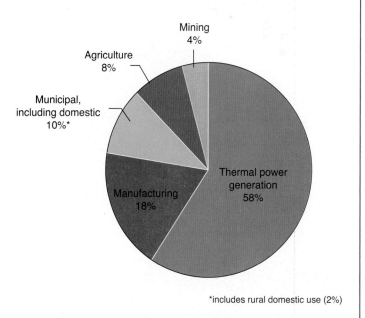

FIGURE 19.2 The five main users of fresh water in Canada. (*Source*: Environment Canada, 2008.)

Expand Your Thinking—What keeps water moving through the hydrologic cycle?

19-2 Runoff Enters Channels that Join Other Channels to Form a Drainage System

LO 19-2 *Describe the concept of drainage systems.*

As we learned in Chapters 7 ("Weathering") and 8 ("Sedimentary Rock"), when water erodes land, it picks up sedimentary particles and carries them away in the flow of streams. A stream is any flowing body of water that follows a **channel** (the physical confines of flowing water consisting of a bed and banks; **FIGURE 19.3**). A *river* is a major branch of a stream system. Every year, 36,000 km³ of water runs off the land worldwide, and it has been calculated that the energy of all the water flowing off the land in a single year is equal to 9 billion kilowatts. Much of this energy is spent eroding the land, and worldwide, streams carry about 16 billion tonnes of sedimentary particles and 2 to 4 billion tonnes of dissolved ions every year. Weathered rock and regolith provides this material, and when streams carry it away grain by grain, the land is worn down. Stream erosion, working over geologic time, has brought many mountain ranges down to the level of flat plains. As is evident to anyone gazing out of an airplane window, much of Earth's landscape is the product of stream erosion.

FIGURE 19.3 Channels, such as this one containing a stream in Waterton National Park, carry runoff across the land to the sea. In the process, the water erodes and transports sediment and dissolved chemicals.

Hal Horwitz/Science Photo Library

 Describe the effects of water as it runs across the land.

Drainage System

Steams do not flow in isolation; they are part of a large network of channels of various sizes known as a **drainage system.** Drainage systems are fed by surface runoff that quickly becomes organized into channels. The total area feeding water to a stream is called the **watershed** or *drainage basin* (**FIGURE 19.4**). Within a watershed, all runoff drains into the same stream. Every point on Earth's surface is part of a watershed, since all the rain falling on land must drain somewhere. For example, the watershed for the Mississippi River drains two-thirds of the continental United States as it flows 4,070 km to the Gulf of Mexico. But nested within the Mississippi watershed are many smaller watersheds that drain every channel from every slope of every hill and ridge in the midwestern states, and in the very southern parts of Alberta and Saskatchewan. The Mackenzie River, which flows into the Arctic Ocean, is the longest river in Canada at 1,738 km, and has the largest watershed, draining parts of British Columbia, Alberta, Saskatchewan, Yukon, and the Northwest Territories.

Watersheds are separated by *drainage divides*. Drainage divides are the topographic highs, such as ridges, that force water to drain in separate directions into different watersheds. For example, in North America, the Continental Divide separates the drainage basins of streams that flow to the Atlantic, the Gulf of Mexico, Hudson Bay, and the Arctic Ocean from those that flow to the Pacific.

Drainage systems can be described using a simple scheme, called *stream ordering*, that is widely used by *hydrologists* (scientists who study water in nature) (**FIGURE 19.5**). The lowest-order

FIGURE 19.4 Smaller watersheds are nested within larger watersheds to create a drainage system.

 Within a watershed the volume of runoff tends to increase from top to bottom. Why is this?

First-order basin   Second-order basin

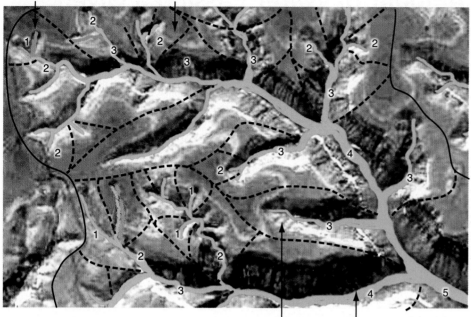

Third-order basin   Fourth-order basin

FIGURE 19.5 The system of stream ordering identifies the rank of a channel within a drainage system.

State a hypothesis about how the topography of the land controls the drainage pattern in a watershed.

stream is defined as one with no **tributaries;** this is a *first-order stream.* First-order streams are also called *headwaters.* A second-order stream is formed by the *confluence* (or juncture) of two first-order streams, and a third-order stream is formed by the confluence of two second-order streams. A drainage basin can be described using similar criteria. A basin is ranked by the highest-order stream it contains. Hence, a *second-order basin* contains first- and second-order streams but no third-order stream. Scientists have discovered that the relationship between the number of stream segments in one order and the next, called the *bifurcation ratio,* is consistently around three. This has been called the *law of stream numbers.*

Drainage Patterns

Hydrologists have noticed that drainage patterns reflect the local topography and nature of the bedrock. The *dendritic* (meaning to have a branching structure like a tree) drainage pattern is the most common (**FIGURE 19.6**). This pattern develops where the underlying rock has a uniform resistance to erosion and lacks features (such as joints or fractures) that could significantly influence drainage. Essentially, the dendritic pattern is the drainage pattern that water establishes for itself in the absence of other factors.

A *trellis* pattern develops where topography, in the form of parallel ridges and valleys, exerts strong control over the orientation of streams. The topography is likely related to the structure of the underlying geology. The major tributaries flow down the valleys, and are joined by minor tributaries approaching at right angles. *Radial* drainage occurs where runoff flows away from a dome or hilltop in all directions. Areas where the crust has parallel faults or repeated sets of joints cause streams to take on a grid-like or *rectangular* pattern. There are several other drainage patterns as well, all depending on the nature of the underlying geology.

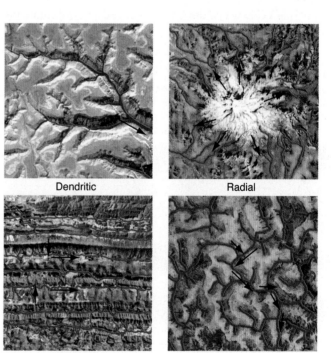

Dendritic   Radial

Trellis   Rectangular

FIGURE 19.6 The character of the underlying geology and topography controls the geometry of a drainage pattern. Geologists learn to recognize drainage patterns as clues to the nature of the underlying crust.

Are bedrock fractures important in drainage geometry? Why or why not?

Expand Your Thinking—How would you use the law of stream numbers to help manage streams?

19-3 Discharge Is the Amount of Water Passing a Given Point in a Measured Period of Time

LO *19-3 Identify how discharge, flow, and channel characteristics vary between headwaters and base level.*

A stream channel is a conduit that allows water and sediment to move from a *source area* to a base level. The **base level** is the theoretical lowest level to which the land will erode—globally, base level is sea level, the average level of the ocean surface. *Local base level* is the lowest point a particular channel will reach—usually the point where it joins with another channel. The local base level also may be a lake, marsh, the ocean, or a reservoir. The ultimate base level is the ocean. If all streams were allowed to flow continuously through time, with no new uplift of the land, eventually they would carve their channels down to sea level.

Discharge

A stream continually adjusts its shape and path as the amount of water passing through the channel changes. The volume of water (per unit of time) passing any point on a stream is called the **discharge.** Discharge is measured in units of volume divided by time (m^3/s) and is defined as the amount of water passing a given point over a measured period of time. The discharge (Q) is equal to the cross-sectional area (A [m^2]) of a channel (width [m] × average depth [m]) times the average velocity (V [m/s]) of the flow.

$$Q = A \times V$$

By measuring stream discharge, it is possible to improve our understanding of flow characteristics that are useful in managing a stream as a resource. During certain seasons of the year, streams tend to flood. In other seasons, they can become dry. By collecting data on stream discharge over a period of years, authorities responsible for managing stream resources learn how much water a healthy stream typically carries at various times of the year. This information improves their ability to make sustainable decisions about activities that may withdraw water from the system (such as irrigation) or contribute water to it (such as runoff from paved areas).

Flow

Water flowing within a channel may be either *laminar,* in which all water molecules travel along parallel and uniform flow paths, or *turbulent,* in which individual water molecules follow irregular paths (**FIGURE 19.7**).

Turbulent flow can keep sediment suspended in the water column longer than laminar flow can, and it increases erosion of the stream bottom and channel walls. The average velocity of laminar flow is generally greater than that of turbulent flow because turbulent flow is characterized by some water movement that is directed upstream in the form of eddies, against the mean flow going downstream. *Turbulent eddies* also direct flow vertically from the bed toward the surface (**FIGURE 19.8**).

FIGURE 19.8 This turbulent flow is created by the high velocity of the water flowing over a ledge of bedrock in the stream channel.

Emmanuel Layani/Getty Images, Inc.

Turbulent

Laminar

FIGURE 19.7 Turbulent flow is characterized by water motion that goes against the mean flow direction or is directed vertically from the channel bed toward the water's surface. Laminar flow occurs when water molecules travel along uniform flow paths.

 Why does turbulent flow cause more erosion?

 What pathways do water particles follow in turbulent flow?

Because eddies contain flow components that are not directed downstream, they tend to reduce the efficiency of the flow and produce locally chaotic water movement. However, the same turbulent action enhances the water's sediment-carrying capacity. Turbulent eddies tend to scour the bottom and suspend sediment, and the chaotic water movement keeps particles suspended in upward-directed currents.

Characteristics of Channels

In a channel, both width and depth increase as the flow moves downstream. This occurs because the amount of water flowing in a channel increases downstream as additional tributaries join the system (FIGURE 19.9).

Stream channels have different characteristics, depending on three primary variables: *gradient, discharge,* and *sediment load.* Channel gradient, or slope, is the angle between the channel bed or floor and a horizontal plane. The slope determines the flow velocity of the water in the channel. A steeper gradient produces an increase in flow velocity, which in turn increases the amount of sediment moved by the stream.

The cross-sectional shape of a channel changes as the stream becomes deeper and wider due to the influx of tributaries. The deepest parts of a channel develop where velocity is highest because of the ability of faster water to erode sediment and deepen the channel. The cross-sectional shape of a channel varies with its location in the stream and with changes in the discharge (FIGURE 19.10). As discharge increases, the width and depth of the channel increase faster than the flow velocity. A typical channel with a large discharge flows across relatively low-gradient slopes. Hence, discharge and channel gradient tend to change inversely; that is, when one increases, the other decreases.

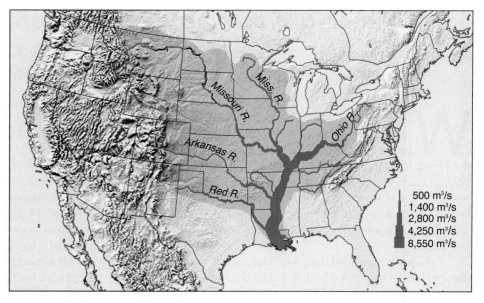

FIGURE 19.9 The amount of water flowing in the Mississippi watershed increases as the water moves downstream because of the addition of tributaries.

(?) How do stream gradient and discharge change with distance downstream?

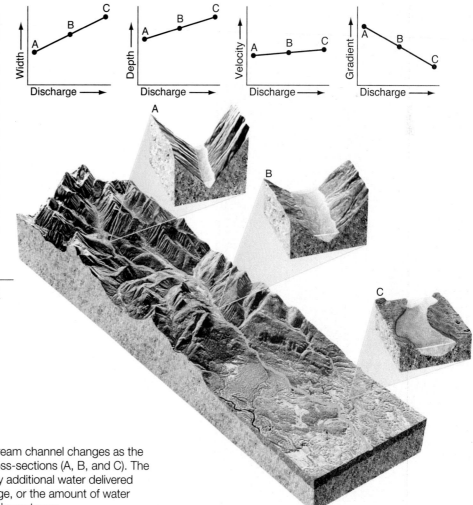

(?) **Expand Your Thinking**—Describe the field methods you would use to obtain the data necessary for calculating the discharge in a stream in your neighbourhood.

(?) How do stream gradient and discharge change with distance downstream?

FIGURE 19.10 The cross-sectional shape of a stream channel changes as the water flows downstream, as shown by the three cross-sections (A, B, and C). The channel becomes deeper and wider in order to carry additional water delivered by a greater number of tributaries. Thus the discharge, or the amount of water passing a particular point in the channel, increases downstream.

19-4 Running Water Erodes Sediment

LO 19-4 *Compare the various types of sediment transport (suspended load, bed load, and dissolved load) with the current velocities they need, as described in the Hjulstrom diagram.*

Within a channel, the point of maximum velocity shifts position depending on the characteristics of the channel. In straight segments, maximum flow is near the surface in the centre, away from the friction caused by the bed and walls. In a sinuous segment, the maximum velocity migrates toward the outer bank and lies somewhat below the surface (**FIGURE 19.11**). This is because *momentum* (the tendency to continue moving forward) directs the water toward the outside of a curve, causing it to collide with the far bank.

A stream carries a sediment load that may be transported as either bed load or suspended load. A stream also transports a large volume of dissolved load (**FIGURE 19.12**). *Suspended load* consists of particles that are carried along suspended in the water. The size of these particles depends on their density and the velocity of the stream. Turbulent eddies in higher-velocity currents keep particles suspended and support larger and denser particles. The suspended load is what gives most streams their muddy appearance and brown or red colour.

Bed load consists of large particles that remain on the stream bed most of the time that they are moved by the water. They move by sliding or rolling—called *traction*—or jumping—known as *saltation*—as a result of collisions between particles. As stream velocity increases, sediment that was carried as bed load tends to become suspended, and sediment that was not transported at lower velocities begins to move as bed load. In general, as stream velocity increases, more sediment is transported and grains move through the stages of traction, then saltation, and then suspension. Bed load moves at only a small fraction of the average flow velocity of the stream. But suspended load moves at a large fraction of the stream velocity; hence, bed load (5 percent to 10 percent) is generally less important than suspended load (90 percent to 95 percent) in terms of total sediment load of a stream.

Dissolved load consists of ions that have entered the water as a result of chemical weathering of rocks. This load is invisible because the ions are dissolved in the water. The dissolved load consists mainly of HCO^{-3} (bicarbonate ions), calcium (Ca^{+2}), sulphate (SO_4^{-2}), chloride (Cl^-), sodium (Na^+), magnesium (Mg^{+2}), and potassium (K^+). Eventually these ions are carried to the oceans and give the oceans their salty taste. Streams fed by groundwater that has flowed thro ugh the crust generally carry a higher dissolved load than do those whose only source is runoff.

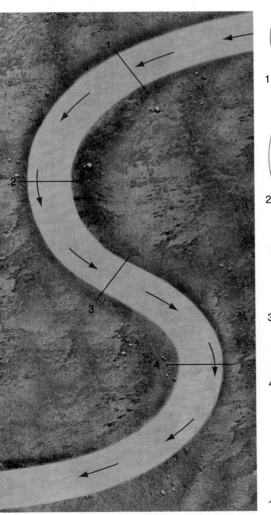

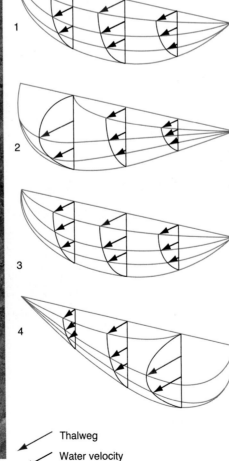

1

2

3

4

Thalweg

Water velocity

(?) How does the outer bank of a channel in a curved segment change with time?

FIGURE 19.11 In curved segments of channel, maximum flow velocity (thalweg: red arrow) migrates toward the outer bank and lies below the surface. The thalweg is discussed in more detail in Section 19-5. The four channel sections shown provide an indication of the depth of the channel and the velocity of the water at different depths. The longer arrows indicate higher flow velocity.

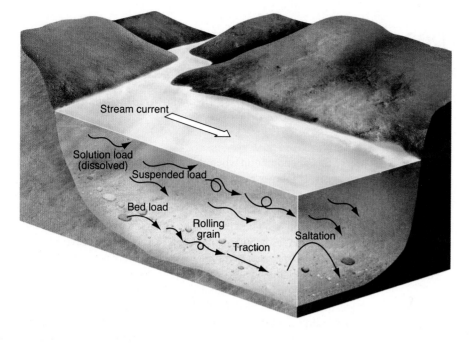

FIGURE 19.12 Sediment load consists of suspended load and bed load. Bed load includes saltation (particles jumping into the water column) and traction (grains moving by sliding and rolling along the bed). Dissolved load consists of dissolved compounds.

(?) Which carries the largest volume of sediment in a stream: bed load or suspended load?

Sediment Erosion

Erosion is of great importance to humans, as it plays a role in the management of natural resource problems such as soil loss, stabilization of hillsides, and unwanted sediment buildup in areas such as lakes, coastlines, and coral reefs. Because of the need to control landscape erosion, geologists have developed a simple model of the nature of erosion. This model is known as the *Hjulstrom diagram* (**FIGURE 19.13**).

The Hjulstrom diagram plots the relationship between water velocity and sediment size, and shows two curves representing (1) the

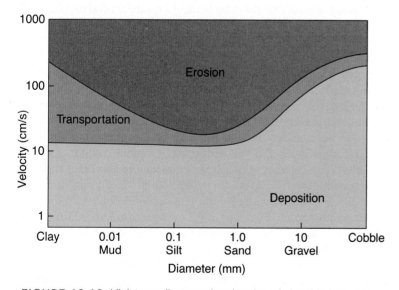

FIGURE 19.13 Hjulstrom diagram showing the relationship between grain size and the velocity of the current needed to erode, transport, and deposit sediment.

 Which is harder to erode: gravel or clay?

approximate stream velocity needed to erode sediments of varying sizes from the stream bed and (2) the approximate velocity required to continue to transport sediments of varying sizes once they are moving in the channel. Figure 19.13 is divided into three regions. The pink region (erosion) shows the water velocity needed to erode sediment of various sizes from the bed. The blue region (transportation) shows water velocities needed to keep sediments moving once they have been eroded; notice that it takes less velocity to keep sediments moving than it does to erode them in the first place. The green region (deposition) depicts the velocities at which sediments typically come to rest on the bed—that is, when they are deposited.

The Hjulstrom diagram reveals some interesting things about erosion and deposition in moving water. Notice that once mud and silt have been eroded, they will stay in motion even if the velocity of the water decreases drastically. However, it appears as though the finer-grained sediment, which often consists of clay minerals, tends to start being deposited at the same velocity, because clay minerals flocculate (link together) as their surfaces are typically negatively and positively charged. It is not as easy to transport coarser sediments (sand and gravel), as they require nearly as much water velocity as is required to erode the grains in the first place. Also notice that it takes as much water energy to erode large pieces of gravel as it does to erode clay. Clay is very cohesive ("sticky") and hence is relatively hard to erode.

One of the surprises offered by the Hjulstrom diagram is the fact that sand is the easiest sediment to erode despite the fact that sand grains are heavier than silt or clay. As it turns out, the reason for this phenomenon is simple. Sand grains, being larger than grains of silt or clay, stick up from the bed into the moving water. As a result, they travel along the bed before silt and clay do. Sand grains are usually rounded, but both clay and silt consist of flat particles that do not stick up into the moving water. It is also thought that turbulent eddies containing strong currents that dig into the bed are needed to erode clay because clay particles tend to stick together.

(?) **Expand Your Thinking**—Describe the research program that must have been conducted in order to develop the Hjulstrom diagram.

19-5 There Are Three Types of Stream Channels: Straight, Meandering, and Braided

LO 19-5 *Describe how meandering channels change with time and what causes the formation of braided channels.*

Three general types of stream channels can be described: straight, meandering, and braided.

Straight channels are rare. They are usually found only as short segments of otherwise meandering channels or where the underlying topography and structure of the bedrock (such as a fault or fracture pattern) force the channel to be straight. The reason that straight channels are rare is that water naturally flows in a sinuous fashion due to minute differences in flow characteristics. You can

observe this phenomenon when a raindrop runs down the windshield of your car. It follows a snaky, curving path. Even in straight channel segments, water meanders or migrates from side to side down the length of the channel (**FIGURE 19.14**).

Flow velocity is highest in the zone overlying the deepest part of the stream. For that reason, sediment there is readily eroded, leaving behind a *pool* or *scour depression* in the channel. If you were to draw a line running along the channel and connecting the deepest parts of the stream, your line would mark the natural direction (the profile) of the watercourse. Such a line is called the **thalweg** ("valley line" in German). The thalweg is almost always the line of fastest flow in a river. Where stream velocity is low, sediment will be deposited to form a *bar*. Alternating erosion and deposition along the channel creates a specific type of channel shape, with regularly spaced pools and bars. The path of highest-velocity flow is a line that connects the pools and goes around the bars.

Meandering channels are the most common characteristic of a stream that is free to roam across a valley floor (**FIGURE 19.15**). Meandering channels develop most readily on low-gradient slopes composed of easily eroded sediment. Like all objects with mass, water in a channel possesses momentum. This means that as water enters a bend in a channel, momentum will keep it moving in a straight line so that it collides with the far bank.

This intersection of water and erodible bank leads to a constant, persistent tendency for the outer bank of a meander bend to erode and thus *migrate outward* even more. For this reason, the outer bank of a meander is known as the *cutbank*. Simultaneously, sediment deposition takes place along the inside bank of a meander bend, where the flow velocity is lowest. The resulting sediment deposit produces an exposed bar, called a *point bar*. Point bars tend to migrate inward at nearly the same pace that the cutbank migrates outward; hence, the width of the channel stays relatively stable even as it migrates across the valley floor.

If you consider the consequences of a perpetually outwardly migrating channel, you will realize that it cannot continue forever. Eventually the stream will become so sinuous and tortuous that the movement of water through the valley becomes inefficient. At its extreme, the channel will loop from one side of a valley to the other, with greatly reduced efficiency in the movement of water through the drainage system. Nature, ever an efficient solver of problems, resolves this potential crisis by cutting off a highly sinuous meander at the neck.

FIGURE 19.16 shows how erosion on the outside of a highly sinuous meander bend eventually will cut across the neck between two meanders and form a new, straighter, and more efficient channel. When this occurs, the cut-off meander bend forms an **oxbow lake.** In time, the oxbow will become a *meander scar* filled with a wetland as it accumulates organic plant material and sediment delivered by runoff and wind.

The third type of channel is known as a **braided channel.** Braided channels are formed when a stream contains more sediment than it can readily transport, for example those draining the end of alpine glaciers. The excess sediment accumulates in the channel and forms bars, or small islands, that split the water into multiple channels. This gives the channel a "braided" appearance (**FIGURE 19.17**).

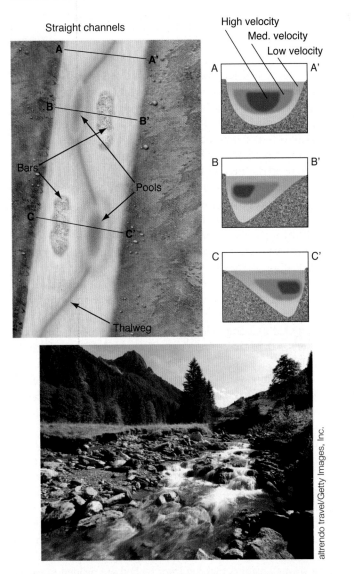

Straight channels

High velocity
Med. velocity
Low velocity

Bars

Pools

Thalweg

altrendo travel/Getty Images, Inc.

FIGURE 19.14 Water tends to flow in a sinuous fashion even in straight channels. This means that the highest-velocity flow migrates from side to side across the channel, as you can see in the three cross-sections, A–A', B–B', and C–C'. As a result, the channel is characterized by alternating pools (erosion) and bars (deposition).

(?) Why are straight channels rare in nature?

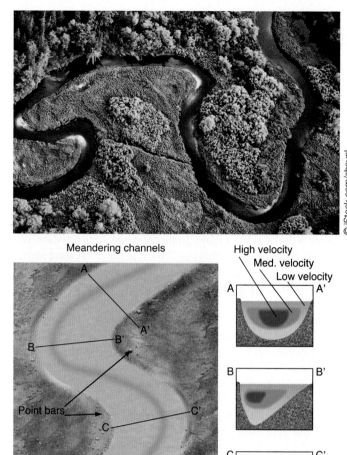

FIGURE 19.15 When a stream channel is able to roam freely, it tends to develop a meandering channel. Meandering channels readily develop on low-gradient slopes composed of easily eroded sediment. The three cross-sections show the variations in velocity across the channel. The areas of low velocity are where the point bars form as sand is deposited. The sandy point bars can be seen on the inside of the meanders on the photograph.

 What is the effect of momentum on water flow?

Braided streams are characterized by highly variable discharge, easily eroded banks, and excessive sediment. The sediment load is carried primarily during periods of high discharge. When the rate of discharge returns to normal, sediment is deposited and forms bars and islands. The water is forced to flow in a braided pattern around the emergent deposits, dividing and reuniting as it moves downstream. During periods of high discharge, the entire channel may contain water and the bars become covered. As they erode and are reshaped under the high waters, sediment is redistributed throughout the channel. Usually, when the water level lowers and deposits re-emerge, the locations and shapes of the bars are entirely different from what they were previously. However, in some cases, bars may become vegetated and thus become more permanent islands in the stream channel.

 Expand Your Thinking—State a hypothesis describing water flow in a channel. Describe how you would test your hypothesis.

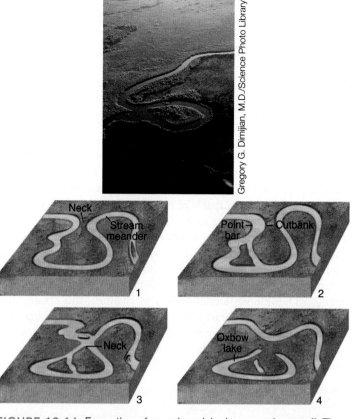

FIGURE 19.16 Formation of an oxbow lake by meander cutoff. The photograph shows the Okavango River in Botswana, and shows the situation after the neck has been breached. Eventually, sediment will be deposited that will isolate the old channel, forming the oxbow lake.

 How does a meander cutoff change with time?

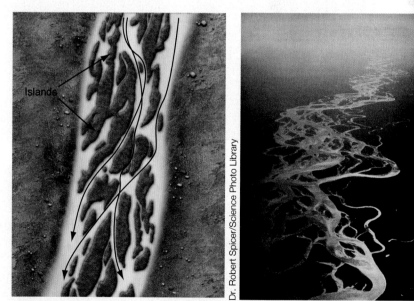

FIGURE 19.17 A braided channel is characterized by bars, islands, and multiple channels, such as exhibited by the Yukon River in central Alaska. Braided streams develop because excess sediment that cannot be transported by the stream flow builds bars and islands that split the water into many interweaving channels.

 Describe the effect of high discharge on a braided channel.

19-6 Flooding Is a Natural Process in Normal Streams

LO 19-6 *Identify the causes of flooding and the hazards associated with building on a flood plain.*

The rate of discharge in a channel can change rapidly, producing a **flood.** Flooding occurs when discharge increases so much that there is too much water for a channel to carry within its banks. Excess water flows outside the channel on the adjoining land, which is known as the **flood plain** (FIGURE 19.18).

Flooding happens for many reasons, but in most cases, it is the result of high discharge in several tributary streams feeding a single river channel that is incapable of handling the excess water without flowing outside its banks. Three common examples include: (1) an intense but short rainstorm in the headwaters of a drainage basin, leading to downstream flooding because multiple tributaries carrying high amounts of discharge feed into a single channel that floods; (2) prolonged rainfall in a drainage basin, saturating the ground and forcing all additional rainfall to flow in channels; this creates high discharge in many tributaries that cause flooding in a river that they feed; and (3) a winter with heavy snowfall ending with a series of very warm weeks so that accumulated snow melts rapidly and many tributaries carry high discharge into a single channel that floods.

All these cases result in a "wave" of high discharge that moves down a drainage system into a single high-order river channel. The wave of high discharge builds over several hours to a day or more and may peak in the space of only an hour or two (FIGURE 19.19).

A flooding channel is observed first as slowly rising water of increasing velocity that contains a great deal of suspended sediment. Within a few hours or in some cases a day or more, water rises to the limit of the stream banks (a condition called *bankful discharge*) and then overruns the banks as a flood. In the following hours or days, water returns to its normal discharge condition as the flood wave moves downstream. The flood may grow downstream if more tributaries continue to add water. On particularly large rivers, the migrating flood may last over a few weeks.

As a stream overtops its banks during a flood, the velocity of the flow is high at first but suddenly decreases as the water flows out over the gentle gradient of the flood plain. Because of the sudden decrease in velocity, coarse-grained suspended sediment such as sand is deposited along the riverbank, eventually building up a *natural levee* (FIGURE 19.20). Natural levees provide some protection from flooding because they increase the relative height of the channel banks.

Flood Plains

Flood plains are flat areas adjacent to a stream channel; they are composed primarily of sediments that were deposited during floods. A number of features are found on a flood plain. Some of these features are depositional, such as natural levees, former point bars that have been abandoned by meander cutoff, and wetlands located in meander scars. Other features, such as oxbow lakes, have been created by past meander processes.

Flood plains develop when streams overtop their levees and spread sediment-laden flood water over the land surface. When the flood waters retreat, stream velocities drop and a layer of *alluvium,* sediment that originates

(a)

(b)

 Why would a major city be located on the flood plain of a river?

FIGURE 19.18 Stream flooding occurs when discharge increases so much that flow must occur outside the channel. *(a)* In 2006, the Red River in Manitoba overflowed its banks. Significant flooding of the river also occurred in 2009 and 2011. *(b)* In June 2013, heavy rainfall and snow melt in the Rocky Mountains led to increased discharge on the Bow River and flooding of downtown Calgary and other low-lying neighbourhoods, crippling the city for days and causing an estimated $1 billion of damage.

CP PHOTO/Winnipeg Free Press-Ken Gigliotti

© Rosanne Tackaberry/Alamy

from a stream, is deposited across the flood plain. Over time, repeated flood cycles result in the deposition of successive layers of alluvium so that the flood plain's elevation rises. Flood plains, then, are composed of a combination of point bar sands and gravels and alluvium deposits of silt and clay. A flood plain is as much a part of a stream system as bark is part of a tree. The flood plain is the location where a stream stores excess water and sediment during a flood. It is the environment where a stream is free to roam and form meanders, oxbows, point bars, and other natural channel features.

 Efforts to decrease the hazards of stream flooding by confining a channel with *artificial levees* (levees built by humans to keep flood water from causing damage) actually damage fertile flood plains by preventing the continued deposition of sediment and the annual cycle of natural wetting. Flood water that is not allowed to flow onto a flood plain is passed downstream to the next community, causing more severe flooding there. Flood plains that are deprived of sediment eventually decrease in elevation and, ironically, experience more frequent flooding, actually increasing the hazard to local communities. The best way to solve flooding problems is to *avoid* them altogether by not locating communities in flood-prone regions such as flood plains. In the long run, avoiding building in flood plains is cheaper and less damaging to the environment than attempting to control the natural flooding process. To learn more about flooding, read the "Critical Thinking" box titled "The 100-Year Flood Plain."

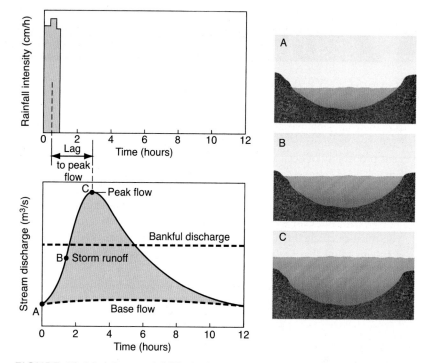

FIGURE 19.19 Intense rainfall in the headwaters of a drainage basin increases the discharge in tributaries feeding a high-order channel. There may be a lag of several hours to a day or more between the time of the rainfall and the time when flooding occurs farther downstream. Normal discharge A will increase to B until bankful discharge is exceeded. The peak of the flood C occurs several hours after the initial increase in discharge and then slowly tails off.

(?) Why is flooding described as a "wave"?

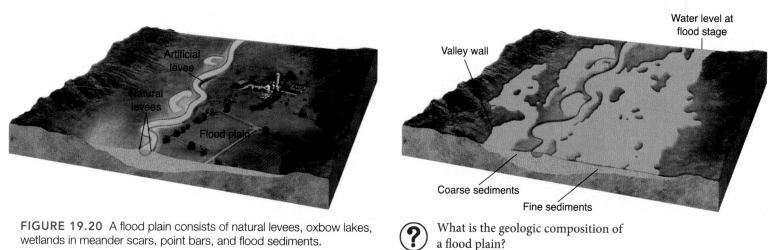

FIGURE 19.20 A flood plain consists of natural levees, oxbow lakes, wetlands in meander scars, point bars, and flood sediments.

(?) What is the geologic composition of a flood plain?

 Expand Your Thinking—Explain why building on a flood plain is bad for both humans and the environment.

CRITICAL THINKING

THE 100-YEAR FLOOD PLAIN

You have been hired to analyze Deer Creek, where developers propose new housing. No new development is allowed in the stream's "100-year flood plain." A 100-year flood is a flood that has a 1 percent probability of occurring in any given year. They are thus rare, but will flood large areas of the flood plain when they do occur. As no development is allowed in the "100-year flood plain," any area with a greater than 1 percent chance of flooding in a single year is off limits.

To determine the legal location for construction of new homes, you must calculate the height, or elevation, of floods measured by a *stream gauge* that has recorded the maximum water level every year since 1935.

Calculate the percent probability (P) of annual recurrence for each of the flood events listed in TABLE 19.1.

$$P = 100 \times M/(n + 1)$$

where M is the flood magnitude shown in column 3 and n is the total number of years, 69.

Use the graph paper (FIGURE 19.21) to plot the data in Table 19.1. Plot the stage or elevation of the flood and the value of P for the same year. Be sure to use the scale at the top of the graph, labelled *Percent probability of recurrence (P)*, to locate the P value.

Draw a straight line through the data points that best represents the trend of the data.

A flood level recurrence interval (RI), or time interval between floods that are similar in size, is calculated using the equation, $RI = (n + 1)$. The recurrence interval is also equal to 100/P. Values for the RI are given on the lower axis of Figure 19.21.

TABLE 19.1 Deer Creek Flood History

Year	Maximum Stage (metres)	Magnitude M	P
1936	6.8	30	
1946	5.5	48	
1959	2.5	67	
1964	4.9	54	
1976	3.4	65	
1984	8.4	8	
1985	7.6	17	
1993	10.2	2	

Source: U.S. Dept. of the Interior, USGS

The elevation of the stream gauge on Deer Creek is 210.31 m. Use your line to determine the elevation (or stage) of the 100-year flood plain.

Add the elevation of the 100-year flood plain to the elevation of the stream gauge, and the answer is the elevation below which it is illegal to develop the land. On the map (FIGURE 19.22), land elevation is contoured in feet using a 10-ft contour interval, and so you will need to convert the contour intervals in feet to metres (10 feet = 3.048 metres). Using a pencil, outline and shade the land area that lies at or below the elevation of the 100-year flood plain.

Please answer the following questions.

1. Why is there a law making it illegal to build in the 100-year flood plain?
2. Why should someone who does not live in the flood plain care if someone else builds there?
3. Why does the government have an obligation to identify geologic hazards for its citizens and to limit people's ability to engage in certain activities, such as building in hazardous locations?
4. You have mapped lands vulnerable to the "100-year flood." What additional hazards may threaten lands in the map area? Be sure to include in your answer considerations of population growth, global warming, stream meandering, potential upstream changes, slope stability, and other possibilities. To accompany your analysis, mark on the map where these threats may be found.
5. Based on your analysis, indicate where you recommend future development of new housing should be located.

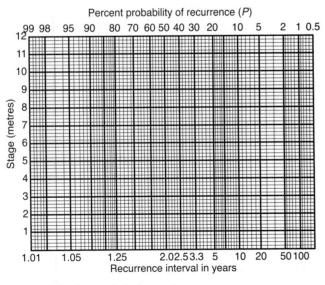

FIGURE 19.21 Probability graph paper.

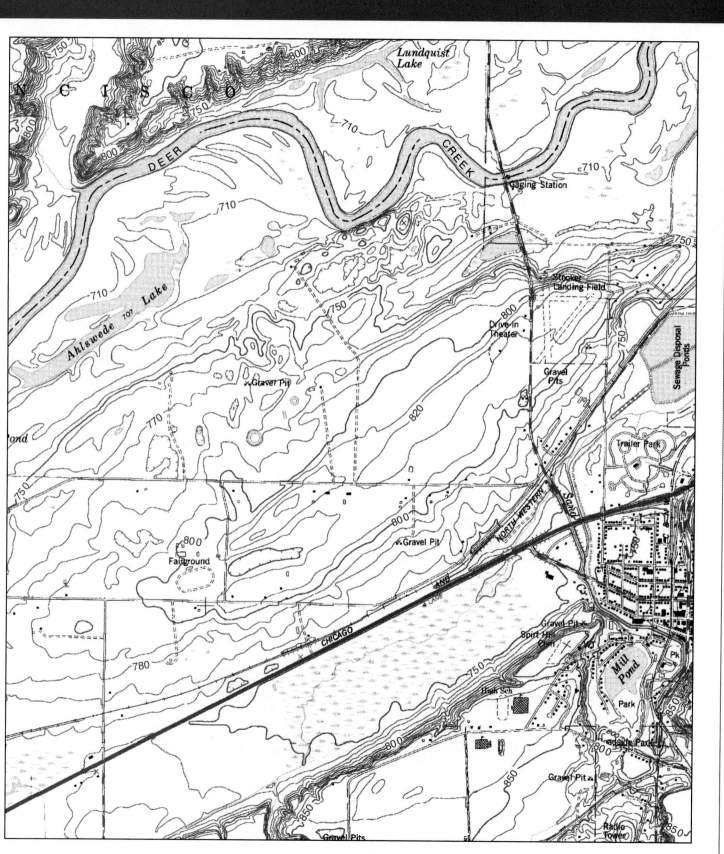

FIGURE 19.22 Deer Creek map. Contour interval 10 feet.
(*Source:* U.S. Dept. of the Interior, USGS.)

19-7 Streams May Develop a Graded Profile

LO 19-7 *Define the concept of a graded stream.*

From its headwaters to where a channel connects with the sea, a stream system has a characteristic *profile* that is gently concave with a flattened lower end (**FIGURE 19.23**). Generally, in their upper reaches, stream valleys tend to be narrow and steep, while in their lower reaches, they are wider and have gentler slopes.

Because of the steep gradient, water in the upper reaches of a stream profile flows with high velocity. This rapid flow promotes downward cutting (channel formation) and *headward erosion* (rather than meandering). The stream valley therefore is typically V-shaped in cross section. Headward erosion is the tendency for a stream channel to lengthen upslope as a result of greater erosion at valley headwaters. The exact shape of the valley (i.e., wide or narrow) is also determined by mass wasting, which depends on hill slope processes and the stability of regolith (recall Chapter 18, "Mass Wasting"). As a channel cuts down and back, the valley it has formed tends to widen. This widening favours meandering, so, with time, meandering takes precedence over headward erosion in sculpting the landscape.

An interesting consequence of headward erosion is stream piracy. **Stream piracy** (or *stream capture*) occurs when headward erosion breaches a drainage divide, intersects another channel, and captures parts of its flow. Piracy increases drainage area by diverting runoff from neighbouring basins. With increased runoff, headward erosion may accelerate and capture additional neighbouring streams (**FIGURE 19.24**).

In the middle reaches of a stream valley, two important changes usually occur: The stream develops an **alluvial channel**, meaning that it flows through its own alluvium (stream sediment), and the channel has a greater tendency to erode laterally (meander), creating a wide, flat-bottomed valley (**FIGURE 19.25**). In the lower reaches, the flood plain grows wider, the meander grows more pronounced, and the gradient becomes very gentle.

The flood plain is an important site of sediment storage within a fluvial system. Continual deposition of sediment in a flood plain is known as *aggradation* (thickening accumulation of flood plain sediment through deposition). But it is possible for a stream to attain a state of *dynamic equilibrium*. That is, if the amount of deposition is balanced by the amount of erosion, such that there is no net gain or loss, then sediment will be transported through the system without a net buildup of the flood plain.

Graded Streams

Scientists who study streams have defined a concept called the **graded stream** (or *graded profile*). This term is applied to streams that have apparently achieved, throughout long segments, a state of dynamic equilibrium between the rate of sediment transport (erosion) and the rate of sediment supply (deposition). Stream grading always is defined over a specific segment of channel. Defining a graded stream over an entire drainage system would have no meaning because, generally speaking, the lower parts of a drainage system always deposit while the uppermost parts always erode. However, over a particular segment, a graded stream has no *net* erosion or deposition; the input of sediments into the segment is the same as the output, and erosion equals deposition over that segment.

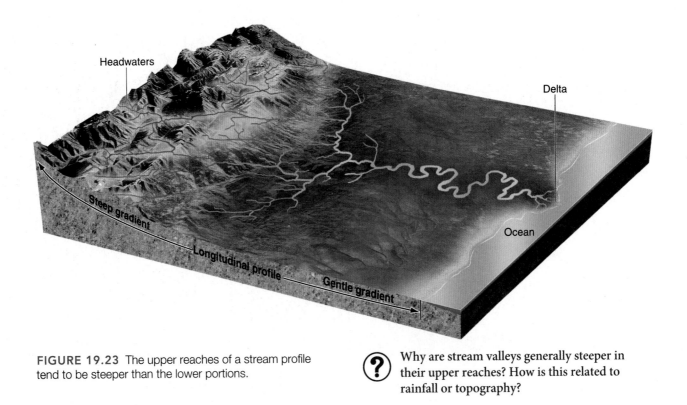

FIGURE 19.23 The upper reaches of a stream profile tend to be steeper than the lower portions.

(?) Why are stream valleys generally steeper in their upper reaches? How is this related to rainfall or topography?

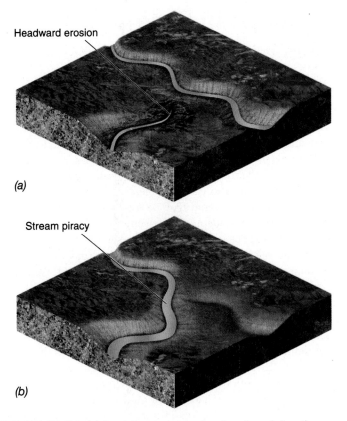

Headward erosion

(a)

Stream piracy

(b)

FIGURE 19.24 *(a)* As a stream erodes in a headward direction, it may *(b)* intersect a neighbouring drainage system and capture its discharge.

? Describe how headward erosion changes drainage basins.

A graded stream maintains an equilibrium between the processes of erosion and deposition and, therefore, between aggradation and *degradation* (downward erosion of a stream bed and flood plain). If excess sediment enters a stream, the stream will store the excess in the form of islands and bars that are propelled downstream during times of high discharge (a braided channel). A stream that has a deficit of sediment will cut downward into its bed. A high gradient generally leads to a stream that erodes its bed. A lower gradient may produce a graded stream that successfully transports all the available sediment and does not erode the bed.

In a very simple way, this concept describes the evolution of the **fluvial** (from the Latin word for "river" or "flowing water") landscape depicted in Figure 19.25. In their early stages, as tectonic activity uplifts an area and streams have steep gradients, channels are characterized by waterfalls and rapids in high-relief topography. These features promote abrasion and sediment production. Gorges and canyons form, and active headward erosion is prevalent. Waterfalls migrate upstream as resistant rock ledges are eroded. Mass wasting widens the valley as slopes are oversteepened by stream undercutting and collapse into the channel. Eventually a valley widens and the landscape is altered so that the stream gradient is lessened and a flood plain develops. The stream reaches equilibrium when its sediment load matches its ability to carry it. The channel does not cut into its bed, nor does the flood plain accumulate sediment deposits. The stream is then said to have a graded profile.

? **Expand Your Thinking**—How would you determine whether a stream has achieved a graded profile? Describe the research you would conduct and the data you would collect.

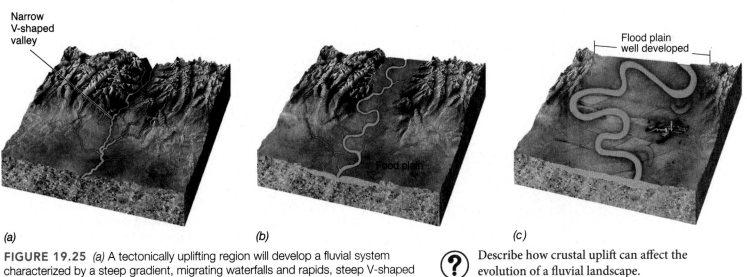

Narrow V-shaped valley

Flood plain

Flood plain well developed

(a)　　　　　(b)　　　　　(c)

FIGURE 19.25 *(a)* A tectonically uplifting region will develop a fluvial system characterized by a steep gradient, migrating waterfalls and rapids, steep V-shaped valleys, and gorges. *(b)* Mass wasting and undercutting of slopes widen valleys, and meandering becomes more pronounced. *(c)* Eventually the gradient decreases as a wide flood plain develops and the stream flows through a flood plain composed of its own alluvium.

? Describe how crustal uplift can affect the evolution of a fluvial landscape.

19-8 Fluvial Processes Adjust to Changes in Base Level

LO 19-8 *Identify the influence of base-level changes, such as dams and waterfalls, on fluvial processes.*

Earlier we introduced the concept of *base level,* the elevation below which a stream can no longer erode the land. As a stream approaches its base level (the ocean, a lake, or another stream), it develops a low gradient. A waterfall formed by a resistant layer of rock can be considered a local base level. Until the rock has been eroded, the stream can erode no lower and will develop a low gradient above the waterfall. Eventually it will erode through the rock, and the waterfall will migrate upstream until it reaches an elevation equal to that of the graded profile of the stream. Once the bed of the graded channel matches the surface of the resistant layer, the waterfall will no longer exist (**FIGURE 19.26**).

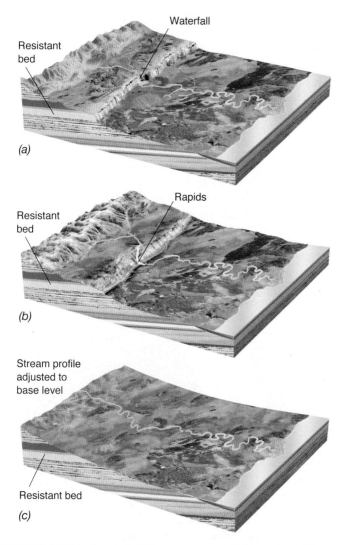

FIGURE 19.26 A waterfall *(a)* represents a local base level. The waterfall will erode and migrate upstream *(b)* until it lies below the gradient of the river *(c)*.

(?) If you were a field geologist in the area of panel *(c)* of Figure 19.26, what evidence would convince you that the area once had a waterfall?

One benefit of running water is its ability to turn *turbines,* which are machines used in the production of electric power. *Hydroelectric dams* provide a vertical drop in order to accelerate the velocity of water and turn turbines to produce electricity. In addition, such dams create *reservoirs* (stored water in the form of a lake) that provide recreational opportunities and drinking water sources for urban areas. However, dams also cause problems. Reservoirs drown river valleys, and valuable farmland may be lost (**FIGURE 19.27**). Dams also destroy ecosystems adapted to a river environment. Flooded archeological sites may be lost forever. Since reservoirs represent a local base level, the velocity of streams immediately drops to zero when they enter a reservoir, causing sediments to be deposited. With time, as streams continue to deliver sediment, a reservoir will fill with sand and mud, and *dredging* will be required in order to extend the life of the reservoir.

Downstream of a dam, a channel will be deprived of its normal load of sediment because of deposition in the reservoir. The stream below a dam will experience an increase in velocity and a decrease in sediment load, both of which cause it to erode (or *incise*) into its channel. What previously had been a graded profile now becomes a highly erosive stream that cuts into its bed. Upstream of the reservoir, the stream experiences a rise in base level, and the profile of the stream valley accumulates sediment. This means that the stream will build a higher flood plain and decrease its gradient as it deposits sediment.

Incised Channels and Terraces

Once a stream has established a graded profile, it is vulnerable to changes in sediment availability, shifts in base level, and tectonic changes in land level. Each of these changes can disrupt the equilibrium of the graded profile. Uplift of the land related to tectonic movement of Earth's crust, such as at a convergent plate boundary, changes a stream's gradient. Even if the land does not tilt as it is uplifting, the gradient must increase because of the stream's higher elevation relative to its base level. An increased gradient results in increased flow velocity within the channel, causing channel erosion. As a channel incises its bed, it cuts into the underlying bedrock and develops steep banks (**FIGURE 19.28**).

Alluvial terraces develop when a graded stream incises its flood plain. Incision can be the result of decreased sediment load, lowered base level, or increased gradient due to uplift. Terraces are former flood-plain deposits that are exposed when a stream incises its channel. A channel is said to become *rejuvenated* when the gradient increases or the sediment load decreases, and the stream cuts through its flood plain.

(?) **Expand Your Thinking**—What are the upstream and downstream problems that may be created by the building of a dam?

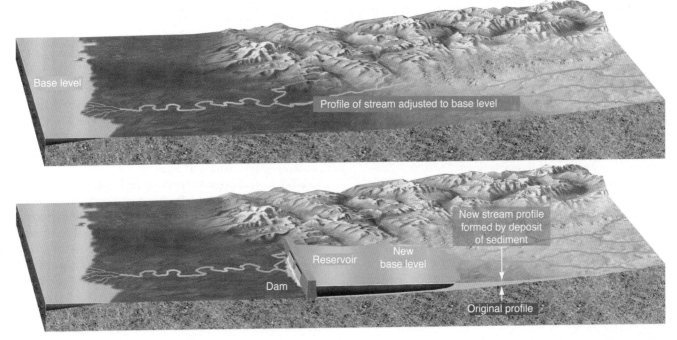

FIGURE 19.27 Below a dam, the stream is deprived of sediment and erodes into its bed. Above the dam, the stream experiences a rise in base level and the stream channel accumulates sediment.

(?) Why do reservoirs require periodic dredging?

(a)

(b)

(c)

© ONTHEBIKE.PL/Alamy

FIGURE 19.28 (a) Graded profile in equilibrium with the sediment load. (b) Decrease in sediment load, land uplift, or base-level fall initiates stream incision. (c) Alluvial terraces develop after prolonged incision. The photograph shows a river, which now has a braided channel that has eroded through its older valley floor, leaving flat terraces high above the present valley floor.

(?) What causes the formation of an alluvial terrace?

19-9 Fluvial Sediment Builds Alluvial Fans and Deltas

LO 19-9 *Describe an alluvial fan and the avulsion process in delta building.*

Alluvial fans are fluvial features that result from the deposition of stream sediment in arid environments. An alluvial fan is a semicircular, gently sloping cone of channel sediment deposited when a high-gradient stream leaves a narrow canyon or gully and enters a flat plain or valley floor. The decrease in gradient causes a drop in flow velocity, and the stream loses its ability to carry sediment. Sediment of all sizes, from boulders to mud, is deposited at the base of the valley, forming a low, conical apron of clastic debris (**FIGURE 19.29**).

Alluvial fans can be large features, as much as 10 km across. The sediment composing the fan is subtly sorted so that coarse material, such as boulders and cobbles, tend to be deposited near the valley mouth, where the stream first experiences a decline in velocity. Finer particles, such as sand and mud, are deposited across the fan surface and can even be carried to the distant edge of the fan and out onto the floor of the adjacent valley. Typically, braided channels drain the sides and centre of the fan surface. Alluvial fans are most likely to develop in areas where intermittent but powerful rainstorms suddenly load stream channels with unsorted alluvium containing intermixed boulders, cobbles, gravel, sand, and mud.

Rainfall in arid regions tends to be related to the buildup of large thunderheads that contain abundant precipitation. Because of the arid climate, there is no groundwater contribution to stream discharge. As a result, streams flow only when there is direct runoff from a rainstorm, and the intense nature of rainfall means that there is little time for runoff to soak into the ground. The lack of vegetation in arid regions also means that runoff is more rapid. Nearly all precipitation becomes runoff into a drainage basin, and a condition known as *flash flooding* occurs.

Major floods do not happen every year. Often several years or decades may pass between flood events. In between these floods, physical and chemical weathering litters mountain slopes with loose sediment. A storm then washes this sediment into dry stream gullies. There, high-gradient, high-velocity flow develops into a raging flood that sweeps sediment of all sizes down the narrow valley. Such flash floods are very hazardous. Hikers report hearing a sound like a locomotive approaching as a flash flood roars down a narrow canyon.

Deltas

A **delta** is a sedimentary deposit that forms when a stream flows into standing water (i.e., reaches its base level) and deposits its load of sediment. With continued deposition over time, a delta can extend out from the coast and become a large geologic feature. The delta is so named because it fans outward from its point of origin at a channel mouth and resembles the Greek letter delta (Δ). Deltas accumulate three types of sedimentary beds: *bottomset beds, foreset beds,* and *topset beds* (**FIGURE 19.30**).

Dan Suzio/Science Source

FIGURE 19.29 Alluvial fan in Death Valley National Park, California.

(?) What conditions lead to the formation of an alluvial fan?

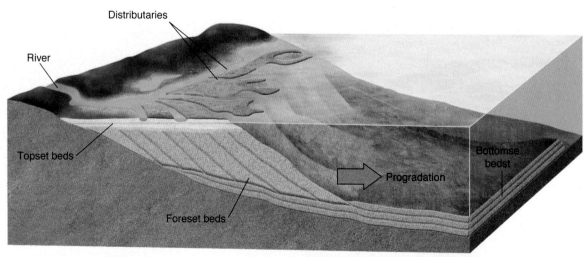

FIGURE 19.30 A simple delta is composed of three types of beds: bottomset beds, foreset beds, and topset beds.

(?) What is the difference between a delta and an alluvial fan?

As a river encounters its base level, silt and clay carried in the channel settle onto the basin floor to form bottomset beds. Coarser grains of sand accumulate closer to the river's mouth and build out a series of angled beds into the basin and bury bottomset beds. These are foreset beds, and they represent the forward-building front of the delta. Above these are topset beds, which consist of flat-lying sand and mud layers. Topset beds constitute the *delta plain* and are host to freshwater and saltwater **wetlands** if the delta is formed in the ocean. One of the most famous deltas in the world is the *Mississippi River Delta* on the Gulf of Mexico coastline of the United States (**FIGURE 19.31**).

The Mississippi River drains over 3 million square kilometres of the United States. Through time, the river has delivered immense amounts of sediment to the ocean through a major channel reaching into the Gulf of Mexico. Approximately seven similar channels have released sediment over the last 8,000 years, each with an average life span of 1,000 years (**FIGURE 19.32**). In time, each channel is abandoned and a new one started through the process of *avulsion*. Avulsion is similar to the meander cutoff process. A flood that has swept down the river arrives at the flat delta plain, overflows its banks, and establishes a new route to the sea; this process results in avulsion, or abandonment of the old channel. The new pathway represents a more efficient route, and the old channel is abandoned. Eventually the old channel fills with sediment and becomes part of the vast wetland on the delta plain. In recent history, the Mississippi River has experienced several major avulsion events, each of which has produced a new channel in the delta system. This process will continue into the future.

Apply your knowledge of surface water to the "Critical Thinking" exercise "Channel Systems."

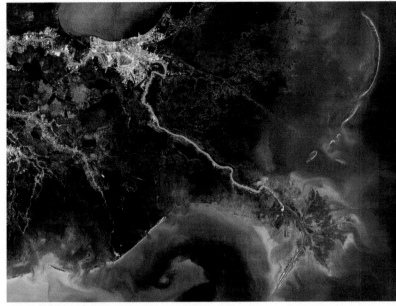

M-Sat Ltd/Science Source

FIGURE 19.31 The Mississippi River drains the interior of the United States. Sediments carried by the river accumulate at the point where the river enters the ocean, its base level, and build the Mississippi River Delta shown in this satellite image.

(?) How is the size of the Mississippi River Delta related to the size of its watershed?

FIGURE 19.32 Avulsion events have resulted in the abandonment of major distributary channels (lobes) and the establishment of new ones in the Mississippi River Delta. Seven former delta lobes, 7 (oldest) through 1 (youngest), are outlined, each of which lasted an average of 1,000 years before being abandoned as a result of avulsion.

(?) What causes avulsion?

Approximate age

1
2
3
4
5
6
7

(thousands of years ago)

Mississippi Delta

Gulf of Mexico

(?) **Expand Your Thinking**—Develop a hypothesis that predicts the impact of building artificial levees along the Mississippi River and delta. How would this affect downstream communities, the delta region, ecosystems, and coastal waters?

? CRITICAL THINKING

CHANNEL SYSTEMS

Please work with a partner and answer the following questions.

1. Describe how the size of eroded sediment grains varies with changes in stream velocity.
2. What is a graded profile, and what would you measure to determine if a stream is graded?
3. Describe the changing nature of sediment deposition in a stream from its headwaters to its base level.
4. What controls the development of straight, meandering, and braided channels?
5. On **FIGURE 19.33**, label components of the hydrologic cycle, fluvial landforms, fluvial processes, and geologic hazards.
6. Pick one watershed in this scene. Combine your knowledge of channel systems with your understanding of weathering and sedimentary environments to build a hypothesis describing the evolution of your watershed.
   a. What role does tectonics play?
   b. What role does climate play?
   c. Where and how are humans exposed to natural hazards?
   d. What role does mass wasting play?

FIGURE 19.33 Typical components of a channel system.

19-10 Water Problems Exist on a Global Scale

LO 19-10 *Describe the freshwater problems that are increasing worldwide.*

Earth has only a finite supply of fresh water. Fresh water is stored in the ground (see Chapter 20, "Groundwater"), in various surface environments such as lakes, glaciers, and streams, and in the atmosphere. Some people think that the salt water of the ocean is available for human use, but the high amount of energy needed to convert salt water into fresh water (known as *desalinization*) makes it prohibitively expensive for all but the wealthiest communities. According to the United Nations, more than 2.8 billion people in 48 countries will face *water stress* or scarcity conditions by 2025. The countries that are predicted to have the most serious problems because they withdraw over 40 percent of the fresh water available are those in northern Africa, the Middle East, India, Pakistan, and South Africa. However, increasing population and overuse in other countries, like China, the United States, Mexico, and western Europe may also result in water scarcity in many areas (**FIGURE 19.34**). By 2050, the number of countries facing water stress or scarcity could rise to 54, with a combined population of 4 billion—about 40 percent of the projected global population of 9.4 billion.

Fresh water is a basic human need that is fundamental to life itself. Restricted access to fresh water inevitably will lead to political disputes and even warfare. The projected level of water stress by the middle of this century represents a very serious issue that has the potential to affect the peaceful coexistence of neighbouring nations around the world. These problems will not go away by themselves. They can be solved, however, through early planning, new technologies for water management, and population control.

Polluted Runoff

Storm water runoff is water that flows overland during a rainstorm. Overland flow during storms is a natural part of the water cycle that serves many beneficial purposes, such as clearing stream channels of blockages and delivering sediment to flood plains during floods. However, human activities that alter the land surface can change the natural character of storm water runoff in ways that can be damaging to both the environment and human communities.

On heavily vegetated land that is in its natural state, storm water runoff moves slowly, allowing for some infiltration into the ground and absorption by vegetation. Vegetated land partially buffers downstream portions of the drainage system from the full impact of a rainstorm. Of course, as we saw earlier, in arid regions that lack vegetation, storm runoff may lead to flash floods.

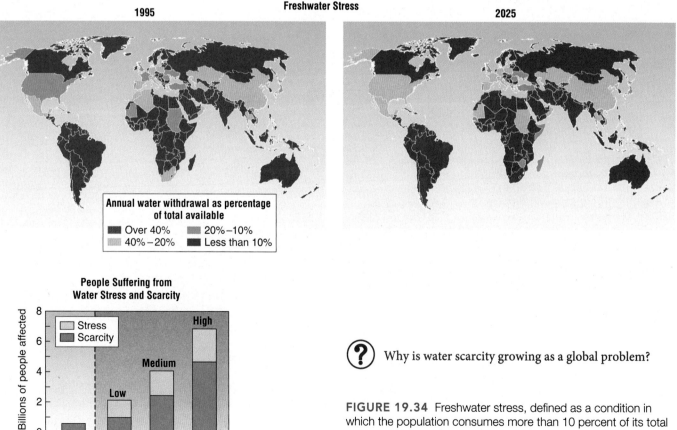

Freshwater Stress

1995   2025

Annual water withdrawal as percentage of total available
- Over 40%
- 40%–20%
- 20%–10%
- Less than 10%

People Suffering from Water Stress and Scarcity

- Stress
- Scarcity

(?) Why is water scarcity growing as a global problem?

FIGURE 19.34 Freshwater stress, defined as a condition in which the population consumes more than 10 percent of its total water supply per year, is projected to increase by 2025 to the point where two out of every three people on Earth will face water scarcity.

After a property has been developed with buildings and the land surface paved, the hydrologic character of the runoff changes. Storm water is no longer buffered by the friction of vegetation and soil and is no longer absorbed by porous regolith. Instead, smooth paved surfaces prevent infiltration and accelerate flow across the ground. *Impervious surfaces,* such as storm sewers, channels, sidewalks, curbs, and gutters, focus runoff downhill into fewer and fewer conduits. Lack of infiltration results in higher runoff discharge flowing at higher velocity across urbanized surfaces. Eventually, with great negative impact, urbanized flow enters a natural environment such as a stream channel, wetland, or seashore estuary. Many of these natural settings lack the capacity to withstand the high discharge and high velocity runoff that occurs under these conditions, and their ecosystem is damaged or destroyed.

Studies have shown that storm water runoff is a significant source of pollution, causing declines in fisheries and restrictions on swimming and limiting our ability to enjoy many of the other benefits that water provides. Storm water runoff contains contaminants that can cause problems in any natural environment receiving the water. Contaminated runoff is referred to as "nonpoint source pollution" (NPS) or **polluted runoff.** Polluted runoff may contain bacteria, oil, detergent, pesticides, heavy metals, animal and human feces, and dirt. These pollutants come from a variety of sources, including roads, gutters, lawns, construction sites, and even the atmosphere. Unlike wastewater entering treatment plants, polluted storm water runoff does not receive any treatment before it is washed into creeks, rivers, and lakes. NPS pollution can have serious impacts on drinking water supplies, recreation, fish, and wildlife. It is the leading cause of most of today's water quality problems (**FIGURE 19.35**).

For many years, efforts to control the discharge of storm water focused on quantity (flood control) more than the quality of the storm water (pollutant and sediment control). In recent years, awareness of the need to improve *water quality* has increased. As a result, federal, provincial, and local governments have established a variety of programs and laws designed to reduce the types and amounts of pollutants contained in storm water discharges into waterways. These programs promote the concept and the practice of *preventing pollution at the source,* before it can cause environmental or health problems.

Anoxic Dead Zones

"Dead zones" are forming off the mouths of several major river systems around the world. A dead zone is a place where water in contact with the sea floor ("bottom water") is *anoxic* (has no dissolved oxygen), meaning that it has such low concentrations of dissolved oxygen that it cannot support most marine life (**FIGURE 19.36**).

The cause of anoxic bottom waters has been traced to nutrients entering the ocean from agriculture products (fertilizer) and human sewage in stream water. These nutrients feed a community of *phytoplankton* (microscopic algae) that float on the surface of the ocean. Because the plankton thrive in the abundant nutrients delivered by the streams, the rapid turnover of their populations means that the amount of dead algae falling to the bottom is very high. Bacteria on the sea floor use dissolved oxygen to decay this organic matter. So much organic matter decays that the water is robbed of all oxygen, depriving other forms of life (crabs, clams, shrimp, zooplankton, and fish) of the ability to breathe. Entire ecosystems can suffocate in the space of one afternoon.

Anoxic zones are spreading worldwide: the number of such sites doubled between 2000 and 2010. An important reason that coastal waters are experiencing such high rates of anoxia is the widespread use of fertile flood plain soils for agriculture. Almost all the flood plains along major river systems throughout the world, but particularly in North America, are used for agriculture. Given the inevitability of regularly flooding, this is not a wise use of the land. Nutrients, fertilizers, insecticides, herbicides, and other chemicals used in farming produce excess chemical residues that enter watersheds and are swept into the ocean. Freshwater aquatic and coastal marine ecosystems are exposed to these residues in ways that are not well understood, but evidence suggests that negative impacts are widespread.

FIGURE 19.35 Polluted runoff collects in streams, lakes, and the ocean.

(?) What are some ways to control polluted runoff?

FIGURE 19.36 Anoxic waters cause the death of thousands of marine organisms, especially fish.

(?) What is the cause of "dead zones"?

 Expand Your Thinking—What are the important water problems where you live?

LET'S REVIEW "GEOLOGY IN OUR LIVES"

Now that you have finished the chapter, "Geology in Our Lives" will have taken on new meaning for you. Let us review it: Water is naturally cleansed and renewed as it moves through the hydrologic cycle between the atmosphere, the ocean, and the crust. The hydrologic cycle includes condensation, precipitation, infiltration, runoff, and evapotranspiration. However, rates of water use by humans may outpace natural rates of water renewal. Water scarcity is increasing in dozens of nations, and polluted runoff transports pathogens and contaminants into natural ecosystems and human drinking water supplies. Another widespread problem is stream flooding that threatens communities built on flood plains. Flood plains are natural extensions of channels, and periodic flooding is a normal process. It is best to avoid building on flood plains because attempts to stop flooding usually result in environmental damage and actually worsen flooding problems in the next community downstream.

Flowing water, or runoff, is responsible for shaping more of the land surface than any other process. Runoff is part of the hydrologic cycle, a geologic concept that describes the movement of water through Earth environments. Runoff becomes channelized flow on the land surface. Water flowing in a channel is known as a stream, and streams are part of a larger drainage system that transports a load of sediment and dissolved compounds to its base level.

Fresh water is a critical resource that is subject to pollution and overuse; this theme is explored further in Chapter 20 ("Groundwater"). Although nature continually refreshes water resources, humans tend to overuse water so it is important to carefully manage water as a valuable resource for our use as well as for that of future generations.

STUDY GUIDE

19-1 The hydrologic cycle moves water between the atmosphere, the ocean, and the crust.

- Water covers 71 percent of Earth's surface and is the dominant agent governing environmental processes. Although nature continually refreshes water resources, rates of human usage outpace natural rates of renewal. For this reason, it is important to carefully manage water as a valuable resource. To do this, it is necessary to understand the **hydrologic cycle** and the nature of channelized water.

- Five natural processes keep water moving through the hydrologic cycle: condensation, precipitation, infiltration, runoff, and evapotranspiration.

19-2 Runoff enters channels that join other channels to form a drainage system.

- Every year, 36,000 km³ of water runs off the land. This water does an enormous amount of work in the form of eroding and transporting sediments and dissolved ions toward the ocean.

- A **stream** is any flowing body of water following a **channel,** and a river is a major branch of a stream system. Worldwide, streams carry about 16 billion tonnes of sediment particles and 2 to 4 billion tonnes of dissolved ions every year. Streams do not flow in isolation; they are part of a large network of channels of various sizes known as a **drainage system.**

19-3 Discharge is the amount of water passing a given point in a measured period of time.

- **Base level** is the theoretical lowest level toward which erosion of Earth's surface constantly progresses but seldom, if ever, reaches.

Local base level is the lowest point a particular channel will reach. The ultimate base level is the ocean. If all streams were allowed to flow continuously through time, with no new uplift of the land, they eventually would carve their channels down to sea level.

- **Discharge** (Q) is equal to the cross-sectional area (A) of a channel (width times average depth [m²]) times the average velocity (V) of the flow [m/s]. Discharge is measured in cubic metres of water per second [m³/s].

19-4 Running water erodes sediment.

- Stream channels have varied characteristics, depending on three primary variables: gradient, discharge, and sediment load. Bed load consists of large particles that remain on the stream bed most of the time. They move by sliding or rolling (also called traction) or by jumping (also known as saltation) as a result of collisions between particles. Sediment moves between bed load and suspended load as the velocity of the stream changes, depending on discharge. Dissolved load consists of ions introduced into the water through the chemical weathering of rocks.

- The Hjulstrom diagram reveals that once mud and silt have been eroded, they stay in motion even if the flow velocity decreases drastically. It is not as easy to transport coarser sediments (sand and gravel); they require nearly as much water velocity as is needed to erode the grains in the first place. It takes as much water energy to erode large pieces of gravel as it does to erode clay.

19-5 There are three types of stream channels: straight, meandering, and braided.

- Three general types of stream channels can be described: **straight, meandering,** and **braided.**

- Meandering channels are the most common characteristic of a stream that is free to roam across a valley floor. Meandering channels develop most readily on low-gradient slopes with an easily eroded bed and banks. As water enters a bend in a channel, momentum keeps it moving in a straight line so that it collides with the far bank. This intersection of water and erodible bank leads to a tendency for the outer bank of a meander bend to erode and thus migrate outward.

- Braided channels are formed when a stream is carrying excess sediment. Because the stream is unable to transport the excess sediment, the sediment accumulates in the channel and forms bars and islands that split the water into multiple channels.

19-6 Flooding is a natural process in normal streams.

- Discharge in a channel can change rapidly, producing a **flood.** Flooding occurs when discharge increases to the point that there is too much water for a channel to carry within its banks. Excess water flows outside the channel onto the adjoining land, known as a **flood plain.**

- A number of features are found on a flood plain. Some of these are depositional, such as natural levees, former point bars that have been abandoned as a result of meander cutoff, and wetlands located in meander scars. Other features, such as oxbow lakes, were created by past erosion. When flood waters retreat, stream velocities decrease and a layer of alluvium, sediment that originates from a stream, is deposited across the flood plain. Repeated flood cycles over time result in the deposition of many successive layers of alluvium, so the elevation of the flood plain rises. Flood plains, then, are composed of point bar sands and gravels and flood deposits of silt and clay.

19-7 Streams may develop a graded profile.

- From its headwaters to the point where a channel connects with the sea, a stream system will have a characteristic profile that is gently concave with a flattened lower end. Generally, in their upper reaches, stream valleys tend to be narrow and steep, while in lower reaches, they are wider and have gentler slopes.

- A **graded stream** is one that has reached, throughout long sections, a state of practical equilibrium between the rate of sediment transport (erosion) and the rate of sediment supply (deposition). Stream grading is always defined over a specific segment of channel. Defining a graded stream over an entire drainage system would have no meaning because the lower parts always deposit while the uppermost parts always erode.

19-8 Fluvial processes adjust to changes in base level.

- A waterfall formed by a resistant layer of rock can be considered a local base level. Until the rock has been eroded, the stream can erode no lower and will develop a low gradient above the waterfall. Eventually the stream will erode through the rock, and the waterfall will migrate upstream until it reaches an elevation equal to the elevation of the stream's graded profile. Once the bed of the graded channel matches the surface of the resistant layer, the waterfall will no longer exist.

- Uplift of the land related to tectonic movement of Earth's crust, such as during continental collision, changes a stream's gradient. An increased gradient results in increased flow velocity within the channel, which in turn causes channel erosion. As a channel erodes its bed, it becomes incised.

- Alluvial terraces develop when a graded stream incises its flood plain. Incision can be the result of decreased sediment load, lowered base level, or increased gradient due to tectonic uplift. Terraces are former flood plain deposits that are exposed when a stream incises its channel. A channel is said to become rejuvenated when the gradient is increased or sediment load is decreased and it cuts through its flood plain.

19-9 Fluvial sediment builds alluvial fans and deltas.

- An **alluvial fan** is a semicircular, gently sloping cone of channel sediment deposited when a high-gradient stream leaves a narrow valley and enters a flat plain or valley floor. The significant change in gradient causes a drop in flow velocity. As a result, the stream loses its ability to carry its former sediment load. Sediment of all sizes, from boulders to mud, is deposited at the base of the valley, forming a low conical apron of clastic debris.

- A **delta** is a sedimentary deposit that forms when a stream flows into standing water (i.e., reaches its base level) and deposits its sediment load. With continued deposition over time, the deposit can extend into the basin and become a large geologic feature with major environmental impacts.

19-10 Water problems exist on a global scale.

- Storm water runoff is water that flows overland during a rainstorm. Overland flow during storms is a natural part of the water cycle that serves many beneficial purposes, such as clearing stream channels of entanglements and blockages and delivering sediment to flood plains during floods. However, alterations to the land surface due to human activities can change the natural character of storm water runoff in ways that can be damaging to both the environment and human habitation.

- According to the United Nations, more than 2.8 billion people in many countries in the Middle East, northern Africa, and the Indian subcontinent will face water stress or scarcity conditions by 2025.

- Contaminated runoff is referred to as "nonpoint source pollution" or **polluted runoff.** Nonpoint source pollution occurs when water runs over the land or through the ground, picking up contaminants such as bacteria, oil, detergent, pesticides, animal feces, and dirt. Polluted runoff can have serious impacts on drinking water supplies, recreation, fish, and wildlife. It is the leading cause of most of today's water quality problems.

KEY TERMS

alluvial channel (p. 524)
alluvial fans (p. 528)
base level (p. 514)
braided channel (p. 518)
channel (p. 512)
delta (p. 528)
discharge (p. 514)
drainage system (p. 512)

flood (p. 520)
flood plain (p. 520)
fluvial (p. 525)
graded stream (p. 524)
hydrologic cycle (p. 510)
meandering channels (p. 518)
oxbow lake (p. 518)
polluted runoff (p. 533)

storm water runoff (p. 532)
straight channels (p. 518)
stream piracy (p. 524)
streams (p. 510)
thalweg (p. 518)
tributaries (p. 513)
watershed (p. 512)
wetlands (p. 529)

ASSESSING YOUR KNOWLEDGE

Please answer these questions before coming to class. Identify the best answer.

1. The hydrologic cycle describes
   a. the movement of water through the environment.
   b. condensation, runoff, evapotranspiration, precipitation, and infiltration.
   c. the water in various natural reservoirs.
   d. the water being exchanged in natural processes.
   e. All of the above.

2. Within a drainage system,
   a. water all flows away from the largest channel.
   b. all runoff flows in only one channel.
   c. water comes only from the same storm.
   d. all runoff drains into the same stream.
   e. all runoff infiltrates.

3. When a stream experiences an increase in gradient, it will
   a. flow faster and erode its channel.
   b. develop a graded profile.
   c. produce more flash floods.
   d. erode all its alluvial fans.
   e. form a delta.

4. A graded profile is one in which
   a. headward erosion is prevalent.
   b. rejuvenation has occurred.
   c. stream piracy is taking place.
   d. erosion is stronger than deposition.
   e. sediment deposition is equal to erosion.

5. Between headwaters and base level, a channel will
   a. widen and deepen.
   b. develop a flatter slope.
   c. tend to accumulate finer-grained sediment.
   d. collect discharge from more tributaries.
   e. All of the above.

6. The Hjulstrom diagram tells us that
   a. clay and gravel require about the same water velocity to erode.
   b. clay is more easily eroded than gravel.
   c. gravel is more easily eroded than clay.

   d. sand is harder to erode than clay.
   e. all sediments erode at the same water velocity.

7. Why is clay difficult to erode but easy to transport?
   a. The particles are so heavy.
   b. The particles stick to the channel bed and then exhibit saltation.
   c. Actually, clay is easy to erode.
   d. Clay sticks to the channel bed but stays in suspension once it has been eroded.
   e. None of the above.

8. The lowest level to which a stream can erode is known as the
   a. Hjulstrom diagram.
   b. topset bed.
   c. graded profile.
   d. base level.
   e. flood plain.

9. Channels generally take one of three forms:
   a. oxbow, meander scar, and alluvial fan.
   b. meandering, braided, and straight.
   c. turbulent, laminar, and graded.
   d. alluvial, aggraded, and erosional.
   e. None of the above.

10. Braided channels form when the
    a. sediment load is small.
    b. ability of the stream to transport sediment is larger than the amount of sediment.
    c. ability of the stream to resist erosion is great.
    d. ability of the stream to transport sediment is exceeded by the amount of sediment.
    e. stream discharge is very high.

11. The discharge of a channel that is 75 m wide and 3.2 m deep, with a flow velocity of 0.34 m/s is
    a. 81.6 m³/s.
    b. 8.16 m³/s.
    c. 24.0 m³/s.
    d. 240 m³/s.
    e. None of the above.

12. Flooding is a result of
    a. prolonged rainfall saturating the ground.
    b. rapid melting of winter snow and ice.
    c. rapid and heavy rainfall.
    d. the presence of upstream artificial levels that pass the flood wave downstream.
    e. All of the above.

13. If the base level lowers,
    a. a stream will rejuvenate.
    b. the flood plain will collect excess sediment.
    c. deposition will increase greater than erosion.
    d. a stream will decrease discharge.
    e. a stream will decrease its gradient.

14. Avulsion is when flooding causes a stream to
    a. decrease discharge.
    b. build an alluvial fan.
    c. rejuvenate.
    d. establish a graded profile.
    e. establish a new channel to reach base level.

15. An important challenge for the world economy in the future will be
    a. a condition in which a stream flows only occasionally because of poor groundwater flow.
    b. a lack of sufficient fresh water to support the natural ecosystem.
    c. a situation in which humans use more water than can be supplied in the next season.
    d. a situation in which a population consumes more than 10 percent of its total water supply a year.
    e. a situation where water pollution has caused a "dead zone."

FURTHER RESEARCH

1. How do deposits in meandering streams compare with the deposits in braided streams?

2. How does the gradient of a stream change as it flows from steep headwaters into the ocean?

3. Why does a stream tend to develop a graded profile?

4. Each year streams and other erosional processes remove about 10.75 km³ of sediment from the continents. The total volume of continents above sea level is approximately 93,000,000 km³.

Thus, continents should erode down to sea level in about 8.6 million years. Of course, if this were true, then the world's continents would all lie at sea level now. What is wrong with this reasoning?

5. List at least three factors that contribute to flooding.

6. Describe the logic behind the statement "Floods today are more frequent than they were 100 years ago."

ONLINE RESOURCES

Explore more about geologic resources on the following websites:

Environment Canada's website on water issues:
www.ec.gc.ca/inre-nwri

U.S. Geological Survey information on surface water and floods:
http://water.usgs.gov/floods

Government of Canada flood preparedness information:
www.getprepared.gc.ca/cnt/hzd/flds-eng.aspx

Additional animations, videos, and other online resources are available at this book's companion website:
www.wiley.com/go/fletchercanada

This companion website also has additional information about WileyPLUS *and other Wiley teaching and learning resources.*

CHAPTER 20
GROUNDWATER

Chapter Contents and Learning Objectives (LO)

20-1 Groundwater is a very important source of fresh water.

LO **20-1** *Describe general features of groundwater.*

20-2 Groundwater is fed by snowmelt and rainfall in areas of recharge.

LO **20-2** *Describe the processes by which the water table gains and loses water.*

20-3 Groundwater moves in response to gravity and hydraulic pressure.

LO **20-3** *Identify the factors that influence groundwater movement.*

20-4 Porous media and fractured aquifers hold groundwater.

LO **20-4** *Compare and contrast different types of aquifers.*

20-5 Groundwater is vulnerable to several sources of pollution.

LO **20-5** *Identify ways in which pollutants enter the groundwater system.*

20-6 Common human activities contaminate groundwater.

LO **20-6** *List human activities that contaminate groundwater.*

20-7 Groundwater remediation includes several types of treatment.

LO **20-7** *Describe methods of cleaning groundwater.*

20-8 Groundwater is responsible for producing springs and karst topography.

LO **20-8** *Identify the major features of karst topography.*

20-9 Hydrothermal activity and cave formation are groundwater processes.

LO **20-9** *Describe the major features of geysers and cave deposits.*

GEOLOGY IN OUR LIVES

Groundwater is one of our most precious natural resources. It is the most important source of drinking water, and it supplies water for irrigation, industry, manufacturing, domestic uses, and most other water needs in human society. Less than 3 percent of the water on Earth is fresh water. About 75 percent of this fresh water is frozen in glaciers, and about 25 percent is groundwater. Only 0.005 percent is surface water (lakes and streams). Groundwater is vulnerable to overuse and pollution. This is because most groundwater moves slowly (usually less than 0.3 m per day), has been in the ground for decades to centuries, and is not quickly renewed by the hydrologic cycle. Rates of human usage generally outpace natural rates of groundwater replenishment; thus conservation measures are in place in nearly every community. Properly conserving groundwater requires careful analysis of where it is located, how it moves, and how it is recharged. The central question in groundwater science is: How shall we use groundwater today so that it is still useful for future generations tomorrow?

(?) This is an aerial photograph of an area in the Northwest Territories that is underlain by limestone. The lake on the surface is draining into the ground via a sinkhole. How does groundwater move, and how do we get it out of the ground? What are the physical characteristics of groundwater?

20-1 Groundwater Is a Very Important Source of Fresh Water

LO 20-1 *Describe general features of groundwater.*

There is water beneath our feet. In fact, about 95 percent of the total supply of fresh water in the world lies within the tiny pore spaces and fractures in the crust. **Groundwater** is our most important source of fresh water.

The Importance of Groundwater

As we saw in Chapter 19 ("Surface Water"), when rain falls to the ground, the water does not stop moving. Some of it flows along the surface as runoff and enters streams or lakes; some of it is used by plants; some evaporates and returns to the atmosphere; and some sinks into the ground. Imagine pouring a glass of water onto a pile of sand. Where does the water go? It moves into the spaces between the particles of sand and becomes groundwater. Overall, precipitation in the form of either rain or snow is the main source of fresh groundwater.

The United States Geological Survey estimates that groundwater in the upper 800 m of continental crust is 3,000 times more plentiful than the water found in all the world's rivers at any given time. This water is not in vast subterranean lakes and pools. It is in pore spaces between grains of sand and gravel and fills the tiniest cracks and fractures within the rocks and sediments of the crust. Most groundwater is in unconsolidated sands and gravels, which account for nearly 90 percent of all groundwater sources. Other sources include porous sandstone, limestone, and highly fractured crystalline and volcanic rock.

Environment Canada estimates that about 30 percent of the population of Canada depends on groundwater for use at home, with 80 percent of people living in rural areas depending on it for drinking water. However, the proportion used at home, compared to groundwater used for agriculture, industry, and municipalities in general, varies widely across the country. Prince Edward Island is completely dependent on groundwater for all uses. In contrast, the agricultural industry is the biggest user of groundwater in the Prairie provinces, and other industries use the most water in British Columbia and Quebec. Most groundwater is pumped out of the ground from water wells, and the Groundwater Information Network maps show that there are thousands of wells across Canada. Because it is critically important to our lives, geologists in Canada and around the world study groundwater in order to develop policies for its sustainable use.

The Water Table

Groundwater is water that is found underground in cracks, pores, and spaces in soil, sediments, and rocks. The area in which water fills these spaces is called the *saturated zone*. The top of this zone is called the **water table.** The top of the water table is the *capillary fringe*, a narrow zone where water seeps upward, pulled by surface tension, to fill empty pore spaces. The water table may be only a metre below the ground's surface, or it may be hundreds of metres down (**FIGURE 20.1**). The portion above the water table where pore spaces and other openings are filled with both air and water is called the *unsaturated zone*.

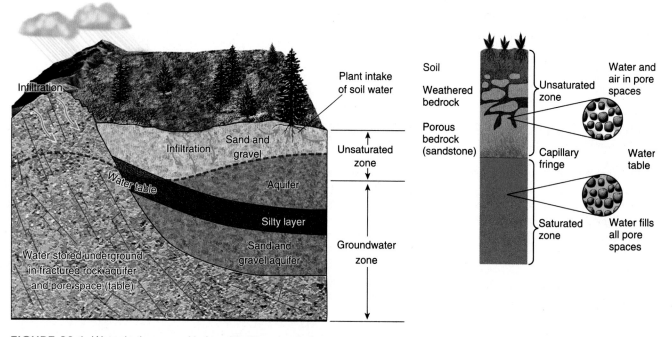

FIGURE 20.1 Water in the ground is found in the saturated zone beneath the water table and in the unsaturated zone above the water table.

? What processes are capable of causing the water table to rise and fall?

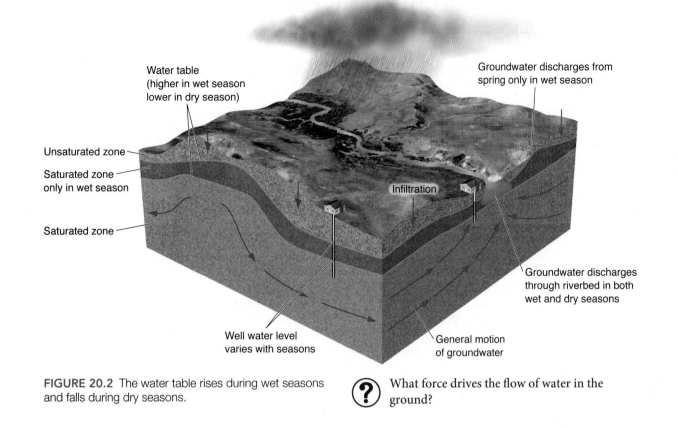

FIGURE 20.2 The water table rises during wet seasons and falls during dry seasons.

(?) What force drives the flow of water in the ground?

Groundwater can be found almost everywhere. As just noted, the water table may be deep or shallow; it may also rise or fall, depending on many factors (e.g., near the coast, the water table rises and falls with the tide and with groups of waves). In humid areas, the water table is a subtle reflection of the ground's topography (**FIGURE 20.2**). That is, where there are hills, the water table is high, and where there are valleys, the water table is low. Heavy rains or melting snow may cause the water table to rise, or an extended period of dry weather may cause it to fall. When the water table falls, streams and wetlands dry up, lake levels drop, and the ground becomes parched and dry.

Water Wells

If you live in or near an urban area, you probably never give much thought to where your water comes from, how it is delivered, or whether it is clean. One of the services provided by local government, paid for by taxes, is the delivery of clean water, and most city governments are so successful at this that there is rarely a reason for the average city dweller to worry about water. All you need to know is how to open the tap at the sink and fill a glass with water. However, that water arrives at your tap as the result of a lot of hard work by many people as well as by the application of modern technology.

In the developed world, most groundwater is extracted using wells, and most such wells are drilled by *truck-mounted drill rigs* (**FIGURE 20.3**), which use various types of cutting heads on the end of a spinning shaft to bore through soil, sediment, sedimentary rock,

and even crystalline igneous and metamorphic rock to access the water table. The *borehole* must penetrate below the water table to account for changes in water table height related to seasonal changes in elevation. Once a well has been drilled, it is lined with a pipe that is perforated to allow the inflow of water but designed to keep sand and other sediments out, and to prevent the hole from caving in. A pump is installed to withdraw the water, a gravel layer is added along the walls of the borehole to keep out fine sediment, and the top of the well is sealed to keep shallow, contaminated water from seeping into the well.

FIGURE 20.3 Water wells are drilled by truck-mounted rigs.

(?) Where does the drinking water in your community come from?

(?) **Expand Your Thinking**—Geologists search for groundwater. Describe the data you would need to develop groundwater resources for a small town.

20-2 Groundwater Is Fed by Snowmelt and Rainfall in Areas of Recharge

LO 20-2 *Describe the processes by which the water table gains and loses water.*

Rainfall and snowmelt add new water to the groundwater system through the process of **recharge.** Recharge may occur in several ways. In most cases, runoff soaks into the ground through percolation and adds to the groundwater. Areas of high rainfall where percolation enters the saturated zone can recharge a groundwater system that extends many kilometres into the surrounding countryside.

High-rainfall areas that recharge the groundwater system are often located at higher elevations, where warm moist air is forced to flow into cooler (higher) levels of the atmosphere. There it condenses as rain or snow, falls to the land surface, and creates a recharge area. Recharge can also occur where streams, lakes, and wetlands act as a source of water for the groundwater system (FIGURE 20.4).

Eventually, groundwater may reappear above the ground. This process is called *groundwater discharge.* Groundwater may flow into streams, wetlands, lakes, and oceans, or it may be discharged in the form of springs and flowing wells. Groundwater discharge can contribute to surface water and runoff. In dry periods, groundwater may supply the entire flow of streams. At all times of the year, in fact, the nature of groundwater flow through the crust and the position of the water table have a profound effect on the volume of surface runoff. Although the rate of discharge determines the volume of water moving from the saturated zone into streams, the rate of recharge determines the volume of water running across the surface. For example, when it rains, the volume of water running into streams and rivers depends on how much rainfall the underground rock and sediment layers can absorb. When there is more water on the surface than can be absorbed into the ground, it runs off into streams and lakes.

Streams that recharge the groundwater system are known as *losing streams* or *influent streams.* In losing streams, the height of the water table is lower than the stream's water surface. Water seeps into the ground through the stream channel bed. In locations where the water table is so low that it is located below the base of the channel, it often bulges upward beneath the recharge area (FIGURE 20.5). In *gaining streams,* or *effluent streams,* discharge from the water table contributes partly or entirely to their flow. In such cases, the water table must be at an elevation above the stream's surface. Water that enters a stream system from the water table through the channel bed is known as **base flow.** A third situation includes both types of water gain and loss along different portions of the same channel. Often a stream gains water during wet seasons of the year and loses water during dry seasons.

Wetlands

Once thought of as useless swamps and wastelands, today **wetlands** are recognized as critically important components of natural ecosystems, which provide habitat for numerous threatened species. It is estimated that about 14 percent of Canada's land area is wetland, which is about 25 percent of all wetlands globally. Wetlands form in places where the water table intersects the land surface. Generally, they are lands where saturation with water is the dominant factor in determining the nature of soil development and the types of plant and animal communities living on and in the soil.

As with streams, groundwater interactions with wetlands can be grouped into three categories that focus on gains and losses of water (FIGURE 20.6). (1) Wetlands may serve as a source of recharge in places where surface water seeps into the ground and feeds the groundwater system. Alternatively, (2) the groundwater system may be a source of water for the wetland. It is also possible (3) for groundwater to feed a wetland from the upslope side, flow through the wetland, and seep back into the ground from the downslope side.

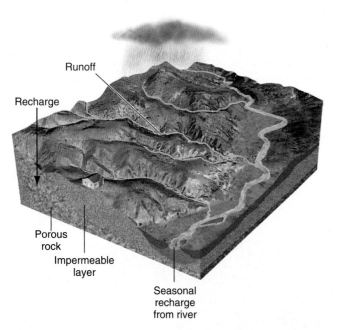

Runoff

Recharge

Porous rock

Impermeable layer

Seasonal recharge from river

FIGURE 20.4 The groundwater system gains water from recharge areas that may be located in regions of high humidity or near lakes, streams, and wetlands. In fact, groundwater is recharged in all locations where rainwater can infiltrate into the ground.

(?) What are the most common ways to recharge groundwater?

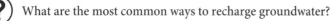

(?) **Expand Your Thinking**—If you were the water manager for the city, how would you determine the recharge area for the groundwater that your city's population drinks?

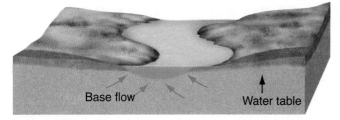

(a) Gaining stream

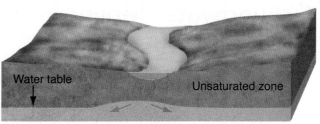

(c) Losing stream (disconnected)

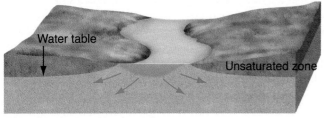

(b) Losing stream (connected)

FIGURE 20.5 (a) Gaining streams experience base flow where groundwater flows into the stream. (b) Losing streams contribute a portion of their discharge to the groundwater table by means of seepage through the channel bed. (c) The water table often bulges upward beneath a losing stream where the water table is not connected to the channel bed.

 Under what conditions would a stream receive base flow during part of the year but provide recharge at other times?

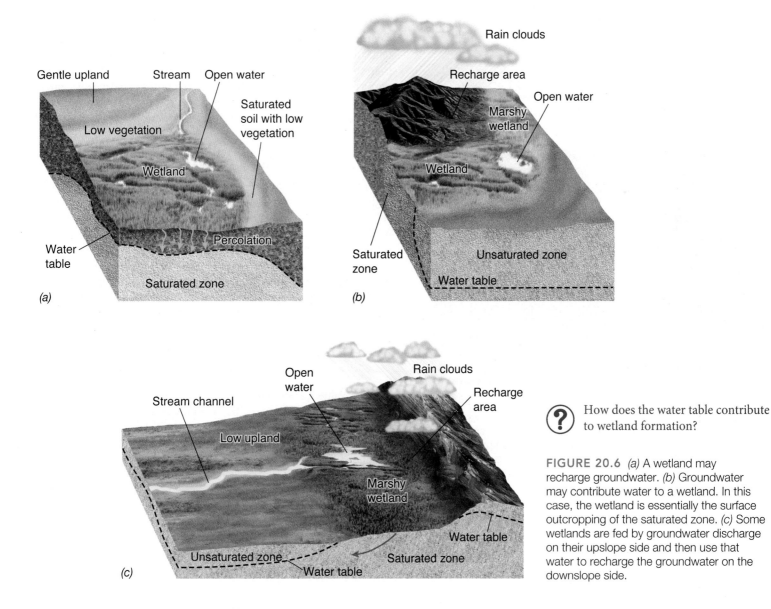

How does the water table contribute to wetland formation?

FIGURE 20.6 (a) A wetland may recharge groundwater. (b) Groundwater may contribute water to a wetland. In this case, the wetland is essentially the surface outcropping of the saturated zone. (c) Some wetlands are fed by groundwater discharge on their upslope side and then use that water to recharge the groundwater on the downslope side.

20-3 Groundwater Moves in Response to Gravity and Hydraulic Pressure

LO 20-3 *Identify the factors that influence groundwater movement.*

Groundwater may move in response to gravity as it seeps downward toward the water table. However, in most cases, groundwater flows in response to **hydraulic pressure.** That is, it migrates from regions of high pressure to regions of low pressure. As a result, it is able to move upward (against gravity) within the crust (**FIGURE 20.7**). Because it has to move along tortuous pathways within rock and sediment, it does not move as fast as it would on the surface.

Although water exists everywhere beneath the water table, some parts of the saturated zone contain more than others. An **aquifer** is

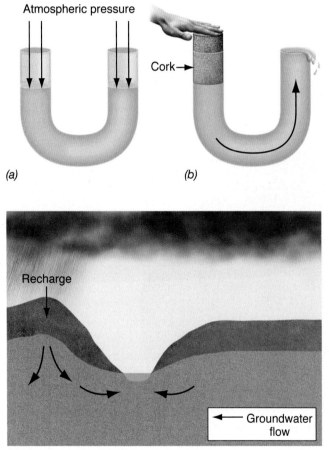

(a)   *(b)*

(c)

FIGURE 20.7 Groundwater moves from areas of high pressure to areas of low pressure. *(a)* Water will not flow if there is an even distribution of pressure acting on the surface. *(b)* When pressure is applied to one part of the surface, the water will flow toward an area of lower pressure. *(c)* Recharge creates an area of high pressure.

(?) What processes may control the length of time that water stays in the ground?

saturated crust that can produce useful quantities of water when tapped by a well. Aquifers may be small, extending under only a few hectares, or very large, underlying thousands of hectares of Earth's surface.

The ability of rock or sediment to transmit groundwater is called **permeability,** and it depends on the size of the spaces in an aquifer and how well the spaces are connected. Aquifers typically consist of highly permeable materials, such as gravel, sand, sandstone, or fractured rock such as limestone. These are permeable because they contain large connected spaces that allow water to flow with relative ease. In some permeable materials, groundwater may move several metres in a day; in others, it moves only a few centimetres in a century (**FIGURE 20.8**).

The amount of empty space within an aquifer is known as **porosity.** In sedimentary rocks, porosity depends on cementation, sorting, and the shape of the grains. Sediments or sedimentary rocks with large, well-rounded grains have high porosity and can store abundant water. Poorly sorted sedimentary rocks have lower porosity. Cements, because they crystallize in the pore spaces between grains, tend to decrease the porosity of the rocks (**FIGURE 20.9**). Porosity is an important characteristic that may control permeability. Highly porous rock or sediment with connected pores is permeable. But if the pores are not connected, permeability is low. Learn more about groundwater with the exercise in the "Critical Thinking" box titled "Groundwater Dynamics."

Hydraulic Gradient

Clay, shale, and other fine-grained rocks and sediments may have high porosity and low permeability because the many pore spaces in clay are extremely small and are not connected. Hence, clay tends to inhibit groundwater flow. In igneous and metamorphic rock, porosity and permeability are typically very low because of the intergrown nature of the minerals. However, if the rock is fractured, the lines of breakage can connect, and high permeability can develop.

The rate at which groundwater moves depends on both the permeability of the rock and the **hydraulic gradient,** which is the difference in elevation between two points in an aquifer divided by their distance (**FIGURE 20.10**). The velocity (V) of groundwater flow is defined as:

$$V = K(h_2 - h_1)/L$$

The value K is the *coefficient of permeability,* which describes the permeability of the aquifer. The hydraulic gradient, $(h_2 - h_1)/L$, is essentially the slope of the water table between two points h_2 and h_1. L is the distance between the two points. The groundwater discharge (Q) is defined by multiplying V times an area (A) through which water flows. This relationship is known as **Darcy's law** and is usually measured in cubic metres per day:

$$Q = AK(h_2 - h_1)/L$$

(?) **Expand Your Thinking**—Can groundwater flow "uphill" underground? Why or why not?

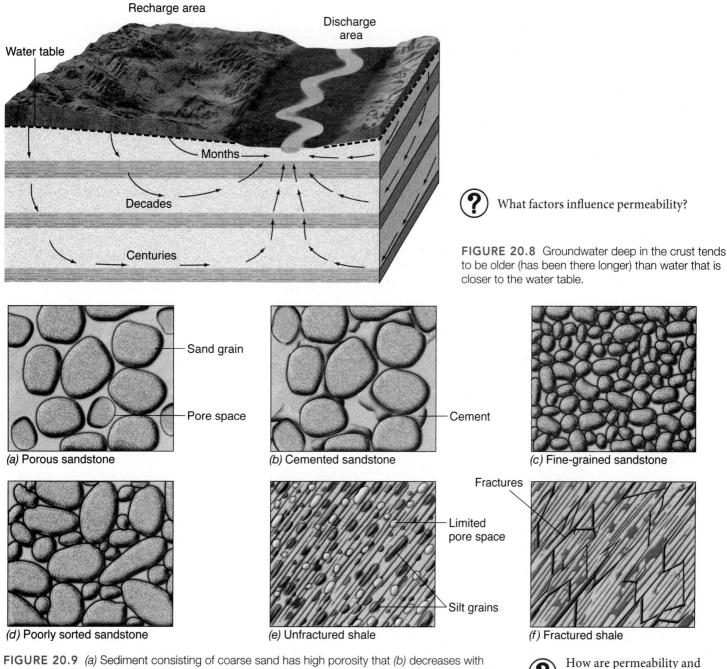

Recharge area

Discharge area

Water table

Months

Decades

Centuries

? What factors influence permeability?

FIGURE 20.8 Groundwater deep in the crust tends to be older (has been there longer) than water that is closer to the water table.

Sand grain

Pore space

(a) Porous sandstone

Cement

(b) Cemented sandstone

(c) Fine-grained sandstone

(d) Poorly sorted sandstone

Fractures

Limited pore space

Silt grains

(e) Unfractured shale

(f) Fractured shale

FIGURE 20.9 (a) Sediment consisting of coarse sand has high porosity that (b) decreases with the addition of cements. (c) Well-sorted, fine-grained sands or sandstones have high porosity and permeability that is (d) decreased when the sorting is reduced. (e) Shale usually has low permeability because the very small size of the grains means that pores are very tiny, and the pores are not connected. Permeability can be increased (f) if the shale is fractured so that the fractures provide new pathways for the movement of groundwater.

? How are permeability and porosity related to one another?

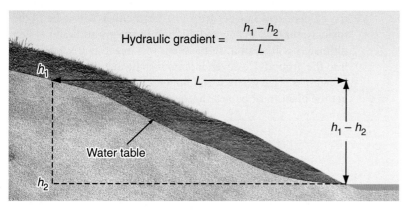

Hydraulic gradient = $\dfrac{h_1 - h_2}{L}$

h_1

L

$h_1 - h_2$

Water table

h_2

? How might Darcy's law be useful in planning a program of groundwater conservation?

FIGURE 20.10 Calculated using Darcy's law, the hydraulic gradient is the difference in elevation between two points on the water table divided by their distance from each other.

? CRITICAL THINKING

GROUNDWATER DYNAMICS

Because groundwater flows from areas of high pressure to areas of low pressure, it is possible to create maps showing the direction of groundwater flow by contouring the surface of the water table. The elevation of the water table can be mapped using wells, springs, lakes, and other evidence of its height. When the surface of the water table is contoured with lines that connect points on the water table that have the same elevation, a line drawn perpendicular to the contours is assumed to indicate the direction of flow. This assumption is based on the hypothesis that water will flow from areas where the water table is high to where it is lower (FIGURE 20.11), and on the water table, different elevations represent differences in pressure. That is, like runoff, groundwater flows across contours rather than along them.

FIGURE 20.12 is a map of an area where *limestone crust* has been heavily *karstified*. That is, chemical weathering has dissolved the rock and created many caverns, sinkholes, and depressions. The climate in the area is humid, so there is high annual rainfall. Groundwater has filled the depressions and caverns with a high

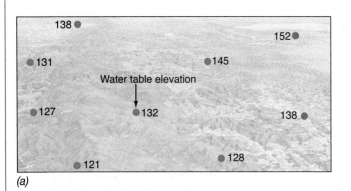

(a)

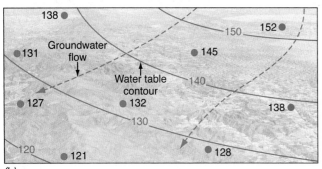

(b)

FIGURE 20.11 Using data from wells, springs, and other sources, it is possible to make a contour map of the water table. In most instances, groundwater will tend to flow perpendicular to the contour lines. The elevations are in metres.

water table, with the result that numerous lakes and ponds follow the shape of the weathered crust. These various bodies of water are connected by the water table.

The map shows several features:

- two roads, CCC Road and Highway 52
- a small town located at the intersection of Highway 52 and CCC Road
- elevations of the water surface in lakes and ponds of the region
- a wetland, shown in dark green
- Cedar Lake and Deep Lake, which are popular locations for recreation and swimming
- Clear Lake, which supplies the town's drinking water
- one contour line of the water table at 20 m elevation

Directions

1. Contour the water table: With a pencil, add additional contours to this map using a 2 m contour interval. The 20 m contour line has been added to help you get started.
2. Draw dashed arrows predicting groundwater flow: Groundwater flow follows the hydraulic gradient; that is, it flows along the shortest path from a higher elevation to a lower elevation of the water table. (See Figure 20.11.)
3. Identify major "drainage" directions in the water table. Place two or three long, bold arrows on the map showing these, and assign them a compass direction at the tip of the arrow (e.g., NW, SSE).
4. Calculate the average gradient along each of your drainages in metres of elevation change per kilometre of distance. Write these on the map. Which direction has the greatest gradient? What are the implications of this with regard to groundwater movement?
5. Imagine that a southbound tanker truck carrying pesticide is rounding the bend on CCC Road. Travelling too fast, it tips over, spilling its contents into the wetland (shown in dark green). Once the poison enters the groundwater system, in which direction will the toxic plume move? The town at the intersection of CCC Road and Highway 52 draws its water from Clear Lake. Is the water supply in danger? As a consultant, propose a method for cleaning up the groundwater and removing this contamination.
6. The town is planning to build a new sewage processing plant. It needs a safe location. The plant will treat liquid waste from the town to the "secondary level." This means that after treatment, the effluent will not be drinkable but neither will it be highly toxic. Natural cleansing and dilution in the groundwater system ultimately will clean it, so the plant will inject its treated sewage into the ground. However, it is important that the sewage be injected in a place where it will not end up in the drinking supply. Yet the plant must be near to the town so that building a sewage

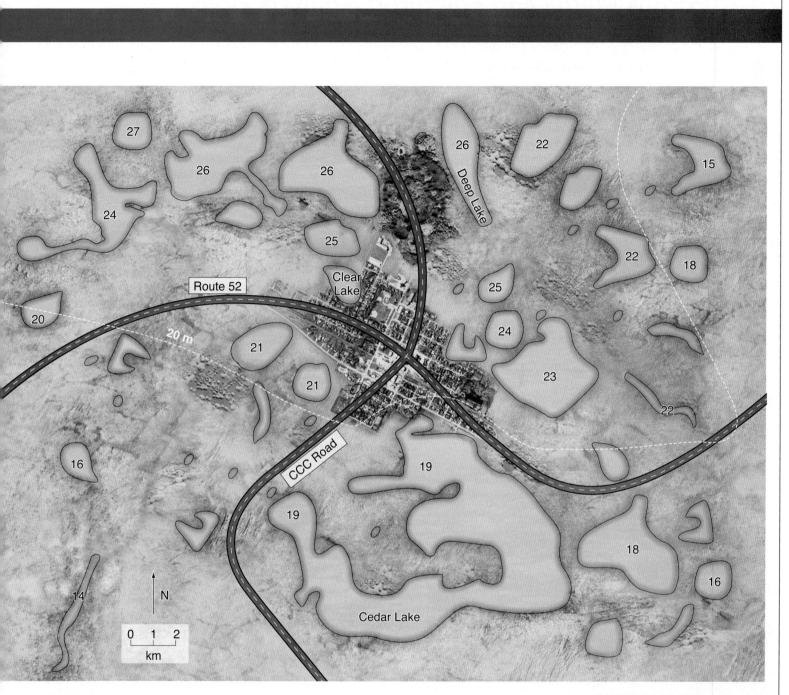

FIGURE 20.12 Water table map.

delivery pipeline to the plant is not too expensive. Where do you recommend that the plant and the injection well should be located?

7. The town is planning to create a landfill for solid waste. The landfill will be lined with clay and an impervious liner. However, it is important that the landfill be located in a place where leaks cannot enter the drinking water or interfere with the recreational use of Deep Lake and Cedar Lake, both of which are popular places for swimming. Where do you recommend they build the landfill? Why is it lined with clay?

20-4 Porous Media and Fractured Aquifers Hold Groundwater

LO 20-4 *Compare and contrast different types of aquifers.*

Two Latin words, *aqua* (water) and *ferre* (to bear or carry) form the modern term "aquifer." Aquifers literally carry water underground. They exist in several forms: a layer of gravel or sand, a layer of sandstone or fractured limestone, or a fractured body of crystalline rock. **Hydrogeologists** (geologists who study groundwater) generally define two types of aquifers in terms of their physical characteristics: *porous media* and *fractured aquifers.*

Porous media (**FIGURE 20.13**) are aquifers that consist of aggregates of particles, such as layers or beds of sand and gravel. These make excellent, high-yielding aquifers because the water is stored within, and moves through, the openings between individual grains. Porous media in which the grains are not connected to each other are considered *unconsolidated,* whereas those with cemented or compacted grains are called *consolidated.* Sandstones are examples of consolidated porous media, while loose sands are examples of unconsolidated porous media. Porous media such as sandstone may become so highly cemented or recrystallized that all of the original pore space is filled. In such cases, the rock is no longer a porous medium. However, if it contains cracks, it can still act as a fractured aquifer.

Fractured aquifers consist of strata, typically of limestone, cemented sandstone, or crystalline rock, in which groundwater moves through cracks, joints, or fractures in otherwise solid rock. Limestone often forms fractured aquifers where the cracks are the result of chemical weathering due to groundwater dissolution; these cracks may develop into large channels or even caverns. Igneous and metamorphic rock that is highly fractured or jointed also can act as a fractured aquifer.

Layers of impermeable rock or sediment that do not allow the easy passage of groundwater are termed either an **aquiclude** or an **aquitard.** An aquiclude prevents (or precludes) the flow of groundwater, whereas the flow through an aquitard is extremely slow (retarded). The presence of aquicludes or aquitards, which are usually formed by layers of shale or clay, is important in determining the movement and location of usable groundwater (**FIGURE 20.14**). A *confined aquifer* is one that lies beneath an aquiclude.

A special kind of confined aquifer, an *artesian aquifer,* is one in which water rises above the upper surface of the aquifer, driven by the hydraulic gradient of the water table. An artesian aquifer has internal pressure because of the confining bed that prevents water flow. This pressure is the result of the recharge area being at a higher level than the rest of the aquifer. The force of gravity pulls the higher water downward, creating a steep hydraulic gradient inside the aquifer. In some cases, the water may rise as far as the ground surface. This will form an **artesian spring** or **well** which will flow by itself; the pressure forces the water out of the ground, and it will rise to a level known as the *potentiometric surface.*

Learn about groundwater contamination by reading "Government-Funded Cleanup" in the "Earth Citizenship" box.

(a)

(b)

FIGURE 20.13 *(a)* Unconsolidated porous media such as these sand and gravel beds may form aquifers that are rich sources of groundwater. *(b)* Fractured aquifers are formed when cracks, joints, and fractures open up zones of permeability in otherwise impermeable rock, such as granite.

 What types of depositional environments would lead to the development of porous media and fractured aquifers?

 Expand Your Thinking—Toxic chemicals have spilled onto the ground. How will you determine whether your drinking water is threatened?

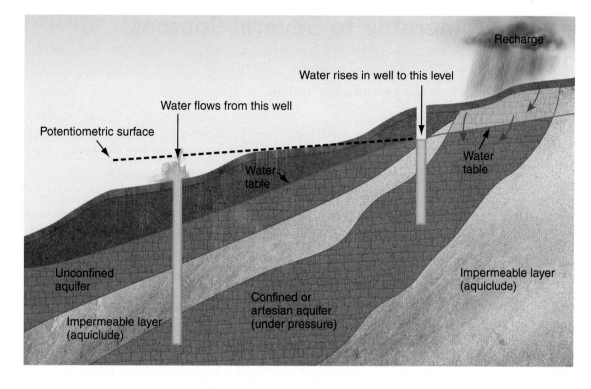

How does an artesian aquifer differ from a normal aquifer?

FIGURE 20.14 Water in a confined aquifer is restricted by an aquiclude. An unconfined aquifer is one in which the water table can freely move up or down. An artesian aquifer forms when hydraulic pressure within an aquifer is high because the recharge area lies at a higher elevation than the aquifer. The high pressure causes the free flow of water through a spring if the confining bed is breached.

EARTH CITIZENSHIP

GOVERNMENT-FUNDED CLEANUP

Throughout most of the 19th century and the early and middle 20th century, people were not aware that dumping chemicals in the ground or disposing of them in shallow pits would lead to groundwater contamination. On tens of thousands of properties across North America, these practices were common, continuous, and intensive. Now we recognize that disposing of toxic waste in the ground threatens human health and the environment. Hazardous waste sites resulting from chemical dumping include abandoned warehouses, landfills, and undocumented disposal sites (**FIGURE 20.15**).

In the 1990s, the Canadian government started to track and identify contaminated sites that might affect groundwater supplies in their areas, and developed the Federal Contaminated Sites Action Plan in 2005. The government has committed about $1 billion to clean up over 1,000 sites that include toxic waste sites, abandoned mines, contaminated military installations, and leaking fuel storage depots.

In the United States, concern over the extent of this problem led Congress to establish the Superfund program in 1980. The goal was to locate, investigate, and clean up the worst toxic waste sites nationwide. The program is administered by the Environmental Protection Agency (EPA) in cooperation with individual states and tribal governments. The cost to clean up sites is enormous, and it is estimated that about US$25 billion has been spent so far. The Superfund program deals only with abandoned waste disposal sites,

Richard Lautens/GetStock.com

FIGURE 20.15 Distant Early Warning (DEW) site at Cape Dyer, Baffin Island, Nunavut, one of 63 United States–built radar installations in Arctic Canada designed to provide early warning of attacks on North America during the Cold War era. The Department of National Defence budget for cleanup of the DEW sites is estimated at about $575 million, with about $77 million being provided by the United States military. The Cape Dyer site is the last to be cleaned up, and was scheduled to be completed in 2013.

and the EPA estimates that there are about 2,500 active disposal sites that will require remediation in the future. Ultimately, total cleanup costs may exceed $1 trillion.

20-5 Groundwater Is Vulnerable to Several Sources of Pollution

LO 20-5 *Identify ways in which pollutants enter the groundwater system.*

Groundwater may move slowly and reside in the crust for long periods of time, which means that water in the ground is vulnerable to contamination. Because water is so effective at dissolving compounds, it is very susceptible to carrying pollutants. For instance, if underground storage tanks at gas stations leak, gasoline and other toxic chemicals may contaminate groundwater.

Groundwater is vulnerable to multiple sources of pollution that may float on the water table, dissolve within the aquifer, or sink to the base of the aquifer, depending on the relative densities of the water and the pollutant (FIGURE 20.16). It does not take much to pollute groundwater. Just 1 L of oil can contaminate up to 1 million litres of drinking water or cause an oil slick extending over more than 8,000 m².

Despite these problems, in Canada, the chemical (pollutants) and biological (pathogens and microbes) quality of groundwater is acceptable for most uses. However, in most urbanized areas, particularly in locations where groundwater is shallow, water quality has deteriorated as a result of human activities. The best way to care for groundwater is to understand the natural and human processes that influence it. With a careful understanding of this precious resource, it can be recovered, sustained, and conserved for the future.

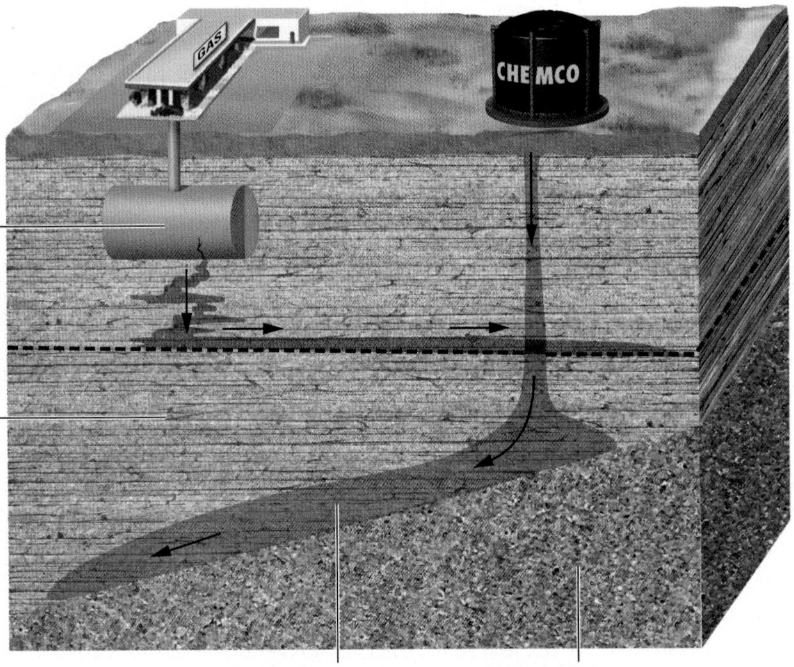

Overpumping and Saltwater Intrusion

At present, 6 out of 10 people live within 60 km of a coast, and more than two-thirds of the population of developing countries lives in the vicinity of an ocean. The increasing concentration of human settlements in coastal areas gives rise to heavy pressure to use coastal groundwater, resulting in *overpumping* of aquifers. When water is pumped from an aquifer at a rate that exceeds the permeability of rock in the area, the water table around the well drops. This process creates a **cone of depression** that increases the hydraulic gradient and rapidly draws water toward the well (FIGURE 20.17). The cone of depression will enlarge as long as the aquifer is unable to restore pumped water at a rate that matches the rate of withdrawal.

Continued pumping at an excessive rate in coastal areas will cause *saltwater intrusion* that damages the quality of the aquifer. Saltwater intrusion occurs when fresh water is withdrawn faster than it can be recharged near a coastline. Salt water generally intrudes upward and landward into an aquifer and around a well, although it can occur "passively" whenever there is a general lowering of the water table near a coastline. The *transition zone* at the base of the aquifer (the interface where fresh water naturally mixes with salt water as it is discharged to the ocean) naturally descends landward, causing the aquifer to be shaped like a wedge at the coastline. Saltwater intrusion is usually first noticed when water wells tap into this transition zone.

Many communities in coastal areas have been or potentially could be affected by saltwater intrusion caused by heavy pumping of groundwater. Saltwater intrusion threatens major sources of fresh water that coastal communities depend on. Sea-level rise due to global warming accelerates this process by raising the transition zone under the aquifer and moving it landward.

Dissolved Minerals

Because water is an effective solvent, it dissolves minerals from the rocks and sediments it encounters. For this reason, most groundwater contains dissolved minerals and gases that give it the slightly tangy taste many people enjoy. Without these dissolved substances, the water would taste flat. The most commonly dissolved constituents include the elements sodium (Na), iron (Fe), calcium (Ca), magnesium (Mg), potassium (K), and chlorine (Cl), and the ions of

Leaky gasoline storage tank

Groundwater in permeable sandstone

Toxic chemicals denser than water

Relatively impermeable rock

FIGURE 20.16 Differences in density between fresh water and chemical contaminants determine where a pollutant is concentrated.

(?) Describe how the density of a pollutant governs its role in contaminating groundwater.

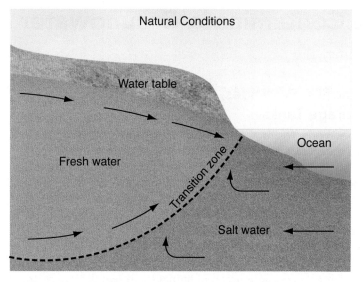

Natural Conditions

FIGURE 20.17 Population growth in coastal areas puts a high demand on groundwater, resulting in saltwater intrusion and deterioration of the water quality.

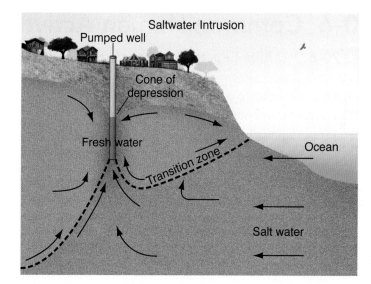

Saltwater Intrusion

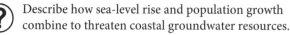

Describe how sea-level rise and population growth combine to threaten coastal groundwater resources.

bicarbonate (HCO_3) and sulphate (SO_4). Most water tastes bad if the level of dissolved constituents exceeds 1,000 mg/L (milligrams per litre). When the level of dissolved constituents reaches a few thousand mg/L, the water tastes salty, but it still may be used in areas where less mineralized water is not available.

Groundwater that has passed through limestone and dolomite contains high levels of dissolved calcium and magnesium ions; it is known as *hard water*. This concentration of ions can cause problems as coatings of mineral precipitates from the water may form on plumbing fixtures or in appliances such as kettles. Some wells and springs contain very large concentrations of dissolved minerals and cannot be tolerated by humans and other animals or plants. For example, the water in aquifers in sedimentary sequences that contain evaporites may become very salty.

Sewage Waste

In general, groundwater is less vulnerable to bacterial pollution than surface water because the soil, sediments, and rocks through which groundwater percolates in the unsaturated and saturated zones tend to filter out most bacteria by a combination of chemical and physical processes. But bacteria do occasionally occur in groundwater, sometimes in dangerously high concentrations.

A common cause of bacterial contamination is direct disposal of liquid sewage from a home or business into the ground. Most liquid sewage generated by human waste enters a community system of buried pipes through which it is pumped to a local sewage treatment plant. However, many rural and suburban homes are not connected to community sewage disposal lines. Instead, these buildings dispose of their liquid waste using *septic systems* and *cesspools (drainage fields)* that rely on the properties of soils and rocks in the unsaturated zone to filter out harmful bacteria (**FIGURE 20.18**). In-ground sewage

disposal systems like these allow liquid sewage to cleanse naturally as it percolates through the unsaturated zone. But if sewage enters an aquifer before it is completely purified, it may contaminate a drinking water source.

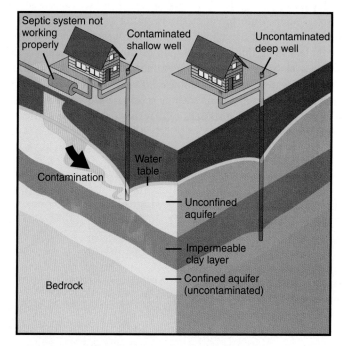

FIGURE 20.18 Domestic sewage in cesspools and septic systems may contaminate drinking water.

Why does a cone of depression tend to draw pollutants toward a well?

Expand Your Thinking—As a consultant you have been asked by a small coastal town to assess its future freshwater availability from groundwater. What must you consider when making this assessment?

20-6 Common Human Activities Contaminate Groundwater

LO 20-6 *List human activities that contaminate groundwater.*

Before the enactment of modern laws controlling waste disposal and monitoring water quality, it was common practice to dispose of liquid and solid wastes in the ground. Even today, illegal dumping of chemicals and contaminants continues, particularly in the developing world, where regulatory controls are less stringent. Illegal disposal practices range from disposal of hundreds of drums of highly toxic chemicals to dumping used engine oil in the backyard (**FIGURE 20.19**).

Often a simple hole in the ground is used to hide disposed chemicals. But eventually containers decay and leak their toxic contents into the ground. All liquids dumped in the ground will percolate through the unsaturated zone, often accelerated by rainfall and melting snow, and form a **contaminant plume** that follows the same hydraulic gradient that drives groundwater flow in aquifer systems. Contaminants include used engine oil, pesticides, paint thinners, cleaning compounds, lubricants, inks, and manufactured toxic substances. A contaminant plume will follow the hydraulic gradient leading to a well if there is a cone of depression in the area.

Salt Contamination

It is common to combat icy conditions by spreading salt on roads and highways. The salt absorbs water and breaks down ice rapidly so that driving is safer. Unfortunately, the resulting salty water turns into polluted runoff that percolates into the unsaturated zone and eventually enters the groundwater system, causing contamination. High chloride levels are found in well water due to this problem, and drinking water may develop a salty, unpleasant taste. The solution is to use alternative de-icing materials such as sand, use salt more sparingly, and better manage salt stockpiles so that they do not pose a threat to drinking water.

Landfills, Pesticides, and Underground Storage Tanks

Since World War II, the growth of industry, technology, population, and water use has increased the stress on both land and water resources. Locally, the quality of groundwater has been degraded. Municipal and industrial wastes disposed in **landfills;** chemical fertilizers, herbicides, and pesticides used in agriculture; and the use of underground storage tanks at gas stations have all resulted in chemical contamination of groundwater resources (**FIGURE 20.20**).

Landfills are excavations in the ground that are meant to hold waste generated by a community. Modern construction techniques use clay barriers and waterproof linings to prevent the movement of any contaminants out of a landfill. But leaks can still develop, and old landfills had few safeguards against contamination moving into the groundwater system. Any of a wide variety of toxic liquids may leach out of a landfill, including bacteria, viruses, organic chemicals that can disrupt the human *endocrine system* (the system of hormone production and use), nitrates, dissolved metals, various cleansing and lubricating agents, and many other potentially dangerous chemicals.

Pesticides, herbicides, and other poisons that kill weeds and insects and other pests may be applied in excessive amounts or during inappropriate times in the crop-growing cycle. This practice can lead to groundwater contamination and consumption of contaminated water by humans as well as to pollution of nearby streams and wetlands. Pesticide use is widespread. The agricultural industry uses the greatest proportion of pesticides and herbicides, although the amount that is being used has been falling for the last 20 years. Pollution by pesticides can be reduced with careful application of these chemicals, restricted use near water wells, and use of alternative pest control methods.

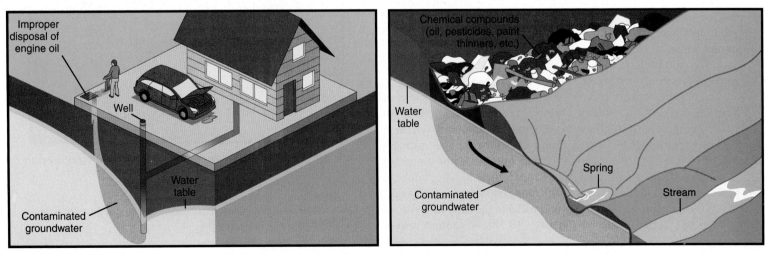

FIGURE 20.19 Dumping chemicals and other unwanted liquids in backyards and shallow pits creates a contaminant plume that may lead to groundwater contamination.

? What is a contaminant plume and how does it move?

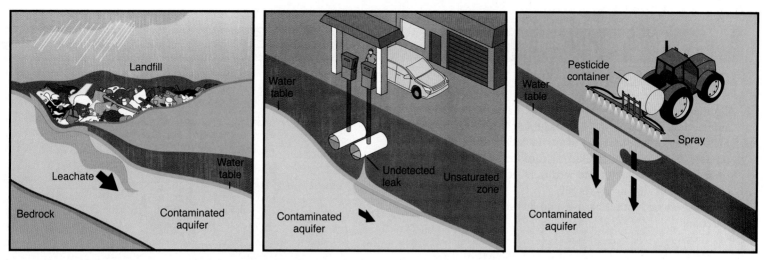

FIGURE 20.20 Chemical substances from landfills, underground storage tanks, and agricultural practices can pollute the groundwater system.

 Describe a method of controlling contamination plumes by pumping groundwater. What aspect of the water table would you be manipulating?

Gasoline service stations store large volumes of gasoline in underground storage tanks. Gasoline is highly toxic to plants and animals, and leakage from tanks is an enormous problem (FIGURE 20.21). Provincial and territorial regulations require that underground storage tanks must be registered, and the preference is for double-walled tanks with monitors and alarms to warn of a potential leak. Older tanks are typically single-walled and eventually rust and leak if not replaced.

© Earth Gallery Environment/Alamy

 What is the density difference between gasoline and water? Where is gasoline likely to collect in an aquifer?

FIGURE 20.21 To reduce groundwater contamination, leaky underground storage tanks at gasoline stations must be replaced.

Expand Your Thinking—A contaminant plume probably has formed under the gas station on the corner. As city water manager, how will you assess the threat to your city's drinking water?

CRITICAL THINKING

YOUR DRINKING WATER

Groundwater can be contaminated by chemical products, animal and human waste, and bacteria (FIGURE 20.22). Major sources for these pollutants include storage tanks, septic systems, animal feedlots, hazardous waste disposal sites, landfills, pesticides and fertilizers, road salts, industrial chemicals, and urban runoff.

Please work with a partner and answer the following questions.

1. Make a list of the sources of contamination in Figure 20.22.
2. How many of these sources are found where you live?
3. List additional sources of groundwater contamination that are not shown in Figure 20.22.
4. The contamination plumes shown have city managers concerned. Describe a method of controlling the plumes and eventually stopping their movement.
5. Do you agree that this area has historically lacked an overarching plan for managing water? Why or why not? Describe the necessary elements of such a plan. Write a list of steps to ensure future safe drinking water in this region.
6. As mayor of the city, you know that just controlling the spread of groundwater contamination (question 4 above) is a short-term fix. What longer-term actions do you propose for sustainably managing the water resource? Write your plan in the form of a press release.
7. Participate in a class forum on managing water.

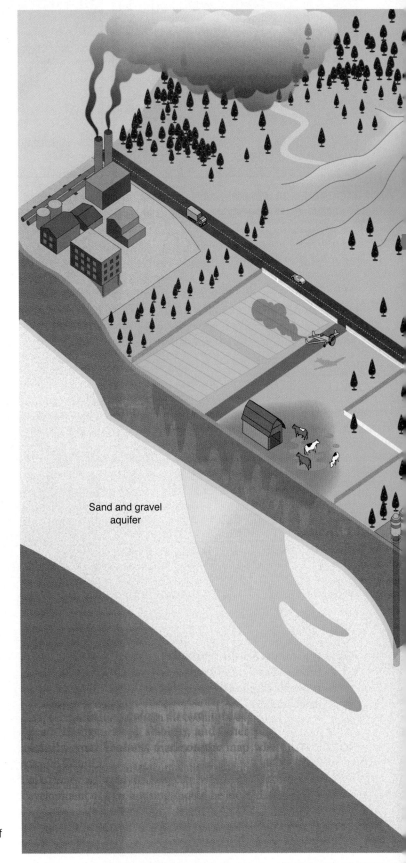

Sand and gravel aquifer

FIGURE 20.22 Potential sources of groundwater contamination.

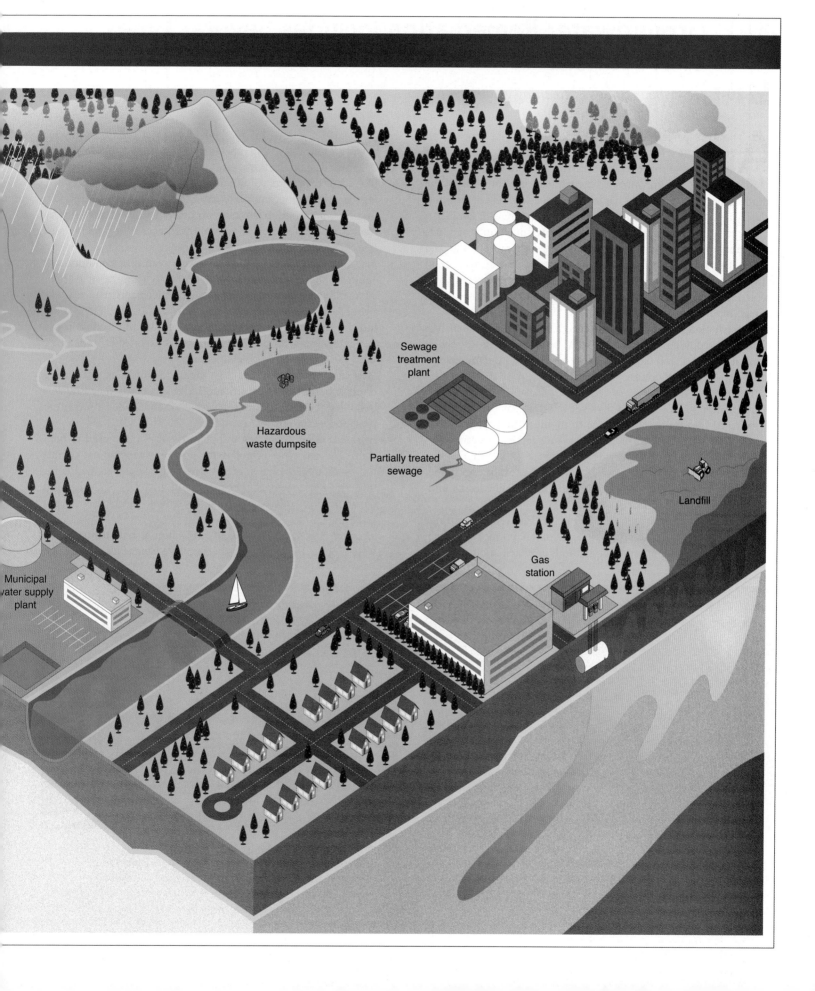

Sewage treatment plant

Hazardous waste dumpsite

Partially treated sewage

Landfill

Municipal water supply plant

Gas station

20-7 Groundwater Remediation Includes Several Types of Treatment

LO 20-7 *Describe methods of cleaning groundwater.*

About 30 percent of the population of Canada depends on groundwater for its drinking water, and groundwater is an important source of water for irrigation and manufacturing. The intense and often uncontrolled contamination of portions of the nation's groundwater supplies has led to the development of an entire industry dedicated to **groundwater remediation (FIGURE 20.23).**

Threats to groundwater are increasing as population grows, cities expand, and the consumer economy demands more goods and services. Estimates of remediation costs at sites under the jurisdiction of the Canadian government alone are in the billions of dollars. Thus, protecting the quality of groundwater supplies is a problem of broad societal importance.

Remediation methods are extremely expensive, and often it is difficult to predict how effective they will be. These methods rely on chemical, physical, and biological processes. Among the more commonly employed methods are:

- *Air sparging.* This method involves injecting gas (usually air or oxygen) into the saturated zone to mobilize contaminants that are volatile or easily stirred into a gaseous phase. "Air stripping" is a similar method of spraying groundwater into the air so that it can vent off gases from pollutants such as gasoline and other petroleum-based pollutants. The water is then allowed to infiltrate back into the ground.
- *Directional wells.* Directional wells are drilled at near-horizontal angles in order to gain access to contaminant plumes and either pump them out or apply other methods of remediation.

- *Recirculation wells.* This method involves circulating contaminated groundwater up to the surface and then back down into the ground again. It allows gases to be vented off, pollutants to be filtered, or contaminated water to be filled with bubbles, converting dissolved compounds into gases that can then be vented off.
- *Aquifer fracturing.* Techniques for enhancing permeability by fracturing aquifer rock allow improved access to pollutants so that they can be removed.
- *Injection wells.* This method involves treating a contaminated aquifer with chemical additives in order to cleanse or strip a specific pollutant. Injection wells of clean water or a chemical additive are also used to create an artificial hydraulic gradient in order to drive a contamination plume in a certain direction so it can be vented, sparged, vacuumed, isolated, or treated in some other way.
- *Permeable reactive barriers.* This method involves using trenches or other barriers treated with a reactive substance to neutralize a pollutant. It often requires that a shallow aquiclude be present to provide access to the entire aquifer.
- *Bioremediation.* Microbes and plant tissues can be used to neutralize various types of pollutants. *Phytoremediation* uses plants to absorb metals onto root systems and take up contaminants into leaves and stems. Other plants and microbes neutralize and decompose organic compounds into simple forms that are not hazardous.
- *Thermal or electrical treatments.* Injection of hot steam, heated water, radio frequency, or electrical resistance can cause the

(a)

(b)

FIGURE 20.23 *(a)* Contaminated groundwater can produce polluted discharge that leads to surface water contamination. *(b)* The remediation of contaminated groundwater involves intensive engineering efforts to remove pollutants.

 Perhaps you are now concerned with the safety of your drinking water. Describe the steps you will take to allay your concern.

breakdown of various pollutants. Steam can mobilize volatile (gas-forming) compounds into a gaseous phase. Electrical probes in the ground can cause ions to migrate toward the probes, thereby breaking down chemical compounds.

Another technique that is frequently used because of the high cost and logistical problems associated with engineered remediation is known as **natural attenuation.** Basically, this approach involves shutting down any source of ongoing pollution, isolating a contamination plume, and allowing natural processes to reduce the toxic potential of contaminants. Although this method may be perceived as "walking away from the problem," often it is a rational approach based on cost and a decision that the aquifer can be abandoned.

GEOLOGY IN OUR LIVES

HYDROGEOLOGICAL REGIONS OF CANADA

The *Groundwater Mapping Program*, along with the new *Groundwater Geoscience Program* organized by Natural Resources Canada, provides a summary of the major hydrogeological regions in Canada (**FIGURE 20.24**) and information on each of the principal aquifers or rock units that will yield usable quantities of water to wells.

The program will provide a comprehensive summary of groundwater resources in Canada and a basic reference on its major aquifers. It will be useful to federal, provincial, and local officials with responsibilities for water allocation, waste disposal, and *wellhead protection* (protecting wells from contamination).

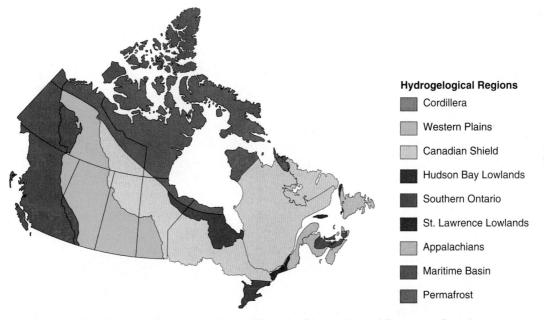

Hydrogelological Regions

- Cordillera
- Western Plains
- Canadian Shield
- Hudson Bay Lowlands
- Southern Ontario
- St. Lawrence Lowlands
- Appalachians
- Maritime Basin
- Permafrost

FIGURE 20.24 The hydrogeological regions of Canada. *Source:* Natural Resources Canada. 2007. Groundwater Mapping Program. Hydrogeological Regions.

 Expand Your Thinking—Describe the factors that should be considered in making a decision to allow natural attenuation of a contaminated aquifer.

20-8 Groundwater Is Responsible for Producing Springs and Karst Topography

LO 20-8 *Identify the major features of karst topography.*

Groundwater has significant implications for the development of geologic features in the crust. The chemistry of water in the ground is capable of causing hydrolysis, dissolution, and oxidation of many minerals. The dissolution mechanism is so strong that it can completely dissolve large volumes of carbonate rock and change the topography of the land and the character of the crust.

Springs

Natural **springs** are places where the water table intersects Earth's surface producing water that flows on the land. Springs were the main source of fresh water for early civilizations and pioneers. As a result, many towns and villages have grown up around springs and, knowing that a groundwater resource existed nearby, communities have drilled water wells supporting continued population growth. Today, in less developed parts of the world, springs are still the principal—sometimes the only—source of water for groups of families and small villages.

The discharge from springs may fluctuate due to seasonal and long-term periods of aridity and humidity that cause variations in the position of the

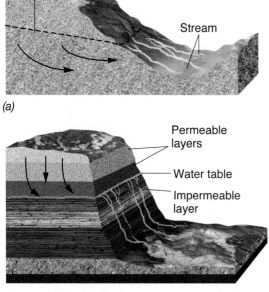

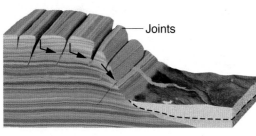

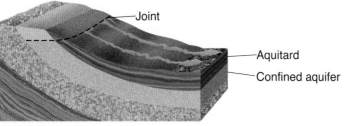

FIGURE 20.25 There are several types of springs: *(a)* A groundwater discharge area will form a spring. *(b)* A spring will form where a water table perched on an aquiclude intersects a slope. *(c)* As groundwater moves along the surface of an aquiclude, it may form a spring. *(d)* Springs form where an aquifer crops out on a hillside. *(e)* If a fracture pattern directs groundwater flow toward a hillside, a spring may develop. *(f)* If aquifer flow is directed toward the surface along a fault plane, a spring may form. *(g)* When water in a confined aquifer migrates along a joint, a spring may form.

water table. It is important to understand the geology of springs so that sustainable use of these resources can be developed. Most springs form at locations where the water table is directed toward the surface by the structure of rock layers that hold groundwater (**FIGURE 20.25**).

Karst Geology

We first explored **karst geology** in Chapter 7 ("Weathering") in our discussion of dissolution as a form of chemical weathering. Karst is a distinctive geologic form that develops when groundwater dissolves *carbonate rock* (i.e., limestone, dolomite, marble). The dissolution process, occurring over many thousands of years, results in unusual surface and subsurface features, including sinkholes, vertical shafts, disappearing streams, subterranean drainage systems, caves, springs, and unusually steep hillsides and slopes.

Over time, as dissolution removes carbonate rock and groundwater flows through dissolved rock with high permeability, aquifers are transformed from *diffuse-flow aquifers* with water moving as laminar flow through small openings, into *conduit-flow aquifers* with water moving primarily as turbulent flow through well-developed permeability channels called *conduit systems*. The flow eventually moves toward discharge points at springs, submarine openings, and surface runoff sites. With time, the process of dissolution lowers the water table as broad cavities open in the aquifer and fill with water. When the water table sinks below the level of surface streams, runoff moves into cave systems below ground level, creating disappearing streams.

In time, more of the surface drainage is diverted underground, and stream valleys virtually disappear to be replaced by closed basins called **sinkholes**. Sinkholes vary from small cylindrical pits to large conical basins that collect and funnel runoff into karst aquifers. Because sinkholes mark the location of underground flow routes, they often form lines on the surface that signal the presence of subsurface caverns. As sinkholes coalesce, they eventually form a steep-walled valley, leaving the surrounding topography as a high-relief remnant of the former land surface.

As we saw in Chapter 7, the process of karst formation involves dissolving of carbon dioxide (CO_2) into the groundwater, creating a mild *carbonic acid* (H_2CO_3). The formation of carbonic acid starts with rain falling through the atmosphere. The rain picks up CO_2 that dissolves in the droplets. After it hits the ground and percolates through the soil, it picks up more CO_2 from the decay of organic tissue produced by vegetation and animals. The CO_2-enriched groundwater (H_2O) in the unsaturated zone forms a weak solution of carbonic acid: $H_2O + CO_2 = H_2CO_3$.

Infiltrating water naturally exploits any cracks or crevices in the carbonate bedrock. Over time, with a continuous supply of CO_2-enriched water, carbonate bedrock begins to dissolve. Openings in the bedrock increase in size, and an underground drainage system develops that is a precursor to a type of cavern growth called *epikarst*. This drainage system allows more water to pass, further accelerating the dissolution process and leading to the formation of caverns and conduit flow systems. Eventually the full development of subsurface caves, sinkholes, and vertical shafts leads to karst topography (**FIGURE 20.26**).

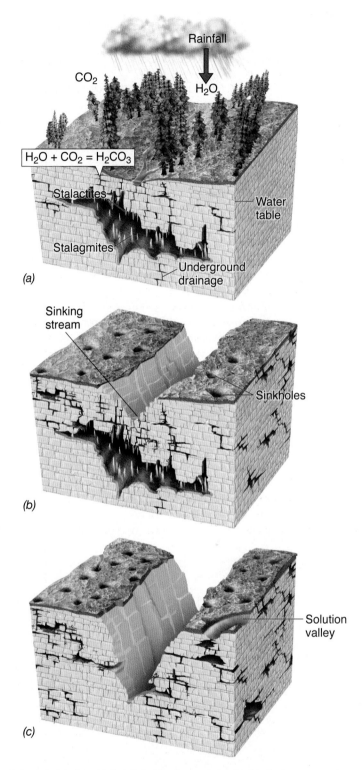

FIGURE 20.26 *(a)* Rainfall absorbs carbon dioxide in the atmosphere and soil and becomes carbonic acid. *(b)* Carbonic acid dissolves carbonate rock and thus contributes to the formation of karst topography *(c)*.

(?) What are the unique characteristics of groundwater in a karst system?

(?) **Expand Your Thinking**—Explain why groundwater in karst is highly vulnerable to contamination.

20-9 Hydrothermal Activity and Cave Formation Are Groundwater Processes

LO 20-9 *Describe the major features of geysers and cave deposits.*

Geysers, *fumaroles* (also called *solfataras*), and *hot springs* are features created by the interaction of groundwater with young (hot) volcanic rocks. These features are particularly common in areas on Earth that have active volcanoes as these are areas where the geothermal gradient is high (temperature increases with depth more rapidly than normal). Typically, runoff percolates through the unsaturated zone and comes into contact with high-temperature rock. As the water gains heat, it loses density and rises toward the surface along cracks and fractures, forming a hot spring. Or it turns into steam and erupts more forcefully as a geyser (from the Icelandic word *geysa,* "to gush") (**FIGURE 20.27**).

Geysers are relatively rare. There are only about 50 known geyser fields on Earth, and approximately two-thirds of those contain five or fewer active geysers. Yellowstone National Park in Wyoming has, by nearly an order of magnitude, more geysers than any other known field. At least three essential conditions must be met for a geyser to form, but many additional factors influence the character of geyser eruptions. The basic elements of a geyser are (1) a supply of hot and cold water, (2) a heat source, and (3) a reservoir and the natural plumbing system associated with it.

Geyser eruptions are of two types: those from *pool geysers* and those from *columnar geysers.* Pool geysers use two separate water sources, one bringing large amounts of shallow, cool water and another bringing a small quantity of boiling water from deeper in the crust. The two waters mix in the reservoir until the entire system gradually increases in temperature and fills the reservoir. Heating continues, and when the system reaches a critical temperature, the hot water rises through the crust and erupts as steam because of the reduction in pressure at the surface. Columnar geysers (**FIGURE 20.28**) originate when hot water in the geyser's plumbing system mixes with cooler water from the surface. When the water reaches the boiling point, steam is caught in a constriction in the geyser's plumbing system. In columnar geysers, constrictions in the channel are essential for eruption to occur. The pressure builds until it lifts overlying water up and out of the channel so that steam eventually escapes in an explosive eruption. The eruption continues until either the reservoir is out of water or the temperature of the system drops below the boiling point.

Fumaroles are vents that allow volcanic gases, such as water vapour and hydrogen sulphide (H_2S) to escape into the atmosphere. Hydrogen sulphide readily oxidizes to become sulphuric acid (H_2SO_4) and native sulphur (S) in the atmosphere, leaving a brilliant yellow coating on rocks around a vent. If the water and heat source are persistent, they may last for decades to centuries. If they occur atop a quickly cooling volcanic deposit, they may disappear within weeks.

Hot springs occur at locations where groundwater comes into contact with a heat source in the crust. The temperature and discharge of hot springs depend on several factors, including the permeability of the aquifer at the spring, the rate of heating, and the nature and quantity of mixing with cool groundwater near the surface.

Caves and Cave Deposits

A **cave** is a natural opening in the crust that extends beyond the range of light and is large enough to permit the entry of humans. Caves occur in a wide range of rock types and for a variety of reasons. They include

(a)

(b)

(c)

FIGURE 20.27 *(a)* A geyser is an eruptive explosion of hot groundwater created by the release of steam. *(b)* Fumaroles form where volcanic gases issue from the ground. *(c)* Hot springs are found in places where groundwater comes into contact with a heat source and the heating causes the water to rise to the surface as it cools.

 What is the difference between a hot spring and a geyser?

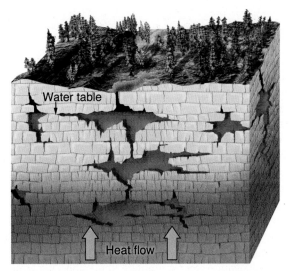

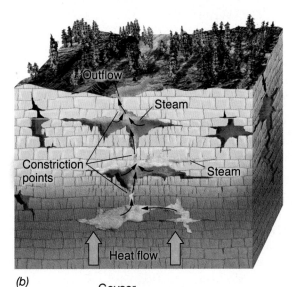

(a)

(b)

Geyser eruption

(c)

FIGURE 20.28 (a) Columnar geysers erupt when groundwater turns into steam above a heat source but is prevented from exiting by a constriction (b) in the fracture system leading to the surface. When the steam eventually is freed, the rapid decrease in pressure lowers the boiling point of the remaining groundwater, causing it to erupt explosively (c) as a columnar geyser.

ice caves, limestone caverns, lava tubes, sea caves, eolian caves made by windblown sand, and caves formed by joints and fracture patterns.

By far the most common type of cave is related to the karst process. Caves created in this way are called *limestone caverns.* The scientific study of caves is known as *speleology* (from the Greek words *spelaion,* "cave," and *logos,* "study") and includes analysis using the techniques of geology, biology, and archeology. Groundwater that dissolves limestone strata creates limestone caverns. These cave systems can extend for hundreds of kilometres, include "rooms" of immense size, and contain fascinating and beautiful features. Mammoth Cave, in the state of Kentucky, is the most extensive in the world, running for 540 km, while Carlsbad Caverns includes a single room that covers an area equivalent to about 14 football fields, with a ceiling 25 m higher than the height of Niagara Falls.

In caves, dissolved calcium carbonate in groundwater precipitates on the floor, the roof, or the walls, creating complex "dripstone" features called *speleothems.* The most familiar of these are **stalactites** and **stalagmites** (**FIGURE 20.29**). Stalactites hang downward from a ceiling and are formed as drop after drop of water slowly trickles through cracks in the cave roof. As each drop of water hangs from the ceiling, it gives off carbon dioxide, causing the precipitation of a thin film of calcite. Stalagmites grow upward from the floor of a cave, generally as a result of water dripping from overhanging stalactites. A *column* forms when a stalactite and a stalagmite grow until they join. A *curtain* or *drapery* begins to form on an inclined ceiling when the drops of water trickle along a sloping roof. Gradually a thin sheet of calcite grows downward from the ceiling and hangs in decorative folds like a drape.

FIGURE 20.29 A complex network of speleothems consisting of stalactites, stalagmites, and columns make up cave deposits.

(?) What is the chemical composition of these features? How do they form?

(?) **Expand Your Thinking**—Are speleothems igneous, sedimentary, or metamorphic? Explain your reasoning.

LET'S REVIEW "GEOLOGY IN OUR LIVES"

Now that you have finished the chapter, "Geology in Our Lives" will have taken on new meaning for you. Let us review it: Groundwater is one of our most precious natural resources. It is the most important source of drinking water, and it supplies water for irrigation, industry, manufacturing, domestic uses, and most other water needs in human society. Less than 3 percent of the water that exists on Earth is fresh water. About 75 percent of this fresh water is frozen in polar ice and glaciers, and about 25 percent is groundwater. Only 0.005 percent is surface water (lakes and streams). Groundwater is vulnerable to overuse and pollution. This is because most groundwater moves slowly (usually less than 0.3 m per day), has been in the ground for decades to centuries, and is not quickly renewed by the hydrologic cycle. Rates of human usage generally outpace natural rates of groundwater replenishment; thus conservation measures are in place in nearly every community. Properly conserving groundwater requires careful analysis of where it is located, how it moves, and how it is recharged. The central question in groundwater science is: How shall we use groundwater today so that it is still useful for future generations tomorrow?

Because it moves slowly and stays in the ground for long periods, groundwater is vulnerable to contamination. Sources of contamination are widespread and usually related to chemical wastes and sewage finding their way to the water table. When a well pumps the water table at a rate that exceeds the permeability of the aquifer, a cone of depression forms that creates a localized hydraulic gradient, pulling water and any contamination toward the well. Sources of groundwater contamination are so prevalent where people live that major efforts are under way to monitor and clean up groundwater pollutants, usually at great expense.

Do not take your water for granted. Remember that groundwater was used to grow your food and manufacture things you need, and it is probably the water you drink.

STUDY GUIDE

20-1 Groundwater is a very important source of fresh water.

- **Groundwater** is water that is underground in cracks, pores, and spaces in soil, sediments, and rocks. The area where water fills these spaces is called the saturated zone. The top of this zone is called the **water table**.

- About 30 percent of Canada's population depends on groundwater for use at home, but in some locations all water is obtained from underground sources.

- In most cases, water is pumped from the ground by a water well. Water wells are drilled by truck-mounted drill rigs, which use various types of cutting heads on the end of a spinning shaft to bore through soil, sediment, sedimentary rock, and even crystalline igneous and metamorphic rock to gain access to the water table.

20-2 Groundwater is fed by snowmelt and rainfall in areas of recharge.

- Rainfall and snowmelt add water to the groundwater system through the process of **recharge**. Recharge may occur in several ways. In most cases, runoff soaks into the ground through percolation and contributes to the water table below.

- Eventually, groundwater reappears above the ground. This process is called groundwater discharge. Groundwater may flow into streams, **wetlands**, lakes, and oceans, or it may be discharged in the form of springs and flowing wells.

- Groundwater supplies many streams, lakes, and wetlands.

- Wetlands are critically important components of natural ecosystems, and it is estimated that Canada has 25 percent of the world's wetlands. Generally, wetlands are lands where saturation with water is the dominant factor determining the nature of soil development and the types of plant and animal communities that live in the soil and on its surface.

20-3 Groundwater moves in response to gravity and hydraulic pressure.

- In most cases in which water moves within the saturated zone, groundwater flows in response to **hydraulic pressure.** That is, it moves from an area of high pressure into areas of lower pressure.

- An **aquifer** is an underground formation of saturated crust that can produce useful quantities of water when tapped by a well.

- A measure of the ability of rock or sediment to transmit groundwater is called **permeability.** It depends on the size of the spaces in the soil or rock of an aquifer and how well the spaces are connected. Aquifers typically consist of highly permeable materials such as gravel, sand, sandstone, or fractured rock like limestone.

- The amount of empty space in an aquifer is its **porosity.** In sediments or sedimentary rocks, porosity depends on the degree of cementation, the sorting, and the shape of the grains.

- The rate at which groundwater moves through an aquifer depends on both the permeability of the rock and the **hydraulic gradient.** This gradient is defined as the difference in elevation between two points in an aquifer divided by their distance from each other.

20-4 Porous media and fractured aquifers hold groundwater.

- **Hydrogeologists** define two types of aquifers in terms of their physical characteristics: porous media and fractured aquifers.

- **Aquicludes** are layers of impermeable rock or sediment that are important in determining the movement and location of usable groundwater. Aquicludes are usually formed by layers of shale or clay that prevent the free flow of groundwater in an aquifer. A confined aquifer is one that lies below an impermeable layer.

- An artesian aquifer is a special kind of confined aquifer found where the escape of water to the surface is driven by the hydraulic gradient of the water table.

20-5 Groundwater is vulnerable to several sources of pollution.

- About one-quarter of Earth's population drinks contaminated water.

- Since water is so effective at dissolving compounds, it often carries toxic compounds that are unhealthy for plants and animals. For instance, leaking underground storage tanks have contributed to contamination in every province. It is estimated that about 70 percent of all underground storage tanks leak gasoline and other toxic chemicals into the ground.

- When water is pumped from an aquifer at a rate that exceeds the permeability of the surrounding soil and rock, the water table around the well will drop. This practice creates a **cone of depression** that increases the hydraulic gradient and draws water toward the well.

- Continued pumping at an excessive rate in coastal areas will cause saltwater intrusion that damages the quality of the aquifer. Saltwater intrusion occurs near a coastline when fresh water is withdrawn faster than it can be recharged.

- One of the most common forms of bacterial contamination occurs through the disposal of liquid sewage from a home or business. Many rural and suburban homes are not connected to community sewage disposal lines. Instead, these buildings dispose of their liquid waste using septic and cesspool systems that rely on the properties of soils and rocks in the unsaturated zone to filter out harmful wastes.

20-6 Common human activities contaminate groundwater.

- There are thousands of known abandoned and uncontrolled hazardous waste sites in Canada and the United States. Illegal disposal practices range from disposal of hundreds of drums of highly toxic chemicals to dumping used engine oil in the backyard.

- **Contaminant plumes** follow the same hydraulic gradient that drives groundwater flow in aquifer systems. These contaminants may include used engine oil and other chemicals, pesticides, paint thinners, cleaning compounds, lubricants, inks, and manufactured toxic substances. A contaminant plume will follow the hydraulic gradient leading to a well if a cone of depression is present.

20-7 Groundwater remediation includes several types of treatment.

- The intense and often uncontrolled contamination of portions of the nation's groundwater supplies has led to the development of an entire industry dedicated to **groundwater remediation.**

- Estimates of remediation costs at Canadian government sites alone is in the order of billions of dollars. Protecting the quality of groundwater supplies is a problem of broad societal importance.

20-8 Groundwater is responsible for producing springs and karst topography.

- The chemistry of water in the ground is capable of causing hydrolysis, dissolution, and even oxidation of many minerals in crustal rocks. The dissolution mechanism is so strong that it can completely dissolve vast tracts of carbonate rock and change the topography of the land and the character of the crust.

- **Springs** and **karst geology** are produced by groundwater.

20-9 Hydrothermal activity and cave formation are groundwater processes.

- **Geysers,** fumaroles, and hot springs are features created by the interaction of groundwater with young volcanic rocks. Typically, runoff percolates through the unsaturated zone and comes into contact with high-temperature rock surrounding a magma body.

- The most common types of **caves** are related to the karst process. These are called limestone caverns. The scientific study of caves is termed speleology and includes the study of the geology, biology, and archeology of caves.

KEY TERMS

aquiclude (p. 548)
aquifer (p. 544)
aquitard (p. 548)
artesian spring (p. 548)
base flow (p. 542)
cave (p. 560)
cone of depression (p. 550)
contaminant plume (p. 552)
Darcy's law (p. 544)
geysers (p. 560)

groundwater (p. 540)
groundwater remediation (p. 556)
hydraulic gradient (p. 544)
hydraulic pressure (p. 544)
hydrogeologists (p. 548)
karst geology (p. 559)
landfills (p. 552)
natural attenuation (p. 557)
permeability (p. 544)
porosity (p. 544)

recharge (p. 542)
sinkholes (p. 559)
springs (p. 558)
stalactites (p. 561)
stalagmites (p. 561)
water table (p. 540)
well (p. 548)
wetlands (p. 542)

ASSESSING YOUR KNOWLEDGE

Please answer these questions before coming to class. Identify the best answer.

1. Groundwater typically is found as
   a. large underground lakes.
   b. large underground rivers.
   c. small pockets under rainforests.
   d. rims of water around individual grains and in fractures.
   e. vapour in pore spaces in the crust.

2. Groundwater is lost by
   a. base flow and groundwater discharge.
   b. evaporation.
   c. capillary flow.
   d. Darcy's law.
   e. all types of streams.

3. Which of the following is a characteristic of wetlands?
   a. Wetlands may recharge groundwater.
   b. Wetlands may be fed by groundwater discharge.
   c. Both a and b.
   d. Neither a nor b.
   e. None of the above.

4. Permeability is
   a. the rate of groundwater flow above the water table.
   b. a measure of the ability of rock or sediment to transmit groundwater.
   c. the size of pore spaces.
   d. the cementation of grains in the capillary fringe.
   e. None of the above.

5. Groundwater flows in response to
   a. gravity and density.
   b. gravity and porosity.
   c. permeability and topography.
   d. gaseous pressure and slope.
   e. gravity and hydraulic pressure.

6. An aquifer is
   a. any water in the ground.
   b. any water at the water table.
   c. a useful source of groundwater.
   d. water that is confined by aquicludes.
   e. groundwater that can be pumped from the ground.

7. Groundwater contamination may result from
   a. saltwater intrusion and illegal chemical disposal.
   b. agricultural waste and pesticides.
   c. landfill seepage and polluted runoff.
   d. human sewage and leaky underground storage tanks.
   e. All of the above.

8. Once human contamination enters the groundwater system, it may
   a. form a plume.
   b. last for centuries.
   c. be stripped off if it forms a gas.
   d. be allowed to naturally attenuate.
   e. All of the above.

9. Groundwater remediation is
   a. the activity of finding new aquifers.
   b. recharge by polluted runoff.
   c. cleaning up sewage before releasing it to a treatment plant.
   d. flow in the direction of hydraulic conductivity.
   e. cleaning polluted groundwater.

10. Geysers
    a. are the reason why caves have speleothems.
    b. may occur as non-violent pools of warm water.
    c. are the main reason why groundwater migrates.
    d. are violent eruptions of hot groundwater.
    e. All of the above.

11. Most liquid fresh water is found
    a. in rivers.
    b. in lakes and streams.
    c. in the ocean.
    d. as groundwater.
    e. during percolation.

12. Porosity is
    a. the percentage of empty space in the crust.
    b. the percentage of connected space for groundwater movement in the crust.
    c. the percentage of crustal space filled with groundwater.
    d. the percentage of pore space in the crust located at the water table.
    e. the water in the capillary fringe.

13. The rule of thumb for mapping the direction of groundwater flow at the water table is:
    a. Draw arrows parallel to the contours of the water table.
    b. Draw arrows at 45° to the contours of the water table.
    c. Draw arrows at 90° to the contours of the water table.
    d. Draw arrows that point downhill on the surface topography above the water table.
    e. None of the above.

14. An aquiclude may lead to formation of an artesian well because
    a. it allows atmospheric pressure to drain an unconfined aquifer.
    b. it confines an aquifer and closes off groundwater recharge.
    c. it confines an aquifer so that hydraulic pressure increases, leading to artesian flow.
    d. it causes karstification of the crust.
    e. All of the above.

15. Rainfall absorbs_____ to become acidic and dissolve limestone.
    a. sulphur dioxide
    b. oxygen
    c. nitrous oxide
    d. carbon dioxide
    e. nitrogen

FURTHER RESEARCH

1. Describe the evolution of a landscape in which groundwater is the dominant agent of change.

2. Where does the water that you drink come from? You may have to contact the local city or provincial water supply office and ask where the water supplied to your address comes from.

3. What is the closest contaminated site to you, as identified by the Canadian government? How is it being remediated?

4. Under what geologic/strata-forming conditions would a confined aquifer be created? (*Hint:* Describe environments of deposition.)

5. What is the difference between porosity and permeability? How are they related?

ONLINE RESOURCES

Explore more about geologic resources on the following websites:

Groundwater on Earth:
www.un-igrac.org

Canadian Ground Water Association:
www.cgwa.org

Natural Resources Canada—Groundwater Geoscience program:
www.nrcan.gc.ca/earth-sciences/about/current-program
groundwater-geoscience/4106

Environment Canada information on groundwater:
www.ec.gc.ca/eau-water/default.asp?lang=En&n=300688DC-1

Additional animations, videos, and other online resources are available at this book's companion website:
www.wiley.com/go/fletchercanada

This companion website also has additional information about WileyPLUS and other Wiley teaching and learning resources.

CHAPTER 21
DESERTS AND WIND

Chapter Contents and Learning Objectives (LO)

21-1 Deserts may be hot or cold, but low precipitation is a common trait.
LO 21-1 *Define the term "desert" and describe a desert environment.*

21-2 Atmospheric moisture circulation determines the location of most deserts.
LO 21-2 *Describe global atmospheric circulation.*

21-3 Not all deserts lie around 30° latitude.
LO 21-3 *Describe the factors that contribute to desert formation.*

21-4 Each desert has unique characteristics.
LO 21-4 *Identify the principal desert types.*

21-5 Wind is an important geological agent.
LO 21-5 *Describe sedimentary processes in deserts.*

21-6 Sand dunes reflect sediment availability and dominant wind direction.
LO 21-6 *Define the primary types of dunes.*

21-7 Arid landforms are shaped by water.
LO 21-7 *Discuss the role of water in desert landscapes.*

21-8 Desertification threatens all six inhabited continents.
LO 21-8 *Describe the conditions leading to desertification.*

GEOLOGY IN OUR LIVES

A desert is a landscape that receives little rainfall, has sparse vegetation, and is unable to support significant populations of animals. Deserts are more widespread than you might think. In fact, they take up one-third of Earth's land surface, and their range is growing. As human populations swell and spread into marginally habitable areas bordering deserts, poor land management (such as overgrazing) and changes in the water cycle due to global warming can permanently damage the native vegetation and soil. These factors lead to desertification. With human populations expanding worldwide, and global warming changing rainfall distribution, desertification is likely to be a problem that continues to grow in the future.

 What factors contribute to the formation of deserts and features like these sand dunes in the Arabian Desert?

21-1 Deserts May Be Hot or Cold, but Low Precipitation Is a Common Trait

LO 21-1 *Define the term "desert" and describe a desert environment.*

A desert is an arid region (**FIGURE 21.1**) that receives less than 25 cm of precipitation per year or an area where the rate of evapotranspiration exceeds the rate of precipitation. Additionally, vegetation may be sparse and thus may not be able to support significant populations of animals. Slightly wetter regions, receiving on average 25 cm to 51 cm of precipitation per year, are considered *semi-arid*.

The 10 largest deserts (**TABLE 21.1**, **FIGURE 21.2**) are spread across the planet. Surprisingly, the two at the top of the list, Antarctica and the Arctic, are covered with water (in the form of ice and snow) but receive little annual precipitation. Of course, the most common image that comes to mind is a place like the Sahara in northern Africa (#3 in Table 21.1). (Read about the Sahara's past in the "Geology in Action" box titled "The Great Sahara.")

Deserts have large daily and seasonal temperature ranges. High daytime temperatures (as high as 45°C) followed by low nighttime temperatures are the result of the low humidity. Water traps infrared radiation coming from both the Sun and the ground. The absence of water means that dry desert air cannot block much sunlight during the day or trap much heat radiating from the ground at night. As a result, deserts heat up strongly during the day as most of the sunlight reaches the ground, but as soon as the Sun sets, heat rapidly escapes from the surface and deserts cool quickly.

Why do we study deserts? Deserts are a major global environmental feature, and they cover approximately one-third of Earth's

TABLE 21.1 Ten Largest Deserts

Rank	Desert	Area (km²)
1	Antarctic Desert (Antarctica)	13,829,430
2	Arctic Desert (regions of permafrost above the Arctic Circle)	13,700,000+
3	Great Sahara (Africa)	9,100,000+
4	Arabian Desert (Middle East)	2,330,000
5	Gobi Desert (Asia)	1,300,000
6	Kalahari Desert (Africa)	900,000
7	Patagonian Desert (South America)	670,000
8	Great Victorian Desert (Australia)	647,000
9	Syrian Desert (Middle East)	520,000
10	Great Basin Desert (North America)	492,000

land surface. They have significant impacts on human populations living both near and far, and shifts in desert boundaries are a cause of great concern to neighbouring populations. There are also many sedimentary rocks that may have formed in desert environments, and so understanding how present-day deserts form and the processes operating in them help us to interpret rocks in the geological record.

The expansion of deserts, a phenomenon known as **desertification**, changes the distribution of available natural resources (especially water and food), which can lead to political unrest as one group tries to maintain its resource level by taking resources from another or by migrating to new resource-rich lands. But changing resources can bring groups of people together as well. Managing water resources is one of the few areas in which unfriendly neighbours in the Middle East have been forced to work together. In other cases, however, the demand for scarce resources in desert regions can cause environmental damage. For instance, because so many people need its water for drinking and irrigation, the Colorado River in the United States has nearly dried up and no longer reaches the Gulf of California. The collapse of agriculture due to a lack of water in many areas of the world has led to famine, political unrest, and civil strife. This is particularly true of nations bordering the Sahara. Niger, Chad, Sudan, and Ethiopia among others seem to be particularly susceptible to the changing environment.

© Tim Whitby/Alamy

FIGURE 21.1 Deserts are characterized by a lack of precipitation and sparse vegetation. The Atacama Desert in Chile is the driest place on Earth.

 State a hypothesis explaining the location of the largest deserts on Earth.

 Expand Your Thinking—Describe how desertification would impact or change specific aspects of the community where you live.

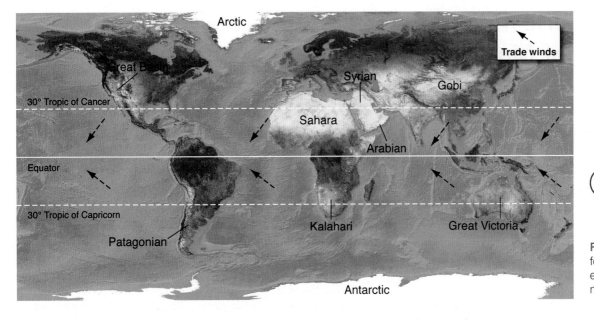

What are the physical characteristics of a desert?

FIGURE 21.2 Deserts are found on every continent. Some examples are shown on this map.

GEOLOGY IN ACTION

THE GREAT SAHARA

Amazingly, the Sahara was a lush and humid place between 5,000 and 10,000 years ago. The region had plenty of rainfall and was teaming with large animals, such as giraffes, gazelles, lions, elephants, and even hippopotamuses and crocodiles. Populations of early farmers and herders lived throughout the area (**FIGURE 21.3**).

The Sahara lies under the influence of the yearly *monsoons* (seasons of high rainfall in the tropics). The monsoon in Africa (and elsewhere) is due to heating during the summer. Air over land becomes warmer and rises, pulling in cool wet air from the ocean, which brings rain. What was different thousands of years ago? Recall the orbital parameters from Chapter 17 ("Glaciers and Paleoclimatology"): Subtle variations in Earth's orbit altered the path of the monsoon and led to changes in rainfall. Approximately 10,000 years ago, a change in *insolation* (sunlight received on Earth's surface) caused the African monsoon to shift roughly 600 km north. This brought summer rains

to what is now the Sahara, creating an environment similar to the grasslands of lush modern east Africa and filling large rivers and lakes. About 5,000 years ago, the monsoons shifted back south, and by 4,000 years ago, the Sahara was as dry as it is today.

We know about the large animals in the Sahara because their fossilized bones have been found at various sites across the desert, often associated with sediment deposited in former lakes and flood plains. This discovery led scientists to conclude that the Sahara must have been wetter in order to support such populations of large animals. However, we did not truly appreciate the amount of water involved until the space shuttle was equipped with downward-looking radar able to penetrate the dry desert sands and image ancient riverbeds. These ancient river channels are not visible on the ground, but their networks (**FIGURE 21.4**) were readily apparent in the images from the high-flying space shuttle.

FIGURE 21.3 Five-thousand-year-old rock art preserves images depicting ancient cultures in the now-uninhabitable central Sahara.

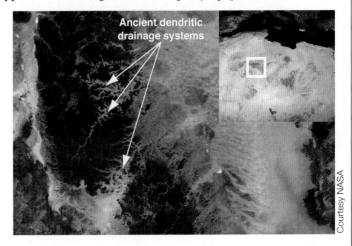

FIGURE 21.4 Ancient dendritic drainage patterns in the Acacus-Amsak region of the central Sahara are evidence of a wetter past.

21-2 Atmospheric Moisture Circulation Determines the Location of Most Deserts

LO 21-2 *Describe global atmospheric circulation.*

Global atmospheric circulation is a primary factor determining variations in rainfall and, hence, the location of deserts. We discussed the atmospheric circulation system in Chapter 16 ("Global Warming," Section 16-2), but it is worth reviewing this system again as it helps to explain the distribution of deserts on Earth. The basic components of the system are the *Hadley cell*, the *Ferrel cell*, and the *Polar cell* (**FIGURE 21.5**). There is one of each cell type in the Northern Hemisphere and one of each type in the Southern Hemisphere.

Atmospheric Circulation

Atmospheric circulation starts with the basic principle that air heated by the Sun rises at the equator (where solar heating is greatest). As the air moves toward the poles, it cools and eventually sinks. Rising air causes low air pressure (at the equator) and sinking air causes high air pressure (at the poles). If Earth were perfectly still and smooth, we might have a single cell in each hemisphere where hot air rises at the equator, moves north or south toward the poles, and then sinks to ground level as it cools at the poles. This air would then flow back to the equator along the ground surface. We would see this pattern expressed in the Northern Hemisphere as a constant north wind and in the Southern Hemisphere as a constant south wind. Fortunately, however, Earth is neither still nor smooth. Earth spins on its axis, causing the changes of day and night, and large mountain ranges deflect the direction of surface winds. Life on Earth is much more interesting this way.

By the time an air mass that has risen at the equator has travelled to about 30° latitude, it has cooled sufficiently to sink back to Earth's surface (forming an area of high pressure). When this air reaches the surface, it must flow away, and it moves back either toward the equator or toward the pole. The air that flows back to the equator is reheated and rises again to repeat the process. This completes the Hadley cell.

At the poles, cold, dense air descends causing a high pressure area. Air flows away from the high pressure and toward the equator. By the time this air nears 60° latitude, it begins to meet the air flowing poleward from the Hadley cell. When these two air masses meet, they have nowhere to go but up. As they rise, they cool and lose moisture, causing high precipitation. Once high in the atmosphere, they must head poleward, where they will cool and sink again, or toward the

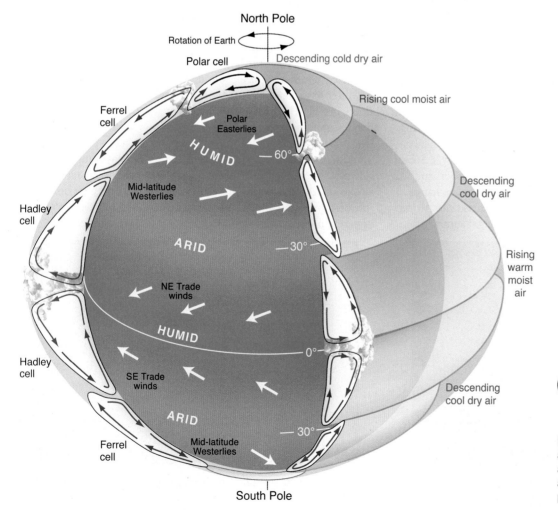

? How are the world's deserts related to Hadley cells?

FIGURE 21.5 The Hadley, Ferrel, and Polar cells are the three major atmospheric convection cells. Deserts are concentrated where these cells produce sinking, dry air masses.

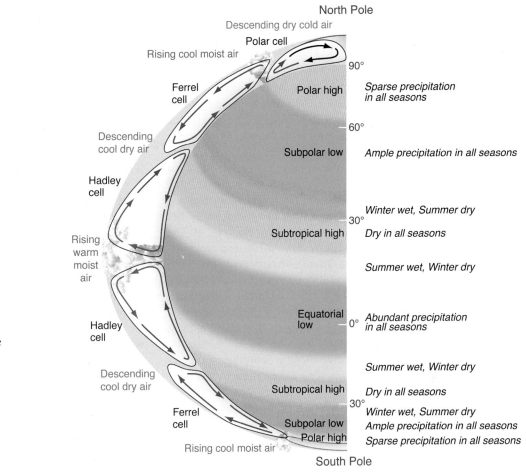

How do surface climates compare or contrast with the atmospheric circulation system?

FIGURE 21.6 Many of Earth's deserts are clustered around 30° latitude due to the sinking air produced where the Hadley and Ferrel cells meet

equator, where they will meet the flow heading poleward from the equator and sink. The circulatory cell sinking at the poles and rising at 60° latitude is called the Polar cell, and the final cell, sinking at 30° latitude and rising at 60° latitude, is named the Ferrel cell.

In 1856, William Ferrel demonstrated that, due to the rotation of Earth, air and water currents moving distances of tens to hundreds of kilometres tend to be deflected to the right in the Northern Hemisphere and to the left in the Southern Hemisphere. This phenomenon is known as the **Coriolis effect.** Because surface winds in a Hadley cell are moving south (in the Northern Hemisphere), when they are deflected to the right, they turn westward and are called the *northeast trade winds.* In the Southern Hemisphere, they turn left to become the *southeast trade winds.* The surface winds in the Northern Hemisphere's Ferrel cell are moving north, and when deflected right, they become the mid-latitude *westerlies.* The surface winds in the northern Polar cell are heading south, and when deflected right, they become the *polar easterlies.* Check a globe to convince yourself of these patterns and determine what part of the global atmospheric circulation system you live in.

Forming Deserts

But what do these patterns have to do with deserts? As air rises, it cools and expands. This is due to the increased distance from the warming effects of Earth's surface and the lower air pressure found at higher altitude. As a rising air mass cools and expands, so does the water vapour contained in it. As the water vapour cools and expands, more water condenses than evaporates, causing water droplets and then clouds to form. Continued condensation produces precipitation, which falls as rain or snow. Therefore, in areas where relatively warm moist air is rising, such as near the equator and around 60° latitude, there will be a lot of precipitation.

The opposite is also true; air warms and contracts as it sinks closer to Earth's surface. This causes evaporation to exceed condensation. No clouds form in locations with lots of sinking air. These areas, such as at the poles and around 30° latitude, have few clouds and little precipitation, thus forming a great belt of arid climate (and deserts) that girdles the globe. Many of the world's deserts are clustered around 30°N and 30°S latitudes for this reason (**FIGURE 21.6**).

 Expand Your Thinking—What are the parts or main features of the Northern Hemisphere atmospheric circulation system?

21-3 Not All Deserts Lie around 30° Latitude

LO 21-3 *Describe the factors that contribute to desert formation.*

Several factors in addition to atmospheric circulation contribute to the location of deserts. The first is called the **orographic effect,** which causes a *rain shadow.* The orographic effect occurs when a mountain range forces the prevailing winds to rise up over them. The rising air cools and expands, and the rate of condensation exceeds the rate of evaporation, causing clouds to form and rain to fall (**FIGURE 21.7**). As a result, the windward side of these mountains can be extremely wet. However, on the leeward side of the mountains, the sinking air warms and compresses—behaving just as it did when sinking as part of the Hadley cell. The resulting increase in evaporation limits cloud formation and rainfall, causing arid and semi-arid conditions. The dry area downwind from the mountains can be very well defined and is called a rain shadow due to the lack of rain. In the western United States, much of the desert in Nevada lies in the rain shadow of the Sierra Nevada Mountains of eastern California. The southern Okanagan Valley in British Columbia is a smaller example of a rain shadow formed by the same process. The mountains trap moisture coming off the Pacific Ocean in winds that blow from west to east, namely the mid-latitude westerlies.

The second factor contributing to desert formation has to do with the *distance moisture is transported in the atmosphere.* Since moisture gets into the atmosphere principally by evaporation from the ocean, distance from the ocean is another cause of low rainfall. The vast mountainous central Asian deserts of Kazakhstan, Afghanistan, Mongolia, and the Tarim Basin and Gobi Desert of northern China are far downwind from any oceans (**FIGURE 21.8**). Their extreme interior location ensures that by the time air masses reach them, the air has lost most of its moisture.

Polar deserts form as a result of a third important factor: the very cold *air descending over polar regions contains little water vapour.* This dry air, and the little water vapour it contains, warms and expands as it descends, further inhibiting cloud formation.

A fourth factor contributing to desert formation is a *cold ocean current next to a tropical coast.* Cold ocean currents flow toward the equator along the west coast of most continents. Cold air above these ocean regions moves onshore over hot land and quickly and dramatically heats up and expands. This process causes high rates of evaporation that produce few clouds and little rain. The deserts along the west coasts of South America (Peru and Chile) and Africa (Namibia and Angola) are particularly dry due in part to the presence of a cold current just offshore.

Another important factor in desert formation is *poor human management* of farmland. Read more in the "Earth Citizenship" box titled "The Dust Bowl."

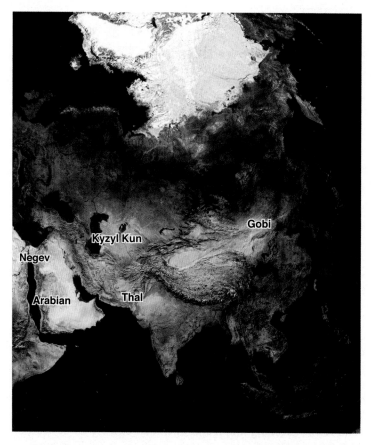

FIGURE 21.8 The great deserts of central Asia are far from any moisture source.

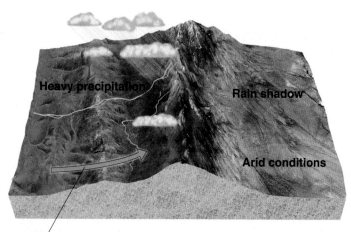

Orographic
uplift

FIGURE 21.7 Moist air is forced upward by orographic uplift due to the presence of mountain ranges. A rain shadow is produced on the downwind side of the mountains.

 Why does a rain shadow develop on the downwind side of large topographic features?

 What factors contribute to the great deserts of central Asia?

Expand Your Thinking—State a hypothesis explaining (accounting for) the climate in your area. How would you test your hypothesis?

THE DUST BOWL

Known to early European explorers as the "Great Desert," the plains of central North America frequently experience prolonged drought. Often a decade or more in length, these droughts greatly reduce the native vegetation, allowing the region's strong winds to produce large dust storms. During the 19th and early 20th centuries, repeated periods of drought resulted in the plains being swept by immense dust clouds. It turned out that these were only a prelude for what was to come.

By the late 1920s, growth in agricultural technology was reshaping the plains to become the bread basket of the world. New tractors pulling disk plows increased the amount of land cultivated for wheat. The increased wheat production depressed world wheat prices; farmers then grew as much wheat as they could in order to avoid losing money. In good years, rain was sufficient to grow wheat on the Great Plains—barely. In 1931, disaster struck in the form of no rain. When spring returned in 1932, the combination of thoroughly tilled, bone-dry soil and strong winds caused the severe erosion of farmland. So much soil was picked up and blown around that the term "black blizzard" was used to describe the dust storms. Choking dust got into everything. Wet sheets were hung over windows in a largely futile effort to keep dust out, children wore dust masks to and from school, and crops blew away with soil that was too dry to be anchored (**FIGURE 21.9**). The "dust bowl" of the 1930s, which affected the Canadian prairies in Saskatchewan and Alberta, and the prairies in the northern United States, had begun.

The crisis in the farming economy caused by the dust bowl, in combination with the economic Great Depression, was very severe. Wheat prices fell to their lowest in history at that time, with the income from farming in Canada catastrophically dropping from $363 million in 1928 to $11 million in 1933. About two-thirds of prairie farmers had to receive monthly financial assistance, and many farmers lost their farms completely. It is thought that at least a quarter of a million people left the prairies during this time. During that time, the central and maritime provinces of Canada sent hundreds of railway carloads of food and clothing to assist the destitute families in Saskatchewan and Alberta.

Toward the end of the dust bowl years, the Canadian government formed the Prairie Farm Rehabilitation Administration, which was designed to help prairie farmers mitigate the effects of drought by providing expertise on farming practices and weather issues. By the time the series of droughts ended in the fall of 1939, the prairie landscape had changed dramatically. Experts estimated that 850 million tonnes of topsoil were blown off the surface of the Great Plains in 1935 alone. Over 1 million square kilometres of agricultural land were significantly affected by topsoil loss. The damage caused by dust storms peaked on April 14, 1935, with Black Sunday when the worst "black blizzard" occurred.

Farming techniques were changed to preserve valuable and irreplaceable topsoil, because the dust bowl was an extreme example of what has happened in many areas when agricultural practices that ignore the potential for soil erosion clash with the environment. This is a North American example of desertification due to a combination of drought and agricultural practices. Droughts are still common in the Canadian prairies, and even though the effects are not as severe, there are still economic ramifications. This is because the prairies are one of the world's major agricultural areas, producing significant amounts of wheat, canola, oats, and barley. For example, the drought of 2001 led to a loss of about $4 billion in grain revenues, which required significant government assistance.

FIGURE 21.9 The dust bowl was an agricultural, ecological, and economic disaster in the Great Plains region of North America.

Science Source

21-4 Each Desert Has Unique Characteristics

LO 21-4 *Identify the principal desert types.*

There are several types of deserts, which can be classified by their geographical location and the dominant weather pattern governing the availability of water. In previous sections, we learned about the factors that lead to desert formation, including global atmospheric circulation that produces areas of dry descending air, regions that fall in the shadow of an orographic barrier, dry polar regions (also related to high atmospheric pressure), remote continental areas located far from an ocean, and tropical coastal areas adjacent to cold ocean currents. The desert types include trade wind deserts, mid-latitude deserts, rain shadow deserts, coastal deserts, monsoon deserts, and polar deserts.

Desert Types

Trade wind deserts are found in the Northern and Southern Hemispheres in the belt of the trade winds. Trade winds originate at the subtropical high-pressure zones near 30°N and 30°S latitude. They flow toward the equator out of the northeast in the Northern Hemisphere or the southeast in the Southern Hemisphere. Trade winds are the product of dry descending air masses at the boundaries of the Hadley and Ferrel circulation cells. As the air masses descend, they heat up and their rate of evaporation increases. Hence they hold more moisture, and areas below them lose moisture rather than gain it. Trade winds are the extension of this dry air. As they sweep toward the equator, they dissipate much of the cloud cover they encounter, evaporating moisture and allowing more sunlight to heat the land. The Kalahari in southern Africa is an example of a trade wind desert (**FIGURE 21.10**).

Mid-latitude deserts occur between 30° and 50° north and south, poleward of the subtropical high-pressure zones. These deserts lie in remote continental regions far from oceans and have a wide range of annual temperatures; they often experience snowfall in the winter and scorching temperatures in the summer. The Gobi Desert of southern Mongolia is a typical mid-latitude desert (**FIGURE 21.11**). These deserts experience a lack of precipitation because the descending air masses at the boundary of the Hadley and Ferrel cells are dry. Winds at mid-latitude deserts tend to be westerlies, originating at the boundary of the two cells. At mid-latitude deserts, the air is flowing poleward from the high-pressure zone rather than toward the equator, as is the case for trade wind deserts.

Rain shadow deserts form where topographic barriers drive winds up to cool high altitudes, where the rate of moisture condensation exceeds the rate of evaporation within the air mass. Since most available moisture goes into cloud formation, the air that descends on the lee side is dry, forming a rain shadow desert. The Great Basin Desert of North America consists of a series of deserts, which formed primarily in the rain shadow of the Sierra Nevada Mountains (**FIGURE 21.12**). The area around Osoyoos in the southern Okanagan Valley of British Columbia is located in a rain shadow. Part of it is called the Nk'mip Desert, and it is considered to be the northernmost extent of the Great Basin Desert. However, it has formed in an area that is semi-arid and not a true desert.

The fourth type of desert, called a *coastal desert,* is found on the western edges of continents in both the Northern and Southern Hemispheres where cold ocean currents come close to shore (**FIGURE 21.13**). The cold air above these currents is drawn over the land by heating, and as the air mass rises, it produces strong evaporation, preventing accumulation of surface water. The high rates of evaporation produce few clouds and little rain. The Atacama Desert in South America, which is the driest place on Earth, is an example of this type of desert.

FIGURE 21.10 The Kalahari is a trade wind desert that lies between 30°S latitude and the equator in southern Africa.

 Explain why trade winds are dry.

FIGURE 21.11 Lying north of the subtropical high-pressure zones, mid-latitude deserts have a wide range of annual temperatures. The Gobi Desert is the fifth-largest desert in the world.

? What is the primary cause of the mid-latitude deserts? Classify each of the deserts in Table 21.1 in terms of the principal desert types.

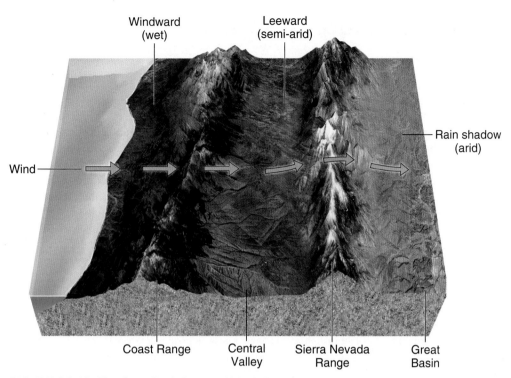

FIGURE 21.12 The Great Basin Desert of North America lies in the rain shadow of the Sierra Nevada Mountains in California.

What causes the predominately western airflow over the Sierra Nevada Mountains?

The Indian *monsoon* is a seasonal wind that flows off the Indian Ocean and Arabian Sea and prevails for several months in the summer. It brings intense rainfall produced by condensation as the warm, moisture-laden air lifts to high altitudes. *Monsoon deserts* are found where the rain is blocked by mountain ranges that produce a rain shadow. India, a nation whose climate is controlled by the annual monsoon, is home to the Rajasthan Desert, and neighbouring Pakistan is home to the Thar Desert. Both of these are monsoon deserts located in the rain shadow of the Aravalli Range of western India.

Polar deserts form in the Arctic and Antarctic where the cold climate prevents air from holding even a small amount of moisture. Rain or snowfall freezes so quickly that the surface water, which would otherwise support plants and animals, is locked up in frozen landscapes of ice. Antarctica is the world's largest polar desert. Precipitation is in the form of snow, but in terms of the equivalent amount of water, the average annual precipitation is only about 50 mm, less than the Sahara receives. Along the coast, the amount of precipitation increases, but is still only about 200 mm. Unlike other deserts, there is little evaporation in Antarctica; hence, the relatively little snow that does fall does not go away again. Instead, it builds up over hundreds and thousands of years into enormously thick ice sheets.

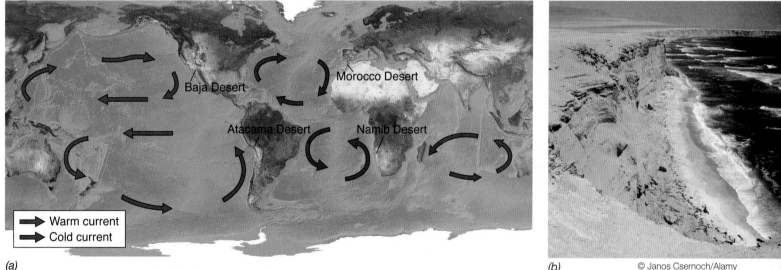

FIGURE 21.13 (a) Where cold ocean currents approach land, cold air heats up and causes intense evaporation. (b) The Atacama Desert, located in coastal Chile, is the driest place on Earth.

© Janos Csernoch/Alamy

Four coastal deserts are shown. What do you think is the likely influence of atmospheric circulation on each? (See Section 21-2.)

Expand Your Thinking—Climate change is warming the atmosphere and oceans. Consider how this will change each of the processes responsible for the principal desert types, and hypothesize how each desert in Table 21.1 is likely to change in the future. How would you monitor the expected changes?

21-5 Wind Is an Important Geological Agent

LO *21-5 Describe sedimentary processes in deserts.*

Due to a lack of water and vegetation, wind is a particularly important agent of erosion and deposition in deserts. Unlike water, wind is not confined to a channel, so it can affect broad tracts of land. Wind is much less dense than water, however, and so it typically moves only grains of sand and silt. Larger grains are too heavy unless the wind speed reaches extreme levels. Many geologic features in the desert are characteristic of wind erosion, deposition, or moving sediment.

There are two methods by which wind transports sediment. Which method prevails is determined by the speed of the wind and the size of the sediment. We have already encountered the first method in our study of water erosion in Chapters 7 ("Weathering") and 19 ("Surface Water"); it is called *saltation*. Saltation occurs close to the ground with grains that are sand size (0.625 mm to 2 mm in diameter). It refers to the process by which grains move short distances above the ground before falling back down. Just as with water erosion, when a grain lands, it may dislodge other grains and cause them to saltate downwind and in turn dislodge more grains when they land. This sequence of events can look as if the grains are leaping downwind and can be compared to the bouncing of a beach ball across the sand. If the grains are small (typically the size of silt), they may not fall back to the ground but may be carried in *suspension* by the wind. This process is enhanced if the wind is blowing at high speed and is turbulent. Suspended grains can reach great altitudes and travel thousands of kilometres. Dust suspended by the wind during storms in the Sahara and Gobi Deserts has crossed the Atlantic and Pacific Oceans.

Once grains begin to accumulate—often around an obstacle of some sort, such as vegetation—**ripples** (centimetre-scale features) and **dunes** (metre-scale features) can form (**FIGURE 21.14**). A simple ripple or dune has two angled slopes, one facing upwind (the *stoss* slope) and another facing downwind (the *lee* slope). Wind will continuously push grains up the stoss slope to the top of the ripple until the sand collapses under its own weight. This occurs by avalanching down the *slip face* on the lee side. The collapsing sand comes to rest when it reaches a slope of about 30° to 34° (the *angle of repose*). Every pile of loose particles has a unique angle of repose, depending on the properties of the particles, such as the grain size and roundness. *Ripples grow into dunes as the wind speed and sand availability increase.* The repeating cycle of sand inching up the windward side to the dune crest, then sliding down the slip face allows the dune (or ripple) to move forward, migrating in the direction the wind blows.

Deposition of windblown sediments leaves highly sorted deposits where the wind slows down. Sediments of a particular size are left in a particular location, and sediments of other sizes are deposited elsewhere. As wind velocity slows, large grains are deposited first, and small grains are transported farther. **Loess** (pronounced "luss") is an important example of a windblown deposit. Loess deposits are composed of windblown silt exposed by repeated advances and retreats of continental glaciers. Loess deposits create particularly rich soil; significant loess deposits are major agriculture areas in Asia (the Yellow River Valley), northern Europe (the plains of Germany, Poland, Ukraine, and western Russia), and North America (the American Midwest) (**FIGURE 21.15**).

If all the fine material on a desert surface is transported away by the wind, what is left is a *lag deposit* composed of coarse sediment (often gravel size) that the wind is unable to transport. When a lag deposit is extensive, it is called a **desert pavement** (**FIGURE 21.16**).

Sand grains are the primary cause of *abrasion* in the desert. Due to their size and weight, sand grains are rarely lifted more than 1 m off the ground. As a result, abrasion is confined close to the ground. Rock outcrops can be sculpted by sand abrasion producing what is known as a *yardang* (**FIGURE 21.17**). In places with a predominant wind direction, rocks are abraded on one side. Pebbles, cobbles, and boulders with such flat wind-abraded surfaces are called *ventifacts*. (Ventifacts can be seen in Figure 21.16.) Shifting winds or a rock that has been turned may cause a ventifact to have more than one flat surface.

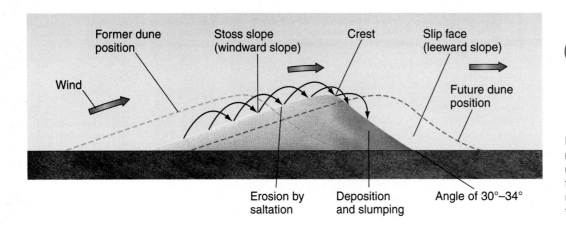

As a ripple or dune migrates with the wind, what would you predict the internal structure (or organization of the grains) to look like?

FIGURE 21.14 Ripples and dunes migrate downwind as sand travels up the windward slope, collects at the top of the pile until it becomes unstable, and then avalanches down the leeward slope.

Expand Your Thinking—Desert pavement is also referred to as "armouring." Why? What is the net effect of armouring on sediment transport?

(a)

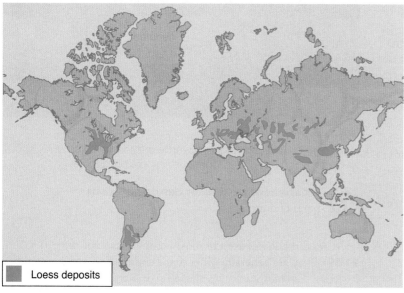

(b)

© Pixtal/Age Fotostock America, Inc.

(c)

FIGURE 21.15 Loess is silt deposited by the wind. *(a)* Wind deposits heavier grains, such as sand, first but may continue to transport silt for great distances, forming loess deposits. *(b)* Many loess deposits are found downwind of formerly glaciated areas. *(c)* Loess landscapes can be massive in scale, dominating the geology of the surface environments. The loess deposits in the Yellow River Valley in China, for example, can be up to 100 m thick. The human-made terraces in this photograph are designed to reduce the amount of erosion.

(?) Why does loess develop in formerly glaciated regions?

© J. Ritterbach/F1 Online/Corbis

FIGURE 21.16 Desert pavement forms when the wind erodes fine sediments and leaves behind a lag deposit of coarse grains.

(?) What conditions promote the development of desert pavement?

© Mike P. Shepherd/Alamy

FIGURE 21.17 A yardang is a wind-abraded ridge found in a desert environment.

(?) How does a wind-abraded ridge form?

21-6 Sand Dunes Reflect Sediment Availability and Dominant Wind Direction

LO *21-6 Define the primary types of dunes.*

Sand dunes immediately come to mind when you think of deserts. Although deserts have many other characteristics, dunes are some of the most spectacular and dynamic landscapes found in arid regions.

Sand Dunes

Four conditions must be met for dunes to form. (1) It is important to have abundant *loose sediment* (usually sand size). Environments with abundant loose sediment include beaches, river channels, and deserts. (2) There must be *sufficient energy* to move the sediment. In the case of deserts, this would be wind energy, but dunes can form underwater in rivers from sand that is transported by running water. (3) There must be some *obstacle around which sand accumulates*. Areas with sparse vegetation commonly form dunes because individual bushes can trap moving sand and start the process of building a dune. Dunes also form around outcrops of rocks. (4) Dunes in the desert require a *dry climate*. Moisture causes sand grains to stick together, and larger grains, or clumps of grains, are more difficult to transport by wind. Moisture also facilitates plant growth, which stabilizes loose sediment and prevents it from moving (**FIGURE 21.18**).

Types of Dunes

Dunes are classified on the basis of shape. There are four principal types (**TABLE 21.2**); however, not all dunes can be classified because many have irregular shapes. The first three dune types are asymmetrical with a gently sloping windward side and a steeply sloping leeward face. The fourth type, the *star dune,* is complex, with arms and ridges formed by wind blowing from several different directions.

FIGURE 21.18 The formation of dunes in a desert requires loose sediment (usually sand), energy to move the sediment (usually wind), an obstacle to trap sand (often a bush), and a dry climate.

 What criteria are used for classifying various types of dunes?

Crescentic dunes are the most common type of dune (**FIGURE 21.19**). They are generally wider than they are long and come in two varieties. *Barchan dunes* are shaped like a crescent moon with the horns pointing downwind. These dunes form where sediment is limited and may be separated from one another by significant distances of bare rock. *Transverse dunes* are long, wavy linear dunes oriented perpendicular to the wind direction. Like barchan dunes, transverse dunes have their steep slip face on the concave side of the dune.

TABLE 21.2 Sand Dunes

Dune Type	Shape	Wind Characteristics	Environment of Deposition	Grain Characteristics
1. Crescentic a. *Barchan*	Crescent moon; horns point downwind	Constant moderate wind from one direction	Limited vegetation and sediment, often on flat bare rock	Well sorted, very fine to medium sand
b. *Transverse*	Long, wavy, linear; oriented perpendicular to wind direction	Constant, moderate wind from one direction	Limited vegetation; more sediment available than barchans	Well sorted, very fine to medium sand
2. Parabolic	Crescent moon; horns point upwind	Variable strength, unidirectional wind	Abundant supply of sand; vegetation common	Well sorted, very fine to medium sand
3. Longitudinal	Long, straight or sinuous sand ridge; generally much longer than wide	Strong, steady winds that blow from two directions	Form in parallel sets of sand ridges	Well sorted, very fine to medium sand
4. Star	Pyramid; three or more arms radiate from a peaked centre	Wind blows from several different directions	Grow taller rather than migrating	Well sorted, very fine to medium sand

Tove, Jan/Johner Images//Getty Images, Inc.

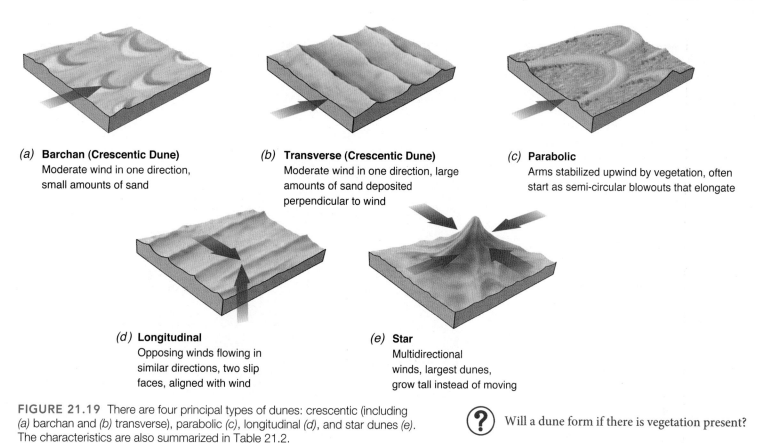

(a) **Barchan (Crescentic Dune)**
Moderate wind in one direction, small amounts of sand

(b) **Transverse (Crescentic Dune)**
Moderate wind in one direction, large amounts of sand deposited perpendicular to wind

(c) **Parabolic**
Arms stabilized upwind by vegetation, often start as semi-circular blowouts that elongate

(d) **Longitudinal**
Opposing winds flowing in similar directions, two slip faces, aligned with wind

(e) **Star**
Multidirectional winds, largest dunes, grow tall instead of moving

FIGURE 21.19 There are four principal types of dunes: crescentic (including (a) barchan and (b) transverse), parabolic (c), longitudinal (d), and star dunes (e). The characteristics are also summarized in Table 21.2.

(?) Will a dune form if there is vegetation present?

Parabolic dunes are also shaped like a crescent moon except that the horns point upwind. These dunes require an abundant supply of sediment and sufficient vegetation. They can be initiated when the wind preferentially erodes the sand creating a depression, called the blowout zone, while the horns of the dune are anchored by vegetation. The *Longitudinal dunes* are long, straight, or slightly sinuous sand ridges generally much longer than they are wide. They usually form in areas with two directions of wind occurring over the year, often associated with separate seasons. Their long axis is parallel to the direction of net sand movement. *Star dunes* are pyramidal sand mounds with steep slip faces on three or more arms that radiate from the elevated centre of the mound. They form in regions where the wind blows from several different directions.

Sand seas are large regions (larger than 125 km²) of windblown sand that contain numerous, very large dunes, where sand covers more than 20 percent of the ground surface (**FIGURE 21.20**). Also known by the Arabic word *erg*, sand seas differ from *dune fields* mainly in size and complexity of dune forms. Sand seas form downwind from sources of large volumes of loose, dry sand. Dry and abandoned riverbeds, flood plains, glacial outwash plains, dry lakes, and beaches are all excellent sources of sand. Dune fields are considered local features and have smaller and simpler dunes than sand seas.

See the "Critical Thinking" box "Interpreting Dunes" to analyze more desert features.

Courtesy of NASA/GSFC/MITI/ERSDAC/JAROS, and U.S./Japan ASTER Science Team

FIGURE 21.20 A sand sea, or erg, is probably what most people imagine when picturing a desert. Although they are not particularly common, they are the archetypical and most dramatic desert landscape. This is the Namib Erg in southern Africa with dunes tens of metres high along the Atlantic Ocean coastline.

(?) What conditions lead to the formation of an erg?

 Expand Your Thinking—What would happen to a parabolic dune if the wind direction changed and precipitation decreased?

? CRITICAL THINKING

INTERPRETING DUNES

Please refer to **FIGURE 21.21**. This is a landscape of dune types. Your job is to analyze the morphology of each of these dunes and interpret the environmental conditions. Please work with a partner and answer the following questions.

1. Name each dune type.
2. Describe the environmental conditions leading to each dune type.
   a. Draw arrows indicating the dominant wind(s) for each dune type.
   b. Describe the apparent sediment abundance related to each dune type.
   c. Indicate sediment sources, the role of vegetation, and transitional dune forms. Describe each.
3. Considering all the above factors, and based on the several deserts we have studied:
   a. State in which desert each of these dune types may be located.
   b. Describe the desert-forming processes active in each of the deserts you have identified.
   c. List the features in this scene that are consistent with the location you have chosen.
   d. List the features in this scene that are not consistent with the location you have chosen.
4. Could this assemblage of dune morphologies in Figure 21.21 be found in this configuration? What criteria will you consider in arriving at your decision?

FIGURE 21.21

21-7 Arid Landforms Are Shaped by Water

LO *21-7 Discuss the role of water in desert landscapes.*

Although desert landscapes may look very different, they are formed by the same geologic forces that shape humid landscapes. The aridity of the desert environment emphasizes different aspects of the same processes and highlights the degree to which the different climates affect those processes.

Chemical weathering is an important agent of change in humid climates. Because of the lack of moisture and organic acids from decaying plants, chemical weathering is not as significant in arid climates. However, it is not entirely absent; over time, clay minerals and soils do form, and iron-rich minerals oxidize to create the red colours typical of some deserts.

Due to low rainfall, permanent streams are very rare in desert environments. **Ephemeral streams,** where flowing water is present only after rainfall events, are the general rule (**FIGURE 21.22**). An ephemeral stream may have water in it for only a few hours or days each year, typically immediately after a rainstorm. Normally desert rains are hard and brief so the water does not soak into the hard-baked soil. Much of the water will form surface flow and, because vegetation is thin, runoff into dry streambeds is rapid. These dry streambeds (called *wash* [English], *arroyo* [Spanish], *wadi* [Arabic], *oued* [French], *vadi* [Hebrew], and *nullah* [Hindi]) quickly fill with fast-moving water causing *flash floods* as a result of the heavy downpour. Unlike floods on perennial streams, which may take days to reach their crests and then subside, flash floods may rise and fall within a few hours. Additionally, the lack of vegetation allows the fast-moving water to erode unanchored soil and rocks at a much greater rate than in a similar-size flood in a humid climate.

Another consequence of low rainfall is that rivers in deserts tend to be small and disappear before they reach the ocean. Low rainfall, coupled with high evaporation and infiltration into the streambed, means that more water is leaving the stream than entering it; such a stream, unless quite large to begin with, will quickly disappear. The Nile and Colorado Rivers are notable examples of large desert rivers that originate outside of a desert environment. In these cases, the rivers start in mountains with snow or rainfall supplying enough water to the river to overcome losses occurring as they traverse the desert regions. However, although the Nile River does reach the sea, the Colorado River no longer does due to water removal by humans for irrigation and drinking.

Despite low rainfall, most erosion in deserts is the result of flowing water. Wind erosion is more significant in deserts than in other environments, but flowing water is still the main erosive agent in deserts.

Landform Evolution

The *Basin and Range* region of the southwestern United States (recall Chapters 11, "Mountain Building,") includes many particularly well-developed examples of desert landscapes and their evolution. This region is characterized by a large number of small north-south trending ranges separated by long, narrow basins. Much of this area is downwind of the Sierra Nevada Mountains and thus lies within its rain shadow. Additionally, the north-south orientation of the ranges produces multiple rain shadow effects against the dominant winds that blow west to east (westerlies). As a result, much of this area is extremely dry and provides an excellent case study of desert landscapes.

Ridge crests in the Basin and Range have sufficiently high elevations to receive snow and rain. As a result, they are chemically and physically weathered, and the resulting sediment is eroded during flash floods and deposited in the basins. The basins have flat floors, and the sediment is quickly deposited at the mouth of discharging streams. This sediment builds up and forms an *alluvial fan*. Coarse sediment is deposited near the top of the fan where the stream emerges from the mountains, and fine sediment is deposited at the base of the fan or even out on the floor of the basin. Through time, alluvial fans grow and spread along the base of the range. When a series of alluvial fans from adjacent stream valleys grows large enough to join together, a *bajada* is formed (**FIGURE 21.23**). Bajadas are broad, gently sloping depositional surfaces lining the entire front of a mountain range where they are crossed by meandering dry streambeds.

After a rain event, the floor of a basin may hold a shallow lake called a **playa lake** (Spanish for "beach"; a lake controlled by rainfall and evaporation). Since these basins often have no drainage outlets, evaporation is the primary means by which water leaves playa lakes. When a playa lake evaporates, it leaves a thin layer of organic-rich mud on the floor of the basin. Remember, the much coarser gravel and sand have been deposited on the alluvial fans, or bajada, and so they do not reach the playa lakes. If the water flowing into the playa lake contains a significant amount of salt, a layer of white salt will be left on the floor of the basin when the water evaporates. This basin floor with its very flat surface composed of hard-baked mud and salt is called a *playa*. Death Valley

FIGURE 21.22 Because of low rainfall in the desert, streams tend to be ephemeral, flowing only after heavy rainstorms.

Dr. Marli Miller

(?) What are the characteristics of flow in an ephemeral stream?

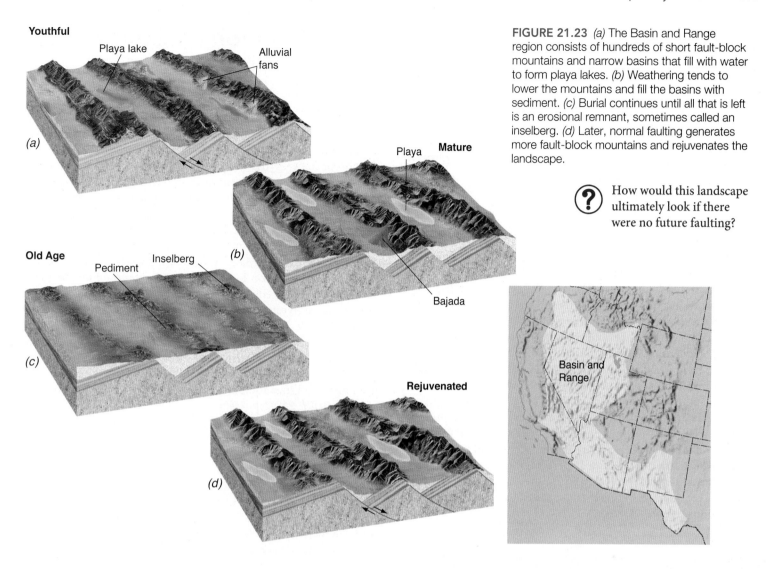

FIGURE 21.23 (a) The Basin and Range region consists of hundreds of short fault-block mountains and narrow basins that fill with water to form playa lakes. (b) Weathering tends to lower the mountains and fill the basins with sediment. (c) Burial continues until all that is left is an erosional remnant, sometimes called an inselberg. (d) Later, normal faulting generates more fault-block mountains and rejuvenates the landscape.

(?) How would this landscape ultimately look if there were no future faulting?

in California and the Bonneville Salt Flats of Utah are excellent examples of playas (**FIGURE 21.24**).

Over time, the ranges erode and the basins fill with their sediments. Eventually, all that may be visible above the flat plain of sediment is a small tip of older rock, sometimes called an inselberg (German for "island mountain"). Perhaps the best-known example is Uluru, or Ayers Rock, in Australia. Where the sediment over the eroded bedrock is thin or discontinuous, a pediment forms. A pediment may look similar to a bajada, but a pediment is a bedrock surface with a thin veneer of alluvial sediment while a bajada is a thick layer of loose sediment. The final step in this landscape evolution is when normal faulting once again results in the formation of fault-block mountains, and the process begins anew.

(?) **Expand Your Thinking**—Why would the layers of organic mud in playa lakes lead to the formation of oil shale? What are the geologic steps involved from silt deposition to making oil shale?

Stephen Marks/The Image Bank/Getty Images, Inc.

(?) Develop a hypothesis that predicts the stratigraphy of a playa. How can you test your hypothesis?

FIGURE 21.24 A playa is a flat, dry lake bed of hard, mud-cracked clay and salt.

21-8 Desertification Threatens All Six Inhabited Continents

LO 21-8 *Describe the conditions leading to desertification.*

Desertification is the process by which land loses its vegetation and turns into a desert. (Remember the second characteristic that defines a desert is a loss of vegetation.) Characteristics of desertification include destruction of native vegetation (**FIGURE 21.25**), unusually high rates of soil erosion, declines in surface water supplies, increasing water and topsoil saltiness, and widespread lowering of groundwater tables. Currently over 250 million people experience the direct consequences of desertification. Many of them are the world's most destitute and vulnerable citizens.

Although it can be a natural process progressing over thousands or millions of years, desertification today frequently has a combination of natural and human causes. Typically, plants native to a semi-arid environment are adapted to occasional drought conditions. But global warming can lead to changes in the distribution of water abundance and hamper the ability of vegetation to deal with already marginal conditions. Where human activities, such as farming and ranching, have weakened, damaged, or cleared much of the native vegetation in an area, a few drought years may kill off the remaining natural plants (most importantly, any that are edible to humans), and the area takes on many characteristics of a desert even if it does not meet the long-term definition of a desert (25 cm of rain per year). Although desertification is a more serious problem in poor, heavily populated countries, more than 35 percent of the semi-arid regions of North America have experienced "severe" desertification. All six inhabited continents have significant areas that are threatened by desertification (**FIGURE 21.26**).

© Tor Eigeland/Alamy

FIGURE 21.25 Desertification is the process by which land loses its vegetation and turns into a desert. The photograph is of the Sahara encroaching on a palm grove near a village in Mauritania, west Africa.

? What conditions lead to an increase in desertification?

The Sahel

Sahara is an Arabic word meaning "desert." **Sahel** means "border" or "margin" in Arabic. The Sahel region of North Africa is indeed the border between the shifting sands of the Sahara in the north and the moist tropical rainforest regions of central Africa. This swath stretches from the Atlantic to the Indian Ocean and includes parts or all of the countries of Senegal, Mauritania, Mali, Burkina Faso, Niger, Nigeria, Chad, Sudan, Ethiopia, Eritrea, Djibouti, and Somalia. More than 300 million people live in this region of dry and unstable weather. Many of them are poor farmers or nomadic herders. They rely on local rains or migrate to follow the rains. The population exceeds the carrying capacity of the land, and their overabundant livestock denude vegetation that is already stressed by low rainfall. Periodic multi-year droughts cause the remaining native vegetation to die and crops to fail. Millions of livestock animals die during these droughts. With the loss of livestock, crop failures, and low water supplies, millions of people also have died from thirst and starvation brought on by the desertification of their lands (**FIGURE 21.27**).

Many of the largest famines and humanitarian crises of the last 50 years occurred in the Sahel during droughts. The crises begun by desertification frequently were compounded by political strife started in no small part by people under stress from desertification. In an effort to combat desertification and thereby limit a root cause of the political unrest, the United Nations formed the *UN Convention to Combat Desertification*. Its purpose is to ensure the long-term productivity of inhabited dry areas. The convention recognizes that human activities contribute to the process of desertification and promotes behaviours that reduce those contributions. It also recognizes that desertification adds to the lethal nature of political unrest and the likelihood of civil war. It is difficult for starving people to get along with their neighbours, and desertification is one of the basic causes of their starvation.

The Sahel is not the only region threatened by desertification. Every continent on Earth has significant areas where desertification is an important hazard. The United Nations is implementing programs in individual countries based on the threats of desertification in specific ecosystems. Local communities are involved to help ensure that the solutions will work in each area. Each plan includes details for management of natural resources such as soil, water, forests, livestock, crops, wildlife and tourist resources, and mineral resources. Human socioeconomic concerns are also examined, such as farm policies, social and economic infrastructure, natural resource access rights, environmental economics, population growth, and local community participation in natural resource management. These programs show the complex web of processes that combine to create conditions for desertification. They also acknowledge that local communities are most familiar with their own regions and that even the best-laid plans are doomed to failure if local communities are not involved in plan development. The experiences in the Sahel mirror the strife that is emerging in other areas of the world where desertification is taking place.

Desertification Vulnerability

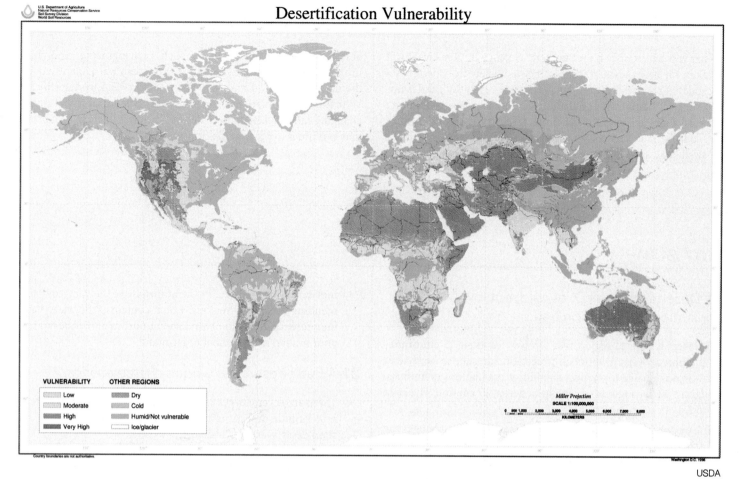

VULNERABILITY
- Low
- Moderate
- High
- Very High

OTHER REGIONS
- Dry
- Cold
- Humid/Not vulnerable
- Ice/glacier

Miller Projection
SCALE 1:100,000,000

KILOMETERS

Country boundaries are not authoritative.

USDA

FIGURE 21.26 All six inhabited continents have significant areas that are threatened by desertification.

 What is the relationship between regions threatened by desertification and the major deserts we studied earlier?

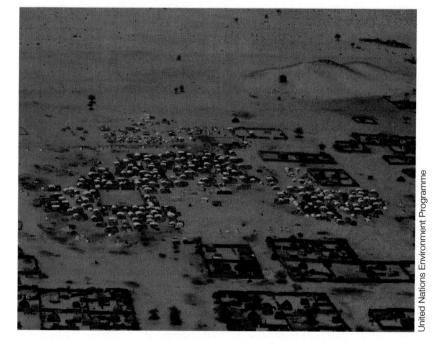

United Nations Environment Programme

 Identify some potential consequences of an influx of environmental refugees to a village already stressed by desertification.

FIGURE 21.27 The advance of desertification in the Sahel threatens food and water security, leading to a population of "environmental refugees." Displaced populations settle on the outskirts of existing towns, as here in El Fasher, northern Sudan. The new settlement is distinguished by white plastic sheeting. These new arrivals add to the environmental burden on the surrounding desert environment.

 Expand Your Thinking—How would desertification in countries that border each other potentially lead to political strife?

LET'S REVIEW "GEOLOGY IN OUR LIVES"

Now that you have finished the chapter, "Geology in Our Lives" will have taken on new meaning for you. Let us review it: A desert is a landscape that receives little rainfall, has sparse vegetation, and is unable to support significant populations of animals. Deserts are more widespread than you might think. In fact, they take up one-third of Earth's land surface, and their range is growing. As human populations swell and spread into marginally habitable areas bordering deserts, poor land management (such as overgrazing) and changes in the water cycle due to global warming can permanently damage the native vegetation and soil. These factors lead to desertification. With human populations expanding worldwide, and global warming changing rainfall distribution, desertification is likely to be a problem that continues to grow in the future.

STUDY GUIDE

21-1 Deserts may be hot or cold, but low precipitation is a common trait.

- A **desert** is an arid region that receives less than 25 cm of precipitation per year. Additionally, vegetation must be so sparse that the area is unable to support significant populations of animals. Slightly wetter regions, receiving between 25 cm and 51 cm of precipitation, are considered semi-arid.

- The expansion of deserts, known as **desertification**, changes the distribution of natural resources (especially water), which can lead to political unrest as one group of humans tries to maintain its resource level by taking resources from another or by migrating to new, resource-rich lands.

- Deserts have large daily and seasonal temperature ranges. High daytime temperatures (as high as 45°C) followed by low nighttime temperatures (down to 0°C) are the result of the low humidity.

21-2 Atmospheric moisture circulation determines the location of most deserts.

- Global atmospheric circulation creates the general characteristics of weather everywhere. The basic components of the system are the Hadley cell, the Ferrel cell, and the Polar cell. There is one of each cell type in the Northern Hemisphere and one of each cell type in the Southern Hemisphere.

- Air warms and contracts as it sinks closer to Earth's surface. This causes evaporation to exceed condensation. No clouds form in locations with a lot of sinking air. These areas, such as at the poles and 30° latitude, have few clouds and little precipitation, thus forming a great belt of deserts that girdles the globe. Many of the world's deserts are clustered around 30°N and 30°S latitudes for this reason.

21-3 Not all deserts lie around 30° latitude.

- In addition to the location of atmospheric circulation cells, several factors contribute to the location of deserts. These include the **orographic effect,** which causes a rain shadow; the distance moisture is transported in the atmosphere; the very cold air descending over polar regions, which contains little water vapour; the presence of a cold ocean current next to a tropical coast; and poor human management of farmland.

21-4 Each desert has unique characteristics.

- There are several types of deserts, including trade wind deserts, mid-latitude deserts, rain shadow deserts, coastal deserts, monsoon deserts, and polar deserts.

21-5 Wind is an important geological agent.

- Due to a lack of water and vegetation, wind is a particularly important agent of erosion and deposition in deserts.

- Saltation occurs close to the ground and is the process by which sand grains move short distances above the ground before falling back down. Just as happens during water erosion, when a grain lands, it may dislodge other grains and cause them to move downwind, where they in turn dislodge more grains where they land. If the grains are small (typically silt size), the grains may not fall back to the ground but may be carried in suspension by the wind.

- If all the fine material on the surface is transported away by the wind, all that is left is a lag deposit composed of coarse sediment that the wind is unable to transport. An extensive lag deposit is often called a **desert pavement.**

21-6 Sand dunes reflect sediment availability and dominant wind direction.

- In order for **sand dunes** to form in a desert, there must be: (1) abundant loose sediment; (2) sufficient energy to move the sediment; (3) some obstacle around which sand accumulates; and (4) a dry climate.

- There are four principal types of dunes: (1) crescentic, consisting of barchan and transverse dunes; (2) parabolic dunes; (3) longitudinal dunes; and (4) star dunes.

21-7 Arid landforms are shaped by water.

- Desert landscapes are formed by the same geologic forces that shape humid landscapes. The desert environment emphasizes different aspects of the same processes highlighting the degree to which physical, chemical, and biological processes are differently affected by contrasting climates.
- Due to low rainfall, permanent streams are very rare in desert environments. **Ephemeral streams,** where flowing water is present only after rainfall events, are the general rule in deserts. As a result of heavy downpours that occur intermittently in deserts, these streams quickly fill with fast-moving water that causes flash floods.
- After a rain event, the floor of a basin may hold a shallow lake called a **playa lake.** If these basins have no drainage outlet, evaporation is the primary means by which water leaves such lakes.
- Over time, the ranges erode and fill the basins with their sediments. Eventually, the ranges erode down and the basins fill up so that all that may be visible above the flat plain of sediment is a small tip of rock called an inselberg.

21-8 Desertification threatens all six inhabited continents.

- Desertification is the process by which land loses its vegetation and turns into a desert environment. Characteristics of desertification include destruction of native vegetation, unusually high rates of soil erosion, declines in surface water supplies, increasing water and topsoil saltiness, and widespread lowering of groundwater tables.
- More than 300 million people live in the **Sahel** region of dry and unstable weather in Africa. Many of them are poor farmers or nomadic herders. They rely on local rains or migrate to follow the rains. The population of people exceeds the carrying capacity of the land, and their overabundant livestock overgraze the vegetation already stressed by low rainfall.

KEY TERMS

Coriolis effect (p. 571)
desert (p. 568)
desertification (p. 568)
desert pavement (p. 576)

dunes (p. 576)
ephemeral streams (p. 582)
loess (p. 576)
orographic effect (p. 572)

playa lake (p. 582)
ripples (p. 576)
Sahel (p. 584)
sand dunes (p. 578)

ASSESSING YOUR KNOWLEDGE

Please answer these questions before coming to class. Identify the best answer.

1. A desert
   a. is hot.
   b. is remote.
   c. is covered by drifting sand.
   d. receives less than 25 cm of rain a year.
   e. is any place covered by sand dunes.

2. Global atmospheric circulation tends to produce arid areas at
   a. 30° latitude.
   b. 20° latitude.
   c. 40° latitude.
   d. the equator.
   e. None of these locations.

3. A Hadley cell
   a. is found only in the Northern Hemisphere.
   b. develops between the equator and about 30° latitude.
   c. develops between 30° and 70° latitude.
   d. is a cell of moist, falling air above the equator.
   e. is found above every desert.

4. The orographic effect is
   a. when winds turn to the right in the Southern Hemisphere.
   b. the evaporation that occurs above a cold coastal current near the tropics.
   c. when a mountain range forces moisture to rise and condense, creating a rain shadow.
   d. when a mountain range causes moisture to rise and evaporate, forming a polar desert.
   e. where the rate of evaporation exceeds the rate of condensation.

5. The principal types of deserts include
   a. continental deserts and sand deserts.
   b. Coriolis deserts and oceanic deserts.

    c. plateau deserts and valley deserts.

    d. high-elevation deserts and plateau deserts.

    e. trade wind deserts and monsoon deserts.

6. A barchan dune is identifiable because
   a. it has a crescent shape and the horns point downwind.
   b. it has a crescent shape and the horns point upwind.
   c. it is a long, linear ridge formed by winds blowing from two different directions.
   d. it is made of loess.
   e. it is the only unvegetated dune form.

7. An erg is
   a. a type of dune.
   b. a region of north Africa near the Sahel.
   c. the name of a dry wind.
   d. a wind-abraded landform.
   e. the Arabic name for a sand sea.

8. Sediment transport by wind
   a. has similar characteristics to transport in running water.
   b. involves particle saltation and suspension.
   c. typically entails turbulent air that moves larger particles.
   d. produces ripples and dunes.
   e. All of the above.

9. Why are there no important loess deposits in Africa?
   a. There were no continental glaciers in Africa.
   b. African continental glaciers did not produce loess.
   c. African glaciers deposited their loess in Europe.
   d. African loess turned into clay.
   e. None of the above.

10. Sand ripples migrate by
    a. sand moving around the outside edge of a pile and accumulating in front.
    b. Sand ripples do not migrate.
    c. wind moving sand as a single pile.
    d. sand moving up the stoss slope and avalanching down the lee slope.
    e. sand saltating into a single form.

11. Star dunes
    a. are shaped like a pyramid, with three or more arms that radiate from a peaked centre.
    b. are formed by wind that blows from several different directions.
    c. grow taller rather than migrating.
    d. require a rich source of sand.
    e. All of the above.

12. Water in a desert is
    a. not an important feature.
    b. mostly used up by the vegetation.
    c. involved in ripple and dune formation.
    d. second only to wind in shaping the desert surface.
    e. the primary agent shaping desert landforms.

13. Loess
    a. is windblown silt that accumulates in thick soil.
    b. is windblown sand that forms ergs.
    c. collects only in arid regions below Hadley cells.
    d. is common primarily in polar deserts.
    e. None of the above.

14. A _____ forms when wind removes finer sediments, leaving behind a layer of coarse sediments.
    a. yardang
    b. barchan
    c. dust bowl
    d. pavement
    e. erg

15. Desertification is
    a. caused by the contraction of a desert and the exposure of unvegetated regions.
    b. primarily caused by warfare.
    c. not related to global warming.
    d. a serious problem in the Arctic.
    e. typically related to climate change and/or poor land management.

FURTHER RESEARCH

1. How can you tell if a dune is a barchan dune or a parabolic dune?

2. How can you tell the direction of the wind that formed a dune?

3. The Atacama Desert of northern Chile and the deserts of the Basin and Range area of the western United States are both created, in part, by the orographic effect of nearby mountain ranges. However, the Atacama lies to the west of its mountains and the Basin and Range area lies to the east of its mountains. Explain how this is possible.

4. The Atacama Desert is usually drier than the Basin and Range. Give a couple of possible reasons for this.

ONLINE RESOURCES

Explore more about deserts on the following websites:

U.S. Geological Survey—"Deserts: Geology and Resources":
http://pubs.usgs.gov/gip/deserts

Are there deserts in Canada? See the following link:
www.thecanadianencyclopedia.com/articles/desert

The United Nations Convention to Combat Desertification:
www.unccd.int

A description of desertification and its impact on human communities:
www.greenfacts.org/en/desertification/

A description of desertification around the world:
http://en.wikipedia.org/wiki/Desertification

Additional animations, videos, and other online resources are available at this book's companion website:
www.wiley.com/go/fletchercanada

This companion website also has additional information about WileyPLUS and other Wiley teaching and learning resources.

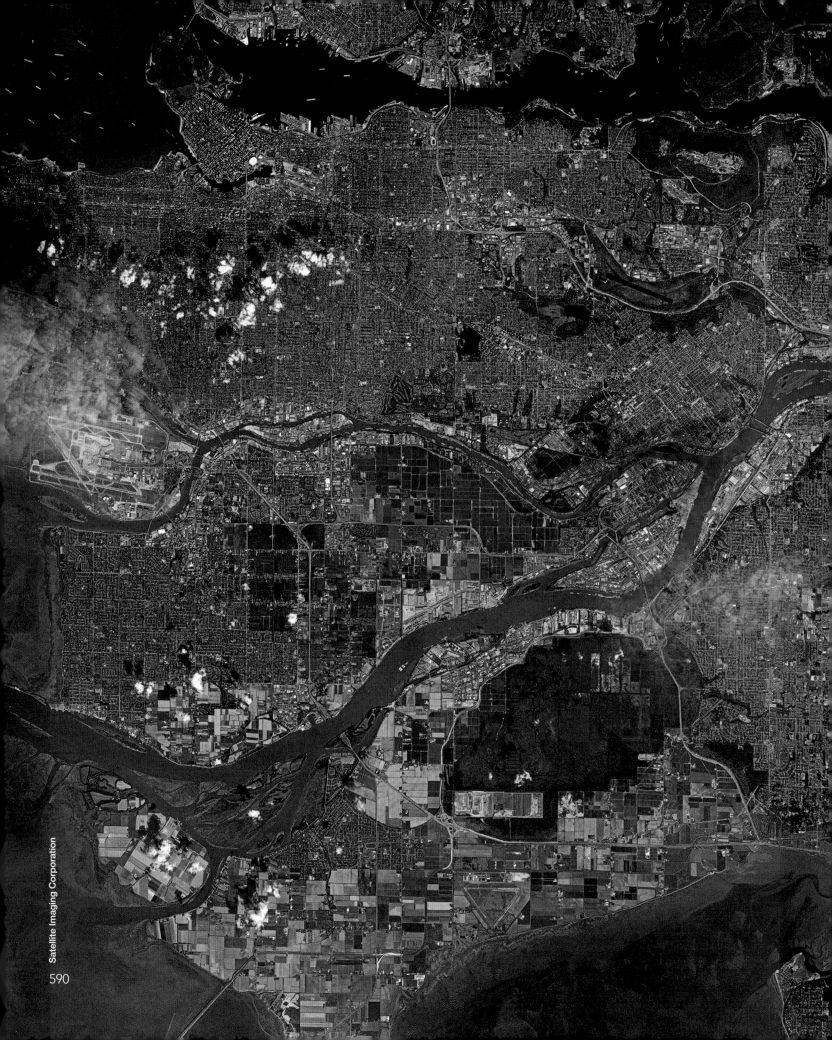

CHAPTER 22
COASTAL GEOLOGY

Chapter Contents and Learning Objectives (LO)

22-1 Change is constantly occurring on the shoreline.
LO 22-1 *Identify the many influences on the world's coastlines.*

22-2 Wave energy is the dominant force driving natural coastal change.
LO 22-2 *Describe the transformation of waves as they travel from deeper to shallower water.*

22-3 Wave refraction and wave-generated currents occur in shallow water.
LO 22-3 *Define "wave refraction," and describe how wave-generated currents form.*

22-4 Longshore currents and rip currents transport sediment in the surf zone.
LO 22-4 *Define and describe "longshore currents" and "rip currents."*

22-5 Gravity and inertia create two tides every day.
LO 22-5 *Compare and contrast the forces that generate tides.*

22-6 Hurricanes and tropical storms cause enormous damage to coastal areas.
LO 22-6 *Describe the nature and impact of hurricanes.*

22-7 Sea-level rise since the last ice age has shaped most coastlines and continues to do so.
LO 22-7 *Describe how sea-level rise affects coastlines.*

22-8 Barrier islands migrate with rising sea level.
LO 22-8 *Define "barrier islands," and explain how they migrate.*

22-9 Rocky shorelines, estuaries, and tidal wetlands are important coastal environments.
LO 22-9 *Identify several types of coastal environments.*

22-10 Coasts may be submergent or emergent, depositional or erosional, or exhibit aspects of all four of these characteristics.
LO 22-10 *Distinguish between submergent and emergent coasts and between depositional and erosional coasts.*

22-11 Coral reefs are home to 25 percent of all marine species.
LO 22-11 *Describe the formation of coral reefs, and define the main types of reefs.*

22-12 Coastal problems are growing as populations increase.
LO 22-12 *Discuss the effects of human activities on coastal environments.*

GEOLOGY IN OUR LIVES

Canada's ocean and Great Lakes coastlines are home to almost 13 million people, about 40 percent of the total Canadian population. In the United States, more than 10 times this amount—almost 153 million people or about 53 percent of the U.S. population—live along coastlines. Surging population growth in coastal zones exposes more people to the dangers of geologic hazards, such as storms, hurricanes, and tsunamis, than in any other geological environment. The world's coasts are home to fragile ecosystems, beautiful vistas, pristine waters, and major growing cities, all coexisting in a narrow and constricted space. Expanding human communities compete for more space at the expense of extraordinary wild lands. There are problems with coastal erosion, waste disposal, a dependency on imported food and water, and rising sea level. Management strategies to ensure the long-term viability of our coasts are largely science based; thus, scientists are important partners in building sustainable coastal communities.

(?) The opening image shows the city of Vancouver, located at the interface between the Coast Mountains and the Pacific Ocean. What will happen in the future as human populations in coastal zones continue to grow and sea level continues to rise?

22-1 Change Is Constantly Occurring on the Shoreline

LO 22-1 *Identify the many influences on the world's coastlines.*

Coastlines are unique and complex environmental systems influenced by geologic, atmospheric, oceanographic, and biologic processes and materials. As a result, coastlines are ever-changing environments. In fact, the only constant on coastlines is change.

Because of their temperate climate (air temperature is moderated by heat stored in the ocean) as well as their beautiful vistas and abundant resources, coastal lands are the most crowded and developed areas in the world. Fifty percent of the population of the industrialized world lives within 100 km of a coast, a population that will grow by 15 percent during the next two decades. In the United States, over 37 million people and 19 million homes were added to coastal areas over the last three decades. This narrow fringe—comprising less than one-fifth of the contiguous land area of the United States—accounts for over half the nation's population and housing supply.

Yet coasts are subject to destructive storms and tsunamis (**FIGURE 22.1**) that sweep inland from the sea, as well as to coastal erosion that slowly eats away at beaches and cliffs and to accelerating sea-level rise due to global warming. Development pressure on coasts has led to pollution of coastal water bodies, depletion of coastal fisheries, and changes in sedimentary processes that support vital ecosystems.

With so much of our culture and economy focused on the shoreline, geologists need to play an active role in advancing the scientific understanding of coastal environments and providing scientific data that can be used to improve management plans and programs. As a result, in recent decades, the new field of **coastal geology** has emerged. Coastal geology is concerned with understanding coastal processes (such as how waves and currents move sediment), the geologic history of coastal areas, and how humans live in coastal environments (**FIGURE 22.2**).

Coastal Processes

Sediments in the *coastal zone* may come from watersheds (typically siliciclastic sediments) as well as from the ocean (typically carbonate sediments). Winds, waves, and currents transport and deposit these sediments, thereby shaping various coastal environments, such as beaches, estuaries, deltas, wetlands, and reefs. Since the last ice age, rising global sea level (over 120 m) has flooded river mouths, drowned low-lying coastal lands, and shifted the coastline landward, in many cases over hundreds of kilometres. Winds, waves, tides, sea-level change, and currents act as agents of change; they are collectively called **coastal processes.** They sculpt the shoreline through erosion and deposition of sediment and by flooding low areas. Understanding how coastal processes and sedimentary materials interact is crucial to interpreting the geologic history of coastal systems and finding ways for people to live in the coastal zone in a sustainable manner.

May 21, 2009

November 5, 2012

(?) Study the damage pattern in Figure 22.1. Are there construction techniques that would allow a home to survive a hurricane?

FIGURE 22.1 Hurricane Sandy made landfall on October 29, 2012, at Brigantine, New Jersey. The storm produced large, damaging waves and strong winds and raised the ocean level more than 3 m to 6 m above normal. Large segments of the U.S. eastern seaboard were heavily damaged, and roads, homes, and businesses in many communities were destroyed.

FIGURE 22.2 Trash washes up on beaches around the world because we have been polluting coastlines for decades.

(?) What are the potential consequences of dumping garbage in coastal environments?

Coastal Terminology

Scientists use a number of specific terms when describing the coastal zone. Here we focus on beaches because that is where we can most readily observe the processes associated with waves. But the concepts that apply to beaches apply to other types of coastal environments as well (FIGURE 22.3).

A **beach** is an unconsolidated accumulation of sand and gravel along the shoreline. Beaches form in freshwater environments such as lakes and rivers, but here we emphasize beaches that form along the shores of oceans. Most beaches are formed and maintained by waves. Many people do not realize that beaches consist of submerged and *subaerial* (exposed to the atmosphere) portions. The submerged portion extends seaward past the *breaker zone,* the region where waves from the deep ocean turn into breakers. After breaking, a wave enters the *surf zone* as a *bore,* which is a wave (or waves) of turbulent water advancing toward the shoreline. It then moves toward the beach and enters the *swash zone.* There the energy of the bore creates horizontal surging of the ocean surface up and down the sloping *foreshore* of the beach.

The landward extent of the foreshore is marked by a *berm,* a ridge of sand built by waves at high tide that deliver sand to the top of the foreshore. Landward of the berm is the *backshore,* which is characterized by a *coastal dune,* an accumulation of wind-blown sand in the shape of a low hill vegetated with salt-tolerant plants.

(?) **Expand Your Thinking**—Sea-level rise has accelerated due to global warming. Describe the ways in which sea-level rise will affect coastal cities.

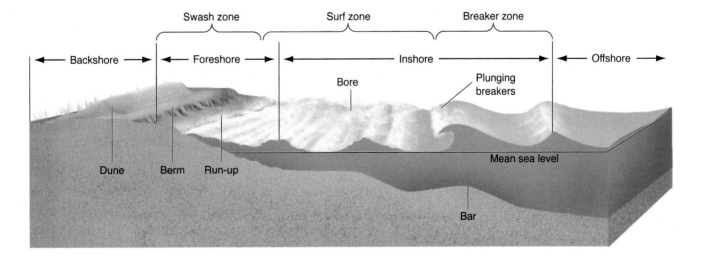

FIGURE 22.3 The coastal zone consists of several environments in which sediments, waves, and currents interact to produce coastal processes.

(?) What kind of sediments collect on beaches? Where do they come from?

22-2 Wave Energy Is the Dominant Force Driving Natural Coastal Change

LO 22-2 *Describe the transformation of waves as they travel from deeper to shallower water.*

Waves on the ocean's surface are generated by the friction of air blowing across the water. Hence, they are called *wind waves*.

Wind Waves

Perhaps you have watched a breeze move across a quiet body of water. As it disturbs the surface and forms small ripples, it is transferring energy from the atmosphere into the water by displacing individual molecules of water. The molecules move in a circle called an *orbital* that is largest at the surface and gets smaller with depth (**FIGURE 22.4**).

A moving wave looks as if it is transporting water, but it is not. Just as music travels through the air without making wind, a wave travels across the ocean surface without making a current. (In the surf zone, however, waves do make currents.) Energy (waves) radiates outward from the point at which the wind interacts with the ocean surface, but water does not. The size of a wave is governed by

wind velocity, fetch (the distance over which the wind blows), and the duration of time that the wind disturbs the water.

Waves are described by their *height,* the vertical distance between the peak of a crest and the bottom of a trough; their *wavelength* (λ), the horizontal distance between two successive waves; and their *period,* the time it takes for two successive wave crests to pass a given point in space. In general, the circular water motion generated by a wave extends to a depth equal to half the wave length ($\lambda/2$).

The ocean's surface in a wave-generating area, called a *sea*, is chaotic and disturbed due to high wind velocities. A sea consists of steep, sharp-crested waves of many different heights, lengths, and periods. As waves move out of the area of generation in the absence of local wind, they become *swells*, which are long, regular, symmetrical waves with periods ranging from 5 to 20 s. Swells are noticeable as the smooth rolling action of the ocean surface. Each swell transfers energy horizontally across the ocean's surface (**FIGURE 22.5**).

The shape and speed of a wave are governed by the displacement of water particles and controlled by two characteristics: wavelength and water depth. In deep water (where the depth is greater than the wavelength), a swell takes the form of a symmetrical sine wave (refer to Figure 22.4). However, the form of a wave changes dramatically when it enters shallow water (where the water depth is less than half the wavelength).

The nautical term for shallow water is *shoal,* and a wave entering shallow water is said to be a *shoaling wave.* Shoal water interferes with the movement of water particles at the base of the wave, slowing its forward motion. Interaction with the bottom changes a water

? What factors govern the size of a wave?

FIGURE 22.4 A wave is described in terms of its height, length, and period. Particles of water disturbed by the wind travel in a circle called an orbital. A shoaling wave encounters the bottom at depths of less than half the wavelength. This slows the forward movement of the wave and causes it to become higher and to develop a steep face.

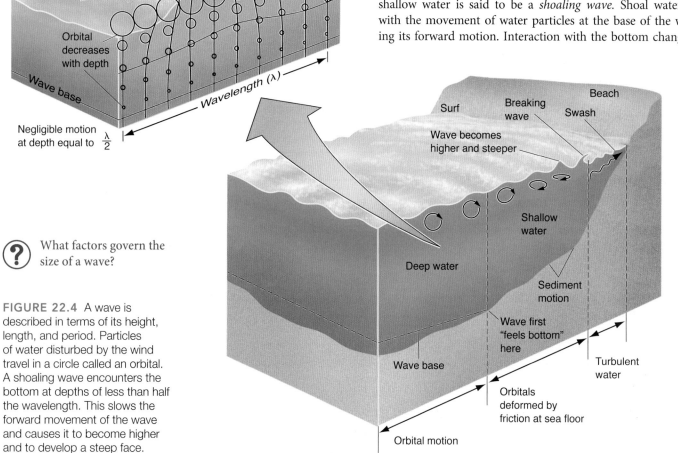

(a)

(b)

FIGURE 22.5 (a) Waves are generated by strong winds in a chaotic environment called a "sea." (b) Beyond the sea, wave energy is organized into swells, which appear as a gently rolling motion of the ocean's surface.

(?) How does a wave change as it enters shallow water?

particle's orbital into an ellipse (a flattened circle), and at the sea floor, water particles experience back-and-forth movement. As a wave enters shallow water, it slows, and the following wave, moving faster, "catches up" with it. The wavelength decreases and the wave's height increases. The crests become narrow and pointed with steep faces, while the troughs become wide, shallow curves. Because the wavelength decreases and the speed slows, the wave's period remains unchanged.

In very shallow water, the energy from sea and swell is released in the surf zone, where waves become *breakers*. When a wave's height is approximately equal to the depth of the water, the wave "breaks." That is, the peak of the crest pitches forward and tumbles down the face of the wave in the direction in which the wave is moving (usually landward) and breaks into a mass of aerated water. A wave breaks when it becomes overly steep. Wave breaking occurs because the velocities of water particles in the crest exceed the velocity of the wave and the crest surges ahead.

Global Wind and Wave Patterns

The biggest waves are generated by strong winds blowing across vast ocean distances. **FIGURE 22.6** shows a map of global wind speeds and wave heights as measured by satellite radar pulses. Wind speed is determined by the strength of a return pulse reflected from the ocean surface. Calm areas serve as good reflectors and return a strong pulse; a rough sea scatters a radar signal and returns a weak pulse. The radar instrument uses these differences to measure wind speed and wave height.

The Southern Ocean, with its intense belt of winds and vast fetch uninterrupted by continents, tends to generate the highest winds and waves on the planet: exceeding 15 m/s and 6 m in height. Strong winds and high waves are also associated with stormy regions in the North Atlantic and North Pacific oceans. These are indicated by the most intense red tones. In general, there is a high degree of correlation between wind speed and wave height. The weakest winds and lowest waves (represented by magenta and dark blue) are found in the western tropical Pacific Ocean, the tropical Atlantic Ocean, and the tropical Indian Ocean. These areas are characterized by trade winds that tend to be persistent but relatively low in velocity.

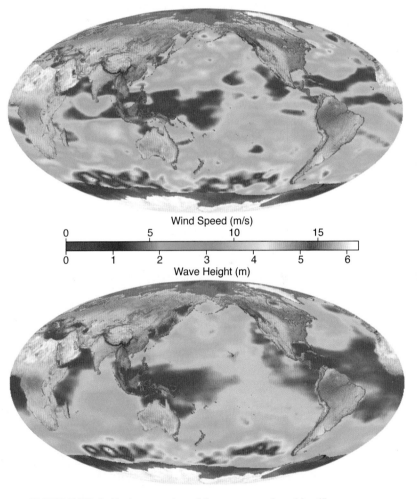

FIGURE 22.6 Radar mapping of the ocean surface identifies areas of greatest wind speed (top) and wave height (bottom). Both the wind speed and wave height were measured simultaneously and averaged over the period of October 3–12, 1992.

(?) Why do the world's largest waves occur in the Southern Ocean?

(?) **Expand Your Thinking**—Referring to Figure 22.6, if you were to sail around the world to visit your favourite places, what safe route would you pick?

22-3 Wave Refraction and Wave-Generated Currents Occur in Shallow Water

LO 22-3 *Define "wave refraction," and describe how wave-generated currents form.*

When a wave enters shallow water, it is subject to **refraction.** Wave refraction occurs when the wave crest bends so that it is aligned parallel to the contours of the sea floor. For straight coasts with parallel contours on the shallow sea floor, refraction tends to orient waves so that the wave crest is parallel to the coastline.

Refraction occurs when a wave approaches a shoreline at an angle to the contours of the sea floor. As the portion of the wave in shallow water experiences shoaling, it slows due to frictional resistance from the bottom. The rest of the wave, still travelling unhindered in deeper water, tends to change its direction like a door slowly swinging on its hinge. The rapidly travelling portion of the wave swings landward and lines up with bottom contours so that the wave crest and bottom contours remain parallel (**FIGURE 22.7**).

As a wave approaches a *rocky headland* that sticks out into the ocean beyond the surrounding shoreline, the wave front begins to refract when it first encounters the sea floor. This leads to *convergence* of wave energy on the headland. The wave crest takes on the orientation of the bottom contours and expends energy breaking against all sides of the headland, causing it to erode. Simultaneously, the portion of the wave crest moving into the adjacent bay also experiences

refraction. In this case, however, the wave energy *diverges* across a broad area of shoreline. The wave energy is effectively reduced in the embayment because the energy spreads out as the wave arcs into the bay (**FIGURE 22.8**).

Wave-Generated Currents

Water particles at the crest of a wave move forward with the wave, while particles in the trough move in the opposite direction (see Figure 22.4). Together these patterns define a not-quite-closed circle. The orbital motion of a water particle during the passage of a wave carves a slightly open curve rather than a perfectly closed circle. The momentum of a wave drives this forward movement of the water (called "Stokes drift"). In deep water, this effect is minor, but in shallow water, it causes water to pile up along a coastline.

Water movement and sediment transport along the coast are highly complex. In the surf zone, there is net shoreward transport of water near the surface associated with the movement of bores. Most of this shoreward movement occurs under the crests of the bores because water particle motion there is directed onshore. This motion produces a gradual advance of the water mass toward a beach. Water piles up against the shore, raising the water level and creating

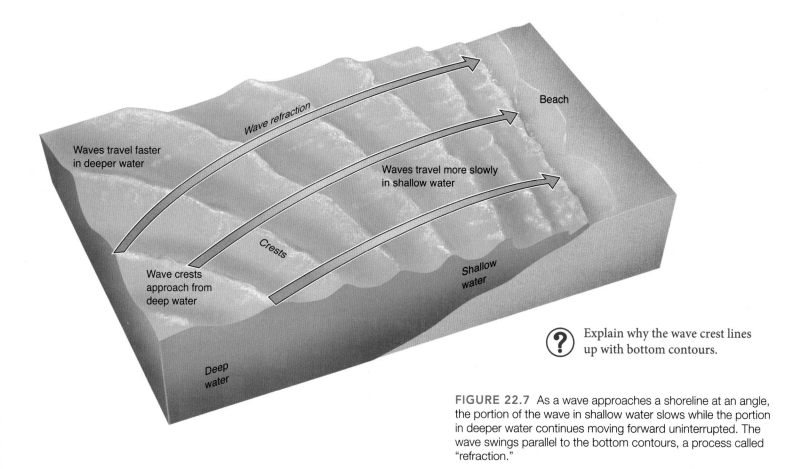

Beach

Wave refraction

Waves travel faster in deeper water

Waves travel more slowly in shallow water

Crests

Shallow water

Wave crests approach from deep water

Deep water

(?) Explain why the wave crest lines up with bottom contours.

FIGURE 22.7 As a wave approaches a shoreline at an angle, the portion of the wave in shallow water slows while the portion in deeper water continues moving forward uninterrupted. The wave swings parallel to the bottom contours, a process called "refraction."

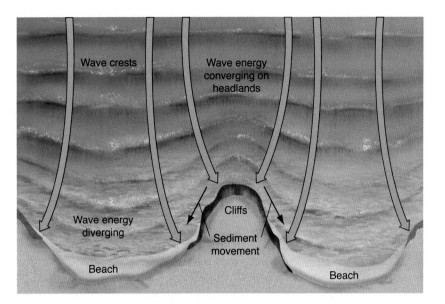

FIGURE 22.8 As a wave approaches a rocky headland, refraction causes the energy to converge on the headland and diverge in the adjoining bay. Hence, the headland erodes and the bay accumulates the resulting sediment as a beach.

(?) How does sand transport in Figure 22.8 compare and contrast with transport in Figure 22.9? What are the implications for the beach?

sea-level set-up, which in turn leads to the formation of **wave-generated currents.**

Of course, water cannot pile up against the coast forever. A current that flows along the bottom in the seaward direction counterbalances the onshore movement of water in the surf zone. This is the *undertow.* FIGURE 22.9 provides a simple two-dimensional model of circulation in the surf zone. This model does not account for the delivery of sand onto a beach. Sand moves landward in the surf zone, which causes beaches to become stable and recover after erosion by large waves. Beach recovery following a storm has been observed along many shorelines. If the net movement of sand along the seabed were always directed offshore in the surf zone (as in Figure 22.9), beaches would not exist.

How does sand move toward a beach? Researchers continue to investigate how beaches function, but the answer may be found in two areas: (1) Three-dimensional surf zone circulation may create some regions with onshore sand movement and some areas with offshore sand movement, and (2) the onshore water velocities under the bore crests may be very important in transporting sediment toward the beach. Undertow appears to operate most strongly during storms. This fact may account for the erosion that is often observed during high waves. A particularly strong offshore flow during intense storm conditions might transport sand so far offshore that it is permanently lost from the beach, resulting in net recession of the shoreline.

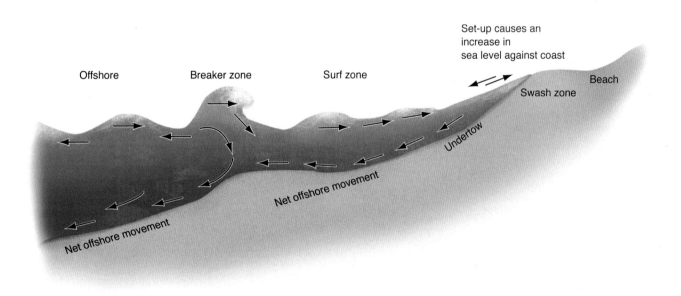

FIGURE 22.9 Breakers and bores cause net movement of water toward the shoreline, raising sea level along the coast and creating sea-level set-up, which causes the water surface to gently slope from the shoreline toward deeper water. Sea-level set-up establishes an offshore-directed pressure gradient that drives a current called "undertow" along the seabed in the seaward direction.

 How do offshore-directed currents affect sand transport in the coastal zone?

 Expand Your Thinking—If rocky headlands experience more wave energy than beaches do, what will eventually become of them? What is the long-term trend of shoreline evolution?

22-4 Longshore Currents and Rip Currents Transport Sediment in the Surf Zone

LO 22-4 *Define and describe "longshore currents" and "rip currents."*

When waves approach a shoreline from an angle, they generate a **longshore current** that moves along the shoreline in the direction of wave movement (**FIGURE 22.10**). Part of this current is due to the periodic release of sea-level set-up as a volume of water that moves along the shoreline, and part of it is due to the momentum of a moving wave that is directed obliquely to the shoreline. The longshore current may transport, along the shoreline, sand that has been suspended in the water column by breaking waves. This movement is important because it establishes a sand-sharing system along a beach whereby one segment of a beach can contribute sand to another, helping to counteract the effects of erosion in which sand is lost offshore. At other times of the year, when waves approach from a different direction, the sand may return.

Grains of sand moving along the foreshore follow an asymmetrical or sawtooth path in a longshore current. They are moved up the beach by the up-rush of a wave in the swash zone at an angle directed along the shoreline, but their return path is perpendicular to the shoreline, following a line directed by gravity. This process of sediment transport is called **longshore drift.**

Another important consequence of sea-level set-up along a shoreline is the formation of **rip currents** (**FIGURE 22.11**). As water piles up along a beach, it is released periodically in the form of offshore-directed currents that surge through the surf zone. These currents carry suspended sand and floating debris. As a result, rip currents can be detected as dirty, rough, or dark water that extends through a surf zone beyond the breakers. Rip currents may be spaced along a beach in a recurring pattern so that a long stretch of beach can develop several rip currents.

As water moves offshore in a rip current, it excavates a channel in the sandy sea floor, like a stream channel, and keeps the channel open for several hours or days. Because the water depth is greater in the rip channel than in the surrounding surf zone (where there are shallow areas called **sand bars),** the water appears darker and moves offshore at a relatively high velocity (making rip currents extremely hazardous to swimmers). Between rip channels, water moves onshore under waves and builds an accumulation of sand in a sand bar. The difference in depth between the sand bar and the rip channel may be greater than the height of a person. For this reason, and because of the strong current, rip channels are hazardous to swimmers and lead to drownings every year on beaches around the country. The way to escape from a rip current is to avoid swimming against it and instead swim out of it sideways (along the shoreline).

Shorelines Straighten over Time

Another form of wave-generated current develops on a shoreline that has a rocky headland and an adjoining bay, as described earlier (see Figure 22.8). On rocky embayed coasts, two areas of refraction occur: one where energy converges on the headland and a second where energy diverges in an embayment. As a result of these two areas of refraction, sand is eroded from the headland and moves into the embayment, where it builds a beach. How does this sand move? It is propelled through the action of wave-generated currents related to set-up at the headland.

The convergence of energy leads to sea-level set-up around headlands. That is, sea level at the headland is higher due to greater wave energy and momentum. In the adjoining embayment, sea level is lower because wave momentum is lower. Hence, the water surface slopes from the headland into the embayment. This gradient drives a longshore current that carries eroded sediment from the headland into the adjoining bay. Over time, as sand accumulates in the bay, a beach builds up and slowly fills the embayed area. When this process is carried to its conclusion, the headland continuously erodes back, the embayment slowly fills with sand, and the shoreline straightens over time (**FIGURE 22.12**). In fact, one of the accepted laws of coastal geology is that *shorelines straighten over time.*

Net movement of sand grains

Path of sand particles

Longshore current

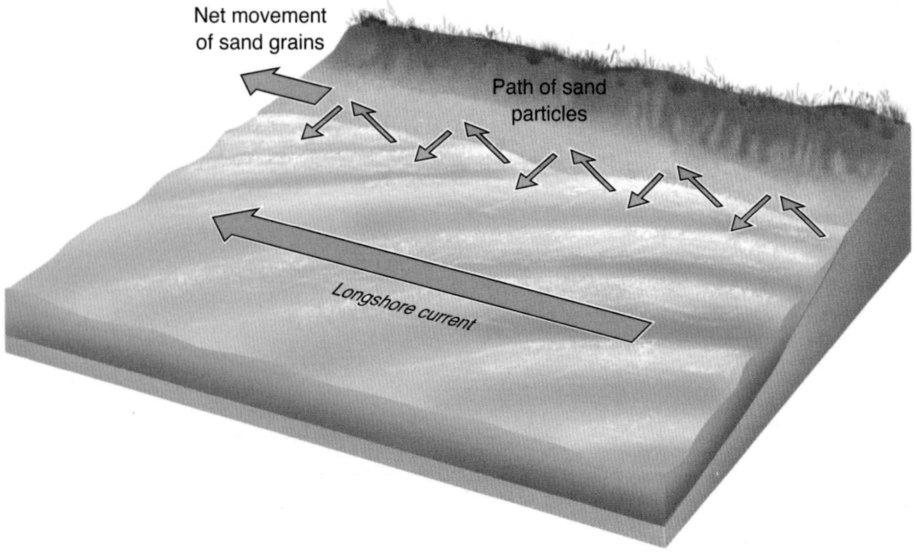

 In this situation, what would the effect be of building a jetty (a solid wall) across the beach and out into the ocean?

FIGURE 22.10 Waves that approach the shoreline at an angle experience refraction but still may intersect the beach obliquely. The momentum of the wave drives water onto the beach in the direction of wave movement. Water travels up the beach at an angle but returns to the ocean, under the force of gravity, in a direction perpendicular to the shoreline.

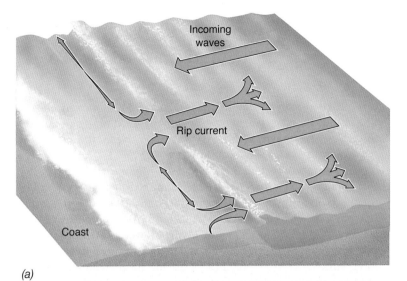

(a)

(b)

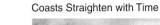
Coasts Straighten with Time

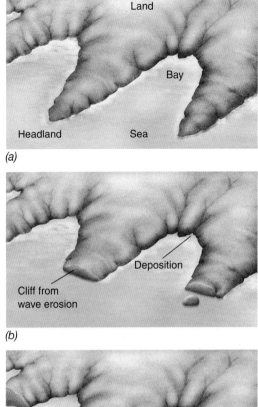

(a)

(b)

(c)

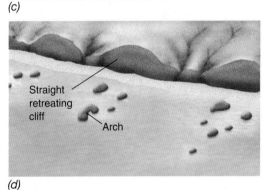

(d)

Peter Chadwick/Gallo Images/Getty Images, Inc.

FIGURE 22.11 Sea-level set-up is released periodically by offshore-directed circulation. *(a)* Rip currents remove this water from the surf zone and carry suspended sediment and debris beyond the breaker zone. *(b)* Dark channels or the presence of turbid water travelling through the breaker zone reveals rip currents. Sand bars may develop between rip currents.

 If rip currents carry excess water offshore, what brings excess water onshore?

FIGURE 22.12 *(a-d)* Wave-generated currents move sand eroded from a rocky headland into the adjoining embayment and build a beach. Over time, the headland is worn down and the bay fills with sand. Eventually the shoreline becomes a long, straight beach. Portions of the headland that are resistant to erosion are stranded offshore to form sea stacks and arches, such as those found in the Hopewell Rocks within the Bay of Fundy *(e)*.

 What is the eventual fate of sea stacks and sea arches?

Tourism New Brunswick

(e)

 Expand Your Thinking—Considering all that you have learned about coastal processes, draw an illustration that captures sediment transport on a beach. Label all components and write an accompanying description.

22-5 Gravity and Inertia Create Two Tides Every Day

LO 22-5 *Compare and contrast the forces that generate tides.*

The oceans rise and fall twice each day due to *tide-raising forces* related to the gravitational attraction between Earth, the Moon, and the Sun (**FIGURE 22.13**). These regular and predictable oscillations in the ocean's surface are called **tides.** Newton's law of universal gravitation states that the force of gravity existing between two bodies is directly proportional to their masses and inversely proportional to the square of the distance between them. This means that the greater the mass of the objects and the closer they are to each other, the greater the gravitational attraction between them. The Moon, although not large on a planetary scale, is near Earth and therefore exerts a significant gravitational attraction on it. The Sun, although not near Earth, is large and also exerts a significant attraction, but its attraction is about half that of the Moon.

Lunar Tide

Gravity is only one of the major tide-raising forces responsible for creating tides. Another is *inertial force,* which counterbalances gravity in the Earth–Moon system. Inertia is the tendency of moving objects to continue moving in a straight line unless acted on by an outside force. Together, gravity and inertia are responsible for the creation of two major tidal bulges on Earth's surface.

The attraction between Earth and the Moon is strongest on the side facing the Moon. Water is able to immediately respond to the Moon's pull. Hence, gravitational attraction causes water on Earth's nearest side to be pulled toward the Moon. As a result, the oceans "bulge out" on the side facing the Moon. However, at the same time that gravitational force draws the water closer to the Moon, inertial force works to keep the water in place. But the gravitational force is stronger, and hence the water is pulled toward the Moon on Earth's near side.

On Earth's opposite side, the gravitational force is less because the Moon is farther away. Here, inertial force exceeds gravitational force, and the water tries to keep going in a straight line, moving away from Earth and also forming a bulge. Many people believe that the

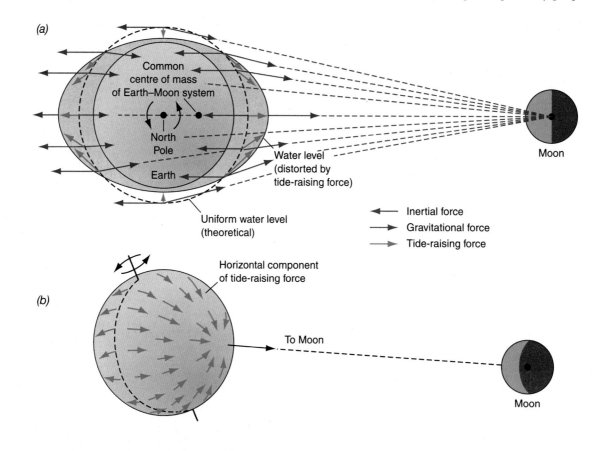

FIGURE 22.13 *(a)* Two tide-raising forces are responsible for the lunar ocean tide: the gravitational attractive force (red arrow) and the inertial force (blue arrow). The balance between these two opposite forces (green arrow) defines the tide-raising force. On the side of Earth facing the Moon, the gravitational attraction force is greater than the inertial force and a tidal bulge is created in the ocean surface. On the side facing away from the Moon, the inertial force is greater and creates a tidal bulge. Earth rotates beneath its watery envelope and encounters these two bulges (high tide) twice every day. *(b)* Green arrows display the theoretical movement of water toward the tidal bulge. As continents rotate into the tidal bulges, this water must flow within, throughout, and around coastlines in the form of a series of tidal currents.

(?) How are tidal currents produced?

tides travel around the planet, but in reality it is Earth that rotates on its axis and slides under the watery bulges created by the Earth–Moon system. Because Earth is rotating while this is happening, two tides normally occur each day: one when Earth slides under the lunar bulge of the oceans and another when Earth slides under the inertial bulge on the opposite side.

Coastlines encounter the tidal bulges as Earth rotates. As a bulge approaches, coastal environments experience *tidal currents,* in which water fills bays, inlets, lagoons, and other environments in the shoreline. Arriving currents are called the *flooding tides.* As a tidal bulge passes a coastal point, the water drains from these environments during *ebbing tides.*

Solar Tide

The Sun also plays an important role in generating tide-raising forces. Two tidal bulges are created by the Earth–Sun tidal system: a gravitational bulge facing the Sun and an inertial bulge on Earth's opposite side. Although these tidal bulges create tidal oscillations in their own right, their more important role is to modulate, or influence, the lunar tide-raising forces. The interaction of the forces generated by the Moon and the Sun can be quite complex (**FIGURE 22.14**).

Because this chapter is just an introduction to tidal theory, we cannot discuss all aspects of the tides here. However, it should be apparent that when the Sun–Earth–Moon system is aligned, the tide range will be higher than usual. This is called *spring tide.* When the solar and lunar components are at 90° to each other, the tide range will be lower than usual. This is called *neap tide.*

(?) How many times each month does spring tide occur?

FIGURE 22.14 Tidal bulges in the ocean surface related to the gravitational and inertial forces of the Sun–Earth system modulate the lunar tide. When the Sun, Earth, and Moon are aligned, the spring tide occurs. When the Moon is at 90° to the Sun–Earth alignment, neap tide occurs.

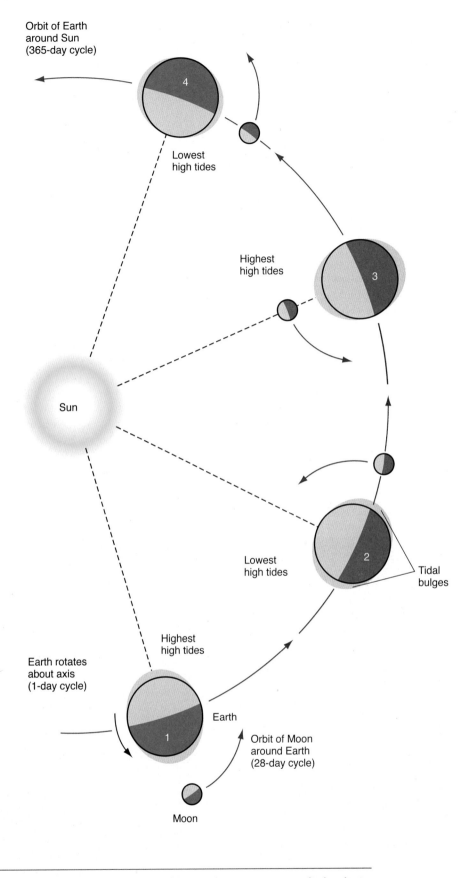

Orbit of Earth around Sun (365-day cycle)

Lowest high tides

Highest high tides

Sun

Lowest high tides

Tidal bulges

Earth rotates about axis (1-day cycle)

Highest high tides

Earth

Orbit of Moon around Earth (28-day cycle)

Moon

(?) **Expand Your Thinking**—What role might tidal currents play in the coastal processes you studied earlier?

22-6 Hurricanes and Tropical Storms Cause Enormous Damage to Coastal Areas

LO 22-6 *Describe the nature and impact of hurricanes.*

Few things in nature compare to the destructive force of a **hurricane** (FIGURE 22.15). Called "the greatest storm on Earth," a hurricane is capable of causing massive damage to coastal areas with sustained winds that can reach well over 200 km per hour, intense rainfall, and flooding ocean waters with huge waves. In fact, during its life cycle, a hurricane can expend as much energy as 10,000 nuclear bombs. The term "hurricane" is derived from *Huracan,* the name of a god of evil recognized by the Tainos, an aboriginal tribe in Central America. It should be noted that the term "hurricane" is applied to the Atlantic Ocean; *cyclone* is used for such storms in the Indian Ocean and parts of the Pacific; and *typhoon* is used in the northwest Pacific region.

The Gulf Coast of the United States and the Atlantic coast of both the United States and Canada are known for their vulnerability to hurricane and tropical storm damage, but any coastline in a region where hurricanes develop and travel is vulnerable. In fact, three regions in the Northern Hemisphere are known for their tendency to spawn hurricanes: the tropical Atlantic, the eastern tropical Pacific, and the western tropical Pacific. In the Southern Hemisphere, hurricanes form in two primary areas: the western tropical Pacific and the Indian Ocean (FIGURE 22.16).

For a storm to be classified as a hurricane, its winds must reach a sustained speed of at least 119 km per hour or higher. Hurricanes tend to form around a pre-existing atmospheric disturbance in warm tropical oceans where there is high humidity in the atmosphere, light winds above the storm, and a high rate of condensation in the atmosphere. If the right conditions last long enough, a hurricane can produce violent winds, incredible waves, torrential rains, and floods.

On average, there are six Atlantic hurricanes each year; over a three-year period, approximately five hurricanes strike the Atlantic coastline of the United States and Canada. When hurricanes move onto land, the heavy rain, strong winds, and large waves can damage buildings, trees, cars, roads, and power lines. But a hurricane weakens on land because its source of heat—the warm ocean water—is gone. The hurricane becomes a *tropical storm* (a storm with wind speeds between 63 km and 117 km per hour). These storms are also highly destructive, as they dump immense quantities of rain on the land and batter the coast with waves.

Accompanying the large waves is a high-sea-level phenomenon called **storm surge** (FIGURE 22.17). Storm surge is a combined effect that results from low atmospheric pressure above the ocean surface causing a bulge of water to travel beneath a storm, wind shear pushing the water in the direction of the storm's forward movement, and sea-level set-up due to wave momentum. The combined processes of low pressure, wind shear, and set-up can raise the water level several metres along a coastline. Add to this waves 6 m to 10 m high (or more) formed by the winds of a storm, and it is easy to see why hurricanes and tropical storms are very dangerous when they reach the shore, especially one with low topography. They cause massive damage and flooding of low-lying coastlines, damage that is compounded if they hit at high tide.

Hurricanes are not the only type of storm that damages coastal areas, but they are among the worst (FIGURE 22.18). High winds and storm surge are a primary cause of hurricane-inflicted loss of life and property damage. When a hurricane makes landfall along low-lying coastal lands, ocean waters sweep across beaches and roads and into adjacent communities. Heavy rains compound the problem by saturating the ground and causing flooding of nearby streams. Coastal erosion is caused by high waves and storm surge that can strip a beach of its sand and undermine homes, roadways, and businesses. How can such damage be mitigated? The simplest and most direct way is to *avoid the hazard* by not developing the shoreline. If we stopped putting communities on the edge of the ocean, we would greatly reduce the suffering, loss of life, and expensive damage caused by hurricanes and other coastal hazards.

NASA

FIGURE 22.15 Hurricane Irene inundated the Atlantic coast of United States and Canada in 2011.

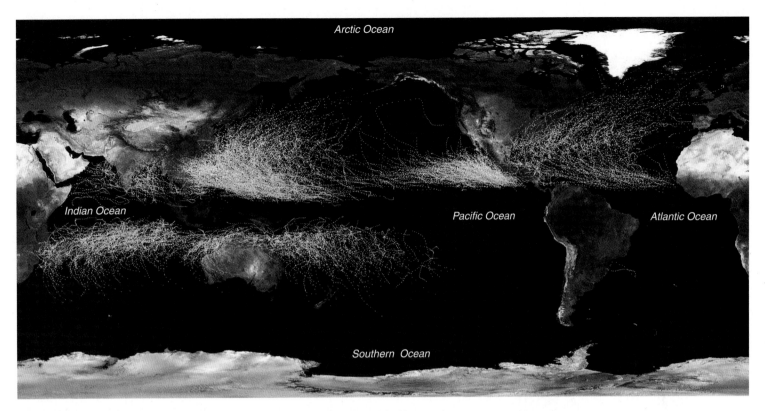

FIGURE 22.16 The higher concentrations of the storm tracks in this global image show that there are three major areas of hurricane formation in the Northern Hemisphere and two major areas in the Southern Hemisphere.

 What do hurricanes do to coastal environments when they strike?

FIGURE 22.17 Hurricane Katrina hit New Orleans, Louisiana, in 2005. It was the costliest and one of the five deadliest hurricanes in the history of the United States. The combination of high waves, wind shear on the water, high tide, sea-level set-up by wave momentum, and low atmospheric pressure on the ocean surface caused a storm surge that flooded 80 percent of the city.

FIGURE 22.18 Hurricanes cause damage to coastal communities with high winds and storm surge. This community in Atlantic City, New Jersey, was flooded by a massive storm surge related to Hurricane Sandy in 2012.

 How does flooding damage a community?

 What preventive measures could have been taken to minimize the flooding caused by Hurricane Sandy?

 Expand Your Thinking—Imagine that you are going to build a beachfront home in a hurricane-prone area. What special design features would you use? Explain their purpose. Find information on the Internet to help you come up with an answer.

22-7 Sea-Level Rise since the Last Ice Age Has Shaped Most Coastlines and Continues to Do So

LO *22-7 Describe how sea-level rise affects coastlines.*

One of the most influential geological processes shaping coastal environments around the world is the rise in sea level since the end of the last ice age. The last ice age reached its peak 20,000 to 30,000 years ago, and rapid global warming began shortly after. Because sea level at that time was more than 120 m below the present level, the world's shallow sea floors and continental shelves were exposed (**FIGURE 22.19**).

The last ice age began about 80,000 years ago and was punctuated by short-lived warming spells, but overall it lasted nearly 60,000 years. Between about 20,000 and 6,000 years ago, global warming caused glaciers around the world to shrink and seas to expand. The rising sea level flooded river mouths and continental shelves on every coastline of the world. This flooding shaped the look and character of modern coastal environments. Barrier islands were born, estuaries widened, tidal wetlands accumulated sediment, and coral reefs flourished in warm seas.

Sea level is rising today and will continue to rise in the centuries ahead. Greenhouse gas-induced global warming leads to warming of the ocean. Warming causes ice to melt and ocean water to expand; these two processes are the main causes of global sea-level rise. Today rising seas threaten coastal wetlands, estuaries, islands, beaches, and all types of coastal environments. Frequent flooding by storms and high tides, accelerated erosion, and saltwater intrusion into streams and aquifers threatens cities, ports, coastal communities, and other types of human development. Because this threat has enormous economic and environmental consequences, it is important to study records of past sea-level change to improve our understanding of how coastal environments evolve during periods of sea-level rise.

Future Sea-Level Rise

Tide gauge data from around the world show that, during the 20th century, global average sea level rose between 10 cm and 20 cm. Models of future climate change suggest that this increase was the precursor to an extended period (many centuries) of accelerated sea-level rise beginning in the 21st century. Indeed, early in the 21st century, researchers identified acceleration in global sea-level rise using a combination of tide gauge and satellite data.

There is 30 percent more carbon dioxide (CO_2) in the atmosphere now than there was 100 years ago. As we saw in Chapter 16 ("Global Warming"), this greenhouse gas is effective at trapping heat in the lower atmosphere. Researchers report that the decade 2000 to 2009 was the warmest decade in over 120 years, with 2005 being second only to 2010 as the warmest year ever measured. This strong warming trend leads to sea-level rise because water expands as it warms, and when the ocean surface expands, the easiest direction is up—hence, sea level rises.

Warming also melts ice. Studies have found that both the Greenland and Antarctic Ice Sheets are experiencing melting that is pouring water at an accelerating rate into the oceans. The rate of melting on Greenland has tripled in the last decade. Alpine glaciers are likewise melting around the world, and that water, too, is flowing into the oceans.

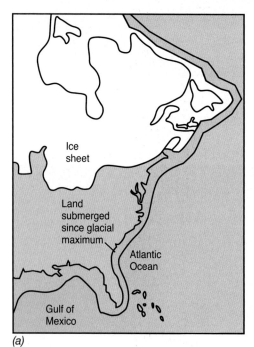

(a)

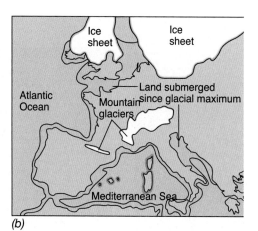

(b)

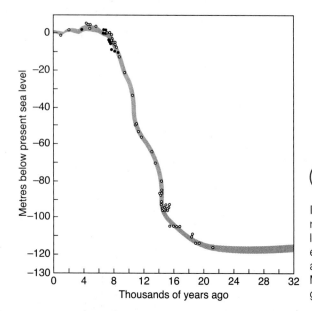

(c)

(?) What effects will rising sea level have on a coast?

FIGURE 22.19 When the last ice age ended, melting of the great ice sheets raised global sea levels 120 m to 130 m. As a result, coastlines retreated landward and modern coastal environments came into existence. *(a)* The eastern coast of Canada and the United States at the peak of the last ice age approximately 20,000 to 30,000 years ago; *(b)* the coast of Europe and the Mediterranean Sea during the last ice age; *(c)* dates of fossil corals (circles), which grow near sea level, provide a history of sea-level change since the last ice age.

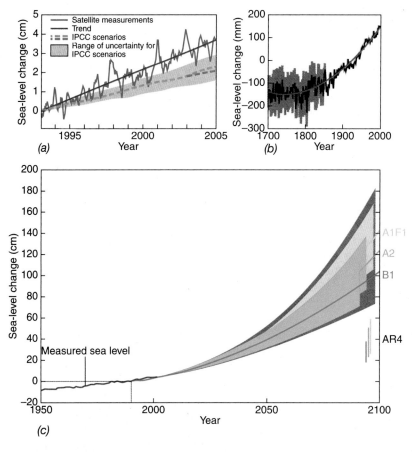

FIGURE 22.20 *(a)* The rate of modern sea-level rise (blue and red lines) is exceeding the worst-case scenario of computer models from the Intergovernmental Panel on Climate Change (IPCC; pink zone and dashed lines). *(b)* Studies have found that the modern acceleration in the rate of sea-level rise may have started over 200 years ago (uncertainty of the study is shown in grey). *(c)* Comparisons of global temperature and sea-level change over the past 130 years have been used to project future sea-level position. Research indicates that sea level may rise between 0.75 m and 1.9 m by 2100, depending on the rate at which we produce greenhouse gases. The future sea-level projection of the 2007 IPCC Assessment Report 4 (AR4) is now considered an underestimate of the threat of sea-level rise. B1 is a best case scenario of greenhouse gas production; A2 and A1F1 are moderate and worst-case scenarios (uncertainty shown in shades).

(?) What options do coastal cities have when faced with sea-level rise?

become a global problem plaguing coastal communities and the tourism industry.

The relationship between rising sea level and coastal erosion is not well understood. A beach cross-section (called a **beach profile**) has a characteristic shape that depends on the size of sand grains and the energy of waves. The beach profile can be described by the distance offshore at which waves first affect the profile (L) and the depth there (D). Typically the ratio L/D can vary between 10 and 200. As sea level rises, the profile shifts landward to regain the L/D ratio that is natural for that setting; it achieves this ratio by eroding the shoreline (**FIGURE 22.21**). If L/D is approximately 100, the change in D due to sea-level rise (usually in millimetres) can translate into horizontal beach erosion two orders of magnitude greater than D. Hence, with global sea-level rise currently at over 3 mm/y, this translates into approximately 30 cm/y of erosion, or 3 m per decade.

The rate of sea-level rise today, measured by satellites mapping the ocean surface and extrapolated forward in time to reflect the estimated rate of rise over a century, is the highest ever observed: 34 cm/century. This is approximately twice the rate recorded in the 20th century. Scientists have projected that, by the end of this century, sea level will rise between 0.75 m and 1.9 m depending on the rate at which we produce greenhouse gases (**FIGURE 22.20**).

The physical impacts of sea-level rise along the coast can be grouped into five categories: inundation of low-lying areas, erosion of beaches and bluffs, salt intrusion into aquifers and surface waters, higher water tables, and increased flooding and storm damage. All of these impacts have important effects, but the first two have had and are continuing to have very dramatic impacts on coastal regions worldwide. Tidal wetlands, mapped by satellite, are shrinking on the shores of estuaries around the world, and beach erosion has

(?) **Expand Your Thinking**—Most of the goods traded between countries travel by ship. How could sea-level rise affect the global shipping industry?

(?) What will happen to houses or roads built next to an eroding beach?

FIGURE 22.21 As sea level rises, the average beach profile will shift landward, eroding the land. The amount of erosion may be approximately 100 times the amount of sea-level rise.

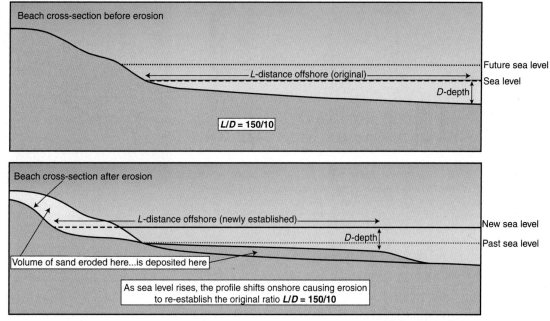

22-8 Barrier Islands Migrate with Rising Sea Level

LO *22-8 Define "barrier islands," and explain how they migrate.*

Coastal processes related to waves, storms, tides, and sea-level rise interact with the underlying geology of the coastline to produce various types of *coastal environments*. Low-elevation coastlines with an abundance of sand tend to form **barrier islands** characterized by sandy beaches, tidal inlets, lagoons, and tidal marshes (FIGURE 22.22).

Barrier Islands

Barrier islands are long, low, narrow sandy islands with one shoreline that faces the open ocean and another that faces a saltwater *tidal lagoon* protected by the barrier. The barrier is separated from the mainland by one or more *tidal inlets* through which flows sea water driven by differences in tide level between the lagoon and the open ocean. Changes in tide lead to the movement of strong tidal currents through the inlet, exchanging water and sediment between the lagoon and the ocean. Sediments carried by a flooding tide from the ocean into the lagoon collect at the landward end of the inlet in a deposit known as a *flood-tide delta*. Those that collect on the seaward end of the inlet, carried by an ebbing tide, form an *ebb-tide delta*. Tidal deltas are important sources of sediment for the adjacent lagoon and the ocean shoreline and sea floor.

Barrier islands are found on coastlines around the world. In the United States, chains of barrier islands are found along the east and south coasts, where they extend from New England down the Atlantic coast, around the Gulf of Mexico, and south to Mexico. They include the Outer Banks of North Carolina (FIGURE 22.23), a long line of sandy barrier islands extending over 280 km that protects the broad waters of Pamlico Sound, a tidal lagoon.

(a)

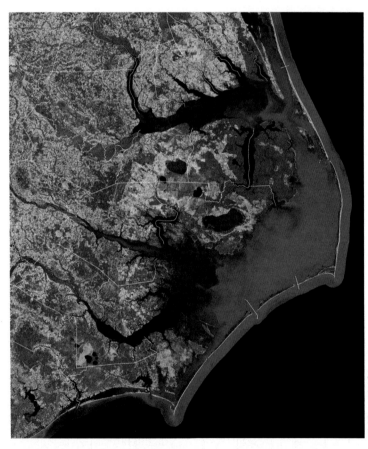

(b)

FIGURE 22.22 *(a)* Many barrier islands such as this one along Miami, Florida, are heavily developed because they are popular destinations for vacationers. *(b)* Differences in water level created by the tides drive circulation between a lagoon and the open ocean. Sediments deposited at the ocean and lagoon ends of the tidal inlet are known as tidal deltas, as shown here for Delaware Bay.

FIGURE 22.23 The long, thin line of barrier islands located off the Atlantic coast of North Carolina is called the Outer Banks. These islands protect a broad tidal lagoon, Pamlico Sound, that exchanges water with the ocean through tidal inlets.

 Describe the role of storms in a barrier-lagoon setting.

 Describe the role of tides in a barrier-lagoon setting.

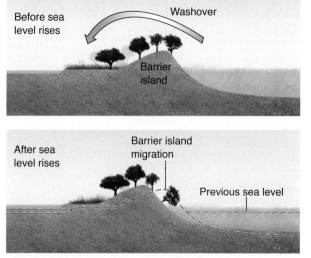

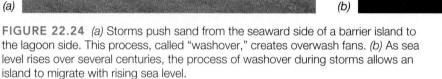

FIGURE 22.24 *(a)* Storms push sand from the seaward side of a barrier island to the lagoon side. This process, called "washover," creates overwash fans. *(b)* As sea level rises over several centuries, the process of washover during storms allows an island to migrate with rising sea level.

Given how barrier islands migrate with rising sea level, are they a safe place to build a community? Explain your answer.

Barrier islands are constantly changing environments that shift their position when they are overrun by storm surge. When a storm such as a hurricane encounters a low barrier island, the flood waters carry sand from the ocean side of the island and deposit it on the lagoon side. This process is called **washover,** and the resulting sandy deposit is called an *overwash fan.* Sand also moves offshore during storms and into the lagoon through tidal inlets under non-storm conditions. Like a giant treadmill, washover causes barrier islands to roll landward by eroding on the ocean side and depositing on the lagoon side while at the same time losing some sand offshore. This process has been referred to as *barrier rollover.* Barrier rollover, driven by storm surge, washover, and sand transport through tidal inlets into lagoons, allows an island to migrate landward under conditions of sea-level rise and thereby not drown in place (**FIGURE 22.24**).

Barrier islands are composed of several habitats, typically oriented parallel to the shore (**FIGURE 22.25**). Immediately landward of the ocean beach is a line of sand dunes that store sand delivered by winds blowing off the sea. Inland lies a wide sand-covered plain vegetated by drought-tolerant grasses or a forest thicket. On the lagoon side, **tidal wetlands** grow on the surface of old overwash fans. Salt-tolerant plants grow in the muddy substrate of these marshes, which are flooded daily by high tide. As these plants die and accumulate over time, they form *peat* between the levels of high and low tide. Eventually barrier rollover buries the peat under overwash fans and migrating wind-blown dunes as the island shifts landward during periods of rising sea level. The emergence of these peat layers on the ocean side of the island after hundreds or thousands of years of burial provides remarkable proof of barrier rollover.

© Jupiter Images/Thinkstock/Getty Images, Inc.

FIGURE 22.25 Barrier islands consist of several zones or habitats. On the seaward side of the island is a beach and surf zone. Landward of the beach is a system of sand dunes and a wide sand-covered plain vegetated by drought-tolerant grasses or a forest thicket; on the lagoon side are tidal wetlands growing on former overwash fans.

What happens to tidal wetlands when the barrier island rolls over them?

Expand Your Thinking—Explain how the process of barrier rollover would differ under slowly rising sea level versus rapidly rising sea level.

22-9 Rocky Shorelines, Estuaries, and Tidal Wetlands Are Important Coastal Environments

LO 22-9 *Identify several types of coastal environments.*

Rocky shorelines are found around the world in places where steep hills or mountains descend directly into the sea (FIGURE 22.26). Types of rock that differ in their ability to withstand wave erosion are a characteristic of rocky shores. Those rocks that succumb to erosion more easily retreat while those that are more resistant protrude into the surf. *Sea stacks,* composed of resistant rock that can withstand the pounding of waves and bioerosion by intertidal organisms, form in this manner. Sea stacks mark the former position of the shoreline before erosion caused the coast to retreat landward.

As a rocky shoreline retreats due to erosion, it leaves behind a flat, wave-scoured platform marking the position between high and low tides. This *wave-cut platform* continues to widen as the shoreline retreats. Platforms widen as the sea erodes the base of a cliff by forming a wave-cut notch (FIGURE 22.27). As the notch undercuts the cliff face, the weight of the unsupported rock above eventually collapses in a rock slide, which is a type of *mass wasting.* The sea removes the slide debris, and the notch-eroding process begins again. As the cliff retreats, the very base of the notch is left behind, leaving a wide bench—a wave-cut platform. Eventually, the platform becomes so wide that the sea expends too much energy reaching the cliff, and the shoreline retreat slows and may end.

FIGURE 22.27 As wave erosion cuts away at the base of a cliff, forming a notch, it creates a broad rocky platform marking the retreat of the rocky shoreline like that shown here below the sandstone cliffs of Kildare Capes, Prince Edward Island.

(?) Explain the role of tides in forming a wave-cut platform.

Estuaries and Tidal Wetlands

As ocean waters rose after the last ice age, they flooded coastlines and moved shoreline environments landward, a process termed **marine transgression.** Sea water invaded and flooded the mouths of stream valleys, widening them to form our modern estuaries. An **estuary** is a partially enclosed body of water that forms in places where fresh water from rivers and streams flows into the ocean, mixing with the salty sea water. Estuaries, and the lands surrounding them, mark the transition from terrestrial/freshwater systems to marine/saltwater systems. They are protected from the full force of marine waves and storms by reefs, barrier islands, or fingers of land *(spits)* that absorb the ocean's energy. Generally thought of as relatively calm bodies of water, estuaries are influenced by the tides, although the restricted opening of an estuary changes the timing and height of tides compared to those in the nearby open ocean.

Also known by other names, such as bays, lagoons, harbours, inlets, or sounds, estuaries come in all different sizes and shapes. However, the defining feature of an estuary, regardless of its name or size, is the mixing of fresh and salt water. Familiar examples of estuaries include Victoria Harbour, Burrard Inlet at the mouth of the Fraser River, San Francisco Bay, Puget Sound, Chesapeake Bay, and the Gulf of St. Lawrence (FIGURE 22.28). Estuarine environments are among the most biologically productive on Earth, creating more organic matter each year than similar-size areas of forest, grassland, or agriculture. The sheltered waters of estuaries are home to unique communities of plants and animals that are specially adapted for life at the margin of the sea. These plants and animals occupy various

FIGURE 22.26 Formed of erosion-resistant rock, sea stacks such as this along the East Coast Trail of Newfoundland mark the locations of former shorelines that have shifted landward as a result of erosion.

(?) How might a sea stack influence coastal processes in the immediate vicinity?

Explain the role of sea-level rise in the evolution of the Gulf of St. Lawrence.

FIGURE 22.28 The Gulf of St. Lawrence is the world's largest estuary, covering an area of about 236,000 km² where the Great Lakes of North America drain into the Atlantic Ocean via the St. Lawrence River.

habitats in and around estuaries, including shallow open waters, tidal wetlands, sandy beaches, mudflats and sand flats, oyster reefs, mangrove forests, river deltas, coral lagoons, and sea grass beds.

Tidal wetlands, also called *salt marshes,* develop along the fringes of many estuaries (FIGURE 22.29). Water draining from the uplands carries sediments, nutrients, and pollutants that are deposited in mud-rich wetlands and prevented from entering estuarine waters. Wetland plants and soils act as a natural buffer between the land and ocean, absorbing flood waters and dissipating storm surges. Salt marsh grasses and other plants help stabilize the shoreline and counteract coastal erosion.

Salt marshes are formed in quiet, low-energy environments that are protected from direct wave action yet still experience the energy of storm surge, weather, and daily tide changes. Marshes are typically flooded daily by high tide, but the spring flood tide reaches farthest inland and covers the entire wetland surface. During flood tide, fine-grained suspended silts and clays settle out of the water and accumulate on the marsh surface. The stalks and leaves of salt-tolerant plants assist this process by capturing grains of silt and clay while they are still suspended in flood waters; rain later washes these onto the marsh surface. As sediment accumulates, the coastal basins and embayments holding marshes are filled with mud.

Estuarine and wetland histories are tied to sea-level rise. Globally, sea-level rise slowed about 5,000 to 3,000 years ago; by 2,000 years ago, the rate of global sea-level rise was only about 1 mm per year or less (less than one-third of its present rate). Under this new environment of slow marine transgression on the U.S. Atlantic coast, a single species of tidal marsh grass, smooth cord-grass (*Spartina alterniflora*), established the first permanent foothold in the low-energy embayments of drowned stream valleys. Grass shoots slowed the movement of the water, trapping and accumulating even more sediment. This enhanced sedimentation, coupled with the

increasing volume of root systems, allowed the newly developing marsh surface to keep up with rising sea level and even expand seaward into the open estuary. As a result, tidal wetlands have become important, widespread estuarine environments that play key roles in coastal sedimentation and biological diversity.

FIGURE 22.29 Tide wetlands are characterized by broad expanses of salt-tolerant grasses and numerous meandering tide channels.

Expand Your Thinking—Look at Figure 22.28; what will happen to the Gulf of St. Lawrence estuary as sea level continues to rise due to global warming? What major impacts can be expected?

22-10 Coasts May Be Submergent or Emergent, Depositional or Erosional, or Exhibit Aspects of All Four of These Characteristics

LO 22-10 *Distinguish between submergent and emergent coasts and between depositional and erosional coasts.*

Sea level has risen on coastlines around the world since the end of the last ice age. The influence of this phenomenon on coastal environments depends in part on the relative rates of vertical change in land elevation in relation to vertical change in sea level. Those coasts where the land is tectonically uplifting at a rate that matches or exceeds sea-level rise are known as **emergent coasts.** Those where sea-level rise exceeds uplift are known as **submergent coasts.**

Another way to classify coastal systems is by the relative magnitude of erosional versus depositional processes. **Erosional coasts** are dominated by erosional processes, and **depositional coasts** are dominated by depositional processes.

Emergent and Submergent Coasts

Emergent coasts are characterized by fossil wave-cut platforms, or *beach ridges,* that are uplifted above the present-day sea level. For example, the coasts of British Columbia and California (**FIGURE 22.30**) are being uplifted by tectonic forces associated with the active boundary between the Pacific Plate and the western edge of the North American Plate. Uplift can be sudden when it is associated with large earthquakes, which is known as *co-seismic uplift.* That is, uplift occurs at the same time as an earthquake and is caused by the same seismic event. This type of uplift produces a cliff at the coastline that is eroded by waves during the interval of time between co-seismic events.

Another type of emergent coast is one that is uplifting because the weight of a glacier has been removed due to climate warming since the end of the last ice age. In these areas, usually located in extreme northern and southern latitudes along continental shores (e.g., Norway; Hudson Bay, Canada), the lithosphere slowly rebounds due to the removal of the weight of a past glacier. This process is called **glacioisostatic uplift.** It occurs because the lithosphere has elastic properties that cause it to recover from the former weight (**FIGURE 22.31**). Often sequences of emerged beach ridges mark the history of the uplift.

Submergent coastlines experience slow flooding by the ocean as a result of sea-level rise, lithospheric subsidence, or both. Any place where the net result of sea-level change and vertical land movement results in marine drowning is a submerging coast. In the current era of accelerated sea-level rise, many coastal systems are submerging. These include areas where estuaries form in the flooded mouths of river systems, such as along the eastern coast of the United States and Canada (e.g., Gulf of St. Lawrence, Figure 22.28).

In places where great volumes of sediment collect, such as major delta regions, significant subsidence occurs due to sediment compaction and the downward flexing of the lithosphere by the enormous weight of the sediment. When a rise in sea level is added to this subsidence, rapid flooding occurs across broad expanses of low-lying shoreline. The Ganges-Brahmaputra River delta region is an example

FIGURE 22.30 The coastlines of central California, British Columbia, and Alaska are dominated by emergent coasts because tectonic forces cause uplift faster than the rate of sea-level rise. In this photo, a broad fossil marine terrace marks the top of the coastal cliff.

 Describe the tectonic setting of the British Columbia coast. Why does it lead to co-seismic uplift?

FIGURE 22.31 When the weight of a glacier is removed, the lithosphere elastically rebounds. In places such as this beach at Bathurst Inlet, Nunavut, a series of gravel beach ridges may form, with each ridge marking a former shoreline that has since been lifted above the wave zone.

 What is the evidence that this coast is emergent?

Today
Total population: 15 million
Total land area: 134,000 km^2

(a)

Sea-level rise: 1.5 m
Total population affected: 17 million (15%)
Total land area affected: 22,000 km^2 (16%)

(b)

FIGURE 22.32 The region around Dhaka, Bangladesh, is one of the most densely populated on Earth. By 2050, continued sea-level rise will displace over 17 million people in Bangladesh and drown 16 percent of the nation's land.

(?) Why are deltas especially vulnerable to sea-level rise?

of a submerging coast. This region on the coast of Bangladesh, one of the world's poorest countries, is expected to undergo a rise of 1.5 m in sea level by 2050 due to the combination of global warming and submergence. The nation will experience a 16 percent loss of land area, and 15 percent of its population will be displaced (**FIGURE 22.32**).

Erosional and Depositional Coasts

Another way to classify coastal systems is by the net result of physical processes. Erosional coasts are those that are experiencing net landward migration due to erosion. Most often they include rocky shorelines undergoing landward retreat. Sea stacks, wave-cut beaches, and undercut cliffs characterize erosional coasts.

Erosional coasts typically have steep topography associated with tectonically active margins, with a narrow continental shelf, steep coastal plains, and youthful mountain systems that have not existed long enough to shed high volumes of sediments. Rivers tend to cut into steep mountainous watersheds with high gradients. Sediments are coarse grained, and wetlands and estuaries tend to be narrow and poorly developed. All these features are typical of shorelines that generally act as sediment sources rather than as sites of sediment accumulation.

Depositional coasts often are associated with older mountain systems that are no longer tectonically active. With long periods of geologic time in which to shed sediments into the coastal zone, a broad

(a)

(b)

FIGURE 22.33 (a) Erosional coasts are characterized by eroding sea cliffs, sea stacks, and short beaches in narrow embayments. (b) Depositional coasts collect large quantities of sediment. Coastal wetlands tend to be found on depositional coasts.

(?) How is the tectonic setting likely to differ between erosional and depositional coasts?

continental shelf to decrease wave energy, and wide coastal plains to collect and store sediments, these coastal systems are characterized by barrier island chains, deltas, broad lagoons, and extensive tidal wetlands. River valleys tend to be wide and flat. In general, depositional coasts are found in mid-plate settings where little active mountain building takes place (**FIGURE 22.33**).

(?) **Expand Your Thinking**—Considering that global sea-level rise has accelerated, which of these four coastal types is most vulnerable to coastal hazards, and which is least vulnerable? Explain your reasoning.

CRITICAL THINKING

COASTAL SYSTEMS

Coastal environments are complex geologic systems that receive sediments, dissolved compounds, nutrients, and a mixture of fresh and salty water from terrestrial and marine environments. Population increase along coasts has led to declines in environmental resources, pollution, increased vulnerability to coastal hazards, and other negative impacts.

Please work with a partner and answer the following questions using FIGURE 22.34.

1. Label the coastal environments.
2. Make a table with five columns labelled (left to right): Environment, Sediment Type, Rock Type, Coastal Processes, and Sea-Level Rise. List each coastal environment in the left column. In the next four columns, fill in the following:
   a. Describe the sediments that collect in each environment.
   b. Name the type of sedimentary rock that will result from each environment.
   c. Describe how waves and tides influence each environment.
   d. Consider sea-level rise. How will it affect each environment?
3. Label the diagram with the *coastal hazards* that will affect each environment.
4. You are the mayor of a small coastal town with a population of 5,000. In the summer of 2013, a hurricane hit the barrier island on which your town is located. Seventy-five percent of the buildings in town were completely destroyed by a combination of high winds and storm surge. The other buildings were damaged to some degree. Roads were torn up by waves; electricity and sewage infrastructure was destroyed. The damage to the community totalled $7 billion. The government asks you to consider abandoning the town and offers you new land on the adjacent mainland. What do you do? How do you involve the whole community in making the decision? Write a scenario in which you lead the community to a decision.

FIGURE 22.34 Coastal environments.

22-11 Coral Reefs Are Home to 25 Percent of All Marine Species

LO 22-11 *Describe the formation of coral reefs, and define the main types of reefs.*

Although a large and interdependent ecosystem relies on **coral reefs** for survival, reefs occupy a mere 0.2 percent of the world's oceans. Not only are these underwater habitats crucial to the ocean's biodiversity, they are home to 25 percent of all marine species.

Coral Reefs

Coral reefs are shallow coastal environments where numerous types of *coral* (an animal) and *algae* (a plant) build massive solid structures of calcium carbonate ($CaCO_3$ [limestone]). Corals secrete this material, which forms their skeletons, as part of their metabolic activity (**FIGURE 22.35**).

Individual coral animals are called *polyps.* A polyp has a single opening that functions both to take in food and expel waste; this opening is surrounded by a fringe of tentacles. Each polyp grows within its own hard cup, called a *calyx,* depositing a solid skeleton of $CaCO_3$ as it grows. As layer upon layer of calcium carbonate builds up over time, the reef framework is created, with individual coral animals, numbering in the hundreds of thousands, living like a skin on the surface. Under ideal conditions, some corals can grow several centimetres per year. The process of secreting calcium carbonate in enormous groups allows corals to grow into huge colonies. Each colony is made up of individual polyps, all of which are linked to their neighbours by connective tissue that includes their stomach. So when one polyp eats, they all eat!

Living among the reef builders are myriad fish, molluscs, echinoderms, and other types of organisms in the world's most biologically diverse marine ecosystem. Up to 3,000 different species of plants and animals may be found living together on a single coral reef. It is estimated that one-quarter of all known fish species live on or around coral reefs. Reefs are valuable natural assets that provide food, jobs, protection from large waves and storm surge, and billions of dollars in revenues each year to local communities and national economies.

Living inside the corals are tiny *symbiotic algae,* a beneficial plant that converts sunlight and nutrients into food for coral growth and calcium carbonate production. Because these algae depend on light for photosynthesis, corals require direct access to sunlight for growth and can be damaged by sediment or other factors that reduce the clarity or quality of the water around them.

Coral reefs are generally found only between 30°N and 30°S of the equator (**FIGURE 22.36**), but even within this belt, they do not occur in areas where fresh water and heavy coastal sedimentation prevent coral growth. Over geologic time, reefs have survived fluctuating sea levels, uplifting land masses, periods of widespread warming and repeated glaciations, as well as recurrent short-term natural disasters, such as cyclones and hurricanes. In short, coral reefs have a remarkable ability to adapt and survive.

David Nardini/The Image Bank/Getty Images, Inc.

? Reefs are threatened around the world. What steps would you take to manage them?

FIGURE 22.35 Composed of diverse marine organisms, the world's coral reefs have been called "the rainforests of the sea."

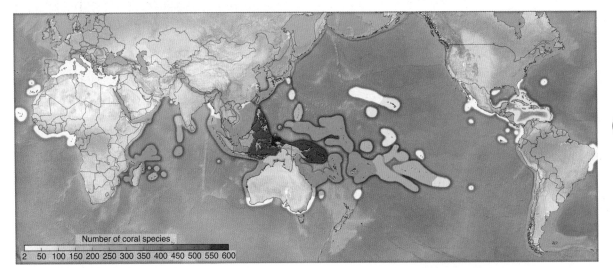

Number of coral species
2 50 100 150 200 250 300 350 400 450 500 550 600

? What types of environmental conditions do corals like?

FIGURE 22.36 Located between 30°N and 30°S of the equator, coral reefs require shallow, warm, sunlit seas.

(a)

(b)

(c)

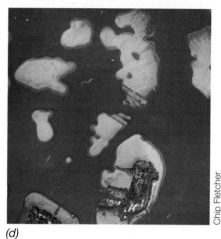
(d)

? Describe how atolls form.

FIGURE 22.37 *(a)* Fringing reefs are attached to a mainland coast. *(b)* Barrier reefs, such as the Great Barrier Reef of northeastern Australia, have a deep lagoon that separates the reef from the coast. *(c)* An atoll forms when the island to which it is attached subsides below sea level. Small sandy islets often develop on the atoll's rim. *(d)* Patch reefs are small, isolated reefs that grow in lagoons or on reef flats.

Types of Reefs

There are four basic kinds of coral reefs: *fringing reefs, barrier reefs, atolls,* and *patch reefs* (**FIGURE 22.37**).

Fringing reefs grow in shallow water and are directly attached to a coastline. They consist of several ecological zones that are defined by their depth, the shape of the reef's surface, and the plant and animal community it supports (**FIGURE 22.38**). These zones include the *shore zone*; the inner *reef* or *reef flat* (located in shallow water with little wave energy); the *reef crest* (the shallowest seaward portion of the reef, which causes waves to break); and the *outer reef* or *fore reef* (the seaward-sloping front below the reef crest). Each zone is characterized by a unique assemblage of coral, algae, and other organisms. All four types of reefs have similar zones.

Barrier reefs are separated from land by a lagoon. Lagoons can be very deep, exceeding 30 m in depth and characterized by sandy floors, prolific coral growth, and strong tidal currents. Barrier reefs can be large, regional-scale features characterized by separate islands, inlets, and patch reefs. The zones found on a fringing reef are also found on a barrier reef.

An atoll is another type of coral reef. Atolls are shaped like irregular doughnuts with a broad, deep lagoon in the middle. Atolls originate from fringing reefs attached to volcanic islands that subside over time. As the island sinks (usually due to plate cooling as it moves away from a hotspot) and erodes (due to stream and wave erosion), the reef continuously grows upward in order to remain in shallow sunlit waters. Eventually a barrier reef may develop when the volcanic island subsides below sea level. By the time the last of the island sinks below the waves, the reef has formed a nearly circular shape with a deep central lagoon.

Patch reefs are small, isolated reefs typically several tens of metres in diametre; they are located in lagoons or on fringing reef flats.

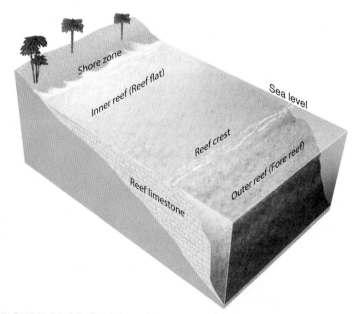

FIGURE 22.38 Reefs have four ecological zones: shore zone, reef flat, reef crest, and fore reef.

? What are the environmental conditions in each zone?

? **Expand Your Thinking**—Explain the various ways in which global warming could affect the world's reefs.

22-12 Coastal Problems Are Growing as Populations Increase

LO 22-12 *Discuss the effects of human activities on coastal environments.*

Because the coastal zone has become the focus of dramatic population growth, development, and investment, humans are increasingly exposed to coastal hazards and damage to coastal resources. This is particularly important for Canada which hosts the longest coastline in the world at about 243,000 km. Every year, Canadians invest about $1 billion on projects related to our coasts. The funds are used to maintain ports and harbours, for erosion protection, pollution control and aquaculture, and for recreation facilities. This cost will undoubtedly increase with time as sea-level rise is bringing coastal hazards to the doorsteps of our buildings. This trend is especially worrisome, as researchers have documented changes in the behaviour of hurricanes due to increased sea surface temperatures.

Coastal Erosion

Coastal erosion is a worldwide problem that threatens beaches, coastal lands, and the economic and environmental systems that rely on them. For instance, according to a U.S. study released by the Federal Emergency Management Agency (FEMA), approximately 25 percent of homes and other structures within 150 m of the U.S. coastline and the shorelines of the Great Lakes will fall victim to the effects of coastal erosion within the next 60 years. The worst-hit areas include the Atlantic and Gulf coastlines, which are expected to account for 60 percent of nationwide losses due to erosion. Costs to U.S. home owners will average more than $500 million per year in the form of higher insurance premiums, replacement of damaged buildings, and engineering measures to counteract erosion (FIGURE 22.39).

As a result of studies such as FEMA's, the U.S. and Canadian governments, along with their state and provincial counterparts, are recommending that coastal communities take steps to avoid erosion. This *avoidance policy* is designed to circumvent the predicted future of high nationwide costs, hardships and financial burdens for home owners, and negative impacts on shoreline environments that suffer when erosion and human land use run afoul of each other. There is really only one good response when coastal erosion strikes: Get out of the way. Seawalls and other types of walls may protect a house or road temporarily, but they usually cause beach loss, wetland loss, and other types of environmental impacts because they set an artificial boundary on the ecosystem preventing it from migrating with changing sea level and other shifting conditions. Seawalls also separate beaches from sand dunes, an important source of sand. Given the inevitability of continued sea-level rise and changes in coastal storms, remaining on an eroding coast is a dangerous choice.

Coastal Pollution

Among the major threats to coastal resources is the increased occurrence of **polluted runoff** in coastal waters adjacent to urban, suburban, industrialized, and agricultural lands. Polluted runoff is the contaminated water that runs off city streets, building sites, parking lots, and farmlands. It now accounts for 75 percent of the pollution in coastal waters. Polluted runoff comes from many sources: used engine oil poured down storm drains; pesticides and fertilizers applied to farms and yards; dirt that washes into the coastal zone from logging, agricultural, or mining sites; industrial solvents and chemicals; and manure from livestock operations. Along Canadian and U.S. coasts, the problem of polluted runoff is especially fierce because there are more people living in coastal communities than ever before. These millions of people produce huge quantities of pollution that is released into coastal waters.

Agriculture is the largest contributor of polluted runoff. Fertilizer and manure, along with more than a billion tonnes of eroded soil, flow from farm fields into coastal waters every year. Over 13,000 tonnes of farming chemicals are applied annually in areas that drain into the coasts. Some of these compounds kill native marine ecosystems while others fuel the growth of noxious forms of algae that lead to significant depletion of oxygen in the water in a process called **eutrophication** (FIGURE 22.40). This causes other forms of life literally to suffocate.

© Brett R. Henry/Alamy

(?) How will this scene change as sea-level rise continues?

FIGURE 22.39 Over the next 60 years within the United States, approximately 25 percent of homes within 150 m of the shoreline will be lost to erosion.

FIGURE 22.40 Eutrophication occurs when algae populations in coastal waters increase rapidly because of high nutrient content. This increase robs the water column of oxygen that is necessary to other forms of life, often leading to massive fish kills in coastal waters and estuaries.

FIGURE 22.41 In cities such as Vancouver, as populations grow in the coastal zone, buildings, roads, homes, and hospitals are built in locations that are vulnerable to natural coastal hazards. Sea-level rise associated with global warming will make these locations even more vulnerable.

Coastal Hazards

Throughout geologic history, high waves, coastal erosion, tsunamis, strong winds, hurricanes, and changing sea levels have come and gone largely unnoticed. Natural coastal processes are a cause for concern only when humans get in the way. When they occur near buildings and roadways, these natural processes become **coastal hazards,** a term that refers to any coastal process that may cause loss of life or property.

Storms have battered our coasts for millennia, but in recent decades they have become stronger and more destructive. Mounting scientific evidence links this trend to global warming. Why? Because hurricanes get their power from warm surface waters underlying the storm. Since ocean temperatures have risen with global warming, fiercer storms are brewing around the globe, which compounds the losses, both human and economic, related to natural disasters.

Worldwide, economic losses caused by natural hazards have more than tripled over the last three decades and amount to $3.5 trillion. In Canada over the last 50 years, losses due to natural disasters exceeded $25 billion. In the United States, annual losses due to natural disasters have climbed from $4.5 billion in 1970 to $15 billion to $20 billion (in 1970 dollars) in 2003; a large portion of this increase is directly related to coastal hazards. In 2005, Hurricane Katrina alone caused over $100 billion in losses to the city of New Orleans. These figures illustrate a trend of escalating costs due to natural hazards that has many government agencies and private insurance companies concerned. The primary explanation for this increase is population growth in hazard-prone locations, including coastal areas. As a result, homes, businesses, schools, highways, and hospitals are built in locations that are particularly vulnerable to chronic or catastrophic coastal hazards, such as hurricanes, severe storms, sea-level rise, coastal erosion, and tsunamis (**FIGURE 22.41**).

Studies show a global trend toward increased wind speed in the strongest hurricanes over the past two or three decades. The strongest trends are in the North Atlantic Ocean and the Indian Ocean. According to the Intergovernmental Panel on Climate Change (IPCC), it is "more likely than not" that there is a human contribution to the observed trend of hurricane intensification since the 1970s. The IPCC predicts that "it is likely that future hurricanes will become more intense, with larger peak wind speeds and heavier precipitation." Researchers have concluded that with global warming, the strongest storms may grow in intensity, and there may be fewer weak ones. The overall strength of storms as measured in wind speed would rise by 2 percent to 11 percent, but there would be between 6 percent and 34 percent fewer storms in number. Essentially, there would be fewer weak and moderate storms and more of the big damaging ones, which also are projected to be stronger due to warming.

The size and scope of these highly damaging events have had a profound effect on public policy and perceptions concerning hazards and what can, or should, be done to minimize their impacts on vulnerable populations. Because of the significant costs of catastrophic natural disasters, agencies in charge of emergency management have expanded their focus in recent years beyond disaster preparedness to include **hazard mitigation.** Hazard mitigation is an effort to minimize individual and community vulnerability to disasters. The primary purpose of mitigation is to ensure that fewer individuals and their property are victims of disaster. The intent of this effort is to reduce injuries and loss of life as well as the overall cost and economic burden of natural hazards.

⁇ Expand Your Thinking—What options do coastal communities have in dealing with the issues of coastal erosion, coastal pollution, and coastal hazards?

LET'S REVIEW "GEOLOGY IN OUR LIVES"

Now that you have finished the chapter, "Geology in Our Lives" will have taken on new meaning for you. Let us review it: Canada's ocean and Great Lakes coastlines are home to almost 13 million people, about 40 percent of the total Canadian population. In the United States, more than 10 times this amount—almost 153 million people or about 53 percent of the U.S. population—live along coastlines. Surging population growth in coastal zones exposes more people to the dangers of geologic hazards, such as storms, hurricanes, and tsunamis, than in any other geological environment. The world's coasts are home to fragile ecosystems, beautiful vistas, pristine waters, and major growing cities, all coexisting in a narrow and constricted space. Expanding human communities compete for more space at the expense of extraordinary wild lands. There are problems with coastal erosion, waste disposal, a dependency on imported food and water, and rising sea level. Management strategies to ensure the long-term viability of our coasts are largely science based; thus, scientists are important partners in building sustainable coastal communities.

As the population grows, coastal problems increase. Growing coastal pollution degrades the natural ecosystem. Because the coastal zone has become the focus of dramatic population growth, development, and investment, exposure to coastal hazards and damage to natural coastal resources have risen equally dramatically. Processes such as sea-level rise, storm surge, and coastal erosion, along with many other coastal hazards, threaten the lives of coastal residents and the billions of dollars of infrastructure that has been built along the edge of the oceans. In the end, as government agencies work to mitigate these threats, the simplest, but perhaps not the easiest, solution is to pull back from the water's edge and avoid the area of greatest hazard. To achieve this goal in the era of global warming will require a citizenry that is educated in the science of coastal geology.

STUDY GUIDE

22-1 Change is constantly occurring on the shoreline.

- Coastlines are unique and complex geologic systems. Influenced by processes and materials from the land, sea, and atmosphere, coastlines are ever-changing environments. Because of their temperate climate (air temperature is moderated by heat stored in the ocean) as well as their beautiful vistas and abundant resources, coastal lands are the most crowded and developed areas in the world. Fifty percent of the population of the industrialized world lives within 100 km of a coast, a population that will grow by 15 percent during the next two decades.

- Winds, waves, tides, sea-level change, and currents act as agents of change, collectively called **coastal processes.** They sculpt the shoreline through erosion and deposition of sediment and by flooding low areas. Understanding how coastal processes and sedimentary materials interact is crucial to interpreting the geologic history of coastal systems and finding ways for people to live in the coastal zone in a sustainable manner.

22-2 Wave energy is the dominant force driving natural coastal change.

- Waves on the ocean's surface are generated by the friction of air blowing across the water. Hence they are called "wind waves." A wave travels through water without making a current. Energy radiates outward from the source but water does not. The size of a wave is governed by wind velocity, the fetch or distance over which the wind blows, and the length of time the wind disturbs the water.

- As a wave enters shallow water, it slows before those following it. The following wave, still moving at its original speed, tends to catch up with the wave in front of it. The wavelength decreases and the wave's height increases. The crests become narrow and pointed with steep faces while the troughs become wide, shallow curves. Because the wavelength decreases and the speed slows, the wave period remains unchanged.

- The strongest winds (over 54 km per hour) and highest waves (over 6 m) are found in the Southern Ocean and are associated with stormy regions in the North Atlantic and the North Pacific.

22-3 Wave refraction and wave-generated currents occur in shallow water.

- Wave **refraction** occurs when the wave crest bends so that it aligns parallel to the contours of the sea floor. For straight coasts with parallel contours on a shallow sea floor, refraction tends to orient waves parallel to the coastline.

- Water piles up against the shore, raising the water level and creating sea-level set-up, which leads to the formation of **wave-generated currents.** The current that flows along the bottom in the seaward direction to counterbalance the onshore movement of water in the surf zone is called an *undertow.*

22-4 Longshore currents and rip currents transport sediment in the surf zone.

- When waves approach a shoreline from an angle, they generate a **longshore current** that moves along the shoreline in the direction of wave movement. Part of this current is due to the periodic release of sea-level set-up and part of it is due to wave momentum directed obliquely to the shoreline.

- An important consequence of sea-level set-up along a shoreline is the formation of **rip currents.** As water piles up along a beach, it is released periodically in the form of offshore-directed currents that surge through the surf zone. These currents carry suspended sand and floating debris. As a result, rip currents can be detected by an observer as dirty or dark water that extends through a surf zone beyond the breakers.

- Wave-generated currents move sand that has been eroded from a rocky headland into the adjoining embayment, thereby building a beach. Over time the headland is worn down, and the bay fills with sand. Eventually the shoreline becomes a long, straight beach.

22-5 Gravity and inertia create two tides every day.

- The oceans rise and fall twice every day in response to **tide**-raising forces related to the gravitational attraction between Earth, the Moon, and the Sun.

- Two tide-raising forces are responsible for the lunar ocean tide: the gravitational attractive force and the inertial force. The balance between these two opposite forces defines the tide-raising force. On the side of Earth facing the Moon, the gravitational attraction force is greater than the inertial force, and a tidal bulge is created in the ocean's surface. On the side of Earth facing away from the Moon, the inertial force is greater and creates a tidal bulge. Earth rotates beneath its watery envelope and encounters these two bulges (high tide) twice every day.

- When the Sun, Earth, and the Moon are aligned, the spring tide occurs. When the Moon is at 90° to the Sun–Earth alignment, the neap tide occurs.

22-6 Hurricanes and tropical storms cause enormous damage to coastal areas.

- Called "the greatest storm on Earth," a **hurricane** is capable of causing massive damage to coastal areas with sustained winds of over 200 km per hour, intense rainfall, and flooding ocean waters with huge waves. In fact, during its life cycle, a hurricane can expend as much energy as 10,000 nuclear bombs.

- **Storm surge** is a combined effect that results from low atmospheric pressure above the ocean surface causing a bulge of water to travel beneath the hurricane; wind shearing the water surface in the direction of the hurricane's forward movement; and sea-level set-up due to wave momentum. The combined processes of low pressure, wind shear, and set-up can raise the water level by several metres. Add to this the waves, 6 m to 10 m high or more, formed by the winds of a hurricane, and it is easy to see why hurricanes are very dangerous when they intersect a coastline, especially one with low topography. They cause massive damage and flooding of low-lying coastlines.

22-7 Sea-level rise since the last ice age has shaped most coastlines and continues to do so.

- One of the most influential geological processes shaping coastal environments around the world is the rise in sea level since the end of the last ice age. The last ice age reached its peak 20,000 to 30,000 years ago, and rapid global warming began shortly after. Because sea level at that time was more than 120 m below the present level, much of the world's shallow sea floor and continental shelves were exposed.

- As glaciers around the world melted, the meltwater flowed to the oceans and raised sea level. Rising sea level flooded river mouths and continental shelves on every coastline in the world. This flooding shaped the look and character of modern coastal environments. Barrier islands formed, estuaries expanded, tidal wetlands accreted, and coral reefs flourished in warm seas.

- Scientists have projected that by the end of this century, sea level will rise between 0.75 m and 1.9 m. The physical impacts of sea-level rise along the coast can be grouped into five categories: inundation of low-lying areas, erosion of beaches and bluffs, salt intrusion into aquifers and surface waters, higher groundwater tables, and increased flooding and storm damage.

22-8 Barrier islands migrate with rising sea level.

- Low-elevation coastlines with an abundance of sand tend to form long, narrow **barrier islands** characterized by sandy beaches, tidal inlets, lagoons, and tidal marshes.

- Barrier islands are fragile, constantly changing geologic environments that shift their position when overrun by storm surge. When a storm such as a hurricane encounters a low barrier island, the flood waters carry sand from the ocean side of the island and deposit it on the lagoon side.

- Like a giant treadmill, the island rolls landward by eroding on the ocean side and building on the lagoon side while at the same time losing some sand offshore. This process is referred to as "barrier rollover." Barrier rollover, driven by storm surge and sand transport through tidal inlets into lagoons, allows an island to migrate landward under conditions of sea-level rise and thereby not drown in place.

22-9 Rocky shorelines, estuaries, and tidal wetlands are important coastal environments.

- **Rocky shorelines** are found around the world in places where steep hills or mountains descend directly into the sea. Those rocks that are more easily eroded retreat while those that are more resistant protrude into the surf. Sea stacks, composed of resistant rock able to withstand the pounding of waves and bioerosion by intertidal organisms, form in this manner. Sea stacks mark the former position of the shoreline before erosion caused the coast to retreat landward.

- As ocean waters rose following the last ice age, they flooded coastal uplands and moved the shoreline inland, a process termed **marine transgression.** Sea water invaded and flooded the mouths of stream valleys, forming estuaries. An **estuary** is a partially enclosed body of water that forms where fresh water from rivers and streams flows into the ocean, mixing with the salty sea water.

- Tidal wetlands, also called salt marshes, develop along the fringes of many estuaries. Water draining from the uplands carries sediments, nutrients, and pollutants that are deposited in mud-rich wetlands and prevented from entering estuarine waters. Wetland plants and soils act as a natural buffer between the land and ocean, absorbing flood waters and dissipating storm surges. Salt marsh grasses and other plants help stabilize the shoreline and counteract coastal erosion.

22-10 Coasts may be submergent or emergent, depositional or erosional, or exhibit aspects of all four of these characteristics.

- Coasts where the land is tectonically uplifting at a rate that matches or exceeds sea-level rise are known as **emergent coasts.**

Those where sea-level rise exceeds uplift are known as **submergent coasts.**

- Another way to classify coastal systems is by the relative magnitude of erosional versus depositional processes. **Erosional coasts** are dominated by erosional processes, and **depositional coasts** are dominated by depositional processes.

22-11 Coral reefs are home to 25 percent of all marine species.

- Reefs are crucial to supporting the ocean's biodiversity; they are home to 25 percent of all marine species. **Coral reefs** are shallow coastal environments where numerous types of coral (an animal) and algae (a plant) build massive solid structures of calcium carbonate ($CaCO_3$). There are four basic kinds of coral reefs: fringing reefs, barrier reefs, atolls, and patch reefs.

22-12 Coastal problems are growing as populations increase.

- Because the coastal zone has become the focus of dramatic population growth, development, and investment, exposure to coastal hazards and damage to natural coastal resources has increased. Sea-level rise brings coastal hazards to the doorsteps of buildings in coastal areas.

- **Coastal erosion** is a worldwide problem that threatens beaches, coastal lands, and the economic and environmental systems that rely on them. Among the major threats to coastal resources is the increased occurrence of **polluted runoff** in coastal waters adjacent to urban, suburban, industrialized, and agricultural coastal lands. When natural coastal processes encounter buildings and roadways, they become **coastal hazards,** a term that refers to any coastal process that may cause loss of life or property.

KEY TERMS

barrier islands (p. 606)
beach (p. 593)
beach profile (p. 605)
coastal erosion (p. 616)
coastal geology (p. 592)
coastal hazards (p. 617)
coastal processes (p. 592)
coral reefs (p. 614)
depositional coasts (p. 610)
emergent coasts (p. 610)

erosional coasts (p. 610)
estuary (p. 608)
eutrophication (p. 616)
glacioisostatic uplift (p. 610)
hazard mitigation (p. 617)
hurricane (p. 602)
longshore current (p. 598)
longshore drift (p. 598)
marine transgression (p. 608)
polluted runoff (p. 616)

refraction (p. 596)
rip currents (p. 598)
rocky shorelines (p. 608)
sand bars (p. 598)
storm surge (p. 602)
submergent coasts (p. 610)
tidal wetlands (p. 607)
tides (p. 600)
washover (p. 607)
wave-generated currents (p. 597)

ASSESSING YOUR KNOWLEDGE

Please answer these questions before coming to class. Identify the best answer.

1. The coastal zone is constantly changing because
   a. waves and currents are constantly shifting the position of sand and other sediments.
   b. storms frequently visit the shore.
   c. sea level rises and falls through time, changing the shoreline position.
   d. humans influence coastal environments.
   e. All of the above.

2. As waves move from deep to shallow water, they
   a. grow smaller.
   b. become more energetic.
   c. travel faster.
   d. grow steeper.
   e. do not change significantly.

3. Wave-generated currents include
   a. rip currents and longshore currents.
   b. hurricanes and storm surge.
   c. refraction and diffraction.
   d. shoaling and water orbitals.
   e. Stokes drift and set-up.

4. Longshore currents and rip currents develop
   a. in deep water.
   b. on the subaerial beach.
   c. in nearshore circulation.
   d. because of global warming.
   e. All of the above.

5. Spring tide occurs when
   a. March and April tides reach their highest point.
   b. the lunar and solar tides are aligned.
   c. the lunar and solar tides are opposed.
   d. the Moon is farthest from Earth.
   e. winter turns to spring.

6. Hurricanes
   a. do not occur outside the tropics.
   b. have decreased their impact on U.S. and Canadian shores.
   c. cause coastal damage by high winds and flooding.
   d. are not strong enough to inflict serious damage on houses and roads.
   e. None of the above.

7. Sea-level rise
   a. has exceeded 120 m since the last ice age.
   b. has formed estuaries in the mouths of major rivers.
   c. is a growing problem on all coastlines.
   d. is a major cause of coastal erosion.
   e. All of the above.

8. Barrier islands may migrate with rising sea level.
   a. This process is called "rollover."
   b. Barrier islands cannot migrate with rising sea level.
   c. Rollover only occurs because sand dunes on the island can move.
   d. Barrier islands cannot migrate if they are damaged by a hurricane.
   e. Barrier islands migrate seaward as sea level rises.

9. Tidal deltas form where
   a. sediments collect in the mouths of rivers.
   b. barrier islands roll over.
   c. storm surge is especially frequent.
   d. tides enter and exit lagoons at inlets.
   e. rocky shorelines accumulate beaches.

10. The Gulf of St. Lawrence is
    a. the world's largest estuary.
    b. the result of co-seismic uplift.
    c. formed by sediment erosion in the mouths of streams.
    d. a major river delta.
    e. an example of an emergent coastline.

11. Shorelines on tectonically stable lands that collect sediment are called
    a. emergent coasts.
    b. submergent coasts.
    c. erosional coasts.
    d. depositional coasts.
    e. estuaries.

12. Corals are
    a. animals that build huge colonies of calcium carbonate.
    b. organisms known as polyps.
    c. builders of reefs, along with plants (algae).
    d. dependent on sunlight.
    e. All of the above.

13. The main types of reefs are
    a. coral, algal, and oyster reefs.
    b. barrier, atoll, fringing, and patch reefs.
    c. submergent, emergent, stable, and unstable reefs.
    d. tropical, arctic, and temperate reefs.
    e. None of the above.

14. Humans impact coastal environments because of
    a. pollution, erosion, and exposure to hazards.
    b. building too close to the ocean.
    c. rising sea level, placing human development at risk.
    d. polluted runoff, which may cause eutrophication.
    e. All of the above.

15. Coastal hazards are costing more in recent years because
    a. hurricanes last longer and are stronger.
    b. there is more vulnerable development on the coastline.
    c. sea level is higher.
    d. the coastal population is growing.
    e. All of the above.

FURTHER RESEARCH

1. What will be the future impact of rising sea level on barrier islands that have been developed with towns and highways?

2. How will future sea-level rise influence coral reefs? (Explain in terms of all four types of reefs.)

3. How do barrier islands migrate landward? Why do tidal wetland deposits become exposed in the ocean beach?

4. Make a chart with two columns labelled "Emergent Coasts" and "Submergent Coasts" and two rows labelled "Erosional Coasts" and "Depositional Coasts." Fill in geographic examples of each type.

5. Name five coastal processes.

6. How do waves change as they approach and finally intersect a shoreline?

ONLINE RESOURCES

Explore more about coasts on the following websites:

Centre for Natural Hazard Research:
http://www.sfu.ca/cnhr/risk.html

Natural Resources Canada—Coastal Research:
www.nrcan.gc.ca/earth-sciences/geography-boundary/coastal-research/about-canada-coastline/8504

U.S. NOAA Ocean and Coastal Resource Management:
http://coastalmanagement.noaa.gov/welcome.html

U.S. NOAA National Hurricane Center:
http://hurricanes.noaa.gov

U.S. Environmental Protection Agency—Climate Change Impacts and Adapting to Change:
http://www.epa.gov/climatechange/impacts-adaptation/coasts.html*U.S. Geological Survey Marine and Coastal Geology Program monthly newsletter:* http://soundwaves.usgs.gov

Additional animations, videos, and other online resources are available at this book's companion website:
www.wiley.com/go/fletchercanada

This companion website also has additional information about WileyPLUS *and other Wiley teaching and learning resources.*

CHAPTER 23
MARINE GEOLOGY

Chapter Contents and Learning Objectives (LO)

23-1 Marine geology is the study of geologic processes within ocean basins.
LO 23-1 *List and describe the world's five oceans.*

23-2 Ocean waters are mixed by a global system of currents.
LO 23-2 *Draw the patterns of ocean circulation.*

23-3 A continental shelf is the submerged border of a continent.
LO 23-3 *Describe the main features of a continental shelf.*

23-4 The continental margin consists of the shelf, the slope, and the rise.
LO 23-4 *Identify the major processes shaping the geologic history of a continental margin.*

23-5 Most ocean sediment is deposited on the continental margin.
LO 23-5 *Describe sediment that collects on a continental margin.*

23-6 Pelagic sediment covers the abyssal plains.
LO 23-6 *Identify the processes that control the composition of pelagic sediments.*

23-7 Pelagic stratigraphy reflects dissolution, dilution, and productivity.
LO 23-7 *Describe the stratigraphy of pelagic sediments.*

23-8 The mid-ocean ridge is the site of seafloor spreading.
LO 23-8 *Identify the principal components of oceanic crust.*

23-9 Oceanic trenches occur at subduction zones.
LO 23-9 *Describe the characteristics of oceanic trenches.*

23-10 Human impacts on the oceans are global in extent.
LO 23-10 *Identify human impacts on the oceans.*

GEOLOGY IN OUR LIVES

The oceans are Earth's largest and most important environment. Geological processes are essential to the character of the oceans. For example, currents in the five major oceanic circulation systems carry heat from the tropics toward the poles and return again to the tropics to absorb more heat. The distribution of heat by these currents controls the climate, weather, winds, precipitation, and other fundamental aspects of Earth's environment. Among the many ecosystems that depend on ocean circulation are plankton communities on the ocean surface. These communities govern the type of sediment that accumulates on the ocean floor and the dissolved compounds (such as carbon dioxide) that are buried with these sediments. This chain of processes regulates and stabilizes the chemistry and temperature of the atmosphere and ocean water. Oceans cover more than 70 percent of Earth's surface, and human impacts on the oceans are global in extent.

(?) What are the most important geologic processes governing the character of the oceans?

23-1 Marine Geology Is the Study of Geologic Processes within Ocean Basins

LO 23-1 *List and describe the world's five oceans.*

Viewed from outer space, Earth is a planet of swirling whites and deep blues (**FIGURE 23.1**). In fact, these colours are evidence of water in its many forms: circulating clouds of water vapour, deep oceans, and perpetual snow and ice. Dominating this scene are the oceans. With an average depth of 3,800 m, they contain roughly 97 percent of the water on Earth. (The other 3 percent is found in the atmosphere, on the land surface, or locked in the lithosphere.)

Five Oceans

The largest single feature on the planet's surface (covering more than 70 percent), our oceans are unique in the Solar System. No other planet has open oceans of liquid water, although the discovery of ice on Mars and the well-developed channels across Martian landscapes indicate that there has been liquid water there in the past. Scientists also speculate that Europa, a moon of Jupiter, has an ocean of liquid water, but it is trapped beneath a smooth veneer of ice that covers the moon's surface. It is highly likely that life on Earth originated in the oceans, and they continue to house an amazingly diverse web of ecosystems. Yet despite this dominant role among Earth's environments, we know more about the surface of the Moon than we do about the floor of the oceans.

Until the year 2000, there were four recognized oceans: the *Pacific, Atlantic, Indian,* and *Arctic.* However, in the spring of that year, the International Hydrographic Organization identified a new ocean, the *Southern Ocean,* which surrounds Antarctica and extends to 60° latitude (**TABLE 23.1, FIGURE 23.2**). There are also many *seas* (smaller branches of an ocean); seas often are partly enclosed by land. The largest seas are the South China Sea, the Caribbean Sea, and the Mediterranean Sea.

FIGURE 23.1 Earth is the water planet, and oceans are the largest single feature on the surface.

 Why do oceans dominate Earth's climate?

TABLE 23.1 The Oceans

Ocean	Area (km²)	Average Depth (m)	Greatest Depth (m)
Southern Ocean	20,327,000	4,000 to 5,000	Southern end of the South Sandwich Trench: 7,235 m deep
Arctic Ocean	13,224,479	1,204	Eurasia Basin: 5,450 m deep
Indian Ocean	73,426,163	3,963	Java Trench: 7,725 m deep
Pacific Ocean	166,240,977	4,637	Mariana Trench: 11,033 m deep
Atlantic Ocean	86,557,402	3,926	Puerto Rico Trench: 8,605 m deep

Because oceans influence the entire planet, many fields of science are devoted to their study. **Marine geology** involves geophysical, geochemical, sedimentological, and paleontological investigations of the ocean floors and coastal margins. Marine geology has strong ties to *physical oceanography* (the study of ocean currents and waves), *chemical oceanography* (the study of chemical processes in the ocean), *biological oceanography* (the study of living organisms in the ocean), and plate tectonics. *Marine geologists* view the ocean in terms of the geological processes that take place within ocean basins. Geologists study and observe sedimentary and geochemical processes on the sea floor and along the coasts, collect and analyze cores of fossilized plankton from the sea floor, interpret how seafloor topography relates to plate tectonic processes, and examine aspects of water circulation and climate that are central to how oceans interact with Earth's other geological systems.

Why Is the Ocean Salty?

When it comes to the oceans, perhaps the first question on everyone's mind is, "Why is the water salty?" The ocean is salty because it contains dissolved ions and compounds that come from the chemical weathering of continental rocks and from hydrothermal processes at spreading centres and subduction zones where sea water interacts with hot crust. Over 90 percent of the dissolved ions in sea water are chloride (Cl) and sodium (Na) (which form halite, NaCl). However, marine saltiness also comes from dissolved sulphur (S), calcium (Ca), magnesium (Mg), bromine (Br), potassium (K), and bicarbonate (HCO_3).

All together, the concentration of these dissolved ions, known as the **salinity,** has a global average of about 35 parts per 1,000. In other words, about 35 of every 1,000 parts (3.5 percent) of the weight of sea water come from dissolved ions; thus, in 4 km³ of sea water, the weight of the sodium chloride (NaCl) would be about 120 million tonnes. The same 4 km³ would also contain up to 25 tonnes of gold

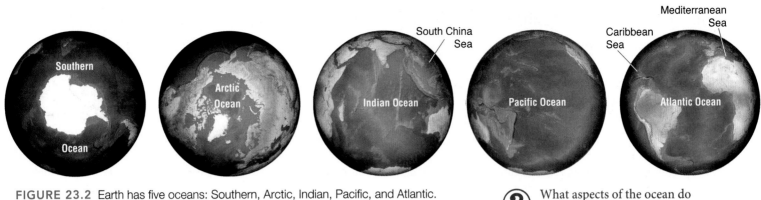

FIGURE 23.2 Earth has five oceans: Southern, Arctic, Indian, Pacific, and Atlantic.

(?) What aspects of the ocean do scientists study?

and 45 tonnes of silver. But, before you go out and try to get rich with sea water, just think about how big 4 km³ is. It has been estimated that if all the dissolved salt in the ocean were removed and spread evenly across the land surface, it would form a layer more than 166 m thick, about the height of a 40-storey office tower.

Although the average salinity of the ocean is relatively stable through time (with some variation due to environmental processes, such as climate, that change through geologic time), it does vary from place to place. Salinity is higher in mid-latitude oceans because evaporation exceeds precipitation, and in restricted areas, such as the Mediterranean and Red Seas. Salinity is lower near the equator because of high rainfall and near major rivers where fresh water enters the ocean (**FIGURE 23.3**).

Ocean water is a complex solution of dissolved ions and compounds as well as decaying organic matter. Most of the dissolved components are derived from chemical weathering of the crust. Others (Cl, S) come from hydrothermal venting at spreading centres and other sites where sea water leaches ions from the crust. There is geologic evidence that the oceans have been as salty as they are today for over 1 billion years. Hence, if dissolved ions and compounds are being perpetually delivered to the oceans by runoff and hydrothermal activity, something must be regulating the mixture such that both the amount of water and the amount of dissolved material are stable.

The oceans' stable salinity is a result of geologic processes that regulate the concentration of the dissolved components; that is, inputs of water and dissolved ions and compounds match outputs. The amount of water in the oceans is regulated by the hydrologic cycle. Over time, the input of water by runoff and precipitation is evenly matched by evaporation. This balance keeps the volume of water in the oceans approximately stable except when it is influenced by climate changes and volcanism.

Likewise, the volume of dissolved ions and compounds is held approximately stable over time. This stability is accomplished by sedimentary processes that remove the dissolved components at about the same rate that they are introduced. Processes that remove dissolved components include (1) *inorganic precipitation* (mostly along the ocean margins), where evaporation concentrates the dissolved components and they form crystalline evaporite deposits; (2) salt storage in the pore water among marine sediments as they are deposited; (3) chemical interactions between sea water and the basalt sea floor that create new minerals, which are stored in the crust; and (4) *organic precipitation* of dissolved components by marine plants and animals to build solid tissue and skeletal parts. These components are eventually buried as sediment when an organism (mostly plankton) dies and falls to the sea floor. All of these processes regulate the volume of dissolved ions and compounds. Hence, ocean salinity has remained essentially stable for at least the past 1 billion years because of geological processes.

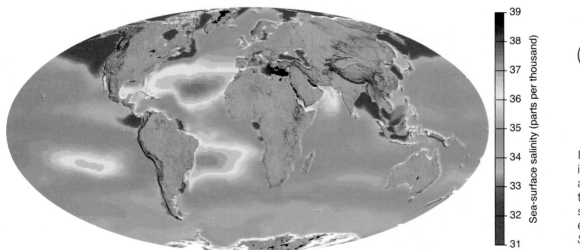

(?) What are the primary factors that influence ocean salinity?

FIGURE 23.3 Ocean salinity is high in mid-latitude oceans and restricted areas, such as the Mediterranean and Red seas. Salinity is low near the equator and near major rivers. *Source:* NODC, World Ocean Atlas, 1999.

(?) **Expand Your Thinking**—Are the dissolved ions and compounds removed from sea water stored permanently in the lithosphere, or do they eventually return to the surface? Explain.

23-2 Ocean Waters Are Mixed by a Global System of Currents

LO 23-2 *Draw the patterns of ocean circulation.*

The oceans influence many other natural systems on the planet, and they especially affect the weather and climate. Ocean water moderates surface temperatures by absorbing heat from the Sun and transporting that heat toward the poles and the sea floor. Restless ocean currents distribute this heat around the globe (**FIGURE 23.4**), warming the land and air during winter and cooling it in summer.

Ocean circulation is the large-scale movement of water by **currents**, which are large masses of continuously moving water. There are basically two types of oceanic circulation: winds drive *surface circulation*, and cool water at the poles that sinks and moves through the lower ocean drives *deep circulation*. The general pattern of circulation consists of surface currents carrying warm salty water away from the tropics toward the poles while releasing heat to the atmosphere during the journey. Winter at the poles further cools this surface water. Because it is now cooler and still salty, the water sinks to the deep ocean, creating currents along the sea floor and at mid-depths in the water column. This process is especially pronounced in the North Atlantic and in the Southern Ocean in the coastal waters of Antarctica, where *deep-water production* is the strongest. As deep ocean water gradually warms, it returns to the surface nearly everywhere in the ocean as new, colder water sinks to depth. Once at the surface, it is carried back to the tropics by surface currents, where it is warmed again and the cycle begins anew. The more efficient the cycle, the more heat is transferred from the tropics to the poles, and the more this heat warms the climate.

Gyres

Due to Earth's rotation, currents are deflected to the right in the Northern Hemisphere and to the left in the Southern Hemisphere. This is known as the **Coriolis effect,** named after the French scientist *Gaspard-Gustave Coriolis* (1792–1843), who described the transfer of energy in rotating systems. These currents eventually come into contact with the continents, which redirect them (again, to the right and left, depending on which hemisphere they are in), creating large-scale circulation systems called **gyres**, which sweep the major ocean basins. There are five major basin-wide gyres: the North Atlantic, South Atlantic, North Pacific, South Pacific, and Indian Ocean Gyres. Each gyre is composed of a strong and narrow *western boundary current* and a weak and broad *eastern boundary current*.

Let us examine the surface circulation of the *North Pacific Gyre* as a typical example of how winds and the Coriolis effect combine with continental deflection to create circulation. In the North Pacific atmosphere, a descending column of dry air that originated at the equator (the northern end of the Hadley cell; see Chapters 16, "Global Warming," and 21, "Deserts and Wind") blows toward the equator but is deflected to the west by the Coriolis effect. This southwest-flowing wind is known as the **trade wind.** All the major ocean basins in both the Northern and the Southern Hemispheres have trade winds. The Pacific trade wind drives the *North Equatorial Current* to the west just north of the equator at about 15°N latitude. This current is deflected north near the Philippines to create the warm western boundary current known as the Japan or *Kuroshio Current*. The Kuroshio Current carries warm water away from the tropics until it turns to the east at approximately 45°N latitude and becomes the *North Pacific Current,* which moves across the basin toward North America. As it approaches the North American continent, the North Pacific Current branches, with one arm moving north to circulate through the Gulf of Alaska and the Bering Sea as the *Alaska Current*. A southern arm becomes the cool, slow-moving eastern boundary current called the *California Current*. The California Current moves from about 60°N to 15°N latitude and merges with the North Equatorial Current. From there it travels, once again, thousands of miles across the basin to Asia. Each of the five major gyres in the oceans has similar systems of currents.

Vertical Circulation

In the North Atlantic basin, the western boundary current is known as the *Gulf Stream*. The Gulf Stream carries warm tropical

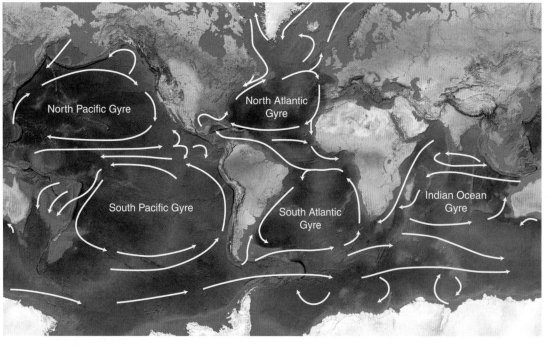

FIGURE 23.4 Currents flow in complex patterns affected by wind, Earth's rotation, and water density.

 Explain the general pattern of ocean circulation.

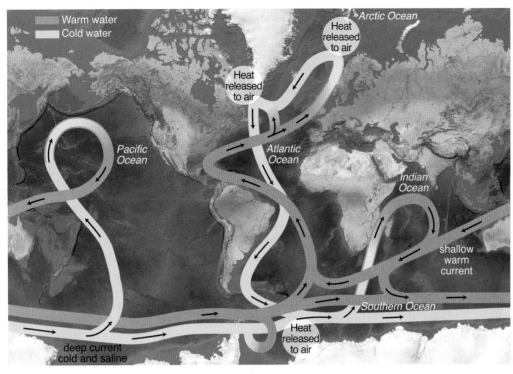

FIGURE 23.5 The thermohaline circulation is a global pattern of currents that carries heat, dissolved gas, and other compounds on a trip that can take up to 1,300 years.

 How does Ekman transport influence coastal waters?

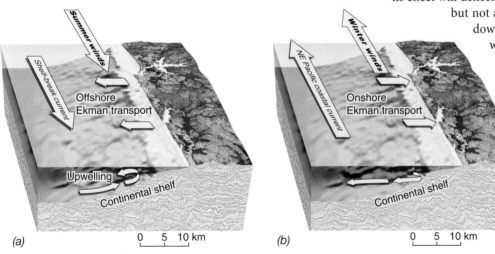

(a)   0   5   10 km

(b)   0   5   10 km

FIGURE 23.6 (*a*) Along the Pacific coast of Canada, summer winds from the north create Ekman transport to the west and stimulate upwelling of nutrient-rich deep water. (*b*) In the winter, winds flow to the north causing Ekman transport to the east, onto the shoreline.

 What drives the oceanic conveyor belt?

water from the Caribbean to the cold waters of the North Atlantic. As it moves, the Gulf Stream cools and evaporates, thus greatly increasing its density. By the time it arrives in the North Atlantic as a cold, salty body of water, it can no longer stay afloat and begins a long descent of 2 km to 4 km, where it becomes a deep current known as the *North Atlantic Deep Water.* The Deep Water travels south through the Atlantic and eventually joins similar deep water that is forming in

the Southern Ocean. There it becomes the *Circumpolar Deep Water,* which journeys throughout the Southern Ocean. An arm of the Circumpolar Deep Water migrates into the North Pacific, and there, after a voyage of approximately 35,000 km, water that originated in the North Atlantic Gulf Stream eventually surfaces. It has been estimated that up to 1,300 years may pass before this water returns to its place of origin. This **thermohaline circulation,** also called the *oceanic conveyor belt* (recall the discussion of thermohaline circulation in Chapter 16, "Global Warming"), travels through all the world's oceans, connecting them in a truly global system that transports both energy (heat) and matter (solids, dissolved compounds, and gases), and thereby influences global climate (FIGURE 23.5).

Ocean currents may cause the water beneath them to flow in a different direction. This unexpected result, called *Ekman transport,* is a consequence of the Coriolis effect interacting with a column of water. Recall that the Coriolis effect causes currents in the Northern Hemisphere to deflect to the right (and currents in the Southern Hemisphere to deflect to the left). As wind blows across the water and creates a current, the Coriolis effect will deflect it slightly. Water in lower layers deflects as well but not as fast as the surface water. This effect translates downward through the water column in the form of a weakening spiral. Theoretically, this spiral results in water transport that is 90° to the right of the wind direction in the Northern Hemisphere.

Ekman transport is especially important in coastal zones, where wind blowing parallel to the shore can generate a current that is perpendicular to the shoreline. For instance, along the Pacific coast of Canada, the wind generally blows to the south in the summer. Ekman transport stimulated by this wind drives coastal water to flow offshore, 90° to the right of the wind. To replace this water, a deep, nutrient-rich current rises along the sea floor into the coastal zone in a phenomenon called *upwelling.* This upwelling sustains a seasonal fishery that is crucial to the economy of western Canada. In the winter, the wind reverses and blows along the shoreline to the north. This wind stimulates Ekman transport that is directed onshore, forcing shallow waters into the coastal zone (FIGURE 23.6). As these waters pile up against the coast, they drive a current along the sea floor that is directed offshore, a process called *downwelling.*

 Expand Your Thinking—Deep water pumped from the sea floor off Hawaii is desalinated and sold as the cleanest drinking water on Earth. Explain the reasoning behind this statement.

23-3 A Continental Shelf Is the Submerged Border of a Continent

LO 23-3 *Describe the main features of a continental shelf.*

If you look closely at a map of the sea floor around most continents you will notice a shallow platform that extends offshore from the coastline (**FIGURE 23.7**). This is the **continental shelf,** an area of the ocean floor that is generally less than 200 m deep surrounding each continental land mass. Continental shelves make up about 8 percent of the entire ocean area. An undersea extension of a continent, the continental shelf ranges from over 100 km wide to only a few kilometres wide, depending on its geologic history. The fact that the shelf happens to be under water is due to its geologic origin as the outer margin of the continent.

The Canadian and U.S. Coastal Plain

In Canada and the United States, the continental shelf is particularly well developed along the eastern coast and Gulf of Mexico. It is part of the geologic province known as the *Coastal Plain.* The Coastal Plain originated with the breakup of the supercontinent Pangaea in the Late Paleozoic to Early Mesozoic Eras about 280 to 230 million years ago (recall Chapter 15, "The Geology of Canada," Section 15-5). A basin developed between the four newly separated continents—Europe,

Africa, North America, and South America—and gradually grew to become the Atlantic Ocean and Gulf of Mexico. As the North American Plate pulled away to the west, the thick continental crust along its eastern edge collapsed into a series of normally faulted blocks caused by the tensional stress of the rift (**FIGURE 23.8**). These blocks now lie under the continental shelf, deeply buried by sediments washing off the Appalachian Highlands. This once-active divergent plate boundary has become the passive margin of westward-moving North America. The Coastal Plain is widely recognized as a classic example of a *passive continental margin.*

Sediments eroded from the Appalachians moved to the continental margin and were deposited in broad deltas, estuaries, barrier islands, and other coastal and marine sedimentary environments. They gradually covered the faulted continental margin, burying it under sedimentary layers thousands of metres thick. These units are Mesozoic to Cenozoic in age and include thick sequences of Cretaceous rocks. In the northern region of the Coastal Plain, boreholes have been drilled to a depth of 5 km without encountering the igneous or metamorphic *basement.* The sediment sequence is even thicker along the Texas portion of the province, with up to 10 km of Jurassic to Quaternary sedimentary beds. These layers are famous as a rich source of petroleum.

As sediments accumulated along the Atlantic coast, the level of the ocean rose and fell with changes in global climate (warm climate causes a rise in sea level, cool climate causes a fall in sea level) and vertical changes in the elevation of the continent due to changing heat flow associated with the rifting process. As a result, the seashore migrated back and forth across the newly forming coastal plain, depositing various types of sediments on the landscape. Over many tens of millions of years, the margin of North America cooled and subsided as it retreated from the hot environment of the Mid-Atlantic spreading centre. During this time, the rate of sediment delivery and deposition from the Appalachians approximately matched the rate of continental subsidence, and a great wedge of sediment, 5 km to 15 km thick, developed into the flat continental shelf we see today.

DATA SIO, NOAA, US Navy, NGA, GEBCO, Image Landsat, Image IBCAO

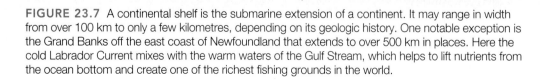

FIGURE 23.7 A continental shelf is the submarine extension of a continent. It may range in width from over 100 km to only a few kilometres, depending on its geologic history. One notable exception is the Grand Banks off the east coast of Newfoundland that extends to over 500 km in places. Here the cold Labrador Current mixes with the warm waters of the Gulf Stream, which helps to lift nutrients from the ocean bottom and create one of the richest fishing grounds in the world.

(?) What is the origin of eastern Canada's continental shelf?

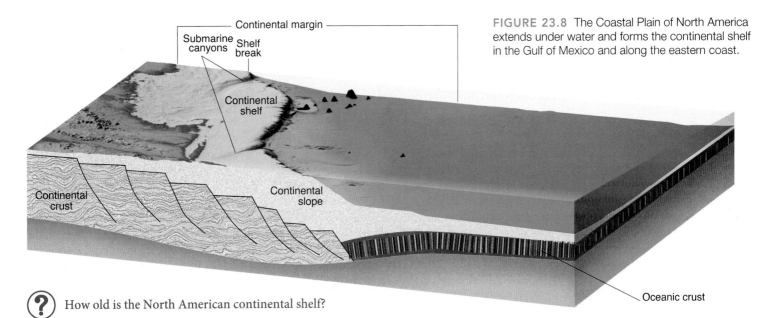

FIGURE 23.8 The Coastal Plain of North America extends under water and forms the continental shelf in the Gulf of Mexico and along the eastern coast.

How old is the North American continental shelf?

Shelf Areas

Continental shelves extend from the shore (at an average slope of about 0.1°) to the *shelf break,* the point at which the sea floor steepens sharply, marking the boundary between the continental shelf and the **continental slope.** The break is found at a depth ranging from 100 m to 200 m. It has been theorized that this depth may mark a former coastline and that global variations in the depth of the shelf break are the result of differences in the vertical movement of continents.

The shelf surface is swept by waves, tidal currents, and storms, which, in places, continually move a layer of loose sediment consisting of sand, silt, and silty mud. In other locations, the shelf is starved of sediment, and the sea floor consists of outcropping sedimentary rocks of (usually) Cenozoic age. The shelf surface exhibits some relief, featuring small hills and ridges that alternate with shallow depressions and troughs. In some cases, steep-walled V-shaped **submarine canyons** cut deeply into both the shelf and the slope below.

Many of the world's continental shelves are rich in natural resources, such as marine life, sand and gravel deposits, natural gas and oil reserves (**FIGURE 23.9**), and phosphorite beds (sedimentary rock that contains high amounts of phosphate often produced by biological activity in the overlying water column). Researching and mapping shelf areas and resources has been an important task of marine geologists over the years.

Most nations have claimed *mineral and land rights* to their shelves and actively manage these resources through systems of laws and rules. For instance, according to the 2003 United Nations Convention on the Law of the Sea, the government of Canada has sovereign rights over the resources (including mineral, oil, and gas) on and below the seabed on their continental shelves that extend 200 nautical miles, or 370 km, beyond the coastline. Nationally, through the Oceans Act of 1996 (amended in 2005), the federal government, in partnership with the provinces, oversees the development of natural resources, such as natural gas and oil reserves, in an orderly and timely manner;

meets the energy needs of the country; protects the human, marine, and coastal environments; and receives a fair and equitable return on the resources of the continental shelf.

Str/Jonathan Hayward/Canadian Press

FIGURE 23.9 Canada's Atlantic margin is rich with oil and gas deposits, which are drilled by almost 330 active wells (red dots). By comparison, the U.S. Gulf of Mexico has almost 4,000 active wells. The inset photo shows the Hibernia Platform located on the Grand Banks east of Newfoundland.

Who manages the natural energy resources of the continental shelf?

Expand Your Thinking—If you were to make a map of a continent, where would you draw the border? Explain.

23-4 The Continental Margin Consists of the Shelf, the Slope, and the Rise

LO 23-4 *Identify the major processes shaping the geologic history of a continental margin.*

The **continental margin** (labelled in Figure 23.8) is the broad zone that includes the continental shelf, the continental slope, and the **continental rise.** The margin separates the thin oceanic crust of greater density from the thick continental crust of lesser density and constitutes about 28 percent of the total oceanic area. The transition from continental to oceanic crust typically occurs at the outer part of the margin under the continental rise. Beyond the seaward edge of the rise lies the **abyssal plain** on fully oceanic crust (studied in Section 23-6).

Active and Passive Margins

The tectonic setting of a continental margin provides the basis for interpreting its history and structure. Margins are classified as "active" or "passive," each having certain fundamental characteristics (**FIGURE 23.10**). Active continental margins occur at plate boundaries (convergent, divergent, or transform) and are tectonically active, with typically less width and sediment input than passive margins.

Active margins located at convergent boundaries where there is a subduction zone may eventually have **accreted terrane** (discussed in Chapter 15, "The Geology of Canada"). Accreted terrane (also known as *exotic terrane*) develops where foreign blocks of crust, carried on a subducting plate toward the margin, collide with the overriding plate. These terranes add to the volume and topography of a continental margin and complicate the structure of the crust. Accreted terranes

may include island arcs, small bits of continental-type crust (think New Zealand or Japan), or hotspot islands that collide with the active margin. Since the active margin is on the overriding plate, it is also likely to have a volcanic arc fed by magma generated by water released from the subducting plate.

Each new accreted terrane that collides with the volcanic arc may cause the subduction site to close down and a new one to open on the seaward side of the collision where there is oceanic crust. Hence, the subduction zone "jumps" back to the advancing front of the active margin, and the accreted terrane becomes part of the active continental margin. The coast-parallel ranges and valleys of the Cascadia margin on the British Columbia, Washington, and Oregon coasts are, in part, a result of accreted terranes that collided with the North American Plate as it migrated to the west with the breakup of Pangaea. These terranes were island arcs and continental fragments in the Pacific that were swept up in the westward movement of the North American Plate and added to its active western margin.

Passive continental margins are located within plates. For instance, the passive margin along the Canadian and U.S. east coast and the Gulf Coast is separated from the Mid-Atlantic Ridge by an expanse of oceanic crust that was generated after Pangaea rifted. Oceanic and continental crusts meet in a region of low tectonic activity beneath the continental rise off the east coast of Canada and the United States. Passive margins are generally wide and may receive a large influx of sediment coming from large and mature watersheds (such as the St. Lawrence and Mississippi Rivers) and intra-basinal carbonate sedimentation associated with plankton deposition or coral reef construction.

Submarine Canyons

A submarine canyon is a steep-sided valley in the shape of a meandering channel that cuts into the shelf and slope of both active and passive continental margins. Many submarine canyons end with a thick sedimentary deposit on the continental rise called a **submarine fan.** Submarine canyons are complex features that form in several different ways, each with its own unique set of circumstances. Here we describe one type that is common on margins around the world.

Submarine canyons have winding valleys with V-shaped cross-sections and a central axis that slopes outward and downward (typical of a stream channel) across the continental slope. They may have numerous tributaries entering from both sides and relief that is equivalent to major canyons on land. For instance, the Monterey Submarine Canyon off the central California coast (**FIGURE 23.11**) and the Grand Canyon of the Colorado River are comparable in size.

The facts that submarine canyons resemble terrestrial stream channels and that many canyons line up with continental watersheds have provided strong arguments that the heads of submarine canyons were originally cut by rivers during periods of low sea level (such as during Pleistocene glaciations). Dendritic drainage patterns (typical of stream-carved watersheds) are well developed in many shallow

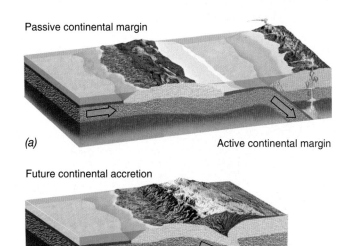

Passive continental margin

(a)

Active continental margin

Future continental accretion

(b)

FIGURE 23.10 *(a)* A passive margin is not located at a plate boundary. *(b)* An active margin is located at a plate boundary. As a passive margin, on a moving plate, approaches and collides with an active margin, an accreted terrane develops.

(?) Where have accreted terranes played a major role in shaping the margins of North America?

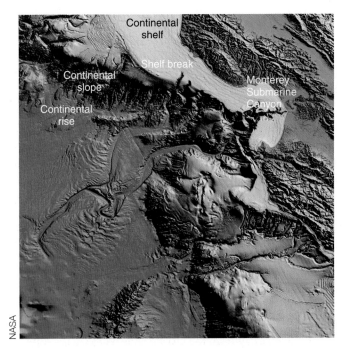

FIGURE 23.11 The origin of Monterey Submarine Canyon off the coast of California is unknown. However, one hypothesis suggests that the shallow portion of the canyon was cut millions of years ago by a large river that no longer flows in the region.

(?) What is the evidence that the shallow portion of submarine canyons was originally carved by streams?

canyon heads but become less prominent in deeper water. Also, various continental margins are known to have undergone subsidence over millions of years, which would have submerged upper canyons originally eroded by streams. This evidence has convinced many marine geologists that the upper portion of submarine canyons are carved by stream erosion during low sea-levels on continental margins that subsided over recent geologic history.

However, even if fluvial erosion processes are important in forming the shallow extent of canyons, they cannot have influenced deep canyon sections at depths of thousands of metres. Hence, a *composite origin* for submarine canyons has been hypothesized, involving the formation of erosive submarine landslides known as **turbidity currents** (FIGURE 23.12). Turbidity currents result from the collapse of a sedimentary deposit on the slope or shelf edge. The collapse may be caused by an earthquake, a storm, or internal waves and currents within the ocean that travel along thermal and salinity layers in the water column. Sediments thrown into suspension by the collapse develop a dense, sediment-laden current that travels down the axis of the canyon at high speed, carving the floor and sides. The density and speed of the turbidity current carries it into deep water (where it scours the lower sections of the canyon) and out onto the continental rise (where it feeds sediment to the submarine fan). Turbidity currents generally are credited with both excavating submarine canyons and transporting great quantities of sediment down the canyons to form the fans that build the continental rise at the base of the continental slope. Continual marine deposition on the continental slope flanking the sides of a canyon can raise the level of the continental slope over time, which in turn causes the canyon to gain greater relief because they continue to be excavated by turbidity currents.

Turbidity currents carry sand and mud that is deposited on the rise when the currents spread out laterally and decelerate at the base of the canyons. These deposits are called *turbidites*. Such deposits normally have graded bedding—their grains become progressively finer upward, reflecting sedimentation as the current velocity gradually decreases.

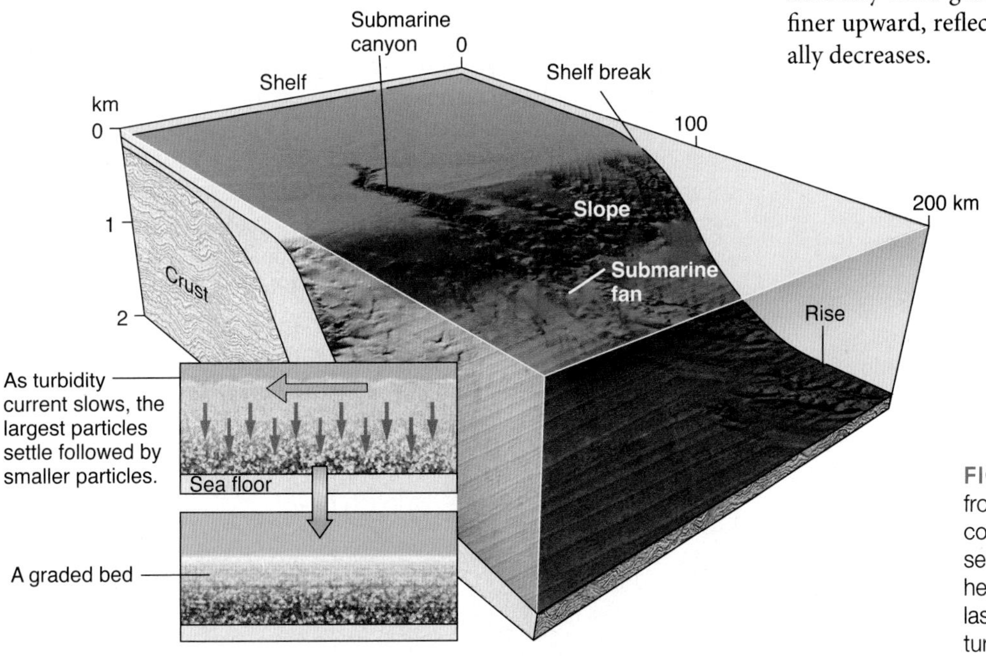

(?) What kind of rock is made by turbidites? In what type of tectonic setting would you find such deposits on land?

FIGURE 23.12 Turbidity currents carry sediment from the continental shelf and slope onto the continental rise. As the current slows and the sediment is deposited in a submarine fan, the largest/heaviest grains settle first and the fine grains settle last. This makes a sedimentary deposit known as a turbidite.

(?) **Expand Your Thinking**—Considered simply, a submarine canyon consists of two types of sedimentary deposits. Turbidity currents deliver one kind. What is the other kind?

23-5 Most Ocean Sediment Is Deposited on the Continental Margin

LO 23-5 *Describe sediment that collects on a continental margin.*

Most of the oceans' sediment (about 75 percent by volume) is deposited on the shelf, slope, and rise, where accumulations can exceed 10 km in thickness. Water characteristics (such as temperature, sediment load, and nutrient content), climate, tectonic setting, depth, history of sea-level change, and other factors govern the type of sediment found on a continental margin. Some margins are glaciated, some are dominated by volcanic arcs, and others are flanked by arid deserts or humid rainforests. These environments, the lithology of the continental crust, and the processes that weather the crust all determine the nature of sediments that enter the watersheds and travel onto the margin.

The Coastal Zone

The coastal zone of most margins receives *detrital sediment* via watersheds that drain neighbouring uplands. Coastal environments include tidal wetlands and mudflats that collect silt and clay delivered by daily tides; lagoons and estuaries that collect muds and sands transported by streams as well as tides from the ocean; sandy barrier islands, beaches, and dunes; tidal inlets that are typically sand rich; and deltas and coral reefs (**FIGURE 23.13**). For a refresher on the geologic character of these environments, refer to Chapters 8 ("Sedimentary Rock") and 22 ("Coastal Geology").

During the last ice age, many rivers flowed directly across the continental shelf and deposited their sediments at the outer shelf or onto the continental slope. But when the ice age ended and sea level rose around the world, the coast retreated—over 100 km in many cases. The rising waters flooded into river channels, driving them back into the watersheds, and formed estuaries that today trap modern sediments. This history and the fact that most large rivers have human-made hydroelectric dams with sediment-trapping reservoirs mean that modern shelves tend to be "sediment starved." The Gulf of St. Lawrence on Canada's east coast and the Fraser River estuary on the west coast are examples of such sediment-storing estuaries. Thus, many shelves do not receive large amounts of modern sediment; instead, they hold *relict sediments* deposited as sea level migrated across the shelf 3,000 to 15,000 years ago, when climate was changing. Relict sediment, rather than modern sediment, composes the sea floor on many shelves.

The Continental Shelf

Across the shelf, away from the coastal zone where many watershed sediments are trapped, the sediment character reflects a different set of influences. Sediment that collects on the sea floor between the shoreline and the edge of the shelf is termed **neritic sediment** (from *nerita*, Latin for "sea mussel"). On some margins, neritic sediment is composed of the skeletal remains of plankton living in the water column that collect on the sea floor after they die, forming *biogenic sediment.* Some shelves get their sediment from winds that blow off the land (**FIGURE 23.14**) or by weathering of the rocks exposed on the sea floor (termed *residual sediment*). Still others collect *authigenic sediment,* a type that inorganically precipitates from the water column or whose mineralogy is changed by interactions with sea water. Authigenic sediment includes *glauconite,* a clay formed when anoxic sea water interacts with sedimentary deposits on the shelf, or *phosphorite,* a rock of phosphates derived from marine invertebrates that secrete calcium phosphate shells and the bones and excrement of vertebrates. On low-latitude shelves, coral reefs may be widespread, producing carbonate sediment. On many shelves, neritic sediment is not modern but is relict, left over from the time when the shoreline was transgressing across the shelf during rising sea level. The resulting sedimentary record of these variations can be highly complex, often containing combinations of many sediment sources (**FIGURE 23.15**).

(?) What is the typical composition of detrital coastal sediments?

FIGURE 23.13 Coastal environments that trap sediments, such as this one showing the mouth of the Murray River in South Australia, include tidal wetlands, lagoons, estuaries, inlets, barrier islands, dune fields, and beaches.

Bill Bachman/Alamy

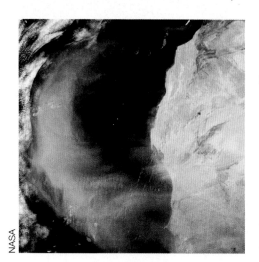

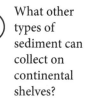

(?) What other types of sediment can collect on continental shelves?

FIGURE 23.14 Off the western coast of Africa, strong winds blow desert silt into the ocean.

NASA

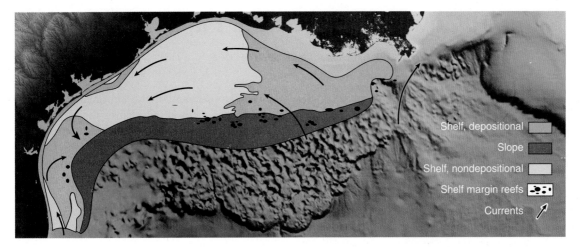

 What is the probable source of modern sediment that is deposited on this shelf?

FIGURE 23.15 In the Gulf of Mexico, the U.S. continental shelf has large areas of relict sediment, left over from the rise of sea level after the last ice age.

Some shelves receive massive amounts of modern sediment because they have large watersheds that drain broad areas of land. Deltas formed by rivers such as the Mississippi (United States), Amazon (Brazil), Ganges-Brahmaputra (India), and the Yangtze (China) contain so much detrital sediment that they bury the topography of the margin. For instance, the Ganges and Brahmaputra Rivers empty into the Bay of Bengal on the north coast of the Indian Ocean (**FIGURE 23.16**). This is the world's largest delta and one of the most fertile regions on the planet. Between 125 and 143 million people live on the delta, despite risks from floods caused by seasonal monsoon rains, heavy runoff from melting glaciers in the Himalayan Mountains, and the threat of storm surge that rolls across the low-lying delta from frequent tropical cyclones sweeping through the bay. A large part of the nation of Bangladesh lies on the *Ganges-Brahmaputra Delta,* and many of the country's people depend on the delta for survival. Offshore lies the *Bengal Fan,* the largest submarine accumulation of sediment on Earth. River sediment is carried from the mouth of the Ganges and Brahmaputra Rivers through several submarine canyons, some of which are over 2,000 km long, and across the sea floor of the Bay of Bengal, finally coming to rest up to 30° of latitude (over 3,000 km) from where it originated.

Hemipelagic Sediments

Along many margins, currents and waves redistribute sediment laterally on the shelf. However, some sediment is moved across the shelf edge and collects on the continental slope and rise. Sediments that drape upper and middle continental slopes around the world are known as **hemipelagic sediments.** They grade from predominantly continental muds (closer to watershed sources) into biogenic sediments (farther from watershed sources). Even where biogenic constituents are dominant, hemipelagic sediments typically have a dark colour, reflecting the detrital mud component. Variations in the composition of terrigenous muds reflect weathering intensity on the adjacent land. Terrigenous sediments are delivered to the ocean by rivers and remain in suspension as they are carried beyond the shelf edge by surface currents or by downwelling currents related to Ekman transport.

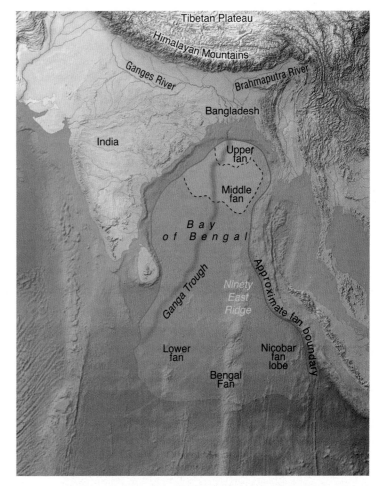

FIGURE 23.16 The Ganges and Brahmaputra Rivers transport sediment from the highest mountains on Earth (the Himalayan Mountains) to the Bengal Fan, the world's largest submarine accumulation of sediment.

 Why is so much sediment available to build the Bengal Fan?

Expand Your Thinking—Describe factors that control the types of sediment that collect on continental margins.

23-6 Pelagic Sediment Covers the Abyssal Plains

LO 23-6 *Identify the processes that control the composition of pelagic sediments.*

Beyond the continental margin lies the abyssal plain (**FIGURE 23.17**). Untouched by sunlight, the abyssal plain is the horizontal, or very gently sloping, sediment-draped region of the deep ocean floor that lies between the continental rise and a mid-ocean ridge. Abyssal plains are among Earth's largest and least explored geologic features.

Generally located at depths of 3,000 m to 6,000 m, abyssal plains are hundreds of kilometres wide and thousands of kilometres long, and they cover approximately 40 percent of the ocean floor. They are, in fact, one of the major topographic provinces on the planet. For example, in the North Atlantic basin between Nova Scotia and the Mid-Atlantic Ridge, the *Sohm Abyssal Plain* is massive. It has an area of approximately 900,000 km². Abyssal plains are globally distributed, but the largest are found in the Atlantic Ocean. In the topographically complex Indian and Pacific Oceans, with their numerous mid-oceanic ridges, island chains, seamounts, and wide spreading centres, abyssal plains occur mainly as the small, flat floors of marginal seas (such as the floor of the *Philippine Sea*) or as the narrow, elongated bottoms of deep sea trenches at subduction zones.

Abyssal Sediment

A prominent aspect of abyssal plains is their sediment, known as **pelagic sediment** (pelagic means "in the water column"). Much of the reason the abyssal sea floor lacks dramatic relief is that the rugged topography created by faulting and fracturing at spreading centres is buried under thick blankets of marine sediment. Abyssal sediment deposits thicken with the age of the oceanic crust, generally increasing away from the spreading centres. Abyssal sediment is typically 100 m to 500 m thick in the central abyssal plains (**FIGURE 23.18**). Sediment accumulation increases dramatically beneath the continental rise and slope where turbidity currents and other sources of clastic sediment (such as river deltas) lead to sediment thicknesses of 3,000 m and more.

The abyssal environment is very calm, far removed from storms that disturb shallow ocean waters, rivers that drain the continents,

and thick deep-sea fans at the base of the continental margins. The low energy of the abyssal plain is reflected in the fine-grained character of pelagic sediment, which is composed predominantly of **biogenic ooze** and **abyssal clay.** Biogenic ooze consists of microscopic plankton remains produced in the food chain in the overlying waters, from which they fall to the sea floor. Abyssal clay is derived from windblown continental sediment that falls from the atmosphere into the sea. Other pelagic sediment components include *volcanogenic particles, cosmogenic particles,* and authigenic sediments. Each of these is discussed below.

Pelagic sediment reflects the nature of the marine environment: water temperature, depth, and proximity to a continent. In equatorial to temperate regions where the water is shallower than 4 km to 5 km, pelagic sediment is composed primarily of calcareous shells of microscopic plankton consisting of *foraminiferans* (an animal) and *coccolithophores* (a plant). These sediments make *calcareous ooze* (any carbonate sediment composed of more than 30 percent microscopic skeletal debris) (**FIGURE 23.19**). Below a depth of 4 km to 5 km, calcareous shells dissolve, and the principal sediment is abyssal clay mixed with the remains of plankton called *radiolarians* (an animal) and *diatoms* (a plant) that are made of silica.

The depth at which calcium carbonate ($CaCO_3$) completely dissolves is known as the **carbonate compensation depth (CCD).** Whether the sea floor is deeper or shallower than the CCD is a major factor governing the composition of pelagic sediment. In the absence of cold water and abundant terrigenous sediment, sea floor that lies above the CCD tends to be rich in calcareous ooze. Sea floor that lies below the CCD tends to be rich in *siliceous ooze* and abyssal clay.

Calcareous sediment dissolves at the CCD because it interacts with a weak acid that is formed by dissolved carbon dioxide in sea water. Cold sea water under high pressure (deep water) holds more dissolved carbon dioxide than warmer water at low pressure (shallow water). The dissolved carbon dioxide (CO_2) combines with water (H_2O) and forms *carbonic acid* (H_2CO_3) (this reaction was important in Chapter 7, "Weathering"), which reacts with calcareous particles falling through the water column. Below the CCD, calcareous particles will dissolve completely. The exact depth of the CCD varies because the temperature, pressure, and chemical composition of sea water varies around the world (and through geologic time) due to variations in water circulation, salinity, and other factors. The CCD can also be driven downward by high rates of calcareous sediment production in the water above (because waters become saturated with respect to carbonate).

Below the CCD, pelagic sediment consists largely of abyssal clay (also termed *red clay*) and siliceous ooze. Abyssal clay is blown off the continents into the ocean, particularly on the western coasts of continents adjacent to major deserts. This aeolian dust is deposited everywhere, but it is abundant only in abyssal regions because low biological productivity and the dissolution of calcareous skeletal debris prevent other, normally dominant, types of sediment from accumulating. As humans degrade the landscape and set large brushfires, and as desertification intensifies, aeolian dust has grown in abundance.

FIGURE 23.17 A tripod fish momentarily comes to rest on the sediment-covered abyssal plain of the Pacific Ocean near Hawaii.

NOAA

(?) What is the source of sediment on the abyssal plains?

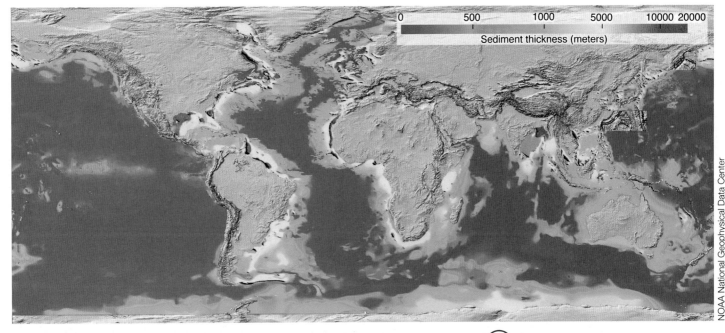

NOAA National Geophysical Data Center

FIGURE 23.18 Sediment thickness on the sea floor varies from a few metres on young oceanic crust to thousands of metres along continental margins.

(?) What is the composition of abyssal sediment?

Although less common, abyssal sediments can also include volcanogenic particles, cosmogenic particles, and authigenic sediments. Volcanogenic particles, typically consisting of fine ash that falls from the atmosphere, can be a rich source of pelagic sediment in localities near volcanically active regions. Cosmogenic sediments accumulate in very low amounts. They come from dust created by asteroid impacts, from debris from comets (that falls to Earth from space), and from meteorites that release particles as they encounter friction in the atmosphere. Authigenic sediments are made up of minerals that precipitate from sea water. These include *metallic sulphides* that form where hydrothermal processes at spreading centres precipitate minerals (see Chapter 10, "Geologic Resources") and *manganese nodules* and *phosphate deposits* that accumulate from precipitated ions dissolved in sea water.

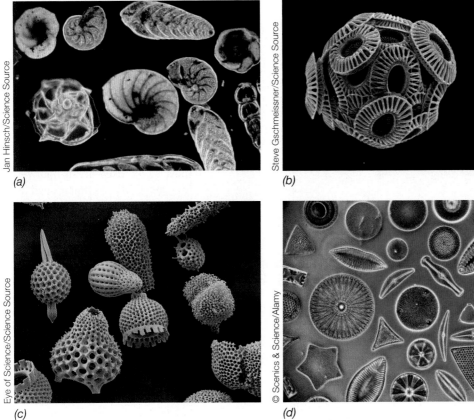

(a) Jan Hinsch/Science Source

(b) Steve Gschmeissner/Science Source

(c) Eye of Science/Science Source

(d) © Scenics & Science/Alamy

FIGURE 23.19 Pelagic sediment usually has a high percentage of microscopic skeletal debris consisting of calcareous ooze, made up of *(a)* foraminifera and *(b)* coccolithophores, or siliceous ooze, made up of *(c)* radiolarians and *(d)* diatoms.

(?) What factors determine the composition of biogenic ooze?

 Expand Your Thinking—Make a table listing the types of sediment that accumulate on the abyssal plain and the major controls on their abundance.

23-7 Pelagic Stratigraphy Reflects Dissolution, Dilution, and Productivity

LO 23-7 *Describe the stratigraphy of pelagic sediments.*

The rate at which pelagic sediment accumulates depends on three processes: *dissolution* (of biogenic particles), *dilution* (mixing with other types of sediment), and *productivity* (production of new organic matter by living organisms in the water).

Dissolution

Typical ocean waters are *undersaturated* with respect to silica. Sea water that is undersaturated with respect to a substance tends to continuously dissolve that substance until the water becomes saturated, at which point no more will dissolve. Hence, sea water tends to dissolve silica plankton shells. However, conditions favouring deposition of silica or calcium carbonate are opposite from one another. As sea water warms and pressure decreases, silica solubility (tendency to dissolve) increases. Conversely, silica's solubility decreases in the cold, pressurized water of the deep ocean. This means that objects made of silica, such as plankton shells, tend to favour cold surface water, and as they fall through cold deep water, their chances of arriving at the sea floor and forming a sedimentary deposit are enhanced. On the other hand, carbonate solubility increases with depth so that bottom waters become less saturated in calcium carbonate. The patterns of carbonate and silica deposits on the sea floor reflect the different processes of formation and preservation, resulting in calcareous oozes that have little biogenic silica and silica oozes that have little biogenic carbonate.

The solubility of calcium carbonate varies from one ocean to another because of variations in temperature and carbon dioxide content. Near the surface of all oceans where the temperature is relatively high, sea water generally is saturated with respect to calcium carbonate (hence calcium carbonate material tends not to dissolve). However, at average depths of 4 km to 5 km, the dissolved carbon dioxide content in sea water is sufficiently high to cause calcium carbonate to dissolve. Consequently, calcareous shells are rarely found on the ocean floor at depths below 5 km; however, this may vary. For instance, the CCD occurs at 6 km in the Atlantic and 3.5 km in parts of the Pacific.

Dilution and Productivity

Generally speaking, two types of sediment cover most of the sea floor away from continents: abyssal clay and biogenic ooze (**FIGURE 23.20**). Abyssal clay accumulates at the very slow rate of about 1 mm per 1,000 years and covers most of the deep-ocean floor below the CCD. As mentioned in Section 23-6, abyssal clay is deposited everywhere in the oceans. But it is found in high concentrations only where other types of sediment (such as biogenic oozes) are absent and hence do not dilute the clay accumulation.

There are two main groups of ooze: siliceous ooze and calcareous ooze. Siliceous ooze is composed mostly of the tiny remains of diatoms (plants) and radiolarians (animals). Both diatoms and radiolarians secrete skeletal hard parts made up of *opaline silica* ("opaline" simply means that there is water inside the silica molecule, which makes it appear iridescent). Calcareous oozes are composed of the remains of coccolithophores (plants) and foraminiferans (animals). They

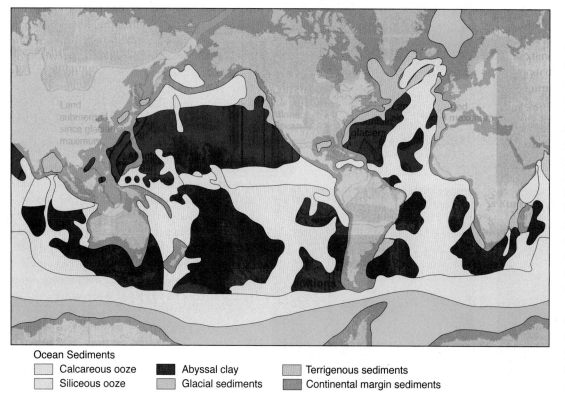

Ocean Sediments
- Calcareous ooze
- Siliceous ooze
- Abyssal clay
- Glacial sediments
- Terrigenous sediments
- Continental margin sediments

FIGURE 23.20 The modern ocean floor is covered with various sediments. Near Antarctica and to the east of Greenland, the sediment is detrital and comes from ice-rafting and glacial deposition. In the cold water of the Southern Ocean and in the north Pacific, the sea floor tends to collect siliceous ooze. Siliceous ooze also collects in the equatorial Pacific and Indian Oceans. Below the CCD, the sediment is mostly abyssal clay. Above the CCD, away from the continents, most of the sea floor is covered in calcareous ooze. Along continental margins, watersheds on the land deliver detrital (terrigenous) sediment in great quantities.

 Point to one area on this map and explain the composition of the pelagic sediment.

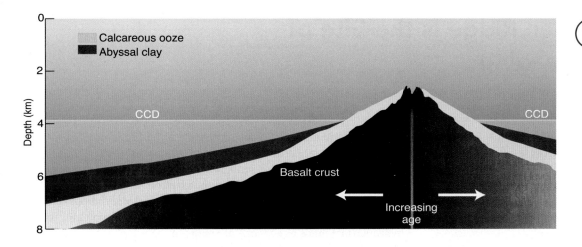

(?) Describe the stratigraphy of the sediments if you were to drill a core into the Atlantic sea floor.

FIGURE 23.21 The Mid-Atlantic Ridge collects calcareous ooze above the CCD. As the sea floor moves away from the ridge into deeper water, it passes down through the CCD, where it collects predominantly abyssal clay. Thus, the resulting stratigraphy moving away from the spreading ridge consists of two layers, with ooze overlain by clay.

secrete hard parts composed of calcium carbonate. Coccoliths are a type of algae; foraminifera may live on the sea floor as well as float in the water column, and they, like radiolaria, feed mainly on algae.

Marine Stratigraphy

The depth of the Mid-Atlantic Ridge is about 2,500 m at the crest and 7,500 m at its base. The CCD intersects the ridge midway down its flanks and controls the nature of pelagic sedimentation. Calcareous ooze collects on the shallow portions of the ridge and forms a thick layer where the slowly spreading sea floor approaches the CCD. But once the sea floor passes through the CCD, the deposition of calcareous ooze halts, and abyssal clay (no longer diluted with biogenic sediment) becomes the dominant component of the pelagic sediment. This pattern produces a relatively simple two-layer stratigraphy, consisting of a layer of calcareous ooze underlying a layer of abyssal clay (**FIGURE 23.21**).

In the Pacific Ocean, the stratigraphy is more complex (**FIGURE 23.22**). The East Pacific Rise, the mid-ocean ridge that is the birthplace of the Pacific Plate, is located in the southeast portion of the Pacific basin. From there the plate moves to the northwest through the Southern Hemisphere, across the equator, and eventually to

subduction zones in the northwestern Pacific. With a typical depth of approximately 2,500 m, the East Pacific Rise collects calcareous ooze along its crest and upper flanks. However, much of the Pacific sea floor lies below the CCD, and abyssal clay collects on it as the plate moves toward the equator.

At the equator, something interesting happens. Because of high biological productivity in the warm, nutrient-rich water, abundant siliceous and calcareous skeletal debris falls through the water column. Dissolution of this large volume of material results in sea water that is only slightly undersaturated with respect to calcium carbonate and silica. Hence, there is little dissolution of biogenic particles, and the sea floor collects a mixture of siliceous and calcareous ooze—essentially the CCD is depressed to the depth of the sea floor. Once the constantly moving plate passes beyond the equatorial waters, the CCD re-forms at a shallower depth, and a second layer of abyssal clay collects that buries the mixed siliceous/calcareous sediments that formed at the equator. Hence, a four-layer model of stratigraphy is produced in the Pacific Ocean. From the base to the top, this model consists of calcareous ooze, abyssal clay, a layer of mixed siliceous and calcareous ooze, and another layer of abyssal clay.

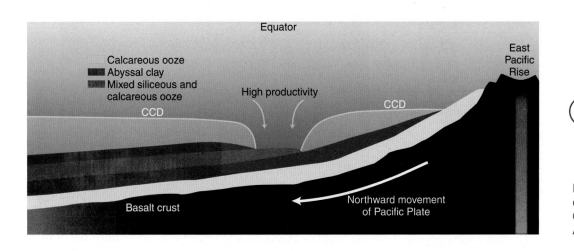

(?) Explain what happens to the CCD in the Pacific Ocean at the equator.

FIGURE 23.22 The stratigraphy of abyssal sediments in the Pacific Ocean is more complex than in the Atlantic Ocean.

(?) **Expand Your Thinking**—If abyssal clay is prominent only when ooze is absent, describe in relative terms the rates of accumulation of the two types of sediment. Explain the differences.

23-8 The Mid-Ocean Ridge Is the Site of Seafloor Spreading

LO 23-8 *Identify the principal components of oceanic crust.*

The world's longest mountain range, about 60,000 km, is the global **mid-ocean ridge** that runs across the floor of all five oceans (FIGURE 23.23). As we have learned, the mid-ocean ridge occurs where lithospheric plates separate; as they gradually move apart, magma rises to fill the gap. This magma, interacting with cold sea water that descends through fractures in the sea floor, drives the eruption of *hydrothermal vents* along the ridges (FIGURE 23.24). Sea water that contacts hot rock or magma heats up to 400°C and dissolves elements such as sulphur, lead (Pb), copper (Cu), and zinc (Zn) from the hot crust. This superheated water rises through the sea floor and gushes from vents, carrying with it these newly acquired elements. Valuable metallic sulphide minerals precipitate when the emerging superheated water is rapidly cooled on contact with the cold (about 4°C) sea water. (Recall the sulphide mineral group from Chapters 4, "Minerals," and 10, "Geologic Resources.")

Researchers descending in deep-sea submersibles to study mid-ocean ridges may encounter the astonishing site of tall, thin "chimneys" of precipitated metallic sulphide minerals spewing clouds of black smoke. These are known as *black smokers,* and they mark the location of hydrothermal vents. Unique *hydrothermal vent communities* consisting of giant tubeworms, clams, crabs, and shrimp have evolved to live around these seafloor geysers.

How do living things survive in such an inhospitable environment? Most life on Earth depends on sunlight as the ultimate source of energy. Green plants use sunlight to make food by the process of photosynthesis, and other organisms then feed on plants. In the darkness of the ocean depths, there is no sunlight for photosynthesis. Instead, a special class of bacteria use dissolved sulphur compounds in hydrothermal fluids as a source of energy to replace photosynthesis. These bacteria are the base of vent food webs, and all other animals ultimately depend on them for food. Thus, hydrothermal vent communities are forever tied to plate tectonics for their living environment.

Oceanic Crust

Spreading centre volcanism occurs at the site of mid-ocean ridges where two lithospheric plates diverge from one another. A rift valley at the crest of the ridge is characterized by extension of oceanic crust, normal faulting, magma intrusion, seafloor volcanic activity, and the creation of a rugged terrain made of faulted blocks of rock. As the plates separate, magma from the asthenosphere rises to fill fissures and fractures and in the process produces new oceanic crust. The rise of this hot mantle gives thermal buoyancy to the sea floor—this is the reason that ridges stand high above the surrounding abyssal plain. Researchers estimate that, every year, approximately 20 seafloor eruptions occur along mid-ocean rift valleys and that 2.5 km^2 of new sea floor is formed. About 4 km^3 of new ocean crust forms every year, with a thickness averaging 1 km to 2 km along the ridge axis.

The lower two-thirds (5 km) of the newly forming oceanic crust consists of slow-cooling intrusive magma composed of phaneritic gabbro and

FIGURE 23.23 The mid-ocean ridge is the world's longest mountain range. Here the Atlantic mid-ocean ridge marks the spreading centre between the African Plate and the South American Plate.

J.R. Delaney and D.S. Kelley, University of Washington, Seattle

FIGURE 23.24 The hot springs at the surface of hydrothermal vents are called "black smokers" because of their billowing clouds of precipitating metals.

 In addition to the minerals in basalt, what other minerals are added to the sea floor at spreading centres?

 What is the basis for the food chain in hydrothermal vent communities?

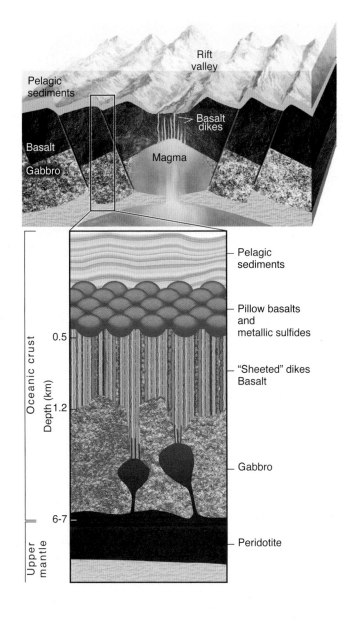

FIGURE 23.25 Oceanic crust consists of a layer of phaneritic gabbro and ultramafic rocks. This layer is overlain by densely spaced basalt dikes that carry lava to the sea floor. Extrusive basalts are interbedded with deposits of metallic sulphide minerals. As the sea floor is carried away from the rift valley, it is slowly buried in pelagic sediment.

(?) How does the igneous rock composition and texture change from the upper mantle to the sea floor?

layered ultramafic rocks. Above this layer is an approximately 1.5-km-thick unit of vertical basalt dikes, often called *sheeted dikes* due to their dense spacing. This layer is buried by a 0.5-km-thick unit of volcanic pillow basalt interbedded with rich metallic sulphide minerals produced by the hydrothermal process. Within rift valleys, the sea floor is composed of pillow basalts and metallic sulphide deposits. After the basalts and deposits form, volcanism and hydrothermal action ceases, and these rocks accumulate a blanket of pelagic sediment. The sedimentary layer thickens and begins to bury the rugged topography of the rift valley as the crust moves away from the spreading centre (**FIGURE 23.25**).

Abyssal Hills

You have probably never heard of **abyssal hills,** but some scientists think they are Earth's most abundant topographic feature (**FIGURE 23.26**). They are small, well-defined hills, ridges, and peaks that rise a few metres to several hundred metres above the abyssal sea floor. Abyssal hills are produced when pelagic sediment buries the rugged topography formed in rift valleys. The new sea floor formed by normal faulting and volcanism at rift zones is very jagged and uneven with high topographic relief. This topography is slowly buried with layers of sediment that express the highs and lows of the relief but not its craggy nature, thus producing abyssal hills.

As the hills grow older and move away from the rift valley, they are buried in pelagic sediment, producing subdued but abundant topographic features. Typically, abyssal hills are most abundant near mid-ocean ridges and decrease away from ridges. This is because they eventually disappear as they are buried in sediment.

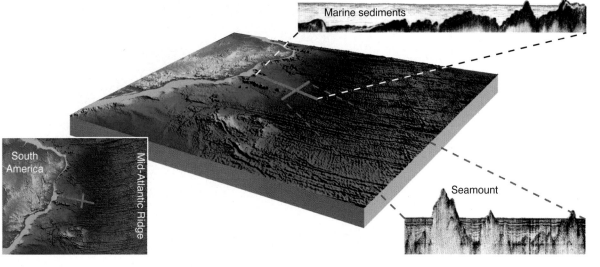

(?) How do abyssal hills change as they grow older?

FIGURE 23.26 Abyssal hills have been described as Earth's most abundant topographic feature. They are sediment-covered terrain produced by rifting at mid-ocean ridges.

 Expand Your Thinking—What principal factors control the distribution and relief of abyssal hills?

? CRITICAL THINKING

STRUCTURE OF THE SEA FLOOR

On the North American Plate, the Atlantic sea floor is created at the Mid-Atlantic Ridge and moves to the west. With a partner, please refer to **FIGURE 23.27** and answer the following questions.

1. Draw an expanded cross-section of the rock and sediment types that form the oceanic crust. Be sure to include igneous as well as sedimentary rock types and label the important features.
2. Figure 23.27 shows several important features. Label as many as you can.
3. Circle the abyssal hills along the flanks of the Mid-Atlantic Ridge. What is the origin of these features?
4. Describe the processes that form submarine canyons. How are they related to submarine fans?
5. Why does the source and composition of sediment change between the coast, shelf, rise, abyssal plain, and oceanic ridge?
6. Classify the types of rocks and minerals that are formed in the rift valley of the Mid-Atlantic Ridge.
7. As a marine scientist, you have a research submersible available for four dives, a rare and expensive opportunity. What research will you conduct using these dives? State a hypothesis motivating each dive, identifying where it will take place and the data you will collect to test your hypothesis.

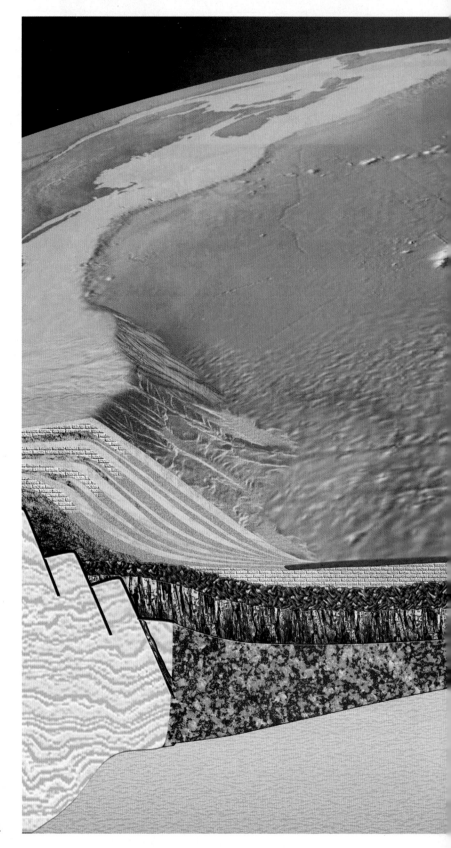

FIGURE 23.27 The Atlantic sea floor.

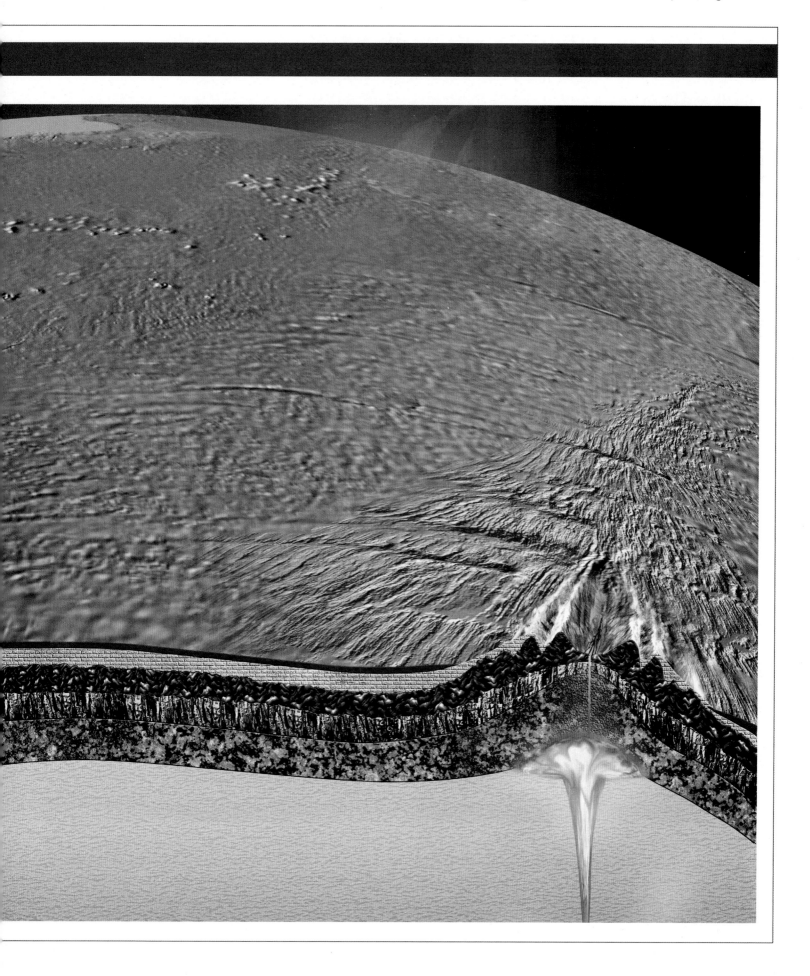

23-9 Oceanic Trenches Occur at Subduction Zones

LO 23-9 *Describe the characteristics of oceanic trenches.*

In our studies of plate tectonics, you learned that convergent margins are characterized by the meeting of two plates with the denser plate subducting beneath the less dense plate. Typically older oceanic lithosphere subducts beneath younger oceanic lithosphere, and oceanic lithosphere (of any age) subducts when it converges with continental lithosphere. This process results in the formation of one of the most dramatic features in the ocean basins, **oceanic trenches.** Trenches range in depth from 6 km to 11 km and help define plate boundaries.

In the case of ocean–ocean convergence, a volcanic island arc forms (**FIGURE 23.28**). In the case of ocean–continent convergence, a *volcanic arc* develops. Both these volcanically active ranges are the product of partial melting beneath the overriding plate where water released from the subducting slab lowers the melting point of rock in the upper mantle. This process generates magma that intrudes the overlying crust and produces volcanism.

The oceanic trenches are the deepest parts of the ocean floor. The floor in the *Challenger Deep* of the *Mariana Trench* in the west Pacific is 11 km deep and sits 3 km to 4 km deeper than the surrounding ocean floor. The Mariana Trench, the world's deepest, is formed where two plates, both of oceanic lithosphere, converge with the older Pacific Plate subducting beneath the younger Philippine Plate. Major world trenches include (**FIGURE 23.29**):

- the *Mariana Trench,* 11 km deep, where the Pacific Plate subducts beneath the smaller Mariana Plate
- the *Aleutian Trench,* 7.6 km deep, where the Pacific Plate subducts beneath an oceanic portion of the North American Plate.
- the *Kuril Trench,* 10.5 km deep, which links the Aleutian Trench with the Japan Trench in the northwest Pacific basin.

- the *Japan Trench,* 9 km deep, where the Pacific Plate subducts beneath the Eurasian Plate.
- the *Tonga Trench,* 10.8 km deep, where the Pacific Plate subducts beneath the Tonga Plate and the Indo-Australian Plate.
- the *Kermadec Trench,* 10 km deep, where the Pacific Plate subducts beneath the Indo-Australian Plate.
- the *Middle America Trench,* 6.6 km deep, where the Pacific Plate subducts beneath the Caribbean Plate.
- the *Peru-Chile Trench,* also known as the *Atacama Trench,* 8 km deep, where the Nazca Plate subducts beneath the South America Plate.
- the *Cayman Trough,* 7 km deep, which is not a proper trench, formed by a transform fault between the North American Plate and the Caribbean Plate.
- the *Puerto Rico Trench,* 8.6 km deep, which marks the boundary of the North American Plate and the Caribbean Plate.

Oceanic trenches generally have a V cross-section, similar to a stream-carved valley, except the side on the overriding plate tends to be steeper than the side on the subducting plate. Trench walls can be as steep as 45° (Tonga Trench), although walls dipping between 4° and 16° are more typical. The floor of a trench is usually a flat abyssal plain, well below the CCD, and blanketed in pelagic sediment, although not in the thicknesses found in mid-ocean areas. Because oceanic lithosphere is being subducted at trenches, there is less time to accumulate sediment than at more stable settings. Abyssal clay, siliceous ooze, volcanic ash and lapilli, layers of turbidites (coarse sediments), and disturbed layers of sediment that have slumped from steep trench walls are the typical material that layer the trench floor.

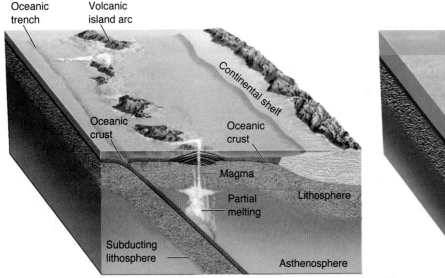

(a)

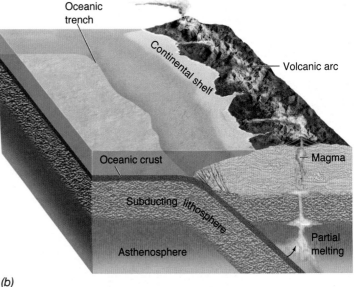

(b)

FIGURE 23.28 Oceanic trenches form where one plate subducts beneath another. *(a)* An island arc is formed by ocean–ocean convergence, and *(b)* a volcanic arc is formed by ocean–continent convergence.

(?) What is the source of magma that produces volcanism on the overlying plate?

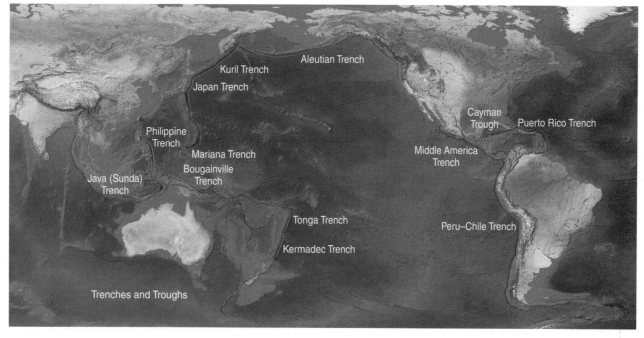

FIGURE 23.29 Oceanic trenches occur at convergent plate margins.

GEOLOGY IN ACTION

THE PUERTO RICO TRENCH

The Puerto Rico Trench marks the convergence of the North American Plate and the Caribbean Plate (**FIGURE 23.30**). South and east of Puerto Rico, the boundary is convergent, but at Puerto Rico and Hispaniola, the plates form a transform boundary with a significant seismic and tsunami risk. This has long been recognized by geologists, and was revealed to the world with fatal consequences on January 12, 2010, during the Haiti earthquake.

Because Puerto Rico lies on an active plate boundary, earthquakes are a constant threat to the over 4 million citizens who live there, and tsunamis threaten the densely populated coastal region, where most communities are located. The region has a history of large earthquakes. At least seven quakes with Richter magnitudes of 6.9 or greater have occurred since 1787. In 1946, a magnitude 8.1 earthquake occurred north of Hispaniola, a large island to the east of Puerto Rico where Haiti is located. This quake generated a tsunami that drowned nearly 1,800 people in northeastern Hispaniola. Other earthquakes in the region are known to have generated tsunamis as well.

Haiti, unfortunately, experienced the most damaging earthquake in the history of the Caribbean region on January 12, 2010. The death toll was unusually high because of the poor construction

FIGURE 23.30 The Puerto Rico Trench marks the boundary between the North American Plate and the Caribbean Plate.

methods used in buildings in the impoverished nation: over 200,000 people died as a result of the 7.0 magnitude earthquake.

 Expand Your Thinking—Why would oceanic trenches be a source of tsunamis?

23-10 Human Impacts on the Oceans Are Global in Extent

LO 23-10 *Identify human impacts on the oceans.*

The science of marine geology is mainly concerned with the history of ocean basins, the nature of sediments that collect there and what they can tell us about past events, and the processes related to plate tectonics that influence the oceanic lithosphere and adjoining continents. However, many geologists also play important roles in helping to manage marine resources and environments associated with coral reefs, beaches, estuaries, barrier islands, the shelf, and the larger ocean.

Human impacts on the marine environment are now known to be global in scale. In February 2008, researchers at the University of California, Santa Barbara in the United States released a map showing areas of the world where ocean environments had been negatively impacted by human activities. The stunning feature of the map is that across the entire 360 million square kilometres covered by oceans (70 percent of Earth's surface), less than 4 percent of the ocean remains unaffected by humans (**FIGURE 23.31**).

Nineteen scientists collaborating from a range of universities and government agencies compiled the map using past studies of human impacts on marine ecosystems, such as coral reefs, sea grass beds, fisheries, continental shelves, and the deep ocean. Most previous work had focused largely on localized impacts. The new study expands the scope to encompass human impacts on marine ecosystems on a global scale. Fishing, fertilizer runoff, pollution, shipping, climate change, and other impacts (a total of 17 ways humans affect oceans) were mapped as individual "geographic information layers" and overlain on one another to produce a composite map of global impacts. These data reveal that the degree of human impact over large areas of the ocean is poorly understood.

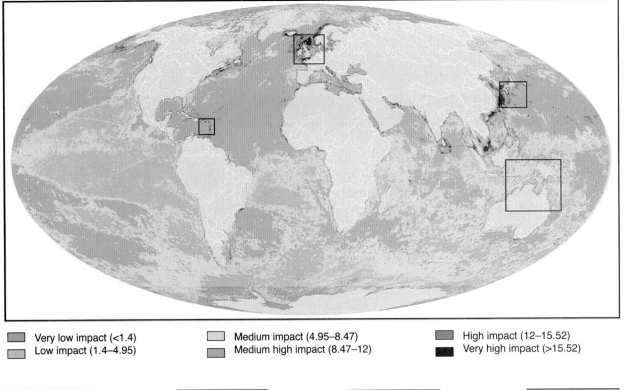

	Very low impact (<1.4)		Medium impact (4.95–8.47)		High impact (12–15.52)
	Low impact (1.4–4.95)		Medium high impact (8.47–12)		Very high impact (>15.52)

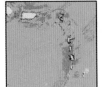

Eastern Caribbean

North Sea

Japanese Waters

Torres Straight

University of California

FIGURE 23.31 Map of human impacts on the global ocean. Colour coding is used to depict the degree of impact based on a scale of 17 ways humans affect the oceans; darker orange and red colours indicate greatest impact.

(?) Where are human impacts on the oceans highest? Where are they lowest?

An important outcome of this study is the realization that the past practice of managing resources by dealing with one type of impact at a time overlooks the combined sum of multiple impacts. Researchers conclude that managers can no longer afford to focus only on fishing or coastal wetland loss or pollution as if they are separate effects. Human impacts overlap in space and time, and in too many cases, the combined magnitude is frighteningly high. Conservation action that cuts across the entire range of human impacts is needed now around the globe if the oceans are to recover.

Expand Your Thinking—Managers of different marine environments often work in different agencies and do not communicate. How would you correct this situation? What barriers would you encounter?

EARTH CITIZENSHIP

LOSS OF CORAL REEFS

Although reefs cover less than 1 percent of the ocean floor, they play an integral role in coastal communities (**FIGURE 23.32**). They provide economic benefits through fisheries and tourism, buffer coastal areas from large waves, produce sand for beaches, and offer a shallow protective environment for numerous marine species. Many coastal communities live off fish caught on coral reefs. When corals die, the economic benefits quickly disappear. Coral disease, predators, rising ocean temperatures due to climate change, nutrient pollution, destructive fishing practices, and sediment runoff from coastal development can all impact reef communities.

A study of corals in the central and western Pacific Ocean and the eastern Indian Ocean has revealed that living coral is dying faster than was previously thought. Coral communities are disappearing at a rate of 1 percent a year, and nearly 1,554 km^2 of living coral have disappeared every year since the late 1960s. Researchers analyzed 6,000 quantitative surveys of reef communities performed between 1968 and 2004 from more than 2,600 coral reefs. They found that living coral declined from 40 percent in the early 1980s to approximately 20 percent by 2003.

The researchers conclude that half of the world's reef-building corals already have been lost and that coral loss is a global phenomenon probably due in part to large-scale impacts such as climate change. Continued decline of Indo-Pacific reefs could mean the loss of millions of dollars of income from fisheries and tourism and has important impacts on marine ecology and diversity.

(a) Darryl Leniuk/Digital Vision/Getty Images, Inc. (b) Michael Patrick O'Neill/Science Source

FIGURE 23.32 (a) Healthy coral reefs provide important economic and environmental benefits and are one of the most ecologically diverse communities in the ocean. (b) Human impacts on reefs include pollution, overfishing, and climate change.

LET'S REVIEW "GEOLOGY IN OUR LIVES"

Now that you have finished the chapter, "Geology in Our Lives" will have taken on new meaning for you. Let us review it: The oceans are Earth's largest and most important environment. Geological processes are essential to the character of the oceans. For example, currents in the five major oceanic circulation systems carry heat from the tropics toward the poles and return again to the tropics to absorb more heat. The distribution of heat by these currents controls the climate, weather, winds, precipitation, and other fundamental aspects of Earth's environment. Among the many ecosystems that depend on ocean circulation are plankton communities on the ocean surface. These communities govern the type of sediment that accumulates on the ocean floor and the dissolved compounds (such as carbon dioxide) that are buried with these sediments. This chain of processes regulates and stabilizes the chemistry and temperature of the atmosphere and ocean water. Oceans cover more than 70 percent of Earth's surface, and human impacts on the oceans are global in extent.

Sediment accumulation and plate tectonics are connected. Mid-ocean ridges and oceanic trenches mark the location of spreading centres and subduction zones. Plates move between these sites beneath waters of varying temperatures, depths, and productivity. These conditions govern the stratigraphy of pelagic sediment. Continental margins contain most of the sediment. Detrital sediment is trapped in the coastal zone and the shelves collect biogenic and authigenic sediment. Some margins pour sediment into the ocean, forming massive deltas and fans that become the dominant topographic features. Sediment moving across the shelf is funnelled through submarine canyons to deep-sea fans composed of turbidite sequences. Overall, the oceans are an integrated geological system where living organisms, plate tectonics, the continents, the atmosphere, and the water column all influence one another.

STUDY GUIDE

23-1 Marine geology is the study of geologic processes within ocean basins.

- There are five recognized oceans: the Pacific, Atlantic, Indian, Southern, and Arctic. There are also many seas, which are smaller branches of an ocean that are partly enclosed by land. The largest seas are the South China Sea, the Caribbean Sea, and the Mediterranean Sea.

- **Marine geology** involves geophysical, geochemical, sedimentological, and paleontological investigations of the ocean floor and coastal margins.

- The ocean is salty because it contains dissolved ions and compounds that come from chemical weathering and hydrothermal processes where sea water interacts with hot crust. Chloride and sodium ions make up over 90 percent of all dissolved ions in sea water. The concentration of dissolved ions in sea water, known as the **salinity,** has a global average of about 35 parts per 1,000. Ocean salinity is a result of geologic processes that regulate the concentration of dissolved components; inputs of water and dissolved ions are matched by outputs.

23-2 Ocean waters are mixed by a global system of currents.

- Ocean circulation is the large-scale movement of water by **currents,** which are large masses of continuously moving ocean water. There are two types of oceanic circulation: surface circulation, which is driven by winds, and deep circulation, which is driven by cool water at the poles that sinks and moves through the lower ocean.

- The general pattern of oceanic circulation is that surface currents carry warm water away from the tropics toward the poles while releasing heat to the atmosphere during the journey. At the poles, surface water cools during winter and sinks to the deep ocean, creating currents along the sea floor and at mid-depths in the water column.

- Due to Earth's rotation, currents are deflected to the right in the Northern Hemisphere and to the left in the Southern Hemisphere. This deflection is known as the **Coriolis effect.** Continents redirect currents and create large-scale circulation systems called **gyres.** There are five major gyres: the North Atlantic, South Atlantic, North Pacific, South Pacific, and Indian Ocean Gyres.

- Ekman transport occurs when the Coriolis effect interacts with a column of water or air. As the Coriolis effect deflects a current, water in lower layers deflects more slowly than surface water. This effect translates down as a weakening spiral resulting in water transport that is 90° to the right of the wind direction in the Northern Hemisphere.

23-3 A continental shelf is the submerged border of a continent.

- A **continental shelf** is the area generally less than 200 m deep surrounding a continental land mass. Shelves make up about 8 percent of the oceanic area. A shelf is the undersea extension of a continent and ranges from over 100 km wide to a few kilometres wide depending on its geologic history.

- The shelf surface is swept by waves, tidal currents, and storms, which, in places, continually transport a layer of sediment on the sea floor consisting of sand, silt, and silty mud. In other locations, the shelf is starved of sediment, and the sea floor consists of outcropping sedimentary rocks of (usually) Cenozoic age.

- Many continental shelves are rich in natural resources, such as marine life, sand and gravel deposits, natural gas and oil reserves, and phosphorite beds. Most nations claim mineral and land rights to their shelves and actively manage these resources.

23-4 The continental margin consists of the shelf, the slope, and the rise.

- The **continental margin** includes the continental shelf, the continental slope, and the **continental rise.** The margin separates thin oceanic crust of greater density from thick continental crust of lesser density and constitutes about 28 percent of the oceanic area. Beyond the rise lies the **abyssal plain** on fully oceanic crust.

- Active margins mark plate boundaries (convergent, divergent, or transform) and typically have less width and sediment input than passive margins. Active margins located at convergent plate boundaries where there is a subduction zone may have **accreted terrane.** Passive margins are located within plates. For instance, the passive margin along the east coast of Canada and the United States is separated from the Mid-Atlantic Ridge by an expanse of oceanic crust.

- A **submarine canyon** is a valley in the shape of a meandering channel that cuts into the shelf and slope. Many canyons end with a thick sedimentary deposit on the continental rise called a **submarine fan.** The upper portion of submarine canyons was probably carved by stream erosion during low sea-levels on continental margins that have subsided. Lower portions of canyons likely were carved by **turbidity currents,** which are dense, sediment-laden flows that travel down the axis of a canyon at high speed, carving the floor and sides. Turbidity currents carry sand and mud that is deposited on the rise when the currents spread out laterally and decelerate at the base of the canyon. These deposits are called turbidites.

23-5 Most ocean sediment is deposited on the continental margin.

- Most of the oceans' sediment (about 75 percent by volume) is deposited on the shelf, slope, and rise, where accumulations can exceed 10 km in thickness. Sediment between the shoreline and the edge of the shelf is termed **neritic sediment.** Neritic sediment may be detrital, biogenic, aeolian, residual, authigenic, or relict. Sediments that drape upper and middle continental slopes around the world are known as **hemipelagic sediments.** They grade from terrigenous mud into biogenic sediments.

- The coastal zone receives detrital sediment via watersheds. When the last ice age ended and sea level rose, rising water flooded river channels, driving them back into the watersheds. This process formed estuaries that today trap modern sediments, causing shelves to be sediment starved. Shelf sediment tends to be relict from the period of sea-level rise.

- Deltas formed by rivers such as the Mississippi (United States), Amazon (Brazil), Ganges-Brahmaputra (India and Bangladesh), and the Yangtze (China) contain so much detrital sediment that they bury the topography of the margin.

23-6 Pelagic sediment covers the abyssal plains.

- The abyssal plain lies between the continental rise and a mid-ocean ridge. Abyssal plains are among Earth's least explored geologic features. Abyssal sediment is composed predominantly of **biogenic ooze** and **abyssal clay.** Biogenic ooze consists of plankton remains, and abyssal clay is derived from continents where sediment is eroded by the wind and falls into the sea. Other abyssal sediments include volcanogenic particles, cosmogenic particles, and authigenic sediments.

- Calcareous shells come from foraminiferans (animals) and coccolithophores (plants). These make calcareous ooze (more than 30 percent skeletal debris). Siliceous ooze comes from plankton with silica shells, such as radiolarians (animals) and diatoms (plants).

- The **carbonate compensation depth (CCD)** is the depth at which calcium carbonate ($CaCO_3$) completely dissolves because the water is acidic. Whether the sea floor is deeper or shallower than the CCD governs the composition of pelagic sediment. Sea floor that lies below the CCD tends to be rich in abyssal clay.

23-7 Pelagic stratigraphy reflects dissolution, dilution, and productivity.

- Pelagic sediment accumulation depends on dissolution of biogenic particles, dilution from mixing with other types of sediment, and production of new organic matter.

- Calcareous ooze collects on the shallow portions of a mid-ocean ridge. Below the CCD, abyssal clay is the dominant sediment. This pattern results in a two-layer stratigraphy consisting of calcareous ooze underlying abyssal clay.

- The East Pacific Rise collects calcareous ooze on its crest and upper slope. Below the CCD, abyssal clay collects as the sea floor moves toward the equator. At the equator, high biological productivity suppresses the CCD, and mixed calcareous/siliceous ooze collects. Beyond equatorial waters, the CCD re-forms and abyssal clay collects that buries the biogenic sediment from the equator. Hence, a four-layer model of stratigraphy is produced in the Pacific Ocean.

23-8 The mid-ocean ridge is the site of seafloor spreading.

- The world's longest mountain range, about 60,000 km, is the global **mid-ocean ridge.** Valuable metallic sulphide minerals precipitate in the rift valley at mid-ocean ridges from hot water expelled at black smokers that mark the location of hydrothermal vents.

- Rock in the upper mantle is ultramafic with the composition of peridotite. This rock partially melts to produce mafic magma, which intrudes the crust to form gabbro. Crystallization yields layered ultramafic rocks among the gabbros. Basaltic magma rising rapidly to the sea floor forms sheeted dikes and fissure eruptions, generating submarine lava fields. Nestled among pillow lava in the rift valley are metallic sulphides formed by hydrothermal activity.

- **Abyssal hills** are small, well-defined hills, ridges, and peaks that rise a few metres to several hundred metres above the abyssal sea floor. They are produced when pelagic sediment blankets the rugged topography formed in rift valleys.

23-9 Oceanic trenches occur at subduction zones.

- Plate convergence results in **oceanic trenches** that range in depth from 6 km to 11 km. The Mariana Trench, the world's deepest, formed where the older Pacific Plate subducts beneath the younger Philippine Plate. The floor of a trench is usually a flat abyssal plain, well below the CCD and blanketed in pelagic sediment, which is not as thick as that in mid-ocean settings. Because oceanic lithosphere is subducted at trenches, there is less time for sediment to accumulate than at more stable settings.

- Ocean–ocean convergence produces a volcanic island arc, and ocean–continent convergence produces a volcanic arc. Convergent boundaries are marked by seismic activity. Earthquake foci outline the position of a subducting oceanic plate.

23-10 Human impacts on the oceans are global in extent.

- Marine geologists play important roles in managing marine resources such as coral reefs and coastal environments.
- Researchers have produced a map showing areas of the world where ocean environments are negatively impacted by human activities. Across the entire 360 million square kilometres covered by the oceans (70 percent of Earth's surface), less than 4 percent remains unaffected by humans.

- Managing resources by dealing with one type of impact at a time overlooks the combined sum of multiple impacts. Managers cannot afford to focus only on fishing or coastal wetland loss or pollution as if they are separate effects. Human impacts overlap in space and time, and in many cases, the combined magnitude is high. Conservation action that cuts across the range of impacts is needed around the globe if the oceans are to recover.

KEY TERMS

abyssal clay (p. 634)
abyssal hills (p. 639)
abyssal plain (p. 630)
accreted terrane (p. 630)
biogenic ooze (p. 634)
carbonate compensation depth (CCD) (p. 634)
continental margin (p. 630)
continental rise (p. 630)

continental shelf (p. 628)
continental slope (p. 629)
Coriolis effect (p. 626)
currents (p. 626)
gyres (p. 626)
hemipelagic sediments (p. 633)
marine geology (p. 624)
mid-ocean ridge (p. 638)
neritic sediment (p. 632)

oceanic trenches (p. 642)
pelagic sediment (p. 634)
salinity (p. 624)
submarine canyon (p. 629)
submarine fan (p. 630)
thermohaline circulation (p. 627)
trade wind (p. 626)
turbidity currents (p. 631)

ASSESSING YOUR KNOWLEDGE

Please answer these questions before coming to class. Identify the best answer.

1. The major oceans are the
   a. Mediterranean, Caribbean, Indian, Atlantic, and Pacific.
   b. Antarctic, Arctic, Pacific, Atlantic, and Mediterranean.
   c. Southern, Arctic, Pacific, Atlantic, and Indian.
   d. Pacific, Atlantic, Arctic, and Indian.
   e. Indian, Red Sea, Arctic, Pacific, and Antarctic.

2. Surface circulation is propelled principally by
   a. thermohaline circulation.
   b. Earth's rotation.
   c. plate tectonics.
   d. the wind.
   e. the tides.

3. Ekman transport occurs when
   a. the Coriolis effect causes surface water to travel to the left in the Northern Hemisphere.
   b. water below the surface lags behind surface water in responding to the Coriolis effect.
   c. winds blow water toward the coast, causing upwelling.
   d. surface water and water that is deeper travel in the same direction due to the Coriolis effect.
   e. currents sink due to density and evaporation.

4. The continental margin consists of the
   a. abyssal plain, slope, rise, and mid-ocean ridge.
   b. shelf, slope, rise, and abyssal plain.
   c. shelf, slope, and rise.
   d. watershed, shelf, and slope.
   e. abyssal hills, submarine fans, and submarine ridges.

5. Submarine canyons are
   a. formed by erosion of the continental slope by streams and turbidity currents.
   b. drowned watersheds.
   c. inactive drainage systems formed by past tectonic processes.
   d. caused by sediment migration across the continental rise.
   e. the result of hurricane storm surge.

6. Active margins are characterized by
   a. wide, sediment-rich shelves.
   b. narrow shelves along tectonic boundaries.
   c. high sediment accumulation due to large watersheds and their deltas.
   d. abyssal clay accumulation because of a shallow CCD.
   e. slow subsidence.

7. Modern shelves tend to be sediment starved because
   a. coastal environments trap detrital sediments.
   b. neritic sediments are buried by pelagic sediments on shelves.
   c. biogenic sediment overwhelms detrital sediment accumulation.
   d. strong currents prevent sediment from accumulating.
   e. Actually, shelves are not sediment starved.

8. Pelagic sediment predominantly consists of
   a. relict, authigenic, and detrital sediments.
   b. biogenic ooze and abyssal clay.
   c. hemipelagic sediment and turbidites.
   d. metallic sulphide deposits.
   e. abyssal hills and turbidites.

9. Calcareous ooze is mostly composed of
   a. foraminifera and coccolithophores.
   b. abyssal clay and diatoms.
   c. diatoms and radiolarians.
   d. partially dissolved abyssal clay.
   e. cosmogenic and volcanogenic sediments.

10. The carbonate compensation depth is
   a. the depth at which calcium carbonate precipitates in water.
   b. generally at 2 km in most oceans.
   c. controlled by the salinity.
   d. the level of silicate precipitation.
   e. None of the above.

11. Pelagic sediment stratigraphy is influenced by
   a. the CCD.
   b. productivity in overlying waters.
   c. proximity to continents.
   d. dissolution and dilution of sediments.
   e. All of the above.

12. Oceanic crust is composed of
   a. gabbro, andesite, and basalt.
   b. gabbro, basalt, hydrothermal deposits, and pelagic sediment.
   c. basalt, pillow lava, peridotite, andesite, and marine rhyolite.
   d. rhyolite lava flows, gabbro, basalt, and hemipelagic sediment.
   e. None of the above.

13. Most oceanic trenches are characterized by
   a. great depth.
   b. mixed types of sediments.
   c. earthquake activity.
   d. nearby active volcanism.
   e. All of the above.

14. Managing marine environments requires
   a. focusing on one problem at a time.
   b. sticking to the coastal zone because that is where most of the problems are.
   c. managing overlapping human impacts simultaneously.
   d. focusing mostly on the surface waters.
   e. In reality, most marine environments do not need management.

15. The ocean is known as "Earth's heat engine" because
   a. surface circulation carries heat from the equator toward the poles and releases it to the atmosphere during the journey, warming the air.
   b. thermohaline circulation carries heat to the deep ocean, releasing it around the globe as the water circulates through the deep sea.
   c. ocean water stores heat during the day and releases it at night, warming the air when the Sun is absent.
   d. water stores heat and moderates global warming of the atmosphere.
   e. All of the above.

FURTHER RESEARCH

1. What type of sediment is most likely to accumulate in the Arctic Ocean?

2. Why do mid-ocean ridges rise above the surrounding sea floor?

3. How do abyssal hills form?

4. Predict the type of sediment you would find at these locations:
   a. on the sea floor at 60 m of water depth offshore of St. John's, Newfoundland
   b. on the abyssal plain around Hawaii
   c. in the Puerto Rico Trench
   d. south of Greenland on the Atlantic sea floor

5. Describe the major topographic provinces of the sea floor.

6. Why are many modern continental shelves "starved" of sediment?

7. Draw the stratigraphy of the Pacific sea floor where it descends into the Aleutian Trench.

8. What is the origin of submarine canyons?

ONLINE RESOURCES

Explore more about oceans on the following websites:

Maps of the sea floor and various types of marine geology information:
www.ngdc.noaa.gov/mgg/mggd.html

Marine geology research being conducted by the Geological Survey of Canada:
GSC Atlantic
http://gsc.nrcan.gc.ca/org/atlantic/index_e.php
GSC Pacific (Sidney)
http://gsc.nrcan.gc.ca/org/sidney/index_e.php

Marine geology research being conducted by the U.S. Geological Survey:
http://marine.usgs.gov

How the U.S. federal government manages oceans and coasts and how the National Ocean Service makes maps of the sea floor:
http://oceanservice.noaa.gov

Additional animations, videos, and other online resources are available at this book's companion website:
www.wiley.com/college/fletcher

This companion website also has additional information about WileyPLUS and other Wiley teaching and learning resources.

Common Rock-Forming Silicate Minerals		
Silicate Mineral	**Composition**	**Physical Properties**
Quartz	Silicon dioxide (silica, SiO_2)	Hardness of 7 (on scale of 1 to 10)*; will not cleave (fractures unevenly); typically colourless, but may exhibit a variety of colours; transparent; specific gravity: 2.65
Potassium feldspar group	Aluminosilicates of potassium	Hardness of 6.0–6.5; cleaves well in two directions; pink or white; specific gravity: 2.5–2.6
Plagioclase feldspar group	Aluminosilicates of sodium and calcium	Hardness of 6.0–6.5; cleaves well in two directions; white or grey; may show striations on cleavage planes; specific gravity: 2.6–2.7
Muscovite mica	Aluminosilicates of potassium with water	Hardness of 2–3; cleaves perfectly in one direction, yielding flexible, thin plates; colourless; transparent in thin sheets; specific gravity: 2.8–3.0
Biotite mica	Aluminosilicates of magnesium, iron, and potassium, with water	Hardness of 2.5–3.0; cleaves perfectly in one direction, yielding flexible, thin plates; black to dark brown; specific gravity: 2.7–3.2
Pyroxene group	Silicates of aluminum, calcium, magnesium, and iron	Hardness of 5–6; cleaves in two directions at 87° and 93°; black to dark green; specific gravity: 3.1–3.5
Amphibole group	Silicates of aluminum, calcium, magnesium, and iron	Hardness of 5–6; cleaves in two directions at 56° and 124°; black to dark green; specific gravity: 3.0–3.3
Olivine	Silicates of magnesium and iron	Hardness of 6.5–7.0; poor cleavage in two directions; light green; transparent to translucent; specific gravity: 3.2–3.6
Garnet group	Aluminosilicates of iron, calcium, magnesium, and manganese	Hardness of 6.5–7.5; uneven fracture; red, brown, green, or yellow; specific gravity: 3.5–4.3

(Biotite mica, Pyroxene group, Amphibole group, and Olivine are bracketed as *Ferromagnesian minerals*)

*The scale of hardness used by geologists was formulated in 1822 by Frederich Mohs. Beginning with diamond as the hardest mineral, he arranged the following table:

10 Diamond	8 Topaz	6 Feldspar	4 Fluorite	2 Gypsum
9 Corundum	7 Quartz	5 Apatite	3 Calcite	1 Talc

The Periodic Table

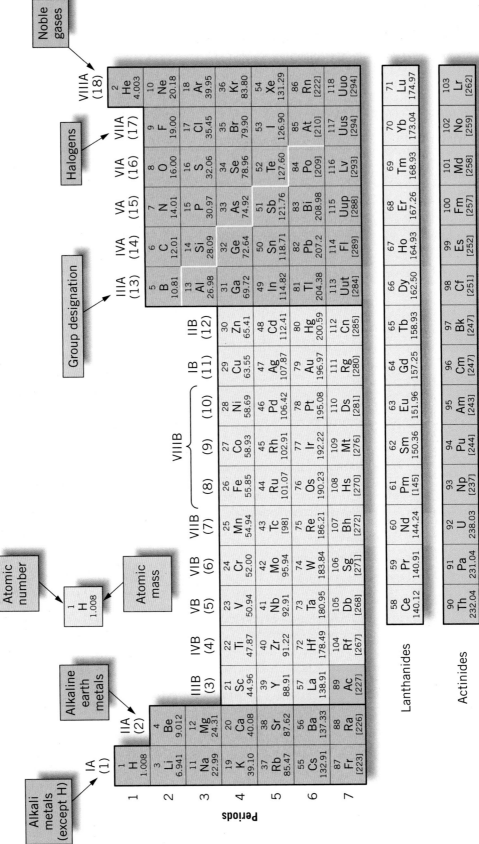

Elements and Their Chemical Symbols

Element	Symbol	Element	Symbol	Element	Symbol	Element	Symbol
Actinium	Ac	Erbium	Er	Mercury	Hg	Seaborgium	Sg
Aluminum	Al	Europium	Eu	Molybdenum	Mo	Selenium	Se
Americium	Am	Fermium	Fm	Neodymium	Nd	Silicon	Si
Antimony	Sb	Flerovium	Fl	Neon	Ne	Silver	Ag
Argon	Ar	Fluorine	F	Neptunium	Np	Sodium	Na
Arsenic	As	Francium	Fr	Nickel	Ni	Strontium	Sr
Astatine	At	Gadolinium	Gd	Niobium	Nb	Sulphur	S
Barium	Ba	Gallium	Ga	Nitrogen	N	Tantalum	Ta
Berkelium	Bk	Germanium	Ge	Nobelium	No	Technetium	Tc
Beryllium	Be	Gold	Au	Osmium	Os	Tellurium	Te
Bismuth	Bi	Hafnium	Hf	Oxygen	O	Terbium	Tb
Bohrium	Bh	Hassium	Hs	Palladium	Pd	Thallium	Tl
Boron	B	Helium	He	Phosphorus	P	Thorium	Th
Bromine	Br	Holmium	Ho	Platinum	Pt	Thulium	Tm
Cadmium	Cd	Hydrogen	H	Plutonium	Pu	Tin	Sn
Calcium	Ca	Indium	In	Polonium	Po	Titanium	Ti
Californium	Cf	Iodine	I	Potassium	K	Tungsten	W
Carbon	C	Iridium	Ir	Praseodymium	Pr	Ununnilium	Uun
Cerium	Ce	Iron	Fe	Promethium	Pm	Ununoctium	Uuo
Cesium	Cs	Krypton	Kr	Protactinium	Pa	Ununpentium	Uup
Chlorine	Cl	Lanthanum	La	Radium	Ra	Ununseptium	Uus
Chromium	Cr	Lawrencium	Lr	Radon	Rn	Ununtrium	Uut
Cobalt	Co	Lead	Pb	Rhenium	Re	Uranium	U
Copernicium	Cn	Lithium	Li	Rhodium	Rh	Vanadium	V
Copper	Cu	Livermorium	Lv	Roentgenium	Rg	Xenon	Xe
Curium	Cm	Lutetium	Lu	Rubidium	Rb	Ytterbium	Yb
Darmstadtium	Ds	Magnesium	Mg	Ruthenium	Ru	Yttrium	Y
Dubnium	Db	Manganese	Mn	Rutherfordium	Rf	Zinc	Zn
Dysprosium	Dy	Meitnerium	Mt	Samarium	Sm	Zirconium	Zr
Einsteinium	Es	Mendelevium	Md	Scandium	Sc		

PHYSICAL REGIONS
This artist-rendered relief map depicts Earth's landforms above and below the surface of the ocean. Major mountain systems are shaded to emphasize their elevation. The Himalayan Mountains tower over India's Ganges plain; the Andes and Rocky Mountains reign over the Americas. All are dwarfed by the Mid-Atlantic Ridge, a submarine mountain range that stretches from Iceland to near Antarctica.

WESTERN HEMISPHERE

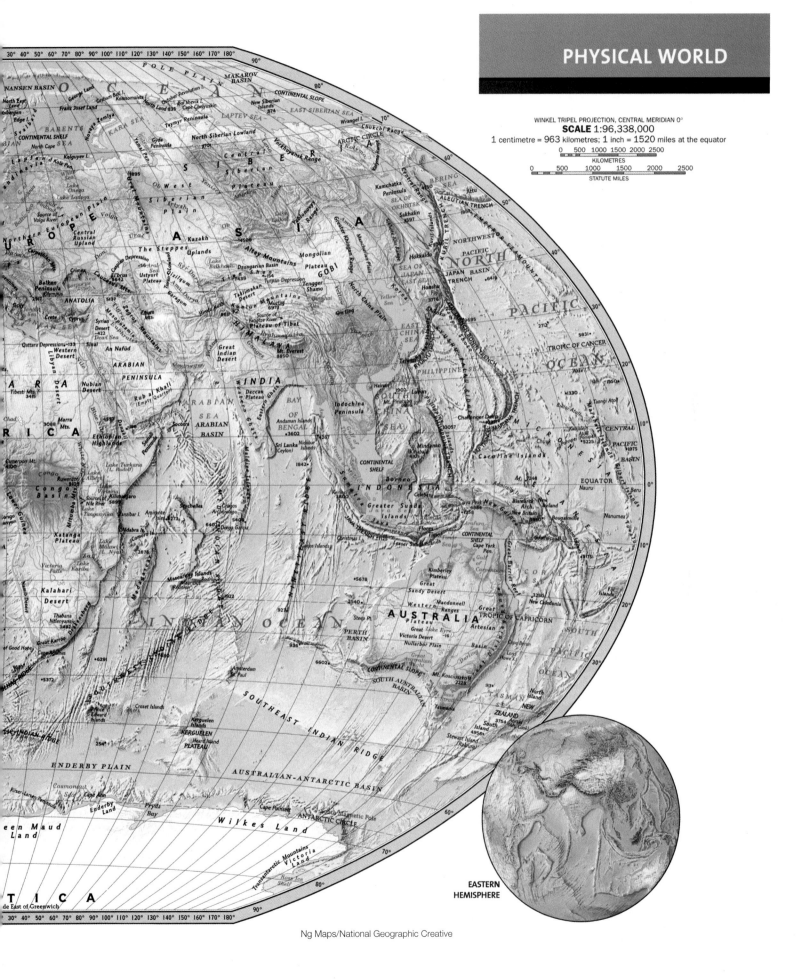

PHYSICAL WORLD

WINKEL TRIPEL PROJECTION, CENTRAL MERIDIAN 0°
SCALE 1:96,338,000
1 centimetre = 963 kilometres; 1 inch = 1520 miles at the equator

0   500 1000 1500 2000 2500
KILOMETRES

0   500   1000   1500   2000   2500
STATUTE MILES

EASTERN HEMISPHERE

Ng Maps/National Geographic Creative

GLOSSARY

Abrasion   Process in which solid rock is worn away by impact; grains become smaller and more spherical the longer they are transported.

Abyssal clay   Component of pelagic sediment (silt and clay) that is typically windblown from a continent.

Abyssal hill   Well-defined hill, ridge, or peak that rises a few metres to several hundred metres above the abyssal sea floor.

Abyssal plain   Large flat area of the deep sea floor having slopes less than about 1 m/km and ranging in depth below sea level from 3 km to 6 km.

Accreted terrane   Land mass that originated as an island arc or a microcontinent that was later added onto a continent, typically by convergent tectonics.

Accretionary prism   Wedge-shaped formation of rock along the front of an overriding plate; formed as a plate subducts and some of the rock and sediment are scraped off and collect as a series of angular rock slabs thrust on top of each other.

Active margin   Evolving margin of a continent on an overriding plate, typically containing a volcanic arc and an accretionary prism.

Aftershock   Earthquake that occurs after a large earthquake, on the same fault or nearby fault; related to the stress environment of the main shock.

Alluvial channel   Channel located in its own alluvium or stream sediment; has a tendency to erode laterally and create a wide, flat-bottomed valley.

Alluvial fan   Fan-shaped depositional feature of stream sediment that forms where a channel in a steep watershed discharges into a flat valley; typically forms in arid environments.

Alpine glacier   Glacier confined to a mountain valley.

Amphibole group   Most common group of silicate minerals consisting of double chains of silicate tetrahedra.

Andesitic magma   Magma that is andesitic (intermediate) in composition.

Anion   Atom that gains one or more electrons and acquires an overall negative charge.

Anthracite   Hard coal formed by metamorphism.

Anticline   Convex upward arch (fold) with an axial plane that is vertical, having the oldest rocks in the centre.

Appalachian Orogen   Ancient mountain belt formed by terrane accretion to the eastern continental margin of North America, starting in the Middle Ordovician (460 million years ago) and culminating by Late Carboniferous (350 million

years ago) when Laurentia, Gondwana, and Baltica collided to form the supercontinent Pangaea.

Aquiclude   Body of impermeable, or distinctly less permeable, rock that prevents groundwater movement.

Aquifer   Underground formation of saturated crust that can produce useful quantities of water when tapped by a well.

Aquitard   Body of distinctly less permeable rock that slows down the movement of groundwater.

Archean Eon   Period of geologic time that followed the Hadean Eon, between 3.8 to 2.5 billion years ago, during which Earth formed a solid crust able to support life.

Artesian spring   Natural spring from which the water flows out of the ground under its own pressure without the need for a pump and draws its supply of water from an artesian aquifer.

Atom   Smallest component in nature that has the properties of a given substance.

Atomic number   Number of protons in an atom's nucleus by which each element is defined.

Avalanche   Fast-flowing, fluidized mass of snow, ice, air, and occasionally some regolith that cascades down a mountainside due to the collapse of a snowfield.

Avoidance   Recognition of the presence and characteristics of a geologic hazard and, as a result, not building or otherwise using the land there.

Banded iron formation (BIF)   Formation with fine layering of cherty silica and iron, generally in the form of hematite, magnetite, or siderite; usually Precambrian in age and a major source of iron ore.

Barrier island   Long, low, narrow sandy island with one shoreline that faces the open ocean and another that faces a saltwater tidal lagoon protected by the barrier.

Basaltic magma   Magma that is basaltic (mafic) in composition.

Base flow   Water that enters a stream system from the water table through the channel bed.

Base level   Theoretical lowest level to which land will erode; globally this is sea level.

Bauxite   Ore of aluminum formed by intense weathering and found as a residual soil; mainly $Al(OH)_3$ (gibbsite).

Beach   Unconsolidated accumulation of sand and gravel along a shoreline.

Beach profile   Topographic cross-section of a beach.

Biogenic ooze   Component of pelagic sediment consisting of microscopic plankton remains that land on the sea floor from the overlying water where the plankton lived.

Biogenic sediments   Sediment produced by organic precipitation of compounds or consisting of the remains of living organisms.

Biological weathering   Weathering that occurs when rock disintegrates due to the chemical and/or physical activity of a living organism.

Bituminous coal   Coal formed by the removal of water and volatile gases from lower grade lignite coal.

Black smoker   Underwater hydrothermal spring that releases dark, billowing clouds of metal-rich particles.

Body wave   Seismic wave that travels into and through Earth's interior.

Bowen's reaction series   Process in which magma evolves to a new chemical state that is relatively enriched in silica as a result of fractionally crystallizing iron, magnesium, and calcium-rich compounds.

Braided channel   Channel with more sediment than it can transport; the sediment accumulates in the channel and forms bars or small islands to give it a "braided" appearance.

Calcite   Most common carbonate mineral, consisting of calcium bonded with a carbonate ion.

Caldera   Roughly circular, steep-walled volcanic basin formed when a volcano collapses because a large volume of magma was removed from the underlying magma chamber during an eruption.

Canadian Cordillera   Canada's youngest mountain belt, formed by terrane accretion to the western continental margin. The main phase of mountain building started in the Jurassic (at least 180 million years ago) and culminated in the Late Cretaceous (90 million years ago). However, mountain building continues there today.

Canadian Shield   Stable continental region not subjected to orogenesis for nearly 1 billion years; composed of Precambrian igneous and metamorphic rocks that have been exposed by erosion over millions of years.

Carbon cycle   Production, storage, and movement of carbon on Earth.

Carbonate compensation depth (CCD)   Depth in the ocean at which calcium carbonate ($CaCO_3$) completely dissolves.

Cation Atom that loses electrons and acquires an overall positive charge.

Cave Natural opening in rock that extends beyond the range of light and is large enough to permit the entry of humans.

Cenozoic Era Youngest era of the Phanerozoic Eon.

Central vent volcano Volcano built around a central vent.

Channel Physical confine of flowing water consisting of a bed and banks.

Charles Darwin (1809–1882) English natural scientist who formulated a theory of evolution by natural selection.

Charles Lyell (1797–1875) Author of *Principles of Geology*, which popularized uniformitarianism—the idea that Earth is shaped by slow-moving forces still in operation today. Lyell was a close and influential friend of Charles Darwin.

Chemical differentiation Process in which chemically distinct magma forms by the incomplete melting of rock or incomplete crystallization of magma.

Chemical sediment Sediment produced by inorganic precipitation of dissolved compounds.

Chemical weathering Chemical decomposition of minerals in rock.

Chondrite Primitive non-metallic meteorite considered to be representative of the most primitive material in the Solar System.

Cinder cone Ash, lapilli, blocks, and bombs of congealed lava built up around the single vent from which they were ejected.

Clastic sediments Broken pieces of crust deposited by water, wind, ice, or some other physical process.

Clay Most abundant and finest grained sediment particle, created by the decomposition of silicate rocks, usually during hydrolysis; composed of fine-grained minerals (phyllosilicates).

Climate feedback Secondary change that occurs within the climate system in response to a primary change, either positive or negative.

Climate proxy Indicator of past climate.

Coal seam Layer of coal deposited by a "coal swamp" and lithified to compressed and concentrated carbon; typically 1 m to 10 m in thickness.

Coastal erosion Erosion of coastlines due to sea-level rise and/or decreased sediment availability to the shoreline.

Coastal geology Study of coastal processes, the geologic framework of coastal areas, and how humans live in coastal environments.

Coastal hazard Coastal process that may cause loss of life or property.

Coastal process Agent of change in coastal areas such as winds, waves, storms, tides, sea-level change, and currents.

Compound Material comprising two or more chemical elements bonded together, usually with a characteristic structure.

Compressional stress Directed stress that squeezes rock.

Cone of depression Conical depression in the water table immediately surrounding a well; develops when the rate of groundwater removed is greater than the rate of natural replacement of that water.

Contact metamorphism Metamorphism produced when an igneous intrusion, such as a pluton, heats the surrounding country rock.

Contaminant plume Combination of toxic liquids flowing in the same hydraulic gradient that drives groundwater flow in an aquifer system.

Continental crust Part of Earth's crust that comprises the continents.

Continental glacier Glacier covering a large area and not confined by topography.

Continental margin Broad zone that includes the continental shelf, the continental slope, and the continental rise.

Continental rise Region of gently changing slope where the abyssal plain meets the continental margin.

Continental shelf Area of the ocean floor that is generally less than 200 m deep surrounding each continental land mass.

Continental slope Pronounced slope beyond the seaward margin of the continental shelf and ending offshore at the continental rise.

Continent-continent convergent boundary Location where two plates composed of continental crust collide, building mountains in the process.

Convection Transfer of heat by the movement of material from areas of high heat to areas of low heat.

Convergent boundary Location within the crust where two or more plates push together.

Coral reef Shallow coastal environment where numerous types of coral and algae build massive, solid structures of calcium carbonate.

Core Largely metallic iron mass at Earth's centre; composed of inner solid and outer liquid regions.

Coriolis effect Effect that, due to Earth's rotation, causes air and water currents to be deflected to the right in the Northern Hemisphere and to the left in the Southern Hemisphere.

Country rock Phrase often used by miners and geologists for the rock in the crust that surrounds an igneous intrusion.

Crater Small, circular depression created primarily by the explosive excavation of rock during an eruption.

Craton Core of ancient Precambrian rock in the continental crust that has attained tectonic and isostatic stability.

Creep Slow, down-slope migration of soil under the influence of gravity.

Critical thinking Use of reasoning to develop hypotheses and theories that can be rigorously tested.

Crust Outermost and thinnest of Earth's compositional layers; consists of brittle rock that is less dense than the mantle.

Crystalline structure Geometric pattern that atoms assume in a solid.

Crystallization Process through which atoms or compounds in a liquid state are arranged into an orderly solid state.

Cumulate Layer of dense minerals forming a concentrated deposit on the floor of a magma chamber.

Current Mass of continuously moving water.

Darcy's law Relationship between discharge, permeability, and hydraulic gradient in groundwater movement.

Debris fall Relatively free fall or collapse of regolith from a cliff or steep slope.

Debris flow Downslope movement of a mass of unconsolidated, water-saturated regolith, more than half of which is coarser than sand.

Debris slide Slow to rapid downslope movement of regolith across an inclined surface.

Decay chain Steps of radioactive decay involving multiple radioactive "daughters."

Decompression melting Melting that occurs when pressure decreases; *also called* pressure-release melting.

Deforestation Removal of trees for human use of the land.

Deformation Process in which stress causes bending, pulling, and fracturing of the solid rock of the crust.

Delta Sedimentary deposit formed when a stream flows into standing water (i.e., reaches its base level) and deposits its load of sediment.

Deposition Process of material, such as sediment, being added to a landform.

Depositional coast Coast dominated by depositional processes.

Desert Arid region that receives less than 25 cm of precipitation per year, or an area where the rate of evapotranspiration exceeds the rate of precipitation.

Desert pavement Surface layer of coarse particles in deserts, concentrated by the removal of finer material by wind.

Desertification Expansion of desert into non-desert areas.

Differentiation Process by which a magma changes composition or, on a larger scale, the process by which Earth's internal layers formed.

Dike Arm of intrusive rock that cuts vertically (or nearly so) across other rocks in the crust.

Dip Angle in degrees between a horizontal plane and an inclined plane, measured down from horizontal in a plane perpendicular to the strike.

Dip-slip fault Fault in which displacement occurs parallel to the dip of the fault plane.

Direct mitigation Practices that reduce vulnerability to mass wasting hazards when it is too late for avoidance.

Directed stress Pressure acting on the crust that is greatest in one direction.

Discharge Volume of water passing any point on a stream in a measured period of time.

Discontinuity Interface marking a change in rock density, causing seismic wave velocity to change suddenly.

Dissolution   Chemical weathering reaction in which a mineral dissolves in water.

Divergent plate boundary   Location within the crust where two or more plates pull away from each other.

Drainage system   Network of linked channels that carry water downhill.

Dune   Mound or ridge of sand deposited by wind.

Earthquake   Sudden shaking of the crust.

Effusive eruption   Volcanic eruption of fluid lava.

Elastic rebound theory   Theory that earthquakes result from the release of stored elastic energy by slippage on faults.

Ellesmerian Orogeny   Late Devonian to Early Carboniferous (about 360 million years ago) orogeny representing the first of two mountain-building episodes responsible for the formation of the Innuitian Orogen in northern Canada.

Emergent coast   Coast where the land is tectonically uplifting at a rate that exceeds sea-level rise.

Energy resources   Non-renewable resources (oil, coal, natural gas, and uranium) and renewable resources (solar energy, wind energy, bioenergy, geothermal energy, fuel cells, hydroelectricity, and ocean energy) that provide energy for human activities.

Environment of deposition   Sedimentary basin where sediment is deposited.

Eon   Unit of geological time that lasts for hundreds of millions to billions of years.

Ephemeral stream   Stream where flowing water is present only seasonally or after rainfall events.

Epicentre   Geographic spot on Earth's surface directly above an earthquake focus.

Epoch   Unit of geologic time, commonly used to subdivide the Cenozoic Era, that lasts for hundreds of thousands to millions of years.

Era   Unit of geologic time that lasts tens of millions to hundreds of millions of years.

Erosion   Complex group of related processes by which rock particles are physically moved.

Erosional coast   Coast dominated by erosional processes.

Eruption column   Massive, high-velocity, billowing cloud of gas, molten rock, and solid particles that is blasted into the air with tremendous force from a volcano.

Estuary   Partially enclosed body of water that forms in places where fresh water from rivers and streams flows into the ocean, mixing with salty sea water.

Eurekan Orogeny   Early Cenozoic (about 50 million years ago) orogeny representing the second of two mountain-building episodes responsible for the formation of the Innuitian Orogen in northern Canada.

Eutrophication   Increase of noxious forms of algae that use up oxygen in the water when decaying and cause other forms of life to suffocate; caused by nutrient contamination.

Evaporite   Mineral formed when evaporation leads to inorganic precipitation of dissolved compounds; a form of chemical sediment.

Evolution   Changing of plant and animal communities through time.

Explosive eruption   Dramatic volcanic eruption that throws volcanic products kilometres into the air; characterized by high gas discharge, an eruption column, ash, and extreme violence.

Extrapolation   Form of critical thinking used to make estimates of phenomena that cannot be directly observed; consists of inferring or estimating an answer by projecting or extending a known value.

Extrusive   Of or pertaining to igneous rock formed from material that has been extruded or ejected onto the Earth's surface

Fault   Zone or plane of breakage across which layers of rock are displaced relative to one another.

Fault-block mountain   Mountain characterized by tensional stress, crustal thinning, and normal faulting.

Faulting   Process that causes rock layers on one side of a break or fracture in the crust to be displaced relative to rock layers on the other side.

Feldspar group   Most common group of silicate minerals, with a three-dimensional framework structure of tetrahedra held together by combinations of calcium, sodium, and potassium.

Felsic   Of or pertaining to igneous rock composed of minerals with the lowest melting temperatures, light in colour, and relatively enriched in sodium, potassium, oxygen, and silicon.

Field geology   Obtaining of data directly from the rocks exposed at Earth's surface or those known to exist at very shallow depths under the soil and sediment.

Flood   Water discharge great enough to cause a stream to overflow its banks.

Flood basalt   Large volume of basaltic lava erupted as part of a large igneous province; may be preserved as very extensive plateaus.

Flood plain   Part of any stream valley that is regularly inundated during floods and whose geologic features are formed by floods.

Fluvial   Of or pertaining to streams or rivers, especially erosional and depositional processes of streams and the sediments and landforms resulting from them.

Fluvial erosion   Active movement of sediment particles by running water.

Focus   Point where an earthquake originates due to slippage on a fault within the crust.

Fold   Individual bend or warp in layered rock.

Fold-and-thrust mountain   Mountain that develops where tectonic plates collide at a convergent margin.

Foliation   Planar texture in metamorphic rock due to the alignment of recrystallizing platy minerals; caused by directed stress.

Foreshock   Small earthquake that precedes a larger earthquake on the same fault or a nearby fault and is related to the same general stress environment.

Fossil   Remains of an animal and plant or traces of its presence that have been preserved in the crust.

Fossil fuels   Concentrated forms of burnable organic carbon contained in rock originating from fossil accumulations.

Fracture   Breakage of brittle material.

Gas giant   One of the outermost planets of the Solar System (Jupiter, Saturn, Uranus, and Neptune); characterized by great mass, low density, and a thick atmosphere consisting primarily of hydrogen and helium.

Genetic mutation   Random change to genetic material, either RNA or DNA.

Genetic variation   Differences in inherited traits.

Geologic event   Notable occurrence of a common geologic process: deposition, deformation, faulting, intrusion, volcanism, and erosion.

Geologic hazard   Dangerous natural process such as a landslide, flood, erosion, volcanic eruption, tsunami, storm, earthquake, and hurricane.

Geologic province   Region with a geologic history that distinguishes it from other provinces.

Geologic resource   Non-living material mined to maintain our system of industry and quality of life.

Geologic time scale   Calendar of events that covers Earth's entire history.

Geological engineering   Study of geological processes and the role they play in engineering projects and safety.

Geological map   Map that shows the topography of an area, as well as the lithology, age, strike, and dip of rock units exposed at the surface or shallowly buried under loose sediment and soil.

Geology   Study of Earth and other planets in the Solar System and beyond.

Geomagnetic field   Magnetic field that surrounds Earth.

Geothermal energy   Heat that originates within Earth largely through decay of radioactive isotopes.

Geyser   Thermal spring having a system of groundwater recharge and heating that causes intermittent eruptions of water and steam.

Glacial landform   Landform created by glaciation.

Glacial moraine   Deposit of till that marks a former position of glacial ice.

Glacial outwash   Poorly sorted, stratified sediment deposited in an outwash plain.

Glacial till   Unstratified and unsorted sediment that has been deposited directly by melting ice.

Glacier   Semi-permanent body of ice, consisting largely of recrystallized snow, that shows evidence of downslope or outward movement, due to the stress of its own weight.

Glacier mass balance   Process in which the terminus of an active glacier advances or retreats, depending on the relative rates of ice delivery and wastage.

Glacioisostatic uplift   Process in which crust is uplifting because the weight of a glacier has been removed.

Global carbon cycle Transfer of carbon between Earth's atmosphere, hydrosphere, lithosphere, and biosphere.

Global circulation models (GCMs) Computer-based mathematical programs that simulate the physics of Earth's oceans, land, and atmosphere.

Global climate change Large-scale change in climate over time, whether natural or as a result of human activity.

Global warming Increase in the average temperature of Earth's surface, including the oceans.

Graded stream Stream that has achieved, throughout long segments, a state of dynamic equilibrium between the rate of sediment transport and the rate of sediment supply.

Greenhouse effect Atmosphere's natural ability to store heat radiated from Earth.

Greenhouse gas Gas in the atmosphere (mainly H_2O, CO_2, CFCs, N_2O, and CH_4) that absorbs thermal radiation from Earth's surface, a process referred to as the greenhouse effect.

Groundwater All the water contained in the pore spaces within soil, sediment, and bedrock.

Groundwater remediation Cleanup of polluted groundwater by a variety of chemical, physical, and biological processes.

Gyre Large-scale ocean circulation system; gyres sweep the major ocean basins.

Half-life Time required for half of a radioisotope species in a sample to decay.

Hazard mitigation Effort to minimize individual and community vulnerability to disasters.

Hemipelagic sediment Sediment that drapes upper and middle continental slopes around the world.

High-grade metamorphism Metamorphism under conditions of high temperature and high pressure.

Historical geology Study of Earth's history.

Hominidae Family of organisms that includes humans.

Homologous structures Similar characteristics of organisms that are due to their shared ancestry.

Hotspot Single stationary source of magma in the mantle that periodically erupts onto the crust, forming an active volcano.

Humus layer Soil layer rich in partially decomposed organic plant debris.

Hurricane Tropical cyclonic storm having winds that exceed 119 km/h.

Hydraulic gradient Slope of the water table.

Hydraulic pressure Difference in pressure between areas of groundwater, which drives the flow of groundwater from areas of high pressure to areas of low pressure.

Hydrogeologist Geologist who studies groundwater.

Hydrologic cycle Transfer of water between the atmosphere, the ocean, and the crust.

Hydrolysis Weathering process involving silicate minerals and their decomposition by acidic water; chemical reaction in which the H^+ or OH^- ions of water replace ions of a mineral.

Hydrosphere Liquid water on Earth, including water in the atmosphere, oceans, rivers, and lakes.

Hydrothermal fluid Chemically enriched hot fluid in the crust that can form when groundwater interacts with hot rock.

Hydrothermal vein filling Crystallized minerals in cracks and fractures, precipitated from hot fluid enriched with dissolved minerals leached from hot rock associated with intrusions.

Hydrothermal vent Location at which hot groundwater is released into sea water.

Hypothesis Testable educated guess that attempts to explain a phenomenon.

Ice age Cool climate period, also called a glacial period, during which as much as 30 percent of Earth's surface is covered by glaciers, sea level lowers, and other global phenomena occur.

Ice wedging Process in which the growth of ice crystals forces a joint to split open due to the increased volume of water when it turns to ice.

Igneous evolution Process in which an igneous rock partially melts and the resulting magma differentiates via fractional crystallization or other processes to produce myriad igneous products, which then re-enter the rock cycle.

Igneous mineral Mineral created by crystallization from cooling magma.

Igneous rock Rock produced by crystallization of magma.

Index mineral Metamorphic mineral whose first appearance marks a specific zone of metamorphism.

Induced fission Caused to decay radioactively.

Industrial Revolution Period beginning in the mid-1800s in which there was rapid population growth and expansion of industrial activities; included deforestation and burning of fossil fuels, affecting the mixture of gases in the atmosphere.

Inner core Rock body at Earth's centre thought to be composed of solid metal alloy consisting mostly of iron and nickel (about 5,000°C).

Innuitian Orogen Canada's northernmost mountain belt formed by two mountain-building episodes; the first occurred during the Late Devonian to Early Carboniferous (about 360 million years ago); the second occurred during the Early Cenozoic (about 50 million years ago).

Insolation Amount of energy Earth receives from the Sun.

Insoluble residue Mineral product remaining after intense chemical weathering of the original rock.

Interglacial Warm interval in long-term global climate.

Intergovernmental Panel on Climate Change (IPCC) International panel created by the United Nations and the World Meteorological Organization to address the complex issue of global warming.

Intermediate Of or pertaining to igneous rock whose composition is between mafic and felsic.

Intrusion Movement of magma through the crust.

Intrusive Of or pertaining to igneous rock formed from magma that has crystallized while being trapped in the crust or mantle.

Ion Atom with a net charge from the gain or loss of electrons.

Iron catastrophe Event in Earth's history in which the planet's temperature passed the melting point of iron (1,538°C), which resulted in the internal layers that characterize Earth today.

Iron oxide Compound composed of oxidized iron (e.g., hematite, Fe_2O_3).

Island arc Chain of insular andesitic stratovolcanoes on oceanic crust, parallel to a seafloor trench, produced by partial melting of the upper mantle at a subduction zone.

Isostasy Process that restores equilibrium when the mass of the crust changes.

Isotope Atom of an element having the same atomic number but a different mass number from other atoms of the same element.

Isotopic dating Process that uses naturally occurring radioactive decay to estimate the age of geologic samples.

James Hutton (1726–1797) Scottish farmer and naturalist, known as the founder of modern geology; originator of uniformitarianism and deep time (extended Earth history).

Joint Opening or "parting" in rock where no lateral displacement has occurred.

Karst geology Geologic features that develop over time as groundwater dissolves carbonate rocks.

Karst topography Assemblage of topographic forms resulting from dissolution of carbonate bedrock and consisting primarily of closely spaced sinkholes.

Lahar Mudflow on a volcano; typically composed of fluidized ash.

Landfill Location used for the mass disposal of municipal and industrial waste.

Large igneous province Vast area (at least 100,000 km²) covered by huge volumes of predominantly mafic extrusive and intrusive rock from fissure eruptions·

Large-scale volcanic terrain Large-volume volcanically derived crust formed by eruptive products coming from a network of sources rather than from a classic central vent volcano.

Laterite Iron oxide-rich soil produced by weathering.

Laurentia Predecessor to the current North American continent that developed during the Precambrian by the amalgamation of Archean cratons and Proterozoic mountain belts.

Lava Magma that reaches Earth's surface through a volcanic vent.

Lineation Linear texture in metamorphic rock due to the preferred alignment of elongate minerals; caused by directed stress.

Lithification Process that converts sediment into sedimentary rock.

Lithosphere Rigid upper mantle and crust.

Loess Wind-deposited silt, sometimes accompanied by some clay and fine sand.

Longshore current Wave-generated current that flows parallel to the coast within the surf zone.

Longshore drift Sediment transport in a longshore current.

Louis Agassiz (1807–1873) Influential paleontologist, glaciologist, and geologist; originator of the theory of the ice ages.

Love wave Seismic surface wave that moves the ground from side to side.

Low-grade metamorphism Metamorphism under conditions of low temperature and low pressure.

Mafic Of or pertaining to igneous rocks composed of minerals with the highest melting temperatures, mostly dark in colour, and relatively enriched in iron, magnesium, and calcium.

Magma Molten rock, together with any minerals, rock fragments, grains, and dissolved gases, that forms when temperatures rise and melting occurs in the mantle or crust.

Magma differentiation Processes that change the composition of magma as it migrates into and through the crust.

Main shock Largest earthquake in a cluster.

Mantle Thickest layer of Earth's interior; made of dense, hot rock that lies directly beneath the crust and surrounds Earth's core.

Marine geology Geophysical, geochemical, sedimentological, and paleontological investigations of the ocean floor and coastal margins.

Marine isotopic stages Numbering system developed to name the alternating warm periods and ice ages of recent geologic history based on isotopic climate proxies.

Marine transgression Shoreline movement toward the land; process in which ocean waters rose after the last ice age, flooding the coast and moving the shoreline inland.

Mass extinction Event in which large numbers of species permanently die out within a very short period.

Mass number Number of neutrons plus protons in the nucleus of an atom.

Mass wasting Process by which gravity pulls soil, debris, sediment, and broken rock down a hillside or cliff.

Meandering channel Channel with loop-like bends that evolves over time.

Meltwater Layer of water below a glacier, formed when increased pressure toward the base of a glacier lowers the melting point of ice and when percolating meltwater accumulates from above.

Mesozoic Era Middle era of the Phanerozoic Eon.

Metamorphic facies Set of metamorphic mineral assemblages repeatedly found together, indicating certain pressure and temperature conditions of metamorphism.

Metamorphic mineral Mineral created by recrystallization of rock under conditions of high heat and pressure within the crust.

Metamorphic rock Rock whose original compounds, textures, or both, have been transformed by reactions in the solid state as a result of a change in temperature, pressure, and/or chemically active fluids.

Metamorphism All changes in mineral assemblage, rock texture, or both, that takes place in rocks in the solid state within Earth's crust as a result of changes in temperature, pressure and/or chemically active fluids.

Methane Colourless, odourless, clean-burning fossil fuel.

Mica group Group of silicate minerals that have their silicate tetrahedra arranged in sheets.

Mid-ocean ridge Ridge that develops along a rift zone in the ocean due to high heat flow and magma upwelling.

Milky Way Galaxy Galaxy that contains billions of stars, including the Sun (and thus our Solar System), held together by mutual gravitation.

Mineral Naturally occurring, inorganic, crystalline solid with a definite chemical composition that can vary within a restricted range.

Mineral resource Useful rock and mineral material that has value in its natural state or in a processed state.

Mohorovicic discontinuity Boundary between the crust and the mantle.

Mohs hardness scale Scale that uses numerical values from 1 to 10 as a relative measure of hardness based on comparison with specific minerals.

Monocline Local steepening in an otherwise uniformly dipping layer of strata.

Monogenetic field Collection of volcanic vents and flows, sometimes numbering in the hundreds or thousands, typically from a single magma source.

Mountain belt Group of several mountain ranges, usually with related histories extending across a broad range of time.

Mud Unconsolidated, water-saturated sediment of silt and clay grains.

Mudflow Flowing mass of water-saturated silt and clay grains.

Natural attenuation Technique for groundwater remediation involving shutting down any source of ongoing pollution, defining and isolating a known contamination plume, and allowing natural processes to reduce the toxic potential of contaminants.

Natural law Successful theory that over time has been shown to be always true for a broad array of natural processes.

Natural resource Material that occurs in nature and is essential or useful to humans, such as water, air, building stone, topsoil, and minerals.

Natural selection Process by which favourable traits that are heritable become more common in successive generations of a population of reproducing organisms.

Neritic sediment Sediment that collects on the sea floor between the shoreline and the edge of the continental shelf.

Nicholas Steno (1638–1686) Considered to be the father of geology and stratigraphy, credited with the law of superposition, the principle of original horizontality, the principle of lateral continuity, and other fundamental geological principles.

Nonfoliated texture Texture found in metamorphic rocks that have undergone contact metamorphism or in rocks that, when viewed with the unaided eye, show no evidence of foliated texture.

Non-renewable Resource whose total amount available on Earth is reduced every time it is used, and what is used is not replenished within a human generation.

Non-renewable energy resource Oil, coal, natural gas, or uranium used for energy, whose total amount available on Earth is reduced every time it is used, and what is used is not replenished within a human generation.

Normal fault Fault, generally steeply inclined, along which the hanging-wall block has moved downward relative to the foot wall.

Nuclear fusion Thermonuclear reaction in which the nuclei of light atoms join to form nuclei of heavier atoms, releasing energy and creating new heavier elements.

Nucleation Initial grouping of a few atoms that starts the process of crystal growth.

Nuna Supercontinent formed during the Early Proterozoic, about 1.8 billion years ago, when Archean cratons collided and were welded together. Part of Nuna became the core of the Canadian Shield.

Ocean acidification Increasing acidity of oceanic waters in the form of carbonic acid.

Ocean-continent convergent boundary Boundary formed where oceanic crust collides with continental crust, causing the denser oceanic crust to subduct below the continental crust and be recycled into the mantle.

Oceanic crust Crust beneath the abyssal oceans, generally basaltic in composition.

Oceanic fracture zone Type of transform boundary consisting of inactive and active portions.

Oceanic trench Arc-linear seafloor depression at a plate boundary where an oceanic plate subducts into the mantle beneath a converging plate.

Ocean-ocean convergent boundary Boundary formed where two plates composed of oceanic crust converge; typically the denser (older) will subduct.

Olivine group Group of silicate minerals consisting of independent silicate tetrahedra held together by varying proportions of magnesium and iron cations.

Orbital parameters   In paleoclimatology, typically three of the six Keplerian parameters derived from the laws of planetary motion: axial tilt, eccentricity, and precession.

Ore   Aggregate of minerals from which one or more minerals can be extracted profitably.

Ore mineral   Mineral that contains metal of interest in a mine and from which the metal can be extracted profitably.

Orogenesis   Mountain building in any form where lithospheric plates converge.

Orogenic belt   Large mountain belt formed by tectonic forces that deform and raise the crust.

Orogeny   Mountain building, commonly by folding and thrusting caused by the collision of lithospheric plates; synonymous with orogenesis.

Orographic effect   Topographic effect in which a mountain range causes prevailing winds to rise over it, leading to rainfall on the windward side and the development of a rain shadow or arid or semi-arid conditions on the leeward side.

Outcrop   Place where layers of rock are exposed.

Outer core   Layer surrounding the inner core; extremely hot and melted because it is under less pressure than the inner core, which is solid.

Oversteepening   Slope failure due to the addition of material to the top of a slope or the removal of material at the base of a slope, increasing the slope angle so that it is unstable.

Oxbow lake   Crescent-shaped shallow lake or marsh occupying the abandoned channel of a meandering stream.

Oxidation   Process in which a chemical element loses electrons by bonding with oxygen.

Paleoclimatology   Study of past climate.

Paleomagnetism   Remnant magnetism in ancient rock recording past conditions of the geomagnetic field.

Paleozoic Era   Oldest era of the Phanerozoic Eon.

Pangaea   Supercontinent in which Europe, North America, Africa, Antarctica, South America, and all continental lands were joined together by about 300 million years ago.

Partial melting   Melting of rock that occurs when some compounds melt while others remain solid.

Passive margin   Tectonically quiet margin of a continent.

Pegmatite   Exceptionally coarse-grained intrusive igneous rock, commonly granitic in composition.

Pelagic sediment   Sediment consisting of the remains of marine organisms in the open ocean.

Period   Unit of geologic time that lasts millions to tens of millions of years.

Permafrost   Soil whose moisture is frozen for more than two years.

Permeability   Degree to which a fluid is able to flow through a material such as rock; related to the interconnectedness of pore space.

Petroleum   Gaseous, liquid, or semi-solid substance consisting of compounds of carbon and hydrogen.

Phreatomagmatic eruption   Explosion caused by rapid expansion of steam when magma comes in contact with groundwater.

Phylogeny   Evolutionary lineage of an organism.

Physical geology   Study of materials that compose Earth; the chemical, biological, and physical processes that create them; and the ways in which they are organized and distributed throughout the planet.

Physical weathering   Type of weathering that occurs when rock is fragmented by physical processes that do not change its chemical composition.

Pillow lava   Submarine deposit of talus and bulbous rock created when basalt erupting onto the sea floor is quickly frozen by the cold sea water.

Placer deposit   Deposit formed when weathered soil erodes into a channel and dense mineral grains are concentrated by the moving water.

Planetesimal accretion   Largely collisional process by which matter gathered to form planets.

Plate boundary   Location within the crust where lithospheric plates meet.

Plate tectonics   Unifying theory that explains Earth's geology in terms of the motions and interactions of lithospheric plates that move around over the asthenosphere.

Playa lake   Shallow lake on the floor of a basin after a rain event, the hydrology of which is controlled solely by precipitation and evaporation.

Plinian eruption   Large volcanic explosion that pushes thick, dark columns of pyroclasts and gas high into the stratosphere.

Plutonic   Of or pertaining to an intrusive igneous rock; typically describes large intrusive bodies at great depth in the crust.

Polluted runoff   Runoff that contains contaminants from a variety of sources, including roads, gutters, lawns, and construction sites.

Porosity   Measure (percentage) of the amount of empty space contained in a rock.

Porphyry copper   Type of hydrothermal deposit containing disseminated copper-bearing minerals associated with a porphyritic intrusive igneous rock.

Primary sedimentary structure   Feature that reflects physical processes that act on sediments when they are being deposited.

Principle of uniformitarianism   Principle stating that Earth is very old, natural processes have been essentially uniform through time, and the study of modern geologic processes is useful in understanding past geologic events.

P-wave   Primary wave; a type of seismic body wave that compresses and stretches the material through which it moves.

Pyroclastic debris   Erupted pieces of rock and lava that may vary from ash to bomb or block size.

Pyroclastic flow   Hot, highly mobile avalanche of tephra and gas that rushes down the flank of a volcano during an eruption.

Pyroxene group   Most common group of silicate minerals, comprising single chains of silicate tetrahedra.

Quartz   Common silicate mineral consisting of a three-dimensional framework of silicate tetrahedra, and thus containing only silicon and oxygen.

Quaternary Period   Most recent period of geologic time (2.6 million years ago to present).

Quick clay   Special type of marine clay, known as Leda clay, that can spontaneously change to a fluid state in a process called liquefaction, which is caused by slight disturbances such as ground shaking.

Radioactive decay   Natural process of transformation by which the nucleus of the atom of an element transforms into the nucleus of an atom of a different element and emits energy in the process.

Radioisotope   Radioactive atom; each radioisotope decays at a unique, fixed rate.

Rayleigh wave   Seismic surface wave that rolls along the ground, moving up and down and forward and backward in the direction of movement.

Reasoning by analogy   Scientific thinking in which one thing is inferred to be similar to another, usually the more simplified thing, in order to promote understanding.

Recharge   Process through which precipitation adds new water to the groundwater system.

Recrystallization   Metamorphic process in which minerals within a rock change in order to be stable under new conditions. This results in the growth of new minerals by recycling old minerals in the solid state.

Refraction   Change in direction when a wave passes from one medium to another.

Regional metamorphism   Metamorphism affecting large volumes of crust; typically involving high pressure conditions.

Regolith   Soil, debris, sediment, and broken rock.

Relative dating   Process for determining the order of formation among geologic events based on the relationship of rocks to one another.

Renewable energy   Energy taken from inexhaustible sources.

Reverse fault   Fault, generally steeply inclined, in which the hanging-wall block has moved upward relative to the foot wall.

Rhyolite caldera complex   Caldera developed as a result of a very explosive eruption of gas-rich, viscous rhyolitic magma.

Rhyolitic magma   Magma that is rhyolitic (felsic) in composition.

Richter scale   Measurement of the largest amplitude on a seismogram, describing an earthquake's magnitude.

Rift valley   Linear, fault-bounded valley along a divergent plate boundary or spreading centre.

Ring of fire   Outer boundary of the Pacific Plate where it meets the surrounding plates, creating the most seismically and volcanically active zone in the world.

Rip current   High-velocity, wave-generated current flowing seaward from the shore.

Ripple   Migrating, small wave of sand deposited by wind or water.

Rock   Solid aggregation of minerals.

Rock cleavage   Property by which rock breaks into plate-like fragments along flat planes.

Rock cycle   Recycling of rock material, in the course of which rock is created, destroyed, and altered through internal and external Earth processes; powered by gravity, geothermal energy, and solar energy.

Rock slide   Sudden and rapid downslope movement of a detached mass of bedrock across an inclined surface.

Rockfall   Free fall of detached bodies of bedrock from a cliff or steep slope.

Rocky shoreline   Shoreline where solid rock rather than unconsolidated sediment characterizes the water's edge.

Rodinia   Late Proterozoic supercontinent that formed about 1.1 to 1.0 billion years ago when many continents on Earth recombined.

Root wedging   Fracturing of rock by the pressure of growing plant roots; a type of physical weathering.

Sahel   Region of North Africa that is the border between the sands of the Sahara in the north and the moist tropical rainforest regions of central Africa.

Salinity   Concentration of dissolved sulphur, calcium, magnesium, bromine, potassium, and bicarbonate in water.

Sand   Rock and mineral particles 0.0625 mm to 2 mm in diameter.

Sand bar   Sand accumulation along a coast, creating an area of shallow water.

Sand dune   Typical feature of deserts that develops in dry climates; requires the availability of loose sand, wind to transport it, and an obstacle around which it can start to develop.

Scientific method   Use of reasoning to interpret evidence in order to construct a testable hypothesis as an explanation of a natural phenomenon; testing of hypotheses.

Seafloor spreading   Lateral movement of oceanic crust away from mid-ocean ridges due to the combined effect of mantle convection, ridge push and slab pull.

Sea-level rise   Increase in the level of the world's oceans, typically at a rate and with variability that is unique to a location.

Sediment   Particles of mineral, broken rock, and organic debris that are unconsolidated; formed by weathering.

Sedimentary mineral   Mineral created by crystallization in Earth's surface environments.

Sedimentary quartz   Weathering product that results from crystallization of silica.

Sedimentary rock   Rock formed by organic or inorganic precipitation or by deposition and cementation of sediment.

Seismic shaking hazards   Damaging, dangerous processes related to seismicity, including building damage and collapse.

Seismic tomography   Method of imaging Earth's structure using differences in seismic wave behaviour.

Seismic wave   Energy produced by an earthquake.

Seismometer   Instrument that detects earthquakes.

Shear stress   Directed stress that shears rock side to side.

Shield volcano   Volcano that emits fluid lava, typically basaltic, and builds up a broad dome-shaped structure with a low surface slope.

Silicate   Type of mineral composed of silicon and oxygen; the building block of all silicate minerals.

Silicate structure   Mineral structure of silica tetrahedra.

Silt   Particle between 0.002 mm and 0.063 mm.

Sinkhole   Depression created when an underground cavern grows and undermines an area until the roof eventually collapses.

Slope stability   Degree to which a slope or cliff is stable and therefore resistant to mass wasting.

Slump   Landform that occurs when regolith suddenly drops (or rotates) a short distance down a slope.

Smelting   Heating or melting a mineral to obtain the metal it contains.

Soil   Unconsolidated mineral and organic material forming the uppermost layer on Earth's surface.

Soil classification   Grouping of soils on the basis of several related factors including mineralogy, texture, organic composition, moisture, and other factors.

Soil profile   Set of regolith layers created when percolation forms and moves dissolved ions that may recombine to form clay minerals.

Solar energy   Radiant light and heat from the Sun.

Solar nebula hypothesis   Hypothesis stating that the Sun, planets, and other objects orbiting the Sun originated at the same time from the same source through the collapse and condensation of a planetary nebula, and have evolved in different ways since that time.

Solar System   Sun, planets, and other objects trapped by the Sun's gravity.

Solar wind   Subatomic particles and radiation originating from the Sun.

Solifluction   Soil creep that occurs in water-saturated regolith on permafrost.

Sorting   Separation of grains by density and size.

Spheroidal weathering   Exfoliation characterized by preferential weathering of edges and corners.

Spring   Groundwater emerging naturally at the ground surface.

Stalactite   Downward-hanging inorganic precipitate in a cave.

Stalagmite   Upward-oriented inorganic precipitate on a cave floor.

Stellar nucleosynthesis   Process by which the chemical elements assemble in the cores of stars, primarily by nuclear fusion.

Stock   Relatively small igneous intrusion.

Storm surge   Coastal flooding resulting from high sea level caused by a storm.

Storm water runoff   Water that flows overland during a rainstorm.

Straight channel   Channel whose underlying topography and bedrock structure force the channel to be straight.

Strain   Stress that changes the shape and volume of a rock.

Stratigraphic principles   Seven basic but important concepts that provide the foundation for relative dating, including the principles of superposition, original horizontality, original lateral continuity, cross-cutting relationships, inclusions, unconformities, and fossil succession.

Stratovolcano   High steep-sided volcano typically formed from intermediate magma, often above a convergent plate margin.

Stream   Flowing body of water that follows a channel.

Stream piracy   Headward stream erosion that breaches a drainage divide, intersects another channel, and captures its flow.

Stress   Magnitude and direction of a deforming force exerted on a surface.

Strike   Compass direction of a horizontal line that marks the intersection of an inclined plane with Earth's surface.

Strike-slip fault   Fault in which displacement has been horizontal and parallel to the strike of the fault.

Stromatolite   Dome-like structure formed by microscopic bacteria that trap sediments between fine strands.

Structural geology   Branch of geology concerned with determining the architecture of the crust and inferring from it the nature of folds and faults in the rock that are usually below ground and therefore not in full view.

Subduction zone   Plate margin where one plate moves into the mantle beneath another.

Submarine canyon   Steep-sided valley on the continental shelf or slope resembling a river-cut canyon on land.

Submarine fan   Thick sedimentary deposit on the continental rise at the end of a submarine canyon.

Submergent coast   Coast where sea-level rise exceeds the rate of land uplift.

Surface wave   Seismic wave that travels along Earth's surface.

Sustainable resource   Any natural product used by humans that meets the needs of the present without compromising the ability of future generations to meet their own needs.

S-wave   Secondary wave; seismic body wave that can move only through solid materials, moving up and down or side to side.

Syncline   Concave upward fold or arch with an axial plane that is vertical, having the youngest rocks in the centre of the fold.

Talus   Slope of fallen rock that collects at the base of cliffs and steep hillsides.

Tensional stress   Directed stress that stretches rock.

Tephra Loose assemblage of airborne pyroclasts.

Terrane Fault-bounded body that consists of fragments of continental crust, island arcs, or oceanic rocks accreted to continental margins during orogenesis at convergent plate boundaries.

Terrestrial planet One of the innermost planets of the Solar System (Mercury, Venus, Earth, and Mars); characterized by high densities and generally silicate and metallic compositions.

Thalweg Line running along a channel connecting the deepest parts of the stream and the fastest flow, marking the natural direction of the watercourse.

Theory Hypothesis that passes repeated and rigorous tests and is applicable to a broad phenomenon.

Theory of evolution Theory stating that changes occur in the inherited traits of a biological population from one generation to the next.

Thermohaline circulation Global pattern of ocean circulation driven by temperature and salinity differences between parts of the oceans.

Thrust fault Low-angle reverse fault with dip less than 20°.

Tidal wetland Marsh flooded by high tide daily and characterized by growth of salt-tolerant plants in the muddy substrate.

Tide Regular and predictable oscillation in the sea surface due to gravitational forces.

Trade wind Flow of air from subtropical regions toward the equator; ships carrying goods relied on these winds for several centuries.

Transform boundary Plate boundary characterized by lateral slip of the plates past each another.

Tributary Stream that joins a larger stream.

Tsunami Series of waves in the ocean caused by rapid movement of the sea floor.

Tuff Rock formed when pyroclastic debris, dominated by ash, accumulates on the ground and solidifies.

Turbidity current Gravity-driven current consisting of a dilute mixture of sediment and water, which has a density greater than the surrounding water and travels down the continental slope and rise.

Ultramafic Of or pertaining to igneous rock with the lowest silica content and enriched in iron and magnesium.

Unconformity Surface that represents a gap in the temporal record of Earth's history; produced by the removal of rock or sediment by erosion or by a period of non-deposition.

Undercutting Process in which rock or regolith is removed from the base of a hillside, such as by stream erosion, thereby increasing the slope.

Uniformitarianism Concept stating that Earth's history is best explained by observations of modern processes, suggesting that geologic principles have been uniform over time.

Uplift Raising of crust by geologic processes.

Vent Opening in the crust through which lava flows.

Ventifact Rock with an unusual shape and a flat face that has been abraded by wind-blown sediment.

Vestigial structure Vestige of a body part that was used by ancestral forms but is now non-essential.

Volcanic arc Line of volcanoes generated above a subducting plate.

Volcanic mountain Mountain formed at a volcanic arc above a subduction zone.

Volcanic outgassing Process in which volatiles, such as water, nitrogen, and carbon, trapped in minerals are released by melting within Earth's interior.

Volcano Landform that releases lava, gas, or ashes to the surface, or has done so in the past.

Volcanologist Scientist who studies volcanoes.

Wadati-Benioff zone Steeply dipping plane defined by earthquake foci that corresponds to a subducting oceanic slab.

Washover Process of marine inundation carrying sand from the ocean side of a barrier island to the lagoon side.

Water table Upper surface of the saturated zone of groundwater.

Watershed Total area feeding water to a stream.

Wave-generated current Current created by the movement of waves.

Wave refraction Process by which the direction of waves, moving into shallow water at an angle to the shoreline, is changed by interaction with the sea floor.

Weathering Set of physical, chemical, and biological processes that break down rocks and minerals in the crust to create sediment, new minerals, soil, and dissolved ions and compounds.

Welded tuff Welded mass of glass shards solidified when pyroclastic flows lose their energy and come to rest as thick beds of ash and lapilli.

Wetland Land whose soil is saturated with moisture either permanently or seasonally.

Zone of accumulation Upper zone on a glacier, covered by snowfall, and representing an area of net gain in mass.

Zone of wastage Area of a glacier where ice is lost not only by melting, but also by sublimation and calving.

INDEX

A

A horizon, 178
aa lava, 132, 135, 135*f*
Abitibi, Quebec, 221
abrasion, 164–165, 196, 197, 454, 576
abundant metals, 246, 247*t*
abyssal clay, 192, 634, 636
abyssal hills, 639, 639*f*
abyssal plains, 209, 630, 634–635
Acadian Orogeny, 295*f*, 296, 407
Acasta gneiss, 229*f*, 338*f*
accreted terranes, 290, 630
accretionary prism, 62, 412
active layer, 453
active margin, 64, 414, 630, 630*f*
Adamant Range, British Columbia, 164, 164*f*, 450*f*
Adirondack Mountain, New York, 295
Aeolian erosion, 182
aerosols, 429
Afghanistan, 572
Africa, 10, 12, 59, 65, 407*f*, 628, 632*f*
African Plate, 52, 132, 638*f*
African Rift Valley, 153*f*
aftershocks, 311, 311*f*
Agassiz, Louis, 452
age
    see also dating
    Earth's age, 340–341, 356–357
    of geologic samples, 346–347, 349
aggradation, 524, 525
agriculture, 553*f*, 616
Agriculture and Agri-Food Canada (AAFC), 184
agriculture minerals, 246, 247*t*
AIDS, 372–374
air sparging, 556
Alaska, 610*f*
Alaska Current, 626
Alaska Range, 413
albedo flip, 464
Alberta, 403
Alberta Badlands, 8*f*, 160*f*
Aleutian Islands, 71, 124
Aleutian Trench, 642
Alleghenian Orogeny, 295*f*, 296
alluvial channel, 524
alluvial fan, 206, 207*f*, 528–529, 582
alluvial terraces, 526, 527*f*
alluvium, 520–521
alpha decay, 346, 346*f*, 347
Alpide Belt, 71, 71*f*
alpine environments, 176–177
alpine glaciers, 454, 455*f*, 464, 464*f*
Alpine-Himalayan Mountain System, 386
the Alps, 286
aluminum, 48, 87*f*, 88*t*, 91*t*, 92, 172, 245*t*, 246, 250*f*

aluminum oxide, 177
Amazon River, 633
ammonia, 32
amphibole, 110, 114, 115, 229
amphibole facies, 233
amphibole group, 93*f*, 94
amphibolite, 227*t*, 237*t*
Andaman-Sumatra quake, 68, 71
Andaman-Sumatra tsunami, 69
Andes Mountains, Peru, 153*f*, 453
Andes Range, 124
andesine, 94
andesite, 115, 116, 117*f*, 119
andesitic magma, 134, 134*t*, 135
angle of repose, 576
anhydrite, 96, 97*f*
anhydrous, 237
Aniakchak Caldera, Alaska, 145*f*
animal respiration, 170
anion, 86
Annapolis Lowlands, 295
Anomalocaris canadensis, 381*f*
anoxic dead zones, 533, 533*f*
anoxic mud, 90
Antarctic Desert, Antarctica, 568*t*
Antarctic Ice Sheet, 441, 465, 604
Antarctic Peninsula, 465
Antarctic Plate, 52
Antarctica, 10, 12, 65, 386, 423*f*, 427*f*, 455, 456, 457, 470, 470*f*, 471, 473
Antarctica Peninsula, 457*f*
anthracite, 227*t*, 230*f*, 231, 260
anthracite coal, 227, 230*t*
anthropogenic emissions, 427
antibodies, 372
anticlines, 257, 282, 283*f*
antiform, 282
aphanitic, 112, 113*t*
aphelion, 474
Apollo, 356
Appalachian belt, 294–296, 294*f*
Appalachian Mountains, 65, 286, 394, 628
Appalachian Orogen, 399, 400, 403, 404–407, 404*f*, 405*f*, 407*f*
Appalachian Plateau, 295
aquiclude, 548
aquifer fracturing, 556
aquifers, 544, 548, 549*f*
aquitard, 548
Arabian Desert, Middle East, 568*t*
Arabian Plate, 52
Aravalli Range, India, 575
arc volcanism, 150, 151
archaea branch, 372
Archaeopteryx, 20*f*, 382
Archean Eon, 21*t*, 48, 377*f*, 378, 378*f*
Arches National Park, Utah, 179

the Arctic, 414, 438
Arctic Desert, 568*t*
Arctic environments, 176–177
Arctic National Wildlife Refuge, 255
Arctic Ocean, 624, 624*t*, 625*f*
Arctic sea ice, 464
arête, 459
argon, 88*t*
arid environments, 176
    see also desert
arkose, 202, 202*t*, 203*f*
artesian aquifer, 548, 549*f*
artesian spring, 548
artificial levees, 521
asbestos, 221, 224, 225
Asbestos Hazard Emergency Response Act, 225
aseismic, 312
ash, 113, 138, 139*f*
asteroid belt, 37
asteroids, 28, 37
asthenosphere, 49
Atacama Desert, Chile, 575*f*
Atacama Trench, 642
Athabasca Sedimentary Basin, Saskatchewan, 250
Athans, Pete, 296
Atlantic City, New Jersey, 603*f*
Atlantic continental shelf, 414
Atlantic Ocean, 52, 295*f*, 382, 383*f*, 394*f*, 595, 624, 624*t*, 625*f*, 637*f*
Atlantic Ocean Basin, 59
Atlantic Uplands, 295
atmosphere, 9, 424
atmospheric moisture circulation, 570–571
atmospheric temperature, 438–439, 466–467
atoll, 615, 615*f*
atomic number, 86, 346
atoms, 86, 86*f*
    atomic number, 86, 346
    bonds, 88–89
    covalent bonds, 89
    electrons, 86–87
    ionic bonds, 88–89
    mass number, 86, 346
    metallic bonds, 89
    neutrons, 86
    protons, 86
    stable electron configuration, 88
    structure of atoms, 86
augite, 93*f*, 95*f*
aureole, 222
Australia, 10, 12, 48, 65, 356*f*, 386, 456, 485*t*
Australia National University, 31
Australopithecines, 386
Australopithecus afarensis, 387, 387*f*
authigenic sediment, 632, 634
avalanches, 492
Avalon terrane, 406

Avalonia, 294
avoidance, 484
avoidance policy, 616
avulsion, 529, 529f
Axel Heiberg Island, Nunavut, 408f
axial plane, 282
axial tilt, 474
Ayers Rock, Australia, 583

B

B horizon, 178
backshore, 593
badlands, 400
Baltica, 405, 405f
Bam quake, Iran, 305
banded iron formations (BIFs), 250–251, 379
bankful discharge, 520
Barazangi, Muawia, 312
barchan dunes, 578, 578t, 579f
barite, 96, 97f
barrier islands, 207, 208f, 606–607, 606f, 607f
barrier reefs, 615, 615f
barrier rollover, 607
basal cleavage, 85f
basal sliding, 454
basalt, 106, 107f, 112, 113, 115, 116–117, 117f,
        119, 122–124, 173t, 227t, 237, 237t
basaltic magma, 134–135, 134t
base flow, 542
base level, 514
base-level changes, 526, 527f
basement, 628
Basin and Range province, 292–293, 292f, 582,
        583f
basins, 282, 283f
batholiths, 116, 125
Bathurst Inlet, Nunavut, 610f
bauxite, 172, 173t, 177, 177f, 245t, 250, 250f
beach, 197t, 207, 208f, 593, 593f
beach profile, 605
beach ridges, 610
Becquerel, Henri, 340
bed load, 182, 516
bedforms, 212, 213f
bedrock, 178, 400–402, 404–406
benchmark glaciers, 468
Bengal Fan, 633, 633f
Bering Strait, 458f
berm, 593
beryllium, 88t
beta decay, 346, 346f, 347
bicarbonate ion, 431
bifurcation ratio, 513
Big Island of Hawaii, 312
bioenergy, 264
biogenic methane, 259
biogenic ooze, 192, 634, 636
biogenic sedimentary rock, 205, 205f, 205t
biogenic sediments, 194, 195f, 198–199, 632
biological diversity, 382–383
biological oceanography, 624
biological weathering, 73, 162, 163f, 170–171
biomass energy, 254
bioremediation, 556
biosphere, 9, 441
biotite, 82, 82f, 93f, 94, 95f, 173t, 226
biotite mica, 110, 115

biotite schist, 229
Bismarck Plate, 52
bituminous coal, 227t, 260
black blizzard, 573
black smokers, 122, 151, 638, 638f
bleaching, 440
blind faults, 323
blue-green cyano-bacteria, 378
Blue Ridge, 295
blue-shifted light, 35
blueschist, 227t, 229
blueschist facies, 233
boa constrictors, 371f
body waves, 310
bogs, 177
Bonneville Salt Flat, Utah, 583
bore, 593, 597f
borehole, 541
bornite, 245t, 248
boron, 88t
bottom waters, 476
bottomset beds, 528
Bow River, Alberta, 520f
Bow Valley, Alberta, 459f
Bowen, Newfoundland, 110
Bowen's reaction series, 110–111, 110f, 114
brachiopod fossils, 369f
braided channel, 518–519, 521f
braided streams, 460
Brazil, 81
breaker zone, 593, 599f
breakers, 595, 597f
breccia, 202, 202t, 203f
Brigantine, New Jersey, 592f
British Columbia, 304, 540, 610, 610f
brittle deformation, 274
Bronze Age, 244
brunisolic soil order, 184t
brunisols, 184, 184f
Burgess Shale, 381f
burial metamorphism, 233
burrowing organisms, 170, 170f
Butterfly Nebula, 30f
Bylot Island, Nunavut, 461f

C

C horizon, 178
calcareous ooze, 634, 636, 637f
calcification, 431, 431f
calcite, 80, 83t, 90, 95, 95f, 97f, 167, 171, 171t,
        173t, 231, 231f
calcium, 48, 82, 88t, 89, 91t, 92, 110, 111
calcium carbonate, 195f
calcium-rich plagioclase feldspar, 110, 173t
calderas, 145, 145f
California, 280, 485t, 610, 610f
California Current, 626
Callisto, 34
calyx, 614
Cambrian Period, 21t, 380
Canada
    Appalachian Orogen, 404f, 407
    Atlanta margin, 629f
    Canadian Shield. See Canadian Shield
    coal, use of, 260
    Coastal Plain, 628
    coastlines, 591, 616

    continental platform, 400–403
    continental shelves, 414, 415f
    Cordilleran Orogen, 410–413
    dominant soil orders, 184–185, 184t, 185f
    energy use in, 254f
    fundamental geologic divisions, 394–395
    geologic components, 397f
    geologic provinces, 394–397, 395f
    geological map, 396–397, 396f
    greenhouse gas emissions, 435t
    hydrogeological regions, 557
    Innuitian Orogen, 408–409
    and Kyoto Protocol, 445
    landslides, 502, 502f
    largest lakes, 400
    last ice age, 399f
    major earthquakes, 313
    margins of Canada, 414
    minerals, mining and processing of, 83f
    natural gas reserves, 259
    natural hazards, 617
    northernmost mountain belt, 408–409
    Phanerozoic Orogens, 403
    physiography of northern Appalachians, 295
    seismic hazard map, 306f
    seismic zones, 70
    slopes, 414
    volcanoes in, 143, 143f
    wells, 540
    youngest mountain belt, 410–413
Canadian Association of Petroleum Producers,
        259
Canadian Cordillera, 275f, 410–411, 410f, 411f,
        412f
Canadian Prairies, 182, 183, 403, 445f
Canadian Precambrian Shield, 106
Canadian Rockies, 270f, 286
Canadian Shield, 123, 395, 397, 398–399, 398f,
        399f
Canadian soil classification system, 184–185,
        184t, 185f
CANDU reactor, 262, 263, 263f
cap rock, 257
capillary fringe, 540
capture hypothesis, 41
carbon, 86, 87f, 88t, 354, 430f, 434f
carbon-14, 354f, 355
carbon cycle, 174
carbon dioxide, 9, 32, 37, 170, 174, 355, 422,
        427–428, 427f, 430, 434–435
carbon monoxide, 37
carbon quotas, 444
carbonate compensation depth (CCD), 634
carbonate reservoirs, 257
carbonates, 96f, 97, 97f
carbonic acid, 440, 559, 634
Carboniferous Period, 339, 377f, 381
Caribbean Plate, 52, 643
Caribbean Sea, 624
Caroline Plate, 52
Carolinia, 294
Carrara, Italy, 221
carrier bed, 257
Cascade Mountains, British Columbia, 290–291
Cascade Range, 107, 124, 151, 290
Cascades Volcanic Belt, 143
Cascadia subduction zone, 68, 291, 291f, 382
cast, 368, 369f
cation, 86

cation substitution, 92–93
cave, 560–561
cave deposits, 560–561
Cayman Trough, 642
cementation, 15, 200
Cenozoic Era, 21t, 377f, 386–387
Central Lowlands, 295
central rift valley, 59f
central vent volcanoes, 132, 140, 140t, 141f,
 142–143
 rhyolite caldera complexes, 145
 shield volcanoes, 142–143
 stratovolcanoes, 144–145
centrifugal force, 29
Ceres, 28–29, 28f
cesspools, 498, 551
Chad, 387
chalcocite, 245t, 247
chalcopyrite, 83t, 96, 97f, 245t, 248
Chaleur Uplands, 295
chalk, 205, 205f, 205t
Challenger Deep, 63, 642
channel, 182, 512, 512f, 515
 see also stream channel
Charlevoix impact structure, 313
Charlevoix Seismic Zone, 70, 313
chemical differentiation, 48
chemical oceanography, 624
chemical sedimentary rock, 204–205, 204f, 204t
chemical sediments, 194–195, 198–199
chemical weathering, 73, 162, 166–168
chemically active fluids, 224
Chernobyl accident, 262
chernozemic, 403
chernozemic soil order, 184t
chernozems, 184, 184f
chert, 171t, 172, 173t, 204–205, 204f
Chic-Choc Mountain, Quebec, 295
Chicago Climate Exchange, 445
Chicxulub Crater, 374, 375f
Chile, 70, 305, 322
China, 305, 435t, 444
 earthquake, 68, 69, 69f, 281, 305, 312–313
 minerals, demand for, 81
china clay, 166
chlorine, 88t
chlorite, 122, 222, 226, 229, 231f
chlorofluorocarbons, 428f
Chomolungma, 296
 see also Mount Everest, Himalayas
chondrites, 357
chrysotile, 221, 225, 225f
cinder cones, 141f, 142, 143f, 148
cinnabar, 96
Circum-Pacific Seismic Belt, 70, 71f
Circumpolar Deep Water, 627
cirque, 458
Clarion fracture zone, 67
classification
 igneous rock, 114–115, 115f
 sedimentary rock, 202–203, 203f
 soil classification, 184–185, 184t, 185f
 volcanoes, 140, 140t, 141f
clastic grains, 198–199
clastic reservoirs, 257
clastic sedimentary rock, 202, 202f
clastic sediments, 194, 194f
clastic wedge, 296
clay, 166, 171, 173, 173f, 192, 200, 220f

clay minerals, 173
claystone, 202, 202t, 203f
cleavage, 84, 84t, 85f
climate, 176, 422, 433f
 cold climates, 176–177
 hot, arid climates, 176
 hot, humid climates, 177
 oxygen isotopes and global climate history,
  466–467
 soil formation, 176
 soils, and general climate conditions, 185
 and weathering, 172
climate change, 5
 see also global warming
 Sun, role of, 433
 and weathering, 174–175
climate feedback, 429, 476–477
climate modelling, 432–433
climate proxy, 466
closed system, 352
clusters, 311
co-seismic uplift, 610
coal, 205, 205f, 205t, 260–261, 260f, 339, 435
coal seams, 261
Coast-Cascade Ranges, 290–291, 290f
Coast Mountains, British Columbia, 106, 411,
 452t
coastal desert, 574
coastal dune, 593
coastal environments, 207, 208f, 606
 see also coastlines
 coral reefs, 614–615
 depositional coasts, 610, 611, 611f
 emergent coasts, 610–611
 erosional coasts, 610, 611, 611f
 human activities, effects of, 616–617
 population increase, effect of, 616–617
 submergent coasts, 610–611
 types of, 608–609
coastal erosion, 616
coastal geology, 592
coastal hazards, 617
Coastal Plain, 295, 629f
coastal pollution, 616
coastal processes, 592
coastal terminology, 593
coastal zone, 592, 593f, 632
coastlines
 see also coastal environments
 change, 592–593
 hurricanes, 602, 602f
 longshore current, 598
 rip currents, 598
 sea-level rise, 604–605
 straightening, over time, 598
 tides, 600–601
 tropical storms, 602
 wave energy, 594–595
 wave-generated currents, 596–597
 wave refraction, 596
cobalt, 245t
cobaltite, 245t
coccolithophores, 634
coccoliths, 205
Cocos Plate, 52, 58, 293
coefficient of permeability, 544
cold climates, 176–177
cold ocean current, 572
collisional orogenesis, 64

Colorado Plateau, 292–293
Colorado River, 582
colour index, 114, 114f
colour (minerals), 84–85
Columbia Plateau, Washington, 140t, 148, 151
Columbia River Gorge, Washington, 149f
Columbia River Plateau, 123
column, 204, 561, 561f
columnar geysers, 560
columnar jointing, 276
coma, 37f
comets, 28
commercial energy, 435
compaction, 15, 200, 200f
compass bearing, 277
complex folds, 282
composite origin, 631
composite volcanoes, 135, 140t
compounds, 88
compressional stress, 272, 273, 273f
conchoidal fracture, 84
condensation, 510
conduit-flow aquifers, 559
conduit systems, 559
cone of depression, 550
confined aquifer, 548, 549f
confining stress, 272, 273f
confluence, 513
conglomerate, 202, 202f, 202t, 203f, 227t, 230
consistency, 357
construction minerals, 246, 247t
contact aureole, 222
contact metamorphism, 222, 223f, 227, 230–231,
 231f, 232f, 233, 234
 see also metamorphism
contaminant plume, 552, 552f
continent-continent convergent boundary, 64
continental crust, 50, 51
continental drift, 10–11
continental environments, 206, 207f
continental freeboard, 344
continental glaciers, 454, 455f
continental ice sheets, 455
continental margin, 208f, 209, 630–631, 632–633
continental platform, 395, 400–403
continental rise, 209, 630
continental shelf, 62, 207–209, 208f, 414, 415f,
 628–629, 628f, 632–633
continental slope, 209, 629
continental volcanic arc, 124
continents
 see also specific continents
 felsic igneous rocks, 119
 reorganization of, 380–383
 supercontinents, 397
controlled breeding, 372
convection, 11, 48, 108
convergent boundaries, 58, 59f, 68t
 continent-continent convergent boundary, 64
 ocean-continent convergent boundary, 62
 ocean-ocean convergent boundary, 63, 63f
 oceanic crust subducts, 62–63
 orogenesis, 64–65
convergent margins, 236
convergent seismicity, 314
conveyor belt hypothesis, 477
cooling aerosols, 429
copper, 82, 96f, 97f, 245t
Copper Age, 244

Coppermine volcanics, 123
coquina, 205, 205f, 205t
coral reefs, 169f, 440, 614–615, 614f, 615f, 645
Cordilleran Ice Sheet, 470
Cordilleran Orogen, 399, 400, 403, 410–413
core, 48, 50
    discontinuities, 329
    inner core, 48, 49f
    outer core, 48, 49f
Coriolis effect, 571, 626
corundum, 97f
cosmogenic particles, 634
cosmogenic radioisotopes, 348
country rock, 125, 222
covalent bonds, 89
Crab Nebula, 29f
craters, 145
craton, 378, 378f, 394
creep, 488, 489f, 496
crescentic dunes, 578, 578t, 579f
Cretaceous-Paleogene extinction, 374, 382
Cretaceous Period, 8f, 21t, 344, 377f, 382, 383,
    383f
Cretaceous-Tertiary extinction, 374
crevasses, 454
critical thinking, 3, 4, 6–7
    extrapolation, 31
    global challenges, solving, 8–9
    inductive and deductive reasoning, 6f
    tools of, and Earth's history, 366–367
cross beds, 212, 212f
cross-cutting relationships, 342f, 343
Crouse, Bill, 296
crust, 5, 48, 49f, 50–51
    continental crust, 50, 51
    eight most common elements in, 91t
    oceanic crust, 50, 51
    outcrop patterns, 284–285
    primordial crust, 48, 356
    strain, 276–277
    stresses in, 272–273
    structure of, 284–285
crustal extension, 292–293, 292f
crustal rifting, 108
crustal uplift, 65
cryosolic soil order, 184t
cryosols, 184f
crystal settling, 108–109, 109f
crystalline structure, 80
crystallization, 7, 90, 90f, 108–109
    continuous branch, 111
    discontinuous branch, 111
    igneous rock, texture of, 112–113, 113f
    of magma, 110–111
    recrystallization, 15, 90, 220, 221, 231
    substitution, 92–93
cumulate, 109, 248
Curie, Marie and Pierre, 340
Curie temperature, 56
currents, 626–627
curtain, 561
cutbank, 518
cyclone, 16, 602
Cynognathus, 10

D

da Vinci, Leonardo, 340

dam, 526, 527f
Darcy's law, 544
Darwin, Charles, 367, 367f, 370
dating, 338
    accuracy of dating, 352–353
    isotopic clocks, 348, 353, 354–355
    isotopic dating, 338, 346–347, 350, 352–353
    radiocarbon dating, 354–355
    radioisotope, 348
    relative dating. See relative dating
    sources of uncertainty, 352–353
daughter isotope, 346
David, 221
dead zones, 533
Death Valley, California, 582–583
debris falls, 490, 491f, 496
debris flow, 490, 496
debris slides, 490, 491f
decay chain, 352–353
Deccan Traps, India, 148
decompression melting, 59, 108
deductive reasoning, 6, 6f
deep burial, 237
deep circulation, 626
deep roots, 287n
deep-sea environments, 208f, 209
deep-sea trenches, 11, 52, 53
deep time, 340, 366
deep-water production, 626
Deer Creek, 522, 523f
deforestation, 436, 488, 497
deformation, 272, 341, 341f
    brittle deformation, 274
    ductile deformation, 274, 282, 283f
    elastic deformation, 274
    plastic deformation, 274, 282, 283f
degradation, 525
delta, 207, 208f, 528–529, 528f, 606, 633
delta plain, 529
dendritic, 513
density, 494
density segregation, 248, 249f
deposition, 341, 341f
depositional coasts, 610, 611, 611f
depositional glacial landforms, 459–461
desalinization, 532
desert, 206, 207f, 567, 568, 569f
    atmospheric moisture circulation, 570–571
    desert environment, 568
    formation of, 571, 572
    low precipitation, 568
    sand dunes, 578–579, 578f, 578t, 579f
    sedimentary processes, 576, 577f
    10 largest deserts, 568t
    types of, 574–575
    unique characteristics, 574–575
    water, role of, 582–583
    wind, 576, 577f
desert pavement, 176, 576
desert soils, 185
desertification, 17, 441, 568, 584, 585f
detrital sediment, 632
Devonian Period, 21t, 377f, 381
Dhaka, Bangladesh, 611f
diamond, 80, 84
diatoms, 256, 634
differential stress, 272–273
differentiation, 106
diffuse-flow aquifers, 559

dikes, 125, 150
dilation, 494, 494f
dilation fractures, 179
dilution, 636–637
Dimetrodon, 374f
Dinosaur Provincial Park, Alberta, 192f
diopside, 231f
diorite, 115, 116, 117f, 119, 172f
dip, 277, 284f
dip-slip faults, 278
direct mitigation, 503
direct solar energy, 254
directed stress, 223, 223f, 228–229, 272
directional wells, 556
discharge, 514–515
discontinuity, 326, 328–329
disseminated minerals, 248
dissolution, 167–168, 636
dissolved load, 516
dissolved minerals, 550–551
distance moisture, 572
Distant Early Warning site, Nunavut, 549f
divergent boundaries, 58–59, 59f, 68t
divergent margins, 236
divergent plate boundaries, 57
divergent seismicity, 314
dolomite, 97f, 204, 227t, 231
dolostone, 204, 204f, 204t
domes, 282, 283f
Dominican Republic, 436f
Doppler, Christian, 35
Doppler effect, 35
Dorman, James, 312
double planet hypothesis, 41
double substitution, 92
drainage basin, 512, 521f
drainage divides, 512
drainage fields, 551
drainage patterns, 513, 513f
drainage system, 512–513
Drake, Edwin L., 256
drapery, 561
drawback, 308
dredges, 251f
dredging, 526
dropstones, 461
droughts, 440
drumlins, 461
ductile deformation, 274, 282, 283f
ductility, 275
dune fields, 579
dunes, 207f, 576, 576f
    see also sand dunes
dunite, 114
Dunnage, 294
Dunnage terrane, 406
dust bowl, 182–183, 572, 573
dust tail, 37, 37f
dwarf planets, 28, 36
dynamic equilibrium, 31, 524

E

Early Mesozoic Era, 628
Earth, 28, 29, 30, 32, 32f
    age of, 340–341, 356–357
    atmosphere, 33, 424
    atmospheric temperature, 438–439, 466–467

consequences of heat buildup, 40–41
continuous changes, 9
core, 48, 50
crust, 5, 48, 49f, 50–51
    see also crust
discontinuities in Earth's interior, 326, 328–329
and Earth system, 9
evidence of past events, 9
extraterrestrial bombardment, 40–41
formation of, 40
geologic change, 5
geologic time scale, 20, 20t
global warming. See global warming
during Hadean Eon, 48–49, 48f
heat budget, 426–427, 426f
heat sources for early Earth, 40–41
history of the Earth. see Earth's history
inner core, 48, 49f
interior, 48–49, 49f
layers of, 48–49
magnetic field, 56f
mantle, 48, 49f, 50, 51f, 108, 328–329
minerals, 81
origin of, 28–29
outer core, 48, 49f
recycling of Earth materials over time, 8–9
seasons, 474f
solar nebula hypothesis, 28, 29
surface temperature, 426, 432f
tectonic plates, distribution of, 394f
Earth citizenship, 4
coal use, problems with, 261
coral reefs, loss of, 645
El Salvador's deadly landslide, 497
global warming and mass wasting, 498
government-funded cleanup, 549
minerals, 81
natural resources, management of, 17
ocean acidification, 168–169
peak oil, 255
power, problem with, 262
water facts, 511
Earth systems, 9
earthquake hazards, types of, 306–309
earthquakes, 16, 52, 304
causes of, 304–305
convergent seismicity, 314
divergent seismicity, 314
earthquake hazards, types of, 306–309
elastic rebound theory, 310
epicentre, 310, 320
fire, 308
frequency of occurrence since 1900, 323t
global distribution of foci, 312f
great earthquake, 322
ground shaking, 306–307
historical quakes, 305t
intensity, 320–321, 322
intraplate earthquakes, 312–313
landslides, 307
locations of, 70–71
magnitude, 322–323
micro-earthquake, 322
moment magnitude, 323
origin of, 310–311
plate boundaries, 68–71, 312
prediction, 313
Richter scale, 322

risk of, 68–69
seismic tomography, 330–331
seismic waves, 318
seismology, 326–327
seismometers, 310, 320–321, 320f
transform seismicity, 314–315
tsunami, 308–309, 309f
Earth's history
    see also specific periods, eons, and eras
from composition of sand, 202–203, 203f, 203t
critical thinking, tools of, 366–367
crystallization history, 112–113
evidence of past environments, 206–209
evolution. See evolution
fossils, 368–369
geologic time scale, 13f, 20, 20t, 376, 377f
ice ages, cycles of, 470–471
interglacials, cycles of, 470–471
retreating ice in North America, 471f
as sequence of geologic events, 340–341
East African Rift Valley, 53, 59, 132
East Antarctica, 465
East Antarctica Ice Sheet, 455
East Coast Trail, Newfoundland, 608f
East Pacific Rise, 12, 67, 140t, 150, 637
eastern boundary current, 626
Eastern System, 410, 411
ebb-tide delta, 606
eccentricity, 474
the ecliptic, 29
ecosystems, 430
Ediacaran Period, 377f
effluent streams, 542
effusive eruptions, 132, 136
Ekati diamond mine, Northwest Territories, 399f
Ekman transport, 627, 627f
El Fasher, Sudan, 585f
El Salvador, 485t, 497
elastic deformation, 274
elastic energy, 310
elastic limit, 274
elastic rebound theory, 310
electrical treatments, 556–557
electron capture, 346, 347
electron patterns of elements, 88t
electrons, 86–87
elements
    see also specific elements
    electron patterns of, 88t
    ions, 89
    most common elements in Earth's crust, 91t
Ellesmere Island, Nunavut, 409
Ellesmerian Orogeny, 408, 409
embryology, 371
emergence from last ice age, 476
emergent coasts, 610–611
emission trading, 445
end moraine, 460
energy resources, 244–245
    bioenergy, 264
    biomass energy, 254
    coal, 260–261
    energy use in Canada, 254f
    fossil fuels, 254
    geoexchange energy, 264
    geothermal energy, 254, 264
    hydroelectric energy, 254, 264
    hydrogen fuel cells, 265
    natural gas, 258–259

non-renewable energy resources, 254
nuclear power plants, 262–263
ocean energy, 265
oil, 254, 256–257, 258–259
renewable energy, 264–265, 265f
solar energy, 254, 264
wind energy, 254, 264
Enriquillo-Plantain Garden fault, Haiti, 278, 280
Environment Canada, 441f, 540
environmental energy, 196
Environmental Protection Agency (EPA), 549
environments of deposition, 162, 196, 210
Eocene Epoch, 377f, 386
eons, 20, 376
ephemeral streams, 582
epicentre, 310, 320
epidote, 122
epikarst, 559
epochs, 20, 376
Equatorial Current, 425
eras, 20, 376
erg, 579f
Eris, 28, 28f
erosion, 9, 14, 162, 285f, 287f, 341, 341f
    coastal erosion, 616
    fluvial erosion, 182
    gully erosion, 182, 183f
    headward erosion, 524, 525f
    by ice, 454–455
    rill erosion, 182
    sediment, 517
    sheet erosion, 182
    soil erosion, 182, 183
    tectonic erosion, 62
    by water, 182, 516–517
    wave erosion, 608f
    wind erosion, 182–183, 183f
erosional coasts, 610, 611, 611f
erosional glacial landforms, 458–459
Erta Ale Volcano, Ethiopia, 137t
eruption column, 138
eskers, 461
Eskola, Pentti, 232
estuary, 608
eukaryote branch, 372
Euramerica, 381
Eurasia, 12, 64f, 407f
Eurasian Plate, 52, 64f, 65, 71, 313
Eurekan Orogeny, 408, 409
Europa, 34, 624
Europe, 10, 137t, 471f, 628
eutrophication, 616, 617f
evapo-transpiration, 510
evaporation, 511
evaporite, 195, 251
evaporite deposits, 194f
Eve Cone, British Columbia, 142, 143f
evolution, 343, 366, 366f
    Appalachian Orogen, 404–406
    Canadian Cordillera, 411–413
    embryology, 371
    homologous structures, 370
    horse evolution, 370, 370f
    igneous evolution, 118–119, 118f
    Innuitian Orogen, 408–409
    lines of evidence for, 370–371
    magma evolution, 119
    mass extinctions, 374–375, 375f

evolution (cont.)
  molecular biology, 372
  phylogeny, 370
  process of evolution, 372
  theory of evolution, 366
  vestigial structures, 371
exfoliation, 164, 164f
exhumation, 287
exotic terrane, 630
expansion of the universe, 35
experiment, 7
Explorer Plate, 290f, 293
explosive eruptions, 132, 136–137, 138
extension stress, 272
external mould, 369, 369f
extinctions, 374–375
extrapolation, 31
extrusive igneous rock, 14, 107, 112, 115
Eyjafjallajökull volcano, Iceland, 137, 137f

F

failed rift, 409
failure surface, 488
Faraday, Michael, 265
Farallon plate, 292, 293, 331
fast-moving mass-wasting events, 490, 491f
fault, 66, 274, 276, 276f
  see also specific faults
  dip-slip faults, 278
  evidence of faulting, 280
  and mountain building, 286
  normal faults, 278
  reverse faults, 278, 279f, 280
  shear fracture or fault, 164
  strike-slip faults, 278
  thrust fault, 278
  transform faults, 66–67, 67f
fault-block mountains, 286, 287f, 288–289f
fault breccia, 280
fault gouge, 280
fault trap, 257
faulting, 341, 341f
fayalite, 167
Federal Contaminated Sites Action Plan, 549
Federal Emergency Management Agency
    (FEMA), 616
feldspar, 83t, 92, 93f, 94, 227
feldspar group, 94
felsic igneous rock, 111, 114, 115, 115f, 118f, 119
felt reports, 321
Fennoscandian Ice Sheet, 470
Ferrel, William, 571
Ferrel cell, 425, 570, 570f
fibrous fracture, 84
field geology, 396
fire, 308
firn, 453
first-order stream, 513
fission, 263
fission hypothesis, 41
fissure eruptions, 122, 140
fjord glacier, 461
fjords, 457
flank eruptions, 140
flash flooding, 528, 582
flint, 205
flood, 16, 498, 520–521, 528

flood basalts, 123, 148
flood plains, 16, 520–521, 521f, 524
flood-tide delta, 606
flow, 514–515
fluorine, 88t, 89
fluorite, 83t, 96f, 97f
fluorocarbons, 429
fluvial, 525
fluvial erosion, 182
fluvial processes, 526, 527f
fluvial sediment, 528–529
focus, 310
fold-and-thrust belts, 294–296
fold-and-thrust mountains, 286, 288–289f
folds, 276, 276f, 282, 282f, 283f, 286
foliated metamorphic rock, 228–229, 228f, 228t
foliation, 221, 223
fool's gold, 84
footwall block, 278
foraminifera, 256, 431, 466
foraminiferans, 634
fore-arc area, 233
fore-arc basin, 62
fore-arc ridge, 62
fore reef, 615
foreland basins, 402
foreset beds, 528
foreshocks, 311, 311f
foreshore, 593
formaldehyde, 80
Fossil Fuel Age, 244
fossil fuels, 17, 17f, 175, 193, 193f, 254, 423, 434
  see also coal; oil
fossil stromatolites, 379f
fossil succession, 343, 343f, 376
fossilization, 368
fossils, 206f, 368–369
400-km discontinuity, 329
fractionation, 119
fracture, 84, 84t, 274
fracture zone, 67
fractured aquifers, 548, 548f
fragmental volcanic rocks, 113
framework silicates, 91
Fraser River estuary, 632
freeze-thaw, 165
Freon, 429
fresh water, 17, 511
  see also groundwater
freshwater problems, 532–533
friction, 494
Friedrich, Carl, 50
fringing reefs, 615, 615f
frost line, 33f
fuel rods, 254, 263
fumaroles, 560, 560f

G

gabbro, 112, 115, 116–117, 117f, 119
gaining streams, 542, 543f
galena, 83t, 96, 97f, 245t
Galilean satellites, 34
Gander terrane, 406
Ganderia, 294
Ganges-Brahmaputra Delta, 633
Ganges-Brahmaputra River, 610–611, 633, 633f
gangue, 247

Ganymede, 34
Garibaldi Volcanic Belt, 143
garnet, 226, 231f
garnet-mica schist, 229
gas chemistry, and volcanoes, 136–137
gas giants, 29, 34–35
gas or ion tail, 37, 37f
gases, 430
gasoline service stations, 553f
Gaspard-Gustave Coriolis, 626
gemstones, 82
general circulation of the atmosphere, 424
genetic mutation, 372
genetic variation, 372
geoexchange energy, 264
geologic change, 5
geologic events, 341
  Earth's history, as sequence of geologic events,
      340–341
  order of geologic events, 344–345
geologic hazards, 16
  earthquakes, 16
    see also earthquakes
  floods, 16
  hurricanes, 16, 602, 602f
  mass wasting, 16–17
    see also mass wasting
  study of, 16–17
  tornadoes, 16
  tsunamis, 16
  volcanic eruptions, 16, 154–155
    see also volcanoes
geologic map. See geological map
geologic provinces, 395–396, 395f
geologic resources, 244
  energy resources, 244–245, 254–265
  meaning of, 244–245
  mineral resources, 244, 246–251
  non-renewable resources, 244
  renewable energy, 245
  study of, 245
geologic time
  dating. See dating
  deep time, 340, 366
  Earth history as sequence of geologic events,
      340–341
  Hadean time, 376
  study of, 339
  time, 338
  walk through time, 339
geologic time scale, 13f, 20, 20t, 376, 377f
geological engineering, 485
geological map, 284–285, 284f, 345, 345f,
      396–397, 396f
geological products, 485
Geological Survey of Canada, 321, 395, 396,
      502
geology, 4
  characteristics of, 8–9
  coastal geology, 592
  geologic hazards, study of, 16–17
  historical geology, 4
  importance of, in daily life, 4–5
  integrative science, 4
  layer cake geology, 342
  marine geology, 624–625
  physical geology, 4
  "science of time," 338–339
    see also geologic time

structural geology, 284
vast range of time and space, 8
geomagnetic field, 5, 50, 50f
Georges Bank, 414, 415f
GEOSAT (Geodectic Satellite), 209
geosphere, 9
geothermal energy, 254, 264
geothermal gradient, 222, 232
Germany, 435t
geysers, 560, 560f, 561f
gibbsite, 250
Giotto, 37
glacial drift, 459
glacial erratics, 459
glacial interglacial cycles, 474–475
glacial landforms, 452
glacial moraine, 460
glacial outwash, 460
glacial periods, 427
glacial striations, 452, 454
glacial till, 459, 461f
glacial troughs, 458
glacial valleys, 458
glaciar mass balance, 456
glaciation, in Canadian Shield, 399
glacier ice, 453, 453f
Glacier Peak, 291
glaciers, 197t, 206, 207f, 452, 452f, 453f
    alpine glaciers, 454, 455f, 464, 464f
    benchmark glaciers, 468
    calving, 456
    continental glaciers, 454, 455f
    depositional glacial landforms, 459–461
    described, 452–453
    equilibrium line, 456
    erosion by, 454–455, 455f
    erosional glacial landforms, 458–459
    fjord glacier, 461
    glacial interglacial cycles, 474–475
    global warming, effect of, 464
    mass balance of a glacier, 468
    movement of ice inside glacier, 456–457, 456f,
        457f
    outlet glaciers, 454–455
    oxygen isotopes and global climate history,
        466–467
    polar glaciers, 454
    sediment, transportation of, 460f
    subpolar glaciers, 454
    temperate glaciers, 454
    temperature at base of, 454f
    terminus, 456
    tributary glacier, 459
    types of, 454–455
    valley glaciers, 454
    vestiges of last ice age, 456–457
glacioisostatic rebound, 274, 457
glacioisostatic uplift, 610
glaciomarine drift, 461
glassy, 112, 113t
glauconite, 632
Glenwood Springs, Colorado, 491f
gleysolic soil order, 184t
global atmospheric circulation, 570–571
global biogeochemical cycling, 430
global carbon cycle, 430–431
global challenges, 8–9
global change, 423
global circulation models (GCMs), 432–433, 472

global climate change, 422–423
global climate history, 466–467
global cooling, 21t, 175
global tectonics, 54
global temperature, 438f, 439f, 440f
global thermohaline circulation, 476
global warming, 422
    cause of, 422–423
    climate modelling, 432–433
    glaciers, effect on, 464
    global carbon cycle, 430–431
    global change, 423
    global climate change, 422–423
    greenhouse effect, 426–429
    greenhouse gases, 423, 427–429, 444f
    heat circulation, 424–425
    human activities, and carbon dioxide,
        434–436
    and ice environments, 464
    impact of, 440–441
    international efforts, 444–445
    Kyoto Protocol, 444–445
    last interglacial, 472–473
    and mass wasting, 498
    ocean acidification, 440
    potential, 429
    significant climate anomalies and events, 2009,
        442–443f
    temperature increases, 438–439
global warming potential (GWP), 429
global water, 425
global wind and wave patterns, 595
global winds, 424–425
Glossopteris, 10
glucose, 80
gneiss, 144, 173t, 220f, 223f, 227, 227t, 228f, 229,
    275
gneissic texture, 229
Gobi Desert, Asia, 441, 568t, 572, 574f
Goderich, Ontario, 16
gold, 89, 97f, 247
Gondwana, 64f, 65, 380, 381, 382f, 404, 405, 405f,
    407
Gorda Plate, 290f, 293
Gorda Ridge, 67
Gould, Stephen J., 366
government-funded cleanup, 549
grabens, 292
graded beds, 212
graded profile, 524–525, 527f
graded stream, 524–525
gradient, 515
Grand Banks, Newfoundland, 414, 415f, 628f,
    629f
Grand Canyon, 190f, 293, 338f, 488f
granite, 14f, 82, 82f, 106, 107f, 112, 112f, 115, 116,
    116f, 119, 173t
granulite, 237t
granulite facies, 233
graphite, 80, 83t, 97f
gravel, 196f
gravity, 494, 544, 600–601
Great Barrier Reef, Australia, 615f
Great Basin Desert, North America, 568t, 575f
Great Bear Lake, Canada, 400
Great Depression, 573
great earthquake, 322
Great Lakes, 399f, 616
Great Plains, 573

Great Red Spot, 34
Great Sahara, Africa, 441, 568t, 569
Great Slave Lake, Canada, 400
Great Victorian Desert, Australia, 568t
Green River Formation, Wyoming, 206f
greenhouse effect, 426–429
greenhouse gases, 174, 423, 427–429, 433, 435t,
    437f, 444f
Greenland, 386, 427f, 455, 456, 457, 470, 470f,
    471, 473, 476
Greenland Ice Sheet, 455f, 457f, 465, 473, 473f,
    604
greenschist, 227t, 229, 237t
greenschist facies, 233
Grenville Orogen, 398, 400
Grenville Orogeny, 379
ground moraines, 460
ground shaking, 306–307
groundmass, 112
groundwater, 224, 250f, 510, 540
    aquifers, types of, 548, 549f
    cave formation, 560–561
    cleaning groundwater, 556–557
    contaminated groundwater, 552–553, 556f
    dynamics, 546
    factors influencing groundwater movement,
        544, 545f
    general features of, 540–541
    gravity, 544
    human activities, contamination by, 552–553
    hydraulic pressure, 544
    hydrothermal activity, 560
    importance of, 540
    karst geology, 559
    karst topography, 558–559
    pollution, vulnerability to, 550–551
    porous media, 548
    recharge, 542, 542f
    remediation, 556–557
    springs, 558–559, 558f
    water wells, 541
groundwater discharge, 542
Groundwater Information Network, 540
groundwater remediation, 556–557
group (mineral), 94
Guatemala, 485f
Gulatee, B.L., 296
Gulf Coast, 602
Gulf of Mexico, 633f
Gulf of St. Lawrence, 608, 609f, 632
Gulf Stream, 425, 425f, 476–477, 626–627, 628f
gully erosion, 182, 183f
gypsum, 78–79f, 82, 83t, 96f, 97f
gyres, 626

H

Hadean Eon, 21t, 40, 48–49, 48f, 377f, 378
Hadean time, 376
Hadley cells, 424, 570, 570f
Haiti, 69, 69f, 280, 280f, 305, 436f
Hale-Bopp comet, 40–41
half-life, 263, 347, 347f, 348t, 352
halides, 96f, 97, 97f
halite, 80, 81f, 83t, 85f, 97f, 173t
Halley's comet, 37
hanging valleys, 459
Hapke, Bruce, 163

hard water, 551
hardness, 84, 84t
Harrison Lake, British Columbia, 9
Harvard School of Public Health, 261
Haumea, 28, 28f
Hawaii, 472f
Hawaii Volcanoes National Park, 132
Hawaiian Islands, 11, 106, 117, 142, 143, 152f, 493
hazard mitigation, 617
hazardous waste sites, 549
headwall scarps, 488
headward erosion, 524, 525f
headwaters, 513
Health Canada, 347
heat, and metamorphism, 222
heat budget, 426–427, 426f
heat circulation, 424–425
heavy minerals, 194
heavy water, 263
Heezen, Bruce, 11
Heiltskuk Icefield, British Columbia, 452f
helium, 88t
helium fusion, 30
hematite, 82, 83t, 90, 96f, 97f, 167, 171, 171t, 173t, 193, 245t
hemipelagic sediments, 633
herbicides, 552
Herculaneum, 145
Herschel Island, Yukon, 495f
Hess, Harry, 11, 72
heterogeneous body, 48
Hibernia Platform, 629f
high-grade metamorphism, 226
Highland Valley copper mine, British Columbia, 413f
Himalayan fold-and-thrust belt, 386
Himalayan Mountains, 64f, 65, 286, 294
Himalayas, 71
hinge line, 282
hinge or hinge point, 282
historical geology, 4, 366
"History of Ocean Basins" (Heezen), 11
history of the Earth. See Earth's history
HIV, 373, 373f
Hjulstrom diagram, 517
Holocene Epoch, 21t, 377f, 458, 470, 472
Hominidae, 386
Homo erectus, 387, 387f
Homo habilis, 244, 244f, 386
Homo sapiens, 339
Homo sapiens, 386, 387, 387f
homogeneous body, 48
homologous structures, 370
hoodoo, 8f
Hooke, Robert, 274
Hooke's law, 274, 274f
Hopewell Rocks, Nova Scotia, 213f, 599f
hornblende, 93f, 94, 95f
hornfels, 227, 227t, 230f, 230t, 231
hornfels facies, 233
horse evolution, 370, 370f
Horseshoe Falls, Ontario, 402f
Horseshoe Canyon, Alberta, 212f
horsts, 292
hot, arid climates, 176
hot, humid climates, 177
hotspots, 11–12, 12f, 90, 90f, 117
    see also mantle plumes

basalt, formation of, 122–124
    and metamorphism, 236–237
Hubble, Edwin P., 35
Hubble Space Telescope, 30f
Huckleberry Ridge eruption, 143
human activities
    and carbon dioxide in atmosphere, 434–436
    coastal environments, effect on, 616–617
    groundwater contamination, 552–553
    mass wasting, cause of, 498
    oceans, impact on, 644–645, 644f
human origins, 386–387
human population growth, 435, 436f, 551f, 616–617
humans, 386–387
Humber Arm allochthon, Newfoundland, 406f
humid environments, 177
humus layer, 178
Hurricane Katrina, 16, 603f, 617
Hurricane Sandy, 441f, 592f
hurricanes, 16, 602, 602f
Hutton, James, 20, 72, 338, 340, 366, 367f
hydrated minerals, 122
hydraulic action, 165
hydraulic gradient, 544
hydraulic pressure, 544
hydrocarbons, 256, 430
hydroelectric dams, 526
hydroelectric energy or power, 254, 264
hydrogen, 88t
hydrogen fuel cells, 265
hydrogeological regions of Canada, 557
hydrogeologists, 548
hydrologic cycle, 510–511, 510f
hydrologists, 512
hydrolysis, 166, 171t
hydrosphere, 9, 425
hydrothermal activity, 560
hydrothermal fluids, 96, 224
hydrothermal mineral deposits, 248–250, 248t
hydrothermal vein fillings, 224
hydrothermal vent communities, 638
hydrothermal vents, 122, 638, 638f
hydrous, 237
hydroxides, 247
hypothesis, 6, 7

I

Iapetan Realm, 406
Iapetus Ocean, 295, 295f, 405f, 406
ice, 97f
    see also glaciers
ice, melting, 441, 465
ice ages, 386, 399f, 433, 452, 456–457, 470–471, 476
ice cores, 466, 470f
ice divides, 457
ice sheets, 455, 465
ice shelves, 457, 464
ice streams, 457
ice wedging, 165, 165f, 177, 454, 495
icebergs, 457
icecaps, 454
igneous crystallization, 90
igneous evolution, 118–119, 118f
igneous fractionation, 119
igneous intrusions, 125

igneous minerals, 90
igneous rock, 14, 106
    see also specific types of igneous rock
    classification, 114–115, 115f
    colour, 114
    colour index, 114, 114f
    common igneous rock-forming minerals, 111t
    common textures, 112–113, 113t
    common types of igneous rock, 116–117
    crystallization, 108–109
    extrusive igneous rock, 14, 107, 112
    formation of, 106–107, 108–109
    igneous evolution, 118–119, 118f
    igneous intrusions, 125
    intrusive igneous rock, 14, 14f, 107, 112
    magma differentiation, 108–109, 118–119, 118f
    naming of, 114–115
    and plate tectonics, 118–119, 119f
    study of, 107
    texture, and crystallization history, 112–113, 113f
illite, 173
ilmenite, 245t
immiscible melts, 248
impact hypothesis, 41, 41f
impervious surfaces, 533
in-ground waste disposal, 498
incised channels, 526
inclusions, 342f, 343
index minerals, 226–227
India, 10, 64f, 65, 81, 435t
Indian-Australian Plate, 52, 65, 294, 313
Indian Ocean, 595, 617, 624, 624t, 625f
Indian Ocean Gyre, 626
Indonesia, 16, 68, 151, 152f, 491f
induced fission, 263
inductive reasoning, 6, 6f, 7, 48, 206
industrial minerals, 246, 247t
Industrial Revolution, 427
inert gases, 88
inertia, 600–601
inertial force, 600
inference, 112
infiltration, 510–511
influent streams, 542
injection wells, 556
inner core, 48, 49f
Innuitian Orogen, 399, 403, 408–409, 408f
inorganic precipitation, 625
insolation, 433, 474
insolation weathering, 165
insoluble residues, 177
integrative science, 4
interglacial periods, 399, 427
interglacials, 386, 458
    glacial interglacial cycles, 474–475
    interglacial cycles, 470–471
    last interglacial, 472–473
Intergovernmental Panel on Climate Change (IPCC), 444, 617
interior, 49f
interior of the Earth, 48–49
Interior System, 411, 413
intermediate igneous rock, 111, 114, 115, 115f, 119
internal mould, 368, 369f
International Astronomical Union, 28
International Hydrographic Organization, 624

interplate slip, 291, 291*f*
intraplate earthquakes, 312–313
intraplate volcanism, 150, 151, 312
intrusion, 341, 341*f*
intrusive igneous rock, 14, 14*f*, 107, 112, 115
invertebrates, 386
Io, 34
ionic bonds, 88–89
ionosphere, 424
ions, 86–87, 89
Iran, 69, 71
iron, 48, 82, 91*t*, 110, 111, 172, 245*t*, 246, 251*f*, 379*f*
Iron Age, 244
iron catastrophe, 48
iron oxidation, 167*f*
iron oxide cemented sandstone, 195*f*
iron oxides, 193
island arc, 63, 119, 124, 144, 151
island arc volcanism, 151*f*
isostasy, 287
isostatic adjustment, 287
isotopes, 86–87, 346
isotopic analysis, 350*t*
isotopic clocks, 348, 353, 354–355
isotopic dating, 338, 346–347, 350, 352–353

J

Jack Hills conglomerate, Australia, 376
Japan, 71, 435*t*
Japan Current, 626
Japan Trench, 642
"Jason," 149*f*
jasper, 205
Java, 71
Johnson's Landing, British Columbia, 485*t*
joints, 164, 276–277, 276*f*
Joly, John, 340
Juan de Fuca Plate, 52, 58, 143, 290*f*, 293, 413, 414
Jupiter, 28, 29, 34, 34*f*, 37, 624
Jurassic Period, 21*t*, 377*f*, 382–383, 382*f*

K

Kalahari Desert, Africa, 568*t*, 574*f*
kame terraces, 460
kame topography, 461
kames, 460
Kant, Immanuel, 340
kaolinite, 166, 171*t*, 173, 173*t*
kaolinite clay, 171*t*
Karijini National Park, Australia, 379*f*
karst geology, 559
karst topography, 168, 168*f*, 559*f*
karstification, 168
Kashmir quake, 305, 308*f*
Kaskawulsh Glacier, Yukon, 461
Katla volcano, 137
Kauai (Hawaii), 12
Kazakhstan, 572
Keeling, Charles, 428
Kenaston, Saskatchewan, 445*f*
Kermadec Trench, 642
kerogen, 256
kettle holes, 460

kettle lakes, 460
kettle topography, 461
Kilauea Volcano, Hawaii, 132, 136, 137*t*, 142, 312
Kildare Capes, Prince Edward Island, 608*f*
kimberlite pipes, 328
Klondike gold rush, 84, 251*f*
Kobe earthquake, Japan, 308, 309*f*
komatiite, 115, 117
Kona, Hawaii, 136
Krakatau Volcano, Indonesia, 68
Kuiper Belt, 28, 37, 40
Kuril Trench, 642
Kuroshio Current, 626
kyanite, 226
Kyoto Protocol, 444–445

L

Labrador Current, 628*f*
Labrador Sea, 409
labradorite, 94
lag deposit, 576
lagoon, 207, 208*f*
Laguna del Maule caldera, Chile, 135*f*
lahar, 155, 155*f*, 490, 492, 493*f*
lake, 206, 207*f*
    see also specific lakes
Lake Huron, Canada, 400
Lake Ontario, Canada, 400
Lake Superior, Canada, 400
Lake Winnipeg, Canada, 400
Laki fissure eruption, Iceland, 141, 141*f*
Lambert Glacier, East Antarctica, 454*f*, 455
laminar, 514
laminations, 200
land rights, 629
land use, 435–436
land-use planning, 498
landfills, 552, 553*f*
landslides, 16–17, 307, 308*f*, 484, 485*f*, 485*t*, 495*f*, 497, 502–503, 502*f*
    see also mass wasting
lapilli, 138, 139*f*
Laplace, Pierre Simon, 340
large igneous provinces (LIPs), 123, 140*t*, 148–150
large-scale volcanic terrains, 132, 140, 140*t*, 141*f*, 148–149
Late Devonian extinction, 374
Late Paleozoic Era, 628
latent heat, 426
lateral moraines, 460
laterite, 177, 177*f*
Laurasia, 65, 382*f*
Laurentia, 399, 400–402, 403, 405, 405*f*, 407
Laurentian, 294
Laurentian Realm, 406
Laurentide Ice Sheet, 470
lava, 14, 106, 134
lava plains, 32
lava plateaus, 148
Lave Creek eruption, 143
law of stream numbers, 513
layer cake geology, 342
leach fields, 498
leaching, 178, 250
lead, 82, 245*t*, 357*f*
lead isotopes, 357

Leclerc, Georges-Louis (Comte de Buffon), **340**
Leda clay, 490, 491*f*
lee slope, 492
left-lateral strike-slip fault, 278
Leyte, Philippine, 490, 490*f*
Libby, Willard F., 354
lidar data, 396
lignite, 260
limbs, 282
limestone, 168*f*, 173*t*, 204, 227*t*, 231*f*, 548
limestone caverns, 561
lineation, 221, 223
Lipari, Sicily, 132
liquefaction, 490
lithic fragments, 194, 194*f*
lithic sandstone, 202, 202*t*, 203*f*
lithification, 200
lithified, 162
lithium, 88*t*
lithology, 202
lithosphere, 49, 53*f*
lithospheric plates, 52–53
lithostatic stress, 222–223, 272
Little Ice Age, 471
living organisms, 9
local base level, 514
loess, 197, 576, 577*f*
logarithm of the maximum seismic wave amplitude, 322
Loma Prieta earthquake, San Francisco, 305*f*
longitudinal dunes, 578*t*, 579, 579*f*
Longmenshan fault, China, 280, 313
longshore current, 598
longshore drift, 598
loose sediment, 192, 192*f*
loose snow avalanches, 492
losing streams, 542, 543*f*
Love, A.E.H., 318
Love waves, 318
low-grade metamorphism, 226
low-velocity zone, 328
Lower St. Lawrence Seismic Zone, 70
lunar highlands, 356
lunar tide, 600–601
lung cancer, 225
lustre, 84, 84*t*
luvisolic soil order, 184*t*
luvisols, 184, 184*f*
Lyell, Charles, 340, 367, 367*f*

M

maar vents, 148
Mackenzie dykes, 123
Mackenzie River, 414, 415*f*
MacMillan Pass, Yukon, 249
Madagascar, 65
mafic igneous rock, 110, 114, 115*f*, 119
magma, 14*f*, 106
    andesitic magma, 134, 134*t*, 135
    assimilation, 109, 109*f*
    basaltic magma, 134–135, 134*t*
    Bowen's reaction series, 110–111, 110*f*
    chemistry of, 134
    crystallization of, 110–111
    decompression melting, 59
    formation of, 106, 108, 108*f*
    genesis, 144–145

magma (cont.)
  high-silica magma, 134
  low-silica magma, 134
  magma evolution, 119
  migration, 109, 109f
  mixing, 109, 109f
  partial melting, 106, 108
  rhyolitic magma, 134, 134t, 135
  types of, 134–135, 134t
  and volcanoes, 106, 106f
magma assimilation, 109, 109f
magma differentiation, 106, 108–109, 109f,
    118–119, 118f
magma foam, 136
magma migration, 109, 109f
magma mixing, 109, 109f
magma supply rates, 12
magmatic segregation mineral deposits, 248, 248t
magnesium, 88t, 89, 91t, 92, 110, 111, 246
magnetic field, 50, 56f
magnetic polarity, 50
magnetic striping, 57
magnetite, 83t, 97f
magnetometers, 56
main shock, 311
Makemake, 28, 28f
malachite, 96f, 97f
mammals, 386–387
manganese, 246
manganese nodules, 635
mantle, 11, 48, 50, 51f, 108, 328–329
mantle convection, 108
mantle plumes, 50, 51, 122–124, 123f, 149
  see also hotspots
mantle transition zone, 329
maps, 277, 277f, 284–285, 284f
marble, 173t, 221, 227, 227t, 230f, 230t
margins of Canada, 414
Mariana Trench, 63, 642
marine environments, 207–209, 208f
marine geologists, 624
marine geology, 624
  continental margin, 630–631, 632–633
  continental shelf, 628–629, 632–633
  currents, 626–627
  human impacts on oceans, 644–645, 644f
  mid-ocean range, 638–639
  ocean circulation, 626–627
  ocean sediment, 632–633
  oceanic trenches, 642
  oceans, 624–625
  pelagic sediment, 634–635
  pelagic stratigraphy, 636–637
  salinity of oceans, 624–625, 625f
marine isotopic stages, 470–471
marine sediment, 124
marine stratigraphy, 637
marine transgression, 608
Maritime Plain, 295
Mars, 28, 29, 32, 32f, 33, 37, 624
mass balance, 468
mass extinctions, 374–375, 375f
mass number, 86, 346
mass spectrometer, 348
mass wasting, 16–17, 182, 307, 483, 484
  avalanches, 492
  in coastal environment, 608
  common types of, 488, 489f
  described, 484–485

  factors contributing to, 494–495
  fast-moving mass-wasting events, 490, 491f
  and global warming, 498
  human activities, as cause of, 498
  human toll, 484
  lahar, 492
  landslides. See landslides
  moisture content, variations in, 496
  research, and mitigation practices, 502–503
  speed, variations in, 496
  submarine landslides, 493
  trigger, 495
matrix, 112
Mauna Kea Volcano, Hawaii, 286, 286f
Mauna Loa Volcano, Hawaii, 140t, 142, 142f,
    286f
Mayon Volcano, Philippines, 139f, 155
Mazama (Crater Lake), 291
McArthur River deposit, 250
McAuley Creek, British Columbia, 491f
Meager, 107
meandering channels, 518
mechanical weathering, 73
medial moraine, 460
Mediterranean, 71
Mediterranean Sea, 624
meguma, 294
Meguma terrane, 406, 407
melting Arctic, 438
melting ice, 441
melting ice sheets, 465
melting points, 108
meltwater, 454
Mercalli, Giuseppe, 320
Mercury, 28, 29, 30, 32, 32f
Mesosaurus, 10
mesosphere, 424
mesothelioma, 225
Mesozoic Era, 21t, 377f, 382–383
Messenger, 32
metaconglomerate, 227, 227t, 230f, 230t
metal ores, 245t, 247t
metal strapping, 503, 503f
metallic bonds, 89
metallic cations, 92–93, 96
metallic mineral resources, 246–247, 247t
metallic sulphide deposits, 122
metallic sulphides, 635
metamorphic facies, 232, 233f
metamorphic grades, 226–227, 226f
metamorphic minerals, 15, 90, 220
metamorphic pathways, 232–233, 232f
metamorphic products, 227t
metamorphic rock, 14, 15, 220
  defined, 220–221
  foliated rock, 228–229, 228f, 228t
  metamorphic grades, 226–227, 226f
  metamorphism. See metamorphism
  nonfoliated rock, 228, 230–231, 230f, 230t
  nonfoliated texture, 228
  texture, 227, 228, 228f, 228t, 230–231
metamorphism, 15, 15f, 220, 220f
  of basalt crust, 237, 237t
  burial metamorphism, 233
  causes of, 222–223
  contact metamorphism, 222, 223f, 227, 230–
    231, 231f, 232f, 233, 234
  heat, 222
  high-grade metamorphism, 226

  and hotspots, 236–237
  low-grade metamorphism, 226
  and plate tectonics, 234, 236–237
  pressure, 222–223
  regional metamorphism, 223, 227, 227f, 228–
    229, 230–231, 234
  thermal metamorphism, 222
  water, role of, 224
metasomatism, 224
meteorites, 356–357, 356t, 357f
methane, 32, 37, 256, 258, 427f, 428
methanogens, 259
mica, 93f
mica group, 94
Michelangelo, 221
micrite, 205
micro-earthquake, 322
micrometeorites, 163
Mid-Atlantic Ridge, 12, 71, 71f, 150, 153f, 414,
    630, 634, 637f, 640
mid-latitude deserts, 574, 574f
mid-ocean range, 638–639, 638f
mid-ocean ridge, 12, 52, 140t, 149
mid-plate volcanism, 150
mid-plate volcanoes, 12f
Middle America Trench, 642
Milankovitch, Milutin, 474, 475
Milky Way Galaxy, 28, 35
mineral cements, 195
mineral chemistry, 172
mineral deposits, 247, 248–251, 252
mineral resources, 244
  common mineral resources, 246f
  metallic mineral resources, 246–247, 247t
  non-metallic mineral resources, 246, 247t
  ore, 248–251
mineral rights, 629
mineral stability, 172, 173t, 177f
mineralogists, 82
mineralogy, 82
minerals, 14, 80, 81f
  see also specific minerals
  agriculture minerals, 246, 247t
  atoms, 86–87
  in Canada, 83f, 399
  Canadian Shield, 399
  chemical composition, 80
  cleavage, 84, 84t
  colour, 84–85
  common igneous rock-forming minerals, 111t
  common minerals and their uses, 83t
  common rock-forming minerals, 94–95
  compounds of atoms bonded together, 88–89
  construction minerals, 246, 247t
  crystal form, 84t
  crystalline solid, 80
  crystallization, 82f
  defined, 80, 94
  disseminated minerals, 248
  dissolved minerals, 550–551
  fracture, 84, 84t
  group name, 94
  hardness, 84, 84t
  heavy minerals, 194
  hydrated minerals, 122
  igneous minerals, 90
  index minerals, 226–227
  industrial minerals, 246, 247t
  inorganic, 80

lustre, 84, 84t
major classes of minerals, 96–97, 97t
management of, 17
metamorphic minerals, 15, 90, 220
mineral environments, 98
mineral identification, 84–85
Mohs hardness scale, 84, 85f
naturally occurring, 80
ore minerals, 247, 248t
physical properties, 84–85, 84t
reaction to HCl, 84t
recrystallized minerals, 15
as resource, 81
rock, as solid aggregate of minerals, 82
sedimentary minerals, 90, 162
species, 94
streak, 84t
variety, 94
Miocene Epoch, 377f, 386
MIS 5, 472
MIS 5e, 472–473, 473f
Mississippi River, 512, 633
Mississippi River Delta, 529, 529f
Mississippian Period, 21t, 377f, 381
Mississippian Rundle Group, 412f
mitigation practices, 502–503
model, 7
modern epoch, 21t
modification of hypothesis, 7
modified Mercalli (MM) intensity scale, 320, 321t
Mohorovicic, Andrija, 328
Mohorovicic discontinuity (Moho), 328–329, 328f
Mohs, Friedrich, 84
Mohs hardness scale, 84, 85f
molecular biology, 372
Molokai, Hawaii, 493f
Molokai fracture zone, 67
molybdenite, 245t
molybdenum, 245t
moment magnitude, 323
momentum, 516
Momotombo Volcano, Nicaragua, 137t
Monashee Mountains, British Columbia, 218–219f, 492f
Mongolia, 572
monocline, 282, 283f
monogenetic fields, 140t, 148
monsoon, 575
monsoon deserts, 575
Monterey Submarine Canyon, 631f
montmorillonite, 173
Montreal Protocol, 429
Moon, 41, 41f, 163, 600, 600f
Moon rocks, 356, 356t
moraines, 460
Morrell, Andre, 141
Morro do Bumba, Brazil, 482f
motion of the sea floor, 11
Mount Assiniboine Provincial Park, 270f
Mount Baker, Washington, 107, 144
Mount Edziza, British Columbia, 142, 143f
Mount Everest, Himalayas, 65, 286f, 296
Mount Fuji, Japan, 144, 144f
Mount Galunggung, Indonesia, 492
Mount Garibaldi, British Columbia, 107, 144
Mount Hood, Oregon, 107
Mount Kidd, Alberta, 412f

Mount Logan, Yukon, 106, 410f, 411, 411f
Mount Meager, British Columbia, 143, 144, 286, 291, 485t
Mount Pelée, Martinique, 138
Mount Pinatubo, Philippines, 16, 140t, 144, 429, 493f
Mount Rainer, Washington, 107, 492
Mount Robson, British Columbia, 410f, 411
Mount Ruapehu, New Zealand, 492
Mount Shasta, California, 107, 291
Mount St. Helens, Washington, 107, 136f, 140t, 143, 144, 291
Mount Stromboli, Italy, 16f
Mount Vesuvius, Italy, 145
Mount Waddington, British Columbia, 410f, 411
Mount Yamnuska, Alberta, 279f
mountain, 286
    see also mountain building; specific mountains
    Canada's northernmost mountain belt, 408–409
    Canada's youngest mountain belt, 410–413
    fault-block mountains, 286, 287f, 288–289f
    fold-and-thrust mountains, 286, 288–289f
    formation of mountains, 52
    rift-shoulder mountains, 409
    volcanic mountains, 286, 288, 289f, 290–291
mountain belts, 286, 294–296
mountain building
    in Canadian Cordillera, 411–413
    crustal extension, 292–293, 292f
    dip-slip faults, 278
    fault-block mountains, 286, 287f, 288–289f
    fold-and-thrust belts, 294–296
    fold-and-thrust mountains, 286, 288–289f
    outcrop patterns, 284–285
    rock folds, 282, 283f
    strain, 274–277
    stress in the crust, 272–273
    strike-slip faults, 278
    types of mountain-building processes, 286–287, 288
    volcanic mountains, 286, 288, 289f, 290–291
mountain range, 286
mud, 192, 200
mud cracks, 213, 213f
mudflows, 490, 496
mudstone, 166, 227t
Murray River, Australia, 632f
muscovite, 93f, 222, 226
muscovite mica, 111, 111t, 115
muskeg, 177
Myrdalsjökull glacier, 137

N

Namib Desert, Namibia, 192f
National Aeronautics and Space Administration (NASA), 32
National Oceanic and Atmospheric Administration, 422
native elements, 96, 96f, 97f
native metals, 247
natural arches, 179
natural attenuation, 557
natural frequency, 307
natural gas, 258–259
natural law, 6
natural levee, 520

natural processes, 485
natural resources, 17
Natural Resources Canada, 306f, 307, 439f, 484, 502
natural selection, 367, 367f
A Naturalist's Voyage on the Beagle (Darwin), 367
Nazca Plate, 52, 58
neap tide, 601
nebula, 29
negative climate feedback, 429, 476–477
Neogene Period, 21t, 377f, 386
Neolithic Age, 244
neon, 88t
Neptune, 28, 29, 34–35, 34f, 37
Neptunism, 366
neritic sediment, 632
Neumann, Frank, 320
neutral compounds, 92–93
neutrons, 86
Nevada del Ruiz Volcano, Columbia, 69
New Brunswick, 295
New Brunswick Highlands, 295
New Madrid earthquakes, 313
New-Quebec Orogen, 398
New Stone Age, 244
New Zealand, 71, 279
Newfoundland, 295
Newfoundland Highlands, 295
nickel, 48
Nile River, 582
Nisga'a people, 142, 143
nitrogen, 88t
nitrous oxide, 429
non-metallic mineral resources, 246, 247t
non-renewable, 244
non-renewable energy resources, 254
nonfoliated metamorphic rock, 230–231, 230f, 230t
nonfoliated texture, 228
nonpoint source pollution, 533
normal faults, 278
normal polarity, 56
normal stress, 494
North America, 10, 12, 16, 393, 407f, 471f, 628, 629f
North American Cordillera, 286
North American Plate, 52, 66, 66f, 67, 143, 330–331, 382, 394, 394f, 407, 413, 414, 610, 630, 640, 643
North Anatolian fault, Turkey, 278, 280, 314, 315f
North Atlantic Basin, 60
North Atlantic Deep Water, 425, 425f, 477, 627
North Atlantic Gyre, 626
North Atlantic Ocean, 407, 595, 617, 626
North Equatorial Current, 626
north magnetic pole, 56
North Pacific Current, 626
North Pacific Gyre, 626
North Pacific Ocean, 595
North Vancouver, British Columbia, 485t
Notre Dame Mountain, Quebec, 295
nova, 30
Nova Scotia, 295
Nova Scotia Highlands, 295
Nova Scotia Uplands, 295
Nubian Plate, 132
nuclear fusion, 30–31

nuclear power plants, 262–263
nucleation, 90
nucleation seed, 90
Nuna, 398
Nuvvuagittuq greenstone belt, Quebec, 338f

O

Oahu, Hawaii, 493f
objects in the Solar System, 36–37
obliquity, 474
observation, 7
obsidian, 114
Occam's razor, 6
ocean acidification, 168–169, 440
ocean basins, 9
ocean circulation, 626–627
ocean-continent convergent boundary, 62, 62f
ocean energy, 265
ocean-ocean convergent boundary, 63, 63f
ocean sediment, 632–633, 636f
    see also pelagic sediment
oceanic conveyor belt, 627
oceanic crust, 50, 51, 638–639, 639f
    basalt, 134
    formation of, 57
    mafic igneous rocks, 119
    origin and recycling of, 52–53
    subducts, at convergent boundaries, 62–63
oceanic fracture zones, 67, 67f
oceanic trench, 63, 124, 642, 642f, 643f
oceans, 624–625, 624t
    see also marine geology
Oceans Act, 629
offshore drilling platforms, 193f
oil, 254
    composition of, 256–257
    origins of, 256
    peak oil, 255
    status of world's oil supply, 258, 259f
oil shale, 292, 293f
Oil Springs, Ontario, 256
oil traps, 257
oil window, 256
Okanagan Valley, British Columbia, 572
Okavango River, Botswana, 519f
Olbers, Heinrich, 35
Olbers' Paradox, 35
oldest rocks in the world, 398–399
Oligocene Epoch, 377f, 386
olivine, 80, 92, 92f, 93f, 94, 95f, 97f, 110, 115, 171t, 173t
olivine group, 94
Omega Centauri, 31f
On the Origin of Species (Darwin), 367
100-year flood plain, 522
Ontong Java Plateau, 148
Oort Cloud, 28, 37, 40
opaline silica, 636
open-system behaviour, 352, 352f
orbital forcing, 475f, 476–477
orbital parameters, 474–475, 475f
Ordovician Period, 21t, 295, 344, 377f, 380, 401f
Ordovician-Silurian extinction, 374
ore, 247, 248–251
ore bodies, 244
ore minerals, 247, 248t
organic decay, 170

organic matter, 430
organic precipitation, 625
organic soil order, 184t
organic soils, 184, 184f
Organization of the Petroleum Exporting
    Countries (OPEC), 258
original horizontality, 284, 342, 342f
original lateral continuity, 342–343, 342f
orogenesis, 64–65, 175f
orogenic belts, 286
orogeny, 286, 379
orographic effect, 572
orthoclase, 93f, 97f, 115
orthoclase feldspar, 82, 94, 95f, 96f, 111t, 114,
    115, 167f, 171t, 173t
orthogneiss, 227n, 227t
Ottoia, 381f
outcrop patterns, 284–285
outcrops, 276
Outer Banks, North Carolina, 606, 606f
outer core, 48, 49f
outer reef, 615
outlet glaciers, 454–455
outwash plain, 460
outwash terraces, 460
overpumping, 550
oversteepening, 494–495
overwash fan, 607
oxbow lake, 518, 519f
oxidation, 167
oxide mineral class, 167
oxides, 96, 96f, 97f, 247
oxidized carbon, 431
oxygen, 9, 48, 50, 82, 88t, 89, 90, 91t, 111
oxygen isotopes, 466–467
ozone, 429

P

P-wave shadow zone, 326, 327f
P waves, 318, 329f
Pacific Ocean, 394f, 595, 624, 624t, 625f, 637,
    637f
Pacific Ocean Basin, 143
Pacific Plate, 52, 58, 63, 66f, 67, 70, 143, 331,
    394f, 610, 637, 642
Pacific Ring of Fire. See Ring of Fire
Pacific Shelf, 414
pahoehoe lava, 132, 135, 135f
Paleocene Epoch, 377f, 386
paleoclimates, 476–477
paleoclimatology, 452
Paleogene Period, 21t, 175, 377f, 382, 386
paleomagnetism, 56–57
paleotologists, 387
Paleozoic Era, 21t, 369f, 377f, 380–381, 380f
palis, 493, 493f
Pangaea, 10, 12, 13f, 60, 65, 229, 295f, 296, 381,
    382f, 400, 403, 405, 405f, 407f, 413, 413f,
    414, 628
Panthalassa, 412, 413f
Papua New Guinea, 485t
parabolic dunes, 578t, 579, 579f
paragneiss, 227n, 227t
parent isotope, 346
parsimony, 6, 7
partial melting, 106, 108, 118–119, 124f
passive continental margin, 628

passive margin, 64, 414, 630, 630f
Patagonia, South America, 368f
Patagonian Desert, South America, 568t
patch reefs, 615, 615f
paternoster lakes, 459
pavements, 197
peak ground acceleration (PGA) seismic hazard
    map, 307f
peak oil, 255
Pearya, 409
peat, 607
pegmatite, 248
pelagic sediment, 634–635, 635f
pelagic stratigraphy, 636–637
peneplain, 287
Pennsylvanian Period, 21t, 381
peri-Gondwanan Realm, 406
peridotite, 115, 117, 117f, 328
periglacial environments, 453
perihelion, 474
periods, 20, 376
permafrost, 453
permafrost zones, 488
permeability, 257, 544, 547f
permeable reactive barriers, 556
Permian Period, 21t, 339, 377f, 381
Permian-Triassic extinction, 374
permineralization, 369
perovskite, 329
Peru-Chile Trench, 642
pesticides, 552
Peterman Glacier, 465f
petroleum, 256, 257f
pH scale, 168
phaneritic, 112, 113t
Phanerozoic basins, 401f
Phanerozoic Eon, 366, 377f, 380
Phanerozoic Orogens, 403
phenocrysts, 112
Philippine Islands, 71
Philippine Plate, 52, 63, 642
Philippine Sea, 634
Philippines, 485t
phosphate deposits, 635
phosphorite, 632
phosphorus, 88t
photosynthesis, 431
phreatomagmatic eruption, 142, 148, 154, 291
phyllite, 227, 227t, 228f, 229
phyllitic texture, 229
phyllosilicates, 173
phylogeny, 370
physical geology, 4
physical oceanography, 624
physical properties of minerals, 84–85, 84t
physical weathering, 162, 163f, 164–165
physiographic provinces, 294
physiography, 395
phytoplankton, 431, 434, 533
Piedmont, 294, 295
Pietà, 221
pillow lava, 122, 123f
Pinian eruption column, 154f
Pioneer fracture zone, 67
Piton de la Fournaise Volcano, 136f
placer deposits, 251
plagioclase, 94, 97f
plagioclase feldspar, 82, 82f, 92, 93f, 94, 95f, 110,
    111, 111t, 114, 115

planetary embryos, 29
planetary nebula, 28, 30, 32
planetesimal accretion, 29, 29f
planetesimals, 29
planets, 28, 29
plastic deformation, 274, 282, 283f
plastic flow, 454
plasticity, 50
plate boundaries, 52, 58–59, 59f
    convergent boundaries, 58, 59f, 62–65, 68t
    divergent boundaries, 58–59, 59f, 68t
    earthquakes, 68–71, 312
    geologic processes at, 68t
    transform boundaries, 58, 59f, 66–67, 66f, 68t, 273
plate collision, 286
plate drag model, 72
plate movement, mechanisms of, 72, 72f
plate tectonics, 9
    Basin and Range province, 292–293
    Canadian Cordillera, 411–413
    continent-continent convergent boundary, 64
    continental drift, 10–11
    convergent boundaries, 58, 59f, 62–65
    and critical thinking, 10–13
    distribution of Earth's tectonic plates, 394f
    divergent boundaries, 58–59, 59f
    earthquakes, and plate boundaries, 68–71
    Earth's layers, 48–51
    global tectonics, 54
    hotspots, 11–12
    igneous evolution, 118–119, 119f
    lithosphere, 53f
    lithospheric plates, 52–53
    and metamorphism, 234, 236–237
    ocean-continent convergent boundary, 62, 62f
    ocean-ocean convergent boundary, 63, 63f
    oceanic crust subducts at convergent boundaries, 62–63
    orogenesis, 64–65
    paleomagnetism, 56–57
    Pangaea, 12
    plate boundaries, 58–59, 59f, 68–71, 68t
    plate movement, mechanisms of, 72, 72f
    rift valley topography, 12
    and rock cycle, 72–73, 73f
    seafloor spreading, 11, 56–57
    spreading centres, 53
    subduction zones, 53
    transform boundaries, 58, 59f, 66–67, 66f
plates, 9, 47, 52, 52f
platinum, 245t, 247
platinum group materials, 245t
playa, 583f
playa lake, 582
Pleistocene Epoch, 21t, 377f, 470
plesiosaur, 383f
Plinian eruption, 145
Pliocene Epoch, 377f, 386
plucking, 454
plunging anticlines, 282
plunging fold, 282, 283f
plunging synclines, 282
Pluto, 28, 28f
plutonic, 107
plutonism, 124f
plutons, 63, 125, 125f
podsols, 185
podzolic soil order, 184t

podzols, 184f
Poe, Edgar Allen, 35
point bar, 518
Polar cell, 425, 570, 570f
polar deserts, 572, 575
polar glaciers, 454
polar soils, 185
polarity events, 57f
polarity reversals, 57f
polarized, 56
polarized molecule, 166
poles, 56
polluted runoff, 532–533, 533f, 616
pollution
    coastal pollution, 616
    of groundwater, 550–551
        groundwater contamination, 552–553
        polluted runoff, 532–533, 616
polyps, 614
Pompeii, 145
pool depression, 518
poorly sorted deposit, 196
population growth, 435, 436f, 551f, 616–617
porosity, 200, 257, 544
porous media, 548
porphyritic, 112, 113t
porphyroblasts, 229, 231
porphyry copper, 248
Porteau Cove, British Columbia, 5
positive climate feedback, 429, 476–477
potash, 403
potassium, 48, 88t, 89, 91t, 111, 115
potassium-argon, 354
potassium feldspar, 82f, 93f, 111
potassium-rich feldspar, 115
potentiometric surface, 548
powder snow avalanche, 492, 492f
Prairie Farm Rehabilitation Administration, 573
Prairie soil, 403
Precambrian Eon, 366, 377f, 380f
precession, 475
precious metals, 82
precipitation, 510
prediction, 7
pressure, and metamorphism, 222–223
pressure release, 164
primary radioisotopes, 348
primary sedimentary structures, 212–213
primordial crust, 48, 356
primordial lead, 357
Prince Edward Island, 295, 540
principle of cross-cutting relationships, 342f, 343
principle of fossil succession, 343, 343f, 376
principle of inclusions, 342f, 343
principle of original horizontality, 284, 342, 342f
principle of original lateral continuity, 342–343, 342f
principle of parsimony, 6, 7
principle of superposition, 338, 342, 342f
principle of unconformities, 342f, 343
principle of uniformitarianism, 20
Principles of Geology (Lyell), 367
processes, 430
productivity, 636–637
Proterozoic Eon, 21t, 295, 295f, 313, 377f, 378–379, 378f, 379f
Proterozoic Orogens, 398–399
protolith, 221, 227t
protons, 86

Public Safety Canada, 502
Puerto Rico Trench, 642, 643
pull-apart stress, 272
pumice, 113, 138, 139f
pyrite, 83t, 84, 96, 96f, 97f
pyroclastic, 113t
pyroclastic debris, 106, 134
pyroclastic flows, 138, 139f
pyroclastic volcanic rocks, 113
pyroxene, 92f, 94, 110, 114, 115, 173t, 329
pyroxene group, 93f, 94

Q

quartz, 80, 82, 82f, 83t, 90, 93f, 94, 95f, 97f, 111, 111t, 114, 115, 171t, 173t, 227, 229
quartz sandstone, 173t, 202, 202t, 203f
quartzite, 173t, 227, 227t, 230f, 230t
Quaternary Period, 21t, 377f, 386, 452
Quebec, 295, 540
Quebec Highlands, 295
Queen Charlotte fault, British Columbia, 66f, 68, 143, 273, 314, 315
quick clay, 490

R

R horizon, 178
radar, 396
radial drainage, 513
radioactive decay, 338, 346–347
radioactivity, 346–347
radiocarbon age, 355f
radiocarbon dating, 354–355
radioisotopes, 346, 348, 348t
radiolarians, 634
radon-222, 347
rain shadow, 292, 572
rain shadow deserts, 574
rainfall, 528
Rajasthan Desert, 575
rare-metal pegmatites, 248
rate of geologic renewal, 244
rate of strain accumulation, 282
ray paths, 326
Rayleigh waves, 318
reasoning, 6–7
reasoning by analogy, 58
rebound, 287
recent epoch, 377f
recharge, 542, 542f
recirculation wells, 556
recrystallization, 15, 90, 220, 221, 231
rectangular pattern, 513
red clay, 634
red giant, 30
Red River, Manitoba, 16, 520f
Red Rock Coulee, Alberta, 195f
Red Sea, 53, 59
red-shifted light, 35
reduced carbon, 431
reef, 207, 208f
reef crest, 615
reef flat, 615
Reelfoot rift, 313
refraction, 596, 597f

regional metamorphism, 223, 227, 227f, 228–229, 230–231, 234
  see also metamorphism
regolith, 163, 484
regosolic soil order, 184t
regrading land, 503
Reid, Harry Fielding, 310
rejection of hypothesis, 7
relative dating, 338, 341, 341f, 350
    example of, 344–345, 344f
    order of geologic events, 344–345
    stratigraphic principles, 342–343
relict sediments, 632
renewable energy, 245, 264–265, 265f
replacement, 369, 369f
reservoir, 257, 430, 526
reservoir bed, 257
residual mineral deposits, 248t, 250
residual sediment, 632
respiration, 431
retaining walls, 503
Revelstoke Dam, British Columbia, 503f
reverse faults, 278, 279f, 280
reversed polarity, 56
Rheic Ocean, 405f, 407
rhyolite, 115, 116, 116f
rhyolite breccia, 116
rhyolite caldera complexes, 145, 152f, 155
rhyolite lava, 116
rhyolitic magma, 134, 134t, 135
Richter, Charles F., 322
Richter magnitude, 322, 322f
Richter scale, 322
ridge push, 72
ridge slide, 72
rift-shoulder mountains, 409
rift valley topography, 12
rift valleys, 12, 52, 59f
rift zone, 53
right-lateral strike-slip fault, 278
rill erosion, 182
Ring of Fire, 58, 68, 70, 124, 143, 150
rip currents, 598
rip-rap, 503
ripple mark, 213, 213f
ripples, 576, 576f
river, 512, 633
Robert-Bourassa hydroelectric power station, 264f
rock bolts, 503
rock cleavage, 228
rock cycle, 8, 72–73, 73f, 201, 366
rock flour, 454
rock folds, 282, 283f
rock gypsum, 173t, 204, 204f, 204t
rock netting, 503f
rock record, 338
rock salt, 173t, 204, 204f, 204t
rock slides, 490, 491f, 608
rock stability, 172, 173t
rockfall sheds, 503
rockfalls, 490, 491f, 496
Rockies. See Canadian Rockies
rocks, 14, 430
    in the crust, 272–273
    defined, 82
    fragmentation of rock, 164–165
    igneous rock, 14
    magnetic properties, 56

metamorphic rock, 14, 15
    see also metamorphic rock
Moon rocks, 356, 356t
oldest rocks in the world, 398–399
sedimentary rock, 14–15
    see also sedimentary rock
as solid aggregate of minerals, 82
stress, 272
texture, 112–113, 113f, 204
rocky headland, 596
Rocky Mountains, 286, 410, 411, 520f
rocky shorelines, 608
Rodinia, 295, 378, 379f, 398, 400
root wedging, 170, 170f
runoff, 510, 511, 512–513, 532–533
Russia, 435t, 445, 485t
rutile, 83t, 245t

S
S-folds, 282, 283f
S-wave shadow zone, 326, 327f
S waves, 318, 329f
Sagarmatha, 296
    see also Mount Everest, Himalayas
Sahara Desert. See Great Sahara, Africa
Sahel, 584, 585f
Sahelanthropus tchadensis, 386–387, 387f
Saint-Jude, Quebec, 485t
salinity, 624–625, 625f
salinization, 17
salt contamination, 552
salt crystals, growth of, 165
salt marshes, 609
saltation, 182, 182f, 516, 576
saltwater intrusion, 550
San Andreas fault, California, 66, 66f, 67, 273, 278, 280, 280f, 314, 315, 341
San Francisco Volcanic Field, Arizona, 148, 148f
sand, 192, 196f, 547f
    history from composition of, 202–203, 203f, 203t
sand bars, 598
sand dunes, 578–579, 578f, 578t, 579f
sand seas, 579, 579f
sandblasting, 164
sandstone, 14, 15f, 195f, 202, 202f, 227t, 231, 548
Saskatchewan, 402, 403
satellite radar, 209
saturated zone, 540
Saturn, 28, 29, 34, 34f
Saudi Arabia, 59, 435t
scarce metals, 246, 247t
scattered disc, 28
schist, 173t, 227, 227t, 228f, 229
schistose texture, 229
scientific method, 7, 7f
scoria, 113, 138
Scotian Shelf, 414, 415f
scour depression, 518
scree, 495
sea floor, 209, 640
sea-level rise, 422, 438–439, 439f, 604f, 605f
    barrier islands, 606–607
    coastlines, effect on, 604–605
    future sea-level rise, 604–605
    during last ice age, 458f
    during last interglacial, 472–473

sea-level set-up, 597, 597f, 599f
sea stacks, 608, 608f
sea-surface temperature (SST), 440
Sea-to-Sky highway, 5
seafloor magnetism, 57
seafloor spreading, 11, 56–57, 59, 296
seamount, 11
seas, 624
seasonal intensity, 433
seawalls, 616
second-order basin, 513
second-order stream, 513
secondary enrichment, 250, 250f
sediment, 14, 192
    see also specific types of sediment
    biogenic sediments, 194, 195f, 198–199, 632
    changes in, during transportation, 196–197
    changing composition of, 199
    chemical sediments, 194–195, 198–199
    clastic grains, 198–199
    clastic sediments, 194, 194f
    components of, 198f
    deposition, 341, 341f
    erosion, 517
    fluvial sediment, 528–529
    hemipelagic sediments, 633
    lithified, 162
    marine sediment, 124
    neritic sediment, 632
    ocean sediment, 632–633, 636f
    particle size, 196
    pelagic sediment, 634–635
    sedimentary cycle, 200–201
    transportation of, by glaciers, 460f, 461f
    types of, 194–195
sediment liquefaction, 306
sediment load, 515, 517f
sedimentary cycle, 200–201
sedimentary environments, 197t
sedimentary mineral deposits, 248t, 250–251
sedimentary minerals, 90, 162
sedimentary products of weathering, 171, 171t
sedimentary quartz, 171, 171t
sedimentary rock, 14–15, 192
    biogenic sedimentary rock, 205, 205f, 205t
    chemical sedimentary rock, 204–205, 204f, 204t
    clastic sedimentary rock, 202, 202f, 202t
    composition of, 200f
    in continental platform, 400–403
    evidence of past environments, 206–209
    formation of, 192–193
    fossils, 206f
    metamorphosis, 73
    within rock cycle, 201
    sedimentary cycle, 200–201
    strata, 212f
    study of, 193
    types of, 202–203, 203f
sedimentary structures, 212–213
sedimentologists, 194
sedimentology, 292
Sego Canyon, Utah, 336f
seismic hazard zone, 70
seismic hazards, 307
seismic tomography, 330–331
seismic waves, 51f, 304, 305f, 318, 318t, 319f, 326–327, 326f, 327f
seismogram, 318, 319f

seismology, 326–327
seismometers, 310, 320–321, 320f
Selkirk Mountains, British Columbia, 164, 495f
semi-arid, 568
sensible heat, 426
septic systems, 551
septic tanks, 498
serpentine, 122
serpentinite, 231f
severe events, 440
sewage waste, 551, 551f
shale, 166, 173t, 202f, 202t, 203f, 220f, 226f, 227f, 227t, 228f, 231
shallow water, 596–597
Shark Bay, Australia, 379f
shear fracture or fault, 164
shear stress, 272, 273, 273f, 279f, 494, 494f
shearing, 66
sheet erosion, 182
sheeted dikes, 122, 150, 639
sheeted joints, 164, 164f, 179f
shelf areas, 629
shelf break, 629
shield volcanoes, 117, 140t, 142–143, 144
shoal, 594
shoaling wave, 594, 594f
shore zone, 615
shorelines. See coastlines
short-wave ultraviolet (UV) radiation, 426
shotcrete, 503, 503f
Siberia, 453
Siberian Traps, Russia, 148
Sichaun quake, China, 281, 305, 312–313
Sierra Nevada Ranges, 286, 292, 572, 575f, 582
Sikdar, Radhanath, 296
silica, 59, 115, 124
silica cemented sandstone, 195f
silicate compound, 90
silicate structures, 90–91, 91f, 92–93
silicate tetrahedra, 92
silicates, 89, 90–91, 96, 96f, 97f
siliceous ooze, 636
silicon, 48, 50, 88t, 89, 91t
Silicon Age, 244
silicon dioxide, 53
sill, 125
sillimanite, 226
silt, 193, 196f, 200
siltstone, 173t, 202, 202f, 202t, 203f, 227t
Silurian Period, 21t, 377f, 380–381
simple folds, 282
single substitution, 92
sinkholes, 168, 559
660-km discontinuity, 329
skarns, 224
skeletal limestone, 205, 205f, 205t
skeletal parts, 192, 193
slab avalanches, 492
slab pull, 72
slab window, 293, 293f
slaking, 165
slate, 221, 227, 228–229, 228f
slaty cleavage, 228
slaty texture, 228–229
slickenlines, 280, 281f
slickensides, 280, 281f
slip face, 576
slope failure, 484, 495, 496f
    see also mass wasting

slope stability, 484
slopes, 414
sloughs, 445f
slump, 488, 489f, 495f
slump complexes, 488
smelting, 247
Smith, William, 340, 345
smooth cord-grass (Spartina alterniflora), 609
Snell's law, 326
snow, 453f
snow line, 452
sodium, 48, 88t, 89, 91t, 92, 111, 115
sodium chloride. See table salt
sodium plagioclase feldspar, 114
sodium-rich plagioclase feldspar, 115
Sohm Abyssal Plain, 634
soil, 162
    A horizon, 178
    B horizon, 178
    bedrock, 178
    C horizon, 178
    Canadian soil classification system, 184–185, 184t, 185f
    components of, 176, 176f
    desert soils, 185
    formation of, 176–177
    and general climate conditions, 185
    humus layer, 178
    management of, 17
    moisture, 170, 511
    polar soils, 185
    Prairie soil, 403
    R horizon, 178
    subsoil, 178
    temperate soils, 185
    topsoil, 178
    tropical soils, 185
    and weathering, 176–177
soil classification, 184–185, 184t
soil creep. See creep
soil erosion, 182, 183
soil horizons, 178
soil profile, 178, 178f
solar constant, 433
solar energy or power, 254, 264, 264f, 433
solar nebula, 33f
solar nebula hypothesis, 28, 29, 40, 340
solar radiation, 474–475
Solar System, 28
    asteroids, 37
    comets, 37, 37f
    dwarf planets, 28, 36
    Earth's origin, 28–29
    gas giants, 29, 34–35
    Milky Way Galaxy, 28, 35
    objects in the Solar System, 36–37
    planets, 28, 29
    solar nebula hypothesis, 28, 29, 340
    Sun, 28, 30–31
    terrestrial planets, 29, 32–33, 32f
solar tide, 601
solar wind, 32, 40, 40f
solfataras, 560
solid-state reaction, 220, 222
solifluction, 488, 489f, 496
solonetzic soil order, 184t
sorting, 196–197
Soufrière Hills, West Indies, 144
source area, 202, 514

source bed, 257
sources of uncertainty, 352–353
Souris River, 16
South America, 10, 12, 65, 407f, 628
South American Plate, 52, 638f
South Atlantic Gyre, 626
South Cascade Glacier, 468, 468f, 469f
South China Sea, 624
South Korea, 435t
south magnetic pole, 56
Southern Ocean, 595, 624, 624t, 625f, 626, 627
space weathering, 163
species (mineral), 94
specific gravity, 110
speleology, 561
speleothems, 561f
sphalerite, 83t, 96, 97f, 245t
spheroidal weathering, 178–179
spinel, 329
spreading centre volcanism, 150–151, 150f, 638
spreading centres, 11, 11f, 53, 66–67, 90, 90f, 122, 124f
spring tide, 601
springs, 558–559, 558f
sputtering, 163
Sri Lanka, 65
St. Elias Mountains, Yukon, 106, 411, 413
St. Lawrence Lowland, 400
St. Lawrence paleo-rift faults, 313
stability, 172
stable electron configuration, 88
stagnant slab, 330
stalactites, 204, 561, 561f
stalagmites, 204, 561, 561f
star, 28, 31
star dune, 578, 578t, 579, 579f
static electricity, 87, 88
Statistics Canada, 255
steam turbine, 265
steel, 82
steep hillside, 197t
stellar nucleosynthesis, 30
Steno, Nicholas, 366, 367f
Stewart, British Columbia, 272f
stocks, 116, 125
Stokes drift, 596
storm surge, 602
storm water runoff, 532–533
storms, 440
straight channels, 518, 518f
strain, 304, 310
    Cascadia subduction zone, 291, 291f
    in the crust, 276–277
    dip, 277
    ductile deformation, 274
    elastic deformation, 274
    fracture, 274
    governing factors, 275
    joints, 276–277
    plastic deformation, 274
    rate of strain accumulation, 282
    stages of, 274–275
    strike, 277
strata, 212f
stratification, 212, 212f
stratified drift, 460
stratigraphic principles, 342–343, 344–345
stratigraphy, 376
stratosphere, 424

stratovolcanoes, 116, 135, 140t, 144–145, 155
stream, 206, 207f, 510, 524–525, 542, 582
stream capture, 524
stream channel, 197t, 515, 515f, 518–519, 518f, 519f
    see also channel
stream gauge, 522
stream ordering, 512–513, 513f
stream piracy, 524
stress, 220, 272–273, 304
    see also specific types of stresses
stretched pebble conglomerate, 230
strike, 277, 277f, 284f
strike-and-dip symbols, 277f, 284
strike-slip faults, 278
stromatolites, 378, 379f
Stromboli, Sicily, 132
structural geology, 284
structural traps, 257
Strutt, John William, 318
subatomic particle, 346
subduction zones, 11, 53, 64, 64f, 68, 69f, 90, 90f, 124f
    basalt, formation of, 124
    Cascadia subduction zone, 68, 291, 291f, 382
    oceanic trenches, 642
    volcanism, 296
subducts, 53, 62–63
sublimation, 454
submarine canyons, 629, 630–631
submarine fan, 630
submarine landslides, 493
submarine volcanism, 149f
submergent coasts, 610–611
subpolar glaciers, 454
subsoil, 178
substitution, 92–93
Sullivan Mine, British Columbia, 249
sulphates, 96, 96f, 97f
sulphides, 96, 96f, 97f, 247
sulphur, 82, 88t, 96, 97f
Sumatra, 71, 311
Sumatra-Andaman tsunami, 16
Sumatra quake, Indonesia, 305
summit crater, 140
Sun, 28, 30–31, 163, 174
    and climate change, 433
    solar wind, 32, 40, 40f
supercontinents, 397
    see also specific supercontinents
supergiant oil fields, 258
supernova, 30
superposition, 338, 342, 342f
supervolcano, 143
surf zone, 593
surface circulation, 626
surface temperature, 426, 432f
surface water. See water
surface waves, 310, 318
suspended load, 182, 516
suspension, 182, 576
sustainability, 4
sustainable resource, 4
Sutton Mountain, Quebec, 295
Sverdrup Basin, 408, 408f, 409
swash zone, 593
swells, 594
symbiotic algae, 614

syncline, 282, 283f
Syrian Desert, Middle East, 568t

T

table salt, 80, 89f
Taconic Orogeny, 295f, 296, 405f, 406, 406f
Tahiti, 117
talc, 83t, 221
talus, 122, 150, 165, 495, 495f
Tangshan, China, 69
tar sands of Alberta, 403
Tarim Basin, 572
tarn, 458
Tōhoku earthquake, Japan, 309, 309f
Technology Age, 244
tectonic settings, types of, 153f
tectonic theory. See plate tectonics
tectonic uplift, 201
tectonics, 9
    see also plate tectonics
temperate glaciers, 454
temperate soils, 185
temperature
    atmospheric temperature, 438–439, 466–467
    at base of glacier, 454f
    changes, 423f
    Curie temperature, 56
    global temperature, 438f, 439f, 440f
    sea-surface temperature (SST), 440
    surface temperature, 426, 432f
tensional stress, 272, 273f, 279f
tephra, 136, 138
terminal moraines, 460
terraces, 526
terranes, 397, 404
terrestrial planets, 29, 32–33, 32f
Tertiary Period, 377f
testing predictions, 7
Tethys Sea, 383
tetrahedron linkage, 91
texture
    foliated texture, 228–229, 228f, 228t
    igneous rock, 112–113, 113t
    metamorphic rock, 227, 228, 228f, 228t, 230–231
    nonfoliated texture, 228, 230–231
    rock, 204
thalweg, 516f, 518
Thar Desert, 575
Tharpe, Marie, 11
Thelon-Taltson Orogen, 398
theory, 6
theory of evolution, 366
theory of the ice ages, 452
thermal metamorphism, 222
thermal treatments, 556–557
thermohaline circulation, 425, 425f, 477, 627
thermosphere, 424
Thingvellir National Park, Iceland, 46–47f
Thingvellir rift, 59
third-order stream, 513
thrust fault, 278
Tibetan-Himalayan region, 175f
Tibetan Plateau, 64f, 65
tidal bulges, 601f
tidal currents, 601
tidal deltas, 606, 606f

tidal inlets, 606, 606f
tidal lagoon, 606
tidal wetland, 207, 208f, 607, 608–609, 609f
tide-raising forces, 600
tides, 600–601
Tiktaalik roseae, 381f
time, 338
tin, 82
titanium, 82, 245t, 246
titanium dioxide, 245t
Titusville, Pennsylvania, 256
toe of the slump, 488
tomography, 330
Tonga Trench, 642
topography, 400
topset beds, 528
topsoil, 178
"Tornado Alley," 16
tornadoes, 16
Torngat Orogen, 398
toxic waste, 244
traction, 516
trade wind, 425, 571, 626
trade wind deserts, 574
Trans-Hudson Orogen, 398
transform boundaries, 58, 59f, 66f, 68t, 273
    connection of two spreading centres, 66–67
    and earthquakes, 68
transform faults, 66–67, 67f
transform seismicity, 314–315
transition zone, 550
transpiration, 511
transverse dunes, 578, 578t, 579f
"trap," 148
travertine, 167, 171t, 204, 204f, 204t
trellis pattern, 513
Triassic-Jurassic extinction, 374
Triassic Period, 21t, 377f, 382
tributaries, 513
tributary glacier, 459
tropical soils, 185
tropical storms, 602
troposphere, 424, 425f
truck-mounted drill rigs, 541
Tseax Cone, British Columbia, 142, 143
tsunami, 16, 291, 291f, 308–309, 309f
tuff, 113, 138
turbidites, 631
turbidity currents, 209, 631, 631f
turbine, 254, 265, 526
turbulent eddies, 514
turbulent flow, 490, 514, 514f
Turkey, 71, 280, 280f
Turtle Mountain, Alberta, 490, 491f
typhoon, 602
Tyrannosaurus rex, 20, 368f

U

Uganda, 485t
Ukinrek Maars, Alaska, 140t
ultramafic igneous rock, 110, 115, 115f, 118f
Uluru, Australia, 583
UN Convention to Combat Desertification, 584
uncertainty, 352–353
unconformities, 341, 342f, 343
unconformity-type uranium deposits, 250
unconsolidated porous media, 548, 548f

undercutting, 307, 485, 495, 497, 503
underground storage tanks, 553, 553f
undersea volcano, 209
undertow, 597, 597f
unglaciated topography, 459f
Unicorn Horn, British Columbia, 164
uniformitarianism, 20, 366
United Kingdom, 435t
United Nations, 435, 436, 444, 511, 532, 584
United Nations Convention on the Law of the
    Sea, 629
United States, 143f, 293f
    carbon dioxide emissions, 444
    Coastal Plain, 628
    coastlines, 591
    energy, consumption of, 435
    greenhouse gas emissions, 435t
    and Kyoto Protocol, 445
    major earthquakes, 313
    natural disasters, 617
    Superfund program, 549
    weather disasters, 441f
universe, nature of, 35
unstable slopes. See mass wasting
uplift, 341
uplift weathering hypothesis, 174–175, 386
upwelling, 627
Ural Mountains, Eurasia, 65, 286
uraninite, 245t
uranium, 245t, 250, 263
uranium decay chain, 352–353
Uranus, 28, 29, 34, 34f
U.S. Environmental Protection Agency, 225
U.S. Geological Survey (USGS), 307, 310, 540
Ussher, James, 340, 340f

V

vaccines, 372
Vaiont Dam, Italy, 484–485, 486
valence electrons, 86
Valley and Ridge, 295
valley glaciers, 454
Vancouver, British Columbia, 617f
varieties (mineral), 94
vegetation cover, 503
vein, 248
vein deposits, 96
vent, 132
ventifacts, 165, 576
Venus, 28, 29, 30, 32–33, 32f, 174
vertical circulation, 626–627
vertisolic soil order, 184t
vesicular, 113, 113t
vestigial structures, 371
vibration, 495
Vinland Map, 355, 355f
viruses, 372–373
volatile gases, 32
volcanic arc, 62f, 63, 144, 290f, 642
    arc volcanism, 150, 151, 151f
    continental volcanic arc, 124
volcanic arc terranes, 413f
volcanic block, 138
volcanic bomb, 138, 139f
volcanic eruptions, 16, 108
volcanic hazard, 154–155
volcanic mountains, 286, 288, 289f, 290–291

volcanic outgassing, 40, 174
volcanic risk, 155
volcanism, 124f, 286
    arc volcanism, 150, 151
    intraplate volcanism, 150, 151, 312
    island arc volcanism, 151f
    mid-plate volcanism, 150
    spreading centre volcanism, 150–151, 150f,
        638
    subduction zones, 296
    submarine volcanism, 149f
volcanoes, 107, 108, 132
    arc volcanism, 150, 151
    in Canada, 143, 143f
    central vent volcanoes, 132, 140, 140t, 141f,
        142–143, 144–145
    cinder cones, 141f, 142, 143f, 148
    classification, 140, 140t, 141f
    composite volcanoes, 135, 140t
    cooling effect, 429
    defined, 132–133
    effusive eruptions, 136
    explosive eruptions, 132, 136–137, 138
    formation of, 52
    high-silica volcanoes, 155
    human communities, threat to, 154–155
    intermediate igneous rocks, 119
    intraplate volcanism, 150, 151
    large igneous provinces (LIPs), 140t, 148–150
    large-scale volcanic terrains, 132, 140, 140t,
        141f, 148–149
    magma, 106, 106f
    map of active volcanoes, 133f
    mid-ocean ridge, 140t
    mid-plate volcanism, 150
    mid-plate volcanoes, 12f
    monogenetic field, 140t
    pyroclastic flows, 138, 139f
    rhyolite caldera complexes, 145, 155
    shield volcanoes, 117, 140t, 142–143, 144
    spreading centre volcanism, 150–151, 150f
    stratovolcanoes, 116, 135, 140t, 144–145, 155
    study of, 132–133
    supervolcano, 143
    types of, 153f
    undersea volcano, 209
    volcanic gases, 136
volcanogenic massive sulphide (VMS), 249, 249f
volcanogenic particles, 634, 635
volcanologists, 137, 138, 154, 155
Vulcano, Sicily, 132

W

Wadati-Benioff zone, 314
Walcott Quarry, 381f
Wasatch Mountain Range, Utah, 281f
Washington State, 485t
washover, 607, 607f
water
    alluvial fan, 528–529
    base-level changes, 526, 527f
    bottom waters, 476
    chemically active fluids, 224
    coastlines. See coastlines
    at convergent boundaries, 237
    delta building, 528–529
    and deserts, 582–583

discharge, 514–515
drainage system, 512–513
and Earth system, 9
erosion by, 182, 516–517
facts, 511
flooding, 520–521
flow, 514–515
fluvial processes, 526, 527f
fresh water, 17, 511
    see also groundwater
freshwater problems, 532–533
global water, 425
global water problems, 532–533
groundwater. See groundwater
hard water, 551
hydrologic cycle, 510–511, 510f
infiltration, 503
and mass wasting, 494
meltwater, 454
and metamorphism, 224
oceanic plate subducts, 63
pH of water, 168
quality, 533
runoff, 512–513
shallow water, 596–597
stream channel. See stream channel
streams. See stream
wells, 541
water molecule, 166, 166f, 167f
water stress, 532, 532f
water table, 540–541, 541f, 543f
water vapour, 429, 511
waterfall, 526f
watershed, 512, 512f
Waterton National Park, 512f
wave, 595f, 598f
wave-cut platform, 608
wave energy, 594–595
wave erosion, 608f
wave-generated currents, 596–597, 599f
wave refraction, 326, 596
waves, 594–595, 594f
weather, 422
weathering, 5, 14, 15f, 73, 162
    biological weathering, 73, 162, 163f, 170–171
    chemical weathering, 73, 162, 166–168
    and climate change, 174–175
    insolation weathering, 165
    mechanical weathering, 73
    mineral stability, 172, 173t
    natural arches, 179
    physical weathering, 162, 163f, 164–165
    rock stability, 172, 173t
    and sediment, 201f
    sedimentary products, 171, 171t
    soil, production of, 176–177
    space weathering, 163
    spheroidal weathering, 178–179
    stability, 172–173
    uplift weathering hypothesis, 174–175
weathering products, 171, 171t
Wegener, Alfred, 10–11, 12, 52
welded tuff, 138, 139f
well, 541, 548, 556
well-sorted deposit, 196
wellhead protection, 557
Wells Gray-Clearwater Volcanic Field, British
    Columbia, 148

Wells Gray Provincial Park, British Columbia, 142
West Antarctica, 465
West Antarctica Ice Sheet, 473
westerlies, 571
Western Canada Sedimentary Basin, 400, 401*f*
Western Interior Seaway, 344, 345*f*, 402–403
Western System, 411, 413
wetland, 197*t*, 206, 207*f*, 511, 529, 542, 543*f*, 608–609, 609*f*
whales, 371*f*
Whipple, Fred, 37
white asbestos, 225, 225*f*
Williams, Hank, 406*f*
Williams, James M., 256
Wilson , J. Tuzo, 11, 66
wind, 576
wind-deposited dune, 197*t*
wind energy or power, 254, 264

wind erosion, 182–183, 183*f*
wind farms, 264, 264*f*
wind wave, 308, 594–595, 594*f*, 595*f*
Wood, Harry, 320
Wopmay Orogen, 398
World Meteorological Organization, 444
World Resources Institute, 435

X

xenoliths, 125

Y

Yangtze River, 633
yardang, 576
Yellow River Valley, China, 577*f*

Yellowstone National Park, Wyoming, 140*t*, 143, 152*f*, 312
Yellowstone Volcano, 148
Yosemite National Park, California, 106
Younger Dryas, 476–477, 477*f*
Yukon River, Alaska, 519*f*

Z

Z-folds, 282
zeolite, 237*t*
zeolite facies, 232
zinc, 82, 245*t*
Zion National Park, Utah, 212*f*
zircon, 353, 376, 398
zone of accumulation, 456
zone of wastage, 456
zoned mineral recrystallization, 231